GEOREF Thesaurus and Guide to Indexing

Third Edition

Sharon J. Riley, Editor

Published by the
AMERICAN GEOLOGICAL INSTITUTE

Earlier Editions:
GeoRef Thesaurus and Guide to Indexing, First Edition

Second Edition, 1978

GeoRef Thesaurus and Guide to Indexing, Third Edition

Library of Congress Catalog Card Number 81-68160
International Standard Book Number 0-913312-53-3
Printed in the United States of America

American Geological Institute
5205 Leesburg Pike
Falls Church, Virginia 22041
703/379-2480

The Thesaurus was photocomposed on a Videocomp
from a tape generated by SAMANTHA programs.
Eterna bold and light fonts were used for the text.

Contents

Introduction

The third edition of the *GeoRef Thesaurus and Guide to Indexing* contains approximately 13,500 terms. There are geographic place names, systematic terms (rocks, fossils, minerals, etc.) and non-systematic terms (geologic features, processes, properties, materials, etc.).

The ANSI Z39 Standard, *Guidelines for Thesaurus Structure, Construction and Use* (1974), has been followed herein.[1]

Source of the Vocabulary

The GeoRef data base consists of five segments:
(1) Bibliography and Index of North American Geology (1961-1970)
(2) Bibliography of Theses in Geology (1965-1966)
(3) Geophysical Abstracts (1966-1971)
(4) Bibliography and Index of Geology Exclusive of North America (1967-1968)
(5) Bibliography and Index of Geology (1969 to present)

The index terms and sets used in the five segments have much in common. They derive from a common source, the early volumes of the *Bibliography and Index of North American Geology*, edited by John M. Nickles. This Thesaurus is based on the indexing found in that Bibliography as it has evolved through the years.

This edition has derived from terms of GeoRef file Segments 4 and 5. Most of the terms in Segments 1-3 are also included, but we have been unable as yet to study the indexing in those Segments in detail in order to incorporate notes and cross-references for them. We intend to do so. Meanwhile, help in searching Segments 1-3 can be found in List A of the Guide to Indexing in the back of the Thesaurus.

Changes in this Edition

We have attempted to minimize changes in the Thesaurus. Where changes have been made we have documented them. A number of significant changes have been made between this edition and the last. These are categorized below and described in detail under the terms involved and in the lists in the Guide to Indexing.

In mid-1981, AGI will begin to exchange references in GeoRef for references from the GEODE and PASCAL data bases of Bureau de Recherches Geologiques et Minieres and Centre National[1] de la Recherche Scientifique.

[1] Available from the American National Standards Institute, 1430 Broadway, New York, N.Y. 10018.

This exchange will eliminate duplicate effort in preparation of citations among these data bases. Implementation of this exchange has caused some changes in the Thesaurus, best seen by referring to the lists in the Guide to Indexing.

The most extensive changes are shown in List O, Geographic terms, where many of the terms followed by (3) are new. These new terms together with the others in List O are used in conjunction with a special set of maps on which oceans and land masses have been systematically divided into regions, usually based on physiography or natural features. Their boundaries appear on the maps and the term for each region is given. The USSR, for example, has been divided into major and minor regions. The oceans have also been divided. At least one of the terms in List O is to be included by the indexer for each paper in which a location is mentioned. The most specific term applicable to the entire location in a paper is to be used.

Some common names for fossils were added and restricted to biostratigraphy citations (see List F). New terms for many commodities have been added while retaining the old terms for noneconomic papers (see the asterisked terms in List C). Plural forms of rock terms were added for use as grouping terms for papers on igneous rocks. In List H, terms labelled (2) are new except volcanic rocks, plutonic rocks, hypabyssal rocks and diabase. A number of pre-coordinated terms have been added for sedimentation environments and processes (see List N). And grouping terms have been added for islands such as Indian Ocean Islands (see List O).

These changes were undertaken to achieve a common set of index terms for the exchange of references. The changes were agreed to as part of the lengthy process of vocabulary comparison in which the parties to the exchange participated. The process began years ago with work on a multilingual thesaurus of geology terms done under the auspices of the International Union of Geological Sciences and the International Council of Scientific Unions-Abstracting Board. That work is still in progress. But much more intensive effort began on the French and English terms in 1979 in preparation for the exchange of references. This work was accomplished in meetings attended by G.N. Rassam for AGI, Jacques Gravesteijn for BRGM and Edith Lagarde for CNRS. Rassam and Gravesteijn had also worked for years previously on the multilingual thesaurus. The exchange will involve not only a common set of index terms but also computer translation between French and English versions of the terms.

Apart from the above changes, 277 new geographic terms and 40 non-geographic terms were added, based on frequency of use as candidate terms. Also added were terms for Deep Sea Drilling Project legs and sites.

Indexing

Prior to 1977 when the first edition of the Thesaurus appeared, the *Guide to Indexing* was the only tool available for GeoRef indexers and searchers. The Guide, which consists of structured lists of terms, has been revised and is included herein.

GeoRef indexing is in three-level sets (see the first page of the Guide), which follow definite rules which can be found under each first-level term in the Thesaurus.

Level-one and level-two terms have been tightly controlled in GeoRef, and a few changes have been made in terms on these levels. Level three has been more open-ended to accommodate specific locations, minerals, etc. Beginning in 1978, level three has been shortened to a single term. The other terms that had been included in level three became supplemental index terms. This change does not affect searching. Its effect on the printed indexes is evident by a comparison between 1978 and earlier publications. Briefly, the 1978 indexes have a single term plus title on level three instead of a term string. Also in 1978 cross-references were added to the index and selected supplemental terms to the citations.

A relatively small number of terms, fewer than 2,000, are valid on levels one and two. All are included herein. Also included are all systematic and geographic level-three terms used over four times in GeoRef and all non-systematic terms used over 24 times.

Autoposting

If a term appears in the indexing of a document, it does not necessarily follow that its broader terms will also appear. Certain broader terms *will* be added because they are called for by the set structures. For example, for most geographic, stratigraphic, and fossil narrower terms, one, but not every, broader term is required by the set structures. Other broader terms are added to facilitate computer searching.

As of January 1978, selected broader terms are autoposted (automatically added) by the computer. For example, each time Kansas is entered by an indexer, the computer adds United States.

Those broader terms which are autoposted are designated as BA or BZ terms. For example:

apatite
    BA phosphates
    BT minerals

Each time apatite is indexed, phosphates is autoposted, but minerals is not autoposted.

BA and BZ broader terms are autoposted. However, the reciprocal autoposting of narrower terms does not occur. In the above example, use of the term apatite causes phosphates to be autoposted. However, use of phosphates does *not* result in apatite being autoposted. Thus, narrower terms do not occur in citations simply because their broader terms were used. This assures they occur only when a concept at least as specific is found in the source document.

Autoposting is used selectively for:

a. Areas (see List O in the Guide)
b. Age and stratigraphic terms, excluding formations (see List E)
c. Fossils (see List F)
d. Igneous, sedimentary, and metamorphic rocks (see Lists H, I, J)
e. Sediments (see List N)
f. Minerals (see List L)
g. Soils (see List M)
h. Non-systematic terms (in a few cases, such as meteorites and faults)

Term Relationships

In term entries, relationships are indicated by the tags CO, UF, BT, BA, BX, BZ, NT, NA, NX, NZ, and SA.

a. **Geographic Coordinates**

In use since September 1977, coordinates (CO) are assigned to the principal geographic area(s) studied in a document.

A fixed-field format of thirty characters is used to represent coordinates. For example, the coordinates for Alabama, lat. N30°-N35°, long. W85°-W88°30′ are given as:

CO N300000N350000
W0850000W0883000

Principal geographic areas and a few others have coordinates herein. Any area is eligible and coordinates will be added for more areas. Coordinates for areas are indexed as needed. Those not appearing herein are looked up in an atlas and added to the citation.

Coordinates are used *in addition to* applicable area terms.

b. **Use/Used For**

The use/used for (UF) relationship indicates synonyms or alternate spellings of a term. For example:

columbium
    use niobium

niobium
    UF columbium

Ezan Cape
    use Esan Cape

Esan Cape
    UF Ezan Cape

Also, inverted forms of multiword terms have been cross-referenced when appropriate. For example:

control, erosion
    use erosion control

erosion control
    UF control, erosion

c. **Broader Term/Narrower Term**

The broader term/narrower term (BT-NT) relationship is that of a group to an individual in the group, e.g., BT minerals to NT silicates. It is a genus-species or class to member-of-the-class relationship. Variants of BT-NT are BA-NA, BX-NX, and BZ-NZ, explained below. A term can have several broader terms. The broader terms are arranged in ascending order, with each successive broader term being more inclusive. An example is:

peridot
    BA olinine group
        nesosilicates
        orthosilicates
        silicates
    BT minerals

In the above example, peridot is one of the specific rocks in the olivine group, the olivine group is one of several rock groups in the nesosilicates, the nesosilicates etc. All of a term's broader terms are shown, except for some area terms for which certain higher level broader terms are omitted. Examples are omission of Eastern U.S. and North America from the broader terms of Hanover County.

There is no reciprocal entry for each broader term. Referring to the above example of peridot, the reciprocals are:

olivine group
    NA peridot

nesosilicates
    NA olivine group

orthosilicates
    NA nesosilicates

silicates
    NA orthosilicates

minerals
    NT silicates

But in the case of narrower terms, not all levels of terms are given—only the next lower level. In the above example, of the five levels of terms between minerals and peridot, only the next lower level, silicates, is shown under minerals. Similarly, only the level containing orthosilicates is shown under silicates, etc.

A part-whole relationship is not allowed for narrower term-broader term pairs, e.g., caldera cannot have the broader term volcanoes. Two important exceptions are geographic and geologic age terms. Here part-whole relationships are the rule. Thus, Turin has the broader term Italy.

Variants of BT-NT are BA-NA, BX-NX, and BZ-NZ. In each of these, the basic relationship is that of broader term-narrower term, as described above, but with an added feature.

BA-NA means BT-NT *and* the BT is automatically posted by the computer each time the NT is used. For example, BA Mammalia under Primates means that if the term Primates is used to index a paper, then as a consequence, the term Mammalia is automatically added to the indexing for that paper, by the computer.

BX-NX means BT-NT *and* the BT is one of two or more BTs which are on the same level in a hierarchy or are from different hierarchies. The most frequent use of BX-NX is for counties of the United States. It is common in the United States that several states have counties of the same name. For example, both Kansas and New York have a Chautauqua County. The Thesaurus entry reads:

Chautauqua County
    Index states as applicable.
        BX Kansas
        BX New York
        BT United States

This entry indicates that *either* Kansas *or* New York may be the BT in any particular case, and the indexer should include the proper alternative.

BZ-NZ means BT-NT *and* the BT is autoposted and there are two or more BTs on the same level or in different hierarchies. Thus, BZ-NZ combines BA-NA and BX-NX. It seldom occurs in the Thesaurus. In the following example, Hawaii has BTs in three different hierarchies:

Hawaii
    BZ Pacific Ocean
        United States
        Polynesia

d. **See also**

The see also (SA) indicates related terms or terms often appearing together in GeoRef indexing other than broader terms and narrower terms. Examples are:

Gallup Sandstone
BT Upper Cretaceous
Cretaceous
SA Mesaverde Group
New Mexico

geochronology
SA absolute age

Notes for Terms

The following kinds of notes are included for terms:

a. Date Term Became Valid—if no date is given, the term is valid for GeoRef Segments 4 and 5. If a date *is* given, this is the year in which the term became officially valid in GeoRef. It may have been used before that date.

b. Set Level—The set level on which the term is currently used may be given. For example:

boudinage
Includes use on level 3 under sedimentary structures (1) soft sediment deformation (2). . .

(Boudinage can be found under soft sediment deformation, which is a second-level term under sedimentary structures.)

All valid index terms can be used on level three. Unless otherwise specified in a note, a term cannot be used on level one or two. List A in the Guide shows the valid first-level terms for all segments of GeoRef.

c. Set Structure—For each valid level-one term, the Thesaurus shows how the sets are to be structured.

d. Usage—The phrase "includes use," which occurs frequently in the notes, suggests an example of current and significant GeoRef usage, but is not meant as a listing of all possibilities. It means "this is an important use," but not "this is the only use."

e. Reference to the Guide—Notes may refer to the lists in the Guide to Indexing in the back of the Thesaurus.

f. Previously Used Term—If a term is no longer valid for indexing, a note to "also search" the term occurs under the preferred term entry.

g. Geography and Stratigraphy—Brief notes are given for most terms to assist the user in locating an area. Directions are abbreviated in these notes as N, E, S, and W.

h. Combined Terms in Sets—Usually only single terms occur on any level of a set. However, a few terms can occur together, separated by commas, such as mineral deposits, genesis. Explanatory notes are provided under each of these terms.

Searching

Much is said in this introduction about sets and levels. These relate to the printed index of the *Bibliography and Index of Geology* in which the level-one terms are the entry points and the entries are three-level sets of terms. Knowledge of the sets is most helpful in using the Bibliography.

In online searching sets and levels can be ignored to the extent that any term can be searched directly. However, a knowledge of the sets is helpful here in that it shows which sets are to be expected for certain subject areas, and which terms are appropriate for those sets.

Information on sets can be found in the explanatory notes which accompany the lists in the Guide in the back of the Thesaurus. The notes under the entries for each level-one term in the Thesaurus indicate how sets are structured for that term. Level-one terms also appear as a group in List A.

Under Igneous Rocks, List H, for example, the Explanatory notes tell the kinds of papers in which the term igneous rocks will be used on level one and under what circumstances the term may *not* be included for papers which include igneous rocks.

Term Validation

The Thesaurus on computer tape is the source of direct access validation files currently in use to check the spelling, capitalization, and level of terms. Index terms which don't validate can be accepted by overriding the validation. These become candidate Thesaurus terms when they have occurred five or more times.

Alphabetization

Terms are sorted word-by-word rather than letter-by-letter. Specifically:

- The sort is on letters, numbers, spaces and hyphens.
- Spaces sort before hyphens; hyphens sort before letters; letters sort before numbers.
- An open parenthesis is sorted as if it were a space.
- All characters, including punctuation marks, other than letters, numbers, hyphens and the open parenthesis, are squeezed out in the sort, i.e., are treated as if they did not exist.
- A few terms are identical except that one begins with a lower-case letter, one with an upper-case letter, e.g., alpine and Alpine. When this happens, the term with the lower-case letter precedes the other term.

Feedback

The computer file from which the Thesaurus was produced will be regularly updated for future editions. Users are encouraged to send in corrections and comments. Let us hear from you!

Authorities

The following publications are consulted as authorities for new GeoRef index terms:

General

Bates, Robert L. and Jackson, Julia A. (eds.), Glossary of Geology, 2nd ed. (Am. Geol. Inst., Falls Church, VA, 1980)

Geography

Seltzer, L.E. (ed.), Columbia Lippincott Gazetteer of the World (Columbia Univ. Press, New York, 1966)

U.S. Dept. of the Interior, Board on Geographic Names, Official Standard Names

Webster's New Geographical Dictionary (G & C Merriam Co., Springfield, MA, 1972)

Meteorites

Hey, Max H., Catalogue of Meteorites (British Museum, London, 1966)

Wasson, John T., Meteorites; Classification and Properties (Springer, Berlin, 1974)

Mineralogy (Silicates)

Strunz, H., Mineralogische Tabellen, 6th ed. (Akad. Verlag., Leipzig, 1977)

Mineralogy (Nonsilicates)

Fleischer, M., 1975 Glossary of Mineral Species (Mineral. Record. Inc., Bowie, MD)

Paleontology, Invertebrate

Moore, R.C. (ed.), Treatise on Invertebrate Paleontology, 1st ed. [2nd ed. for Part E (Archaeocyatha), Part V (Graptolithina), Part W (Trace Fossils)] (Geol. Soc. Am./Univ. Kansas Press)

Paleontology, Vertebrate

Romer, A.S., Vertebrate Paleontology, 3rd ed. (Univ. Chicago Press, 1974)

Gregory, Joseph T., et al., Bibliography of Fossil Vertebrates 1978 (Am. Geol. Inst., Falls Church, VA, 1981)

Stratigraphy

Lexique Stratigraphique International (Cent. Natl. Rech. Sci., Paris, 1950s-1960s [for United States, 1967 ed.])

Abbreviations

For term relationships:

- **CO** Geographical coordinates
- **BT** Broader term
- **BA** Autoposted broader term
- **BX** Broader term on same level with another broader term
- **BZ** Autoposted broader term on same level with another broader term
- **NT** Narrower term
- **NA** Autoposting narrower term
- **NX** Narrower term on same level with another narrower term
- **NZ** Autoposting narrower term on same level with another narrower term
- **SA** See also
- **UF** Used for

(Explanations of these relationships are in the Introduction.)

A

aa
use aa lava

aa lava
Term introduced in 1978. Before 1978, also search aa AND lava.
UF aa
BT lava
SA pahoehoe
volcanism

Aachen
City near the intersection of the Belgium and Netherland borders.
BT North Rhine-Westphalia
Germany

Aaland
use Aland

Aalenian
Europe. Includes use on level 3 under age terms(1). See list E.
BA Middle Jurassic
Jurassic
BT Mesozoic
SA Dogger
Lower Jurassic

Aar Massif
BT Switzerland

Aar Valley
River valley in central and N Switzerland. Also search Aare Valley.
UF Aare Valley
BT Switzerland

Aare Valley
use Aar Valley

Aargau
Canton in N.
BT Switzerland

Abakan
Town on Yenisei River SW of Krasnoyarsk in Khakass Autonomous Oblast.
BT Russian Republic
USSR

Abashiri
City in NE Hokkaido.
UF Abasiri
BT Hokkaido
Japan

Abasiri
use Abashiri

Abee
Village in central.
BT Alberta
Canada

Aberdeen
City on North Sea.
BT Scotland
Great Britain
United Kingdom

Aberdeenshire
County in NE.
CO N565000N574300
W0014500W0034800
BT Scotland
Great Britain
United Kingdom

Aberystwyth
Town on Saint Georges Channel.
BT Wales
Great Britain
United Kingdom

Aberystwyth Grits
Upper Llandoverian. Forms part of the Ystwyth Stage. N and central Wales and NW England.
BT Silurian
SA England
United Kingdom
Wales

Abilene
Arch in central Kansas.
UF Abilene Arch
BT Kansas
United States

Abilene Arch
use Abilene

Abitibi
County on James Bay.
BT Quebec
Canada

Abitibi Belt
Term introduced in 1981. Structural subprovince in SW Quebec and E Ontario.
CO N460000N503000
W0735000W0850000
BT Superior Province
Canadian Shield
SA Ontario
Quebec

Abkhasia
use Abkhazia

Abkhazia
Abkhaz Autonomous Soviet Socialist Republic. Also search Abkhasia.
CO N422000N432000
E0421500E0400000
UF Abkhasia
BT Georgian Republic
USSR

ablation
Includes use on level 3 under sedimentation(1).
SA glaciers
mass balance
sedimentation
wind erosion
wind transport

Abo
use Turku

abrasion
UF mechanical erosion
SA detritus
erosion
glaciation
grinding
planation

abrasives
General. Includes use as level 3 commodity term under industrial minerals(1). See list C.
SA corundum deposits
diamonds
diatomite
garnet deposits
industrial minerals
pumice deposits
silica

Abruzzi
Autonomous region on the Adriatic Sea.
CO N414000N425500
E0144800E0130000
BT Italy
NT Chieti
Marsica

Absaroka Mountains
use Absaroka Range

Absaroka Range
Range of the Rocky Mountains. Index states as applicable. Also search Absaroka Mountains.
CO N433000N450000
W1090000W1101500
UF Absaroka Mountains
BT United States
SA Beartooth Mountains
Montana
Rocky Mountains
Wyoming

absarokite
As of 1981, includes use on level 3 under igneous rocks(1) basalts(2). See list H. Before 1981, included use on level 3 under igneous rocks(1) basalt family(2).
BA basalts
volcanic rocks
BT igneous rocks

absolute age
For radiometric or radiogenic (isotopic) dating; for non-isotopic dating, see geochronology. Includes use on level 1; on level 2 under orogeny(1). If 1, term set options are:
dates
material [rock group, rock type, mineral name (e.g. charcoal, granite, metamorphic rocks, shells, sediments)]
methods
name of method [Ar/Ar, C-14, H-3, He-4/He-3, Io/Th, K/Ar, Pb-alpha, Pb/Pb, Pb/Th, Pb-210, Re/Os, Sr/Rb, Th/Th, Th/U, U/He, U/Pb, U-238/Pb-206, U-235/Pb-207, U/Th/Pb, uranium disequilibrium]
techniques
subtopic [e.g. instruments, models, sample preparation, sampling]
topic [applications, bibliography, catalogs, interpretation, philosophy]
subtopic
UF actual age (absolute age)
SA age
Ar/Ar
C-14
changes of level
charcoal
dates
diffusion
geochemistry
geochronology
He-4/He-3
Io/Th
Io/U
isochrons
isotopes
K/Ar
new methods
orogeny
Pb-210
Pb/Pb
Pb/Th
radioactive decay
relative age
Sr/Rb
standard materials
Th/Th
Th/U
U-238/Pb-206
U/He
U/Pb
U/Th/Pb
uranium disequilibrium
whole rock

absorbent materials
Term introduced in 1981.
SA industrial minerals
materials

absorption
Includes use on level 2 under aurora(1); on level 3 under spectroscopy(1) methods(2); on level 3 under geochemistry(1) processes(2).
SA absorption and scattering
adsorption
atomic absorption
aurora
emission spectroscopy
geochemistry
sorption
spectroscopy

absorption and scattering
Includes use on level 2 under aeronomy(1).
SA absorption
aeronomy
aurora
scattering

absorption spectroscopy
Not a valid index term. Use absorption under spectroscopy(1).

Abu Dhabi
Sheikdom. One of federation of 7 states at S end of Persian Gulf.
BT United Arab Emirates
Arabian Peninsula

Abukuma Mountains
NE of Utsunomiya in E central Honshu. Also search Abukuma Plateau.
UF Abukuma Plateau
BT Honshu
Japan

Abukuma Plateau
use Abukuma Mountains

abundance
Includes use on level 2 under chemical elements (list D) and under isotopes(1). Before 1981, also included use on level 2 under commodity terms. As of 1981, use only in the chemical sense.
SA organic materials

abyssal cones
use submarine fans

abyssal environment
As of 1981, no longer a valid term for GeoRef. Use deep-sea environment. Included use on level 3.

abyssal fans
use submarine fans

abyssal plains
Includes use on level 3 under oceanography(1).
SA continental rise
ocean basins
ocean floors
oceanography
plains

Abyssinian Rift valley
use Ethiopian Rift

abyssolith
use batholiths

Ac
use actinium

Acadian
Provincial series, Canada. Includes use on level 3 under age terms(1). See list E.
BA Middle Cambrian
Cambrian
BT Paleozoic
NT Lancara Formation
SA Acadian Phase
Canada

Acadian Orogeny
use Acadian Phase

Acadian Phase
Use Acadian for the age. Before 1978, also search Acadian Orogeny. Also search orogeny AND Acadian.
UF Acadian Orogeny
BT Devonian
SA Acadian
Antler Orogeny
orogeny

Acantharina
Suborder. Includes use on level 2 under Radiolaria(1). See list F.

BA Porulosida
Radiolaria
BT Protista
Invertebrata

acanthite
BA sulfides
BT minerals
SA argentite

Acanthodes
Genus. Includes use on level 3 under Pisces(1) Osteichthyes(2).
BA Acanthodii
Osteichthyes
Pisces
BT Vertebrata

Acanthodii
As of 1981, includes use on level 3 under Pisces(1) Osteichthyes(2). See list F. Before 1981, included use on level 3 under Pisces(1) Placodermi(2). Before 1981, also search acanthodians. As of 1981, search acanthodians only for biostratigraphy.
BA Osteichthyes
Pisces
BT Vertebrata
NA Acanthodes

acanthopores
SA Bryozoa

Acceglio
Village in S.
BT Italy
SA Piemonte

accelerograms
Includes use on level 3 under earthquakes(1). Also search acceleration; accelerographs; accelerometers.
SA earthquakes
engineering geology
seismology

accelerometers
Term introduced in 1981.
SA engineering geology
instruments
seismology

accessory minerals
Includes use on level 3 under petrology(1). Before 1978, search minerals AND accessory.
BT minerals
SA heavy minerals

Accomac County
use Accomack County

Accomack County
On the Delmarva Peninsula. Also search Accomac County.
UF Accomac County
BT Virginia
United States

Accra
City on the Gulf of Guinea.
BT Ghana

accretion
Term used for sedimentation through 1977. After 1977, term includes use under planetology(1) as genetic concept for the Moon and planets.
SA deposition
Moon
planetology
sedimentation

accumulation
As of 1981, no longer a valid term for GeoRef. Before 1981, used as a general term.

accuracy
Used as a general term and for precision. Also search precision.
SA calibration
corrections
efficiency
errors
reliability

Acer
Genus. Includes use on level 3 under angiosperms(1) Dicotyledoneae(2).
BA Dicotyledoneae
angiosperms
BT Spermatophyta
Plantae

Acheulean
use Acheulian

Acheulian
Archaeologic classification.
UF Acheulean
Acheullian
BA Paleolithic
BT Cenozoic

Acheullian
use Acheulian

achondrites
Includes use on level 3 under meteorites(1).
BT meteorites
NA angrite
chassignite
chladnite
diogenite
howardite
nakhlite
ureilite
SA chondrites
eucrite
iron meteorites
stony irons

acid mine drainage
Before 1978, also search drainage AND mines.
SA drainage
environmental geology
mine drainage
mines
pollution
sulfuric acid

acidic
A valid level 2 index term through 1977. After 1977, use acidic composition on level 2 under igneous rocks(1).

acidic composition
Term introduced in 1978. Includes use on level 2 under igneous rocks(1). Before 1978, also search acidic.
SA acidic magmas
acids
composition
igneous rocks
pH

acidic magmas
Term introduced in 1981.
SA acidic composition
magmas

acidity
use pH

acids
SA acidic composition
amino acids
compounds
fatty acids
fulvic acids
humic acids
inorganic acids
organic acids
pH
sulfuric acid

acmite
BA clinopyroxene
pyroxene group
chain silicates
silicates
BT minerals
SA aegirine

Acoela
Term introduced in 1981. Includes use on level 2 under Mollusca(1). See list F.
BA Gastropoda
Mollusca
BT Invertebrata

Aconcagua Province
Central Chile. Also search Aconcagua.
BT Chile
NT La Ligua

acoustic logging
use acoustical logging

acoustic methods
use acoustical methods

acoustic surveys
use acoustical surveys

acoustic waves
use acoustical waves

acoustical
A valid level 2 index term through 1977. After 1977, use acoustical logging on level 2 under well-logging(1).

acoustical logging
Not a valid index term from 1975 to 1977. After 1977, includes use on level 2 under well-logging(1). Before 1978, also search well-logging AND acoustic; acoustic logging.
UF acoustic logging
logging, acoustical
BA well-logging
SA acoustical methods
acoustical surveys

acoustical methods
Includes use on level 2 under geophysical methods(1). Also search acoustic methods; acoustic.
UF acoustic methods
BA geophysical methods
SA acoustical logging
acoustical surveys
deep-tow methods
echo sounding
methods
side-scanning methods
sonar methods

acoustical properties
Includes use on level 3 under geophysical methods(1).
SA geophysical surveys
physical properties
properties
seismology

acoustical surveys
Includes use on level 3 under area terms(1) geophysical surveys(2). Also search acoustic surveys; acoustic.
UF acoustic surveys
BA geophysical surveys
BT surveys
SA acoustical logging
acoustical methods
sonar methods

acoustical waves
Also search acoustic AND waves; acoustical AND waves; sonic waves; sound waves.
UF acoustic waves
sonic waves
sound waves
BA elastic waves
SA seismology
waves

Acqui
Town in SE.
BT Italy
SA Piemonte

acquisition, data
use data acquisition

Acre
Term introduced in 1981. State in W Brazil.
CO S110000S070000
W0663000W0740000
BT Brazil

Acreodi
Term introduced in 1981. Includes use on level 2 under Mammalia(1). See list F.
BA Mammalia
BT Tetrapoda
Vertebrata

acritarch flora
Term introduced in 1981. To be used for acritarchs when they are used as tools in stratigraphy. See list F. Before 1981, search acritarchs AND biostratigraphy.
SA acritarchs
biostratigraphy

acritarchs
Hystrichosphaeridae is included here and under Dinoflagellata. Includes use on level 2 under palynomorphs(1). See list F.
BA palynomorphs
NA Baltisphaeridium
NZ Hystrichosphaeridae
SA acritarch flora
Dinoflagellata

Actiniaria
Includes use on level 2 under Coelenterata(1). See list F.
BA Anthozoa
Coelenterata
BT Invertebrata

Actinistia
Term introduced in 1981. Includes use on level 2 under Pisces(1). See list F.
BA Osteichthyes
Pisces
BT Vertebrata

actinium
Includes use on level 1 and 2 as a chemical element (list D).
UF Ac
SA elements

Actinodontida
Term introduced in 1981. Includes use on level 2 under Mollusca(1). See list F.
BA Bivalvia
Mollusca
BT Invertebrata

actinolite
BA clinoamphibole
amphibole group
chain silicates
silicates
BT minerals
SA actinolite facies
asbestos

actinolite facies
Term introduced in 1978.
BT facies
SA actinolite
metamorphic rocks

Actinopterygii
Includes use on level 3 under Pisces(1) Osteichthyes(2). See list F.
BA Osteichthyes
Pisces
BT Vertebrata
NA Chondrostei
Holostei
Teleostei

action, frost
use frost action

activation analysis
Including field applications and instruments. Includes use on level 3 under chemical analysis(1) methods(2). Also search activation.
UF radioactivation analysis
SA analysis
chemical analysis
isotopes
neutron activation analysis

activation energy
SA energy
particles

active faults
BA faults

active layer
As of 1978, term is used on level 2 under permafrost(1).
UF annually thawed layer
layer, active
mollisol
SA permafrost
soils

activity
Includes use on level 3 under geochemistry(1).
SA geochemistry

activity, igneous
use igneous activity

actual age (absolute age)
use absolute age

Acungui Group
BT Precambrian
SA Brazil
Parana

Ada County
SW Idaho.
BT Idaho
United States

Adak Island
In central part of Aleutian Islands SW of Alaska Peninsula.
BT Alaska
United States

Adalia
use Antalya

Adamawa
Administrative region in N.
BT Cameroon

adamellite
As of 1981, includes use on level 3 under igneous rocks(1) granites(2). See list H. Before 1981, included use on level 3 under igneous rocks(1) granite-granodiorite family(2).
BA granites
BT igneous rocks

Adamello Massif
In Rhaetian Alps in N.
BT Italy

adamite
BA arsenates
BT minerals

Adamow Mine
In W central part of country.
BT Poznan
Poland

Adams County
Index states as applicable.
BX Colorado
Idaho
Illinois
Indiana
Iowa
Mississippi
Nebraska
North Dakota
Ohio
Pennsylvania
Washington
Wisconsin
BT United States

Adana
Province in S Anatolia. Also a city.
BT Turkey
Middle East

adaptation
Includes use on level 3 under ecology(1) or paleoecology(1).
SA ecology
paleoecology
paleontology

adaptive filters
Term introduced in 1981.
BT filters

Adavale Basin
S central.
BT Queensland
Australia

Addis Ababa
City in central part of country.
BT Ethiopia

Addison County
Term introduced in 1981. W Vermont.
CO N434500N441800
W0724400W0732500
BT Vermont
United States

Adelaide
City on Gulf Saint Vincent in SE.
BT South Australia
Australia

Adelaide Geosyncline
Also search Adelaide.
BT South Australia
Australia
SA geosynclines

Adelaide Island
Off W coast of Antarctic Peninsula S of Cape Horn.
BT Antarctica

Adelaidean
SE Australia. See list E.
UF Adelaidean system
BA Precambrian
SA upper Precambrian
upper Proterozoic

Adelaidean system
use Adelaidean

Adelie Coast
In the French Sector on the Indian Ocean side S of Australia. Also search Adelie Land.
UF Adelie Land
BT Antarctica

Adelie Land
use Adelie Coast

Aden
City on Gulf of Aden. Former British Colony.
BT Southern Yemen
Arabian Peninsula

adiabatic demagnetization
Term introduced in 1981. Includes use on level 3 under paleomagnetism(1) and geophysics(1).
BT demagnetization
SA geophysics
magnetization
paleomagnetism
remanent magnetization

Adirondack Anorthosite
NE New York.
BT Precambrian
SA New York

Adirondack Mountains
NE New York. Also search Adirondacks.
CO N430000N443000
W0733000W0750000
UF Adirondacks
BT New York
United States

Adirondacks
use Adirondack Mountains

Admire
Village in Lyon County in E central.
BT Kansas
United States

Adour Basin
River basin in extreme SW France. Also search Adour.
BT France

Adrar
Interior region in W central near Western Sahara border.
BT Mauritania

Adriatic Coast
use Adriatic region

Adriatic region
Index countries as applicable. Also search Adriatic Coast.
UF Adriatic Coast
BT Europe
SA Albania
Italy
Yugoslavia

Adriatic Sea
Between Italy on the W, and Albania and Yugoslavia on the E. Includes use on level 1 as an area term (list O). For term set options see list B.
CO N395000N454500
E0200000E0121500
BA East Mediterranean
Mediterranean Sea
SA Mediterranean region

adsorption
Includes use on level 3 under geochemistry(1) processes(2).
SA absorption
chromatography
clay mineralogy
desorption
geochemistry
processes
solution
sorption

adularia
BA feldspar group
framework silicates
silicates
BT minerals
SA aluminosilicates
K-feldspar
orthoclase

Adzhar
use Adzharistan

Adzharia
use Adzharistan

Adzharistan
Adzhar Autonomous Soviet Socialist Republic in SW Georgian Republic. Also search Adzhar; Adzharia.
CO N410000N411500
E0423000E0411500
UF Adzhar
Adzharia
BT Georgian Republic
USSR

Aegean Islands
BT Mediterranean region
SA Greek Aegean Islands

Aegean Sea
Between Greece on the W and Turkey on the E. As of 1977 includes use on level 1 as an area term (list O). See list B for term set options.
CO N360000N410000
E0282000E0230000
BA East Mediterranean
Mediterranean Sea

aegerine
use aegirine

Aegina
Town and island in Saronic Gulf. Central Greece and Euboea.
BT Greece

aegirine
UF aegerine
BA clinopyroxene
pyroxene group
chain silicates
silicates
BT minerals
SA acmite

aenigmatite
UF enigmatite
BA chain silicates
silicates
BT minerals
SA rhonite

aeolian environment
use eolian environment

Aeolian Islands
use Lipari Islands

aeolotropy
use anisotropy

aerial photography
As of 1978, term is used on level 2 under remote sensing(1). Also search aerial; aerial photographs.
SA cartography
geomorphology
landform description
maps
photogeology
photogrammetry
photography
remote sensing

aerobic environment
Term introduced in 1978. Before 1978, search aerobic.
SA anaerobic environment
environment
paleoecology

aeromagnetic surveys
A valid term through 1973. Use magnetic surveys and airborne methods. Also search aeromagnetic.

aeronomy
Usually out-of-scope for GeoRef. Includes use on level 1. Term set options are:
absorption and scattering
subtopic
composition
element or compound
densities and temperatures
element or compound
diffusion
mechanism
instruments
name of instrument
ionization
latitude
source of energy [e.g. cosmic rays, X-rays]
techniques
name of technique
tides
subtopic
turbulence
type
waves
type
winds
altitude by region [e.g. stratosphere, troposphere]
SA absorption and scattering
atmosphere
aurora
cosmic rays
densities and temperatures
diffusion

interplanetary space
ionization
ionosphere
magnetosphere
meteorology
stratosphere
tides
troposphere
turbulence
waves
winds

aerosols
Includes use on level 2 under meteorology(1).
SA air-sea interface
atmosphere
convection
meteorology
particles
radioactive tracers
turbidity

aeschynite
As of 1981, includes use on level 3 under minerals(1) niobates(2). See list L. Before 1981, included use on level 3 under minerals(1) oxides(2).
UF aschynite
eschinite
eschynite
BA niobates
oxides
BT minerals

Afar
Large desert region mostly in Ethiopia. Also search Danakil. Index Djibouti and/or Ethiopia as applicable.
UF Danakil
Dankalia
BT Africa
SA Djibouti
Ethiopia

Afar Depression
Part of Afar lying between the "Danakil Alps" and the Ethiopian highlands in NE Ethiopia. Section of the Great Rift Valley. Also search Afar; Afar Rift; Danakil; Danakil Depression.
UF Afar Rift
Danakil Depression
BT Ethiopia
SA East African Rift
Great Rift Valley

Afar Rift
use Afar Depression

Afars and Issas Territory
A valid level 1 term through 1977. After 1977, use Djibouti.

affinities
Includes use on level 2 or 3 under commodity terms (list C); used mostly on level 3 under fossil groups(1). See list F. Also search geochemical affinities.

Afghan-Tadzhik Basin
use Afghan-Tadzhik Depression

Afghan-Tadzhik Depression
Index Afghanistan and/or Tadzhikistan as applicable. Also search Afghan-Tadzhik; Afghan-Tadzhik Basin.
UF Afghan-Tadzhik Basin
BT Asia
SA Afghanistan
Tadzhik Depression
Tadzhikistan

Afghanistan
Includes use on level 1 as an area term (list O). For term set options see list B.
CO N290000N381500
E0750000E0600000
BA Indian Peninsula
Asia
NT Altimur Mountains
Bamian
Dasht-i-Nawar
Ghazni
Kabul
Logar
Paktia
Wardak
SA Afghan-Tadzhik Depression
Amu Darya
Badakhshan
Hindu Kush
Murgab Basin
Turkestan

Africa
Includes use on level 1 or 2 as an area term (list O). For term set options see list B. To retrieve all documents, individual countries and physiographic regions should also be searched (see list O).
CO S350000N370000
E0510000W0180000
NT Afar
African Platform
Atlas Mountains
Benue Valley
Blue Nile
Cap Blanc
NZ Cape Verde Islands
NA Central Africa
NT Chad Basin
Congo Basin
Congo River
NA East Africa
NT East African Lakes
East African Rift
Gregory Rift
Kalahari Desert
Kasai River
Lake Chad
Lebombo Mountains
Libyan Desert
Limpopo Basin
Limpopo Belt
Logone River
NZ Malagasy Republic
NT Mali-Niger Syneclise
Namaqualand
Niger River
Niger Valley
Nile River
Nile Valley
Nimba Mountains
NA North Africa
NT Nubia
Nubian Shield
Orange River
Reguibat Ridge
NA Sahara
NT Sahel
Senegal Basin
Senegal River
NA Southern Africa
NT Tanezrouft
Taourirt
Umba
Volta Basin
NA West Africa
NT West African Shield
Zambezi Valley
SA African Plate
Dwyka Series
Gondwana
Karroo System
Katangan Orogeny
Nubian Sandstone
Tethys

African Plate
Includes the continent of Africa, adjoining areas of the Atlantic and Indian oceans, the Malagasy Republic, and much of E Mediterranean Sea.
SA Africa
plate tectonics
plates

African Platform
A vast platform of complex Precambrian and Paleozoic rocks, partly covered by Mesozoic and Tertiary sedimentary rocks, which underlies the entire continent.
BT Africa

aftershocks
Includes use on level 2 under earthquakes(1).
SA earthquakes
focus
foreshocks
seismology

Aftonian
Interglacial interval. North America.
BA lower Pleistocene
Pleistocene
Quaternary
BT Cenozoic

Ag
use silver

Agades
Town in W central Niger. Also search Agadez.
UF Agadez
BT Niger

Agadez
use Agades

Agarak
Mining region in central.
BT Armenia
USSR

agate
BA silica minerals
framework silicates
silicates
BT minerals
SA chalcedony
gems
quartz

age
Includes use as level 2 or 3 term appropriate to a large number of topics, e.g. on level 2 under ocean basins(1) and under ground water(1). See list G.
SA absolute age
exposure age
relative age

Ager Formation
Catalonia and the Pyrenees.
BT lower Eocene
Eocene
SA Spain

agglomerate
Includes use on level 3 under igneous rocks(1) or sedimentary rocks(1).
SA breccia
conglomerate
igneous rocks
pyroclastics
sedimentary rocks

agglutinates
SA fines
Moon
particles

aggradation
SA geomorphology
processes

aggregate
Includes use as level 3 commodity term under construction materials(1). See list C.
UF lightweight aggregate
BA construction materials
SA gravel deposits
pumice deposits
rocks
sand
sands

aggregation
A valid term through 1975. After 1975, use sedimentation.

Agly Massif
S part of country.
BT Pyrenees-Orientales
France

agmatite
Includes use on level 3 under metamorphic rocks(1) migmatites(2). See list J.
BA migmatites
BT metamorphic rocks

Agnatha
As of 1981, includes use on level 1 and 2 as a fossil term (list F). Before 1981, included use on level 2 under Pisces(1).
BT Vertebrata
NA Heterostraci
Ostracodermi
SA Pisces

Agnes
A storm in 1972 on the East Coast of United States.
SA hurricanes

Agnostida
Includes use on level 2 under Trilobita(1). See list F.
BA Trilobita
BT Trolobitomorpha
Arthropoda
Invertebrata

Agnotozoic
use Proterozoic

agpaite
As of 1981, includes use on level 3 under igneous rocks(1) syenites(2). See list H. Before 1981, included use on level 3 under igneous rocks(1) syenite family(2).
BA syenites
BT igneous rocks

Agricola Lake
BT Northwest Territories
Canada

agricultural waste
Term introduced in 1978. Before 1978, search agricultural.
UF waste, agricultural
SA agriculture
waste disposal

agriculture
As of 1978, term is used on level 2 under land use(1).
SA agricultural waste
fertilization
fertilizers
land use
soils
utilization

Agto
Fishing settlement on small island in Davis Strait.
BT Greenland
Arctic region

Agua Blanca Fault
Index Baja California and/or California.
SA Baja California
California

Agulhas Bank
South of Cape of Good Hope.
BT Indian Ocean

Ahaggar
Volcanic upland in SE Algeria. Also search Ahaggar Mountains.
CO N194000N264800
E0110000E0005900
UF Ahaggar Mountains

Hoggar Mountains
BT Algeria
SA Sahara

Ahaggar Mountains
use Ahaggar

ahermatypic taxa
Term introduced in 1978 on level 3 under fossil groups (list F). Before 1978, search ahermatypic.
UF taxa, ahermatypic
SA Coelenterata
hermatypic taxa

Ahnet
Region in S central.
BT Algeria

Aichi
Prefecture in central Honshu. Capital is Nagoya.
BT Honshu
Japan

aid projects
Term introduced in 1981.
SA economic agreements
economics
international cooperation

Aigoual Massif
In Cevennes Mountains in S.
BT France

Aiguilles Rouges
Alpine range in SE.
BT France

Aiken
City in Aiken County in W.
BT South Carolina
United States

Aiken County
W South Carolina.
BT South Carolina
United States

aikinite
BA sulfides
BT minerals

Ain
Department in E France.
CO N453500N463000
E0061000E0044000
BT France

air
As of 1978, term is used on level 2 under pollution(1).
SA atmosphere
pollution

air-sea interface
Includes use on level 3 under oceanography(1) or meteorology(1) aerosols(2).
UF interface, air-sea
SA aerosols
atmosphere
interfaces
oceanography

Aira Caldera
BT Kyushu
Japan

airborne
A valid index term through 1977. After 1977, use airborne methods.

airborne methods
Term introduced in 1978. Includes use on level 3 under geophysical methods(1) magnetic methods(2). Before 1978, also search airborne.
BA geophysical methods
SA geophysical surveys
magnetic methods
magnetic surveys
methods
remote sensing

airfields
Term introduced in 1981.
SA engineering geology
environmental geology
land use

airglow
As of 1981, no longer a valid term for GeoRef. Before 1981, included use on level 2 under ionosphere(1).

Airy waves
Term introduced in 1978. Includes use on level 3 under seismology(1).
BA elastic waves
SA seismology
surface waves
waves

Aisne
Department in N.
CO N484500N501000
E0041000E0010000
BT France
SA Oise River valley
Somme River valley

Aix-en-Provence
City in S part of country.
BT Bouches-du-Rhone
France

Aizu Basin
Fukushima Prefecture in N central Honshu. Also search Aizu.
BT Honshu
Japan

Ajay River
Index Indian states as applicable.
BT India
SA Bihar
West Bengal

Ajman
Sheikdom. One of federation of 7 states at S end of Persian Gulf.
BT United Arab Emirates
Arabian Peninsula

Ajmer
City and district in central.
BT Rajasthan
India

akaganeite
BA oxides
BT minerals

Akbastau
Village NE of Chimkent in S central.
BT Kazakhstan
USSR

Akchagylian
Europe. See list C.
BA upper Pliocene
Pliocene
Neogene
Tertiary
BT Cenozoic
SA upper Tertiary

Akchatau
Town in SE Karaganda Oblast in E central.
BT Kazakhstan
USSR

Akenobe Mine
Hyogo Prefecture of which Kobe is the capital.
BT Honshu
Japan

Akera
River in SW.
BT Azerbaidzhan
USSR

akermanite
BA melilite group
sorosilicates
orthosilicates
silicates
BT minerals

Akhaltsikh
use Akhaltsikhe

Akhaltsikhe
City.
UF Akhaltsikh
BT Georgian Republic
USSR

Akhtala
Town in N.
BT Armenia
USSR

Akita
City on Japan Sea in N.
BT Honshu
Japan

Akiyoshi
Village in extreme SW.
BT Honshu
Japan

Akiyoshi Limestone
SW Honshu.
BT Permian
SA Honshu
Japan

Akjoujt
Village in W central.
BT Mauritania

Akron
City in Summit County in NE.
BT Ohio
United States

Aksu
Village in Taldy Kurgan Oblast in SE.
BT Kazakhstan
USSR

Aktyubinsk
City and oblast in NW.
BT Kazakhstan
USSR

Akzhal
Town in E Semipalatinsk Oblast in E.
BT Kazakhstan
USSR

Al
use aluminum

Al-Kufrah
use Kufra Basin

Al-Qusayr
use Kosseir

Al-Quseir
use Kosseir

Al-26
Includes use on level 3 under isotopes(1).
SA aluminum
isotopes

Ala-Kul Lake
use Alakol

Alabama
Includes use on level 1 as an area term (list O). For term set options see list B.
CO N300000N350000
W0850000W0883000
BA Southern U.S.
United States
NT Alabama River
NX Birmingham
Blount County
Butler County
Calhoun County
Cherokee County
NT Chilton County
NX Clarke County
Clay County
Coffee County
NT Coosa County
NX Dallas County
NT Escambia County
Etowah County
NX Fayette County
Franklin County
Greene County
NT Highland Rim
NX Jackson County
Jefferson County
Lawrence County
Lee County
Limestone County
Macon County
Madison County
Marion County
Marshall County
NT Martin Lake
Mobile
Mobile County
NX Monroe County
Montgomery County
Morgan County
Perry County
Pickens County
Pike County
Randolph County
NT Red Mountain
NX Russell County
Saint Clair County
Shelby County
NT Talladega Front
Tallapoosa County
Tuscaloosa County
NX Washington County
SA Appalachian Basin
Appalachian Plateau
Bangor Limestone
Black Warrior Basin
Brevard Zone
Chattahoochee River
Chattanooga Shale
Chesterian
Chickamauga Group
Chipola Formation
Citronelle Formation
Claiborne Group
Clayton Formation
Cumberland Plateau
Fernvale Formation
Fort Payne Formation
Gulf Coastal Plain
Jackson Group
Marianna Limestone
Midway Group
Mississippi Embayment
Monteagle Limestone
Ocala Group
Pennington Formation
Piedmont
Pottsville Group
Red Mountain Formation
Ripley Formation
Rome Formation
Saint Louis Limestone
Sainte Genevieve Limestone
Shady Dolomite
Smackover Formation
Talladega Group
Tennessee River
Tennessee Valley
Tuscaloosa Formation
Valley and Ridge Province
Vicksburg Group
Warsaw Formation
Yazoo Clay

Alabama River
Central and SW.
BT Alabama
United States

alabandine
use alabandite

alabandite
Also search alabandine.
UF alabandine
manganblende
BA sulfides
BT minerals

alabaster
UF onyx marble
BA sulfates
BT minerals
SA aragonite

calcite
gypsum

Alacran Reef
Off north Yucatan coast.
BT Gulf of Mexico

Alae Crater
Kilauea volcano on island of Hawaii.
BT Hawaii
United States
SA Kilauea

Alagoas
State in NE Brazil.
CO S103000S084500
W0350000W0382000
BT Brazil
SA Sao Francisco Basin
Sergipe-Alagoas Basin

Alai Range
One of the W ranges of the Tien Shan in SW Kirghizia. Also search Alai.
UF Alay Range
BT Kirghizia
USSR
SA Darvaza Range
Hissar Range

Alaia
use Alanya

Alais
use Ales

Alaiye
use Alanya

Alakol
Lake in Taldy Kurgan Oblast E of Lake Balkhash in E Kazakhstan. Also search Ala-Kul Lake.
UF Ala-Kul Lake
BT Kazakhstan
USSR

Alameda County
On San Francisco Bay.
CO N373000N375000
W1213000W1222500
BT California
United States

Alaminos Canyon
BT Gulf of Mexico

Aland
Islands belonging to Finland. Also search Aland Islands; Aland Archipelago.
UF Aaland
Aland Archipelago
Aland Islands
BT Baltic Sea
Atlantic Ocean

Aland Archipelago
use Aland

Aland Islands
use Aland

Alandroal
Town in S central part of country.
BT Evora
Portugal

Alanya
Town on the Gulf of Adalia in S Anatolia.
UF Alaia
Alaiye
Alaya
BT Turkey
Middle East

Alaska
Includes use on level 1 as an area term (list O). For term set options see list B.
CO N510000N720000
W1300000E1730000
BA Western U.S.
United States
NT Adak Island
Alaska Peninsula
Alaska Range
Aleutian Islands
Amchitka Island
Anchorage
Augustine
Barrow
Birch Creek
Blying Sound Quadrangle
Brooks Range
Cannikin
Chandalar Quadrangle
Chignik Quadrangle
Chugach Mountains
Colville River
Colville River delta
Cook Inlet
Delta River
Fairbanks
Fairweather Fault
Glacier Bay
Goodnews Quadrangle
Hagemeister Island Quadrangle
Juneau
Juneau ice field
Katmai National Monument
Kenai Peninsula
Ketchikan Quadrangle
Knik Arm
Kodiak Island
Martin River Glacier
Matanuska Valley
Meade Basin
Montague Island
Nome
North Slope
Nunivak Island
Philip Smith Mountains Quadrangle
Point Barrow
Prince Rupert Quadrangle
Prince William Sound
Prudhoe Bay
Rat Island
Rat Islands
Seward Peninsula
Seward Quadrangle
Shemya Island
Sitka Sound
Susitna River basin
Sutwik Island Quadrangle
Talkeetna Mountains Quadrangle
Talkeetna Quadrangle
Trans-Alaska Pipeline
Valdez
Valley of Ten Thousand Smokes
Wrangell Mountains
Yakutat Bay
Yukon-Tanana Upland
SA Arctic region
Bowser Formation
Chitistone Pass
Coast Mountains
Coast Ranges
Denali Fault
Kenai Group
Lisburne Group
Prince of Wales Island
Rampart Group
Rocky Mountains
Sadlerochit Formation
Saint Elias Mountains
Tintina fault zone
White Mountain
White River
Yukon River

Alaska Peninsula
SW Alaska.
BT Alaska
United States

Alaska Range
S central.
BT Alaska
United States
SA North American Cordillera

alaskite
As of 1981, includes use on level 3 under igneous rocks(1) granites(2). See list H. Before 1981, included use on level 3 under igneous rocks(1) granite-granodiorite family(2).
BA granites
BT igneous rocks

Alava
One of the Basque Provinces in N.
CO N422900N431200
W0021400W0031500
BT Basque Provinces
Spain

Alaverdi
City in N Armenia.
UF Allaverdy
BT Armenia
USSR

Alay Range
use Alai Range

Alaya
use Alanya

Alazani
Mining and oil producing area in E.
BT Georgian Republic
USSR

Alba-Iulia
Town in Alba County in SW.
BT Transylvania
Romania

Albacete
Province in SE.
CO N380200N392700
W0005500W0025500
BT Murcia region
Spain

Alban Hills
Part of the Lower Apennines SE of Rome.
BT Latium
Italy
SA Apennines

Albania
Includes use on level 1 as an area term (list O). For term set options see list B.
CO N393500N424000
E0210500E0191500
BA Southern Europe
Europe
NT Albanides
Ionian Zone
SA Adriatic region
Balkan Peninsula
Mediterranean region

Albanides
Mountain area in W and SW.
BT Albania
SA Alps

Albany
City in Albany County in E.
BT New York
United States

Albany County
Index states as applicable.
BX New York
Wyoming
BT United States

Albatross Cordillera
use East Pacific Rise

albedo
Includes use on level 2 under ionosphere(1) for neutron albedo, magnetic albedo, and albedo of electromagnetic waves; on level 3 under interplanetary space(1) cosmic rays(2) and under magnetosphere(1) cosmic rays(2).
UF Bond albedo
SA cosmic rays
interplanetary space
ionosphere
magnetosphere
Moon
reflectivity
remote sensing

Albert Canyon
Village between Mount Revelstone and Glacier National Park in SE.
BT British Columbia
Canada

Alberta
Includes use on level 1 as an area term (list O). For term set options see list B.
CO N490000N600000
W1100000W1200000
BA Western Canada
Canada
NT Abee
Athabasca Glacier
Athabasca Oil Sands
Athabasca River
Banff
Banff National Park
Bow River Valley
Bruderheim
Calgary
Drumheller
Edmonton
Fort McMurray
Jasper National Park
Lake Louise
Leduc
Medicine Hat
Red Deer River
Smoky River
Strathcona Mine
Swan Hills
Wabasca region
SA Altyn Limestone
Bearpaw Formation
Belly River Formation
Blairmore Group
Canadian Cordillera
Canadian Rocky Mountains
Canadian Shield
Churchill Province
Cold Lake
Edmonton Formation
Elk Point Basin
Elk Point Group
Great Plains
Hector Formation
Keg River Formation
Kicking Horse River valley
Mannville Formation
Miette Group
Milk River Formation
Missouri River basin
North Saskatchewan River
Oldman Formation
Palliser Formation
Paskapoo Formation
Peace River
Souris River basin
South Saskatchewan River
Sweetgrass Arch
Viking Formation
Waterways Formation

Albian
Europe. Above Aptian, below Cenomanian. Includes use on level 3 under age terms(1). See list E.
BA Lower Cretaceous
Cretaceous
BT Mesozoic
SA Middle Cretaceous

Albion Range
Cassia County in S.
BT Idaho
United States

albite
As of 1981, includes use on level 3 under minerals(1) framework silicates, plagioclase(2). See list L. Before 1981, included use on level 3 under minerals(1) framework silicates, feldspar group(2).
BA plagioclase
feldspar group
framework silicates
silicates
BT minerals
SA alkali feldspar
andesine
granite

albitite
As of 1981, includes use on level 3 under igneous rocks(1) syenites(2). See list H. Before 1981, included use on level 3 under igneous rocks(1) trachyte-phonolite family(2).
BA syenites
BT igneous rocks

albitization
Includes use on level 3 under metasomatism(1) processes(2).
BA metasomatism
SA autometamorphism
processes
spilitization

albitophyre
As of 1981, includes use on level 3 under igneous rocks(1) syenites(2). See list H. Before 1981, included use on level 3 under igneous rocks(1) syenite family(2).
BA syenites
BT igneous rocks

Alboran Sea
Between S Spain and Morocco in W.
CO N350500N364500
W0010000W0053000
BA West Mediterranean
BT Mediterranean Sea

Alborz Mountains
use Elburz

Albuquerque
City in Bernalillo County in central.
BT New Mexico
United States

Alcoy
City in SE part of country.
BT Alicante Province
Valencia region
Spain

Alcudia Valley
S central part of country.
BT Ciudad Real
New Castile
Spain

Aldabra Island
Atoll 250 miles NW of Malagasy Republic.
CO S092400S092400
E0462000E0462000
BT Seychelles

Aldan
City in SE Yakutia.
BT Russian Republic
USSR

Aldan Plateau
SE Yakutia.
CO N562000N581000
E1303000E1230000
BT Russian Republic
USSR

Aldan River
SE Yakutia. Flows into the Lena River.
BT Russian Republic
USSR

Aldan Shield
The Aldan Plateau area of SE Yakutia. Part of greater Angara Shield.
CO N553000N613000
E1370000E1163000
BA Siberian Platform
BT Russian Republic
USSR

Aldanian
Europe. See list E.
BA Lower Cambrian
Cambrian
BT Paleozoic

Aleksinac
Town in E Serbia.
UF Alexinats
Alexinatz
BT Serbia
Yugoslavia

Aleksod
Region in Ahaggar in SE.
BT Algeria

Alentejo
Former province in central and S.
BT Portugal

Ales
City in S part of country. Also search Alais.
UF Alais
BT Gard
France

Alessandria
City and province in SE.
BT Italy
SA Piemonte

Aleutian Arc
use Aleutian Islands

Aleutian Islands
Chain of volcanic islands extending in great curve from Alaska Peninsula in SW. Also search Aleutian Arc and Aleutians.
CO N510000N553000
W1610000E1720000
UF Aleutian Arc
Aleutians
BT Alaska
United States
SA Aleutian Trench
Amchitka Island
Rat Island
Rat Islands

Aleutian Ridge
Along S side of Aleutian chain.
BT Pacific Ocean
SA Aleutian Trench

Aleutian Trench
S of Aleutian Ridge and extending along the entire Aleutian chain.
BT Pacific Ocean
SA Aleutian Islands
Aleutian Ridge

Aleutians
use Aleutian Islands

Alexander Island
Large island off W coast of Antarctic Peninsula S of Cape Horn.
BT Antarctica

Alexandria
Governate. Also a city on the Mediterranean Sea.
BT Egypt

Alexandrian
North American provincial series. Lower Silurian (above Cincinnatian of Ordovician, below Niagaran). See list E.
UF Alexandrian Series
BA Lower Silurian
Silurian
BT Paleozoic

Alexandrian Series
use Alexandrian

alexandrite
BA oxides
BT minerals
SA chrysoberyl

Alexinats
use Aleksinac

Alexinatz
use Aleksinac

Alfisols
Includes use on level 3 under soils(1). See list M.
BA soils

Alfold
Plain in central and E. Also search Great Hungarian Plain; Hungarian Basin; Hungarian Great Plain; Hungarian Plain.
CO N455500N482700
E0225500E0184500
UF Great Alfold
Great Hungarian Plain
Hungarian Basin
Hungarian Great Plain
Hungarian Plain
BT Hungary

algae
Includes use on level 1 and 2 as a fossil term (list F).
BT thallophytes
Plantae
NA calcareous algae
Chlorophyta
Chrysophyta
Coccolithophoraceae
Cyanophyta
diatoms
Microcodium
nannofossils
Phaeophyta
Pyrrhophyta
Rhodophyta
NZ stromatolites
SA algal biscuits
algal flora
algal mats
algal mounds
algal structures
bioherms
fungi
lichens
microorganisms
phytoplankton
plankton
Protista
reef builders
zooxanthellae

algal banks
See list K.
BA algal structures
biogenic structures
sedimentary structures
SA banks

algal biscuits
Includes use on level 3 under sedimentary structures(1) biogenic structures(2). See list K.
UF lake biscuits
water biscuits
BA algal structures
biogenic structures
sedimentary structures
SA algae
girvanella

algal flora
Term introduced in 1981 as the common name for algae. See list F. To be used for algae when they are used as tools in stratigraphy. Includes use on level 1. Before 1981, search algae AND biostratigraphy. If 1, term set options are:
biostratigraphy
age [list E]
SA algae
biostratigraphy

algal limestone
Common in early Paleozoic, especially in the Cambro-Ordovician of the Appalachian region of the United States. Before 1978, also search algal AND limestone.
BA limestone
carbonate rocks
BT sedimentary rocks

algal mats
Includes use on level 3 under sedimentary structures(1) biogenic structures(2). See list K.
BA algal structures
biogenic structures
sedimentary structures
SA algae
stromatolites

algal mounds
Includes use on level 3 under sedimentary structures(1) biogenic structures(2). See list K.
UF mounds, algal
BA algal structures
biogenic structures
sedimentary structures
SA algae
calcilutite
limestone

algal structures
Term introduced in 1981. Includes use on level 3 under sedimentary structures(1) biogenic structures(2). See list K.
BA biogenic structures
sedimentary structures
NA algal banks
algal biscuits
algal mats
algal mounds
SA algae

Algarve
Region in extreme S Portugal. Former province.
BT Portugal

Algau Alps
use Allgau Alps

Algeria
Includes use on level 1 as an area term (list O). For term set options see list B.
CO N190000N370000
E0120000W0090000
BA North Africa
Africa
NT Ahaggar
Ahnet
Aleksod
Algiers region
Annaba
Aures
Babor Range
Constantine region
Dellys
Gour Oumelalen
Great Kabylia
Hassi Messaoud Field
Kabylia
Laghouat
Mouydir
Oasis region
Oran region
Ougarta
Saoura region
Tassili n'Ajjer
Tell
Tindouf
Tindouf Basin
SA Atlas Mountains
Mediterranean region
Sahara

Tanezrouft
Taourirt

Algerian Plain
use Balearic Basin

Algiers
No longer a valid term for GeoRef as of 1981. Use Algiers region.

Algiers region
Term introduced in 1981. Before 1981, search Algiers.
CO N335000N365500
E0050200E0004200
BT Algeria

Algoma
District on W Lake Superior.
BT Ontario
Canada

Algonkian
Includes use on level 3 as an age term (list E).
BA Precambrian
NT Baraboo Quartzite
SA Proterozoic
upper Precambrian

algorithms
SA automatic data processing
computer programs
equations
mathematical geology

Alicante
As of 1981, refers to only the city. Before 1981, also included the province. In SE Spain.
BT Alicante Province
Valencia region
Spain

Alicante Province
Term introduced in 1981. Before 1981, search Alicante. In SE Spain.
CO N375200N385300
E0001500W0010500
BT Valencia region
Spain
NT Alcoy
Alicante

Alice Arm
Village on Observatory Inlet near Alaska border NE of Prince Rupert.
BT British Columbia
Canada

Alice Springs
Town in S central.
BT Northern Territory
Australia

Alismidae
Term introduced in 1981. Includes use on level 2 under angiosperms(1). See list F.
BA Monocotyledoneae
angiosperms
BT Spermatophyta
Plantae

alkali amphibole
use alkalic amphibole

alkali basalt
Includes use on level 3. As of 1981, not to be used in the igneous rocks set. See list H. Before 1981, included use on level 3 under igneous rocks(1) alkali basalt family(2). Also search alkalic basalt; alkaline basalt.
UF alkalic basalt
alkaline basalt
BA alkali basalts
basalts
volcanic rocks
BT igneous rocks
SA basalt
tholeiite

alkali basalt family
As of 1981, no longer a valid term for GeoRef. Use alkali basalts or trachybasalts as applicable. See list H. Before 1981, included use on level 2 under igneous rocks(1).

alkali basalts
Term introduced in 1981. Includes use on level 2 under igneous rocks(1). See list H. Before 1981, search alkali basalt family.
BA basalts
volcanic rocks
BT igneous rocks
NA alkali basalt
ankaratrite
basanite
crinanite
hawaiite
mugearite
nepheline basalt
spilite
tephrite
trachybasalts
trachydolerite

alkali diorites
Term introduced in 1981. Includes use on level 2 under igneous rocks(1). See list H.
BA diorites
BT igneous rocks

alkali feldspar
As of 1981, includes use in combination with framework silicates (i.e. framework silicates, alkali feldspar) on level 2 under minerals(1). See list L. Before 1981, included use on level 3 under minerals(1) framework silicates, feldspar group(2).
BA feldspar group
framework silicates
silicates
BT minerals
SA albite
K-feldspar

alkali gabbro family
As of 1981, no longer a valid term for GeoRef. Use alkali gabbros. See list H. Before 1981, included use on level 2 under igneous rocks(1).

alkali gabbros
Term introduced in 1981. Includes use on level 2 under igneous rocks(1). See list H. Before 1981, search alkali gabbro family.
BA gabbros
BT igneous rocks
NA essexite
ijolite
leucitite
melteigite
nephelinite
olivine nephelinite
teschenite
urtite
SA foyaite
shonkinite

alkali granite
Term introduced in 1981. Includes use on level 3 and is not to be used in the igneous rocks set. See list H. Before 1981, search alkalic granite.
BA alkali granites
granites
BT igneous rocks

alkali granites
Term introduced in 1981. Includes use on level 2 under igneous rocks(1). See list H. Before 1981, search alkalic granite.
BA granites
BT igneous rocks
NA alkali granite

alkali metals
UF alkaline metals
BT metals
SA cesium
lithium
potassium
rubidium
sodium

alkali olivine basalt
As of 1981, includes use on level 3 under igneous rocks(1) basalts(2). See list H. Before 1981, included use on level 3 under igneous rocks(1) basalt family(2).
BA basalts
volcanic rocks
BT igneous rocks
SA basalt
olivine basalt

alkali syenite
Includes use on level 3. As of 1981, not to be used in the igneous rocks set. See list H. Before 1981, included use on level 3 under igneous rocks(1) syenite family(2).
UF alkalic syenite
BA alkali syenites
syenites
BT igneous rocks

alkali syenites
Term introduced in 1981. Includes use on level 2 under igneous rocks(1). See list H. Before 1981, search alkali syenite.
BA syenites
BT igneous rocks
NA alkali syenite

alkalic
Valid level 2 term through 1977. After 1977, use alkalic composition on level 2 under igneous rocks(1).

alkalic amphibole
Term introduced in 1981. Includes use in combination with chain silicates (i.e. chain silicates, alkalic amphibole) on level 2 under minerals(1). See list L. Before 1981, also search alkali amphibole.
UF alkali amphibole
BA amphibole group
chain silicates
silicates
BT minerals

alkalic basalt
use alkali basalt

alkalic composition
Term introduced in 1978. Includes use on level 2 under igneous rocks(1). Before 1978, also search alkalic AND igneous rocks.
SA calc-alkalic composition
composition
igneous rocks

alkalic granite
As of 1981, no longer a valid term for GeoRef. Use alkali granite. Before 1981, included use on level 3 under igneous rocks(1) granite-granodiorite family(2).

alkalic pyroxene
Term introduced in 1981. Includes use in combination with chain silicates (i.e. chain silicates, alkalic pyroxene) on level 2 under minerals(1). See list L.
BA pyroxene group
chain silicates
silicates
BT minerals

alkalic syenite
use alkali syenite

alkaline basalt
use alkali basalt

alkaline metals
use alkali metals

alkalinity
Includes use on level 3 under geochemistry(1).
SA geochemistry
pH

Allan Hills
Term introduced in 1981. Victoria Land, Antarctica.
CO S770000S763000
E1600000E1590000
BT Victoria Land
Antarctica

Allan Hills Meteorites
Term introduced in 1981. Due to numbering of these meteorites, such as Allan Hills 76-01 and Allan Hills 77005, also truncate when searching. Before 1981, also search Allan Hills AND meteorites.
BT meteorites
SA Antarctica
Victoria Land

allanite
Also search orthite.
UF orthite
BA epidote group
sorosilicates
orthosilicates
silicates
BT minerals

Allarechenskiy
Ore field in the Kola Peninsula in extreme NW European USSR. Also search Allarechenskoye.
UF Allarechenskoye
BT Russian Republic
USSR
SA Kola Peninsula

Allarechenskoye
use Allarechenskiy

Allaverdy
use Alaverdi

Allegany County
Index states as applicable.
BX Maryland
New York
BT United States

Alleghany County
Index states as applicable.
BX North Carolina
Virginia
BT United States

Alleghany Mountains
use Allegheny Mountains

Alleghenies
use Allegheny Mountains

Allegheny County
W Pennsylvania.
BT Pennsylvania
United States

Allegheny Front
Eastern slope of the Allegheny Mountains. Index states as applicable.
BT United States
SA Allegheny Mountains
Maryland
Pennsylvania
Virginia
West Virginia

Allegheny Group
Subdivided into three formations: Clarion, Kittanning, and Freeport. E Kentucky, W Maryland, E Ohio, W Pennsylvania, W Virginia, and West Virginia.
BT Middle Pennsylvanian
Pennsylvanian
SA Kentucky
Maryland
Ohio
Pennsylvania

Virginia
West Virginia

Allegheny Mountains
Ranges of the Appalachians constituting a part of the Allegheny Plateau. Index States as applicable.
UF Alleghany Mountains
Alleghenies
BT United States
SA Allegheny Front
Allegheny Plateau
Appalachians
Maryland
Pennsylvania
Virginia
West Virginia

Allegheny Orogeny
Term introduced in 1978. An event which deformed the rocks of the Valley and Ridge province, and those of the adjacent Allegheny Plateau in the central and S Appalachians. Before 1978, also search Allegheny AND orogeny.
SA orogeny
Pennsylvanian
Permian

Allegheny Plateau
W section of the Appalachians extending from Cumberland Plateau on S to Mohawk Valley in New York on N and including the Allegheny Mountains. Index states as applicable.
BT United States
SA Allegheny Mountains
Appalachian Plateau
Appalachians
Kentucky
New York
Ohio
Pennsylvania
Virginia
West Virginia

Allen County
Index states as applicable.
BX Kansas
Kentucky
Ohio
BT United States

Allende Meteorite
Term introduced in 1981. Impact in SE Chihuahua. Before 1981, also search Allende AND meteorites.
BA carbonaceous chondrites
chondrites
BT meteorites
SA Chihuahua
Mexico

Allerod
Europe. An interval of late-glacial time following the Older Dryas and preceding the Younger Dryas.
BA Weichselian
Quaternary
BT Cenozoic

allevardite
use rectorite

Allgaeu Alps
use Allgau Alps

Allgau Alps
Index Bavaria and/or Tyrol as applicable.
CO N471000N473000
E0105000E0100000
UF Algau Alps
Allgaeu Alps
BT Alps
Europe
SA Austria
Bavaria
Germany
Tyrol

Allier
Department in central.
CO N455000N464500
E0040000E0023000
BT France
SA Limagne

allochthons
Includes use on level 3 under tectonics(1).
SA autochthons
faults
klippen
nappes
overthrust faults
structural geology
tectonics
windows

alloclasite
BA sulfides
BT minerals

Allogromiina
Includes use on level 2 under foraminifera(1). See list F.
BA foraminifera
BT Protista
Invertebrata

allometry
SA fossils

allophane
BA clay minerals
sheet silicates
silicates
BT minerals

alloys
As of 1981, includes use on level 2 under minerals(1). See list L. Before 1981, also search native elements and alloys.
BT minerals
NA awaruite
carbides
NZ hedleyite
NA josephinite
kamacite
niggliite
nitrides
phosphides
plessite
silicides
taenite
SA base metals
construction materials
metallurgy
metals
native elements

alluvial deposits
use alluvium

alluvial fans
Includes use on level 3 under geomorphology(1) fluvial features(2).
UF detrital fan
dry delta
talus fan
SA Alluvial soils
alluvium
deposition
fluvial features

alluvial plains
Includes use on level 3 under geomorphology(1).
UF river plain
SA deposition
geomorphology
paleogeography
plains

Alluvial soils
Includes use on level 3 under soils(1). See list M.
BA soils
SA alluvial fans
alluvium

alluvion
use alluvium

alluvium
Includes use on level 3 under sediments(1) clastic sediments(2). See list N.
UF alluvial deposits
alluvion
BA clastic sediments
BT sediments
SA alluvial fans
Alluvial soils
clay
colluvium
eluvium
floodplains
gravel
sand
silt
soil mechanics
soils

Alma Dag
use Amanos Mountains

Alma River
In Crimea. Also search Alma.
BT Ukraine
USSR

Alma-Ata
City in SE Kazakhstan.
UF Vernyi
Vyernyi
BT Kazakhstan
USSR

Almaden
Village in S central part of country.
BT Ciudad Real
New Castile
Spain

Almalyk
Town SE of Tashkent.
BT Uzbekistan
USSR

almandine
UF almandite
BA garnet group
nesosilicates
orthosilicates
silicates
BT minerals

almandite
use almandine

Almas Valley basin
River basin in W central.
BT Transylvania
Romania

Almeria
As of 1981, refers to only the city. Before 1981, also included the province. On the Mediterranean Sea.
BT Almeria Province
Andalusia
Spain

Almeria Province
Term introduced in 1981. Before 1981, search Almeria. In SE Spain.
CO N364000N375700
W0013500W0031000
BT Andalusia
Spain
NT Almeria
Cerro del Hoyazo
Sierra de Gador
Sierra de los Filabres
Velez Rubio

Almond Formation
In Mesaverde Group. Overlies Ericson Sandstone and underlies Lewis Shale. SW Wyoming.
BT Upper Cretaceous
Cretaceous
SA Mesaverde Group
Wyoming

Almora
Town and district in N.
BT Uttar Pradesh
India

Alno
Island in Gulf of Bothnia.
CO N623000N623000
E0173000E0173000
BT Sweden
SA Gulf of Bothnia

Alnus
Genus.
BA Dicotyledoneae
angiosperms
BT Spermatophyta
Plantae

Alpago Basin
Belluno Province in N.
BT Veneto
Italy

Alpena County
On Lake Huron in Lower Peninsula.
BT Michigan
United States

Alpenvorland
Term introduced in 1981.
CO N471500N490400
E0133000E0083500
BT Germany
SA West Germany

Alpes-de-Haute Provence
Department in SE France. Before 1977, also search Basses-Alpes.
CO N433000N450000
E0070000E0053000
UF Basses-Alpes
BT France
NT Castellane
Digne
Verdon Valley
SA Eoulx Basin
Provence
Provence Alps

Alpes-Maritimes
Department in extreme SE.
CO N433000N442000
E0074500E0064000
BT France
NT Antibes
Nice
Saint-Vallier
SA Esterel
Provence
Provence Alps

alpha chalcocite
use digenite

Alpha Cordillera
Undersea feature N and NW of Queen Elizabeth Islands of Northwest Territories, Canada.
BT Arctic Ocean

alpha rays
Use when referring to the rays themselves. For application or methodology, use alpha-ray spectroscopy. Includes use on level 3 under spectroscopy(1) and under chemical element(1).
UF rays, alpha
SA alpha-ray spectroscopy
cosmic rays
spectroscopy

alpha-ray spectroscopy
Term introduced in 1978. Includes use on level 3 under spectroscopy (1) and under chemical element(1) analysis(2). Before 1978, also search alpha ray AND spectroscopy.
BT spectroscopy
SA alpha rays
analysis
gamma-ray spectroscopy
radioactivity

alpine
A valid index term through 1977. Use alpine environment.

Alpine
A valid term used for Alpine orogeny through 1977. After 1977, use Alpine Orogeny.

alpine environment
Term introduced in 1978. Includes use on level 3 under ecology(1). Before 1978, also search alpine.
SA Alps
climate
ecology
environment

Alpine Fault
BT New Zealand

Alpine Foreland
use Prealps

Alpine Geosyncline
Term introduced in 1978. Before 1978, search geosynclines AND Alpine.
SA Alpine Orogeny
geosynclines

Alpine Orogeny
Term introduced in 1978. Includes use on level 3 under orogeny(1) and tectonics(1). Before 1978, also search orogeny AND Alpine; Alpine structure.
UF Alpine structure
SA Alpine Geosyncline
alpine-type
Cenozoic
Mesozoic
orogeny

Alpine structure
use Alpine Orogeny

alpine-type
Used mostly as a general term on level 3 under mineral deposits, genesis(1).
SA Alpine Orogeny
Alps
ecology
mineral deposits
tectonics

Alps
Great mountain system of S central Europe. Index countries as applicable. Includes use on level 1 as an area term (list O). For term set options see list B.
BA Europe
NT Allgau Alps
NZ Bavarian Alps
NT Carnic Alps
Central Alps
NZ Central Austrian Alps
NT Cottian Alps
Eastern Alps
NZ French Alps
NT Julian Alps
Karawanken
Lepontine Alps
Maritime Alps
NZ North Austrian Alps
NT Otztal Alps
Pennine Alps
Rhaetian Alps
NZ South Austrian Alps
Swiss Alps
NT Western Alps
Zillertal Alps
SA Albanides
alpine environment
alpine-type
Austria
Bernese Alps
Brianconnais Zone
Dauphine Alps
Dinaric Alps
Dolomites
France
Gailtal Alps
Germany
Glarus Alps
Hellenides
Hohe Tauern
Italy
Ligurian Alps
Limestone Alps
Northern Limestone Alps
Oberalp Pass
Piedmont Alps
Prealps
Provence Alps
Sau Alps
Savoy Alps
Sesia-Lanzo Zone
Southern Alps
Stubai Alps
Switzerland
Tavetsch
Vanoise
Yugoslavia

Alpujarras
Mountainous area in S.
BT Spain

Alsace
Region and former province in NE.
BT France

alstonite
UF bromlite
BA carbonates
BT minerals

Alta
Village in extreme N part of country.
BT Finnmark
Norway

Altai
Administrative Territory or Kray on the Kazakhstan and Mongolian borders.
UF Altay
BT Russian Republic
USSR

Altai Mountains
Mountain system at meeting of borders of Altai Kray, Kazakhstan, Mongolia, and Sinkiang Uighur. Index countries as applicable. Also search Altai Range.
CO N443000N500000
E1030000E0870000
UF Altai Range
Altay Mountains
BT Asia
NZ Soviet Altai Mountains
SA China
Chuya Alps
Gorny Altai
Kalba Range
Kuznetsk Alatau
Mongolia
Mongolian Altai
USSR

Altai Range
use Altai Mountains

Altai-Sayan region
Region including adjoining Altai and Sayan mountains. Index Mongolia and/or USSR as applicable. Also search Altai-Sayan.
CO N500000N550000
E1000000E0800000
BT Asia
SA Mongolia
USSR

altaite
As of 1981, includes use on level 3 under minerals(1) tellurides(2). See list L. Before 1981, included use on level 3 under minerals(1) sulfides(2).
BA tellurides
sulfides
BT minerals

Altamaha River
SE Georgia.
BT Georgia
United States

Altan Teeli
Village in W Mongolia. Also search Altan Teli.
UF Altan Teli
Dzereg
BT Mongolia

Altan Teli
use Altan Teeli

Altay
use Altai

Altay Mountains
use Altai Mountains

Altenberg
Town near Czechoslovak Border in S part of country.
BT Karl-Marx-Stadt Bezirk
East Germany

alteration
Formerly a general term appropriate to a large number of topics. As of 1976, use was restricted to any process having to do with changes in temperature and pressure, i.e. diagenesis, hydrothermal alteration, metasomatism, weathering, etc. The term is used to denote the whole spectrum of chemical and physical change. Includes use on level 2 under igneous rocks(1).
SA autometamorphism
changes
diagenesis
fenitization
hydrothermal alteration
hydrothermal processes
igneous rocks
kaolinization
leaching
metasomatic rocks
metasomatism
processes
serpentinization
thermal alteration
wallrock alteration
weathering

Altimur Mountains
E part of country.
BT Afghanistan

Altin Tepe
use Altin-Tepe

Altin-Tepe
Ore bearing region. Also search Altin Tepe.
UF Altin Tepe
BT Dobruja
Romania

altitude
Used as a general term. Also search elevation.
UF elevation

Alto Adige
Along with Trento comprises Trentino-Alto Adige autonomous region in NE.
BT Italy

Alto Alentejo
Former province in central.
BT Portugal

Altonian
Provincial series, New Zealand. See list E.
BA lower Miocene
Miocene
Neogene
Tertiary
BT Cenozoic

Altoona
City in Blair County in W central.
BT Pennsylvania
United States

Altyn Formation
use Altyn Limestone

Altyn Limestone
In Ravalli Group. Precambrian (Belt series). SW Alberta, SE British Columbia and NW Montana. Also search Altyn Formation.
UF Altyn Formation
BT Precambrian
SA Alberta
British Columbia
Montana
Ravalli Group

Altyn-Topkan
Village in N.
BT Tadzhikistan
USSR

alum
Also search potassium alum.
UF potash alum
potassium alum
BA sulfates
BT minerals

alum rock
use alunite

alumina
Includes use on level 3 under geochemistry(1) or under commodities (list C). Also search aluminum oxide.
UF aluminum oxide
SA aluminum
corundum
geochemistry
oxides

aluminosilicates
As of 1981, includes use on level 2 under minerals(1). See list L. Before 1981, included use on level 3.
BA silicates
BT minerals
SA adularia
amazonite
bavenite
berthierine
bytownite
carpholite
chloritoid
clinozoisite
cookeite
cordierite
corrensite
cymrite
epidote
euclase
eucryptite
glaucophane
hydrobiotite
hydrosodalite
kaliophilite
kalsilite
lawsonite
leifite
maskelynite
milarite
mordenite
offretite
osumilite
petalite
prehnite
pumpellyite
sapphirine
spodumene
stilpnomelane
sudoite

aluminum
Includes use on level 1 and 2 as a chemical element (list D). As of 1981, use aluminum ores for aluminum as a commodity. Before 1981, included use on level 1 as a commodity term.
UF Al

SA Al-26
alumina
aluminum ores
bauxite
elements

aluminum ores
Term introduced in 1981. Includes use on level 1 and 2 as a commodity term (list C). Before 1981, search aluminum AND ore deposits.
SA aluminum
bauxite

aluminum oxide
use alumina

alumite
use alunite

alumohydrocalcite
BA carbonates
BT minerals

alumstone
use alunite

alunite
UF alum rock
alumite
alumstone
BA sulfates
BT minerals
SA jarosite

Alveolina
Includes use on level 3 under foraminifera(1) Alveolinellidae(2).
BA Alveolinellidae
Miliolina
foraminifera
BT Protista
Invertebrata

Alveolinellidae
Includes use as level 2 term under foraminifera(1). See list F.
BA Miliolina
foraminifera
BT Protista
Invertebrata
NA Alveolina

Alwar
City in NE Rajasthan. Formerly a princely state.
UF Alwur
BT Rajasthan
India

Alwur
use Alwar

Am
use americium

Amadeus Basin
SW Northern Territory and W Western Australia. Index state and/or territory as applicable.
CO S260000S230000
E1340000E1290000
BT Australia
SA Northern Territory
Western Australia

Amador County
E California.
BT California
United States

Amagase
Village S of Osaka.
BT Honshu
Japan

Amami
use Oshima

Amami-O-shima
Largest island in Amami-gunto Island group.
BT Ryukyu Islands
Japan
SA O-shima

Amanos Mountains
Along SE Gulf of Alexandretta in SE Anatolia.
UF Alma Dag
Gavur Mountains
BT Turkey
Middle East

Amapa
Federal Territory on French Guiana border.
CO S011000N043000
W0495000W0550000
BT Brazil
NT Serra do Navio

Amargosa Desert
Index California and/or Nevada as applicable.
BT United States
SA California
Nevada

amargosite
use bentonite

Amarjola
Region in the Rajmahal Hills in E.
BT Bihar
India

Amasra
Town on the Black Sea in NW Anatolia.
BT Zonguldak
Turkey

Amasra Basin
Coal basin in NW Anatolia.
BT Turkey
Middle East

Amazon Basin
River drainage basin. Index countries as applicable. Also search Amazon; Amazon region.
UF Amazon region
BT South America
SA Bolivia
Brazil
Colombia
Ecuador
Peru
Venezuela

Amazon region
use Amazon Basin

Amazon River
Largest river in the world if measured by volume of water carried. Rises in 2 major headstreams in the Peruvian Andes flowing N and then E across N Brazil into the Atlantic Ocean. Index countries as applicable. Also search Amazon.
BT South America
SA Brazil
Colombia
Peru

Amazonas
State in NW Brazil.
CO S090000N020000
W0562000W0735000
BT Brazil

Amazonian Shield
use Brazilian Shield

amazonite
UF amazonstone
BA feldspar group
framework silicates
silicates
BT minerals
SA aluminosilicates
microcline

amazonstone
use amazonite

Amba Dongar
Region.
BT Gujarat
India

amber
UF bernstein
BA organic compounds
BT minerals
SA resins

amber mica
use phlogopite

Ambin Massif
SE part of country.
BT Savoie
France

amblygonite
UF hebronite
BA phosphates
BT minerals
SA lithium
montebrasite

Amblypoda
As of 1981, includes use on level 3 under Mammalia(1). See list F. Before 1981, included use on level 2 under Mammalia(1).
BA Mammalia
BT Tetrapoda
Vertebrata

Ambre Mountain
N part of country.
BT Malagasy Republic

Ambrosia Lake
BT New Mexico
United States

Ambrosia Lake District
Term introduced in 1981. Uranium-mining district in McKinley County, New Mexico.
BT McKinley County
New Mexico
United States
SA Grants mineral belt

Amchitka Island
In center of Aleutian Islands SW of Alaska Peninsula. Also search Amchitka.
BT Alaska
United States
SA Aleutian Islands

Amelia
Mineral bearing region in Amelia County in S central.
BT Virginia
United States

America
Term introduced in 1981. Includes all of the Americas.
SA Central America
Latin America
North America
South America
West Indies

americium
Includes use on level 1 and 2 as a chemical element (list D).
UF Am
SA elements

Americus Limestone
use Americus Limestone Member

Americus Limestone Member
Of Foraker Limestone. E Kansas, SE Nebraska and central N Oklahoma. Also search Americus Limestone.
UF Americus Limestone
BT Permian
SA Kansas
Nebraska
Oklahoma

Ames Limestone
Of Conemaugh Formation. E Kentucky, SW Maryland, E Ohio, SW Pennsylvania, and N West Virginia.
BT Upper Pennsylvanian
Pennsylvanian
SA Kentucky
Maryland
Pennsylvania
West Virginia

amethyst
BA silica minerals
framework silicates
silicates
BT minerals
SA corundum
gems
quartz

Amiidae
As of 1981, includes use on level 3 under Pisces(1) Halecostomi(2). See list F. Before 1981, included use on level 3 under Pisces(1) Osteichthyes(2).
BA Halecostomi
Holostei
Actinopterygii
Osteichthyes
Pisces
BT Vertebrata

amino acids
Includes use on level 2 under organic materials(1).
BA organic materials
SA acids
fatty acids
proteins
racemization

Amisk Group
Considered as oldest rocks in Flin Flon Belt. Overlain by Missi Group.
BT Archean
SA Manitoba

Amisk Lake
Near Manitoba border in E.
BT Saskatchewan
Canada

Amite River
Rises in SW Mississippi and flows S and E into Lake Maurepas in SE Louisiana. Index states as applicable.
BT United States
SA Louisiana
Mississippi

Ammobaculites
Genus. Includes use on level 3 under foraminifera(1) Lituolacea(2).
BA Lituolacea
Textulariina
foraminifera
BT Protista
Invertebrata

Ammodiscacea
Includes use on level 2 under foraminifera(1). See list F.
BA Textulariina
foraminifera
BT Protista
Invertebrata
NA Ammodiscidae
Astrorhizidae

Ammodiscidae
Family. Includes use on level 3 under foraminifera(1) Ammodiscacea(2).
BA Ammodiscacea
Textulariina
foraminifera
BT Protista
Invertebrata

Ammonia
Genus. Includes use on level 3 under foraminifera(1) Rotaliacea(2).
BA Rotaliacea
Rotaliina
foraminifera
BT Protista
Invertebrata
NA Ammonia beccarii

SA ammonia compound

Ammonia beccarii
Includes use on level 3 under foraminifera(1) Rotaliacea(2).
BA Ammonia
Rotaliacea
Rotaliina
foraminifera
BT Protista
Invertebrata

ammonia compound
Term introduced in 1978 to distinguish from fossil genus, Ammonia. Includes use on level 3 under geochemistry(1). Before 1978, also search ammonia.
SA Ammonia
ammonium
geochemistry

Ammonites
As of 1981, includes use on level 3 under Mollusca(1) Ammonoidea(2). See list F. Before 1981, included use on level 3 under Mollusca(1) Cephalopoda(2).
BA Ammonoidea
Tetrabranchiata
Cephalopoda
Mollusca
BT Invertebrata
SA aptychi
Ceratites

ammonium
Includes use on level 3 under geochemistry(1).
SA ammonia compound
geochemistry

Ammonoidea
As of 1981, includes use on level 2 under Mollusca(1). Before 1981, included use on level 3 under Mollusca(1) Cephalopoda(2). See list F.
BA Tetrabranchiata
Cephalopoda
Mollusca
BT Invertebrata
NA Ammonites
Anarcestida
Bactritida
Baculites
Bouleiceras
Ceratitida
Clymeniida
Dactylioceratidae
Desmoceratida
Goniatitida
Hildocerataceae
Hildoceratidae
Lytoceratida
Perisphinctida
Phylloceratida
Prolecanitida
Psiloceratida
Scaphites
SA ammonoids

ammonoids
As of 1981, to be used for Ammonoidea when they are used as tools in stratigraphy. See list F. Before 1981, search (ammonoids OR Ammonoidea) AND biostratigraphy. Before 1981, included use on level 3 under Mollusca(1) Cephalopoda(2).
SA Ammonoidea
biostratigraphy

amorphous materials
General term introduced in 1978 for all materials not crystalline. Descriptive, not geologic term. Before 1978, also search amorphous AND materials.
SA materials

amosite
Commercial term.
BA amphibole group
chain silicates
silicates
BT minerals
SA anthophyllite
asbestos
clinoamphibole
gedrite
orthoamphibole

Amparo
City in E.
BT Sao Paulo
Brazil

Amphibia
Includes use on level 1 and 2 as a fossil term (list F).
BT Tetrapoda
Vertebrata
NA Labyrinthodontia
Lepospondyli
Lissamphibia
SA amphibians

amphibians
Term introduced in 1981 as the common name for Amphibia. See list F. To be used for Amphibia when they are used as tools in stratigraphy. Includes use on level 1. Before 1981, search Amphibia AND biostratigraphy. If 1, term set options are:
biostratigraphy
age [List E]
SA Amphibia
biostratigraphy

amphibole
As of 1981, no longer a valid term for GeoRef. See amphibole group. Before 1981, included use on level 3 under minerals(1) chain silicates, amphibole group(2).

amphibole group
Includes use in combination with chain silicates (i.e. chain silicates, amphibole group) on level 2 under minerals(1). See list L.
BA chain silicates
silicates
BT minerals
NA alkalic amphibole
amosite
clinoamphibole
crocidolite
orthoamphibole
SA jade

amphibolite
As of 1981, no longer a valid term for GeoRef. Use amphibolites. Before 1981, included use on level 3 under metamorphic rocks(1) amphibolites(2).
SA para-amphibolite

amphibolite facies
Includes use on level 3 under metamorphic rocks(1).
BT facies
SA epidote-amphibolite facies
metamorphic rocks

amphibolites
Includes use on level 2 under metamorphic rocks(1). See list J. Before 1981, also search amphibolite.
BT metamorphic rocks
NA orthoamphibolite
para-amphibolite

Amphineura
Not a valid term for GeoRef. Use Aplacophora or Polyplacophora under Mollusca(1). See list F.

Amphistegina
Genus. Includes use on level 3 under foraminifera(1) Orbitoidacea(2).
BA Orbitoidacea
Rotaliina
foraminifera
BT Protista
Invertebrata

amplitude
Not to be used to indicate abundance. Strictly for use in seismology. Includes use on level 3 under seismology(1) elastic waves(2).
SA elastic waves
seismic intensity
seismology

amplitude distortion
Term introduced in 1981.
SA distortion
elastic waves
seismology

Ampurdan
Coastal plain in Catalonia.
BT Gerona Province
Catalonia
Spain

Amsterdam
City on IJsselmeer (former Zuider Zee). North Holland.
BT Netherlands

Amsterdam Island
Dependency of Malagasy Republic 2800 miles off SE African coast.
BT Indian Ocean

Amu Darya
River which rises in Pamirs and flows into Aral Sea. Index Afghanistan and Soviet republics as applicable.
UF Oxus
BT Asia
SA Afghanistan
Tadzhikistan
Turkmenia
Uzbekistan

Amund Ringnes Island
One of the Queen Elizabeth Islands in District of Franklin.
BT Northwest Territories
Canada
SA Queen Elizabeth Islands

Amur Basin
River basin covering an area including parts of Amur and Chita oblasts, and Khabarovsk Kray in the Russian Republic; parts of Inner Mongolia, and Manchuria in China; and E Mongolia. Index Manchuria, Mongolia, or Russian Republic as applicable. Before 1981, also search Amur; Amur region.
CO N470000N540000
E1410000E1210000
BT Asia
SA Inner Mongolia
Manchuria
Mongolia
Russian Republic
Soviet Far East

Amur region
Term introduced in 1981.
CO N473000N553000
E1390000E1213000
BA Northeastern USSR
BT Russian Republic
USSR

Amur River
Formed at junction of the Shilka and Argun rivers. After serving as the boundary between Manchuria and 2 oblasts of the Russian Republic, it flows NE across Khabarovsk Kray into the N end of the Tatar Strait. Index Manchuria and/or Russian Republic as applicable. Also search Amur.
BT Asia
SA Argun River
Manchuria
Russian Republic

Anabar Anteclise
use Anabar Shield

Anabar Bay
On the Laptev Sea in NW Yakutia. Also search Anabar.
BT Russian Republic
USSR

Anabar Massif
use Anabar Shield

Anabar River
Flows into Anabar Bay in NW Yakutia. Also search Anabar.
BT Russian Republic
USSR

Anabar Shield
Part of the Siberian Platform in NW central Yakutia. Also search Anabar; Anabar Anteclise; Anabar Massif.
CO N650000N720000
E1230000E0960000
UF Anabar Anteclise
Anabar Massif
BA Siberian Platform
BT Russian Republic
USSR

Anacostia River basin
Index Maryland and/or District of Columbia as applicable. Also search Anacostia River.
BT United States
SA District of Columbia
Maryland

Anadarko Basin
Index states as applicable.
BT United States
SA Kansas
Oklahoma
Texas

Anadyr Basin
River basin in Chukchi National Okrug in NE Siberia. Also search Anadyr.
BT Russian Republic
USSR

Anadyr Range
Extends SE from Chaun Bay into W Chukchi Peninsula in extreme NE Siberia. Also search Anadyr and Chukchi Range.
UF Chukchi Range
BT Russian Republic
USSR

anaerobic environment
Term introduced in 1978. Before 1978, search anaerobic.
SA aerobic environment
environment
paleoecology
sapropel

anaerobic taxa
Term introduced in 1978. Before 1978, search anaerobic.
UF taxa, anaerobic

Anagalida
Term introduced in 1981. Includes use on level 2 under Mammalia(1). See list F.
BA Mammalia
BT Tetrapoda
Vertebrata

analbite
BA feldspar group
framework silicates
silicates
BT minerals

analcime
Also search analcite.
UF analcite

BA zeolite group
framework silicates
silicates
BT minerals

analcite
use analcime

analog filters
Term introduced in 1981.
BT filters

analog simulation
Term introduced in 1978. Also search analog AND automatic data processing; analog techniques.
SA automatic data processing
mathematical geology
simulation

analog techniques
As of 1981, no longer a valid term for GeoRef. Use analog simulation.

analyses
A valid term through 1974. After 1974, use analysis.

analysis
Includes use on level 2 under name of element(1), under isotopes(1) and paleoecology(1). See list D. For methods, see chemical analysis(1), spectroscopy(1), X-ray analysis(1), and thermal analysis(1). For data, see material name. Before 1975, also search analyses.
SA activation analysis
alpha-ray spectroscopy
analytical methods
atomic absorption
autoradiography
Bacukus-Gilbert analysis
canonical analysis
chemical analysis
chemical composition
chromatography
cluster analysis
colorimetry
correspondence analysis
differential thermal analysis
discriminant analysis
electron diffraction analysis
electron paramagnetic resonance
electron probe
emission spectroscopy
environmental analysis
factor analysis
finite difference analysis
finite element analysis
flame photometry
fluid inclusions
Fourier analysis
gamma-ray spectroscopy
infrared spectroscopy
ion probe
isotopes
least-squares analysis
major-element analyses
microwave spectroscopy
minor-element analyses
modal analysis
Mossbauer spectroscopy
multispectral analysis
multivariate analysis
neutron activation analysis
neutron diffraction analysis
nuclear magnetic resonance
numerical analysis
optical spectroscopy
polarography
pollen analysis
qualitative analysis
quantitative analysis
radio-frequency spectroscopy
Raman spectroscopy
regression analysis
sampling
shape analysis
spectral analysis
spectroscopy
spherical harmonic analysis
statistical analysis
structural analysis
thermal analysis
thermogravimetric analysis
thermomagnetic analysis
Thomson-Haskell analysis
time domain analysis
trace-element analyses
trend-surface analysis
ultraviolet spectroscopy
vacuum fusion analysis
variance analysis
wet methods
Wiener-Hopf analysis
X-ray analysis
X-ray diffraction analysis
X-ray fluorescence
X-ray spectroscopy

analytical
A valid level 2 index term through 1977. After 1977, use analytical methods on level 3 under level 1 terms.

analytical data
A valid term through 1975. After 1975, use data.

analytical methods
Not a valid term from 1975 to 1977. After 1977, includes use on level 3. Before 1978, also search analytical AND methods.
SA analysis
methods

Anantapur
City and district in SW.
BT Andhra Pradesh
India

Anantnag
Town and district in S central.
BT Jammu and Kashmir
India

Anapa
Town in Krasnodar Kray in the Northern Caucasus.
BT Russian Republic
USSR

Anapsida
Includes use on level 2 under Reptilia(1). See list F.
BA Reptilia
BT Tetrapoda
Vertebrata
NA Captorhinomorpha
Chelonia
Cotylosauria
Pareiasauria

Anarak
Town in central.
BT Iran

Anarcestida
Term introduced in 1981. Includes use on level 2 under Mollusca(1). See list F.
BA Ammonoidea
Tetrabranchiata
Cephalopoda
Mollusca
BT Invertebrata

Anasco Bay
On Mona Passage on the W coast.
BT Puerto Rico
West Indies

anastomosing streams
use braided streams

anatase
UF octahedrite (mineral)
BA oxides
BT minerals
SA brookite
rutile

anatectite
use anatexite

anatexis
Includes use on level 3 under metamorphism(1).
SA assimilation
contamination
magmas
metamorphism
palingenesis

anatexite
Includes use on level 3 under metamorphic rocks(1) migmatites(2). See list J.
UF anatectite
BA migmatites
BT metamorphic rocks

Anatolia
In E and central Turkey.
CO N363700N413800
E0445000E0300000
BT Turkey

Ancenis
Town on the Loire River in W part of country.
BT Loire-Atlantique
France

Anchorage
City on Cook Inlet in S.
BT Alaska
United States

anchors
SA engineering geology

ancient
A valid term through 1977 used with ice ages (i.e. ice ages, ancient) on level 2 under glacial geology(1). After 1977, use ancient ice ages on level 2.

ancient ice ages
Term introduced in 1978 on level 2 under glacial geology (1). Before 1978, search ice ages AND ancient.
SA glacial geology

Ancona
City and province on the Adriatic Sea.
BT Marches
Italy

Andalusia
Eight provinces in S and SW Spain.
CO N360000N385000
W0004000W0073000
BT Spain
NT Almeria Province
Cadiz Province
Cordoba Province
Granada Province
Huelva Province
Jaen Province
Malaga Province
Seville Province
SA Andalusian

Andalusian
BA Triassic
BT Mesozoic
SA Andalusia
Betic Cordillera
Spain

andalusite
Includes use on level 3 under minerals(1). See list L. As of 1981, use andalusite deposits for andalusite as a commodity. Before 1981, included use on level 3 as a commodity term under ceramic materials(1).
BA orthosilicates
silicates
BT minerals
SA andalusite deposits
kyanite
sillimanite

andalusite deposits
Term introduced in 1981. Includes use on level 3 as a commodity term (list C) under ceramic materials(1). Before 1981, search andalusite AND deposits.
SA andalusite
ceramic materials

Andaman Basin
Between the Andaman and the Nicobar islands.
BT Bay of Bengal
Indian Ocean

Andaman Islands
E Bay of Bengal, N of Nicobar Islands. Andaman and Nicobar Territory.
CO N103000N134500
E0930000E0922000
BA Bengal Islands
BT India

Andaman Sea
W of Burmese and Thai section of Malay Peninsula.
CO N050000N170000
E0980000E0923000
BT Indian Ocean

Andean Geosyncline
Term introduced in 1978. Includes use on level 3 under geosynclines(1) evolution(2). Before 1978, search geosynclines AND Andean.
SA Andean Orogeny
Andes
geosynclines

Andean Orogeny
Term introduced in 1978. Before 1978, search orogeny AND Andean.
SA Andean Geosyncline
Andes
orogeny
tectonics

Anderson County
Index states as applicable.
BX Kansas
Kentucky
Tennessee
South Carolina
Texas
BT United States
NT Walker Branch watershed

Andes
Mountain system extending along W coast from Venezuela and Colombia to Tierra del Fuego. Index countries as applicable. Includes use on level 1 as an area term (list O). For term set options see list B.
UF Andes Mountains
BA South America
NZ Argentine Andes
NA Central Andes
NT Eastern Cordillera
NA Northern Andes
Southern Andes
NZ Venezuelan Andes
NT Western Cordillera
SA Andean Geosyncline
Andean Orogeny
Argentina
Bolivia
Chile
Colombia
Ecuador
Peru
Venezuela

Andes Mountains
use Andes

andesine
As of 1981, includes use on level 3 under minerals(1) framework silicates, plagioclase(2). See list L. Before 1981, included use on level

3 under minerals(1) framework silicates, feldspar group(2).
BA plagioclase
feldspar group
framework silicates
silicates
BT minerals
SA albite
anorthite

andesite
Includes use on level 3. As of 1981, not to be used in the igneous rocks set. See list H. Before 1981, included use on level 3 under igneous rocks(1) andesite-rhyolite family(2).
BA andesites
volcanic rocks
BT igneous rocks
SA andesitic composition
meta-andesite
propylite
pyroxene andesite

andesite basalt
Not a valid term for GeoRef. See basalt.

andesite porphyry
As of 1981, includes use on level 3 under igneous rocks(1) andesites(2). See list H. Before 1981, included use on level 3 under igneous rocks(1) andesite-rhyolite family(2).
BA andesites
volcanic rocks
BT igneous rocks
SA porphyry

andesite tuff
As of 1981, includes use on level 3 under igneous rocks(1) pyroclastics(2). See list H. Before 1981, included use on level 3 under igneous rocks(1) pyroclastics and glasses(2). Also search andesitic tuff.
UF andesitic tuff
BA pyroclastics
volcanic rocks
BT igneous rocks
SA tuff

andesite-rhyolite family
As of 1981, no longer a valid term for GeoRef. Use andesites, trachyandesites, dacites, rhyodacites, or rhyolites as applicable. See list H. Before 1981, included use on level 2 under igneous rocks(1).

andesites
Term introduced in 1981. Includes use on level 2 under igneous rocks(1). See list H. Before 1981, search andesite-rhyolite family.
BA volcanic rocks
BT igneous rocks
NA andesite
andesite porphyry
pyroxene andesite
SA propylite

andesitic composition
Term introduced in 1978. Before 1978, search andesitic AND composition.
SA andesite
composition

andesitic tuff
use andesite tuff

Andhra Pradesh
State in central.
CO N130000N200000
E0845000E0764000
BT India
NT Anantapur
Cuddapah
Cuddapah Basin
Guntur
Khammam
Kurnool
Nellore
Nellore mica belt
Rajahmundry
Srikakulam
Visakhapatnam
SA Aravalli System
Cuddapah System
Deccan Plateau
Godavari River
Godavari Valley
Krishna
Kurnool System
Pranhita-Godavari Valley
Raghavapuram Shales

Ando soils
use Andosols

Andorra
A principality in E Pyrenees between France and Spain. Includes use on level 1 as an area term (list O). For term set options see list B.
CO N422500N424000
E0014500E0012500
BA Western Europe
Europe
SA Pyrenees

Andosols
UF Ando soils
BA soils

andradite
BA garnet group
nesosilicates
orthosilicates
silicates
BT minerals

Andriamena
Village in N central.
BT Malagasy Republic

Andros Island
Large island in W.
CO N233000N250000
W0773000W0783000
BT Bahamas
West Indies

Andrychow
Town in S part of country.
BT Krakow
Poland

Anegada Island
Northernmost island. British Virgin Islands.
BA British Virgin Islands
Virgin Islands
Leeward Islands
BT Lesser Antilles

anelastic materials
Term introduced in 1981. Includes use on level 3 under seismology(1). Before 1981, also search anelastic media; anelastic.
SA materials
seismology

anelastic media
As of 1981, no longer a valid term for GeoRef. Use anelastic materials.

anelasticity
SA deformation
elasticity
seismology

Angara
use Angara River

Angara River
Flows from Lake Baikal into Yenisei River. Formerly called Upper Tunguska in its lower course. Also search Angara.
UF Angara
Upper Tunguska
BT Russian Republic
USSR
SA Tunguska
Tunguska River

Angara-Lena Basin
Adjoining basin of two rivers in Irkutsk Oblast W and NW of Lake Baikal.
CO N520000N630000
E1150000E0950000
BA Siberian Platform
BT Russian Republic
USSR
SA Lena Basin

angiosperms
Includes use on level 1 and 2 as a fossil term (list F).
BT Spermatophyta
Plantae
NA Dicotyledoneae
Monocotyledoneae
SA Dryas
Florinites
fossil wood
gymnosperms
Pteropsida
sporangia

angle of dip
use dip

Anglesea
use Anglesey

Anglesey
Island and County in N Wales. Also search Anglesea.
CO N531000N533000
W0040000W0044000
UF Anglesea
BT Wales
Great Britain
United Kingdom

anglesite
BA sulfates
BT minerals

Angola
Peoples Republic of Angola. Includes use on level 1 as an area term (list O). For term set options see list B.
CO S180000S060000
E0230000E0120000
UF Portuguese West Africa
BA Central Africa
Africa
NT Cabinda
Catanda
Cuanza Basin
Cuanza-Sul
Luanda
Mossamedes
SA Congo Basin
Congo River
Kasai River
Zambezi Valley

Angora
use Ankara

Angouleme
City in W part of country.
BT Charente
France

Angra dos Reis
City in SW.
BT Rio de Janeiro
Brazil

Angren
City in Tashkent Oblast.
BT Uzbekistan
USSR

angrite
Term introduced in 1981. Includes use on level 3 under meteorites(1).
BA achondrites
BT meteorites

Angus
County on the North Sea. Formerly Forfar or Forfarshire.
CO N563000N570000
W0023000W0032500
UF Forfar
Forfarshire
BT Scotland
Great Britain
United Kingdom

anhydrite
Includes use on level 3 under sedimentary rocks(1) chemically precipitated rocks(2) and on level 3 under minerals(1) sulfates(2). See list I (sed. rocks) and list L (minerals). Before 1981, also included use on level 3 as a commodity term (list C) under gypsum(1) and under sediments(1) chemically precipitated sediments(2). As of 1981, use anhydrite deposits for anhydrite as a commodity.
UF cube spar
BA sulfates
BT minerals
SA anhydrite deposits
cap rocks
chemically precipitated rocks
evaporites
gypsum
sedimentary rocks
sediments

anhydrite deposits
Term introduced in 1981. Includes use on level 3 as a commodity term (list C) under gypsum deposits(1). Before 1981, search anhydrite AND deposits.
SA anhydrite
evaporite deposits
gypsum deposits

anhysteretic remanent magnetization
BA remanent magnetization
SA paleomagnetism

anilite
BA sulfides
BT minerals

animals
As of 1981, no longer a valid term for GeoRef. Before 1981, used as a general term.

Animas Mountains
Hidalgo County in extreme SW.
BT New Mexico
United States

Animikie Group
Includes Pokegama Quartzite, Biwabik Iron Formation Series, and Virginia Slate. N Michigan, NE Minnesota, Ontario, and N Wisconsin.
BT Precambrian
SA Biwabik Iron Formation
Gunflint Iron Formation
Michigan
Minnesota
Ontario
Wisconsin

anions
SA cations
electrolytes
geochemistry
ion exchange
ions

Anisian
Europe. Above Scythian, below Ladinian. Includes use on level 3 under age terms(1). See list E.
BA Middle Triassic
Triassic
BT Mesozoic

anisotropic materials
Term introduced in 1978. Before 1978, search anisotropic.
SA anisotropy
materials

anisotropy
Includes use on level 3 under seismology(1).
UF aeolotropy
SA anisotropic materials
magnetic susceptibility
mechanical properties
seismology

Anjou
Old province in W.
BT France

Anjouan Island
One of the French Comoro Islands in the Mozambique Channel.
UF Johanna Island
BT Indian Ocean
SA Comoro Islands

Ankara
Province in W central Anatolia. Also a city formerly called Angora.
UF Angora
BT Turkey
Middle East

ankaramite
As of 1981, includes use on level 3 under igneous rocks(1) basalts(2). See list H. Before 1981, included use on level 3 under igneous rocks(1) basalt family(2) and ultramafic family(2).
BA basalts
volcanic rocks
BT igneous rocks
SA ultramafics

ankaratrite
As of 1981, includes use on level 3 under igneous rocks(1) alkali basalts(2). See list H. Before 1981, included use on level 3 under igneous rocks(1) alkali basalt family(2).
BA alkali basalts
basalts
volcanic rocks
BT igneous rocks
SA nepheline basalt
olivine nephelinite

ankerite
UF cleat spar
ferroan dolomite
BA carbonates
BT minerals
SA dolomite

Ann Arbor
City in Washtenaw County in SE Lower Peninsula.
BT Michigan
United States

Annaba
Department on the Mediterranean Sea. Also a city. Formerly called Bone.
BT Algeria

Annapolis Valley
River valley in SW Nova Scotia. Also search Annapolis.
BT Nova Scotia
Canada

Anne Arundel County
Term introduced in 1981. Central Maryland.
CO N384300N391400
W0762400W0765100
BT Maryland
United States

Annecy
Town S of Lake Geneva in W.
BT Haute-Savoie
France

Annelida
As of 1981, includes use on level 2 under worms(1). See list F. Before 1981, included use on level 1.
BA worms
BT Invertebrata
SA scolecodonts

Annonaceae
Family.
BA Dicotyledoneae
angiosperms
BT Spermatophyta
Plantae

Annot Sandstone
Priabonian (Bartonian). In the Alps of SE France.
BT Eocene
SA France

annual growth rings
use tree rings

annual meeting
As of 1981, no longer a valid term for GeoRef. See symposia.

annual report
From 1978-1980, included use on level 2 under surveys(1). As of 1981, includes use on level 2 under survey organizations(1).
SA associations
current research
industry
progress report
report
survey organizations
symposia

annually thawed layer
use active layer

anomalies
Includes use as a level 2 or 3 term appropriate to a large number of topics, e.g. on level 2 under atmosphere(1), ecology(1), heat flow(1), and isostasy(1). See list G.
SA Bouguer anomalies
electrical anomalies
free-air anomalies
gravity anomalies
leakage anomalies
magnetic anomalies
regional anomalies
residual anomalies

Anomalinidae
Family. Includes use on level 3 under foraminifera(1) Cassidulinacea(2).
BA Cassidulinacea
Rotaliina
foraminifera
BT Protista
Invertebrata

anorthite
As of 1981, includes use on level 3 under minerals(1) framework silicates, plagioclase(2). See list L. Before 1981, included use on level 3 under minerals(1) framework silicates, feldspar group(2).
BA plagioclase
feldspar group
framework silicates
silicates
BT minerals
SA andesine

anorthoclase
BA feldspar group
framework silicates
silicates
BT minerals
SA orthoclase

anorthosite
As of 1981, includes use on level 3 under igneous rocks(1) gabbros(2). See list H. Before 1981, included use on level 3 under igneous rocks(1) gabbro family(2).
BA gabbros
BT igneous rocks
SA gabbroic anorthosite

anorthositic gabbro
As of 1981, includes use on level 3 under igneous rocks(1) gabbros(2). See list H. Before 1981, included use on level 3 under igneous rocks(1) gabbro family(2).
BA gabbros
BT igneous rocks
SA gabbro

Ansbach
City in W.
BT Bavaria
West Germany

Antalya
Province in S Anatolia on Gulf of Adalia. Also a city formerly called Adalia.
UF Adalia
BT Turkey
Middle East

Antarctic Continent
use Antarctica

Antarctic Ocean
An arbitrary definition, considered the equivalent of Southern Ocean. Waters about Antarctica are sometimes called the Antarctic Ocean, but actually are only the parts of the Atlantic, Pacific and Indian Oceans south of approximately 55° S. Also search Antarctic. Index oceans as applicable. Includes use on level 1 or 2 as an area term (list O). If 1, see term set options under list B.
NT Bellingshausen Sea
Drake Passage
McMurdo Sound
Mid-Antarctic Ridge
Ross Sea
Scotia Sea
Weddell Sea
SA Antarctica
Atlantic Ocean
DSDP Site 270
DSDP Site 274
DSDP Site 322
DSDP Site 323
DSDP Site 325
Indian Ocean
Leg 28
Leg 29
Leg 35
Leg 36
Pacific Ocean
Pacific-Antarctic Ridge
subantarctic regions

Antarctic Peninsula
Peninsula jutting northward toward Cape Horn. Also search Palmer Peninsula.
UF Palmer Peninsula
BT Antarctica

Antarctic Plate
SA Antarctica
plate tectonics
plates

Antarctic Platform
Central area which constitutes a high plateau. Also search Antarctic.
BT Antarctica

Antarctic Polar Cap
Term introduced in 1981. Before 1981, also search Antarctica AND Polar Cap.
BT Antarctica
SA South Pole

Antarctica
Continent centering on the South Pole. Includes use on level 1 or 2 as an area term (list O). If 1, see term set options under list B. To retrieve all documents, individual countries and physiographic regions should also be searched (see list O). Also search Antarctic.
CO S900000S610000
W0400000W0395900
UF Antarctic Continent
BA Polar regions
NT Adelaide Island
Adelie Coast
Alexander Island
Antarctic Peninsula
Antarctic Platform
Antarctic Polar Cap
Anvers Island
Arthur Harbor
Beacon Valley
Beardmore Glacier
Byrd Station
Coalsack Bluff
Don Juan Pond
NA Dry Valley Drilling Project
NT Dufek Intrusion
East Antarctica
East Ongul Island
Ellsworth Land
Ellsworth Mountains
Enderby Land
Fosdick Mountains
Graham Land
Horlick Mountains
Hut Point Peninsula
Jones Mountains
Kerguelen Islands
Lutzow-Holm Bay
Marguerite Bay
Marie Byrd Land
McMurdo Ice Shelf
Mirnyy Station
Molodezhnaya Station
Ohio Range
Pensacola Mountains
Polar Continental Shelf
Prince Charles Mountains
Queen Alexandra Range
Queen Maude Land
Queen Maude Range
Ross Dependency
Ross Ice Shelf
Ross Island
Scotia Sea Islands
Shackleton Glacier
Shackleton Range
Sor-Rondane Mountains
South Pole
Taylor Glacier
Taylor Valley
Transantarctic Mountains
Victoria Land
Vostok Station
Wilkes Land
Wisconsin Range
Wright Valley
SA Allan Hills Meteorites
Antarctic Ocean
Antarctic Plate
Beacon Supergroup
Circum-Antarctic region
Falkland Islands
Ferrar Group
Fremouw Formation
Gondwana
subantarctic regions
Yamato Meteorites

Antelope Valley
Term introduced in 1981. W Mojave Desert, N of Los Angeles.
CO N343000N352000
W1174000W1183000
BT Mojave Desert
California
United States

Antelope Valley Limestone
In Pogonip Group. Central Nevada.
BT Ordovician
SA Lower Ordovician
Middle Ordovician
Nevada
Pogonip Group

anthophyllite
BA orthoamphibole
amphibole group
chain silicates
silicates
BT minerals
SA amosite
asbestos
cummingtonite
gedrite

Anthozoa
Includes use on level 2 under Coelenterata(1). See list F.
BA Coelenterata
BT Invertebrata
NA Actiniaria
Ceriantipatharia
Corallimorpharia
Heterocorallia
Hexactiniaria
Octocorallia
Rugosa
Scleractinia
Tabulata
Tetracorallia
Zoantharia
Zoanthiniaria

anthracite
Includes use on level 3 under sedimentary rocks(1) organic residues(2). See list I.
BA coal
organic residues
BT sedimentary rocks
SA lignite
organic materials

Anthraconaia
Genus. Includes use on level 3 under Mollusca(1). See list F.
BA Bivalvia
Mollusca
BT Invertebrata
SA Anthraconauta

Anthraconauta
Genus. Includes use on level 3 under Mollusca(1). See list F.
BA Bivalvia
Mollusca
BT Invertebrata
SA Anthraconaia

anthraxolite
SA bitumens
coal
oil shale
organic materials

anthropogenic (activity)
use human activity

anthropology
Includes use on level 3.
SA fossil man

Anti-Atlas
Range of the Atlas Mountains SW of High Atlas Mountains in SW Morocco. Also search Anti-Atlas Mountains.
CO N274000N320500
W0020000W0131500
UF Anti-Atlas Mountains
BT Morocco
SA Atlas Mountains
Moroccan Atlas Mountains

Anti-Atlas Mountains
use Anti-Atlas

Antibes
Town on the Mediterranean Sea in extreme SE part of country.
BT Alpes-Maritimes
France

anticlinal
A valid term through 1977. After 1977, use anticlines.

anticlines
Not a valid index term from 1971 to 1977. Now use on level 3 under folds(1) style(2). Before 1978, also search folds AND anticlinal.
BA folds
SA anticlinoria
antiform folds
arches
diapirs
domes
geanticlines
synclines

anticlinoria
Includes use on level 3 under folds(1) systems(2).
BA folds
SA anticlines
antiform folds
geanticlines
geosynclines
synclinoria
systems

Anticosti Island
In Gulf of Saint Lawrence.
BT Quebec
Canada

antidunes
Includes use on level 3 under sedimentary structures(1) bedding plane irregularities(2). See list K.
BA bedding plane irregularities
sedimentary structures
SA dunes
flame structures
sand waves

Antietam Formation
In Chilhowee Group. Maryland, SE Pennsylvania, Virginia, West Virginia.
BT Lower Cambrian
Cambrian
SA Chilhowee Group
Maryland
Pennsylvania
Virginia
West Virginia

antiform folds
Term introduced in 1978. Includes use on level 3 under folds(1) style(2). Before 1978, also search folds AND antiform.
BA folds
SA anticlines
anticlinoria
synform folds

Antigonish
Town in Antigonish County, NE Nova Scotia.
BT Antigonish County
Nova Scotia
Canada

Antigonish County
Term introduced in 1981. N Nova Scotia. Before 1981, also search Antigonish.
CO N452200N455400
W0612600W0621500
BT Nova Scotia
Canada
NT Antigonish

antigorite
BA serpentine group
sheet silicates
silicates
BT minerals

Antilles
Index Greater Antilles and/or Lesser Antilles.
BT West Indies
SA Caribbean region
Greater Antilles
Lesser Antilles

antimonates
Term introduced in 1981. Includes use on level 2 under minerals(1). See list L. Before 1981, search antimonates and antimonites.
BT minerals

antimonates and antimonites
As of 1981, no longer a valid term for GeoRef. Use antimonates or antimonites as applicable. See list L. Before 1981, included use on level 2 under minerals(1).

antimonides
As of 1981, includes use on level 2 under minerals(1). See list L. Before 1981, included use on level 3 under minerals(1) sulfides(2).
BA sulfides
BT minerals
NA dyscrasite

antimonite
use stibnite

antimonites
Term introduced in 1981. Includes use on level 2 under minerals(1). See list L. Before 1981, search antimonates and antimonites.
BT minerals

antimony
Includes use on level 1 and 2 as a chemical element (list D). As of 1981, use antimony ores for antimony as a commodity. Before 1981, included use on level 1 as a commodity term.
UF Sb
SA antimony ores
elements
native elements

antimony ores
Term introduced in 1981. Includes use on level 1 and 2 as a commodity term (list C). Before 1981, search antimony AND ore deposits.
SA antimony
native elements

Antioquia
Department in NW Colombia. Also a city.
BT Colombia

antiperthite
BA feldspar group
framework silicates
silicates
BT minerals
SA perthite

antitrust legislation
Term introduced in 1981.
SA economics
legislation

Antler orogenic belt
use Antler Orogeny

Antler Orogeny
An orogeny which extensively deformed Paleozoic rocks of the Great Basin in Nevada. Named for relations in the Antler Peak Quadrangle near Battle Mountain. Before 1978, also search Antler; Antler orogenic belt.
UF Antler orogenic belt
BT Paleozoic
SA Acadian Phase
Devonian
Mississippian
Nevada
orogeny

Antlers Sands
In Trinity Group. SW Oklahoma and NE Texas.
BT Lower Cretaceous
Cretaceous
SA Comanchean
Oklahoma
Texas
Trinity Group

Antofagasta
Province in N Chile. Also a seaport N of Santiago.
BT Chile
NT Chuquicamata
SA Atacama Desert

Antong Java Rise
use Ontong Java Plateau

Antrim
County. Also a town.
CO N543000N551500
W0054500W0063000
BT Northern Ireland
United Kingdom

Antwerp
Province in N Belgium. Also a city at the mouth of the Western Scheldt River.
BT Belgium
NT Malines

Anura
Order. Includes use on level 3 under Amphibia(1) Lissamphibia(2).
BA Lissamphibia
Amphibia
BT Tetrapoda
Vertebrata

Anvers Island
Just off NW coast of the Antarctic Peninsula.
BT Antarctica

Anza Desert State Park
Occupies most of E San Diego County. Also search Anza.
BT California
United States

Anzoategui
State in N extending from the Caribbean Sea to the Orinoco River.
BT Venezuela
NT Oficina

Aomori
City and prefecture in N.
BT Honshu
Japan

Aosta Valley
use Valle d'Aosta

Aouelloul
A meteor crater near Chinguetti in W central Mauritania. Also search Aouelloul Crater.
UF Aouelloul Crater
BT Mauritania

Aouelloul Crater
use Aouelloul

Apache County
NE Arizona.
BT Arizona
United States

Apalachicola
City at mouth of Apalachicola River on Apalachicola Bay in Franklin County in panhandle.
BT Florida
United States

apatite
BA phosphates
BT minerals
SA carbonate apatite
chlorapatite
fluorapatite
francolite
heavy minerals
hydroxylapatite
pyromorphite

apatite ores
use phosphate deposits

Apennine Front
Linear outer slopes of the Apennines. Also search Apennine AND front. Also used for the Moon.
BX Moon
Italy
SA Apennines

Apennine Mountains
use Apennines

Apennine Range
use Apennines

Apennines
Mountain range extending the full length of the Italian peninsula. Includes use on level 1 as an area term (list O). For term set options see list B.
CO N380000N450000
E0170000E0082000
UF Apennine Mountains
Apennine Range
BA Italy
Southern Europe
Europe
NT Northern Apennines
Southern Apennines
SA Alban Hills
Apennine Front
Apuane Alps
Ligurian Apennines

aphanitic texture
Term introduced in 1978. Before 1978, search aphanitic.
SA igneous rocks
textures

Aphebian
North America, middle Precambrian. See list E.
BA lower Proterozoic
Proterozoic
BT Precambrian
NT Hurwitz Group
SA middle Precambrian

Aplacophora
Includes use on level 2 under Mollusca(1). See list F.
BA Mollusca
BT Invertebrata

aplite
As of 1981, includes use on level 3 under igneous rocks(1) granites(2). See list H. Before 1981, included use on level 3 under igneous rocks(1) granite-granodiorite family(2).
BA granites
BT igneous rocks

apogranite
As of 1981, includes use on level 3 under igneous rocks(1) granites(2). See list H. Before 1981, included use on level 3 under igneous rocks(1) granite-granodiorite family(2).
BA granites
BT igneous rocks
SA granite

Apollo
As of 1981, use Apollo Program. Included use on level 3 under Moon(1).

Apollo Program
Term introduced in 1981. Includes use on level 3 under Moon(1) and remote sensing(1). Before 1981, search Apollo.
SA extraterrestrial geology
landing sites
Moon
planetology
remote sensing

Apollo 11
Includes use on level 3 under Moon(1).
SA Moon
remote sensing

Apollo 12
Includes use on level 3 under Moon(1).
SA Moon
remote sensing

Apollo 13
Includes use on level 3 under remote sensing(1).
SA remote sensing

Apollo 14
Includes use on level 3 under Moon(1).
SA Moon
remote sensing

Apollo 15
Includes use on level 3 under Moon(1).
SA Moon
remote sensing

Apollo 16
Includes use on level 3 under Moon(1).
SA Moon
remote sensing

Apollo 17
Includes use on level 3 under Moon(1).
SA Moon
remote sensing

Apollo 9
Includes use on level 3 under remote sensing(1).
SA Moon
remote sensing

Apollo-Soyuz Program
Term introduced in 1981.
SA extraterrestrial geology
planetology
remote sensing

apophyllite
BA sheet silicates
silicates
BT minerals

Appalachian Basin
Also search Appalachian Plateau. Index states as applicable.
BT United States
SA Alabama
Appalachian Plateau
Kentucky
Maryland
New York
North Carolina
Ohio
Pennsylvania
Tennessee
Virginia
West Virginia

Appalachian Mountains
use Appalachians

Appalachian Orogeny
use Appalachian Phase

Appalachian Phase
Term introduced in 1978. Includes use on level 3 under tectonics(1) evolution(2). Before 1978, also search Appalachian structure; orogeny AND Appalachian; Appalachian Orogeny.
UF Appalachian Orogeny
Appalachian structure
BT Permian
SA orogeny
tectonics

Appalachian Plateau
Westernmost part of the Appalachians including the Allegheny and Cumberland plateaus. Index States as applicable. Also search Appalachian AND plateau.
BT United States
SA Alabama
Allegheny Plateau
Appalachian Basin
Appalachians
Cumberland Plateau
Kentucky
Maryland
New York
North Carolina
Ohio
Pennsylvania
Tennessee
Virginia
West Virginia

Appalachian structure
use Appalachian Phase

Appalachians
Mountain system of eastern North America extending from Canadian maritime provinces to Alabama. Index countries as applicable. Includes use on level 1 as an area term (list O). For term set options see list B.
CO N330000N473000
W0670000W0870000
UF Appalachian Mountains
BA North America
NT Blue Ridge Province
Brevard Zone
Carolina slate belt
NA Central Appalachians
Northern Appalachians
Southern Appalachians
SA Allegheny Mountains
Allegheny Plateau
Appalachian Plateau
Blue Ridge Mountains
Canada
Catskill Mountains
Cumberland Plateau
Great Appalachian Valley
Great Smoky Mountains
Green Mountains
Hudson Highlands
United States

apparatus
Do not use for instruments. As of 1981, includes use on level 3 under Conodonta(1). See list F. Before 1981, included use under conodonts(1).
SA Conodonta

appinite
As of 1981, includes use on level 3 under igneous rocks(1) syenites(2) and diorites(2). See list H. Before 1981, included use on level 3 under igneous rocks(1) diorite family(2) and syenite family(2).
BT igneous rocks
SA diorites
syenites

applications
Includes use as level 2 or 3 term appropriate to a large number of topics, e.g. on level 2 under mineral exploration(1). See list G. Also search application.

Apsheron Peninsula
Peninsula jutting into the Caspian Sea on E. Also search Apsheron.
CO N402000N404000
E0503000E0493000
BT Azerbaidzhan
USSR

Apt
Town in SE part of country.
BT Vaucluse
France

Aptian
Europe. Above Barremian below Albian. Includes use on level 3 under age terms(1). See list E.
BA Lower Cretaceous
Cretaceous
BT Mesozoic
SA Areado Formation
Gargasian
Middle Cretaceous
Zubair Formation

aptychi
Also search aptychus.
UF aptychus
SA Ammonites

aptychus
use aptychi

Apuane Alps
Division of Etruscan Apennines in E.
BT Tuscany
Italy
SA Apennines

Apulia
Autonomous region comprising approximately the southern third of east coast.
CO N394500N415800
E0183200E0145700
UF Puglia
BT Italy
NT Bari
Brindisi
Foggia
Gargano
Lecce
Murge
Otranto
Salentina Peninsula
Taranto

Apure Guarico Plain
Term introduced in 1981.
CO N061000N095000
W0644000W0715500
BT Venezuela

Apuseni Mountains
A large mountain massif in WSW Transylvania. Also search Apuseni.
BT Romania
SA Muntii Metalici

aqueous solutions
Includes use on level 3 under geochemistry(1).
SA geochemistry
solution
solutions
water

Aquia Formation
In Pamunkey Group. Delaware, Maryland and E Virginia.
BT lower Eocene
Eocene
SA Delaware
Maryland
Virginia

aquifer properties
A valid term through 1978. After 1978, see aquifers or specific property, e.g. transmissivity.

aquifers
Includes use on level 2 under ground water(1); on level 3 under hydrogeology(1).
NT unconfined aquifers
SA artesian waters
Darcy's law
ground water
hydrogeology
recharge
transmissivity
water balance
water resources
water supply

Aquitaine
A region of SW.
BT France

Aquitaine Basin
Large lowland in SW bounded by Pyrenees on S, Central Massif on N and NE, and Bay of Biscay on W.
CO N431500N460000
E0020000W0013000
BT France

Aquitanian
Europe. Above Chattian (Oligocene), below Burdigalian. Includes use on level 3 under age terms(1). See list E.
BA lower Miocene
Miocene
Neogene
Tertiary
BT Cenozoic

Ar
use argon

Ar-36
Includes use on level 3 under isotopes(1).
SA Ar-40/Ar-36
argon
isotopes

Ar-36/Ar-40
use Ar-40/Ar-36

Ar-37
Artificial. Includes use on level 3 under isotopes(1).
SA argon
isotopes

Ar-38
Includes use on level 3 under isotopes(1).
SA argon
isotopes

Ar-39
Artificial. Includes use on level 3 under isotopes(1).
SA Ar-40/Ar-39
argon
isotopes

Ar-39/Ar-40
use Ar-40/Ar-39

Ar-40
Includes use on level 3 under isotopes(1).
SA Ar-40/Ar-36
Ar-40/Ar-39
argon
isotopes

Ar-40/Ar-36
Includes use on level 3 under isotopes(1). Also search Ar-36/Ar-40; Ar-36 AND Ar-40.
UF Ar-36/Ar-40
SA Ar-36
Ar-40
argon

Ar-40/Ar-39
Includes use on level 3 under isotopes(1). Also search Ar-39/Ar-40; Ar-39 AND Ar-40.
UF Ar-39/Ar-40
SA Ar-39
Ar-40
argon

Ar/Ar
Includes use on level 3 under absolute age(1) methods(2). Also search argon argon.
SA absolute age
argon
methods

Arabia
use Arabian Peninsula

Arabian Desert
use Eastern Desert

Arabian Gulf
use Persian Gulf

Arabian Peninsula
Between Red Sea on the W; and the Persian Gulf, Gulf of Oman, and Arabian Sea on the E. Also search Arabia. Index countries as applicable. Includes use on level 1 and 2 as an area term (list O). For term set options see list B.
CO N123000N304000
E0594000E0343000
UF Arabia
BA Asia
NT Arabian Shield
NA Bahrain
Kuwait
Oman
Qatar
Saudi Arabia
Southern Yemen
NT Timna
NA United Arab Emirates
Yemen
SA Middle East

Arabian Plate
SA plate tectonics
plates
Saudi Arabia

Arabian Ridge
use Carlsberg Ridge

Arabian Sea
Between the Arabian Peninsula on the W and Pakistan and India in the E. As of 1977, includes use as a level 1 area term (list 0). For term set options see list B.
CO S010000N303000
E0770000E0473000
BA Indian Ocean
NA Gulf of Aden
NT Gulf of Cambay
Gulf of Oman
Gulf of Tadjoura
NA Persian Gulf
SA Leg 23
Leg 24

Arabian Shield
NE part of the Ethiopian Shield. Index countries as applicable.
BT Arabian Peninsula
SA Saudi Arabia
Southern Yemen
Yemen

Arabian-Indian Midoceanic Ridge
use Carlsberg Ridge

Arabian-Indian Ridge
use Carlsberg Ridge

Arabic
Includes use on level 3 to indicate language of a document.

Aracaju
City on the Atlantic Ocean in NE part of country.
BT Sergipe
Brazil

Aracena
Town near the Gulf of Cadiz.
BT Huelva Province
Andalusia
Spain

Arachnida
Class. Includes use on level 2 under Arthropoda(1). See list F.
BA Chelicerata
Arthropoda
BT Invertebrata

Arad
City and county near the Hungarian border.
BT Banat
Romania

Araeoscelidia
Term introduced in 1981. Includes use on level 2 under Reptilia(1). See list F.
BA Euryapsida
Reptilia
BT Tetrapoda
Vertebrata
NA Trilophosauria

Arafura Sea
Sea between N Australia and Indonesia; 800 miles long by about 350 miles wide.
BA North Australian Seas
West Pacific
BT Pacific Ocean
SA Australia
Indonesia

Aragats
A mountain in NW.
BT Armenia
USSR

Aragon
Region and ancient kingdom in NE.
CO N395200N425700
E0004500W0021000
BT Spain
NT Huesca Province
Saragossa Province
Teruel

aragonite
BA carbonates
BT minerals
SA alabaster
calcite
calcium carbonate
pearls
witherite

Aragua
State in N.
BT Venezuela

Aragvi
River in N central.
BT Georgian Republic
USSR

Aral region
Index Soviet republics as applicable. Also search Aral; Aral Sea region.
UF Aral Sea region
BT USSR
SA Kazakhstan
Turkmenia
Uzbekistan

Aral Sea
Between Kazakhstan and Uzbekistan. Also search Aral.
CO N440000N460000
E0620000E0580000
BT USSR

Aral Sea region
use Aral region

Arapahoe County
S and SE of Denver.
BT Colorado
United States

Ararat
Mountain near Iranian and USSR borders.
BT Turkey
Middle East

Arauco
Province in N.
BT Chile

Aravalli Range
Central.
BT Rajasthan
India

Aravalli System
Includes Dabari, Matoon, and Udaipur formations.
BT Archean
SA Andhra Pradesh
Gujarat
India
Rajasthan

Araxa
Town in W.
BT Minas Gerais
Brazil

Araya Peninsula
W Sucre.
BT Venezuela
SA Sucre

Arbarastakh
Carbonatite complex in Yakutia.
BT Russian Republic
USSR

Arbuckle Group
Subdivided into Fort Sill Limestone, Royer Marble, Signal Mountain Limestone, Chapman Ranch Dolomite, Wolf Creek Dolomite, Cool Creek Limestone, Alden Limestone, and West Spring Creek Formation. S Oklahoma.
BT Paleozoic
SA Cambrian
Lower Ordovician
Oklahoma
Ordovician
Upper Cambrian

Arbuckle Mountains
S central.
CO N340000N343000
W0964500W0973000
BT Oklahoma
United States

Arcachon Basin
Inlet of Bay of Biscay in SW part of country.
BT Gironde
France
SA Bay of Biscay

arch dams
UF archdams
BT dams

Archaean
use Archean

Archaediscidae
Family. Includes use on level 3 under foraminifera(1) Fusulinina(2).
BA Fusulinina
foraminifera
BT Protista
Invertebrata

Archaeocyatha
Includes use on level 1 and 2 as a fossil term (list F).
BT Invertebrata

Archaeogastropoda
As of 1981, includes use on level 2 under Mollusca(1). See list F. Before 1981, included use on level 3 under Mollusca(1) Gastropoda(2).
BA Gastropoda
Mollusca
BT Invertebrata

Archaeolithothamnium
As of 1981, includes use on level 3 under algae(1) Corallinaceae(2). Before 1981, included use on level 3 under algae(1) Rhodophyta(2).
BA Corallinaceae
Rhodophyta
algae
BT thallophytes
Plantae

archaeological sites
UF sites, archaeological
SA archaeology
artifacts
fossil man

archaeology
Includes use on level 3 under area terms(1) stratigraphy(2) and under paleoecology(1).
UF archeology

SA archaeological sites
artifacts
fossil man
paleoecology

Archaeomonadaceae
BA ebridians
Protista
BT Invertebrata

Archaeopteryx
Genus. Includes use on level 3 under Aves(1) Archaeornithes(2).
BA Archaeornithes
Aves
BT Tetrapoda
Vertebrata

Archaeornithes
Includes use on level 2 under Aves(1). See list F.
BA Aves
BT Tetrapoda
Vertebrata
NA Archaeopteryx
SA Neornithes

Archaeozoic
use Archean

Archangel
use Arkhangelsk

archdams
use arch dams

Archean
Includes use on level 1 and 2 as an age term (list E) as of 1978. As of 1981, used for lower Precambrian. Also search lower Precambrian.
UF Archaean
Archaeozoic
Archeozoic
BA Precambrian
NT Amisk Group
Aravalli System
Arunta Complex
Bulawayan Group
Gangpur Series
Kalgoorlie System
Krivoy Rog Series
Rice Lake Group
Sausar Series
Singhbhum Granite
Willyama Complex
Yellowknife Group
SA Moldanubian

Archeocopida
Order. Includes use on level 2 under Ostracoda(1).
BA Ostracoda
BT Crustacea
Mandibulata
Arthropoda
Invertebrata

archeology
use archaeology

Archeozoic
use Archean

arches
SA anticlines
domes
folds

Arches National Park
Term introduced in 1981. In SE Utah. Before 1981, also search Arches.
BT Utah
United States

Archosauria
Includes use on level 2 under Reptilia(1). See list F.
BA Reptilia
BT Tetrapoda
Vertebrata
NA Crocodilia
Ornithischia
Pterosauria
Saurischia
Thecodontia
SA dinosaurs

Archostemata
Includes use on level 3 under Insecta(1).
BA Insecta
BT Mandibulata
Arthropoda
Invertebrata

Arcina
Term introduced in 1981. Includes use on level 2 under Mollusca(1). See list F.
BA Cyrtodontida
Bivalvia
Mollusca
BT Invertebrata

arcs
Includes use on level 3 under aurora(1) electrical field(2).
SA aurora
electrical field

arcs, island
use island arcs

Arctic Archipelago
Large group of islands in the Arctic Ocean nearly coextensive with District of Franklin, Northwest Territories, Canada. Used as level 1 area term from 1977 to 1978. Now used only on level 3. Also search Arctic Islands.
UF Arctic Islands
BT District of Franklin
Northwest Territories
Canada
SA Arctic region

Arctic Coastal Plain
SA Arctic region

arctic environment
Term introduced in 1978. Includes use on level 3 under ecology(1). Before 1978, also search arctic.
SA climate
ecology
environment
paleoclimatology
paleoecology

Arctic Islands
use Arctic Archipelago

Arctic Ocean
Extends from North Pole southward to approximately 70° N and is bounded by Alaska, Canada, Greenland, Norway, and the USSR. Includes use on level 1 or 2 as an area term (list O). If 1, see term set options under list B.
CO N700000N900000
W0400000W0395900
UF Arctic Sea
North Polar Sea
NT Alpha Cordillera
NX Baffin Bay
NT Barents Sea
Beaufort Sea
Canada Basin
Chukchi Sea
East Siberian Sea
Eurasia Basin
Greenland Sea
Kara Sea
Laptev Sea
Lomonosov Ridge
Mendeleyev Ridge
Mid-Arctic Ocean Ridge
Mohns Ridge
Norwegian Sea
White Sea
SA Arctic region
Bering Strait
Jan Mayen
Leg 38
Svalbard

Arctic region
The Arctic Ocean and islands in it and adjacent to it plus mainland surfaces N of the Arctic circle. Index Alaska, Arctic Archipelago, Arctic Ocean, Greenland, and countries as applicable. Includes use on level 1 or 2 as an area term (list O). If 1, see term set options under list B.
BA Polar regions
NA Greenland
Jan Mayen
NT Svalbard
SA Alaska
Arctic Archipelago
Arctic Coastal Plain
Arctic Ocean
Canada
Finland
Hecla Hoek Formation
Norway
Soviet Arctic
subarctic regions
Sweden
USSR

Arctic Sea
use Arctic Ocean

Arctocyonia
Term introduced in 1981. Includes use on level 2 under Mammalia(1). See list F.
BA Mammalia
BT Tetrapoda
Vertebrata

arcuate faults
Before 1978, also search arcuate AND faults.
BA faults

Ardara Pluton
An igneous intrusion in NW part of country.
BT Donegal
Ireland

Ardeche
Department in S.
CO N441500N453000
E0050000E0034500
BT France
NT Largentiere
Privas
Vivarais

Ardennes
Wooded plateau region. Index countries as applicable.
CO N493000N503000
E0053000E0041000
UF Forest of Ardennes
BT Europe
NT Givet
Rocroi
Sedan
SA Belgium
France
Luxembourg

Ardennes Department
Term introduced in 1981.
CO N491200N501000
E0052500E0040200
BT France

Ardnamurchan
Peninsula and parish in Argyll County in W.
BT Scotland
Great Britain
United Kingdom

arduinite
use mordenite

Areado Formation
BT Lower Cretaceous
Cretaceous
SA Aptian
Brazil
Minas Gerais

areal geology
Used for entries that might properly be placed under three or more of the second-order headings such as geomorphology, stratigraphy, structural geology. Includes use on level 2 under area terms(1). See list B.
SA bibliography
explanatory text
geology
guidebook
maps

areal studies
Includes use on level 2 under clay mineralogy(1).
UF studies, areal
SA clay mineralogy
mineral data

Arecibo
Town on the Atlantic Ocean W of San Juan.
BT Puerto Rico
West Indies

Arecidae
Term introduced in 1981. Includes use on level 2 under angiosperms(1). See list F.
BA Monocotyledoneae
angiosperms
BT Spermatophyta
Plantae

arenaceous texture
Term introduced in 1978. Includes use on level 3 under sedimentary rocks(1) and sediments(1). Before 1978, also search arenaceous AND specific sediment type or sedimentary rock.
UF arenarious texture
psammitic texture
sabulous texture
sandy texture
SA sand
sandstone
sedimentary rocks
sediments
textures

Arenal
Extinct volcano in NW.
BT Costa Rica

arenarious texture
use arenaceous texture

Arendal
A city on the Skagerrak.
BT Norway

arendalite
use epidote

Arenig
Two mountains in N Merioneth in N Wales. One is called Arenig Fach and the other Arenig Fawr.
BT Wales
Great Britain
United Kingdom

Arenigian
Europe. Above Tremadocian, below Llanvirnian. Includes use on level 3 under age terms(1). See list E.
BA Lower Ordovician
Ordovician
BT Paleozoic

arenite
Includes use on level 3 under sedimentary rocks(1) clastic rocks(2). See list I.
BA clastic rocks
BT sedimentary rocks
SA terrigenous materials

Arequipa
Department in S Peru. Also a city.

BT Peru

arfvedsonite
BA clinoamphibole
amphibole group
chain silicates
silicates
BT minerals

Argentera
Village in S.
BT Italy
SA Piemonte

Argentera Massif
S Piemonte.
BT Italy
SA Piemonte

Argentiere Glacier
In the Mount Blanc Massif of the Pennine Alps.
BT Haute-Savoie
France

Argentina
Includes use on level 1 as an area term (list O). For term set options see list B.
CO S550000S220000
W0533000W0730000
BA South America
NX Argentine Andes
NT Barreal
Buenos Aires Province
Catamarca
Chubut
Cordoba Department
Corrientes
NX Federal District
NT Jujuy
La Pampa
La Rioja
Mendoza
Neuquen
Neuquen Basin
Pampas
Pampean Mountains
Patagonia
Precordillera
Puerto Deseado
Rio Negro Province
Salado Basin
Salta
San Juan Province
San Luis
Santiago del Estero
Sierras Pampeanas
Talacasto
Tucuman
SA Andes
Baquero Formation
Campo del Cielo Meteorite
Chaco
Chanares Formation
Magdalena
Paganzo Group
Parana Basin
Parana River
Patagonian Andes
Rio de la Plata
Santa Cruz
Santa Fe
Tierra del Fuego
Yacoraite Formation

Argentine Andes
Term introduced in 1981. Region in Argentina.
CO S450000S213000
W0644000W0734000
BZ Andes
BX Argentina

Argentine Basin
E of Argentina.
BT Atlantic Ocean

argentite
UF argyrite
BA sulfides
BT minerals
SA acanthite
silver

Arges River
In S part of country.
BT Romania

argillaceous rocks
use argillaceous texture

argillaceous texture
Term introduced in 1978. Includes use on level 3 under sediments(1) and sedimentary rocks(1). Before 1978, also search argillaceous AND specific sediment type or sedimentary rock. Also search argillaceous rocks.
UF argillaceous rocks
SA argillite
clay
clay minerals
marl
sedimentary rocks
sediments
shale
textures

argillite
Includes use on level 3 under sedimentary rocks(1) clastic rocks(2). See list I.
UF dropstone
BA clastic rocks
BT sedimentary rocks
SA argillaceous texture
terrigenous materials

Argo abyssal plain
Eastern.
BT Indian Ocean

Argolis
Department in E.
BT Peloponnesus
Greece

argon
Includes use on level 1 and 2 as a chemical element (list D). Also search Ar.
UF Ar
BT noble gases
SA Ar-36
Ar-37
Ar-38
Ar-39
Ar-40
Ar-40/Ar-36
Ar-40/Ar-39
Ar/Ar
elements
helium
isotopes
K/Ar
krypton
neon
radon
xenon

Argovian
Europe. Substage in Great Britain: upper Jurassic (lower Lusitanian; above Oxfordian Stage, below Rauracian Substage). Includes use on level 3 under age terms(1). See list E.
BA Upper Jurassic
Jurassic
BT Mesozoic

Argun
Village in central Tomsk Oblast in W central Siberia.
BT Russian Republic
USSR

Argun River
Rises in NW Manchuria and serves as part of boundary between NE China and USSR. Joins with Shilka River to form the Amur River. Index Manchuria and/or Russian Republic as applicable. Also search Argun.
BT Asia
SA Amur River
Manchuria
Russian Republic
Shilka Valley

Argyll (County)
use Argyllshire

Argyllshire
County in W Scotland. Also search Argyll.
UF Argyll (County)
BT Scotland
Great Britain
United Kingdom

argyrite
use argentite

Ariake Bay
On Pacific Ocean in SE.
BT Kyushu
Japan

arid environment
Term introduced in 1978. Includes use on level 2 under land use(1). Before 1978, also search arid; before 1975 search arid regions.
SA climate
deserts
ecology
environment
geomorphology
land use
paleoclimatology
semi-arid environment

arid regions
A valid term through 1974. After 1974, arid was used as an index term. After 1977, use arid environment.

Aridisols
Includes use on level 3 under soils(1). See list M.
BA soils

Ariege
Department in S.
CO N423500N432500
E0021000E0005000
BT France
NT Castillon Massif
Querigut Massif
Saint-Girons
SA Salat Valley

ariegite
As of 1981, includes use on level 3 under igneous rocks(1) ultramafics(2). See list H. Before 1981, included use on level 3 under igneous rocks(1) ultramafic family(2).
BA ultramafics
BT igneous rocks

Aries Valley
River valley in W central Transylvania. Also search Aries.
BT Transylvania
Romania

Arikaree Group
Comprises Sharps, Monroe Creek, and Harrison Formations. Colorado, S South Dakota, SE Montana, and SE Wyoming.
BT Miocene
SA Colorado
Montana
South Dakota
Wyoming

Arikareean
Provincial series, North America.
BA Tertiary
BT Cenozoic
SA lower Miocene
Neogene
Paleogene
upper Oligocene

Arita River
W Kii Peninsula S of Osaka. Also search Aritagawa.
UF Aritagawa
BT Honshu
Japan

Aritagawa
use Arita River

Ariyalur
Town in S.
BT Tamil Nadu
India

Ariyalur Stage
S India.
BT Upper Cretaceous
Cretaceous
SA India

Arize Massif
N of the central Pyrenees.
BT France

Arizona
Includes use on level 1 as an area term (list O). For term set options see list B.
CO N311500N370000
W1090000W1150000
BA Southwestern U.S.
United States
NT Apache County
Bisbee
Black Mesa
Black Mesa Basin
Cochise County
Coconino County
Flagstaff
Gila County
NX Graham County
NT Grand Canyon
Greenlee County
Holbrook
Hopi Buttes Field
NX Jerome
NT Marble Canyon
Maricopa County
Mogollon Plateau
Mohave County
Naha test well
Navajo County
O'Leary Peak
Petrified Forest National Park
Phoenix
Pima County
Pinal County
Salt River
San Carlos Indian Reservation
San Francisco Mountain
San Francisco Peaks
San Manuel
Santa Catalina Mountains
NX Santa Cruz County
NT Santa Rita Mountains
Sierra Ancha
Sierrita Mountains
Superstition Mountains
Tombstone
Tonto Basin
Tucson
Tucson Basin
Tucson Mountains
University of Arizona
Verde Valley
Yavapai County
Yuma
NX Yuma County
SA Basin and Range Province
Bidahochi Formation
Canyon Diablo Meteorites
Chinle Formation
Chuar Group
Coconino Sandstone
Colorado Plateau
Colorado River
Colorado River basin
Cutler Formation

Four Corners
Gila River
Kaibab Formation
Kayenta Formation
Lake Mead
Lake Powell
Mancos Shale
Martin Formation
Mesaverde Group
Moenkopi Formation
Morrison Formation
Navajo Indian Reservation
Navajo Sandstone
Paradox Basin
Pedregosa Basin
Redwall Limestone
San Francisco Mountains
San Juan Basin
Sonoran Desert
Supai Formation
Todilto Formation
Toroweap Formation
Unkar Group
Verde Formation
Virgin River valley
White Mountains

Arkansas
Includes use on level 1 as an area term (list O). For term set options see list B.
CO N330000N363000
W0894000W0944000
BA Southern U.S.
United States
NT Barber County
Batesville
NX Benton County
Boone County
Calhoun County
Carroll County
Clark County
Clay County
Columbia County
Crawford County
Crittenden County
Dallas County
NT Fayetteville
NX Franklin County
Fulton County
Grant County
Greene County
Howard County
Jackson County
Jefferson County
Johnson County
Lafayette County
Lawrence County
Lee County
Lincoln County
Logan County
Madison County
Marion County
Monroe County
Montgomery County
Nevada County
Newton County
Perry County
Phillips County
Pike County
Polk County
Pope County
Pulaski County
Randolph County
Saline County
Scott County
Sevier County
Union County
Washington County
White County
SA Arkansas River
Arkansas River valley
Arkoma Basin
Atoka Formation
Bloyd Formation
Bossier Formation
Brassfield Formation
Buckner Formation
Chattanooga Shale
Cotton Valley Group
Desmoinesian
Everton Formation
Fayetteville Formation
Fernvale Formation
Jackfork Group
Johns Valley Formation
Midcontinent
Midway Group
Mississippi Embayment
Mississippi River
Mississippi Valley
Morrow Formation
New Madrid region
Norphlet Formation
Ouachita Mountains
Ozark Mountains
Paluxy Formation
Pitkin Limestone
Red River
Red River valley
Saint Peter Sandstone
Saratoga Chalk
Savanna Formation
Schuler Formation
Smackover Formation
Smithville Formation
Stanley Group
Stuttgart
Tennessee Sandstone
Trinity Group
Washita Group
White River
Woodbine Formation

Arkansas City
Cowley County in S.
BT Kansas
United States

Arkansas River
S central U.S., flows 1,450 mi. generally ESE from the Rocky Mts. of central Colorado, through Kansas, Oklahoma, and Arkansas, to Mississippi River N of Greenville, Mississippi. Index states as applicable. Also search Arkansas River valley.
BT United States
SA Arkansas
Colorado
Kansas
Mississippi
Oklahoma

Arkansas River valley
Index states as applicable.
BT United States
SA Arkansas
Colorado
Kansas
Oklahoma

Arkhangelsk
City and oblast on the White Sea in NW Russian Republic. Also search Archangel.
UF Archangel
BT Russian Republic
USSR

Arkoma Basin
Index states as applicable.
BT United States
SA Arkansas
Oklahoma

arkose
Includes use on level 3 under sedimentary rocks(1) clastic rocks(2). See list I.
BA clastic rocks
BT sedimentary rocks
SA arkosic composition
meta-arkose
sandstone
subarkose
terrigenous materials

arkosic composition
Term introduced in 1978. Includes use on level 3 under sediments(1) and sedimentary rocks(1). Before 1978, also search arkosic AND specific sediment type or sedimentary rock.
SA arkose
composition
sedimentary rocks
sediments

Arlan
Oil bearing region in Bashkiria in S Urals.
BT Russian Republic
USSR

armalcolite
BA oxides
BT minerals
SA pseudobrookite

Armenia
Armenian Soviet Socialist Republic.
CO N390000N410000
E0460000E0430000
BT USSR
NT Agarak
Akhtala
Alaverdi
Aragats
Arteni
Dastakert
Kadzharan
Kafan
Lake Sevan
Megri
Megrinskiy Pluton
Pambak
Razdan
Sevan
Shamlug
Sisian
Yerevan
Zod
SA Caucasus
European USSR
Lesser Caucasus
Oktemberyan Series
Shakhdag Range
Transcaucasia
Zangezur

Armidale
Town in NE.
BT New South Wales
Australia

armored mud balls
Includes use on level 3 under sedimentary structures(1) secondary structures(2). See list K.
UF mud balls, armored
BA secondary structures
sedimentary structures
SA clay

Armorica
use Armorican Massif

Armorican Massif
NW part of country.
CO N463000N494500
W0000000W0045000
UF Armorica
BT France

Arno River Basin
N central.
BT Tuscany
Italy

aromatic hydrocarbons
Term introduced in 1978. Before 1978, search aromatic or aromatics AND hydrocarbons.
BA organic materials
SA hydrocarbons

Aroostook County
Term introduced in 1981. N Maine.
CO N453500N473000
W0674700W0700200
BT Maine
United States

Arosa Bay
On the Atlantic Ocean in W between provinces of La Coruna and Pontevedra.
BT Galicia
Spain

Arran
Island in Firth of Clyde in SW.
CO N552500N554500
W0050500W0053000
BT Scotland
Great Britain
United Kingdom

arrays
As of 1981, restricted to seismology. Includes use on level 3 under seismology(1). Before 1981, also used under geophysical methods(1).
SA LASA
networks
seismology

arrival time
SA earthquakes
elastic waves
seismology

Arrow Canyon Range
BT Nevada
United States

arroyos
SA geomorphology

arsenates
Includes use on level 2 under minerals(1). See list L.
BT minerals
NA adamite
austinite
NZ beudantite
NA conichalcite
haidingerite
legrandite
NZ mimetite
NA pharmacolite
pharmacosiderite
scorodite

arsenic
Includes use on level 1 and 2 as a chemical element (list D). As of 1981, use arsenic ores for arsenic as a commodity. Before 1981, included use on level 1 as a commodity term.
UF As
SA arsenic ores
arsenopyrite
elements
native elements

arsenic ores
Term introduced in 1981. Includes use on level 1 and 2 as a commodity term (list C). Before 1981, search arsenic AND ore deposits.
SA arsenic
native elements

arsenical pyrites
use arsenopyrite

arsenides
As of 1981, includes use on level 2 under minerals(1). See list L. Before 1981, included use on level 3 under minerals(1) sulfides(2).
BA sulfides
BT minerals
NA arsenopyrite
cobaltite
lollingite
maucherite
niccolite
pararammelsbergite
rammelsbergite
safflorite

skutterudite
sperrylite

arsenites
Includes use on level 2 under minerals(1). See list L.
BT minerals
NA mixite

arsenopyrite
As of 1981, includes use on level 3 under minerals(1) arsenides(2). See list L. Before 1981, included use on level 3 under minerals(1) sulfides(2).
UF arsenical pyrites
white pyrites
BA arsenides
sulfides
BT minerals
SA arsenic
lollingite

arsenosulfides
BA sulfides
BT minerals

Artemisia
Genus. Includes use on level 3 under angiosperms(1) Dicotyledoneae(2) or under palynomorphs(1) miospores(2).
BA Dicotyledoneae
angiosperms
BT Spermatophyta
Plantae
SA miospores
palynomorphs

Artemovsk
City in the Donets Basin.
UF Bakhmut
BT Ukraine
USSR

Arteni
Region.
BT Armenia
USSR

artesian waters
Includes use on level 2 under ground water(1). Before 1976, also search artesian waters and wells.
SA aquifers
ground water
hydrogeology
water
water resources

artesian waters and wells
A valid term through 1971. After 1972, use artesian waters.

Arthrodira
Includes use on level 3 under Pisces(1) Placodermi(2).
BA Placodermi
Pisces
BT Vertebrata

Arthropoda
Includes use on level 1 and 2 as a fossil term (list F).
BT Invertebrata
NA Chelicerata
Mandibulata
Trilobitomorpha

Arthur Harbor
On Anvers Island. Off NW Antarctic Peninsula.
BT Antarctica

Articulata
Includes use on level 2 under Brachiopoda(1). See list F.
BA Brachiopoda
BT Invertebrata
NA Dictyonellidina
Orthida
Pentamerida
Rhynchonellida
Spiriferida
Strophomenida
Terebratulida
Thecideidina
SA Inarticulata

Articulatae
Includes use on level 3 under pteridophytes(1). See list F.
BA Sphenopsida
pteridophytes
BT Plantae
SA Sphenopteris

artifacts
Includes use on level 3.
SA archaeological sites
archaeology
fossil man

artificial
A valid term through 1977. After 1977, use artificial recharge.

artificial minerals
Not a valid index term for GeoRef. Use synthetic materials on level 3 under minerals(1).

artificial recharge
Also search artificial AND recharge.
SA ground water
recharge

artinite
BA carbonates
BT minerals

Artinskian
Europe. Above Sakmarian, below Kungurian. Includes use on level 3 under age terms(1). See list E.
BA Lower Permian
Permian
BT Paleozoic

Artiodactyla
Includes use on level 2 under Mammalia(1). See list F.
BA Mammalia
BT Tetrapoda
Vertebrata
NA Hippopotamus
Ruminantia
Suiformes
Tylopoda

Artois
Historic region in extreme N to the S and SW of Flanders.
BT France

Arunta Complex
BT Archean
Precambrian
SA Australia
Northern Territory

Arve Valley
River Valley. Index countries as applicable.
BT Europe
SA France
Switzerland

Arvicolidae
BA Myomorpha
Rodentia
Mammalia
BT Tetrapoda
Vertebrata

As
use arsenic

Asama
Active volcano 90 miles NW of Tokyo. Also search Asamayama.
UF Asama Volcano
Asamayama
BT Honshu
Japan

Asama Volcano
use Asama

Asamayama
use Asama

Asaphidae
Family. Includes use on level 3 under Trilobita(1) Ptychopariida(2).
BA Ptychopariida
Trilobita
BT Trilobitomorpha
Arthropoda
Invertebrata

asbestos
As of 1981, use asbestos deposits for asbestos as a commodity. Includes use on level 3 under minerals(1). Before 1981, included use on level 1 as a commodity term.
BA silicates
BT minerals
SA actinolite
amosite
anthophyllite
asbestos deposits
chain silicates
chrysotile
sheet silicates

asbestos deposits
Term introduced in 1981. Includes use on level 1 and 2 as a commodity term (list C). Before 1981, search asbestos AND deposits.
SA asbestos

Ascension Island
British posession 700 miles NW of Saint Helena.
BT Atlantic Ocean

ascharite
use szaibelyite

Aschersleben
City in N.
BT Halle Bezirk
East Germany

aschynite
use aeschynite

Ascoli Piceno
Town and province in S.
BT Marches
Italy

ASEAN
Term introduced in 1981.
SA international cooperation

aseismic design
Term introduced in 1981.
SA design
engineering geology

aseismic margins
Term introduced in 1981.
SA continental margin
passive margins
plate tectonics

aseismic ridges
Term introduced in 1978. Before 1978, search aseismic.
SA mid-ocean ridges

ash
As of 1978 term is used to indicate "artificial" ash. Also search fly ash.
UF fly ash
industrial ash
SA tuff
volcanic ash

ash falls
SA pyroclastics
volcanic ash
volcanism
volcanoes
volcanology

ash flows
As of 1981, includes use on level 3 under igneous rocks(1) pyroclastics(2). See list H. Before 1981, included use on level 3 under igneous rocks(1) pyroclastics and glasses(2).
UF glowing avalanche
BA pyroclastics
volcanic rocks
BT igneous rocks
SA ash-flow tuff
flows
ignimbrite
tuff
volcanic ash
volcanism

Ash Hollow Formation
In Ogallala Group. W Kansas, W Nebraska, South Dakota, and SE Wyoming.
BT Pliocene
SA Kansas
Nebraska
Ogallala Formation
South Dakota
Wyoming

ash-flow tuff
As of 1981, includes use on level 3 under igneous rocks(1) pyroclastics(2). See list H. Before 1981, included use on level 3 under igneous rocks(1) pyroclastics and glasses(2).
BA pyroclastics
volcanic rocks
BT igneous rocks
SA ash flows
ignimbrite
tuff

Ashe Formation
NW North Carolina, E Tennessee, and SW Virginia.
BT lower Paleozoic
Paleozoic
SA North Carolina
Tennessee
Virginia

Ashgillian
Europe. Upper Caradocian, below Llandoverian (Silurian). Includes use on level 3 under age terms(1). See list E.
BA Upper Ordovician
Ordovician
BT Paleozoic
SA Bala

Ashio
Town WNW of Utsunomiya in central.
BT Honshu
Japan

Ashkhabad
City in S near the Iranian border.
BT Turkmenia
USSR

Ashmore Meteorite
Term introduced in 1981. Impact in Gaines County, W Texas. Before 1981, also search Ashmore AND meteorites.
BA H chondrites
chondrites
BT meteorites
SA Texas

ashtonite
use mordenite

Asia
Includes use on level 1 or 2 as an area term (list O). If 1, see term set options under list B. To retrieve all documents, individual countries and physiographic regions should also be searched (see list O).
NT Afghan-Tadzhik Depression
Altai Mountains
Altai-Sayan region
Amu Darya
Amur Basin
Amur River
NA Arabian Peninsula
NT Argun River
Badakhshan
Bengal

Brahmaputra River
Dzhungarian Alatau
Euphrates River
NA Far East
NT Ganges River
Gobi Desert
NA Himalayas
NT Hindu Kush
Hyderabad
Ili Basin
Indian Peninsula
Indian Shield
Indus River
Irtysh River
Kangwon
Karakoram
Kerulen River
Khanka Lake
Kosi Basin
Kumaun Himalayas
Ladakh
Lesser Himalayas
Murgab Basin
Ramnagar
Selenga River valley
Siwalik Range
Southeast Asia
Strait of Malacca
Tannu-Ola Range
Tarbagatay Range
Turkestan
Vindhyan Basin
Yenisei Basin
Yenisei-Khatanga basin
Zagros
SA Eurasia
Pacific mobile belt
Phuket Group
Tethys

Asiatic
A valid term through 1977 used in combination with USSR. After 1977, use Asiatic USSR.

Asiatic USSR
Term introduced in 1978. Includes the Russian Republic E of the Urals and the republics of Soviet Central Asia. Index Soviet republics as applicable. Before 1978, search Asiatic AND USSR.
BT USSR
SA Kazakhstan
Kirghizia
Russian Republic
Tadzhikistan
Turkmenia
Uzbekistan

Askania gravimeters
Term introduced in 1981.
BT gravimeters

Aso
Central Kyushu. Also search Aso Volcano and Aso-san.
UF Aso Volcano
Aso-san
BT Kyushu
Japan
SA Aso Caldera

Aso Caldera
Central.
BT Kyushu
Japan
SA Aso

Aso Volcano
use Aso

Aso-san
use Aso

asphalt
Includes use on level 3 under organic materials(1); on level 3 as a commodity term under bitumens(1). See list C.
BA bitumens
organic materials

Assam
State NE of Bangladesh.
BA Northeastern India
BT India
NT Shillong
Shillong Plateau

Asselian
Stage. See list E.
BA Permian
BT Paleozoic

assemblages
Includes use on level 3 under fossil groups (list F). Also search fossil assemblages; faunal assemblages.
UF fossil assemblages
SA biofacies
communities
faunal provinces
fossils
index fossils

assimilation
Includes use on level 3 under magmas(1). As of 1981, restricted to magmas. Not to be used as a general term.
SA anatexis
contamination
country rocks
emplacement
hybridization
inclusions
intrusions
magmas
palingenesis

Assiniboine River
S Canada; 590 mi. long. Rises in SE Saskatchewan, flows S and E across S Manitoba into the Red River of the North at Winnipeg.
BT Canada
SA Manitoba
Saskatchewan

Assiniboine River valley
Important wheat growing region. Index provinces as applicable. Also search Assiniboine Valley.
UF Assiniboine Valley
BT Canada

Assiniboine Valley
use Assiniboine River valley

associations
Includes use on level 1 (list A). Reserved for organizations not included under museums or survey organizations. Before 1976, also search meetings. Term set options are:
topic [List B (except areal geology), or extraterrestrial geology, general, geophysics]
name of group, conference, meeting, symposium [conventional abbreviations for well known organizations may be used, e.g. AAPG, AGI, CNRS, COMECON, CSIRO, EURATOM, GSA, INQUA, IUGS, NASA, NATO, NOAA, UNESCO, VINITI]
SA annual report
bibliography
catalogs
collections
current research
education
expeditions
geology
guidebook
INQUA
institutions
international cooperation
International Geological Congress
IUGS
museums
organization
planning
programs
progress report
publications
report
research
survey organizations
symposia
United Nations

Assyntian Orogeny
use Assyntic Orogeny

Assyntic Orogeny
Term introduced in 1978. Orogenic event at the end of the Precambrian. Before 1978, search Assyntian.
UF Assyntian Orogeny
BT Precambrian
SA Cambrian
orogeny

Astarte
Genus. Includes use on level 3 under Mollusca(1). See list F.
BA Bivalvia
Mollusca
BT Invertebrata

Astartida
Term introduced in 1981. Includes use on level 2 under Mollusca(1). See list F.
BA Bivalvia
Mollusca
BT Invertebrata

Asteridae
Term introduced in 1981. Includes use on level 2 under angiosperms(1). See list F.
BA Dicotyledoneae
angiosperms
BT Spermatophyta
Plantae

Asterocyclina
Genus. Includes use on level 3 under foraminifera(1) Orbitoidacea(2).
BA Orbitoidacea
Rotaliina
foraminifera
BT Protista
Invertebrata

Asteroidea
As of 1981, includes use on level 2 under Echinodermata(1). See list F. Before 1981, included use on level 3 under Echinodermata(1) Stelleroidea(2).
BA Stelleroidea
Asterozoa
Echinodermata
BT Invertebrata

asteroids
Includes use on level 1 (list A). Term set options are:
topic [composition, cosmochemistry, distribution, evolution, genesis, observations, properties, theoretical studies]
(name of asteroid) or subtopic
UF minor planets
SA cosmic dust
cosmic rays
cosmochemistry
interplanetary space
meteorites
meteoroids
Moon
planetology
planets
solar system
tektites

Asterozoa
As of 1981, includes use on level 3 under Echinodermata(1). See list F. Before 1981, included use on level 2 under Echinodermata(1).
BA Echinodermata
BT Invertebrata
NA Stelleroidea

asthenosphere
Includes use on level 3 under tectonophysics(1) and under plate tectonics(1).
UF zone of mobility
SA crust
Earth
elastic waves
isostasy
lithosphere
lower crust
magmas
mantle
Mohorovicic discontinuity
plate tectonics
tectonophysics
upper mantle

Asti
City and province in central.
BT Italy
SA Piemonte

Astian
Europe. Above the Piacenzian. Includes use on level 3 under age terms(1). See list E.
BA upper Pliocene
Pliocene
Neogene
Tertiary
BT Cenozoic

Astoria Canyon
Just off Astoria, Oregon and the mouth of the Columbia River.
BT Pacific Ocean

Astrakhan
City and oblast on NW Caspian Sea.
BT Russian Republic
USSR

Astrapotheria
Includes use on level 2 under Mammalia(1). See list F.
BA Mammalia
BT Tetrapoda
Vertebrata

astroblemes
Includes use on level 3 under meteor craters(1) or geomorphology(1) impact features(2). Also search fossil meteor craters.
UF fossil meteorite craters
SA craters
cryptoexplosion features
geomorphology
impact features
meteor craters

astronomy
SA astrophysics and solar physics
planetology

astrophyllite
BA sheet silicates
silicates
BT minerals

astrophysics and solar physics
Usually out-of-scope for GeoRef. Includes use on level 1. Term set options are:
corona
subtopic
electromagnetic radiation
type of radiation or subtopic
solar flares
type of radiation or subtopic
instruments
name of instrument
phenomena measured
magnetic field
subtopic
particle radiation
name of particle or subtopic
surface phenomena

name of phenomena or subtopic
UF solar physics, astrophysics and
SA astronomy
aurora
corona
electromagnetic radiation
extraterrestrial geology
geophysics
interplanetary space
magnetic field
magnetosphere
particle radiation
planetology
solar cycles
solar flares
Sun
sunspots
surface phenomena

Astrorhizidae
Family. Includes use on level 3 under foraminifera(1) Ammodiscacea(2).
BA Ammodiscacea
Textulariina
foraminifera
BT Protista
Invertebrata

Asturian Arc
Term introduced in 1981.
BT Spain

Asturian Massif
In NW part of country.
BT Asturias
Spain

Asturian Orogenic Phase
use Asturian Orogeny

Asturian Orogeny
Term introduced in 1978. Before 1978, search Asturian AND orogeny.
UF Asturian Orogenic Phase
BT Carboniferous
SA orogeny

Asturias
Region and ancient kingdom in NW. Before 1981, also search Oviedo.
CO N425400N434000
W0043000W0071500
UF Oviedo
BT Spain
NT Asturian Massif
Aviles
La Caridad
SA Cantabrian Basin

Asturreta
Region in N part of country.
BT Navarra
Spain

Aswan
Governate in SE Egypt. Also a city on the Nile and name of dam.
BT Egypt
NT Kom Ombo

asymmetric folds
Term introduced in 1978. Includes use on level 3 under folds(1) style(2). Before 1978, also search folds AND asymmetric.
BA folds
SA symmetric folds

asymmetry
use symmetry

Ata Caldera
On uninhabited island of Ata in S Tonga in S Pacific Ocean.
BT Tonga
Polynesia

Atacama
Province in N.
BT Chile
NT Copiapo
SA Atacama Desert

Atacama Desert
N central part of country. Index provinces as applicable.
BT Chile
SA Antofagasta
Atacama

atacamite
As of 1981, includes use on level 3 under minerals(1) chlorides(2). See list L. Before 1981, included use on level 3 under minerals(1) halides(2).
BA chlorides
halides
BT minerals
SA paratacamite

Atascosa County
Term introduced in 1981. SW Texas.
CO N283800N291500
W0981000W0985000
BT Texas
United States

Atasu
Village in central, NW of Lake Balkhash.
BT Kazakhstan
USSR

ataxite
Term introduced in 1981.
BA iron meteorites
BT meteorites
NA Campo del Cielo Meteorite
Warburton Meteorite

Ataxophragmiidae
Family. Includes use on level 3 under foraminifera(1) Lituolacea(2).
BA Lituolacea
Textulariina
foraminifera
BT Protista
Invertebrata

Atchafalaya Bay
On the Gulf of Mexico off Saint Mary and Terrebonne parishes.
BT Louisiana
United States
SA Gulf Coastal Plain

Atchison
City in Atchison County in NE.
BT Kansas
United States

Atchison County
NE Kansas.
BT Kansas
United States

atectonic processes
Term introduced in 1981.
UF nontectonics
SA processes
sedimentation
tectonics

Atera Fault
Gifu Prefecture in central.
BT Honshu
Japan

Athabasca District
Term introduced in 1981. Uranium-mining district in Saskatchewan.
BT Saskatchewan
Canada

Athabasca Formation
Term introduced in 1981. Primarily an orthoquartzite. Outcrops over a large area S of Lake Athabasca in Churchill Province.
BT Proterozoic
Precambrian
SA Churchill Province
Saskatchewan

Athabasca Glacier
In Rocky Mountains between Jasper and Banff national parks near the British Columbia border.
BT Alberta
Canada

Athabasca Oil Sands
Term introduced in 1981. In E Alberta. One of the largest petroleum reservoirs in the world. Oil is found in the McMurray, Clearwater, and Grand Rapids formations.
BT Alberta
Canada

Athabasca River
Flows into Lake Athabasca.
UF Athabaska River
BT Alberta
Canada

Athabaska River
use Athabasca River

Athens
City in Ohio; city in Georgia; and city in Central Greece and Euboea. Index Greece/Georgia/or Ohio as applicable.
SA Georgia
Greece
Ohio

Athgarh Sandstone
Sandstones of this formation are also grouped as part of Cuttack Stage. Part of Upper Gondwana sequence of the E coast.
BT Jurassic
SA India

Atlanta
City in Fulton County in N central.
BT Georgia
United States

Atlantic
Used primarily in Europe for an interval of post/glacial time following the Boreal and preceding the Subboreal. It corresponds to most of the Altithermal and the middle part of the Hypsithermal.
BA Holocene
Quaternary
BT Cenozoic

Atlantic Coastal Plain
Index states as applicable. Includes use on level 1 as an area term (list O). For term set options see List B.
BA United States
North America
NT Central Atlantic Coastal Plain
Chesapeake Bay
Southern Atlantic Coastal Plain
SA Carolina Bays
Connecticut
Delaware
Florida
Georgia
Maryland
Massachusetts
New Jersey
New York
New York Bight
North Carolina
Rhode Island
South Carolina
Virginia

Atlantic County
Term introduced in 1981. SE New Jersey.
CO N391700N394400
W0741900W0745900
BT New Jersey
United States

Atlantic Ocean
Includes use on level 1 or 2 as an area term (list O). If 1, see term set options under list B.
CO S550000N750000
E0200000W0800000
NT Argentine Basin
Ascension Island
Atlantis fracture zone
Atlantis II
Atlantis Seamount
Barbados Ridge
Barracuda Ridge
Bay of Fundy
Bermuda Platform
Bermuda Rise
Blake Plateau
Blake-Bahama Basin
Blake-Bahama Outer Ridge
Block Island Sound
Bouvet Island
Cape Basin
Cape Verde Basin
Cape Verde Rise
Celtic Sea
Davis Strait
NA East Atlantic
NZ Falkland Islands
NT Falkland Plateau
Flemish Cap
Georges Bank
Gibbs fracture zone
Grand Banks
Great Bahama Bank
Great Meteor Seamount
Gulf of Maine
Gulf Stream
Hare Bay
Hatteras abyssal plain
NX Hudson Bay
NT Iceland-Faeroe Ridge
Labrador Basin
Long Island Sound
Maury Channel
Mid-Atlantic Ridge
NA North Atlantic
Northeast Atlantic
Northwest Atlantic
NT Puerto Rico Trench
Reykjanes Ridge
Rockall Bank
Rockall Plateau
Rockall Trough
Romanche fracture zone
Romanche Trench
Sable Island Bank
Saint Paul Rocks
Sargasso Sea
Scotia Ridge
Scotian Shelf
NZ Shetland Islands
NT Sierra Leone Rise
NA South Atlantic
Southwest Atlantic
NT Strait of Gibraltar
Straits of Florida
Tristan da Cunha
Trou Sans Fond
Vema fracture zone
Walvis Ridge
NA West Atlantic
NT Wilkinson Basin
Wilmington Canyon
Yucatan Channel
SA Antarctic Ocean
Atlantic Ocean Islands
Atlantic region
Bering Strait
DSDP Site 105
DSDP Site 327
DSDP Site 329
DSDP Site 330
DSDP Site 332
DSDP Site 333
DSDP Site 334
DSDP Site 335
DSDP Site 356
DSDP Site 357
DSDP Site 361
DSDP Site 362
DSDP Site 363
DSDP Site 364
DSDP Site 367

DSDP Site 368
DSDP Site 369
DSDP Site 370
DSDP Site 391
DSDP Site 395
DSDP Site 396
DSDP Site 417
DSDP Site 418
FAMOUS
Iapetus
Jan Mayen
Leg 1
Leg 11
Leg 12
Leg 14
Leg 2
Leg 3
Leg 36
Leg 37
Leg 38
Leg 39
Leg 4
Leg 40
Leg 41
Leg 43
Leg 44
Leg 45
Leg 46
Leg 47
Leg 48
Leg 49
Leg 50
Leg 51
Leg 52
Leg 53
Scotia Sea
South Sandwich Islands

Atlantic Ocean Islands
Term introduced in 1981. Includes use on level 1 as an area term (list O). For term set options, see list B.
NA Azores
Bermuda
Canary Islands
NZ Cape Verde Islands
NA Madeira
NT Saint Helena
Sao Tome
SA Atlantic Ocean
Falkland Islands

Atlantic region
Term introduced in 1976. An artificial term used to indicate the coastal region immediately adjacent to the Atlantic Ocean. Includes land and the immediate littoral zone. Includes use on level 1 and 2 as an area term (list O). For term set options see list B.
SA Atlantic Ocean

Atlantic-type margins
use passive margins

Atlantis
Term introduced in 1981. Legendary island or continent said to have been destroyed by a cataclysm.
SA popular geology

Atlantis fracture zone
Across the Mid-Atlantic Ridge in N.
BT Atlantic Ocean
SA Mid-Atlantic Ridge

Atlantis II
Seamounts off U.S. E coast roughly midway between Nova Scotia and Bermuda.
BT Atlantic Ocean

Atlantis II Deep
BT Red Sea

Atlantis Seamount
Term introduced in 1981. SW of the Azores. Before 1981, search Atlantis.
BT Atlantic Ocean

atlas
Includes use on level 3 under glossaries(1).
SA glossaries
maps
report

Atlas
use Atlas Mountains

Atlas Mountains
Mountain system of NW and N Africa extending from SW Morocco to NE Tunisia. Index countries as applicable. Also search Atlas AND appropriate area.
UF Atlas
BT Africa
NZ Moroccan Atlas Mountains
SA Algeria
Anti-Atlas
High Atlas
Middle Atlas
Morocco
Tunisia

Atlin
Village near Alaska border NE of Juneau in NW.
BT British Columbia
Canada

atmosphere
Includes use on level 1(only for Earth); on level 2 or 3 under Moon(1) or other planets; on level 3 under environmental geology(1) pollution(2). For studies dealing with geochemical changes (short-term or during geologic time), observations related to interactions with phenomena of the Earth's surface, and other geophysical aspects. For changes in climate, see paleoclimatology. If 1, term set options are:
topic [age, changes, circulation, composition, evolution, general, genesis, instruments, observations, processes, properties, temperature]
subtopic
NA upper atmosphere
SA aeronomy
aerosols
air
air-sea interface
atmospheric precipitation
atmospheric pressure
biosphere
boundary interactions
circulation
climate
convection
convection currents
cycles
degassing
Earth
Ekman spiral
environmental geology
fallout
geochemistry
geophysics
glacial geology
humidity
hydrology
hydrosphere
ionosphere
magnetosphere
meteorology
Moon
ozone
paleoclimatology
planetology
pollution
sea water
storms
stratosphere
sulfur dioxide
troposphere
volcanology

atmospheric precipitation
As of 1978, term is used on level 2 under hydrology(1). Before 1978, also search hydrology AND precipitation.
SA atmosphere
climate
evapotranspiration
hydrologic cycle
hydrology
meteorology
moisture
precipitation
raindrops
rainfall
retention
snow
water

atmospheric pressure
BT pressure
SA atmosphere

Atoka Formation
Sandstone members: Coody (Coata), Pope Chapel, Georges Fork, Dirty Creek, Webbers Falls, and Blackjack School. Subdivided, in Arkansas, to include Greenland Sandstone Member at base.
BT Atokan
Middle Pennsylvanian
Pennsylvanian
SA Arkansas
Oklahoma

Atokan
Provincial series, North America: lower Middle Pennsylvanian. Above Morrowan, below Desmoinesian. Includes use on level 3 under age terms(1). See list E.
BA Middle Pennsylvanian
Pennsylvanian
Carboniferous
BT Paleozoic
NT Atoka Formation

atolls
Includes use on level 3 under reefs(1).
BT reefs
SA bioherms
Coelenterata
islands
lagoons

atomic absorption
Includes use on level 3 under spectroscopy(1) methods(2); under chemical element(1) analysis(2). See list D.
SA absorption
analysis
spectroscopy

atomic packing
Term introduced in 1978 for use under crystal structure(1), crystallography(1), or mineralogy(1). Before 1978, also search packing.
SA crystal structure
crystallography
mineralogy
packing

Atrypa
Genus. As of 1981, includes use on level 3 under Brachiopoda(1) Spiriferida(2). See list F. Before 1981, included use on level 3 under Brachiopoda(1) Articulata(2).
BA Atrypidae
Spiriferida
Articulata
Brachiopoda
BT Invertebrata

Atrypidae
As of 1981, includes use on level 3 under Brachiopoda(1) Spiriferida(2). See list F. Before 1981, included use on level 3 under Brachiopoda(1) Articulata(2).
BA Spiriferida
Articulata
Brachiopoda
BT Invertebrata
NA Atrypa

Atsumi Peninsula
S of Nagoya.
BT Honshu
Japan

attapulgite
use palygorskite

attenuation
Includes use on level 3 under seismology(1). Also search damping.
SA elastic waves
seismology
wave dispersion

Atterberg limits
UF consistency limits
limits, Atterberg
SA deformation
plasticity
rock mechanics
soil mechanics

Attica
Ancient division and state. Currently a department. Central Greece and Euboea.
BT Greece

Attock
Town in NW.
BT Punjab Province
Pakistan

Aturia
Genus. As of 1981, includes use on level 3 under Mollusca(1) Nautiloidea(2). See list F. Before 1981, included use on level 3 under Mollusca(1) Cephalopoda(2).
BA Nautiloidea
Tetrabranchiata
Cephalopoda
Mollusca
BT Invertebrata

Au
use gold

Aube
Department in N central.
CO N475500N484500
0E004500E0032000
BT France

Aubrac
Mountain range in S central.
BT France

aubrite
use chladnite

Auckland
City in NW.
BT North Island
New Zealand

Auckland Islands
200 miles south of New Zealand.
BT Pacific Ocean

Aude
Department in S.
CO N423000N434000
E0031500E0014500
BT France
NT Corbieres
Mouthoumet Massif
Narbonne
Portel
Quillan
Salsigne Mine
SA Montagne Noire

augen gneiss
Includes use on level 3 under metamorphic rocks(1) gneisses(2). See list J.

BA gneisses
BT metamorphic rocks

augite
BA clinopyroxene
pyroxene group
chain silicates
silicates
BT minerals
SA basalt
fassaite
omphacite
titanaugite

augitite
As of 1981, includes use on level 3 under igneous rocks(1) ultramafics(2). See list H. Before 1981, included use on level 3 under igneous rocks(1) ultramafic family(2).
BA ultramafics
BT igneous rocks

Augusta
City in Kennebec County in S.
BT Maine
United States

Augusta County
Western Virginia.
BT Virginia
United States

Augustine
Volcano on Augustine Island in Kamishak Bay at mouth of Cook Inlet in S Alaska. Also search Augustine Volcano.
UF Augustine Volcano
BT Alaska
United States

Augustine Volcano
use Augustine

Augustow
Town in NE part of country.
BT Bialystok
Poland

aulacogens
Includes use on level 3 under tectonics(1).
SA grabens
tectonics
troughs

aureoles
Includes use on level 3 under metamorphism(1) and intrusions(1). Also search contact aureole; contact zone; metamorphic aureoles; metamorphic zone; thermal aureoles.
UF contact aureole
contact zone
exomorphic zone
metamorphic aureoles
metamorphic zone
thermal aureole
SA contact metamorphism
country rocks
haloes
intrusions
metamorphism

Aures
Mountain massif in NE.
BT Algeria

aurora
Usually out-of-scope for GeoRef. Includes use on level 1 for special bibliographies; on level 2 under ionosphere(1). If 1, term set options are:
absorption
subtopic
auroral zone (including auroral oval)
subtopic
electrical field
topic [arcs, currents, electrojet, particle precipitation]
magnetic field
topic [currents, electrojet, magnetic storms, micropulsations, type of aurora]
morphology
motions or subtopic
orientation or subtopic
synoptic studies or subtopic
time variations or subtopic
observations
type of aurora
theoretical studies
topic
SA absorption
absorption and scattering
aeronomy
arcs
astrophysics and solar physics
auroral oval
auroral zone
currents
electrical field
electrojet
intensity
ionosphere
magnetic field
magnetic storms
magnetosphere
meteorology
micropulsations
morphology
motions
solar wind
time variations

auroral oval
Includes use on level 3 under aurora(1).
SA aurora
auroral zone

auroral zone
Including auroral oval. Includes use on level 2 under aurora(1).
UF zone, auroral
SA aurora
auroral oval

Austin
City in Travis County in central.
BT Texas
United States

Austin Group
Comprises Ector Tongue, Bonham Clay, Blossom Sand, Brownstone Marl, Gober Tongue, Austin Chalk, and Burditt Marl. E Texas. Also search Austin Chalk.
BT Gulfian
Upper Cretaceous
Cretaceous
SA Texas

austinite
BA arsenates
BT minerals

Austral Islands
Group in French Oceania 330 miles S of Society Islands. Also search Austral.
UF Tubuai Islands
BT French Polynesia
Polynesia

Australasia
Islands of the South Pacific including Australia, New Zealand, and Papua New Guinea. Index countries as applicable. Includes use on level 1 or 2 as an area term (list O). If 1, see term set options under list B.
NA Australia
New Zealand
Papua New Guinea
SA Oceania
Solomon Islands

Australia
Includes use on level 1 or 2 as an area term (list O). If 1, see term set options under list B. To retrieve all documents, individual countries and physiographic regions should also be searched (see list O).
CO S440000S100000
E1540000E1130000
BA Australasia
NT Amadeus Basin
Australian Shield
Bass Strait
Canberra
Carpentaria Basin
Cooper Basin
Eucla Basin
Gambier Embayment
Georgina Basin
Giles Complex
Great Artesian Basin
Joseph Bonaparte Gulf
Lachlan fold belt
Murray Basin
Murray River
Musgrave Ranges
New England Batholith
NA New South Wales
Northern Territory
NT Nullarbor Plain
Officer Basin
Otway Basin
NA Queensland
NT Simpson Desert
Snowy Mountains
NA South Australia
NT Tasman Geosyncline
NA Tasmania
Victoria
Western Australia
SA Arafura Sea
Arunta Complex
Baralaba Coal Measures
Bitter Springs Formation
Brockman Iron Formation
Broken River Formation
Bulli Seam
Gondwana
Hamersley Group
Hawkesbury Sandstone
Heemskirk Granite
Henbury Meteorite
Illawarra Coal Measures
Indian Plate
Ipswich Coal Measures
Lilydale Limestone
Mersey River
Mersey Valley
Murchison Meteorite
Narrabeen Group
Oceania
Singleton Coal Measures
Warburton Meteorite
Weekeroo Station Meteorite
Wolf Creek Meteorite

Australian Capital Territory
SE Australia, within state of New South Wales. Formerly called Federal Capital Territory.
UF Federal Capital Territory
BT New South Wales
Australia

Australian Plate
use Indian Plate

Australian Shield
SW and central.
BT Australia

australite
BA tektites
SA meteorites

Australopithecinae
Subfamily. Includes use on level 3 under Mammalia(1) Primates(2).
BA Primates
Mammalia
BT Tetrapoda
Vertebrata

Australopithecus
Genus. As of 1981, includes use on level 3 under Mammalia(1) Hominidae(2). See list F. Before 1981, included use on level 3 under Mammalia(1) Primates(2).
BA Hominidae
Primates
Mammalia
BT Tetrapoda
Vertebrata

Austria
Includes use on level 1 as an area term (list O). For term set options see list B.
CO N461500N490000
E0171500E0093000
BA Central Europe
Europe
NX Austrian Vienna Basin
NT Burgenland
Carinthia
NX Central Austrian Alps
NT Enns Valley
Gailtal Alps
Hohe Tauern
Koralpe Range
Limestone Alps
Lower Austria
NX North Austrian Alps
NT North Austrian Crystallines
North Austrian Molasse
Salzburg State
Salzkammergut
Semmering
NX South Austrian Alps
NT South Austrian Molasse
Styria
Tauern Tunnel
Totes Gebirge
Tyrol
Upper Austria
Venediger Group
Vorarlberg
Wechsel
SA Allgau Alps
Alps
Carnic Alps
Central Alps
Danube River
Danube Valley
Eastern Alps
Gosau Formation
Hallstatt Limestone
Inn Valley
Karawanken
Lake Constance
Molasse Basin
Morava River valley
Otztal Alps
Pannonia
Pannonian Basin
Rhaetian Alps
Rhine Basin
Rhine River
Vienna Basin
Wetterstein Limestone
Zillertal Alps

Austrian Vienna Basin
Term introduced in 1981.
CO N474000N485000
E0171000E0154000
BZ Vienna Basin
BX Austria

authigenesis
Used for mineral genesis under diagenesis(1).
SA authigenic minerals
diagenesis
minerals
recrystallization
sedimentary rocks

authigenetic minerals
use authigenic minerals

authigenic minerals
Also search authigenic AND minerals.
UF authigenetic minerals
BT minerals

SA authigenesis

autochthons
SA allochthons
overthrust faults
tectonics
windows

autocorrelation
BA statistical analysis
SA mathematical methods
statistical analysis

automatic cartography
Term introduced in 1981.
SA automatic data processing
cartography
maps

automatic data processing
Includes use on level 1 and 2 (list A) for computer applications and machine modeling of geological data. Before 1972, also search computer methods; computer techniques. Term set options are:
methods
name of method or application
topic [List B (except areal geology), or extraterrestrial geology, general, geophysics]
subtopic related to 2nd level term
UF data processing, automatic
SA algorithms
analog simulation
automatic cartography
computer programs
computers
data
data acquisition
data bases
data handling
data retrieval
data storage
digital simulation
Fortran
Fortran IV
Fourier analysis
graphic display
hand calculators
holography
information systems
mathematical geology
mathematical models
microcomputers
minicomputers
models
nomograms
punch cards
simulation
statistical analysis
trend-surface analysis
well-logging

autometamorphism
Includes use on level 2 under metamorphism(1). Also search autometasomatism.
BA metamorphism
SA albitization
alteration
hydrothermal alteration
igneous rocks
mineral assemblages
processes

autometasomatism
As of 1981, no longer a valid term for GeoRef. Use autometamorphism. Included use on level 3 under metasomatism(1).

autoradiography
Includes use on level 3 under geophysical methods(1) methods(2). Also search radiography.
UF radiography
SA analysis
geophysical methods
methods
radioactivity

Autun
Town in E central part of country.
BT Saone-et-Loire
France

Autunian
Europe. Above Stephanian, below Saxonian. Includes use on level 3 under age terms(1). See list E.
BA Lower Permian
Permian
BT Paleozoic
SA Rotliegendes

autunite
BA phosphates
BT minerals

Auvergne
Historical region in S central.
CO N443000N461500
E0044500E0020000
BT France

Auversian
Europe. Above Lutetian, below Bartonian. Includes use on level 3 under age terms(1). See list E.
BA upper Eocene
Eocene
Paleogene
Tertiary
BT Cenozoic
SA Priabonian

Auxerre
City in N central part of country.
BT Yonne
France

Avacha
Active volcano on S Kamchatka Peninsula. Also search Avacha Volcano.
UF Avacha Volcano
BT Russian Republic
USSR

Avacha Volcano
use Avacha

avalanches
As of 1978, term is used on level 2 under geologic hazards(1).
SA geologic hazards
landslides
slope stability

Avallon
Town in N central part of country.
BT Yonne
France

Avalon Peninsula
SW of Saint John's.
CO N463000N481000
W0523000W0541500
BT Newfoundland
Canada

Avalonian Orogenic Phase
use Avalonian Orogeny

Avalonian Orogeny
Term introduced in 1978. Used in connection with orogenic event near the end of Precambrian time along E border of North America. Named for the Avalon Peninsula in SE Newfoundland. Before 1978, search Avalonian AND orogeny.
UF Avalonian Orogenic Phase
BT Precambrian
SA orogeny

Aves
Includes use on level 1 and 2 as a fossil term (list F).
BT Tetrapoda
Vertebrata
NA Archaeornithes
Neornithes
SA birds
eggs

Aves Island
Uninhabited Venezuelan island just East of Bonaire, Netherlands Antilles. Also search Aves in combination with Caribbean Sea.
UF Bird Island
BT Caribbean Sea

Aves Ridge
S of Aves Island in W.
BT Caribbean Sea

Aveyron
Department in S central.
CO N434000N450000
E0033000E0014500
BT France
NT Decazeville
SA Central Massif
Rodez Trough
Rouergue

Avila
As of 1981, refers to only the city. Before 1981, also included the province. In W central Spain.
BT Avila Province
Old Castile
Spain

Avila Province
Term introduced in 1981. Before 1981, search Avila. In W central Spain.
CO N400500N411000
W0041000W0054500
BT Old Castile
Spain
NT Avila

Aviles
City on the Bay of Biscay.
BT Asturias
Spain

Avonian
Europe. Includes use on level 3 under age terms(1). See list E.
BA Carboniferous
BT Paleozoic
SA Culm
Dinantian

Awamoan
Stage. New Zealand.
UF Awamoan Stage
BA lower Miocene
Miocene
Neogene
Tertiary
BT Cenozoic

Awamoan Stage
use Awamoan

awaruite
As of 1981, includes use on level 3 under minerals(1) alloys(2). See list L. Before 1981, included use on level 3 under minerals(1) native elements and alloys(2).
BA alloys
BT minerals
SA iron
nickel

Awash Valley
River valley in E.
BT Ethiopia

Awashima
Island in Sea of Japan.
BT Honshu
Japan

Axel Heiberg Island
Largest of Sverdrup Islands of District of Franklin.
CO N780000N820000
W0850000W0970000
BT Northwest Territories
Canada

axes, fold
use fold axes

axial surface
use axial-plane structures

axial-plane structures
Term introduced in 1978. Used for foliation or cleavage. Includes use on level 3 under foliation(1) style(2) and under structural analysis(1) theoretical studies(2). Before 1978, also search axial-plane; axial plane; axial plane foliation; axial planes.
UF axial surface
SA cleavage
deformation
flow cleavage
folds
foliation
schistosity
slaty cleavage
structural analysis
structures

axinite
BA ring silicates
silicates
BT minerals
SA borosilicates

Ayacucho
Department in SW.
BT Peru

Ayrshire
County in SW.
CO N550000N555000
W0033000W0050000
BT Scotland
Great Britain
United Kingdom

Azalea Field
Midland County in W.
BT Texas
United States

Azerbaidzhan
Azerbaidzhan Soviet Socialist Republic in Transcaucasia. When referring to Iran, use Azerbaijan.
CO N383000N420000
E0500000E0450000
BT USSR
NT Akera
Apsheron Peninsula
Baku
Baku Archipelago
Bank
Dashkesan
Filizchay
Karabakh
Kedabek
Kirovabad
Kobystan
Lenkoran
Mugan Steppe
Nakhichevan
Neftechala
Neftyanyye Kamni
Ordubad
Peschanyy
Shemakha
Surakhany
SA Azerbaijan
Caspian Basin
Caucasus
European USSR
Greater Caucasus
Kura Lowland
Kura River
Lesser Caucasus
Shakhdag Range
Transcaucasia
Zangezur

Azerbaijan
Used when referring to Iran. For USSR, use Azerbaidzhan.
BT Iran
SA Azerbaidzhan

Azof
use Azov

Azolla
Genus.
BA Filicopsida
pteridophytes
BT Plantae

Azores
Island group belonging to Portugal in E central North Atlantic Ocean. Includes use on level 1 as an area term (list O). For term set options see list B.
CO N370000N400000
W0250000W0310000
BA Atlantic Ocean Islands
NT Sao Miguel Island
Terceira Island

Azores-Gibraltar Ridge
Term introduced in 1981.
BA North Atlantic
BT Atlantic Ocean

Azov
Town near NE arm of Azov Sea in SW Rostov Oblast. Azov has also been used to index Azov Sea.
UF Azof
BT Russian Republic
USSR
SA Azov Sea

Azov region
Includes Azov Sea, the Crimea, the southern Ukraine bordering Azov Sea, and western section of the Northern Caucasus. Index Azov Sea and Soviet republics as applicable. Also search Azov; Azov-Kuban Basin.
CO N443000N470000
E0390000E0350000
UF Azov-Kuban Basin
BT USSR
SA Azov Sea
Russian Republic
Ukraine

Azov Sea
Shallow north arm of the Black Sea bounded on the SW by the Crimea, N by the Ukraine, and E by the Northern Caucasus. Also search Sea of Azov, and Azov.
CO N450000N470000
E0390000E0340000
UF Sea of Azof
Sea of Azov
BT USSR
SA Azov
Azov region
Black Sea

Azov-Kuban Basin
use Azov region

azurite
BA carbonates
BT minerals
SA smithsonite

B

b
use magnitude-frequency ratio

B
use boron

Ba
use barium

Babadag Lake basin
N Dobruja.
BT Dobruja
Romania

babingtonite
BA chain silicates
silicates
BT minerals

Babor Range
Coastal range of the Little Atlas Mountains in N Algeria. Also search Babor; Babors.
UF Babors
BT Algeria

Babors
use Babor Range

Bacau
City and county in central.
BT Moldavia
Romania

bacteria
Schizomycetes included here. Includes use on level 1 and 2 as a fossil term (list F); on level 3 under soils(1) biota(2) and under geochemistry(1).
UF schizomycetes
BT thallophytes
Plantae
SA biology
biota
fungi
microorganisms
paleobotany
parasites
Protista
soils

Bactritida
Term introduced in 1981. Includes use on level 2 under Mollusca(1). See list F.
BA Ammonoidea
Tetrabranchiata
Cephalopoda
Mollusca
BT Invertebrata

Bacukus-Gilbert analysis
Term introduced in 1981.
SA analysis
seismology

Baculites
Genus. As of 1981, includes use on level 3 under Mollusca(1) Ammonoidea(2). See list F. Before 1981, included use on level 3 under Mollusca(1) Cephalopoda(2).
BA Ammonoidea
Tetrabranchiata
Cephalopoda
Mollusca
BT Invertebrata

Bad Deutsch Altenburg
Village on the Danube in E.
BT Lower Austria
Austria

Bad Gastein
Town in W central part of country.
BT Salzburg State
Austria

Bad Pyrmont
Town in S.
BT Lower Saxony
West Germany

Badajoz
As of 1981, refers to only the city. Before 1981, also included the province. In W Spain.
BT Badajoz Province
Extremadura
Spain

Badajoz Province
Term introduced in 1981. Before 1981, search Badajoz. In W Spain on Portuguese border.
CO N375700N393000
W0044000W0072000
BT Extremadura
Spain
NT Badajoz

Badakhshan
Frontier province in NE Afghanistan and part of Gorno-Badakshan Autonomous Oblast in Tadzhikistan. Index Afghanistan and/or Tadzhikistan as applicable.
BT Asia
SA Afghanistan
Tadzhikistan

Badami Series
The Badami Group lies over the Kaladgi Group and is overlain by thick cover of Deccan Traps. Belgaum District in SW part of India.
BT Proterozoic
Precambrian
SA India
Mysore

baddeleyite
BA oxides
BT minerals

Baden
Former German state.
BT Baden-Wurttemberg
West Germany

Baden-Baden
City in West Germany. Also search Baden.
BT Baden-Wurttemberg
West Germany

Baden-Wuerttemberg
use Baden-Wurttemberg

Baden-Wurttemberg
A state of West Germany in SW. Also search Baden-Wuerttemberg.
CO N473000N494800
E0103000E0073000
UF Baden-Wuerttemberg
BT West Germany
NT Baden
Baden-Baden
Black Forest
Freiburg
Hegau
Heidelberg
Kaiserstuhl
Karlsruhe
Katzenbuckel
Neckar River
Swabian Alb
Weinheim
Wurttemberg
Wutach Valley
SA Franconia
Main River
Odenwald
Palatinate
Swabia

Badenian
Provincial series, Europe. See list E.
BA Miocene
Neogene
Tertiary
BT Cenozoic
SA lower Miocene
middle Miocene

Badkhyz
Steppe region in extreme S near Afghanistan and Iranian borders.
BT Turkmenia
USSR

badlands
Generic term for a type of topography.
SA erosion features
geomorphology

Badlands National Monument
Badlands area in Theodore Roosevelt National Memorial Park in SW. Also search Badlands AND South Dakota.
BT South Dakota
United States

Baffin Bay
Part of Arctic Ocean between Greenland and Baffin Island, and inlet of Laguna Madre, S Texas.
BX Arctic Ocean
Texas
SA Northwest Territories

Baffin Island
Largest and most easterly of the Canadian Arctic Archipelago in District of Franklin.
CO N620000N733000
W0613000W0900000
BT Northwest Territories
Canada

Bagalkot
Town in NW.
BT Mysore
India

Bagh Beds
Includes Nimar Sandstone, Nodular Limestone, Deola Marl, Coralline Limestone.
BT Upper Cretaceous
Cretaceous
SA Gujarat
India
Madhya Pradesh

Bahama Islands
use Bahamas

Bahama Platform
use Blake Plateau

Bahamas
A chain of islands SE of Florida and N of Cuba. Includes use on level 1 as an area term (list O). For term set options see list B.
CO N204500N280000
W0720000W0793000
UF Bahama Islands
BA West Indies
NT Andros Island
Bimini
Exuma Sound
Northeast Providence Channel
Tongue of the Ocean
SA San Salvador

Bahariya Oasis
In the Libyan Desert in W.
BT Egypt

Bahia
State in E.
CO S182000S082000
W0372000W0464000
BT Brazil
NT Boquira Mine
Curaca River basin
Reconcavo Basin
Salvador
Tucano Basin
Uaua
SA Barreiras Formation
Brazilian Shield
Sao Francisco Basin

Bahrain
An archipelago in the W Persian Gulf. Includes use on level 1 as an area term (list O). For term set options see list B.
CO N253000N263000
E0505500E0501000
UF Bahrein
BA Arabian Peninsula
Asia

Bahrein
use Bahrain

Baia Mare
Town in Maramures County in N Transylvania. Also search Baia-Mare.
UF Baia-Mare
BT Transylvania
Romania

Baia-Mare
use Baia Mare

Baikal (Lake)
use Lake Baikal

Baikal Mountains
West shore of Lake Baikal mostly in Irkutsk Oblast. Also search Baikal; Baikal Range.
UF Baikal Range
BT Russian Republic
USSR

Baikal Range
use Baikal Mountains

Baikal region
Lake Baikal, Baikal Mountains, NE Irkutsk Oblast, and N Buryat A.S.S.R. Also search Baikal.
CO N510000N610000
E1183000E1040000
BX Russian Republic
BZ Siberian fold belt
BT USSR
SA Lake Baikal

Baikal rift zone
Term introduced in 1981. S central Siberia.
CO N513000N580000
E1200000E1000000
BT Russian Republic
USSR
SA Buryat
Irkutsk
Lake Baikal

Baikalian Orogenic Phase
use Baikalian Phase

Baikalian Orogeny
use Baikalian Phase

Baikalian Phase
Term introduced in 1978. Widely used throughout the USSR for orogenies late in Precambrian time. Named after Lake Baikal in Siberia. Before 1978, search Baikalian AND orogeny; Baikalian Orogeny.
UF Baikalian Orogenic Phase
Baikalian Orogeny
BT Cambrian
SA orogeny
Precambrian
Russian Republic
USSR

Bainbridge Formation
Characterized by two lithologically well-differentiated formations, St. Clair below, and Moccasin Springs above. SW Illinois and E Missouri.
BT Niagaran
Middle Silurian
Silurian
SA Illinois
Missouri

Bairdia
Genus. Includes use on level 3 under Ostracoda(1). See list F.
BA Bairdiidae
Bairdiacea
Podocopida
Ostracoda
BT Crustacea
Mandibulata
Arthropoda
Invertebrata

Bairdiacea
Includes use on level 3 under Ostracoda(1). See list F.
BA Podocopida
Ostracoda
BT Crustacea
Mandibulata
Arthropoda
Invertebrata
NA Bairdiidae

Bairdiidae
Includes use on level 3 under Ostracoda(1). See list F.
BA Bairdiacea
Podocopida
Ostracoda
BT Crustacea
Mandibulata
Arthropoda
Invertebrata
NA Bairdia

Bairdiomorpha
Term introduced in 1981. Includes use on level 2 under Ostracoda(1). See list F.
BA Ostracoda
BT Crustacea
Mandibulata
Arthropoda
Invertebrata

Baita
Village in Maramures County, in N.
BT Transylvania
Romania

Baixo Alentejo
A region in S.
BT Portugal

Baja California
A peninsula. Northern half is state of Baja California. Southern half is territory of Baja California Sur.
CO N230000N323000
W1090000W1170000
BT Mexico
NT Baja California Sur
Cerro Prieto
Mexicali
Ojo de Liebre Lagoon
Todos Santos Bay
SA Agua Blanca Fault
Colorado River
Colorado River delta
Rosario Formation

Baja California Sur
Term introduced in 1981. Territory in NW Mexico on peninsula of Baja California.
CO N225000N280000
W1091000W1150000
BT Baja California
Mexico

Bajocian
Europe. Includes use on level 3 under age terms(1). See list E.
BA Middle Jurassic
Jurassic
BT Mesozoic
SA Dogger

Bakal
City in Chelyabinsk Oblast in Southern Urals.
BT Russian Republic
USSR

Baker County
Term introduced in 1981. Index states as applicable.
BX Florida
Georgia
Oregon
BT United States

Bakersfield
City in Kern County in S.
BT California
United States

Bakhchisarai
Town in S Crimea. Also search Bakhchisaray.
UF Bakhchisaray
BT Ukraine
USSR

Bakhchisaray
use Bakhchisarai

Bakhmut
use Artemovsk

Bakhtiari Formation
Synonomous with Bakhtiari Group. Comprises Upper and Lower Bakhtiari Formations.
BT Pliocene
SA Iran
Iraq

Bakony Mountains
In NW central Hungary. Also search Bakony, and Transdanubian Central Mountains.
UF Transdanubian Central Mountains
BT Hungary

Baku
City on Apsheron Peninsula on Caspian Sea.
BT Azerbaidzhan
USSR

Baku Archipelago
Also search Baku.
BT Azerbaidzhan
USSR

Bala
England: Middle and Upper Ordovician. Comprises the Caradocian and Ashgillian. Includes use on level 3 under age terms(1). See list E.
BA Ordovician
BT Paleozoic
SA Ashgillian
Caradocian
Middle Ordovician
Upper Ordovician

Bala District
Urban district in NE Marioneth in NW Wales.
BT Wales
Great Britain
United Kingdom

Balaghat
Town and district in central.
BT Madhya Pradesh
India

balance, mass
use mass balance

balance, water
use water balance

Balasore
Town and district on Bay of Bengal in N.
BT Orissa
India

Balaton (Lake)
use Lake Balaton

Balaton region
Area in W around Lake Balaton. Also search Balaton.
BT Hungary
SA Lake Balaton

Balcanoona
Village in E central.
BT South Australia
Australia

Bald Mountain
Name of peaks which occur in several states: central Colorado; central and SW central Idaho; E central Oregon; W South Dakota; NE Utah; N Wyoming and W central Wyoming. Index states as applicable.
BT United States
SA Colorado
Idaho
Oregon
South Dakota
Utah
Wyoming

Balearic abyssal plain
use Balearic Basin

Balearic Basin
Between Balearic Islands and Sardinia.
UF Algerian Plain
Balearic abyssal plain
BT Mediterranean Sea

Balearic Islands
Off E coast of Spain. Forms Spanish province of Baleares. Includes use on level 1 as an area term (list O). For term set options see list B.
CO N383500N400500
E0042500E0010500
BA Spain
NT Ibiza
Majorca
Minorca
Palma

Balei
City in S central Chita Oblast E of Lake Baikal. Also search Baley.
UF Baley
BT Russian Republic
USSR

Baley
use Balei

Bali
Island off E end of Java.
BT Lesser Sunda Islands
Indonesia

Balkan Foreland
A stable area marginal to the Balkan Mountains of central.
BT Bulgaria

Balkan Mountains
Range extending E and W across central.
CO N423500N440000
E0280000E0221500
BT Bulgaria

Balkan Peninsula
SE Europe between the Adriatic and Ionian Seas on the W; the Mediterranean Sea on the S; and the Aegean and Black seas on the E. Includes use on level 1 as an area term (list O). For term set options see list B. Before 1978, also search Balkans.
UF Balkan States
Balkans
BA Europe
SA Albania
Bulgaria
Greece
Romania
Turkey
Yugoslavia

Balkan States
use Balkan Peninsula

Balkans
use Balkan Peninsula

Balkhan
Two mountain ranges in Krasnovodsk Oblast in W: the Greater Balkhan Range and the Lesser Balkhan Range.
BT Turkmenia
USSR

Balkhash
City on N shore of Lake Balkhash.
BT Kazakhstan
USSR

Balkhash (Lake)
use Lake Balkhash

Balkhash region
Area around Lake Balkhash. Also search Balkhash.
BT Kazakhstan
USSR
SA Lake Balkhash

ball and pillow
use ball-and-pillow

ball-and-pillow
Includes use on level 3 under sedimentary structures(1) soft sediment deformation(2). See list K. Also search ball and pillow.
UF ball and pillow
BA soft sediment deformation
sedimentary structures
SA pillow structure
primary structures

balls, coal
use coal balls

Balmat
No longer a valid term for GeoRef as of 1981. Use Balmat-Edwards District if applicable.

Balmat-Edwards District
Term introduced in 1981. Mining district in N New York. Before 1981, also search Balmat; Edwards.
CO N441500N442000
W0751500W0752500
BT Saint Lawrence County
New York
United States

Baltic Basin
use Baltic region

Baltic Coast
use Baltic region

Baltic Glaciation
Comprises Baltic region plus Norway which was completely covered by Pleistocene glaciation. Also search Baltic AND glaciation.
BT Pleistocene
SA Europe

Baltic Plain
Term introduced in 1981.
CO N540000N600000
E0313000E0190000
BA Russian Platform
BT USSR
NA Estonia
Latvia

Baltic region
Index Baltic Sea, countries and Soviet republics as applicable. Also search Baltic; Baltic Basin; Baltic Coast. Includes use on level 1 or 2 as an area term (list O). If 1, see term set options under list B.
CO N520000N680000
E0310000E0100000
UF Baltic Basin
Baltic Coast
BA Eurasia
SA Baltic Sea
Denmark
Estonia
Europe
Finland
Germany
Latvia
Lithuania
Peribaltic Syneclise
Russian Republic
Sweden

Baltic Sea
Enclosed by Denmark, Sweden, Finland, Soviet Union, Poland, and Germany. Includes use on level 1 as an area term (list O). For term set options see list B.
CO N540000N660000
E0300000E0100000
BA European Atlantic
North Atlantic
Atlantic Ocean
NT Aland
Gulf of Bothnia
Gulf of Finland
Gulf of Riga
Kiel Bay
SA Baltic region

Baltic Shield
Index countries and Soviet republics as applicable.
BT Europe
SA Estonia
Finland
Ladoga Series
Lake Ladoga region
Latvia
Norway
Russian Republic
Sweden

Baltic Syneclise
use Peribaltic Syneclise

Baltimore
City on Chesapeake Bay in Baltimore County.
BT Maryland
United States

Baltimore County
Nearly surrounds city of Baltimore in N.
BT Maryland
United States

Baltimore Gneiss
Unconformable below Glenarm Series. W Maryland, SE Pennsylvania, and central N Virginia.
BT Precambrian
SA Maryland
Pennsylvania
Virginia

Baltisphaeridium
BA acritarchs
palynomorphs

Baluchistan
Province in W.
BT Pakistan
SA Sulaiman Range

Bambak
use Pambak

Bamble
Village on the Skagerrak.
BT Telemark
Norway

Bambu
Village in NE Zaire.
UF Kilo
Kilo-Mines
BT Zaire

Bambui Group
BT Proterozoic
Precambrian
SA Brazil
Minas Gerais

Bamian
Province in N central Afghanistan. Also a town.
BT Afghanistan

Banat
Region. Part of Voyvodina in Serbia and the Banat region of Romania. Index countries as applicable.
UF Banat of Temesvar
BT Europe
NT Arad
Delinesti
Mehadia
Moldova Noua
Resita
Rusca Montana
Sasca-Montana
Semenic Mountains
Severin
Svinita
SA Getic Nappe
Mehedinti Plateau
Romania
Yugoslavia

Banat of Temesvar
use Banat

Bancroft
Village between Ottowa and Georgian Bay in SE.
BT Ontario
Canada

Band-e-Amir
A settlement in S part of country.
BT Fars
Iran

Bandama River
Rises in N and flows into Gulf of Guinea.
BT Ivory Coast

banded gneiss
See list J.
BA gneisses
BT metamorphic rocks

banded materials
Term introduced in 1978. Before 1978, also search banded.
SA materials

banded structures
Includes use on level 3 under igneous rocks(1) or metamorphic rocks(1). Also search banding; banded.
UF banding
SA igneous rocks
metamorphic rocks
petrology
structures

banding
use banded structures

Bandundu
Term introduced in 1981.
CO S081000S004000
E0210000E0121000
BT Zaire

Banff
Town in Banff National Park near the British Columbia border in SW.
CO N570500N574500
W0021500W0033000
BT Alberta
Canada

Banff National Park
On the British Columbia border in SW.
BT Alberta
Canada

Banffshire
County on Moray Firth in NE.
BT Scotland
Great Britain
United Kingdom

Bangalore
City and district in SE.
BT Mysore
India

Bangka
An island between SE Sumatra and SW Borneo.
BT Indonesia

Bangkok
City on the Gulf of Siam in S.
BT Thailand

Bangla Desh
use Bangladesh

Bangladesh
Formerly East Pakistan. Includes use on level 1 as an area term (list O). For term set options see list B. Before 1973, also search East Pakistan or Pakistan AND east.
CO N203000N263200
E0924000E0880000
UF Bangla Desh
BA Indian Peninsula
Asia
NT Magura
SA Bengal
Brahmaputra River
Ganges River
Indian Plate
Vindhyan Basin

Bangor Limestone
Includes Burgess Oolite and Rockwood Oolite, and Spout Spring Oolite near base. Alabama, central and E Tennessee, and NW Georgia.
BT Upper Mississippian
Mississippian
SA Alabama
Georgia
Tennessee

Banja Luka
Town in N Bosnia and Herzegovina.
BT Yugoslavia

Bank
Town in SE.
BT Azerbaidzhan
USSR

banks
See list K.
BA biogenic structures
sedimentary structures
SA algal banks

Banks Island
W District of Franklin.
CO N720000N743000
W1150000W1250000
BT Northwest Territories
Canada

Bankura
Town and district in W.
BT West Bengal
India

Bannock County
SE Idaho.
BT Idaho
United States

Banska Stiavnica
Town in S central.
BT Slovakia
Czechoslovakia

Baoha
A town in N North Vietnam.
BT Vietnam

Baquero Formation
BT Lower Cretaceous
Cretaceous
SA Argentina
Santa Cruz

Bar
Town SW of Kiev in W Vinnitsa Oblast.
BT Ukraine
USSR

Barabash Suite
Primorye Kray in Soviet Far East.
BT Upper Permian
Permian
SA Russian Republic
USSR

Baraboo
City in Sauk County in S central. Also a river.
BT Wisconsin
United States

Baraboo Quartzite
Underlies Seeley Slate. Sauk and Columbia counties in S central Wisconsin.
BT Algonkian
Precambrian
SA Wisconsin

Barakar
Town in W.
BT West Bengal
India

Barakar Formation
use Barakar Stage

Barakar Stage
Lowest stage of the Damuda Series. Named for Barakar River in Raniganj coalfields of West Bengal. In N, NE, and E India. Also search Barakar Formation.
UF Barakar Formation
BT Lower Permian
Permian
SA Damuda Series
India

Baralaba Coal Measures
Permo-Carboniferous.
BT Paleozoic
SA Australia
Carboniferous
Pennsylvanian
Permian
Queensland

Baranya Mountains
use Mecsek Mountains

Baraolt
Town in SE.
BT Transylvania
Romania

Baraolt Basin
SE Transylvania. Also search Baraolt.
BT Transylvania
Romania

Barataria Bay
Inlet of Gulf of Mexico between Lafourche and Plaquemines parishes.
BT Louisiana
United States
SA Gulf Coastal Plain

Barbados
Easternmost island of the Lesser Antilles. Includes use on level 1 as an area term (list O). For term set options see list B.
CO N130000N132000
W0592000W0594500
BA Lesser Antilles
West Indies

Barbados Ridge
Extends N and S of Barbados with the island in the center.
BT Atlantic Ocean

Barber County
S on the Oklahoma border.
BT Arkansas
United States

Barberton
Town in SE near Swaziland border.
BT Transvaal
South Africa

Barberton Mountain Land
SE Transvaal.
BT Transvaal
South Africa

barbosalite
BA phosphates
BT minerals

Barcelona
As of 1981, refers to only the city. Before 1981, also included the province. On the Mediterranean Sea.
BT Barcelona Province
Catalonia
Spain

Barcelona Province
Term introduced in 1981. Before 1981, search Barcelona. On the Mediterranean Sea.
CO N411000N422000
E0025000E0012200
BT Catalonia
Spain
NT Barcelona
Llogrebat River basin
Vich
SA Ter River basin

Bardo Mountains
In SW part of country.
BT Wroclaw
Poland

Barents Sea
Part of Arctic Ocean N of Norway and USSR, and between Spitsbergen and Novaya Zemlya.
CO N700000N800000
E0670000E0100000
BT Arctic Ocean

Bari
City and province on the Adriatic Sea.
BT Apulia
Italy

Barinas
State in SW Venezuela. Also a city.
BT Venezuela

Baringo Lake
use Lake Baringo

barite
As of 1981, use barite deposits for barite as a commodity. Includes use on level 3 under minerals(1) sulfates(2). Before 1981, included use on level 1 as a commodity term.
BA sulfates
BT minerals
SA barite deposits

barite deposits
Term introduced in 1981. Includes use on level 1 and 2 as a commodity term (list C). Before 1981, search barite AND deposits.
SA barite

barium
Includes use on level 1 and 2 as a chemical element (list D). Also search Ba.
UF Ba
SA barylite
elements

barium feldspar
Term introduced in 1981. Includes use in combination with framework silicates (i.e. framework silicates, barium feldspar) on level 2 under minerals(1). See list L. Before 1981, included use on level 3 under minerals(1) framework silicates, feldspar group(2).
BA feldspar group
framework silicates
silicates
BT minerals
NA celsian
paracelsian

Barmer
Town in W.
BT Rajasthan
India

Barnes ice cap
Central Baffin Island in District of Franklin.
BT Northwest Territories
Canada

Barnstable County
Coextensive with Cape Cod.
BT Massachusetts
United States

Barnwell County
Term introduced in 1981. W South Carolina.
CO N330700N333000
W0811300W0814700
BT South Carolina
United States

Baroda
City and former state in E.
BT Gujarat
India

barometry, geologic
use geologic barometry

Baronnies
Hilly region in SE part of country.
BT Drome
France

Barracuda Ridge
In the Anegada Passage area between the Virgin Islands and the N Leeward Islands.
BT Atlantic Ocean

Barrandian
Stage. See list E.
BA Middle Cambrian
Cambrian
BT Paleozoic

Barrandian area
use Barrandian Basin

Barrandian Basin
UF Barrandian area
BT Czechoslovakia

Barre Granite
An elongated, oval-shaped body that crops out in northwestern and west-central parts of East Barre quadrangle. Washington County, Vermont.
BT Devonian
SA Vermont

Barreal
Town in SW.
BT Argentina
SA San Juan Province

Barreiras Formation
BT Tertiary
SA Bahia
Brazil
Espirito Santo
Para
Pernambuco

Barreirinhas Basin
NE Maranhao.
BT Maranhao
Brazil

Barremian
Europe. Above Hauterivian, below Aptian. Includes use on level 3 under age terms(1). See list E.
BA Lower Cretaceous
Cretaceous
BT Mesozoic
SA Speeton Clay

barrier bars
use longshore bars

barrier islands
As of 1978, term is used on level 2 under shorelines(1). Before 1981, also search barriers.
SA bars
islands
longshore bars
shorelines

barrier reefs
BT reefs
SA fringing reefs
lagoons

barriers
As of 1981, no longer a valid term for GeoRef. See barrier islands; longshore bars. Included use on level 3 under sedimentation(1) and geomorphology(1).

Barrow
City at Point Barrow on the Arctic Ocean.
BT Alaska
United States

Barrow Island
Off NW coast.
BT Western Australia
Australia

bars
Includes use on level 3 under geomorphology(1) fluvial features(2) and shore features(2); on level 3 under sedimentary structures(1) planar bedding structures(2). See list K.
NT longshore bars
SA barrier islands
bedding plane irregularities
fluvial features
geomorphology
planar bedding structures
sedimentary structures
shore features

Barsakelmes
Region in Ust Urt Plateau in SW.
BT Kazakhstan
USSR

Barstovian
North America. See list E.
BA Miocene
Neogene
Tertiary
BT Cenozoic

Barstow
City in San Bernardino County in S.
BT California
United States

Barth
Town on the Baltic Sea.
BT East Germany

Bartlesville Sand
In Kansas and Oklahoma. Also search Bartlesville Sandstone.
BT Carboniferous
SA Kansas
Mississippian
Oklahoma
Pennsylvanian

Barton Beds
Overlain by the lower Headon Beds and rest on the upper Bracklesham Beds. Divided into lower, middle, and upper Barton Beds. Hampshire County in S England. Also search Barton.
BT Bartonian
Eocene
SA England
United Kingdom

Barton County
Index states as applicable.
BX Kansas
Missouri
BT United States

Bartonian
Europe. Above Auversian, below Ludian. Includes use on level 3 under age terms(1). See list E.
BA Eocene
Paleogene
Tertiary
BT Cenozoic
NT Barton Beds
SA Priabonian

Barwell Meteorite
Term introduced in 1981. Impact in Leicestershire, England. Before 1981, also search Barwell.
BA L chondrites
chondrites
BT meteorites
SA England
Leicestershire

barylite
BA ring silicates
silicates
BT minerals
SA barium

Barytherioidea
Term introduced in 1981. Includes use on level 2 under Mammalia(1). See list F.
BA Proboscidea
Mammalia
BT Tetrapoda
Vertebrata

Bas-Dauphine
Region. Part of the historic province of Dauphine in SE France. Associated with the Dauphine Alps.
BT France
SA Dauphine Alps

Bas-Rhin
Department in Alsace in extreme NE.
CO N480500N490500
E0081200E0070000
BT France
NT Bouxwiller
Strasbourg
SA Saar Basin

basalt
Includes use on level 3. As of 1981, not to be used in the igneous rocks set. See list H. Before 1981, included use on level 3 under igneous rocks(1) basalt family(2). Also search basaltic rocks.
BA basalts
volcanic rocks
BT igneous rocks
SA alkali basalt
alkali olivine basalt
augite
basaltic composition
columnar basalt
lava
melaphyre
metabasalt
nepheline basalt
pillow lava
tholeiitic basalt
tholeiitic composition
trap rocks

basalt family
As of 1981, no longer a valid term for GeoRef. Use basalts, alkali basalts, trachybasalts, or diabase as applicable. See list H. Before 1981, included use on level 2 under igneous rocks(1).

basalt glass
Not a valid term for GeoRef. Use volcanic glass and basalt.

basaltic
Not a valid term for GeoRef. Use basaltic composition or basaltic layer.

basaltic composition
Term introduced in 1978. Before 1978, search basaltic.
SA basalt
composition
igneous rocks

basaltic domes
use shield volcanoes

basaltic layer
Also search sima.
UF layer, basaltic
sima
SA granitic layer
lower crust

basaltic rocks
Not a valid term for GeoRef. Use basalts or basaltic composition.

basalts
Term introduced in 1981. Includes use on level 2 under igneous rocks(1). See list H. Before 1981, search basalt family.
BA volcanic rocks
BT igneous rocks
NA absarokite
alkali basalts
alkali olivine basalt
ankaramite
basalt
columnar basalt
melaphyre
oceanite
olivine basalt
olivine tholeiite
shoshonite
tholeiite
tholeiitic basalt
trap rocks
SA picrite
picrite porphyry

basanite
As of 1981, includes use on level 3 under igneous rocks(1) alkali basalts(2). See list H. Before 1981, included use on level 3 under igneous rocks(1) alkali basalt family(2).
BA alkali basalts
basalts
volcanic rocks
BT igneous rocks

basculating faults
use wrench faults

base exchange
use ion exchange

base metals
As of 1978, term includes use on level 1 and 2 as a commodity (list C).
SA alloys
metal ores
metals

base surges
Used mostly in sedimentation.
SA sedimentation
surges

base-level peneplain
use peneplains

Basel
Canton composed of demicantons of Basel-Land and Basel Stadt. Also a city.
UF Basle
BT Switzerland

basement
Includes use on level 3 under tectonics(1). Also search basement rocks; basement structure.
UF basement rocks
basement structure
SA crust
igneous rocks
metamorphic rocks
Mohorovicic discontinuity
sedimentary cover
shields
tectonics

basement rocks
use basement

basement structure
use basement

Bashkiria
Bashkir Autonomous Soviet Socialist Republic in SE European USSR.
CO N520000N570000
E0600000E0530000
BT Russian Republic
USSR

Bashkirian
Europe. See list E.
UF Bashkirian Stage
BA Upper Carboniferous
Carboniferous
BT Paleozoic

Bashkirian Stage
use Bashkirian

basic rocks
Not a valid term for GeoRef. Use mafic composition.

Basilicata
Autonomous region in S.
CO N395500N411000
E0165000E0152200
BT Italy
NT Matera
Potenza
SA Lucania

Basin and Range Province
Level 1 area term as of 1978. Physiographic province in W and SW United States characterized by a series of tilted fault blocks forming longitudinal, asymmetric ridges of mountains and broad intervening basins. Also search Basin and Range. Index states as applicable.
CO N290000N433000
W1023000W1220000
BA United States
North America
SA Arizona
Battle Mountain High
California
Great Basin
Idaho
Nevada
New Mexico
Oregon
Texas
Utah

basin range structure
Includes use on level 3 under tectonics(1). Also search basin range; basin-range; basin-range structure.
UF range, basin
SA basins
faults
tectonics

basins
Includes use on level 3 under folds(1) style(2), under tectonics(1), and under paleogeography(1).
UF structural basins
SA basin range structure
depressions
drainage basins
folds
intermontane basins
intracratonic basins
marginal basins
ocean basins
paleogeography
playas
sedimentary basins
sedimentation
tectonics

Baskunchak Series
S Urals.
SA Permian
Russian Republic
Triassic
USSR

Basle
use Basel

Basommatophora
Term introduced in 1981. Includes use on level 2 under Mollusca(1). See list F.
BA Gastropoda
Mollusca
BT Invertebrata

Basque Provinces
On the Bay of Biscay near the French border. Also search Basque. Index provinces as applicable.
CO N422900N432800
W0014500W0033300
BT Spain
NT Alava
Guipuzcoa
Vizcaya
SA Cantabrian Basin

Bass Basin
use Bass Strait

Bass Strait
Large channel separating Victoria from Tasmania and the Indian Ocean from the Tasman Sea. Also search Bass Basin.
UF Bass Basin
BT Australia
NX Western Port

bassanite
BA sulfates
BT minerals

Basses-Alpes
use Alpes-de-Haute Provence

Basses-Pyrenees
use Pyrenees-Atlantiques

Bastar
Former Indian State. Now a district in SE.
BT Madhya Pradesh
India

bastnaesite
UF bastnasite
BZ fluorides
carbonates
BT minerals

bastnasite
use bastnaesite

Bastogne
Town, E Luxembourg province.
BT Belgium

Basutoland
use Lesotho

Batchawana Bay
On Lake Superior N of Sault Sainte Marie.
BT Ontario
Canada

Batenev Ridge
In the Kuznetsk Alatau of the Khakass Autonomous Oblast SW of Krasnoyarsk.
BT Russian Republic
USSR

Batesville
City in Independence County in NE central.
BT Arkansas
United States

Bath
City and county borough near Bristol in W.
BT England
Great Britain
United Kingdom

batholiths
Includes use on level 2 under intrusions(1).
UF abyssolith
central granite
intrusive mountain
BA intrusions
SA dikes
domes
igneous rocks
laccoliths
lopoliths
plutons
stocks

Bathonian
Europe. Above Bajocian, below Callovian. Includes use on level 3 under age terms(1). See list E.
BA Middle Jurassic
Jurassic
BT Mesozoic
NT Great Oolite Series
SA Dogger

Bathurst
Region in NE.

BT New Brunswick
Canada

Bathurst Island
NW District of Franklin.
CO N750000N763000
W0973000W1023000
BT Northwest Territories
Canada

bathymetric maps
Term introduced in 1978. Before 1978, search maps AND bathymetric.
BA maps
SA bathymetry
ocean floors

bathymetric surveys
use bathymetry

bathymetry
Includes use on level 3 under marine geology(1) or geophysical methods(1). Also search bathymetric surveys.
UF bathymetric surveys
SA bathymetric maps
echo sounding
geophysical surveys
marine geology
ocean floors
oceanography
paleo-oceanography
paleobathymetry

Baton Rouge
City on the Mississippi River in East Baton Rouge Parish.
BT Louisiana
United States

Battambang
Town in W.
BT Cambodia

Battle Mountain High
Term introduced in 1981. Heat flow anomaly in region including parts of N Nevada, SE Oregon, S Idaho, SW Montana, and NW Wyoming.
BT United States
SA Basin and Range Province
Great Basin
heat flow
Idaho
Montana
Nevada
Oregon
Western U.S.
Wyoming

Batum
use Batumi

Batumi
City on the Black Sea near the Turkish border. Also search Batum.
UF Batum
BT Georgian Republic
USSR

Bauchi Plateau
use Jos Plateau

Bauer Deep
SE Pacific Ocean, E of the East Pacific Rise.
BT Pacific Ocean

Baumkirchen
Region in Inn Valley.
BT Tyrol
Austria

Bauru Formation
SA Brazil
Cretaceous
lower Tertiary
Minas Gerais
Sao Paulo
Tertiary
Upper Cretaceous

bauxite
Includes use on level 1 and 2 as a commodity term (list C). As of 1981, also includes use on level 3 under sedimentary rocks(1) clastic rocks(2). See list I.
SA aluminum
aluminum ores
bauxitization
clastic rocks
clay minerals
laterites
sedimentary rocks

bauxitization
SA bauxite
laterites
laterization
mineral deposits
processes
weathering

Bavaria
State in SE.
CO N474000N503500
E0134500E0093000
BT West Germany
NT Ansbach
Bavarian Forest
Bayreuth
Berchtesgaden
Coburg
Eichstatt
Erlangen
Fichtelgebirge
Franconian Jura
Hagendorf
Kelheim
Kreuth
Middle Franconia
Munchberg Gneiss Massif
Munich
Nabburg
Neuburg
Nuremberg
Passau
Regensburg
Ries Crater
Sandelzhausen
Solnhofen
Spessart
Upper Bavaria
Upper Franconia
Upper Palatinate
Weissenburg
SA Allgau Alps
Bavarian Alps
Franconia
Inn Valley
Isar Valley
Main River
Odenwald
Palatinate
Rhon Mountains
Salzach River
Solnhofen Limestone
Swabia

Bavarian Alps
Range of the Central Alps along the Austro-German border.
CO N472200N475000
E0130700E0104500
BZ Alps
BX Germany
SA Bavaria
Central Alps
West Germany

Bavarian Forest
A subsidiary forested mountain range of the Bohemian Forest in E Bavaria.
BT Bavaria
West Germany

Bavarian Massif
Term introduced in 1981.
CO N483000N502000
E0135000E0113500
BT Germany
SA West Germany

bavenite
BA sheet silicates
silicates
BT minerals
SA aluminosilicates

Bay of Bengal
Part of Indian Ocean between India and the W coast of Burma.
CO N050000N230000
E0940000E0800000
BT Indian Ocean
NT Andaman Basin

Bay of Biscay
Large inlet of the Atlantic Ocean between the W tip of Brittany and NW Spain. Also search Biscay; Biscay Bay; Gulf of Gascony.
CO N432000N480000
W0011000W0080000
UF Biscay (Bay)
Biscay Bay
BA European Atlantic
North Atlantic
BT Atlantic Ocean
NT Cantabria Seamount
SA Arcachon Basin

Bay of Bourgneuf
On Bay of Biscay off Vendee and Loire-Atlantique departments.
BT France

Bay of Fundy
Inlet between Maine and New Brunswick on the W and Nova Scotia on the E.
BT Atlantic Ocean
SA Minas Basin

Bay of Islands
Inlet of Gulf of Saint Lawrence off W coast.
BT Newfoundland
Canada

Bay of Mont-Saint-Michel
use Bay of Saint-Michel

Bay of Naples
Inlet of Tyrrhenian Sea off the city of Naples.
BT Campania
Italy

Bay of Plenty
Large inlet off NE coast of North Island, New Zealand.
BT Pacific Ocean

Bay of Saint Michel
use Bay of Saint-Michel

Bay of Saint-Michel
On English channel off SW corner of Normandy. Location of Mont-Saint-Michel. Also search Bay of Saint-Michel.
UF Bay of Mont-Saint-Michel
Bay of Saint Michel
Mont-Saint-Michel Bay
BT France

Bay of the Seine
Inlet of the English Channel off Normandy between Cotentin Peninsula on W to mouth of Seine River on E.
BT France

Bay Park
Town in Nassau County on Long Island.
BT New York
United States

Bayan-Nuurin-khotnor Basin
Term introduced in 1981. Part of the Bayan-khongor fault valley, central Mongolia. Between the Hangay Mountains to the NE and the South Hangay Upland to the SW. Contained a pluvial lake during the Pleistocene.
CO N463000N473000
0E101000E0990000
BT Mongolia
SA Hangay Mountains

bayerite
Artificial.
BA oxides
BT minerals
SA gibbsite

Baylor County
SW of Wichita Falls in N.
BT Texas
United States

Bayn Dzak
Region in the Gobi Desert.
BT Mongolia

Bayreuth
City in N.
BT Bavaria
West Germany

bays
Includes use on level 3 under shorelines(1) or geomorphology(1) shore features(2).
SA fjords
geomorphology
inlets
shore features
shorelines

Be
use beryllium

Be-10
Includes use on level 3 under isotopes(1).
SA beryllium
isotopes

Be-7
Includes use on level 3 under isotopes(1).
SA beryllium
isotopes

beach erosion
Not a valid index term for GeoRef. Use beaches and erosion.

beach ridges
Includes use on level 3 under shorelines(1) or geomorphology(1) shore features(2).
SA cheniers
geomorphology
shore features
shorelines

beach rock
use beachrock

beaches
As of 1978, term is used on level 2 under shorelines(1).
SA coastal environment
erosion
geomorphology
lacustrine features
lagoons
landform evolution
paleoecology
sedimentation
shore features
shorelines
spits

beachrock
Includes use on level 3 under sedimentary rocks(1) carbonate rocks(2). See list I.
UF beach rock
BA carbonate rocks
BT sedimentary rocks

Beacon Group
use Beacon Supergroup

Beacon Supergroup
In Queen Maud Range along SW Ross Ice Shelf and in Victoria Land W of Ross Sea, both in Ross Dependency S of New Zealand. Also search

Beacon Group.
UF Beacon Group
SA Antarctica
Carboniferous
Devonian
Jurassic
Mesozoic
Paleozoic
Permian
Triassic

Beacon Valley
In S Victoria Land near SE Ross Sea in Ross Dependency S of New Zealand.
BT Antarctica

Bear Gulch Limestone Member
In Tyler Formation. Central N Montana. Also search Bear Gulch Limestone.
BT Mississippian
SA Montana

Bear Island
240 miles N of Norway. Along with Spitsbergen forms Svalbard island group.
BT Svalbard
Arctic region

Bear Lake
Index states as applicable.
BT United States
SA Idaho
Utah

Bear Province
Structural province of the Canadian Shield.
CO N630000N733000
W1060000W1240000
BT Canadian Shield
SA Northwest Territories

Bear River basin
Index states as applicable.
BT United States
SA Idaho
Utah

Bear River Range
Index states as applicable.
BT United States
SA Idaho
Utah

Bear Valley
Central California.
BT California
United States

Bear-Slave Operation
Geochemical operation in Great Bear and Great Slave lakes region of District of Mackenzie.
BT Northwest Territories
Canada

Beardmore Glacier
At NW end of Queen Maude Range just off Ross Ice Shelf in Ross Dependency S of New Zealand.
BT Antarctica
SA Queen Maude Range
Ross Ice Shelf

bearing capacity
Includes use on level 3 under foundations(1).
SA compaction
foundations
physical properties
soil mechanics
stress

Bearn
An historical region of SW.
BT France

Bearpaw Formation
In Montana Group. S Alberta; N, E, and S Montana; and Elk Basin region of central N Wyoming. Also search Bearpaw Shale.
UF Bearpaw Shale
BT Upper Cretaceous
Cretaceous
SA Alberta
Edmonton Formation
Montana
Montana Group
Wyoming

Bearpaw Shale
use Bearpaw Formation

Beartooth Mountains
Index states as applicable. NE spur of Absaroka Range.
CO N441000N455000
W1090000W1110000
BT United States
SA Absaroka Range
Montana
Wyoming

Beas River
One of the "Five Rivers" of the Punjab. Rises in Himachal Pradesh and flows into the Sutlej River. Index states as applicable.
BT India
SA Himachal Pradesh

Beata Ridge
Off south central coast of Hispaniola.
BT Caribbean Sea

Beatty
Underground nuclear explosion site. NW of Las Vegas.
BT Nevada
United States

Beauce
Ancient region in N central. Index departments as applicable.
BT France
SA Eure-et-Loir
Loir-et-Cher

Beaufort
City in Carteret County in E.
BT North Carolina
United States

Beaufort County
On both sides of the Pamlico River in E.
BT North Carolina
United States

Beaufort Formation
Underlies Castle Hayne Limestone or Yorktown Formation, or locally, an unnamed middle(?) Miocene unit; overlies Upper Cretaceous Pendee Formation. In E North Carolina. Also search Beaufort Series.
BT Paleocene
SA North Carolina

Beaufort Sea
Between N Alaska and the Arctic Archipelago of Canada.
CO N690000N760000
W1230000W1560000
BT Arctic Ocean

Beaumont
City in Jefferson County in E.
BT Texas
United States

Beaver County
Index states as applicable.
BX Oklahoma
Pennsylvania
Utah
BT United States

Beaver Creek
Two Beaver Creeks. One in NE and one in W central.
BT Wyoming
United States

Beaver Lake
Garden County in western panhandle.
BT Nebraska
United States

Beaver River
Extreme SE.
BT Yukon Territory
Canada

Beaverhead County
Extreme SW.
BT Montana
United States

Beaverhead Formation
Unconformably overlies Kootenai, Dinwoody, Thaynes, Phosphoria, Quadrant and Madison. E central Idaho and SW Montana.
SA Cretaceous
Eocene
Idaho
Montana
Paleocene
Upper Cretaceous

Beaverhead Mountains
S part of Bitterroot Range. Index states as applicable.
BT United States
SA Bitterroot Range
Idaho
Montana

Beaverlodge
Lake N of Lake Athabaska.
BT Saskatchewan
Canada

Bechar
Town about 440 miles SW of Algiers.
BT Saoura region
Algeria

Becraft Mountain
Columbia County in E.
BT New York
United States

bed forms
use bedforms

bed load
use bedload

bed-load
use bedload

Bedarieux
Town in S part of country.
BT Herault
France

bedded volcano
use stratovolcanoes

bedding
Includes use on level 3 under sedimentary structures(1) planar bedding structures(2) and under fractures(1) style(2). See list K.
BA planar bedding structures
sedimentary structures
SA bedding plane irregularities
columnar joints
cross-bedding
flaser bedding
fractures
graded bedding
laminations
lineation
massive bedding
primary structures
rhythmic bedding
stratification
strike

bedding faults
Term introduced in 1978. Before 1978, search faults AND bedding.
BA faults
SA bedding joints

bedding joints
Term introduced in 1981.
SA bedding faults
joints

bedding plane irregularities
Includes use on level 2 under sedimentary structures(1). See list K.
BA sedimentary structures
NA antidunes
current markings
flute casts
fossil ice wedges
frost features
groove casts
grooves
megaripples
mounds
mud lumps
mudcracks
parting lineation
ripple marks
sand waves
scour casts
scour marks
striations
tool marks
SA bars
bedding
dunes
ice wedges
imbrication
load casts
sand bodies
sole marks

bedding structures, planar
use planar bedding structures

bedding-plane slip
use flexural-slip

Bedford County
Index states as applicable.
BX Pennsylvania
Tennessee
Virginia
BT United States

Bedford Shale
Devonian and Mississippian(?): NE Kentucky, E Ohio, and SW Pennsylvania.
BT Paleozoic
SA Devonian
Kentucky
Mississippian
Ohio
Pennsylvania

Bedfordshire
County in SE.
CO N514500N522000
W0001000W0004000
BT England
Great Britain
United Kingdom

bedforms
Also search bed forms.
UF bed forms
SA fluvial features
hydrology

bedload
Also search bottom load.
UF bed load
bed-load
bottom load
SA stream transport
streams

bedrock
Used as a general term.
SA outcrops
rocks
soils

beds, convoluted
use convoluted beds

beds, key
use key beds

beds, marker
use marker beds

Beemerville
Village in Sussex County in extreme N.
BT New Jersey
United States

Beerenberg
Extinct volcano on Norway's Jan Mayen Island.
BT Jan Mayen
Arctic region

Behar
use Bihar

behavior
Used as a general term.

beidellite
BA clay minerals
sheet silicates
silicates
BT minerals
SA montmorillonite

Beil Limestone
use Beil Limestone Member

Beil Limestone Member
Of Lecompton Limestone. SW Iowa, E Kansas, SE Nebraska. Also search Beil Limestone.
UF Beil Limestone
BT Virgilian
Pennsylvanian
SA Iowa
Kansas
Lecompton Limestone
Nebraska

Beira Baixa
Former province in N central.
BT Portugal

Beirut
City on the Mediterranean Sea.
BT Lebanon
Middle East

Beius Basin
Bihor County in NW.
BT Transylvania
Romania

Bek-Budi
use Karshi

Belaya Gora
Village S of Arkhangelsk in Arkhangelsk Oblast in NW European USSR.
BT Russian Republic
USSR

Belchatow
Town in central part of country.
BT Lodz
Poland

Belcher Islands
Group in SE Hudson Bay.
BT Northwest Territories
Canada

belemnites
As of 1981, restricted to use as the common name for Belemnitidae. See list F. To be used for Belemnitidae when they are used as tools in stratigraphy. Before 1981, included use on level 3 under Mollusca(1) Cephalopoda(2). Before 1981, search (Belemnitidae OR belemnites) AND biostratigraphy.
SA Belemnitidae
biostratigraphy

Belemnitidae
Family. Includes use on level 3 under Mollusca(1) Cephalopoda(2). Before 1981, also search belemnites. As of 1981, search belemnites only for biostratigraphy.
BA Cephalopoda
Mollusca
BT Invertebrata
SA belemnites
Belemnopsis

Belemnoidea
As of 1981, includes use on level 2 under Mollusca(1). See list F. Before 1981, included use on level 3 under Mollusca(1) Cephalopoda(2).
BA Dibranchiata
Cephalopoda
Mollusca
BT Invertebrata
SA Belemnopsis

Belemnopsis
Includes use on level 3 under Mollusca(1) Cephalopoda(2).
BA Cephalopoda
Mollusca
BT Invertebrata
SA Belemnitidae
Belemnoidea

Belfast
City in Co. Antrim and Co. Down.
BT Northern Ireland
United Kingdom

Belfort
Term introduced in 1981. Territory in NE.
CO N472500N475000
E0071000E0064000
BT France

Belgaum
Town and district in W.
BT Mysore
India

Belgian Congo
use Zaire

Belgium
Includes use on level 1 as an area term (list O). For term set options see list B.
CO N493000N513000
E0063000E0023000
BA Western Europe
Europe
NT Antwerp
Bastogne
Brabant
Brabant Massif
Hainaut
Liege
Luxembourg Province
Namur
SA Ardennes
Campine
European Platform
Grimmertingen
Limburg
Luxembourg Sandstone
Meuse River
Meuse Valley
North Sea Coast
Scheldt River
Stavelot-Venn Massif

Belgorod
City N of Kharkov in SE Russian Republic. Also an oblast.
BT Russian Republic
USSR

Belgrade
City on Danube River in N central Serbia.
UF Beograd
BT Serbia
Yugoslavia

Beli Isvor
use Beli Izvor

Beli Izvor
Village in E Bulgaria. Also search Beli Isvor.
UF Beli Isvor
BT Bulgaria

Belitung
use Billiton

Belize
Formerly British Honduras. Includes use on level 1 as an area term (list O). For term set options see list B. Before 1975, also search British Honduras.
CO N155500N183000
W0872000W0892000
BA Central America
NT Maya Mountains

Beljak
use Villach

Bell Canyon Formation
In Delaware Mountain Group. Includes Hegler Limestone, Pinery Limestone, Radar Limestone, Lamar Limestone and McCombs Limestone members. S New Mexico, and W Texas.
BT Guadalupian
Permian
SA New Mexico
Texas

Bell Creek Field
Oil and gas field.
BT Montana
United States
SA oil and gas fields

Bell Island
In Conception Bay W of Saint John's.
BT Newfoundland
Canada

Bellary
Town in E.
BT Mysore
India

Belle Isle
Island constituting a salt dome on N shore of Atchafalaya Bay in Saint Mary Parish in S.
BT Louisiana
United States

Belledonne
use Belledonne Massif

Belledonne Massif
Part of mountain range in SE France. Also search Belledonne; Belledonne Range.
UF Belledonne
Belledonne Mountain Range
Belledonne Range
BT France

Belledonne Mountain Range
use Belledonne Massif

Belledonne Range
use Belledonne Massif

Bellerophontina
Term introduced in 1981. Includes use on level 2 under Mollusca(1). See list F.
BA Gastropoda
Mollusca
BT Invertebrata

Bellingshausen Sea
Inlet of South Pacific Ocean extending westward from Alexander Island off the W coast of Antarctic Peninsula to Thurston Island.
CO S730000S660000
W0750000W0980000
BT Antarctic Ocean

Belluno
Town and province in N.
BT Veneto
Italy

Belly River Formation
Has been considered equivalent to Two Medicine Formation and Virgelle Sandstone of Blackfoot Indian Reservation, N Montana, and Judith River Formation, Claggett Formation, and Eagle Sandstone of central Montana.
BT Upper Cretaceous
Cretaceous
SA Alberta
Saskatchewan

Belomorsk
Town in Karelian ASSR in NW European USSR.
UF Soroka
BT Russian Republic
USSR

Belorussia
use Byelorussia

Belozerka
Town in Kherson Oblast N of the Crimea.
BT Ukraine
USSR

Belozero
Village W of Arkhangelsk in Arkhangelsk Oblast.
BT Russian Republic
USSR

Belsk
Town in in NE part of country. Also search Bielsk Podlaski.
UF Bielsk
Bielsk Podlaski
BT Bialystok
Poland

Belt Basin
Cascade County in W central.
BT Montana
United States

Belt Supergroup
Comprises Piegan Group, Missoula Group, and Ravalli Group. Also search Belt Series.
BT Precambrian
SA British Columbia
Idaho
Missoula Group
Montana
Ravalli Group
Washington

Beltana
Village in E central.
BT South Australia
Australia

belts
A valid term through 1978. After 1978, see specific terms, e.g. fold belts, metamorphic belts, etc.

belts, fold
use fold belts

belts, greenstone
use greenstone belts

belts, metamorphic
use metamorphic belts

belts, mobile
use mobile belts

belts, volcanic
use volcanic belts

Bembridge Marls
Defined to include Bembridge Oyster Beds, Lower Bembridge Marls, and Upper Bembridge Marls. Isle of Wight in English Channel S of Southampton.
BT Oligocene
SA England
United Kingdom

benches
Includes use on level 3 under geomorphology(1) shore features(2).
SA erosion
geomorphology

shore features
terraces

beneficiation
Used for commodity production technology.
SA flotation
purification

Bengal
A region and former Indian province. Index Bangladesh and Indian states as applicable. Also search Bengal Basin.
UF Bengal Basin
BT Asia
SA Bangladesh
Tripura
West Bengal

Bengal Basin
use Bengal

Bengal Islands
Term introduced in 1981. Includes Andaman and Nicobar Islands.
CO N063000N135000
E0943000E0913000
BT India
NA Andaman Islands
Nicobar Islands
SA Indian Ocean Islands

Benham
Village in Harlan County in SE.
BT Kentucky
United States

Beni Bouchera
Region in the Rif in NE.
BT Morocco

Benin
Was established in 1960 as Dahomey. Includes use on level 1 as an area term (List O). For term set options see list B. Before 1976, also search Dahomey.
CO N061000N123000
E0035000E0004000
BA West Africa
Africa
SA Mali-Niger Syneclise
Niger River
Niger Valley

Benioff seismic zone
use Benioff zone

Benioff zone
Includes use on level 3 under plate tectonics(1) subduction(2); on level 3 under seismology(1).
UF Benioff seismic zone
Benioff zones
zone, Benioff
SA lithosphere
plate tectonics
plates
sea-floor spreading
seismology
subduction
subduction zones
trenches

Benioff zones
use Benioff zone

Bennettitales
Includes use on level 2 under gymnosperms(1). See list F.
BA gymnosperms
BT Spermatophyta
Plantae
SA Ptilophyllum

Benoue Valley
use Benue Valley

Benson Mines
Also search Benson AND New York.
BT New York
United States

benthic taxa
use benthonic taxa

benthonic
A valid index term through 1977. After 1977, use benthonic taxa.

benthonic taxa
Term introduced in 1977. Includes use on level 3 under fossil group(1). Before 1978, search benthonic; benthic.
UF benthic taxa
taxa, benthonic
SA bottom features
fossils
marine geology
paleoecology

Benton County
BX Arkansas
Indiana
Iowa
Minnesota
Mississippi
Missouri
Oregon
Tennessee
Washington
BT United States

Benton Formation
In Colorado Group. Comprising Graneros Shale, Greenhorn Limestone and Carlile Shale as members. Contains Codell Sandstone Member. E Colorado, South Dakota, Kansas, S Minnesota, SE Montana, Nebraska, NE New Mexico, E Wyoming.
BT Cretaceous
SA Carlile Shale
Colorado
Colorado Group
Graneros Shale
Greenhorn Limestone
Kansas
Lower Cretaceous
Minnesota
Montana
Nebraska
New Mexico
South Dakota
Upper Cretaceous
Wyoming

bentonite
Includes use on level 3 under sedimentary rocks(1) clastic rocks(2). See list I. As of 1981, use bentonite deposits for bentonite as a commodity. Before 1981, included use on level 1 as a commodity term.
UF amargosite
mineral soap
soap clay
volcanic clay
BA clastic rocks
BT sedimentary rocks
SA bentonite deposits
clastic sediments
clay
clay mineralogy
clay minerals
metabentonite
pyroclastics
sheet silicates
terrigenous materials
tuff
volcanic ash

bentonite deposits
Term introduced in 1981. Includes use on level 1 and 2 as a commodity term (list C). Before 1981, search bentonite AND deposits.
SA bentonite
clays

Benue Trough
use Benue Valley

Benue Valley
River valley. Index countries as applicable. Also search Benue Trough.
UF Benoue Valley
Benue Trough
Binue Valley
BT Africa
SA Cameroon
Nigeria

Beograd
use Belgrade

Beppu
City in NE.
BT Kyushu
Japan

beraunite
BA phosphates
BT minerals

Berchtesgaden
Town in extreme SE.
BT Bavaria
West Germany

Berea Sandstone
Devonian or Mississippian. NE Kentucky, S Michigan, Ohio, W Pennsylvania, and N West Virginia.
BT Paleozoic
SA Devonian
Kentucky
Michigan
Mississippian
Ohio
Pennsylvania
West Virginia

Beregovo
Town in Transcarpathian Oblast in extreme SE.
BT Ukraine
USSR

Beresford Lake
In SW Manitoba.
BT Manitoba
Canada

beresite
As of 1981, includes use on level 3 under metasomatic rocks(1) greisen(2). Before 1981, included use on level 3 under igneous rocks(1).
BA greisen
BT metasomatic rocks
SA hypabyssal rocks
quartz porphyry

berg crystal
use quartz crystal

Bergamo
City in Bergamo Province in central.
BT Lombardy
Italy

Bergell Massif
In SE part of country.
CO N461000N462000
E0094500E0093000
BT Switzerland

Bergen
City on the North Sea in SW part of country.
BT Norway

Bergisch Gladbach
City in SW central.
BT North Rhine-Westphalia
West Germany

Bergisches Land
Region on right bank of the Rhine River.
BT West Germany

Bergstrasse
Name applied to the W slope of the Odenwald hills of SW.
BT West Germany

Berici Hills
Range of volcanic hills in Vicenza Province in W central.
BT Veneto
Italy

Bering land bridge
As of 1981, no longer a valid term for GeoRef. Use Beringia.

Bering Sea
Between NE Siberia and Alaska with the Aleutian Islands on the S and Bering Strait on the N. Includes use on level 1 as an area term (list O). For term set options see list B.
CO N513000N653000
W1573000E1620000
BA West Pacific
Pacific Ocean
NT Bowers Ridge
Bristol Bay
Norton Sound
Saint Paul Island
SA Bering Strait
Beringia

Bering Strait
Term introduced in 1981. Connects Chukchi Sea and Bering Sea between Asia and North America.
CO N650000N670000
W1670000W1700000
SA Arctic Ocean
Atlantic Ocean
Bering Sea
Beringia
Chukchi Sea

Beringia
Term introduced in 1981. NE Siberia, NW North America, and intervening continental shelves. Early man entered North America via this region. Before 1981, also search Bering land bridge.
SA Bering Sea
Bering Strait
Chukchi Sea
North America
Siberia

Berkeley
City in Alameda County across San Francisco Bay from San Francisco.
BT California
United States

Berks County
SE central.
BT Pennsylvania
United States

Berkshire
County in S England.
CO N512000N514500
W0003000W0014000
BT England
Great Britain
United Kingdom

Berkshire Anticlinorium
use Berkshire Hills

Berkshire County
Term introduced in 1981. W Massachusetts.
CO N420300N424500
W0725600W0733100
BT Massachusetts
United States

Berkshire Highlands
use Berkshire Hills

Berkshire Hills
Hills in Berkshire County in W Massachusetts. Also search Berkshire Anticlinorium; Berkshire Massif.
UF Berkshire Anticlinorium
Berkshire Highlands
Berkshire Massif
Berkshires
BT Massachusetts
United States

Berkshire Massif
use Berkshire Hills

Berkshires
use Berkshire Hills

Berlin
City partitioned into East Berlin and West Berlin since 1945. Index countries as applicable.
BT Germany
SA East Germany
West Germany

Bermuda
British colony comprising over 300 coral islands about 640 miles ESE of Cape Hatteras. Includes use on level 1 as an area term (list O). For term set options see list B. Also search Bermuda Platform; Bermuda Rise.
CO N321500N322500
W0644000W0645000
UF Bermuda Island
Bermudas
BA Atlantic Ocean Islands
NT Harrington Sound

Bermuda Island
use Bermuda

Bermuda Platform
Part of Bermuda Rise.
BT Atlantic Ocean
SA Bermuda Rise

Bermuda Rise
Runs in SW-NE direction to the west of Bermuda.
BT Atlantic Ocean
SA Bermuda Platform

Bermudas
use Bermuda

Bern
Canton in W central Switzerland. Also a city. Also search Berne.
UF Berne
BT Switzerland
NT Biel Lake
SA Grindelwald

Bernalillo County
Central.
BT New Mexico
United States

berndtite
BA sulfides
BT minerals

Berne
use Bern

Bernese Alps
N division of Central Alps in Bern and Valais cantons. Also search Bernese Oberland.
UF Bernese Oberland
BT Switzerland
NT Grindelwald
SA Alps
Central Alps

Bernese Oberland
use Bernese Alps

Bernic Lake
BT Manitoba
Canada

bernstein
use amber

Berounka System
Term introduced in 1981.
CO N492000N501500
E0145000E0125000
BT Czechoslovakia

Berriasian
Europe. Lowermost Lower Cretaceous, above Portlandian (Jurassic), below Valanginian. Includes use on level 3 under age terms(1). See list E.
BA Lower Cretaceous
Cretaceous
BT Mesozoic
SA Neocomian

Berrien County
Index states as applicable.
BX Georgia
Michigan
BT United States

berryite
BA sulfosalts
BT minerals

berthierine
BA sheet silicates
silicates
BT minerals
SA aluminosilicates
chamosite

berthonite
use bournonite

bertrandite
BA ring silicates
silicates
BT minerals

Berwickshire
County in SE.
BT Scotland
Great Britain
United Kingdom

beryl
Includes use on level 3 under minerals(1) and on level 3 as a commodity term. See list C.
BA ring silicates
silicates
BT minerals
SA beryllium
beryllium ores
emerald
gems

beryllium
Includes use on level 1 and 2 as a chemical element (list D). As of 1981, use beryllium ores for beryllium as a commodity. Before 1981, included use on level 1 as a commodity term.
UF Be
SA Be-10
Be-7
beryl
beryllium ores
elements

beryllium ores
Term introduced in 1981. Includes use on level 1 and 2 as a commodity term (list C). Before 1981, search beryllium AND ore deposits.
SA beryl
beryllium

Besancon
City near Swiss border.
BT Doubs
France

Beshchady Mountains
Section of the Carpathians. Index countries and Ukraine as applicable.
BT Europe
SA Carpathians
Czechoslovakia
Poland
Ukraine

Beskid Mountains
Both West and East Beskid mountains comprise mountain group of the Carpathians on NE border. Also search Beskids.
UF Beskids
BT Czechoslovakia
SA Carpathians

Beskids
use Beskid Mountains

Bessarabian
Pertains to Bessarabia, a former Romanian province which is now coextensive with Moldavia, USSR.
BT Moldavia
USSR

Bet-Pak-Dala
Desert steppe W of Lake Balkhash. Also search Golodnaya Step.
CO N450000N470000
E0730000E0673000
UF Golodnaya Step
Hunger Steppe
BT Kazakhstan
USSR

betafite
Also search blomstrandite, ellsworthite, and hatchettolite.
UF blomstrandite
ellsworthite
hatchettolite
BZ niobates
tantalates
BA oxides
BT minerals
SA pyrochlore

betekhtinite
BA sulfides
BT minerals

Betic Cordillera
Extends along the N extremity of Andalusia between the Guadiana and Guadalaquivir rivers. It is a southern series of the Alpine System. Before 1978, also search Sierra Morena.
CO N380000N390000
W0030000W0060000
UF Cordillera Betica
Cordillera Marianica
BT Spain
SA Andalusian
Betic Zone
Prebetic Zone
Subbetic Zone

Betic Zone
Used in association with the Betic Cordillera in SW.
BT Spain
SA Betic Cordillera

Betsiboka Basin
Around town of Majunga on Bombetoka Bay. Before 1978, also search Majunga; Majunga Basin.
UF Majunga Basin
BT Malagasy Republic

Betts Cove
BT Newfoundland
Canada

Betula
Genus.
BA Dicotyledoneae
angiosperms
BT Spermatophyta
Plantae

beudantite
BZ arsenates
sulfates
BT minerals

Bevier Coal
Bevier Formation comprises Bevier Coal with underlying gray clay and overlying black slate. SE Kansas.
BT Pennsylvanian
SA Desmoinesian
Kansas

Bewcastle
Village and parish in Cumberland County in NE.
BT England
Great Britain
United Kingdom

Beyrichicopina
Term introduced in 1981. Includes use on level 2 under Ostracoda(1). See list F.
BA Ostracoda
BT Crustacea
Mandibulata
Arthropoda
Invertebrata

Bhandara
Town in extreme NE.
BT Maharashtra
India

Bhander Group
Bhander Series includes upper Bhander Sandstone, Sirbu shales, Middle Bhander Sandstone and Ganurgarh shales. Named after range of hills N of Narmada Valley. Also search Bhander Series; Bhander Limestone.
UF Bhander Limestone
Bhander Series
SA Cambrian
India
Madhya Pradesh
Precambrian

Bhander Limestone
use Bhander Group

Bhander Series
use Bhander Group

Bhilwara
Town in S.
BT Rajasthan
India

Bhuj
Town in Cutch in NW.
BT Gujarat
India

Bhuj Series
Consists of Zamia beds, Ptilophyllum beds, and Palmoxylon beds. Named after town of Bhuj in Cutch.
BT Middle Cretaceous
Cretaceous
SA Gujarat
India

Bhutan
Kingdom in E Himalayas between Tibet and NE India. Includes use on level 1 as an area term (list O). For term set options see list B.
CO N263000N283000
E0921000E0884000
BA Indian Peninsula
Asia
SA Himalayas
Lesser Himalayas

Bi
use bismuth

Bialowieza
National park in E.
BT Poland

Bialystok
Province in extreme NE Poland. Also a city.
BT Poland
NT Augustow
Belsk
Suwalki

Biarritz
Town on the Bay of Biscay near Spanish border.
BT Pyrenees-Atlantiques
France

Biarritzian
Europe: middle to upper Eocene. Includes use on level 3 under age terms(1). See list E.
BA Eocene
Paleogene
Tertiary
BT Cenozoic
SA middle Eocene
upper Eocene

bibliography
Includes use on level 1 (list A); on

level 2 under major disciplines; as a level 2 or 3 term appropriate to a large number of topics (list G). Covers major bibliographies on topics, areas or persons. If 1., term set options are:
topic [List B or extraterrestrial geology, general, geophysics]
subtopic
SA areal geology
associations
biography
catalogs
geology
glossaries
lexicons
mineralogy
oceanography
paleobotany
paleontology
petrology
research
review
sedimentary petrology
stratigraphy
structural geology
survey organizations
tectonophysics
volcanology

Bicaz Valley
River valley in Bacau County.
BT Moldavia
Romania

Bida
Region in W central.
BT Nigeria

Bidahochi Formation
Divided into six members for mapping purposes. Fifth member informally referred to as White Cone Member. In NE Arizona.
BT Pliocene
SA Arizona
New Mexico

bideauxite
BZ chlorides
fluorides
BA halides
BT minerals

Biel Lake
In NW Bern.
UF Bienne Lake
Lake of Bienne
BT Bern
Switzerland

Biella
Town in Vercelli Province in N central.
BT Italy
SA Piemonte

Bielsk
use Belsk

Bielsk Podlaski
use Belsk

Bielsko
City in S part of country. Also search Bielsko-Biala.
UF Bielsko-Biala
BT Katowice
Poland

Bielsko-Biala
use Bielsko

Bienne Lake
use Biel Lake

Big Belt Mountains
Range of Rocky Mountains in W central.
BT Montana
United States
SA Rocky Mountains

Big Bend National Park
In Brewster County on the Mexican border in SW Texas. Also search Big Bend.
BT Texas
United States

Big Horn Basin
use Bighorn Basin

Big Horn County
Index states as applicable.
BX Montana
Wyoming
BT United States

Big Horn Mountains
use Bighorn Mountains

Big Pine Key
One of the principal islands in the Florida Keys.
BT Florida
United States
SA Florida Keys

Big Snowy Mountains
N central.
BT Montana
United States

Bigby-Cannon Limestone
Contemporary facies comprising this unit are Bigby Limestone and Cannon Limestone. Overlies Hermitage Formation and underlies Catheys Formation. Central Tennessee.
BT Middle Ordovician
Ordovician
SA Tennessee

Bighorn Basin
River basin. Index states as applicable. Also search Big Horn Basin.
UF Big Horn Basin
BT United States
SA Montana
Wyoming

Bighorn Mountains
Range of Rocky Mountains. Index states as applicable. Also search Big Horn Mountains.
CO N433000N453000
W1063000W1083000
UF Big Horn Mountains
BT United States
SA Montana
Rocky Mountains
Wyoming

Bihar
State in E India.
CO N220000N273000
E0882000E0831000
UF Behar
BT India
NT Amarjola
Bokaro
Dhanbad
Gaya
Hazaribagh
Jharia
Jharia coal field
Karanpura
Karanpura coal field
Karharbari
Monghyr
Mosaboni copper mines
Palamau
Rajgir
Rajmahal
Rajmahal Hills
Ramgarh coal field
Ranchi
Singhbhum
Sini
SA Ajay River
Chainpur Meteorite
Damodar Valley
Damuda Series
Karharbari Stage
Kosi Basin
Panchet Series
Semri Series
Singhbhum Granite
Singhbhum shear zone
Son Valley

Bihar Mountains
use Bihor Mountains

Bihor
County in W.
BT Transylvania
Romania

Bihor Mountains
In W central Transylvania.
UF Bihar Mountains
BT Transylvania
Romania
SA Vladeasa Mountain

Biikzhal
Drill hole in Emba River region of Caspian Basin.
BT Kazakhstan
USSR

Bijapur
Town in NW.
BT Mysore
India

Bijawar System
Includes Majhauli Group, Bhitri Group, Lora Group, and Chanderdip Group.
BT Precambrian
SA India
Madhya Pradesh

Bikaner
A city and former state in NW.
BT Rajasthan
India

Bilaspur
Former Indian state. Now a district in N part of country.
BT Himachal Pradesh
India

Bilbao
City on the Bay of Biscay in the Basque country.
BT Vizcaya
Basque Provinces
Spain

Billings
City in Yellowstone county in S central.
BT Montana
United States

Billings County
In W North Dakota.
CO N464000N472000
W1030500W1034000
BT North Dakota
United States

Billiton
Island between Sumatra and SW Borneo.
UF Belitung
BT Indonesia

Bimini
Two small islands off S Florida coast.
UF Bimini Islands
BT Bahamas
West Indies

Bimini Islands
use Bimini

Bingham
Mining town in Salt Lake County in N central Utah.
BT Utah
United States

Bingham County
Before 1978, also search Bingham for the county.
BT Idaho
United States

Binghamton
City in Broome County on the Pennsylvania border.
BT New York
United States

Binn Valley
use Binnental

Binnatal
use Binnental

Binnen Tal
use Binnental

Binnental
Valley in E Valais. Also search Binn Valley, and Binnatal.
UF Binn Valley
Binnatal
Binnen Tal
BT Valais
Switzerland

Binue Valley
use Benue Valley

biocalcarenite
BA carbonate rocks
BT sedimentary rocks
SA calcarenite

biocenoses
Includes use on level 2 and 3 under fossil group(1). See list F. Also search biocenosis.
UF biocenosis
biocoenoses
life assemblage
SA communities
paleontology
thanatocenoses

biocenosis
use biocenoses

biochemical sedimentation
Term introduced in 1981.
SA sedimentation

biochemistry
Includes use on level 2 under fossil group(1). See list F. Also search biogeochemistry.
SA biology
fossils
geochemistry
paleobiology

biocirculation
Includes use on level 2 under ocean circulation(1).
SA ocean circulation

bioclastic sedimentation
Term introduced in 1981.
SA sedimentation

biocoenoses
use biocenoses

biofacies
Includes use on level 3 under ecology(1), paleoecology(1), sedimentation(1), or sediments(1).
UF biologic facies
SA assemblages
biostratigraphy
biota
ecology
facies
sedimentation
sediments

biogenic
A valid term through 1977. See biogenic structures; biogenic effects.

biogenic effects
Term introduced in 1981. Used for the effects of organisms on their surroundings.
SA ecology

biogenic structures
Includes use on level 2 under sedimentary structures(1). See list K.
BA sedimentary structures
NA algal structures

banks
bioturbation
borings
burrows
coprolites
girvanella
lebensspuren
oncolites
NZ stromatolites
NA tracks
trails
SA coal
peat
reefs
structures

biogeochemical methods
Includes use on level 2 under mineral exploration(1). Before 1972, also search biogeochemical prospecting; biogeochemistry.
BT geochemical methods
SA geobotanical methods
geochemistry
methods
mineral exploration

biogeochemical prospecting
A valid term through 1971. After 1971, use biogeochemical methods.

biogeochemistry
A valid term through 1978. After 1978, see biochemistry for fossils (list F), or biogeochemical methods under mineral exploration(1).

biogeography
Includes use on level 1 (list A); on level 2 under fossil groups (list F). To be used for descriptions of the geographic distribution of both fossil and modern organisms. If 1, term set options are:
age (List E; only where one age is discussed)
area (or global)
fossil group [List F]
age
topic [catalogs, concepts, distribution, evolution, interpretation, nomenclature, observations, patterns]
subtopic
UF chorology
phytogeography
zoogeography
SA Boreal
continental drift
ecology
faunal provinces
fossils
geography
paleobotany
paleoecology
paleogeography
paleontology
palynology
range
stratigraphy

biography
Includes use on level 1 (list A). This set is used only where there is an extensive bibliography. Term set options are:
general
name of individual (last, first).
SA bibliography

bioherms
Includes use on level 3 under reefs(1).
UF organic mound
SA algae
atolls
biostromes
calcareous composition
Coelenterata
reefs

biologic facies
use biofacies

biological processes
As of 1981, no longer a valid term for GeoRef. Use biology.

biology
Before 1981, also search biological processes.
SA bacteria
biochemistry
biosphere
biotypes
ecology
epibiotism
exobiology
paleobiology

biomass
Includes use on level 3 under ecology(1).
SA ecology
new energy sources

biometry
Includes use on level 3 under fossil group(1) and under paleontology(1). See list F.
SA fossils
mathematical geology
paleontology
statistical analysis

biomicrite
Includes use on level 3 under sedimentary rocks(1) carbonate rocks(2).
BA carbonate rocks
BT sedimentary rocks
SA micrite

biopelite
use black shale

biorhexistasy
Term introduced in 1981.
SA laterites
soils
weathering

biosparite
BA carbonate rocks
BT sedimentary rocks
SA limestone

biosphere
SA atmosphere
biology
Earth
hydrosphere
lithosphere
tectonosphere

biostratigraphy
As of 1981, includes use on level 2 under common names of fossil groups(1). See list F. Before 1981, included use on level 2 under scientific names of fossil groups(1).
SA acritarch flora
algal flora
ammonoids
amphibians
belemnites
biofacies
birds
bivalves
brachiopods
bryozoans
cephalopods
coccoliths
conodonts
corals
crustaceans
desmids
diachronism
diatom flora
dinoflagellates
echinoderms
ferns
fish
foraminifers
fossils
gastropods
graptolites
historical geology
index fossils
invertebrates
mammals
mollusks
nannoconids
ostracods
paleoecology
paleontology
palynomorphs
radiolarians
range
reptiles
rodents
rudists
stratigraphy
trilobites
vertebrates

biostromes
Includes use on level 3 under reefs(1).
SA bioherms
Coelenterata
reefs

biota
Includes use on level 2 under soils(1). Also search organisms.
SA bacteria
biofacies
ecology
faunal studies
floral studies
Invertebrata
living taxa
microorganisms
soils
Vertebrata

biotite
BA mica group
sheet silicates
silicates
BT minerals
SA heavy minerals

biotite gneiss
Includes use on level 3 under metamorphic rocks(1) gneisses(2). See list J.
BA gneisses
BT metamorphic rocks

biotite granite
As of 1981, includes use on level 3 under igneous rocks(1) granites(2). See list H. Before 1981, included use on level 3 under igneous rocks(1) granite-granodiorite family(2).
BA granites
BT igneous rocks
SA granite

biotite schist
Includes use on level 3 under metamorphic rocks(1) schists(2). See list J.
BA schists
BT metamorphic rocks

bioturbation
Includes use on level 3 under sedimentary structures(1) biogenic structures(2). See list K.
BA biogenic structures
sedimentary structures
SA borings
burrows
sedimentation

biotypes
Term introduced in 1978. Includes use on level 3 under ecology(1) or paleoecology(1).
SA biology
ecology
paleoecology

Birch Creek
Village in E Alaska. Also a creek.
BT Alaska
United States

Bird Island
use Aves Island

Bird River
Village in SE.
BT Manitoba
Canada

Bird Spring Formation
In Spring Mountain area, Nevada, underlies Spring Mountain Formation and overlies Illipah Formation. Also search Bird Spring Group.
BT Paleozoic
SA California
Lower Permian
Mississippian
Nevada
Pennsylvanian
Permian
Upper Mississippian
Utah

birds
Term introduced in 1981 as the common name for Aves. See list F. To be used for Aves when they are used as tools in stratigraphy. Includes use on level 1. Before 1981, search Aves AND biostratigraphy. If 1, term set options are:
biostratigraphy
age [List E]
SA Aves
biostratigraphy

birefraction
use birefringence

birefringence
UF birefraction
double refraction
SA minerals
optical properties
polarization
refractive index

Birmingham
City in Jefferson County in central Alabama, and city in W central England. Index Alabama and/or England as applicable.
BX Alabama
England

Birmingham University
In Birmingham in W central.
BT England
Great Britain
United Kingdom

birnessite
BA oxides
BT minerals

Biryusa
River SW of Tayshet in W Irkutsk Oblast.
BT Russian Republic
USSR

Bisbee
City in Cochise County in SE.
BT Arizona
United States

Biscay (Bay)
use Bay of Biscay

Biscay Bay
use Bay of Biscay

Biscayne Bay
S of Miami on E coast of Dade County.
BT Florida
United States

Bishop Tuff
E central California. Bishop Tuff was erupted a short time after Sherwin

glacial stage.
BT Pleistocene
Quaternary
SA California

Bismarck Archipelago
Volcanic group of islands N and NE of E New Guinea.
BT Papua New Guinea
Australasia
NT Rabaul Caldera
SA Melanesia
New Britain
New Ireland

Bismarck Sea
Part of W Pacific Ocean NE of New Guinea and NW of New Britain.
BT Pacific Ocean

bismuth
Includes use on level 1 and 2 as a chemical element (list D). As of 1981, use bismuth ores for bismuth as a commodity. Before 1981, included use on level 1 as a commodity term.
UF Bi
SA bismuth ores
elements
heavy metals
native elements

bismuth ores
Term introduced in 1981. Includes use on level 1 and 2 as a commodity term (list C). Before 1981, search bismuth AND ore deposits.
SA bismuth
native elements

bismuthides
As of 1981, includes use on level 2 under minerals(1). See list L. Before 1981, included use on level 3 under minerals(1) sulfides(2).
BA sulfides
BT minerals
NA froodite
michenerite

bismuthinite
BA sulfides
BT minerals

Bison
Genus. As of 1981, includes use on level 3 under Mammalia(1) Ruminantia(2). See list F. Before 1981, included use on level 3 under Mammalia(1) Artiodactyla(2).
BA Bovidae
Ruminantia
Artiodactyla
Mammalia
BT Tetrapoda
Vertebrata
NA Bison latifrons
Bison occidentalis

Bison latifrons
As of 1981, includes use on level 3 under Mammalia(1) Ruminantia(2). See list F. Before 1981, included use on level 3 under Mammalia(1) Artiodactyla(2).
BA Bison
Bovidae
Ruminantia
Artiodactyla
Mammalia
BT Tetrapoda
Vertebrata

Bison occidentalis
As of 1981, includes use on level 3 under Mammalia(1) Ruminantia(2). See list F. Before 1981, included use on level 3 under Mammalia(1) Artiodactyla(2).
BA Bison
Bovidae
Ruminantia
Artiodactyla
Mammalia
BT Tetrapoda
Vertebrata

Bistrita
City in Bistrita-Nasaud County in central.
BT Transylvania
Romania

Bistrita Mountains
Range of the Carpathians in NE Romania.
UF Bistritei Mountains
BT Romania
SA Carpathians

Bistrita Valley
River valley in W central.
BT Moldavia
Romania

Bistritei Mountains
use Bistrita Mountains

Bitkov
In W Ukraine.
UF Bitkow
Bytkov
BT Ukraine
USSR

Bitkow
use Bitkov

Bitlis
Term introduced in 1981. Town and province in SE.
BT Turkey

bitter spar
use dolomite

Bitter Springs Formation
Underlies Middle Cambrian sediments and overlies Heavitree Quartzite. SW and S central Northern Territory.
BT Proterozoic
Precambrian
SA Australia
Northern Territory

Bitterfeld
City in W.
BT Halle Bezirk
East Germany

Bitterroot Range
Range of the Rocky Mountains. Index states as applicable.
BT United States
SA Beaverhead Mountains
Idaho
Montana
Rocky Mountains

bitumenite
use torbanite

bitumens
Includes use on level 1 as a commodity term (list C); on level 2 under organic materials(1).
BA organic materials
NA asphalt
SA anthraxolite
hydrocarbons
kerogen
oil seeps
petroleum

bituminous coal
Before 1978, also search coal AND bituminous.
UF soft coal
BA coal
organic residues
BT sedimentary rocks
SA coal

bituminous shale
As of 1981, no longer a valid term for GeoRef. Use black shale.

bivalves
Term introduced in 1981 as the common name for Bivalvia. See list F. To be used for Bivalvia when they are used as tools in stratigraphy. Before 1981, search Bivalvia AND biostratigraphy.
SA biostratigraphy
Bivalvia

Bivalvia
Includes use on level 2 under Mollusca(1). See list F. Before 1976, also search Pelecypoda.
BA Mollusca
BT Invertebrata
NA Actinodontida
Anthraconaia
Anthraconauta
Astarte
Astartida
Carbonicola
Cardiidae
Carditida
Crassostrea virginica
Ctenodontida
Cyrtodontida
Daonella
Didacna
Edmondia
Glycymeris
Gryphaea
Halobia
Hippuritacea
Inoceramidae
Inoceramus
Limnocardium
Lucinidae
Mactra
Mercenaria
Monotis
Myophoria
Mytilus
Nuculanidae
Nuculidae
Ostreidae
Pectinacea
Pectinidae
Pholadomyida
Praecardiida
Rudistae
Septibranchia
Solemyida
Tridacna
Trigoniidae
Unionidae
Venerida
SA bivalves

Biwa Lake
use Lake Biwa

Biwabik Iron Formation
In Animikie Group. In St. Louis County, NE Minnesota. Part of Mesabi Iron Range. Also search Biwabik Iron-Formation.
UF Biwabik Iron-Formation
BT Precambrian
SA Animikie Group
Minnesota

Biwabik Iron-Formation
use Biwabik Iron Formation

bixbyite
UF partridgeite
BA oxides
BT minerals

Bjerkrem-Sogndal Massif
S part of country.
BT Norway

Bjurbole Meteorite
Term introduced in 1981. Impact ENE of Helsinki. Before 1981, also search Bjurbole AND meteorites.
BA L chondrites
chondrites
BT meteorites
SA Finland

Black earth
use Chernozems

Black Forest
Mountainous region in W.
CO N473000N490000
E0083000E0073000
BT Baden-Wurttemberg
West Germany

Black Hills
Group of mountains primarily in SW South Dakota. Index states as applicable.
CO N431500N443500
W1023000W1040500
BT United States
SA South Dakota
Wyoming

black lead
use graphite

Black Mesa
Tableland region in NE.
CO N360000N370000
W1090000W1110000
BT Arizona
United States

Black Mesa Basin
In NE Arizona.
BT Arizona
United States

Black Range
Extends N-S through parts of Grant and Sierra counties in SW.
BT New Mexico
United States

Black River Group
Comprises Pamelia, Lowville, and Chaumont formations.
BT Middle Ordovician
Ordovician
SA Blackriverian
New York
Pennsylvania

Black Rock
Term introduced in 1981. In Millard County, Utah. Also search Black Rock Desert AND Utah.
BT Millard County
Utah
United States

Black Rock Desert
In Pershing and Humboldt counties in NW.
BT Nevada
United States

Black Sea
Large inland sea bounded on the N and NE by the USSR, S by Turkey, and W by Bulgaria and Romania. Includes use on level 1 as an area term (list O). For term set options see list B.
CO N410000N470000
E0420000E0280000
BA East Mediterranean
Mediterranean Sea
SA Azov Sea
DSDP Site 379
DSDP Site 380
DSDP Site 381
Leg 42B

Black Sea Basin
use Black Sea region

Black Sea Coast
use Black Sea region

Black Sea Lowland
use Black Sea region

Black Sea region
Index countries as applicable. Also search Black Sea Coast, Black Sea Lowland and Black Sea Basin.

CO N423000N473000
E0400000E0300000
UF Black Sea Basin
Black Sea Coast
Black Sea Lowland
BT Eurasia
SA Bulgaria
Romania
Turkey
USSR

black shale
Includes use on level 3 under sedimentary rocks(1) clastic rocks(2). Also search carbonaceous shale; bituminous shale.
UF biopelite
carbonaceous shale
BA clastic rocks
BT sedimentary rocks
SA shale
terrigenous materials

Black Warrior Basin
Primarily in Alabama. Index states as applicable.
BT United States
SA Alabama
Mississippi

Blackford County
E central.
BT Indiana
United States

Blackhawk Formation
In Mesaverde Group. Six principal members are Desert, Grassy, Sunnyside, Kenilworth, Aberdeen, and Spring Canyon. Central E Utah.
BT Upper Cretaceous
Cretaceous
SA Mesaverde Group
Utah

Blackriverian
North America. Lower Mohawkian, below Trentonian. Includes use on level 3 as an age term (list E).
BA Middle Ordovician
Ordovician
BT Paleozoic
SA Black River Group

Blacksburg
Town in Montgomery County in SW.
BT Virginia
United States

Blackwater
River and falls in Tucker County in E.
BT West Virginia
United States

Blaine County
Index states as applicable.
BX Idaho
Montana
Nebraska
Oklahoma
BT United States

Blaine Formation
In Nippewalla, El Reno, or Pease River Group. W Oklahoma, central North and panhandle of Texas.
BT Permian
SA Oklahoma
Texas

Blairmore Group
Includes the Gladstone Formation, Beaver Mines Formation, and Mill Creek Formation in S Alberta.
BT Cretaceous
SA Alberta
Lower Cretaceous
Upper Cretaceous

Blake Basin
use Blake-Bahama Basin

Blake Nose
Term introduced in 1981. NE-jutting spur of the Blake Plateau.
CO N293000N303000
W0760000W0770000
BT Blake Plateau
Atlantic Ocean

Blake Outer Ridge
use Blake-Bahama Outer Ridge

Blake Plateau
E of N Florida and N of Grand Bahama Island. Also search Bahama Platform.
UF Bahama Platform
BT Atlantic Ocean
NT Blake Nose

Blake Ridge
use Blake-Bahama Outer Ridge

Blake-Bahama Basin
Just to the E of the Blake Plateau E of North Florida. Also search Blake Basin.
UF Blake Basin
BT Atlantic Ocean

Blake-Bahama Outer Ridge
E of the Blake-Bahama Basin which is E of the Blake Plateau off N Florida.
UF Blake Outer Ridge
Blake Ridge
BT Atlantic Ocean

Blancan
North America. Above Hemphillian.
BT Cenozoic
SA Pleistocene
Pliocene
Quaternary
Tertiary
Texas

Blanco fracture zone
Off coast of Oregon. NE of E end of Mendocino Escarpment.
BT Pacific Ocean

blasting
Includes use on level 3 under engineering geology(1) or foundations(1).
SA engineering geology
excavations
explosions
foundations

Blastoidea
Includes use on level 2 under Echinodermata(1). See list F.
BA Crinozoa
Echinodermata
BT Invertebrata

blastomylonite
Includes use on level 3 under metamorphic rocks(1) mylonites(2). See list J.
BA mylonites
BT metamorphic rocks

Blattopteroida
Term introduced in 1981. Includes use on level 2 under Insecta(1). See list F.
BA Insecta
BT Mandibulata
Arthropoda
Invertebrata

Bleiberg
Town in S part of country.
BT Carinthia
Austria

Bleikvassli
Ore bearing region.
BT Nordland
Norway

Blekinge
County on the Baltic Sea in extreme SE.
BT Sweden

blind deposits
Term introduced in 1981.
SA mineral deposits

Blind River
Town on Lake Huron.
BT Ontario
Canada

bloating shale
use shale

block
A valid index term through 1977. Now use block structures.

block clay
use melange

block faults
use block structures

Block Island Sound
Between Washington County, Rhode Island and Block Island.
BT Atlantic Ocean

block mountains
use block structures

block structures
Term introduced in 1978. Includes use on level 3 under faults(1) systems(2). Before 1978, also search faults AND block; block faults; block structure; block mountains.
UF block faults
block mountains
fault blocks
SA faults
horsts
normal faults
systems

blomstrandite
use betafite

Bloomsburg Formation
In Cayuga Group. Central and S Pennsylvania, W Maryland, W Virginia, and N West Virginia.
BT Upper Silurian
Silurian
SA Maryland
Pennsylvania
Virginia
West Virginia

Blount County
Index states as applicable.
BX Alabama
Tennessee
BT United States

Blow River
Flows into Mackenzie Bay in the extreme N.
BT Yukon Territory
Canada

Bloyd Formation
Shale consists of (ascending) Brentwood Limestone Member, Woolsey Member and an unnamed shale division. NW Arkansas and E Oklahoma.
BT Morrowan
Lower Pennsylvanian
Pennsylvanian
SA Arkansas
Oklahoma

blue algae
use Cyanophyta

blue chalcocite
use digenite

Blue Glacier
Mount Olympus, Olympic National Park in NW.
BT Washington
United States

blue lead
use galena

Blue Mountain Lake
In the Adirondack Mountains.
BT New York
United States

Blue Mountains
In W Idaho, NE Oregon, and SE Washington, and also a range in E Jamaica. Index Jamaica and states as applicable.
SA Idaho
Jamaica
Oregon
Washington

Blue Nile
Right headstream of the Nile River. Index countries as applicable.
BT Africa
SA Ethiopia
Nile River
Sudan

Blue Ridge Mountains
E and SE range of the Appalachians. Index states as applicable. Also search Blue Ridge.
BT United States
SA Appalachians
Georgia
Grandfather Mountain
North Carolina
Virginia
West Virginia

Blue Ridge Province
Embraces the Appalachians E of the Great Appalachian Valley from South Mountain, Pa., southward including the Piedmont. Composed of some of the most ancient rocks of the Appalachians. Index states as applicable. Also search Blue Ridge.
BT Appalachians
North America
SA Georgia
Maryland
North Carolina
Pennsylvania
South Carolina
Virginia
West Virginia

blue-green algae
use Cyanophyta

Bluebell Mine
E shore of Kootenay Lake in SE.
BT British Columbia
Canada

blueschist
Includes use on level 3 under metamorphic rocks(1) schists(2). See list J.
BA schists
BT metamorphic rocks
SA glaucophane schist

blueschist facies
BT facies
SA greenschist facies
metamorphic rocks

bluff formation
use loess

Blyava
A section of the City of Mednogorsk in the S Urals near the Kazakh border.
BT Russian Republic
USSR

Blying Sound Quadrangle
Term introduced in 1981. S central Alaska.
CO N590000N600000
W1470000W1500000
BT Alaska
United States

Bochnia
A mining town in S part of country.
BT Krakow
Poland

Bochum
City in the Ruhr Valley.
BT North Rhine-Westphalia
West Germany

Bodaibo
Town in NE Irkutsk Oblast.
BT Russian Republic
USSR

Bodega Head
Tip of peninsula sheltering Bodega Bay from the Pacific Ocean in Sonoma County.
BT California
United States

Bodensee
use Lake Constance

bodies, ore
use ore bodies

bodily tide
use Earth tides

bodily waves
use body waves

body waves
Includes use on level 3 under seismology(1) elastic waves(2).
UF bodily waves
BA elastic waves
NA P-waves
PcP-waves
PKiP-waves
PKP-waves
PKS-waves
PPP-waves
PPS-waves
S-waves
ScS-waves
SKP-waves
SKS-waves
SA seismology
surface waves
waves

boehmite
BA oxides
BT minerals

Boeotia
A district in Attica Department. Central Greece and Euboea.
BT Greece

Boerzsoeny Mountains
use Borzsony Mountains

Bog soils
Includes use on level 3 under soils(1). See list M.
BA soils
SA Histosols
Hydromorphic soils
Intrazonal soils
peat
soil group

Bogota
City in central part of country.
BT Colombia

bogs
Include use on level 3 under geomorphology(1).
NT peat bogs
SA geomorphology
marshes
swamps

Boguszow
City in SW part of country.
BT Wroclaw
Poland

bohdanowiczite
BA sulfides
BT minerals

Bohemia
Region in westernmost part of country.
CO N483200N510400
E0165000E0120600
BT Czechoslovakia
NT Bohemian Massif
Central Bohemian Pluton
Chvaletice
Jachymov
Karlovy Vary
Kladno
Kutna Hora
Pilsen Basin
Plana
Prague
Pribram
Slany
Sobotka
Sokolov
Sokolov Basin
Teplice
SA Karkonosze Mountains
Sudeten
Sudeten Mountains
Weisse Elster Basin
White Mountain

Bohemia-Moravia System
Term introduced in 1981.
CO N484500N500000
E0171000E0132500
BT Czechoslovakia

Bohemian Basin
Term introduced in 1981.
CO N494000N504500
E0163000E0134000
BT Czechoslovakia

Bohemian Massif
A dissected quadrangular plateau bounded by the Bohemian Forest, Erzgebirge, the Sudeten Mountains and the Bohemian-Moravian Heights.
CO N483000N501000
E0163000E0123000
BT Bohemia
Czechoslovakia
SA Central Bohemian Pluton

Boise
City in Ada County in WSW.
BT Idaho
United States

Bojnice
Village in W central.
BT Slovakia
Czechoslovakia

Bokaro
Coal field in central Bihar. Also search Bokaro coal field, Bokaro Coalfield, and Bokaro Field.
UF Bokaro coal field
Bokaro Coalfield
Bokaro Field
BT Bihar
India

Bokaro coal field
use Bokaro

Bokaro Coalfield
use Bokaro

Bokaro Field
use Bokaro

Bolangir
Town and district in W.
BT Orissa
India

Boleslawiec
Town is SW part of country.
BT Wroclaw
Poland

Boliden
Village in NE.
BT Vasterbotten
Sweden

Bolinas Lagoon
Off Bolinas Bay in SW Marin County.
BT California
United States

Bolivar
State in SE.
BT Venezuela
NT Cartagena
Imataca Complex
Magdalena Delta

Bolivia
Includes use on level 1 as an area term (list O). For term set options see list B.
CO S230000S093000
W0583000W0693000
BA South America
NT Cochabamba
La Paz
Oruro
NX Sucre
SA Amazon Basin
Andes
Chaco
Eastern Cordillera
Lake Titicaca
Magdalena
Santa Cruz

Bolivina
Genus. Includes use on level 3 under foraminifera(1) Buliminacea(2).
BA Buliminacea
Rotaliina
foraminifera
BT Protista
Invertebrata

Bolivinitidae
Includes use on level 3 under foraminifera(1) Buliminacea(2).
BA Buliminacea
Rotaliina
foraminifera
BT Protista
Invertebrata

Bolivinoides
Genus. Includes use on level 3 under foraminifera(1) Buliminacea(2).
BA Buliminacea
Rotaliina
foraminifera
BT Protista
Invertebrata

Bologna
City and province in E central.
BT Emilia-Romagna
Italy
NT Farneto Cave

Bolshezemelskaya Tundra
In Nenets National Okrug of Arkhangelsk Oblast. Also search Bol'shezemel'skaya Tundra.
BT Russian Republic
USSR

Boltyshka Depression
Central Ukraine. Also search Boltyshka.
BT Ukraine
USSR

Bolzano
City and province in N central.
BT Trentino-Alto Adige
Italy

Bombay
City on the Arabian Sea.
BT Maharashtra
India

Bonaire
Island off coast of Venezuela, 30 miles E of Curacao.
BT Netherlands Antilles

Bonaparte Basin
use Bonaparte Gulf basin

Bonaparte Gulf basin
N of Joseph Bonaparte Gulf and W of Northern Territory. Also search Bonaparte Basin.
UF Bonaparte Basin
BT Timor Sea
Pacific Ocean
SA Joseph Bonaparte Gulf

Bond albedo
use albedo

bonding
As of 1978, term is used on level 2 under crystal chemistry(1) and crystal structure(1).
SA cell dimensions
crystal chemistry
crystal structure
order-disorder

Bondoc Peninsula
Between Ragay Gulf and Mompog Pass of S Quezon Province in S.
BT Luzon
Philippine Islands

bone beds
Term introduced in 1981. Includes use on level 3 under sedimentary rocks(1) chemically precipitated rocks(2). See list I. Before 1981, also search bone bed.
BA phosphate rocks
chemically precipitated rocks
BT sedimentary rocks

Bone Valley Formation
Unconformably overlain by Pleistocene terrace deposits ranging in age from Sunderland to Pamlico. S central Florida.
BT middle Pliocene
Pliocene
SA Florida

bones
Includes use on level 3 under fossil groups(1) for type of material. See list F.
SA fossils
jaws
osteology
paleontology
skeletons
skulls
teeth
Vertebrata

Bong Range
In N part of country.
BT Liberia

Bonin Islands
Group of 27 volcanic islands 600 miles S of Tokyo. Part of Tokyo Prefecture. Also search Ogasawara Islands.
UF Ogasawara Islands
BT Pacific Ocean

Bonner Formation
In Missoula Group. Precambrian (Belt series): In W Montana.
UF Bonner Quartzite
BT Precambrian
SA Missoula Group
Montana

Bonner Quartzite
use Bonner Formation

Bonneterre Dolomite
use Bonneterre Formation

Bonneterre Formation
Includes Tom Sank limestone member at base. Also search Bonneterre Dolomite.
UF Bonneterre Dolomite
BT Upper Cambrian
Cambrian
SA Missouri

Bonneville Salt Flats
Tooele County in NW.
BT Utah
United States

book reviews
Includes use on level 1. Term set

options are:
topic [List B, extraterrestrial geology, general, geophysics]
title
SA catalogs
review
textbooks

Boone County
Index states as applicable.
BX Arkansas
Illinois
Indiana
Iowa
Kentucky
Missouri
Nebraska
West Virginia
BT United States

Boothia Felix Peninsula
use Boothia Peninsula

Boothia Peninsula
N extremity of mainland Canada W of Gulf of Boothia in District of Franklin.
CO N690000N720000
W0920000W0963000
UF Boothia Felix Peninsula
BT Northwest Territories
Canada

Boquira Mine
Also search Boquira.
BT Bahia
Brazil

Bor
Town in E.
BT Serbia
Yugoslavia

boracite
BZ borates
chlorides
BT minerals

borate deposits
Term introduced in 1981. Includes use on level 3 as a commodity term (list C) under boron deposits(1). Before 1981, search borates AND deposits.
SA borates
boron deposits
evaporite deposits

borates
Includes use on level 2 under minerals(1). See list L. As of 1981, use borate deposits for borates as a commodity. Before 1981, included use on level 3 as a commodity term.
BT minerals
NZ boracite
NA borax
colemanite
NZ fluoborite
NA hulsite
hydroboracite
kaliborite
kernite
kotoite
kurchatovite
kurnakovite
ludwigite
preobrazhenskite
NZ sakhaite
NA solongoite
suanite
szaibelyite
ulexite
vimsite
vonsenite
SA borate deposits

borax
BA borates
BT minerals

Borax Lake
BT California
United States

Borborema
A plateau in NE part of country. Index states as applicable.
BT Brazil
SA Paraiba
Rio Grande do Norte

Bordeaux
City in SW part of country.
BT Gironde
France

Bordelais
Region in SW part of country.
BT Gironde
France

Borden Group
In Kentucky. Comprises New Providence Formation, Brodhead Formation, Floyds Knob Formation, and Muldraugh Formation. In S Indiana, comprises New Providence Shale, Locust Point, Carwood, Floyds Knob, and Edwardsville formations. S Indiana and E central Kentucky.
BT Mississippian
SA Indiana
Kentucky
Lower Mississippian
Upper Mississippian

borderland, continental
use continental borderland

bore
use boreholes

Boreal
Includes use on level 3 as an age unit. Used primarily in Europe for an interval of postglacial time following the Preboreal and preceding the Atlantic, during which the inferred climate was warm and dry. See list E.
BA Holocene
Quaternary
BT Cenozoic
SA biogeography
paleoclimatology
Preboreal

boreal environment
Term introduced in 1978. Includes use on level 3 under ecology(1). Before 1978, also search boreal; before 1981, also search boreal region (Boreal is used as an age unit).
SA ecology
environment
paleoclimatology
paleoecology

boreal region
As of 1981, no longer a valid term for GeoRef. Use boreal environment.

borehole sections
Term introduced in 1981.
SA boreholes
sections

boreholes
Includes use on level 3 under well-logging(1). Before 1976, also search drill holes.
UF bore
SA borehole sections
cores
cuttings
dipmeter logging
drilling
electrical logging
heat flow
well-logging
wells

borings
As of 1978, restricted to worm borings. For well-logging, use boreholes or cores. Includes use on level 3 under ichnofossils(1) and sedimentary structures(1).
UF worm borings
BA biogenic structures
sedimentary structures
SA bioturbation
burrows
cores
ichnofossils
lebensspuren

Borislav
City in Lvov Oblast in W.
BT Ukraine
USSR

Borkut Deposit
Region of mineral deposits in Transcarpathian Oblast in W Ukraine. Also search Borkut.
BT Ukraine
USSR

Borneo
An island, the S two-thirds of which is known as Kalimantan, Indonesia, with the remainder comprising the two states of East Malaysia and the British protected sultanate of Brunei. Index Brunei, Kalimantan, and the states of East Malaysia as applicable.
CO S040000N073000
E1192000E1085000
BX Far East
Malay Archipelago
SA Brunei
East Malaysia
Indonesia
Kalimantan
Malaysia
Sabah
Sarawak

Borneo-Java Shelf
use Sunda Shelf

Bornholm
Island in W Baltic Sea S of Sweden. Also search Bornholm Island.
UF Bornholm Island
BT Denmark

Bornholm Island
use Bornholm

bornite
UF erubescite
horseflesh ore
purple copper ore
BA sulfides
BT minerals
SA copper

boron
Includes use on level 1 and 2 as a chemical element (list D). As of 1981, use boron deposits for boron as a commodity. Before 1981, included use on level 1 as a commodity term.
UF B
SA boron deposits
elements

boron deposits
Term introduced in 1981. Includes use on level 1 and 2 as a commodity term (list C). Before 1981, search boron AND deposits.
SA borate deposits
boron

boronatrocalcite
use ulexite

borosilicates
BA silicates
BT minerals
SA axinite
dravite

Borrego Mountain
In S California.
BT California
United States

Borrowdale Volcanic Series
Llanvirn and ? Llandeilo Series. Underlain by Skiddaw Slates and overlain by the Coniston Limestone Group. Lake District in NW England.
BT Ordovician
SA England
Llandeilian
Llanvirnian

borrowing
Term introduced in 1981.
SA economics

Borshchev
City in Ternopol Oblast in W Ukraine. Also search Borshchov.
UF Borshchov
Borszczow
BT Ukraine
USSR

Borshchov
use Borshchev

Borsod Basin
Coal basin in NW Hungary. Also search Borsod.
BT Hungary

Borszczow
use Borshchev

Borzeta
Boreholes in W Carpathians.
BT Poland

Borzsony Mountains
Near Slovakian border in NW Nograd, Hungary. Also search Boerzsoeny Mountains.
UF Boerzsoeny Mountains
BT Hungary

Bos
Genus. As of 1981, includes use on level 3 under Mammalia(1) Ruminantia(2). See list F. Before 1981, included use on level 3 under Mammalia(1) Artiodactyla(2).
BA Bovidae
Ruminantia
Artiodactyla
Mammalia
BT Tetrapoda
Vertebrata
NA Bos primigenius

Bos primigenius
As of 1981, includes use on level 3 under Mammalia(1) Ruminantia(2). See list F. Before 1981, included use on level 3 under Mammalia(1) Artiodactyla(2).
BA Bos
Bovidae
Ruminantia
Artiodactyla
Mammalia
BT Tetrapoda
Vertebrata

Bosano
Region in NW.
BT Sardinia
Italy

Boshchekul
Village in Pavlodar Oblast in NE.
BT Kazakhstan
USSR

Bosnia
Region. Northern part of Republic of Bosnia and Herzegovina.
BT Yugoslavia

Bosnia-Herzegovina
Term introduced in 1981. Region in

W central.
CO N423500N452000
E0194000E0153800
BT Yugoslavia

Boso Peninsula
use Chiba Peninsula

Bosphorus
use Bosporus

Bosporus
Strait connecting Black Sea with Sea of Marmara. Also search Bosphorus.
UF Bosphorus
BT Turkey
Middle East

Bossier Formation
In Cotton Valley Group. S Arkansas, W Louisiana, and E Texas.
BT Upper Jurassic
Jurassic
SA Arkansas
Cotton Valley Group
Louisiana
Texas

Boston
City in Suffolk County on Massachusetts Bay.
BT Massachusetts
United States

bostonite
As of 1981, includes use on level 3 under igneous rocks(1) syenites(2). See list H. Before 1981, included use on level 3 under igneous rocks(1) syenite family(2).
BA syenites
BT igneous rocks

Bosumtwi Crater
S central.
BT Ghana

botany, paleo-
use paleobotany

Bothnian Sea
use Gulf of Bothnia

Botryococcus
Includes use on level 3 under algae(1) Chlorophyta(2).
BA Chlorophyta
algae
BT thallophytes
Plantae

Botswana
Before 1966 was British protectorate of Bechuanaland. Includes use on level 1 as an area term (list O). For term set options see list B.
CO S270000S174000
E0293000E0200000
BA Southern Africa
Africa
SA Dwyka Series
Kalahari Desert
Karroo System
Limpopo Basin
Limpopo Belt
Zambezi Valley

bottling
Term introduced in 1981.
SA water
water resources

bottom currents
Papers are usually out of scope for GeoRef. Includes use on level 3 under ocean circulation.
BT currents
SA ocean circulation

bottom features
Includes use on level 2 under ocean floors(1) and marine geology(1); on level 3 under oceanography(1).
UF features, bottom
SA benthonic taxa
marginal basins
marine geology
mid-ocean ridges
ocean floors
submarine canyons
trenches
troughs

bottom load
use bedload

bottom water
SA oceanography
water

Botucatu Formation
Also search Botucatu Sandstone.
UF Botucatu Sandstone
SA Brazil
Mato Grosso
Mesozoic
Paleozoic
Paraguay
Permian
Sao Paulo
South America
Triassic

Botucatu Sandstone
use Botucatu Formation

Bou Azzer
Locality in Marrakesh region in W central Morocco. Also search Bou-Azzer.
UF Bou-Azzer
BT Morocco

Bou-Azzer
use Bou Azzer

Bouches-du-Rhone
Department on the Mediterranean Sea.
CO N431500N435500
E0054500E0041000
BT France
NT Aix-en-Provence
Carry-le-Rouet
Marseilles
Rhone Delta
Sainte-Victoire Mountain
SA Provence

boudinage
Includes use on level 3 under sedimentary structures(1) soft sediment deformation(2); on level 3 under deformation(1), and lineation(1) style(2); on level 3 under structural analysis(1). See list K. Before 1981, also search boudins.
UF pull apart structures
sausage structure
BA soft sediment deformation
sedimentary structures
SA deformation
lineation
melange
structural analysis

boudins
As of 1981, no longer a valid term for GeoRef. Use boudinage.

Bougainville
Largest of the Solomon Islands. ESE of the Bismarck Archipelago in the SW Pacific.
BT Papua New Guinea
Australasia
SA Solomon Islands

Bouguer
A valid term through 1976. After 1976, use Bouguer anomalies.

Bouguer anomalies
Includes use on level 3 under geophysical methods(1). Also search Bouguer.
BA gravity anomalies
SA anomalies
free-air anomalies
geophysical methods
geophysical surveys

boulangerite
BA sulfantimonites
sulfosalts
BT minerals

Boulder
City in Boulder County in N central.
BT Colorado
United States

Boulder Batholith
BT Montana
United States

boulder clay
Term introduced in 1981. Includes use on level 3 under sediments(1) clastic sediments(2). See list N.
BA clastic sediments
BT sediments
SA glacial geology
till

Boulder County
N central.
CO N395000N402000
W1050500W1053500
BT Colorado
United States

boulders
Includes use on level 3 under sediments(1) clastic sediments(2); on level 3 under geomorphology(1), glacial geology(1). Included use as sedimentary rock until 1977.
UF bowlder
BA clastic sediments
BT sediments
SA cobbles
erratics
geomorphology
glacial geology
gravel
rock glaciers
sedimentary rocks
terrigenous materials

Bouleiceras
Genus. As of 1981, includes use on level 3 under Mollusca(1) Ammonoidea(2). See list F. Before 1981, included use on level 3 under Mollusca(1) Cephalopoda(2).
BA Ammonoidea
Tetrabranchiata
Cephalopoda
Mollusca
BT Invertebrata

Boulogne
City on the English Channel. Also search Boulogne-sur-Mer.
UF Boulogne-sur-Mer
BT Pas-de-Calais
France

Boulogne-sur-Mer
use Boulogne

Boulonnais
Old district in extreme N part of country.
BT Pas-de-Calais
France

boundary
Used for stratigraphic boundaries only.
SA stratigraphy
stratotypes

boundary interactions
Includes use on level 2 under meteorology(1).
UF interactions, boundary
SA atmosphere
meteorology

boundary layer
Includes use on level 2 under ocean circulation(1). Before 1976, also search boundary layer processes.
UF layer, boundary
SA ocean circulation
oceanography
turbulence

boundary layer processes
A valid term through 1975. After 1975, use boundary layer.

boundstone
BA carbonate rocks
BT sedimentary rocks

Bourges
City in central part of country.
BT Cher
France

Bourgogne
use Burgundy

bournonite
UF berthonite
endellionite
wheel ore
BA sulfantimonites
sulfosalts
BT minerals

Bouvet Island
Norwegian island 1600 miles SSW of Cape of Good Hope.
BT Atlantic Ocean

Bouxwiller
Town in extreme NE part of country.
BT Bas-Rhin
France

Bovidae
As of 1981, includes use on level 3 under Mammalia(1) Ruminantia(2). See list F. Before 1981, included use on level 3 under Mammalia(1) Artiodactyla(2).
BA Ruminantia
Artiodactyla
Mammalia
BT Tetrapoda
Vertebrata
NA Bison
Bos

Bow River Valley
In S Alberta.
BT Alberta
Canada

bow shock waves
Includes use on level 2 under magnetosphere(1). Before 1976, also search bow shock.
SA intensity
magnetopause
magnetosheath
magnetosphere
shock waves
waves

Bowen
Town on the Coral Sea in the S Great Barrier Reef area.
BT Queensland
Australia

Bowen Basin
In E Queensland. Also search Bowen.
CO S250000S210000
E1500000E1480000
BT Queensland
Australia

Bowers Ridge
N of the Aleutian Islands.
BT Bering Sea
Pacific Ocean

Bowie Seamount
Just W of the Queen Charlotte Islands off central British Columbia.
BT Pacific Ocean

bowlder
use boulders

Bowser Formation
In Tuxedni Group, Middle(?) and Upper Jurassic: Central S Alaska.
BT Jurassic
SA Alaska
Middle Jurassic
Upper Jurassic

Box Elder County
Extreme NW.
BT Utah
United States

Boza Wola
Region in S.
BT Poland

Br
use bromine

Braarudosphaeridae
BA nannofossils
algae
BT thallophytes
Plantae

Brabant
Province in central.
BT Belgium
NT Louvain
University of Louvain

Brabant Massif
Also search Brabant.
BT Belgium

Brachiopoda
Includes use on level 1 and 2 as a fossil term (list F).
BT Invertebrata
NA Articulata
Inarticulata
SA brachiopods
reef builders

brachiopods
Term introduced in 1981 as the common name for Brachiopoda. See list F. To be used for Brachiopoda when they are used as tools in stratigraphy. Includes use on level 1. Before 1981, search Brachiopoda AND biostratigraphy. If 1, term set options are:
biostratigraphy
age [List E]
SA biostratigraphy
Brachiopoda

Brachiopterygii
Term introduced in 1981. Includes use on level 2 under Pisces(1). See list F.
BA Osteichthyes
Pisces
BT Vertebrata

Brachyura
Suborder. Includes use on level 3 under Arthropoda(1) Malacostraca(2).
BA Malacostraca
Crustacea
Mandibulata
Arthropoda
BT Invertebrata

brackish water
Includes use on level 3 under ecology(1). Before 1978, also search brackish.
SA brackish-water environment
ecology
paleosalinity
salinity
salt lakes
salt water
sea water
water

brackish-water environment
Term introduced in 1978. Includes use on level 3 under ecology(1) name of fossil group(2). Before 1978, also search brackish or brackish water.
SA brackish water
ecology
environment
salinity
salt marshes
swamps
tidal flats

Brad
Town in Hunedoara County in SW.
BT Transylvania
Romania

Bradyodonti
As of 1981, includes use on level 3 under Pisces(1) Elasmobranchii(2). See list F. Before 1981, included use on level 3 under Pisces(1) Chondrichthyes(2).
BA Elasmobranchii
Chondrichthyes
Pisces
BT Vertebrata

Braganca
District in NE Portugal. Also a city.
BT Portugal
SA Tras-os-Montes

Brahmaputra River
Rises in the Kailas Range of SW Tibet; it flows E across S Tibet, then S, SW, and again S through Bangladesh where it becomes the Jamuna River which merges with the Ganges River in the Ganges-Brhamaputra Delta on the Bay of Bengal. Index Tibet and countries as applcable.
BT Asia
SA Bangladesh
China
India
Tibet

braided streams
Includes use on level 3 under geomorphology(1) fluvial features(2). Also search braided channels.
UF anastomosing streams
BA streams
SA channels
fluvial features
geomorphology
sedimentation

Braila
City and county in NE.
BT Walachia
Romania

Bramsche
Town in SW.
BT Lower Saxony
West Germany

Branchiopoda
Includes use on level 2 under Arthropoda(1). See list F. As of 1981, used for Conchostraca. Before 1981, also search Conchostraca.
BA Crustacea
Mandibulata
Arthropoda
BT Invertebrata

Brandenburg
Region and former state in central East Germany. Also a town on the Havel River.
BT East Germany

Brandenburg Stade
Europe. Term introduced in 1978. Before 1978, also search Brandenburg.
BA Weichselian
upper Pleistocene
Pleistocene
Quaternary
BT Cenozoic

Brandon
Region in Warwick County in central.
BT England
Great Britain
United Kingdom

Brandon Stade
Europe. Term introduced in 1978. Before 1978, also search Brandon.
BA Weichselian
upper Pleistocene
Pleistocene
Quaternary
BT Cenozoic

brannerite
BA oxides
BT minerals

Brasov
City and county in SE Transylvania. Also search Stalin.
UF Stalin
BT Transylvania
Romania

Brassfield Formation
Considered to consist of lower dolomitic limestone, including Whitfieldella bed, and Plum Creek Clay Member above. In central N Arkansas, S Indiana, SW Illinois, central Kentucky, SW Ohio, and S Tennessee. Also search Brassfield Limestone.
BT Lower Silurian
Silurian
SA Arkansas
Illinois
Indiana
Kentucky
Ohio
Tennessee

Bratislava
City on the Danube River in SW.
BT Slovakia
Czechoslovakia

braunite
BA orthosilicates
silicates
BT minerals

Bravais lattice
use lattice

bravoite
BA sulfides
BT minerals
SA pyrite

Brazil
Includes use on level 1 as an area term (list O). For term set options see list B.
CO S340000N051500
W0340000W0740000
BA South America
NT Acre
Alagoas
Amapa
Amazonas
Bahia
Borborema
Brazilian Shield
Ceara
Espirito Santo
Goias
Maranhao
Mato Grosso
Minas Gerais
Para
Paraiba
Parana
Parnaiba Basin
Pelotas Basin
Pernambuco
Piaui
Recife-Joao Pessoa
Rio de Janeiro
Rio Grande do Norte
Rio Grande do Sul
Rondonia
Roraima
Santa Catarina
Sao Francisco Basin
Sao Paulo
Sergipe
Sergipe-Alagoas Basin
Serra do Mar
Tocantins River region
SA Acungui Group
Amazon Basin
Amazon River
Areado Formation
Bambui Group
Barreiras Formation
Bauru Formation
Botucatu Formation
Estrada Nova Formation
Guiana Basin
Guyana Shield
Irati Formation
Minas Series
Parana Basin
Parana River
Passa Dois Group
Pirabas Formation
Rio Bonito Formation
Rio Negro
Rio Negro Meteorite
Roraima Formation
Santana Formation
Serra Geral Formation
Tubarao Group
Vitoria

Brazilian Shield
The Brazilian uplands of N central, E central, and S. Index states as applicable.
UF Amazonian Shield
BT Brazil
SA Bahia
Goias
Maranhao
Mato Grosso
Minas Gerais
Para
Parana
Santa Catarina
Sao Paulo

Brazos River
Central Texas.
BT Texas
United States

Brazzaville
City on the Congo River across from Kinshasa.
BT Congo

breakers
use breaking waves

breaking strength
use fracture strength

breaking waves
Term introduced in 1978. Includes use on level 2 under ocean waves(1). Before 1978, also search waves AND breaking; breakers.
UF breakers
BT ocean waves
SA waves

breakwaters
UF water-break
BT marine installations
SA engineering geology
shorelines

Breathitt County
Term introduced in 1981. E central Kentucky.
CO N372000N374000
W0825700W0833400
BT Kentucky
United States

Breathitt Formation
In Pottsville Group. In SE Kentucky.

BT Middle Pennsylvanian
Pennsylvanian
SA Kentucky
Pottsville Group

breccia
Includes use on level 3 under sedimentary rocks(1) clastic rocks(2) and on level 3 under faults(1) effects(2). See list I. As of 1981, use lunar breccia for breccia on the Moon. Before 1981, also included use on level 3 under Moon(1).
UF rubblerock
BA clastic rocks
BT sedimentary rocks
NA tectonic breccia
volcanic breccia
SA agglomerate
breccia pipes
brecciation
conglomerate
faults
gouge
gravel
lunar breccia
meteor craters
microbreccia
Moon
sediments
shear zones
slickensides
structural analysis
suevite
tectonics
terrigenous materials

breccia pipes
Includes use on level 3.
BT pipes
SA breccia
diatremes
intrusions
plugs
volcanic rocks
volcanism
volcanoes

brecciation
SA breccia
processes

Brecknock (shire)
use Brecknockshire

Brecknockshire
County in SE Wales. Also search Brecknock.
UF Brecknock (shire)
Breconshire
BT Wales
Great Britain
United Kingdom

Breconshire
use Brecknockshire

Breidamerkurjokull
Glacier.
BT Iceland

Brenham
City in Washington County in SE central.
BT Texas
United States

Brent Crater
Meteor crater at Brent in Algonquin Provincial Park E of Georgian Bay.
BT Ontario
Canada

Brescia
City and province in E.
BT Lombardy
Italy

Breslau
use Wroclaw

Brest
City on the Atlantic Ocean at W end of Brittany in Finistere, and a city in SW Byelorussia. Index Byelorussia and/or France.
SA Byelorussia
France

Brest Basin
Named after city of Brest in SW Byelorussia. Index Poland and Soviet republics as applicable. Also search Brest.
BT Europe
SA Byelorussia
Poland
Ukraine

Brevard County
Term introduced in 1981. Central Florida, on the Atlantic Ocean.
CO N275000N284800
W0802500W0805800
BT Florida
United States

Brevard fault zone
use Brevard Zone

Brevard Zone
Index states as applicable. Also search Brevard fault zone.
UF Brevard fault zone
BT Appalachians
SA Alabama
Georgia
North Carolina

Brewster County
In SW Texas.
BT Texas
United States

Brezno
Town in central.
BT Slovakia
Czechoslovakia

Briancon
Town in SE part of country.
BT Hautes-Alpes
France

Briancon Zone
use Brianconnais Zone

Brianconnais Zone
SE France and NW Italy. Index countries as applicable. Also search Brianconnaise Zone.
UF Briancon Zone
Brianconnaise Zone
BT Europe
SA Alps
Cretaceous
Eocene
France
Italy
Jurassic
Mesozoic
Paleocene
Permian
Tertiary

Brianconnaise Zone
use Brianconnais Zone

briartite
BA sulfides
BT minerals

brick clays
Not a valid term for GeoRef. Use clays.

Bridgerian
North America.
BA Eocene
Paleogene
Tertiary
BT Cenozoic

bridges
As of 1978, term is used on level 2 under foundations(1).
SA engineering geology
foundations
geomorphology
highways

brief report
A valid term through 1978. After 1978, use report, e.g. under associations(1) or symposia(1).

Brindisi
City and province on the S Adriatic Sea.
BT Apulia
Italy

brines
Includes use on level 1 and 2 as a commodity term (list C).
SA bromine
bromine deposits
desalinization
iodine
iodine deposits
mineral waters
salt
salt lakes
salt water
salt-water intrusion
sea water

Brioverian
Europe. See list E.
BA upper Proterozoic
Proterozoic
Precambian

Brisbane
City in extreme SE.
BT Queensland
Australia

Bristol
City on the Bristol Channel.
BT England
Great Britain
United Kingdom

Bristol Bay
Between SW Alaskan mainland and Alaska Peninsula.
BT Bering Sea
Pacific Ocean

Bristol Channel
Arm of Atlantic Ocean. Index England and/or Wales.
BT United Kingdom
SA England
Wales

britholite
BZ phosphates
silicates
BT minerals

British Columbia
Includes use on level 1 as an area term (list O). For term set options see list B.
CO N490000N600000
W1140000W1390000
BA Western Canada
Canada
NT Albert Canyon
Alice Arm
Atlin
Bluebell Mine
Cariboo Mountains
Copper Mountain
Crowsnest Pass
Dogtooth Mountains
Endako
Esplanade Range
Fraser River
Guichon Creek Batholith
Harrison Lake
Hope
Kamloops
Kootenay Lake
Manning Park
Omineca Mountains
Pinchi Lake
Queen Charlotte Islands
Quesnel Lake
Revelstoke
Rossland
Saanich Inlet
Salmo
Selkirk Mountains
Shuswap Complex
Sifton Basin
Skeena Mountains
Slocan mining camp
Smithers
Sustut Basin
Tofino Basin
Topley Intrusions
Tulameen coal area
Vancouver Island
SA Altyn Limestone
Belt Supergroup
Burgess Shale
Canadian Cordillera
Canadian Rocky Mountains
Coast Mountains
Coast Ranges
Columbia River
Columbia River basin
Juan de Fuca Strait
Kicking Horse River valley
Kootenay Formation
Lardeau Group
Liard River
Miette Group
Nahanni Formation
Nicola Group
Okanagan Valley
Pasayten Group
Peace River
Purcell System
Rocky Mountain Trench
Skagit Valley
Strait of Georgia
Vancouver
Windermere System

British Guiana
use Guyana

British Honduras
Not a valid index term for GeoRef since 1974. Use Belize.

British North Borneo
use Sabah

British Virgin Islands
Term introduced in 1978 for Virgin Islands excluding those of the United States. Before 1978, also search Virgin Islands for British Virgin Islands.
BA Virgin Islands
Leeward Islands
BT Lesser Antilles
NA Anegada Island
SA United Kingdom

Brittany
Region and former province occupying peninsula between English Channel and Bay of Biscay.
BT France

Brive
Basin in S central part of country. Also search Brive-la-Gaillarde.
UF Brive-la-Gaillarde
BT Correze
France

Brive-la-Gaillarde
use Brive

Brno
City in S central.
BT Moravia
Czechoslovakia

Broach
City in S.
BT Gujarat
India

Broadlands
Geothermal field.
BT New Zealand

Broadwater County
SW central.
BT Montana

United States

brochantite
BA sulfates
BT minerals

Brocken Massif
In Harz Mountains in SE.
BT East Germany

Brockman Iron Formation
Wittenoom Gorge area in NW Western Australia.
BT Precambrian
SA Australia
Western Australia

Broegger Peninsula
use Brogger Peninsula

Brogger Peninsula
On the Island of Spitsbergen. Also search Broggerhalvoya.
UF Broegger Peninsula
Broggerhalvoya
BT Spitsbergen
Svalbard
Arctic region

Broggerhalvoya
use Brogger Peninsula

Broken Bay
Inlet of Pacific Ocean.
BT New South Wales
Australia

Broken Hill
City in W.
BT New South Wales
Australia

Broken River Formation
NE Queensland. Also search Broken River Group.
BT Paleozoic
SA Australia
Devonian
Queensland
Silurian

bromellite
BA oxides
BT minerals

bromides
Term introduced in 1981. Includes use on level 2 under minerals(1). See list L. Before 1981, also search bromide.
BA halides
BT minerals

bromine
Includes use on level 1 and 2 as a chemical element (list D). As of 1981, use bromine deposits for bromine as a commodity. Before 1981, included use on level 1 as a commodity term.
UF Br
SA brines
bromine deposits
elements
salt

bromine deposits
Term introduced in 1981. Includes use on level 1 and 2 as a commodity term (list C). Before 1981, search bromine AND deposits.
SA brines
bromine
salt

bromlite
use alstonite

Bronson Hill Anticlinorium
In S New Hampshire and N central Massachusetts.
BT United States
SA Massachusetts
New Hampshire

bronzite
BA orthopyroxene
pyroxene group
chain silicates
silicates
BT minerals
SA enstatite

bronzitite
As of 1981, includes use on level 3 under igneous rocks(1) ultramafics(2). See list H. Before 1981, included use on level 3 under igneous rocks(1) ultramafic family(2).
BA ultramafics
BT igneous rocks
SA pyroxenite

brookite
BA oxides
BT minerals
SA anatase
rutile

Brooks Range
Between Yukon River and Arctic Ocean in N central.
BT Alaska
United States
SA Rocky Mountains

Broward County
Term introduced in 1981. S Florida, on the Atlantic Ocean.
CO N255700N262100
W0800400W0805400
BT Florida
United States

brown algae
use Phaeophyta

brown coal
As of 1981, no longer a valid term for GeoRef. Use lignite. Before 1981, included use on level 3 as a commodity term under lignite(1) and on level 3 under sedimentary rocks(1) organic residues(2).

Brown County
Index states as applicable.
BX Indiana
Kansas
BT United States

Brown forest soils
Includes use on level 3 under soils(1). See list M.
BA soils
SA Intrazonal soils
soil group

brown mica
use phlogopite

Brown soils
Includes use on level 3 under soils(1). See list M.
BA soils
SA Zonal soils

Bruce Peninsula
Term introduced in 1981. Peninsula of SE Ontario, Canada. Extends N between Lake Huron and Georgian Bay.
CO N444000N451500
W0805500W0814500
UF Saugeen Peninsula
BT Ontario
Canada
SA Niagara Escarpment

brucite
BA oxides
BT minerals

brucite marble
Includes use on level 3 under metamorphic rocks(1) marbles(2). See list J.
BA marbles
BT metamorphic rocks

Bruderheim
Village NE of Edmonton in central Alberta.
UF Bruederheim
BT Alberta
Canada

Bruederheim
use Bruderheim

Brule Formation
In White River Group. In NE Colorado, W Nebraska, W South Dakota, and E Wyoming.
BT Oligocene
SA Colorado
middle Oligocene
Nebraska
South Dakota
upper Oligocene
White River Group
Wyoming

Brunei
Sultanate under British protection in NE part of Borneo.
CO N040000N051000
E1153000E1140000
BT Far East
SA Borneo
East Malaysia

Brunhes Epoch
Geomagnetic epoch.
UF Brunhes Normal
BA Quaternary
BT Cenozoic
SA Holocene
paleomagnetism
Pleistocene
reversals
upper Pleistocene

Brunhes Normal
use Brunhes Epoch

Brunswick
City in Lower Saxony, West Germany, and in Glynn County, Georgia. Index Georgia and/or West Germany as applicable.
SA Georgia
West Germany

Brunswick County
Term introduced in 1981. Index states as applicable.
BX North Carolina
Virginia
BT United States

brushite
BA phosphates
BT minerals

Bryansk
City and oblast SW of Moscow.
BT Russian Republic
USSR

bryophytes
Mosses. Includes use on level 1 and 2 as a fossil term (list F).
BT Plantae
NA Hepaticae
Musci
SA paleobotany
thallophytes

Bryozoa
Includes use on level 1 and 2 as a fossil term (list F).
UF Polyzoa
BT Invertebrata
NA Cheilostomata
Cryptostomata
Ctenostomata
Cyclostomata
Cystoporata
Ectoprocta
Trepostomata
SA acanthopores
bryozoans
reef builders

bryozoans
Term introduced in 1981 as the common name for Bryozoa. See list F. To be used for Bryozoa when they are used as tools in stratigraphy. Includes use on level 1. Before 1981, search Bryozoa AND biostratigraphy. If 1, term set options are: biostratigraphy
age [List E]
SA biostratigraphy
Bryozoa

Bucegi Mountains
Group in the E Transylvanian Alps in S central.
BT Romania
SA Transylvanian Alps

Buchanan County
Index states as applicable.
BX Iowa
Missouri
Virginia
BT United States

Bucharest
City and administrative district in S central.
BT Walachia
Romania

Buckingham County
S central.
BT Virginia
United States

Buckner Formation
Subsurface. Overlies Smackover Limestone; underlies Cotton Valley Formation.
BT Upper Jurassic
Jurassic
SA Arkansas
Louisiana
Mississippi
Texas

Budapest
City on both sides of the Danube River in N central part of country.
BT Hungary

Budapest region
Term introduced in 1981.
CO N472000N474500
E0193000E0184500
BT Hungary

Budennovsk
use Prikumsk

Buenos Aires
City on the Rio de la Plata.
BT Federal District
Argentina

Buenos Aires Province
In E part of country.
CO S410000S331500
W0564000W0633000
BT Argentina
NT Sauce Grande River valley

buergerite
BA ring silicates
BT minerals

Buffalo River
E of the Tennessee River in W central.
BT Tennessee
United States

Bug region
Area including both the Southern Bug and the Western Bug rivers in SW European USSR and E Poland. Index Poland and Soviet republics as applicable.
BT Europe
SA Byelorussia
Poland
Ukraine

Bug River
The Southern Bug River in S Ukraine and the Western Bug River of the Ukraine, Byelorussia and E Poland.

Index Poland and Soviet republics as applicable.
BT Europe
SA Byelorussia
Poland
Ukraine

building stone
Includes use on level 3 as a commodity term. See list C.
UF stone, building
BA construction materials
SA dimension stone
granite deposits
limestone deposits
marble deposits
sandstone deposits

buildings
As of 1978, term is used on level 2 under foundations(1).
SA foundations
structures

Bukantau
Mountains SE of Aral Sea in N.
BT Uzbekistan
USSR

Bukhara
City and oblast in S near border of Turkmenia.
BT Uzbekistan
USSR

Bukhara-Khiva
A region comprising neighboring former Khanates SE of the Aral Sea.
BT Uzbekistan
USSR

Bukk Mountains
Southern spur of the Carpathians in NE.
BT Hungary
SA Carpathians

Bukovina
Region in E central Europe. Index Ukraine and/or Romania as applicable.
BT Europe
SA Romania
Ukraine

Bukusu
Carbonatite complex.
BT Uganda

Bulawayan Group
Strong unconformities separate it from the overlying and underlying systems (Shamvaian and Sebakwian).
BT Archean
Precambrian
SA Zimbabwe

Bulgaria
Includes use on level 1 as an area term (list O). For term set options see list B.
CO N413000N441500
E0284500E0221500
BA Southern Europe
Europe
NT Balkan Foreland
Balkan Mountains
Beli Izvor
NX Bulgarian Dobruja
Bulgarian Rhodope Mountains
NT Burgas
Central Bulgaria
Glavatsi
Isker River
Khaskovo
North Bulgarian Hills
Panagyurishte
Pernik coal basin
Pleven
Plovdiv
Rila Mountains
Ruse
Sakar Mountains
Sofia
Sredna Gora
Varna
Vidin
Vratsa
West Bulgarian Hills
SA Balkan Peninsula
Black Sea region
Danube Plain
Danube River
Danube Valley
Dobruja
Dobruja Basin
Istranca Mountains
Krajiste
Macedonia
Maritsa River
Moesia
Moesian Platform
Osogovo Mountains
Rhodope Mountains
Struma River valley

Bulgarian Dobruja
Term introduced in 1981.
CO N431100N440000
E0284000E0273000
BZ Dobruja Basin
BX Bulgaria

Bulgarian Rhodope Mountains
Term introduced in 1981.
CO N411200N421800
E0261500E0230500
BZ Rhodope Mountains
BX Bulgaria

Bulimina
Genus. Includes use on level 3 under foraminifera(1) Buliminacea(2).
BA Buliminacea
Rotaliina
foraminifera
BT Protista
Invertebrata

Buliminacea
Includes use on level 2 under foraminifera(1). See list F.
BA Rotaliina
foraminifera
BT Protista
Invertebrata
NA Bolivina
Bolivinitidae
Bolivinoides
Bulimina
Gabonella
Uvigerina
Uvigerinidae

bulk modulus
UF incompressibility modulus
modulus, bulk
modulus of incompressibility
volume elasticity
BA elastic constants
SA deformation
elasticity
shear modulus

Bulli Seam
Top of the Illawarra Coal Measures is marked by the Bulli Coal Seam. Coal seam in Sydney Basin in E New South Wales.
BT Permian
SA Australia
Illawarra Coal Measures
New South Wales

Buncombe County
In W North Carolina.
BT North Carolina
United States

Bundelkhand
Region S of the Jumna River. Index states as applicable.
BT India
SA Madhya Pradesh
Uttar Pradesh

Bundenbach
Area in Westerwald region in W.
BT Hesse
West Germany

Bungonia Caves
In SE New South Wales.
BT New South Wales
Australia

Bunol
Town in E part of country.
BT Valencia Province
Valencia region
Spain

Bunter
Europe. Above Permian, below Muschelkalk. Includes use on level 3 under age terms(1). See list E. Also search Buntsandstein; Bunter Sandstone.
UF Bunter Sandstone
Buntsandstein
BA Lower Triassic
Triassic
BT Mesozoic
NT Voltzia Sandstone

Bunter Sandstone
use Bunter

Buntsandstein
use Bunter

Burbank Sand
Subsurface. Osage and Kay counties in N Oklahoma. Also search Burbank Sands.
UF Burbank Sands
BT Carboniferous
SA Kansas
Oklahoma

Burbank Sands
use Burbank Sand

Burdekin Delta
On Upstart Bay on Pacific Ocean in Great Barrier Reef area in E.
BT Queensland
Australia

Burdekin River
Flows into Upstart Bay on Pacific Ocean in Great Barrier Reef area in E.
BT Queensland
Australia

Burdigalian
Europe. Includes use on level 3 under age terms(1). See list E.
BA lower Miocene
Miocene
Neogene
Tertiary
BT Cenozoic

Burdwan
A city, district and division in central.
BT West Bengal
India

Bureya
Town in SE Amur Oblast near the Manchurian border.
CO N480000N520000
E1333000E1260000
BT Russian Republic
USSR

Burgas
Province on the Black Sea in SE Bulgaria. Also a city.
BT Bulgaria

Burgenland
State on the Hungarian border in E.
CO N464600N480800
E0171000E0160000
BT Austria

Burgess Shale
BT Middle Cambrian
Cambrian
SA British Columbia
Canada

Burgos
As of 1981, refers to only the city. Before 1981, also included the province. In N central Spain.
BT Burgos Province
Old Castile
Spain

Burgos Province
Term introduced in 1981. Before 1981, search Burgos. In N central Spain.
CO N413000N431500
W0023000W0042000
BT Old Castile
Spain
NT Burgos

Burgundy
Region and former province of E central and E France. Also search Bourgogne.
UF Bourgogne
BT France

Buriadia
Includes use on level 3 under gymnosperms(1) Coniferales(2).
BA Coniferales
gymnosperms
BT Spermatophyta
Plantae

burial metamorphism
Term introduced in 1978. Includes use on level 2 under metamorphism(1). Before 1978, also search metamorphism AND burial.
BA metamorphism
SA geosynclines
regional metamorphism

buried channels
Includes use on level 3 under geomorphology(1) fluvial features(2).
SA buried features
buried valleys
channels
drift
fluvial features
geomorphology
glacial geology
paleogeography
planar bedding structures

buried features
Term introduced in 1978. Before 1978, search buried.
UF features, buried
SA buried channels
buried valleys

buried valleys
Includes use on level 3 under geomorphology(1), paleogeography(1), and under glacial geology(1).
SA buried channels
buried features
drift
geomorphology
glacial geology
paleogeography
valleys

Burke County
Index states as applicable.
BX Georgia
North Carolina
North Dakota
BT United States

Burleson County
SE central.
BT Texas
United States

Burlington Peninsula
Term introduced in 1981. On N coast of Newfoundland.
CO N490000N501000

W0553000W0570000
BT Newfoundland
Canada

Burma
Includes use on level 1 as an area term (list O). For term set options see list B.
CO N090000N284500
E1014500E0920000
BA Far East
Asia
NT Shan State
SA Indochina
Kama
Phuket Group
Siwalik System

Burnet County
Central.
BT Texas
United States

Burnie
Town on Bass Strait in N.
BT Tasmania
Australia

Burono
use Vourinos

Burro Mountain
BT California
United States

burrows
Includes use on level 3 under sedimentary structures(1) and ichnofossils(1).
BA biogenic structures
sedimentary structures
SA bioturbation
borings
ichnofossils
lebensspuren

bursts, rock
use rock bursts

Burundi
Formerly Urundi which was part of Belgian trust territory of Uranda-Urundi. Includes use on level 1 as an area term (list O). For term set options see list B.
CO S043000S022000
E0305000E0285500
BA Central Africa
Africa
SA East African Rift
Lake Tanganyika
Nile River

Buryat
Buryat Autonomous Soviet Socialist Republic. S and SE of Lake Baikal. Also search Buryatia; Buryat-Mongol; Buryat-Mongolia.
UF Buryat-Mongol
Buryat-Mongolia
Buryatia
BT Russian Republic
USSR
SA Baikal rift zone

Buryat-Mongol
use Buryat

Buryat-Mongolia
use Buryat

Buryatia
use Buryat

Bushveld Complex
Igneous complex in E Transvaal. Also search Bushveld.
BT Transvaal
South Africa

bustamite
BA chain silicates
silicates
BT minerals

Butler County
Index states applicable.
BX Alabama
Iowa
Kansas
Kentucky
Missouri
Nebraska
Ohio
Pennsylvania
BT United States

Butte
City in Silver Bow County in SW.
BT Montana
United States

Butte County
Index states as applicable.
BX California
Idaho
South Dakota
BT United States
NT Oroville

Butte District
Mining district in the Butte area in SW Montana. Also search Butte.
BT Montana
United States

Buzau
City and county in NE.
BT Walachia
Romania

Buzau River
In NE Walachia.
BT Walachia
Romania

Buzzards Bay
Inlet of Atlantic Ocean W of Cape Cod.
BT Massachusetts
United States

Bydgoszcz
Province in N central Poland. Also a city.
BT Poland
NT Dobrzyn
Grudziadz
Inowroclaw
Mogilno
Torun

Byelorussia
Byelorussian Soviet Socialist Republic. Also search Belorussia.
CO N510000N560000
E0330000E0240000
UF Belorussia
White Russia
BA Russian Platform
BT USSR
NT Byelorussian Massif
Kletno
Mezhrechye
Minsk
Mogilev
Orsha
Ostashkovichi
Rechitsa
Starobin
Ustron
Zhitkovichi
SA Brest
Brest Basin
Bug region
Bug River
Dnieper Basin
Dnieper River
Dvina River
European USSR
Neman River basin
Polesye
Pripet Basin
Russian Plain
Vyshkovo

Byelorussian Massif
S part of Lithuanian-Byelorussian Upland.
BT Byelorussia
USSR

Byrd Station
In Marie Byrd Land SW of the Amundsen Sea on the Pacific Ocean side.
CO S795900S795900
W1200100W1200100
BT Antarctica

Bystrzyca
River in SW part of country.
BT Wroclaw
Poland

Bytkov
use Bitkov

Bytom
City in central.
BT Katowice
Poland

bytownite
As of 1981, includes use on level 3 under minerals(1) framework silicates, plagioclase(2). See list L. Before 1981, included use on level 3 under minerals(1) framework silicates, feldspar group(2).
BA plagioclase
feldspar group
framework silicates
silicates
BT minerals
SA aluminosilicates

C

C
use carbon

C-12
Includes use on level 3 under isotopes(1).
UF carbon-12
SA C-13/C-12
carbon
isotopes

C-12/C-13
use C-13/C-12

C-13
Includes use on level 3 under isotopes(1).
UF carbon-13
SA C-13/C-12
carbon
isotopes

C-13/C-12
Includes use on level 3 under isotopes(1). Also search C-12/C-13; C-12 AND C-13.
UF C-12/C-13
SA C-12
C-13
carbon
geologic thermometry
isotopes
paleoclimatology
stable isotopes

C-14
Includes use on level 3 under absolute age(1) methods(2). Before 1976, also search carbon-14.
SA absolute age
carbon
isotopes

Ca
use calcium

Cabaniss Formation
In Cherokee Group, SE Kansas, W Missouri, central and NE Oklahoma.
BT Desmoinesian
Middle Pennsylvanian
Pennsylvanian
SA Cherokee Group
Kansas
Missouri
Oklahoma

Cabinda
A district and an enclave N of the mouth of the Congo River.
UF Kabinda
BT Angola

Cabo Ortegal
Cape in NW part of country in N.
BT La Coruna Province
Galicia
Spain

Cabo Rojo
Town in SW.
BT Puerto Rico
West Indies

Cabrieres
Location in the Montagne Noire in S part of country.
BT Herault
France

Caceres
As of 1981, refers to only the city. Before 1981, also included the province. In W Spain.
BT Caceres Province
Extremadura
Spain

Caceres Province
Term introduced in 1981. Before 1981, search Caceres. In W Spain on Portuguese border.
CO N390300N403000
W0045800W0073000
BT Extremadura
Spain
NT Caceres

Cache County
Term introduced in 1981. N Utah.
CO N412200N420000
W1112200W1121000
BT Utah
United States

Cache La Poudre River
N central.
BT Colorado
United States

Cache Valley
Index Idaho and/or Utah as applicable.
BT United States
SA Idaho
Utah

cacoxenite
BA phosphates
BT minerals

Caddo County
SW of Oklahoma City.
BT Oklahoma
United States

Cadiz
As of 1981, refers to only the city. Before 1981, also included the province. In SW on the Atlantic Ocean.
BT Cadiz Province
Andalusia
Spain

Cadiz Province
Term introduced in 1981. Before 1981, search Cadiz. In SW on the Atlantic Ocean.
CO N360000N370500
W0050500W0063000
BT Andalusia
Spain

NT Cadiz
Vejer de la Frontera
SA Serrania de Ronda

cadmium
Includes use on level 1 and 2 as a chemical element (list D).
UF Cd
SA elements
geochemistry
heavy metals

cadmium blende
use greenockite

cadmium ocher
use greenockite

Cadomian Orogeny
Term introduced in 1978. Before 1978, search Cadomian AND orogeny.
SA Cambrian
orogeny
Precambrian

Caen
City in Normandy.
BT Calvados
France

Caernarvon County
use Caernarvonshire

Caernarvonshire
County in NW Wales.
CO N524500N532000
W0034500W0044500
UF Caernarvon County
BT Wales
Great Britain
United Kingdom

Cairngorm Mountains
Range of the Grampian Hills in highlands of central.
BT Scotland
Great Britain
United Kingdom

Cairo
Governorate in NE Egypt. Also a city on the Nile River.
BT Egypt

Caithness
County in extreme NE.
CO N581000N584000
W0030000W0035000
BT Scotland
Great Britain
United Kingdom

Calabria
An autonomous region in S Italy. The toe of the Italian boot.
CO N380000N401000
E0170000E0154500
BT Italy
NT Catanzaro
Cosenza
Reggio di Calabria
Sila Massif

Calabrian
Europe. Above Astian, below middle Pleistocene. Includes use on level 3 under age terms(1). See list E.
BA lower Pleistocene
Pleistocene
Quaternary
BT Cenozoic
SA Villafranchian

Calamites
Genus. Includes use on level 3 under pteridophytes(1) Sphenopsida(2).
BA Equisetales
Sphenopsida
pteridophytes
BT Plantae

Calatayud-Teruel Basin
NE Spain. Index provinces as applicable.
BT Spain
SA Saragossa Province
Teruel

Calaveras County
Central.
BT California
United States

Calaveras Fault
Central.
BT California
United States

calaverite
As of 1981, includes use on level 3 under minerals(1) tellurides(2). See list L. Before 1981, included use on level 3 under minerals(1) sulfides(2).
BA tellurides
sulfides
BT minerals
SA gold

calc-alkalic
A valid term through 1977. After 1977, use calc-alkalic composition(2) under igneous rocks(1).

calc-alkalic composition
Term introduced in 1978. Includes use on level 2 under igneous rocks(1).
SA alkalic composition
composition
igneous rocks

calc-schist
Includes use on level 3 under metamorphic rocks(1) schists(2). See list J.
BA schists
BT metamorphic rocks

calc-sinter
use travertine

calc-tufa
use tufa

calcarenite
Includes use on level 3 under sedimentary rocks(1) carbonate rocks(2). See list I.
BA carbonate rocks
BT sedimentary rocks
SA biocalcarenite
limestone

calcareous
Valid index term through 1977. After 1977, use calcareous composition.

calcareous algae
Includes use on level 3 under algae(1). See list F.
BA algae
BT thallophytes
Plantae
SA stromatolites

calcareous clay
use marl

calcareous composition
Term introduced in 1978. Includes use on level 3 under sediments(1) and sedimentary rocks(1). Before 1978, also search calcareous AND specific sediment type or sedimentary rock.
SA bioherms
composition
ooze
sedimentary rocks
sediments

calcareous nannofossils
use nannofossils

calcareous ooze
Not a valid term for GeoRef. Use calcareous composition and ooze.

calcareous sinter
use travertine

Calcareous soils
Before 1978, also search calcareous.
BA soils

calcareous tufa
use tufa

Calcasieu Parish
On the Texas border in SW.
BT Louisiana
United States

calcic composition
Term introduced in 1978. Includes use on level 2 under igneous rocks(1). Before 1978, also search igneous rocks AND calcic.
SA composition
igneous rocks

calcicrete
use calcrete

calcification
Term refers to fossils or soils.
SA diagenesis
fossils
processes
soils

calcilutite
Includes use on level 3 under sedimentary rocks(1) carbonate rocks(2). See list I.
BA carbonate rocks
BT sedimentary rocks
SA algal mounds
limestone

calciphyre
Includes use on level 3 under metamorphic rocks(1) marbles(2). See list J.
BA marbles
BT metamorphic rocks

Calcisphaerulidae
Family. Also search Calcispherulidae.
BA problematic fossils

Calcispongea
Includes use on level 2 under Porifera(1). See list F.
BA Porifera
BT Invertebrata
NA Sphinctozoa

calcite
Includes use on level 3 under minerals(1). See list L. As of 1981, use calcite deposits for calcite as a commodity. Before 1981, included use on level 1 as a commodity term.
BA carbonates
BT minerals
SA alabaster
aragonite
calcite deposits
calcitization
calcium carbonate
chalk
dolomite
Iceland spar
light minerals
limestone
magnesian calcite
travertine
tufa

calcite deposits
Term introduced in 1981. Includes use on level 1 and 2 as a commodity term (list C) for optical calcite. Before 1981, search calcite AND deposits.
SA calcite
marble deposits

calcitization
SA calcite
dedolomitization
diagenesis
dolomitization
processes

calcium
Includes use on level 1 and 2 as a chemical element (list D). Also search Ca.
UF Ca
SA elements

calcium carbonate
SA aragonite
calcite
carbonates
geochemistry
lime

calcrete
As of 1981, includes use on level 3 under sedimentary rocks(1) carbonate rocks(2). See list I. Before 1981, included use on level 3 under sediments(1) chemically precipitated sediments(2).
UF calcicrete
BA carbonate rocks
BT sedimentary rocks
SA caliche
conglomerate
duricrust
soils
weathering crust

calculation
A valid term through 1978. After 1978, see mathematical methods or mathematical models.

Calcutta
City on the Hooghly River near Bay of Bengal in S.
BT West Bengal
India

calderas
Includes use on level 3 under volcanology(1), and under geomorphology(1) volcanic features(2).
BT volcanic features
SA cauldrons
craters
geomorphology
volcanism
volcanoes
volcanology

Caldwell County
Term introduced in 1981. Index states as applicable.
BX Kentucky
Missouri
North Carolina
Texas
BT United States

Caledonia County
NE Vermont. Also search Caledonia.
BT Vermont
United States

Caledonian
A valid term through 1977. After 1977, use Caledonian Orogeny.

Caledonian Geosyncline
As of 1981, no longer a valid term for GeoRef. Use Caledonian Orogeny. Term was introduced in 1978.

Caledonian Orogeny
Name commonly used for the early Paleozoic deformation in Europe which created an orogenic belt extending from Ireland and Scotland northwestward through Scandinavia. Before 1978, also search Caledonian. Before 1981, also search Caledonian Geosyncline.
BT Paleozoic
SA Caledonides
orogeny

Caledonides
Orogenic belt extending from Ireland and Scotland northwestward through Scandinavia, formed by the Caledonian Orogeny.

SA Caledonian Orogeny
Europe
Great Britain
Scandinavia

Calgary
City in SW.
BT Alberta
Canada

Calhoun County
Index states as applicable.
BX Alabama
Arkansas
Florida
Georgia
Illinois
Iowa
Michigan
Mississippi
South Carolina
Texas
West Virginia
BT United States

calibration
Used as a general term.
SA accuracy

caliche
As of 1981, includes use on level 3 under sedimentary rocks(1) carbonate rocks(2). See list I. Before 1981, included use on level 3 under sediments(1) chemically precipitated sediments(2). Also includes use on level 3 under soils(1) and weathering(1).
BA carbonate rocks
BT sedimentary rocks
SA calcrete
duricrust
soils
weathering crust

Calico Mountains
Small range in Mojave Desert.
BT California
United States

Caliente Range
BT California
United States

California
Includes use on level 1 as an area term (list O). For term set options see list B.
CO N323000N420000
W1141500W1243000
BA Pacific Coast
Western U.S.
United States
NT Alameda County
Amador County
Anza Desert State Park
Bakersfield
Barstow
Bear Valley
Berkeley
Bodega Head
Bolinas Lagoon
Borax Lake
Borrego Mountain
Burro Mountain
NX Butte County
NT Calaveras County
Calaveras Fault
Calico Mountains
Caliente Range
Calistoga
Cape Mendocino
Carmel
Carmel Bay
Central California
Channel Islands
Cholame
Clear Lake
Coalinga
Colusa County
Contra Costa County
Coyote Creek Fault
Crestmore
Death Valley
Del Monte Beach
Del Norte County
Diablo Range
Eel River
El Centro
El Dorado County
Elk Hills Field
Elsinore Fault
NX Eureka
NT Feather River
Fortuna
Fresno
Fresno County
Furnace Creek
Gabilan Range
Garlock Fault
Hayward
Hayward Fault
Hollister
NX Humboldt County
NT Imperial County
Imperial Valley
Inglewood
Inyo County
Inyo Mountains
Kern County
Kettleman Hills
NX Kings County
NT La Jolla
NX Lake County
NT Lassen Peak
Lassen Volcanic National Park
Lompoc
Long Beach
Los Angeles
Los Angeles Basin
Los Angeles County
Madera County
Marin County
Mariposa County
Melones Fault
Mendocino County
Merced County
Mill Creek
Modoc County
Modoc Plateau
Mojave Desert
Mono Basin
Mono County
Mono Craters
Mono Lake
Monterey
Monterey Bay
Monterey County
Mount Lyell
Napa County
Needles
NX Nevada County
NT Newport Bay
Newport-Inglewood Fault
Nicolaus
Northern California
NX Orange County
NT Oroville Dam
Owens Valley
Oxnard
Pacoima Dam
Pajaro Valley
Palm Springs
Palo Alto
Palos Verdes Hills
Panamint Range
Parkfield
Pasadena
Peninsular Ranges
Perris
Pinnacles National Monument
Pisgah Crater
Placer County
Plumas County
Point Arena
Point Loma
Point Mugu
Point Reyes
Portola Valley
Preston Peak
Puente Hills
Rand Mountains
Riverside
Riverside County
Sacramento
Sacramento County
Sacramento Valley
Salinas
Salinas Valley
NX Salinian Block
NT Salton Sea
Salton Trough
San Andreas Fault
San Benito County
San Bernardino
San Bernardino County
San Bernardino Mountains
San Clemente Island
San Diego
San Diego County
San Fernando
San Fernando Valley
San Francisco
San Francisco Bay
San Francisco County
San Francisco Peninsula
San Gabriel Mountains
San Giorgio Mountain
San Gregorio Fault
San Jacinto Fault
San Joaquin County
San Joaquin Valley
San Jose
San Juan Bautista
San Luis Obispo
San Luis Obispo County
San Mateo County
San Miguel Island
San Nicolas Island
San Pedro
San Pedro Valley
Santa Ana
Santa Ana Mountains
Santa Barbara
Santa Barbara Channel
Santa Barbara County
Santa Catalina Island
Santa Clara
Santa Clara County
Santa Clara Valley
NX Santa Cruz County
NT Santa Cruz Island
Santa Cruz Mountains
Santa Lucia Range
Santa Maria
Santa Monica Mountains
Santa Rosa
Santa Rosa Mountains
Santa Ynez Mountains
Sargent
Scripps Institution of Oceanography
Searles Lake
Shasta County
NX Sierra County
NT Sierra Nevada
Sierra Nevada Batholith
Simi Hills
Siskiyou County
Solano County
Sonoma County
Southern California
Southern California Batholith
Stanislaus County
Sur fault zone
Sylmar Fault
Tehachapi Mountains
Tehama County
Temblor Range
The Geysers
Tomales Bay
Transverse Ranges
NX Trinity County
NT Trona Village
Tulare County
Tuolumne County
University of California
Ventura
Ventura Basin
Ventura County
Whittier Fault
Wilmington oil field
Yosemite National Park
Yuba County
SA Agua Blanca Fault
Amargosa Desert
Basin and Range Province
Bird Spring Formation
Bishop Tuff
Capistrano Formation
Cascade Range
Central Valley
Coast Ranges
Colorado River
Colorado River basin
Franciscan Formation
Great Basin
Great Valley Sequence
Hat Creek Basalt
Hidden Valley Dolomite
Johnnie Formation
Kaibab Formation
Klamath Mountains
Lake Tahoe
Lost Burro Formation
Marca Shale Member
Moenkopi Formation
Monte Cristo Limestone
Monterey Formation
Moreno Formation
Noonday Dolomite
Pelona Schist
Pico Formation
Pogonip Group
Poway Conglomerate
Puente Formation
Punchbowl Formation
Rose Canyon Formation
Saint George Formation
San Diego Formation
San Lorenzo Formation
San Onofre Breccia
Santa Margarita Formation
Santa Maria Formation
Sespe Formation
Shadow Mountains
Sonoran Desert
Stirling Quartzite
Supai Formation
Tejon Formation
Truckee River
Tulare Formation
Twin Lakes
Umpqua Formation
White Mountain
White Mountains
Wilmington
Wood Canyon Formation
Wyman Formation

California City
Term introduced in 1981. Also search California AND Colombia.
BT Santander Department
Colombia

californium
includes use on level 1 and 2 as chemical element (list D).
UF Cf
SA elements

Caliman (Mountains)
use Calimani Mountains

Calimani Mountains
Range in the Eastern Carpathians in E Transylvania. Also search Caliman, Calimani, and Calimanului Mountains.
UF Caliman (Mountains)
Calimanului Mountains
BT Transylvania
Romania
SA Carpathians

Eastern Carpathians

Calimanului Mountains
use Calimani Mountains

caliper logging
As of 1978, term is used on level 2 under well-logging(1). Before 1978, also search well-logging AND caliper.
UF logging, caliper
BA well-logging

Calistoga
City in Napa County NE of San Francisco Bay.
BT California
United States

Callander Bay
W edge of Lake Nipissing near city of North Bay E of Georgian Bay.
BT Ontario
Canada

Callianassa
Genus. Includes use on level 3 under Arthropoda(1) Malacostraca(2).
BA Malacostraca
Crustacea
Mandibulata
Arthropoda
BT Invertebrata

Callipteris
Genus. As of 1981, includes use on level 3 under gymnosperms(1). Before 1981, included use on level 3 under pteridophytes(1) Filicopsida(2).
BA Pteridospermae
gymnosperms
BT Spermatophyta
Plantae

Callovian
Europe. Above Bathonian, below Oxfordian. Includes use on level 3 under age terms(1). See list E.
BA Middle Jurassic
Jurassic
BT Mesozoic
SA Dogger

Calpionella
Genus.
BA Tintinnidae
Protista
BT Invertebrata

Calpionellidae
Includes use on level 3 under Protista(1) Tintinnidae(2). See list F.
BA Tintinnidae
Protista
BT Invertebrata

Calpionellites
Genus.
BA Tintinnidae
Protista
BT Invertebrata

Caltanissetta
City in central.
BT Sicily
Italy

Calvados
Department in NW.
CO N485000N493500
E0004000W0011500
BT France
NT Caen
May-sur-Orne
SA Normandy
Seine Estuary

Calvert County
Term introduced in 1981. S Maryland.
CO N381800N384700
W0762230W0764200
BT Maryland
United States

Calvert Formation
In Chesapeake Group. Delaware, E Maryland, and Virginia.
BT middle Miocene
Miocene
SA Delaware
Maryland
Virginia

Camaguey
Province in E central Cuba. Also a city.
BT Cuba
West Indies

Camarines Norte
Province in SE.
BT Luzon
Philippines

Camaroidea
Includes use on level 2 under Graptolithina(1). See list F.
BA Graptolithina
BX Hemichordata
Invertebrata

Cambay
City and former Indian state at N end of Gulf of Cambay.
BT Gujarat
India

Cambay Basin
Gulf of Cambay area.
BT Gujarat
India

Cambodia
Khmer Republic or Kampuchea. Includes use on level 1 as an area term (list O). For term set options see list B.
CO N103000N144000
E1074000E1020000
UF Kampuchea
Khmer Republic
BA Far East
Asia
NT Battambang
SA Indochina

Cambrian
Includes use on level 1 as an age term (list E); on level 2 under paleoterms, e.g. paleoecology, paleomagnetism, paleogeography. Above Precambrian, below Ordovician.
BT Paleozoic
NT Baikalian Phase
Lardeau Group
NA Lower Cambrian
Middle Cambrian
NT Nama System
Phuket Group
Semri Series
Tintic Quartzite
Tons Member
NA Upper Cambrian
SA Arbuckle Group
Assyntic Orogeny
Bhander Group
Cadomian Orogeny
Dalradian
Eocambrian
Jacobsville Sandstone
Kaimur Sandstone
Knox Group
Kurnool System
Porsanger Dolomite Formation
Talladega Group
Vindhyan
Wood Canyon Formation

Cambridge
City in Middlesex County in NE Massachusetts, and a city NNE of London. Also a former county in E England now a part of Cambridgeshire and Isle of Ely. Index England and/or Massachusetts as applicable.
CO N520000N524500
E0003000W0001500
BX England
Massachusetts

Cambridge Arch
Index states as applicable.
BT United States
SA Kansas
Nebraska

Cambridge University
City of Cambridge NNE of London.
UF University of Cambridge
BT England
Great Britain
United Kingdom

Camelidae
As of 1981, includes use on level 3 under Mammalia(1) Ruminantia(2). See list F. Before 1981, included use on level 3 under Mammalia(1) Artiodactyla(2).
BA Ruminantia
Artiodactyla
Mammalia
BT Tetrapoda
Vertebrata

Cameron Parish
Extreme SW.
BT Louisiana
United States

Cameroon
Includes former French trust territory and British trust territory of Southern Cameroons. Includes use on level 1 as an area term (list O). For term set options see list B.
CO N020000N130000
E0170000E0080000
BA West Africa
Africa
NT Adamawa
SA Benue Valley
Chad Basin
Lake Chad
Logone River

Camp Century
On the Greenland Ice Sheet NE of Thule Air Base in NW.
BT Greenland
Arctic region

Campania
An autonomous region in S.
CO N400000N413000
E0154500E0134500
BT Italy
NT Bay of Naples
Caserta
Cilento
Ischia
Monte Somma
Naples
Phlegraean Fields
Pozzuoli
Roccamonfina
Salerno
Sorrento Peninsula
Vesuvius
SA Lucania

Campanian
Europe. Above Santonian, below Maestrichtian. Includes use on level 3 under age terms(1). See list E.
BA Senonian
Upper Cretaceous
Cretaceous
BT Mesozoic
SA Rosario Formation

Campbell County
Index states as applicable.
BX Kentucky
South Dakota
Tennessee
Virginia
Wyoming
BT United States

Campbell Island
Term introduced in 1981. Island in S Pacific Ocean, S of New Zealand.
CO S530000S520000
E1693000E1683000
BT Pacific Ocean

Campeche
State on W part of Yucatan Peninsula.
BT Mexico
SA Gulf Coastal Plain
Yucatan Peninsula

Campeche Bank
N of the Yucatan Peninsula.
BT Gulf of Mexico
SA Yucatan Shelf

Campine
Heathland area. Index countries as applicable.
BT Europe
SA Belgium
Netherlands

Campo del Cielo Meteorite
Term introduced in 1981. Point of impact N of village of Campo del Cielo in S Argentina. Before 1981, also search Campo del Cielo AND meteorites.
BA ataxite
iron meteorites
BT meteorites
SA Argentina
Chaco

Campobasso
Town in S central part of country.
BT Italy

camptonite
As of 1981, includes use on level 3 under igneous rocks(1) lamprophyres(2). See list H. Before 1981, included use on level 3 under igneous rocks(1) lamprophyre and carbonatite family(2).
BA lamprophyres
BT igneous rocks

Camptostromatoidea
Includes use on level 2 under Echinodermata(1). See list F.
BA Echinozoa
Echinodermata
BT Invertebrata

Canada
Includes use on level 1 or 2 as an area term (list O). If 1, see list B for term set options.
CO N420000N840000
W0520000W1410000
BT North America
NT Assiniboine River
Assiniboine River valley
NX Canadian Cordillera
Canadian Rocky Mountains
NT Carswell Structure
Chaleur Bay
Cold Lake
Dunnage Melange
NA Eastern Canada
NT Elk Point Basin
Grenville Front
NX Hudson Bay
NT Hudson Bay Lowlands
James Bay
Kicking Horse River valley
Labrador Trough
Lake Timiskaming
Liard River
Mackenzie Mountains
North Saskatchewan River
Northumberland Strait
NA Northwest Territories
NT Ottawa Valley
Peace River
Peel River
Reindeer Lake

Restigouche Estuary
Rice Lake
Richardson Mountains
Saskatchewan River
Selwyn Mountains
South Saskatchewan River
Timiskaming
Ungava
NA Western Canada
SA Acadian
Appalachians
Arctic region
Burgess Shale
Christopher Formation
Cloridorme Formation
Cypress Hills Formation
Davidsville Group
Espanola Formation
Flinton Group
Gander Lake Group
Geological Survey of Canada
Gowganda Formation
Great Lakes
Great Lakes region
Great Plains
Harbour Main Group
Hudsonian Orogeny
Huronian
Keg River Formation
Kenoran Orogeny
Kootenay Formation
Lardeau Group
Leda Clay
Mannville Formation
Meguma Group
Michelle Formation
Miette Complex
Miette Group
Milk River Formation
Missi Group
Mississippi River basin
Missouri River basin
Nahanni Formation
Nicola Group
Nipissing Diabase
North American Cordillera
Onaping Formation
Osler Series
Palliser Formation
Paskapoo Formation
Peel Sound Formation
Prince Albert Group
Ramparts Formation
Ravenscrag Formation
Read Bay Formation
Rice Lake Group
Rocky Mountains
Saint Lawrence River
Shaunavon Formation
Signal Hill Formation
Sokoman Formation
Taconic Allochthon
Taconic Orogeny
Western Interior
Windsor Group
Yellowknife Group

Canada Basin
N of Alaska and WNW of Canada's Queen Elizabeth Islands. Also search Canadian Basin.
UF Canadian Basin
BT Arctic Ocean

Canadian Basin
use Canada Basin

Canadian Cordillera
Index provinces as applicable.
BX Canada
North American Cordillera
SA Alberta
British Columbia
Coast Mountains
Rocky Mountains

Canadian River
Index states as applicable.
UF South Canadian River
BT United States
SA Colorado
New Mexico
Oklahoma
Texas

Canadian Rocky Mountains
Term introduced in 1981. Before 1981, search Canada AND Rocky Mountains. Index provinces as applicable.
BX Canada
Rocky Mountains
SA Alberta
British Columbia
Cariboo Mountains
North American Cordillera
Northern Rocky Mountains

Canadian Series
Provincial series, North America. Above Croixian (Cambrian), below Champlainian. Includes use on level 3 under age terms(1). See list E. Before 1981, also search Canadian AND Ordovician.
BA Lower Ordovician
Ordovician
BT Paleozoic

Canadian Shield
Index states and provinces as applicable. Introduced as a level 1 area term in 1978. Also search Canadian AND Shield.
CO N430000N790000
W0560000W1240000
UF Laurentian Highlands
Laurentian Plateau
Precambrian Shield
BA North America
NT Bear Province
Churchill Province
Grenville Province
Slave Province
Southern Province
Superior Province
SA Alberta
Manitoba
Michigan
New York
Northwest Territories
Ontario
Quebec
Saskatchewan
Wisconsin

canals
As of 1978, term is used on level 2 under waterways(1).
SA channels
streams
waterways

Cananea
City in N.
BT Sonora
Mexico

Canaries
use Canary Islands

Canary Islands
Spanish controlled island group off NW Africa. Includes use on level 1 as an area term (list O). For term set options see list B.
CO N274500N291500
W0130000W0180000
UF Canaries
BA Atlantic Ocean Islands
NT Fuerteventura
Gomera
Grand Canary
Hierro
Lanzarote
Tenerife
SA Spain

Canavese Zone
Mountain region N of Turin.
BT Italy
SA Piemonte

Canberra
City between Sydney and Melbourne. Australian Capital Territory.
BT Australia

cancrinite
BZ carbonates
framework silicates
BT minerals

Candona
Genus. Includes use on level 3 under Ostracoda(1). See list F.
BA Cyprididae
Podocopida
Ostracoda
BT Crustacea
Mandibulata
Arthropoda
Invertebrata

canfieldite
BZ sulfogermanates
sulfostannates
BA sulfosalts
BT minerals

Canidae
Includes use on level 3 under Mammalia(1). See list F.
BA Carnivora
Mammalia
BT Tetrapoda
Vertebrata
NA Canis

Canis
Genus. Includes use on level 3 under Mammalia(1). See list F.
BA Canidae
Carnivora
Mammalia
BT Tetrapoda
Vertebrata

Cannikin
Name of site of nuclear explosion on Amchitka Island in the Aleutian Islands.
BT Alaska
United States

Canning Basin
Arid region in northern Western Australia, about 790 miles NNE of Perth.
CO S240000S180000
E1280000E1200000
UF Desert Basin
BT Western Australia
Australia

Cannon County
Central.
BT Tennessee
United States

Canon City
Fremont County in central.
BT Colorado
United States

canonical analysis
BA statistical analysis
SA analysis
mathematical geology

Cantabria Knoll
use Cantabria Seamount

Cantabria Seamount
W Bay of Biscay. Also search Cantabria.
UF Cantabria Knoll
BT Bay of Biscay
Atlantic Ocean

Cantabrian Basin
Region in the Cantabrian Mountain area in N and NW Spain. Index Asturias, Basque Provinces, and Santander Province as applicable.
BT Spain
SA Asturias
Basque Provinces
Santander Province

Cantabrian Mountains
Range extending westward from the Pyrenees along the Bay of Biscay to the Galician Mountains.
CO N423000N433000
W0024000W0072000
BT Spain

Cantabrian region
Term introduced in 1981.
BT Spain

Cantal
Department in S central.
CO N443000N453000
E0031500E0020500
BT France
NT Cantal Massif
Massiac
SA Central Massif

Cantal Massif
Part of Central Massif covering central.
BT Cantal
France
SA Central Massif

Canterbury
District in E central.
BT South Island
New Zealand

Canyon Diablo Meteorites
Term introduced in 1981. Refers to three meteorites found in Coconino County, Arizona in 1891, 1936, and 1949. (Used for Canon Diablo, Canyon Diablo no. 2, and Canyon Diablo no. 3.) Before 1981, also search Canon Diablo AND meteorites; Canyon Diablo AND meteorites; Canyon Diablo Meteorite. Due to numbering of these meteorites, also truncate when searching.
BA octahedrite
iron meteorites
BT meteorites
SA Arizona
Coconino County

Canyon Group
Includes Graford Formation, Winchell Limestone, Brad Formation, and Caddo Creek Formation. Central and N central Texas.
BT Upper Pennsylvanian
Pennsylvanian
SA Texas
United States

canyons
Includes use on level 3 under geomorphology(1). As of 1976, usage restricted to land canyons. Before 1976, also refers to submarine canyons.
UF land canyons
SA geomorphology
submarine canyons

Cap Blanc
Index countries as applicable.
BT Africa
SA Mauritania

cap rocks
Restricted to use for salt tectonics: an impervious layer or body of rock over a salt dome.
SA anhydrite
diapirs
gypsum
rocks
salt domes
salt tectonics

capacity
As of 1981, no longer a valid term for GeoRef. Before 1981, included use as a general term.

C

Capbreton
Town in SW.
BT Landes
France

Cape Basin
Term introduced in 1981. SE Atlantic Ocean.
CO S410000S260000
E0100000E0010000
UF Walvis Basin
BT Atlantic Ocean

Cape Bojador
Term introduced in 1981. Headland extending into Atlantic Ocean on S coast of Morocco.
CO N260700N260800
W0142900W0143000
BT Morocco
Africa

Cape Breton Island
N part of province.
CO N453000N470000
W0593000W0613000
BT Nova Scotia
Canada

Cape Cod
SE Massachusetts. Coextensive with Barnstable County.
BT Massachusetts
United States

Cape Dyer
Village and cape on Davis Strait on E Baffin Island in District of Keewatin.
BT Northwest Territories
Canada

Cape Fear Arch
Extends NW from Cape Fear.
BT North Carolina
United States

Cape Hatteras
Promontory on Hatteras Island between Pamlico Sound and the Atlantic Ocean.
BT North Carolina
United States

Cape Kennedy
Formerly Cape Canaveral in Brevard County on the Atlantic Ocean. Site of John F. Kennedy Space Center.
BT Florida
United States

Cape Lookout
Headland at S end off Carteret County.
BT North Carolina
United States

Cape Mendocino
On W coast of Humboldt County. Extreme W part of state. Also search Mendocino.
UF Mendocino (Cape)
BT California
United States

Cape Province
Province in S.
CO S345000S244500
E0301500E0162500
BT South Africa
NT Cape Town
False Bay
Karroo Basin
Kimberley
Knysna
Langebaanweg
Molteno
Port Elizabeth
Postmasburg
SA Ecca Series
Namaqualand
Orange River
Vaal River

Cape Sable
S point of Cape Sable Island in extreme S.
BT Nova Scotia
Canada

Cape Town
City in extreme SW Cape Province.
UF Capetown
BT Cape Province
South Africa

Cape Valse
use False Cape

Cape Verde Atlantic
Term introduced in 1981 to indicate a region of the Atlantic Ocean.
BA North Atlantic
BT Atlantic Ocean

Cape Verde Basin
Term introduced in 1981. E Atlantic Ocean between the Mid-Atlantic Ridge and West Africa.
CO N080000N350000
W0200000W0420000
BT Atlantic Ocean

Cape Verde Islands
Former Portuguese overseas province W of Dakar, Senegal. Includes use on level 1 as an area term (list O). For term set options see list B.
CO N144000N172000
W0223000W0253000
BZ Atlantic Ocean Islands
Africa

Cape Verde Rise
Term introduced in 1981. Between Cape Verde Islands and West Africa.
CO N140000N210000
W0180000W0260000
BT Atlantic Ocean

Cape York Peninsula
Between Gulf of Carpentaria and Coral Sea in N.
BT Queensland
Australia

Capetown
use Cape Town

capillarity
SA capillary pressure
capillary water
soils

capillary pressure
BT pressure
SA capillarity
capillary water

capillary water
Term introduced in 1981. Before 1981, also search capillarity.
SA capillarity
capillary pressure
soils
unsaturated zone
water table

Capistrano Formation
Subdivided to include Osa Member. Unconformably underlies Niguel Formation. Southern California between Santa Ana and Oceanside.
BT Neogene
Tertiary
SA California
lower Pliocene
upper Miocene

Capitan Formation
Massive reef limestone which grades basinward into upper Delaware Mountain Sandstone and lagoonward into Carlsbad Limestone. SE New Mexico and W Texas.
BT Guadalupian
Permian
SA New Mexico
Texas

Capnic
use Cavnic

Captorhinomorpha
Term introduced in 1981. Includes use on level 2 under Reptilia(1). See list F.
BA Anapsida
Reptilia
BT Tetrapoda
Vertebrata

capture
use stream capture

Carabobo
State on Caribbean Sea. Also a village.
BT Venezuela
NT Puerto Cabello

Caracas
City in N.
BT Federal District
Venezuela

Caracas Group
Middle and upper Mesozoic. Includes Las Brisas Formation, Antimano Formation, Tacagua Formation. In Cordillera de la Costa of Venezuela.
UF Caracas Series
BT Mesozoic
SA Federal District
middle Mesozoic
upper Mesozoic
Venezuela

Caracas Series
use Caracas Group

Caradocian
Europe. Below Ashgillian. Includes use on level 3 under age terms(1). See list E.
BA Upper Ordovician
Ordovician
BT Paleozoic
SA Bala

Caravaca
City in SE part of country.
BT Murcia Province
Murcia region
Spain

carbargilite
use coal

carbides
As of 1981, includes use on level 2 under minerals(1). See list L. Before 1981, included use on level 3 under minerals(1) native elements and alloys(2).
BA alloys
BT minerals
NA cohenite
moissanite

carbohydrates
Includes use on level 2 under organic materials(1).
BA organic materials
NA cellulose

carbon
Includes use on level 1 and 2 as a chemical element (list D). Also search C.
UF C
SA C-12
C-13
C-13/C-12
C-14
carbonaceous composition
carbonado
chaoite
charcoal
diamond
diamonds
elements
graphite
hydrocarbons
isotopes
lonsdaleite
moissanite
organic carbon
organic materials
stable isotopes

Carbon County
Index states as applicable.
BX Montana
Pennsylvania
Utah
Wyoming
BT United States

carbon dioxide
Includes use on level 3 under geochemistry(1), fluid inclusions(1), or phase equilibria(1). Also search CO2.
UF CO2
SA carbon monoxide
fluid inclusions
geochemistry
phase equilibria

carbon monoxide
UF CO
SA carbon dioxide

carbon-12
use C-12

carbon-13
use C-13

carbon-14
A valid term through 1976. After 1976, use C-14.

carbonaceous
A valid index term through 1977. After 1977, use carbonaceous chondrites under meteorites(1), and use carbonaceous composition as a general term referring to carbon and coal.

carbonaceous chondrites
Includes use on level 3 under meteorites(1). Before 1978, also search chondrites AND carbonaceous.
BA chondrites
BT meteorites
NA Allende Meteorite
Murchison Meteorite
Murray Meteorite
Orgueil Meteorite
Vigarano Meteorite

carbonaceous composition
A general term used with carbon and coal. Before 1978, also search carbonaceous.
SA carbon
coal
composition
organic materials

carbonaceous shale
use black shale

carbonado
As of 1981, includes use on level 3 under minerals(1) native elements(2). See list L. Before 1981, included use on level 3 under minerals(1) native elements and alloys(2).
BA native elements
BT minerals
SA carbon
diamonds
industrial minerals

carbonate
As of 1981, no longer a valid term for GeoRef. See carbonates; carbonate ion; carbonate rocks; carbonate sediments.

carbonate apatite
Also search carbonate-apatites; tavistockite.
UF carbonate-apatites
tavistockite

BA phosphates
BT minerals
SA apatite
dahllite

carbonate compensation
use carbonate compensation depth

carbonate compensation depth
Also search calcite compensation depth.
UF carbonate compensation
depth of compensation
SA lysoclines

carbonate composition
As of 1981, no longer a valid term for GeoRef. Before 1981, included use on level 2 under nodules(1).

carbonate ion
Term introduced in 1981.
SA carbonates
geochemistry
ions

carbonate rocks
Includes use on level 2 under sedimentary rocks(1). See list I.
BT sedimentary rocks
NA beachrock
biocalcarenite
biomicrite
biosparite
boundstone
calcarenite
calcilutite
calcrete
caliche
chalk
coquina
dolostone
grainstone
limestone
micrite
packstone
travertine
wackestone
SA carbonate sediments
carbonates
carbonatite
dolomite
intraclasts
marl
oolite
rocks
tufa

carbonate sediments
Includes use on level 2 under sediments(1). See list N.
BT sediments
SA carbonate rocks
carbonates
coquina
dolomite
dolostone
gravel
limestone
oolite
sand
travertine
tufa

carbonate-apatites
use carbonate apatite

carbonates
Includes use on level 2 under minerals(1). See list L.
BT minerals
NA alstonite
alumohydrocalcite
ankerite
aragonite
artinite
azurite
NZ bastnaesite
NA calcite
NZ cancrinite
NA cerussite
chalybite
NZ dahllite
NA dawsonite
dolomite
gaylussite
huntite
hydromagnesite
Iceland spar
magnesian calcite
magnesite
malachite
NZ meionite
NA monohydrocalcite
nahcolite
nesquehonite
norsethite
NZ phosgenite
NA protodolomite
pyroaurite
rhodochrosite
NZ sakhaite
NA shortite
siderite
sjogrenite
smithsonite
NZ spurrite
NA strontianite
NZ synchisite
thaumasite
NA thermonatrite
trona
vaterite
weddellite
weloganite
whewellite
witherite
SA calcium carbonate
carbonate ion
carbonate rocks
carbonate sediments
sodium carbonate

carbonatite
Includes use on level 3. As of 1981, not to be used in the igneous rocks set. See list H. Before 1981, included use on level 3 under igneous rocks(1) lamprophyre and carbonatite family(2).
BA carbonatites
BT igneous rocks
SA carbonate rocks

carbonatites
Term introduced in 1981. Includes use on level 2 under igneous rocks(1). See list H. Before 1981, search lamprophyre and carbonatite family.
BT igneous rocks
NA carbonatite
sovite

carbonatization
As of 1981, no longer a valid term for GeoRef. See carbonates.

Carbondale
City in Jackson County in S.
BT Illinois
United States

Carbondale Formation
In Kewanee Group. Illinois and W Kentucky.
BT Middle Pennsylvanian
Pennsylvanian
SA Illinois
Kentucky
Springfield Coal Member

Carbonicola
Genus. Includes use on level 3 under Mollusca(1). See list F.
BA Bivalvia
Mollusca
BT Invertebrata

Carboniferous
Includes use on level 1 as an age term (list E); on level 2 under paleoterms, e.g. paleoecology, paleogeography, paleomagnetism. Above Devonian, below Permian.
BT Paleozoic
NT Asturian Orogeny
NA Avonian
NT Bartlesville Sand
Burbank Sand
NA Culm
Dinantian
NT Essen Beds
Jackfork Group
Johns Valley Formation
NA Middle Carboniferous
Mississippian
Pennsylvanian
Silesian
NT Springer Formation
Tubarao Group
NA Upper Carboniferous
SA Baralaba Coal Measures
Beacon Supergroup
Dwyka Series
Geirud Formation
Gondwana System
Karroo System
Paganzo Group
Singleton Coal Measures
Talchir Formation

carbonification
use coalification

Cardiff
City in Glamorgan County on the Bristol Channel.
BT Wales
Great Britain
United Kingdom

Cardigan Bay
Large inlet of Saint George's Channel.
BT Wales
Great Britain
United Kingdom
SA Irish Sea

Cardigan County
use Cardiganshire

Cardiganshire
County on Cardigan Bay.
CO N520000N524000
W0034000W0044000
UF Cardigan County
BT Wales
Great Britain
United Kingdom

Cardiidae
Includes use on level 3 under Mollusca(1). See list F.
BA Bivalvia
Mollusca
BT Invertebrata
NA Cardium

Carditida
Term introduced in 1981. Includes use on level 2 under Mollusca(1). See list F.
BA Bivalvia
Mollusca
BT Invertebrata

Cardium
Genus. Includes use on level 3 under Mollusca(1). See list F.
BA Cardiidae
Bivalvia
Mollusca
BT Invertebrata
NA Cardium edule

Cardium edule
Includes use on level 3 under Mollusca(1). See list F.
BA Cardium
Cardiidae
Bivalvia
Mollusca
BT Invertebrata

Cariaco Basin
W of Cariaco Bay and between Isla la Tortuga and Venezuelan coast. Also search Cariaco Trench.
UF Cariaco Trench
BT Caribbean Sea

Cariaco Trench
use Cariaco Basin

Caribbean Mountain Range
Islands comprising the Antilles stretching from Florida to the coast of Venezuela constituting the remainder of several submerged Andean spurs meeting in Puerto Rico.
BT West Indies

Caribbean Plate
The Caribbean Sea including the Caribbean islands and across Central America to the Cocos plate just to the W.
SA Caribbean region
plate tectonics
plates

Caribbean region
Index Antilles, Caribbean Sea, Central America, Colombia, and Venezuela as applicable. Also search Caribbean. Includes use on level 1 or 2 as an area term (list O). If 1, see list B for term set options.
CO N070000N230000
W0580000W0900000
SA Antilles
Caribbean Plate
Caribbean Sea
Central America
Colombia
Greater Antilles
Hispaniola
Lesser Antilles
Venezuela
West Indies
Yucatan Peninsula

Caribbean Sea
Bounded by the West Indies on the N and E, South America on the S, and Central America on the W. Also search Caribbean. Includes use on level 1 or 2 as an area term (list O). If 1, see list B for term set options.
CO N090000N220000
W0600000W0780000
BA North American Atlantic
North Atlantic
Atlantic Ocean
NT Aves Island
Aves Ridge
Beata Ridge
Cariaco Basin
Cayman Trough
Colombian Basin
Gulf of Cariaco
Los Testigos
Mid-Cayman Rise
Nicaragua Rise
Venezuelan Basin
Yucatan Basin
SA Caribbean region
Leg 15
Leg 4

Cariboo Mountains
Range of the Rocky Mountains in E central.
BT British Columbia
Canada
SA Canadian Rocky Mountains
Rocky Mountains

Caribou County
Term introduced in 1981. SE Idaho.
CO N422300N430200
W1110300W1121000
BT Idaho
United States

Carinthia
State in S.
CO N463000N470000
E0150000E0123000
BT Austria
NT Bleiberg
Krappfeld
Sau Alps
Villach
Wolfsberg

Carixian
Europe. See list E.
BA Lower Jurassic
Jurassic
BT Mesozoic

Carlile Shale
In Colorado Group. E Colorado, Kansas, SE Montana, Nebraska, NE New Mexico, South Dakota, E Wyoming.
BT Upper Cretaceous
Cretaceous
SA Benton Formation
Colorado
Colorado Group
Kansas
Montana
Nebraska
New Mexico
South Dakota
Wyoming

Carlin
Town in Elko County in NE.
BT Nevada
United States

Carlin Mine
NE Nevada. Also search Carlin.
BT Nevada
United States

Carlsbad
City in Eddy County in SE. Use Karlovy Vary for town in Czechoslovakia.
BT New Mexico
United States

Carlsbad Caverns
In SW Eddy County just N of Texas border.
BT New Mexico
United States

Carlsberg Ridge
Between Somalia and the Seychelles in the W and the Laccadive and Maldive islands in the E. Also search Arabian-Indian Midoceanic Ridge and Arabian-Indian Ridge.
CO S090000N100000
E0700000E0570000
UF Arabian Ridge
Arabian-Indian Midoceanic Ridge
Arabian-Indian Ridge
BT Indian Ocean
SA mid-ocean ridges

Carmarthenshire
County in W on Bristol Channel.
UF Carmethen County
BT Wales
Great Britain
United Kingdom

Carmel
City in Monterey County.
UF Carmel-by-the-Sea
BT California
United States

Carmel Bay
NW Monterey County.
BT California
United States

Carmel-by-the-Sea
use Carmel

Carmethen County
use Carmarthenshire

Carmona
City in SW part of country.
BT Seville Province
Andalusia
Spain

carnallite
As of 1981, includes use on level 3 under minerals(1) chlorides(2). See list L. Before 1981, included use on level 3 under minerals(1) halides(2).
BA chlorides
halides
BT minerals

Carnarvon Basin
S of Lake McCleod in Carnarvon region in W.
CO S270000S210000
E1160000E1143000
BT Western Australia
Australia

Carnegie Ridge
Just S of the equator between Ecuador and Galapagos Islands.
CO S030000N010000
W0810000W0920000
BA South American Pacific
BT Pacific Ocean

Carnian
Europe. Above Ladinian, below Norian. Includes use on level 3 under age terms(1). See list E.
BA Upper Triassic
Triassic
BT Mesozoic
SA Hallstatt Limestone

Carnic Alps
Range of Eastern Alps. Index countries as applicable.
BT Alps
Europe
SA Austria
Eastern Alps
Italy
Karawanken

Carnivora
Includes use on level 2 under Mammalia(1). See list F.
BA Mammalia
BT Tetrapoda
Vertebrata
NA Canidae
Felidae
Fissipeda
Hyaenidae
Mustelidae
Pinnipedia
Ursidae

carnotite
BA vanadates
BT minerals

Carolina Bays
Term introduced in 1981. Depressions on the E coast of the U.S. from Florida to New Jersey that are of controversial origin.
BT United States
SA Atlantic Coastal Plain
Delaware
depressions
Florida
geomorphology
Georgia
Maryland
New Jersey
North Carolina
South Carolina
Virginia

Carolina slate belt
W North Carolina and NW South Carolina. Index states as applicable.
BT Appalachians
SA North Carolina
South Carolina

Caroline Islands
Central and W island group of U.S. Trust Territory of the Pacific Islands.
CO N050000N100000
E1660000E1300000
UF Carolines
BT Micronesia
SA Pacific Ocean

Carolines
use Caroline Islands

carotenoids
BA organic materials
SA pigments

Carpathian Foredeep
Elongated depressions bordering Carpathians. Index countries and Ukraine as applicable. Also search Carpathian AND foredeep.
BT Europe
SA Carpathians
Czechoslovakia
Poland
Romania
Ukraine

Carpathian Foreland
Stable area marginal to Carpathians. Index countries and Ukraine as applicable. Also search Carpathian AND foreland.
BT Europe
SA Carpathians
Czechoslovakia
Poland
Romania
Subcarpathians
Ukraine

Carpathian Mountains
use Carpathians

Carpathians
Mountain system of E and central Europe enclosing the Alfold. Index Ukraine and countries as applicable. Includes use on level 1 as an area term (list O). For term set options see list B. Before 1973, also search Carpathian Mountains.
CO N460000N495000
E0280000E0180000
UF Carpathian Mountains
BA Europe
NT Eastern Carpathians
NZ Polish Carpathians
Slovakian Carpathians
Soviet Carpathians
NT Tatra Mountains
Western Carpathians
SA Beshchady Mountains
Beskid Mountains
Bistrita Mountains
Bukk Mountains
Calimani Mountains
Carpathian Foredeep
Carpathian Foreland
Czechoslovakia
Hungary
Low Tatra Mountains
Matra Mountains
Poland
Rodna Mountains
Romania
Spis-Gemer
Subcarpathians
Tokaj-Eperjes Mountains
Transylvanian Alps
Ukraine
Ukrainian Carpathians
Vrancea

Carpentaria Basin
Region around Gulf of Carpentaria. Index Northern Territory and/or Queensland as applicable.
CO S190000S150000
E1430000E1400000
BT Australia
SA Northern Territory
Queensland

Carpentarian
SE Australia. See list E.
BA middle Precambrian
Precambrian

carpholite
BA chain silicates
silicates
BT minerals
SA aluminosilicates
ferrocarpholite

Carriacou
Largest island of the Grenadines in the Windward Islands.
BA Windward Islands
BT Lesser Antilles
SA Grenada

Carroll County
Index states as applicable.
BX Arkansas
Georgia
Illinois
Indiana
Iowa
Kentucky
Maryland
Mississippi
Missouri
New Hampshire
Ohio
Tennessee
Virginia
BT United States

carrollite
UF sychnodymite
BA sulfides
BT minerals
SA linnaeite

Carry-le-Rouet
A section in S.
BT Bouches-du-Rhone
France

carst
use karst

Carswell Structure
Lake Athabaska area of NW Saskatchewan, and the Baker Lake area in District of Keewatin of the Northwest Territories. Index province and territory as applicable.
BT Canada
SA Northwest Territories
Saskatchewan

Cartagena
City on the Caribbean Sea.
BT Bolivar
Colombia

Carter County
Index states as applicable.
BX Kentucky
Missouri
Montana
Oklahoma
Tennessee
BT United States

Carteret County
Term introduced in 1981. E North Carolina, on the Atlantic Ocean.
CO N343500N350500
W0760200W0771000
BT North Carolina
United States

Carterinacea
Includes use on level 2 under foraminifera(1). See list F.
BA Rotaliina
foraminifera
BT Protista
Invertebrata

Cartersville
City in Bartow County in NW.

BT Georgia
United States

cartography
For methods, instruments, programs. Includes use on level 2 under maps(1). Before 1972, also search geological exploration; before 1976 also search geologic mapping; before 1978 also search mapping.
UF chartology
SA aerial photography
automatic cartography
coordinates
geodesy
geomorphology
legends
leveling
maps
photogeology
photogrammetry
topography
triangulation

Caryophyllidae
Term introduced in 1981. Includes use on level 2 under angiosperms(1). See list F.
BA Dicotyledoneae
angiosperms
BT Spermatophyta
Plantae

Casablanca
City on the Atlantic Ocean.
BT Morocco

Casapalca
Town NE of city of Lima.
BT Lima
Peru

Cascade County
Term introduced in 1981. W central Montana.
CO N464900N474100
W1103900W1120300
BT Montana
United States

Cascade Mountains
use Cascade Range

Cascade Range
Northern continuation of the Sierra Nevada Mountains. Index states as applicable. Also search Cascade Mountains and Cascades.
CO N400000N510000
W1203000W1230000
UF Cascade Mountains
Cascades
BT United States
NX Golden Horn Batholith
SA California
Mount Rainier
Oregon
Sierra Nevada
Washington

cascades
Includes use on level 3 under hydrology(1) or geomorphology(1) fluvial features(2).
SA fluvial features
geomorphology
hydrology
streams
waterfalls

Cascades
use Cascade Range

Cascadia Basin
Off Washington coast.
BT Pacific Ocean

Cascadia Channel
Off coast of Oregon SW of Cascadia Basin.
BT Pacific Ocean

case histories
As of 1981, no longer a valid term for GeoRef. Use case studies.

case studies
As of 1978, term is used on level 2 under pollution(1), rock mechanics(1), and soil mechanics(1). Before 1981, also search case histories.
UF studies, case
SA pollution
rock mechanics
soil mechanics

Caserta
City and province NE of Naples.
BT Campania
Italy

Caseyville Formation
In McCormick Group. SE Illinois and W Kentucky.
BT Lower Pennsylvanian
Pennsylvanian
SA Illinois
Kentucky

cash transactions
Term introduced in 1981.
SA economics

Cashel
Town at the base of the Rock of Cashel in S central part of country.
BT Tipperary
Ireland

Casper
City in Natrona County in central.
BT Wyoming
United States

Casper Formation
Includes all beds from top of Madison Limestone to base of Permian (Opeche) red shales. In SE Wyoming.
BT Paleozoic
SA Pennsylvanian
Permian
Wyoming

Caspian Basin
Region including the Caspian Sea and surrounding areas. Index Caspian Sea, Soviet republics and Iran as applicable. Also search Caspian; Caspian region.
CO N363000N490000
E0553000E0450000
UF Caspian region
BT Eurasia
SA Azerbaidzhan
Caspian Depression
Caspian Sea
Iran
Kazakhstan
Russian Republic
Turkmenia

Caspian Depression
That part of the Caspian Basin lying N of the Caspian Sea; it is a large lowland area which was submerged by the Caspian Sea at one time. Index Soviet republics as applicable. Also search Caspian; Caspian Lowland.
CO N450000N520000
E0570000E0423000
UF Caspian Lowland
Caspian Syneclise
BA Central Asia
BT USSR
SA Caspian Basin
Kazakhstan
Russian Republic

Caspian Lowland
use Caspian Depression

Caspian region
use Caspian Basin

Caspian Sea
Largest inland sea in the world. Index Iran and/or the USSR as applicable. Also search Caspian. Includes use on level 1 as an area term (list O). For term set options see list B.
CO N360000N480000
E0530000E0470000
BA Eurasia
SA Caspian Basin
Iran
Kara-Bogaz-Gol
Neftyanyye Kamni
USSR

Caspian Syneclise
use Caspian Depression

Cass County
Index states as applicable.
BX Illinois
Indiana
Iowa
Michigan
Minnesota
Missouri
Nebraska
North Dakota
Texas
BT United States

Cassidulinacea
Includes use on level 2 under foraminifera(1). See list F.
BA Rotaliina
foraminifera
BT Protista
Invertebrata
NA Anomalinidae

cassiterite
BA oxides
BT minerals
SA heavy minerals

Castellane
Village in SE.
BT Alpes-de-Haute Provence
France

Castellon
use Castellon de la Plana Province

Castellon de la Plana
As of 1981, refers to only the city. Before 1981, also included the province. In E Spain.
BT Castellon de la Plana Province
Valencia region
Spain

Castellon de la Plana Province
Term introduced in 1981. Before 1981, search Castellon de la Plana. Also search Castellon/IT. In E Spain.
CO N394200N404800
E0003000W0005300
UF Castellon
BT Valencia region
Spain
NT Castellon de la Plana
SA Maestrazgo

Castile
Region and ancient kingdom of central and N central part of country consisting of 13 modern provinces.
BT Spain
NA New Castile
Old Castile

Castile Formation
Formation, as now defined, includes beds in lower part of Ochoa Series that are confined in extent to Delaware basin and overlap sloping surface of Capitan Limestone along its margins. SE New Mexico, and W Texas.
BT Permian
SA New Mexico
Texas

Castilla la Nueva
use New Castile

Castilla la Vieja
use Old Castile

Castillon Massif
In Ariege Pyrenees in S.
BT Ariege
France

Castle Hayne
Village in New Hanover County in SE.
BT North Carolina
United States

Castle Hayne Limestone
Underlies Yorktown Formation near Pollicksville, North Carolina. Coastal Plain of North Carolina and South Carolina.
BT Eocene
SA middle Eocene
North Carolina
South Carolina
upper Eocene

Castleton
Town in Rutland County in S central.
BT Vermont
United States

Castoridae
Family. As of 1981, includes use on level 3 under Mammalia(1) Sciuromorpha(2). See list F. Before 1981, included use on level 3 under Mammalia(1) Rodentia(2).
BA Sciuromorpha
Rodentia
Mammalia
BT Tetrapoda
Vertebrata

casts, flute
use flute casts

casts, groove
use groove casts

casts, load
use load casts

cataclasis
Term introduced in 1981.
SA deformation
metamorphism

cataclasites
Includes use on level 2 under metamorphic rocks(1). See list J. Also search cataclasite.
BT metamorphic rocks

catagenesis
Also search katagenesis.
UF katagenesis
SA sedimentary rocks

catalogs
Includes use on level 1 (list A); as a level 2 or 3 term appropriate to a large number of topics, e.g. on level 2 under paleontology(1). See list G. Also search catalog. If 1, term set options are:
topic [list B, extraterrestrial geology, general, geophysics]
subtopic (e.g. language if not English)
SA associations
bibliography
book reviews
collections
education
geology
glossaries
lexicons
maps
mineralogy
museums
oceanography
paleobotany
paleontology

petrology
sedimentary petrology
stratigraphy
structural geology
survey organizations
volcanology

Catalonia
Historical region in NE Spain. Index provinces as applicable.
CO N403000N423000
E0033000E0000000
BT Spain
NT Barcelona Province
Gerona Province
Lerida
Tarragona
Ter River basin
SA Catalonian Coastal Ranges

Catalonian Coastal Ranges
Term introduced in 1981.
BT Spain
SA Catalonia

catalysis
SA chemical analysis
geochemistry

Catamarca
Province in NW Argentina. Also a city.
CO S300000S250000
W0650000W0690000
BT Argentina
SA Sierras Pampeanas

Catanda
Village in E.
BT Angola

Catanzaro
City in central.
BT Calabria
Italy

catastrophes
Term introduced in 1978. Includes use on level 2 under geologic hazards(1).
SA destruction
geologic hazards

catastrophic
A valid term through 1977. After 1977, use catastrophic waves on level 2 under ocean waves(1).

catastrophic advance
use glacier surges

catastrophic waves
Term introduced in 1978. Includes use on level 2 under ocean waves(1). Before 1978, also search catastrophic AND ocean waves.
BT ocean waves
SA tsunamis
waves

catastrophism
SA uniformitarianism

catchment hydrodynamics
Term introduced in 1981.
SA drainage basins
ground water
hydrodynamics
hydrology

catchments
use drainage basins

catenas
Soil association of a given area developed from common parent material.
SA soils

cathodoluminescence
SA luminescence

cation exchange
A valid term through 1978. After 1978, use ion exchange or cation exchange capacity, e.g. under geochemistry(1) or soils(1).

cation exchange capacity
Includes use on level 3 under geochemistry(1) properties(2). Also search cation exchange.
SA clay mineralogy
geochemistry
ion exchange
ions

cations
Term reintroduced in 1981. Was valid through May 1978. Also search ions.
SA anions
geochemistry
ions

Catoctin Formation
Divided into Warrenton Agglomerate Member at base and basalt flows now altered to metabasalt or "greenstone" schist. W Maryland, N Virginia, and NE West Virginia.
BT Precambrian
SA Maryland
Virginia
West Virginia

Catron County
In W on Arizona border.
BT New Mexico
United States
NT Mogollon Mountains

Catskill Delta
Devonian age complex. Index states as applicable.
BT United States
SA Maryland
New York
Pennsylvania
West Virginia

Catskill Formation
In Susquehanna Group.
BT Paleozoic
SA Devonian
Lower Mississippian
Maryland
Middle Devonian
Mississippian
New York
Pennsylvania
Upper Devonian
Virginia

Catskill Mountains
Group of the Appalachians in SE.
BT New York
United States
SA Appalachians

Cattaraugus County
Term introduced in 1981. SW New York.
CO N420000N423300
W0781900W0790400
BT New York
United States
NT West Valley

cattierite
BA sulfides
BT minerals

Caucasian Foreland
use Caucasus Foreland

Caucasus
Mountain system between the Black and Caspian Seas. Also search Caucasus Mountains. Index Greater Caucasus, Lesser Caucasus, Northern Caucasus, and Soviet republics as applicable.
CO N410000N440000
E0480000E0400000
UF Caucasus Mountains
BT USSR
NA Greater Caucasus
Lesser Caucasus
NZ Northern Caucasus
SA Armenia
Azerbaidzhan
Georgian Republic
Russian Republic

Caucasus Foreland
The Northern Caucasus or Ciscaucasia area which includes the northern slopes and foothills of the Caucasus plus the Kuban Steppes to the north. Also search Caucasian Foreland.
UF Caucasian Foreland
BT Russian Republic
USSR
SA Northern Caucasus

Caucasus Mountains
use Caucasus

cauldrons
SA calderas
volcanic features
volcanism
volcanology

causes
Includes use as level 2 or 3 term appropriate to a large number of topics, e.g. on level 2 under orogeny(1). See list G.

Causses
An arid Jurassic limestone plateau of S Central Massif in S and SW France. Also search South Massif.
UF South Massif
BT France
SA Central Massif

caustobiolith
Includes use on level 3 under sedimentary rocks(1) organic residues(2). See list I. Also search caustobioliths.
BA organic residues
BT sedimentary rocks
SA organic materials

Cauvery Basin
River basin in S part of country. Index states as applicable.
UF Kaveri Basin
BT India
SA Mysore
Tamil Nadu

Caux
Chalky tableland in N.
BT Seine-Maritime
France

cave maps
Term introduced in 1981. Before 1981, also search caves AND maps.
BA maps
SA caverns
caves

Cavellinidae
Includes use on level 3 under Ostracoda(1). See list F.
BA Podocopida
Ostracoda
BT Crustacea
Mandibulata
Arthropoda
Invertebrata
SA Healdiidae

caverns
Icludes use on level 3 under geomorphology(1) solution features(2).
SA cave maps
caves
geomorphology
solution features

caves
Includes use on level 3 under geomorphology(1) solution features(2) and volcanic features(2).
SA cave maps
caverns
exsurgence
geomorphology
karst
solution
solution cavities
solution features
speleology
speleothems
volcanic features

cavities
Used as a general term through 1978. After 1978, see specific terms, e.g. underground space or solution cavities.

cavities, solution
use solution cavities

Cavnic
Village in Maramures County in N Transylvania.
UF Capnic
BT Transylvania
Romania

Cayman Islands
Term introduced in 1981. Includes use on level 1 as an area term (list O).
BA West Indies
NT Grand Cayman Island

Cayman Trench
use Cayman Trough

Cayman Trough
South of the Cayman Islands and Cuba. Also search Cayman Trench.
UF Cayman Trench
BT Caribbean Sea

Caytoniales
Includes use on level 2 under gymnosperms(1). See list F.
BA gymnosperms
BT Spermatophyta
Plantae
NA Pachypteris

Cayugan
Provincial series, North America. Above Niagaran, below Ulsterian (Devonian). Includes use on level 3 under age terms(1). See list E.
BA Upper Silurian
Silurian
BT Paleozoic
NT Williamsport Sandstone

Cd
use cadmium

CDP method
use common-depth-point method

Ce
use cerium

Ceara
State in NE.
CO S075000S025000
W0371000W0412000
BT Brazil
NT Jaguaribe

Cebu
Island in Visayan Islands in central.
BT Philippine Islands

Cedar City Formation
BT Devonian
SA Missouri

Cedar Mountain Formation
Together with Buckhorn Conglomerate comprises Cedar Mountain Group. NW Colorado and central Utah.
BT Lower Cretaceous
Cretaceous
SA Colorado
Utah

Cedar Valley Formation
Includes Linwood, Littleton, Coralville, Solon and Rapid members. E Iowa, SW Illinois, and SE Minnesota.
BT Middle Devonian

Devonian
SA Illinois
Iowa
Minnesota

celadonite
BA mica group
sheet silicates
silicates
BT minerals
SA glauconite

Celebes
Island E of Borneo. Also search Sulawesi, the official Indonesian name.
CO S053700N014500
E1250500E1184500
UF Sulawesi
BT Indonesia

Celebes Sea
Bounded on N by Philippines, W by Borneo, and S by island of Celebes. Also search Celebes. Includes use on level 1 as an area term (list O). For term set options see list B.
BA Indonesian Seas
West Pacific
Pacific Ocean
SA Darvel Bay
Malay Archipelago

celestite
BA sulfates
BT minerals

Celje
Town in NW part of country.
BT Slovenia
Yugoslavia

cell dimensions
As of 1978, term is used on level 2 under crystal structure(1).
SA bonding
crystal structure
dimensions
lattice
minerals

cell, unit
use unit cell

cellulose
BA carbohydrates
organic materials

celsian
As of 1981, includes use on level 3 under minerals(1) framework silicates, barium feldspar(2). See list L. Before 1981, included use on level 3 under minerals(1) framework silicates, feldspar group(2).
BA barium feldspar
feldspar group
framework silicates
silicates
BT minerals
SA paracelsian

Celtic Sea
Body of water over the continental shelf between the S coast of Ireland and Brittany. It is arbitralily divided from the English Channel by a line from Land's End to Ushant Island off Brittany.
BT Atlantic Ocean

cement
Term restricted to use as the material. Use cementation for the process under sedimentary rocks(1).
SA cement materials
cementation
cementation deposits
sedimentary rocks
sediments

cement materials
Includes use on level 3 as a commodity term (list C).
BA construction materials
SA cement
concrete
limestone deposits
materials

cementation
Includes use on level 3 under sedimentation(1) diagenesis(2); on level 3 under sedimentary rocks(1) diagenesis(2). See list I.
BA diagenesis
SA cement
cementation deposits
clastic sediments
consolidation
deposition
lithification
sedimentation

cementation deposits
Term introduced in 1981.
SA cement
cementation
mineral deposits

Cenari Zone
use Ceneri Zone

Ceneri Zone
In Southern Alps. Index countries as applicable.
UF Cenari Zone
BT Europe
SA Italy
Paleozoic
Switzerland

Cenomanian
Europe. Above Albian, below Turonian. Includes use on level 3 under age terms(1). See list E.
BA Upper Cretaceous
Cretaceous
BT Mesozoic
SA Middle Cretaceous

Cenozoic
Includes use on level 1 as an age term (list E); on level 2 under paleoterms, e.g. paleoecology, paleogeography, paleomagnetism.
BT Phanerozoic
NT Blancan
Leda Clay
Loup Fork Group
NA lower Cenozoic
Matuyama Epoch
NT Osaka Group
NA Paleolithic
NT Quaternary
San Juan Formation
Santa Maria Formation
Siwalik System
Tertiary
Tulare Formation
Uonuma Group
NA upper Cenozoic
NT Verde Formation
SA Alpine Orogeny
Himalayan Orogeny

centers, color
use color centers

centers, spreading
use spreading centers

Central Africa
Term introduced in 1978. Before 1978, search Africa AND central.
CO S180000N113000
E0313000E0083000
BA Africa
NA Angola
Burundi
Central African Republic
Congo
Equatorial Guinea
Gabon
Rwanda
Zaire
SA East Africa
North Africa
Southern Africa
West Africa

Central African Republic
Formerly French territory of Ubangi-Shari. Became independent in 1960. Includes use on level 1 as an area term (list O). For term set options see list B.
CO N030000N120000
E0273000E0140000
BA Central Africa
Africa
NT Koto
SA Congo Basin

Central Alps
Ranges of the Alps extending from the Great Saint Bernard Pass in the W to Innsbruck and the Brenner Pass in the E. Index countries as applicable.
CO N460000N470000
E0113000E0090000
BT Alps
Europe
SA Austria
Bavarian Alps
Bernese Alps
Glarus Alps
Italy
Lepontine Alps
Pennine Alps
Rhaetian Alps
Switzerland
West Germany

Central America
Index countries as applicable. Includes use on level 1 or 2 as an area term (list O). If 1, see list B for term set options.
CO N071000N183000
W0771000W0921500
NA Belize
Costa Rica
El Salvador
Guatemala
Honduras
Nicaragua
Panama
NT Panama Canal Zone
SA America
Caribbean region

Central Andes
Term introduced in 1978. Before 1978, also search Andes AND central.
BA Andes
BT South America
SA Northern Andes
Southern Andes

Central Appalachians
Term introduced in 1978. Before 1978, also search Appalachians AND central.
BA Appalachians
BT North America
SA Northern Appalachians
Southern Appalachians

Central Asia
Region known as Soviet Central Asia. Index Soviet republics as applicable.
CO N370000N530000
E0860000E0430000
UF Soviet Central Asia
BT USSR
NA Caspian Depression
NZ Central Kazakhstan
Eastern Kazakhstan
NA Fergana Basin
NZ Karakum
Kopet-Dag Range
NA Kyzylkum
Pamirs
Tien Shan
Turgay Basin
Ustyurt
SA Kazakhstan
Kirghizia
Tadzhikistan
Turkmenia
Uzbekistan

Central Atlantic Coastal Plain
Term introduced in 1981. Before 1981, also search Atlantic Coastal Plain AND Central.
CO N340000N373000
W0752000W0800000
BT Atlantic Coastal Plain

Central Austrian Alps
Term introduced in 1981.
CO N463500N474600
E0163500E0095200
BZ Alps
BX Austria

Central Basin
Central.
BT Tennessee
United States

Central Bohemian Pluton
Part of the Bohemian Massif in central.
BT Bohemia
Czechoslovakia
SA Bohemian Massif

Central Bulgaria
Term introduced in 1981.
CO N414000N431000
E0280000E0224500
BT Bulgaria

Central California
Term introduced in 1981. Before 1981, also search California AND Central.
CO N354800N390000
W1154000W1234000
BT California
United States

Central Europe
CO N453000N550000
E0240000E0060000
BA Europe
NA Austria
Czechoslovakia
Germany
Hungary
Liechtenstein
Poland
Switzerland

Central Florida
Term introduced in 1981. Before 1981, also search Florida AND Central.
CO N270000N290000
W0800500W0825000
BT Florida
United States

central granite
use batholiths

Central Indian Ridge
Term introduced in 1981.
CO S400000S050000
E0800000E0630000
BA Mid-Indian Ridge
BT Indian Ocean
SA mid-ocean ridges

Central Kazakhstan
Term introduced in 1981.
CO N450000N550000
E0790000E0650000
BZ Central Asia
BX Kazakhstan
BT USSR

Central Massif
An old, eroded plateau region in S central part of country. Index departments as applicable. Also search Massif Central.
CO N440000N470000

E0050000E0004500
UF Massif Central
BT France
NT Millevaches Plateau
SA Aveyron
Cantal
Cantal Massif
Causses
Haute-Loire
Montagne Noire
Morvan
Sidobre Massif

Central Pacific
Term introduced in 1981.
BA Pacific Ocean

Central Polish Glaciation
The Elster and Saale stages of glaciation both of which covered central Poland. They were respectively the oldest and the middle stage of 3 stages which covered northern Europe.
BT Poland
SA Saalian

Central Rocky Mountains
Term introduced in 1978. Before 1978, also search Rocky Mountains AND central.
BA Rocky Mountains
BT North America
SA Northern Rocky Mountains
Southern Rocky Mountains
U. S. Rocky Mountains

Central Siberian Plateau
use Siberian Platform

Central Swiss Alps
Term introduced in 1981.
BA Swiss Alps
BT Switzerland

Central Transdanubia
Term introduced in 1981.
CO N464200N474900
E0191000E0171000
BZ Transdanubia
BX Hungary

Central Urals
Term introduced in 1981.
CO N530000N600000
E0623000E0563000
BZ Urals
BX Russian Republic
BT USSR

Central Valley
Valley of Sacramento and San Joaquin rivers in California, and a 600 mile long valley between the Andes and the Coastal Range in central Chile. Index California and/or Chile as applicable. Before 1978, also search Great Valley for the valley in California.
SA California
Chile
Sacramento Valley
San Joaquin Valley

centrallasite
use gyrolite

Centre County
Central.
BT Pennsylvania
United States

centrum (seismology)
use focus

Cephalodiscida
Includes use on level 2 under Pterobranchia(1). See list F.
BA Pterobranchia
BX Hemichordata
Invertebrata

Cephalonia
Island in center of Ionian Islands in Ionian Sea W of Greek mainland. Also search Kefallinia.
UF Kefallinia
BT Greek Ionian Islands
Greece

Cephalopoda
Includes use on level 2 under Mollusca(1). See list F.
BA Mollusca
BT Invertebrata
NA Belemnitidae
Belemnopsis
Ceratites
Coleoidea
Dibranchiata
Tetrabranchiata
SA cephalopods

cephalopods
Term introduced in 1981 as the common name for Cephalopoda. See list F. To be used for Cephalopoda when they are used as tools in stratigraphy. Before 1981, search Cephalopoda AND biostratigraphy.
SA biostratigraphy
Cephalopoda

ceramic materials
Includes use on level 1 as a commodity term (list C).
SA andalusite deposits
clay mineralogy
kyanite deposits
materials
refractory materials
sillimanite deposits

Ceratites
Genus. Includes use on level 3 under Mollusca(1) Cephalopoda(2).
BA Cephalopoda
Mollusca
BT Invertebrata
SA Ammonites

Ceratitida
Term introduced in 1981. Includes use on level 2 under Mollusca(1). See list F.
BA Ammonoidea
Tetrabranchiata
Cephalopoda
Mollusca
BT Invertebrata

Ceratolithus
BA Coccolithophoraceae
algae
BT thallophytes
Plantae

Ceratomorpha
Term introduced in 1981. Includes use on level 2 under Mammalia(1). See list F.
BA Perissodactyla
Mammalia
BT Tetrapoda
Vertebrata

Ceriantipatharia
Includes use on level 2 under Coelenterata(1). See list F.
BA Anthozoa
Coelenterata
BT Invertebrata

Cerithiidae
Includes use on level 3 under Mollusca(1). See list F.
BA Gastropoda
Mollusca
BT Invertebrata
NA Cerithium

Cerithium
Includes use on level 3 under Mollusca(1). See list F.
BA Cerithiidae
Gastropoda
Mollusca
BT Invertebrata

cerium
Includes use on level 1 and 2 as a chemical element (list D).
UF Ce
SA elements
lanthanum
rare earths

cerolite
BA serpentine group
sheet silicates
silicates
BT minerals

Cerozem
use Sierozems

Cerro de Pasco
Town in W central part of country.
BT Peru

Cerro del Hoyazo
In the Sierra Alhamilla area in SE part of country.
BT Almeria Province
Andalusia
Spain

Cerro Prieto
Geothermal area near Mexicali in N.
BT Baja California
Mexico

cerussite
BA carbonates
BT minerals

Cervidae
As of 1981, includes use on level 3 under Mammalia(1) Ruminantia(2). See list F. Before 1981, included use on level 3 under Mammalia(1) Artiodactyla(2).
BA Ruminantia
Artiodactyla
Mammalia
BT Tetrapoda
Vertebrata
NA Cervus

Cervus
As of 1981, includes use on level 3 under Mammalia(1) Ruminantia(2). See list F. Before 1981, included use on level 3 under Mammalia(1) Artiodactyla(2).
BA Cervidae
Ruminantia
Artiodactyla
Mammalia
BT Tetrapoda
Vertebrata

cesium
Includes use on level 1 and 2 as a chemical element (list D). Also search Cs.
UF Cs
SA alkali metals
Cs-137
elements

Cetacea
Includes use on level 2 under Mammalia(1). See list F.
BA Mammalia
BT Tetrapoda
Vertebrata

Cette
use Sete

Ceuta
Term introduced in 1981.
BT Morocco

Cevennes
Mountain Range in S.
BT France

Ceylon
A valid term through 1973. After 1973, use Sri Lanka.

Cf
use californium

chabazite
BA zeolite group
framework silicates
silicates
BT minerals

Chablais
Limestone massif of Savoy Alps and region in SE part of country.
BT Haute-Savoie
France

Chaco
Region. Index countries as applicable.
UF Gran Chaco
BT South America
SA Argentina
Bolivia
Campo del Cielo Meteorite
Paraguay

Chad
Includes use on level 1 as an area term (list O). For term set options see list B.
CO N072000N233000
E0240000E0183000
BA West Africa
Africa
NT Kanem
Tibesti Massif
SA Chad Basin
Lake Chad
Logone River
Sahara
Sahel

Chad Basin
Index countries as applicable.
BT Africa
SA Cameroon
Chad
Lake Chad
Niger
Nigeria

Chadak
Village in E.
BT Uzbekistan
USSR

Chadobets Uplift
SE Krasnoyarsk Kray in S Siberia.
BT Russian Republic
USSR

Chadron Arch
NW Nebraska and SW South Dakota. Index states as applicable.
BT United States
SA Nebraska
South Dakota

Chadron Formation
In White River Group. NE Colorado, W Nebraska, W South Dakota, and E Wyoming.
BT lower Oligocene
Oligocene
SA Chadronian
Colorado
Nebraska
South Dakota
White River Group
Wyoming

Chadronian
North America. See list E.
BA lower Oligocene
Oligocene
Paleogene
Tertiary
BT Cenozoic
SA Chadron Formation

Chaetetidae
Family.
BA Tabulata
Anthozoa
Coelenterata
BT Invertebrata

Chaetognatha
Includes use on level 2 under worms(1). See list F.
BA worms
BT Invertebrata

Chaetopoda
Includes use on level 3.
BA worms
BT Invertebrata
SA Polychaetia

Chaffee County
Central.
BT Colorado
United States

Chain Deep
BT Red Sea

chain silicates
Includes use on level 2 under minerals(1); in combination with amphibole group, alkalic amphibole, clinoamphibole, orthoamphibole, pyroxene group, alkalic pyroxene, clinopyroxene, and orthopyroxene (i.e. chain silicates, amphibole group) to form terms on level 2 under minerals(1). See list L. Also search inosilicates.
UF inosilicates
BA silicates
BT minerals
NA aenigmatite
amphibole group
babingtonite
bustamite
carpholite
deerite
ferrocarpholite
howieite
jade
neptunite
pectolite
plancheite
pyroxene group
pyroxmangite
rhodonite
rhonite
scawtite
shattuckite
tobermorite
wollastonite
xonotlite
SA asbestos

Chaine des Puys
Term introduced in 1981. Division of Auvergne Mountains in S central part of country. Before 1981, also search Monts Dome.
BT Puy-de-Dome
France

Chainpur Meteorite
Term introduced in 1981. Impact in W Bihar. Before 1981, also search Chainpur AND meteorites.
BA LL chondrites
chondrites
BT meteorites
SA Bihar
India

chalcanthite
BA sulfates
BT minerals

chalcedony
BA silica minerals
framework silicates
silicates
BT minerals
SA agate
chert
gems
jasper
jasperoid
quartz

chalcocite
BA sulfides
BT minerals
SA copper
copper sulfides
digenite

chalcopyrite
UF copper pyrites
BA sulfides
BT minerals
SA copper

chalcostibite
BA sulfantimonites
sulfosalts
BT minerals

Chaleur Bay
Inlet of Gulf of Saint Lawrence between New Brunswick and the Gaspe Peninsula. Index provinces as applicable.
BT Canada
SA New Brunswick
Quebec

chalk
Includes use on level 3 under sedimentary rocks(1) carbonate rocks(2). See list I. As of 1981, use chalk deposits for chalk as a commodity. Before 1981, included use on level 3 as a commodity term.
BA carbonate rocks
BT sedimentary rocks
SA calcite
chalk deposits
limestone

Chalk
As of 1981, no longer a valid term for GeoRef. Use Upper Cretaceous.

chalk deposits
Term introduced in 1981. Includes use on level 3 as a commodity term (list C) under limestone deposits(1). Before 1981, search chalk AND deposits.
SA chalk
limestone deposits

Challenger Knoll
BT Gulf of Mexico

chalmersite
use cubanite

chalybite
BA carbonates
BT minerals
SA siderite

Chama Basin
River basin in N.
BT New Mexico
United States

Chamba
Town and district in N part of country.
BT Himachal Pradesh
India

chambers, magma
use magma chambers

chamosite
BA chlorite group
sheet silicates
silicates
BT minerals
SA berthierine

Champlain Valley
Lake Champlain area between Adirondack Mountains and Green Mountains and extending into S Quebec. Index Quebec and states as applicable.
SA Lake Champlain
New York
Quebec
Vermont

Champlainian
Provincial series, North America. Above Canadian. Includes use on level 3 under age terms(1). See list E.
BA Middle Ordovician
Ordovician
BT Paleozoic
SA Trentonian

Chanares
Village in central.
BT Cordoba Department
Argentina

Chanares Formation
BT Triassic
SA Argentina
La Rioja

Chandalar Quadrangle
Term introduced in 1981. Central Alaska.
CO N670000N680000
W1470000W1500000
BT Alaska
United States

Chandigarh
City in E.
BT Punjab State
India

Chandler wobble
Includes use on level 3 under Earth(1) motions(2).
SA Earth
motions

Changai Mountains
use Hangay Mountains

changes
Includes use as level 2 or 3 term appropriate to a large number of topics, e.g. on level 2 under paleoecology(1). See list G.
SA alteration
conversion
transformations

changes of level
Used mostly under stratigraphy for sea-level (sometimes lake-level) changes of the upper Cenozoic. Includes use on level 1(list A). Term set options are:
topic [age, causes, concepts, correlation, detection, evolution, genesis, interpretation, mechanism, observations, patterns, processes, rates]
subtopic (no area term)
UF level, changes of
SA absolute age
continental shelf
eustacy
geodesy
geomorphology
glacial geology
glaciation
isostasy
neotectonics
paleoecology
paleogeography
reefs
regression
sedimentation
shorelines
stratigraphy
submergence
tectonics
tectonophysics
terraces
transgression

Changwat Nakhon Phanom
Province in NE Thailand. Nakhon Phanom is also a city.
BT Thailand

channel geometry
For channel cross-section only. Includes use on level 3 under geomorphology(1) and hydrology(1).
SA channels
fluvial features
geometry
geomorphology
hydrology
rivers
rivers and streams

Channel Islands
Used for the Santa Barbara Islands off SW California coast between Santa Barbara and San Diego. Before 1981, term referred to the Channel Islands in the English Channel as well, and was indexed under California or England as applicable. Beginning in 1981, use English Channel Islands for the British islands.
CO N324000N341000
W1181500W1203000
UF Santa Barbara Islands
BT California
SA English Channel Islands

channel order
use stream order

channels
As of 1978, term is used on level 2 under waterways(1).
SA braided streams
buried channels
canals
channel geometry
continental slope
fluvial features
geomorphology
grooves
gullies
lava channels
levees
planar bedding structures
rivers
sedimentary structures
sedimentation
sinuosity
streams
tidal channels
troughs
underground channels
waterways

chaoite
As of 1981, includes use on level 3 under minerals(1) native elements(2). See list L. Before 1981, included use on level 3 under minerals(1) native elements and alloys(2).
BA native elements
BT minerals
SA carbon
diamond
graphite
lonsdaleite
meteorites

Char
Rock belt in the Lake Zaysan and Kalba Mountain areas of NE.
BT Kazakhstan
USSR

Characeae
Family.
BA Charophyta
Chlorophyta
algae
BT thallophytes
Plantae

characterization
Includes use on level 3 under soils(1) water regimes(2).
SA soils
water regimes

charcoal
Includes use on level 3 under coal(1) as a commodity term.
SA absolute age
carbon
coal

C

Chardzhov
City and oblast in E.
BT Turkmenia
USSR

Charente
Department in W France. Also a river.
CO N451000N461000
E0005500W0003000
BT France
NT Angouleme
Saint-Severin

Charente-Inferieure
use Charente-Maritime

Charente-Maritime
Department in W France.
CO N450500N462000
E0000200W0013500
UF Charente-Inferieure
BT France

Charentes
Region in W and W central.
BT France

Charles County
Term introduced in 1981. S Maryland.
CO N381500N384200
W0764000W0771700
BT Maryland
United States

Charleston
City in Charleston County in SE South Carolina, and a city in Kanawha County, West Virginia. Index states as applicable.
BX South Carolina
West Virginia
BT United States

Charleston County
Along the Atlantic Ocean in SE.
BT South Carolina
United States

Charlevoix
Cryptoexplosion or impact feature in Quebec. Also a city on Lake Michigan. Index Quebec or Michigan as applicable. Also search Charlevoix structure.
SA cryptoexplosion features
impact features
Michigan
Quebec

Charlevoix County
N Lower Peninsula.
BT Michigan
United States

Charlotte
Term introduced in 1981. City in Mecklenburg County, S North Carolina.
CO N350300N350300
W0805000W0805000
BT North Carolina
United States

Charlotte County
In extreme S New Brunswick. Also counties in Florida and Virginia, U.S. Index New Brunswick and states as applicable.
BX Florida
New Brunswick
Virginia

charnockite
As of 1981, includes use on level 3 under igneous rocks(1) granites(2). See list H. Before 1981, included use on level 3 under igneous rocks(1) granite-granodiorite family(2).
BA granites
BT igneous rocks

Charophyta
Includes use on level 2 under algae(1). See list F. Before 1981, also search charophytes. As of 1981, search charophytes only for biostratigraphy.
BA Chlorophyta
algae
BT thallophytes
Plantae
NA Characeae

Charters Towers
City just E of the Great Dividing Range in NE central.
BT Queensland
Australia

chartology
use cartography

Chase County
Index states as applicable.
BX Kansas
Nebraska
BT United States

Chase Group
Includes Odell Formation, Nolans Formation, Wreford Limestone, Matfield Shale, Doyle Shale, Winfield Limestone, Herrington Limestone, Fort Riley Limestone and Barneston Limestone. E Kansas, N central Oklahoma, and SE Nebraska.
BT Permian
SA Fort Riley Limestone
Kansas
Nebraska
Oklahoma
Wreford Limestone

chassignite
Term introduced in 1981. Includes use on level 3 under meteorites(1).
BA achondrites
BT meteorites

Chateaulin Basin
In extreme W Brittany.
BT Finistere
France

Chatham County
Index states as applicable.
BX Georgia
North Carolina
BT United States

Chatham Rise
500 miles E of New Zealand.
CO S443000S430000
W1770000E1743000
BT Pacific Ocean

Chatkal Range
Branch of NW Tien Shan. Also search Chatkal.
BT Kirghizia
USSR
SA Tien Shan

Chattahoochee River
Term introduced in 1981. In Georgia, Alabama, and Florida.
CO N304000N345000
W0833000W0851500
BT United States
SA Alabama
Florida
Georgia

Chattanooga Shale
Consists of lower Dowelltown Member and upper Gassaway Member. In some areas, includes Hardin Sandstone Member below the Dowelltown. N Alabama, Arkansas, NW Georgia, Illinois, E and W Kentucky, NE Mississippi, Missouri, Oklahoma, and Tennessee.
BT Paleozoic
SA Alabama
Arkansas
Devonian
Georgia
Illinois
Kentucky
Mississippi
Mississippian
Missouri
Oklahoma
Tennessee
Upper Devonian

Chattian
Europe. Above Rupelian, below Aquitanian (Miocene). Includes use on level 3 under age terms(1). See list E.
BA upper Oligocene
Oligocene
Paleogene
Tertiary
BT Cenozoic

Chaturi
use Chiatura

Chautauqua County
Index states as applicable.
BX Kansas
New York
BT United States

Chaves County
In SE New Mexico.
BT New Mexico
United States

Chaves Island
Term introduced in 1981 for one of the Galapagos Islands. Not to be confused with Santa Cruz Island which is one of the Channel Islands. Also search Santa Cruz Island AND Galapagos.
UF Indefatigable Island
BT Galapagos Islands
SA Santa Cruz Island

Chaya Massif
A gabbro-peridotite massif in northern Baikal region.
BT Russian Republic
USSR

Chazyan
North America. Below Mohawkian, above lower Ordovician. Includes use on level 3 under age terms(1). See list E.
BA Middle Ordovician
Ordovician
BT Paleozoic

Chechen-Ingush
Autonomous Soviet Socialist Republic on N slopes of Caucasus.
CO N423000N440000
E0463000E0443000
UF Checheno-Ingush
BT Russian Republic
USSR

Checheno-Ingush
use Chechen-Ingush

Chedabucto Bay
Inlet of Atlantic Ocean in NE just S of Cape Breton Island.
BT Nova Scotia
Canada

Cheilostomata
Includes use on level 2 under Bryozoa(1). See list F.
BA Bryozoa
BT Invertebrata

Cheju Island
In East China Sea S of mainland.
BT Korea

Chekalin
Town in W Tula Oblast in SW European USSR. Also search Likhvin.
UF Likhvin
BT Russian Republic
USSR

Chelan County
Term introduced in 1981. Central Washington.
CO N471700N483500
W1195000W1211000
BT Washington
United States

Cheleken
Town on the Caspian Sea in Krasnovodsk Oblast.
BT Turkmenia
USSR

Cheleken Peninsula
In Krasnovodsk Oblast on Caspian Sea.
BT Turkmenia
USSR

Chelicerata
Subphylum. Includes use on level 2 under Arthropoda(1). See list F.
BA Arthropoda
BT Invertebrata
NA Arachnida
Merostomata

Chelonia
As of 1981, includes use on level 2 under Reptilia(1). See list F. Before 1981, included use on level 3 under Reptilia(1) Anapsida(2).
BA Anapsida
Reptilia
BT Tetrapoda
Vertebrata
NA Pelomedusidae

Chelyabinsk
City and oblast in Southern Urals.
BT Russian Republic
USSR

chemical
A valid term through 1977. After 1977, use more specific term, e.g. chemical composition, or chemical properties, or chemical analysis, or chemical weathering.

chemical analysis
Use only for methodology. Includes use on level 2 under minerals(1). For data, see appropriate material. Includes use on level 1 (list A). Also search chemical analyses; analysis AND chemical. Before 1981, also search chemical methods. Term set options are:
methods
name of method [activation analysis, chromatography, colorimetry, electrolytic analysis, infrared spectroscopy, major-element analyses, minor-element analyses, polarography, spectroscopy, trace-element analyses, vacuum fusion analysis, volumetric analysis, wet methods, X-ray fluorescence]
techniques
subtopic [e.g. sample preparation, titration, material, reagent, etc.]
SA activation analysis
analysis
catalysis
chemical composition
chromatography
clay mineralogy
color
colorimetry
crystal chemistry
crystallography
decrepitation
differential thermal analysis
diffusion
electron microscopy
electron probe

emission spectroscopy
flame photometry
geochemical methods
geochemistry
geochronology
igneous rocks
inclusions
infrared spectra
ion probe
ion probe data
isotopes
major-element analyses
minerals
minor-element analyses
neutron activation analysis
organic materials
photometry
polarography
qualitative analysis
quantitative analysis
sample preparation
solubility
spectroscopy
standard materials
thermal analysis
thermogravimetric analysis
trace-element analyses
vacuum fusion analysis
wet methods
X-ray analysis
X-ray diffraction analysis
X-ray fluorescence

chemical composition
To discuss data and results of methods. Includes use on level 3 under chemical analysis(1) methods(2) and under soils(1) composition(2). Also search chemical AND composition.
SA analysis
chemical analysis
composition
gas chromatography
geochemistry
soils

chemical demagnetization
Term introduced in 1981. Includes use on level 3 under paleomagnetism(1) and geophysics(1).
BT demagnetization
SA geophysics
magnetization
paleomagnetism
remanent magnetization

chemical elements
use elements

chemical explosions
As of 1978, term is used on level 2 under explosions(1).
BT explosions

chemical fossils
SA fossils
ichnofossils
problematic fossils

chemical industry
Term introduced in 1981.
SA economic geology
industry

chemical magnetization
use chemical remanent magnetization

chemical methods
As of 1981, no longer a valid term for GeoRef. See geochemical methods; chemical analysis. Before 1981, included use on level 3 under soils(1) analysis(2).

chemical properties
Includes use under commodity terms (list C).
SA coal
geochemistry
properties
thermochemical properties
valency

chemical reactions
use reactions

chemical remanence
use chemical remanent magnetization

chemical remanent magnetization
Includes use on level 3 under paleomagnetism(1). Also search chemical magnetization.
UF chemical magnetization
chemical remanence
CRM
crystallization magnetization
crystallization remanent magnetization
BA remanent magnetization
SA magnetization
paleomagnetism

chemical sedimentation
Term introduced in 1981.
SA sedimentation

chemical weathering
Includes use on level 3 under weathering(1). Also search decomposition.
UF decomposition
BT weathering
SA corrosion

chemically precipitated rocks
Includes use on level 2 under sedimentary rocks(1). See list I.
BT sedimentary rocks
NA chert
duricrust
evaporites
ferricrete
flint
iron formations
ironstone
jaspilite
phosphate rocks
silcrete
siliceous sinter
taconite
tufa
weathering crust
SA anhydrite
dolomite
gypsum
halite
iron-rich composition
novaculite
rocks
travertine

chemically precipitated sediments
As of 1981, no longer a valid term for GeoRef. Use chemically precipitated rocks. Before 1981, included use on level 2 under sediments(1).

chemistry
As of 1981, no longer a valid term for GeoRef. See geochemistry. Before 1981, included use on level 2 under soils(1).

Chenango River
In S central.
BT New York
United States

cheniers
Term introduced in 1981. Includes use on level 3 under geomorphology(1) shore features(2).
SA beach ridges
geomorphology
shore features
shorelines

Cher
Department in central France.
CO N462500N474000
E0030700E0014500
BT France
NT Bourges

Cheremkhovo Basin
use Irkutsk Basin

Cheremkovo coal basin
use Irkutsk Basin

Cherkasi
use Cherkassy

Cherkassy
City and oblast in central Ukraine.
UF Cherkasi
BT Ukraine
USSR

Chernigov
City in Chernigov Oblast in N.
BT Ukraine
USSR

Chernozems
Includes use on level 3 under soils(1). See list M. Also search Black earth.
UF Black earth
Chernozyom
Tchornozem
Tschernosem
Tschernosiom
BA soils
SA Chestnut soils
soil group
Zonal soils

Chernozyom
use Chernozems

Cherokee County
Index states as applicable.
BX Alabama
Georgia
Iowa
Kansas
North Carolina
Oklahoma
South Carolina
Texas
BT United States

Cherokee Group
Divided into Riverton, Warner, Rowe, Dry Wood, Bluejacket, Seville, Weir, Tebo, Scammon, Mineral, Robinson, Branch, Fleming, Croweburg, Verdigris, Bevier, Lagonda, Mulky, and Excello formations. SW Iowa, E Kansas, W Missouri, and SE Nebraska.
BT Desmoinesian
Middle Pennsylvanian
Pennsylvanian
SA Cabaniss Formation
Iowa
Kansas
Missouri
Nebraska

chert
Includes use on level 3 under sedimentary rocks(1) chemically precipitated rocks(2). See list I. Also search hornstone.
UF hornstein
BA chemically precipitated rocks
BT sedimentary rocks
NA jasperoid
SA chalcedony
flint
jasper

chertification
BA diagenesis
SA silicification

Chesapeake Bay
Inlet of Atlantic Ocean. Index states as applicable.
CO N365500N393700
W0754000W0763300
BT Atlantic Coastal Plain
SA Maryland
Virginia

Cheshire
County in NW.
CO N530000N533000
W0020000W0032000
BT England
Great Britain
United Kingdom

Chesil Beach
On the English Channel in Dorset.
BT England
Great Britain
United Kingdom

Chester County
Index states as applicable.
BX Pennsylvania
South Carolina
Tennessee
BT United States

Chester Series
use Chesterian

Chesterian
Provincial series, North America. Formations included in standard section: Aux Vases Sandstone, Renault Limestone, Bethel Sandstone, Point Creek Formation. N Alabama, Illinois, S Indiana, Kentucky, E Missouri, and Tennessee. Includes use on level 3 under age terms(1). See list E.
UF Chester Series
BA Upper Mississippian
Mississippian
Carboniferous
BT Paleozoic
SA Alabama
Illinois
Indiana
Kentucky
Meramecian
Missouri
Morrowan
Springer Formation
Tennessee

Chestnut soils
Includes use on level 3 under soils(1). See list M.
BA soils
SA Chernozems
soil group
Zonal soils

Cheviot Hills
Range of hills along the English-Scottish border. Index England and/or Scotland as applicable.
BT Great Britain
United Kingdom
SA England
Scotland

chevkinite
BA orthosilicates
silicates
BT minerals

chevron folds
Term introduced in 1978. Includes use on level 3 under folds(1). Before 1978, also search folds AND chevron.
BA folds
SA kink folds

Cheyenne County
Index states as applicable.
BX Colorado
Kansas
Nebraska
BT United States

Cheyenne Sandstone
Underlies Kiowa Shale Member; overlies Morrison Formation. SW

Kansas, SE Colorado, and W Oklahoma.
BT Lower Cretaceous
Cretaceous
SA Colorado
Kansas
Oklahoma

Chhindwara
Town in central.
BT Madhya Pradesh
India

Chhindwara District
Central.
BT Madhya Pradesh
India

Chiang Mai
Term introduced in 1981. City and province in NW.
BT Thailand

Chiapas
State in SE.
BT Mexico

Chiatura
City in central Georgian Republic, USSR. Also search Chiaturi.
UF Chaturi
BT Georgian Republic
USSR

Chiayi-Hsinying area
BT Taiwan

Chiba
City and prefecture SE of Tokyo.
BT Honshu
Japan

Chiba Peninsula
Between Tokyo Bay and Sagami Sea. Also search Boso Peninsula, and Chiba.
UF Boso Peninsula
BT Honshu
Japan

Chibougamau
Village SE of Lake Mistassini in central.
BT Quebec
Canada

Chicago
City on Lake Michigan in Cook County.
BT Illinois
United States

Chichibu
City NW of Tokyo in central.
BT Honshu
Japan

Chichibu Belt
BT Honshu
Japan

Chichibu Mine
Central.
BT Honshu
Japan

Chickamauga Group
Includes limestone strata of Middle and Upper Ordovician age. N Alabama, NW Georgia, E Tennessee, and SW Virginia. Also search Chickamauga Limestone.
BT Ordovician
SA Alabama
Georgia
Middle Ordovician
Tennessee
Upper Ordovician
Virginia

Chickasawhay Formation
In Limestone Creek Group. In SE Mississippi.
BT upper Oligocene
Oligocene
SA Mississippi

Chickmagalur
Village in SW.
BT Mysore
India

Chieti
Town in E.
BT Abruzzi
Italy

Chignik Quadrangle
Term introduced in 1981. S central Alaska.
CO N560000N570000
W1580000W1603000
BT Alaska
United States

Chigoku Sammayuku Mountains
use Chugoku

Chihuahua
State in northern Mexico. Also a city.
CO N254500N320000
W1030000W1080000
BT Mexico
SA Allende Meteorite
Chihuahua tectonic belt
Pedregosa Basin
Rio Grande
Rio Grande Rift
Rio Grande Valley
Sierra Madre Occidental

Chihuahua tectonic belt
Index Mexican and U. S. states as applicable.
SA Chihuahua
Coahuila
New Mexico
Rio Grande Rift
Texas

Chile
Includes use on level 1 as an area term (list O). For term set options see list B.
CO S560000S174500
W0670000W0760000
BA South America
NT Aconcagua Province
Antofagasta
Arauco
Atacama
Atacama Desert
Coquimbo
Magallanes
Santiago
Tarapaca
Valparaiso
SA Andes
Central Valley
Patagonia
Patagonian Andes
Tierra del Fuego
Western Cordillera

Chile Ridge
Runs SE-NW off Chilean coast between 80° W-105° W. Also search Chile Rise.
CO S453000S350000
W0800000W1050000
UF Chile Rise
BA South American Pacific
BT Pacific Ocean
SA mid-ocean ridges

Chile Rise
use Chile Ridge

Chilhowee Group
Includes Loudoun Formation, Weverton Sandstone, Harpers Shale, Antietam Sandstone, Unicoi Formation, Hampton Shale, Erwin Quartzite, Vann Quartzite, Sandsuck Shale, Cochran Quartzite, Nebo Quartzite, Murry Shale, Hesse Quartzite, Shady Dolomite, Rome Formation, Nichols Shale, Helenmode Formation.
BT Lower Cambrian
Cambrian
SA Antietam Formation
Maryland
North Carolina
Shady Dolomite
Tennessee
Virginia

Chilton County
Central.
BT Alabama
United States

China
Peoples Republic of China. Includes use on level 1 as an area term (list O). For term set options see list B.
CO N200000N530000
E1350000E0740000
BA Far East
Asia
NT Han River basin
Hebei
Heilungkiang
Inner Mongolia
Manchuria
Shandong
Shansi
Shensi
Sinkiang Uighur
Szechwan
Tibet
Yunnan
SA Altai Mountains
Brahmaputra River
Khanka Lake

China Sea
Bordering on China and divided by Taiwan into East China Sea to N and South China Sea extending to Malaysia in the S. Index seas as applicable.
BT Pacific Ocean
SA East China Sea
South China Sea

Chingis-Tau
Mountain range in Semipalatinsk Oblast in E Kazakhstan. Also search Chingiz Range.
UF Chingiz Range
BT Kazakhstan
USSR

Chingiz Range
use Chingis-Tau

Chinkuashih
Village in N Taiwan.
UF Kinnashih
BT Taiwan

Chinkuashih Mine
N Taiwan. Also search Chinkuashih.
UF Kinwashih Mine
BT Taiwan

Chinle Formation
In Dockum Group. N Arizona, SW Colorado, SE Nevada, N New Mexico, S Utah.
BT Upper Triassic
Triassic
SA Arizona
Colorado
Dockum Group
Nevada
New Mexico
Utah

Chios
City and island in Aegean Sea off W coast of Turkey.
UF Khios
BT Greek Aegean Islands
Greece

Chipola Formation
In Alum Bluff Group. SE Alabama and NW Florida.
BT lower Miocene
Miocene
SA Alabama
Florida

Chirchik
City in Tashkent Oblast in E.
BT Uzbekistan
USSR

Chirchik River
Flows into the Syr Darya at Chinaz. Index Soviet republics as applicable. Also search Chirchik.
BT USSR
SA Kazakhstan
Uzbekistan

Chiroptera
Includes use on level 2 under Mammalia(1). See list F.
BA Mammalia
BT Tetrapoda
Vertebrata

Chita
City and oblast E of Lake Baikal on the Mongolian and Manchurian borders.
BT Russian Republic
USSR

Chitaldroog
use Chitradurga

Chitaldrug
use Chitradurga

Chitaldrug schist belt
Crystalline rocks in this area are considered to be Dharwars. Also search Chitaldrug.
BT Mysore
India
SA Dharwars

Chitinozoa
Includes use on level 2 under palynomorphs(1). See list F. Before 1981, also search chitinozoans. As of 1981, search chitinozoans only for biostratigraphy.
BA palynomorphs

Chitistone Pass
In the St. Elias Mountains. Index Alaska and/or Yukon Territory as applicable.
SA Alaska
Yukon Territory

Chitradurga
Town in E central Mysore. Also search Chitaldrug.
UF Chitaldroog
Chitaldrug
BT Mysore
India

Chitral
Town and river.
BT North-West Frontier Province
Pakistan

Chittenden County
On Lake Champlain in NW.
BT Vermont
United States

Chiuzbaia
Region in Maramures County in N.
BT Transylvania
Romania

Chkalov
use Orenburg

chkalovite
BA framework silicates
silicates
BT minerals

chladnite
Term introduced in 1981. Used for enstatite achondrites. Includes use on level 3 under meteorites(1). Before 1981, also search aubrite.
UF aubrite
BA achondrites
BT meteorites

NA Norton County Meteorite

chlorapatite
BA phosphates
BT minerals
SA apatite

chloride ion
Term introduced in 1978. Before 1978, search chloride.
SA chlorides
ions

chlorides
As of 1981, includes use on level 2 under minerals(1). See list L. Before 1981, included use on level 3 under minerals(1) halides(2).
BA halides
BT minerals
NA atacamite
NZ bideauxite
boracite
NA carnallite
NZ eudialyte
NA halite
NZ kainite
mimetite
NA paratacamite
NZ phosgenite
pyromorphite
NA rinneite
sylvite
NZ vanadinite
SA chloride ion
meionite
sodium chloride

chlorine
Includes use on level 1 and 2 as a chemical element (list D). Also search Cl.
UF Cl
SA elements

chlorite
BA chlorite group
sheet silicates
silicates
BT minerals
SA chloritization

chlorite group
Includes use in combination with sheet silicates (i.e. sheet silicates, chlorite group) on level 2 under minerals(1). See list L.
BA sheet silicates
silicates
BT minerals
NA chamosite
chlorite
clinochlore
cookeite
kammererite
sudoite

chlorite schist
Includes use on level 3 under metamorphic rocks(1) schists(2). See list J.
BA schists
BT metamorphic rocks

chloritization
Includes use on level 3 under metasomatism(1).
BA metasomatism
SA chlorite
processes

chloritoid
BA orthosilicates
silicates
BT minerals
SA aluminosilicates

chloropal
use nontronite

chlorophaeite
BA silicates
BT minerals

Chlorophyceae
Term introduced in 1981. Includes use on level 2 under algae(1). See list F.
BA Chlorophyta
algae
BT thallophytes
Plantae
NA Codiaceae
Dasycladaceae
Receptaculitaceae

chlorophyll
BA organic materials

Chlorophyta
Includes use on level 2 under algae(1). See list F.
UF green algae
BA algae
BT thallophytes
Plantae
NA Botryococcus
Charophyta
Chlorophyceae
Desmidiales
Tasmanites

Choctawhatchee Bay
Inlet of Gulf of Mexico in the Florida panhandle in NW.
BT Florida
United States

Cholame
Town in San Luis Obispo County on the Pacific in S.
BT California
United States

Chondrichthyes
Includes use on level 2 under Pisces(1). See list F.
BA Pisces
BT Vertebrata
NA Elasmobranchii
Eubradyodonti
Holocephali

chondrites
Includes use on level 3 under meteorites(1). Search chondrites AND meteorites to distinguish from the ichnofossil.
BT meteorites
NA carbonaceous chondrites
enstatite chondrites
H chondrites
HL chondrites
L chondrites
LL chondrites
Sharps Meteorite
SA achondrites
Chondrites ichnofossils
chondrules
Coldwater Meteorites
iron meteorites
Odessa Meteorite
stony irons

Chondrites ichnofossils
Term introduced in 1981 to distinguish the trace fossil from the meteorite. Includes use on level 3 under ichnofossils(1). See list F. Before 1981, search Chondrites AND NOT meteorites.
SA chondrites
ichnofossils

Chondrostei
Term introduced in 1981. Includes use on level 2 under Pisces(1). See list F.
BA Actinopterygii
Osteichthyes
Pisces
BT Vertebrata

chondrules
Includes use on level 3 under meteorites(1).
SA chondrites
meteorites

Choptank Formation
In Chesapeake Group. E Maryland and Virginia.
BT middle Miocene
Miocene
SA Maryland
Virginia

Choptank River
Index states as applicable.
BT United States
SA Delaware
Maryland

Chor
Village in SE part of country.
BT Sind
Pakistan

Chordata
Includes use on level 1 and 2 as a fossil term (list F). Term is to be used only when more specific terms do not apply.
NT Protochordata
Vertebrata
SA Hemichordata

chorology
use biogeography

Chorzow
City in S part of country.
BT Katowice
Poland

Chouteau County
Term introduced in 1981. N central Montana.
CO N472400N482000
W1093000W1112500
BT Montana
United States

Choybalsan
Town on Kerulen River in E Mongolia. Also search Kerulen.
UF Kerulen
BT Mongolia
Asia

Christchurch
City on Pacific Ocean in E.
BT South Island
New Zealand

Christian County
Term introduced in 1981. Index states as applicable.
BX Illinois
Kentucky
Missouri
BT United States

Christiansund
use Kristiansund

Christmas Island
An external territory of Australia, about 225 miles S of W end of Java.
BT Indian Ocean

Christmas Mountains
In Big Bend region in SW.
BT Texas
United States

Christopher Formation
Ellef Ringnes Island in Franklin District.
BT Lower Cretaceous
Cretaceous
SA Canada
Northwest Territories

chromates
Includes use on level 2 under minerals(1). See list L.
BT minerals
NA crocoite
hemihedrite

chromatograms
Term introduced in 1981.
SA chromatography

chromatography
Includes use on level 3 under chemical analysis(1) methods(2).
SA adsorption
analysis
chemical analysis
chromatograms
gas chromatography
spectroscopy

C

chrome diopside
UF chrome-diopside
BA clinopyroxene
pyroxene group
chain silicates
silicates
BT minerals
SA diopside

chrome spinel
Also search picotite.
UF picotite
BA oxides
BT minerals
SA spinel

chrome-diopside
use chrome diopside

chromite
Includes use on level 3 under minerals(1) oxides(2). See list L. As of 1981, use chromite ores for chromite as a commodity. Before 1981, included use on level 1 as a commodity term.
BA oxides
BT minerals
SA chromite ores
chromitite

chromite ores
Term introduced in 1981. Includes use on level 1 and 2 as a commodity term (list C). Before 1981, search chromite AND ore deposits.
SA chromite

chromitite
As of 1981, includes use on level 3 under igneous rocks(1) ultramafics(2). See list H. Before 1981, included use on level 3 under igneous rocks(1) ultramafic family(2).
BA ultramafics
BT igneous rocks
SA chromite

chromium
Includes use on level 1 and 2 as a chemical element (list D). Also search Cr.
UF Cr
SA elements

chronology
Used as a general term.
SA geochronology

chronostratigraphy
Includes use under stratigraphy(1) or individual ages (list E).
SA stratigraphy

chrysoberyl
BA oxides
BT minerals
SA alexandrite

chrysocolla
BA sheet silicates
silicates
BT minerals

Chrysophyta
Including golden-brown algae. Includes use on level 2 under algae(1). See list F.
UF golden-brown algae
BA algae
BT thallophytes
Plantae

chrysotile
Also search chrysotile asbestos.
BA serpentine group
sheet silicates
silicates
BT minerals
SA asbestos

Chu
Town and river in Dzhambul Oblast in S.
BT Kazakhstan
USSR

Chu-Ili Mountains
Branch of the Tien Shan in Dzhambul Oblast in S.
BT Kazakhstan
USSR
SA Tien Shan

Chuar Group
Grand Canyon Series. Grand Canyon region in N Arizona.
BT Precambrian
SA Arizona

Chubut
Province in S.
CO S460000S420000
W0640000W0730000
BT Argentina
SA Patagonia

Chuckanut Formation
Overlies schists and igneous rocks. San Juan Islands in NW Washington.
BT Eocene
SA Washington

Chudleigh
Town in Devon in SW.
BT England
Great Britain
United Kingdom

Chugach Mountains
Along S coast.
BT Alaska
United States
SA Coast Ranges

Chugoku
Mountains in SW Honshu.
UF Chigoku Sammayuku Mountains
BT Honshu
Japan

Chukchi Peninsula
NE extremity of Siberia in Chukchi National Okrug of Magadan Oblast. Also search Chukchi; Chukotka. As of 1981, term includes Wrangel Island.
CO N640000N720000
W1700000E1530000
UF Chukot Peninsula
Chukotka (Peninsula)
Chukotsk Peninsula
Chukotski Peninsula
Chukotskiy Peninsula
BA Northeastern USSR
BT Russian Republic
USSR
SA Okhotsk-Chukchi

Chukchi Range
use Anadyr Range

Chukchi Sea
N of Bering strait. Between Alaska and Siberia. Also search Chukchi.
CO N660000N713000
W1560000E1770000
BT Arctic Ocean
SA Bering Strait
Beringia

Chukot Peninsula
use Chukchi Peninsula

Chukotka (Peninsula)
use Chukchi Peninsula

Chukotsk Peninsula
use Chukchi Peninsula

Chukotski Peninsula
use Chukchi Peninsula

Chukotskiy Peninsula
use Chukchi Peninsula

Chuma River
use Uda River

Chuquicamata
Mining settlement in N part of country.
BT Antofagasta
Chile

Church Stretton
Urban district in S central Shropshire in W.
BT England
Great Britain
United Kingdom

Churchill
Town on Hudson Bay.
BT Manitoba
Canada

Churchill County
In W Nevada.
BT Nevada
United States

Churchill Falls
On Churchill River in SW and S Labrador. Also search Grand Falls.
UF Grand Falls
BT Labrador
Canada

Churchill Province
Index Labrador and provinces as applicable.
CO N533000N790000
W0640000W1140000
BT Canadian Shield
SA Alberta
Athabasca Formation
Labrador
Manitoba
Northwest Territories
Ontario
Quebec
Saskatchewan

Chuya
Village in NE Irkutsk Oblast near the SW Yakut ASSR border.
BT Russian Republic
USSR

Chuya Alps
Range in Altai Mountains near the Mongolian and Sinkiang-Uighur borders.
BT Russian Republic
USSR
SA Altai Mountains

Chuya Basin
River basin in Altai Kray near the Mongolian and Sinkiang-Uighur borders.
BT Russian Republic
USSR

Chvaletice
Mine site in E.
BT Bohemia
Czechoslovakia

Cibicides
Includes use on level 3 under foraminifera(1) Orbitoidacea(2).
BA Orbitoidacea
Rotaliina
foraminifera
BT Protista
Invertebrata

Cicatricosisporites
BA miospores
palynomorphs

Cieszyn
City in S part of country.
BT Katowice
Poland

Cilento
Area in SW part of country.
BT Campania
Italy

Cima d'Asta
Peak in the Dolomites in E.
BT Trentino-Alto Adige
Italy
SA Dolomites

Ciminna Basin
Term introduced in 1981. Structural basin in NW Sicily.
CO N375200N375600
E0133700E0132700
BT Sicily
Italy

Cimmaron County
In extreme W of panhandle.
BT Oklahoma
United States

Cimmerian
Black Sea area. Above Meotian, below Kuyalnikian. Includes use on level 3 under age terms(1). See list E.
UF Kimmerian
BA Pliocene
Neogene
Tertiary
BT Cenozoic

Cincinnati
City on the Ohio River in Hamilton County in SW.
BT Ohio
United States

Cincinnati Arch
Index states as applicable.
BT United States
SA Indiana
Kentucky
Ohio

Cincinnatian
Provincial series, North America. Above Champlainian, below Alexandrian (Silurian). Includes use on level 3 under age terms(1). See list E.
BA Upper Ordovician
Ordovician
BT Paleozoic

cinder cones
UF cones, cinder
BT volcanic features
SA volcanism
volcanoes

cinematics
use kinematics

cinerite
Includes use on level 3 under sedimentary rocks(1) clastic rocks(2). See list I.
BA clastic rocks
BT sedimentary rocks
SA terrigenous materials

cinnabar
BA sulfides
BT minerals
SA mercury

circulation
A valid level 1 term through 1977 used in combination with oceans (i.e. oceans, circulation). After 1977, includes use under meteorology(1).
SA atmosphere
climate-induced circulation
meteorology
ocean circulation
ocean waves
paleocirculation
permeability
springs
thermal circulation
thermohaline circulation
winds

Circum-Antarctic region
SA Antarctica

Circum-Pacific Belt
use Circum-Pacific region

Circum-Pacific region
Term introduced in 1978. The great-circle belt that borders the Pacific Ocean along the continental margin of Asia and the Americas. Before 1978, also search Circum-Pacific Belt; Circum-Pacific.
UF Circum-Pacific Belt
BT Pacific Ocean
SA Pacific region

cirques
Includes use on level 3 under glacial geology(1) glacial features(2).
BT glacial features
SA erosion
eskers
glacial geology
glacial lakes
kames

Cirripedia
Includes use on level 2 under Arthropoda(1). See list F.
BA Crustacea
Mandibulata
Arthropoda
BT Invertebrata

Ciscaucasia
use Northern Caucasus

cistern rock
use laccoliths

Citronelle Formation
Recognized to be equivalent of Willis Formation of Texas and Louisiana. Graham Ferry Formation of Pliocene and Pleistocene age disconformably underlies Citronelle Formation. Gulf Coastal Plain from W Florida and S Georgia to E Texas inclusive.
BT Pliocene
SA Alabama
Florida
Georgia
Louisiana
Mississippi
Texas

Ciudad de Valles
use Valles

Ciudad Real
Province in S central.
CO N382100N393700
W0024000W0050500
BT New Castile
Spain
NT Alcudia Valley
Almaden
SA Montes de Toledo

Ciudad Trujillo
use Santo Domingo

civil engineering
UF engineering, civil
SA engineering geology

Cl
use chlorine

Cladocopina
Term introduced in 1981. Includes use on level 2 under Ostracoda(1). See list F.
BA Ostracoda
BT Crustacea
Mandibulata
Arthropoda
Invertebrata

Cladophlebis
BA miospores
palynomorphs

Claiborne Group
Includes Carrizo Sand, Reklaw Formation, Queen City Sand, Weches Glauconitic Marl, Viesca Glauconitic Marl, Sparta Sand, Stone City Beds, Crockett Formation, Landrum Shale, Spiller Sand, Yegua Formation, Gosport Sand, Tallahatta Formation, Zilpha Clay, Winona Sand, Wautubbee Formation, Cockfield Formation, Avon Park Limestone, Tallahassee Limestone, Lake City Limestone, Lisbon Formation, Neshoba Sand, Cook Mountain Formation, Meridian Formation, Kosciusko Formation, Cane River Formation. Gulf Coastal Plain from Georgia to S Texas.
BT middle Eocene
Eocene
SA Alabama
Cook Mountain Formation
Florida
Georgia
Louisiana
Mississippi
Queen City Formation
Texas

Claiborne Parish
In NW Louisiana.
BT Louisiana
United States

Clallam County
Term introduced in 1981. NW Washington.
CO N475100N482500
W1225500W1244500
BT Washington
United States

Clare
County on the Atlantic in W.
BT Ireland

Clark County
Index states as applicable.
BX Arkansas
Idaho
Illinois
Indiana
Kansas
Kentucky
Missouri
Nevada
Ohio
South Dakota
Washington
Wisconsin
BT United States

Clarke County
Index states as applicable.
BX Alabama
Georgia
Iowa
Mississippi
Virginia
BT United States
NT Las Vegas Valley

Clarno
Village in Wheeler County in N central.
BT Oregon
United States
SA Clarno Formation

Clarno Formation
Subdivided into four units. N central Oregon.
BT Eocene
SA Clarno
Oregon

classification
Includes use as level 2 or 3 term appropriate to a large number of topics, e.g. on level 2 under geochemistry(1) and under lava(1). See list G.
SA definition
identification
nomenclature
terrain classification

Classopollis
BA miospores
palynomorphs

clastic dikes
Includes use on level 3 under sedimentary structures(1) soft sediment deformation(2). See list K.
BA soft sediment deformation
sedimentary structures
SA dikes
sandstone dikes

clastic rocks
Valid index term for GeoRef since 1976. Before 1976, search clastics AND terrigenous or clastics AND nonterrigenous in combination with sedimentary rocks. From 1976 through 1977, search clastic rocks AND terrigenous or nonterrigenous. After 1977, use clastic rocks on level 2 under sedimentary rocks(1). Also search clastic.
BT sedimentary rocks
NA arenite
argillite
arkose
bentonite
black shale
breccia
cinerite
claystone
conglomerate
contourite
diamictite
diatomaceous earth
eolianite
fanglomerate
flysch
gaize
graywacke
marl
microbreccia
mixtite
molasse
mudstone
novaculite
orthoquartzite
porcellanite
radiolarite
red beds
sandstone
saprolite
shale
siltstone
sparagmite
spongolite
subarkose
subgraywacke
tillite
tilloid
tonstein
turbidite
SA bauxite
clastic sediments
fragments
greensand
pyroclastics
siliceous composition
terrigenous materials
wildflysch

clastic sediments
Valid index term for GeoRef since 1976. Before 1976 search clastics AND terrigenous or clastics AND nonterrigenous. Before 1978 included use in combination with terrigenous or nonterrigenous (i.e. clastic sediments, terrigenous) on level 2 under sediments(1). After 1978, use clastic sediments on level 2. Also search clastic.
BT sediments
NA alluvium
boulder clay
boulders
clay
cobbles
colluvium
diamicton
drift
dust
eluvium
erratics
flint clay
gravel
kaolin
loess
mud
ooze
outwash
pebbles
proluvium
quartz sand
residual clays
residuum
sand
shingle
silt
till
SA bentonite
cementation
clastic rocks
flysch
gaize
greensand
particulate materials
pyroclastics
terrigenous materials
volcanic ash
wildflysch

clastics
A level 2 term through 1975 used in combination with terrigenous or nonterrigenous to indicate sedimentary rocks or sediments (i.e. clastics, terrigenous or clastics, nonterrigenous). After 1975 through 1977, clastic rocks and clastic sediments were used on level 2 in combination with terrigenous or nonterrigenous (i.e. clastic rocks, terrigenous). After 1977, use clastic rocks or clastic sediments on level 2 under respective level 1 term.

clasts
Includes use on level 3 under sedimentary rocks(1) or sediments(1).
SA fragments
sedimentary rocks
sediments

Clatsop County
Extreme NW.
CO N454500N461500
W1232000W1240200
BT Oregon
United States

clay
Includes use on level 3 under sediments(1) clastic sediments(2). See list N. As of 1981, includes use as a material in engineering geology; before 1981, also search clays for engineering applications. When of economic value, use clays.
BA clastic sediments
BT sediments
SA alluvium
argillaceous texture
armored mud balls
bentonite
clay mineralogy
clays
claystone
fireclay
flint clay
fuller's earth
kaolin
loam
marl
mud
mud lumps
quick clay
residual clays
sensitive clays
silt
soft clays
stiff clays
terrigenous materials
underclay

clay (soil)
use Clay soils

Clay County
Index states as applicable.
BX Alabama
Arkansas
Florida
Georgia
Illinois
Iowa
Kansas
Kentucky
Minnesota
Mississippi
Missouri
Nebraska
North Carolina
South Dakota
Tennessee
Texas
West Virginia
BT United States

clay mineralogy
Includes use on level 1 (list A). Term set options are:
areal studies
area
experimental studies
chemical property
material
mineral
physical property
subtopic
mineral data
mineral name
SA adsorption
areal studies
bentonite
cation exchange capacity
ceramic materials
chemical analysis
clay
clay minerals
clays
crystal chemistry
crystal growth
crystal structure
crystallinity
crystallography
differential thermal analysis
electron microscopy
flocculation
fuller's earth
glauconite
ion exchange
kaolin
mineral data
mineralogy
minerals
paragenesis
sedimentary petrology
sedimentary rocks
sediments
sheet silicates
silicates
soils
spectroscopy
thermal analysis
transformations
vermiculite
weathering
X-ray analysis

clay minerals
Includes use in combination with sheet silicates (i.e. sheet silicates, clay minerals) on level 2 under minerals(1). See list L.
BA sheet silicates
silicates
BT minerals
NA allophane
beidellite
corrensite
dickite
halloysite
hectorite
illite
imogolite
kaolinite
montmorillonite
nacrite
nontronite
palygorskite
rectorite
saponite
sepiolite
smectite
stevensite
tosudite
vermiculite
SA argillaceous texture
bauxite
bentonite
clay mineralogy
differential thermal analysis
hisingerite
hydromica
hydromuscovite
kaolin
kaolinization
metabentonite
metakaolin
mixed-layer minerals

Clay soils
UF clay (soil)
BA soils

clay stone
use claystone

clays
Includes use as a commodity term on level 1 (list C). Before 1981, included use when of economic value or of engineering application. As of 1981, use only when of economic value; use clay for material in engineering geology. Before 1981, search clays AND deposits.
SA bentonite deposits
clay
clay mineralogy
engineering geology
flint clay
fuller's earth
kaolin
kaolin deposits
quick clay
refractory materials
residual clays
sensitive clays
shale
soft clays
soil mechanics
stiff clays
vermiculite deposits

claystone
Includes use on level 3 under sedimentary rocks(1) clastic rocks(2). See list I.
UF clay stone
BA clastic rocks
BT sedimentary rocks
SA clay
mudstone
terrigenous materials

Clayton Formation
In Midway Group. S Alabama, SW Georgia, NE Mississippi, SE Missouri, and S Tennessee.
BT Paleocene
SA Alabama
Georgia
Mississippi
Missouri
Tennessee

Clear Creek
River in N central.
BT Colorado
United States

Clear Creek County
N central.
BT Colorado
United States
NT Henderson Mine

Clear Lake
Lake County in NW central.
BT California
United States

Clearwater County
Index states as applicable.
BX Idaho
Minnesota
BT United States

cleat spar
use ankerite

cleavage
As of 1981, use only for cleavage of rocks produced by deformation or metamorphism. Use mineral cleavage for minerals. Before 1978, term also referred to cleavage folds.
SA axial-plane structures
cleavage folds
flow cleavage
folds
foliation
fracture cleavage
lineation
mineral cleavage
schistosity
slaty cleavage
slip cleavage
structural analysis

cleavage folds
Term introduced in 1978. Includes use on level 3 under folds(1). Before 1978, also search cleavage AND folds.
UF shear-cleavage folds
BA folds
SA cleavage

Clermont
City in N part of country.
BT Oise
France

Clermont-Ferrand
City in S central part of country.
BT Puy-de-Dome
France

Cleveland
City on Lake Erie in Cuyahoga County in N.
BT Ohio
United States

cliff of displacement
use fault scarps

cliffs
Includes use on level 3 under geomorphology(1) erosion features(2), fluvial features(2), and shore features(2).
SA erosion features
fault scarps
fluvial features
geomorphology
gullies
scarps
shore features
slopes
talus slopes

climate
Includes use on level 3 under soils(1) genesis(2) as type of factor.
SA alpine environment
arctic environment
arid environment
atmosphere
atmospheric precipitation
climate effects
factors
humid environment
humidity
length of day
meteorology
obliquity of the ecliptic
paleoclimatology
semi-arid environment
soils
storms
subtropical environment
taiga environment
temperate environment
temperature
tropical environment
winds

climate effects
Term introduced in 1981. Not to be used for climates of the past (paleoclimatology).
SA climate
effects

climate-induced circulation
Term introduced in 1978. Includes use on level 2 under ocean circulation(1). Before 1978, also search oceans AND circulation AND climate induced or climate-induced.
SA circulation
ocean circulation

climatology, paleo-
use paleoclimatology

Climax
Village in Lake County in central.
BT Colorado
United States

clinoamphibole
As of 1981, includes use in combination with chain silicates (i.e. chain silicates, clinoamphibole) on level 2 under minerals(1). See list L. Before 1981, included use on level 3 under minerals(1) chain silicates, amphibole group(2).
BA amphibole group
chain silicates
silicates
BT minerals
NA actinolite
arfvedsonite
crossite
cummingtonite
ferrohastingsite
glaucophane
grunerite
hastingsite
hornblende
kaersutite
magnesioriebeckite
nephrite
pargasite
richterite
riebeckite
tirodite
tremolite
tschermakite
SA amosite

clinochlore
BA chlorite group
sheet silicates
silicates
BT minerals

clinoenstatite
BA clinopyroxene
pyroxene group
chain silicates
silicates
BT minerals

clinohumite
BZ fluorides
humite group
BT minerals
SA titanoclinohumite

clinohypersthene
BA clinopyroxene
pyroxene group
chain silicates
silicates
BT minerals

clinoptilolite
BA zeolite group
framework silicates
silicates
BT minerals
SA heulandite

clinopyroxene
As of 1981, includes use in combination with chain silicates (i.e. chain silicates, clinopyroxene) on level 2 under minerals(1). See list L. Before 1981, included use on level 3 under minerals(1) chain silicates, pyroxene group(2).
BA pyroxene group
chain silicates
silicates
BT minerals
NA acmite
aegirine
augite
chrome diopside
clinoenstatite
clinohypersthene
diopside
fassaite
hedenbergite
jadeite
johannsenite
kunzite
omphacite
pigeonite
spodumene
titanaugite

clinopyroxenite
As of 1981, includes use on level 3 under igneous rocks(1) ultramafics(2). See list H. Before 1981, included use on level 3 under igneous rocks(1) ultramafic family(2).
BA ultramafics
BT igneous rocks
SA pyroxenite

clinozoisite
BA epidote group
sorosilicates
orthosilicates
silicates
BT minerals
SA aluminosilicates
zoisite

Clinton County
Term introduced in 1981. Index states as applicable.
BX Illinois
Indiana
Iowa
Kentucky
Michigan
Missouri
New York
Ohio
Pennsylvania
BT United States

Clinton Group
Includes Willowvale Shale, Rose Hill Formation, Keefer Sandstone, Rochester Shale, Osgood Formation, Laurel Formation, Waldron For-

mation, Reynales Formation, Irondequoit Formation, Decew Formation, Neahga Formation, Thorold Formation, Maplewood Formation, Furnaceville Formation, Sodus Formation, Williamson Formation. New York to NE Tennessee; also Michigan.
BT Middle Silurian
Silurian
SA Maryland
Michigan
New York
Pennsylvania
Virginia
West Virginia

clintonite
Also search xanthophyllite.
UF xanthophyllite
BA mica group
sheet silicates
silicates
BT minerals

Clitheroe
City E of Liverpool in E Lancashire.
BT England
United Kingdom

Cloridorme Formation
Gaspe Peninsula.
BT Middle Ordovician
Ordovician
SA Canada
Quebec

closed fractures
Term introduced in 1978. Includes use on level 3 under fractures(1). Before 1978, also search fractures AND closed.
BA fractures

closed systems
Used as a general term.
SA open systems
systems

Cloud County
N central.
BT Kansas
United States

clouds
Includes use on level 3 under meteorology(1) water(2).
SA droplets
meteorology
raindrops
water

Clovis
City in Curry County on the Texas border in E.
BT New Mexico
United States

Cluj
City in Cluj County in central.
BT Transylvania
Romania

cluster analysis
BA statistical analysis
SA analysis
mathematical geology

Clymeniida
Term introduced in 1981. Includes use on level 2 under Mollusca(1). See list F.
BA Ammonoidea
Tetrabranchiata
Cephalopoda
Mollusca
BT Invertebrata

Cm
use curium

Co
use cobalt

CO
use carbon monoxide

Coahuila
State in NE.
CO N243000N295000
W1000000W1040000
BT Mexico
NT Parras Basin
SA Chihuahua tectonic belt
Difunta Group
Rio Grande
Rio Grande Rift
Rio Grande Valley
Sierra Madre Oriental
Zuloaga Limestone

coal
Includes use on level 1 as a commodity term (list C); on level 3 under sedimentary rocks(1) organic residues(2). Also search carbargilite.
UF carbargilite
BA organic residues
BT sedimentary rocks
NA anthracite
bituminous coal
coke coal
NT humic coal
NA lignite
macerals
sapropelite
steam coal
SA anthraxolite
biogenic structures
bituminous coal
carbonaceous composition
charcoal
chemical properties
coal balls
coal fields
coal seams
coalification
energy sources
gasification
lignite
lithification
organic materials
peat
rank
reflectance
strip mining
volatiles

coal balls
Includes use on level 3 under coal(1) as a commodity term.
UF balls, coal
SA coal
coal seams
concretions

coal basins
As of 1981, no longer a valid term for GeoRef. See coal. Before 1981, included use on level 3.

coal beds
use coal seams

coal clay
use underclay

Coal County
SE central.
BT Oklahoma
United States

coal fields
Includes use on level 3 under coal(1) as a commodity term. Also search coalfields.
UF coalfields
fields, coal
SA coal

coal measures
As of 1981, no longer a valid term for GeoRef. See coal. Before 1981, included use on level 3.

Coal Measures
Europe. Includes use on level 3 under age terms(1). See list E.
BA Upper Carboniferous
Carboniferous
BT Paleozoic

coal seams
Includes use on level 3 under coal(1) as a commodity term. Also search coal beds; coal deposits; coal strata.
UF coal beds
seams, coal
SA coal
coal balls
underclay

coalfields
use coal fields

coalification
Includes use on level 3 under coal(1) as a commodity term.
UF carbonification
incarbonization
incoalation
SA coal
diagenesis
processes

Coalinga
City in Fresno County in S central.
BT California
United States

Coalsack Bluff
Site of paleontological investigations in Beardmore Glacier area around NW end of Queen Maud Range off the Ross Ice Shelf in Ross Dependency.
BT Antarctica

coast
A valid term through 1978. After 1978, use coastal environment.

Coast Mountains
Extend from Yukon Territory to the Fraser River serving as border between NW British Columbia and the Alaskan panhandle. Index Alaska, British Columbia, and Yukon Territory as applicable.
BT North America
SA Alaska
British Columbia
Canadian Cordillera
Coast Ranges
North American Cordillera
Yukon Territory

Coast Ranges
Mountains extending along the Pacific Ocean from California to Alaska. Index states and British Columbia as applicable.
BT North America
NX Salinian Block
SA Alaska
British Columbia
California
Chugach Mountains
Coast Mountains
Diablo Range
Gabilan Range
Kenai Peninsula
Klamath Mountains
Kodiak Island
North American Cordillera
Olympic Mountains
Oregon
Peninsular Ranges
Queen Charlotte Islands
Saint Elias Mountains
San Bernardino Mountains
San Gabriel Mountains
Santa Ana Mountains
Santa Cruz Mountains
Santa Lucia Range
Santa Monica Mountains
Santa Ynez Mountains
Vancouver Island
Washington

coastal
A valid term through 1977. After 1977, use coastal environment.

coastal dunes
Term introduced in 1981. Includes use on level 3 under geomorphology(1) eolian features(2) and shore features(2).
BA dunes

coastal effects
Term introduced in 1981 for use in relation to the Earth's magnetic field.
SA effects
magnetic field

coastal environment
Term introduced in 1978. Includes use on level 3 under sedimentation(1). Before 1978, also search coastal; search coast.
SA beaches
coastal plains
ecology
environment
inlets
lagoons
playas
sedimentation
shorelines
spits
tidal flats

coastal plains
Includes use on level 3 under geomorphology(1).
SA coastal environment
geomorphology
plains

coastlines
use shorelines

cobalt
Includes use on level 1 and 2 as a chemical element (list D). As of 1981, use cobalt ores for cobalt as a commodity. Before 1981, included use on level 1 as a commodity term.
UF Co
SA cobalt ores
cobaltite
elements
heavy metals

Cobalt
Mining town in SE Ontario.
BT Ontario
Canada

cobalt ores
Term introduced in 1981. Includes use on level 1 and 2 as a commodity term (list C). Before 1981, search cobalt AND ore deposits.
SA cobalt

cobalt pyrites
use linnaeite

cobaltite
As of 1981, includes use on level 3 under minerals(1) arsenides(2). See list L. Before 1981, included use on level 3 under minerals(1) sulfides(2).
BA arsenides
sulfides
BT minerals
SA cobalt

Cobb Seamount
Aproximately 550 miles W of Washington.
BT Pacific Ocean

cobbles
Includes use on level 3 under sediments(1) clastic sediments(2). See list N.
BA clastic sediments
BT sediments
SA boulders
gravel
pebbles

shingle
terrigenous materials

Cobequid Bay
Eastern arm of Minas Basin off Bay of Fundy.
BT Nova Scotia
Canada
SA Minas Basin

Coburg
City in N.
BT Bavaria
West Germany

Coccolithophoraceae
Includes use on level 2 under algae(1). See list F. Before 1981, also search coccoliths. As of 1981, search coccoliths only for biostratigraphy.
BA algae
BT thallophytes
Plantae
NA Ceratolithus
Coccolithus
SA coccoliths
discoasters
nannofossils

coccoliths
As of 1981, used as the common name for Coccolithophoraceae. See list F. To be used for Coccolithophoraceae when they are used as tools in stratigraphy. Before 1981, search (coccoliths OR Coccolithophoraceae) AND biostratigraphy.
SA biostratigraphy
Coccolithophoraceae

Coccolithus
BA Coccolithophoraceae
algae
BT thallophytes
Plantae

Cochabamba
Department in central Bolivia. Also a city.
BT Bolivia

Cochise County
Extreme SE.
BT Arizona
United States

Cochrane
Town and district S of James Bay.
BT Ontario
Canada

Coconino County
N central.
CO N341500N370000
W1104500W1132000
BT Arizona
United States
SA Canyon Diablo Meteorites

Coconino Sandstone
In Aubrey Group. N Arizona, S Utah, and SE Nevada.
BT Permian
SA Arizona
Nevada
Utah

Cocos Plate
In the Pacific Ocean W of the Caribbean Plate, E and S of the Pacific Plate, and N of the Nazca Plate which is S of the Galapagos Islands.
BT Pacific Ocean
SA Galapagos Rift
plate tectonics
plates

Cocos Ridge
Between Costa Rica and the Galapagos Islands.
CO N000000N090000
W0830000W0923000
BA South American Pacific
BT Pacific Ocean

coda waves
Includes use on level 3 under seismology(1).
BA surface waves
elastic waves
SA seismograms
seismology
waves

Codiaceae
As of 1981, includes use on level 2 under algae(1). See list F. Before 1981, included use on level 3 under algae(1) Chlorophyta(2).
BA Chlorophyceae
Chlorophyta
algae
BT thallophytes
Plantae
NA Halimeda

Cody Shale
In Colorado Group or Montana Group. Bighorn Basin in N Wyoming.
BT Upper Cretaceous
Cretaceous
SA Colorado Group
Montana Group
Wyoming

coefficient, correlation
use correlation coefficient

coefficient of permeability
use hydraulic conductivity

coefficient of transmissibility
use transmissivity

coefficients, partition
use partition coefficients

Coelenterata
Includes use on level 1 and 2 as a fossil term (list F).
BT Invertebrata
NA Anthozoa
Hydrozoa
Scyphozoa
Stromatoporoidea
SA ahermatypic taxa
atolls
bioherms
biostromes
corals
hermatypic taxa
Receptaculitaceae
reef builders
reefs

Coelodonta antiquitatis
Includes use on level 3 under Mammalia(1) Perissodactyla(2).
BA Perissodactyla
Mammalia
BT Tetrapoda
Vertebrata

coercive force
use coercivity

coercivity
Before 1978, also search coercive force.
UF coercive force
SA magnetic field
paleomagnetism
remanent magnetization

coesite
BA silica minerals
framework silicates
silicates
BT minerals

Coeur d'Alene
City in Kootenai County in N.
BT Idaho
United States

Coeur d'Alene River
Flows into Coeur d'Alene Lake in Kootenai County in N.
BT Idaho
United States

coexisting
A valid term through 1977. After 1977, see coexisting minerals; coexisting materials.

coexisting materials
As of 1981, no longer a valid term for GeoRef. Included use on level 3. Term was introduced in 1978.

coexisting minerals
Term introduced in 1978. Before 1978, also search coexisting AND minerals.
BT minerals

Coeymans Formation
Includes Stormville Sandstone Member, Elbow Ridge Sandstone Member. W Maryland, E New York, E Pennsylvania, W Virginia, N West Virginia, and New Jersey.
UF Coeymans Limestone
BT Lower Devonian
Devonian
SA Maryland
New York
Pennsylvania
Virginia
West Virginia

Coeymans Limestone
use Coeymans Formation

Coffee County
Index states as applicable.
BX Alabama
Georgia
Tennessee
BT United States

coffinite
BA orthosilicates
silicates
BT minerals

cohenite
As of 1981, includes use on level 3 under minerals(1) carbides(2). See list L. Before 1981, included use on level 3 under minerals(1) native elements and alloys(2).
BA carbides
alloys
BT minerals
SA meteorites

cohesion
use shear strength

cohesionless materials
Term introduced in 1981.
SA engineering geology
materials

cohesive materials
Introduced in 1978 as general term mostly used in engineering geology. Before 1978, also search cohesive.
SA engineering geology
materials
rock mechanics
soil mechanics

coiling
SA fossils
morphology

Coimbatore
City in E.
BT Tamil Nadu
India

Coimbra
District in W Portugal. Also a city.
BT Portugal

Cojedes
State in NW central.
BT Venezuela
NT Tinaquillo

coke coal
Term introduced in 1981. Includes use on level 3 under sedimentary rocks(1) organic residues(2). See list I. Also search coke.
UF natural coke
BA coal
organic residues
BT sedimentary rocks

Colchis
Ancient country on the Black Sea S of the Caucasus.
BT Georgian Republic
USSR

Cold Lake
Term introduced in 1981. On Alberta-Saskatchewan border.
CO N542500N544000
W1095000W1101500
BT Canada
SA Alberta
Saskatchewan

Coldwater Meteorites
Term introduced in 1981. Refers to two meteorites found in Comanche County, S Kansas. One is an iron meteorite (octahedrite); the other is a chondrite (H chondrite). Before 1981, also search Coldwater AND meteorites.
BT meteorites
SA chondrites
Comanche County
H chondrites
iron meteorites
Kansas
octahedrite

colemanite
BA borates
BT minerals

Coleoidea
Subclass. Includes use on level 3 under Mollusca(1) Cephalopoda(2).
BA Cephalopoda
Mollusca
BT Invertebrata

Coleoptera
As of 1981, includes use on level 3 under Insecta(1) Coleopteroida(2). Before 1981, included use on level 2 under Insecta(1). See list F.
BA Coleopteroida
Insecta
BT Mandibulata
Arthropoda
Invertebrata

Coleopteroida
Term introduced in 1981. Includes use on level 2 under Insecta(1). See list F.
BA Insecta
BT Mandibulata
Arthropoda
Invertebrata
NA Coleoptera

Coles Bay
BT Tasmania
Australia

Colfax County
Index states as applicable.
BX Nebraska
New Mexico
BT United States

collagen
BA organic materials
SA proteins

collapse structures
SA land subsidence
slump structures
structures

collecting
Includes use on level 3 under minerals(1).
SA collections
fossil localities
mineral localities

mineralogy
popular geology

collections
Usage restricted to collections in museums, etc. Includes use on level 2 under meteorites(1); on level 3 under fossil names (list F), mineral names (list L), and paleontology(1). Also use collecting on level 3 under minerals(1).
SA associations
catalogs
collecting
geology
museums
paleontology

college level
A valid term through 1977. After 1977, use college-level education.

college-level education
Term introduced in 1978 on level 3 under education(1). Before 1978, search college level.
SA education
elementary school
high school
junior high school

Collenia
Genus.
BA stromatolites
algae
BT thallophytes
Plantae
SA Cyanophyta
ichnofossils

collinite
Term introduced in 1981.
BA vitrinite
macerals
coal
organic residues
BT sedimentary rocks

collision
No longer a valid term for GeoRef as of 1981. Use plate collision if applicable.

colloidal materials
Term introduced in 1978. Before 1978, search colloidal; colloids.
UF colloids
NT gels
SA flocculation
materials

colloids
use colloidal materials

colloquia
use symposia

colluvium
Includes use on level 3 under sediments(1) clastic sediments(2). See list N.
BA clastic sediments
BT sediments
SA alluvium
erosion features
geomorphology
soils
talus slopes
terrigenous materials

Cologne
City on the left bank of Rhine River in W.
BT North Rhine-Westphalia
West Germany

Colombia
Includes use on level 1 as an area term (list O). For term set options see list B.
CO S040000N121500
W0670000W0790000
BA South America
NT Antioquia
Bogota
Magdalena River
Magdalena Valley
Sabana de Bogota
Santa Marta
Santander Department
Sierra Nevada de Santa Marta
NX Sucre
SA Amazon Basin
Amazon River
Andes
Caribbean region
Eastern Cordillera
Guajira Peninsula
Llanos
Magdalena
Orinoco River
Rio Negro
Western Cordillera

Colombia Basin
use Colombian Basin

Colombian Basin
N of Colombia in S Caribbean. Also search Colombia Basin.
UF Colombia Basin
Colombian Plain
BT Caribbean Sea

Colombian Plain
use Colombian Basin

Colon Archipelago
use Galapagos Islands

colonial taxa
Term introduced in 1978. Before 1978, also search colonial; colonies.
UF colonies
taxa, colonial

colonies
use colonial taxa

color
Includes use on level 3 under soils(1) morphology(2).
SA chemical analysis
colorimetry
soils

color centers
UF centers, color
SA crystal structure
crystallography
minerals
optical properties

color imagery
Term introduced in 1981.
BT imagery

Colorado
Includes use on level 1 as an area term (list O). For term set options see list B.
CO N370000N410000
W1020000W1090000
BA Western U.S.
United States
NX Adams County
NT Arapahoe County
Boulder
Boulder County
Cache La Poudre River
Canon City
Chaffee County
NX Cheyenne County
NT Clear Creek
Clear Creek County
Climax
Colorado Springs
Cortez
Crested Butte
Cripple Creek
NX Custer County
NT Denver
Dillon
NX Douglas County
NT Eagle County
Eaton
NX El Paso County
NT Elk Mountains
NX Fremont County
NT Front Range
Front Range urban corridor
NX Garfield County
NT Gilpin County
Golden
Gore Range
NX Grand County
NT Grand Junction
Greeley
Gunnison County
NX Jackson County
Jefferson County
Kiowa County
Lake County
NT Larimer County
Las Animas County
Leadville
NX Lincoln County
Logan County
NT Loveland
Mesa County
NX Mineral County
NT Montezuma County
NX Morgan County
NT Mosquito Range
NX Otero County
NT Ouray County
NX Park County
Phillips County
NT Piceance Creek basin
Pikes Peak
Pikes Peak Batholith
Pitkin County
Pueblo County
Rangely
Rangely Anticline
Rangely Field
Rifle
Rio Blanco County
Saguache County
Salida
NX San Juan County
NT San Juan District
San Juan Mountains
San Juan volcanic field
NX Sand Wash Basin
NT Sawatch Range
NX Sedgwick County
NT Silverton Caldera
Spanish Peaks
NX Summit County
NT Vail Pass
NX Washington County
NT Weld County
Wet Mountains
White River Plateau
NX Yuma County
SA Arikaree Group
Arkansas River
Arkansas River valley
Bald Mountain
Benton Formation
Brule Formation
Canadian River
Carlile Shale
Cedar Mountain Formation
Chadron Formation
Cheyenne Sandstone
Chinle Formation
Colorado Group
Colorado Plateau
Colorado River
Colorado River basin
Cutler Formation
Dakota Formation
Denver Basin
Dockum Group
Duchesne River Formation
Fort Hays Limestone Member
Fort Union Formation
Four Corners
Fox Hills Formation
Frontier Formation
Graneros Shale
Great Plains
Green River
Green River basin
Green River Formation
Greenhorn Limestone
Kayenta Formation
Kiowa Formation
Lance Formation
Laney Shale Member
Laramie Formation
Laramie Mountains
Leadville Formation
Lewis Shale
Loup Fork Group
Madison Group
Mancos Shale
Manitou Formation
Medicine Bow Mountains
Menefee Formation
Mesaverde Group
Midcontinent
Minturn Formation
Missouri River basin
Moenkopi Formation
Montana Group
Morrison Formation
Navajo Sandstone
Niobrara Formation
Ogallala Formation
Paradox Basin
Paradox Member
Pierre Shale
Pinedale Glaciation
Raton Basin
Rio Grande
Rio Grande Valley
San Jose Formation
San Juan Basin
San Juan Formation
San Juan River
San Luis Valley
Sangre de Cristo Mountains
Smoky Hill Chalk Member
South Platte River valley
Sundance Formation
Sunlight
Todilto Formation
Twilight Gneiss
Twin Lakes
U. S. Rocky Mountains
Uinta Basin
Wasatch Formation
Washakie Basin
Weber Sandstone
White Limestone
White River Group
Yule Marble

Colorado Group
Includes Thermopolis Shale, Mowry Shale, Warm Creek Shale, Graneros Shale, Greenhorn Limestone, Carlile Shale, Niobrara Formation, Belle Fourche Shale, Marias River Formation, Skull Creek Shale, Fall River Sandstone, Newcastle Sandstone, and Telegraph Shale.
BT Cretaceous
SA Benton Formation
Carlile Shale
Cody Shale
Colorado
Frontier Formation
Graneros Shale
Greenhorn Limestone
Idaho
Iowa
Kansas
Lower Cretaceous
Montana
Mowry Shale
Nebraska
New Mexico
Niobrara Formation
North Dakota
South Dakota
Upper Cretaceous
Wyoming

Colorado Plateau
Index states as applicable. Introduced as level 1 area term in 1978.
CO N334000N403000
W1062000W1140000
BA United States
North America
SA Arizona
Colorado
New Mexico
Utah

Colorado River
Rises in N Colorado and flows SW, then W through the Grand Canyon, and finally S into the Gulf of California. Index Mexican and U. S. states as applicable.
SA Arizona
Baja California
California
Colorado
Nevada
Sonora
Utah

Colorado River basin
Term introduced in 1981. Drainage basin.
CO N310000N440000
W1053000W1163000
BT North America
SA Arizona
California
Colorado
Mexico
Nevada
New Mexico
Utah
Wyoming

Colorado River delta
At N tip of Gulf of California. Index states as applicable.
BT Mexico
SA Baja California
Sonora

Colorado Springs
Term introduced in 1981. City in central Colorado at junction of Fountain and Monument creeks, near base of Pikes Peak.
CO N385000N385000
W1044900W1044900
BT Colorado
United States

coloradoite
As of 1981, includes use on level 3 under minerals(1) tellurides(2). See list L. Before 1981, included use on level 3 under minerals(1) sulfides(2).
BA tellurides
sulfides
BT minerals

colorimetry
Includes use on level 3 under chemical element(1) analysis(2) and under chemical analysis(1) methods(2). See list D.
SA analysis
chemical analysis
color
spectroscopy

columbates
use niobates

Columbia
Name of cities in Missouri and South Carolina. Also cities in Illinois, Kentucky, Louisiana, and Tennessee. Town in Connecticut. Index states as applicable.
BX Connecticut
Illinois
Kentucky
Louisiana
South Carolina
Tennessee
BT United States

Columbia Basin
use Columbia River basin

Columbia County
Index states as applicable.
BX Arkansas
Florida
Georgia
New York
Oregon
Pennsylvania
Washington
Wisconsin
BT United States

Columbia Plateau
A level 1 area term as of 1978. Lava basin. Index states as applicable. Also search Columbia River plateau.
CO N440000N483000
W1150000W1200000
UF Columbia River plateau
BA United States
North America
SA Idaho
Oregon
Washington

Columbia River
Rises in SE British Columbia and flows S and then W into the Pacific Ocean on the Oregon-Washington border. Also search Columbia AND appropriate area. Index British Columbia and states as applicable.
CO N453000N481500
W1185400W1241000
SA British Columbia
Oregon
Washington

Columbia River Basalt
Miocene and Pliocene (?). N Idaho, Oregon, and Washington. Also search Columbia River Group.
BT Tertiary
SA Idaho
Miocene
Oregon
Pliocene
Washington

Columbia River basin
Drainage basin. Index British Columbia, and states as applicable. Also search Columbia Basin.
UF Columbia Basin
SA British Columbia
Idaho
Montana
Nevada
Oregon
Utah
Washington
Wyoming

Columbia River estuary
Tidal mouth of the Columbia River. Index states as applicable.
BT United States
SA Oregon
Washington

Columbia River plateau
use Columbia Plateau

columbite
BZ tantalates
niobates
BA oxides
BT minerals

columbium
use niobium

Columbus
City in Franklin County in central Ohio; a city in Muscogee County in W Georgia. Index states as applicable.
BX Georgia
Ohio
BT United States

Columbus Limestone
Comprises Bellepoint, Eversole (?), Delhi, Marblehead, and Venice members. Named for exposure at Columbus in Franklin County, Ohio.
BT Middle Devonian
Devonian
SA Ohio

columnar basalt
As of 1981, includes use on level 3 under igneous rocks(1) basalts(2). See list H. Before 1981, included use on level 3 under igneous rocks(1) basalt family(2). Before 1978, also search columnar AND basalt.
BA basalts
volcanic rocks
BT igneous rocks
SA basalt

columnar jointing
use columnar joints

columnar joints
Term introduced in 1978. Includes use on level 3 under fractures(1) style(2). Before 1978, also search fractures AND columnar.
UF columnar jointing
BA fractures
SA bedding
joints

Colusa County
N central.
BT California
United States

Colville River
Flows E along N slope of Brooks Range and then N into Arctic Ocean.
BT Alaska
United States

Colville River delta
On Arctic Ocean in N.
BT Alaska
United States

Comanche County
Index states as applicable.
BX Kansas
Oklahoma
Texas
BT United States
SA Coldwater Meteorites

Comanchean
Provincial series, North America. Below Gulfian. Includes use on level 3 under age terms(1). See list E.
UF Comanchian
BA Cretaceous
BT Mesozoic
NT Edwards Formation
Fredericksburg Group
Mentor Beds
Paluxy Formation
Trinity Group
SA Antlers Sands
Lower Cretaceous
Upper Cretaceous

Comanchian
use Comanchean

combustion
SA geochemistry
processes

comendite
As of 1981, includes use on level 3 under igneous rocks(1) rhyolites(2). See list H. Before 1981, included use on level 3 under igneous rocks(1) andesite-rhyolite family(2).
BA rhyolites
volcanic rocks
BT igneous rocks
SA pantellerite

comets
Includes use on level 2 under interplanetary space(1).
SA interplanetary space
meteorites
meteors
solar system
solar wind
tektites

Commelinidae
Term introduced in 1981. Includes use on level 2 under angiosperms(1). See list F.
BA Monocotyledoneae
angiosperms
BT Spermatophyta
Plantae

commencements, sudden
use sudden commencements

commodity exchange
Term introduced in 1981.
SA economics

common mica
use muscovite

common salt
use halite

common-depth-point method
UF CDP method
SA methods
seismic methods

communist countries
Term introduced in 1981.
SA Western World

communities
Includes use on level 3 under ecology(1) and paleoecology(1).
SA assemblages
biocenoses
ecology
paleoecology

Como
City, lake and province in NW.
BT Lombardy
Italy

Comoro Islands
Group of volcanic islands in N Mozambique Channel between Mozambique and Malagasy Republic. An overseas territory of France. As of 1977 includes use as a level 1 area term (list O). See list B for term set options.
CO S130000S110000
E0450000E0430000
BA Indian Ocean Islands
SA Anjouan Island

compaction
Includes use on level 3 under diagenesis(1) and soil mechanics(1).
SA bearing capacity
consolidation
diagenesis
lithification
porosity
processes
soil mechanics

company diversification
Term introduced in 1981.
SA economics

company integration
Term introduced in 1981.
SA economics

company mergers
Term introduced in 1981.
SA economics

comparison
Used as a general term through May 1978. No longer a valid term for GeoRef.

compensation
Includes use on level 2 under isosta-

sy(1).
SA crust
isostasy

competence
As of 1981, no longer a valid term for GeoRef. See sedimentation, transport, hydrology, and hydraulics. From 1978-1980, was restricted to water transport of detritus. Prior to 1978, wind transport was included.

competent materials
Term introduced in 1978.
SA deformation
materials
rock mechanics
soil mechanics

complexes
Includes use on level 3 under igneous rocks(1) and metamorphic rocks(1). Before 1981, also search structural complexes.
SA igneous rocks
metamorphic rocks
metamorphism
ring complexes

complexing
SA geochemistry
metals
organic materials
processes

components
Used as a general term.

composite cone
use stratovolcanoes

composite volcano
use stratovolcanoes

composition
Includes use as level 2 or 3 term appropriate to a large number of topics, e.g. on level 2 under meteorology(1). See list G. Also search content.
UF content
SA acidic composition
alkalic composition
andesitic composition
arkosic composition
basaltic composition
calc-alkalic composition
calcareous composition
calcic composition
carbonaceous composition
chemical composition
dolomitic composition
ferromanganese composition
ferruginous composition
gabbroic composition
glauconitic composition
granitic composition
iron-rich composition
mafic composition
mineral composition
phosphate composition
potassic composition
rhyolitic composition
saline composition
siliceous composition
tholeiitic composition
ultramafic composition

compounds
Used as a general term.
SA acids
organic compounds

compressibility
Includes use on level 3 under soil mechanics(1).
UF modulus of compression
SA compression
elasticity
physical properties
plasticity
soil mechanics

compression
Includes use on level 3 under deformation(1) field studies(2).
SA compressibility
deformation
flow cleavage
lithification
pressure
tension
yield strength

compressional wave
use P-waves

compressive strength
SA deformation
rock mechanics
soil mechanics
strength

computer methods
A valid term through 1971. After 1971, use automatic data processing.

computer programs
NT Fortran
Fortran IV
SA algorithms
automatic data processing
programs

computers
For documents discussing the machine itself.
SA automatic data processing
hand calculators
microcomputers
minicomputers
programs

concentration
Used as a general term.
SA localization
saturation
solubility
solution

concentric folds
Term introduced in 1978. Includes use on level 3 under folds(1). Before 1978, also search concentric AND folds; search parallel AND folds.
UF parallel folds
BA folds
SA similar folds

concentric fractures
Term introduced in 1978. Includes use on level 3 under fractures(1) style(2). Before 1978, also search fractures AND concentric.
BA fractures

concepts
Includes use as level 2 or 3 term appropriate to a large number of topics, e.g. on level 2 under geochemistry(1) and mineral exploration(1). See list G.

conchiolin
BA organic materials
SA proteins

Conchostraca
As of 1981, no longer a valid term for GeoRef. Use Branchiopoda. Before 1981, included use on level 3 under Arthropoda(1) Branchiopoda(2).

concrete
Includes use on level 3 under engineering geology(1).
BA construction materials
SA cement materials
engineering geology

concretions
Includes use on level 3 under sedimentary structures(1) secondary structures(2). See list K. As of 1981, used for nodules other than those on the ocean floor.
BA secondary structures
sedimentary structures
SA coal balls
cone-in-cone
geodes
nodules
sedimentary rocks
sediments
septaria

condensates
Liquid hydrocarbon that emanates from a gas well or from the gas-cap of an oil well. Not to be used for other condensed materials.
UF gas condensates
SA heavy oil
natural gas
petroleum

condensation
Includes use on level 3 under Moon(1), planetology(1), and geochemistry(1).
SA geochemistry
Moon
planetology
planets

conditions, P-T
use P-T conditions

conductive materials
Term introduced in 1981.
SA engineering geology
geophysical methods
materials

conductivity
Includes use on level 3 under heat flow(1). A level 2 term under heat flow(1) until 1976; now use thermal conductivity(2).
SA electrical conductivity
electrical logging
geothermal gradient
heat flow
hydraulic conductivity
measurement
resistivity
thermal conductivity

Condylarthra
Includes use on level 2 under Mammalia(1). See list F.
BA Mammalia
BT Tetrapoda
Vertebrata

cone-in-cone
Includes use on level 3 under sedimentary structures(1) secondary structures(2). See list K.
BA secondary structures
sedimentary structures
SA concretions
septaria

Conemaugh
River in SW.
BT Pennsylvania
United States

Conemaugh Group
Divided into nine parts: Mahoning, Buffalo, Saltsburg, Grafron, Barton, Morgantown, Lonaconing, Connellsville, and one not yet named. Also includes Nadine Limestone, Woods Run Limestone, Carnahan Run Shale, and Brush Creek Limestone. W Maryland, E Ohio, Pennsylvania, N Virginia, and W west Virginia. Also search Conemaugh Formation.
BT Pennsylvanian
SA Maryland
Ohio
Pennsylvania
Virginia
West Virginia

cones, cinder
use cinder cones

cones, shatter
use shatter cones

conferences
Not a valid term for GeoRef. See under associations and under symposia.

configuration
As of 1981, no longer a valid term for GeoRef. Included use on level 2 under magnetosphere(1).

congelifraction
Also search frost shattering; frost splitting; frost weathering; frost wedging; gelifraction.
UF frost bursting
frost shattering
frost splitting
frost weathering
frost wedging
gelifraction
gelivation
SA cryoturbation
frost action
frost heaving
periglacial features
rock mechanics
soil mechanics

congeliturbation
use cryoturbation

conglomerate
Includes use on level 3 under sedimentary rocks(1) clastic rocks(2). See list I.
UF conglomerite
BA clastic rocks
BT sedimentary rocks
SA agglomerate
breccia
calcrete
ferricrete
metaconglomerate
terrigenous materials

conglomerite
use conglomerate

Congo
This term is now limited to the present Congo (People's Republic of the Congo). Formerly it also included what is now Zaire. Includes use on level 1 as an area term (list O). For term set options see list B.
CO S050000N034000
E0184000E0110000
UF People's Republic of the Congo
BA Central Africa
Africa
NT Brazzaville
SA Congo Basin
Congo River
Zaire

Congo (River)
use Congo River

Congo Basin
River drainage basin. Index countries as applicable.
BT Africa
SA Angola
Central African Republic
Congo
Zaire
Zambia

Congo River
Rises as the Lualaba River in SE Zaire and flows N, then W, and finally SW into the Atlantic Ocean. Index countries as applicable.
UF Congo (River)
Kongo River
Zaire River
BT Africa
SA Angola
Congo
Zaire

C

congresses
Not a valid term for GeoRef. See under associations and under symposia.

Coniacian
Europe. Above Turonian, below Santonian. Includes use on level 3 under age terms(1). See list E.
BA Senonian
Upper Cretaceous
Cretaceous
BT Mesozoic

conical fractures
Term introduced in 1978. Includes use on level 3 under fractures(1) style(2). Before 1978, also search fractures AND conical.
BA fractures

conichalcite
BA arsenates
BT minerals

Coniferae
Includes use on level 3 under gymnosperms(1). See list F.
BA gymnosperms
BT Spermatophyta
Plantae

Coniferales
Includes use on level 2 under gymnosperms(1). See list F.
BA gymnosperms
BT Spermatophyta
Plantae
NA Buriadia
Cupressaceae
Picea
Pinaceae
Taxodiaceae
Taxodium
SA Dadoxylon

conjugate folds
Term introduced in 1978. Includes use on level 3 under folds(1) style(2). Before 1978, search folds AND conjugate.
BA folds
SA joints

connate waters
Includes use on level 2 under ground water(1). Also search formation water; formation waters.
SA fossil waters
ground water
pore water
water

Connecticut
Includes use on level 1 as an area term (list O). For term set options see list B.
CO N410000N420300
W0714800W0734400
BA New England
Eastern U.S.
United States
NX Columbia
NT Hartford County
Litchfield County
NX Middlesex County
NT Roaring Brook Valley
Suffield
Tolland County
NX Windham County
Woodbury
SA Atlantic Coastal Plain
Connecticut River
Connecticut Valley
Manhattan Formation
New York City Group
Newark Group
Westerly Granite

Connecticut River
Rises in N New Hampshire and flows S emptying into Long Island Sound. Index states as applicable.
BT United States
SA Connecticut
Massachusetts
New Hampshire
Vermont

Connecticut River valley
use Connecticut Valley

Connecticut Valley
River valley. Index states as applicable. Also search Connecticut; Connecticut River valley.
CO N413000N452000
W0721500W0724500
UF Connecticut River valley
BT United States
SA Connecticut
Massachusetts
New Hampshire
Vermont

Connemara
Barren, mountainous coastal region in County Galway on the Atlantic Ocean.
CO N531500N534000
W0090000W0102000
BT Ireland

Conodonta
Term introduced in 1981. Includes use on level 1 and 2 as a fossil term (list F). Before 1981, search conodonts. As of 1981, search conodonts only for biostratigraphy.
NA Hindeodella
Icriodus
Ozarkodina
Palmatolepis
Panderodus
Polygnathus
Spathognathodus
SA apparatus
conodonts

conodonts
As of 1981, restricted to use as the common name for Conodonta. See list F. To be used for Conodonta when they are used as tools in stratigraphy. Includes use on level 1. Before 1981, search conodonts AND biostratigraphy. Before 1981, included use on level 1 and 2 as a fossil term. If 1, term set options are:
biostratigraphy
age [List E]
SA biostratigraphy
Conodonta

Conrad discontinuity
Includes use on level 3 under crust(1).
SA crust
discontinuities
Earth
seismology
velocity

conservation
Level 1 term as of 1978. Includes use on level 2 under impact statements(1) and land use(1). Used for environmental conservation of natural resources. If 1, term set options are:
environment
beaches
bogs
caves
ecosystems
marshes
reefs
shorelines
natural resources
coal
energy sources
forests
ground water
metal ores
mineral resources
underground space
water resources
wetlands
topic [experimental studies, impact statements, methods, programs]
subtopic
SA deforestation
depletion
energy conservation
environmental geology
erosion control
impact statements
land leases
land use
natural resources
pollution
RARE II regions
reclamation
resources
soils
urbanization
waste disposal

consistency
As of 1981, no longer a valid term for GeoRef. Use consistent materials if applicable. Included use under soils(1) morphology(2).

consistency limits
use Atterberg limits

consistent materials
Term introduced in 1981.
SA engineering geology
materials

consolidation
Includes use on level 3 under diagenesis(1). Before 1978, also search solidification.
UF solidification
SA cementation
compaction
diagenesis
lithification
overconsolidated materials
soil mechanics

consolidometer tests
Term introduced in 1981.
SA engineering geology
soil mechanics

Constance Lake
use Lake Constance

constant, dielectric
use dielectric constant

Constanta
County. Also a city on the Black Sea.
UF Constantsa
BT Dobruja
Romania

Constantine
No longer a valid index term for GeoRef as of 1981. Use Constantine region.

Constantine region
Term introduced in 1981.
CO N340500N370700
E0084000E0040000
BT Algeria
NT Guelma
Hodna Basin

constants, elastic
use elastic constants

Constantsa
use Constanta

construction
As of 1978, term is used on level 2.
SA construction materials
dams
dredging
foundations
highways
marine installations
reservoirs
shorelines
tunnels
underground installations

construction materials
Includes use on level 1 as a commodity term (list C).
NA aggregate
building stone
cement materials
concrete
dimension stone
SA alloys
construction
granite deposits
gravel deposits
limestone deposits
marble deposits
materials
perlite
refractory materials
sand
sands
sandstone deposits
shale
trap rocks

contact
As of 1981, no longer a valid term for GeoRef. See boundary; contact metamorphism. Before 1981, included use on level 2 under intrusions(1). Before 1977, used to indicate type of metamorphism.

contact aureole
use aureoles

contact metamorphism
Not a valid term from 1973 through 1977. After 1977, includes use on level 2 under metamorphism. Also search metamorphism AND contact. As of 1981, also includes use on level 2 under intrusions(1).
BA metamorphism
SA aureoles
country rocks
pneumatolysis

contact zone
use aureoles

contamination
As of 1981, restricted to use on level 2 under magmas(1) and is not to be confused with pollution. Before 1981, included use on level 2 under ground water(1).
SA anatexis
assimilation
country rocks
emplacement
hybridization
intrusions
magmas
palingenesis
pollution

contemporaneous faults
use growth faults

content
use composition

conterminous regions
Used to indicate regions with a common boundary. Used especially for the United States.
SA United States

continental
A valid term through 1977. After 1977, use continental borderland, continental type (for the crust), or continental rise.

continental borderland
The area of the continental margin between the shoreline and the continental slope which is topographically more complex than the continental shelf.
UF borderland, continental

SA continental margin
continental shelf
continental slope

continental crust
use continental type

continental displacement
use continental drift

continental drift
For general discussions of "classical" drift concepts as well as plate tectonic reconstructions. Includes use on level 1 (list A); on level 3 under plate tectonics(1). Also search continental AND drift. Term set options are:
area [use terms like Gondwana, Iapetus, Laurasia, Mesogaea, Paleoafrica, Pangaea, Paratethys, Tethys]
locality
topic [causes, concepts, evolution, general, genesis, interpretation, mechanism, paleobotany, paleoclimatology, paleomagnetism, paleontology, patterns]
subtopic (no area term)
UF continental displacement
continental migration
displacement theory
epeirophoresis theory
Wegener hypothesis
SA biogeography
continents
geosynclines
Gondwana
Laurasia
mantle
Mohorovicic discontinuity
ocean basins
paleogeography
paleomagnetism
paleontology
Pangaea
Paratethys
plate tectonics
polar wandering
reconstruction
sea-floor spreading
stratigraphy
tectonics
tectonophysics
Tethys
volcanology

continental dunes
Term introduced in 1981. Includes use on level 3 under geomorphology(1) eolian features(2).
BA dunes

continental genesis
use continental type

continental margin
Use term to denote both shelf and slope. Includes use on level 3 under area sets (list B) and under plate tectonics(1) or tectonics(1). Also search continental margins; continental AND margin.
UF margin, continental
SA aseismic margins
continental borderland
continental rise
continental shelf
continental slope
continents
cratons
island arcs
ocean floors
passive margins
plate tectonics
submarine canyons
tectonics
tectonophysics

continental migration
use continental drift

continental platform
use continental shelf

continental rise
Includes use on level 3 under continental slope(1) or ocean floors(1).
UF rise, continental
SA abyssal plains
continental margin
continental slope
marine geology
nepheloid layer
ocean floors

continental shelf
For studies of geological processes taking place on the shelf. Restricted to the modern shelf area. Includes use on level 1 (list A). Also search continental AND shelf. Also used on level 3 in area sets under oceanography(2) where the level 1 term refers to the adjacent land area. A term for the ocean is also indexed for studies on specific continental shelves; however, the first level term is usually the land area. Term set options are:
topic [List B except for areal geology]
subtopic (no area term)
UF continental platform
shelf, continental
NA inner shelf
outer shelf
SA changes of level
continental borderland
continental margin
continental slope
continents
currents
marine geology
ocean circulation
ocean floors
oceanography
offshore
Polar Continental Shelf
reefs
sedimentation
shelf environment
slope environment
slumping
submarine canyons

continental slope
Includes use on level 1 (list A). For modern slope area. Also search continental AND slope. Term set options are:
topic [List B except for areal geology]
subtopic (no area term)
UF slope, continental
NA inner slope
outer slope
SA channels
continental borderland
continental margin
continental rise
continental shelf
continents
marginal basins
marine geology
ocean circulation
ocean floors
oceanography
offshore
sedimentation
slope environment
slopes
slumping
submarine canyons

continental type
Used in relation to crust. Before 1978, search continental; continental crust; genesis AND continental; search continental AND crust.
UF continental crust
continental genesis
SA crust
plate tectonics

continents
Includes use on level 3 when discussing continents as units (e.g. geometry of a continent). See continental drift(1) or topic under area terms, list B.
SA continental drift
continental margin
continental shelf
continental slope
cratons
Earth
epeirogeny
island arcs
islands
isostasy
marginal seas
tectonic platforms

continuous filters
Term introduced in 1981.
BT filters

continuous materials
Term introduced in 1981.
SA engineering geology
materials

contour maps
BA maps
SA structure contour maps
topographic maps

contourite
Includes use on level 3 under sedimentary rocks(1) clastic rocks(2). See list I.
BA clastic rocks
BT sedimentary rocks
SA terrigenous materials

Contra Costa County
On San Francisco Bay.
BT California
United States

contraction
Used as a general term.
SA processes

control, erosion
use erosion control

control, production
use production control

controls
Includes use on level 2 under mineral deposits, genesis(1) and sedimentation(1).
SA geochemical controls
hydrogeological controls
lithologic controls
mechanical controls
mineral deposits
paleogeographic controls
sedimentation
slope stability
stratigraphic controls
structural controls
tectonic controls
waste disposal
waterways

Conularida
Term introduced in 1981. Includes use on level 2 under Coelenterata(1). See list F.
BA Scyphozoa
Coelenterata
BT Invertebrata

convection
Includes use on level 2 and 3 under meteorology(1) and under tectonophysics(1); on level 3 under plate tectonics(1).
SA aerosols
atmosphere
convection currents
heat flow
mantle
meteorology
oceanography
plate tectonics
tectonophysics

convection cells
As of 1981, no longer a valid term for GeoRef. Use convection currents.

convection currents
Includes use on level 2 under tectonophysics(1); on level 3 under Earth(1) and plate tectonics(1). Before 1981, also search convection cells.
BT currents
SA atmosphere
convection
density
mantle
plate tectonics
tectonophysics

convergence
As of 1981, no longer a valid term for GeoRef. Use plate convergence if applicable.

Converse County
E central.
BT Wyoming
United States

conversion
Used as a general term.
SA changes

convoluted beds
Includes use on level 3 under sedimentary structures(1) soft sediment deformation(2) and turbidity current structures(2). See list K.
UF beds, convoluted
convoluted deformation
BA soft sediment deformation
sedimentary structures
SA slump structures
turbidity current structures

convoluted deformation
use convoluted beds

Coo
use Kos

Cooch Behar
City just N of N Bangladesh. A former state.
BT West Bengal
India

Cook County
Index states as applicable.
BX Georgia
Illinois
Minnesota
BT United States

Cook Inlet
Arm of the Pacific Ocean W of Kenai Peninsula in S.
BT Alaska
United States

Cook Islands
Group of 15 islands W of French Polynesia and E of Samoa and Tonga islands.
CO S230000S180000
W1560000W1650000
UF Southern Cook Islands
BT Polynesia
SA Pacific Ocean

Cook Mountain Formation
In Claiborne Group. NW Louisiana, S and E Texas.
BT middle Eocene
Eocene
SA Claiborne Group
Louisiana
Texas

cookeite
BA chlorite group

sheet silicates
silicates
BT minerals
SA aluminosilicates
lepidolite

cooling
Used as a general term.
SA temperature

Cooma
Town in S.
BT New South Wales
Australia

Cooper Basin
Also search Cooper's Creek basin. Index states as applicable.
CO S300000S220000
E1430000E1370000
UF Cooper's Creek basin
BT Australia
SA Queensland
South Australia

cooperation, international
use international cooperation

Cooper's Creek basin
use Cooper Basin

coordinate systems
Includes use on level 2 under magnetosphere(1).
SA magnetosphere
systems

coordinates
Includes use on level 3 under Moon(1) or maps(1). Use geodetic coordinates under geodesy(1). For geographic coordinates, see specific area.
SA cartography
geodesy
geodetic coordinates
maps
Moon

coordination
As of 1978, term is used on level 2 under crystal chemistry(1).
SA crystal chemistry

Coorong Lagoon
Off Lacepede Bay of Indian Ocean in SE South Australia. Also search Coorong.
UF The Coorong
BT South Australia
Australia

Coos Bay
Inlet on coast of Coos County on the Pacific Ocean in SW.
CO N431500N433500
W1241000W1243000
BT Oregon
United States

Coos County
On the Pacific Ocean in SW.
CO N424000N434000
W1234000W1243500
BT Oregon
United States

Coosa County
Central.
BT Alabama
United States

Copenhagen
County on Zealand Island in E Denmark. Also a city.
UF Kobenhavn
BT Denmark

Copepoda
Includes use on level 2 under Arthropoda(1). See list F.
BA Crustacea
Mandibulata
Arthropoda
BT Invertebrata

copiapite
BA sulfates
BT minerals

Copiapo
Town in N part of country. Also a river.
UF San Francisco de la Selva
Selva
BT Atacama
Chile

copper
Includes use on level 1 and 2 as a chemical element (list D). As of 1981, use copper ores for copper as a commodity. Before 1981, included use on level 1 as a commodity term.
UF Cu
SA bornite
chalcocite
chalcopyrite
copper ores
elements
enargite
heavy metals
native elements
porphyry copper

Copper Canyon
Canyon in Lander County in N central.
BT Nevada
United States

Copper Harbor Conglomerate
Upper Keweenawan. Keweenaw County on the Upper Peninsula, NW Michigan.
BT Precambrian
SA Michigan

Copper Mountain
Location of copper deposit in British Columbia.
BT British Columbia
Canada

copper ores
Term introduced in 1981. Includes use on level 1 and 2 as a commodity term (list C). Before 1981, search copper AND ore deposits.
SA copper
enargite
native elements
porphyry copper

copper pyrites
use chalcopyrite

copper sulfides
BA sulfides
BT minerals
SA chalcocite
covellite

Copperbelt
Area.
CO S180000S170000
E0283000E0270000
BT Zambia

Coppermine River
Flows into Coronation Gulf in N District of Mackenzie.
BT Northwest Territories
Canada

coprolites
Includes use on level 1 and 2 as a fossil term(list F); on level 3 under sedimentary structures(1) biogenic structures(2). See list K.
BA biogenic structures
sedimentary structures
SA Vertebrata

coquimbite
BA sulfates
BT minerals

Coquimbo
Province in central Chile. Also a city.
BT Chile

coquina
As of 1981, includes use on level 3 under sedimentary rocks(1) carbonate rocks(2). See list I. Before 1981, included use on level 3 under sediments(1) clastic sediments(2).
BA carbonate rocks
BT sedimentary rocks
SA carbonate sediments
terrigenous materials

coral pinnacle
use pinnacle reefs

coral reefs
A valid term through 1974. After 1974, use reefs.

Coral Sea
Between Queensland, Australia, on the W and New Hebrides and New Caledonia on the E. As of 1977, includes use on level 1 as an area term (list O). For term set options see list B.
CO S300000S040000
E1700000E1423000
BA West Pacific
Pacific Ocean
NT Coral Sea Basin
Great Barrier Reef
SA Leg 21

Coral Sea Basin
Between NE Queensland and Solomon Islands.
BT Coral Sea
Pacific Ocean

Corallimorpharia
Includes use on level 2 under Coelenterata(1). See list F.
BA Anthozoa
Coelenterata
BT Invertebrata

Corallinaceae
As of 1981, includes use on level 2 under algae(1). Before 1981, included use on level 3 under algae(1) Rhodophyta(2). See list F.
BA Rhodophyta
algae
BT thallophytes
Plantae
NA Archaeolithothamnium
Lithophyllum
Lithothamnium

corals
As of 1981, restricted to use as the common name for Coelenterata. See list F. To be used for Coelenterata when they are used as tools in stratigraphy. Includes use on level 1. Before 1981, search (corals OR Coelenterata) AND biostratigraphy. Before 1981, included use on level 3 under Coelenterata(1) Anthozoa(2), and on level 3 as a reef constituent. If 1, term set options are:
biostratigraphy
age [List E]
SA biostratigraphy
Coelenterata

Corbieres
Outliers of the E Pyrenees in S Aude. Also search Corbieres Mountains.
UF Corbieres Mountains
BT Aude
France

Corbieres Mountains
use Corbieres

Cordaitales
Includes use on level 2 under gymnosperms(1). See list F.
BA gymnosperms
BT Spermatophyta
Plantae
NA Cordaites
SA Dadoxylon

Cordaites
Genus.
BA Cordaitales
gymnosperms
BT Spermatophyta
Plantae

cordierite
BA ring silicates
silicates
BT minerals
SA aluminosilicates

Cordillera
As of 1981, no longer a valid term for GeoRef. See North American Cordillera.

Cordillera Betica
use Betic Cordillera

Cordillera de la Costa
Along Caribbean Sea in central N.
BT Venezuela

Cordillera Marianica
use Betic Cordillera

Cordillera Occidental
use Western Cordillera

Cordillera Oriental
use Eastern Cordillera

Cordilleran Geosyncline
Term introduced in 1978. Before 1978, search geosynclines AND Cordilleran.
SA Cordilleran Orogeny
geosynclines

Cordilleran Orogeny
Term introduced in 1978. Includes use on level 3 under orogeny(1). Before 1978, search Cordilleran AND orogeny.
SA Cordilleran Geosyncline
Laramide Orogeny
orogeny
tectonics

Cordoba
As of 1981, refers to only the city in Spain. Before 1981, also referred to the province in Spain, and the city and department in Argentina. In S central part of country.
BT Cordoba Province
Andalusia
Spain

Cordoba Department
Term introduced in 1981 for the province in central Argentina. Before 1981, also search Cordoba AND Argentina.
CO S350000S292000
W0613000W0654500
BT Argentina
NT Chanares
SA Sierras Pampeanas

Cordoba Province
Term introduced in 1981. Before 1981, also search Cordoba AND Spain. In S central part of country. For province in Argentina, use Cordoba Department.
CO N371000N385000
W0035700W0053500
BT Andalusia
Spain
NT Cordoba
Los Pedroches
Valsequillo

core
Used for the core of the Earth. For that of other planets and Moon, see appropriate set. Includes use on level 1 (list A); on level 2 under Earth(1), and under seismology(1).

Term set options are:
topic [composition, concepts, evolution, general, genesis, interpretation, processes, properties, temperature, theoretical studies]
subtopic
NA inner core
outer core
SA crust
Earth
geophysics
heat flow
heat sources
interior
lithosphere
magnetic field
magnetohydrodynamics
mantle
Mohorovicic discontinuity
phase transitions
seismology
tectonophysics
transition zones

cores
Includes use on level 3 under well-logging(1). Before 1978, also search borings.
UF drill cores
SA boreholes
borings
cuttings
well-logging
wells

Corinth
City on Gulf of Corinth in Corinth Department in NE.
BT Peloponnesus
Greece

Coriolis force
Includes use on level 2 under ocean circulation(1).
UF force, Coriolis
geostrophic force
SA Earth
ocean circulation

Cork
County on the Atlantic Ocean. Also a city. Search County Cork.
CO N513000N522000
W0081500W0103000
UF County Cork
BT Ireland

Cormeilles-en-Parisis
Town just N of Paris.
BT Val-d'Oise
France

Cornwall
County in extreme SW.
CO N500000N505000
W0041500W0063000
BT England
Great Britain
United Kingdom

Cornwallis Island
One of the Parry Islands in central District of Franklin.
CO N743000N753000
W0933000W0970000
BT Northwest Territories
Canada

Coromandel Coast
SE coast from Point Calimere in the S to mouths of the Krishna River in Andhra Pradesh. Also search Coromandel.
BT India

Coromandel Peninsula
ENE of Auckland. Also search Coromandel.
BT North Island
New Zealand

corona
Includes use on level 2 under astrophysics and solar physics(1).
SA astrophysics and solar physics
plasma instabilities
solar wind
Sun

coronadite
BA oxides
BT minerals
SA hollandite

Coronation Mine
E central.
BT Saskatchewan
Canada

Corophioides
BA ichnofossils

Corpus Christi Bay
Inlet of Gulf of Mexico between Neuces and Patricio counties in SE.
BT Texas
United States

corrections
Used as a general term.
SA accuracy
errors

correlation
Formerly a general term appropriate to a large number of topics (list G). As of 1976, the term was restricted to stratigraphical correlation. Includes use on level 2 under stratigraphy(1).
SA stratigraphy

correlation coefficient
UF coefficient, correlation
BA statistical analysis

corrensite
BA clay minerals
sheet silicates
silicates
BT minerals
SA aluminosilicates

correspondence analysis
BA statistical analysis
SA analysis

Correze
Department in S central.
CO N445000N454500
E0023000E0011000
BT France
NT Brive
SA Millevaches Plateau

Corrientes
Province between the Parana and the Uruguay rivers in NE Argentina. Also a city.
CO S300000S270000
W0560000W0600000
BT Argentina

corrosion
Includes use under geomorphology(1). Chemical meaning also used.
SA chemical weathering
destruction
erosion
geomorphology

Corsica
Island in the Mediterranean Sea and a department of France. Includes use on level 1 as an area term (list O). For term set options see list B.
CO N411500N430000
E0093000E0083000
BA Western Europe
France
Europe
SA Mediterranean region

Cortez
City in Montezuma County in extreme SW.
BT Colorado
United States

Cortez Mountains
Eureka County in central Nevada. Also search Cortez.
BT Nevada
United States

Cortina
use Cortina D'Ampezzo

Cortina D'Ampezzo
Town in Dolomites in N Veneto.
UF Cortina
BT Veneto
Italy

corundum
Includes use on level 3 under minerals(1) and gems(1). As of 1981, use corundum deposits for corundum as a commodity (abrasives). Before 1981, included use on level 1 as a commodity term (abrasives).
BA oxides
BT minerals
SA alumina
amethyst
corundum deposits
gems
sapphire

corundum deposits
Term introduced in 1981. Includes use on level 1 and 2 as a commodity term (list C). Before 1981, search corundum AND deposits. For corundum as a gem, index corundum and gems.
SA abrasives
corundum
industrial minerals

Corynexochida
Includes use on level 2 under Trilobita(1). See list F.
BA Trilobita
BT Trilobitomorpha
Arthropoda
Invertebrata

Cos
use Kos

cosalite
BA sulfobismuthites
sulfosalts
BT minerals

Cosenza
Town and province in W central.
BT Calabria
Italy

cosmic dust
Includes use on level 2 under planetology(1) and (as of 1981) extraterrestrial geology(1).
UF zodiacal dust
SA asteroids
dust
extraterrestrial geology
interplanetary dust
interplanetary space
meteoroids
particles
planetology

cosmic rays
Includes use on level 2 under interplanetary space(1), magnetosphere(1), planetology(1), and (as of 1981) extraterrestrial geology(1); on level 3 under aeronomy(1) ionization(2).
UF rays, cosmic
SA aeronomy
albedo
alpha rays
asteroids
cutoff rigidities
electromagnetic radiation
exposure age
extraterrestrial geology
gamma rays
intensity
interplanetary space
ionization
magnetic field
magnetosphere
Moon
particle-track dating
particles
protons
solar cycles
solar flares
spallation

cosmochemistry
For "geochemistry" of extraterrestrial materials. Includes use on level 2 under extraterrestrial geology(1), planetology(1), and asteroids(1) as of 1981. Before 1981, included use on level 3.
SA asteroids
cosmogenic elements
elements
extraterrestrial geology
geochemistry
planetology

cosmogenic elements
Term introduced in 1978. Before 1978, search cosmogenic; cosmogenic AND isotopes.
SA cosmochemistry
elements
isotopes

cosmolites
use meteorites

Coso Hot Springs
use Coso Hot Springs KGRA

Coso Hot Springs KGRA
Term introduced in 1981. Known Geothermal Resources Area. Before 1981, also search Coso Hot Springs.
CO N360200N360300
W1174500W1174700
UF Coso Hot Springs
BT Inyo County
California
United States

cost
Term introduced in 1981.
SA economics

Costa Rica
Includes use on level 1 as an area term (list O). For term set options see list B.
CO N080000N111500
W0823000W0860000
BA Central America
NT Arenal

Cote d'Or
use Cote-d'Or

Cote-d'Or
Department in E central France. Also search Cote d'Or.
CO N465500N480500
E0053000E0040200
UF Cote d'Or
BT France
NT Dijon
Pouilly-en-Auxois
SA Morvan

Cotentin Peninsula
Peninsula in Normandy jutting into the English Channel in N.
BT Manche
France

Cotes-du-Nord
Department on the English Channel in N Brittany.
CO N480000N490000
W0020000W0034500
BT France

Cotswold Hills
Range of hills in Gloucestershire in W central England. Also search Cotswolds.
CO N514500N521500
W0011500W0014500
UF Cotswolds
BT England
Great Britain
United Kingdom

Cotswolds
use Cotswold Hills

Cottbus
As of 1981, refers to only the city. Before 1981, also included the district.
BT Cottbus Bezirk
East Germany

Cottbus Bezirk
Term introduced in 1981. District in SE. Before 1981, search Cottbus.
CO N511600N520900
E0150000E0124800
BT East Germany
NT Cottbus
Doberlug
SA Saxony
Upper Lusatia

Cottian Alps
Division of Western Alps. Index countries as applicable.
BT Alps
Europe
SA Dauphine Alps
France
Italy
Piedmont Alps
Western Alps

Cotton
County on the Texas border in SW.
BT Oklahoma
United States

Cotton Valley Group
Includes Schuler Formation, and Bossier Formation. S Arkansas, N Louisiana, W Mississippi, and E Texas.
BT Upper Jurassic
Jurassic
SA Arkansas
Bossier Formation
Louisiana
Mississippi
Schuler Formation
Texas

Cottonwood Limestone
In Council Grove Group. E Kansas, SE Nebraska, and central northern Oklahoma. Also search Cottonwood Limestone Member.
UF Cottonwood Limestone Member
BT Permian
SA Kansas
Nebraska
Oklahoma

Cottonwood Limestone Member
use Cottonwood Limestone

Cotylosauria
Term introduced in 1981. Includes use on level 2 under Reptilia(1). See list F.
BA Anapsida
Reptilia
BT Tetrapoda
Vertebrata

Coulombs' modulus
use shear modulus

country rocks
Includes use on level 3 under intrusions(1).
SA assimilation
aureoles
contact metamorphism
contamination
host rocks
hybridization
igneous rocks
inclusions
intrusions
magmas
rocks
wallrock alteration

County Cork
use Cork

Coupon Bight
Florida keys area.
BT Florida
United States

Courland Spit
Narrow sandspit on the Baltic Sea in both Kaliningrad Oblast and Lithuania. Index Soviet republics as applicable.
UF Kurland Spit
BT USSR
SA Lithuania
Russian Republic

Couvinian
Europe. Above Emsian, below Givetian. Includes use on level 3 under age terms(1). See list E.
BA Middle Devonian
Devonian
BT Paleozoic
SA Eifelian

Covasna
County in SE.
BT Transylvania
Romania

covelline
use covellite

covellite
UF covelline
BA sulfides
BT minerals
SA copper sulfides

cover, sedimentary
use sedimentary cover

Cowley County
On the Oklahoma border in S.
BT Kansas
United States

Cowlitz County
On the Columbia River in SW.
BT Washington
United States

Coyote Creek Fault
W California.
BT California
United States

CO2
use carbon dioxide

Cr
use chromium

cracks
Includes use on level 3 under rock mechanics(1).
NT microcracks
SA fissures
fractures
open fractures
rock mechanics

Cracow
use Krakow

Crai
Forest in Apuseni Mountains in WSW.
BT Transylvania
Romania

crandallite
BA phosphates
BT minerals

Crassostrea virginica
Includes use on level 3 under Mollusca(1). See list F.
BA Bivalvia
Mollusca
BT Invertebrata

Crater Lake
In Cascade Mountains in Klamath County in SW.
BT Oregon
United States

cratering
Includes use on level 3 under earthquakes(1), geomorphology(1), Moon(1), seismology(1), and volcanology(1).
SA craters
cryptoexplosion features
geomorphology
impact features

craters
Includes use on level 3 under geomorphology(1) impact features(2) and landform description(2); on level 3 under volcanology(1) and meteor craters(1).
SA astroblemes
calderas
cratering
cryptoexplosion features
impact features
impacts
lunar craters
maars
meteor craters
microcraters
Moon
volcanology

cratons
Includes use on level 3 under tectonics(1).
SA continental margin
continents
crust
intracratonic basins
shields
tectonic platforms
tectonics

Crawford County
Term introduced in 1981. Index states as applicable.
BX Arkansas
Georgia
Illinois
Indiana
Iowa
Kansas
Michigan
Missouri
Ohio
Pennsylvania
Wisconsin
BT United States
SA Sandusky River basin

Cree Lake
Term introduced in 1981. N Saskatchewan.
CO N572000N574000
W1060000W1071000
BT Saskatchewan
Canada

creep
As of 1978, term is used on level 2 under permafrost(1) and slope stability(1).
UF creeping
SA deformation
elastic limit
elastic strain
etching
landslides
mass movements
mechanical properties
permafrost
slope stability
slumping
soils
solifluction
stress

creeping
use creep

Creil
Town N of Paris.
BT Oise
France

crenulation cleavage
use slip cleavage

Creodonta
Order. Includes on level 2 under Mammalia(1).
BA Mammalia
BT Tetrapoda
Vertebrata

Crested Butte
Town and Peak in Gunnison County in W central.
BT Colorado
United States

Crestmore
Area.
BT California
United States

Cretaceous
Includes use on level 1 as an age term (list E); on level 2 under paleoterms, e.g. paleoecology, paleogeography, paleomagnetism. Above Jurassic, below Tertiary.
BT Mesozoic
NT Benton Formation
Blairmore Group
Colorado Group
NA Comanchean
NT Dakota Formation
Eureka Sound Formation
Gosau Formation
Graneros Shale
NA Lower Cretaceous
NT Lower Greensand
Mancos Shale
NA Middle Cretaceous
NT Mishash Formation
Nubian Sandstone
Potomac Group
Robles Formation
San Felipe Formation
Santana Formation
Shiranish Formation
NA Upper Cretaceous
NT Viking Formation
NA Vraconian
NT Washita Group
Weald Clay
Whitemud Formation
Yacoraite Formation
SA Bauru Formation
Beaverhead Formation
Brianconnais Zone
Deccan Traps
Difunta Group
Ewekoro Formation
Fort Union Formation
Franciscan Formation
Gondwana System
Great Valley Sequence
Intertrappean Beds
Ionian Zone
Kootenay Formation
Laramide Orogeny
Maiolica Limestone
Mardin Formation
Muro Group
Niniyur Group
Paskapoo Formation
Scaglia Formation
Serra Geral Formation
Shimanto Group
Tal Formation

Tertiary

Crete
Administrative region and island in E Mediterranean Sea.
CO N345500N354500
E0263000E0231500
BT Greece
SA Hellenides

Creuse
Department in central France.
CO N453500N462800
E0023500E0012000
BT France
SA Millevaches Plateau

Cricetidae
Family. As of 1981, includes use on level 3 under Mammalia(1) Myomorpha(2). See list F. Before 1981, included use on level 3 under Mammalia(1) Rodentia(2).
BA Myomorpha
Rodentia
Mammalia
BT Tetrapoda
Vertebrata
NA Pliomys

Crimea
Region jutting into the Black Sea.
CO N440000N460000
E0370000E0320000
BZ Russian Platform
BX Ukraine
BT USSR

Crimean Mountains
Range in S Crimea along Black Sea Coast.
BT Ukraine
USSR

Crimean Plain
Dry, level steppe covering the northern 80% of Crimean Peninsula.
BT Ukraine
USSR

crinanite
As of 1981, includes use on level 3 under igneous rocks(1) alkali basalts(2). See list H. Before 1981, included use on level 3 under igneous rocks(1) alkali basalt family(2).
BA alkali basalts
basalts
volcanic rocks
BT igneous rocks

Crinoidea
Includes use on level 2 under Echinodermata(1). See list F.
BA Crinozoa
Echinodermata
BT Invertebrata
NA Delocrinus
Inadunata

Crinozoa
As of 1981, used on level 3 under Echinodermata(1). See list F. Before 1981, used on level 2 under Echinodermata(1).
BA Echinodermata
BT Invertebrata
NA Blastoidea
Crinoidea
Cystoidea
Edrioblastoidea
Eocrinoidea
Lepidocystoidea
Parablastoidea
Paracrinoidea

Cripple Creek
City in Teller County in central.
BT Colorado
United States

Crisana-Maramures
Historical province in NW.
BT Transylvania
Romania

cristobalite
BA silica minerals
framework silicates
silicates
BT minerals
SA quartz
tridymite

critical review
Term introduced in 1981.
SA review

Crittenden County
Index states as applicable.
BX Arkansas
Kentucky
BT United States

CRM
use chemical remanent magnetization

Croatia
Constituent republic in N.
CO N423000N464500
E0191500E0153000
BT Yugoslavia
NT Limski Channel
Northern Limestone Alps
Split
Velebit Mountains
SA Istria

crocidolite
BA amphibole group
chain silicates
silicates
BT minerals
SA riebeckite

Crocodile River basin
use Limpopo Basin

Crocodilia
As of 1981, includes use on level 2 under Reptilia(1). See list F. Before 1981, included use on level 3 under Reptilia(1) Archosauria(2).
BA Archosauria
Reptilia
BT Tetrapoda
Vertebrata
NA Teleosauridae

crocoite
BA chromates
BT minerals

Cromerian
Glaciation. Europe. Includes use on level 3 as an age term (list F)
BA upper Pleistocene
Pleistocene
BT Quaternary
NT South Polish Glaciation

Crook County
Index states as applicable.
BX Oregon
Wyoming
BT United States

Crooked Creek Formation
In Meade Group. Southwestern Kansas and northwestern Oklahoma.
BT Pleistocene
SA Kansas
Oklahoma

cross folds
use superposed folds

cross fractures
Term introduced in 1978. Before 1978, search cross AND fractures.
BA fractures
SA joints

cross joints
As of 1981, no longer a valid term for GeoRef. Use cross fractures.

cross-bedding
Includes use on level 3 under sedimentary structures(1) planar bedding structures(2). See list K.
BA planar bedding structures
sedimentary structures
SA bedding
cross-laminations
cross-stratification
laminations

cross-laminations
Includes use on level 3 under sedimentary structures(1) planar bedding structures(2). See list K.
UF diagonal lamination
BA planar bedding structures
sedimentary structures
SA cross-bedding
cross-stratification
laminations
ripple drift-cross laminations

cross-stratification
Includes use on level 3 under sedimentary structures(1) planar bedding structures(2). See list K.
BA planar bedding structures
sedimentary structures
SA cross-bedding
cross-laminations
stratification

crossite
BA clinoamphibole
amphibole group
chain silicates
silicates
BT minerals
SA glaucophane

Crossopterygii
Order. Includes use on level 3 under Pisces(1) Osteichthyes(2).
BA Osteichthyes
Pisces
BT Vertebrata

Crowsnest Pass
In SE British Columbia.
BT British Columbia
Canada

Crozet Islands
Five small French islands SE of South Africa.
BT Indian Ocean

Crozon Peninsula
S of Brest in Brittany in W.
BT Finistere
France

crust
Used for the Earth's crust, and the crust of the planets and their satellites. As of 1981, for the crust of the Moon, use lunar crust. When used on level 1, or on level 2 or 3 in sets such as seismology(1) or plate tectonics(1), it refers to the crust of the Earth. When used in a planet set (when level 1 term is planet name), it refers to the crust of that planet. Also search crusts. If 1, term set options are:
topic [age, anomalies, composition, concepts, evolution, genesis, interpretation, observations, processes, properties, theoretical studies, thickness]
subtopic (no area term)
NA lower crust
upper crust
SA asthenosphere
basement
compensation
Conrad discontinuity
continental type
core
cratons
discontinuities
Earth
earthquakes
epeirogeny
geophysics
geosynclines
geothermal gradient
heat flow
heat sources
horizontal discontinuities
isostasy
lithosphere
low-velocity zones
lunar crust
magmas
mantle
mobile belts
Mohorovicic discontinuity
neotectonics
ocean basins
oceanic crust
paleomagnetism
phase transitions
plate tectonics
plates
sea-floor spreading
seismic surveys
seismology
shields
tangential discontinuities
tectonics
tectonophysics
thickness
undation
velocity structure
vertical discontinuities
volcanology

crust, weathering
use weathering crust

Crustacea
Includes use on level 2 under Arthropoda(1). See list F.
BA Mandibulata
Arthropoda
BT Invertebrata
NA Branchiopoda
Cirripedia
Copepoda
Malacostraca
NT Ostracoda
SA crustaceans
Trilobita

crustaceans
Term introduced in 1981 as the common name for Crustacea. See list F. To be used for Crustacea when they are used as tools in stratigraphy. Before 1981, search Crustacea AND biostratigraphy.
SA biostratigraphy
Crustacea

crustal shortening
Before 1978, also search shortening AND crustal.
UF shortening, crustal
SA gravity sliding
plate tectonics
tectonics
thrust faults

crustal structure
A valid term through 1978. See crust; structure.

crustal studies
A valid term through 1978. After 1978, see crust(1).

Cruziana
BA ichnofossils
SA Trilobita

cryokarst
use thermokarst

cryolite
As of 1981, includes use on level 3 under minerals(1) fluorides(2). See list L. Before 1981, included use on level 3 under minerals(1) halides(2) and on level 3 as a commodity term under fluorspar(1). As of 1981, use cryolite deposits for cryolite as a

commodity.
BA fluorides
halides
BT minerals
SA cryolite deposits

cryolite deposits
Term introduced in 1981. Includes use on level 3 as a commodity term (list C) under fluorspar(1). Before 1981, search cryolite AND deposits.
SA cryolite
fluorspar

cryopedology
Includes use on level 3 under glacial geology(1) periglacial features(2).
SA engineering geology
frost action
glacial geology
periglacial features
permafrost

cryotectonics
use glaciotectonics

cryoturbation
UF congeliturbation
frost churning
frost stirring
geliturbation
SA congelifraction
frost action
frost heaving
periglacial features
rock mechanics
soil mechanics

cryptoexplosion features
Includes use on level 2 under geomorphology(1). Before 1973, also search cryptoexplosion structures. Used for structures formed by the explosive release of energy, that exhibit intense rock deformation, and that show no obvious genetic relation to volcanism, tectonics, or meteorite impact.
UF features, cryptoexplosion
SA astroblemes
Charlevoix
cratering
craters
explosions
geomorphology
impact features
impactite
impacts
Manicouagan Crater
meteor craters
shatter cones
shock metamorphism
suevite

cryptoexplosion structures
A valid term through 1972. After 1972, use cryptoexplosion features.

Cryptolithus
Genus. Includes use on level 3 under Trilobita (1) Ptychopariida(2).
BA Ptychopariida
Trilobita
BT Trilobitomorpha
Arthropoda
Invertebrata

cryptomelane
BA oxides
BT minerals

cryptoperthite
BA feldspar group
framework silicates
silicates
BT minerals
SA perthite

Cryptostomata
Includes use on level 2 under Bryozoa(1). See list F.
BA Bryozoa
BT Invertebrata
NA Fenestellidae
Rhabdomesidae

crystal chemistry
Used for the relations among chemical composition, structure and properties of crystals. Includes use on level 1 (list A) and on level 2 under minerals(1). Term set options are:
mineral group (List L)
mineral species
topic (if more than one single specific mineral; e.g. bonding, coordination, ion exchange, order-disorder, partitioning, phase equilibria)
subtopic
UF stereochemistry
SA bonding
chemical analysis
clay mineralogy
coordination
crystal growth
crystal structure
crystallography
electron microscopy
geochemistry
ion exchange
isomorphism
lattice parameters
mineralogy
minerals
order-disorder
partition coefficients
partitioning
phase equilibria
substitution
transformations

crystal dislocations
Term introduced in 1981. Includes use on level 3 under crystal structure(1). Before 1981, also search dislocations AND minerals.
UF line defects
SA crystal growth
crystal structure
crystals
minerals

crystal field
Includes use on level 3 under minerals(1).
UF field, crystal
SA minerals

crystal form
UF form, crystal
SA crystal growth
habit

crystal growth
Used for studies on natural or artificial growth of crystals. Includes use on level 1 (list A). Term set options are:
mineral group (list L)
mineral species
topic (if more than one single specific mineral; e.g. crystal form, mechanism, phase equilibria, synthesis, twinning)
subtopic
SA clay mineralogy
crystal chemistry
crystal dislocations
crystal form
crystal structure
crystallography
electron microscopy
epitaxy
fluid inclusions
geochemistry
growth
growth spirals
habit
inclusions
intergrowths
lattice parameters
mineralogy
minerals
nucleation
overgrowths
phase equilibria
single-crystal method
synthesis
twinning
zoning

crystal habit
use habit

crystal lattice
use lattice

crystal, quartz
use quartz crystal

Crystal River Formation
In Ocala Group. N and W Florida.
BT Eocene
SA Florida
Ocala Group

crystal structure
Used for the internal structure of the crystal. Includes use on level 1 (list A) and level 2 under minerals(1). Also search atomic structure. Term set options are:
mineral group (list L)
mineral species
topic (if more than one single specific mineral; e.g. bonding, cell dimensions, defects, refinement)
subtopic
UF crystalline structure
SA atomic packing
bonding
cell dimensions
clay mineralogy
color centers
crystal chemistry
crystal dislocations
crystal growth
crystallinity
crystallography
defects
etching
exsolution
growth spirals
inclusions
ion exchange
lattice
lattice parameters
metamict minerals
metamictization
mineralogy
minerals
molecular structure
perovskite structure
polyhedra
polymorphism
polytypism
refinement
single-crystal method
space groups
superstructure
unit cell
X-ray analysis

crystalline rocks
Used when rocks cannot be distinguished as igneous or metamorphic.
SA igneous rocks
metamorphic rocks
rocks

crystalline structure
use crystal structure

crystallinity
Degree to which a clay mineral, such as illite, is crystalline.
SA clay mineralogy
crystal structure
diagenesis
illite

crystallites
SA magmas
volcanic glass

crystallization
Includes use on level 3 under intrusions(1) or magmas(1).
SA crystals
differentiation
fractional crystallization
genesis
granitization
intergrowths
intrusions
lithification
magmas
precipitation
recrystallization
water of crystallization

crystallization magnetization
use chemical remanent magnetization

crystallization remanent magnetization
use chemical remanent magnetization

crystallography
Treated as a whole. Used for general studies on the discipline of crystallography. Includes use on level 1 (list A). Term set options are:
topic [automatic data processing, bibliography, catalogs, classification, concepts, education, experimental studies, general, history, instruments, methods, nomenclature, objectives, observations, philosophy, practice, principles, symposia, textbooks, theoretical studies]
subtopic
SA atomic packing
chemical analysis
clay mineralogy
color centers
crystal chemistry
crystal growth
crystal structure
crystals
electron microscopy
geochemistry
growth spirals
holography
intergrowths
lattice parameters
mineralogy
minerals
monoclinic system
paragenesis
petrology
phase equilibria
pleochroism
reflectivity
spectroscopy
standard materials
X-ray analysis

crystals
Used as a general term. Includes use on level 3 under minerals(1).
SA crystal dislocations
crystallization
crystallography
intergrowths
minerals
phenocrysts

Cs
use cesium

Cs-137
Includes use on level 3 under isotopes(1).
SA cesium
isotopes

Ctenodontida
Term introduced in 1981. Includes use on level 2 under Mollusca(1). See list F.
BA Bivalvia
Mollusca
BT Invertebrata

Ctenostomata
Includes use on level 2 under Bryozoa(1). See list F.
BA Bryozoa
BT Invertebrata

Cu
use copper

Cuanza Basin
River basin in W central Angola.
UF Kwanza Basin
BT Angola

Cuanza-Sul
District on the South Atlantic in W.
BT Angola

Cuba
Includes use on level 1 as an area term (list O). For term set options see list B.
CO N195000N231500
W0740000W0850000
BA Greater Antilles
West Indies
NT Camaguey
Isle of Pines
La Habana
Las Villas
Oriente
Pinar del Rio

cubanite
UF chalmersite
BA sulfides
BT minerals

cube spar
use anhydrite

Cuddalore Sandstone
use Cuddalore Series

Cuddalore Sandstones
use Cuddalore Series

Cuddalore Series
Late Miocene-Pliocene. Also search Cuddalore Sandstone, and Cuddalore Sandstones.
UF Cuddalore Sandstone
Cuddalore Sandstones
BT Tertiary
SA India
Miocene
Pliocene
Tamil Nadu
upper Miocene

Cuddapah
Town and district in central Madras State, S India.
BT Andhra Pradesh
India

Cuddapah Basin
S central Andhra Pradesh. Also search Cuddapah.
BT Andhra Pradesh
India

Cuddapah System
Now included in Purana Group. Subdivided into Kistna Series, Nallamalai Series, Cheyair Series and Papaghni Series. Also search Cuddapah.
BT Precambrian
SA Andhra Pradesh
India

Cue
Town in W central.
BT Western Australia
Australia

Cuenca
As of 1981, refers to only the city. Before 1981, also included the province. In E central Spain.
BT Cuenca Province
New Castile
Spain

Cuenca Province
Term introduced in 1981. Before 1981, search Cuenca. In E central Spain.
CO N391300N403800
W0011000W0031300
BT New Castile
Spain
NT Cuenca
SA Serrania de Cuenca

Cuisian
Europe. Above Ypresian, below Lutetian. Includes use on level 3 under age terms(1). See list E.
BA lower Eocene
Eocene
Paleogene
Tertiary
BT Cenozoic

Culberson County
In W Texas.
BT Texas
United States

Culm
Provincial series, Europe.
BA Carboniferous
BT Paleozoic
SA Avonian
Dinantian

Cumana
City on the Caribbean Sea.
BT Venezuela
SA Sucre

Cumberland
County in NW England. Cities in Kentucky and Maryland. Town in Rhode Island. Village in central Virginia. Index countries and states as applicable.
BX England
Maryland
Rhode Island
Kentucky
Virginia

Cumberland County
Index states as applicable.
BX Illinois
Kentucky
Maine
New Jersey
North Carolina
Pennsylvania
Tennessee
Virginia
BT United States

Cumberland Mountains
use Cumberland Plateau

Cumberland Peninsula
On easternmost Baffin Island on Davis Strait in District of Franklin. Also search Cumberland AND peninsula.
BT Northwest Territories
Canada

Cumberland Plateau
Southwesternmost division of the Appalachians. Also search Cumberland AND plateau. Index states as applicable.
UF Cumberland Mountains
BT United States
SA Alabama
Appalachian Plateau
Appalachians
Kentucky
Tennessee
Virginia
West Virginia

cummingtonite
BA clinoamphibole
amphibole group
chain silicates
silicates
BT minerals
SA anthophyllite

cumulates
SA igneous rocks
magmas

Cupressaceae
Family.
BA Coniferales
gymnosperms
BT Spermatophyta
Plantae

cuprite
BA oxides
BT minerals

Curaca River basin
Just S of the Sao Francisco River in NE.
BT Bahia
Brazil

Curacao
Island in the Netherlands Antilles just N of W Venezuela.
BT Netherlands Antilles

Curie point
Includes use on level 3 under heat flow(1) or paleomagnetism(1). Also search Curie temperature.
UF Curie temperature
SA heat flow
magnetic properties
magnetic susceptibility
paleomagnetism
thermoremanent magnetization

Curie temperature
use Curie point

curium
Includes use on level 1 and 2 as chemical element (list D).
UF Cm
SA elements

current directions
Includes use on level 3 under ocean circulation(1) or sedimentation(1).
UF directions, current
SA currents
ocean circulation
sedimentation

current lineations
use parting lineation

current markings
Includes use on level 3 under sedimentary structures(1) bedding plane irregularities(2). See list K.
UF markings, current
BA bedding plane irregularities
sedimentary structures
SA flute casts
scour marks
tool marks

current partings
use parting lineation

current research
From 1978-1980, included use on level 2 under surveys(1). As of 1981, includes use on level 2 under survey organizations(1).
SA annual report
associations
geology
progress report
report
research
survey organizations

currents
Includes use on level 2 under ionosphere(1) and ocean circulation(1); on level 3 under aurora(1) electrical field(2) and magnetic field(2); on level 3 under meteorology(1) electrical phenomena(2).
NT bottom currents
convection currents
density currents
longshore currents
turbidity currents
SA aurora
continental shelf
current directions
electrical field
electrical phenomena
electrojet
ionosphere
lightning
magnetic field
meteorology
ocean circulation
paleocurrents
upwelling

curricula
Includes use on level 3 under education(1).
SA education

Curry County
Index states as applicable.
BX New Mexico
Oregon
BT United States

curved seismic interface
Term introduced in 1981.
SA discontinuities
inclined seismic interface
seismology

curves, traveltime
use traveltime curves

Custer County
Index states as applicable.
BX Colorado
Idaho
Montana
Nebraska
Oklahoma
South Dakota
BT United States

customs duty
Term introduced in 1981.
SA economics

cut and fill
Includes use on level 3 under sedimentary structures(1) planar bedding structures(2). See list K.
BA planar bedding structures
sedimentary structures

Cutch
District in E Gujarat. Also search Kutch. Rann of Cutch is to the N and NE.
UF Kutch
BT Gujarat
India

cutinite
Term introduced in 1981.
BA exinite
macerals
coal
organic residues
BT sedimentary rocks

Cutler Formation
Includes Halgaito Tongue, Cedar Mesa Sandstone Tongue, Organ Rock Tongue, White Rim Sandstone Member, DeChelly Sandstone Member, Hoskinnini Tongue, White River Sandstone, Rico transition facies. NE Arizona, SW Colorado, NW New Mexico, and SE Utah.
BT Permian
SA Arizona
Colorado
New Mexico
Utah

cutoff rigidities
Includes use on level 3 under interplanetary space(1) cosmic rays(2) and under magnetosphere(1) cos-

C

mic rays(2).
UF rigidities, cutoff
SA cosmic rays
interplanetary space
magnetosphere

Cuttack
City in E.
BT Orissa
India

cuttings
Includes use on level 3 under well-logging(1).
UF drill cuttings
SA boreholes
cores
well-logging
wells

Cuvier abyssal plain
Off Western Australia. Also search Cuvier.
BT Indian Ocean

Cyanophyta
Blue algae, blue-green algae and Schizophyta are included here. Includes use on level 2 under algae(1). See list F.
UF blue algae
blue-green algae
Schizophyta
BA algae
BT thallophytes
Plantae
SA Collenia

Cycadales
Includes use on level 2 under gymnosperms(1). See list F.
BA gymnosperms
BT Spermatophyta
Plantae
NA Ptilophyllum
SA Pachypteris
Taeniopteris

Cycadofilicales
Term introduced in 1981. Includes use on level 2 under gymnosperms(1). See list F.
BA gymnosperms
BT Spermatophyta
Plantae

Cyclades
Group of islands in S Aegean Sea.
BT Greek Aegean Islands
Greece
NT Naxos
Thera

cycles
Includes use on level 2 under geochemistry(1) and paleoclimatology(1).
SA atmosphere
cyclic processes
cyclothems
geochemistry
hydrologic cycle
hydrology
paleoclimatology
solar cycles

cyclic
A valid level 2 term through 1977. After 1977, use cyclic processes on level 2 under sedimentation(1).

cyclic loading
Includes use on level 3 under engineering geology(1).
SA engineering geology
loading

cyclic processes
Term introduced in 1978. Includes use on level 2 under sedimentation(1). Before 1978, also search sedimentation AND cyclic.
SA cycles
cyclothems
processes
sedimentation

Cyclocystoidea
Includes use on level 2 under Echinodermata(1). See list F.
BA Echinozoa
Echinodermata
BT Invertebrata

cyclosilicates
use ring silicates

Cyclostomata
Includes use on level 2 under Bryozoa(1). See list F.
BA Bryozoa
BT Invertebrata
SA Cystoporata

cyclothems
Includes use on level 3 under sedimentary structures(1) planar bedding structures(2). See list K.
BA planar bedding structures
sedimentary structures
NA megacyclothems
SA cycles
cyclic processes
rhythmite
sedimentation

cylindrical folds
Term introduced in 1978. Before 1978, search folds AND cylindrical.
UF cylindroidal fold
BA folds

cylindrical structures
Includes use on level 2 under sedimentary structures(1). See list K.
BA sedimentary structures
SA structures

cylindrite
BA sulfantimonates
sulfosalts
BT minerals

cylindroidal fold
use cylindrical folds

cymrite
BA sheet silicates
silicates
BT minerals
SA aluminosilicates

Cynodontia
As of 1981, includes use on level 3 under Reptilia(1) Therapsida(2). See list F. Before 1981, included use on level 3 under Reptilia(1) Synapsida(2).
BA Therapsida
Synapsida
Reptilia
BT Tetrapoda
Vertebrata

Cynomorpha
Term introduced in 1981. Includes use on level 2 under Mammalia(1). See list F.
BA simians
Primates
Mammalia
BT Tetrapoda
Vertebrata

Cypress Hills Formation
BT Oligocene
SA Canada
Saskatchewan

Cyprididae
Family. Includes use on level 3 under Ostracoda(1). See list F.
BA Podocopida
Ostracoda
BT Crustacea
Mandibulata
Arthropoda
Invertebrata
NA Candona

Cypridocopina
Term introduced in 1981. Includes use on level 2 under Ostracoda(1). See list F.
BA Ostracoda
BT Crustacea
Mandibulata
Arthropoda
Invertebrata

Cyprinidae
Family. As of 1981, includes use on level 3 under Pisces(1) Teleostei(2). See list F. Before 1981, included use on level 3 under Pisces(1) Osteichthyes(2).
BA Teleostei
Actinopterygii
Osteichthyes
Pisces
BT Vertebrata

Cyprus
Island nation in the Mediterranean Sea. Includes use on level 1 as an area term (list O). For term set options see list B.
CO N343000N354000
E0343000E0323000
BA Middle East
Asia
NT Troodos Massif
SA Mediterranean region
Near East

Cyrenaica
Easternmost part of country. Former province under the Italians.
BT Libya

Cyrtodontida
Term introduced in 1981. Includes use on level 2 under Mollusca(1). See list F.
BA Bivalvia
Mollusca
BT Invertebrata
NA Arcina
Ostreacea
Pteriina

cyrtolite
BA orthosilicates
silicates
BT minerals
SA zircon

Cyrtospirifer
As of 1981, includes use on level 3 under Brachiopoda(1) Spiriferida(2). See list F. Before 1981, included use on level 3 under Brachiopoda(1) Articulata(2).
BA Spiriferida
Articulata
Brachiopoda
BT Invertebrata

Cystoidea
Includes use on level 2 under Echinodermata(1). See list F.
BA Crinozoa
Echinodermata
BT Invertebrata

Cystoporata
BA Bryozoa
BT Invertebrata
SA Cyclostomata

Cytheracea
Includes use on level 3 under Ostracoda(1). See list F.
BA Podocopida
Ostracoda
BT Crustacea
Mandibulata
Arthropoda
Invertebrata
NA Cytheridae
Leptocythere
Trachyleberididae

Cytherella
Genus. Includes use on level 3 under Ostracoda(1). See list F.
BA Cytherellidae
Podocopida
Ostracoda
BT Crustacea
Mandibulata
Arthropoda
Invertebrata

Cytherellidae
Includes use on level 3 under Ostracoda(1). See list F.
BA Podocopida
Ostracoda
BT Crustacea
Mandibulata
Arthropoda
Invertebrata
NA Cytherella
Cytherelloidea

Cytherelloidea
Genus. Includes use on level 3 under Ostracoda(1). See list F.
BA Cytherellidae
Podocopida
Ostracoda
BT Crustacea
Mandibulata
Arthropoda
Invertebrata

Cytheridae
Includes use on level 3 under Ostracoda(1). See list F.
BA Cytheracea
Podocopida
Ostracoda
BT Crustacea
Mandibulata
Arthropoda
Invertebrata

Cytherocopina
Term introduced in 1981. Includes use on level 2 under Ostracoda(1). See list F.
BA Ostracoda
BT Crustacea
Mandibulata
Arthropoda
Invertebrata

Czech Bohemian Forest
Term introduced in 1981. Mountain range in W part of country. Indicates that part of the Bohemian Forest that lies in Czechoslovakia. Also search Czechoslovakia AND Bohemian Forest.
CO N483000N500300
E0150000E0123000
BT Czechoslovakia

Czech Erzgebirge
Term introduced in 1981.
CO N494500N505000
E0143500E0121000
BZ Erzgebirge
BX Czechoslovakia

Czech Sudeten Mountains
Term introduced in 1981.
CO N493000N510000
E0181500E0142000
BZ Sudeten Mountains
BX Czechoslovakia

Czechoslovakia
Includes use on level 1 as an area term (list O). For term set options see list B.
CO N473000N510000
E0224500E0120000
BA Central Europe
Europe
NT Barrandian Basin
Berounka System
Beskid Mountains
Bohemia

Bohemia-Moravia System
Bohemian Basin
Czech Bohemian Forest
NX Czech Erzgebirge
Czech Sudeten Mountains
NT Moravia
Slovakia
NX Slovakian Carpathians
Slovakian Pannonian Basin
NT Sudeten
SA Beshchady Mountains
Carpathian Foredeep
Carpathian Foreland
Carpathians
Danube River
Danube Valley
Elbe River
Elbe Valley
Erzgebirge
Isergebirge
Izera Mountains
Moldanubian
North Sudetic Basin
Oder Valley
Pannonian Basin
Silesia
Snieznik
Subcarpathians
Sudetic Basin
Tatra Mountains
Vienna Basin
Western Carpathians

Czekanowskiales
BA Ginkgoales
gymnosperms
BT Spermatophyta
Plantae

Czestochowa
City in N.
BT Katowice
Poland

Czestochowa-Zawierce Basin
In NE Katowice.
BT Katowice
Poland

D

D-region
Includes use on level 2 under ionosphere(1).
SA E-region
F-region
ionosphere

D/H
SA deuterium
hydrogen
isotopes
stable isotopes
tracers

Dacht-e-Nawar
use Dasht-i-Nawar

Dacian
Europe. Includes use on level 3 under age terms(1). See list E.
BA Pliocene
Neogene
Tertiary
BT Cenozoic

Dacian Basin
Region roughly covering modern Romania.
BT Romania

dacite
Includes use on level 3. As of 1981, not to be used in the igneous rocks set. See list H. Before 1981, included use on level 3 under igneous rocks(1) andesite-rhyolite family(2).
BA dacites
volcanic rocks
BT igneous rocks

dacite porphyry
As of 1981, includes use on level 3 under igneous rocks(1) dacites(2). See list H. Before 1981, included use on level 3 under igneous rocks(1) andesite-rhyolite family(2).
BA dacites
volcanic rocks
BT igneous rocks
SA porphyry

dacites
Term introduced in 1981. Includes use on level 2 under igneous rocks(1). See list H. Before 1981, search dacite.
BA volcanic rocks
BT igneous rocks
NA dacite
dacite porphyry

Dacryoconarida
Order.
BA problematic fossils
SA Mollusca

Dactylioceratidae
As of 1981, includes use under Mollusca(1) Ammonoidea(2). See list F. Before 1981, included use on level 3 under Mollusca(1) Cephalopoda(2).
BA Ammonoidea
Tetrabranchiata
Cephalopoda
Mollusca
BT Invertebrata

Dade County
Index states as applicable.
BX Florida
Georgia
Missouri
BT United States

Dadoxylon
Genus. Includes use on level 3 under gymnosperms(1).
BA gymnosperms
BT Spermatophyta
Plantae
SA Coniferales
Cordaitales
Ginkgoales
Glossopteris

Dagestan
Autonomous Soviet Socialist Republic on W shore of Caspian Sea. Also search Daghestan.
CO N410000N450000
E0470000E0460000
UF Daghestan
BT Russian Republic
USSR

Daghestan
use Dagestan

dahllite
BZ carbonates
phosphates
BT minerals
SA carbonate apatite

Dahomey
A valid index term through 1976. After 1976, use Benin.

daily variation
use diurnal variations

Daito Ridge
Term introduced in 1981. NW Philippine Sea.
CO N215000N250000
E1360000E1313000
BT Philippine Sea
Pacific Ocean

Dakar
City on the Atlantic Ocean.
BT Senegal

Dakota Formation
E Colorado, NE New Mexico, North Dakota, Kansas, Minnesota, SE Montana, Nebraska, South Dakota, W Oklahoma, E Wyoming. Also search Dakota Sandstone; Dakota Group.
UF Dakota Group
Dakota Sandstone
BT Cretaceous
SA Colorado
Kansas
Lower Cretaceous
Minnesota
Montana
Nebraska
New Mexico
North Dakota
Oklahoma
South Dakota
Upper Cretaceous
Wyoming

Dakota Group
use Dakota Formation

Dakota Sandstone
use Dakota Formation

Dalarna
Region in W central Sweden. Also search Dalecarlia.
UF Dalarne
Dalecarlia
BT Sweden

Dalarne
use Dalarna

Dalat
Town NE of Saigon.
BT Vietnam

Dalecarlia
use Dalarna

Dalgaranga
Region.
BT Western Australia
Australia

Dallas
City in Dallas County in NE central.
BT Texas
United States

Dallas County
Index states as applicable.
BX Alabama
Arkansas
Iowa
Missouri
Texas
BT United States

Dalmatia
Region, including many islands, which extends along the Adriatic Sea from Zadar on the N to near the Albanian border on the S. Sometimes the name is applied to most of the Yugoslav coast.
BT Yugoslavia

Dalradian
Great Britain, especially Scotland.
SA Cambrian
Precambrian

damage
Term introduced in 1981. Used for damage resulting from natural phenomena.
SA earthquakes
effects
engineering geology
geologic hazards

damage, radiation
use radiation damage

Damara System
Includes Khomas Series, Marble Series, and Quartzite Series, Chuos Tillite, and Otavi Series. In W Damaraland region. NW of Walvis Bay.
BT Proterozoic
SA Namibia

Damascus
City in SW.
BT Syria

Damodar Valley
River valley. Index states as applicable.
BT India
SA Bihar
West Bengal

Dampier Ridge
NE of Sidney in N.
BT Tasman Sea
Pacific Ocean

damping
As of 1981, no longer a valid term for GeoRef. Use attenuation.

dams
Level 1 term as of 1978 used for geologic studies on dams. Includes use on level 2 under foundations(1). If level 1, term set options are:
construction
subtopic
design
subtopic
foundations
subtopic
site exploration
subtopic
NT arch dams
SA construction
design
earth dams
embankments
engineering geology
foundations
grouting
land subsidence
Oroville Dam
reservoirs
rock mechanics
site exploration
slope stability
soil mechanics
water storage

dams and damsites
Not a valid index term for GeoRef. Use dams.

Damuda Series
Forms middle part of lower Gondwanas. Subdivided into three stages: Raniganj (series), Ironstone shales, and lower Damudas, afterwards called Barakars.
BT Permian
SA Barakar Stage
Bihar
India
Karharbari Stage
Madhya Pradesh
Raniganj Stage

Dan River basin
Index states as applicable.
BT United States
SA North Carolina
Virginia

Danakil
use Afar

Danakil Depression
use Afar Depression

danburite
BA framework silicates
silicates
BT minerals

Dane County
Term introduced in 1981. S Wisconsin.
CO N425000N431700

D

W0890000W0895100
BT Wisconsin
United States

Dangerous Islands
use Tuamotu Islands

Danian
Europe. Above Maestrichtian (Cretaceous), below Montian. Includes use on level 3 under age terms(1). See list E.
BA lower Paleocene
Paleocene
Paleogene
Tertiary
BT Cenozoic
NT Niniyur Group

Daniel's Harbour
Village on upper Gulf of Saint Lawrence in NW.
BT Newfoundland
Canada

Dankalia
use Afar

Danube Delta
On the Black Sea primarily in Romania. Index Romania and/or Ukraine as applicable. Also search Danube River delta.
UF Danube River delta
BT Europe
SA Romania
Ukraine

Danube Plain
Covers the Walachian Plain and the N Bulgaria lowlands. Also search Danube. Index countries as applicable.
BT Europe
SA Bulgaria
Romania

Danube River
Flows E from the Black Forest and turns S just N of Budapest. It then flows S, SE, N, and finally E into the Black Sea. Index Ukraine and countries as applicable. Also Search Danube.
UF Donau River
Duna River
Dunai River
Dunarea River
Dunau River
SA Austria
Bulgaria
Czechoslovakia
Germany
Hungary
Romania
Ukraine
Yugoslavia

Danube River delta
use Danube Delta

Danube Stage
Term introduced in 1978. Europe.
BA lower Pleistocene
Pleistocene
Quaternary
BT Cenozoic

Danube Valley
River valley. Index Ukraine and countries as applicable. Also search Danube.
BT Europe
SA Austria
Bulgaria
Czechoslovakia
Germany
Hungary
Romania
Ukraine
Yugoslavia

Danville
City in Pittsylvania County on the North Carolina border.
BT Virginia
United States

Danzig
use Gdansk

Daonella
Includes use on level 3 under Mollusca(1). See list F.
BA Bivalvia
Mollusca
BT Invertebrata

Darasun
Town in S central Chita Oblast. N of Mongolia.
BT Russian Republic
USSR

Darcy's law
SA aquifers
fluid phase
gases
ground water
hydrodynamics
permeability
petroleum
porosity
reservoir properties

Dare County
In E between Albermarle and Pamlico Sounds.
BT North Carolina
United States

Darfur
Province in W.
BT Sudan

Darien
E part of isthmus between Gulf of Darien on E and Gulf of San Miguel on W.
BT Panama
Central America

Darjeeling
Town, hill station, and district in extreme N near Sikkim border.
UF Darjiling
BT West Bengal
India

Darjiling
use Darjeeling

Darling Downs
Tableland in SE.
BT Queensland
Australia

Darling Range
In SW.
BT Western Australia
Australia

Darmstadt
City in S.
BT Hesse
West Germany

Dartmoor
Tableland in Devonshire in SW.
BT England
Great Britain
United Kingdom

Darvaz
Peak in the W Pamirs in central Tadzhikistan.
UF Kaganovich Peak
BT Tadzhikistan
USSR

Darvaza
Town in Ashkhabad Oblast in S.
BT Turkmenia
USSR

Darvaza Range
Branch of Alai Range N of Afghanistan in S central.
BT Tadzhikistan
USSR
SA Alai Range

Darvel Bay
Inlet of Celebes Sea in Sabah in East Malaysia.
BT Malaysia
SA Celebes Sea

Darwin
City on the Timor Sea in N.
BT Northern Territory
Australia

Darwin glass
Includes use on level 3 under tektites(1).
UF queenstownite
BA tektites
SA glasses
impactite

Darwinula
Includes use on level 3 under Ostracoda(1). See list F.
BA Podocopida
Ostracoda
BT Crustacea
Mandibulata
Arthropoda
Invertebrata

Dashkesan
City in W.
BT Azerbaidzhan
USSR

Dasht-e Bayaz
use Dasht-e-Bayaz

Dasht-e-Bayaz
Region in NE part of country. Locale of earthquake in 1968. Also search Dasht-e Bayaz.
UF Dasht-e Bayaz
BT Iran

Dasht-e-Lut Basin
use Lut Desert

Dasht-e-Nawar
use Dasht-i-Nawar

Dasht-i-Lut
use Lut Desert

Dasht-i-Nawar
Volcanic region in central Afghanistan. Also search Dacht-e-Nawar.
UF Dacht-e-Nawar
Dasht-e-Nawar
BT Afghanistan

Dastakert
Village in S.
BT Armenia
USSR

Dasycladaceae
Includes use on level 2 under algae(1). See list F.
BA Chlorophyceae
Chlorophyta
algae
BT thallophytes
Plantae

data
Used as a general term. Before 1975 also search analytical data.
SA automatic data processing
data acquisition
data bases
data retrieval
electron microscopy data
electron probe data
ion probe data
mineral data
new data
SEM data
TEM data
X-ray data

data acquisition
Term introduced in 1978.
UF acquisition, data
SA automatic data processing
data
data storage

data analysis
As of 1981, no longer a valid term for GeoRef. See automatic data processing.

data bases
Also search data systems.
UF data systems
SA automatic data processing
data
data handling
data retrieval
data storage
information systems

data handling
Term introduced in 1978.
UF handling, data
SA automatic data processing
data bases
data retrieval
information systems

data processing
As of 1981, no longer a valid term for GeoRef. Use automatic data processing.

data processing, automatic
use automatic data processing

data retrieval
UF retrieval, data
SA automatic data processing
data
data bases
data handling
data storage
information systems

data storage
SA automatic data processing
data acquisition
data bases
data retrieval
information systems
storage

data systems
use data bases

dates
Includes use on level 2 under absolute age(1).
SA absolute age
geochronology

Datil-Mogollon volcanic field
Term introduced in 1981. SW New Mexico.
CO N321000N343000
W1063000W1090300
BT New Mexico
United States
SA Mogollon Mountains

dating, fission-track
use fission-track dating

dating, particle-track
use particle-track dating

datolite
BA orthosilicates
silicates
BT minerals

Dauphine
Historical region and former province in SE.
BT France

Dauphine Alps
W offshoot of Cottian Alps in SE.
BT France
SA Alps
Bas-Dauphine
Cottian Alps
Pelvoux Massif
Vercors

Davenport
City on the Mississippi River in Scott County.
BT Iowa
United States

Davidson
Town in Mecklenburg County on the South Carolina border in S.
BT North Carolina
United States

Davidson County
Index states as applicable.
BX North Carolina
Tennessee
BT United States
NT Nashville

Davidsville Group
In NE Newfoundland.
BT Ordovician
SA Canada
Newfoundland

Davis Mountains
Jeff Davis County in W.
BT Texas
United States

Davis Strait
Between SW Greenland and Baffin Island.
BT Atlantic Ocean

Davos
Town consisting of Davos Platz and Davos Dorf in N central.
BT Graubunden
Switzerland

Dawson
City in W central.
BT Yukon Territory
Canada

dawsonite
BA carbonates
BT minerals

De Soto County
Term introduced in 1981. Index states as applicable.
BX Florida
Mississippi
BT United States

De Twente
use Twente

Dead Sea
A salt lake lying in part of the Great Rift Valley and constituting the lowest point on the Earth's surface. Also search Dead Sea Rift. Index countries as applicable.
BT Middle East
SA Dead Sea Rift
Great Rift Valley
Israel
Jordan

Dead Sea Rift
BT Middle East
SA Dead Sea
Israel
Jordan

Deadwood Formation
In N Black Hills, conformably underlies Aladdin Sandstone. SE Montana, W South Dakota (Black Hills), and E Wyoming.
BT Paleozoic
SA Lower Ordovician
Montana
South Dakota
Upper Cambrian
Wyoming

death assemblages
use thanatocenoses

Death Valley
Valley in San Bernardino and Inyo counties, California. If applicable, also index/search Death Valley National Monument which includes a larger area extending into Nevada.
CO N353500N370000
W1161500W1173000
BT California
United States

Debrecen
City in E part of country.
BT Hungary

debris
Used as a general term.
SA detritus
geomorphology
regolith

debris flows
As of 1978, term is used on level 2 under slope stability(1).
SA flows
slope stability

debris slopes
use talus slopes

Decapoda
As of 1981, no longer a valid term for GeoRef. Use Malacostraca. Before 1981, included use on level 3 under Arthropoda(1) Crustacea(2).

Decatur County
In NW Kansas.
BT Kansas
United States

Decaturville
Village in Camden County in central.
BT Missouri
United States

decay, radioactive
use radioactive decay

Decazeville
Town in S central part of country.
BT Aveyron
France

Deccan Intertrappean
use Intertrappean Beds

Deccan Intertrappean Beds
use Intertrappean Beds

Deccan Intertrappean Series
use Intertrappean Beds

Deccan Plateau
Triangular tableland between the Eastern and Western Ghats and between the Narmada River on the N and the Krishna River on the S. Also search Deccan. Index states as applicable.
UF Dekkan Plateau
BT India
NX Lonar Crater
SA Andhra Pradesh
Indian Shield
Madhya Pradesh
Maharashtra
Mysore
Orissa

Deccan Trap
use Deccan Traps

Deccan Traps
Upper Cretaceous to lower Eocene. Named after the step-like aspect of the flat topped hills of The Deccan Plateau. Also search Deccan Trap.
UF Deccan Trap
SA Cretaceous
Eocene
India
Intertrappean Beds
lower Eocene
Upper Cretaceous

Deception Island
One of Britain's South Shetland Islands off N Antarctic Peninsula.
BA South Shetland Islands
Scotia Sea Islands

Dechenellidae
Includes use on level 3 under Trilobita(1).
BA Trilobita
BT Trilobitomorpha
Arthropoda
Invertebrata

decision-making
Term introduced in 1981.
SA legislation
policy

declination
Includes use on level 3 under Earth(1), geochronology(1), or paleomagnetism(1).
SA geochronology
magnetic field
paleomagnetism

decollement
Includes use on level 3 under folds(1) mechanics(2).
UF detachment
SA deformation
disharmonic folds
faults
flexural-slip
folds
mechanics
tectonics

decomposition
use chemical weathering

deconvolution
SA distortion
elastic waves
filters
seismology

decrepitation
Includes use on level 3 under chemical analysis(1) or mineralogy(1).
SA chemical analysis
geologic thermometry
mineralogy
sample preparation

dedolomitization
SA calcitization
diagenesis
dolomitization
processes

deep earthquakes
use deep-focus earthquakes

deep faults
use deep-seated structures

deep focus
A valid term through 1977. After 1977, use deep-focus earthquakes.

deep magnetic sounding
Term introduced in 1981. Includes use on level 3 under geophysical methods(1).
SA geophysical methods
geophysical surveys
magnetic methods
magnetic surveys

Deep Sea Drilling Project
Also search DSDP.
UF DSDP
SA DSDP Site 105
DSDP Site 125
DSDP Site 132
DSDP Site 157
DSDP Site 158
DSDP Site 163
DSDP Site 167
DSDP Site 214
DSDP Site 225
DSDP Site 227
DSDP Site 228
DSDP Site 253
DSDP Site 254
DSDP Site 259
DSDP Site 260
DSDP Site 261
DSDP Site 262
DSDP Site 263
DSDP Site 264
DSDP Site 270
DSDP Site 274
DSDP Site 284
DSDP Site 292
DSDP Site 296
DSDP Site 297
DSDP Site 315
DSDP Site 319
DSDP Site 320
DSDP Site 321
DSDP Site 322
DSDP Site 323
DSDP Site 325
DSDP Site 327
DSDP Site 329
DSDP Site 330
DSDP Site 332
DSDP Site 333
DSDP Site 334
DSDP Site 335
DSDP Site 356
DSDP Site 357
DSDP Site 361
DSDP Site 362
DSDP Site 363
DSDP Site 364
DSDP Site 367
DSDP Site 368
DSDP Site 369
DSDP Site 370
DSDP Site 372
DSDP Site 374
DSDP Site 379
DSDP Site 380
DSDP Site 381
DSDP Site 391
DSDP Site 395
DSDP Site 396
DSDP Site 417
DSDP Site 418
IPOD
Leg 1
Leg 10
Leg 11
Leg 12
Leg 13
Leg 14
Leg 15
Leg 16
Leg 17
Leg 18
Leg 19
Leg 2
Leg 20
Leg 21
Leg 22
Leg 23
Leg 24
Leg 25
Leg 26
Leg 27
Leg 28
Leg 29
Leg 3
Leg 30
Leg 31
Leg 32
Leg 33
Leg 34
Leg 35
Leg 36
Leg 37
Leg 38
Leg 39
Leg 4
Leg 40
Leg 41
Leg 42A
Leg 42B
Leg 43
Leg 44
Leg 45
Leg 46
Leg 47
Leg 48
Leg 49
Leg 5
Leg 50
Leg 51

Leg 52
Leg 53
Leg 54
Leg 55
Leg 56
Leg 57
Leg 58
Leg 59
Leg 6
Leg 60
Leg 61
Leg 62
Leg 63
Leg 64
Leg 65
Leg 66
Leg 7
Leg 8
Leg 9
marine geology
ocean basins
ocean floors
oceanography

deep seismic sounding
Includes use on level 3 under seismology(1) and geophysical methods(1). Also search DSS; seismic sounding.
UF DSS
seismic sounding
SA deep sounding
explosions
geophysical methods
geophysical surveys
geophysics
seismic methods
seismology
sounding

deep sounding
Includes use on level 3 under geophysical methods(1).
SA deep seismic sounding
echo sounding
geophysical methods
geophysical surveys
sounding

deep structure
use deep-seated structures

deep-focus earthquakes
Term introduced in 1978. Includes use on level 3 under earthquakes(1). Before 1978, search earthquakes AND deep focus.
UF deep earthquakes
SA earthquakes
focus
intermediate-focus earthquakes
shallow-focus earthquakes

deep-sea
A valid term through 1977. After 1977, use deep-sea environment.

deep-sea environment
Term introduced in 1978. Before 1978, search deep-sea or deep sea. As of 1981, used for abyssal environment. Also search abyssal environment.
SA deep-sea sedimentation
environment

deep-sea fans
use submarine fans

deep-sea sedimentation
Term introduced in 1981.
SA deep-sea environment
sedimentation

deep-seated structures
Term introduced in 1978. Includes use on level 3 under tectonics(1). Before 1978, search deep structure; deep-seated. Also search deep structure; deep faults.
UF deep faults
deep structure
SA structures
tectonics

deep-tow methods
Term introduced in 1978. Before 1978, search deep-tow. Includes use on level 3 under geophysical methods(1).
BA geophysical methods
SA acoustical methods
methods

deepening
use dredging

deerite
BA chain silicates
silicates
BT minerals

defects
As of 1978, term is used on level 2 under crystal structure(1).
SA crystal structure
deformation
destruction

definition
Used as a general term.
SA classification
identification
nomenclature
photogrammetry

Deflandrea
BA Dinoflagellata
palynomorphs

deforestation
SA conservation
ecology
forestry
forests
reclamation

deformation
Used for the process of folding, faulting, shearing, compression, or extension of the rocks as a result of various forces. Only small-scale deformations considered here. See tectonics(1) for large scale. Includes use on level 1 (list A) and level 2 under rock mechanics(1) and soil mechanics(1). Term set options are:
experimental studies
effects (e.g. creep, elastic strain, flow lines, fractures, kink-band structures, plastic flow, recrystallization, relaxation, twin-gliding) or name of modulus or limit (e.g. compression, elastic limit, fracture strength, shock metamorphism, tension, torsion, yield strength) or theory (e.g. elasticity, plasticity, viscoelasticity, viscosity) or material
field studies
effects or name of modulus or limit or material
theoretical studies
effects or name of modulus or limit or material
SA anelasticity
Atterberg limits
axial-plane structures
boudinage
bulk modulus
cataclasis
competent materials
compression
compressive strength
creep
decollement
defects
dilatancy
distortion
ductility
elastic constants
elastic limit
elastic materials
elastic properties
elastic strain
elastic waves
elasticity
engineering geology
enterolithic folds
extension
extensometers
fabric
failures
faults
field studies
finite strain analysis
flexure
flow lines
folds
foliation
fracture strength
fractures
geodesy
geophysics
Hooke's law
isoclinal folds
kink-band structures
lamellae
lineation
mechanical properties
melange
metamorphic rocks
metamorphism
meteor craters
microcracks
mobilization
mylonitization
orogeny
petrofabrics
plastic flow
plasticity
plate tectonics
pressure
recrystallization
rheology
rock mechanics
salt tectonics
seismology
shatter cones
shear
shear stress
shock waves
soil mechanics
strain
strainmeters
strength
stress
structural analysis
tectonics
tectonite
tectonophysics
tensile strength
tension
torsion
triaxial tests
viscoelasticity
viscosity
yield strength

degasification
use degassing

degassing
Includes use on level 3 under atmosphere(1) or mantle(1). Also search degasification; outgassing.
UF degasification
SA atmosphere
mantle

deglaciation
Includes use on level 3 under glacial geology(1) glaciation(2).
SA glaciation
glaciers

degradation
Includes use on level 3 under geomorphology(1), soils(1), or weathering(1). Also search deterioration.
UF deterioration
SA denudation
erosion
geomorphology
leaching
soils
transport
weathering

Dehra Dun
City in N Uttar Pradesh. Also search Dehradun.
UF Dehradun
BT Uttar Pradesh
India

Dehradun
use Dehra Dun

dehydration
Includes use on level 3 under geochemistry(1), minerals(1), or sediments(1).
UF drying
SA geochemistry
hydration
minerals
sediments
water of dehydration

Deimos
SA Mars
Phobos
satellites

Deinotherioidea
Term introduced in 1981. Includes use on level 2 under Mammalia(1). See list F.
BA Proboscidea
Mammalia
BT Tetrapoda
Vertebrata

Dekkan Plateau
use Deccan Plateau

Del Monte Beach
On Monterey Peninsula in Monterey County in W.
BT California
United States

Del Norte County
Extreme NW.
BT California
United States

Del Rio
City in Val Verde County on the Rio Grande in SW.
BT Texas
United States

delafossite
BA oxides
BT minerals

Delaware
Includes use on level 1 as an area term (list O). For term set options see list B.
CO N382700N395000
W0750500W0754600
BA Eastern U.S.
United States
NX Kent County
Sussex County
SA Aquia Formation
Atlantic Coastal Plain
Calvert Formation
Carolina Bays
Choptank River
Delaware Bay
Delaware River
Delaware River basin
Delmarva Peninsula
Magothy Formation
Marshalltown Formation
Monmouth Group
Navesink Formation
Newark Group
Potomac Group
Raritan Formation
Saint Marys Formation
Wicomico Formation
Wilmington
Wissahickon Formation

Delaware Basin
SE New Mexico and SW Texas. Index states as applicable.
BT Permian Basin
SA New Mexico
Texas

Delaware Bay
Arm of Atlantic Ocean. Index states as applicable.
BT United States
SA Delaware
New Jersey

Delaware County
Index states as applicable.
BX Indiana
Iowa
New York
Ohio
Oklahoma
Pennsylvania
BT United States

Delaware Limestone
Overlies Columbus Limestone; underlies Olentangy Shale. Central and N Ohio.
BT Middle Devonian
Devonian
SA Ohio

Delaware River
Flows from S New York SE into Delaware Bay. Index states as applicable.
BT United States
SA Delaware
New Jersey
New York
Pennsylvania

Delaware River basin
Index states as applicable.
BT United States
SA Delaware
New Jersey
New York
Pennsylvania

Delhi
Territory in N central part of country. Also a city NNE of New Delhi.
BT India
NT New Delhi

Delhi System
Included Ajabgarh Series, Hornstone Breccia, Kushalgarh Limestone, Alwar Series, Raialo limestones and quartzes. Hornstone Breccia and Kushalgarh Limestone later confined to NE Rajputana, and Raialos excluded from Delhis to form separate unit. Named after city of Delhi.
BT Precambrian
SA Haryana
India

Delinesti
Village in Caras-Severin County in W part of country.
BT Banat
Romania

dellenite
As of 1981, includes use on level 3 under igneous rocks(1) rhyodacites(2). See list H. Before 1981, included use on level 3 under igneous rocks(1) andesite-rhyolite family(2).
BA rhyodacites
volcanic rocks
BT igneous rocks
SA rhyodacite

Dellys
City on Mediterranean Sea E of Algiers.
BT Algeria

Delmarva Peninsula
Between Chesapeake and Delaware bays. Index states as applicable.
BT United States
SA Delaware
Maryland
Virginia

Delmas
Village in W central.
BT Saskatchewan
Canada

Delocrinus
Genus. Includes use on level 3 under Echinodermata(1) Crinoidea(2).
BA Crinoidea
Crinozoa
Echinodermata
BT Invertebrata

Delta Area
BT Netherlands

Delta River
In E Alaska.
BT Alaska
United States

deltaic
A valid term through 1977. After 1977, use deltas or (as of 1981) deltaic environment.

deltaic environment
Term introduced in 1981.
SA deltaic sedimentation
deltas
ecology
environment
paleoecology
sedimentation

deltaic sedimentation
Term introduced in 1981.
SA deltaic environment
deltas
sedimentation

deltas
Includes use on level 3 under geomorphology(1) fluvial features(2) and shore features(2). As of 1981, use deltaic environment for type of environment. Before 1981, included use on level 3 under ecology(1), paleoecology(1), and sedimentation(1) as type of environment.
SA deltaic environment
deltaic sedimentation
ecology
fluvial features
geomorphology
paleoecology
sedimentation
shore features

demagnetization
Includes use on level 3 under geophysics(1) or paleomagnetism(1).
NT adiabatic demagnetization
chemical demagnetization
thermal demagnetization
SA geophysics
magnetic hysteresis
magnetization
paleomagnetism
remagnetization
remanent magnetization

demand
Term introduced in 1981.
SA economics

Demospongea
Includes use on level 2 under Porifera(1). See list F.
BA Porifera
BT Invertebrata

Denali Fault
Mount McKinley area in S central. Extends into Yukon Territory.
BT North America
SA Alaska
Yukon Territory

Denbighshire
County in N.
BT Wales
Great Britain
United Kingdom

dendrograms
BA statistical analysis

Dendroidea
Includes use on level 2 under Graptolithina(1). See list F.
BA Graptolithina
BX Hemichordata
Invertebrata

Dengizkul
Lake SW of Bukhara.
BT Uzbekistan
USSR

denitrification
Includes use on level 3 under geochemistry(1).
SA geochemistry

Denmark
Includes use on level 1 as an area term (list O). For term set options see list B.
CO N541500N580000
E0130000E0080000
BA Scandinavia
Western Europe
Europe
NT Bornholm
Copenhagen
Faeroe Islands
Helsingor
Vendsyssel
Zealand
SA Baltic region
European Platform
Fennoscandia
Jutland
North Sea Coast
Oresund
Zechstein

densities and temperatures
Includes use on level 2 under aeronomy(1).
SA aeronomy
density
ion densities and temperatures
temperature

density
Includes use on level 3 under Earth(1). Also search densities.
SA convection currents
densities and temperatures
ion densities and temperatures
isostasy
sea water
specific gravity

density currents
BT currents
SA turbidity currents

Dentalium
Genus. Includes use on level 3 under Mollusca(1) Scaphopoda(2).
BA Scaphopoda
Mollusca
BT Invertebrata

denudation
Includes use on level 3 under geomorphology(1) processes(2).
SA degradation
erosion
erosion features
geomorphology
water erosion
weathering

Denver
City in Denver County in NE central.
BT Colorado
United States

Denver Basin
Primarily in central and NE Colorado. Index states as applicable.
UF Denver Julesberg Basin
BT United States
SA Colorado
Nebraska
Wyoming

Denver Julesberg Basin
use Denver Basin

depletion
As of 1981, restricted to resources. Before 1981, used as a general term.
SA conservation
mineral resources
natural resources
petroleum
reclamation

deposition
Includes use on level 2 under sedimentation(1); on level 3 under glacial geology(1) glaciation(2).
SA accretion
alluvial fans
alluvial plains
cementation
diagenesis
glacial geology
glaciation
ice movement
reservoirs
reworking
sedimentary processes
sedimentary structures
sedimentation
sediments
stratification
uniformitarianism

depositional fault
use growth faults

depositional remanent magnetization
Also search detrital remanent magnetization.
UF detrital remanent magnetization
BA remanent magnetization
SA paleomagnetism

deposits
As of 1981, no longer a valid term for GeoRef. See list C. Before 1981, included use for all papers on nonmetal commodities.

depressions
Includes use on level 3 under geomorphology(1).
SA basins
Carolina Bays
folds
geomorphology

depth
Includes use on level 2 under Mohorovicic discontinuity(1).
SA depth indicators
dimensions
Mohorovicic discontinuity

depth indicators
Term introduced in 1981.
SA depth
indicators

depth of compensation
use carbonate compensation depth

Derby
City in Derbyshire in central.
BT England
Great Britain
United Kingdom

Derbyshire
County in central.
CO N524500N533000
W0011500W0020000
BT England

Great Britain
United Kingdom

Dermapteroida
Term introduced in 1981. Includes use on level 2 under Insecta(1). See list F.
BA Insecta
BT Mandibulata
Arthropoda
Invertebrata

Dermoptera
Term introduced in 1981. Includes use on level 2 under Mammalia(1). See list F.
BA Mammalia
BT Tetrapoda
Vertebrata

Des Moines Series
use Desmoinesian

desalination
use desalinization

desalinization
Includes use on level 3 under ground water(1) or water resources(1). Also search desalination.
UF desalination
SA brines
ground water
leaching
salinity
salt
sea water
soils
treatment

Descartes
Village in NW part of country.
BT Oran region
Algeria

descloizite
BA vanadates
BT minerals

description
Used as a general term.
SA landform description
morphology

Desert Basin
use Canning Basin

desert pediplains
use pediplains

desert plains
use pediplains

Desert soils
Includes use on level 3 under soils(1). See list M.
BA soils
SA soil group
Zonal soils

desertification
SA ecology

deserts
Includes use on level 3 under geomorphology(1) eolian features(2).
SA arid environment
ecology
eolian features
geomorphology
playas

desiccation
SA ecology
moisture
pore water
soils

design
As of 1978, includes use on level 2 under dams(1), foundations(1), marine installations(1), nuclear facilities(1), reservoirs(1), shorelines(1), tunnels(1), underground installations(1), and waterways(1).
SA aseismic design
dams
development
engineering geology
foundations
instruments
marine installations
nuclear facilities
reservoirs
shorelines
tunnels
underground installations
waterways

Desmidiales
Term introduced in 1981. Includes use on level 2 under algae(1). See list F. Before 1981, search desmids. As of 1981, search desmids only for biostratigraphy.
BA Chlorophyta
algae
BT thallophytes
Plantae
SA desmids

desmids
As of 1981, to be used for Desmidiales when they are used as tools in stratigraphy. See list F. Before 1981, search desmids AND biostratigraphy. Before 1981, included use on level 3 under algae(1) Chlorophyta(2).
SA biostratigraphy
Desmidiales

desmine
use stilbite

Desmoceratida
Term introduced in 1981. Includes use on level 2 under Mollusca(1). See list F.
BA Ammonoidea
Tetrabranchiata
Cephalopoda
Mollusca
BT Invertebrata

Desmoinesian
Comprises Cherokee Group, Marmaton Group, Cabaniss Group, and Krebs Group. Subdivided into Venteran and Cygnian substages. Also search Des Moines Series; Desmoinesian Series.
UF Des Moines Series
Desmoinesian Series
BA Middle Pennsylvanian
Pennsylvanian
Carboniferous
BT Paleozoic
NT Cabaniss Formation
Cherokee Group
Labette Shale
Marmaton Group
Savanna Formation
Wewoka Formation
SA Arkansas
Bevier Coal
Francis Creek Shale
Iowa
Kansas
Missouri
Nebraska
Oklahoma

Desmoinesian Series
use Desmoinesian

Desmostylia
Includes use on level 2 under Mammalia(1). See list F.
BA Mammalia
BT Tetrapoda
Vertebrata

Desolation Islands
use Kerguelen Islands

desorption
SA adsorption
geochemistry

destruction
Used as a general term.
SA catastrophes
corrosion
defects
geologic hazards

detachment
use decollement

detection
Includes use as a level 2 or 3 term appropriate to a large number of topics, e.g. on level 2 under earthquakes(1). See list G.

detergents
Includes use on level 3 under environmental geology(1) waste disposal(2).
SA environmental geology
liquid waste
waste disposal
waste water

deterioration
use degradation

detrital deposits
Term introduced in 1981.
SA detritus
mineral deposits

detrital fan
use alluvial fans

detrital remanent magnetization
use depositional remanent magnetization

detrital sedimentation
Term introduced in 1981.
SA sedimentation

detritus
Includes use on level 3 under sediments(1) or sedimentary rocks(1). Also search detrital; detrital minerals.
SA abrasion
debris
detrital deposits
erosion
weathering

Detroit River Group
Includes Sylvania Sandstone, Amherstburg Formation, Lucas Formation. SE Michigan, N Ohio, and W Ontario.
BT Middle Devonian
Devonian
SA Michigan
Ohio
Ontario

deuterium
Includes use on level 1 and 2. See list D (chemical elements). Also search H-2.
UF H-2
SA D/H
elements
hydrogen
isotopes
stable isotopes

Deux-Sevres
Department in W.
CO N460000N471000
E0001500W0010000
BT France
SA Poitou

developing countries
use Third World

development
Used as a general term.
SA design
economic geology
exploration
ground water

Devensian
Europe. See list E.
BA upper Pleistocene
Pleistocene
Quaternary
BT Cenozoic

deviation, standard
use standard deviation

Devils Icebox
Region in Boone County in central.
BT Missouri
United States

devitrification
SA geochemistry
glasses
volcanic glass

Devoluy
Limestone range of the Dauphine Pre-Alps in SE.
BT France

Devon Island
N of Baffin Island off Baffin Bay in District of Franklin. Also search Devon.
CO N743000N763000
W0793000W0970000
UF North Devon Island
BT Northwest Territories
Canada

Devonian
Includes use on level 1 as an age term (list E); on level 2 under paleoterms, e.g. paleoecology, paleogeography, paleomagnetism. Above Silurian, below Carboniferous.
BT Paleozoic
NT Acadian Phase
Barre Granite
Cedar City Formation
Genesee Group
Heemskirk Granite
Hunsruck Shale
Keg River Formation
Levis Shale
Lilydale Limestone
Lost Burro Formation
NA Lower Devonian
NT Martin Formation
NA Middle Devonian
NT Miette Complex
Muth Quartzite
Old Red Sandstone
Ramparts Formation
Tor Formation
Traverse Group
NA Upper Devonian
NT Upper Old Red Sandstone
Waterways Formation
SA Antler Orogeny
Beacon Supergroup
Bedford Shale
Berea Sandstone
Broken River Formation
Catskill Formation
Chattanooga Shale
Galway Granite
Geirud Formation
Hercynian Orogeny
Hidden Valley Dolomite
Hunton Group
Peel Sound Formation
Read Bay Formation
Ringerike Sandstone
Rondout Formation
Talladega Group
Woodford Shale

Devonshire
County in SW England.
CO N501500N511500
W0030000W0043000
BT England
Great Britain
United Kingdom

dewatering
As of 1981, no longer a valid term for GeoRef. See drawdown. Included

use on level 3 under ground water(1).

Dewey County
Index states as applicable.
BX Oklahoma
South Dakota
BT United States

dextral faults
use right-lateral faults

Dhanbad
Town in SE.
BT Bihar
India

Dharwar
City in NW Mysore. Now part of the joint municipality of Hubli-Dharaar.
BT Mysore
India

Dharwars
"Sub metamorphic" crystalline rocks of the Dharwar System, named after city of Dharwar in Mysore, but occurring throughout the country.
SA Chitaldrug schist belt
India

Dhenkanal
Town and district in E central.
BT Orissa
India

diabase
As of 1981, includes use on level 2 under igneous rocks(1). See list H. Before 1981, included use on level 3 under igneous rocks(1) basalt family(2). Before 1981, also search dolerite.
BT igneous rocks
NA diabase porphyry
olivine diabase
quartz diabase
tholeiitic dolerite
SA metadiabase
trap rocks

diabase porphyry
As of 1981, includes use on level 3 under igneous rocks(1) diabase(2). See list H. Before 1981, included use on level 3 under igneous rocks(1) basalt family(2).
BA diabase
BT igneous rocks
SA porphyry

Diablo Bolson
use Diablo Platform

Diablo Platform
Extreme W Texas.
UF Diablo Bolson
BT Texas
United States

Diablo Range
One of the Coast Ranges in W central California.
BT California
United States
SA Coast Ranges

diachronism
SA biostratigraphy
stratigraphy

diagenesis
Level 1 term as of 1978, used for all the chemical, physical and biological changes, modifications, or transformations undergone by a sediment after its initial deposition, and during and after its lithification, exclusive of weathering and metamorphism. Includes use on level 2 under sedimentary rocks(1) and sediments(1). If level 1, term set options are:
controls
subtopic
dolomitization
subtopic
effects
subtopic
geochemistry
subtopic
indicators
subtopic
materials
subtopic
mechanism
subtopic
processes
subtopic
sedimentation
subtopic
topic [e.g. environment, experimental studies, models, theoretical studies]
subtopic
NA cementation
chertification
early diagenesis
late diagenesis
syngenesis
SA alteration
authigenesis
calcification
calcitization
coalification
compaction
consolidation
crystallinity
dedolomitization
deposition
dolomitization
halmyrolysis
indicators
lithification
materials
metamorphism
metasomatism
phosphatization
pressure solution
processes
pyritization
sedimentary petrology
sedimentary rocks
sedimentary structures
sedimentation
sediments
silicification
weathering

diagonal lamination
use cross-laminations

diagonal-slip faults
Term introduced in 1978. Before 1978, search faults AND diagonal-slip.
BA faults

diagrams
As of 1981, no longer a valid term for GeoRef. See graphic methods.

diallagite
As of 1981, includes use on level 3 under igneous rocks(1) ultramafics(2). See list H. Before 1981, included use on level 3 under igneous rocks(1) ultramafic family(2).
BA ultramafics
BT igneous rocks

dialogite
use rhodochrosite

diamictite
Includes use on level 3 under sedimentary rocks(1) clastic rocks(2). See list I. Also search mixtite.
BA clastic rocks
BT sedimentary rocks
SA mixtite
terrigenous materials

diamicton
UF symmicton
BA clastic sediments
BT sediments
SA till

diamond
When of economic value, use diamonds. As of 1981, includes use on level 3 under minerals(1) native elements(2). See list L. Before 1981, included use on level 3 under minerals(1) native elements and alloys(2).
BA native elements
BT minerals
SA carbon
chaoite
diamonds
gems
graphite
kimberlite
lonsdaleite

diamonds
Use only when of economic value. Includes use on level 1 as a commodity term (list C);on level 2 under placers(1).
SA abrasives
carbon
carbonado
diamond
gems
kimberlite
placers

diaphorite
use ultrabasite

diaphthoresis
use retrograde metamorphism

diapiric fold
use diapirs

diapirism
SA diapirs
intrusions
salt domes
salt tectonics

diapirs
Includes use on level 3 under folds(1) style(2) and under salt tectonics(1).
UF diapiric fold
piercement
piercing fold
SA anticlines
cap rocks
diapirism
domes
folds
salt
salt domes
salt tectonics

diaspore
BA oxides
BT minerals

diastrophism
As of 1981, no longer a valid term for GeoRef. See orogeny; epeirogeny.

diatom flora
Term introduced in 1981 as the common name for diatoms. See list F. To be used for diatoms when they are used as tools in stratigraphy. Before 1981, search diatoms AND biostratigraphy.
SA biostratigraphy
diatoms

diatomaceous earth
Includes use on level 3 under sedimentary rocks(1) clastic rocks(2). See list I. Also search tripolite. Before 1981, also search diatomite AND sedimentary rocks.
UF earth, diatomaceous
tripolite
BA clastic rocks
BT sedimentary rocks
SA diatomite
diatoms

diatomite
Includes use on level 1 as a commodity term (list C). As of 1981, to be used only if of economic value; otherwise, use diatomaceous earth. Before 1981, included use on level 3 under sedimentary rocks(1) clastic rocks(2).
SA abrasives
diatomaceous earth

diatoms
Includes use on level 2 under algae(1). See list F.
BA algae
BT thallophytes
Plantae
NA Melosira
Navicula
Nitzschia
SA diatom flora
diatomaceous earth
frustules
phytoplankton
plankton

diatremes
Includes use on level 2 under intrusions(1).
BA intrusions
SA breccia pipes
pipes
volcanic features

Dibranchiata
Term introduced in 1981. Includes use on level 2 under Mollusca(1). See list F.
BA Cephalopoda
Mollusca
BT Invertebrata
NA Belemnoidea

Dickenson County
In SW Virginia.
BT Virginia
United States

Dickinson County
Index states as applicable.
BX Iowa
Kansas
Michigan
BT United States

dickite
BA clay minerals
sheet silicates
silicates
BT minerals

Dicotyledoneae
Includes use on level 2 under angiosperms(1). See list F. Before 1981, also search dicotyledons. As of 1981, search dicotyledons only for biostratigraphy.
BA angiosperms
BT Spermatophyta
Plantae
NA Acer
Alnus
Annonaceae
Artemisia
Asteridae
Betula
Caryophyllidae
Dilleniidae
Ericaceae
Euphorbiaceae
Fagus
Hamamelididae
Juglandaceae
Juglans
Lauraceae
Leguminosae
Magnoliidae
Quercus
Rosidae
Ulmaceae

SA Monocotyledoneae

Dicroidium
Includes use on level 3 under pteridophytes(1) Filicopsida(2).
BA Filicopsida
pteridophytes
BT Plantae
SA gymnosperms
Pteridospermae

dictionaries
Includes use on level 3 under lexicons(1).
SA glossaries
lexicons

Dictyocha
BA Silicoflagellata
Protista
BT Invertebrata

Dictyonellidina
Term introduced in 1981. Includes use on level 2 under Brachiopoda(1). See list F.
BA Articulata
Brachiopoda
BT Invertebrata

Didacna
Genus. Includes use on level 3 under Mollusca(1). See list F.
BA Bivalvia
Mollusca
BT Invertebrata

Didymograptina
Includes use on level 2 under Graptolithina(1). See list F.
BA Graptoloidea
Graptolithina
BX Hemichordata
Invertebrata
NA Didymograptus
Isograptus

Didymograptus
Genus.
BA Didymograptina
Graptoloidea
Graptolithina
BX Hemichordata
Invertebrata

dielectric constant
UF constant, dielectric
SA dielectric properties
electrical field
electrical methods

dielectric properties
SA dielectric constant
electrical field
electrical methods
properties

Dieppe
City on English Channel.
BT Seine-Maritime
France

differential thermal analysis
Used on level 1 for methodology through 1977. After 1977, includes use on level 2 under thermal analysis(1).
UF thermography
SA analysis
chemical analysis
clay mineralogy
clay minerals
DTA data
electron microscopy
sample preparation
spectroscopy
thermal analysis
thermogravimetric analysis
X-ray diffraction analysis

differential thermal analysis data
use DTA data

differential weathering
Before 1978, also search weathering AND differential.
BT weathering

differentiation
Includes use on level 2 under magmas(1) and on level 3 under intrusions(1), igneous rocks(1), and Earth(1).
SA crystallization
Earth
fractional crystallization
horizon differentiation
hybridization
igneous rocks
intrusions
magmas

diffraction
Includes use on level 3 under seismology(1).
SA ocean waves
reflection
refraction
seismology

diffusion
Includes use on level 2 under aeronomy(1), meteorology(1), and ocean circulation(1); on level 3 under geochemistry(1) processes(2). See also under specific elements. Included use on level 2 under oceanography(1) until 1976.
SA absolute age
aeronomy
chemical analysis
geochemistry
ion exchange
meteorology
ocean circulation
oceanography
permeability
polarography
processes

diffusivity, thermal
use thermal diffusivity

Difunta Group
SA Coahuila
Cretaceous
lower Paleocene
Mexico
Paleocene
Upper Cretaceous

digenite
UF alpha chalcocite
blue chalcocite
BA sulfides
BT minerals
SA chalcocite

Digha
Coastal region along the Bay of Bengal.
BT West Bengal
India

digital simulation
Term introduced in 1978. Before 1978, search digital AND simulation.
SA automatic data processing
simulation

digital techniques
As of 1981, no longer a valid term for GeoRef. See automatic data processing.

Digne
City in SE part of country.
BT Alpes-de-Haute Provence
France

Dijon
City in E central part of country.
BT Cote-d'Or
France

dike swarms
Includes use on level 3 under intrusions(1).
SA dikes
igneous rocks
intrusions
ring dikes
swarms

dikes
Includes use on level 2 under intrusions(1).
BA intrusions
NT ring dikes
SA batholiths
clastic dikes
dike swarms
igneous rocks
laccoliths
ring complexes
ring structures
sills

dilatancy
Includes use on level 3 under deformation(1), seismology(1), earthquakes(1), and geophysics(1).
SA deformation
earthquakes
physical properties
seismology

dilatation
use dilation

dilatational wave
use P-waves

dilation
Used as a general term.
UF dilatation

Dilleniidae
Term introduced in 1981. Includes use on level 2 under angiosperms(1). See list F.
BA Dicotyledoneae
angiosperms
BT Spermatophyta
Plantae

Dillon
Village in Summit County in N central.
BT Colorado
United States

dimension stone
As of 1981, includes use on level 3 as a commodity term (list C) under construction materials(1), granite deposits(1), sandstone deposits(1), limestone deposits(1), and marble deposits(1). Before 1981, included use on level 3 as a commodity term under construction materials(1), granite(1), sandstone(1), limestone(1), and marble(1).
UF stone, dimension
BA construction materials
SA building stone
granite deposits
limestone deposits
marble deposits
sandstone deposits

dimensions
SA cell dimensions
depth
grain size
grains

Dimitrovo coal basin
use Pernik coal basin

Dimmit County
Term introduced in 1981. SW Texas.
CO N281000N284000
W0992500W1001000
BT Texas
United States

dimorphism
As of 1981, no longer a valid term for GeoRef. Use polymorphism. Before 1981, included use on level 3 under crystal structure(1) and fossil groups (list F).

Dinant
Town in S part of country.
BT Namur
Belgium

Dinant Basin
S part of country.
BT Namur
Belgium

Dinantian
Europe. Includes Tournaisian and Visean. Includes use on level 3 under age terms(1). See list E. As of 1981, used for Lower Carboniferous. Before 1981, also search Lower Carboniferous.
BA Carboniferous
BT Paleozoic
NA Tournaisian
Visean
SA Avonian
Culm
Mississippian
Upper Carboniferous

Dinaric Alps
SE division of Eastern Alps along Adriatic Sea. Also search Dinaric.
CO N430000N450000
E0180000E0150000
BT Europe
SA Alps
Eastern Alps
Northern Limestone Alps
Velebit Mountains
Yugoslavia
Zlatibor Mountains

Dingle Peninsula
Juts into Atlantic in W County Kerry.
CO N520500N521500
W0094500W0103000
BT Ireland

Dinocerata
Term introduced in 1981. Includes use on level 2 under Mammalia(1). See list F.
BA Mammalia
BT Tetrapoda
Vertebrata

Dinoflagellata
Hystrichosphaeridae is included here and under acritarchs. Includes use on level 2 under palynomorphs(1). See list F. Also search Dinophyceae. Before 1981, also search dinoflagellates. As of 1981, search dinoflagellates only for biostratigraphy.
UF Dinophyceae
BA palynomorphs
NA Deflandrea
NZ Hystrichosphaeridae
NA Peridinium
SA acritarchs
dinoflagellates
plankton
Pyrrhophyta

dinoflagellates
Term introduced in 1981 as the common name for Dinoflagellata. See list F. To be used for Dinoflagellata when they are used as tools in stratigraphy. Before 1981, search Dinoflagellata AND biostratigraphy.
SA biostratigraphy
Dinoflagellata

Dinophyceae
use Dinoflagellata

Dinosauria
use dinosaurs

dinosaurs
As of 1981, includes use on level 2 under Reptilia(1). See list F. Before 1981, included use on level 3 under Reptilia(1).
UF Dinosauria
BA Reptilia
BT Tetrapoda
Vertebrata
SA Archosauria
Hadrosauridae

diogenite
Term introduced in 1981. Includes use on level 3 under meteorites(1).
BA achondrites
BT meteorites

Diois
Massif in SE part of country.
BT Drome
France

diopside
BA clinopyroxene
pyroxene group
chain silicates
silicates
BT minerals
SA chrome diopside

diorite
Includes use on level 3. As of 1981, not to be used in the igneous rocks set. See list H. Before 1981, included use on level 3 under igneous rocks(1) diorite family(2).
BA diorites
BT igneous rocks
SA metadiorite
microdiorite
quartz diorite

diorite family
As of 1981, no longer a valid term for GeoRef. Use diorites, alkali diorites, or quartz diorites as applicable. See list H. Before 1981, included use on level 2 under igneous rocks(1).

diorite porphyry
As of 1981, includes use on level 3 under igneous rocks(1) diorites(2). See list H. Before 1981, included use on level 3 under igneous rocks(1) volcanic rocks(2) and diorite family(2).
BA diorites
BT igneous rocks
SA porphyry

diorites
Term introduced in 1981. Includes use on level 2 under igneous rocks(1). See list H. Before 1981, search diorite family.
BT igneous rocks
NA alkali diorites
diorite
diorite porphyry
ferrodiorite
mangerite
microdiorite
monzodiorite
plagiogranite
quartz diorites
syenodiorite
tonalite
trondhjemite
SA appinite

dip
As of 1978, term is used on level 3 as a noun. Before 1978, included use for dip faults and dip fractures.
UF angle of dip
SA dip-slip faults
dipmeter logging
orientation
strike
structural analysis

dip faults
Term introduced in 1978. Before 1978, search faults AND dip.
BA faults

dip fractures
Term introduced in 1978. Before 1978, search dip AND fractures.
BA fractures

dip-slip faults
Term introduced in 1978. Includes use on level 3 under faults(1) displacements(2). Before 1978, search faults AND dip-slip.
BA faults
SA dip
displacements

Diplograptina
Includes use on level 2 under Graptolithina(1). See list F.
BA Graptoloidea
Graptolithina
BX Hemichordata
Invertebrata

dipmeter logging
Term introduced in 1978. Includes use on level 2 under well-logging(1). Before 1978, also search well-logging AND dipmeter or dip-meter.
UF logging, dipmeter
BA well-logging
SA boreholes
dip

Dipnoi
Term introduced in 1981. Includes use on level 2 under Pisces(1). See list F.
BA Osteichthyes
Pisces
BT Vertebrata

dipole moment
SA magnetic field
paleomagnetism

Diprotodonta
Term introduced in 1981. Includes use on level 2 under Mammalia(1). See list F.
BA Mammalia
BT Tetrapoda
Vertebrata

Diptera
As of 1981, includes use on level 3 under Insecta(1). Before 1981, included use on level 2 under Insecta(1). See list F.
BA Insecta
BT Mandibulata
Arthropoda
Invertebrata
SA Raphidiodea

dipyrite
use pyrrhotite

direct magnetization
Term introduced in 1981.
BA magnetization

direct problem
Term introduced in 1981.
SA geophysics
inverse problem
seismology

directions, current
use current directions

directory
Includes use on level 3 under commodity terms(1), e.g. under mineral resources(1). See list C.
SA guidebook

discharge
Includes use on level 3 under hydrology(1) and ground water(1). Before 1981, also used for effluents.
SA effluents
ground water
hydrology
recharge
runoff
water recovery

Discoasteridae
use discoasters

discoasters
Includes use on level 3 under algae(1) nannofossils(2). See list F. Also search discoaster; Discoasteridae
UF Discoasteridae
BA nannofossils
algae
BT thallophytes
Plantae
NA Ethmodiscus
SA Coccolithophoraceae
Sphenolithus

Discocyclina
Genus. Includes use on level 3 under foraminifera(1) Orbitoidacea(2).
BA Orbitoidacea
Rotaliina
foraminifera
BT Protista
Invertebrata

discontinuities
Includes use on level 3 under crust(1), mantle(1), or seismology(1). Use for interface in seismology.
NT horizontal discontinuities
tangential discontinuities
vertical discontinuities
SA Conrad discontinuity
crust
curved seismic interface
inclined seismic interface
mantle
Mohorovicic discontinuity
seismology

discontinuous materials
Term introduced in 1981.
SA engineering geology
materials

Discorbacea
Includes use on level 2 under foraminifera(1). See list F.
BA Rotaliina
foraminifera
BT Protista
Invertebrata

discordant folds
Term introduced in 1978. Includes use on level 3 under folds(1) orientation(2). Before 1978, also search folds AND discordant.
BA folds
SA orientation

discoveries
As of 1981, restricted to economic geology. Before 1981, used as a general term.
SA mineral deposits
placers

Discovery Bay
Village on central N coast.
BT Jamaica
West Indies

Discovery Deep
Also search Discovery AND deeps.
BT Red Sea

discrete filters
Term introduced in 1981.
BT filters

discriminant analysis
UF discriminant function analysis
BA statistical analysis
SA analysis
mathematical geology

discriminant function analysis
use discriminant analysis

disharmonic folds
Term introduced in 1978. Includes use on level 3 under folds(1) style(2). Before 1978, also search folds AND disharmonic.
BA folds
SA decollement
flexural-slip
harmonic folds

Disko Island
Davis Strait just off SW Greenland. Also search Disko.
CO N683000N702000
W0520000W0550000
BT Greenland

dislocations
As of 1981, no longer a valid term for GeoRef. For dislocations within crystals, use crystal dislocations. Before 1981, included use on level 3 under faults(1), seismology(1), and crystal structure(1).

dispersion
As of 1981, to be used for the chemical process of dispersion. Before 1981, included use as a general term for dispersal; also search dispersal.
SA fresh water
geochemistry
ground water
salt water
salt-water intrusion
sea water

dispersion patterns
Includes use on level 3 under mineral exploration(1) geochemical methods(2).
SA geochemical methods
haloes
leakage anomalies
mineral exploration
patterns
primary dispersion
secondary dispersion

dispersive filters
Term introduced in 1981.
BT filters

displacement theory
use continental drift

displacements
Includes use on level 2 under faults(1).
SA dip-slip faults
drag folds
faults
nappes
overthrust faults
reactivation
reverse faults
seismic moment
transcurrent faults
transform faults
wrench faults

display, graphic
use graphic display

disposal, waste
use waste disposal

disseminated deposits
Term introduced in 1978. Before 1978, search disseminated AND ore deposits; search deposits AND disseminated.
SA mineral deposits

dissipation
As of 1981, no longer a valid term for GeoRef. Before 1981, used as a general term.

dissociation
Includes use on level 3 under geochemistry(1) processes(2).
SA electrolysis
  geochemistry
  processes

dissolution
A valid term through May 1978. After May 1978, use solution or solutions, e.g. under geochemistry(1).

dissolved materials
Term introduced in 1978. Before 1978, search dissolved.
SA materials
  suspended materials

Distephanus
BA Silicoflagellata
  Protista
BT Invertebrata

distortion
Includes use on level 3 under deformation(1) and seismology(1).
SA amplitude distortion
  deconvolution
  deformation
  elastic waves
  filters
  linear distortion
  non-linear distortion
  phase distortion
  seismology
  signal distortion
  wave dispersion

distribution
Includes use as a level 2 or 3 term appropriate to a large number of topics, e.g. on level 2 under faults(1), folds(1), fractures(1) and under ocean circulation(1). See list G.
SA size distribution
  spatial distribution
  statistical distribution
  temporal distribution

District of Columbia
Includes use on level 1 as an area term (list O). For term set options see list B.
CO N384700N390000
  W0765400W0771000
BA Eastern U.S.
  United States
NT Rock Creek Park
  Smithsonian Institution
SA Anacostia River basin
  Potomac River
  Potomac River basin
  Wicomico Formation

District of Franklin
The northernmost district.
CO N630000N850000
  W0600000W1250000
UF Franklin District
BT Northwest Territories
  Canada
NT Arctic Archipelago
SA Eureka Sound Formation

District of Keewatin
In SE Northwest Territories. Before 1978, also search Keewatin.
CO N600000N690000
  W0830000W1020000
UF Keewatin District
BT Northwest Territories
  Canada
SA Keewatin

District of Kenora
Extreme N, NW, and W.
BT Ontario
  Canada

District of Mackenzie
Central and W Northwest Territories. Also search Mackenzie; Mackenzie District.
CO N600000N710000
  W1020000W1370000
UF Mackenzie District
BT Northwest Territories
  Canada

Distrito Federal
use Federal District

disturbances
Term is restricted to extraterrestrial phenomena. Includes use on level 2 under ionosphere(1).
SA intensity
  ionosphere
  magnetic storms
  magnetosphere
  solar activity
  solar cycles
  solar flares
  substorms
  sudden commencements
  variations

Ditrau
Town in Mures County in SW central.
BT Transylvania
  Romania

Dittonian
In Old Red Sandstone, England: upper Gedinnian; above Downtonian. Includes use on level 3 under age terms (1). See list E.
BA Lower Devonian
  Devonian
BT Paleozoic
SA Old Red Sandstone

diurnal variations
Also search diurnal.
UF daily variation
SA magnetic field
  variations

diversity
Includes use on level 3 under fossil groups (list F).
SA species diversity

Dixie Valley
Churchill County in W.
BT Nevada
  United States

Djadjerud Valley
BT Iran

Djadokhta Formation
In the Gobi Desert.
BT Upper Cretaceous
  Cretaceous
SA Mongolia

Djibouti
Became independent June 27, 1977. Formerly French Territory of the Afars and Issas. Capital city also named Djibouti. As of 1978, includes use on level 1 as an area term. Before 1978, also search Afars and Issas Territory.
CO N105000N125000
  E0433000E0414000
UF French Somaliland
BA East Africa
  Africa
SA Afar
  Red Sea Basin

djurleite
BA sulfides
BT minerals

Djursland
Peninsula of Arhus County jutting into the Kattegat in E.
BT Jutland
  Denmark

Dnepropetrovsk
City and oblast in E central.
BT Ukraine
  USSR

Dnieper Basin
River drainage basin. Index Soviet republics as applicable. Also search Dnieper; Dnieper region.
UF Dnieper region
BT USSR
SA Byelorussia
  Dnieper-Donets Basin
  Russian Republic
  Ukraine

Dnieper region
use Dnieper Basin

Dnieper River
Rises in S Valdai Hills and flows S, then SE into a big bend at Dnepropetrovsk, and finally into the Black Sea near Kherson. Index Soviet republics as applicable. Also search Dnieper.
BT USSR
SA Byelorussia
  Russian Republic
  Ukraine

Dnieper-Donets Basin
Region comprising the Donets Basin and that part of the Dnieper Basin in the nearby Dnepropetrovsk area. Index Soviet republics as applicable. Also search Dnieper-Donets; Dnieper-Donets Depression; Dnieper-Donets region.
CO N460000N520000
  E0420000E0300000
UF Dnieper-Donets Depression
  Dnieper-Donets region
BA Russian Platform
BT USSR
SA Dnieper Basin
  Donets Basin
  Russian Republic
  Ukraine

Dnieper-Donets Depression
use Dnieper-Donets Basin

Dnieper-Donets region
use Dnieper-Donets Basin

Dniester River
Flows into Black Sea SW of Odessa. Index Soviet republics as applicable. Also search Dniester, Dniester region, Dniester Valley.
BT USSR
SA Moldavia
  Ukraine

Dniester-Prut Interfluve
Area between the two rivers. Includes all of Moldavia which formerly was Bessarabia. Index Soviet republics as applicable.
BT USSR
SA Moldavia
  Ukraine

Doberlug
Town in SE part of country.
BT Cottbus Bezirk
  East Germany

Doboy Sound
McIntosh County on the Atlantic Ocean.
BT Georgia
  United States

Dobrogea
use Dobruja

Dobrudja
use Dobruja

Dobrudzha
use Dobruja

Dobruja
Region on the Black Sea. Also search Dobrogea which is the Romanian name.
UF Dobrogea
  Dobrudja
  Dobrudzha
BT Romania
NT Altin-Tepe
  Babadag Lake basin
  Constanta
  Mangalia
  Sulina
  Tulcea
SA Bulgaria

Dobruja Basin
Black Sea coastal strip S of the Danube river including NE Bulgaria. Index countries as applicable. Also search Dobruja.
BT Europe
NZ Bulgarian Dobruja
SA Bulgaria
  Romania

Dobrzyn
Town in N central part of country.
BT Bydgoszcz
  Poland

Dockum Group
Includes Sloan Canyon Formation, Sheep Pen Canyon Formation, Tecovas Shale, Santa Rosa Sandstone, Chinle Formation, Pierce Canyon Redbeds, Baldy Hill Formation, Travesser Formation. Colorado, Kansas, Oklahoma, New Mexico, and W Texas.
BT Upper Triassic
  Triassic
SA Chinle Formation
  Colorado
  Kansas
  New Mexico
  Oklahoma
  Texas

Docodonta
Includes use on level 2 under Mammalia(1). See list F.
BA Mammalia
BT Tetrapoda
  Vertebrata

Doda
Village in SW.
BT Jammu and Kashmir
  India

Dodecanese
Group of islands in SE Aegean Sea.
BT Greek Aegean Islands
  Greece
NT Rhodes

Dogger
Europe. Above Lias, below Malm. Includes use on level 3 under age terms(1). See list E.
BA Middle Jurassic
  Jurassic
BT Mesozoic
SA Aalenian
  Bajocian
  Bathonian
  Callovian

Dogger Bank
Submerged sand bank.
BT North Sea
  Atlantic Ocean

Dogo
Largest of Oki Islands off W coast of Honshu in Japan Sea.
BT Honshu
  Japan
SA Oki Islands

Dogtooth Mountains
In SE British Columbia.
BT British Columbia
  Canada

dolerite
As of 1981, no longer a valid term for GeoRef. Use diabase. Before

1981, included use on level 3 under igneous rocks(1) basalt family(2).

Dolina
City in Lvov Oblast in W.
BT Ukraine
USSR

dolinen
use dolines

dolines
UF dolinen
SA hydrogeology
karst
sinkholes

dolomite
Includes use on level 3 under minerals(1) carbonates(2). For sediments, use carbonate sediments; dolomitic composition. To indicate the rock made of dolomite, use dolostone. Before 1976, also search dolomite for the commodity, the rock, and the sediment. See list I (sed. rocks), list L (minerals), and list N (sediments).
UF bitter spar
magnesian spar
BA carbonates
BT minerals
SA ankerite
calcite
carbonate rocks
carbonate sediments
chemically precipitated rocks
dolomitic composition
dolomitic limestone
dolomitization
dolostone
evaporites
light minerals
marbles

Dolomite Alps
use Dolomites

dolomite marble
A valid metamorphic rock term through 1977. After 1977, use marbles AND dolomite.

Dolomites
Mountain range of the Eastern Alps between the Adige and Piave river valleys in NE part of country. Index autonomous regions as applicable.
CO N460000N464500
E0121500E0113000
UF Dolomite Alps
BT Italy
SA Alps
Cima d'Asta
Eastern Alps
Latemar Massif
Lessini Mountains
Trentino-Alto Adige
Veneto

dolomitic composition
Term introduced in 1978. Includes use on level 3 under sediments(1) and sedimentary rocks(1). Before 1978, also search composition AND dolomitic.
SA composition
dolomite
dolomitization
dolostone
sedimentary rocks
sediments

dolomitic limestone
Term introduced in 1981. Includes use on level 3 under sedimentary rocks(1) carbonate rocks(2). See list I. Before 1981, also search magnesian limestone.
BA limestone
carbonate rocks
BT sedimentary rocks
SA dolomite
dolomitization
dolostone

dolomitite
use dolostone

dolomitization
As of 1978, term is used on level 2 under diagenesis(1).
SA calcitization
dedolomitization
diagenesis
dolomite
dolomitic composition
dolomitic limestone
dolostone
limestone
metasomatism
processes
sedimentary rocks
sedimentation
sediments

dolostone
Includes use on level 3 under sedimentary rocks(1) carbonate rocks(2). See list I. To be used for rocks made of dolomite (mineral). As of 1981, use dolostone deposits for dolostone as a commodity. Before 1981, included use on level 1 as a commodity term.
UF dolomitite
BA carbonate rocks
BT sedimentary rocks
SA carbonate sediments
dolomite
dolomitic composition
dolomitic limestone
dolomitization
dolostone deposits

dolostone deposits
Term introduced in 1981. Includes use on level 1 and 2 as a commodity term (list C). Before 1981, search dolostone AND deposits.
SA dolostone
limestone deposits
marble deposits

domain structure
As of 1981, no longer a valid term for GeoRef. Use magnetic domains.

domains
Used as a general term.
SA environment
facies
magnetic domains
sedimentary petrology

Domerian
Europe. Includes use on level 3 under age terms(1). See list E.
BA middle Liassic
Lower Jurassic
Jurassic
BT Mesozoic

domes
Includes use on level 3 under folds(1) style(2).
BA folds
SA anticlines
arches
batholiths
diapirs
geomorphology
intrusions
laccoliths
salt domes
uplifts

Dominica
Term introduced in 1981. Island between Guadeloupe and Martinique in the Lesser Antilles.
BT Lesser Antilles

Dominican Republic
Occupies eastern two thirds of island of Hispaniola in the Greater Antilles. Includes use on level 1 as an area term (list O). For term set options see list B.
CO N173000N200000
W0682000W0720000
BA Hispaniola
Greater Antilles
West Indies
NT Santo Domingo

Don Basin
River basin in central S and S Russian Republic. Also search Don.
BT Russian Republic
USSR
SA Don River

Don Juan Pond
In Wright Valley on W shore of Ross Sea near Ross Ice Sheet in N Victoria Land.
BT Antarctica

Don River
Rises SE of Tula and flows SE into a big bend near the Volga River then flows SW into the Azov Sea at Rostov. Also search Don.
BT Russian Republic
USSR
SA Don Basin

Dona Ana County
On the Mexican border.
BT New Mexico
United States
NT Kilbourne Hole

Donau River
use Danube River

Donbas
use Donets Basin

Donbass
use Donets Basin

Donegal
County. W of Northern Ireland.
BT Ireland
NT Ardara Pluton
Rosses Granite

Donets Basin
Industrial region in plain of Donets River primarily in E Ukraine. Also search Donbas, Donbass, and Donets.
CO N470000N483000
E0400000E0373000
UF Donbas
Donbass
BT USSR
SA Dnieper-Donets Basin
Novorayskoe Formation
Russian Republic
Ukraine

Doniphan County
Extreme NE.
BT Kansas
United States

Donjek Glacier
In SW Yukon Territory.
BT Yukon Territory
Canada

Donjek River
In SW Yukon Territory.
BT Yukon Territory
Canada

Doornik
use Tournai

Dordogne
Department in SW central.
CO N443000N454500
E0021500W0010000
BT France

Dore Lake Complex
BT Quebec
Canada
SA Precambrian

Dorog
Town NW of Budapest.
BT Hungary

Dorog Basin
Also search Dorog.
BT Hungary

Dorpat
use Tartu

Dorset
County on the English Channel. Also search Dorsetshire.
CO N503000N511000
W0014500W0030000
UF Dorsetshire
BT England
Great Britain
United Kingdom

Dorsetshire
use Dorset

double refraction
use birefringence

double-refracting spar
use Iceland spar

Doubs
Department in E France.
CO N463500N473700
E0070000E0054000
BT France
NT Besancon

Douglas County
Index states as applicable.
BX Colorado
Georgia
Illinois
Kansas
Minnesota
Missouri
Nebraska
Nevada
Oregon
South Dakota
Washington
Wisconsin
BT United States

Douglas Group
Includes Stranger Formation, and Lawrence Formation. SW Iowa, E Kansas, NW Missouri, and SE Nebraska.
BT Virgilian
Upper Pennsylvanian
Pennsylvanian
SA Iowa
Kansas
Lawrence Formation
Missouri
Nebraska

Douro River
Term introduced in 1981. River in Spain and Portugal. Rises in N central Spain, flows W to NE Portugal, S along border, and W to Atlantic Ocean.
BT Europe
SA Portugal
Spain

Downtonian
In Old Red Sandstone, England. Uppermost Silurian or lowermost Devonian; Schmidt, 1960–lowermost Gedinnian (Lower Devonian). Includes use on level 3 under age terms(1). See list E.
BA Old Red Sandstone
Devonian
BT Paleozoic
SA Lower Devonian
Upper Silurian

Drac Valley
River valley in SE.
BT France

D

drag folds
Term introduced in 1978. Includes use on level 3 under folds(1) orientation(2). Before 1978, also search drag AND folds.
BA folds
SA displacements
faults
orientation

drainage
Includes use on level 3 under soils(1) water regimes(2).
SA acid mine drainage
drainage basins
drainage patterns
dredging
lakes
mine drainage
seepage
soils
water
water regimes

drainage basins
Includes use on level 3 under geomorphology(1) fluvial features(2). Also search catchments.
UF catchments
feeding ground
SA basins
catchment hydrodynamics
drainage
drainage patterns
fluvial features
geomorphology
precipitation
representative basins
rivers
rivers and streams
storage
valleys
water balance
water yield
watersheds

drainage changes
A valid term through 1974. After 1974, see landform evolution and fluvial features under geomorphology(1).

drainage networks
use drainage patterns

drainage patterns
Includes use on level 3 under geomorphology(1) fluvial features(2).
UF drainage networks
SA drainage
drainage basins
fluvial features
geomorphology
patterns
rivers
stream order

Drake Passage
Strait between Cape Horn and South Shetland Islands N of Antarctica.
UF Drake Strait
BT Antarctic Ocean
SA Leg 36

Drake Strait
use Drake Passage

Drammen
City on a branch of Oslo Fjord.
BT Norway

drape folding
use drape folds

drape folds
Term introduced in 1978. Also search drape folding.
UF drape folding
BA folds

dravite
BA ring silicates
silicates
BT minerals
SA borosilicates
tourmaline

drawdown
Includes use on level 3 under ground water(1) levels(2). Also search pumping.
SA ground water
hydraulics
levels
reservoirs

dredging
Includes use on level 3 under engineering geology(1) or mining geology(1).
UF deepening
SA construction
drainage
engineering geology
excavations
mining geology
ocean floors
waterways

Dresden
As of 1981, refers to only the city. Before 1981, also included the district.
BT Dresden Bezirk
East Germany

Dresden Bezirk
Term introduced in 1981. District in SE. Before 1981, search Dresden.
CO N504200N512700
E0150300E0131000
BT East Germany
NT Dresden
SA Saxony
Upper Lusatia

drift
Includes use on level 3 under sediments(1) clastic sediments(2). See list N.
UF glacial drift
BA clastic sediments
BT sediments
SA buried channels
buried valleys
glacial geology
glacial transport
glaciofluvial features
glaciofluvial sedimentation
loess
moraines
terrigenous materials
till

drill cores
use cores

drill cuttings
use cuttings

drill holes
A valid term through 1975. After 1975, use boreholes.

drilling
Used for the process. Includes use as level 3 term under commodities (list C), e.g. petroleum(1) or under well-logging(1) or engineering geology(1).
SA boreholes
engineering geology
petroleum
sand equivalent tests
well-logging
wells

Drome
Department in SE France.
CO N441500N452500
E0055000E0043500
BT France
NT Baronnies
Diois
Valence
SA Vercors

Dronning Maud Land
use Queen Maude Land

droplets
Includes use on level 3 under meteorology(1) water(2).
SA clouds
meteorology
raindrops
water

dropstone
use argillite

Drumheller
City in central.
BT Alberta
Canada

drumlins
Includes use on level 3 under glacial geology(1) glacial features(2).
BT glacial features
SA glacial geology
moraines
till

dry delta
use alluvial fans

dry hot rocks
use hot dry rocks

Dry Valley Drilling Project
In different locations such as Lake Vanda, Lake Vida, New Harbor, Ross Island, and McMurdo Sound off SW Ross Ice Sheet in the Victoria Land area of Ross Dependency. Also search DVDP.
BA Antarctica

Dryas
Genus. Before 1978, term also included use for the intervals of late-glacial time.
BA miospores
palynomorphs
SA angiosperms
Older Dryas
Oldest Dryas
Younger Dryas

drying
use dehydration

DSDP
use Deep Sea Drilling Project

DSDP Site 105
Term introduced in 1981. NW Atlantic Ocean, Leg 11. Before 1981, search Site 105 AND Deep Sea Drilling Project.
CO N345343N345343
W0691024W0691024
SA Atlantic Ocean
Deep Sea Drilling Project
Leg 11

DSDP Site 125
Term introduced in 1981. E Mediterranean Sea, Leg 13. Before 1981, search Site 125 AND Deep Sea Drilling Project.
CO N343719N343719
E0202541E0202541
SA Deep Sea Drilling Project
Leg 13
Mediterranean Sea

DSDP Site 132
Term introduced in 1981. Tyrrhenian Basin, Mediterranean Sea. Leg 13. Before 1981, search Site 132 AND Deep Sea Drilling Project.
CO N401540N401540
E0112628E0112628
SA Deep Sea Drilling Project
Leg 13
Mediterranean Sea
Tyrrhenian Sea

DSDP Site 157
Term introduced in 1981. E equatorial Pacific Ocean, Leg 16. Before 1981, search Site 157 AND Deep Sea Drilling Project.
CO S014570S014542
W0855420W0855412
SA Deep Sea Drilling Project
Leg 16
Pacific Ocean

DSDP Site 158
Term introduced in 1981. E equatorial Pacific Ocean, Leg 16. Before 1981, search Site 158 AND Deep Sea Drilling Project.
CO N063724N063724
W0851412W0851412
SA Deep Sea Drilling Project
Leg 16
Pacific Ocean

DSDP Site 163
Term introduced in 1981. E equatorial Pacific Ocean, Leg 16. Before 1981, search Site 163 AND Deep Sea Drilling Project.
CO N111442N111442
W1501730W1501730
SA Deep Sea Drilling Project
Leg 16
Pacific Ocean

DSDP Site 167
Term introduced in 1981. Central equatorial Pacific Ocean. Before 1981, search Site 167 AND Deep Sea Drilling Project.
CO N070406N070406
W1764930W1764930
SA Deep Sea Drilling Project
Leg 17
Pacific Ocean

DSDP Site 214
Term introduced in 1981. On crest of Ninetyeast Ridge, Indian Ocean. Leg 22. Before 1981, search Site 214 AND Deep Sea Drilling Project.
CO S112013S112013
E0884305E0884305
SA Deep Sea Drilling Project
Indian Ocean
Leg 22

DSDP Site 225
Term introduced in 1981. Red Sea, Leg 23. Before 1981, search Site 225 AND Deep Sea Drilling Project.
CO N211835N211835
E0381507E0381507
SA Deep Sea Drilling Project
Leg 23
Red Sea

DSDP Site 227
Term introduced in 1981. Red Sea, Leg 23. Before 1981, search Site 227 AND Deep Sea Drilling Project.
CO N211952N211952
E0380758E0380758
SA Deep Sea Drilling Project
Leg 23
Red Sea

DSDP Site 228
Term introduced in 1981. Red Sea, Leg 23. Before 1981, search Site 228 AND Deep Sea Drilling Project.
CO N190510N190510
E0390012E0390012
SA Deep Sea Drilling Project
Leg 23
Red Sea

DSDP Site 253
Term introduced in 1981. SE Indian Ocean, Leg 26. Before 1981, search Site 253 AND Deep Sea Drilling Project.
CO S245239S245239
E0872158E0872158
SA Deep Sea Drilling Project
Indian Ocean
Leg 26

DSDP Site 254
Term introduced in 1981. SE Indian Ocean, Leg 26. Before 1981, search Site 254 AND Deep Sea Drilling Project.
CO S305809S305809
E0875343E0875343
SA Deep Sea Drilling Project
Indian Ocean
Leg 26

DSDP Site 259
Term introduced in 1981. E Indian Ocean, Leg 27. Before 1981, search Site 259 AND Deep Sea Drilling Project.
CO S293700S293700
E1124200E1124200
SA Deep Sea Drilling Project
Indian Ocean
Leg 27

DSDP Site 260
Term introduced in 1981. E Indian Ocean, Leg 27. Before 1981, search Site 260 AND Deep Sea Drilling Project.
CO S160900S160900
E1101800E1101800
SA Deep Sea Drilling Project
Indian Ocean
Leg 27

DSDP Site 261
Term introduced in 1981. E Indian Ocean, Leg 27. Before 1981, search Site 261 AND Deep Sea Drilling Project.
CO S125700S125700
E1175400E1175400
SA Deep Sea Drilling Project
Indian Ocean
Leg 27

DSDP Site 262
Term introduced in 1981. Timor Trough, E Indian Ocean. Leg 27. Before 1981, search Site 262 AND Deep Sea Drilling Project.
CO S105200S105200
E1235100E1235100
SA Deep Sea Drilling Project
Indian Ocean
Leg 27

DSDP Site 263
Term introduced in 1981. E Indian Ocean, Leg 27. Before 1981, search Site 263 AND Deep Sea Drilling Project.
CO S232000S232000
E1105800E1105800
SA Deep Sea Drilling Project
Indian Ocean
Leg 27

DSDP Site 264
Term introduced in 1981. E Indian Ocean, Leg 28. Before 1981, search Site 264 AND Deep Sea Drilling Project.
CO S0345808S03458
E1120241E1120241
SA Deep Sea Drilling Project
Indian Ocean
Leg 28

DSDP Site 270
Term introduced in 1981. SE Ross Sea, Antarctic Ocean. Leg 28. Before 1981, search Deep Sea Drilling Project AND Site 270.
CO S772629S772629
W1783011W1783011
SA Antarctic Ocean
Deep Sea Drilling Project
Leg 28
Ross Sea

DSDP Site 274
Term introduced in 1981. Ross Sea, Antarctic Ocean. Leg 28. Before 1981, search Site 274 AND Deep Sea Drilling Project.
CO S685949S685949
E1732538E1732538
SA Antarctic Ocean
Deep Sea Drilling Project
Leg 28
Ross Sea

DSDP Site 284
Term introduced in 1981. W of New Zealand in the Tasman Sea. Leg 29. Before 1981, search Site 284 AND Deep Sea Drilling Project.
CO S403029S403029
E1674049E1674049
SA Deep Sea Drilling Project
Leg 29
Pacific Ocean
Tasman Sea

DSDP Site 292
Term introduced in 1981. W Philippine Sea, Leg 31. Before 1981, search Site 292 AND Deep Sea Drilling Project.
CO N154907N154907
E1243903E1243903
SA Deep Sea Drilling Project
Leg 31
Pacific Ocean
Philippine Sea

DSDP Site 296
Term introduced in 1981. N Philippine Sea, Leg 31. Before 1981, search Site 296 AND Deep Sea Drilling Project.
CO N292025N292025
E1333131E1333131
SA Deep Sea Drilling Project
Leg 31
Pacific Ocean
Philippine Sea

DSDP Site 297
Term introduced in 1981. N Philippine Sea, Leg 31. Before 1981, search Site 297 AND Deep Sea Drilling Project.
CO N305222N305222
E1340953E1340953
SA Deep Sea Drilling Project
Leg 31
Pacific Ocean
Philippine Sea

DSDP Site 315
Term introduced in 1981. Central equatorial Pacific Ocean near the Line Islands. Leg 33. Before 1981, search Site 315 AND Deep Sea Drilling Project.
CO N041016N041016
W1583132W1583132
SA Deep Sea Drilling Project
Leg 33
Pacific Ocean

DSDP Site 319
Term introduced in 1981. E of the East Pacific Rise, SE Pacific Ocean. Leg 34. Before 1981, search Site 319 AND Deep Sea Drilling Project.
CO S130102S130102
W1013128W1013128
SA Deep Sea Drilling Project
Leg 34
Pacific Ocean

DSDP Site 320
Term introduced in 1981. Off the coast of Peru, SE Pacific Ocean. Leg 34. Before 1981, search Site 320 AND Deep Sea Drilling Project.
CO S090024S090024
W0833148W0833148
SA Deep Sea Drilling Project
Leg 34
Pacific Ocean

DSDP Site 321
Term introduced in 1981. Off the coast of Peru, SE Pacific Ocean. Leg 34. Before 1981, search Site 321 AND Deep Sea Drilling Project.
CO S120117S120117
W0815414W0815414
SA Deep Sea Drilling Project
Leg 34
Pacific Ocean

DSDP Site 322
Term introduced in 1981. Antarctic Ocean between S tip of South America and the Antarctic Peninsula. Leg 35. Before 1981, search Site 322 AND Deep Sea Drilling Project.
CO S600127S600127
W0792529W0792529
SA Antarctic Ocean
Deep Sea Drilling Project
Leg 35

DSDP Site 323
Term introduced in 1981. Antarctic Ocean, SW of South America. Leg 35. Before 1981, search Site 323 AND Deep Sea Drilling Project.
CO S634050S634050
W0975941W0975941
SA Antarctic Ocean
Deep Sea Drilling Project
Leg 35

DSDP Site 325
Term introduced in 1981. Antarctic Ocean, NW of the Antarctic Peninsula. Before 1981, search Site 325 AND Deep Sea Drilling Project.
CO S650247S650247
W0734024W0734024
SA Antarctic Ocean
Deep Sea Drilling Project
Leg 35

DSDP Site 327
Term introduced in 1981. SW Atlantic Ocean, N of the Scotia Sea. Leg 36. Before 1981, search Site 327 AND Deep Sea Drilling Project.
CO S505223S505223
W0464701W0464701
SA Atlantic Ocean
Deep Sea Drilling Project
Leg 36

DSDP Site 329
Term introduced in 1981. SW Atlantic Ocean, N of the Scotia Sea. Leg 36. Before 1981, search Site 329 AND Deep Sea Drilling Project.
CO S503919S503919
W0460544W0460544
SA Atlantic Ocean
Deep Sea Drilling Project
Leg 36

DSDP Site 330
Term introduced in 1981. SW Atlantic Ocean, N of the Scotia Sea. Leg 36. Before 1981, search Site 330 AND Deep Sea Drilling Project.
CO S505511S505511
W0465300W0465300
SA Atlantic Ocean
Deep Sea Drilling Project
Leg 36

DSDP Site 332
Term introduced in 1981. Near crest of Mid-Atlantic Ridge. Leg 37. Before 1981, search Site 332 AND Deep Sea Drilling Project.
CO N365243N365246
W0333828W0333834
SA Atlantic Ocean
Deep Sea Drilling Project
Leg 37

DSDP Site 333
Term introduced in 1981. Near crest of Mid-Atlantic Ridge. Leg 37. Before 1981, search Site 333 AND Deep Sea Drilling Project.
CO N365027N365027
W0334003W0334003
SA Atlantic Ocean
Deep Sea Drilling Project
Leg 37

DSDP Site 334
Term introduced in 1981. W flank of Mid-Atlantic Ridge. Leg 37. Before 1981, search Site 334 AND Deep Sea Drilling Project.
CO N370208N370208
W0342452W0342452
SA Atlantic Ocean
Deep Sea Drilling Project
Leg 37

DSDP Site 335
Term introduced in 1981. W flank of the Mid-Atlantic Ridge. Leg 37. Before 1981, search Site 335 AND Deep Sea Drilling Project.
CO N371744N371744
W0351155W0351155
SA Atlantic Ocean
Deep Sea Drilling Project
Leg 37

DSDP Site 356
Term introduced in 1981. Sao Paulo Plateau, SW Atlantic Ocean. Leg 39. Before 1981, search Site 356 AND Deep Sea Drilling Project.
CO S281713S281713
W0410517W0410517
SA Atlantic Ocean
Deep Sea Drilling Project
Leg 39

DSDP Site 357
Term introduced in 1981. Rio Grande Rise, SW Atlantic Ocean. Leg 39. Before 1981, search Site 357 AND Deep Sea Drilling Project.
CO S300015S300015
W0353335W0353335
SA Atlantic Ocean
Deep Sea Drilling Project
Leg 39

DSDP Site 361
Term introduced in 1981. Off Cape of Good Hope, SE Atlantic Ocean. Leg 40. Before 1981, search Site 361 AND Deep Sea Drilling Project.
CO S350358S350358
E0152655E0152655
SA Atlantic Ocean
Deep Sea Drilling Project
Leg 40

DSDP Site 362
Term introduced in 1981. Walvis Ridge, SE Atlantic Ocean. Leg 40. Before 1981, search Site 362 AND Deep Sea Drilling Project.
CO S194527S194527
E0103157E0103157
SA Atlantic Ocean
Deep Sea Drilling Project
Leg 40

DSDP Site 363
Term introduced in 1981. Walvis Ridge, SE Atlantic Ocean. Leg 40. Before 1981, search Site 363 AND Deep Sea Drilling Project.
CO S193845S193845
E0090248E0090248
SA Atlantic Ocean
Deep Sea Drilling Project
Leg 40

DSDP Site 364
Term introduced in 1981. Angola Basin, SE Atlantic Ocean. Leg 40. Before 1981, search Site 364 AND Deep Sea Drilling Project.
CO S113419S113419

E0115818E0115818
SA Atlantic Ocean
Deep Sea Drilling Project
Leg 40

DSDP Site 367
Term introduced in 1981. Cape Verde Basin, NE Atlantic Ocean. Leg 41. Before 1981, search Site 367 AND Deep Sea Drilling Project.
CO N122913N122913
W0200250W0200250
SA Atlantic Ocean
Deep Sea Drilling Project
Leg 41

DSDP Site 368
Term introduced in 1981. Cape Verde Rise, NE Atlantic Ocean. Leg 41. Before 1981, search Site 368 AND Deep Sea Drilling Project.
CO N173026N173026
W0212114W0212114
SA Atlantic Ocean
Deep Sea Drilling Project
Leg 41

DSDP Site 369
Term introduced in 1981. SE of Canary Islands, NE Atlantic Ocean. Leg 41. Before 1981, search Site 369 AND Deep Sea Drilling Project.
CO N263533N263533
W0145955W0145955
SA Atlantic Ocean
Deep Sea Drilling Project
Leg 41

DSDP Site 370
Term introduced in 1981. Off Morocco, NE Atlantic Ocean. Leg 41. Before 1981, search Site 370 AND Deep Sea Drilling Project.
CO N325015N325015
W0104634W0104634
SA Atlantic Ocean
Deep Sea Drilling Project
Leg 41

DSDP Site 372
Term introduced in 1981. Balearic Basin, Mediterranean Sea. Leg 42A. Before 1981, search Site 372 AND Deep Sea Drilling Project.
CO N400152N400154
E0044747E0044747
SA Deep Sea Drilling Project
Leg 42A
Mediterranean Sea

DSDP Site 374
Term introduced in 1981. Ionian Sea. Leg 42A. Before 1981, search Site 374 AND Deep Sea Drilling Project.
CO N355052N355052
E0181147E0181147
SA Deep Sea Drilling Project
Ionian Sea
Leg 42A
Mediterranean Sea

DSDP Site 379
Term introduced in 1981. E Black Sea. Leg 42B. Before 1981, search Site 379 AND Deep Sea Drilling Project.
CO N430017N430017
E0360041E0360041
SA Black Sea
Deep Sea Drilling Project
Leg 42B

DSDP Site 380
Term introduced in 1981. W Black Sea. Leg 42B. Before 1981, search Site 380 AND Deep Sea Drilling Project.
CO N420559N420559
E0293654E0293654
SA Black Sea
Deep Sea Drilling Project
Leg 42B

DSDP Site 381
Term introduced in 1981. W Black Sea. Leg 42B. Before 1981, search Site 381 AND Deep Sea Drilling Project.
CO N414015N414015
E0292458E0292458
SA Black Sea
Deep Sea Drilling Project
Leg 42B

DSDP Site 391
Term introduced in 1981. Blake-Bahama Basin, NW Atlantic Ocean. Leg 44. Before 1981, search Site 391 AND Deep Sea Drilling Project.
CO N281340N281340
W0753653W0753653
SA Atlantic Ocean
Deep Sea Drilling Project
Leg 44

DSDP Site 395
Term introduced in 1981. N central Atlantic Ocean near the Mid-Atlantic Ridge axis. Leg 45. Before 1981, search Site 395 AND Deep Sea Drilling Project.
CO N224521N224521
W0460454W0460454
SA Atlantic Ocean
Deep Sea Drilling Project
IPOD
Leg 45

DSDP Site 396
Term introduced in 1981. N central Atlantic Ocean near the Mid-Atlantic Ridge axis. Leg 45. Before 1981, search Site 396 AND Deep Sea Drilling Project.
CO N225853N225853
W0433057W0433057
SA Atlantic Ocean
Deep Sea Drilling Project
IPOD
Leg 45

DSDP Site 417
Term introduced in 1981. Near S end of Bermuda Rise, NW Atlantic Ocean. Legs 51 and 52. Before 1981, search Site 417 AND Deep Sea Drilling Project.
CO N250000N250000
W0680000W0680000
SA Atlantic Ocean
Deep Sea Drilling Project
IPOD
Leg 51
Leg 52

DSDP Site 418
Term introduced in 1981. Near S end of Bermuda Rise, NW Atlantic Ocean. Legs 52 and 53. Before 1981, search Site 418 AND Deep Sea Drilling Project.
CO N250205N250205
W0680327W0680327
SA Atlantic Ocean
Deep Sea Drilling Project
IPOD
Leg 52
Leg 53

DSS
use deep seismic sounding

DTA data
Term introduced on level 3 under thermal analysis(1).
UF differential thermal analysis data
SA differential thermal analysis
endothermic reactions
thermal analysis

Dubai
Emirate. One of federation of 7 states at S end of Persian Gulf.
BT United Arab Emirates
Arabian Peninsula

Dublin
County on Irish Sea. Also a city.
BT Ireland

Duchesne County
In NE Utah.
BT Utah
United States

Duchesne River Formation
Unconformably underlies Bishop Conglomerate; unconformably overlies Uinta and Uinta (?) Formation. NW Colorado and NE Utah.
BT Paleogene
Tertiary
SA Colorado
Eocene
Oligocene
Utah

Ducktown
Town in Polk County in extreme SE.
BT Tennessee
United States

ductility
SA deformation
rock mechanics

Dudley
Town SW of Birmingham in central.
BT England
Great Britain
United Kingdom

Duero Basin
Term introduced in 1981.
BT Spain

Dufek Intrusion
In the Dufek Massif area of Forrestal Range in the Pensacola Mountains of Edith Ronne Land S of the Weddell Sea.
BT Antarctica

dufrenoysite
BA sulfarsenites
sulfosalts
BT minerals

Dukla
Town in SE part of country.
BT Rzeszow
Poland

Duluth
City at extreme W end of Lake Superior in Saint Louis County.
BT Minnesota
United States

Duluth Complex
In NE Minnesota.
BT Minnesota
United States
SA Precambrian

dumortierite
BA orthosilicates
silicates
BT minerals

Duna River
use Danube River

Dunai River
use Danube River

Dunantul
use Transdanubia

Dunarea River
use Danube River

Dunau River
use Danube River

Dundee
City on Firth of Tay on North Sea.
BT Scotland
Great Britain
United Kingdom

dune rock
use eolianite

Dunedin
City on Pacific Ocean in SE.
BT South Island
New Zealand

dunes
Includes use on level 3 under geomorphology(1) eolian features(2). Before 1981, also included use on level 3 under sedimentary structures(1) bedding plane irregularities(2).
NA coastal dunes
continental dunes
submarine dunes
SA antidunes
bedding plane irregularities
eluvium
eolian features
geomorphology
sand waves
sedimentary structures

dunite
As of 1981, includes use on level 3 under igneous rocks(1) peridotites(2). See list H. Before 1981, included use on level 3 under igneous rocks(1) ultramafic family(2).
BA peridotites
ultramafics
BT igneous rocks
SA peridotite

Dunkard Basin
In West Virginia, Ohio, and Pennsylvania. Structural basin.
BT United States
SA Ohio
Pennsylvania
West Virginia

Dunkard Group
Includes Washington and Greene formations. W Maryland, E Ohio, SW Pennsylvania, and N West Virginia. Also search Dunkard Series.
BT Paleozoic
SA Maryland
Ohio
Pennsylvania
Pennsylvanian
Permian
West Virginia

Dunkerque
use Dunkirk

Dunkirk
City at S end of North Sea.
UF Dunkerque
BT Nord Department
France

Dunn County
Term introduced in 1981. Index states as applicable.
BX North Dakota
Wisconsin
BT United States

Dunnage Melange
BT Canada

Duplin Formation
Base of Duplin is an unconformity. Florida, E Georgia, E North Carolina, and E South Carolina.
BT upper Miocene
Miocene
SA Florida
Georgia
North Carolina
South Carolina

durain
Includes use on level 3 under sedimentary rocks(1) organic residues(2). See list I.
BA macerals
coal

organic residues
BT sedimentary rocks

Durance
Village in SW part of country.
BT Lot-et-Garonne
France

Durance Basin
River basin in SE France. Also search Durance.
BT France

Durango
State in W central Mexico. Also a city.
BT Mexico
NT Mapimi
SA Sierra Madre Occidental

Durban
City on the Indian Ocean.
BT Natal
South Africa

Durham
County in NE.
CO N543000N545700
W0011500W0022200
BT England
Great Britain
United Kingdom

Durham County
Term introduced in 1981. N central North Carolina.
CO N355000N361400
W0784100W0790100
BT North Carolina
United States

duricrust
As of 1981, includes use on level 3 under sedimentary rocks(1) chemically precipitated rocks(2). See list I. Before 1981, included use on level 3 under sediments(1) chemically precipitated sediments(2). Also includes use on level 3 under soils(1) and weathering(1).
BA chemically precipitated rocks
BT sedimentary rocks
SA calcrete
caliche
ferricrete
soils
weathering
weathering crust

Dushambe
use Dushanbe

Dushanbe
City in W Tadzhikistan.
UF Dushambe
Stalinabad
BT Tadzhikistan
USSR

dust
Includes use on level 3 under sediments(1) clastic sediments(2). See list N.
BA clastic sediments
BT sediments
SA cosmic dust
dust storms
interplanetary dust
terrigenous materials
volcanic ash

dust storms
Also search dust clouds.
SA dust
storms

Dutch Guiana
use Surinam

Dutch New Guinea
use Irian Jaya

Dutchess County
On Connecticut border in SE.
BT New York
United States

Duval County
In NE Florida and S Texas. Index states as applicable.
BX Florida
Texas
BT United States

Dvina River
The Western Dvina River which flows into the Gulf of Riga and the Northern Dvina in the White Sea Basin. Index Soviet republics as applicable.
BT USSR
SA Byelorussia
Latvia
Russian Republic

Dwyka Series
Of the Karroo System. Includes Upper Shales, Conglomerate (now Tillite), and Lower Shales. Also search Dwyka Tillite.
BT upper Paleozoic
Paleozoic
SA Africa
Botswana
Carboniferous
Karroo System
Namibia
Permian
South Africa

Dy
use dysprosium

dynamic
A valid term through 1977. After 1977, use dynamic metamorphism on level 2 under metamorphism(1).

dynamic metamorphism
Term introduced in 1978. Includes use on level 2 under metamorphism(1). Before 1978, also search metamorphism AND dynamic.
BA metamorphism
SA regional metamorphism
retrograde metamorphism

dynamic properties
Term introduced in 1981.
SA dynamics
properties

dynamics
As of 1978, term is used on level 2 under shorelines(1).
SA dynamic properties
shorelines
soil dynamics

dyscrasite
As of 1981, includes use on level 3 under minerals(1) antimonides(2). See list L. Before 1981, included use on level 3 under minerals(1) sulfides(2).
BA antimonides
sulfides
BT minerals

dysprosium
Includes use on level 1 and 2 as chemical element (list D).
UF Dy
SA elements
rare earths

Dzereg
use Altan Teeli

Dzhagdy Range
NW of Komsomolsk in Khabarovsk Kray in Soviet Far East.
BT Russian Republic
USSR

Dzhezkazgan
Town in Karaganda Oblast in E central.
BT Kazakhstan
USSR

Dzhida River
S of Lake Baikal in Buryat ASSR.
BT Russian Republic
USSR

Dzhugdzhur region
Term introduced in 1981. Borders the Okhotsk Sea and includes the Dzhugdzhur Range.
CO N540000N620000
E1563000E1330000
BA Northeastern USSR
BT Russian Republic
USSR

Dzhungaria
Region in N Sinkiang Uighur. Also search Dzungaria.
UF Dzungaria
Zungaria
BT Sinkiang Uighur
China

Dzhungarian Alatau
Northernmost branch of the Tien Shan. Index Kazakhstan and/or Sinkiang Uighur as applicable. Also search Dzungar Alatau, Dzungarian Ala-Tau and Dzungarian Alatau.
CO N445000N453000
E0823000E0784500
UF Dzungar Alatau
Dzungarian Ala-Tau
Dzungarian Alatau
BT Asia
SA Kazakhstan
Sinkiang Uighur

Dzirula Massif
Central.
BT Georgian Republic
USSR

Dzungar Alatau
use Dzhungarian Alatau

Dzungaria
use Dzhungaria

Dzungarian Ala-Tau
use Dzhungarian Alatau

Dzungarian Alatau
use Dzhungarian Alatau

E

E-region
Includes use on level 2 under ionosphere(1).
SA D-region
F-region
ionosphere

Eagle County
W central.
BT Colorado
United States

early diagenesis
Term introduced in 1978. Includes use on level 3 under diagenesis(1).
BA diagenesis
SA late diagenesis

Earth
Treated as a whole. For general concepts or studies. Includes use on level 1(list A). Term set options are: topic [age, composition, concepts, evolution, general, genesis, gravity field, interior, magnetic field, motions, observations, processes, properties, temperature, theoretical studies]
subtopic [e.g. Chandler wobble, convection currents, density, differentiation, Earth tides, expansion, free oscillations, melting, rotation]
SA asthenosphere
atmosphere
biosphere
Chandler wobble
Conrad discontinuity
continents
core
Coriolis force
crust
differentiation
Earth tides
expansion
figure of Earth
free oscillations
geodesy
geodynamics
geoid
geophysics
geothermal gradient
gravity field
heat flow
hydrosphere
interior
isostasy
lithosphere
magnetic field
magnetosphere
mantle
Mohorovicic discontinuity
Moon
motions
obliquity of the ecliptic
oscillations
paleomagnetism
phase transitions
planetology
planets
plate tectonics
pulsations
reversals
rotation
seismology
solar system
spherical harmonic analysis
tectonophysics
terrestrial planets

earth dams
Term introduced in 1981. Also search earthdams.
SA dams
earthworks
engineering geology

earth, diatomaceous
use diatomaceous earth

earth flows
use earthflows

earth pressure
As of 1978, term is used on level 2 under soil mechanics(1).
SA pressure
soil mechanics

Earth tides
Includes use on level 3 under Earth(1) processes(2) and under geophysics(1).
UF bodily tide
SA Earth
geodesy
geophysics
tectonophysics
tides

Earth-current methods
Includes use on level 2 under geophysical methods(1).
UF telluric methods
BA geophysical methods
SA Earth-current surveys
electrical field
magnetic field
magnetotelluric methods
methods

Earth-current surveys
Includes use on level 3 under area terms(1) geophysical surveys(2).
UF telluric surveys
BA geophysical surveys
BT surveys
SA Earth-current methods
magnetotelluric surveys

earthflows
As of 1978, term is used on level 2 under slope stability(1). Also search earth flows.
UF earth flows
SA mudslides
slope stability

earthquake magnitude
use magnitude

earthquake record
use seismograms

earthquake sea wave
use tsunamis

earthquakes
Used on level 1 and 2 for studies emphasizing individual earthquakes. Includes use on level 2 under geologic hazards(1), mantle(1) and seismology(1). If level 1, term set options are:
topic [aftershocks, causes, classification, detection, effects, epicenters, focal mechanism, focus, genesis, magnitude, prediction, seismic intensity]
subtopic [e.g. deep-focus earthquakes, intermediate-focus earthquakes, shallow-focus earthquakes, or year, for specific earthquake] (no area term)
SA accelerograms
aftershocks
arrival time
crust
damage
deep-focus earthquakes
dilatancy
elastic waves
engineering geology
epicenters
explosions
faults
fluid injection
focal mechanism
focus
foreshocks
foundations
geologic hazards
ground motion
icequakes
induced earthquakes
intermediate-focus earthquakes
isoseismic maps
magnitude
magnitude-frequency ratio
mantle
marsquakes
microearthquakes
microseisms
modified Mercalli scale
Mohorovicic discontinuity
moonquakes
nuclear facilities
periodicity
plate tectonics
prediction
Q
reservoirs
Richter Scale
rock mechanics
seismic energy
seismic intensity
seismic moment
seismic response
seismic sources
seismic zoning
seismograms
seismographs
seismology
shallow-focus earthquakes
slope stability
soil mechanics
spatial distribution
stick-slip
surface waves
swarms
teleseismic signals
temporal distribution
tiltmeters
tsunamis
vibration
volcanic earthquakes

earthworks
As of 1978, term is used on level 2 under foundations(1).
SA earth dams
foundations

East Africa
CO S271000N223000
E0513000E0220000
BA Africa
NA Djibouti
Ethiopia
Kenya
Malawi
Mozambique
Somali Republic
Sudan
Tanzania
Uganda
Zambia
SA Central Africa
North Africa
Southern Africa
West Africa

East African Lakes
Term introduced in 1981 as a grouper for all lakes in E Africa.
BT Africa
NA Lake Albert
NZ Lake Baringo
NA Lake Edward
Lake Kariba
Lake Kivu
NZ Lake Magadi
NA Lake Malawi
Lake Natron
Lake Tanganyika
Lake Turkana
Lake Victoria
SA East African Rift

East African Rift
Part of Great Rift Valley. Index countries and lakes as applicable. Also search East African Rift system; East African Rift valley; East African Rift zone.
UF East African Rift system
East African Rift valley
BT Africa
SA Afar Depression
Burundi
East African Lakes
Ethiopia
Ethiopian Rift
Great Rift Valley
Gregory Rift
Kenya
Lake Albert
Lake Edward
Lake Kivu
Lake Magadi
Lake Malawi
Lake Natron
Lake Tanganyika
Lake Turkana
Malawi
Mozambique
Rwanda
Tanzania
Uganda

East African Rift system
use East African Rift

East African Rift valley
use East African Rift

East Anglia
Region including counties of Norfolk and Suffolk on the North Sea in extreme E.
BT England
Great Britain
United Kingdom

East Antarctica
Term introduced in 1981. E of Transantarctic Mountains.
CO S000000S650000
E1710000W0400000
BT Antarctica

East Atlantic
Term introduced in 1978. Before 1978, also search Atlantic Ocean AND east or eastern.
BA Atlantic Ocean
SA Northeast Atlantic
Northwest Atlantic
South Atlantic
Southeast Atlantic
Southwest Atlantic
West Atlantic

East Bay
Term introduced in 1981. In the Gulf of Mexico off the Mississippi Delta.
CO N285500N295000
W0890900W0892300
BT Louisiana
United States

East Carpathians
use Eastern Carpathians

East China Sea
Enclosed by China, Korea, Japan, and Taiwan. As of 1977, includes use on level 1 as an area term (list O).
CO N240000N333000
E1303000E1200000
BA West Pacific
Pacific Ocean
SA China Sea

East European Plain
use Russian Plain

East European Platform
Not a valid term for GeoRef. Use Russian Platform.

East Frisian Islands
In North Sea off NW.
BT Lower Saxony
West Germany

East Germany
Officially German Democratic Republic or Deutsche Demokratische Republik. In N central Europe, bounded on N by the Baltic Sea, on E by Poland, on S by Czechoslovakia and West Germany, and on W by West Germany. Introduced as a level 1 area term in 1978.
CO N503000N543000
E0150000E0100000
BA Germany
Central Europe
Europe
NT Barth
Brandenburg
Brocken Massif
Cottbus Bezirk
Dresden Bezirk
Erfurt Bezirk
Frankfurt Bezirk
Gera Bezirk
Halle Bezirk
Karl-Marx-Stadt Bezirk
Kyffhauser Range
Leipzig Bezirk
Magdeburg Bezirk
Mecklenburg
Neubrandenburg Bezirk
Potsdam Bezirk
Rostock Bezirk
Rugen Island
Saxony
Saxony-Anhalt
Saxony-Thuringia
Schwerin Bezirk
Steinach
Suhl Bezirk
Taubach
Thuringia
Thuringian Basin
Thuringian Forest
Unstrut River
Upper Lusatia
Weissenberg
SA Berlin
Elbe River
Elbe Valley
Erzgebirge
Frankfurt
Harz Foreland
Harz Mountains
Harz region
Lusatia
North German Plain
Northeastern German Plain
Oder Valley
Pomerania
Saale River
Saxonian Massif
Silesia
Thuringian Hills
Thuringian Massif
Weisse Elster Basin
Werra
Zechstein

East Greenland
Term introduced in 1978. Before 1978, search Greenland AND east.
BA Greenland
SA South Greenland

East Indian Ocean
Term introduced in 1978. Before 1978, also search Indian Ocean AND east or eastern.
BA Indian Ocean

East Lothian
County on S bank of Firth of Forth.
CO N555000N560500
W0022000W0030000
UF Hardington County
Hardingtonshire
BT Scotland
Great Britain
United Kingdom

East Malaysia
That part of the Federation of Malaysia which is comprised of the two states of Sabah and Sarawak on the island of Borneo.
BT Malaysia
SA Borneo
Brunei
Sabah
Sarawak

East Mediterranean
Term introduced in 1978. Before 1978, search Mediterranean Sea AND east. As of 1981, includes the Black Sea.
CO N301000N464000
E0420000E0100000
BT Mediterranean Sea
NA Adriatic Sea
Aegean Sea
Black Sea
Ionian Sea
SA West Mediterranean

East Mesa
use East Mesa KGRA

East Mesa KGRA
Term introduced in 1981. Known Geothermal Resources Area. Before 1981, also search East Mesa.
CO N324500N325000 W1151000W1151500
UF East Mesa
BT Imperial Valley
Imperial County
California
United States

East Midlands
Eastern part of the midlands of central.
BT England
Great Britain
United Kingdom

East Ongul Island
In E Lutzow-Holm Bay off Prince Olav Coast in Norwegian Sector on Indian Ocean side.
BT Antarctica

East Pacific
Term introduced in 1978. Before 1978, search Pacific Ocean AND east or eastern.
BA Pacific Ocean
SA Equatorial Pacific
North Pacific
Northeast Pacific
Northwest Pacific
South Pacific
Southeast Pacific
Southwest Pacific
West Pacific

East Pacific Ocean Islands
Term introduced in 1981. Includes use on level 1 as an area term (list O). For term set options see list B.
NT Easter Island
NA Galapagos Islands
SA Pacific Ocean

East Pacific Rise
In S Pacific. Also search Albatross Cordillera. Indicates the mid-ocean ridge extending from the Pacific-Antarctic Ridge to the Gulf of California.
UF Albatross Cordillera
Easter Island Cordillera
BT Pacific Ocean
SA mid-ocean ridges
Pacific-Antarctic Ridge

East Pakistan
A valid term through 1972. After 1972, use Bangladesh. Also search Pakistan AND east.

East Rudolf
Basin in N.
BT Kenya

East Sayan
use Eastern Sayan

East Siberian Sea
Between New Siberian Islands and Wrangel Island N of Magadan Oblast and NE Yakutia.
CO N690000N770000 E1800000E1390000
BT Arctic Ocean

East Texas Field
Oil field in E.
BT Texas
United States

East-Central Nigeria
Term introduced in 1981.
CO N041500N070500 E0080000E0053000
BT Nigeria

Easter Island
About 2,000 miles W of the Chilean coast.
CO S280000S270000 W1090000W1100000
BT East Pacific Ocean Islands

Easter Island Cordillera
use East Pacific Rise

Eastern Alps
Ranges of the Alps extending from Innsbruck in the E to near Vienna and including the Dinaric Alps of Yugoslavia. Index countries as applicable.
BT Alps
Europe
SA Austria
Carnic Alps
Dinaric Alps
Dolomites
Gosau Formation
Hohe Tauern
Julian Alps
Karawanken
Otztal Alps
Salzkammergut
Wechsel
Wiener Wald
Yugoslavia
Zillertal Alps

Eastern Canada
Term introduced in 1981.
CO N420000N640000 W0523000W0950000
BA Canada
NA Labrador
Maritime Provinces
Newfoundland
Ontario
Quebec

Eastern Carpathians
Ranges of the Carpathians in SW Ukraine and Moldavia. Index Ukraine and/or Romania as applicable. Also search East Carpathians.
UF East Carpathians
BT Carpathians
Europe
SA Calimani Mountains
Haghimas Syncline
Romania
Soviet Carpathians
Tulghes Series
Ukraine

Eastern Cordillera
Eastern range of the Andes. Index countries as applicable. Also search Cordillera Oriental.
UF Cordillera Oriental
BT Andes
SA Bolivia
Colombia
Ecuador
Peru

Eastern Desert
Between Nile on W and Gulf of Suez and Red Sea on E. Also search Arabian Desert.
CO N220000N302000 E0365500E0305000
UF Arabian Desert
BT Egypt

Eastern Ghats
Low mountain range extending about 500 miles along SE and E coast as far N as Mahanadi River.
BT India
SA Ghats

Eastern Goldfields
BT Western Australia
Australia

Eastern Hemisphere
Used when discussing many large areas too numerous to mention. Includes use on level 1 and 2 as an area term (list O). If 1, see term set options under list B.
SA Northern Hemisphere
Southern Hemisphere
Western Hemisphere

Eastern Kazakhstan
Term introduced in 1981.
CO N400000N500000 E0860000E0723000
BZ Central Asia
BX Kazakhstan
BT USSR

Eastern Sayan
Term introduced in 1978. Before 1978, use Sayan AND east or eastern.
CO N523000N560000 E1040000E0920000
UF East Sayan
BA Sayan
Siberian fold belt
BT USSR

Eastern Sea
Term introduced in 1981. Before 1981, also search Mare Orientale.
UF Mare Orientale
BT Moon

Eastern Swiss Alps
Term introduced in 1981.
BA Swiss Alps
BT Switzerland

Eastern U.S.
Term introduced in 1978. Includes use as a level 1 area term. Before 1978, also search United States AND east or eastern.
BA United States
NA Delaware
District of Columbia
Florida
Georgia
Maryland
New England
New Jersey
New York
North Carolina
Pennsylvania
South Carolina
Virginia
West Virginia

Eastern Venezuela
Term introduced in 1981.
CO N074000N105000 W0595000W0654000
BT Venezuela

Eaton
Town in Weld County in NE.
BT Colorado
United States

Ebino
Region.
BT Kyushu
Japan

ebridians
Includes use on level 2 under Protista(1). See list F.
BA Protista
BT Invertebrata
NA Archaeomonadaceae

Ebro Basin
River basin in NE extending from Santander Province to Catalonia. Also search Ebro River basin.
UF Ebro River basin
BT Spain

Ebro River
Rises in the Cantabrian Mountains in N and flows SE into the Mediterranean Sea SW of Barcelona.
BT Spain

Ebro River basin
use Ebro Basin

Ecca Series
Karroo Basin.
BT Permian
SA Cape Province
South Africa

eccentricity
Used as a general term.
SA geometry

Echinodermata
Includes use on level 1 and 2 as a fossil term (list F).
BT Invertebrata
NA Asterozoa
Crinozoa
Echinozoa
Homalozoa
Pelmatozoa
SA echinoderms
Receptaculitaceae
reef builders
sclerites

echinoderms
Term introduced in 1981 as the common name for Echinodermata. See list F. To be used for Echinodermata when they are used as tools in stratigraphy. Before 1981, search Echinodermata AND biostratigraphy. Includes use on level 1. If 1, term set options are:
biostratigraphy
age [List E]
SA biostratigraphy
Echinodermata

Echinoidea
Includes use on level 2 under Echinodermata(1). See list F.
BA Echinozoa
Echinodermata
BT Invertebrata

Echinozoa
As of 1981, includes use on level 3 under Echinodermata(1). See list F. Before 1981, included use on level 2 under Echinodermata(1).
BA Echinodermata
BT Invertebrata
NA Camptostromatoidea
Cyclocystoidea
Echinoidea
Edrioasteroidea
Helicoplacoidea
Holothuroidea
Ophiocistioidea

Echiurida
Term introduced in 1981. Includes use on level 2 under worms(1). See list F.
BA worms
BT Invertebrata

Echo Bay
Village just E of Sault Ste. Marie in S Ontario.
BT Ontario
Canada

echo sounding
Includes use on level 3 under geophysical methods(1) acoustical methods(2).
BA geophysical methods
SA acoustical methods
bathymetry
deep sounding
geophysical surveys
oceanography
sonar methods
sounding

eclogite
As of 1981, includes use on level 2 under metamorphic rocks(1). See list J. Before 1981, included use on level 3 under metamorphic rocks(1) granulites(2).
BT metamorphic rocks

ecology
Includes use on level 1 (list A); on level 2 under fossil groups (list F).

Used for the study of relationships between organisms and their environments, including the study of communities, patterns of life, natural cycles, relationships of organisms to each other, biogeography, and population changes. If 1, term set options are:
name of fossil group (List F) (for Recent fauna and flora)
type of environment [e.g. arctic environment, alpine environment, boreal environment, brackish-water environment, coastal environment, deltaic environment, eolian environment, estuarine environment, glacial environment, intertidal environment, lacustrine environment, lagoonal environment, marine environment, paludal environment, reefs, terrestrial environment]
topic [analysis, changes, concepts, evolution, interpretation, observations, processes]
subtopic (no area term)
SA adaptation
alpine environment
alpine-type
arctic environment
arid environment
biofacies
biogenic effects
biogeography
biology
biomass
biota
biotypes
boreal environment
brackish water
brackish-water environment
coastal environment
communities
deforestation
deltaic environment
deltas
desertification
deserts
desiccation
ecosystems
environment
environmental geology
epibiotism
estuaries
estuarine environment
eutrophication
fresh water
fresh-water environment
geomorphology
glacial environment
grasslands
habitat
human ecology
humid environment
hydrology
impact statements
intertidal environment
lacustrine environment
lagoonal environment
lagoons
lakes
living taxa
mangrove swamps
marine environment
paleoecology
paludal environment
peat bogs
ponds
reefs
semi-arid environment
shallow-water environment
soils
subtropical environment
supratidal environment
swamps
taiga environment
temperate environment
terrestrial environment
tropical environment
tundra
vegetation
weathering
wetlands

economic agreements
Term introduced in 1981.
SA aid projects
economics

economic evaluation
Term introduced in 1981. Used for evaluation of mineral deposits.
SA economics
evaluation
mineral deposits

economic geology
For the discipline as a whole. See also specific commodities (list C). Includes use on level 1 (list A); on level 2 under area terms (list B). If 1, term set options are:
topic [applications, bibliography, catalogs, classification, concepts, education, experimental studies, general, history, instruments, interpretation, methods, nomenclature, objectives, philosophy, practice, principles, symposia, textbooks, theoretical studies]
subtopic
SA chemical industry
development
economics
energy sources
gems
geology
industrial minerals
industry
legislation
metallogenic provinces
metallogeny
metals
mineral deposits
mineral economics
mineral exploration
mineral resources
mining geology
new energy sources
nonmetals
oil and gas fields
ore minerals
paragenesis
recycling
water resources

economic geology maps
Term introduced in 1978. Before 1978, search maps AND economic geology.
BA maps

economics
Includes use on level 2 under commodity terms (list C) for general treatments not tied to specific areas.
SA aid projects
antitrust legislation
borrowing
cash transactions
commodity exchange
company diversification
company integration
company mergers
cost
customs duty
demand
economic agreements
economic evaluation
economic geology
European Community
exchange rate
exploitation
export
export value
financing
futures market
import
import value
industrial patents
International Monetary Fund
investment
investment legislation
loans
Lome Convention
London Metal Exchange
manpower
market stabilization
mine reactivation
mineral economics
negotiations
N.Y. Commodity Exchange
pilot plants
prediction maps
price
production value
productive capacity
productivity
profitability
purchases
recycling
sales
subsidies
supply
taxes
technology transfer
Third World
World Bank

ecosystems
Includes use on level 3 under ecology(1).
SA ecology
environment

ecoulement
use gravity sliding

Ecrins-Pelvoux Massif
use Pelvoux Massif

ectinite
Includes use on level 3 under metamorphic rocks(1) migmatites(2). See list J.
BA migmatites
BT metamorphic rocks

Ectoprocta
BA Bryozoa
BT Invertebrata

Ector County
In W Texas.
BT Texas
United States

Ectotropha
Includes use on level 2 under Insecta(1). See list F.
BA Insecta
BT Mandibulata
Arthropoda
Invertebrata

Ecuador
Includes use on level 1 as an area term (list O). For term set options see list B.
CO S050000N013000
W0750000W0810000
BA South America
SA Amazon Basin
Andes
Eastern Cordillera
Galapagos Islands
Western Cordillera

Eddy County
Index states as applicable.
BX New Mexico
North Dakota
BT United States

Eden Shale
Includes Economy Formation, Southgate Formation, McMicken Formation. S Indiana, central Kentucky, and SW Ohio.
BT Edenian
Upper Ordovician
Ordovician
SA Indiana
Kentucky
Ohio

Edenian
North America. Above Mohawkian, below Maysvillian. Includes use on level 3 under age terms(1). See list E.
BA Upper Ordovician
Ordovician
BT Paleozoic
NT Eden Shale

Edentata
Includes use on level 2 under Mammalia(1). See list F.
BA Mammalia
BT Tetrapoda
Vertebrata

Edinburgh
City near the Firth of Forth in SE.
BT Scotland
Great Britain
United Kingdom

edingtonite
BA zeolite group
framework silicates
silicates
BT minerals

Edmondia
BA Bivalvia
Mollusca
BT Invertebrata

Edmonson County
Term introduced in 1981. Central Kentucky.
CO N370300N372200
W0860400W0863000
BT Kentucky
United States

Edmonton
City in central.
BT Alberta
Canada

Edmonton Formation
Now considered equivalent of either Bearpaw Shale or Fox Hills Sandstone of Montana.
BT Upper Cretaceous
Cretaceous
SA Alberta
Bearpaw Formation
Fox Hills Formation
Northwest Territories
Saskatchewan

Edo
use Tokyo

Edrioasteroidea
Includes use on level 2 under Echinodermata(1). See list F.
BA Echinozoa
Echinodermata
BT Invertebrata

Edrioblastoidea
Includes use on level 2 under Echinodermata(1). See list F.
BA Crinozoa
Echinodermata
BT Invertebrata

education
Includes use on level 1 (list A) for general discussions of methodology and curricula in education; on level 2 under major disciplines; as a level 2 or 3 term appropriate to a large number of topics (list G). If 1, term set options are:
topic [List B (except for areal geology), or extraterrestrial geology, general, geophysics]
subtopic (college-level educa-

tion, curricula, elementary geology, elementary school, graduate-level education, high school, junior high school, materials, methods, objectives, popular geology, vocational school)
SA associations
catalogs
college-level education
curricula
elementary geology
elementary school
general
geology
glossaries
graduate-level education
high school
historical geology
junior high school
materials
museums
oceanography
petrology
popular geology
programs
stratigraphy
textbooks

Edwards County
Index states as applicable.
BX Illinois
Kansas
Texas
BT United States

Edwards Formation
Overlies Walnut Clay and underlies Georgetown Limestone of Washita Group. In S Texas. Also search Edwards Limestone.
BT Comanchean
Cretaceous
SA Lower Cretaceous
Texas

Edwards Plateau
Highland region of W central Texas.
CO N300000N313000
W0990000W1020000
BT Texas
United States

Eel River
Rises in N Mendocino County in NW and flows NW into the Pacific Ocean.
BT California
United States

Eemian
Europe. See list E.
BA upper Pleistocene
Pleistocene
Quaternary
BT Cenozoic

effects
Includes use as a level 2 or 3 term appropriate to a large number of topics, e.g. on level 2 under earthquakes(1) and under faults(1). See list G.
SA climate effects
coastal effects
damage
geomorphologic effects
influence
relief effects
thermal effects

efficiency
Used as a general term.
SA accuracy
streams

effluents
Term introduced in 1981 for liquids discharged as waste. Before 1981, also search discharge AND pollution; effluent.
SA discharge
liquid waste
pollutants
pollution
waste water

Eger
City NE of Budapest.
BT Hungary

Egerian
Europe. See list E.
BA upper Oligocene
Oligocene
Paleogene
Tertiary
BT Cenozoic

Eggenburgian
Europe. See list E.
BA lower Miocene
Miocene
Neogene
Tertiary
BT Cenozoic

eggs
Includes use on level 3 under Aves(1) or Reptilia(1).
SA Aves
Reptilia

eggstone
use oolite

Egypt
Includes use on level 1 as an area term (list O). For term set options see list B.
CO N220000N320000
E0353000E0250000
BA North Africa
Africa
NT Alexandria
Aswan
Bahariya Oasis
Cairo
Eastern Desert
Fayum
Fayum Depression
Kharga Oasis
Kosseir
NX Middle Nile Valley
NT Nile Delta
Safaga
Sinai
Suez Canal
Western Desert
SA Libyan Desert
Mediterranean region
Middle East
Near East
Nile River
Nile Valley
Nubia
Nubian Sandstone
Nubian Shield
Red Sea Basin
Sahara

Eh
Includes use on level 3 under geochemistry(1). Also search oxidation-reduction; oxidation-reduction potential; redox; redox potential.
UF oxidation-reduction potential
redox
redox potential
SA geochemistry
oxidation
oxidation zone
reduction

Ehime
Prefecture.
BT Shikoku
Japan

Eichkogel
Region in NE part of country.
BT Lower Austria
Austria

Eichstatt
Town in central.
BT Bavaria
West Germany

Eifel
Hilly region in W between Rhine and Moselle rivers. Also search Eifel Mountains.
UF Eifel Mountains
BT Rhineland-Palatinate
West Germany

Eifel Mountains
use Eifel

Eifelian
Europe. Above Emsian, below Givetian. Includes use on level 3 under age terms(1). See list E.
BA Middle Devonian
Devonian
BT Paleozoic
SA Couvinian

eigenvalues
SA mathematical methods

Eilat
use Elath

Eilenriede
Forest area E of Hanover.
BT Lower Saxony
West Germany

Eire
use Ireland

ejecta
Includes use on level 3 under volcanology(1) or mud volcanoes(1).
SA lapilli
mud volcanoes
pyroclastics
volcanic ash
volcanology

ekanite
BA ring silicates
silicates
BT minerals

Ekaterinburg
use Sverdlovsk

Ekibastuz
Town in Pavlodar Oblast in NE.
BT Kazakhstan
USSR

Ekman spiral
Includes use on level 2 under ocean circulation(1).
SA atmosphere
ocean circulation
patterns
winds

El Alboran Island
Spanish islet 135 miles E of Gibraltar in SW.
BT Mediterranean Sea

El Centro
City in Imperial County in Imperial Valley in S.
BT California
United States

El Dorado County
SW of Lake Tahoe in N.
BT California
United States

El Kharga
use Kharga Oasis

El Paso
City in El Paso County in extreme W.
BT Texas
United States

El Paso County
Index states as applicable.
BX Colorado
Texas
BT United States

El Salvador
Includes use on level 1 as an area term (list O). For term set options see list B.
CO N131000N142800
W0873800W0901000
BA Central America
SA San Salvador

Elasmobranchii
As of 1981, includes use on level 2 under Pisces(1). Before 1981, included use on level 3 under Pisces(1) Chondrichthyes(2). See list F.
BA Chondrichthyes
Pisces
BT Vertebrata
NA Bradyodonti
Selachii

elastic constants
Includes use on level 3 under deformation(1) or geophysics(1). Also search elastic moduli; modulus of elasticity; elastic modulus.
UF constants, elastic
elastic moduli
modulus of elasticity
NA bulk modulus
Poisson's ratio
shear modulus
Young's modulus
SA deformation
elastic properties
elasticity
geophysics
rock mechanics
soil mechanics

elastic limit
Includes use on level 3 under deformation(1) field studies(2).
UF limit, elastic
SA creep
deformation
elastic properties
elastic strain
elasticity
fracture strength
Hooke's law
mechanical properties
strain
viscoelasticity
yield strength

elastic materials
Introduced in 1978 as a general term, mostly under deformation(1). Before 1978, also search elastic AND materials.
SA deformation
elasticity
explosions
geophysics
materials
plastic materials
rock mechanics

elastic moduli
use elastic constants

elastic properties
Includes use on level 3 under deformation(1) and under material name; on level 3 under seismology(1).
SA deformation
elastic constants
elastic limit
elastic strain
elasticity
geophysics
Hooke's law
mechanical properties
properties
seismology
tensile strength

elastic strain
Includes use on level 3 under deformation(1) and under engineering geology(1).

SA creep
deformation
elastic limit
elastic properties
elasticity
engineering geology
Hooke's law
strain
viscoelasticity

elastic waves
Do not use seismic waves. Includes use on level 2 under seismology(1) and under mantle (1); on level 3 under earthquakes(1). Before 1976, also search seismic waves.
NA acoustical waves
Airy waves
body waves
long-period waves
short-period waves
surface waves
SA amplitude
amplitude distortion
arrival time
asthenosphere
attenuation
deconvolution
deformation
distortion
earthquakes
explosions
filters
focus
foreshocks
group velocity
linear distortion
mantle
Mohorovicic discontinuity
non-linear distortion
phase distortion
phase velocity
propagation
Q
scattering
seismic methods
seismic sources
seismology
signal distortion
traveltime
traveltime curves
Vibroseis
wave dispersion
waves

elasticity
As of 1978, term is used on level 2 under rock mechanics(1) and soil mechanics(1).
SA anelasticity
bulk modulus
compressibility
deformation
elastic constants
elastic limit
elastic materials
elastic properties
elastic strain
engineering geology
mechanical properties
plasticity
Poisson's ratio
rigidity
rock mechanics
shear modulus
soil mechanics
strain
stress
viscoelasticity
yield strength
Young's modulus

elastoplastic materials
Term introduced in 1981.
SA engineering geology
materials

Elat
use Elath

Elath
City at head of Gulf of Aqaba. Also search Eilat; Elat.
UF Eilat
Elat
BT Israel

Elba
Island in Mediterranean Sea between Corsica and mainland of Italy.
BT Tuscany
Italy

elbaite
BA ring silicates
silicates
BT minerals
SA tourmaline

Elbe River
Rises in N Bohemia and flows NW through Hamburg into the North Sea. Index countries as applicable. Also search Elbe.
BT Europe
SA Czechoslovakia
East Germany
West Germany

Elbe Valley
Index countries as applicable. Also search Elbe.
BT Europe
SA Czechoslovakia
East Germany
West Germany

Elbingerode
Town in W part of country.
BT Magdeburg Bezirk
East Germany

Elbistan
Town in S central part of Anatolia.
BT Turkey

Elborus
use Elbrus

Elbrus
Peak in a N subsidiary spur of the main range of the Caucasus. Highest peak in Europe.
UF Elborus
Elbrus Mountain
Mount Elbrus
Mt. Elbrus
BT Russian Republic
USSR

Elbrus Mountain
use Elbrus

Elburz
Range in N parallel to S shore of Caspian Sea. Also search Elburz Mountains.
UF Alborz Mountains
Elburz Mountains
BT Iran

Elburz Mountains
use Elburz

Eldorado
Village in NE.
BT Saskatchewan
Canada

electric fields
A valid term through 1974. After 1974, use electrical field.

electrical
A valid term through 1977. After 1977, use electrical logging on level 2 under well-logging(1) or electrical conductivity under geophysical methods(1).

electrical anomalies
Term introduced in 1981. Includes use on level 3 under geophysical methods(1).
SA anomalies
electrical methods
electrical surveys
geophysical methods
geophysical surveys

electrical conductivity
Includes use on level 3 under geophysical methods(1).
SA conductivity
electrical field
geophysical methods
resistivity
thermal conductivity

electrical field
Includes use on level 2 under aurora(1), ionosphere(1), and magnetosphere(1). Before 1974, also search electric fields.
UF field, electrical
SA arcs
aurora
currents
dielectric constant
dielectric properties
Earth-current methods
electrical conductivity
electromagnetic field
electromagnetic induction
electromagnetic waves
induced polarization
ionosphere
magnetic field
magnetosphere
magnetotelluric methods

electrical logging
Not a valid term from 1971 through 1977. Now includes use on level 2 under well-logging(1). Before 1978, also search well-logging AND electrical.
UF logging, electrical
BA well-logging
SA boreholes
conductivity
resistivity

electrical methods
Includes use on level 2 under geophysical methods(1).
BA geophysical methods
SA dielectric constant
dielectric properties
electrical anomalies
electrical properties
electrical surveys
electromagnetic methods
induced polarization
methods
resistivity
self-potential methods

electrical phenomena
Includes use on level 2 under meteorology(1).
UF phenomena, electrical
SA currents
lightning
meteorology
raindrops
thunderstorms

electrical properties
Includes use on level 3 under material name.
SA electrical methods
electrical surveys
properties
surface properties

electrical resistivity
use resistivity

electrical surveys
Includes use on level 3 under area terms(1) geophysical surveys(2).
BA geophysical surveys
BT surveys
SA electrical anomalies
electrical methods
electrical properties
electromagnetic logging
electromagnetic methods
induced polarization
resistivity
self-potential methods

electro-osmosis
Term introduced in 1981. Before 1981, also search electroosmosis.
SA ground water

electrochemical properties
Includes use on level 3 under geochemistry(1) properties(2).
SA geochemistry
properties

electrodes
Used as a general term.
SA electrolysis

electrojet
Includes use on level 2 under ionosphere(1); on level 3 under aurora(1) electrical field(2) and magnetic field(2).
SA aurora
currents
ionosphere
magnetic field
polar regions

electrolysis
Includes use on level 3 under geochemistry(1) processes(2).
SA dissociation
electrodes
geochemistry
polarography
processes
solution

electrolytes
SA anions
geochemistry
ion exchange
ions

electromagnetic
A valid term through 1977. After 1977, use electromagnetic logging on level 2 under well-logging(1).

electromagnetic field
Includes use on level 3 under geophysical methods(1).
UF field, electromagnetic
SA electrical field
electromagnetic induction
geophysical methods
geophysical surveys
magnetic field
magnetotelluric methods

electromagnetic induction
Term introduced in 1981. Also search induction AND electromagnetic.
SA electrical field
electromagnetic field
electromagnetic logging
electromagnetic methods
magnetic field

electromagnetic logging
Term introduced in 1978. Includes use on level 2 under well-logging(1). Before 1978, also search well-logging AND electromagnetic.
UF logging, electromagnetic
BA well-logging
SA electrical surveys
electromagnetic induction
electromagnetic methods
geophysical methods
geophysical surveys

electromagnetic methods
Includes use on level 2 under geophysical methods(1).
UF electromagnetic prospecting
BA geophysical methods
SA electrical methods
electrical surveys
electromagnetic induction

electromagnetic logging
electromagnetic surveys
induction
magnetotelluric methods
magnetotelluric surveys
methods
mineral exploration

electromagnetic prospecting
use electromagnetic methods

electromagnetic radiation
Includes use on level 2 under astrophysics and solar physics(1) and under interplanetary space(1).
SA astrophysics and solar physics
cosmic rays
electromagnetic waves
gamma rays
interplanetary space
luminescence
magnetic field
particles
radiation
radio-wave methods
radioactivity
solar wind
whistlers
X-rays

electromagnetic seismographs
Term introduced in 1981.
BT seismographs

electromagnetic surveys
Includes use on level 3 under area terms(1) geophysical surveys(2).
BA geophysical surveys
BT surveys
SA electromagnetic methods
magnetic field
magnetotelluric surveys

electromagnetic waves
Includes use on level 2 under meteorology(1).
SA electrical field
electromagnetic radiation
magnetic field
meteorology
propagation
radio-wave methods
waves

electron content
Includes use on level 2 under ionosphere(1).
SA ion densities and temperatures
ionosphere

electron diffraction analysis
Includes use on level 3 under chemical analysis(1) and under chemical element(1) analysis(2).
SA analysis
electrons
neutron diffraction analysis
X-ray diffraction analysis

electron microprobe
use electron probe

electron microscopy
Includes use on level 1 (list A) for methodology, not for data. If 1, term set options are:
applications
subtopic
instruments
subtopic
methods
subtopic
techniques
subtopic
UF microscopy, electron
SA chemical analysis
clay mineralogy
crystal chemistry
crystal growth
crystallography
differential thermal analysis
electron microscopy data
scanning method
scattering
SEM data
spectroscopy
structural analysis
TEM data
thermal analysis
thin sections
transmission method
X-ray analysis

electron microscopy data
Term introduced in 1978. Used to distinguish electron microscopy as a method from the data.
SA data
electron microscopy

electron paramagnetic resonance
As of 1978, term is used only for methodology. For data, see EPR spectra. Includes use on level 3 under spectroscopy(1) methods(2); on level 3 under chemical element(1) analysis(2). See list D. Also search electron spin resonance.
UF electron spin resonance
SA analysis
EPR spectra
methods
nuclear magnetic resonance
resonance
spectroscopy

electron paramagnetic resonance spectra
use EPR spectra

electron probe
As of 1978, term is restricted to methodology. For data, see electron probe data. Includes use on level 3 under spectroscopy(1) methods(2); chemical analysis(1) methods(2); on level 3 under chemical element(1) analysis(2). See list D. Also search electron microprobe.
UF electron microprobe
electron probe analysis
probe, electron
SA analysis
chemical analysis
electron probe data
ion probe
methods
spectroscopy

electron probe analysis
use electron probe

electron probe data
Term introduced in 1978. Before 1978, search electron probe.
UF electron probe spectra
SA data
electron probe

electron probe spectra
use electron probe data

electron spin resonance
use electron paramagnetic resonance

electrons
Used as a general term.
SA electron diffraction analysis
neutrons
particles
protons

elementary geology
Term introduced in 1978. Includes use on level 3 under education(1). Before 1978, search popular and elementary geology.
SA education
geology
popular geology

elementary school
Includes use on level 3 under education(1).
SA college-level education
education
high school
junior high school

elements
Includes use on level 3 under geochemistry(1). See list D. Also search chemical elements.
UF chemical elements
SA actinium
aluminum
americium
antimony
argon
arsenic
barium
beryllium
bismuth
boron
bromine
cadmium
calcium
californium
carbon
cerium
cesium
chlorine
chromium
cobalt
copper
cosmochemistry
cosmogenic elements
curium
deuterium
dysprosium
erbium
europium
fluorine
gadolinium
gallium
geochemistry
germanium
gold
hafnium
helium
holmium
indium
iodine
iridium
iron
isotopes
krypton
lanthanum
lead
lithium
lithophile elements
lutetium
magnesium
major elements
major-element analyses
manganese
mercury
metals
migration of elements
minor elements
minor-element analyses
molybdenum
neodymium
neon
neptunium
nickel
niobium
nitrogen
nonmetals
osmium
oxygen
palladium
phosphorus
platinum
plutonium
polonium
potassium
praseodymium
promethium
protactinium
radium
radon
rare earths
rhenium
rhodium
rubidium
ruthenium
samarium
scandium
selenium
siderophile elements
silicon
silver
sodium
strontium
sulfur
tantalum
technetium
tellurium
terbium
thallium
thorium
thulium
tin
titanium
trace elements
trace-element analyses
tritium
uranium
vanadium
volatile elements
xenon
ytterbium
yttrium
zinc
zirconium

Elephantidae
As of 1981, includes use on level 3 under Mammalia(1) Elephantoidea(2). See list F. Before 1981, included use on level 3 under Mammalia(1) Proboscidea(2).
BA Elephantoidea
Proboscidea
Mammalia
BT Tetrapoda
Vertebrata
NA Elephas
Mammuthus
Palaeoloxodon naumanni
Stegodon

Elephantoidea
Term introduced in 1981. Includes use on level 2 under Mammalia(1). See list F.
BA Proboscidea
Mammalia
BT Tetrapoda
Vertebrata
NA Elephantidae

Elephas
Genus. As of 1981, includes use on level 3 under Mammalia(1) Elephantoidea(2). See list F. Before 1981, included use on level 3 under Mammalia(1) Proboscidea(2).
BA Elephantidae
Elephantoidea
Proboscidea
Mammalia
BT Tetrapoda
Vertebrata
NA Elephas antiquus

Elephas antiquus
As of 1981, includes use on level 3 under Mammalia(1) Elephantoidea(2). Before 1981, included use on level 3 under Mammalia(1) Proboscidea(2). See list F.
BA Elephas
Elephantidae
Elephantoidea
Proboscidea
Mammalia
BT Tetrapoda
Vertebrata

elevation
use altitude

ELF
As of 1981, no longer a valid term for GeoRef. See electromagnetic waves. Before 1981, included use on level 3 under ionosphere(1) wave propagation(2).

Elk Hills Field
U. S. naval oil reservation in Kern County in S California. Also search Elk Hills; Elk Hills Oil Field.
UF Elk Hills oil field
BT California
United States

Elk Hills oil field
use Elk Hills Field

Elk Mountains
Range of Rocky Mountains in W central.
BT Colorado
United States
SA Rocky Mountains

Elk Point Basin
Index provinces as applicable.
BT Canada
SA Alberta
Manitoba
Saskatchewan

Elk Point Group
Includes Ashern Formation, Elm Point Formation, Winnipegosis Formation, and Prairie Formation.
BT Middle Devonian
Devonian
SA Alberta
Manitoba
Montana
North Dakota
Prairie Evaporite
Saskatchewan
South Dakota
Winnipegosis Formation

Elko County
In NE Nevada.
BT Nevada
United States

Ellef Ringnes Island
One of the Sverdrup Islands in District of Franklin.
CO N773000N790000
W0980000W1060000
BT Northwest Territories
Canada

Ellesmere Island
W of NW Greenland in District of Franklin.
CO N760000N840000
W0600000W0920000
BT Northwest Territories
Canada

Ellice Islands
use Tuvalu

Elliot Lake
Town in S central.
BT Ontario
Canada

Elliott County
In E Kentucky.
BT Kentucky
United States

ellipticity
General property.
SA geodesy
geometry

Ellis County
Index states as applicable.
BX Kansas
Oklahoma
Texas
BT United States

Ellsworth County
Central.
BT Kansas
United States

Ellsworth Highland
use Ellsworth Land

Ellsworth Land
High plateau just S and SW of the Antarctic Peninsula. Also search Ellsworth Highland.
UF Ellsworth Highland
James W. Ellsworth Land
BT Antarctica

Ellsworth Mountains
Range S of Ellsworth Land and SW of the Antarctic Peninsula.
BT Antarctica

ellsworthite
use betafite

elongate minerals
Includes use on level 3 under lineation(1) style(2).
BT minerals
SA lineation
structural analysis

Elphidium
Genus. Includes use on level 3 under foraminifera(1) Rotaliacea(2).
BA Rotaliacea
Rotaliina
foraminifera
BT Protista
Invertebrata

Elsinore Fault
Extending from a point E of Los Angeles to a point E of San Diego.
BT California
United States

eluvium
Includes use on level 3 under sediments(1) clastic sediments(2). See list N.
BA clastic sediments
BT sediments
SA alluvium
dunes
soils
terrigenous materials

Elvas
City near the Spanish border in S part of country.
BT Portugal

Ely
Town in White Pine County, Nevada, and Saint Louis County, Minnesota. Index states as applicable.
BX Minnesota
Nevada
BT United States

Emba
Town in Aktyubinsk Oblast in NW.
BT Kazakhstan
USSR

Emba Field
Term introduced in 1981. Oil and gas field in Aktyubinsk Oblast in NW.
CO N470000N490000
E0560000E0523000
BT Kazakhstan
USSR

Emba River
Flows into NE Caspian Sea. There are oil fields in its lower course. Also search Emba.
BT Kazakhstan
USSR

embankments
As of 1978, term is used on level 2 under highways(1) and slope stability(1).
SA dams
highways
levees
reservoirs
slope stability
slopes

embrechite
Term introduced in 1981. Includes use on level 3 under metamorphic rocks(1) migmatites(2). See list J.
BA migmatites
BT metamorphic rocks

Embrithopoda
Includes use on level 2 under Mammalia(1). See list F.
BA Mammalia
BT Tetrapoda
Vertebrata

emerald
Variety of beryl.
UF smaragd
BA ring silicates
silicates
BT minerals
SA beryl
gems

Emery
Town in Emery County in central.
BT Utah
United States

Emery County
Central.
BT Utah
United States

Emilia-Romagna
Autonomous region in N Italy. Also search Emilia.
CO N434500N451000
E0125000E0910000
BT Italy
NT Bologna
Modena
Parma
Piacenza
Romagna
Taro Valley
SA Po River
Po Valley
Vigarano Meteorite

emission
A valid term through 1977 used to denote the method. After 1977, use emission spectroscopy.

emission spectroscopy
Not a valid term from 1971 through 1977. After 1977, includes use on level 3 under spectroscopy(1) and under chemical element(1) analysis(2). Before 1978, also search spectroscopy AND emission.
BT spectroscopy
SA absorption
analysis
chemical analysis

emission, thermal
use thermal emission

emissions
As of 1981, no longer a valid term for GeoRef. Before 1981, included use on level 2 and 3 under aurora(1).

Emperor Seamount Chain
use Emperor Seamounts

Emperor Seamounts
SE of Kamchatka Peninsula. Also search Emperor Seamount Chain.
UF Emperor Seamount Chain
BT Pacific Ocean

emplacement
Includes use on level 2 under intrusions(1).
SA assimilation
contamination
hybridization
igneous rocks
intrusions
magmas
plutons
salt tectonics

emplectite
BA sulfobismuthites
sulfosalts
BT minerals

Emporia
City in Lyon County in E central.
BT Kansas
United States

Ems River
Flows into the North Sea. Index West German states as applicable. Also search Ems.
BT West Germany
SA Lower Saxony
North Rhine-Westphalia
Weser-Ems

Emsian
Europe. Above Siegenian, below Eifelian. Includes use on level 3 under age terms(1). See list E.
BA Lower Devonian
Devonian
BT Paleozoic

Emsland
Swampy region between Ems River and Netherlands border.
BT Lower Saxony
West Germany

en echelon faults
Term introduced in 1978. Includes use on level 3 under faults(1) patterns(2) and folds(1). Before 1978, also search faults AND en echelon.
BA faults
SA systems

en echelon folds
Term introduced in 1978. Before 1978, search folds AND en echelon.
BA folds

ENADIMSA
Term introduced in 1981.
SA Spain
survey organizations

enargite
BA sulfarsenates
sulfosalts
BT minerals
SA copper
copper ores

Enari
use Inari

Encounter Bay
Inlet of Indian Ocean SE of Adelaide.
BT South Australia
Australia

encroachment (ground water)
use salt-water intrusion

Encruzilhada
use Encruzilhada do Sul

Encruzilhada do Sul
City in SE Rio Grande do Sul. Also search Encruzilhada.
UF Encruzilhada
BT Rio Grande do Sul
Brazil

Endako
Village in central.
BT British Columbia
Canada

endellionite
use bournonite

endellite
use halloysite

endemic taxa
Includes use on level 3 under fossil groups (list F) as applicable. Also

search endemism.
UF endemism
taxa, endemic

endemism
use endemic taxa

enderbite
As of 1981, includes use on level 3 under igneous rocks(1) granites(2). See list H. Before 1981, included use on level 3 under igneous rocks(1) granite-granodiorite family(2).
BA granites
BT igneous rocks

Enderby Land
Semicircular projection of land on the Indian Ocean side claimed by Australia.
BT Antarctica

endogene processes
Includes use on level 3 under mineral deposits, genesis(1) processes(2). Before 1981, also search hypogene processes.
SA mineral deposits
processes

endothermic reactions
Term introduced in 1978. Includes use on level 3 under thermal analysis(1). Before 1978, also search endothermic.
SA DTA data
reactions
thermal analysis

Endothyra
Genus. Includes use on level 3 under foraminifera(1) Fusulinina(2).
BA Fusulinina
foraminifera
BT Protista
Invertebrata

energy
Used as a general term.
SA activation energy
energy sources
fission
free energy
geothermal energy
kinetics
new energy sources
nuclear energy
seismic energy
solar energy

energy conservation
Term introduced in 1981.
SA conservation
environmental geology

energy relaxation
Term introduced in 1981.
SA seismology

energy sources
Includes use on level 1 as a commodity term (list C). See also names of appropriate commodities.
UF sources, energy
SA coal
economic geology
energy
fuel resources
geothermal energy
heat flow
natural gas
new energy sources
nuclear energy
petroleum
power plants
resources
solar energy
water resources

Enewetak
use Eniwetok Atoll

engineering, civil
use civil engineering

engineering geology
Used for geology as applied to engineering practice, especially mining and civil engineering. See appropriate features and processes under geomorphology(1). Includes use on level 1 (list A); on level 2 under area terms (list B). If 1, term set options are:
field studies
subtopic
materials, properties
(for material not covered under rock mechanics and soil mechanics)
material or property (e.g. elastic strain)
methods
type of method (cartography, photogeology, photogrammetry)
topic [bibliography, education, experimental studies, feasibility studies, frost action, maps, petroleum engineering, practice, seepage, site exploration, symposia, techniques]
subtopic
UF geologic engineering
SA accelerograms
accelerometers
airfields
anchors
aseismic design
blasting
breakwaters
bridges
civil engineering
clays
cohesionless materials
cohesive materials
concrete
conductive materials
consistent materials
consolidometer tests
continuous materials
cryopedology
cyclic loading
damage
dams
deformation
design
discontinuous materials
dredging
drilling
earth dams
earthquakes
elastic strain
elasticity
elastoplastic materials
environmental geology
excavations
explosions
foundations
fractures
frost action
frozen ground
gas storage
geodesy
geologic hazards
geology
geomorphology
geophysical methods
geophysical surveys
geophysics
granular materials
ground water
grouting
highways
hydraulics
hydrology
impact statements
induced earthquakes
industry
injection
inventory
land subsidence
land use
landfills
landslides
layered materials
legislation
linear materials
liquefaction potential
liquid waste
load pressure
maps
marine installations
materials
mechanical properties
microcracks
mining geology
non-linear materials
nuclear explosions
nuclear facilities
overconsolidated materials
penetration
penetration tests
permafrost
permeability
petroleum engineering
photogeology
photogrammetry
plastic materials
Poisson's ratio
pore pressure
quick clay
radioactive waste
radioactivity
remote sensing
reservoirs
road tests
rock mechanics
sealing
sedimentation
seepage
seismic response
seismic risk
seismic zoning
settlement
shear stress
shorelines
site exploration
slope stability
soil mechanics
soils
solid waste
solifluction
stability
strain
structural analysis
submarine installations
surface waves
tensile strength
testing
thermal properties
thermal regime
tunnels
underground installations
uniaxial tests
viscoelastic materials
viscoplastic materials
waste disposal
water storage
waterways
weathering
wells

engineering, petroleum
use petroleum engineering

engineering properties
Includes use on level 3 under soils(1).
SA foundations
highways
permafrost
properties
sediments
soil mechanics
soils

England
Includes use on level 1 as an area term (list O). For term set options see list B.
CO N500000N554500
0E001300W0063000
BA Great Britain
United Kingdom
Western Europe
Europe
NT Bath
Bedfordshire
Berkshire
Bewcastle
NX Birmingham
NT Birmingham University
Brandon
Bristol
NX Cambridge
NT Cambridge University
Cheshire
Chesil Beach
Chudleigh
Church Stretton
Clitheroe
Cornwall
Cotswold Hills
NX Cumberland
NT Dartmoor
Derby
Derbyshire
Devonshire
Dorset
Dudley
Durham
East Anglia
East Midlands
Essex
Exeter
Fawley
Folkestone
Forest of Dean
Freshwater
Gibraltar Point
Gloucestershire
Hampshire
Hampshire Basin
Hertfordshire
Holderness
Horsham
Howgill Fells
Humber Estuary
Ingleborough
Ingleton
Isle of Wight
Kent
Lake District
Lancashire
Land's End
Leicestershire
Lincolnshire
Liverpool Bay
London
London Basin
Malvern
Malvern Hills
Meldon Aplite
Mendip Hills
Midlands
Morecambe Bay
Newcastle
Norfolk
Northamptonshire
Northumberland
Norwich
Nottingham
Nottinghamshire
Okehampton
Oxford
Oxfordshire
Padstow
Peak District
Pennines
Piltdown
Sheffield
Shropshire
Somerset
South-West England
Staffordshire
Stonehenge

Suffolk
Surrey
Sussex
Tamar Valley
Teign Valley
Thames Estuary
Thames River
The Weald
Torquay
Tyne River
Warwickshire
Welsh Borderland
Wenlock Edge
Westmorland
Wiltshire
Wolverhampton
Worcestershire
Yorkshire
SA Aberystwyth Grits
Barton Beds
Barwell Meteorite
Bembridge Marls
Borrowdale Volcanic Series
Bristol Channel
Cheviot Hills
Esk Trough
Great Oolite Series
Great Scar Limestone
London Clay
Lower Greensand
Mersey River
Mersey Valley
North Sea Coast
Oxford Clay
Severn Valley
Shap Granite
Skiddaw Slates
Speeton Clay
Trent Valley
Weald Clay
Wenlock Limestone
Wye Valley
York

English
Includes use on level 3 to indicate language of document.

English Channel
Strait between England and France connecting Atlantic Ocean with the North Sea. Includes use on level 1 as an area term (list O). For term set options see list B.
CO N480000N510000
E0020000W0060000
BA European Atlantic
North Atlantic
Atlantic Ocean

English Channel Islands
Term introduced in 1981 for the British group of islands in the English Channel off the W coast of Normandy. Before 1981, "Channel Islands" was used for both these and the Santa Barbara Islands off California. Before 1981, search Channel Islands AND England. Includes use on level 1 as an area term (list O). For term set options see list B.
CO N491000N494500
W0020000W0024500
BA United Kingdom
Western Europe
Europe
SA Channel Islands

English River Belt
Term introduced in 1981. Subprovince of Superior Province, Canadian Shield. E Manitoba and W Ontario. Before 1981, also search English River Subprovince.
CO N500000N520000
W0860000W0970000
UF English River Subprovince
BT Superior Province
Canadian Shield
SA Manitoba
Ontario

English River Subprovince
use English River Belt

enigmatite
use aenigmatite

Enisei Ridge
use Yenisei Ridge

Enisei River
use Yenisei River

Eniwetok
use Eniwetok Atoll

Eniwetok Atoll
Part of the U. S. Trust Territory of the Pacific Islands. Also search Eniwetok.
UF Enewetak
Eniwetok
BT Marshall Islands
Micronesia

Enna
City and province in central.
BT Sicily
Italy

Enns Valley
River Valley. Index Austrian states as applicable. Also search Enns.
BT Austria
SA Lower Austria
Upper Austria

enrichment
Includes use on level 3 under geochemistry(1), metasomatism(1) or weathering(1).
SA geochemistry
metasomatism
migration of elements
mineral deposits
supergene processes
weathering

enstatite
BA orthopyroxene
pyroxene group
chain silicates
silicates
BT minerals
SA bronzite

enstatite chondrites
BA chondrites
BT meteorites

Enteletacea
Superfamily. Includes use on level 3 under Brachiopoda(1) Orthida(2).
BA Orthida
Articulata
Brachiopoda
BT Invertebrata
NA Schizophoria

enterolithic folding
use enterolithic folds

enterolithic folds
Term introduced in 1978. Also search enterolithic; enterolithic folding.
UF enterolithic folding
BA folds
SA deformation

Enteropneusta
Term introduced in 1981. Includes use on level 2 under Hemichordata(1). See list F.
BX Hemichordata
Invertebrata

enthalpy
Includes use on level 3 under geochemistry(1). Also search heat content.
UF heat content
SA entropy
geochemistry
thermodynamic properties

Entisols
Includes use on level 3 under soils(1). See list M.
BA soils

Entomotaeniata
Term introduced in 1981. Includes use on level 2 under Mollusca(1). See list F.
BA Gastropoda
Mollusca
BT Invertebrata

Entomozocopina
Term introduced in 1981. Includes use on level 2 under Ostracoda(1). See list F.
BA Ostracoda
BT Crustacea
Mandibulata
Arthropoda
Invertebrata

Entotropha
Term introduced in 1981. Includes use on level 2 under Insecta(1). See list F.
BA Insecta
BT Mandibulata
Arthropoda
Invertebrata

entropy
Includes use on level 3 under phase equilibria(1).
SA enthalpy
phase equilibria
sediments
thermodynamic properties

environment
Includes use as a level 2 or 3 term appropriate to a large number of topics, e.g. on level 2 under sedimentation(1). See list G.
SA aerobic environment
alpine environment
anaerobic environment
arctic environment
arid environment
boreal environment
brackish-water environment
coastal environment
deep-sea environment
deltaic environment
domains
ecology
ecosystems
environmental analysis
environmental geology
eolian environment
estuarine environment
fluvial environment
fresh-water environment
geomorphology
glacial environment
glaciofluvial environment
glaciolacustrine environment
glaciomarine environment
habitat
high-energy environment
humid environment
interglacial environment
intertidal environment
lacustrine environment
lagoonal environment
low-energy environment
marine environment
nearshore environment
paleoecology
paleoenvironment
paludal environment
pelagic environment
periglacial environment
postglacial environment
reclamation
sebkha environment
sedimentation
semi-arid environment
shallow-water environment
shelf environment
slope environment
subaerial environment
sublittoral environment
submarine environment
subtidal environment
subtropical environment
supratidal environment
taiga environment
temperate environment
terrestrial environment
tropical environment

environmental analysis
For papers discussing the paleoenvironmental implications of several rock types and for papers whose main purpose is to determine environment. Includes use on level 2 under sedimentary rocks(1), sediments(1), and sedimentary structures(1). Also search environmental analyses.
SA analysis
environment
environmental geology
paleoenvironment
sedimentary rocks
sedimentary structures
sedimentation
sediments

environmental geology
Includes use on level 1 (list A); on level 2 under area terms (list B). Used for studies involving the collection, analysis, and application of geologic data and principles to problems created by human occupancy and use of the physical environment. If 1, term set options are:
topic [concepts, education, maps, symposia]
subtopic [controls, floods, human ecology, regional planning, urban planning]
SA acid mine drainage
airfields
atmosphere
conservation
detergents
ecology
energy conservation
engineering geology
environment
environmental analysis
floods
geochemistry
geologic hazards
geology
geomorphology
ground water
human ecology
hydrology
impact statements
industrial waste
industry
injection
inventory
land use
landfills
legislation
liquid waste
maps
marine geology
medical geology
mine drainage
mineral resources
mining geology
natural resources
pesticides
pollution
preventive measures
radioactive waste
reclamation
recycling
regional planning
remote sensing

seismic zoning
soils
solid waste
spoils
strip mining
surface water
toxic materials
toxicity
trace metals
urban planning
urbanization
waste disposal
waste water
water
water resources
waterways

enzymes
BA organic materials

Eocambrian
Uppermost Precambrian(?)/lowermost Cambrian(?).
BA Precambrian
SA Cambrian
Infracambrian
Riphean
Vendian

Eocene
World. Above Paleocene, below Oligocene. Includes use on level 1 as an age term (list E); on level 2 under paleo- terms, e.g. paleoecology, paleogeography, paleomagnetism.
BA Paleogene
Tertiary
BT Cenozoic
NT Annot Sandstone
NA Bartonian
Biarritzian
Bridgerian
NT Castle Hayne Limestone
Chuckanut Formation
Clarno Formation
Crystal River Formation
Golden Valley Formation
Green River Formation
NA lower Eocene
Lutetian
middle Eocene
NT Mirador Formation
Nanjemoy Formation
NA Narizian
NT Punta Mosquito Formation
Pyrenean Orogeny
NA Refugian
NT Rose Canyon Formation
Swauk Formation
Umpqua Formation
NA upper Eocene
NT Werillup Formation
Yamhill Formation
SA Beaverhead Formation
Brianconnais Zone
Deccan Traps
Duchesne River Formation
Floridan Aquifer
Intertrappean Beds
Paskapoo Formation
San Lorenzo Formation
Scaglia Formation
Sespe Formation
Sinjar Formation
Twin River Formation
Wasatch Formation

Eocrinoidea
Class. Includes use on level 2 under Echinodermata(1). See list F.
BA Crinozoa
Echinodermata
BT Invertebrata

eolian
A valid term through 1977 for type of environment. After 1977, see wind transport; wind erosion; and (as of 1981) eolian environment.

eolian environment
Term introduced in 1981.
UF aeolian environment
SA environment
eolian features
eolian sedimentation
sedimentation
wind erosion
wind transport

eolian features
Includes use on level 2 under geomorphology(1).
UF features, eolian
SA deserts
dunes
eolian environment
geomorphology
loess
wind erosion
wind transport
winds

eolian sedimentation
Term introduced in 1981.
SA eolian environment
sedimentation

eolianite
Includes use on level 3 under sedimentary rocks(1) clastic rocks(2). See list I. Also search dune rocks.
UF dune rock
BA clastic rocks
BT sedimentary rocks
SA terrigenous materials

eosphorite
BA phosphates
BT minerals

Eosuchia
Term introduced in 1981. Includes use on level 2 under Reptilia(1). See list F.
BA Lepidosauria
Reptilia
BT Tetrapoda
Vertebrata

Eoulx Basin
Index departments as applicable.
BT France
SA Alpes-de-Haute Provence
Hautes-Alpes

epeirogenesis
A valid term through 1974. After 1974, use epeirogeny.

epeirogeny
Used for specific epeirogenic activity. For general treatment, see tectonics. Includes use on level 1 (list A). Before 1975, also search epeirogenesis. Term set options are:
age [list E]
area
SA continents
crust
eustacy
isostasy
neotectonics
orogeny
tectonics
uplifts

epeirophoresis theory
use continental drift

Ephedrales
Term introduced in 1981. Includes use on level 2 under gymnosperms(1). See list F.
BA gymnosperms
BT Spermatophyta
Plantae

ephemeral streams
BA streams
SA fluvial features

Ephemeropteroida
Term introduced in 1981. Includes use on level 2 under Insecta(1). See list F.
BA Insecta
BT Mandibulata
Arthropoda
Invertebrata

epibiotism
Term introduced in 1981.
SA biology
ecology
parasites
symbiosis

epicenters
Includes use on level 2 under earthquakes(1).
SA earthquakes
focus
seismology

epidiorite
Includes use on level 3 under metamorphic rocks(1) schists(2). See list J.
BA schists
BT metamorphic rocks

epidote
UF arendalite
pistacite
BA epidote group
sorosilicates
orthosilicates
silicates
BT minerals
SA aluminosilicates

epidote group
Includes use in combination with orthosilicates (i.e. orthosilicates, epidote group) on level 2 under minerals(1). See list L.
BA sorosilicates
orthosilicates
silicates
BT minerals
NA allanite
clinozoisite
epidote
piemontite
tanzanite
zoisite

epidote-amphibolite facies
BT facies
SA amphibolite facies
metamorphic rocks

epigene processes
Includes use on level 3 under mineral deposits, genesis(1) processes(2). Also search epigenesis; epigenic processes.
UF epigenic processes
SA mineral deposits
processes

epigenic processes
use epigene processes

Epirus
Administrative region in NW.
CO N385500N402200
E0213000E0204000
BT Greece
SA Pindus Mountains

epistilbite
BA zeolite group
framework silicates
silicates
BT minerals

epitaxy
Includes use on level 3 under crystal growth(1).
SA crystal growth
twinning

epithermal processes
Also search epithermal, or epithermal AND processes.
SA hydrothermal processes
mineral deposits
processes

EPR spectra
Term introduced in 1978. Before 1978, also search electron paramagnetic resonance for data. After 1978, use electron paramagnetic resonance only for methodology.
UF electron paramagnetic resonance spectra
SA electron paramagnetic resonance
spectra

epsomite
BA sulfates
BT minerals

equations
Used as a general term.
SA algorithms
equations of state
mathematical geology
statistical analysis

equations of state
Used as a general term, e.g. on level 3 under geochemistry(1) or phase equilibria(1).
SA equations
geochemistry
phase equilibria

equatorial
A valid term through 1977. See equatorial region.

Equatorial Guinea
Formerly Spanish Guinea. Comprises province of Rio Muni on the mainland and the province of Fernando Po consisting of the islands of Fernando Po and Annobon. Includes use on level 1 as an area term (list O). For term set options see list B. Before 1976, also search Spanish Guinea.
CO N010000N022000
E0113000E0092000
BA Central Africa
Africa
NT Fernando Po

Equatorial Pacific
Term introduced in 1978. Before 1978, search Pacific Ocean AND equatorial.
BA Pacific Ocean
SA East Pacific
North Pacific
Northeast Pacific
Northwest Pacific
South Pacific
Southeast Pacific
Southwest Pacific
West Pacific

equatorial region
As of 1981, no longer a valid term for GeoRef. Term was introduced in 1978.

Equatorial Zaire
Term introduced in 1981.
CO S023000N051000
E0243000E0162000
BT Zaire

Equidae
Family. Includes use on level 3 under Mammalia(1) Perissodactyla(2).
BA Perissodactyla
Mammalia
BT Tetrapoda
Vertebrata
NA Equus
Hipparion

equilibrium
Used as a general term.
SA phase equilibria

Equisetales
Order.
BA Sphenopsida
pteridophytes

BT Plantae
NA Calamites

Equus
Genus. Includes use on level 3 under Mammalia(1) Perissodactyla(2).
BA Equidae
Perissodactyla
Mammalia
BT Tetrapoda
Vertebrata

Er
use erbium

Er Rif
use Rif

Er Riff
use Rif

Eratosthenian
Term introduced in 1978. Lunar stratigraphy. Eratosthenian rocks are older than those of the Copernican system but younger than those of the Imbrian System.
SA Moon

erbium
Includes use on level 1 and 2 as a chemical element (list D).
UF Er
SA elements
rare earths

Erevan
use Yerevan

Erfurt
As of 1981, refers to only the city. Before 1981, also included the district.
BT Erfurt Bezirk
East Germany

Erfurt Bezirk
Term introduced in 1981. District in SW. Before 1981, search Erfurt.
CO N504000N513900
E0114000E0095500
BT East Germany
NT Erfurt
Ruhla
Weimar
SA Thuringia
Thuringian Forest
Unstrut River

Ericaceae
Family.
BA Dicotyledoneae
angiosperms
BT Spermatophyta
Plantae

Eridostraca
Term introduced in 1981. Includes use on level 2 under Ostracoda(1). See list F.
BA Ostracoda
BT Crustacea
Mandibulata
Arthropoda
Invertebrata

Erie County
Index states as applicable.
BX New York
Ohio
Pennsylvania
BT United States

Erimo Seamount
Off S coast of Hokkaido, Japan.
BT Pacific Ocean

erionite
BA zeolite group
framework silicates
silicates
BT minerals

Erivan
use Yerevan

Erlangen
City in N central.
BT Bavaria
West Germany

erodability
use erodibility

erodibility
Rate of erosion.
UF erodability
erosibility
SA erosion

Eromanga Basin
In SW Queensland.
BT Queensland
Australia

erosibility
use erodibility

erosion
Includes use on level 2 under soils(1); on level 3 under geomorphology(1) processes(2) and under glacial geology(1) glaciation(2).
NT grinding
nivation
water erosion
wind erosion
SA abrasion
beaches
benches
cirques
corrosion
degradation
denudation
detritus
erodibility
erosion control
erosion features
erosion surfaces
etching
exfoliation
geologic hazards
geomorphology
glacial geology
glaciation
ice movement
landform evolution
landslides
mobility
mudflows
planation
potholes
processes
sediment yield
sheet erosion
shorelines
slope stability
slumping
soils
talus slopes
valleys
waterways
weathering

erosion control
Includes use on level 3 under soils(1) conservation(2).
UF control, erosion
SA conservation
erosion
soils
water erosion
wind erosion

erosion features
Includes use on level 2 under geomorphology(1).
UF features, erosion
NT erosion surfaces
potholes
SA badlands
cliffs
colluvium
denudation
erosion
geomorphology
gullies
highlands
inselbergs
klippen
pediments
peneplains
talus slopes
terraces
valleys
water erosion
wave-cut platforms
windows

erosion surfaces
Includes use on level 3 under geomorphology(1) erosion features(2). Also search planation surfaces.
UF planation surfaces
surfaces, erosion
BT erosion features
SA erosion
geomorphology
pediments
pediplains
peneplains
planation
wave-cut platforms

erratic blocks
use erratics

erratic boulders
use erratics

erratics
Includes use on level 3 under glacial geology(1). Also search erratic blocks; erratic boulder(s).
UF erratic blocks
erratic boulders
BA clastic sediments
BT sediments
SA boulders
glacial geology
glacial transport
glaciers

errors
Used as a general term.
SA accuracy
corrections
reliability

ERTS
A valid term through 1977. After 1977, use Landsat.

ERTS-1
A valid term through 1977. After 1977, use Landsat.

erubescite
use bornite

eruptions
Includes use on level 3 under volcanology(1) volcanism(2).
SA explosive eruptions
hawaiian-type eruptions
lava
pyroclastics
volcanic earthquakes
volcanism
volcanoes
volcanology

Ervine Creek Limestone
Of Deer Creek Limestone. SW Iowa, NE Kansas, NW Missouri and SE Nebraska.
UF Ervine Creek Limestone Member
BT Virgilian
Upper Pennsylvanian
Pennsylvanian
SA Iowa
Kansas
Missouri
Nebraska

Ervine Creek Limestone Member
use Ervine Creek Limestone

Erzgebirge
Mountain range on the border of East Germany and Bohemia. Index countries as applicable. Also search Krusne Hory and Krusny Hory Mountains.
CO N501000N510000
E0141000E0120000
UF Krusne Hory
Krusnehory Mountains
Krusny Hory Mountains
BT Europe
NZ Czech Erzgebirge
SA Czechoslovakia
East Germany

Esan Cape
At E entrance of Tsugaru Strait in SW Hokkaido. Also search Esan.
UF Ezan Cape
BT Hokkaido
Japan

Escambia County
In S Alabama.
BT Alabama
United States

escarpments
use scarps

eschinite
use aeschynite

eschynite
use aeschynite

Escondido Formation
In Navarro Group. In S Texas.
BT Gulfian
Upper Cretaceous
Cretaceous
SA Texas

Esk Trough
Index England and/or Scotland as applicable. Also search Esk.
BT United Kingdom
SA England
Scotland

Eskdale
Village in Kanawha County in W.
BT West Virginia
United States

eskebornite
As of 1981, includes use on level 3 under minerals(1) selenides(2). See list L. Before 1981, included use on level 3 under minerals(1) sulfides(2).
BA selenides
sulfides
BT minerals

eskers
Includes use on level 3 under glacial geology(1) glacial features(2).
UF os
BT glaciofluvial features
glacial features
SA cirques
glacial geology
kames

Eskisehir
Province in NW central Anatolia. Also a city.
BT Turkey
Middle East

Eskridge Shale
In Council Grove Group. E Kansas, SE Nebraska, and central N Oklahoma.
BT Permian
SA Kansas
Nebraska
Oklahoma

Esmeralda County
In SE Nevada.
BT Nevada
United States

Espanola Formation
Of the Quirke Lake Group. Divisible into three members. S central Ontario.

BT Precambrian
SA Canada
Ontario

Espirito Santo
State on the Atlantic Ocean NE of Rio de Janeiro.
CO S212500S175000
W0393000W0421000
BT Brazil
SA Barreiras Formation

Esplanade Range
Range of Selkirk Mountains in SE.
BT British Columbia
Canada
SA Selkirk Mountains

Essei
use Yessey

Essen Beds
Ruhr District. Also search Essen.
BT Carboniferous
SA North Rhine-Westphalia
West Germany
Westphalian

Essex
County on the North Sea NE of London.
CO N513000N521000
E0011500E0000000
BT England
Great Britain
United Kingdom

Essex County
Index states as applicable.
BX Massachusetts
New Jersey
New York
Vermont
Virginia
BT United States

essexite
As of 1981, includes use on level 3 under igneous rocks(1) alkali gabbros(2). See list H. Before 1981, included use on level 3 under igneous rocks(1) alkali gabbro family(2).
BA alkali gabbros
gabbros
BT igneous rocks

Essonne
Term introduced in 1981. Department in N.
CO N481700N485000
E0023000E0015500
BT France

Esterel
Mountainous forested region. Index departments as applicable.
BT France
SA Alpes-Maritimes
Var

Esterhazy
Town near the Manitoba border in SE.
BT Saskatchewan
Canada

Estevan
Town near the North Dakota border in SE.
BT Saskatchewan
Canada

estimation
Used as a general term through 1977. After 1977, see evaluation.

Estonia
Soviet Socialist Republic.
CO N580000N593000
E0270000E0220000
BA Baltic Plain
Russian Platform
BT USSR
NT Tallin
Tartu
SA Baltic region
Baltic Shield
European USSR
Russian Plain

Estrada Nova Formation
Formation in Brazil: Upper Permian, Permian, and Triassic.
SA Brazil
Goias
Parana
Permian
Santa Catarina
Sao Paulo
Triassic
Upper Permian

Estremadura
Province in S Portugal. Before 1981, search Estremadura AND Portugal.
BT Portugal
SA Extremadura

estuaries
Includes use on level 3 under geomorphology(1) fluvial features(2) or shore features(2). As of 1981, use estuarine environment for type of environment. Before 1981, included use on level 3 under ecology(1), paleoecology(1), and sedimentation(1) for type of environment.
SA ecology
estuarine environment
estuarine sedimentation
fjords
fluvial features
geomorphology
paleoecology
rivers
rivers and streams
sedimentation
shore features
waterways

estuarine
A valid term through 1977. After 1977, use estuaries or (as of 1981) estuarine environment.

estuarine environment
Term introduced in 1981.
SA ecology
environment
estuaries
estuarine sedimentation
paleoecology
sedimentation

estuarine sedimentation
Term introduced in 1981.
SA estuaries
estuarine environment
sedimentation

etching
Includes use on level 3 under crystal structure(1) or geomorphology(1).
SA creep
crystal structure
erosion
geomorphology
landform evolution
weathering

ethane
BA organic materials
SA hydrocarbons
methane

Ethiopia
Includes use on level 1 as an area term (list O). For term set options see list B.
CO N080000N180000
E0480000E0330000
BA East Africa
Africa
NT Addis Ababa
Afar Depression
Awash Valley
Ethiopian Rift
Fantale
Harar
Omo
Omo River
Omo Valley
Tigre
SA Afar
Blue Nile
East African Rift
Lake Turkana
Nile River
Nubian Shield
Red Sea Basin

Ethiopian Rift
That part of the East African Rift extending NE-SW from the Red Sea to Lake Turkana. Also search Ethiopian Rift system and Ethiopian Rift valley.
UF Abyssinian Rift valley
Ethiopian Rift system
Ethiopian Rift valley
BT Ethiopia
SA East African Rift

Ethiopian Rift system
use Ethiopian Rift

Ethiopian Rift valley
use Ethiopian Rift

Ethmodiscus
Includes use on level 3 under algae(1) nannofossils(2).
BA discoasters
nannofossils
algae
BT thallophytes
Plantae

Etna
use Mount Etna

Etowah County
In NE Alabama.
BT Alabama
United States

ettringite
BA sulfates
BT minerals

Eu
use europium

Eua Island
One of the Tongapu group in S Tonga. Also search Eua.
BT Tonga
Polynesia

Euboea
Island in the Aegean Sea.
CO N375500N390000
E0244000E0225000
UF Evvoia
BT Greece

Eubradyodonti
Term introduced in 1981. Includes use on level 2 under Pisces(1). See list F.
BA Chondrichthyes
Pisces
BT Vertebrata

Eucla Basin
Index states as applicable. Also search Eucla.
CO S320000S300000
E1300000E1250000
BT Australia
SA South Australia
Western Australia

euclase
BA orthosilicates
silicates
BT minerals
SA aluminosilicates

eucrite
Includes use on level 3 under meteorites(1) and igneous rocks(1). See list H.
BX igneous rocks
meteorites
NA Moore County Meteorite
Pasamonte Meteorite
SA achondrites
gabbros

eucryptite
BA orthosilicates
silicates
BT minerals
SA aluminosilicates

eudialite
use eudialyte

eudialyte
Also search eudialite.
UF eudialite
BZ chlorides
ring silicates
BT minerals

Euganean Hills
Range of hills in Padova Province in S central.
BT Veneto
Italy

eugeosynclines
Includes use on level 3 under geosynclines(1) or tectonics(1).
UF pliomagmatic zone
SA geosynclines
miogeosynclines
tectonics

Eulengebirge
use Sowie Mountains

eulite
BA orthopyroxene
pyroxene group
chain silicates
silicates
BT minerals

eulysite
As of 1981, includes use on level 3 under igneous rocks(1) ultramafics(2). See list H. Before 1981, included use on level 3 under igneous rocks(1) ultramafic family(2).
BA ultramafics
BT igneous rocks

Euphorbiaceae
Family.
BA Dicotyledoneae
angiosperms
BT Spermatophyta
Plantae

Euphrates River
Rises in E central Turkey and at its confluence with the Tigris River it forms the Shatt-al-Arab which flows into the Persian Gulf. Index countries as applicable.
BT Asia
SA Iraq
Syria
Turkey

Eurasia
Land mass comprising the continents of Europe and Asia. Index continents as applicable. Includes use on level 1 or 2 as an area term (list O). If 1, see term set options under list B.
NA Baltic region
NT Black Sea region
Caspian Basin
NA Caspian Sea
NT Galitsiya
SA Asia
Eurasian Plate
Europe
Laurasia
USSR

Eurasia Basin
Between the North Pole and the major islands lying N of the Eurasian

land mass. Also search Eurasian Basin.
UF Eurasian Basin
BT Arctic Ocean

Eurasian Basin
use Eurasia Basin

Eurasian Plate
That part of the Earth's crust comprising Eurasia, including the British Isles.
SA Eurasia
plate tectonics
plates

Euratom
Term introduced in 1981.
SA Europe

Eure
Department in NW.
CO N484000N493000
E0014500E0001500
BT France
SA Eure Valley
Normandy
Seine Estuary

Eure Valley
River valley in NW France. Index departments as applicable.
BT France
SA Eure
Eure-et-Loir
Orne

Eure-et-Loir
Department in NW central.
CO N475700N490500
E0020000E0004500
BT France
SA Beauce
Eure Valley

Eureka
Village in Nevada. Cities in California, Kansas, and Utah. Index states as applicable.
BX California
Kansas
Utah
Nevada
BT United States

Eureka County
N central Nevada. Also search Eureka AND Nevada.
CO N391000N410000
W1154500W1163500
BT Nevada
United States

Eureka Sound Formation
BT Cretaceous
SA District of Franklin
Northwest Territories

Europe
Includes use on level 1 or 2 as an area term (list O). If 1, see term set options under list B. To retrieve all documents, individual countries and physiographic regions should also be searched (see list O).
CO N350000N710000
E0750000W0250000
NT Adriatic region
NA Alps
NT Ardennes
Arve Valley
NA Balkan Peninsula
NT Baltic Shield
Banat
Beshchady Mountains
Brest Basin
Brianconnais Zone
Bug region
Bug River
Bukovina
Campine
Carpathian Foredeep
Carpathian Foreland
NA Carpathians
NT Ceneri Zone
NA Central Europe
NT Danube Delta
Danube Plain
Danube Valley
Dinaric Alps
Dobruja Basin
Douro River
Elbe River
Elbe Valley
Erzgebirge
European Platform
Fennoscandia
Frankfurt
Gibraltar
Harz Foreland
Harz region
Inn Valley
Isar Valley
Isergebirge
Izera Mountains
Jura Mountains
Jutland
Karkonosze Mountains
Karst region
Krajiste
Lago Maggiore
Lake Constance
Lake Geneva
Lake Ladoga region
Lapland
Limburg
Lower Rhine Basin
Lusatia
Macedonia
Maritsa River
Meuse River
Meuse Valley
Moesia
Moesian Platform
Molasse Basin
Moldavia
Monte Rosa
Morava River valley
Moselle River
Moselle Valley
North German Plain
North Sea Coast
North Sudetic Basin
Oder Valley
Oresund
Osogovo Mountains
Pannonia
Pannonian Basin
Peribaltic Syneclise
Pomerania
Prut River
NA Pyrenees
NT Rhine Basin
Rhine Graben
Rhine River
Rhine Valley
Rhodope Mountains
Rhone River
Rhone Valley
Roztocze
Saale River
Saar Basin
Salzach River
Savoy
Scheldt River
Silesia
Simplon region
Siret River
Snieznik
Somes Basin
NA Southern Europe
NT Spis
Stavelot-Venn Massif
Struma River valley
Subcarpathians
Sudeten Mountains
Sudetic Basin
Sulitjelma
Tagus Basin
Tagus River
Thrace
Tisza River
Tokaj-Eperjes Mountains
Vardar River
Variscides
Vienna Basin
Weisse Elster Basin
Werra
NA Western Europe
SA Baltic Glaciation
Baltic region
Caledonides
Eurasia
Euratom
Faeroe Islands
Tethys

European
A valid term through 1977 used in combination with USSR. After 1977, use European USSR.

European Atlantic
Term introduced in 1981 to indicate a region of the Atlantic Ocean.
BA North Atlantic
BT Atlantic Ocean
NA Baltic Sea
Bay of Biscay
English Channel
Irish Sea
North Sea

European Community
Term introduced in 1981.
SA economics

European Platform
Ancient platform of Precambrian crystalline rocks overlain by sedimentary deposits of the North European Plain extending from the Low Countries to the Urals. Index countries as applicable.
BT Europe
SA Belgium
Denmark
Germany
Netherlands
Poland
USSR

European USSR
Term introduced in 1978. Index Soviet republics as applicable. Before 1978, also search USSR AND European.
CO N400000N700000
E0750000E0200000
BT USSR
SA Armenia
Azerbaidzhan
Byelorussia
Estonia
Georgian Republic
Latvia
Lithuania
Moldavia
Russian Republic
Ukraine

europium
Includes use on level 1 and 2 as a chemical element (list D).
UF Eu
SA elements
rare earths

Euryapsida
Includes use on level 2 under Reptilia(1). See list F.
BA Reptilia
BT Tetrapoda
Vertebrata
NA Araeoscelidia
Sauropterygia

Eurypterida
Includes use on level 3 under Arthropoda(1) Merostomata(2). See list F.
BA Merostomata
Chelicerata
Arthropoda
BT Invertebrata

eustacy
Includes use on level 3 under isostasy(1). Before 1976, also search eustatism. Also search eustasy.
UF eustasy
SA changes of level
epeirogeny
isostasy

eustasy
use eustacy

eutrophication
Includes use on level 3 under ecology(1) or hydrology(1).
SA ecology
hydrogen sulfide
hydrology
lakes

euxenite
BZ niobates
tantalates
BA oxides
BT minerals

evaluation
Includes use on level 2 under mining geology(1).
SA economic evaluation
mining geology

Evangeline Parish
S central.
BT Louisiana
United States

evaporation
Includes use on level 3 under hydrology(1). Also search vaporization.
UF vaporization
SA evapotranspiration
hydrology
sublimation
water vapor

evaporite deposits
Term introduced in 1981. Includes use on level 1 and 2 as a commodity term (list C). Before 1981, search evaporites AND deposits.
SA anhydrite deposits
borate deposits
evaporites
gypsum deposits
salt

evaporites
Includes use on level 3 under sedimentary rocks(1) chemically precipitated rocks(2). See list I. As of 1981, use evaporite deposits for evaporites as a commodity. Before 1981, also included use on level 1 as a commodity term and on level 3 under sediments(1) chemically precipitated sediments(2).
BA chemically precipitated rocks
BT sedimentary rocks
NA potash
salt
SA anhydrite
dolomite
evaporite deposits
gypsum
halite
sediments
sodium chloride

evapotranspiration
Includes use on level 3 under hydrology(1).
SA atmospheric precipitation
evaporation
hydrology

Everglades
Vast swampy region in six counties lying S of Lake Okeechobee in S.
BT Florida

United States

Everton Formation
In Buffalo River Group. Includes Calico Rock Sandstone Member, Kings River Member, Newton Sandstone Member. N Arkansas, and S Missouri.
BT Middle Ordovician
Ordovician
SA Arkansas
Missouri

evolution
Includes use as a level 2 or 3 term appropriate to a large number of topics, e.g. on level 2 under paleontology(1), tectonics(1) and under fossil groups (list F). See list G (general terms).
SA landform evolution
natural selection
species diversity

Evora
District in SE central.
BT Portugal
NT Alandroal

Evros River
use Maritsa River

Evvoia
use Euboea

Ewekoro
Region in SW.
BT Nigeria

Ewekoro Formation
NW, S, SW, and W Nigeria.
SA Cretaceous
lower Tertiary
Nigeria
Tertiary
Upper Cretaceous

excavations
As of 1978, includes use on level 2 under tunnels(1), rock mechanics(1), and underground installations(1). Used under engineering geology(1) through 1977.
SA blasting
dredging
engineering geology
explosions
rock mechanics
slope stability
tunnels
underground installations

Excello Shale
SE Kansas, W and N Missouri, and NE Oklahoma.
BT Pennsylvanian
SA Kansas
Missouri
Oklahoma

exchange capacity
As of 1981, no longer a valid term for GeoRef. Use cation exchange capacity. Before 1981, included use on level 3 under geochemistry(1) properties(2).

exchange rate
Term introduced in 1981.
SA economics

Exeter
City in Devonshire in SW.
BT England
Great Britain
United Kingdom

exfoliation
SA erosion
geomorphology
weathering

exhalative processes
Includes use on level 3 under mineral deposits, genesis(1).
SA mineral deposits
processes

exine
SA miospores
palynomorphs
pollen
spores
sporopollenin

exinite
Includes use on level 3 under sedimentary rocks(1) organic residues(2). See list I.
BA macerals
coal
organic residues
BT sedimentary rocks
NA cutinite
resinite
sporinite

exobiology
SA biology
extraterrestrial geology
life origin
planetology

exogene processes
Includes use on level 3 under mineral deposits, genesis(1) processes(2).
SA mineral deposits
processes

exogenous inclusions
use xenoliths

exomorphic zone
use aureoles

expanding universe theory
Term introduced in 1981. Includes use on level 3 under extraterrestrial geology(1) evolution(2).
SA extraterrestrial geology

expansion
As of 1981, to be used only for the Earth as a whole. Before 1981, used also as a general term.
SA Earth

expansive materials
Term introduced in 1978. Before 1978, also search expansive soils.
UF expansive soils
swelling soils
SA materials
soils

expansive soils
use expansive materials

expeditions
Used as a general term. As of 1981, use only when the organization or progress of an expedition is discussed.
SA associations
oceanography

experimental studies
Includes use as a level 2 or 3 term appropriate to a large number of topics, e.g. on level 2 under meteorology(1), folds(1), fractures(1), heat flow(1), and under oceanography(1). See list G.
UF studies, experimental
SA laboratory studies
theoretical studies

explanatory text
Includes use on level 3 under areal geology(1); on level 3 under maps(1). See list B.
UF text, explanatory
SA areal geology
guidebook
maps

exploitation
A valid term through 1978. Reintroduced in 1981. Used in relation to commodity terms (list C).
SA economics
production

exploration
Includes use on level 2 or 3 under commodity terms(1), e.g. on level 2 under energy sources(1) and placers(1). See list C. Before 1972, also search geological exploration. Before 1981, also search prospecting.
SA development
geochemical methods
geodesy
geological methods
geophysical methods
mineral exploration
mineral resources
Moon
photogeology
site exploration
speleology
surveys

Explorer Program
Term introduced in 1981. Includes use on level 3 under Moon(1).
SA extraterrestrial geology
Moon
planetology
remote sensing

Explorer 35
SA Moon
remote sensing

explosion phenomena
A valid term through 1972. After 1972, see cryptoexplosion features(2) and impact features(2) under geomorphology(1). See explosions(1) and nuclear explosions(3) under engineering geology(1).

explosions
A level 1 term as of 1978. Includes use on level 2 under geologic hazards(1) and seismology(1). Used for geotechnical and seismological studies on the effects of explosions. Before 1973, also search explosion phenomena. If 1, term set options are:
applications
subtopic
chemical explosions
subtopic
detection
subtopic
effects
subtopic
elastic waves
subtopic
excavations
subtopic
experimental studies
subtopic
materials, properties
subtopic
nuclear explosions
subtopic
site exploration
subtopic
theoretical studies
subtopic
NT chemical explosions
nuclear explosions
SA blasting
cryptoexplosion features
deep seismic sounding
earthquakes
elastic materials
elastic waves
engineering geology
excavations
foundations
geologic hazards
land subsidence
marine installations
mining geology
rock mechanics
seismic sources
seismic surveys
seismology
shatter cones
site exploration
slope stability
soil mechanics

explosive eruptions
Term introduced in 1981.
SA eruptions
volcanism
volcanoes
volcanology

export
Term introduced in 1981.
SA economics

export value
Term introduced in 1981.
SA economics

exposure age
Includes use on level 2 under geochronology(1). Before 1975, also search exposure ages.
SA age
cosmic rays
geochronology

exsolution
Includes use on level 3 under crystal structure(1). Also search unmixing.
SA crystal structure
solution

exsurgence
Term introduced in 1981.
SA caves
springs
streams

extension
Used to indicate the mechanics of deformation. Before 1978, also used to indicate style of structure. Now use extension faults or extension fractures for style of structure.
SA deformation
extension faults
extension fractures

extension faults
Before 1978, also search extension AND faults.
BA faults
SA extension

extension fractures
Term introduced in 1978. Before 1978, search extension AND fractures.
BA fractures
SA extension

extensometers
SA deformation
instruments
rock mechanics
strainmeters
stress

extent
As of 1981, no longer a valid term for GeoRef. See distribution. Before 1981, included use on level 2 and 3 appropriate to a large number of topics, e.g. on level 2 under orogeny(1).

extinct
A valid term through 1977 used in combination with lakes (i.e. lakes, extinct). After 1977, use extinct lakes. When referring to taxa, use extinct taxa.

extinct lakes
Includes use on level 3 under glacial geology(1) glacial features(2) and periglacial features(2); on level 3 under geomorphology(1) lacustrine features(2). Before 1976, also search extinct AND lakes.
SA geomorphology
glacial features

E

glacial geology
glacial lakes
lacustrine features
lakes
periglacial features

extinct taxa
Term introduced in 1978. Includes use on level 3 under fossil groups(1). Before 1978, also search extinct.
UF taxa, extinct

extinction
As of 1981, restricted to paleontology. Before 1981, also used for optical extinction. Also search mass extinction. Includes use on level 3 under fossil groups (list F).
SA optical extinction

extraction
Used for mining geology through May 1978. After May 1978, use production under mining geology(1).

extraterrestrial geology
Broad topic for studies about materials and processes outside the Earth. Includes use on level 2 under automatic data processing(1), associations(1), and symposia(1). When used on level 1 (beginning in 1981), the universe as a whole may be implied. If 1, term set options are:
theoretical studies
subtopic
topic [concepts, cosmic dust, cosmic rays, cosmochemistry, evolution, observations]
subtopic
SA Apollo Program
Apollo-Soyuz Program
astrophysics and solar physics
cosmic dust
cosmic rays
cosmochemistry
exobiology
expanding universe theory
Explorer Program
geology
interplanetary comparison
interplanetary space
life origin
Luna Program
Mariner Program
meteorites
Pioneer Program
planetology
planets
Skylab
solar system
Surveyor Program
Viking Program

Extremadura
Term introduced in 1981. Region in W central Spain. Before 1981, also search Estremadura AND Spain. For the province in Portugal, use Estremadura.
CO N375700N403000
W0044000W0073000
BT Spain
NT Badajoz Province
Caceres Province
SA Estremadura

extremities
Term introduced in 1981.
SA morphology
Vertebrata

extrusive rocks
Not a valid index term through 1977. After 1977, includes use on level 3 under lava (1) and volcanology (1). Before 1978, also search extrusive.
SA igneous rocks
lava
rocks
volcanology

Exuma Sound
Body of water SE of Nassau.
BT Bahamas
West Indies

Eyre Peninsula
Between Great Australian Bight and Spencer Gulf.
BT South Australia
Australia

Ezan Cape
use Esan Cape

F

F
use fluorine

F-region
Includes use on level 2 under ionosphere(1).
SA D-region
E-region
ionosphere

fabric
Includes use on level 3 under igneous rocks(1), metamorphic rocks(1), sediments(1), and sedimentary rocks(1) textures(2).
SA deformation
igneous rocks
lineation
metamorphic rocks
orientation
petrofabrics
preferred orientation
sedimentary rocks
sediments
structural analysis
tectonite

fabric analysis
A valid term through 1972. After 1972, use structural analysis or fabric.

facies
Includes use on level 2 under metamorphic rocks(1).
NT actinolite facies
amphibolite facies
blueschist facies
epidote-amphibolite facies
granulite facies
greenschist facies
prehnite-pumpellyite facies
zeolite facies
SA biofacies
domains
grade
isograds
lithofacies
metamorphic rocks
microfacies

facilities, nuclear
use nuclear facilities

factor analysis
BA statistical analysis
SA analysis
mathematical geology

factors
Includes use on level 3 under soils(1) genesis(2).
SA climate
parent materials
soils
time factor

Faeroe Islands
Self governing Danish island group between Iceland and the Shetland Islands. Includes use on level 1 as an area term (list O). For term set options see list B.
CO N612000N623000
W0060000W0080000
UF Faeroes
BT Denmark
NT Kuno
SA Europe

Faeroe-Iceland Ridge
use Iceland-Faeroe Ridge

Faeroes
use Faeroe Islands

Fagaras Mountains
Highest range in Transylvanian Alps. Index Transylvania and/or Walachia as applicable.
BT Romania
SA Transylvania
Transylvanian Alps
Walachia

Fagus
Genus.
BA Dicotyledoneae
angiosperms
BT Spermatophyta
Plantae
SA gymnosperms

failure
As of 1981, no longer a valid term for GeoRef. Use failures. Included use on level 2 under rock mechanics(1) and slope stability(1).

failures
Term introduced in 1981. Includes use on level 2 under rock mechanics(1) and slope stability(1). Before 1981, search failure; rock failure.
UF rock failure
SA deformation
faults
fractures
rock mechanics
slope stability
stress
tensile strength

Fair Isle
Most southerly of the Shetland Islands.
BT Scotland
Great Britain
United Kingdom
SA Shetland Islands

Fairbanks
City in E central.
BT Alaska
United States

Fairfield
Village in Utah County in N central.
BT Utah
United States

Fairview Formation
In Maysville Group. Includes Mount Hope Member, Fairmount Member. SE Indiana, N central Kentucky, and SW Ohio.
BT Upper Ordovician
Ordovician
SA Indiana
Kentucky
Ohio

Fairview Peak
Location of earthquakes and microearthquakes in central Nevada.
BT Nevada
United States

Fairweather Fault
In the Fairweather Range area near the British Columbia border in SE.
BT Alaska
United States

Faiyum
use Fayum

Falcon
State on the Caribbean Sea in NW.
BT Venezuela
NT Paraguana Peninsula

Falkenau
use Sokolov

Falkland Islands
British colony 300 miles E of Straits of Magellan. Islands are claimed by Argentina. Includes use on level 1 as an area term (list O). For term set options see list B.
UF Islas Malvinas
BZ South America
Atlantic Ocean
SA Antarctica
Atlantic Ocean Islands
South Georgia
South Orkney Islands
South Sandwich Islands

Falkland Plateau
Term introduced in 1981. E and NE of Falkland Islands.
CO S523000S473000
W0400000W0580000
BT Atlantic Ocean

Falknov
use Sokolov

Fall River County
Extreme SW.
BT South Dakota
United States

Fall River Formation
In Inyan Kara Group. Includes Keyhole Sandstone Member. W South Dakota and NE Wyoming.
BT Lower Cretaceous
Cretaceous
SA South Dakota
Wyoming

fallout
Restricted to use for radioactive fallout.
SA atmosphere
isotopes
pollution
radioactivity

False Bay
E of Cape Good Hope.
BT Cape Province
South Africa

False Cape
SW tip of Dolak Island off S coast.
UF Cape Valse
Kaap Valsch
BT Irian Jaya
Indonesia

famatinite
BA sulfantimonates
sulfosalts
BT minerals
SA luzonite

Famennian
Europe. Above Frasnian, below Tournaisian (Carboniferous). Includes use on level 3 under age terms(1). See list E.
DA Upper Devonian
Devonian
BT Paleozoic

FAMOUS
French-American expedition to the Mid-Atlantic Ridge, using submersibles.
SA Atlantic Ocean
Mid-Atlantic Ridge
oceanography
submersibles

fanglomerate
Includes use on level 3 under sedimentary rocks(1) clastic rocks(2). See list I.
BA clastic rocks
BT sedimentary rocks
SA terrigenous materials

Fanning Island
One of the Line Islands S of Hawaii. Included in British colony of Gilbert and Ellice Islands.
BT Line Islands
Polynesia

fans
As of 1981, no longer a valid term for GeoRef. See alluvial fans; submarine fans; deltas. Included use on level 3 under glacial geology(1) periglacial features(2).

Fantale
In Ethiopian Rift valley in W Ethiopia. Also search Fantale Volcano.
UF Fantale Volcano
BT Ethiopia

Fantale Volcano
use Fantale

Far East
Easternmost Asia along the Pacific Ocean. Index countries and regions as applicable. Includes use on level 1 as an area term (list O). For term set options see list B.
BA Asia
NX Borneo
NT Brunei
NA Burma
Cambodia
China
Hong Kong
Indochina
Indonesia
Japan
Korea
Laos
NT Malay Peninsula
NA Malaysia
Mongolia
Philippine Islands
Singapore
Taiwan
Thailand
Vietnam
SA Malay Archipelago
Pacific region
Soviet Far East

Faristan
use Fars

Farneto Cave
Term introduced in 1981. In province of Bologna, Italy.
CO N442500N443100
E0113000E0112000
BT Bologna
Emilia-Romagna
Italy

Fars
Province in SW Iran.
UF Faristan
BT Iran
NT Band-e-Amir
Shiraz

farside
SA Moon

farsundite
As of 1981, includes use on level 3 under igneous rocks(1) granites(2). See list H. Before 1981, included use on level 3 under igneous rocks(1) granite-granodiorite family(2).
BA granites
BT igneous rocks

Farther India
use Indochina

fassaite
BA clinopyroxene
pyroxene group
chain silicates
silicates
BT minerals
SA augite

fatty acids
Includes use on level 2 under organic materials(1).
BA organic materials
SA acids
amino acids

faujasite
BA zeolite group
framework silicates
silicates
BT minerals

fault blocks
use block structures

fault escarpment
use fault scarps

fault ledge
use fault scarps

fault planes
UF planes, fault
SA faults

fault scarps
Includes use on level 3 under faults(1).
UF cliff of displacement
fault escarpment
fault ledge
SA cliffs
faults
scarps
slopes

fault zones
Includes use on level 3.
UF zones, fault
SA faults
lineaments
plate tectonics
rift zones

faults
To be used for studies primarily stressing individual faults or systems of faults. For relationships with other structures, see tectonics. Includes use on level 1 (list A); on level 2 under structural analysis(1), geologic hazards(1), and nuclear facilities(1). Also search fault. If level 1, term set options are:
displacements
subtopic [diagonal-slip faults, dip-slip faults, gravity faults, normal faults, overthrust faults, reverse faults, strike-slip faults, thrust faults, transcurrent faults, transform faults, wrench faults]
distribution
subtopic (e.g. topic or area)
effects
subtopic [breccia, gouge, mullions, mylonite, shear zones, slickensides, ultramylonite]
mechanics
subtopic [e.g. compression, flexure, shear, stick-slip]
orientation
subtopic [arcuate faults, bedding faults, dip faults, longitudinal orientation, oblique orientation, strike faults, transverse faults]
patterns
subtopic [en echelon faults, parallel faults, peripheral faults, radial faults]
systems
subtopic [block structures, grabens, horsts, rift zones, step faults]
theoretical studies
subtopic
NA active faults
arcuate faults
bedding faults
diagonal-slip faults
dip faults
dip-slip faults
en echelon faults
extension faults
growth faults
hinge faults
low-angle faults
normal faults
overthrust faults
parallel faults
peripheral faults
radial faults
reverse faults
right-lateral faults
step faults
strike faults
strike-slip faults
thrust faults
transcurrent faults
transform faults
transverse faults
underthrust faults
wrench faults
SA allochthons
basin range structure
block structures
breccia
decollement
deformation
displacements
drag folds
earthquakes
failures
fault planes
fault scarps
fault zones
folds
foliation
fracture zones
fractures
geologic hazards
gouge
grabens
horsts
klippen
lineaments
mechanics
mylonitization
nappes
neotectonics
nuclear facilities
oblique orientation
orientation
orogeny
plate tectonics
reactivation
salt tectonics
scarps
seismic moment
seismology
shear zones
slickensides
stick-slip
strike
structural analysis
structural geology
systems
tectonic elements
tectonic wedges
tectonics
tectonophysics

fauna
As of 1981, no longer a valid term for GeoRef. See faunal studies.

faunal list
UF list, faunal
SA floral list
fossils

faunal provinces
Includes use on level 3 under biogeography(1).
UF provinces, faunal
SA assemblages
biogeography
paleontology

faunal studies
Includes use on level 2 under fossil group(1) for many classes or orders, or for genera. See list F.
UF studies, faunal
SA biota
floral studies
fossils
living taxa
microorganisms

Favosites
Genus. Includes use on level 3 under Coelenterata(1) Tabulata(2). See list F.
BA Favositidae
Tabulata
Anthozoa
Coelenterata
BT Invertebrata
SA Palaeofavosites

Favositidae
Includes use on level 3 under Coelenterata(1) Tabulata(2). See list F.
BA Tabulata
Anthozoa
Coelenterata
BT Invertebrata
NA Favosites
Palaeofavosites

Fawley
Town and parish SSE of Southampton in S.
BT England
Great Britain
United Kingdom

fayalite
BA olivine group
nesosilicates
orthosilicates
silicates
BT minerals

Fayette County
Index states as applicable.
BX Alabama
Georgia
Illinois
Indiana
Iowa
Kentucky
Ohio
Pennsylvania
Tennessee
Texas
West Virginia
BT United States

Fayetteville
City in Washington County in NW.
BT Arkansas
United States

Fayetteville Formation
Includes Mayes Limestone Member. N Arkansas, S Missouri; and NE, central, and E Oklahoma.
BT Upper Mississippian
Mississippian
SA Arkansas
Missouri
Oklahoma

Fayum
Province in Upper Egypt, SW of Cairo.
UF Faiyum
Fayyum
BT Egypt

Fayum Depression
Bed of ancient Lake Moeris in N Upper Egypt SW of Cairo. Also search Fayum.
BT Egypt

Fayyum
use Fayum

Fe
use iron

Fe-55
SA iron
isotopes

Fe-57
SA iron
isotopes

feasibility studies
As of 1978, term is used on level 2. Also search feasibility.
UF studies, feasibility
SA highways
marine installations
nuclear facilities
reservoirs
tunnels
underground installations

Feather River
Term introduced in 1981. N central California.
CO N384500N394000
W1212000W1214000
BT California
United States

features, bottom
use bottom features

features, buried
use buried features

features, cryptoexplosion
use cryptoexplosion features

features, eolian
use eolian features

features, erosion
use erosion features

features, fluvial
use fluvial features

features, frost
use frost features

features, glacial
use glacial features

features, impact
use impact features

features, lacustrine
use lacustrine features

features, periglacial
use periglacial features

features, shore
use shore features

features, solution
use solution features

features, surface
use surface features

features, volcanic
use volcanic features

Federal Capital Territory
use Australian Capital Territory

Federal District
Region surrounding a capital city. Index countries and cities as applicable.
UF Distrito Federal
BX Argentina
Venezuela
Mexico
NT Buenos Aires
Caracas
Mexico City
SA Caracas Group
Valley of Mexico

Federated Shan States
use Shan State

Fedorov stage
use universal stage

feeding ground
use drainage basins

feldspar
As of 1981, no longer a valid term for GeoRef. See feldspar group. Use feldspar deposits for feldspar as a commodity. Before 1981, included use on level 1 as a commodity term (list C) and on level 3 under minerals(1) framework silicates, feldspar group(2).

feldspar deposits
Term introduced in 1981. Includes use on level 1 and 2 as a commodity term (list C). Before 1981, search feldspar AND deposits.
SA feldspar group
feldspathization

feldspar group
Includes use in combination with framework silicates (i.e. framework silicates, feldspar group) on level 2 under minerals(1). See list L.
BA framework silicates
silicates
BT minerals
NA adularia
alkali feldspar
amazonite
analbite
anorthoclase
antiperthite
barium feldspar
cryptoperthite
hyalophane
K-feldspar
microcline
microperthite
moonstone
orthoclase
peristerite
perthite
plagioclase
sanidine
SA feldspar deposits

feldspathization
Includes use on level 3 under metasomatism(1).
BA metasomatism
SA feldspar deposits
granitization
metamorphism

Felidae
BA Carnivora
Mammalia
BT Tetrapoda
Vertebrata

felsite
As of 1981, includes use on level 3 under igneous rocks(1) granites(2). See list H. Before 1981, included use on level 3 under igneous rocks(1) granite-granodiorite family(2).
BA granites
BT igneous rocks

Fenestellidae
Family. Includes use on level 3 under Bryozoa(1) Cryptostomata(2).
BA Cryptostomata
Bryozoa
BT Invertebrata

fenite
Includes use on level 3 under metasomatic rocks(1). Before 1978, included use on level 3 under metamorphic rocks(1).
BA metasomatic rocks
SA fenitization
metasomatism

fenitization
Includes use on level 3 under metasomatism(1).
BA metasomatism
SA alteration
fenite

Fennoscandia
Geological usage for that part of N Europe consisting of Denmark, Finland and adjacent part of the USSR, Norway, and Sweden. Index countries as applicable.
BT Europe
NZ Soviet Fennoscandia
SA Denmark
Finland
Norway
Sweden
USSR

fensters
use windows

Fenton Hill
Term introduced in 1981. Los Alamos Scientific Laboratory's site for hot-dry-rock geothermal energy experiments. Adjacent to Valles Caldera and Nacimiento Mountains.
CO N354300N360000
W1063700W1065700
BT Sandoval County
New Mexico
United States
SA Nacimiento Mountains
Valle Grande Mountains
Valles Caldera

ferberite
BA tungstates
BT minerals

Fergana
Not a valid term for GeoRef. Use Fergana Basin.

Fergana Basin
Mountain enclosed steppe and desert region. Index Soviet republics as applicable. Also search Fergana; Fergana Valley.
CO N394500N413000
E0730000E0693000
UF Fergana Valley
BA Central Asia
BT USSR
SA Kirghizia
Tadzhikistan
Uzbekistan

Fergana Valley
use Fergana Basin

Fergus County
Central.
BT Montana
United States

fergusonite
BZ niobates
tantalates
BA oxides
BT minerals

Fermanagh
County in W.
BT Northern Ireland
United Kingdom

Fernando Po
Province in Bight of Biafra. Also an island.
UF Fernando Poo
BT Equatorial Guinea

Fernando Poo
use Fernando Po

ferns
As of 1981, includes use on level 1. See list F. To be used for pteridophytes when they are used as tools in stratigraphy. Before 1981, search (pteridophytes OR ferns) AND biostratigraphy. Before 1981, included use on level 3. If 1, term set options are:
biostratigraphy
age [List E]
SA biostratigraphy
pteridophytes

Fernvale Formation
In Richmond Group and Patterson Ranch Group. N Arkansas, SW Illinois, SE Missouri, and W Tennessee, NW Alabama, central E and NE Oklahoma.
BT Upper Ordovician
Ordovician
SA Alabama
Arkansas
Illinois
Missouri
Oklahoma
Richmond Group
Tennessee

Ferrar Group
In Victoria Land W of Ross Sea on Pacific Ocean side.
BT Jurassic
SA Antarctica
Mesozoic

ferric iron
Before 1978, search iron AND ferric.
SA iron

ferricrete
As of 1981, includes use on level 3 under sedimentary rocks(1) chemically precipitated rocks(2). See list I. Before 1981, included use on level 3 under sediments(1) chemically precipitated sediments(2).
BA chemically precipitated rocks
BT sedimentary rocks
SA conglomerate
duricrust

ferrierite
BA zeolite group
framework silicates
silicates
BT minerals

ferrimolybdite
BA molybdates
BT minerals

ferroan dolomite
use ankerite

ferrocarpholite
BA chain silicates
silicates
BT minerals
SA carpholite

ferrodiorite
As of 1981, includes use on level 3 under igneous rocks(1) diorites(2). See list H. Before 1981, included use on level 3 under igneous rocks(1) diorite family(2). Also search ferrogabbro.
UF ferrogabbro
BA diorites
BT igneous rocks

ferrogabbro
use ferrodiorite

ferrohastingsite
BA clinoamphibole
amphibole group
chain silicates
silicates
BT minerals
SA hastingsite

ferromanganese
A valid term through 1977. After 1977, use ferromanganese composition on level 2 under nodules(1).

ferromanganese composition
Term introduced in 1978. Includes use on level 2 under nodules(1). Before 1978, also search ferromanganese.
SA composition
iron
manganese
nodules

Ferron Sandstone
use Ferron Sandstone Member

Ferron Sandstone Member
Of Mancos Shale. Central E Utah.
UF Ferron Sandstone
BT Upper Cretaceous
Cretaceous
SA Mancos Shale
Utah

ferropseudobrookite
BA oxides
BT minerals
SA pseudobrookite

ferrosilicon
As of 1981, includes use on level 3 under minerals(1) silicides(2). See list L. Before 1981, included use on level 3 under minerals(1) native elements and alloys(2).
BA silicides
alloys
BT minerals
SA iron
silicon

ferrosilite
BA pyroxene group
chain silicates
silicates
BT minerals

ferrospinel
use hercynite

ferrous iron
Before 1978, search iron AND ferrous.
SA iron

ferruginous composition
Term introduced in 1978. Includes use on level 3 under sediments(1) and sedimentary rocks(1). Before 1978, also search ferruginous. As of 1981, use only for sedimentary formations.
SA composition
iron
iron formations
iron-rich composition
red beds
sedimentary rocks
sediments

ferruginous quartzite
SA orthoquartzite
quartzites

fertilization
Includes use on level 3 under soils(1) utilization(2).
SA agriculture
fertilizers
nutrients
soils
treatment
utilization

fertilizers
Includes use on level 2 under soils(1).
SA agriculture
fertilization
nitrogen
nutrients
phosphorus
soil management
soils
yields

Fezzan
Desert region in SW.
BT Libya

fibrolite
use sillimanite

Fichtelgebirge
Mountain range in NE.
BT Bavaria
West Germany

field capacity
Term introduced in 1981.
SA ground water
soils

field crops
As of 1981, no longer a valid term for GeoRef. Before 1981, included use on level 3 under soils(1) utilization(2) and yields(2).

field, crystal
use crystal field

field, electrical
use electrical field

field, electromagnetic
use electromagnetic field

field, gravity
use gravity field

field, magnetic
use magnetic field

field studies
Includes use on level 2 under deformation(1), and soils(1).
UF studies, field
SA deformation
sampling
soils

field trips
Term introduced in 1981.
SA guidebook
road log

fields, coal
use coal fields

fields, geothermal
use geothermal fields

fields, giant
use giant fields

fields, lava
use lava fields

fields, oil and gas
use oil and gas fields

Fife
County N of Firth of Forth in E.
BT Scotland
Great Britain
United Kingdom

Fig Tree Group
use Fig Tree Series

Fig Tree Series
In Barberton Mountain Land in SE Transvaal. Also search Fig Tree Group.
UF Fig Tree Group
BT Precambrian
SA South Africa
Transvaal

figure of Earth
Includes use on level 2 under geodesy(1).
SA Earth
geodesy
geoid

Fiji
Independent state consisting of island group between New Caledonia and Samoa in S Pacific Ocean. Includes use on level 1 as an area term (list O). For term set options see list B.
CO S192000S160000
W1780000E1770000
UF Fiji Islands
BA Melanesia
NT Suva
Tavua
Vatukoula
Viti Levu
SA Pacific Ocean

Fiji Islands
use Fiji

Fiji Plateau
Just E of Fiji in S.
BT Pacific Ocean

Filicales
use Filicopsida

Filicopsida
Including Filicales. Includes use on level 2 under pteridophytes(1). See list F. Also search Filicales.
UF Filicales
BA pteridophytes
BT Plantae
NA Azolla
Dicroidium
Gleicheniaceae
Pecopteris
NZ Sphenopteris
NA Taeniopteris
SA Juglans

Filizchay
Ore bearing region.
BT Azerbaidzhan
USSR

Fillmore Formation
Overlies House Limestone; underlies Wahwah Limestone. E Nevada and W central Utah. Also search Fillmore Limestone.
BT Lower Ordovician
Ordovician
SA Nevada
Utah

filtering
use filters

filters
Includes use on level 3 under seismology(1). Also search filtering.
UF filtering
NT adaptive filters
analog filters
continuous filters
discrete filters
dispersive filters
multichannel filters
numerical filters
optimal filters
passband filters
recursive filters
spatial frequency filters
weiner filters
SA deconvolution
distortion
elastic waves
noise
seismic methods
seismology
signals

filtration
As of 1978, term is restricted to use with sediments, i.e. for process of removing suspended material from a liquid.
SA processes
sediments

financing
Term introduced in 1981.
SA economics

fine-grained materials
use fines

fines
Also search fine-grained materials.
UF fine-grained materials
SA agglutinates
grain size
mining
particles
sediments

Finger Lakes
Group of long narrow lakes in W.
BT New York
United States

Finistere
Department on W tip of Brittany.
CO N474000N484000
W0034000W0050000
BT France
NT Chateaulin Basin
Crozon Peninsula
Morlaix

finite difference analysis
Before 1978, also search finite difference method; finite difference.
UF finite difference method
BA statistical analysis
SA analysis
mathematical geology
mathematical methods

finite difference method
use finite difference analysis

finite element analysis
Also search finite-element analysis.
UF finite-element analysis
BA statistical analysis
SA analysis
mathematical geology

finite strain
As of 1981, no longer a valid term for GeoRef. Use finite strain analysis.

finite strain analysis
Term introduced in 1981. Before 1981, search finite strain.
SA deformation
rock mechanics
soil mechanics
strain
stress

finite-element analysis
use finite element analysis

Finland
Includes use on level 1 as an area term (list O). For term set options see list B.
CO N594500N700000
E0314500E0190000
BA Scandinavia
Western Europe
Europe
NT Inari
Lake Lappajarvi
Outokumpu
Tampere
Turku
Ylojarvi
SA Arctic region
Baltic region
Baltic Shield
Bjurbole Meteorite
Fennoscandia
Gulf of Bothnia
Gulf of Finland
Haveroe Meteorite
Ladoga Series
Lake Ladoga region
Lapland

Finney County
W central.
BT Kansas
United States

Finnmark
County in N.
BT Norway
NT Alta
Porsang Fjord
Seiland
Tana Fjord
Varanger Fjord
Varanger Peninsula

SA Porsanger Dolomite Formation

fiord
use fjords

Fiordland National Park
SW South Island. Also search Fiordland.
UF Sounds National Park
BT South Island
New Zealand

fire clay
use fireclay

Fire Island
Long, narrow sand spit off S Long Island.
BT New York
United States

fire-clay
use fireclay

fireclay
Also search fire clay; fire-clay; refractory clay.
UF fire clay
fire-clay
refractory clay
SA clay
refractory materials
underclay

Firenze
use Florence

firn
Includes use on level 3 under glacial geology(1) glacial features(2), glaciers(2) and periglacial features(2).
SA glacial features
glacial geology
glaciers
periglacial features
snow

Firth of Clyde
An estuary of the Clyde River in SW Scotland. Also search Clyde River.
BT Scotland
Great Britain
United Kingdom

Firth of Tay
use Tay Estuary

fish
As of 1981, includes use on level 1. See list F. To be used for Pisces when they are used as tools in stratigraphy. Before 1981, search (Pisces OR fish) AND biostratigraphy. Before 1981, included use on level 3. If 1, term set options are:
biostratigraphy
age [List E]
SA biostratigraphy
Pisces

Fiskenaesset
Settlement on SW coast.
BT Greenland
Arctic region

fission
UF nuclear fission
SA energy
fission tracks
fission-track dating
fusion
particles

fission tracks
Used for the phenomena. For method, use fission-track dating.
SA fission
fission-track dating
radiation damage
radioactivity

fission-track
A valid term through 1977. After 1977, use fission-track dating.

fission-track dating
Term introduced in 1978. Includes use on level 2 under geochronology(1). Before 1978, also search fission-track; fission tracks; fission-track method.
UF dating, fission-track
fission-track method
spontaneous fission-track dating
SA fission
fission tracks
geochronology
methods
radiation damage

fission-track method
use fission-track dating

Fissipeda
Term introduced in 1981. Includes use on level 2 under Mammalia(1). See list F.
BA Carnivora
Mammalia
BT Tetrapoda
Vertebrata

fissures
Includes use on level 3 under fractures(1).
SA cracks
fractures
joints

fixation
Used in geochemistry, e.g. for fixation of ions.
SA geochemistry
ions

fjords
Includes use on level 3 under geomorphology(1) shore features(2); on level 3 under glacial geology(1) glacial features(2).
UF fiord
fyord
SA bays
estuaries
geomorphology
glacial features
glacial geology
glaciation
shore features

Flack Lake
BT Ontario
Canada

Flagstaff
City in Coconino County in N central.
BT Arizona
United States

flame emission spectrometry
use flame photometry

flame photometry
Includes use on level 3 under spectroscopy(1) methods(2); on level 3 under chemical element(1) analysis(2). See list D.
UF flame emission spectrometry
SA analysis
chemical analysis
photometry
quantitative analysis
spectroscopy

flame structures
Includes use on level 3 under sedimentary structures(1) soft sediment deformation(2). See list F.
BA soft sediment deformation
sedimentary structures
SA antidunes
load casts
structures

Flandrian
Europe. See list E.
BA Holocene
Quaternary
BT Cenozoic

flares
Term used on level 2 through 1978. Now use solar flares.

flaser bedding
Includes use on level 3 under sedimentary structures(1) planar bedding structures(2). See list K.
BA planar bedding structures
sedimentary structures
SA bedding

Flathead County
Term introduced in 1981. NW Montana.
CO N473500N490000
W1125200W1150100
BT Montana
United States

Flathead Lake
NW Montana.
BT Montana
United States

Flathead Sandstone
In Wyoming, unconformably overlies Precambrian granite and underlies Gallatin Limestone and Gros Ventre Formation. Montana and NW Wyoming.
BT Middle Cambrian
Cambrian
SA Montana
Wyoming

Fleming Formation
In Grand Gulf Group. Includes Lena, Carnahan Bayou, Dough Hills, Williamson Creek, Castor Creek, and Blounts Creek members. Also considered a group name including Oakville and Cuero formations. E Texas, and W Louisiana.
BT Miocene
SA Louisiana
Texas

Flemish Cap
A marine bank NE of Grand Banks off Newfoundland.
BT Atlantic Ocean

Fleurieu Peninsula
S of Adelaide between Gulf Saint Vincent and Indian Ocean.
BT South Australia
Australia

flexural-slip
Includes use on level 3 under folds(1) mechanics(2).
UF bedding-plane slip
SA decollement
disharmonic folds
folds
mechanics

flexure
Used only to indicate the mechanics of deformation. Before 1978, also used to indicate style of structure. Now use flexure folds or growth faults for style of structure.
SA deformation
flexure folds
growth faults

flexure faults
use growth faults

flexure folds
Before 1978, also search flexure AND folds.
BA folds
SA flexure
monoclines

Flin Flon
Town near Saskatchewan border in W.
BT Manitoba
Canada

Flinders Island
Largest island of the Furneaux Islands off NE Tasmania.
BT Tasmania
Australia

Flinders Range
use Flinders Ranges

Flinders Ranges
Between Lake Frome and Lake Torrens in E central South Australia. Also search Flinders Range.
CO S323000S293000
E1393000E1380000
UF Flinders Range
BT South Australia
Australia

flint
Includes use on level 3 under sedimentary rocks(1) chemically precipitated rocks(2). See list I.
BA chemically precipitated rocks
BT sedimentary rocks
SA chert

flint clay
Includes use on level 3 under sediments(1) clastic sediments(2). See list N.
BA clastic sediments
BT sediments
SA clay
clays
refractory materials
terrigenous materials

Flint Creek Range
Range of Rocky Mountains in SW.
BT Montana
United States
SA Rocky Mountains

Flint Hills
Index states as applicable.
CO N370000N383000
W0963000W0965000
BT United States
SA Kansas
Oklahoma

Flinton Group
E Ontario.
BT Precambrian
SA Canada
Ontario

flocculation
SA clay mineralogy
colloidal materials
geochemistry
processes
suspended materials

flokite
use mordenite

flood plains
use floodplains

flood tuff
use ignimbrite

floodplains
Includes use on level 3 under geomorphology(1), fluvial features(2). Also search flood plains.
UF flood plains
SA alluvium
floods
fluvial features
geomorphology
meanders
plains
rivers
terraces

floods
As of 1978, term is used on level 2 under geologic hazards(1) and waterways(1).
SA environmental geology
floodplains
geologic hazards

reservoirs
rivers and streams
surface water
watersheds
waterways

flora
As of 1981, no longer a valid term for GeoRef. See floral studies.

floral list
SA faunal list

floral provinces
UF provinces, floral

floral studies
Includes use on level 2 under fossil group(1) for many classes or orders, or for genera. See list F.
UF studies, floral
SA biota
faunal studies
fossils
living taxa
microorganisms
paleobotany
palynomorphs
Plantae
vegetation

Florence
City in E central Tuscany. Also search Firenze.
UF Firenze
BT Tuscany
Italy

Florence Rise
Term introduced in 1981. Ridge in the Mediterranean Sea. A submarine extension of the Cyprus Arc. Partially separates the Antalya Basin to the N from the Levantine Basin to the S.
CO N344000N351000
E0320000E0313000
BT Mediterranean Sea

florencite
BA phosphates
BT minerals

Florida
Includes use on level 1 as an area term (list O). For term set options see list B.
CO N243000N310000
W0800000W0873000
BA Eastern U.S.
United States
NT Apalachicola
NX Baker County
NT Big Pine Key
Biscayne Bay
Brevard County
Broward County
NX Calhoun County
NT Cape Kennedy
Central Florida
NX Charlotte County
NT Choctawhatchee Bay
NX Clay County
Columbia County
NT Coupon Bight
NX Dade County
De Soto County
Duval County
NT Everglades
Florida Keys
Floridan Aquifer
NX Franklin County
Hamilton County
Hillsborough County
Jackson County
Jefferson County
NT Key Largo
NX Lafayette County
Lake County
NT Lake Okeechobee
NX Lee County
Madison County
Marion County
NT Miami
NX Monroe County
Nassau County
Orange County
NT Palm Beach County
Pensacola
Pinellas County
NX Polk County
Putnam County
NT Saint Johns River basin
Tampa Bay
NX Taylor County
Union County
Walton County
Washington County
SA Atlantic Coastal Plain
Bone Valley Formation
Carolina Bays
Chattahoochee River
Chipola Formation
Citronelle Formation
Claiborne Group
Crystal River Formation
Duplin Formation
Gulf Coastal Plain
Hawthorn Formation
Jackson Group
Key Largo Limestone
Marianna Limestone
Miami Limestone
Midway Group
Ocala Group
Okefenokee Swamp
Panhandle
Suwannee Limestone
Vicksburg Group
Wicomico Formation
Wilcox Group

Florida Bay
Body of water between S tip of Florida and Florida Keys.
BT Gulf of Mexico

Florida Keys
150 mile chain of coral limestone islands extending SW off S tip of Florida.
CO N243000N252000
W0815000W0830000
BT Florida
United States
SA Big Pine Key

Florida Mountains
Luna County in SW.
BT New Mexico
United States

Florida Strait
use Straits of Florida

Florida Straits
use Straits of Florida

Floridan Aquifer
N and central Florida. Below the confining bed consists from top to bottom, of limestone in the bottom part of the Hawthorn Formation and limestone, dolomite, and dolomitic limestone in formations of Eocene age that include the Ocala Group, the Avon Park Limestone, the Lake City Limestone, and, in part, the Oldsmar Limestone.
BT Florida
SA Eocene
Ocala Group

Florinites
BA miospores
palynomorphs
SA angiosperms

flotation
Includes use on level 3 under commodities (list C).
SA beneficiation

flow
As of 1981, no longer a valid term for GeoRef. Before 1981, used to indicate the mechanics of deformation. Before 1978, also used to indicate style of structure. Before 1977, also related to movement of ground water and surface water. See flow structures; flow folds; flow lines; flow mechanism; flow cleavage.

flow (volcanic)
use lava flows

flow cleavage
Includes use on level 3 under foliation(1) genesis(2).
BA foliation
SA axial-plane structures
cleavage
compression
schistosity
slaty cleavage
structural analysis

flow folds
Before 1978, also search flow AND folds.
UF flowage fold
BA folds

flow lines
Includes use on level 3 under deformation(1) field studies(2).
UF lines, flow
SA deformation
lineation
plastic flow
rheology

flow mechanism
Includes use on level 2 under lava(1).
SA lava
lava flows
mechanism
rheology
viscosity

flow regime
As of 1981, no longer a valid term for GeoRef. Before 1981, included use on level 2 under sedimentation(1) and on level 3 under ground water(1) movement(2).

flow structures
Includes use on level 3 under sedimentary structures(1) soft sediment deformation(2) and turbidity current structures(2). See list K.
BA soft sediment deformation
sedimentary structures
SA sedimentation
structures
turbidity current structures

flowage fold
use flow folds

flows
When discussing lava, use lava flows.
SA ash flows
debris flows
hydraulics
lava flows
mass movements
velocity

Floyd County
Index states as applicable.
BX Georgia
Indiana
Iowa
Kentucky
Texas
Virginia
BT United States

fluctuations
Used as a general term. Includes use on level 3 under paleoclimatology(1).
SA paleoclimatology

fluid dynamics
Term introduced in 1981.
SA hydraulics
hydrodynamics

fluid inclusions
A level 1 term as of 1978. Used for gaseous or liquid inclusions found in crystals. Before 1978, included use on level 2 under inclusions(1). If 1, term set options are:
topic [analysis, changes, composition, detection, experimental studies, genesis, geochemistry, geologic barometry, geologic thermometry, interpretation, P-T conditions, paleosalinity, temperature, theoretical studies]
type of inclusion [e.g. ammonia, carbon dioxide, gases, methane] or subtopic
SA analysis
carbon dioxide
crystal growth
gases
geochemistry
geologic barometry
geologic thermometry
host rocks
igneous rocks
inclusions
mineral deposits
mineral inclusions
minerals
P-T conditions
paleosalinity
petrology
rocks
temperature
xenoliths

fluid injection
Includes use on level 3 under earthquakes(1) or waste disposal(1). Also search injection; injection wells.
SA earthquakes
injection
waste disposal

fluid phase
Term introduced in 1978. Includes both gaseous phase and liquid phase. Includes use as a general term under geochemistry(1). Also search fluids; fluid; liquid.
UF fluids
SA Darcy's law
gaseous phase
geochemistry
liquid phase
solid phase

fluid pressure
SA pressure

fluids
use fluid phase

fluids, ore-forming
use ore-forming fluids

fluoborates
As of 1981, includes use on level 2 under minerals(1). See list L. Before 1981, included use on level 3 under minerals(1) halides(2).
BA halides
BT minerals

fluoborite
BZ borates
fluorides
BT minerals

fluor-phlogopite
BA mica group
sheet silicates
silicates
BT minerals
SA phlogopite

fluorapatite
BA phosphates

BT minerals
SA apatite

fluorescence
SA luminescence
optical properties
thermoluminescence
X-ray fluorescence

fluoride ion
Term introduced in 1978. Before 1978, search fluoride.
SA fluorides
fluorine
ions

fluorides
As of 1981, includes use on level 2 under minerals(1). See list L. Before 1981, included use on level 3 under minerals(1) halides(2).
BA halides
BT minerals
NZ bastnaesite
bideauxite
clinohumite
NA cryolite
NZ fluoborite
NA fluorite
NZ humite
leifite
lepidolite
microlite
minyulite
NA neighborite
NZ norbergite
NA prosopite
NZ pyrochlore
synchisite
NA thomsenolite
NZ topaz
triplite
NA villiaumite
weberite
NZ zinnwaldite
SA fluoride ion

fluorine
Includes use on level 1 and 2 as a chemical element (list D).
UF F
SA elements
fluoride ion

fluorite
As of 1981, includes use on level 3 under minerals(1) fluorides(2). See list L. Before 1981, included use on level 3 under minerals(1) halides(2). When of economic value, use fluorspar.
BA fluorides
halides
BT minerals
SA fluorite structure
fluorspar

fluorite structure
Term introduced in 1978.
SA fluorite

fluorspar
Use only when of economic value; otherwise use fluorite. Includes use on level 1 and 2 as a commodity term (list C).
SA cryolite deposits
fluorite

fluosilicates
As of 1981, includes use on level 2 under minerals(1). See list L. Before 1981, included use on level 3 under minerals(1) halides(2).
BA halides
BT minerals

flute casts
Includes use on level 3 under sedimentary structures(1) bedding plane irregularities(2). See list K.
UF casts, flute
BA bedding plane irregularities
sedimentary structures
SA current markings
load casts
scour casts
scour marks
soft sediment deformation
sole marks
turbidity current structures

fluvial
A valid term through 1977. After 1977, use fluvial features or (as of 1981) fluvial environment.

fluvial environment
Term introduced in 1981.
SA environment
fluvial features
fluvial sedimentation
paleoecology
rivers
rivers and streams
stream sediments
streams

fluvial features
Includes use on level 2 under geomorphology(1).
UF features, fluvial
NT meanders
waterfalls
SA alluvial fans
bars
bedforms
braided streams
buried channels
cascades
channel geometry
channels
cliffs
deltas
drainage basins
drainage patterns
ephemeral streams
estuaries
floodplains
fluvial environment
geomorphology
glaciofluvial features
mud banks
rivers
rivers and streams
sedimentary structures
shoals
sinuosity
stream order
stream sediments
streams
terraces
valleys

fluvial sedimentation
Term introduced in 1981.
SA fluvial environment
sedimentation

fluvial transport
use stream transport

fluvioglacial environment
use glaciofluvial environment

flux
As of 1981, no longer a valid term for GeoRef. See meteorite flux; heat flux. Before 1981, included use on level 2 under meteorites(1).

fly ash
use ash

flysch
Includes use on level 3 under sedimentary rocks(1) clastic rocks(2). See list I. Before 1981, also included use on level 3 under sediments(1) clastic sediments(2).
BA clastic rocks
BT sedimentary rocks
SA clastic sediments
terrigenous materials
wildflysch

Foaming Sea
UF Mare Spumans
BT Moon

focal mechanism
As of 1981, includes use on level 2 under earthquakes(1). Before 1981, included use on level 3. Also search focus AND mechanism.
SA earthquakes
focus
mechanism

foci
use focus

focus
Includes use on level 2 under earthquakes(1). Also search hypocenters; foci.
UF centrum (seismology)
foci
hypocenter
seismic focus
SA aftershocks
deep-focus earthquakes
earthquakes
elastic waves
epicenters
focal mechanism
intermediate-focus earthquakes
seismology
shallow-focus earthquakes

Foggia
City and province in N.
BT Apulia
Italy

fold axes
Includes use on level 3 under structural analysis(1) theoretical studies(2).
UF axes, fold
SA folds
structural analysis

fold belts
Includes use on level 3 under tectonics(1). Also search belts; foldbelts.
UF belts, fold
foldbelts
SA mobile belts
orogenic belts
tectonics

foldbelts
use fold belts

folding
use folds

folds
Used for studies on folds and not those on several types of structure. Includes use on level 1 (list A); on level 2 and 3 under structural analysis(1). Also search folding. If 1, term set options are:
distribution
subtopic
experimental studies
subtopic
geometry
subtopic [cylindrical folds, plane cylindrical folds, plane noncylindrical folds]
mechanics
subtopic [compaction, decollement, flexural-slip, flexure, kink-band structures, shear]
orientation (attitude of fold elements with respect to external coordinates)
a. (folds defined on the basis of orientation in relation to the geographic horizontal plane): [horizontal orientation, inclined folds, nappes, normal folds, overturned folds, plunging folds, recumbent folds, vertical orientation]
b. (orientation of folds relative to spatially associated macroscopic structures such as large folds, fold systems, and orogenic zones): [discordant folds, drag folds, longitudinal orientation, oblique orientation, superposed folds]
style
subtopic [anticlines, antiform folds, asymmetric folds, basins, chevron folds, cleavage folds, concentric folds, conjugate folds, convolute folds, cross folds, diapirs, disharmonic folds, domes, harmonic folds, intrafolial folds, isoclinal folds, kink folds, monoclines, polyclinal folds, ptygmatic folds, similar folds, symmetric folds, synclines, synform folds, troughs]
systems
subtopic [anticlinoria, en echelon folds, synclinoria]
theoretical studies
subtopic
UF folding
NA anticlines
anticlinoria
antiform folds
asymmetric folds
chevron folds
cleavage folds
concentric folds
conjugate folds
cylindrical folds
discordant folds
disharmonic folds
domes
drag folds
drape folds
en echelon folds
enterolithic folds
flexure folds
flow folds
harmonic folds
inclined folds
intrafolial folds
isoclinal folds
kink folds
monoclines
normal folds
overturned folds
plunging folds
ptygmatic folds
recumbent folds
similar folds
superposed folds
symmetric folds
synclines
synclinoria
synform folds
SA arches
axial-plane structures
basins
cleavage
decollement
deformation
depressions
diapirs
faults
flexural-slip
fold axes
foliation
fractures
geometry
horizontal orientation
kink-band structures
lineation
mechanics
metamorphic rocks
metamorphism
nappes
neotectonics
oblique orientation
orientation
orogeny

salt domes
salt tectonics
strike
structural analysis
structural geology
style
systems
tectonic elements
tectonics
tectonophysics
vertical orientation

foliation
Includes use on level 1 (list A); on level 2 and 3 under structural analysis(1). Used for planar arrangements of textural or structural features in any type of rock. If 1, term set options are:
genesis
subtopic [flow cleavage, shear cleavage]
interpretation
subtopic
style
subtopic [axial-plane structures, cleavage, fracture cleavage, lamination, schistosity, slaty cleavage, slip cleavage]
NA flow cleavage
fracture cleavage
schistosity
slaty cleavage
slip cleavage
SA axial-plane structures
cleavage
deformation
faults
folds
fractures
intrusions
laminations
lineation
metamorphic rocks
metamorphism
neotectonics
structural analysis
structural geology
style
tectonics
tectonophysics

Folkestone
City on Strait of Dover in SE.
BT England
Great Britain
United Kingdom

Folldal
Village in SE part of country.
BT Norway

Fontainebleau
Town in N central part of country.
BT Seine-et-Marne
France

foraminifera
Includes use on level 1 and 2 as a fossil term (list F).
UF Foraminiferida
BT Protista
Invertebrata
NA Allogromiina
Fusulinina
Lagenidae
Miliolina
Rotaliina
Textulariina
SA foraminifers
micropaleontology
plankton
Receptaculitaceae

Foraminiferida
use foraminifera

foraminifers
Term introduced in 1981 as the common name for foraminifera. See list F. To be used for foraminifera when they are used as tools in stratigraphy. Includes use on level 1. Before 1981, search foraminifera AND biostratigraphy. If 1, term set options are:
biostratigraphy
age [List E]
SA biostratigraphy
foraminifera

force, Coriolis
use Coriolis force

foreshocks
SA aftershocks
earthquakes
elastic waves
shock waves

Forest City Basin
Index states as applicable.
BT United States
SA Iowa
Kansas
Missouri
Nebraska

Forest of Ardennes
use Ardennes

Forest of Dean
Gloucestershire in W central.
BT England
Great Britain
United Kingdom

forestry
Includes use on level 3 under soils(1) utilization(2).
SA deforestation
forests
soils
utilization

forests
Includes use on level 3 under soils(1) yields(2).
SA deforestation
forestry
soils
yields

Foresudetic Monocline
Term introduced in 1981.
BT Poland

Forfar
use Angus

Forfarshire
use Angus

form, crystal
use crystal form

formations, iron
use iron formations

Formosa
Not a valid index term for GeoRef since 1976. Use Taiwan.

Formosa Strait
Channel between Fukien Province of China and Taiwan. Also search Taiwan Strait.
UF Taiwan Strait
BT Pacific Ocean

formula
Term restricted to use for chemical formulas, e.g. for minerals.
SA minerals

forsterite
BA olivine group
nesosilicates
orthosilicates
silicates
BT minerals

Fort Churchill
Destroyed wooden fort at town of Churchill on Hudson Bay in NE.
BT Manitoba
Canada

Fort Good Hope
Trading station on Mackenzie River in NW District of Mackenzie.
UF Good Hope
BT Northwest Territories
Canada

Fort Gouraud
Village near Western Sahara border in NW central Mauritania. Also search Idjil.
UF Idjil
BT Mauritania

Fort Hays Limestone
use Fort Hays Limestone Member

Fort Hays Limestone Member
Of Niobrara Formation. E Colorado, W Kansas, NE New Mexico, and SE South Dakota. Also search Fort Hayes Limestone.
UF Fort Hays Limestone
BT Upper Cretaceous
Cretaceous
SA Colorado
Kansas
New Mexico
Niobrara Formation
South Dakota

Fort McMurray
Town on Athabaska River in NE central.
BT Alberta
Canada

Fort Payne Chert
use Fort Payne Formation

Fort Payne Formation
Includes Greasy Creek facies, Short Mountain facies. N and E Alabama, NW Georgia, Kentucky, NE Mississippi, and Tennessee. Also search Fort Payne Chert.
UF Fort Payne Chert
BT Lower Mississippian
Mississippian
SA Alabama
Georgia
Kentucky
Mississippi
Tennessee

Fort Riley Limestone
In Chase Group. E Kansas, SE Nebraska, and central N Oklahoma. Also search Fort Riley Limestone Member.
UF Fort Riley Limestone Member
BT Permian
SA Chase Group
Kansas
Oklahoma

Fort Riley Limestone Member
use Fort Riley Limestone

Fort Ross
Trading post on Somerset Island, W of Baffin Island, in District of Franklin.
BT Northwest Territories
Canada

Fort Ternan
Village in SW part of country. Nyanza province.
BT Kenya

Fort Union Formation
Includes Lebo Shale Member, Tongue River Member, Tullock Member, Ludlow Member, Sentinel Butte Member. NW Colorado, Montana, North Dakota, NW South Dakota, and Wyoming. Also search Fort Union Group.
SA Colorado
Cretaceous
Montana
North Dakota
Paleocene
South Dakota
Upper Cretaceous
Wyoming

Fort Wingate
Term introduced in 1981. In McKinley County, NW New Mexico.
CO N352500N353500
W1082800W1083500
BT McKinley County
New Mexico
United States

Fort Worth Basin
N central.
BT Texas
United States

Forth Valley
River valley in S central.
BT Scotland
Great Britain
United Kingdom

Forties Field
SA North Sea
oil and gas fields

Fortran
BT computer programs
SA automatic data processing

Fortran IV
BT computer programs
SA automatic data processing
mathematical geology

Fortuna
Town in Humboldt County in NW.
BT California
United States

Fosdick Mountains
N of Edsel Ford Range in W Marie Byrd Land near the Pacific Ocean.
BT Antarctica

Fossa Magna
Structural trench crossing the mountain ranges of Honshu.
BT Honshu
Japan

fossil
A valid index term through 1977 used in combination with man (i.e. man, fossil) on level 1 and 2. After 1977, use fossil man on level 1 and 2.

fossil assemblages
use assemblages

fossil ice wedges
UF ice wedges, fossil
ice-wedge cast
ice-wedge fill
ice-wedge pseudomorph
wedges, fossil ice
BA bedding plane irregularities
sedimentary structures
SA frost action
frost features
ice wedges
periglacial features

fossil localities
Term introduced in 1981. Use only for collecting.
SA collecting
popular geology

fossil man
Term introduced in 1978. Includes use on level 1 and 2 as a fossil term (list F). Used mainly for archaeological and anthropological studies, but as of 1981, is restricted to papers about actual fossil remains. If taxonomy is given, use Mammalia.
SA anthropology
archaeological sites
archaeology
artifacts
Mammalia
Neanderthal

Primates
Vertebrata

fossil meteorite craters
use astroblemes

fossil soils
use Paleosols

fossil waters
Term introduced in 1981.
SA connate waters
pore water

fossil wood
Includes silicified wood and petrified wood. Includes use on level 3 under Plantae(1), gymnosperms(1), and angiosperms(1). See list F. Before 1981, also search silicified wood; petrified wood.
UF petrified wood
silicified wood
SA angiosperms
gymnosperms
Plantae
wood

fossiliferous materials
Term introduced in 1978. Includes use on level 3 under sediments(1) and sedimentary rocks(1). Before 1978, also search fossiliferous AND specific sediment type or sedimentary rocks.
SA fossilization
fossils
materials
sedimentary rocks
sediments

fossilization
Includes taphonomy. Includes use on level 2 under fossil group(1); on level 2 under paleontology(1). See list F.
SA fossiliferous materials
fossils
ichnofossils
paleobotany
paleontology
problematic fossils
silicification
taphonomy

fossils
A valid level 1 term through 1977 used in combination with problematic (i.e. fossils, problematic). After 1978, includes use on level 3. Use problematic fossils on level 1 and 2.
SA allometry
assemblages
benthonic taxa
biochemistry
biogeography
biometry
biostratigraphy
bones
calcification
chemical fossils
coiling
faunal list
faunal studies
floral studies
fossiliferous materials
fossilization
functional morphology
habitat
histology
holotypes
ichnofossils
index fossils
jaws
living fossils
living taxa
locomotion
microfossils
micropaleontology
nannofossils
neotypes
ontogeny
otoliths
paleobiology
paleobotany
paleoclimatology
paleontology
problematic fossils
provinciality
range
revision
reworking
sclerites
septa
shells
skeletons
skulls
species diversity
sutures
typomorphism

foundations
A level 1 term as of 1978. Includes use on level 2 under dams(1). Used for geological studies on engineering foundations. If 1, term set options are:
bridges
subtopic
buildings
subtopic
construction
subtopic
dams
subtopic
design
subtopic
earthworks
subtopic
experimental studies
subtopic
highways
subtopic
instruments
subtopic
land subsidence
subtopic
materials, properties
subtopic
piles
subtopic
seepage
subtopic
settlement
subtopic
site exploration
subtopic
stability
subtopic
structures
subtopic
theoretical studies
subtopic
SA bearing capacity
blasting
bridges
buildings
construction
dams
design
earthquakes
earthworks
engineering geology
engineering properties
explosions
grouting
highways
land subsidence
marine installations
nuclear facilities
overconsolidated materials
piles
rock mechanics
seepage
settlement
site exploration
slope stability
soil mechanics
soils
stability
structures
tunnels
underground installations

Four Corners
Location where boundaries of four states meet. Index states as applicable.
BT United States
SA Arizona
Colorado
New Mexico
Utah

Fourier analysis
Also search harmonic analysis; Fourier transforms.
UF Fourier transformations
Fourier transforms
harmonic analysis
SA analysis
automatic data processing
mathematical geology

Fourier transformations
use Fourier analysis

Fourier transforms
use Fourier analysis

Foveaux Strait
Channel between S South Island and Stewart Island.
BT New Zealand

Fox Glacier
In the Southern Alps of Tasman National Park in E.
BT South Island
New Zealand

Fox Hills Formation
Includes Milliken Sandstone Member, Colgate Member, Stoneville Member, Trail City Sandstone Member, Timber Lake Sandstone Member, Bullhead Member. Also search Fox Hills Sandstone.
BT Upper Cretaceous
Cretaceous
SA Colorado
Edmonton Formation
Montana
North Dakota
South Dakota
Wyoming

foyaite
As of 1981, includes use on level 3 under igneous rocks(1) syenites(2). See list H. Before 1981, included use on level 3 under igneous rocks(1) syenite family(2) and alkali gabbro family(2).
BA syenites
BT igneous rocks
SA alkali gabbros
nepheline syenite

Fra Mauro
Includes use on level 3 under Moon(1). Also search Fra Mauro crater; Fra Mauro formation.
UF Fra Mauro crater
Fra Mauro Formation
BT Moon

Fra Mauro crater
use Fra Mauro

Fra Mauro Formation
use Fra Mauro

fractional crystallization
Includes use on level 3 under magmas(1). Before 1978, search crystallization AND fractional.
SA crystallization
differentiation
magmas
salt
segregation

fractionation
Includes use on level 2 under isotopes(1); on level 3 under geochemistry(1) processes(2). Use fractional crystallization for fractionation of magmas.
SA geochemistry
isotopes
processes
segregation

fracture cleavage
Includes use on level 3 under foliation(1) style(2).
BA foliation
SA cleavage
fractures
joints

fracture strength
Includes use on level 3 under deformation(1) field studies(2).
UF breaking strength
fracture stress
SA deformation
elastic limit
strength
yield strength

fracture stress
use fracture strength

fracture zones
Includes use on level 3 under plate tectonics(1).
UF zones, fracture
SA faults
mid-ocean ridges
ocean floors
plate tectonics
sea-floor spreading
transform faults

fractures
Includes use on level 1 (list A); on level 3 under deformation(1). Used for a break in a rock usually without displacement. If 1, term set options are:
distribution
subtopic
experimental studies
subtopic
genesis
subtopic [release, shear, tension]
patterns
subtopic
style
subtopic [bedding, closed fractures, columnar joints, concentric fractures, conical fractures, cross fractures, dip fractures, extension fractures, feather fractures, joints, latent fractures, oblique fractures, polygonal fractures, radiating fractures, ring sheeting, strike fractures]
systems
subtopic
theoretical studies
subtopic
NA closed fractures
columnar joints
concentric fractures
conical fractures
cross fractures
dip fractures
extension fractures
joints
open fractures
polygonal fractures
release fractures
SA bedding
cracks
deformation
engineering geology
failures
faults

fissures
folds
foliation
fracture cleavage
lineation
microcracks
neotectonics
oblique orientation
shear zones
strike
structural analysis
structural geology
style
systems
tectonic elements
tectonics
tectonophysics
tension
thermal waters
veins

fracturing
Includes use on level 3 under plate tectonics(1) effects(2).
SA hydraulic fracturing
plate tectonics

fragmentation
Used as a general term.
SA fragments
processes

fragments
SA clastic rocks
clasts
fragmentation
lapilli
particles
regolith

framboidal texture
Term introduced in 1978. Before 1978, search framboidal. Also search framboidal pyrite; search framboids.
SA pyrite
textures

framework silicates
Includes use on level 2 under minerals(1); in combination with feldspar group, alkali feldspar, plagioclase, barium feldspar, nepheline group, scapolite group, silica minerals, sodalite group, and zeolite group (i.e. framework silicates, feldspar group) to form terms on level 2 under minerals(1). See list L. Also search tectosilicates.
UF tectosilicates
BA silicates
BT minerals
NZ cancrinite
NA chkalovite
danburite
feldspar group
NZ genthelvite
NA karpinskyite
NZ leifite
NA leucite
nepheline group
pollucite
pseudoleucite
reedmergnerite
scapolite group
silica minerals
sodalite group
NZ wenkite
NA zeolite group

France
Includes use on level 1 as an area term (list O). For term set options see list B.
CO N423000N510000
E0083000W0050000
BA Western Europe
Europe
NT Adour Basin
Aigoual Massif
Aiguilles Rouges
Ain
Aisne
Allier
Alpes-de-Haute Provence
Alpes-Maritimes
Alsace
Anjou
Aquitaine
Aquitaine Basin
Ardeche
Ardennes Department
Ariege
Arize Massif
Armorican Massif
Artois
Aube
Aubrac
Aude
Auvergne
Aveyron
Bas-Dauphine
Bas-Rhin
Bay of Bourgneuf
Bay of Saint-Michel
Bay of the Seine
Bearn
Beauce
Belfort
Belledonne Massif
Bouches-du-Rhone
Brittany
Burgundy
Calvados
Cantal
Causses
Central Massif
Cevennes
Charente
Charente-Maritime
Charentes
Cher
Correze
Cote-d'Or
Cotes-du-Nord
Creuse
Dauphine
Dauphine Alps
Deux-Sevres
Devoluy
Dordogne
Doubs
Drac Valley
Drome
Durance Basin
Eoulx Basin
Essonne
Esterel
Eure
Eure Valley
Eure-et-Loir
Finistere
Franche-Comte
NX French Alps
NT French Coast
NX French Pyrenees
NT Gard
Garonne River
Gers
Gironde
Gironde Estuary
Haut-Rhin
Haute-Garonne
Haute-Loire
Haute-Marne
Haute-Saone
Haute-Savoie
Haute-Vienne
Hautes-Alpes
Hautes-Pyrenees
Hauts-de-Seine
Herault
Ille-et-Vilaine
Indre
Indre-et-Loire
Isere
Isere Valley
Jura
Landes
Languedoc
Laval Basin
Lherz
Limagne
Limousin
Loir-et-Cher
Loire
Loire River
Loire Valley
Loire-Atlantique
Loiret
Lorraine
Lot
Lot-et-Garonne
Lozere
Maine-et-Loire
Manche
Marne
Marne Valley
Marseilles Basin
Mayenne
Meurthe-et-Moselle
Meuse
Montagne Noire
Monts du Lyonnais
Morbihan
Morvan
Moselle
Navarre
Nievre
Nord Department
Nord-Pas-de-Calais Basin
Normandy
Oise
Oise River valley
Orleans
Orne
Paris
Paris Basin
Pas-de-Calais
Pelvoux Massif
Picardy
Poitou
Provence
Provence Alps
Puy-de-Dome
Pyrenees-Atlantiques
Pyrenees-Orientales
Quercy
Redon
Rennes
Rhone
Rodez Trough
Rouergue
Roussillon
Salat Valley
Saone Valley
Saone-et-Loire
Saone-Rhone Basin
Sarthe
Savoie
Seine
Seine Estuary
Seine River
Seine Valley
Seine-et-Marne
Seine-et-Oise
Seine-Maritime
Seine-Saint-Denis
Somme
Somme River valley
Tarn
Tarn-et-Garonne
Touraine
Val-de-Marne
Val-d'Oise
Var
Vaucluse
Vendee
Vercors
Vienne
Vocontian Trough
Vosges
Vosges Mountains
Yonne
Yonne Valley
Yvelines
SA Alps
Annot Sandstone
Ardennes
Arve Valley
Brest
Brianconnais Zone
Cottian Alps
Jura Mountains
Lake Geneva
Maritime Alps
Mediterranean region
Meuse River
Meuse Valley
Moselle River
Moselle Valley
North Sea Coast
Orgueil Meteorite
Pyrenees
Rhine Basin
Rhine River
Rhine Valley
Rhone River
Rhone Valley
Savoy
Scheldt River
Versailles
Voltzia Sandstone
Western Alps

Franche-Comte
Historical region on the Swiss border S and SW of Alsace.
BT France

Francis Creek Shale
W and central Illinois.
BT Pennsylvanian
SA Desmoinesian
Illinois

Franciscan Formation
Includes Corral Hollow Shales, Oak Ridge Sandstone. W California. Also search Franciscan Complex; Franciscan Group.
BT Mesozoic
SA California
Cretaceous
Jurassic

francolite
BA phosphates
BT minerals
SA apatite

Franconia
Former duchy of S central West Germany. Index states as applicable.
UF Franken
BT West Germany
SA Baden-Wurttemberg
Bavaria
Hesse
Upper Franconia

Franconia Formation
Includes Woodhill, Birkmose, Tomah, Reno and Mazomanie members. N Illinois, SE Minnesota, and SW Wisconsin. Index states as applicable.
BT Upper Cambrian
Cambrian
SA Illinois
Minnesota
Wisconsin

Franconian Alb
use Franconian Jura

Franconian Forest
S outlier of the Thuringian Forest.
BT West Germany

Franconian Jura
Plateau in central Bavaria. A northern continuation of the Swabian Jura. Also search Franconian Alb.
CO N484500N500000
E0120000E0104500
UF Franconian Alb
BT Bavaria

West Germany

Franken
use Franconia

Frankenberg
City in SE part of country.
BT Karl-Marx-Stadt Bezirk
East Germany

Frankfurt
City. Frankfurt am Main in West Germany and Frankfurt an der Oder in East Germany. Index countries as applicable.
BT Europe
SA East Germany
West Germany

Frankfurt Bezirk
Term introduced in 1981. District in E. Before 1981, search Frankfurt AND East Germany.
CO N515600N532000
E0144500E0132200
BT East Germany

Frankfurt Stade
Europe. Term introduced in 1978. Before 1978, search Frankfurt.
BA Weichselian
upper Pleistocene
Pleistocene
Quaternary
BT Cenozoic

Franklin County
Index states as applicable.
BX Alabama
Arkansas
Florida
Georgia
Idaho
Illinois
Indiana
Iowa
Kansas
Kentucky
Maine
Massachusetts
Mississippi
Missouri
Nebraska
New York
North Carolina
Ohio
Pennsylvania
Tennessee
Texas
Vermont
Virginia
Washington
BT United States

Franklin District
use District of Franklin

Franklin Mountains
District of Mackenzie in Canada, and N of El Paso in the United States. Index Northwest Territories and/or Texas as applicable.
SA Northwest Territories
Rocky Mountains
Texas

franklinite
BA oxides
BT minerals

Franz Josef Land
Archipelago in Arctic Ocean. Part of Arkhangelsk Oblast.
CO N790000N830000
E0660000E0450000
BX Russian Republic
BZ Russian Platform
BT USSR

Fraser Range
SE of Kalgoorlie in S central.
BT Western Australia
Australia

Fraser River
S central and SW British Columbia. Flows into Georgia Strait S of Vancouver.
BT British Columbia
Canada

Frasnian
Europe. Above Givetian, below Famennian. Includes use on level 3 under age terms(1). See list E.
BA Upper Devonian
Devonian
BT Paleozoic
SA Greenland Gap Group

Fredericksburg Group
Comprises Finlay Limestone, Kiamichi Formation, Edwards Limestone, Paluxy Sand, Walnut Formation, Comanche Peak Limestone, Goodland Formation. S Oklahoma and Texas.
BT Comanchean
Cretaceous
SA Lower Cretaceous
Oklahoma
Texas

Fredericton
SW New Brunswick.
BT New Brunswick
Canada

Frederikshaab
use Frederikshab

Frederikshab
Settlement on Davis Strait in extreme SW Greenland. Also search Frederikshaab.
UF Frederikshaab
BT Greenland
Arctic region

free energy
Includes use on level 3 under geochemistry(1) or phase equilibria(1). Also search Gibbs free energy.
UF Gibbs free energy
Helmholtz free energy
SA energy
geochemistry
phase equilibria
thermodynamic properties

free oscillations
Includes use on level 3 under Earth(1).
UF free vibrations
SA Earth
oscillations

free vibrations
use free oscillations

free-air anomalies
Term introduced in 1978. Before 1978, search free-air.
SA anomalies
Bouguer anomalies
gravity methods

freeze-and-thaw action
use frost action

freeze-thaw action
use frost action

freezing
Term introduced in 1981.
SA frost action
permafrost
thawing

Freiberg
City in SE part of country.
BT Karl-Marx-Stadt Bezirk
East Germany

freibergite
Steel gray variety of tetrahedrite containing silver.
BA sulfantimonites
sulfosalts
BT minerals
SA tetrahedrite

Freiburg
City in SW part of country. Also search Freiburg im Breisgau.
UF Freiburg im Breisgau
BT Baden-Wurttemberg
West Germany

Freiburg im Breisgau
use Freiburg

freieslebenite
BA sulfantimonites
sulfosalts
BT minerals

Fremont County
Index states as applicable.
BX Colorado
Idaho
Iowa
Wyoming
BT United States

Fremouw Formation
Beardmore Glacier area near the Ross Ice Shelf at N end of Queen Maud Range in Ross Dependency.
BT Triassic
SA Antarctica

French
Includes use on level 3 to indicate language of a document.

French Alps
Term introduced in 1981.
BZ Alps
BX France

French Coast
Along British Channel and Bay of Biscay on the Atlantic Ocean side and the Mediterranean Sea on the S.
BT France

French Guiana
French overseas department on NE coast. Includes use on level 1 as an area term (list O). For term set options see list B.
CO N015500N055000
W0513500W0544000
BA South America
SA Guiana Basin
Guianas
Guyana Shield

French Polynesia
Term introduced in 1981. French overseas territory in S Pacific Ocean. Includes Marquesas, Society, Gambier, Tubuai, and Tuamotu islands.
CO S280000S073000
W1320000W1560000
BT Polynesia
NT Austral Islands
Marquesas Islands
NA Society Islands
NT Tuamotu Islands
SA Pacific Ocean

French Pyrenees
Term introduced in 1981.
BZ Pyrenees
BX France

French Somaliland
use Djibouti

frequency
Used as a general term.
SA periodicity

fresh water
Includes use on level 3 under ground water(1), water resources(1), ecology(1), paleoecology(1), and sedimentation(1). Also search freshwater.
UF freshwater
SA dispersion
ecology
fresh-water environment
ground water
paleoecology
salt water
sedimentation
water
water quality
water resources

fresh-water environment
Term introduced in 1978. Before 1978, search fresh-water or fresh water.
SA ecology
environment
fresh water
fresh-water sedimentation
paleoecology

fresh-water sedimentation
Term introduced in 1981.
SA fresh-water environment
sedimentation

freshwater
use fresh water

Freshwater
Town on Isle of Wight in the English Channel.
BT England
Great Britain
United Kingdom

Fresno
City in Fresno County in the San Joaquin Valley.
BT California
United States

Fresno County
In the San Joaquin Valley of central.
CO N355500N373000
W1182000W1205000
BT California
United States

friction
Used as a general term.
SA mechanical properties
physical properties

Friendly Islands
use Tonga

fringing reefs
UF shore reef
BT reefs
SA barrier reefs

Frio County
Term introduced in 1981. SW Texas.
CO N283900N290500
W0985000W0992500
BT Texas
United States

Frio Formation
Consists of dark to very dark, varicolored shales and silty shales and massive to thin-bedded strata of sand and silty sand. Subsurface.
BT Oligocene
SA Louisiana
Texas

Friuli-Venezia Giulia
Autonomous region in NE.
CO N453500N464000
E0134500E0122200
BT Italy
NT Tarvisio
Trieste
SA Tagliamento Valley
Venetia

Front Range
A range of Rocky Mountains in N central.
CO N383000N410000
W1050000W1060000
BT Colorado
United States
SA Rocky Mountains

Front Range urban corridor
BT Colorado
United States

Frontenac County
Index provinces as applicable.
BX Ontario
Quebec
BT Canada

Frontier Formation
In Colorado Group or Mancos Group. S Montana, and W Wyoming, Colorado, Idaho, Utah.
BT Upper Cretaceous
Cretaceous
SA Colorado
Colorado Group
Idaho
Montana
Utah
Wyoming

froodite
As of 1981, includes use on level 3 under minerals(1) bismuthides(2). See list L. Before 1981, included use on level 3 under minerals(1) sulfides(2).
BA bismuthides
sulfides
BT minerals

Frosinone
Town and province in the Apennines SE of Rome.
BT Latium
Italy

frost action
Includes use on level 2 under engineering geology(1) and geomorphology(1).
UF action, frost
freeze-and-thaw action
freeze-thaw action
SA congelifraction
cryopedology
cryoturbation
engineering geology
fossil ice wedges
freezing
geomorphology
glacial geology
highways
patterned ground
periglacial environment
permafrost
polygons
rock mechanics
soil mechanics

frost bursting
use congelifraction

frost churning
use cryoturbation

frost features
Includes use on level 3 under sedimentary structures(1) bedding plane irregularities(2). See list K.
UF features, frost
BA bedding plane irregularities
sedimentary structures
SA fossil ice wedges

frost heaving
As of 1978, term is used on level 2 under permafrost(1).
UF heaving, frost
SA congelifraction
cryoturbation
permafrost

frost shattering
use congelifraction

frost splitting
use congelifraction

frost stirring
use cryoturbation

frost weathering
use congelifraction

frost wedging
use congelifraction

frozen ground
Includes use in engineering geology.
UF gelisol
ground, frozen
merzlota
taele
tele
tjaele
SA engineering geology
ice
periglacial features
permafrost
soil mechanics
soils

fruits
Includes use on level 3 under soils(1) yields(2); includes use on level 3 as a fossil term.
SA soils
yields

Frunze
City in N Kirghizia.
UF Pishpek
BT Kirghizia
USSR

frustules
Refers to morphology of diatoms.
SA diatoms

Fuego
Term introduced in 1981. Active volcano in S central Guatemala.
CO N142900N142900
W0905300W0905300
BT Guatemala

fuel resources
Term introduced in 1981 as a grouper for petroleum, oil shale, oil sands, and natural gas. Includes use on level 1 and 2 as a commodity term (list C).
SA energy sources
natural gas
oil sands
oil shale
petroleum

Fuerteventura
Island in Spain's Las Palmas Province.
BT Canary Islands

fugacity
Includes use on level 3 under geochemistry(1).
SA geochemistry
partial pressure
thermodynamic properties

Fujairah
Emirate. One of federation of 7 states at S end of Persian Gulf.
BT United Arab Emirates
Arabian Peninsula

Fuji
use Fujiyama

Fujiyama
Sacred mountain and quiescent volcano W of Yokohama. Also search Mount Fuji; Mt. Fuji; Mt. Fuji volcano.
UF Fuji
Mount Fuji
Mt. Fuji
Mt. Fuji volcano
BT Honshu
Japan

Fukui
City and prefecture on the Japan Sea in central.
BT Honshu
Japan

Fukuoka
City and prefecture in N.
BT Kyushu
Japan

Fukushima
City in Fukushima Prefecture N central.
BT Honshu
Japan

Fukushima Prefecture
N central.
BT Honshu
Japan

fuller's earth
Includes use as level 3 commodity term under clays(1). See list C.
SA clay
clay mineralogy
clays
kaolin
kaolin deposits
smectite
soils

Fulton County
Index states as applicable.
BX Arkansas
Georgia
Illinois
Indiana
Kentucky
New York
Ohio
Pennsylvania
BT United States

fulvic acids
Includes use on level 3 under organic materials(1).
BA organic materials
SA acids

fumaroles
As of 1978, term is used on level 2 under thermal waters(1).
NT solfataras
SA gases
geothermal energy
geysers
heat flow
springs
sublimates
thermal waters
volcanism
volcanology

functional morphology
BT morphology
SA fossils

functions
Used as a general term in a mathematical sense.
SA mathematical geology
mathematical methods

fungi
Includes use on level 1 and 2 as a fossil term (list F); on level 3 under soils(1) biota(2).
BT thallophytes
Plantae
NA Myxomycetes
SA algae
bacteria
lichens
microorganisms
paleobotany
parasites

Furnace Creek
Small creek which sinks into valley floor in Death Valley National Monument near Nevada border in SE.
BT California
United States

fusinite
Includes use on level 3 under sedimentary rocks(1) organic residues(2). See list I.
BA inertinite
macerals
coal
organic residues
BT sedimentary rocks

fusion
Also search nuclear fusion.
SA fission
melting

Fusulinidae
Includes use on level 2 under foraminifera(1). See list F.
BA Fusulinina
foraminifera
BT Protista
Invertebrata
NA Schwagerina
Triticites

Fusulinina
Includes use on level 2 under foraminifera(1). See list F.
BA foraminifera
BT Protista
Invertebrata
NA Archaediscidae
Endothyra
Fusulinidae

Futaba Group
Underlain by the Paleozoic and granitic rocks with a distinct unconformity and covered disconformably by the Paleogene. N part of E border of Abukama Mountains in E central Honshu.
BT Senonian
Upper Cretaceous
Cretaceous
SA Honshu
Japan

future
Used as a general term.

futures market
Term introduced in 1981.
SA economics

fyord
use fjords

G

Ga
use gallium

gabbro
Includes use on level 3. As of 1981, not to be used in the igneous rocks set. See list H. Before 1981, included use on level 3 under igneous rocks(1) gabbro family(2).
BA gabbros
BT igneous rocks
SA anorthositic gabbro
gabbroic composition
metagabbro
microgabbro
olivine gabbro

gabbro anorthosite
use gabbroic anorthosite

gabbro family
As of 1981, no longer a valid term for GeoRef. Use gabbros or alkali gabbros as applicable. See list H. Before 1981, included use on level 2 under igneous rocks(1).

gabbroic anorthosite
As of 1981, includes use on level 3 under igneous rocks(1) gabbros(2). See list H. Before 1981, included use on level 3 under igneous rocks(1) gabbro family(2). Also search gabbro anorthosite.
UF gabbro anorthosite
BA gabbros

BT igneous rocks
SA anorthosite

gabbroic composition
Term introduced in 1978. Before 1978, search gabbroic; gabbroic rocks.
SA composition
gabbro

gabbros
Term introduced in 1981. Includes use on level 2 under igneous rocks(1). See list H. Before 1981, search gabbro family.
BT igneous rocks
NA alkali gabbros
anorthosite
anorthositic gabbro
gabbro
gabbroic anorthosite
labradoritite
microgabbro
norite
olivine gabbro
rodingite
troctolite
SA eucrite

Gabilan Range
One of the Coast Ranges in W.
BT California
United States
SA Coast Ranges

Gabon
Formerly part of French Equatorial Africa. Includes use on level 1 as an area term (list O). For term set options see list B.
CO S040000N023000
E0143000E0083000
BA Central Africa
Africa
NT Oklo

Gabonella
Genus. Includes use on level 3 under foraminifera(1) Buliminacea(2).
BA Buliminacea
Rotaliina
foraminifera
BT Protista
Invertebrata

gadolinite
BA orthosilicates
silicates
BT minerals

gadolinium
Includes use on level 1 and 2 as a chemical element (list D).
UF Gd
SA elements
rare earths

gahnite
UF zinc spinel
BA oxides
BT minerals
SA spinel group

Gailtal Alps
S Austria.
BT Austria
SA Alps

Gairloch
Town and parish in NW.
BT Scotland
Great Britain
United Kingdom

gaize
See list I.
BA clastic rocks
BT sedimentary rocks
SA clastic sediments
sandstone

Galaico Massif
Term introduced in 1981.
BT Spain

Galapagos Islands
A territory of Ecuador about 600 miles W of mainland. Includes use on level 1 as an area term (list O). For term set options see list B.
CO S013000N010000
W0890000W0920000
UF Colon Archipelago
BA East Pacific Ocean Islands
NT Chaves Island
SA Ecuador
South America

Galapagos Rift
Term introduced in 1981. Separates the Cocos and Nazca tectonic plates. Forms triple junction with the East Pacific Rise. Before 1981, also search Galapagos spreading center.
CO N000000N050000
W0800000W1020000
UF Galapagos spreading center
BT Pacific Ocean
SA Cocos Plate
mid-ocean ridges
Nazca Plate
plate tectonics

Galapagos spreading center
use Galapagos Rift

galena
Also search galenite.
UF blue lead
galenite
lead glance
BA sulfides
BT minerals

galenite
use galena

galenobismutite
BA sulfobismuthites
sulfosalts
BT minerals

Galezice
Syncline in the Swiety Krzyz Mountains between Vistula and Pilica rivers in SE central.
BT Poland
SA Swiety Krzyz Mountains

Galicia
As of 1981, refers to only the region and ancient kingdom in NW Spain. Before 1981, also referred to the former Austrian crownland now in SE Poland and NW Ukraine. Before 1981, search Galicia AND Spain.
CO N414800N434800
W0064500W0092000
BT Spain
NT Arosa Bay
La Coruna Province
Lugo
Orense Province
Pontevedra Province
SA Galitsiya

Galilean satellites
Includes Callisto, Europa, Ganymede, and Io.
SA Jupiter
satellites

Galilee
Hilly region of N.
BT Israel
Middle East

Galilee Basin
Term introduced in 1981. Structural basin in central Queensland.
CO S263000S195000
E1473000E1410000
BT Queensland
Australia

Galitsiya
Term introduced in 1981 for the former Austrian crownland now in SE Poland and NW Ukraine. Before 1981, also search Galicia AND Poland; Galicia AND Ukraine.
BT Eurasia
SA Galicia
Poland
Ukraine

Gallatin County
SW Montana.
BT Montana
United States

Gallatin Range
Index states as applicable.
BT United States
SA Montana
Wyoming

gallium
Includes use on level 1 and 2 as a chemical element (list D). Also search Ga.
UF Ga
SA elements

Gallivare
Village N of the Arctic Circle.
UF Gellnare
BT Norrbotten
Sweden

Galloway
District in SW.
BT Scotland
Great Britain
United Kingdom

Gallup Sandstone
In Mesaverde Group. NW New Mexico.
BT Upper Cretaceous
Cretaceous
SA Mesaverde Group
New Mexico

Galveston
City on Galveston Island on the Gulf of Mexico in Galveston County.
BT Texas
United States

Galveston Bay
Inlet of Gulf of Mexico protected from gulf by Galveston Island and Bolivar Peninsula.
BT Texas
United States

Galveston County
SE Texas.
BT Texas
United States

Galveston Island
30 mile long island at the entrance of Galveston Bay.
BT Texas
United States

Galway
County in W Ireland. Also a city.
CO N525500N534000
W0080000W0101500
BT Ireland
NT Galway Granite

Galway Granite
Connemara District in W.
BT Galway
Ireland
SA Devonian

Gambia
Includes use on level 1 as an area term (list O). For term set options see list B.
CO N131200N135000
W0135000W0165000
BA West Africa
Africa
SA Georgetown

Gambier Embayment
Index states as applicable.
BT Australia
SA South Australia
Victoria

gamma ray
A valid term through 1977. After 1977 use gamma-ray spectroscopy or gamma-ray methods.

gamma rays
Includes use on level 3 under spectroscopy(1), well-logging(1) and under chemical element(1).
UF rays, gamma
SA cosmic rays
electromagnetic radiation
gamma-gamma methods
gamma-ray methods
gamma-ray spectroscopy
radioactivity
scintillations
X-rays

gamma-gamma methods
Term introduced in 1978. Includes use on level 3 under well-logging(1) and under geophysical methods(1). Before 1978, search gamma-gamma.
BA geophysical methods
SA gamma rays
gamma-ray methods
methods
radioactivity
well-logging

gamma-ray methods
Term introduced in 1978. Includes use on level 3 under well-logging. Before 1978, search gamma ray or gamma-ray.
BA geophysical methods
SA gamma rays
gamma-gamma methods
gamma-ray spectroscopy
methods
radioactivity
well-logging

gamma-ray spectrometry
A valid term through 1978. After 1978, see spectroscopy(1).

gamma-ray spectroscopy
Term introduced in 1978. Includes use on level 3 under spectroscopy(1), under well-logging(1), and under chemical element(1) analysis(2). Before 1978, also search spectroscopy AND gamma ray.
BT spectroscopy
SA alpha-ray spectroscopy
analysis
gamma rays
gamma-ray methods
radioactivity
well-logging

Gander Lake Group
UF Gander Lake Series
BT Ordovician
SA Canada
Newfoundland

Gander Lake Series
use Gander Lake Group

Gandja
use Kirovabad

Gandzha
use Kirovabad

Ganga River
use Ganges River

Gangamopteris
Genus.
BA Glossopteridales
gymnosperms
BT Spermatophyta
Plantae

Ganges River
Sacred river which rises in the Himalayas and flows into the Bay of Bengal. Index countries as applicable.

UF Ganga River
BT Asia
SA Bangladesh
India

Gangpur Group
use Gangpur Series

Gangpur Series
Subdivided into Ghoriajor Stage, Kumarmunda Stage, Birmitrapur Stage, and Laingar Stage.
UF Gangpur Group
BT Archean
SA India
Orissa

gangue
Includes use on level 3 under mineral deposits, genesis(1) or commodities (list C).
SA mineral deposits

Gard
Department in S.
CO N433000N443000
E0050000E0031500
BT France
NT Ales

Garfield County
Index states as applicable.
BX Colorado
Montana
Nebraska
Oklahoma
Utah
Washington
BT United States

Gargano
Mountain promontory extending into Adriatic Sea in NE Apulia. Also search Monte Gargano.
UF Monte Gargano
BT Apulia
Italy

Gargasian
Substage in Switzerland: upper Aptian; above Bedoulian Substage. Includes use on level 3 under age terms(1). See list E.
BA Lower Cretaceous
Cretaceous
BT Mesozoic
SA Aptian

Garhwal
District in N Uttar Pradesh.
UF Gurhwal
BT Uttar Pradesh
India

Garian
Town in extreme NW part of country.
UF Gharian
BT Tripolitania
Libya

Garlock Fault
Runs E-W in S central California. Also search Garlock fault zone.
BT California
United States

Garm
Town and oblast in N central.
BT Tadzhikistan
USSR

garnet
As of 1981, no longer a valid term for GeoRef. See garnet group. Use garnet deposits for garnet as a commodity. Included use on level 3 under minerals(1) orthosilicates, garnet group(2) and on level 3 under industrial minerals(1).

garnet deposits
Term introduced in 1981. Includes use on level 3 as a commodity term (list C) under industrial minerals(1). Before 1981, search garnet AND deposits.
SA abrasives
garnet group
heavy mineral deposits
industrial minerals

garnet group
Includes use in combination with orthosilicates (i.e. orthosilicates, garnet group) on level 2 under minerals(1). See list L.
BA nesosilicates
orthosilicates
silicates
BT minerals
NA almandine
andradite
grossular
hydrogrossular
melanite
pyrope
schorlomite
spessartine
uvarovite
vanadium garnet
SA garnet deposits
gems

garnet lherzolite
As of 1981, includes use on level 3 under igneous rocks(1) peridotites(2). See list H. Before 1981, included use on level 3 under igneous rocks(1) ultramafic family(2).
BA peridotites
ultramafics
BT igneous rocks
SA lherzolite

garnet peridotite
As of 1981, includes use on level 3 under igneous rocks(1) peridotites(2). See list H. Before 1981, included use on level 3 under igneous rocks(1) ultramafic family(2).
BA peridotites
ultramafics
BT igneous rocks
SA peridotite

garnet pyroxenite
As of 1981, includes use on level 3 under igneous rocks(1) ultramafics(2). See list H. Before 1981, included use on level 3 under igneous rocks(1) ultramafic family(2).
BA ultramafics
BT igneous rocks
SA pyroxenite

garnetite
Term introduced in 1981. Includes use on level 2 under metamorphic rocks(1). See list J.
BT metamorphic rocks

Garnett
City in Anderson County in E.
BT Kansas
United States

garnierite
UF nepouite
BA serpentine group
sheet silicates
silicates
BT minerals

Garo Hills
In W Assam state, at bend of Brahmaputra River in NE part of country.
CO N252000N254500
E0904500E0900500
BT India

Garonne River
Southwest.
BT France

garrelsite
BA orthosilicates
silicates
BT minerals

garronite
BA zeolite group
framework silicates
silicates
BT minerals

gas
A valid level 1 term through 1977 used in combination with natural (i.e. gas, natural). After 1977, use natural gas as level 1 term.

gas chromatography
Not a valid term from 1975 through 1977. After 1977, includes use as a valid index term on level 3.
SA chemical composition
chromatography
gases

gas condensates
use condensates

gas fields, oil and
use oil and gas fields

Gas Hills
Fremont County in central.
BT Wyoming
United States

gas phase
use gaseous phase

gas storage
Includes use on level 3 under engineering geology(1) petroleum engineering(2).
SA engineering geology
natural gas
reservoirs
storage
underground installations

Gascoyne abyssal plain
Off western Australia.
BT Indian Ocean

gaseous phase
Term introduced in 1978. Includes use as general term under geochemistry(1). Also search gas phase.
UF gas phase
phase, gaseous
SA fluid phase
gases
gasification
geochemistry
liquid phase
solid phase

gases
Includes use on level 3 for other than natural gas (the commodity).
SA Darcy's law
fluid inclusions
fumaroles
gas chromatography
gaseous phase
gasification
hydrogen sulfide
magmas
mining geology
natural gas
noble gases
oil-gas interface
phase equilibria
pneumatolysis
solfataras
sublimates

gasification
Includes use on level 3 under coal(1).
SA coal
gaseous phase
gases
natural gas

Gaspe
Village in E Quebec near E extremity of Gaspe Peninsula.
BT Quebec

Gaspe Peninsula
Extends into Gulf of Saint Lawrence. Also search Gaspe.
CO N475000N491500
W0641500W0680000
BT Quebec

gastaldite
use glaucophane

gastroliths
SA Pisces
Reptilia

Gastropoda
Includes use on level 2 under Mollusca(1). See list F.
BA Mollusca
BT Invertebrata
NA Acoela
Archaeogastropoda
Basommatophora
Bellerophontina
Cerithiidae
Entomotaeniata
Harpidae
Mesogastropoda
Muricacea
Neogastropoda
Planorbis
Prosobranchia
Pteropoda
Pupillidae
Sacoglossa
Stylommatophora
Turritellidae
Viviparus
SA gastropods

gastropods
Term introduced in 1981 as the common name for Gastropoda. See list F. To be used for Gastropoda when they are used as tools in stratigraphy. Before 1981, search Gastropoda AND biostratigraphy.
SA biostratigraphy
Gastropoda

Gatineau Valley
River valley in SW.
BT Quebec
Canada

gauges
use gauging

gauging
Term introduced in 1978. Also search gauges.
UF gauges
SA hydrology
water

Gault
Middle Europe. A lower Cretaceous clay formation in Great Britain. Includes use on level 3 under age terms (1). See list E. Also search Gault Clay and Gault Formation.
UF Gault Clay
BA Lower Cretaceous
Cretaceous
BT Mesozoic
SA Great Britain

Gault Clay
use Gault

Gaurdak
Town in Chardzhou Oblast in NE.
BT Turkmenia
USSR

Gauss Epoch
Geomagnetic epoch.
UF Gauss Normal
BA Pliocene
Neogene
Tertiary
BT Cenozoic
SA paleomagnetism
reversals

G

upper Tertiary

Gauss Normal
use Gauss Epoch

Gavarnie
Village in S part of country.
BT Hautes-Pyrenees
France

Gavrovo Zone
Central Peloponnesus. Also search Gavrovo.
UF Gavrovo-Tripolis Zone
BT Peloponnesus
Greece

Gavrovo-Tripolis Zone
use Gavrovo Zone

Gavur Mountains
use Amanos Mountains

Gay
Ore fields in the Southern Urals.
BT Russian Republic
USSR

Gaya
City in central.
BT Bihar
India

gaylussite
BA carbonates
BT minerals

Gd
use gadolinium

Gdansk
Province in N Poland. Also a city on the Gulf of Danzig.
UF Danzig
BT Poland
NT Leba
Puck Bay

Ge
use germanium

geanticlines
Includes use on level 3 under tectonics(1).
SA anticlines
anticlinoria
geosynclines
synclinoria
tectonics

Geary County
NE central.
BT Kansas
United States

Geauga County
NE Ohio.
BT Ohio
United States

Gedinnian
Europe. Above Ludlovian (Silurian), below Siegenian. Includes use on level 3 under age terms(1). See list E.
BA Lower Devonian
Devonian
BT Paleozoic

gedrite
BA orthoamphibole
amphibole group
chain silicates
silicates
BT minerals
SA amosite
anthophyllite

gehlenite
BA melilite group
sorosilicates
orthosilicates
silicates
BT minerals

Gehrden
Town in SE central.
BT Lower Saxony
West Germany

geikielite
BA oxides
BT minerals

Geirud Formation
Central Elburz Mountains in N Iran.
BT Paleozoic
SA Carboniferous
Devonian
Iran
Permian

Geisel River valley
use Geisel Valley

Geisel Valley
River valley in S central part of country. Also search Geisel River valley.
UF Geisel River valley
BT Halle Bezirk
East Germany

gelifraction
use congelifraction

gelisol
use frozen ground

geliturbation
use cryoturbation

gelivation
use congelifraction

Gellnare
use Gallivare

gels
BT colloidal materials

Gemer
Village in S central.
BT Slovakia
Czechoslovakia

gems
Includes use on level 1 as a commodity term (list C). See also specific gems.
SA agate
amethyst
beryl
chalcedony
corundum
diamond
diamonds
economic geology
emerald
garnet group
minerals
moonstone
pearls
peridot
sapphire
topaz
turquoise
zircon

general
Includes use as a level 2 or 3 term appropriate to a large number of topics, e.g. on level 2 under Mohorovicic discontinuity(1). As of 1976, the term is to be used on level 2 only when referring to geology. See list G.
SA education
geology

generation
As of 1981, no longer a valid term for GeoRef. See genesis. Before 1981, included use on level 2 under ocean waves(1).

Genesee County
BX Michigan
New York
BT United States

Genesee Group
Comprises Geneseo Shale with Genundewa Limestone Lentil, Standish Flagstone, and West River Shale, Penn Yan Shale, Sherburne Flagstone, Renwick Shale, Ithaca Member. Maryland, New York, Pennsylvania, Western Virginia and northern West Virginia.
BT Devonian
SA Maryland
Middle Devonian
New York
Pennsylvania
Upper Devonian
Virginia
West Virginia

genesis
Includes use on level 1 in combination with mineral deposits (i.e. mineral deposits, genesis); as a level 2 or 3 term appropriate to a large number of topics, e.g. on level 2 under foliation(1), fractures(1), and organic materials(1). See list C (commodities) and list G (general terms). As of 1981, also includes use on level 2 under ocean waves(1).
SA crystallization
mineral deposits
paragenesis

Geneva
Canton surrounding W end of Lake Geneva. Also a city.
BT Switzerland
SA Lake Geneva

Genoa
City on the Ligurian Sea. Also search Genova.
UF Genova
BT Liguria
Italy

Genova
use Genoa

genthelvite
BZ framework silicates
sulfides
BT minerals
SA helvite

geobotanical methods
Includes use on level 2 under mineral exploration(1).
SA biogeochemical methods
geochemical methods
methods
mineral exploration

geochemical
A valid term through 1977. See geochemical maps, geochemical methods, and geochemical prospecting.

geochemical controls
Includes use on level 3 under mineral deposits, genesis(1) controls(2). Also search geochemical.
SA controls
mineral deposits
sedimentation

geochemical exploration
A valid term through 1975. After 1975, use geochemical methods or geochemical prospecting. After 1980, use geochemical methods.

geochemical indicators
Term introduced in 1981.
SA geochemistry
indicators

geochemical maps
Term introduced in 1978. Before 1978, search maps AND geochemical.
BA maps

geochemical methods
Includes use on level 2 under mineral exploration(1). Also search geochemical exploration; geochemical prospecting.
NT biogeochemical methods
SA chemical analysis
dispersion patterns
exploration
geobotanical methods
geochemistry
haloes
methods
mineral exploration

Geochemical Ocean Sections Program
use GEOSECS

geochemical profiles
Term introduced in 1981. Before 1981, search profiles AND geochemistry.
SA geochemistry

geochemical prospecting
As of 1981, no longer a valid term for GeoRef. Use geochemical methods. Before 1981, included use on level 3 under mineral exploration(1) geochemical methods(2). Before 1976, also search geochemical exploration. Term was not used for petroleum.

geochemical surveys
A valid term through 1978. After 1978, see surveys(2) under geochemistry(1).

geochemistry
Used for broad treatments as well as specific experiments. Includes use on level 1 (list A); on level 2 under major disciplines, area terms (list B), commodities (list C), intrusions(1), lava(1), organic materials(1), and rocks (list H, I, J). If 1, term set options are:
processes
name of process [absorption, adsorption, chemical fractionation, chlorination, differentiation, diffusion, dissociation, electrolysis, endothermic reactions, fluorination, fractionation, hydration, hydrogenation, hydrolysis, ionization, ion exchange, nitrification, oxidation, photochemical reactions, photosynthesis, polymerization, pyrolysis, reduction, solution, substitution]
properties
name of property [alkalinity, cation exchange capacity, Eh, electrochemical properties, pH, salinity, solubility, thermochemical properties, thermodynamic properties, etc.]
topic [e.g. bibliography, classification, concepts, cycles, education, experimental studies, instruments, methods, practice, surveys, symposia, textbooks]
subtopic
SA absolute age
absorption
activity
adsorption
alkalinity
alumina
ammonia compound
ammonium
anions
aqueous solutions
atmosphere
biochemistry
biogeochemical methods
cadmium
calcium carbonate
carbon dioxide
carbonate ion
catalysis
cation exchange capacity
cations

chemical analysis
chemical composition
chemical properties
combustion
complexing
condensation
cosmochemistry
crystal chemistry
crystal growth
crystallography
cycles
dehydration
denitrification
desorption
devitrification
diffusion
dispersion
dissociation
Eh
electrochemical properties
electrolysis
electrolytes
elements
enrichment
enthalpy
environmental geology
equations of state
fixation
flocculation
fluid inclusions
fluid phase
fractionation
free energy
fugacity
gaseous phase
geochemical indicators
geochemical methods
geochemical profiles
geochronology
geologic barometry
geologic thermometry
GEOSECS
ground water
halmyrolysis
hydration
hydrochemistry
hydrolysis
inclusions
ion exchange
ions
iron-rich composition
isotopes
kinetics
leaching
leakage anomalies
luminescence
mass transfer
metals
metamorphic rocks
metamorphism
metasomatism
meteorites
migration of elements
mineral deposits
mineral exploration
mobilization
modal analysis
organic materials
organo-metallics
oxidation
paleomagnetism
paleosalinity
partitioning
pH
phase equilibria
photosynthesis
pollution
polymerization
precipitation
pyrolysis
reactions
reagents
reduction
respiration
saturation
sea water
segregation
sinks
sodium chloride
solid phase
solubility
solutes
solution
solutions
sorption
specific heat
spectra
spectroscopy
stability
standard materials
stoichiometry
sublimation
sulfate ion
surveys
suspension
thermochemical properties
thermodynamic properties
trace metals
valency
volatiles
water
water of crystallization
water of dehydration
water vapor
weathering

geochronology
For methods and topics other than those treated under absolute age, mainly for non-isotope methods. Includes use on level 1 (list A); on level 2 under age sets (list E), and area terms (list B). If 1, term set options are:
methods
subtopic
time scales
age (list E)
type of method [exposure age, fission-track dating, hydration of glass, lichenometry, optical mineralogy, paleomagnetism, particle-track dating (e.g. cosmic-ray track), racemization, radiation damage, tephrochronology, thermoluminescence, tree rings, varves]
subtopic
UF geologic chronology
SA absolute age
chemical analysis
chronology
dates
declination
exposure age
fission-track dating
geochemistry
hydration of glass
isochrons
lichenometry
metamorphic rocks
metamorphism
meteor craters
optical mineralogy
paleomagnetism
paleontology
palynology
particle-track dating
racemization
radiation damage
relative age
stratigraphy
tephrochronology
thermoluminescence
time scales
tree rings
varves

geocronite
BZ sulfantimonites
sulfarsenites
BA sulfosalts
BT minerals

geodes
Includes use on level 3 under sedimentary structures(1) secondary structures(2). See list K.
BA secondary structures
sedimentary structures
SA concretions

geodesy
Used infrequently and for geologic applications only. Includes use on level 1 (list A); on level 3 under Earth(1) gravity field(2). If 1, term set options are:
topic [figure of Earth, geodetic coordinates, geoid, harmonics, methods, satellite measurements, surveys]
subtopic
SA cartography
changes of level
coordinates
deformation
Earth
Earth tides
ellipticity
engineering geology
exploration
figure of Earth
geodetic coordinates
geoid
geomorphology
geophysical methods
geophysical surveys
geophysics
gravity field
harmonics
isostasy
Laplace transformations
leveling
maps
measurement
neotectonics
networks
satellite measurements
seismology
selenodesy
strainmeters
surveys
tilt
topography
triangulation

geodetic coordinates
Includes use on level 2 under geodesy(1).
SA coordinates
geodesy
geometry
satellite measurements

geodynamics
SA Earth
interior
plate tectonics
tectonics

geographic distribution
A valid term through 1978. For fossils, see distribution or biogeography.

geography
Also search physical geography.
UF physical geography
SA biogeography
geomorphology
paleogeography

geoid
Includes use on level 2 under geodesy(1).
SA Earth
figure of Earth
geodesy

geologic
A valid term through 1977. After 1977, use geologic maps.

geologic barometry
As of 1978, includes use on level 2 under fluid inclusions(1).
UF barometry, geologic
SA fluid inclusions
geochemistry
geologic thermometry
inclusions
P-T conditions

geologic chronology
use geochronology

geologic engineering
use engineering geology

geologic hazards
A level 1 term as of 1978. Includes use on level 2 under impact statements(1) and nuclear facilities(1). Used for studies on the initiation, controls and effects of various geological phenomena which may be hazardous to human ecology. If 1, term set options are:
avalanches
subtopic
catastrophes
subtopic
causes
subtopic
controls
subtopic
earthquakes
subtopic
effects
subtopic
explosions
subtopic
human ecology
subtopic
faults
subtopic
floods
subtopic
land subsidence
subtopic
landslides
subtopic
mudflows
subtopic
observations
subtopic
prediction
subtopic
rock bursts
subtopic
site exploration
subtopic
storms
subtopic
sunspots
subtopic
tsunamis
subtopic
volcanoes
subtopic
UF hazards, geologic
SA avalanches
catastrophes
damage
destruction
earthquakes
engineering geology
environmental geology
erosion
explosions
faults
floods
ground motion
human ecology
hurricanes
impact statements
land subsidence
land use
landslides
liquefaction
liquefaction potential
marine installations
mudflows
nuclear facilities
pollution
reservoirs
rock bursts
rock mechanics
seismic risk

site exploration
slope stability
soil mechanics
storms
sunspots
tsunamis
tunnels
underground installations
volcanic earthquakes
volcanoes
waste disposal

geologic mapping
A valid term through 1975. To search, see note under cartography. See maps(1).

geologic maps
Term introduced in 1978. Includes use on level 3 under maps(1). Before 1978, also search maps AND geologic.
BA maps

geologic surveys
use survey organizations

geologic thermometry
As of 1978, includes use on level 2 under fluid inclusions(1). Before 1974, also search paleotemperature; geothermometry.
UF geothermometry
SA C-13/C-12
decrepitation
fluid inclusions
geochemistry
geologic barometry
inclusions
O-18/O-16
P-T conditions
S-34/S-32
temperature

geological exploration
A valid term through 1971. After 1971, use exploration under field of study. See maps(1) cartography(2).

geological methods
Includes use on level 2 under mineral exploration(1).
SA exploration
geomorphological methods
methods
mineral exploration

geological oceanography
use marine geology

Geological Survey of Canada
SA Canada
survey organizations

geologists
Before 1975, also search geology as a profession.
SA geology

geology
For general treatments stressing the profession and the discipline. Includes use on level 1 (list A) and 2. If 1, term set options are:
topic [education, history, nomenclature, philosophy, practice, principles, research, textbooks]
subtopic
SA areal geology
associations
bibliography
catalogs
collections
current research
economic geology
education
elementary geology
engineering geology
environmental geology
extraterrestrial geology
general
geologists
geophysics
glossaries
historical geology
marine geology
mathematical geology
medical geology
military geology
mining geology
museums
physical geology
popular geology
research
structural geology
surficial geology
symposia
uniformitarianism

geology as a profession
A valid term through 1975. After 1975, use practice. See geology(1) and education(1).

geomagnetic secular variation
use secular variations

geomagnetism
A valid term through 1975. After 1975, see magnetic field under Earth (and other planets) and Moon; see paleomagnetism.

geometry
Includes use on level 2 under folds(1).
SA channel geometry
eccentricity
ellipticity
folds
geodetic coordinates
mathematical geology
plate geometry
quantitative geomorphology
statistical analysis
structural analysis
tortuosity
waterways

geomorphic geology
use geomorphology

geomorphologic
A valid term through 1977. After 1977, use geomorphologic maps.

geomorphologic effects
Includes use on level 2 under isostasy(1).
SA effects
isostasy

geomorphologic maps
Term introduced in 1978. Includes use on level 3 under maps(1). Before 1978, also search maps AND geomorphologic.
BA maps
SA geomorphology

geomorphological methods
Includes use on level 2 under mineral exploration(1).
SA geological methods
geophysical methods
methods
mineral exploration

geomorphology
Includes use on level 1 (list A); on level 2 under area terms (list B), Moon (and other planets), and age terms (list E). Before 1976, also search physiography. If 1, term set options are:
features [cryptoexplosion features, eolian features, erosion features, fluvial features, frost action, impact features, lacustrine features, mass movements, shore features, solution features, volcanic features]
type of feature or topic [e.g. alluvial fans, beaches, caves, craters, deltas, drainage basins, drainage patterns, estuaries, evolution, karst, lagoons, meteor craters, patterned ground, quantitative geomorphology, reefs, rivers, terraces]
topic [applications, bibliography, concepts, education, environment, history, interpretation, landform description, landform evolution, maps, methods, practice, processes, symposia, textbooks]
subtopic
UF geomorphic geology
SA aerial photography
aggradation
alluvial plains
arid environment
arroyos
astroblemes
badlands
bars
bays
beach ridges
beaches
benches
bogs
boulders
braided streams
bridges
buried channels
buried valleys
calderas
canyons
Carolina Bays
cartography
cascades
caverns
caves
changes of level
channel geometry
channels
cheniers
cliffs
coastal plains
colluvium
corrosion
cratering
cryptoexplosion features
debris
degradation
deltas
denudation
depressions
deserts
domes
drainage basins
drainage patterns
dunes
ecology
engineering geology
environment
environmental geology
eolian features
erosion
erosion features
erosion surfaces
estuaries
etching
exfoliation
extinct lakes
fjords
floodplains
fluvial features
frost action
geodesy
geography
geomorphologic maps
gilgai
glacial geology
glacial lakes
glaciotectonics
glacis
gullies
highlands
hills
hummocks
hydrology
impact features
impacts
inselbergs
isostasy
karren
karst
lacustrine features
lagoons
landform description
landform evolution
landforms
landscapes
landslides
limnology
loess
maps
marshes
mass movements
meanders
meteor craters
morphometry
morphostructures
mountains
mud volcanoes
neotectonics
paleogeography
patterned ground
pediments
peneplains
permafrost
physiographic provinces
piedmonts
plains
plateaus
playas
quantitative geomorphology
reefs
relief
relief exposure
relief inversion
rivers
scarps
sedimentology
semi-arid environment
shoals
shore features
shorelines
sinkholes
slope stability
slopes
slumping
soils
solution cavities
solution features
speleology
spits
steppes
stream order
surface features
surficial geology
swamps
taiga environment
talus slopes
tectonic controls
terrain classification
terrains
topography
tors
tundra
valleys
volcanic features
waterways
wave-cut platforms
weathering
wetlands

geophysical
A valid term through 1977. After 1977, use geophysical maps, or geophysical methods, or geophysical surveys.

geophysical maps
Term introduced in 1978. Before 1978, search maps AND geophysical.
BA maps

geophysical methods
Includes use on level 1 (list A); on level 2 under mineral exploration(1).

Used for discussions which stress methodology of applied geophysics. If 1, term set options are: type of method [acoustical methods, Earth-current methods, electrical methods, electromagnetic methods, gravity methods, infrared methods, magnetic methods, magnetotelluric methods, methods (for more than one method), radioactivity methods, remote sensing, seismic methods]
topic [applications, instruments, interpretation, techniques] or kind of platform [airborne methods, deep-tow methods, ground methods, marine methods, satellite methods], kind of method [induced polarization, resistivity, side-scanning methods, sonar methods]
NA acoustical methods
airborne methods
deep-tow methods
Earth-current methods
echo sounding
electrical methods
electromagnetic methods
gamma-gamma methods
gamma-ray methods
gravity methods
ground methods
helicopter methods
induced polarization
infrared methods
magnetic methods
magnetotelluric methods
marine methods
microwave methods
radar methods
radioactivity methods
satellite methods
seismic methods
sonar methods
SA autoradiography
Bouguer anomalies
conductive materials
deep magnetic sounding
deep seismic sounding
deep sounding
electrical anomalies
electrical conductivity
electromagnetic field
electromagnetic logging
engineering geology
exploration
geodesy
geomorphological methods
geophysical profiles
geophysical surveys
geophysics
gravimeters
gravity anomalies
heat flow
high-resolution methods
induced polarization
induction
inverse problem
isostasy
Landsat
magnetic anomalies
magnetic field
maps
marine geology
methods
mineral exploration
mining geology
networks
noise
paleomagnetism
raypaths
remote sensing
resistivity
seismology
side-scanning methods
signal-to-noise ratio
signals
sonobuoys
sounding
time domain analysis
transient methods
well-logging

geophysical profiles
Term introduced in 1981. Before 1981, search geophysical AND profiles.
SA geophysical methods
geophysical surveys

geophysical surveys
For geophysical methods applied to specific areas. The methods are generally those of exploration and primarily concerned with the shallow structure of the Earth. Includes use on level 2 under area terms(1). See list B.
BT surveys
NA acoustical surveys
Earth-current surveys
electrical surveys
electromagnetic surveys
gravity surveys
infrared surveys
magnetic surveys
magnetotelluric surveys
radioactivity surveys
seismic surveys
SA acoustical properties
airborne methods
bathymetry
Bouguer anomalies
deep magnetic sounding
deep seismic sounding
deep sounding
echo sounding
electrical anomalies
electromagnetic field
electromagnetic logging
engineering geology
geodesy
geophysical methods
geophysical profiles
geophysics
gravity anomalies
gravity field
gravity survey maps
ground methods
ground water
heat flow
helicopter methods
high-resolution methods
imagery
induced polarization
inverse problem
isostasy
Landsat
magnetic anomalies
magnetic field
magnetic survey maps
magnetic susceptibility
maps
marine methods
mineral exploration
mining geology
monitoring
networks
radar methods
reflection
refraction
remote sensing
resistivity
satellite methods
self-potential methods
signal-to-noise ratio
sonar methods
sonobuoys
sounding
well-logging

geophysics
For general treatments stressing the profession and the discipline plus experimental studies on minerals and other materials. Includes use on level 1 (list A);on level 2 under bibliography(1), associations(1), education(1), symposia(1). If 1, term set options are:
topic [applications, bibliography, catalogs, classification, concepts, education, experimental studies, general, history, instruments, methods, nomenclature, objectives, observations, philosophy, practice, principles, symposia, textbooks, theoretical studies]
subtopic
SA adiabatic demagnetization
astrophysics and solar physics
atmosphere
chemical demagnetization
core
crust
deep seismic sounding
deformation
demagnetization
direct problem
Earth
Earth tides
elastic constants
elastic materials
elastic properties
engineering geology
geodesy
geology
geophysical methods
geophysical surveys
heat flow
inverse problem
kinetics
magnetosphere
mathematical geology
meteorology
observatories
paleomagnetism
raypaths
seismology
structural analysis
tectonophysics
thermal demagnetization

geopressure
Related to reservoir rocks. Associated with areas of great thickness of sedimentation. Includes use on level 3 under inclusions(1), under commodity terms (list C), or material.
SA petroleum
pressure
reservoir rocks
sedimentation

Georges Bank
E of Massachusetts and S of Nova Scotia. Also search Georges.
CO N400000N412000
W0664500W0700000
BT Atlantic Ocean
SA Gulf of Maine

Georgetown
City in Georgetown County in South Carolina, a city on the Atlantic Ocean in Guyana, and a city on the Atlantic Ocean in Gambia. Index South Carolina and countries as applicable.
SA Gambia
Guyana
South Carolina

Georgetown County
On the Atlantic Ocean.
BT South Carolina
United States

Georgia
Term is used on level 1 as an area term (list O) only when referring to the United States. Included use for the Georgian Soviet Socialist Republic in Transcaucasia through 1977, but after 1977 use Georgian Republic for that.
CO N302000N350000
W0804500W0853500
BA Eastern U.S.
United States
NT Altamaha River
Atlanta
NX Baker County
Berrien County
Burke County
Calhoun County
Carroll County
NT Cartersville
NX Chatham County
Cherokee County
Clarke County
Clay County
Coffee County
Columbia County
Columbus
Cook County
Crawford County
Dade County
NT Doboy Sound
NX Douglas County
Fayette County
Floyd County
Franklin County
Fulton County
NT Glynn County
NX Greene County
Hancock County
Harris County
Jackson County
Jasper County
Jefferson County
Johnson County
Jones County
Lee County
Lincoln County
NT Macon
NX Macon County
Madison County
Marion County
McIntosh County
Mitchell County
Monroe County
Montgomery County
Morgan County
Murray County
Newton County
Oconee County
Pickens County
Pierce County
Pike County
Polk County
Pulaski County
Putnam County
Randolph County
NT Sapelo Island
Stone Mountain
NX Taylor County
NT Towns County
Twiggs County
NX Union County
Walton County
Warren County
Washington County
Wayne County
Webster County
Wheeler County
White County
Wilkinson County
SA Athens
Atlantic Coastal Plain
Bangor Limestone
Blue Ridge Mountains
Blue Ridge Province
Brevard Zone
Brunswick
Carolina Bays
Chattahoochee River
Chattanooga Shale
Chickamauga Group
Citronelle Formation
Claiborne Group
Clayton Formation
Duplin Formation

G

Fort Payne Formation
Hawthorn Formation
Jackson Group
Knox Group
Lake Chatuge
Midway Group
Murphy Marble
Ocala Group
Okefenokee Swamp
Pennington Formation
Piedmont
Red Mountain Formation
Ripley Formation
Roan Supergroup
Rockwood Formation
Rome Formation
Saint Louis Limestone
Sainte Genevieve Limestone
Sand Hills
Savannah River
Shady Dolomite
Suwannee Limestone
Talladega Group
Trenton Group
Tuscaloosa Formation
Twiggs Clay
Valley and Ridge Province
Wicomico Formation
Wilcox Group

Georgian Bay
Inlet of E Lake Huron.
BT Ontario
Canada

Georgian Republic
Term introduced in 1978 for the Georgian Soviet Socialist Republic in Transcancasia. Before 1978, also search USSR AND Georgia.
CO N420000N430000
E0470000E0400000
BT USSR
NT Abkhazia
Adzharistan
Akhaltsikhe
Alazani
Aragvi
Batumi
Chiatura
Colchis
Dzirula Massif
Inguri River
Kakhetia
Kartlia
Kodor River basin
Rioni Basin
Shiraki Steppe
Svanetia
Tbilisi
Tkibuli
Trialet Range
SA Caucasus
European USSR
Greater Caucasus
Kuban River
Kuban Valley
Kura River
Lesser Caucasus
Ossetia
Terek River
Transcaucasia

Georgina Basin
River drainage basin. Index Northern Territory and/or Queensland as applicable.
BT Australia
SA Northern Territory
Queensland

GEOSECS
Acronym.
UF Geochemical Ocean Sections Program
SA geochemistry
oceanography

geostrophic force
use Coriolis force

geosynclines
Includes use on level 1 (list A). Used for large, mobile downwarping of the crust, which subsides as rocks and sediments accumulate to thicknesses of thousands of meters. Term set options are:
evolution
(name of geosyncline e.g. Alpine Geosyncline, Andean Geosyncline, Cordilleran Geosyncline, Hercynian Geosyncline, Mesogaea, Nevadan Geosyncline, Tethys) or subtopic
genesis
(name of geosyncline) or subtopic
processes
(name of geosyncline) or subtopic
SA Adelaide Geosyncline
Alpine Geosyncline
Andean Geosyncline
anticlinoria
burial metamorphism
continental drift
Cordilleran Geosyncline
crust
eugeosynclines
geanticlines
Hercynian Geosyncline
igneous activity
mantle
metallogeny
miogeosynclines
mobile belts
ocean basins
ophiolite
orogenic belts
orogeny
paleogeography
plate tectonics
Propria Geosyncline
sea-floor spreading
synclinoria
Tasman Geosyncline
tectonics
tectonophysics
thermal history

geotectonics
use tectonics

geothermal energy
Includes use on level 1 as a commodity term (list C).
SA energy
energy sources
fumaroles
geothermal systems
geysers
heat flow
heat sources
hot dry rocks
hot springs
hydraulic fracturing
reservoir rocks
thermal waters

geothermal fields
Before 1978, search geothermal.
UF fields, geothermal
SA geothermal systems
hot dry rocks
oil and gas fields

geothermal gradient
Includes use on level 2 under heat flow(1).
SA conductivity
crust
Earth
heat flow
mantle
measurement
regional patterns
thermal conductivity
thermal waters

geothermal phenomena
use geothermal systems

geothermal processes
As of 1981, no longer a valid term for GeoRef. See hydrothermal processes. Term was introduced in 1978.

geothermal regimes
use geothermal systems

geothermal systems
Term introduced in 1978. Before 1978, also search geothermal; geothermal regions; geothermal phenomena.
UF geothermal phenomena
geothermal regimes
SA geothermal energy
geothermal fields
hot dry rocks
systems
thermal regime

geothermometry
use geologic thermometry

Gera
As of 1981, refers to only the city. Before 1981, also included the district.
BT Gera Bezirk
East Germany

Gera Bezirk
Term introduced in 1981. District in SW. Before 1981, search Gera.
CO N502200N510500
E0121800E0110000
BT East Germany
NT Gera
Schwarzburg Anticlinorium
SA Thuringia
Thuringian Basin
Thuringian Forest

German
Includes use on level 3 to indicate language of a document.

germanates
BA oxides
BT minerals

germanite
BA sulfogermanates
sulfosalts
BT minerals

germanium
Includes use on level 1 and 2 as a chemical element (list D). Also search Ge.
UF Ge
SA elements

Germany
Country in Central Europe. Divided in 1949 into West Germany and East Germany. Includes use on level 1 as an area term (list O). See list B for term set options.
BA Central Europe
Europe
NT Alpenvorland
NX Bavarian Alps
NT Bavarian Massif
Berlin
NA East Germany
NT Harz Mountains
Hesse Basin
Northeastern German Plain
Northern German Hills
Northwestern German Plain
Odenwald
Palatinate
Rhenish Schiefergebirge
Rhine Westphalian Basin
Saar-Nahe Basin
Saxonian Massif
Southwestern German Hills
Thuringian Hills
Thuringian Massif
Upper Rhine Valley
NA West Germany
SA Allgau Alps
Alps
Baltic region
Danube River
Danube Valley
European Platform
Hunsruck Shale
Lake Constance
Molasse Basin
Moselle Valley
North Sea Coast
Posidonia Shale
Poznan Clays
Rhine Graben

Gerona
As of 1981, refers to only the city. Before 1981, also included the province. In NE Spain.
BT Gerona Province
Catalonia
Spain

Gerona Province
Term introduced in 1981. Before 1981, search Gerona. In NE Spain on the Mediterranean Sea.
CO N414000N423000
E0032000E0014500
BT Catalonia
Spain
NT Ampurdan
Gerona
Osor
SA Ter River basin

Gerrei
Region in SE.
BT Sardinia
Italy

Gers
Department in SW France.
CO N431500N440500
E0011000W0002000
BT France

gersdorffite
UF nickel glance
BA sulfides
BT minerals

Getchell Mine
Humboldt County in NW.
BT Nevada
United States

Getic Basin
use Getic Nappe

Getic Nappe
NW part of Transylvanian Alps in Hunedoara County in SW Transylvania and E Banat. Also search Getic Basin. Index regions as applicable.
UF Getic Basin
BT Romania
SA Banat
Transylvania

Geula Cave
use Geula Caves

Geula Caves
Mount Carmel. Also search Geula Cave.
UF Geula Cave
BT Haifa
Israel

geysers
As of 1978, term is used on level 2 under thermal waters(1).
UF gusher
pulsating spring
SA fumaroles
geothermal energy
ground water
hot springs
springs
thermal waters

Geysers, The
use The Geysers

Gezira
Region between Blue Nile and White Nile rivers in central part of country.
BT Sudan

Ghana
Britain's former Gold Coast colony combined with British trust territory of Togo. Includes use on level 1 as an area term (list O). For term set options see list B.
CO N043000N120000
E0013000W0030000
BA West Africa
Africa
NT Accra
Bosumtwi Crater
SA Volta Basin

Gharian
use Garian

Ghats
Two mountain ranges in S India forming E and W edges of Deccan Plateau.
BT India
SA Eastern Ghats
Western Ghats

Ghazni
Province. Also a city in E central.
BT Afghanistan

Ghost Ranch
Region in N central.
BT New Mexico
United States

giant fields
UF fields, giant
BT oil and gas fields
SA natural gas
petroleum

Giant's Causeway
Formation of prismatic basaltic columns extending into sea off N coast of County Antrim.
BT Northern Ireland
United Kingdom

Gibbs fault zone
use Gibbs fracture zone

Gibbs fracture zone
S of Reykjanes Ridge which is SE of Iceland. Also search Gibbs Fault; Gibbs Fracture.
UF Gibbs fault zone
BT Atlantic Ocean

Gibbs free energy
use free energy

Gibbs phase rule
use phase rule

gibbsite
Also search hydrargillite.
UF hydrargillite
BA oxides
BT minerals
SA bayerite

Gibeon Meteorite
Term introduced in 1981. Impact near village of Gibeon in S central Namibia. Before 1981, also search Gibeon AND meteorites.
BA octahedrite
iron meteorites
BT meteorites
SA Namibia

Gibraltar
Peninsula and British colony at S tip of Spain.
BT Europe
SA Strait of Gibraltar

Gibraltar Point
Cape on N side of entrance to the Wash on the North Sea.
BT England
Great Britain
United Kingdom

Gifu
City and prefecture in central.
BT Honshu
Japan

Gila County
E central.
BT Arizona
United States

Gila River
Rises in SW New Mexico and empties into the Colorado River near Yuma, Arizona. Index states as applicable.
BT United States
SA Arizona
New Mexico

Gilau
Village in Cluj County in central.
BT Transylvania
Romania

Gilbert Epoch
Geomagnetic epoch.
UF Gilbert Reversed
BA lower Pliocene
Pliocene
Neogene
Tertiary
BT Cenozoic
SA paleomagnetism
reversals
upper Tertiary

Gilbert Islands
Island group SSE of Marshall Islands and NE of Solomon Islands. Forms main part of British Gilbert and Ellice Islands Colony.
CO S040000N050000
E1770000E1710000
BT Micronesia
SA Pacific Ocean

Gilbert Reversed
use Gilbert Epoch

Giles Complex
Index states as applicable.
BT Australia
SA South Australia
Western Australia

Giles County
Term introduced in 1981. Index states as applicable.
BX Tennessee
Virginia
BT United States

gilgai
Term introduced in 1981. Microrelief of mounds and depressions developed on shrinking and swelling clays.
SA geomorphology
mudcracks
soils

gillespite
BA sheet silicates
silicates
BT minerals

Gilpin County
N central.
CO N394500N395000
W1052000W1053500
BT Colorado
United States

Ginkgo
Genus.
BA Ginkgoales
gymnosperms
BT Spermatophyta
Plantae

Ginkgoales
Includes use on level 2 under gymnosperms(1). See list F.
BA gymnosperms
BT Spermatophyta
Plantae
NA Czekanowskiales
Ginkgo
SA Dadoxylon

Gippsland
Region in SE Victoria. Includes part of the Gippsland Basin.
UF Gippsland region
BT Victoria
Australia
SA Gippsland Basin

Gippsland Basin
Similar to area covered by Gippsland, however the Basin also includes an area offshore to S of Victoria.
CO S390000S380000
E1490000E1470000
BT Victoria
Australia
SA Gippsland

Gippsland region
use Gippsland

Girnar Hills
S central Kathiawar Peninsula in W Gujarat. Also search Girnar.
BT Gujarat
India

Gironde
Department on Bay of Biscay.
CO N441500N453000
E0003000W0013000
BT France
NT Arcachon Basin
Bordeaux
Bordelais
Medoc
SA Gironde Estuary

Gironde Estuary
On Bay of Biscay. Formed by confluence of Garonne and Dordogne rivers.
BT France
SA Gironde

Girvan
Town on Firth of Clyde in Ayrshire in SW.
BT Scotland
Great Britain
United Kingdom

girvanella
See list K.
BA biogenic structures
sedimentary structures
SA algal biscuits

Gisborne
Seaport city in E.
BT North Island
New Zealand

Gissar Range
use Hissar Range

Givet
Town in NE part of country.
BT Ardennes
France

Givetian
Europe. Above Eifelian, below Frasnian. Includes use on level 3 under age terms(1). See list E.
BA Middle Devonian
Devonian
BT Paleozoic
NT Horn Plateau Formation

glacial
A valid term through 1977. After 1977, use glacial environment.

glacial drift
use drift

glacial environment
Term introduced in 1978. Before 1978, search glacial.
SA ecology
environment
glacial geology
glacial sedimentation
glaciofluvial environment
glaciolacustrine environment
glaciomarine environment
interglacial environment
periglacial environment
postglacial environment

glacial features
Includes use on level 2 under glacial geology(1).
UF features, glacial
NT cirques
drumlins
glaciofluvial features
SA extinct lakes
firn
fjords
glacial geology
glacial lakes
glaciation
glaciers
ice
ice caps
icebergs
lakes
moraines
outwash
periglacial features
striations
till
tillite
varves

glacial geology
Includes use on level 1 (list A) for geologic features and effects resulting from the action of glaciers and ice sheets. Term set options are:
ancient ice ages (older than Quaternary)
age
glacial features
name of feature [e.g. drumlins, eskers, fjords, fluting, glacial lakes, glacial polish, hanging valleys, kames, moraines, nunataks, sole marks, striations]
glaciation
topic [e.g. deglaciation, deposition, erosion, ice movement]
glaciers (for present-day glaciers)
subtopic
periglacial features
name of feature [e.g. extinct lakes, ice wedges, patterned ground, permafrost, pingos, solifluction]
topic [applications, bibliography, catalogs, classification, concepts, experimental studies, history, instruments, methods, nomenclature, philosophy, practice, symposia, textbooks, theoretical studies]
subtopic
SA ancient ice ages
atmosphere
boulder clay
boulders
buried channels
buried valleys
changes of level
cirques
cryopedology
deposition
drift
drumlins
erosion
erratics
eskers

G

extinct lakes
firn
fjords
frost action
geomorphology
glacial environment
glacial features
glacial lakes
glacial transport
glaciation
glacier surges
glaciers
glaciofluvial environment
glaciolacustrine environment
glaciomarine environment
glaciotectonics
hydrology
ice movement
ice sheets
ice shelves
ice wedges
interglacial environment
isostasy
kames
mass balance
meltwater
moraines
outwash
paleoclimatology
patterned ground
periglacial environment
periglacial features
permafrost
pingos
postglacial environment
sedimentary structures
sedimentation
sediments
snow
soils
solifluction
stratigraphy
striations
thawing
till

glacial lakes
Includes use on level 3 under glacial geology(1) glacial features(2) and under geomorphology(1) lacustrine features(2).
SA cirques
extinct lakes
geomorphology
glacial features
glacial geology
glaciolacustrine environment
lacustrine features
lakes
limnology
paleolimnology
varves

glacial outwash
use outwash

glacial sedimentation
Term introduced in 1981.
SA glacial environment
sedimentation

glacial transport
includes use on level 3 under sedimentation(1) transport(2).
SA drift
erratics
glacial geology
sedimentation
till
transport

glaciation
Includes use on level 2 under glacial geology(1); on level 3 under changes of level(1) and paleoclimatology(1).
SA abrasion
changes of level
deglaciation
deposition
erosion
fjords
glacial features
glacial geology
glaciers
ice movement
ice sheets
ice shelves
paleoclimatology

Glacier Bay
Narrow inlet of Pacific Ocean in N part of SE Alaska. In center of Glacier Bay National Monument.
BT Alaska
United States

Glacier County
Term introduced in 1981. N Montana.
CO N481900N490000
W1121100W1140500
BT Montana
United States

Glacier National Park
NW Montana.
BT Montana
United States

glacier surges
Term introduced in 1978.
UF catastrophic advance
SA glacial geology
glaciers
surges

glaciers
Includes use on level 2 under glacial geology(1) for present day glaciers; on level 2 under hydrology(1).
SA ablation
deglaciation
erratics
firn
glacial features
glacial geology
glaciation
glacier surges
glaciofluvial environment
glaciolacustrine environment
glaciomarine environment
hydrology
hydrosphere
ice
ice caps
ice movement
ice sheets
ice shelves
icebergs
icequakes
kames
mass balance
meltwater
moraines
nivation
periglacial environment
periglacial features
postglacial environment
rock glaciers
snow

glaciofluvial environment
Term introduced in 1978. Before 1978 search glaciofluvial; fluvioglacial; fluvioglacial environment.
UF fluvioglacial environment
SA environment
glacial environment
glacial geology
glaciers
glaciofluvial features
glaciofluvial sedimentation

glaciofluvial features
Term introduced in 1981. Includes use on level 3 under glacial geology(1).
BT glacial features
NT eskers
kames
outwash plains
SA drift
fluvial features
glaciofluvial environment
outwash

glaciofluvial sedimentation
Term introduced in 1981.
SA drift
glaciofluvial environment
outwash
sedimentation

glaciolacustrine environment
Term introduced in 1981.
SA environment
glacial environment
glacial geology
glacial lakes
glaciers
lacustrine features
lakes
periglacial environment

glaciomarine environment
Term introduced in 1981.
SA environment
glacial environment
glacial geology
glaciers
glaciomarine sedimentation
marine environment

glaciomarine sedimentation
Term introduced in 1981.
SA glaciomarine environment
sedimentation

glaciotectonics
Term introduced in 1981.
UF cryotectonics
SA geomorphology
glacial geology
icequakes
tectonics

glacis
SA geomorphology
slopes

Glamorgan
County on Bristol Channel in S Wales.
CO N513000N514500
W0030500W0041500
UF Glamorganshire
BT Wales
Great Britain
United Kingdom

Glamorganshire
use Glamorgan

Glarus
Canton in E central.
BT Switzerland

Glarus Alps
N division of Central Alps. Chiefly in Glarus Canton. Also search Glarus.
BT Switzerland
SA Alps
Central Alps

Glasgow
City on Clyde River in S central.
BT Scotland
Great Britain
United Kingdom

glass
use glasses

glass, hydration of
use hydration of glass

Glass Mountains
Brewster County in SW.
BT Texas
United States

glass, volcanic
use volcanic glass

glasses
As of 1981, includes use on level 2 under igneous rocks(1). See list H. Before 1981, included use on level 3 under igneous rocks(1) pyroclastics and glasses(2). Also search glass. Before 1981, search pyroclastics and glasses.
UF glass
BA volcanic rocks
BT igneous rocks
NA obsidian
palagonite
pitchstone
volcanic glass
zirkelite
SA Darwin glass
devitrification
maskelynite
pyroclastics
tektites

glassy feldspar
use sanidine

glauberite
BA sulfates
BT minerals

glauconite
As of 1981, use glauconite deposits for glauconite as a commodity. Before 1981, included use on level 1 as a commodity term. Includes use on level 3 under minerals(1) sheet silicates, mica group(2). See list L.
BA mica group
sheet silicates
silicates
BT minerals
SA celadonite
clay mineralogy
glauconite deposits
glauconitic composition
glauconitization
greensand

glauconite deposits
Term introduced in 1981. Includes use on level 1 and 2 as a commodity term (list C). Before 1981, search glauconite AND deposits.
SA glauconite
glauconitization
mica deposits

glauconitic composition
Term introduced in 1978. Includes use on level 3 under sediments(1) and sedimentary rocks(1). Before 1978, also search glauconitic AND specific sediment type or sedimentary rock.
SA composition
glauconite
greensand
sedimentary rocks
sediments

glauconitic sand
A valid term through 1974. After 1974, use greensand.

glauconitic sandstone
use greensand

glauconitization
BT metamorphism
SA glauconite
glauconite deposits

glaucophane
Also search gastaldite.
UF gastaldite
BA clinoamphibole
amphibole group
chain silicates
silicates
BT minerals
SA aluminosilicates
crossite

glaucophane schist
Includes use on level 3 under metamorphic rocks(1) schists(2). See list J.

BA schists
BT metamorphic rocks
SA blueschist

Glavatsi
Region in NW.
BT Bulgaria
Europe

Gleicheniaceae
Family.
BA Filicopsida
pteridophytes
BT Plantae

Glen Coe
Glen in Argyll County in W Scotland.
UF Glencoe
BT Scotland
Great Britain
United Kingdom

Glen Rose Formation
In Trinity Group. Includes Lower Glen Rose Formation, Glen Rose Anhydrite, upper Glen Rose Formation. Also search Glen Rose Limestone.
BT Lower Cretaceous
Cretaceous
SA Texas
Trinity Group

Glenarm Series
Includes Wakefield Marble, Silver Run Limestone, Ijamsville Phyllite, Urbana Phyllite, Marburg Schist, Setters Quartzite, Cockeysville Marble, Wissahickon Schist, Peters Creek Schist, Cardiff Conglomerate, Peach Bottom Slate. Lower Paleozoic (?).
BT lower Paleozoic
Paleozoic
SA Maryland
New Jersey
Pennsylvania
Virginia
Wissahickon Formation

Glencoe
use Glen Coe

Glennie
Village in Alcona County in NE Lower Peninsula.
BT Michigan
United States

Gley soils
use Gleys

Gleys
See list M. Also search Gley soils.
UF Gley soils
BA soils
SA Pseudogleys

gliding (tectonics)
use gravity sliding

Glinsk
Village in SW Suny Oblast in N.
BT Ukraine
USSR

Gliridae
Family. As of 1981, includes use on level 3 under Mammalia(1) Myomorpha(2). Before 1981, included use on level 3 under Mammalia(1) Rodentia(2).
BA Myomorpha
Rodentia
Mammalia
BT Tetrapoda
Vertebrata

global
General term used in a geographic sense. Also search world.
UF world
SA world ocean

Globigerina
Genus. Includes use on level 3 under foraminifera(1) Globigerinacea(2).
BA Globigerinidae
Globigerinacea
Rotaliina
foraminifera
BT Protista
Invertebrata
NA Globigerina bulloides
Globigerina pachyderma

Globigerina bulloides
Includes use on level 3 under foraminifera(1) Globigerinacea(2).
BA Globigerina
Globigerinidae
Globigerinacea
Rotaliina
foraminifera
BT Protista
Invertebrata

Globigerina pachyderma
Includes use on level 3 under foraminifera(1) Globigerinacea(2).
BA Globigerina
Globigerinidae
Globigerinacea
Rotaliina
foraminifera
BT Protista
Invertebrata

Globigerinacea
Includes use on level 2 under foraminifera(1). See list F.
BA Rotaliina
foraminifera
BT Protista
Invertebrata
NA Globigerinidae
Globorotaliidae
Globotruncanidae
Hantkenina
Hedbergella
Heterohelicidae
Rotalipora

Globigerinidae
Family. Includes use on level 3 under foraminifera(1) Globigerinacea(2).
BA Globigerinacea
Rotaliina
foraminifera
BT Protista
Invertebrata
NA Globigerina
Globigerinoides
Orbulina
Sphaeroidinella dehiscens

Globigerinoides
Genus. Includes use on level 3 under foraminifera(1) Globigerinacea(2).
BA Globigerinidae
Globigerinacea
Rotaliina
foraminifera
BT Protista
Invertebrata
NA Globigerinoides ruber
Globigerinoides trilobus

Globigerinoides ruber
Includes use on level 3 under foraminifera(1) Globigerinacea(2).
BA Globigerinoides
Globigerinidae
Globigerinacea
Rotaliina
foraminifera
BT Protista
Invertebrata

Globigerinoides trilobus
Includes use on level 3 under foraminifera(1) Globigerinacea(2).
BA Globigerinoides
Globigerinidae
Globigerinacea
Rotaliina
foraminifera
BT Protista
Invertebrata

Globorotalia
Genus. Includes use on level 3 under foraminifera(1) Globigerinacea(2).
BA Globorotaliidae
Globigerinacea
Rotaliina
foraminifera
BT Protista
Invertebrata
NA Globorotalia menardii
Globorotalia pachyderma
Globorotalia truncatulinoides

Globorotalia menardii
Includes use on level 3 under foraminifera(1) Globigerinacea(2).
BA Globorotalia
Globorotaliidae
Globigerinacea
Rotaliina
foraminifera
BT Protista
Invertebrata

Globorotalia pachyderma
Includes use on level 3 under foraminifera(1) Globigerinacea(2).
BA Globorotalia
Globorotaliidae
Globigerinacea
Rotaliina
foraminifera
BT Protista
Invertebrata

Globorotalia truncatulinoides
Includes use on level 3 under foraminifera(1) Globigerinacea(2).
BA Globorotalia
Globorotaliidae
Globigerinacea
Rotaliina
foraminifera
BT Protista
Invertebrata

Globorotaliidae
Family. Includes use on level 3 under foraminifera(1) Globigerinacea(2).
BA Globigerinacea
Rotaliina
foraminifera
BT Protista
Invertebrata
NA Globorotalia

Globotruncana
Genus. Includes use on level 3 under foraminifera(1) Globigerinacea(2).
BA Globotruncanidae
Globigerinacea
Rotaliina
foraminifera
BT Protista
Invertebrata

Globotruncanidae
Family. Includes use on level 3 under foraminifera(1) Globigerinacea(2).
BA Globigerinacea
Rotaliina
foraminifera
BT Protista
Invertebrata
NA Globotruncana

glossaries
Includes use on level 1 (list A) for glossaries and atlases. Term set options are:
topic [List B (except areal geology), or extraterrestrial geology, general, geophysics]
subtopic
SA atlas
bibliography
catalogs
dictionaries
education
geology
lexicons
paleontology
sedimentary petrology
stratigraphy
structural geology
textbooks

Glossograptina
Includes use on level 2 under Graptolithina(1). See list F.
BA Graptoloidea
Graptolithina
BX Hemichordata
Invertebrata

Glossopteridales
Term introduced in 1981. Includes use on level 2 under gymnosperms(1). See list F.
BA gymnosperms
BT Spermatophyta
Plantae
NA Gangamopteris
Glossopteris

Glossopteris
Genus. As of 1981, includes use on level 3 under gymnosperms(1). See list F. Before 1981, included use on level 2 under gymnosperms(1).
BA Glossopteridales
gymnosperms
BT Spermatophyta
Plantae
NA Glossopteris flora
SA Dadoxylon
Pecopteris
Taeniopteris

Glossopteris flora
As of 1981, includes use on level 3 under gymnosperms(1) Glossopteridales(2). Before 1981, included use on level 3 under gymnosperms(1) Glossopteris(2). See list F.
BA Glossopteris
Glossopteridales
gymnosperms
BT Spermatophyta
Plantae

Gloucestershire
County in W England. Also search Gloucester.
CO N513000N521000
W0013000W0024000
BT England
Great Britain
United Kingdom

glowing avalanche
use ash flows

Glycymeris
Includes use on level 3 under Mollusca(1). See list F.
BA Bivalvia
Mollusca
BT Invertebrata

Glynn County
On the Atlantic Ocean in SE.
BT Georgia
United States

gmelinite
BA zeolite group
framework silicates
silicates
BT minerals

gneiss
As of 1981, no longer a valid term for GeoRef. Use gneisses. Before 1981, included use on level 3 under metamorphic rocks(1) gneisses(2).

G

gneisses
Includes use on level 2 under metamorphic rocks(1). See list J. Before 1981, also search gneiss.
BT metamorphic rocks
NA augen gneiss
banded gneiss
biotite gneiss
granite gneiss
orthogneiss
NZ paragneiss
NA sillimanite gneiss
tonalite gneiss

gneissic texture
Term introduced in 1978. Before 1978, search gneissic.
SA metamorphic rocks
textures

Gnetales
Includes use on level 2 under gymnosperms(1). See list F.
BA gymnosperms
BT Spermatophyta
Plantae

Goa
Former Portuguese possession on the Arabian Sea annexed by India in 1962. Territory of Goa, Daman. and Diu.
BT India

Gobi Desert
Index Mongolia and/or Inner Mongolia as applicable. Also search Gobi.
CO N420000N460000
E1170000E1020000
BT Asia
SA Inner Mongolia
Mongolia

Goczalkowice
Borehole region in Upper Silesian coal basin in S part of country.
BT Katowice
Poland

Godavari River
Rises in NW Maharashtra and flows SE into the Bay of Bengal. Index states as applicable. Also search Godavari.
BT India
SA Andhra Pradesh
Maharashtra

Godavari Valley
River valley. Index states as applicable. Also search Godavari.
BT India
SA Andhra Pradesh
Maharashtra
Pranhita-Godavari Valley

Godthaab
Town on SW coast.
UF Godthab
BT Greenland
Arctic region

Godthab
use Godthaab

goethite
UF gothite
xanthosiderite
BA oxides
BT minerals

Goias
State in central Brazil.
CO S193000S050500
W0460000W0532000
BT Brazil
NT Maranhao Basin
SA Brazilian Shield
Estrada Nova Formation
Irati Formation
Tocantins River region

Golan Heights
Hilly region in SW.
BT Syria
Middle East

gold
Includes use on level 1 and 2 as a chemical element (list D). As of 1981, use gold ores for gold as a commodity. Before 1981, included use on level 1 as a commodity term.
UF Au
SA calaverite
elements
gold ores
heavy metals
native elements

gold ores
Term introduced in 1981. Includes use on level 1 and 2 as a commodity term (list C). Before 1981, search gold AND ore deposits.
SA gold
native elements
placers

Golden
City in Jefferson County in N central.
BT Colorado
United States

Golden Horn Batholith
In the Cascade Range, Washington.
BX Cascade Range
Washington
BT United States

Golden Valley Formation
Includes White Earth, South Ross, Lakeside, and East Tioga Clay Beds. SW North Dakota.
BT Eocene
SA North Dakota

golden-brown algae
use Chrysophyta

Goldfield
Village in Esmeralda County in SW.
BT Nevada
United States

Golodnaya Step
use Bet-Pak-Dala

Golovanevsk
Village in Voronezh Oblast in S central European USSR.
BT Russian Republic
USSR

Gomera
Island belonging to Spain.
BT Canary Islands

gondite
See list J.
BA metamorphic rocks

Gondwana
Theoretical ancient continent including India, Australia, Antarctica; and parts of southern Africa and South America. It is supposed to have fragmented and drifted apart in Post-Carboniferous time. Index countries and continents as applicable. Includes use on level 2 under continental drift (1). Before 1974, also search Gondwanaland. Also search lower Gondwana.
SA Africa
Antarctica
Australia
continental drift
India
Laurasia
Pangaea
South America
Southern Hemisphere

Gondwana System
Upper Carboniferous to Lower Cretaceous. Peninsular.
SA Carboniferous
Cretaceous
India
Jurassic
Lower Cretaceous
Permian
Triassic
Upper Carboniferous

Gondwanaland
A valid term through 1973. After 1973, use Gondwana.

Goniatites
As of 1981, includes use on level 3 under Mollusca(1) Goniatitida(2). See list F. Before 1981, included use on level 3 under Mollusca(1) Cephalopoda(2).
BA Goniatitidae
Goniatitida
Ammonoidea
Tetrabranchiata
Cephalopoda
Mollusca
BT Invertebrata

Goniatitida
As of 1981, includes use on level 2 under Mollusca(1). See list F. Before 1981, included use on level 3 under Mollusca(1) Cephalopoda(2).
BA Ammonoidea
Tetrabranchiata
Cephalopoda
Mollusca
BT Invertebrata
NA Goniatitidae

Goniatitidae
As of 1981, includes use on level 3 under Mollusca(1) Goniatitida(2). See list F. Before 1981, included use on level 3 under Mollusca(1) Cephalopoda(2).
BA Goniatitida
Ammonoidea
Tetrabranchiata
Cephalopoda
Mollusca
BT Invertebrata
NA Goniatites

Gonzales County
Term introduced in 1981. S central Texas.
CO N290500N294500
W0970500W0975000
BT Texas
United States

Good Hope
use Fort Good Hope

Goodnews Quadrangle
Term introduced in 1981. Central Alaska.
CO N590000N600000
W1590000W1620000
BT Alaska
United States

Gorda Ridge
use Gorda Rise

Gorda Ridges
use Gorda Rise

Gorda Rise
Off N California E of the Mendocino Escarpment. Also search Gorda Ridge; Gorda Ridges.
UF Gorda Ridge
Gorda Ridges
BT Pacific Ocean

Gore Range
Part of Park Range in N central.
BT Colorado
United States

Gorki
City and oblast. Also search Gorky.
UF Gorkiy
Gorky
Nizhni Novgorod
BT Russian Republic
USSR

Gorkiy
use Gorki

Gorky
use Gorki

Gorlovka Basin
Industrial basin in the Donbas in E.
BT Ukraine
USSR

Gornaya Shoriya
Mountainous region in Kemerovo Oblast just S of Kuznetsk Basin.
BT Russian Republic
USSR

Gorny Altai
Mountain range in Altai Mountains of Gorno-Altai Autonomous Oblast.
CO N484500N523000
E0900000E0840000
BT Russian Republic
USSR
SA Altai Mountains

Goryachiy Plyazh
Beach on Kunashir Island of the Kuril Islands NE of Japan.
BT Russian Republic
USSR
SA Kunashir Island

Gosau
Village in NW part of country.
BT Upper Austria
Austria

Gosau Formation
Term introduced in 1981. Before 1981, also search Gosau Beds.
BT Cretaceous
SA Austria
Eastern Alps

Gosford
Town N of Sydney on the Tasman Sea.
BT New South Wales
Australia

Goshen County
On the Nebraska border in SE.
BT Wyoming
United States

Gosses Bluff
Meteor crater. Also search Gosses Bluff Meteor Crater.
UF Gosses Bluff Meteor Crater
BT Northern Territory
Australia

Gosses Bluff Meteor Crater
use Gosses Bluff

Gota Valley
River valley between Lake Vanern and Kattegat in SW.
BT Sweden

Goteborg
City on the Kattegat in SW Sweden.
UF Gothenburg
BT Sweden

Gothard
use Gotthard Massif

Gothenburg
use Goteborg

gothite
use goethite

Gothland
use Gotland

Gothlandian
use Silurian

Gotland
Island and county in Baltic Sea off SE coast.
CO N570000N580000
W0193000W0180000

UF Gothland
Gottland
BT Sweden

Gotlandian
use Silurian

Goto Islands
Island chain extending about 100 miles SW from NW Kyushu.
BT Kyushu
Japan

Gotthard Massif
Mountain range of the Lepontine Alps in SE central Switzerland. Also search Gotthard; Saint Gotthard; St. Gotthard.
UF Gothard
Saint Gotthard
St. Gotthard
BT Switzerland
SA Lepontine Alps

Gottingen
City in SE.
BT Lower Saxony
West Germany

Gottland
use Gotland

gouge
Includes use on level 3 under faults(1) effects(2).
SA breccia
faults

Gour Oumelalen
Region in NE Ahaggar in SE.
BT Algeria

Gouverneur
Village in Saint Lawrence County in N.
BT New York
United States

Gove County
NW central.
BT Kansas
United States

Gower Peninsula
Extends S into Bristol Channel from S.
BT Wales
Great Britain
United Kingdom

Gowganda
Village N of Georgian Bay.
BT Ontario
Canada

Gowganda Formation
Consists of paraconglomerate, argillite, siltstone, subarkose, and greywacke. W Ontario.
BT Huronian
Proterozoic
Precambrian
SA Canada
Ontario

grabens
Includes use on level 3 under faults(1) systems(2).
SA aulacogens
faults
horsts
systems

grade
Includes use on level 2 under metamorphism(1).
UF metamorphic grade
SA facies
high-grade metamorphism
isograds
low-grade metamorphism
metamorphism
P-T conditions

graded bedding
Includes use on level 3 under sedimentary structures(1) turbidity current structures(2) and planar bedding structures(2). See list K.
BA turbidity current structures
sedimentary structures
SA bedding
planar bedding structures
varves

gradient
A valid term through 1978. After 1978 see geothermal gradient, or stream gradient for streams.

graduate-level education
Term introduced in 1978. Before 1978, also search graduate level.
SA education

Grafton County
W New Hampshire.
BT New Hampshire
United States

graftonite
BA phosphates
BT minerals

Graham Coast
use Graham Land

Graham County
Index states as applicable.
BX Arizona
Kansas
North Carolina
BT United States

Graham Land
Part of N Antarctic Peninsula S of Cape Horn.
UF Graham Coast
BT Antarctica

grahamite
use mesosiderites

grain size
Includes use on level 3, e.g. under sediments(1). Before 1981, also search size.
SA dimensions
fines
granulometry
sediments
size distribution

grain-size analysis
use granulometry

grains
Includes use on level 3 under sedimentary rocks(1).
SA dimensions
granulometry
particles
recrystallization
sand
sedimentary rocks
sediments
shape analysis
sorting

grainstone
See list I.
BA carbonate rocks
BT sedimentary rocks

Graisivaudan
use Gresivaudan

gramenite
use nontronite

Gramineae
Family.
BA Monocotyledoneae
angiosperms
BT Spermatophyta
Plantae

Grampian Highlands
Term introduced in 1981.
CO N551500N574500
W0014500W0055000
BT Scotland
SA Northern Highlands

Gran Canaria
use Grand Canary

Gran Chaco
use Chaco

Granada
As of 1981, refers to only the city. Before 1981, also included the province. In S Spain.
BT Granada Province
Andalusia
Spain

Granada Depression
S part of country.
BT Granada Province
Andalusia
Spain

Granada Province
Term introduced in 1981. Before 1981, search Granada. In S Spain.
CO N364000N380500
W0021100W0041800
BT Andalusia
Spain
NT Granada
Granada Depression

Grand Atlas Mountains
use High Atlas

Grand Banks
Shoal or banks E and S of Newfoundland.
BT Atlantic Ocean

Grand Canary
One of Spain's Canary Islands NW of Western Sahara. Also search Gran Canaria.
UF Gran Canaria
BT Canary Islands

Grand Canyon
Gorge in Colorado River in NW Arizona. Also search Grand Canyon region.
UF Grand Canyon region
BT Arizona
United States

Grand Canyon region
use Grand Canyon

Grand Cayman Island
Largest of the Cayman Islands. About 200 miles NW of Jamaica. Also search Grand Cayman.
BT Cayman Islands

Grand County
Index states as applicable.
BX Colorado
Utah
BT United States

Grand Falls
use Churchill Falls

Grand Forks
City on the Red River of the North in Grand Forks County in E.
BT North Dakota
United States

Grand Isle
Island in Jefferson Parish off SE coast.
BT Louisiana
United States

Grand Junction
City in Mesa County in W.
BT Colorado
United States

Grand Manitoulin
use Manitoulin Island

Grand Rapids
City in Kent County in W Lower Peninsula.
BT Michigan
United States

Grand River
River in numerous locations. Index Ontario and states as applicable.
SA Louisiana
Michigan
Missouri
Oklahoma
Ontario
South Dakota

Grand Traverse County
NW Lower Peninsula.
BT Michigan
United States

Grandfather Mountain
In the Blue Ridge Mountains in NW.
BT North Carolina
United States
SA Blue Ridge Mountains

grandidierite
BA orthosilicates
silicates
BT minerals

Graneros Shale
In Colorado Group. E Colorado, Kansas, SE Montana, Nebraska, NE New Mexico, South Dakota, and E Wyoming.
BT Cretaceous
SA Benton Formation
Colorado
Colorado Group
Kansas
Lower Cretaceous
Montana
Nebraska
New Mexico
South Dakota
Upper Cretaceous
Wyoming

granite
Includes use on level 3. As of 1981, not to be used in the igneous rocks set. See list H. Before 1981, included use on level 3 under igneous rocks(1) granite-granodiorite family(2) and on level 1 as a commodity term (list C). As of 1981, use granite deposits for granite as a commodity. Also search granitic rocks.
BA granites
BT igneous rocks
SA albite
apogranite
biotite granite
granite deposits
granitic composition
granitization
leucogranite
metagranite
microgranite
muscovite granite

Granite County
W Montana.
BT Montana
United States

granite deposits
Term introduced in 1981. Includes use on level 1 and 2 as a commodity term (list C). Before 1981, search granite AND deposits.
SA building stone
construction materials
dimension stone
granite
granitization

granite gneiss
Includes use on level 3 under metamorphic rocks(1) gneisses(2). See list J. Also search granitic gneiss.
UF granitic gneiss
BA gneisses
BT metamorphic rocks

Granite Mountains
Central.

BT Wyoming
United States

granite porphyry
As of 1981, includes use on level 3 under igneous rocks(1) granites(2). See list H. Before 1981, included use on level 3 under igneous rocks(1) granite-granodiorite family(2).
BA granites
BT igneous rocks
SA porphyry

granite-granodiorite family
As of 1981, no longer a valid term for GeoRef. Use granites, alkali granites, or granodiorites as applicable. See list H. Before 1981, included use on level 2 under igneous rocks(1).

granites
Term introduced in 1981. Includes use on level 2 under igneous rocks(1). See list H. Before 1981, search granite-granodiorite family.
BT igneous rocks
NA adamellite
alaskite
alkali granites
aplite
apogranite
biotite granite
charnockite
enderbite
farsundite
felsite
granite
granite porphyry
granodiorites
NZ granosyenite
NA graphic granite
leucogranite
microgranite
micropegmatite
muscovite granite
pegmatite
quartz monzonite
rapakivi

granitic
A valid term through 1977. After 1977, use granitic composition or granitic layer.

granitic composition
Term introduced in 1978. Before 1978, also search granitic; granitic rocks.
SA composition
granite
granitic layer

granitic gneiss
use granite gneiss

granitic layer
Term introduced in 1978. Before 1978, also search granitic; sial.
UF layer, granitic
sial
SA basaltic layer
granitic composition
granitic layer
lower crust

granitic rocks
Not a valid term. Use granite or granitic composition.

granitification
use granitization

granitization
Includes use on level 3 under metasomatism(1) processes(2) and under metamorphism(1).
UF granitification
BA metamorphism
SA crystallization
feldspathization
granite
granite deposits
metasomatism
processes

granitoid
A valid term through 1978. After 1978, use granite or granitic composition.

granodiorite
Includes use on level 3. As of 1981, not to be used in the igneous rocks set. See list H. Before 1981, included use on level 3 under igneous rocks(1) granite-granodiorite family(2).
BA granodiorites
granites
BT igneous rocks
SA plutonic rocks

granodiorite porphyry
As of 1981, includes use on level 3 under igneous rocks(1) granodiorites(2). See list H. Before 1981, included use on level 3 under igneous rocks(1) granite-granodiorite family(2).
BA granodiorites
granites
BT igneous rocks
SA porphyry

granodiorites
Term introduced in 1981. Includes use on level 2 under igneous rocks(1). See list H. Before 1981, search granite-granodiorite family.
BA granites
BT igneous rocks
NA granodiorite
granodiorite porphyry

granophyre
See list H.
BA volcanic rocks
BT igneous rocks

granosyenite
As of 1981, includes use on level 3 under igneous rocks(1) syenites(2) and granites(2). See list H. Before 1981, included use on level 3 under igneous rocks(1) granite-granodiorite family(2) and syenite family(2).
BZ granites
syenites
BT igneous rocks

Grant County
Index states as applicable.
BX Arkansas
Indiana
Kansas
Kentucky
Minnesota
Nebraska
New Mexico
North Dakota
Oklahoma
Oregon
South Dakota
Washington
West Virginia
Wisconsin
BT United States
SA Hillsboro Quadrangle
San Lorenzo Quadrangle

Grant Range
NE Nye County in central.
BT Nevada
United States

Grants
Town in Valencia County in W. New Mexico.
BT New Mexico
United States

Grants District
Term introduced in 1981. Uranium-mining district in McKinley County, New Mexico.
BT McKinley County
New Mexico
SA Grants mineral belt

Grants mineral belt
Term introduced in 1981. Mineralized region in NW New Mexico.
BT New Mexico
United States
SA Ambrosia Lake District
Grants District

granular materials
Includes use on level 3 under soil mechanics(1). Also search granular.
SA engineering geology
granulometry
materials
particulate materials
soil mechanics

granulite
As of 1981, no longer a valid term for GeoRef. Use granulites. Before 1981, included use on level 3 under metamorphic rocks(1) granulites(2).

granulite facies
Includes use on level 3 under metamorphic rocks(1). See list J.
BT facies
SA metamorphic rocks

granulites
Includes use on level 2 under metamorphic rocks(1). See list J. Also search granulite.
BT metamorphic rocks
NA kinzigite
leptite
pyroxene granulite

granulometry
UF grain-size analysis
SA grain size
grains
granular materials
particles
sedimentation
sediments
shape analysis
size distribution

grapestone
Used to indicate a texture.
BA sedimentary rocks

graphic display
Before 1978, also search graphic AND display.
UF display, graphic
SA automatic data processing

graphic granite
As of 1981, includes use on level 3 under igneous rocks(1) granites(2). See list H. Before 1981, included use on level 3 under igneous rocks(1) granite-granodiorite family(2).
BA granites
BT igneous rocks
SA micropegmatite
pegmatite

graphic methods
Before 1978, also search graphic.
SA histograms
methods

graphic texture
Term introduced in 1978. Before 1978, search graphic.
SA igneous rocks
textures

graphite
As of 1981, use graphite deposits for graphite as a commodity. As of 1981, includes use on level 3 under minerals(1) native elements(2). See list L. Before 1981, included use on level 3 under minerals(1) native elements and alloys(2) and on level 1 as a commodity term.
UF black lead
BA native elements
BT minerals
SA carbon
chaoite
diamond
graphite deposits
lonsdaleite

graphite deposits
Term introduced in 1981. Includes use on level 1 and 2 as a commodity term (list C). Before 1981, search graphite AND deposits.
SA graphite
industrial minerals
refractory materials

graptolites
Term introduced in 1981 as the common name for Graptolithina. See list F. To be used for Graptolithina when they are used as tools in stratigraphy. Includes use on level 1. Before 1981, search Graptolithina AND biostratigraphy. If 1, term set options are:
biostratigraphy
age [List E]
SA biostratigraphy
Graptolithina

Graptolithina
Includes use on level 1 and 2 as a fossil term (list F).
BX Hemichordata
Invertebrata
NA Camaroidea
Dendroidea
Graptoloidea
Stolonoidea
Tuboidea
SA graptolites

Graptoloidea
Includes use on level 2 under Graptolithina(1). See list F.
BA Graptolithina
BX Hemichordata
Invertebrata
NA Didymograptina
Diplograptina
Glossograptina
Monograptina

grasslands
SA ecology
vegetation
wetlands

Gratiot County
Term introduced in 1981. Central Michigan.
CO N430600N432900
W0842200W0845100
BT Michigan
United States

gratonite
BA sulfarsenites
sulfosalts
BT minerals

Gratz
use Graz

Graubunden
Canton in E Switzerland. Also search Grisons.
UF Grisons
BT Switzerland
NT Davos
Oberhalbstein
Tavetsch
SA Oberalp Pass

grauwacke
use graywacke

gravel
Includes use on level 3 under sediments(1) clastic sediments(2). See

list N. As of 1981, use gravel deposits for gravel as a commodity. Before 1981, included use on level 1 as a commodity term.
BA clastic sediments
BT sediments
SA alluvium
boulders
breccia
carbonate sediments
cobbles
gravel deposits
pebbles
sand
shingle
terrigenous materials

gravel deposits
Term introduced in 1981. Includes use on level 1 and 2 as a commodity term (list C). Includes sand when sand is used as a construction material. Before 1981, search gravel AND deposits.
SA aggregate
construction materials
gravel
sands

gravimeters
Includes use on level 3 under geophysical methods(1). Also search gravity meters.
UF gravity meters
NT Askania gravimeters
LaCoste-Romberg gravimeters
marine gravimeters
pendulum gravimeters
quartz-fiber gravimeters
spring gravimeters
vibrating-string gravimeters
Western gravimeters
Worden gravimeters
SA geophysical methods
gravity methods
instruments

gravitational gliding
use gravity sliding

gravitational sliding
use gravity sliding

gravity
A valid term through mid-1978. Use specific term, e.g. gravity field or gravity platforms. For gravity faults, use the term normal faults.

gravity anomalies
Includes use on level 3 under geophysical methods(1). Also search gravity surveys AND anomalies; gravity methods AND anomalies.
NA Bouguer anomalies
SA anomalies
geophysical methods
geophysical surveys
gravity methods
gravity surveys
magnetic anomalies

gravity faults
use normal faults

gravity field
Includes use on level 2 under Earth(1) and Moon(1).
UF field, gravity
SA Earth
geodesy
geophysical surveys
Moon

gravity gliding
use gravity sliding

gravity meters
use gravimeters

gravity methods
Includes use on level 2 under geophysical methods(1).
BA geophysical methods
SA free-air anomalies
gravimeters
gravity anomalies
gravity survey maps
gravity surveys
methods

gravity platforms
Term introduced in 1978 used in engineering geology(1). Before 1978, search gravity AND platforms.
BT marine installations
SA marine platforms
platforms

gravity sliding
Includes use on level 2 under tectonics(1). Also search sliding.
UF ecoulement
gliding (tectonics)
gravitational gliding
gravitational sliding
gravity gliding
sliding, gravity
SA crustal shortening
tectonics

gravity survey maps
Term introduced in 1978. Also search maps AND gravity surveys.
BA maps
SA geophysical surveys
gravity methods
gravity surveys

gravity surveys
Includes use on level 3 under area terms(1) geophysical surveys(2).
BA geophysical surveys
BT surveys
SA gravity anomalies
gravity methods
gravity survey maps

gray antimony
use jamesonite

Gray desert soil
use Sierozems

Gray earth
use Sierozems

Gray forest soils
Also search Grey forest soils.
UF Grey forest soils
BA soils

gray hematite
use specularite

Grays Harbor
Inlet of Pacific Ocean on SW coast of Grays Harbor County.
BT Washington
United States

graywacke
Includes use on level 3 under sedimentary rocks(1) clastic rocks(2). See list I. Also search greywacke.
UF grauwacke
greywacke
BA clastic rocks
BT sedimentary rocks
SA metagraywacke
sandstone
subgraywacke
terrigenous materials

Graz
City in Styrian Alps in SE part of country.
UF Gratz
BT Styria
Austria

Great Alfold
use Alfold

Great Appalachian Valley
Term introduced in 1977. Before 1977, also search Great Valley. A longitudinal chain of lowlands of the Appalachians extending from Canada on NE to Alabama on SW.
BT North America
SA Appalachians

Great Artesian Basin
Index Northern Territory and states as applicable.
CO S200000S180000
E1490000E1370000
BT Australia
SA New South Wales
Northern Territory
Queensland
South Australia

Great Australian Bight
Bay on S coast of Australia.
BT Indian Ocean

Great Bahama Bank
Large shoal in the Bahamas SE of Miami and extending between Cuba and Andros Island.
BT Atlantic Ocean

Great Barrier Reef
Largest coral reef in the world extending 1250 miles off NE Queensland, Australia.
CO S230000S100000
E1530000E1430000
BT Coral Sea
Pacific Ocean

Great Basin
Interior region between Sierra Nevada Mountains and S Cascade Range on W, and Wasatch Range and W face of Colorado Plateau on E. Introduced as a level 1 area term in 1978. Index states as applicable.
CO N350000N430000
W1110000W1200000
BA United States
North America
SA Basin and Range Province
Battle Mountain High
California
Idaho
Nevada
Oregon
Utah

Great Bay
Inland tidal bay in SE New Hampshire and inlet of the Atlantic Ocean in SE New Jersey.
BX New Hampshire
New Jersey
BT United States

Great Bear Lake
NW central District of Mackenzie.
BT Northwest Territories
Canada

Great Britain
Index political divisions as applicable. Includes use on level 1 as an area term (list O). For term set options see list B.
CO N500000N580000
E0013000W0080000
BA United Kingdom
Western Europe
Europe
NT Cheviot Hills
NA England
Scotland
Wales
SA Caledonides
Gault
Upper Old Red Sandstone

Great Dyke
Part of great South African plateau.
BT Zimbabwe

Great Glen Fault
Depression extending along ancient fault NE-SW across Scotland from Moray Firth to Loch Linnhe. Part of Caledonian Canal.
BT Scotland
Great Britain
United Kingdom

Great Hungarian Plain
use Alfold

Great Kabylia
W part of Kabylia which is a mountainous coastal region.
BT Algeria
SA Kabylia

Great Khingan Range
use Khingan Range

Great Lake
Largest lake on the island in central.
BT Tasmania
Australia

Great Lakes
Largest group of fresh-water lakes in the world. Index lakes as applicable. Includes use on level 1 as an area term (list O). For term set options see list B.
CO N414000N490000
W0760000W0922000
BA North America
NA Lake Erie
Lake Huron
Lake Michigan
Lake Ontario
Lake Superior
SA Canada
Great Lakes region
United States

Great Lakes region
Index Great Lakes, states, and provinces as applicable. Also search Great Lakes. Includes use on level 1 as an area term (list O). For term set options see list B.
CO N410000N500000
W0750000W0930000
BA North America
SA Canada
Great Lakes
Illinois
Indiana
Michigan
Minnesota
New York
Niagara Escarpment
Ohio
Ontario
Pennsylvania
United States
Wisconsin

Great Meteor Seamount
About 800 miles S of Azores.
BT Atlantic Ocean

Great Oolite Series
Consists of upper Estuarine Series, Great Oolite Limestone, Great Oolite Clays and Cornbrash. Oxfordshire and Gloucestershire in W central England.
BT Bathonian
Middle Jurassic
Jurassic
SA England

Great Plains
A level 1 area term as of 1978. Sloping plateau extending from Rocky Mountains in W to the margin of the Central Plains in the U. S. and to the margin of the Laurentian Highlands in Canada. Index states and provinces as applicable. Until 1978, High Plains was used for the Great Plains from Nebraska southward. Now use Great Plains, but also search High Plains.
CO N290000N560000
W0970000W1150000
BA North America
NA Northern Great Plains

Southern Great Plains
SA Alberta
Canada
Colorado
Kansas
Montana
Nebraska
New Mexico
North Dakota
Oklahoma
Saskatchewan
South Dakota
Texas
United States
Wyoming

Great Rift Valley
A great fault system extending from the Sea of Galilee in SW Asia to Mozambique. Index East African Rift, and SW Asian countries, lakes, gulfs and seas. Also search Rift Valley.
UF Rift Valley
SA Afar Depression
Dead Sea
East African Rift
Gulf of Aden
Gulf of Aqaba
Israel
Jordan
Jordan Valley
Kenya Rift valley
Lake Baringo
Red Sea
Sea of Galilee
Syria
Wadi Araba

Great Salt Lake
In NW near Salt Lake City.
CO N404000N414000
W1121000W1125500
BT Utah
United States

Great Scar Limestone
Overlain by the Yoredale Series. Yorkshire in N England.
BT Visean
Dinantian
Carboniferous
SA England

Great Slave Lake
S Mackenzie District.
CO N610000N630000
W1100000W1170000
BT Northwest Territories
Canada

Great Smokies
use Great Smoky Mountains

Great Smoky Mountains
Range of the Appalachians. Index states as applicable.
UF Great Smokies
Smokies
BT United States
SA Appalachians
North Carolina
Tennessee

Great South Bay
Long narrow inlet of Atlantic Ocean between Fire Island and S shore of Long Island.
BT New York
United States

Great Valley
A valid index term through 1977. After 1977, see entry for Central Valley in California, and see entry for Great Appalachian Valley.

Great Valley Sequence
In California.
BT Mesozoic
SA California
Cretaceous
Jurassic
Upper Cretaceous
Valley and Ridge Province

Greater Antilles
Includes the major islands of the Antilles. As of 1977, includes use as a level 1 area term (list O). See list B for term set options. Index islands as applicable.
CO N173000N273000
W0650000W0850000
BA West Indies
NA Cuba
NT Hispaniola
NA Jamaica
Puerto Rico
SA Antilles
Caribbean region

Greater Caucasus
The main range of the Caucasus separating the Northern Caucasus from the Colchis and Kura Lowlands. Index Soviet republics as applicable.
CO N390000N443000
E0503000E0370000
BA Caucasus
BT USSR
SA Azerbaidzhan
Georgian Republic
Russian Republic
Sunzha

Greater Khingan Mountains
use Khingan Range

Greater Khingan Range
use Khingan Range

Greece
Includes use on level 1 as an area term (list O). For term set options see list B.
CO N345500N414500
E0284500E0193000
BA Southern Europe
Europe
NT Aegina
Attica
Boeotia
Crete
Epirus
Euboea
Greek Aegean Islands
Greek Ionian Islands
NX Greek Macedonia
Greek Thrace
NT Hellenides
Kremasta
Laurion
Othrys
Parnassus
Peloponnesus
Pindus Mountains
Rhodope
Salonika
Sterea Ellas
Thessaly
Vourinos
SA Athens
Balkan Peninsula
Macedonia
Maritsa River
Mediterranean region
Near East
Rhodope Mountains
Struma River valley
Thrace
Vardar River

Greek Aegean Islands
Term introduced in 1981 for the Greek islands in the Aegean Sea.
CO N352000N405000
E0283000E0230000
BT Greece
NT Chios
Cyclades
Dodecanese
Karpathos
Kos
Lesbos
Samos
Santorin
SA Aegean Islands

Greek Ionian Islands
Term introduced in 1981 for the islands off the W coast of Greece, including Kithira (Cerigo).
CO N360500N395500
E0231000E0192000
BT Greece
NT Cephalonia
SA Ionian Islands
Ithaca

Greek Macedonia
Term introduced in 1981. Before 1981, search Macedonia AND Greece.
CO N395000N413000
E0244000E0204500
BZ Macedonia
BX Greece

Greek Thrace
Term introduced in 1981.
CO N404000N414500
E0264000E0243000
BZ Thrace
BX Greece

Greeley
City in Weld County in N.
BT Colorado
United States

green algae
use Chlorophyta

Green Bay
Inlet of NW Lake Michigan. Index Michigan and/or Wisconsin as applicable. Also a city in Brown County, Wisconsin.
BT United States
SA Michigan
Wisconsin

Green Gully
Near Keilor just NE of Melbourne.
BT Victoria
Australia

Green Lake
Green Lake County in S central.
BT Wisconsin
United States

Green Mountains
Range of the Appalachians. Index Quebec and U.S. states as applicable.
SA Appalachians
Massachusetts
Quebec
Vermont

Green River
Rises in the Wind River Range in W central Wyoming and flows S into the Colorado River N of Lake Powell. Index states as applicable.
BT United States
SA Colorado
Utah
Wyoming

Green River basin
Index states as applicable.
BT United States
NX Sand Wash Basin
SA Colorado
Utah
Wyoming

Green River Formation
Subdivided into Tipton Tongue, Laney Shale Member, and Morrow Creek Member. NW and central W Colorado, E Utah, and SW Wyoming. Also search Green River Shale.
BT Eocene
SA Colorado
Lake Gosiute
Laney Shale Member
lower Eocene
middle Eocene
Utah
Wyoming

green tuff
As of 1981, includes use on level 3 under igneous rocks(1) pyroclastics(2). See list H. Before 1981, included use on level 3 under igneous rocks(1) pyroclastics and glasses(2).
BA pyroclastics
volcanic rocks
BT igneous rocks
SA tuff

Green Tuff Formation
Underlies Daijima Formation.
BT Miocene
SA Hokkaido
Honshu
Japan
Kyushu

Green tuff region
BT Japan

Greenbrier County
SE West Virginia.
BT West Virginia
United States

Greene County
Index states as applicable.
BX Alabama
Arkansas
Georgia
Illinois
Indiana
Iowa
Mississippi
Missouri
New York
North Carolina
Ohio
Pennsylvania
Tennessee
Virginia
BT United States

Greenhorn Limestone
Of Colorado Group. Includes Lincoln Limestone, Jetmore Chalk, Hartland Shale, Pfeifer Shale, Bridge Creek Limestone Member, Orman Lake Limestone Member.
BT Upper Cretaceous
Cretaceous
SA Benton Formation
Colorado
Colorado Group
Kansas
Montana
Nebraska
New Mexico
South Dakota
Wyoming

Greenland
Largest island in the world and a Danish province. Includes use on level 1 as an area term (list O). For term set options see list B.
CO N600000N840000
W0200000W0700000
BA Arctic region
NT Agto
Camp Century
Disko Island
NA East Greenland
NT Fiskenaesset
Frederikshab
Godthaab
Greenland ice sheet
Igaliko
Ivigtut
Jameson Land
Kuhn Island
Milne Land
Nugssuaq

Peary Land
Scoresby Land
Scoresby Sound
Skaergaard Intrusion
Sondre Strom Fjord
NA South Greenland
NT Thule
Umanak
NA West Greenland
SA Ilimaussaq

Greenland Gap Group
BT Upper Devonian
Devonian
SA Frasnian
Maryland
Virginia
West Virginia

Greenland ice sheet
Located in basin surrounded by coastal mountains covering roughly 85% of the island. Also search Greenland.
BT Greenland
Arctic region

Greenland Sea
Section of Arctic Ocean between NE coast of Greenland and Spitsbergen.
BT Arctic Ocean
NT Icelandic Plateau
SA Jan Mayen

Greenlee County
Term introduced in 1981. SE Arizona.
CO N322500N334800
W1090300W1093000
BT Arizona
United States

greenockite
UF cadmium blende
cadmium ocher
xanthochroite
BA sulfides
BT minerals

greensand
Includes use on level 3 under sedimentary rocks(1) clastic rocks(2) and under sediments(1) carbonate sediments(2) and clastic sediments(2). Before 1975, also search glauconitic sand.
UF glauconitic sandstone
SA clastic rocks
clastic sediments
glauconite
glauconitic composition
sand
sandstone
sedimentary rocks

greenschist
Includes use on level 3 under metamorphic rocks(1) schists(2). See list J. Also search prasinite.
UF prasinite
BA schists
BT metamorphic rocks
SA greenschist facies

greenschist facies
Includes use on level 3 under metamorphic rocks(1).
BT facies
SA blueschist facies
greenschist
metamorphic rocks

greenstone
Includes use on level 3 under metamorphic rocks(1) schists(2). See list J. Also used as mineral.
BA schists
BT metamorphic rocks
SA greenstone belts
nephrite

greenstone belts
UF belts, greenstone
SA greenstone

Greenwood County
Index states as applicable.
BX Kansas
South Carolina
BT United States

Gregory Rift
That part of the East African Rift extending SW from Lake Turkana into NW central Tanzania. Index countries as applicable. Also search Gregory Rift valley.
UF Gregory Rift valley
BT Africa
SA East African Rift
Kenya
Tanzania

Gregory Rift valley
use Gregory Rift

greigite
Also search melnikovite.
UF melnikovite
BA sulfides
BT minerals

greisen
Compositional term. As of 1978, term is used on level 2 under metasomatic rocks(1).
BA metasomatic rocks
NA beresite
SA greisenization

greisenization
Includes use on level 3 under metasomatism(1).
BA metasomatism
SA greisen
hydrothermal alteration

Grenada
Island, southernmost of Windward Islands in British West Indies.
BA Windward Islands
BT Lesser Antilles
SA Carriacou

Grenoble
City in SE part of country.
BT Isere
France

Grenville
County in SE Ontario and town on E coast of Grenada Island in West Indies. Index Ontario or West Indies as applicable. Until 1977, used to indicate provincial series. Now see Grenvillian Orogeny for the orogeny.
SA Grenvillian Orogeny
Ontario
West Indies

Grenville Front
Index provinces as applicable.
BT Canada
SA Ontario
Quebec

Grenville Orogeny
use Grenvillian Orogeny

Grenville Province
Index New York, Labrador, Newfoundland, and provinces as applicable.
CO N430000N550000
W0560000W0820000
BT Canadian Shield
SA Labrador
New York
Newfoundland
Ontario
Quebec

Grenvillian
use Grenvillian Orogeny

Grenvillian Orogeny
Before 1978, also search Grenvillian; Grenville AND orogeny. Now Grenville is used as a place name.
UF Grenville Orogeny
Grenvillian
BT Precambrian
SA Grenville
orogeny

Gresivaudan
Alpine glacial trough in SE part of country.
UF Graisivaudan
BT Isere
France

Grey forest soils
use Gray forest soils

greywacke
use graywacke

Grimmertingen
Sands.
BT Oligocene
SA Belgium

Grindelwald
Term introduced in 1981. Valley in the Bernese Alps, including the town of Grindelwald, S central Switzerland.
CO N463700N464000
E0080500E0080000
BT Bernese Alps
Switzerland
SA Bern

grinding
BT erosion
SA abrasion

griquaite
As of 1981, includes use on level 3 under igneous rocks(1) ultramafics(2). See list H. Before 1981, included use on level 3 under igneous rocks(1) ultramafic family(2).
BA ultramafics
BT igneous rocks

Grisons
use Graubunden

groins
BT marine installations
SA jetties

Groix
Groix Island was used through 1978. In Bay of Biscay off Brittany coast. Before 1978, also search Groix Island.
UF Groix Island
BT Morbihan
France

Groix Island
use Groix

Groningen
Province in NE Netherlands. Also a city.
BT Netherlands

groove casts
Includes use on level 3 under sedimentary structures(1) bedding plane irregularities(2). See list K.
UF casts, groove
BA bedding plane irregularities
sedimentary structures
SA grooves

grooves
Includes use on level 3 under sedimentary structures(1) bedding plane irregularities(2). See list K.
BA bedding plane irregularities
sedimentary structures
SA channels
groove casts
striations

Grosseto
City and province in S.
BT Tuscany
Italy

grossular
Also search grossularite.
UF grossularite
BA garnet group
nesosilicates
orthosilicates
silicates
BT minerals
SA hydrogrossular
vanadium garnet

grossularite
use grossular

ground
A valid term through 1977. After 1977, use ground methods.

ground, frozen
use frozen ground

ground ice
As of 1978, term is used on level 2 under permafrost(1).
SA ice
permafrost
thermokarst

ground mass
use matrix

ground methods
Term introduced in 1978. Includes use on level 3 under geophysical methods(1). Before 1978, also search ground AND methods.
BA geophysical methods
SA geophysical surveys
methods

ground motion
Includes use on level 3 under geologic hazards(1) earthquakes(2) or under earthquakes(1).
UF motion, ground
SA earthquakes
geologic hazards
seismology
strong motion
surface waves

ground, patterned
use patterned ground

ground truth
SA remote sensing

ground water
For economically oriented papers, see water resources. This term used for subsurface waters, especially for studies on specific areas. Includes use on level 1 (list A) and 2. If 1, term set options are:
topic [age, aquifers, artesian waters, composition, connate waters, genesis, geochemistry, hydrodynamics, levels, models, movement, pollution, recharge, salt-water intrusion, surveys, systems analogs]
(area)
subtopic (if no area is given)
SA aquifers
artesian waters
artificial recharge
catchment hydrodynamics
connate waters
Darcy's law
desalinization
development
discharge
dispersion
drawdown
electro-osmosis
engineering geology
environmental geology
field capacity
fresh water
geochemistry
geophysical surveys
geysers
hot springs
hydraulics

G

hydrochemistry
hydrodynamics
hydrogeology
hydrology
hydrosphere
impurities
infiltration
infiltration galleries
isotopes
juvenile water
levels
liquid waste
mineral waters
movement
multiphase flow
percolation
permeability
pollution
porosity
potentiometric surface
pump tests
purification
recharge
representative basins
saline composition
salt water
salt-water intrusion
saturated zone
seepage
springs
storage
surface water
surveys
systems analogs
thermal waters
transmissivity
unconfined aquifers
unsaturated zone
waste disposal
water
water hardness
water quality
water recovery
water regimes
water resources
water table
well screens

ground-water increment
use recharge

ground-water movement
A valid term through 1972. After 1972, see movement(2) under ground water(1).

ground-water recharge
use recharge

ground-water replenishment
use recharge

groundmass
use matrix

group velocity
Term introduced in 1981.
SA elastic waves
seismology

grout
use grouting

grouting
Includes use on level 3 under foundations(1). Also search grout.
UF grout
SA dams
engineering geology
foundations

groutite
BA oxides
BT minerals

growth
As of 1978, term is restricted to use for biological growth. For crystals, use crystal growth.
SA crystal growth

growth faults
Before 1978, also search flexure AND faults; contemporaneous faults; slump faults.
UF contemporaneous faults
depositional fault
flexure faults
sedimentary fault
BA faults
SA flexure

growth spirals
Includes use on level 3 under crystal growth(1).
UF spirals, growth
SA crystal growth
crystal structure
crystallography

Grozny
City in Chechen-Ingush A.S.S.R. in Northern Caucasus.
BT Russian Republic
USSR

Grudziadz
City in N central part of country.
BT Bydgoszcz
Poland

Grundy County
Index states as applicable.
BX Illinois
Iowa
Missouri
Tennessee
BT United States

Gruneisen parameters
Term introduced in 1981.
SA seismology

grunerite
BA clinoamphibole
amphibole group
chain silicates
silicates
BT minerals

Gryphaea
Genus. Includes use on level 3 under Mollusca(1). See list F.
BA Bivalvia
Mollusca
BT Invertebrata

Grzybow
Sulfur mining region in S.
BT Poland

Guadalajara
As of 1981, refers to only the city in Spain. Before 1981, also included the province in Spain and the city in Mexico. In central Spain.
BT Guadalajara Province
New Castile
Spain

Guadalajara City
Term introduced in 1981 for the city in Mexico. Before 1981, also search Guadalajara AND Mexico.
BT Jalisco
Mexico

Guadalajara Province
Term introduced in 1981. Before 1981, search Guadalajara AND Spain. In central Spain.
CO N401000N412000
W0013200W0033500
BT New Castile
Spain
NT Guadalajara
SA Serrania de Cuenca

Guadalquivir
River in S Spain. Also search Guadalquivir Basin.
UF Guadalquivir River
BT Spain
SA Guadalquivir Basin

Guadalquivir Basin
River basin in S and SW Spain. Also search Guadalquivir; Gualdalquivir River; Guadalquivir River basin.
UF Guadalquivir River basin
BT Spain
SA Guadalquivir

Guadalquivir River
use Guadalquivir

Guadalquivir River basin
use Guadalquivir Basin

Guadalupe County
Term introduced in 1981. Index states as applicable.
BX New Mexico
Texas
BT United States

Guadalupe Island
Mexican island 180 miles off W central Baja California.
BT Pacific Ocean

Guadalupe Mountains
Index New Mexico and/or Texas as applicable.
BT United States
SA New Mexico
Texas

Guadalupe River
SE Texas.
BT Texas
United States

Guadalupian
Provincial series, North America. Above Leonardian, below Ochoan. Includes use on level 3 under age terms(1). See list E.
BA Permian
BT Paleozoic
NT Bell Canyon Formation
Capitan Formation
SA Lower Permian
San Andres Formation
Upper Permian

Guadeloupe
Combined islands of Basse-Terre and Grande-Terre which constitute an overseas department of France. Includes use on level 1 as an area term (list O). For term set options see list B.
BA Leeward Islands
Lesser Antilles
West Indies
NT La Grande Soufriere

Guajira Peninsula
N Colombia and NW Venezuela.
BT South America
SA Colombia
Venezuela

Guam
Island and unincorporated U. S. territory.
BT Mariana Islands
Micronesia

Guanabara
CO S231000S224000
W0431000W0434500
BT Rio de Janeiro
Brazil

Guanajuato
State in central.
BT Mexico
NT Leon City
Sierra Gorda

guano
As of 1981, includes use on level 3 under sediments(1) organic residues(2). See list N. Before 1981, included use on level 3 under sediments(1) organic sediments(2).
BA organic residues
BT sediments

Guapore
use Rondonia

Guarico
State in N central.
BT Venezuela

Guatemala
Includes use on level 1 as an area term (list O). For term set options see list B.
CO N135500N174500
W0881500W0921500
BA Central America
NT Fuego
Guatemala City
Huehuetenango
Motagua Fault
Pacaya
Santiaguito

Guatemala Basin
Term introduced in 1981. Off the coast of Central America in the Pacific Ocean.
CO N100000N140000
W0920000W0990000
BT Pacific Ocean

Guatemala City
City in S central.
BT Guatemala

Gubbio
Term introduced in 1981. Commune in Perugia province, Umbria, central Italy.
CO N431500N432500
E0134200E0133500
BT Perugia
Umbria
Italy

Gudbrandsdal
use Gudbrandsdalen

Gudbrandsdalen
Valley in S central part of country.
UF Gudbrandsdal
BT Norway

gudmundite
BA sulfides
BT minerals

Guelma
Town in Constantine Mountains in NE part of country.
BT Constantine region
Algeria

Guelph
City in SE.
BT Ontario
Canada

Guerrero
State in SW.
BT Mexico
SA Sierra Madre del Sur

guest element
use trace elements

Guiana (Shield)
use Guyana Shield

Guiana Basin
Region between the Orinoco, Negro, and Amazon rivers and the Atlantic Ocean. Index countries as applicable. Also search Guiana.
BT South America
SA Brazil
French Guiana
Guyana
Surinam

Guiana Highland
use Guyana Shield

Guiana Massif
use Guyana Shield

Guianas
Index countries as applicable.
BT South America
SA French Guiana
Guyana
Surinam

Guichon Batholith
use Guichon Creek Batholith

Guichon Creek Batholith
S central British Columbia. Also search Guichon Batholith.
UF Guichon Batholith
BT British Columbia
Canada

guide fossils
use index fossils

guidebook
Includes use on level 3 under area terms(1) areal geology(2) and under associations(1).
SA areal geology
associations
directory
explanatory text
field trips
manuals
road log

guides, ore
use ore guides

Guinea
Formerly French Guinea. Includes use on level 1 as an area term (list O). For term set options see list B.
CO N071500N124000
W0074000W0150000
BA West Africa
Africa
NT Los Islands
SA Niger River
Niger Valley
Nimba Mountains
Senegal Basin
Senegal River

Guinea-Bissau
Formerly Portuguese Guinea. Includes use on level 1 as an area term (list O). For term set options see list B. Before 1975, also search Portuguese Guinea.
CO N110000N130000
W0130000W0170000
BA West Africa
Africa

Guipuzcoa
One of the Basque Provinces in N.
CO N425600N432500
W0014500W0023200
BT Basque Provinces
Spain
NT San Sebastian

Gujarat
State in W.
CO N200000N245000
E0743000E0680000
BT India
NT Amba Dongar
Baroda
Bhuj
Broach
Cambay
Cambay Basin
Cutch
Girnar Hills
Kathiawar
Mount Girnar
Panch Mahals
Saurashtra
Wadia
SA Aravalli System
Bagh Beds
Bhuj Series
Nari Series
Narmada River
Narmada Valley

Gulbarga
Town in NE.
BT Mysore
India

Gulf Coast
use Gulf Coastal Plain

Gulf Coastal Plain
Index Mexican and U.S. states and Quintana Roo as applicable. Also search Gulf Coast. Includes use on level 1 or 2 as an area term (list O). If 1, see term set options under list B.
CO N170000N370000
W0810000W1000000
UF Gulf Coast
BA North America
SA Alabama
Atchafalaya Bay
Barataria Bay
Campeche
Florida
Louisiana
Mississippi
Quintana Roo
Tabasco
Tamaulipas
Texas
United States
Veracruz
Yucatan

Gulf of Aden
Between S coast of Arabian Peninsula and Somalia, E Africa. Includes use on level 1 or 2 as an area term (list O). If 1, see term set options under list B.
CO N103000N150000
E0520000E0430000
BA Arabian Sea
Indian Ocean
SA Great Rift Valley
Gulf of Tadjoura
Leg 24

Gulf of Alaska
Between Alaska Peninsula and Alaskan panhandle.
CO N540000N611500
W1350000W1630000
BA North American Pacific
BT Pacific Ocean
NT Icy Bay
SA Yakutat Bay

Gulf of Aqaba
Between NW Saudi Arabia and the Sinai Peninsula. Also search Gulf of Eilat.
CO N280000N293000
E0350000E0342000
UF Gulf of Eilat
BT Red Sea
SA Great Rift Valley

Gulf of Bothnia
Between Finland and Sweden. Also search Bothnian Sea.
CO N600000N660000
E0253000E0170000
UF Bothnian Sea
BT Baltic Sea
SA Alno
Finland

Gulf of California
Between peninsula of Baja California and the Mexican states of Sonora and Sinaloa. Includes use on level 1 as an area term (list O). For term set options see list B.
CO N224500N320000
W1060000W1150000
BA North American Pacific
Pacific Ocean
SA Leg 64
Leg 65

Gulf of Cambay
Inlet on W coast of India SE of Kathiawar Peninsula of Gujarat State.
BT Arabian Sea

Gulf of Cariaco
Inlet on NE coast of Venezuela S of Araya Peninsula in Sucre State.
BT Caribbean Sea

Gulf of Eilat
use Gulf of Aqaba

Gulf of Finland
Between Finland and Estonia.
CO N590000N603000
E0302000E0223000
BT Baltic Sea
SA Finland

Gulf of Gascony
No longer a valid term for GeoRef as of 1981. Use Bay of Biscay.

Gulf of Guinea
Wide inlet on W coast of Africa just south of the continent's great bulge.
BA North Atlantic
BT Atlantic Ocean

Gulf of Lion
Wide bay extending from French-Spanish border to Toulon. Also search Gulf of Lions.
CO N420000N433500
E0061000E0030500
UF Gulf of Lions
BA West Mediterranean
BT Mediterranean Sea

Gulf of Lions
use Gulf of Lion

Gulf of Maine
Off New England and S of Nova Scotia in Georges Bank area.
BT Atlantic Ocean
SA Georges Bank

Gulf of Mexico
Relatively shallow oceanic-type basin encircled by Cuba, Mexico, and the United States. Also search Gulf Basin. Includes use on level 1 or 2 as an area term (list O). If 1, see term set options under list B.
CO N180000N300400
W0803000W0980000
BA North American Atlantic
North Atlantic
Atlantic Ocean
NT Alacran Reef
Alaminos Canyon
Campeche Bank
Challenger Knoll
Florida Bay
Orca Basin
Sigsbee Deep
Yucatan Shelf
SA Gulf Stream
Leg 1
Leg 10

Gulf of Oman
Extends between N Oman on the Arabian Peninsula and SE Iran.
CO N223000N260000
E0620000E0560000
BT Arabian Sea

Gulf of Panama
On S coast of Panama.
BT Pacific Ocean

Gulf of Pozzuoli
NW inlet of Bay of Naples.
BT Mediterranean Sea

Gulf of Riga
Between Estonia and Latvia.
CO N570000N583000
E0243000E0220000
BT Baltic Sea

Gulf of Saint Lawrence
Off E coast of Canada encircled by Newfoundland, New Brunswick, Nova Scotia, and Quebec.
CO N450000N513000
W0570000W0650000
BA North American Atlantic
North Atlantic
BT Atlantic Ocean

Gulf of Siam
Between S extension of Thailand on the Malay Peninsula on the W and Cambodia and Vietnam on the E. Also search Gulf of Thailand.
CO N060000N133000
E1044000E0991000
UF Gulf of Thailand
BT South China Sea

Gulf of Suez
Between the Sinai Peninsula and the Arabian Desert of Egypt. Joined to the Mediterranean Sea by the Suez Canal.
CO N274000N300000
E0341000E0322000
BT Red Sea

Gulf of Tadjoura
Inlet of the Gulf of Aden on coast of Afars and Issas Territory, E Africa. Also search Gulf of Tadjura.
UF Gulf of Tadjura
BT Arabian Sea
SA Gulf of Aden

Gulf of Tadjura
use Gulf of Tadjoura

Gulf of Thailand
use Gulf of Siam

Gulf of Tonkin
Term introduced in 1981. Arm of the South China Sea, E of North Vietnam.
CO N170000N220000
E1100000E1054000
UF Tonkin Gulf
BT South China Sea

Gulf Stream
Warm ocean current flowing out of Gulf of Mexico along E U.S. to mid-Atlantic Ocean where it merges with North Atlantic drift current and influences climate of Europe as far N as Norway.
BT Atlantic Ocean
SA Gulf of Mexico

Gulfian
North America, provincial series. Includes use on level 3 under age terms (list E).
BA Upper Cretaceous
Cretaceous
BT Mesozoic
NT Austin Group
Escondido Formation
Woodbine Formation

gullies
Includes use on level 3 under geomorphology(1) erosion features(2).
SA channels
cliffs
erosion features
geomorphology

Gumma
Prefecture in central.
BT Honshu
Japan

Gunflint Formation
use Gunflint Iron Formation

Gunflint Iron Formation
In Animikie Group. Gunflint Lake region and Vermilion District in NE Minnesota. Also search Gunflint Formation and Gunflint Iron-Formation.
UF Gunflint Formation
Gunflint Iron-Formation
BT Precambrian
SA Animikie Group
Minnesota

Gunflint Iron-Formation
use Gunflint Iron Formation

Gunnison County
W central.
CO N380900N391600
W1060900W1073000
BT Colorado
United States

Guntur
City and district in E central.
BT Andhra Pradesh
India

Gunz
Europe. Above Astian (Pliocene), below Mindel. Includes use on level 3 under age terms(1). See list E.
BA Pleistocene
Quaternary
BT Cenozoic

Gurev
use Guryev

Gurghiu
Village in Mures County in E central.
BT Transylvania
Romania

Gurghiu Mountains
In SE Transylvania.
UF Gurghiului Mountains
BT Transylvania
Romania

Gurghiului Mountains
use Gurghiu Mountains

Gurhwal
use Garhwal

Guryev
City and oblast at mouth of Ural River at N end of Caspian Sea.
UF Gurev
BT Kazakhstan
USSR

gusher
use geysers

gustavite
BA sulfosalts
BT minerals

Gutii Mountains
BT Moldavia
Romania

Guyana
Formerly British Guiana. Gained independence in 1966. Also search British Guiana. Includes use on level 1 as an area term (list O). For term set options see list B.
CO N011000N083000
W0562500W0612500
UF British Guiana
BA South America
SA Georgetown
Guiana Basin
Guianas
Guyana Shield
Roraima Formation

Guyana Shield
Highland area extending from E Venezuela across N Brazil and the Guianas. Index countries as applicable. Also search Guiana; Guiana Shield.
UF Guiana (Shield)
Guiana Highland
Guiana Massif
BT South America
SA Brazil
French Guiana
Guyana
Surinam

Guyanese Atlantic
Term introduced in 1981.
BA North Atlantic
BT Atlantic Ocean

guyots
use seamounts

Gymnocodiaceae
Term introduced in 1981. Includes use on level 2 under algae(1). See list F.
BA Rhodophyta
algae
BT thallophytes
Plantae

gymnosperms
Includes use on level 1 and 2 as a fossil term (list F).
BT Spermatophyta
Plantae
NA Bennettitales
Caytoniales
Coniferae
Coniferales
Cordaitales
Cycadales
Cycadofilicales
Dadoxylon
Ephedrales
Ginkgoales
Glossopteridales
Gnetales
Nilssoniales
Pentoxylales
Pteridospermae
Welwitschiales
SA angiosperms
Dicroidium
Fagus
fossil wood
Juglans
Pecopteris
Pteropsida
sporangia
Taeniopteris

gypsite
use gypsum

gypsum
Includes use on level 3 under minerals(1) sulfates(2) and under sedimentary rocks(1) chemically precipitated rocks(2). See lists L and I. As of 1981, use gypsum deposits for gypsum as a commodity. Before 1981, included use on level 1 as a commodity term and on level 3 under sediments(1) chemically precipitated sediments(2).
UF gypsite
plaster of paris
plaster stone
BA sulfates
BT minerals
SA alabaster
anhydrite
cap rocks
chemically precipitated rocks
evaporites
gypsum deposits
halite
sedimentary rocks
sediments
selenite

gypsum deposits
Term introduced in 1981. Includes use on level 1 and 2 as a commodity term (list C). Before 1981, search gypsum AND deposits.
SA anhydrite deposits
evaporite deposits
gypsum

gyrolite
UF centrallasite
BA sheet silicates
silicates
BT minerals

gyttja
As of 1981, includes use on level 3 under sediments(1) organic residues(2). See list N. Before 1981, included use on level 3 under sediments(1) organic sediments(2).
BA organic residues
BT sediments
SA sapropel
soils

Gzhelian
Term introduced in 1981. Includes use on level 3 under age terms(1). See list E.
BA Upper Carboniferous
Carboniferous
BT Paleozoic

H

H
use hydrogen

H chondrites
Term introduced in 1981. Includes use on level 3 under meteorites(1).
UF H-group chondrites
BA chondrites
BT meteorites
NA Ashmore Meteorite
Haviland Meteorite
SA Coldwater Meteorites

H-group chondrites
use H chondrites

H-2
use deuterium

H-3
use tritium

Haast River
W South Island.
BT South Island
New Zealand

habit
Includes use on level 3 under crystal growth(1). Also search crystal habit.
UF crystal habit
SA crystal form
crystal growth
lattice
minerals

habitat
Includes use on level 2 and 3 under fossil group(1). See list F.
SA ecology
environment
fossils

Hachijo-jima
Second largest island in Izu-shichito group about 180 miles S of Tokyo.
BT Pacific Ocean
SA Izu-shichito

Hachinohe
City in N.
BT Honshu
Japan

Hadley Rill
use Hadley Rille

Hadley Rille
UF Hadley Rill
BT Moon

Hadrosauridae
As of 1981, includes use on level 3 under Reptilia(1) Ornithischia(2). See list F. Before 1981, included use on level 3 under Reptilia(1) Archosauria(2).
BA Ornithischia
Archosauria
Reptilia
BT Tetrapoda
Vertebrata
SA dinosaurs

Hadrynian
BA upper Precambrian
Precambrian

hafnium
Includes use on level 1 and 2 as a chemical element (list D). Also search Hf.
SA elements

Hagemeister Island Quadrangle
Term introduced in 1981. Central Alaska.
CO N580000N590000
W1600000W1623000
BT Alaska
United States

Hagendorf
Region in E.
BT Bavaria
West Germany

Haghimas Syncline
BT Romania
SA Eastern Carpathians

haidingerite
BA arsenates
BT minerals

Haifa
District in NE Israel. Also a city on the Mediterranean Sea.
BT Israel
Middle East
NT Geula Caves

Hail
Town and oasis.
UF Hayel
BT Saudi Arabia

Hainaut
Province in SW.
BT Belgium
NT Mons Basin
Tournai

Haiti
Occupies western third of island of Hispaniola. Includes use on level 1 as an area term (list O). For term set options see list B.
CO N180000N202000
W0714000W0743000
BA Hispaniola
Greater Antilles
West Indies

Hakodate
City on Tsugaru Strait in S.
BT Hokkaido
Japan

Hakone
Village and mountain resort in SE Honsu.
BT Honshu
Japan
SA Mount Hakone

Hakone Volcano
use Mount Hakone

Hakusan
Extinct volcano in W Honshu. Also search Hakusan Volcano.
UF Hakusan Volcano
BT Honshu
Japan

Hakusan Volcano
use Hakusan

Halberstadt
City in W part of country.
BT Magdeburg Bezirk
East Germany

Halecostomi
Term introduced in 1981. Includes use on level 2 under Pisces(1). See list F.
BA Holostei
Actinopterygii

Osteichthyes
Pisces
BT Vertebrata
NA Amiidae

half space
use half-space

half-space
Includes use on level 3 under seismology(1). Also search half space.
UF half space
SA seismology
velocity

Haliburton County
Term introduced in 1981. Before 1981, also search Haliburton.
CO N444500N453500
W0775600W0790100
BT Ontario
Canada

halides
Includes use on level 2 under minerals(1). See list L.
UF halogenide
BT minerals
NA bromides
chlorides
fluoborates
fluorides
fluosilicates
iodides
NZ zunyite
SA sodium chloride

Halifax
City on the Atlantic Ocean. Before 1981, also included the county.
BT Halifax County
Nova Scotia
Canada

Halifax County
Term introduced in 1981. Central Nova Scotia on the Atlantic Ocean. Before 1981, search Halifax.
CO N442500N451600
W0620900W0641500
BT Nova Scotia
Canada
NT Halifax

Halimeda
As of 1981, includes use on level 3 under algae(1) Codiaceae(2). See list F. Before 1981, included use on level 3 under algae(1) Chlorophyta(2).
BA Codiaceae
Chlorophyceae
Chlorophyta
algae
BT thallophytes
Plantae

halite
As of 1981, includes use on level 3 under minerals(1) chlorides(2). See list L. Before 1981, included use on level 3 under minerals(1) halides(2).
UF common salt
rock salt
BA chlorides
halides
BT minerals
SA chemically precipitated rocks
evaporites
gypsum
salt
sediments

Halland
County in SW.
BT Sweden

Halle
As of 1981, refers to only the city. Before 1981, also included the district.
BT Halle Bezirk
East Germany

Halle Bezirk
Term introduced in 1981. District in SW central. Before 1981, search Halle.
CO N505600N520300
E0125000E0105500
BT East Germany
NT Aschersleben
Bitterfeld
Geisel Valley
Halle
Mansfeld Syncline
Merseburg
Rottleberode
Schlotheim
SA Saxony-Anhalt
Unstrut River

halloysite
Also search endellite; metahalloysite.
UF endellite
metahalloysite
BA clay minerals
sheet silicates
silicates
BT minerals

Hallstatt Limestone
BT Triassic
SA Austria
Carnian
Ladinian
Upper Triassic

halmyrolysis
UF halmyrosis
submarine weathering
SA diagenesis
geochemistry
sea water

halmyrosis
use halmyrolysis

Halobia
Includes use on level 3 under Mollusca(1). See list F.
BA Bivalvia
Mollusca
BT Invertebrata

haloes
Includes use on level 3 under mineral exploration(1). For intrusions and metamorphism, use aureoles. Also search halos.
UF halos
SA aureoles
dispersion patterns
geochemical methods
mineral deposits
mineral exploration

Halog
Settlement in S.
BT Himachal Pradesh
India

halogenide
use halides

halokinesis
use salt tectonics

halos
use haloes

halotrichite
UF iron alum
BA sulfates
BT minerals

Halsingborg
City on Oresund in SW part of country. Also search Helsingborg.
UF Helsingborg
BT Malmohus
Sweden

Hamamelididae
Term introduced in 1981. Includes use on level 2 under angiosperms(1). See list F.
BA Dicotyledoneae
angiosperms
BT Spermatophyta
Plantae

Hamburg
City state on both sides of the Elbe River in N part of country.
BT West Germany

Hamersley Basin
BT Western Australia
Australia

Hamersley Group
BT Precambrian
SA Australia
Western Australia

Hamersley Range
In NW Western Australia.
BT Western Australia
Australia

Hamilton County
Index states as applicable.
BX Florida
Illinois
Indiana
Iowa
Kansas
Nebraska
New York
Ohio
Tennessee
Texas
BT United States

Hamilton Group
Divisions include Onondaga Limestone, Marcellus Formation, Mahantango Formation, Speeds Formation, Deputy Formation, Silver Creek Formation, Swanville Formation, Beachwood Formation, Backoven (Bakoven) Shale, Mount Marion Formation, Ashokan Formation, Kiskatom Formation, Skaneateles Shale, Ludlowville Shale, Moscow Shale, Montebello Formation, Sherman Ridge Formation.
BT Middle Devonian
Devonian
SA Mahantango Formation
Maryland
New York
Pennsylvania
West Virginia

hammarite
BA sulfobismuthites
sulfosalts
BT minerals

Hampden County
Term introduced in 1981. SW Massachusetts.
CO N420000N422200
W0720800W0730500
BT Massachusetts
United States

Hampshire
Term introduced in 1981. County in S England.
CO N504200N512300
W0004500W0015500
BT England
Great Britain
United Kingdom

Hampshire Basin
BT England
Great Britain
United Kingdom

Hampshire County
Index states as applicable.
BX Massachusetts
West Virginia
BT United States

Han River basin
Also search Han River.
BT China
SA Shensi

Hanawa Mine
Akita Prefecture in N.
BT Honshu
Japan

Hancock County
Index states as applicable.
BX Georgia
Illinois
Indiana
Iowa
Kentucky
Maine
Mississippi
Ohio
Tennessee
West Virginia
BT United States

hand calculators
Term introduced in 1981.
UF pocket calculators
SA automatic data processing
computers
microcomputers
minicomputers

handling, data
use data handling

Hanford Reservation
Atomic energy facility in Benton, Franklin, and Grant counties in SE part of state.
BT Washington
United States

Hangay
use Hangay Mountains

Hangay Mountains
W central Mongolia. Also search Hangay; Hangay Range; Khangai; Khangai Mountains.
CO N460000N481500
E1023000E0970000
UF Changai Mountains
Hangay
Hangay Range
Hangayn Mountains
Khangai
Khangai Mountains
BT Mongolia
SA Bayan-Nuurin-khotnor Basin

Hangay Range
use Hangay Mountains

Hangayn Mountains
use Hangay Mountains

Hanna Basin
S central Wyoming. Also search Hanna.
BT Wyoming
United States

Hannover
use Hanover

Hanover
City in Lower Saxony. Also search Hannover.
UF Hannover
BT West Germany

Hanover County
E central.
BT Virginia
United States

Hanson Lake
Near Manitoba border in E.
BT Saskatchewan
Canada

Hantkenina
Genus. Includes use on level 3 under foraminifera(1) Globigerinacea(2).
BA Globigerinacea
Rotaliina
foraminifera
BT Protista
Invertebrata

Haplophragmoides
Genus. Includes use on level 3 under foraminifera(1) Lituolacea(2).
BA Lituolidae
Lituolacea
Textulariina
foraminifera
BT Protista
Invertebrata

Harar
Province in E Ethiopia. Also a city. Also search Harrar.
UF Harer
Harrar
BT Ethiopia

harbors
As of 1978, term is used on level 2 under waterways(1).
SA waterways

Harbour Main Group
Overlain with angular unconformity by Lower Cambrian strata. The Conception Group conformably overlies the Harbour Main Group except near the Holyrood Granite where it is unconformable. E Newfoundland.
BT Precambrian
SA Canada
Newfoundland

Hardanger Plateau
use Hardangervidda

Hardangervidda
Plateau in Hordaland, SW Norway. Also search Hardanger Plateau.
UF Hardanger Plateau
Vidda
BT Norway

Hardeman County
Index states as applicable.
BX Tennessee
Texas
BT United States

Hardin County
Index states as applicable.
BX Illinois
Iowa
Kentucky
Ohio
Tennessee
Texas
BT United States

Harding County
Index states as applicable.
BX New Mexico
South Dakota
BT United States

Hardington County
use East Lothian

Hardingtonshire
use East Lothian

hardness
Includes use on level 3 under minerals(1).
SA mechanical properties
microhardness
minerals
physical properties
water hardness

Hare Bay
Inlet near N tip of Newfoundland.
BT Atlantic Ocean

Harebell Formation
Unconformable contacts with overlying Pinyon Conglomerate and underlying Bacon Ridge Sandstone. Teton County in NW Wyoming.
BT Upper Cretaceous
Cretaceous
SA Wyoming

Harer
use Harar

Harford County
NE Maryland.
BT Maryland
United States

Harghita
County in E.
BT Transylvania
Romania

Harghita Mountains
E Transylvania.
UF Harghitei Mountains
BT Transylvania
Romania

Harghitei Mountains
use Harghita Mountains

Hariana
use Haryana

harkerite
BA orthosilicates
silicates
BT minerals

Harlech
Village in W.
BT Wales
Great Britain
United Kingdom

Harlech Stage
Term introduced in 1978. Before 1978, search Harlech.
BA Lower Cambrian
Cambrian
BT Paleozoic

harmonic analysis
use Fourier analysis

harmonic folds
Term introduced in 1978. Before 1978, search folds AND harmonic.
BA folds
SA disharmonic folds

harmonics
Includes use on level 2 under geodesy(1).
SA geodesy

harmotome
BA zeolite group
framework silicates
silicates
BT minerals

Harney County
SE central.
CO N420000N440500
W1181000W1195500
BT Oregon
United States

Harper County
Index states as applicable.
BX Kansas
Oklahoma
BT United States

Harpidae
Includes use on level 3 under Mollusca(1). See list F.
BA Gastropoda
Mollusca
BT Invertebrata

Harrar
use Harar

Harrington Sound
BT Bermuda

Harris
Southern section of the island of Lewis in the Outer Hebrides off NW.
BT Scotland
Great Britain
United Kingdom

Harris County
Index states as applicable.
BX Georgia
Texas
BT United States

Harrisburg
City in Dauphin County in SE central.
BT Pennsylvania
United States

Harrison County
Term introduced in 1981. Index states as applicable.
BX Indiana
Iowa
Kentucky
Mississippi
Missouri
Ohio
Texas
West Virginia
BT United States

Harrison Lake
In S British Columbia.
BT British Columbia
Canada

Hartberg
Town in SE part of country.
BT Styria
Austria

Hartford County
N central.
BT Connecticut
United States

Haruna
Gumma Prefecture in N central Honshu. Also search Haruna Volcano.
UF Haruna Volcano
BT Honshu
Japan

Haruna Volcano
use Haruna

Harvey County
S central.
BT Kansas
United States

Haryana
State in N central India. Also search Hariana.
CO N274000N310000
E0773000E0743000
UF Hariana
BT India
NT Hissar
Narnaul
SA Delhi System

Harz Foreland
S Lower Saxony, West Germany, and W East Germany. Index countries as applicable. Also search Harz.
BT Europe
SA East Germany
West Germany

Harz Mountains
Mountain group between Elbe and Weser rivers in SE Lower Saxony, West Germany, and SW East Germany. Index countries as applicable. Also search Harz.
CO N512500N515700
E0114000E0101000
BT Germany
SA East Germany
West Germany

Harz region
Harz Foreland and Harz Mountain area of S Lower Saxony, West Germany, and W East Germany. Index countries as applicable. Also search Harz.
BT Europe
SA East Germany
West Germany

harzburgite
As of 1981, includes use on level 3 under igneous rocks(1) peridotites(2). See list H. Before 1981, included use on level 3 under igneous rocks(1) ultramafic family(2).
BA peridotites
ultramafics
BT igneous rocks
SA peridotite

Hassan
Town in S.
BT Mysore
India

Hassi Messaoud
use Hassi Messaoud Field

Hassi Messaoud Field
Oil field near Great Eastern Erg in NE Algeria. Also search Hassi-Messaoud; Hassi Messaoud; Hassi Messaoud oil field.
UF Hassi Messaoud
Hassi Messaoud oil field
Hassi-Messaoud
BT Algeria
SA oil and gas fields

Hassi Messaoud oil field
use Hassi Messaoud Field

Hassi-Messaoud
use Hassi Messaoud Field

hastingsite
BA clinoamphibole
amphibole group
chain silicates
silicates
BT minerals
SA ferrohastingsite

Hat Creek Basalt
Recent. Hat Creek Valley 10 miles NNE of Lassen Peak. Also search Hat Creek.
BT Holocene
SA California

Hatay
Province in S Anatolia on the NW Syrian border.
BT Turkey
Middle East

hatchettolite
use betafite

Hateg
Town in Hunedoara county in SW.
BT Transylvania
Romania

Hatteras abyssal plain
Off SE coast of U. S.
BT Atlantic Ocean

Hattiesburg
City in Forrest County in S.
BT Mississippi
United States

hausmannite
BA oxides
BT minerals

Haut-Rhin
Department in NE.
CO N473000N481500
E0074500E0070000
BT France
NT Sainte-Marie-aux-Mines
SA Vosges Mountains

Haute-Garonne
Department in S.
CO N424000N435300
E0020000E0002500
BT France
NT Toulouse
SA Rodez Trough
Salat Valley

Haute-Loire
Department in S central.
CO N445500N451500
E0043000E0030000
BT France
NT Velay
SA Central Massif

Haute-Marne
Department in NE.
CO N473500N484000
E0055200E0043500
BT France

Haute-Saone
Department in E.
CO N471500N480200
E0065000E0052000
BT France

Haute-Savoie
Department in E.
CO N454500N463000
E0070000E0054500
BT France
NT Annecy
Argentiere Glacier
Chablais
Mont Blanc
Rumilly
Savoy Alps

Haute-Vienne
Department of W central.
CO N460300N471000
E0011000W0000800
BT France
NT Rochechouart
Saint-Sylvestre Massif

Hauterivian
Europe. Above Valanginian, below Barremian. Includes use on level 3 under age terms(1). See list E.
BA Lower Cretaceous
Cretaceous
BT Mesozoic
SA Neocomian
Zubair Formation

Hautes-Alpes
Department in SE.
CO N441000N451000
E0070200E0052500
BT France
NT Briancon
SA Eoulx Basin
Pelvoux Massif

Hautes-Pyrenees
Department in SW.
CO N423500N434000
E0005000W0001500
BT France
NT Gavarnie
Lourdes

Hauts-de-Seine
Term introduced in 1981. Department in N.
CO N484200N490000
E0022000E0020800
BT France

hauyne
Also search hauynite.
UF hauynite
BA sodalite group
framework silicates
silicates
BT minerals
SA sulfates

hauynite
use hauyne

Havana
City on Straits of Florida in NW part of country.
BT La Habana
Cuba

Haveroe Meteorite
Term introduced in 1981. Impact in Turku area in SW Finland. Before 1981, also search Haveroe AND meteorites.
BA ureilite
achondrites
BT meteorites
SA Finland

Haviland
Village in Kiowa County in S central.
BT Kansas
United States

Haviland Meteorite
Term introduced in 1981. Impact in Kiowa County, Kansas. Before 1981, also search Haviland AND meteorites.
BA H chondrites
chondrites
BT meteorites
SA Kansas
Kiowa County

Hawaii
State including all of the Hawaiian Islands. Also southernmost and largest of the islands. Includes use on level 1 as an area term (list O). For term set options see list B. Also index Western U.S.
CO N190000N283000
W1550000W1790000
BZ Pacific Ocean
United States
Polynesia
NT Alae Crater
Hawaii County
Hilo
Honolulu
Kaneohe Bay
Kauai
Kilauea
Kona
Koolau Range
Maui
Mauna Kea
Mauna Loa
Molokai
Oahu
Pearl Harbor
Red Hill
Waianae
Waimanalo

Hawaii County
Entire island of Hawaii.
BT Hawaii
United States

Hawaii Ridge
use Hawaiian Ridge

Hawaiian Arch
use Hawaiian Ridge

Hawaiian Ridge
NW of Hawaiian Islands. Also search Hawaiian Arch; Hawaii Ridge.
UF Hawaii Ridge
Hawaiian Arch
BT Pacific Ocean

hawaiian-type eruptions
Term introduced in 1981.
SA eruptions
volcanism
volcanology

hawaiite
As of 1981, includes use on level 3 under igneous rocks(1) alkali basalts(2). See list H. Before 1981, included use on level 3 under igneous rocks(1) alkali basalt family(2).
BA alkali basalts
basalts
volcanic rocks
BT igneous rocks

Hawke's Bay
Provincial district on E coast.
BT North Island
New Zealand

Hawkesbury Sandstone
BT Triassic
SA Australia
New South Wales

Hawthorn Formation
In Alum Bluff Group. Central northern, N and S Florida; S and SE Georgia; South Carolina. Also, search Hawthorne Formation.
UF Hawthorne Formation
BT Miocene
SA Florida
Georgia
lower Miocene
middle Miocene
South Carolina

Hawthorne Formation
use Hawthorn Formation

Hayel
use Hail

Haymond Formation
Overlies Dimple Limestone; underlies Gaptank Formation. Marathon region in Brewster County in SW Texas.
BT Lower Pennsylvanian
Pennsylvanian
SA Texas

Hayward
City in Alameda County in W.
BT California
United States

Hayward Fault
NE, E, and SE of San Francisco Bay.
BT California
United States

Hazara
District.
BT North-West Frontier Province
Pakistan

hazards, geologic
use geologic hazards

Hazaribagh
Town and district in central.
BT Bihar
India

He
use helium

He-3
Includes use on level 3 under isotopes(1).
SA He-4/He-3
helium
isotopes

He-3/He-4
use He-4/He-3

He-4
Includes use on level 3 under isotopes(1).
SA He-4/He-3
helium
isotopes

He-4/He-3
Includes use on level 3 under isotopes(1). Also search He-3/He-4; search He-3 AND He-4.
UF He-3/He-4
SA absolute age
He-3
He-4
helium
isotopes

Healdiidae
Includes use on level 3 under Ostracoda(1). See list F.
BA Podocopida
Ostracoda
BT Crustacea
Mandibulata
Arthropoda
Invertebrata
SA Cavellinidae

health
use medical geology

Heart Mountain Fault
NW Wyoming.
BT Wyoming
United States

heat capacity
SA specific heat
thermal properties

heat content
use enthalpy

heat flow
Includes use on level 1 (list A) and 2. Used for the product of the thermal conductivity of a substance and the thermal gradient in the direction of the flow of heat. If 1, term set options are:
topic [anomalies, causes, changes, detection, distribution, experimental studies, genesis, geothermal gradient, heat sources, interpretation, measurement, observations, patterns, rates, regional patterns, temperature, theoretical studies, thermal conductivity]
subtopic (no area term)
NT heat flux
SA Battle Mountain High
boreholes
conductivity
convection
core
crust
Curie point
Earth
energy sources
fumaroles
geophysical methods
geophysical surveys
geophysics
geothermal energy
geothermal gradient
heat sources
heat transfer
hot springs
intrusions
mantle
measurement
plate tectonics
radioactivity
regional patterns
sea-floor spreading
tectonophysics
thermal conductivity
thermal diffusivity
thermal regime
thermal waters

heat flux
Heat flow per unit of time.
BT heat flow

heat sources
Includes use on level 2 under heat flow(1).
UF sources, heat
SA core
crust
geothermal energy
heat flow
mantle
regional patterns
thermal waters
volcanism

heat, specific
use specific heat

heat transfer
UF transfer, heat
SA heat flow

heaving, frost
use frost heaving

heavy metals
Also search individual heavy metals.
BT metals
SA bismuth
cadmium
cobalt
copper
gold
iron

lead
manganese
nickel
platinum
silver
tantalum
tellurium
zinc

heavy mineral deposits
Term introduced in 1981. Includes use on level 1 and 2 as a commodity term (list C). Before 1981, search heavy minerals AND deposits.
SA garnet deposits
heavy minerals
kyanite deposits
monazite deposits
placers
zircon deposits

heavy minerals
Includes use on level 3 under sediments(1), sedimentation(1), and sedimentary rocks(1). See also individual heavy minerals. As of 1981, use heavy mineral deposits for heavy minerals as a commodity. Before 1981, included use on level 1 as a commodity term.
BT minerals
SA accessory minerals
apatite
biotite
cassiterite
heavy mineral deposits
ilmenite
kyanite
light minerals
magnetite
monazite
rutile
sedimentation
sediments
titanite
tourmaline
zircon

heavy oil
SA condensates
petroleum

heazlewoodite
BA sulfides
BT minerals

Hebei
Term introduced in 1981. Province in NE China. Also search Hopei.
CO N360000N424000
E1195000E1131000
UF Hopei
BT China

Hebgen Lake
W of Yellowstone National Park in Gallatin County.
BT Montana
United States

Hebrides
Islands in Atlantic Ocean W of Scotland.
CO N553000N583000
W0053000W0084000
UF Western Islands
BT Scotland
SA Inner Hebrides
Outer Hebrides

hebronite
use amblygonite

Hecla
use Hekla

Hecla Hoek Formation
Spitsbergen Island.
BT Precambrian
SA Arctic region
Spitsbergen

Hector Formation
Hector and Corral Creek formations are lithologically similar to the Miette and Horsethief Creek Groups. SW Alberta.
BT Precambrian
SA Alberta

hectorite
BA clay minerals
sheet silicates
silicates
BT minerals

Hedbergella
Genus. Includes use on level 3 under foraminifera(1) Globigerinacea(1).
BA Globigerinacea
Rotaliina
foraminifera
BT Protista
Invertebrata

hedenbergite
BA clinopyroxene
pyroxene group
chain silicates
silicates
BT minerals

hedleyite
BZ tellurides
alloys
BT minerals

Heemskirk Granite
W Tasmania.
BT Devonian
SA Australia
Tasmania

Hegau
Region in S.
BT Baden-Wurttemberg
West Germany

Heidelberg
City on Neckar River in N.
BT Baden-Wurttemberg
West Germany

Heilungkiang
Province in E Manchuria in NE.
BT China
NT Lesser Khingan Mountains

Heimaey
Island a few miles off S Iceland. Largest of Vestmannaeyjar group.
BT Iceland

heintzite
use kaliborite

Hekla
Volcano in SW Iceland. Also search Hecla.
UF Hecla
BT Iceland

Helderberg Group
Includes Coeymans Formation, New Scotland Limestone, Mandata Shale, Becraft Limestone, Port Ewen Shale, Keyser Limestone, Port Jervis Limestone, Licking Creek Limestone, Alsen, Kalkberg, Manlius. New York, E Pennsylvania, W Maryland, Western Virginia, and N West Virginia.
BT Lower Devonian
Devonian
SA Maryland
New York
Pennsylvania
Virginia
West Virginia

Helderbergian
North America. Above upper Silurian, below Deerparkian. Includes use on level 3 under age terms(1). See list E.
BA Lower Devonian
Devonian
BT Paleozoic

Helena
City in Lewis and Clark County in W central.
BT Montana
United States

Helgoland
Island in North Sea 28 miles off mainland.
BT Schleswig-Holstein
West Germany

Helicoplacoidea
Includes use on level 2 under Echinodermata(1). See list F.
BA Echinozoa
Echinodermata
BT Invertebrata

Helicopontosphaera
BA nannofossils
algae
BT thallophytes
Plantae

helicopter methods
Term introduced in 1981. Includes use on level 3 under geophysical methods(1).
BA geophysical methods
SA geophysical surveys
methods
remote sensing

Helikian
North America. See list E.
BA Precambrian
SA Proterozoic

Heliolites
Genus. Includes use on level 3 under Coelenterata(1) Tabulata(2).
BA Heliolitidae
Tabulata
Anthozoa
Coelenterata
BT Invertebrata

Heliolitidae
Family. Includes use on level 3 under Coelenterata(1) Tabulata(1).
BA Tabulata
Anthozoa
Coelenterata
BT Invertebrata
NA Heliolites

helium
Includes use on level 1 and 2 as a chemical element (list D). Also search He. As of 1981, use helium gas for helium as a commodity. Before 1981, included use on level 1 as a commodity term.
UF He
BT noble gases
SA argon
elements
He-3
He-4
He-4/He-3
helium gas
isotopes
krypton
neon
U/He

helium gas
Term introduced in 1981. Includes use on level 1 and 2 as a commodity term (list C). Before 1981, search helium AND deposits.
SA helium

Hell Creek Formation
Includes Bull Creek Sand, Isabel-Firesteal Coal Member. E, N, and central S Montana; SW North Dakota; and NW and N South Dakota.
BT Upper Cretaceous
Cretaceous
SA Montana
North Dakota
South Dakota

Hellenides
Deep rooted mountain chain of Greece which is part of the Alpine orogenic belt. Bordered by Rhodope Mountains in N.
BT Greece
SA Alps
Crete
Peloponnesus
Pindus Mountains

Hell's Canyon
use Snake River canyon

Helmholtz free energy
use free energy

Helmstedt
City near East German border in W.
BT Lower Saxony
West Germany

Helsingborg
use Halsingborg

Helsingor
City on the Oresund N of Copenhagen.
BT Denmark

Helvetian
Europe. Includes use on level 3 under age terms(1). See list E.
BA middle Miocene
Miocene
Neogene
Tertiary
BT Cenozoic
SA Tortonian

helvine
use helvite

helvite
Also search helvine.
UF helvine
BZ silicates
sulfides
BT minerals
SA genthelvite

hematite
As of 1981, includes use on level 3 as a commodity term (list C) under iron ores(1). Before 1981, included use as a commodity term under iron(1).
BA oxides
BT minerals
SA iron
iron ores
martite
specularite

Hemichordata
Includes use on level 1 and 2 as a fossil term (list F). Term is to be used only when more specific terms do not apply.
UF Stomachorda
NX Camaroidea
Cephalodiscida
Dendroidea
Didymograptina
Didymograptus
Diplograptina
Enteropneusta
Glossograptina
Graptolithina
Graptoloidea
Isograptus
Monograptina
Monograptus
Monograptus uniformis
Pterobranchia
Rhapdopleurida
Stolonoidea
Tuboidea
SA Chordata

hemihedrite
BA chromates
BT minerals

hemimorphite
BA orthosilicates
silicates
BT minerals

Hemingfordian
See list E.
BA lower Miocene
Miocene
Neogene
Tertiary
BT Cenozoic

Hemipteroida
Term introduced in 1981. Includes use on level 2 under Insecta(1). See list F.
BA Insecta
BT Mandibulata
Arthropoda
Invertebrata

Hemphillian
See list E.
BA Neogene
Tertiary
BT Cenozoic
SA Miocene
Pliocene

Henbury Meteorite
Term introduced in 1981. Impact in Northern Territory. Before 1981, also search Henbury AND meteorites.
BA octahedrite
iron meteorites
BT meteorites
SA Australia
Northern Territory

Henderson Mine
Term introduced in 1981. In Clear Creek County, Colorado. Molybdenite deposit.
BT Clear Creek County
Colorado
United States

Hennessey Formation
Overlies Wichita Formation and underlies Duncan Sandstone, or Flowerpot Shale, where the Duncan is absent. Central and SW Oklahoma.
BT Permian
SA Oklahoma

Henry County
N Tennessee.
BT Tennessee
United States

Henry Mountains
Garfield County in S central.
BT Utah
United States

Hepaticae
Includes use on level 2 under bryophytes(1). See list F.
BA bryophytes
BT Plantae

Herault
Department in S.
CO N431500N440000
E0041500E0023000
BT France
NT Bedarieux
Cabrieres
Lodeve
Lodeve Basin
Montpellier
Saint-Chinian
Sete

Hercegovina
use Herzegovina

Hercinico Centro
Term introduced in 1981.
BT Spain

Hercinico Sur
Term introduced in 1981.
BT Spain

Hercoglossa danica
As of 1981, includes use on level 3 under Mollusca(1) Nautiloidea(2). See list F. Before 1981, included use on level 3 under Mollusca(1) Cephalopoda(2).
BA Nautiloidea
Tetrabranchiata
Cephalopoda
Mollusca
BT Invertebrata

Hercynian
A valid index term through 1977. After 1977, use Hercynian Orogeny.

Hercynian Geosyncline
Term introduced in 1978. Before 1978, search geosynclines AND Hercynian.
SA geosynclines
Hercynian Orogeny

Hercynian Orogeny
Term introduced in 1978. Late Paleozoic (Devonian-Lower Triassic) of Europe, extending through the Carboniferous and Permian. Also search orogeny AND Hercynian; Variscan; Variscan Orogeny.
UF Variscan Orogeny
BT Paleozoic
SA Devonian
Hercynian Geosyncline
Lower Triassic
Permian
Triassic
Variscides

hercynite
UF ferrospinel
iron spinel
BA oxides
BT minerals
SA spinel group

herderite
BA phosphates
BT minerals

Herja
Mining region in Maramures County in N.
BT Transylvania
Romania

Herkimer County
N central.
BT New York
United States

hermatypic taxa
Term introduced in 1978 on level 3 under fossil groups (list F). Before 1978, also search hermatypic.
UF taxa, hermatypic
SA ahermatypic taxa
Coelenterata
reefs

Heron Island
Largest of Capricorn Islands in Coral Sea 40 miles off SE coast.
BT Queensland
Australia

Hertfordshire
SE England.
CO N513000N521000
E0001000W0004500
BT England
Great Britain
United Kingdom

hervidero
use mud volcanoes

Herzegovina
Region in N Bosnia and Herzegovina.
UF Hercegovina
BT Yugoslavia

herzenbergite
UF kolbeckine
BA sulfides
BT minerals

Hess Mountains
E central.
BT Yukon Territory
Canada

Hesse
Region in SW Germany, comprising the state of Hesse and the former Prussian province of Hesse-Nassau. Also search Hessen.
CO N492300N514000
E0101500E0074500
UF Hessen
BT West Germany
NT Bundenbach
Darmstadt
Kassel
Marburg
Messel
Rheingau
Taunus
Vogelsberg
Waldeck
Wetterau
Wiesbaden
SA Franconia
Lahn River
Lahn River valley
Main River
Mainz Basin
Odenwald
Palatinate
Rhenish Schiefergebirge
Rhon Mountains
Steinheim
Swabia
Weser River
Westerwald
Westphalia

Hesse Basin
Term introduced in 1981.
CO N501500N514000
E0103500E0083700
BT Germany

Hessen
use Hesse

hessite
As of 1981, includes use on level 3 under minerals(1) tellurides(2). See list L. Before 1981, included use on level 3 under minerals(1) sulfides(2).
BA tellurides
sulfides
BT minerals

Heterocorallia
Includes use on level 2 under Coelenterata(1). See list F.
BA Anthozoa
Coelenterata
BT Invertebrata

heterogeneity
Term introduced in 1981. Also search inhomogeneity.
SA homogeneity
seismology

heterogeneous materials
Term introduced in 1978. Before 1978, search heterogeneous.
SA homogeneous materials
materials

heterogenite
Also search stainierite.
UF stainierite
BA oxides
BT minerals

Heterohelicidae
Family. Includes use on level 3 under foraminifera(1) Globigerinacea(2).
BA Globigerinacea
Rotaliina
foraminifera
BT Protista
Invertebrata

heteromorphic taxa
Term introduced in 1978. Before 1978, search heteromorphs.
UF heteromorphs
taxa, heteromorphic

heteromorphite
BA sulfantimonites
sulfosalts
BT minerals

heteromorphs
use heteromorphic taxa

Heteroptera
As of 1981, used on level 3 under Insecta(1). See list F. Before 1981, included use on level 2 under Insecta(1).
BA Insecta
BT Mandibulata
Arthropoda
Invertebrata

Heterostegina
Includes use on level 3 under foraminifera(1) Rotaliacea(2).
BA Rotaliacea
Rotaliina
foraminifera
BT Protista
Invertebrata
NA Heterostegina depressa
SA Nummulitidae

Heterostegina depressa
Includes use on level 3 under foraminifera(1) Rotaliacea(2).
BA Heterostegina
Rotaliacea
Rotaliina
foraminifera
BT Protista
Invertebrata

Heterostraci
As of 1981, includes use on level 3 under Agnatha(1). See list F. Before 1981, included use on level 3 under Pisces(1) Agnatha(2).
BA Agnatha
BT Vertebrata

Hettangian
Europe. Above Rhaetian (Triassic), below Sinemurian. Includes use on level 3 under age terms(1). See list E.
BA lower Liassic
Lower Jurassic
Jurassic
BT Mesozoic

heulandite
BA zeolite group
framework silicates
silicates
BT minerals
SA clinoptilolite

Hexacorallia
use Scleractinia

Hexactiniaria
Includes use on level 2 under Coelenterata(1). See list F.
BA Anthozoa
Coelenterata
BT Invertebrata

hexahedrite
Term introduced in 1981. Before 1981, also search hexahedrites.
BA iron meteorites
BT meteorites

hexahedrites
No longer a valid term for GeoRef as of 1981. Use hexahedrite.

hexahydrite
BA sulfates
BT minerals

HF
As of 1981, no longer a valid term for GeoRef. See electromagnetic waves. Before 1981, included use on level 3 under ionosphere(1) wave propagation(2).

Hg
use mercury

hibschite
use hydrogrossular

Hida
Former province in central Honshu. Now part of Gifu Prefecture.
BT Honshu
Japan

Hida Belt
use Hida metamorphic belt

Hida metamorphic belt
Central Honshu. Also search Hida Belt; Hida metamorphic rocks.
UF Hida Belt
Hida metamorphic rocks
BT Honshu
Japan

Hida metamorphic rocks
use Hida metamorphic belt

Hida Mountains
Central.
BT Honshu
Japan

Hidaka
Town in S.
BT Honshu
Japan

Hidaka Belt
use Hidaka metamorphic belt

Hidaka metamorphic belt
In S Hokkaido. Also search Hidaka Belt.
UF Hidaka Belt
BT Hokkaido
Japan

Hidaka Mountains
In S Hokkaido. Also search Hidaka.
BT Hokkaido
Japan

Hidalgo
State in E central.
BT Mexico
NT Pachuca
SA Sierra Madre Oriental

Hidalgo County
Index states as applicable.
BX New Mexico
Texas
BT United States

Hidas
Town in S. In Baranya.
BT Hungary

Hidden Valley Dolomite
Underlies Devonian Lost Burro Formation; overlies Ely Springs Dolomite, S California. Also search Hidden Valley.
BT Paleozoic
SA California
Devonian
Lower Devonian
Silurian

Hierro
Spanish island off Western Sahara. Westernmost of the Canary Islands.
BT Canary Islands

High Atlas
A range of the Atlas Mountains containing the highest peaks in the entire mountain system. In W and S part of country.
UF Grand Atlas Mountains
High Atlas Mountains
BT Morocco
SA Atlas Mountains
Moroccan Atlas Mountains

High Atlas Mountains
use High Atlas

High Plains
A valid term through mid-1978 used to indicate the Great Plains from Nebraska southward. Now use Great Plains.

high pressure
Includes use on level 3 under metamorphism(1).
BT pressure
SA low pressure
metamorphism
pore pressure

high school
Includes use on level 3 under education(1).
SA college-level education
education
elementary school
junior high school

High Tatra
use Tatra Mountains

High Tatra Mountains
use Tatra Mountains

High Tatras
use Tatra Mountains

high temperature
Includes use on level 3 under metamorphism(1). Also search high-temperature.
BT temperature
SA metamorphism

high-energy environment
Term introduced in 1978. Before 1978, also search high-energy or high energy.
SA environment
low-energy environment
sedimentation

high-grade metamorphism
Term introduced in 1978. Includes use on level 3 under metamorphism(1). Before 1978, also search high-grade AND metamorphism; high grade AND metamorphism.
BA metamorphism
SA grade

high-level mode
Term introduced in 1981.
SA oscillations
seismographs
seismology

high-resolution methods
Term introduced in 1981.
SA geophysical methods
geophysical surveys
methods
paleomagnetism
remote sensing
seismology

Highland County
Index states as applicable.
BX Ohio
Virginia
BT United States

Highland Rim
BT Alabama
United States

Highland Series
BT Precambrian
SA Sri Lanka

highlands
Includes use on level 3 under geomorphology(1) erosion features(2), or under Moon(1). Before 1981, search highlands AND NOT Scotland. See Highlands.
SA erosion features
geomorphology
maria
Moon
mountains
plateaus

Highlands
As of 1981, no longer a valid term for GeoRef. Use Grampian Highlands and/or Northern Highlands. Was used for that part of Scotland lying NW of a line from Loch Lomond to just S of Aberdeen.

highways
Used for geological studies on highways and materials. A level 1 term as of 1978. Includes use on level 2 under foundations(1). If 1, term set options are:
construction
subtopic
embankments
subtopic
feasibility studies
subtopic
foundations
subtopic
frost action
subtopic
materials, properties
subtopic
planning
subtopic
site exploration
subtopic
SA bridges
construction
embankments
engineering geology
engineering properties
feasibility studies
foundations
frost action
land subsidence
planning
road tests
rock mechanics
site exploration
slope stability
soil mechanics
soils
tunnels

Hildesheim
City in S central.
BT Lower Saxony
West Germany

Hildocerataceae
Includes use on level 3 under Mollusca(1). See list F.
BA Ammonoidea
Tetrabranchiata
Cephalopoda
Mollusca
BT Invertebrata

Hildoceratidae
Includes use on level 3 under Mollusca(1). See list F.
BA Ammonoidea
Tetrabranchiata
Cephalopoda
Mollusca
BT Invertebrata

hills
Includes use on level 3 under geomorphology(1).
SA geomorphology
landforms
mountains
topography
tors

Hillsboro Quadrangle
Term introduced in 1981. SW New Mexico.
CO N324500N330000
W1073000W1074500
BT New Mexico
United States
SA Grant County
Sierra County

Hillsborough County
Term introduced in 1981. Index states as applicable.
BX Florida
New Hampshire
BT United States

Hilo
City in Hawaii county on E coast of island of Hawaii.
BT Hawaii
United States

Himachal Pradesh
State in the Himalayas.
CO N302000N332000
E0790000E0754000
BT India
NT Bilaspur
Chamba
Halog
Kangra
Kinnaur
Mahasu
Mandi
Simla
Simla Hills
SA Beas River
Jumna River
Kasauli Series
Kumaun Himalayas

Himalaya
use Himalayas

Himalaya Mountains
use Himalayas

Himalayan Orogeny
Term introduced in 1981. Includes use on level 3 under orogeny(1) and tectonics(1).
SA Cenozoic
Himalayas
Mesozoic
orogeny
tectonics

Himalayas
Mountain system extending from Jammu and Kashmir in the W to Assam in the E. Index Tibet and countries as applicable. Includes use on level 1 as an area term (list O). For term set options see list B. Also search Himalaya; Himalayan; Himalaya.
CO N270000N370000
E0970000E0720000
UF Himalaya
Himalaya Mountains
The Himalaya
BA Asia
SA Bhutan
Himalayan Orogeny
India
Kumaun Himalayas
Lesser Himalayas
Nepal
Sikkim
Tibet

Hindenburg
use Zabrze

Hindenburg in Oberschlesien
use Zabrze

Hindeodella
Genus.
BA Conodonta

Hindu Kush
Mountain range in central Asia. In-

dex Tadzhikistan and countries as applicable.
BT Asia
SA Afghanistan
Pakistan
Tadzhikistan

hinge faults
Term introduced in 1978. Before 1978, search faults AND hinge.
BA faults

hinsdalite
BZ phosphates
sulfates
BT minerals

hintzeite
use kaliborite

Hipparion
Genus. Includes use on level 3 under Mammalia(1) Perissodactyla(2).
BA Equidae
Perissodactyla
Mammalia
BT Tetrapoda
Vertebrata

Hippomorpha
Term introduced in 1981. Includes use on level 2 under Mammalia(1). See list F.
BA Perissodactyla
Mammalia
BT Tetrapoda
Vertebrata

Hippopotamus
Genus. Includes use on level 3 under Mammalia(1) Artiodactyla(2).
BA Artiodactyla
Mammalia
BT Tetrapoda
Vertebrata

Hippuritacea
Includes use on level 3 under Mollusca(1). See list F.
BA Bivalvia
Mollusca
BT Invertebrata

Hiroshima
City and prefecture on the Inland Sea in W.
BT Honshu
Japan

hisingerite
BA sheet silicates
silicates
BT minerals
SA clay minerals

Hispaniola
Island of the Greater Antilles between Cuba and Puerto Rico. Index countries as applicable.
BT Greater Antilles
West Indies
NA Dominican Republic
Haiti
SA Caribbean region

Hissar
City WNW of Delhi.
BT Haryana
India

Hissar Range
A branch of the Alai Range in NW Tadzhikistan. Also search Gissar Range; Hissar.
CO N380000N381000
E0700000E0673000
UF Gissar Range
BT Tadzhikistan
USSR
SA Alai Range
Zeravshan-Hissar

histograms
BA statistical analysis
SA graphic methods

kurtosis

histology
SA fossils

historical geology
Includes use on level 3 under geology(1), paleontology(1) and stratigraphy (1). See list E (age terms).
SA biostratigraphy
education
geology
paleoenvironment
paleontology
physical geology
stratigraphy

history
Use term under various disciplines. Includes use as a level 2 or 3 term appropriate to a large number of topics, e.g. on level 2 under paleontology(1), geology(1), stratigraphy(1), and mineral exploration(1). See list G.
SA survey organizations
uniformitarianism

Histosols
Includes use on level 3 under soils(1). See list M.
BA soils
SA Bog soils

Hitachi
City on the Pacific Ocean in E.
BT Honshu
Japan

Hiyoshi
An area of Mizunami city in Gifu Prefecture in central.
BT Honshu
Japan

HL chondrites
Term introduced in 1981. Includes use on level 3 under meteorites(1).
UF HL-group chondrites
BA chondrites
BT meteorites

HL-group chondrites
use HL chondrites

Ho
use holmium

Hobart
Coastal city in SE.
BT Tasmania
Australia

Hobdo
use Kobdo

Hobsogol
use Khubsugul

Hodgeman County
SW central.
BT Kansas
United States

Hodna Basin
Interior drainage basin in NE part of country.
BT Constantine region
Algeria

hodographs
use traveltime curves

Hodrusa
Ore bearing region.
BT Slovakia
Czechoslovakia

Hoggar Mountains
use Ahaggar

Hoghiz
Village in Brasov County in SE.
BT Transylvania
Romania

Hohe Tauern
Range of Eastern Alps in SW.
CO N470000N471500
E0131500E0120000
BT Austria
SA Alps
Eastern Alps
Tauern Tunnel

Hohe Venn
Range of low mountains in E part of country.
BT Liege
Belgium

Hokkaido
Northernmost of the four main islands.
CO N412500N453000
E1455500E1394000
BT Japan
NT Abashiri
Esan Cape
Hakodate
Hidaka metamorphic belt
Hidaka Mountains
Ikushumbetsu
Ishikari Bay
Ishikari Plain
Kamikawa
Kitami
Koma-ga-take 1
Kushiro coal field
Nemuro Peninsula
Okushiri Island
Oshima Peninsula
Shikotsu
Sorachi
Tarumae
Tokachi
Tokachi Plain
Usu
SA Green Tuff Formation
Tsugaru Strait

Hoko Gunto
use Penghu Islands

Hoko Shoto
use Penghu Islands

Hokuriku
District.
BT Honshu
Japan

Hokuroku
Term introduced in 1981. District, N Honshu.
BT Honshu
Japan

Holbrook
Town in Navajo County in E central.
BT Arizona
United States

Holderness
Peninsula in Yorkshire in NE.
BT England
Great Britain
United Kingdom

Holland
use Netherlands

hollandite
BA oxides
BT minerals
SA coronadite

Hollister
City in San Benito County in W central.
BT California
United States

holmium
Includes use on level 1 and 2 as a chemical element (list D).
UF Ho
SA elements
rare earths

holmquistite
BA orthoamphibole
amphibole group
chain silicates
silicates
BT minerals

Holocene
Recent. Late Quaternary, from Pleistocene to present time. Includes use on level 1 as an age term (list E); on level 2 under paleo- terms, e.g. paleoecology, paleogeography, paleomagnetism; on level 3 under age terms(1). Above Pleistocene. Also search Recent; Postglacial; Post-glacial.
BA Quaternary
BT Cenozoic
NA Atlantic
Boreal
Flandrian
NT Hat Creek Basalt
NA lower Holocene
Mesolithic
middle Holocene
Neolithic
Preboreal
Subboreal
upper Holocene
Versilian
SA Brunhes Epoch
modern

Holocephali
Term introduced in 1981. Includes use on level 2 under Pisces(1). See list F.
BA Chondrichthyes
Pisces
BT Vertebrata

holography
Used as a general term in crystallography and geophysics.
SA automatic data processing
crystallography

Holostei
As of 1981, includes use on level 2 under Pisces(1). See list F. Before 1981, included use on level 3 under Pisces(1) Osteichthyes(2).
BA Actinopterygii
Osteichthyes
Pisces
BT Vertebrata
NA Halecostomi

Holothuroidea
Includes use on level 2 under Echinodermata(1). See list F.
BA Echinozoa
Echinodermata
BT Invertebrata
SA sclerites

holotypes
Term introduced in 1981.
SA fossils
paleobotany
paleontology
taxonomy

Holstein
Region in S.
BT Schleswig-Holstein
West Germany

Holsteinian
Europe. See list E.
BA upper Pleistocene
Pleistocene
Quaternary
BT Cenozoic

Holy Cross Mountains
use Swiety Krzyz Mountains

Homalozoa
As of 1981, includes use on level 3 under Echinodermata(1). See list F. Before 1981, included use on level 2 under Echinodermata(1).
BA Echinodermata
BT Invertebrata
NA Homoiostelea

Homostelea
Machaeridia
Stylophora

Homestake Mine
Largest gold mine in U. S. In Lead, Lawrence County, in W.
BT South Dakota
United States

Hominidae
As of 1981, includes use on level 2 under Mammalia(1). See list F. Before 1981, included use on level 3 under Mammalia(1) Primates(2).
BA Primates
Mammalia
BT Tetrapoda
Vertebrata
NA Australopithecus
Homo
Neanderthal
Paranthropus
Pithecanthropus
pre-Neanderthal
Ramapithecus

Homo
Genus. As of 1981, includes use on level 3 under Mammalia(1) Hominidae(2). See list F. Before 1981, included use on level 3 under Mammalia(1) Primates(2).
BA Hominidae
Primates
Mammalia
BT Tetrapoda
Vertebrata
NA Homo erectus
Homo habilis
Homo sapiens

Homo erectus
As of 1981, includes use on level 3 under Mammalia(1) Hominidae(2). See list F. Before 1981, included use on level 3 under Mammalia(1) Primates(2).
BA Homo
Hominidae
Primates
Mammalia
BT Tetrapoda
Vertebrata

Homo habilis
As of 1981, includes use on level 3 under Mammalia(1) Hominidae(2). See list F. Before 1981, included use on level 3 under Mammalia(1) Primates(2).
BA Homo
Hominidae
Primates
Mammalia
BT Tetrapoda
Vertebrata

Homo neanderthalensis
use Neanderthal

Homo sapiens
As of 1981, includes use on level 2 under Mammalia(1). See list F. Before 1981, included use on level 3 under Mammalia(1) Primates(2).
BA Homo
Hominidae
Primates
Mammalia
BT Tetrapoda
Vertebrata

homogeneity
Includes use on level 3 under geophysical methods(1).
SA heterogeneity
homogeneous materials
seismology

homogeneous materials
Term introduced in 1981.
SA heterogeneous materials
homogeneity

homogenization
Used as a general term.
SA petrology

Homoiostelea
Includes use on level 2 under Echinodermata(1). See list F.
BA Homalozoa
Echinodermata
BT Invertebrata

homonymy
SA nomenclature
taxonomy

Homostelea
Includes use on level 2 under Echinodermata(1). See list F.
BA Homalozoa
Echinodermata
BT Invertebrata

Honaker Trail Formation
In Hermosa Group. SE Utah.
BT Paleozoic
SA Pennsylvanian
Permian
Utah

Honduras
Includes use on level 1 as an area term (list O). For term set options see list B.
CO N130000N160000
W0831500W0891500
BA Central America

Hong Kong
British crown colony which includes the island of Hong Kong; and Kowloon Peninsula and New Territories on mainland. Includes use on level 1 as an area term (list O). For term set options see list B.
CO N220000N230000
E1150000E1130000
BA Far East
Asia

Honolulu
City and county on Oahu Island.
BT Hawaii
United States

Honshu
Largest of the Japanese islands.
CO N332000N413000
E1420000E1305000
BT Japan
NT Abukuma Mountains
Aichi
Aizu Basin
Akenobe Mine
Akita
Akiyoshi
Amagase
Aomori
Arita River
Asama
Ashio
Atera Fault
Atsumi Peninsula
Awashima
Chiba
Chiba Peninsula
Chichibu
Chichibu Belt
Chichibu Mine
Chugoku
Dogo
Fossa Magna
Fujiyama
Fukui
Fukushima
Fukushima Prefecture
Gifu
Gumma
Hachinohe
Hakone
Hakusan
Hanawa Mine
Haruna
Hida
Hida metamorphic belt
Hida Mountains
Hidaka
Hiroshima
Hitachi
Hiyoshi
Hokuriku
Hokuroku
Ibaragi Complex
Ibaraki
Ikuno Mine
Ise Bay
Ishikawa
Ito
Iwaki
Iwate
Izu Peninsula
Izu-shichito
Izumi
Japanese Alps
Joban coal field
Kakioka
Kanagawa
Kanazawa
Kashiwazaki
Kii Peninsula
Kinki District
Kiso Mountains
Kitakami Mountains
Koma-ga-take 2
Kosaka Mine
Kumamoto
Kwanto Plain
Kyoto
Lake Biwa
Lake Suwa
Maizuru
Matsukawa
Matsushiro
Mie
Miho Bay
Mikawa
Mitsuishi
Miura Peninsula
Miyagi
Miyake-Jima
Mizunami
Mizusawa
Mount Asama
Mount Hakone
Mount Mihara
Musashino
Myoko Mountain
Nagano
Nagano Prefecture
Naka-no-umi
Nanao
Nara
Narugo
Nasudake
Niigata
Nohi Rhyolite
Nojiri Lake
Noto Peninsula
Odate
Oga Peninsula
Ogasawara
Oguni
Ojika Peninsula
Okayama
Oki Islands
Omine Mine
Onikobe
Osaka
Otake
Rokko Mountains
Ryoke
Ryoke Belt
Ryujima
Sado Island
Sagami Bay
Saitama
Sangun Zone
Sendai
Shiga
Shimane
Shimokita Peninsula
Shinano River
Shinji Lake
Shiobara
Shirasu
Shizukuishi
Shizuoka
Shonai River
Suruga Bay
Takanuki
Tamagawa
Tamba Plateau
Tanzawa Mountains
Tochigi
Tohoku
Toki
Tokyo
Tokyo Bay
Tottori
Towada
Toyama
Toyoma
Ube coal field
University of Tokyo
Wakayama
Yahagi River
Yake-Dake
Yamagata
Yamaguchi
Yamanashi
Yatsugatake
Yokohama
Yonezawa
Zao
SA Akiyoshi Limestone
Futaba Group
Green Tuff Formation
Izumi Group
Mikabu System
Motojuku Formation
Nankaido
Narita Formation
Osaka Group
Sambagawa Belt
Tono
Tsugaru Strait
Uonuma Group
Usuginu Conglomerate

Hooke's law
Term introduced in 1978.
SA deformation
elastic limit
elastic properties
elastic strain
strain
stress

Hope
Town in SW.
BT British Columbia
Canada

Hopei
use Hebei

Hopi Buttes
use Hopi Buttes Field

Hopi Buttes Field
In Hopi and Navajo Indian reservations in Navajo County. Also search Hopi Buttes volcanic field; Hopi Buttes.
UF Hopi Buttes
Hopi Buttes volcanic field
BT Arizona
United States

Hopi Buttes volcanic field
use Hopi Buttes Field

horizon differentiation
Includes use on level 3 under soils(1) genesis(2).
SA differentiation
horizons
morphology
soils

horizons
Includes use on level 3 under soils(1) morphology(2).
UF soil horizon
soil zone
SA horizon differentiation
morphology
parent materials
soils

horizontal
A valid term through 1977. After 1977, use horizontal orientation.

horizontal discontinuities
Term introduced in 1981. Includes use on level 3 under crust(1), mantle(1), and seismology(1). Used for a type of interface in seismology.
BT discontinuities
SA crust
mantle
seismology

horizontal movements
Term introduced in 1981. Includes use on level 3 under neotectonics(1) and tectonics(1).
SA neotectonics
tectonics
vertical movements

horizontal orientation
Term introduced in 1978. Reference to folds defined on the basis of orientation in relation to the geographic horizontal plane. Before 1978, also search folds AND horizontal.
SA folds
orientation

horizontal-component seismographs
Term introduced in 1981.
BT seismographs

Horlick Mountains
SE of S tip of Ross Ice Shelf.
BT Antarctica

Horn Plateau Formation
S District of Mackenzie.
BT Givetian
Middle Devonian
Devonian
SA Northwest Territories

hornblende
Also search horn stone.
BA clinoamphibole
amphibole group
chain silicates
silicates
BT minerals
SA kaersutite

hornblende gneiss
A valid metamorphic rock term through 1977. Use gneisses and hornblende.

hornblende schist
A valid metamorphic rock term through 1977. Use schists and hornblende.

hornblendite
As of 1981, includes use on level 3 under igneous rocks(1) ultramafics(2). See list H. Before 1981, included use on level 3 under igneous rocks(1) ultramafic family(2).
BA ultramafics
BT igneous rocks

Horne Mine
Noranda area in SW.
BT Quebec
Canada

hornfels
Includes use on level 2 and 3 under metamorphic rocks(1). See list J.
BT metamorphic rocks

hornstein
use chert

Hornsund
Inlet on SW coast.
BT Spitsbergen
Svalbard
Arctic region

Horry County
Extreme E.
BT South Carolina
United States

horseflesh ore
use bornite

Horsham
City in W Sussex in S.
BT England
Great Britain
United Kingdom

horsts
Includes use on level 3 under faults(1) systems(2).
SA block structures
faults
grabens
systems

host materials
As of 1981, no longer a valid term for GeoRef. Use host rocks or country rocks.

host rocks
Includes use on level 3 under inclusions(1). Before 1981, also search host materials. Used for rock serving as a host for mineral deposits.
SA country rocks
fluid inclusions
inclusions
mineral deposits
mineral inclusions
rocks
veins
xenoliths

hot dry rocks
Term introduced in 1981. Before 1981, also search dry hot rocks.
UF dry hot rocks
SA geothermal energy
geothermal fields
geothermal systems

hot spots
Includes use on level 3 under plate tectonics(1).
SA magmas
mantle
plate tectonics
plumes

hot springs
Includes use on level 2 under springs(1); on level 3 under thermal waters(1).
BA springs
SA geothermal energy
geysers
ground water
heat flow
hydrothermal processes
thermal waters

Houghton County
NW Upper Peninsula.
BT Michigan
United States

House Range
Millard County in W.
BT Utah
United States

Houston
City in Harris County in E.
BT Texas
United States

Howard County
Term introduced in 1981. Index states as applicable.
BX Arkansas
Indiana
Iowa
Maryland
Missouri
Nebraska
Texas
BT United States

howardite
Term introduced in 1981. Before 1981, also search howardites. Includes use on level 3 under meteorites(1).
BA achondrites
BT meteorites
NA Kapoeta Meteorite

howardites
As of 1981, no longer a valid term for GeoRef. Use howardite.

Howgill Fells
Westmoreland and Yorkshire counties in N.
BT England
Great Britain
United Kingdom

howieite
BA chain silicates
silicates
BT minerals

Hoxnian
Europe. See list E.
BA upper Pleistocene
Pleistocene
Quaternary
BT Cenozoic

Hsinchu
Region in N.
BT Taiwan

Hubsugul
use Khubsugul

Hudson Bay
Administratively a part of District of Keewatin.
CO N550000N650000
W0760000W0950000
BX Canada
Atlantic Ocean
SA Hudson Bay Lowlands
Northwest Territories

Hudson Bay Lowland
use Hudson Bay Lowlands

Hudson Bay Lowlands
Index Northwest Territories and provinces as applicable. Also search Hudson Bay Lowland.
UF Hudson Bay Lowland
BT Canada
SA Hudson Bay
Manitoba
Northwest Territories
Ontario
Quebec

Hudson Highlands
Part of Appalachians lying along both banks of the Hudson River S of Newburgh.
BT New York
United States
SA Appalachians

Hudson River
In New York and New Jersey.
BT United States
SA New Jersey
New York

Hudson Valley
River valley primarily in New York. Index states as applicable.
BT United States
SA New Jersey
New York

Hudsonian Orogeny
Name proposed for a time of plutonism, metamorphism, and deformation during the Precambrian in the Canadian Shield. Also search Hudsonian.
BT Precambrian
SA Canada
orogeny

Hudspeth County
W Texas.
BT Texas
United States

Huehuetenango
Department in Guatemala. Also a city.
BT Guatemala
Central America

Huelva
As of 1981, refers to only the city. Before 1981, also included the province. In SW Spain on the Gulf of Cadiz.
BT Huelva Province
Andalusia
Spain

Huelva Province
Term introduced in 1981. Before 1981, search Huelva. In SW Spain.
CO N364800N381000
W0060800W0073200
BT Andalusia
Spain
NT Aracena
Huelva
Puebla de Guzman
Rio Tinto

Huesca
As of 1981, refers to only the city. Before 1981, also included the province. In NE Spain.
BT Huesca Province
Aragon
Spain

Huesca Province
Term introduced in 1981. Before 1981, search Huesca. In NE Spain.
CO N411700N425700
E0004500W0005700
BT Aragon
Spain
NT Huesca

Hughes County
E central.
BT Oklahoma
United States

Hughes Creek Shale
Of Foraker Limestone. NE Kansas and SE Nebraska. Also search Hughes Creek Shale Member.
UF Hughes Creek Shale Member
BT Permian
SA Kansas
Nebraska

Hughes Creek Shale Member
use Hughes Creek Shale

Hugoton
City in Stevens County in SW.
BT Kansas
United States

Hugoton Embayment
SW Kansas and Oklahoma panhandle. Index states as applicable. Also search Hugoton.
BT United States
SA Kansas
Oklahoma

Hugoton Field
SW Kansas. Also search Hugoton; Hugoton gas field; Hugoton-Panhandle Field.
UF Hugoton gas field
Hugoton-Panhandle Field

H

BT Kansas
United States
SA oil and gas fields

Hugoton gas field
use Hugoton Field

Hugoton-Panhandle Field
use Hugoton Field

Hull
City across Ottawa River from Ottawa, Ontario, in S.
BT Quebec
Canada

hulsite
BA borates
BT minerals

human activities
use human activity

human activity
Also search anthropogenic; human activities; human impact; human influence; human interference.
UF anthropogenic (activity)
human activities
human impact
human influence
human interference
SA human ecology

human ecology
As of 1978, term is used on level 2 under geologic hazards(1), land use(1), and pollution(1).
SA ecology
environmental geology
geologic hazards
human activity
human waste
land use
medical geology
pollution

human impact
use human activity

human influence
use human activity

human interference
use human activity

human waste
Term introduced in 1978.
UF waste, human
SA human ecology
waste disposal

humates
Includes use on level 2 under organic materials(1).
BA organic materials
SA humic acids

Humber Estuary
On E coast opening on North Sea.
BT England
Great Britain
United Kingdom

Humboldt County
Index states as applicable.
BX California
Iowa
Nevada
BT United States

Humboldt River valley
N central.
BT Nevada
United States

humic acids
Includes use on level 2 under organic materials(1).
BA organic materials
SA acids
humates
humus

humic coal
Also search humite AND coal.
UF humite (coal)
BT coal

humid environment
Term introduced in 1978. Before 1978, search humid.
SA climate
ecology
environment
humidity
paleoclimatology

humidity
Includes use on level 3 under meteorology(1) water(2).
SA atmosphere
climate
humid environment
meteorology
moisture
raindrops
water
water vapor

humite
BZ fluorides
humite group
BT minerals

humite (coal)
use humic coal

humite group
Includes use in combination with orthosilicates (i.e. orthosilicates, humite group) on level 2 under minerals(1). See list L.
BA nesosilicates
orthosilicates
silicates
BT minerals
NZ clinohumite
humite
norbergite
NA titanoclinohumite

hummocks
SA geomorphology
periglacial features
permafrost

Humphreys Peak
use San Francisco Mountain

humus
Includes use on level 3 under organic materials(1).
UF soil ulmin
BA organic materials
SA humic acids
leaves
Mor
Mull
peat
soils

Hungarian Basin
use Alfold

Hungarian Great Plain
use Alfold

Hungarian Plain
use Alfold

Hungary
Includes use on level 1 as an area term (list O). For term set options see list B.
CO N454500N483000
E0230000E0161000
BA Central Europe
Europe
NT Alfold
Bakony Mountains
Balaton region
Borsod Basin
Borzsony Mountains
Budapest
Budapest region
Bukk Mountains
NX Central Transdanubia
NT Debrecen
Dorog
Dorog Basin
Eger
Hidas
Lake Balaton
Matra Mountains
Mecsek Mountains
Northeastern Hungarian Hills
NX Northwestern Transdanubia
Southeastern Transdanubia
Southern Transdanubia
NT Szolnok
Tokaj
Tokaj Mountains
Transdanubia
Vertes
Vertes Mountains
Veszprem
Villany Mountains
SA Carpathians
Danube River
Danube Valley
Pannonia
Pannonian Basin
Somes Basin
Subcarpathians
Tisza River
Tokaj-Eperjes Mountains

Hunger Steppe
use Bet-Pak-Dala

Hunsruck
Mountain region S of the Mosel River between the Rhine and Saar rivers.
BT Rhineland-Palatinate
West Germany

Hunsruck Shale
Also search Hunsruck shales; Hunsrueck Shale.
UF Hunsruck shales
Hunsrueck Shale
BT Devonian
SA Germany
Rhineland-Palatinate

Hunsruck shales
use Hunsruck Shale

Hunsrueck Shale
use Hunsruck Shale

Hunter Valley
River valley in E central.
BT New South Wales
Australia

huntite
BA carbonates
BT minerals

Huntly
Town in Aberdeen County in E.
BT Scotland
Great Britain
United Kingdom

Hunton Group
Includes Chimneyhill Limestone, Henryhouse Shale, Haragan Shale, Bois d'Arc Limestone, Frisco Limestone, Kite Group with Haragan and Cravatt formations. SE Oklahoma. Also search Hunton Limestone.
BT Paleozoic
SA Devonian
Oklahoma
Silurian

Huon Peninsula
On E coast between Astrolabe Bay and Huon Gulf.
BT Papua New Guinea
Australasia

hureaulite
BA phosphates
BT minerals

Huronian
Great Lakes Region. Division of the Proterozoic of the Canadian Shield. Also search Huronian Supergroup. See list E.
UF Huronian Supergroup
BA Proterozoic
Precambrian
NT Gowganda Formation
Onaping Formation
Signal Hill Formation
SA Canada
United States

Huronian Supergroup
use Huronian

Hurricane Ridge Syncline
Southwest Virginia and SE West Virginia. Index states as applicable.
BT United States
SA Virginia
West Virginia

hurricanes
Includes use on level 3 under geologic hazards(1).
SA Agnes
geologic hazards
meteorology
monsoons
storms
winds

Hurwitz Group
Keewatin District.
BT Aphebian
lower Proterozoic
Proterozoic
Precambrian
SA Northwest Territories

Hut Point Peninsula
On Ross Island which is just off NW edge of Ross Ice Shelf in Ross Dependency.
BT Antarctica

Hutchinson
City in Reno County in S central.
BT Kansas
United States

Hutchinson Salt Member
Of Wellington Formation. Salt beds largely within Hutchinson city limits in Reno County in S central Kansas.
BT Permian
SA Kansas
Wellington Formation

Hyaenidae
Includes use on level 3 under Mammalia(1). See list F.
BA Carnivora
Mammalia
BT Tetrapoda
Vertebrata

hyalite
UF water opal
BA silica minerals
framework silicates
silicates
BT minerals
SA opal

hyalophane
BA feldspar group
framework silicates
silicates
BT minerals

Hyalospongea
Includes use on level 2 under Porifera(1). See list F.
BA Porifera
BT Invertebrata

hybridization
Used in relation to magmas.
SA assimilation
contamination
country rocks
differentiation
emplacement
intrusions
magmas

Hyderabad
Princeley state prior to Indian independence. A city in Sind Province, Pakistan, and a city in Andhra Pra-

desh State in India. Index countries as applicable.
BT Asia
SA India
Pakistan

hydrargillite
use gibbsite

hydration
Includes use on level 3 under geochemistry(1) processes(2).
SA dehydration
geochemistry
processes
water

hydration of glass
Includes use on level 2 under geochronology(1).
UF glass, hydration of
SA geochronology

hydraulic conductivity
Also search permeability coefficient.
UF coefficient of permeability
permeability coefficient
SA conductivity
hydraulics
permeability
transmissivity

hydraulic fracturing
Includes use on level 3.
SA fracturing
geothermal energy
massive hydraulic fracturing
mining geology
petroleum engineering
rock mechanics

hydraulic maps
Term introduced in 1978. Before 1978, search maps AND hydraulic.
BA maps

hydraulic pressure
Term introduced in 1978. Before 1978, search hydraulic AND pressure.
BT pressure

hydraulics
Used when discussing channels, or engineering and practical applications. For mathematical aspects of movement, use hydrodynamics. As of 1978, term is used on level 2 under shorelines(1), and waterways(1).
SA drawdown
engineering geology
flows
fluid dynamics
ground water
hydraulic conductivity
hydrodynamics
hydrographs
hydrostatic pressure
multiphase flow
roughness
shorelines
soil mechanics
waterways

hydrobiotite
BA sheet silicates
silicates
BT minerals
SA aluminosilicates
mica group

hydroboracite
BA borates
BT minerals

hydrocarbons
Includes use on level 2 under organic materials(1).
BA organic materials
SA aromatic hydrocarbons
bitumens
carbon
ethane
methane
natural gas
petroleum
sapropel

hydrochemistry
SA geochemistry
ground water
hydrogeology
hydrology
springs
surface water
thermal waters

hydrodynamics
Includes use on level 2 under ground water(1). Used when discussing mathematical aspects of ground water movement. For discussion of channels, engineering and practical applications, see hydraulics.
SA catchment hydrodynamics
Darcy's law
fluid dynamics
ground water
hydraulics

hydrogen
Includes use on level 1 and 2 as a chemical element (list D). Also search H.
UF H
SA D/H
deuterium
isotopes
stable isotopes
tritium

hydrogen sulfide
SA eutrophication
gases
lakes
natural gas
sulfides

hydrogeologic
A valid term through 1977. After 1977, use hydrogeologic maps.

hydrogeologic maps
Term introduced in 1978. Before 1978, search hydrogeologic AND maps.
BA maps
SA hydrogeology
potentiometric surface

hydrogeological controls
Includes use on level 3 under mineral deposits, genesis(1) controls(2).
SA controls
hydrogeology
mineral deposits

hydrogeology
Used for general studies. For the science that deals with subsurface waters and related geologic aspects of surface waters. To search for the hydrology of a particular river search for the name of the river in coordination with hydrology or hydrogeology. Includes use on level 1 (list A); on level 2 under area terms (list B). If 1, term set options are:
topic [applications, bibliography, catalogs, classification, concepts, education, experimental studies, general, history, instruments, methods, nomenclature, objectives, philosophy, practice, symposia, textbooks, theoretical studies]
subtopic
SA aquifers
artesian waters
dolines
ground water
hydrochemistry
hydrogeologic maps
hydrogeological controls
hydrologic cycle
hydrological methods
hydrology
hydrosphere
representative basins
reservoirs
rivers
springs
thermal regime
thermal waters
transmissivity
water resources
water table

hydrogoethite
BA oxides
BT minerals
SA limonite

hydrographs
SA hydraulics
hydrology

hydrogrossular
Also search hibschite.
UF hibschite
BA garnet group
nesosilicates
orthosilicates
silicates
BT minerals
SA grossular

hydrologic budget
use water balance

hydrologic cycle
Includes use on level 3 under hydrology(1). Also search water cycle.
UF water cycle
SA atmospheric precipitation
cycles
hydrogeology
hydrology
hydrosphere
meteorology
water

hydrologic maps
Term introduced in 1978. Before 1978, search maps AND hydrologic.
BA maps
SA hydrology

hydrological methods
Includes use on level 2 under mineral exploration(1).
SA hydrogeology
hydrology
methods
mineral exploration
water

hydrology
Used for continental surface water. In cases where the interchange between ground and surface water is intimate, this set may be used with cycles on level 2. This covers cases of modern sediment transport. To search for the hydrology of a particular river search for the name of the river in coordination with hydrology or hydrogeology. Includes use on level 1. Term set options are:
cycles
subtopic
glaciers
subtopic
ice
subtopic
instruments
name of instrument
subtopic
limnology
subtopic
methods
subtopic
atmospheric precipitation
subtopic
rivers and streams
subtopic
seepage
subtopic
snow
subtopic
surveys
(area)
techniques
name of technique or application
SA atmosphere
atmospheric precipitation
bedforms
cascades
catchment hydrodynamics
channel geometry
cycles
discharge
ecology
engineering geology
environmental geology
eutrophication
evaporation
evapotranspiration
gauging
geomorphology
glacial geology
glaciers
ground water
hydrochemistry
hydrogeology
hydrographs
hydrologic cycle
hydrologic maps
hydrological methods
hydrosphere
hydrostatic pressure
ice
infiltration
lakes
limnology
meltwater
meteorology
multiphase flow
percolation
pollution
precipitation
purification
rainfall
representative basins
reservoirs
retention
rivers
rivers and streams
runoff
sedimentation
sediments
snow
springs
surface water
surveys
suspension
transmissivity
water
water balance
water quality
water resources
water supply
water table
waterlogging
watersheds
waterways

hydrolysis
Includes use on level 3 under geochemistry(1) processes(2); on level 3 under weathering(1).
SA geochemistry
oxidation
processes
weathering

hydromagnesite
BA carbonates
BT minerals

hydrometallurgy
Term introduced in 1981.
SA metallurgy

hydromica
BA sheet silicates
silicates
BT minerals
SA clay minerals
illite
mica group

hydromica schist
Not a valid term for GeoRef. Use schists and hydromica.

Hydromorphic soils
See list M.
BA soils
SA Bog soils
Intrazonal soils

hydromuscovite
BA sheet silicates
silicates
BT minerals
SA clay minerals
mica group
muscovite

hydrosodalite
Artificial mineral.
BA sodalite group
framework silicates
silicates
BT minerals
SA aluminosilicates

hydrosphere
Includes use on level 3 under hydrogeology(1).
SA atmosphere
biosphere
Earth
glaciers
ground water
hydrogeology
hydrologic cycle
hydrology
ice
lakes
rivers
rivers and streams
snow
streams
surface water
tectonosphere
water

hydrostatic pressure
BT pressure
SA hydraulics
hydrology

hydrothermal
A valid term through 1977. See hydrothermal conditions; ore-forming fluids; hydrothermal alteration; hydrothermal processes.

hydrothermal alteration
Includes use on level 3 under metasomatism(1).
BA metasomatism
SA alteration
autometamorphism
greisenization
hydrothermal conditions
hydrothermal processes
leakage anomalies
ore-forming fluids
propylitization
pyritization
pyrometasomatism
sericitization
serpentinization
thermal alteration
thermal waters
uralitization
wallrock alteration
zeolitization

hydrothermal conditions
Term introduced in 1981.
SA hydrothermal alteration
hydrothermal processes
metasomatism
phase equilibria

hydrothermal fluids
Not a valid term for GeoRef. As of 1981, use ore-forming fluids if applicable. Before 1981, hydrothermal solutions was used for this term.

hydrothermal processes
Includes use on level 3 under mineral deposits, genesis(1) processes(2).
SA alteration
epithermal processes
hot springs
hydrothermal alteration
hydrothermal conditions
igneous processes
mineral deposits
ore-forming fluids
processes
water

hydrothermal solutions
As of 1981, no longer a valid term for GeoRef. Use ore-forming fluids if applicable.

hydroxides
BA oxides
BT minerals

hydroxyapatite
use hydroxylapatite

hydroxylapatite
Also search hydroxyapatite.
UF hydroxyapatite
BA phosphates
BT minerals
SA apatite

Hydrozoa
Includes use on level 2 under Coelenterata(1). See list F.
BA Coelenterata
BT Invertebrata

Hymenoptera
As of 1981, includes use on level 3 under Insecta(1) Hymenopteroida(2). See list F. Before 1981, included use on level 2 under Insecta(1).
BA Hymenopteroida
Insecta
BT Mandibulata
Arthropoda
Invertebrata
SA Raphidiodea

Hymenopteroida
Term introduced in 1981. Includes use on level 2 under Insecta(1). See list F.
BA Insecta
BT Mandibulata
Arthropoda
Invertebrata
NA Hymenoptera

Hyolithes
Term introduced in 1981. Includes use on level 2 under Mollusca(1). See list F. Also search Hyolithidae; Hyolithida; Hyolithimorpha.
BA Mollusca
BT Invertebrata
SA problematic fossils

Hyolithidae
As of 1981, no longer a valid term for GeoRef. Use Hyolithes.

hypabyssal
A valid term through 1977. After 1977, use hypabyssal rocks on level 2 under igneous rocks(1).

hypabyssal processes
Term introduced in 1981.
SA hypabyssal rocks
mineral deposits
processes

hypabyssal rocks
Term introduced in 1978. Includes use on level 2 under igneous rocks(1). Before 1978, also search hypabyssal AND igneous rocks.
BT igneous rocks
SA beresite
hypabyssal processes
intrusive rocks
plutonic rocks
rocks

hypersthene
BA orthopyroxene
pyroxene group
chain silicates
silicates
BT minerals

hypocenter
use focus

hypogene processes
As of 1981, no longer a valid term for GeoRef. Use endogene processes.

hypothermal processes
Term introduced in 1981.
UF katathermal processes
SA mineral deposits
processes

Hyracoidea
Term introduced in 1981. Includes use on level 2 under Mammalia(1). See list F.
BA Mammalia
BT Tetrapoda
Vertebrata

Hystrichosphaeridae
Includes use on level 3 under palynomorphs(1) acritarchs(2) or Dinoflagellata(2). Before 1981, also search hystrichospheres; hystrichosphaerids.
BZ acritarchs
Dinoflagellata
BA palynomorphs

Hystricomorpha
Term introduced in 1981. Includes use on level 2 under Mammalia(1). See list F.
BA Rodentia
Mammalia
BT Tetrapoda
Vertebrata

H2SO4
use sulfuric acid

I

I
use iodine

I-129
Includes use on level 3 under isotopes(1).
SA iodine
isotopes

I-131
Includes use on level 3 under isotopes(1).
SA iodine
isotopes

Iacobeni
Village in Suceava County in N.
BT Moldavia
Romania

Iapetus
Proto-Atlantic Ocean. For satellite of Saturn, use Iapetus Satellite.
SA Atlantic Ocean
Iapetus Satellite

Iapetus Satellite
Term introduced in 1978. Distinguish from Iapetus, proto-Atlantic Ocean.
SA Iapetus
Saturn

Ibadan
City in SW part of country. In Western Provinces.
BT Nigeria

Ibaragi Complex
Granitic complex in Osaka Prefecture.
UF Ibaragi Granitic Complex
BT Honshu
Japan

Ibaragi Granitic Complex
use Ibaragi Complex

Ibaraki
City in Osaka Prefecture. Also a prefecture in N central.
BT Honshu
Japan

Ibbenbueren
use Ibbenburen

Ibbenburen
Town in N North Rhine-Westphalia. Also search Ibbenbueren.
UF Ibbenbueren
BT North Rhine-Westphalia
West Germany

Iberia
use Iberian Peninsula

Iberian Cordillera
use Iberian Mountains

Iberian Mountains
Mountain system on E edge of great central plateau. Also search Iberian Cordillera.
UF Iberian Cordillera
BT Spain
Europe

Iberian Peninsula
Extreme SW Europe S of the Pyrenees and the Bay of Biscay. Also search Iberia. Index countries as applicable.
CO N360000N435000
E0032000W0093000
UF Iberia
BA Southern Europe
Europe
NA Portugal
Spain

Iberica
Term introduced in 1981.
BT Spain

Ibiza
Third largest of Spain's Balearic Islands off E coast of Spanish mainland.
UF Iviza
BT Balearic Islands
Spain

ice
Includes use on level 2 under hydrology(1).
SA frozen ground
glacial features
glaciers
ground ice
hydrology
hydrosphere
ice rafting
meltwater
periglacial features
permafrost
sea ice
snow
thawing
water

ice ages
A valid term through 1977 used in combination with ancient (i.e. ice ages, ancient) on level 2 under glacial geology(1). After 1978, use ancient ice ages on level 2.

ice caps
SA glacial features
glaciers
ice sheets

ice mantle
use ice sheets

ice movement
Includes use on level 3 under glacial geology(1) glaciation(2). Before 1976, also search ice movements.
SA deposition
erosion
glacial geology
glaciation
glaciers
movement

ice rafting
Includes use on level 3 under sedimentation(1). Also search ice-rafting.
UF rafting, ice
SA ice
sedimentation
transport

ice sheets
Includes use on level 3 under glacial geology(1) glaciers(2).
UF ice mantle
SA glacial geology
glaciation
glaciers
ice caps

ice shelves
UF shelves, ice
SA glacial geology
glaciation
glaciers
marine environment

ice veins
use ice wedges

ice wedges
Includes use on level 3 under glacial geology(1) periglacial features(2) and under sedimentary structures(1) bedding plane irregularities(2). See list K.
UF ice veins
wedges, ice
SA bedding plane irregularities
fossil ice wedges
glacial geology
periglacial features
permafrost
sedimentary structures

ice wedges, fossil
use fossil ice wedges

ice-wedge cast
use fossil ice wedges

ice-wedge fill
use fossil ice wedges

ice-wedge pseudomorph
use fossil ice wedges

icebergs
SA glacial features
glaciers

Iceland
Independent country and island 155 miles SE of Greenland. Includes use on level 1 as an area term (list O). For term set options see list B. As of 1981, Atlantic Ocean is no longer autoposted.
CO N634000N663000
W0133000W0244500
BA Western Europe
Europe
NT Breidamerkurjokull
Heimaey
Hekla
Kolbeinsey Island
Reykjanes Peninsula
Skeidararjokull
Snaefellsnes Peninsula
Surtsey
Vatnajokull
Vestmannaeyjar

Iceland crystal
use Iceland spar

Iceland spar
UF double-refracting spar
Iceland crystal
BA carbonates
BT minerals
SA calcite

Iceland-Faeroe Ridge
Between Iceland and Faeroe Islands. Also search Iceland-Faeroe Rise; Iceland-Faeroes Ridge.
UF Faeroe-Iceland Ridge
Iceland-Faeroe Rise
Iceland-Faeroes Ridge
BT Atlantic Ocean

Iceland-Faeroe Rise
use Iceland-Faeroe Ridge

Iceland-Faeroes Ridge
use Iceland-Faeroe Ridge

Icelandic Plateau
Term introduced in 1981. W of Jan Mayen Ridge, N of Iceland.
CO N670000N700000
W0093000W0160000
BT Greenland Sea
Arctic Ocean

icequakes
Term introduced in 1981.
SA earthquakes
glaciers
glaciotectonics

ichnofossils
Includes use on level 1 and 2 as a fossil term (list F). As of 1981, is not to be used under sedimentary structures(1), and refers only to studies in which the emphasis is paleontology, not sedimentary petrology. For ichnofossils as components of sediments or sedimentary rocks, use lebensspuren.
NA Corophioides
Cruziana
Nereites
Planolites
Rusophycus
Skolithos
Thalassinoides
SA borings
burrows
chemical fossils
Chondrites ichnofossils
Collenia
fossilization
fossils
Invertebrata
lebensspuren
morphology
Ophiomorpha
Ophiomorpha nodosa
tracks
trails

Ichthyopterygia
Includes use on level 2 under Reptilia(1). See list F.
BA Reptilia
BT Tetrapoda
Vertebrata
NA Ichthyosauria

Ichthyosauria
Term introduced in 1981. Includes use on level 2 under Reptilia(1). See list F.
BA Ichthyopterygia
Reptilia
BT Tetrapoda
Vertebrata
NA Ichthyosaurus

Ichthyosaurus
As of 1981, includes use on level 3 under Reptilia(1) Ichthyosauria(2). See list F. Before 1981, included use on level 3 under Reptilia(1) Ichthyopterygia(2).
BA Ichthyosauria
Ichthyopterygia
Reptilia
BT Tetrapoda
Vertebrata

Icriodus
Genus.
BA Conodonta

Icy Bay
Term introduced in 1981. Inlet of Gulf of Alaska, SE Alaska, W of Yakutat Bay.
CO N594500N610500
W1410500W1415000
BT Gulf of Alaska
Pacific Ocean

Idaho
Includes use on level 1 as an area term (list O). For term set options see list B.
CO N420000N490000
W1110500W1171500
BA Western U.S.
United States
NT Ada County
NX Adams County
NT Albion Range
Bannock County
Bingham County
NX Blaine County
NT Boise
NX Butte County
NT Caribou County
NX Clark County
Clearwater County
NT Coeur d'Alene
Coeur d'Alene River
NX Custer County
Franklin County
Fremont County
NT Idaho County
NX Jefferson County
Jerome
NT Kootenai County
Latah County
Lemhi County
Lemhi Range
NX Lewis County
Lincoln County
NT Mackay
NX Madison County
Oneida County
NT Owyhee County
Pioneer Mountains
Rainbow Mountain
Reynolds Creek
Salmon
Salmon River
Salmon River breaks
Sawtooth Range
Seven Devils Mountains
Shoshone County
Snake River plain
NX Teton County
Valley County
Washington County
NT Wilson Lake
SA Bald Mountain
Basin and Range Province
Battle Mountain High
Bear Lake
Bear River basin
Bear River Range
Beaverhead Formation
Beaverhead Mountains
Belt Supergroup
Bitterroot Range
Blue Mountains
Cache Valley
Colorado Group
Columbia Plateau
Columbia River Basalt
Columbia River basin
Frontier Formation
Great Basin
Idaho Batholith
Lake Bonneville
Madison Group
Missoula Group
Phosphoria Formation
Pocatello Formation
Prichard Formation
Purcell System
Raft River
Raft River basin
Ravalli Group
Snake River
Snake River basin
Snake River canyon
Stump Formation
Swan Peak Formation
Thaynes Formation
U. S. Rocky Mountains
Wallace Formation
Wasatch Front
Wasatch Range
Western Interior
Western Overthrust Belt
Yellowstone National Park

Idaho Batholith
In Bitterroot Range. Index states as applicable.
BT United States
SA Idaho
Montana

Idaho County
N central.
BT Idaho
United States

idaite
BA sulfides
BT minerals

ideal waves
Term introduced in 1978. Includes use on level 2 under ocean waves(1). Before 1978, also search ocean waves AND ideal.
BT ocean waves
SA waves

identification
Includes use as a level 2 or 3 term appropriate to a large number of topics, e.g. on level 2 under igneous rocks(1) and under sedimentary rocks(1). See list G.
SA classification
definition
nomenclature

Idjil
use Fort Gouraud

idocrase
use vesuvianite

Igaliko
Settlement 25 miles NE of Julianenhaab near S tip of island.
BT Greenland
Arctic region

Igarka
City on the Yenisei River north of Arctic Circle in Krasnoyarsk Kray.
BT Russian Republic
USSR

IGC
use International Geological Congress

Iglesiente
Region in SW.

BT Sardinia
Italy

igneous activity
Includes use on level 3 under geosynclines(1) processes(2) and under orogeny(1).
UF activity, igneous
SA geosynclines
igneous processes
magmas
orogeny
volcanology

igneous intrusions
Not a valid index term for GeoRef. Use intrusions and igneous rocks.

igneous processes
Includes use on level 3 under mineral deposits, genesis(1) processes(2).
SA hydrothermal processes
igneous activity
magmas
mineral deposits
plutonic rocks
processes

igneous rocks
Used to describe crystalline rocks of primary origin (list H). Includes use on level 1 (list A); on level 2 under weathering(1), phase equilibria(1), and paragenesis(1). If 1, term set options are:
rock group [acidic composition, alkali basalts, alkali diorites, alkali gabbros, alkali granites, alkali syenites, alkalic composition, andesites, basalts, calc-alkalic composition, calcic composition, carbonatites, dacites, diabase, diorites, gabbros, glasses, granites, granodiorites, hypabyssal rocks, lamprophyres, mafic composition, monzonites, peridotites, phonolites, plutonic rocks, pyroclastics, quartz diorites, rhyodacites, rhyolites, syenites, trachyandesites, trachybasalts, trachytes, ultramafics, volcanic rocks]
rock name
topic [alteration, classification, composition, distribution, experimental studies, genesis, geochemistry, identification, melting, nomenclature, occurrence, petrography, petrology, properties, textures]
subtopic [e.g. chemical composition, mineral composition, physical properties]
NT appinite
carbonatites
diabase
diorites
NX eucrite
NT gabbros
granites
hypabyssal rocks
lamprophyres
picrite
picrite porphyry
plutonic rocks
NA porphyry
NX pyroclastics
NT syenites
ultramafics
volcanic rocks
NX wehrlite
SA acidic composition
agglomerate
alkalic composition
alteration
aphanitic texture
autometamorphism
banded structures
basaltic composition
basement
batholiths
calc-alkalic composition
calcic composition
chemical analysis
complexes
country rocks
crystalline rocks
cumulates
differentiation
dike swarms
dikes
emplacement
extrusive rocks
fabric
fluid inclusions
graphic texture
inclusions
intrusions
intrusive rocks
labradorite
lava
layered intrusions
lineation
lopoliths
mafic composition
magmas
matrix
megacrysts
metaigneous rocks
metamorphism
metasomatism
paragenesis
petrography
petrology
phase equilibria
phenocrysts
pipes
plutons
porphyritic texture
protoliths
schlieren
spinifex texture
standard materials
stocks
ultramafic composition
volcanology

ignimbrite
As of 1981, includes use on level 3 under igneous rocks(1) pyroclastics(2). See list H. Before 1981, included use on level 3 under igneous rocks(1) pyroclastics and glasses(2).
UF flood tuff
BA pyroclastics
volcanic rocks
BT igneous rocks
SA ash flows
ash-flow tuff
tuff
volcanic glass
welded tuff

ijolite
As of 1981, includes use on level 3 under igneous rocks(1) alkali gabbros(2). See list H. Before 1981, included use on level 3 under igneous rocks(1) alkali gabbro family(2).
BA alkali gabbros
gabbros
BT igneous rocks
SA melteigite
plutonic rocks
urtite

Ikh-Khayrkhan
Ore bearing region in central.
BT Mongolia

Iki Island
In Tsushima Strait off NW coast of Kyushu. Also search Iki-shima.
UF Iki-shima
BT Kyushu
Japan

Iki-shima
use Iki Island

Ikuno Mine
Hyogo Prefecture in central.
BT Honshu
Japan

Ikushumbetsu
Coal mine within town area of Mikasa in W central.
BT Hokkaido
Japan

Ilerdian
Europe. See list E.
BA Paleogene
Tertiary
BT Cenozoic

Iles de Loos
use Los Islands

Ili
Town on Ili River in E.
BT Kazakhstan
USSR

Ili Basin
River basin. Index Kazakhstan and/or Sinkiang Uighur as applicable.
BT Asia
SA Kazakhstan
Sinkiang Uighur

Ili Maussaq
use Ilimaussaq

Ilimaussaq
Alkaline intrusion in S Greenland, and Lovozero area of the Kola Peninsula in NW European USSR. Index Greenland and/or USSR. Also search Ilmaussaq and Ilimaussaq Intrusion.
UF Ili Maussaq
Ilimaussaq Intrusion
Ilmaussaq
SA Greenland
USSR

Ilimaussaq Intrusion
use Ilimaussaq

Il'Intsa
Paleovolcano in Vinnitsa area of the Ukrainian Shield in W central.
BT Ukraine
USSR

Ilion
use Troy

Ilium
use Troy

Illawarra Coal Measures
Top of the measures is marked by Bulli Coal Seam, which is conformably overlain by rocks of the Triassic Narrabeen Group. Illawarra District in SE New South Wales. Also search Illawarra.
BT Permian
SA Australia
Bulli Seam
New South Wales

Ille-et-Vilaine
Department in Brittany.
CO N473500N484500
W0010300W0022000
BT France

Illinoian
Pertaining to third glacial stage of Pleistocene Epoch in North America. See list E.
UF Illinoisan
BA upper Pleistocene
Pleistocene
Quaternary
BT Cenozoic

Illinois
Includes use on level 1 as an area term (list O). For term set options see list B.
CO N370000N423000
W0873000W0913000
BA Midwest
United States
NX Adams County
Boone County
Calhoun County
NT Carbondale
NX Carroll County
Cass County
NT Chicago
NX Christian County
Clark County
Clay County
Clinton County
Columbia
Cook County
Crawford County
Cumberland County
Douglas County
Edwards County
Fayette County
Franklin County
Fulton County
Greene County
Grundy County
Hamilton County
Hancock County
Hardin County
NT Iola Field
Iroquois County
NX Jackson County
Jasper County
Jefferson County
Johnson County
Kane County
NT Knox County
NX La Salle County
Lake County
Lawrence County
Lee County
Livingston County
Logan County
Macon County
Madison County
Marion County
Marshall County
Mason County
NT Mazon Creek
NX McHenry County
McLean County
Mercer County
Monroe County
Montgomery County
Morgan County
Perry County
Pike County
Pope County
Pulaski County
Putnam County
Randolph County
Richland County
NT Rock Island
NX Saint Clair County
Saline County
NT Sangamon
NX Scott County
Shelby County
Stark County
Union County
Urbana
Warren County
Washington County
Wayne County
White County
NT Will County
NX Williamson County
Winnebago County
NT Wood River
SA Bainbridge Formation
Brassfield Formation
Carbondale Formation
Caseyville Formation
Cedar Valley Formation
Chattanooga Shale
Chesterian
Fernvale Formation
Francis Creek Shale
Franconia Formation
Great Lakes region
Illinois Basin
Illinois State Geological Survey
Keokuk Limestone

La Salle Limestone
Lake Michigan
Maquoketa Formation
Mattoon Formation
Midcontinent
Midway Group
Mississippi Embayment
Mississippi River
Mississippi Valley
New Madrid region
Niagara Escarpment
Ohio River
Ohio River valley
Peoria Loess
Platteville Formation
Pottsville Group
Richmond Group
Ripley Formation
Saint Laurent Limestone
Saint Louis Limestone
Saint Peter Sandstone
Sainte Genevieve Limestone
Salem Limestone
Shakopee Formation
Springfield
Springfield Coal Member
Upper Mississippi Valley
Versailles
Warsaw Formation
Wedron Formation
Wilcox Group
Woodfordian

Illinois Basin
Index states as applicable.
BT United States
SA Illinois
Indiana
Kentucky

Illinois State Geological Survey
SA Illinois
survey organizations

Illinoisan
use Illinoian

illite
BA clay minerals
sheet silicates
silicates
BT minerals
SA crystallinity
hydromica
muscovite

Ilmaussaq
use Ilimaussaq

Ilmen Mountains
Near Chelyabinsk in the Central Urals.
BT Russian Republic
USSR

ilmenite
Also search menaccanite.
UF menaccanite
BA oxides
BT minerals
SA heavy minerals
magnetic minerals

Iloilo Basin
SW Panay.
BT Philippine Islands

ilvaite
BA orthosilicates
silicates
BT minerals

imagery
As of 1978, term is used on level 2 under remote sensing(1).
NT color imagery
SA geophysical surveys
Landsat
remote sensing

Imandra
Lake on W Kola Peninsula, Murmansk Oblast.
BT Russian Republic
USSR

Imataca Complex
In Serrania de Imataca Mountains S of Orinoco Delta.
BT Bolivar
Venezuela

imbrication
Includes use on level 3 under sedimentary structures(1) planar bedding structures(2). See list K.
BA planar bedding structures
sedimentary structures
SA bedding plane irregularities
tectonic wedges
tectonics

Imbrium Basin
use Sea of Rains

immiscibility
Includes use on level 3 under magmas(1). Also search miscibility.
UF miscibility
SA magmas
phase equilibria
solubility

imogolite
BA clay minerals
sheet silicates
silicates
BT minerals

impact features
Includes use on level 2 under geomorphology(1).
UF features, impact
SA astroblemes
Charlevoix
cratering
craters
cryptoexplosion features
geomorphology
impactite
impacts
meteor craters
meteorites
shatter cones
suevite

impact phenomena
A valid term through 1972. From 1972 through 1978 impact was used. After 1978, use impacts.

impact statements
A level 1 term as of 1978. For official documents detailing the impact of contemplated works or land use on the environment. Includes use on level 2 under land use(1), nuclear facilities(1), and pollution(1). If 1, term set options are:
conservation
subtopic
geologic hazards
subtopic
land use
subtopic
pollution
subtopic
reclamation
subtopic
waste disposal
subtopic
SA conservation
ecology
engineering geology
environmental geology
geologic hazards
land use
nuclear facilities
pollution
reclamation
waste disposal

impactite
SA cryptoexplosion features
Darwin glass
impact features
metamorphic rocks
meteorites
suevite

impacts
Includes use on level 3 under seismology(1) and volcanology(1). Before 1973, also search impact phenomena; impact.
SA craters
cryptoexplosion features
geomorphology
impact features
meteor craters
meteorites
planetology

Imperial County
BT California
United States

Imperial Valley
Imperial County in extreme SE.
CO N324000N333000
W1151500W1161500
BT California
United States
NT East Mesa KGRA

import
Term introduced in 1981.
SA economics

import value
Term introduced in 1981.
SA economics

impregnated deposits
Term introduced in 1981.
SA mineral deposits

impurities
A general term used mostly in reference to water.
SA ground water
pollution
purification
water

In
use indium

in situ
Applicable to many fields, especially rocks, soils, or fossils when in the situation in which they were originally formed or deposited.

Inadunata
Subclass. Includes use on level 3 under Echinodermata(1) Crinoidea(2).
BA Crinoidea
Crinozoa
Echinodermata
BT Invertebrata

Inangahua
Village in NW.
BT South Island
New Zealand

Inari
Town on Lake Inari in N part of country.
UF Enari
BT Finland

Inarticulata
Includes use on level 2 under Brachiopoda(1). See list F.
BA Brachiopoda
BT Invertebrata
NA Lingula
SA Articulata

incarbonization
use coalification

Inceptisols
Includes use on level 3 under soils(1). See list M.
BA soils

inclined folds
Term introduced in 1978. Includes use on level 3 under folds(1) orientation(2). Before 1978, also search folds AND inclined.
BA folds
SA orientation

inclined seismic interface
Term introduced in 1981.
SA curved seismic interface
discontinuities
seismology

inclusions
Includes use on level 1 (list A) and 2 for studies on enclosures or enclaves in rocks or minerals. If 1, term set options are:
mineral inclusions
type of inclusion or host materials [e.g. name of mineral, name of rock, minerals, rocks] or topic
topic [analysis, changes, composition, detection, experimental studies, genesis, geologic barometry, geologic thermometry, P-T conditions, observations, theoretical studies]
subtopic
xenoliths
type of xenolith or host materials or topic
NT mineral inclusions
xenoliths
SA assimilation
chemical analysis
country rocks
crystal growth
crystal structure
fluid inclusions
geochemistry
geologic barometry
geologic thermometry
host rocks
igneous rocks
isotopes
lava
magmas
mineral deposits
minerals
P-T conditions
petrology
phase equilibria
rocks

incoalation
use coalification

incompressibility modulus
use bulk modulus

increment
use recharge

Indefatigable Island
use Chaves Island

index beds
use key beds

index fossils
Includes use on level 3 under paleontology(1). Also search guide fossils.
UF guide fossils
SA assemblages
biostratigraphy
fossils
microfossils
nannofossils
stratigraphy

index maps
BA maps

index of refraction
use refractive index

India
Includes use on level 1 as an area term (list O). For term set options see list B.

CO N070000N370000
E0970000E0680000
BA Indian Peninsula
Asia
NT Ajay River
Andhra Pradesh
Beas River
Bengal Islands
Bihar
Bundelkhand
Cauvery Basin
Coromandel Coast
Damodar Valley
Deccan Plateau
Delhi
Eastern Ghats
Garo Hills
Ghats
Goa
Godavari River
Godavari Valley
Gujarat
Haryana
Himachal Pradesh
Jammu and Kashmir
Jumna River
Karnataka
Kerala
Krishna
Kunavaram Series
Laccadive Islands
Madhya Pradesh
Mahanadi Valley
Maharashtra
Mysore
Narmada River
Narmada Valley
Neill Island
Northeastern India
Orissa
Pondicherry
Pranhita-Godavari Valley
Punjab State
Rajasthan
Satpura Range
Sikkim
Singhbhum shear zone
Son Valley
Tamil Nadu
Trans-Aravalli Vindhyan Basin
Uttar Pradesh
West Bengal
Western Ghats
SA Aravalli System
Ariyalur Stage
Athgarh Sandstone
Badami Series
Bagh Beds
Barakar Stage
Bhander Group
Bhuj Series
Bijawar System
Brahmaputra River
Chainpur Meteorite
Cuddalore Series
Cuddapah System
Damuda Series
Deccan Traps
Delhi System
Dharwars
Ganges River
Gangpur Series
Gondwana
Gondwana System
Himalayas
Hyderabad
Indian Plate
Indian Shield
Intertrappean Beds
Jabalpur Series
Kaimur Sandstone
Kaladgi System
Karakoram
Karharbari Stage
Kasauli Series
Kurnool System
Ladakh
Lathi Formation
Lesser Himalayas
Muth Quartzite
Nari Series
Neyveli Lignite
Niniyur Group
Panchet Series
Raghavapuram Shales
Ramnagar
Raniganj Stage
Sausar Series
Semri Series
Singhbhum Granite
Siwalik Range
Siwalik System
Tal Formation
Talchir Formation
Talchir Series
Tons Member
Vindhyan
Vindhyan Basin
Warkalli Formation

Indian Ocean
Includes use on level 1 or 2 as an area term (list O). If 1, see term set options under list B.
CO S700000N250000
E1200000E0200000
NT Agulhas Bank
Amsterdam Island
Andaman Sea
Anjouan Island
NA Arabian Sea
NT Argo abyssal plain
Bay of Bengal
Carlsberg Ridge
Christmas Island
Crozet Islands
Cuvier abyssal plain
NA East Indian Ocean
NT Gascoyne abyssal plain
Great Australian Bight
Madagascar Basin
Marion Island
Mascarene Basin
Mascarene Islands
Mid-Indian Ridge
Mozambique Channel
Ninetyeast Ridge
Perth abyssal plain
NA Red Sea
NT Somali Basin
Southwest Indian Ridge
NA West Indian Ocean
NT Wharton Basin
SA Antarctic Ocean
DSDP Site 214
DSDP Site 253
DSDP Site 254
DSDP Site 259
DSDP Site 260
DSDP Site 261
DSDP Site 262
DSDP Site 263
DSDP Site 264
Indian Ocean Islands
Indian Plate
Leg 22
Leg 23
Leg 24
Leg 25
Leg 26
Leg 27
Leg 28

Indian Ocean Islands
Term introduced in 1981. Includes use on level 1 as an area term (list O). For term set options see list B.
NA Comoro Islands
NZ Malagasy Republic
NA Maldive Islands
Mauritius
Reunion
Seychelles
SA Bengal Islands
Indian Ocean
Laccadive Islands

Indian Ocean Ridge
use Mid-Indian Ridge

Indian Peninsula
Term introduced in 1981.
BT Asia
NA Afghanistan
Bangladesh
Bhutan
India
Nepal
Pakistan
Sri Lanka

Indian Plate
Tectonic plate extending from the Himalayas to the Tonga Trench in the Pacific Ocean. Includes Australia as well as most of India. Also search Australian Plate.
UF Australian Plate
SA Australia
Bangladesh
India
Indian Ocean
New Guinea
New Zealand
Pacific Ocean
Pakistan
plate tectonics
plates
Tibet

Indian Ridge
use Mid-Indian Ridge

Indian Shield
Plateau region in S and central India plus Sri Lanka. Index countries as applicable. Also search Peninsular Shield.
UF Peninsular Shield
BT Asia
SA Deccan Plateau
India
Sri Lanka

Indian-Antarctic Ridge
use Southeast Indian Ridge

Indiana
Includes use on level 1 as an area term (list O). For term set options see list B.
CO N374500N414500
W0844500W0881000
BA Midwest
United States
NX Adams County
Benton County
NT Blackford County
NX Boone County
Brown County
Carroll County
Cass County
Clark County
Clinton County
Crawford County
Delaware County
Fayette County
Floyd County
Franklin County
Fulton County
Grant County
Greene County
Hamilton County
Hancock County
Harrison County
Howard County
Jackson County
Jasper County
Jefferson County
Johnson County
NT Kentland
NX Lake County
Lawrence County
Madison County
Marion County
Marshall County
Monroe County
Montgomery County
Morgan County
Newton County
Orange County
Perry County
Pike County
Pulaski County
Putnam County
Randolph County
Rush County
Scott County
Shelby County
Sullivan County
NT Tippecanoe County
NX Union County
Warren County
Washington County
Wayne County
White County
SA Borden Group
Brassfield Formation
Chesterian
Cincinnati Arch
Eden Shale
Fairview Formation
Great Lakes region
Illinois Basin
Lake Michigan
Mansfield Formation
Maumee River valley
Midcontinent
New Albany Shale
New Madrid region
Ohio River
Ohio River valley
Richmond Group
Saint Louis Limestone
Saint Peter Sandstone
Sainte Genevieve Limestone
Salem Limestone
Staunton Formation
Versailles
Wabash Formation
Waldron Shale
Warsaw Formation
Wedron Formation
White River
Whitewater River valley

indicators
Includes use on level 2 under paleoclimatology(1), under paleoecology(1), and under plate tectonics(1).
SA depth indicators
diagenesis
geochemical indicators
paleoclimatology
paleoecology
plate tectonics

Indigirka River
NE Yakutia. Flows into East Siberian Sea. Also search Indigirka.
BT Russian Republic
USSR

indium
Includes use on level 1 and 2 as a chemical element (list D).
UF In
SA elements

Indo-Antarctic Ridge
use Southeast Indian Ridge

Indochina
The SE peninsula of Asia including the Malay Peninsula. (French Indochina consisted of current Cambodia, Laos, and Vietnam). Index countries as applicable. Includes use on level 1 as an area term (list O). For term set options see list B.
UF Farther India
BA Far East
Asia
SA Burma
Cambodia

Laos
Malay Peninsula
Malaysia
Thailand
Vietnam
West Malaysia

indochinite
BA tektites

Indonesia
Includes use on level 1 as an area term (list O). For term set options see list B.
CO S113000N063000
E1413000E0950000
BA Far East
Asia
NT Bangka
Billiton
Celebes
Irian Jaya
Java
Kalimantan
Krakatoa
Lesser Sunda Islands
Moluccas
Sumatra
SA Arafura Sea
Borneo
Indonesian Seas
Malay Archipelago
Timor

Indonesian Seas
Term introduced in 1981. Includes small seas of the Pacific Ocean within Indonesia.
BA West Pacific
BT Pacific Ocean
NA Celebes Sea
Java Sea
SA Indonesia

Indre
Department in W central.
CO N462000N471700
E0021000E0005000
BT France
SA Touraine

Indre-et-Loire
Department in N central.
CO N464500N474500
E0012000E0000000
BT France
NT Tours
SA Touraine

induced earthquakes
Term introduced in 1981. Includes use on level 3 under earthquakes(1).
SA earthquakes
engineering geology
seismology

induced magnetization
Term introduced in 1981.
BA magnetization

induced polarization
Includes use on level 3 under geophysical methods(1).
BA geophysical methods
SA electrical field
electrical methods
electrical surveys
geophysical methods
geophysical surveys
polarization
well-logging

induction
Used for geophysical properties.
SA electromagnetic methods
geophysical methods
magnetic field
paleomagnetism
well-logging

Indus River
Rises in SW Tibet and flows into Arabian Sea. Index Jammu and Kashmir, Pakistan, and Tibet as applicable.
BT Asia
SA Jammu and Kashmir
Pakistan
Tibet

industrial ash
use ash

industrial minerals
Includes use on level 1 as a commodity term (list C).
BT minerals
SA abrasives
absorbent materials
carbonado
corundum deposits
economic geology
garnet deposits
graphite deposits
nonmetal deposits
refractory materials

industrial patents
Term introduced in 1981.
SA economics
industry

industrial waste
As of 1978, term is used on level 2 under waste disposal(1).
UF waste, industrial
SA environmental geology
liquid waste
pollution
radioactive waste
solid waste
waste disposal
waste water

industry
Used as a general term.
SA annual report
chemical industry
economic geology
engineering geology
environmental geology
industrial patents

inert gases
use noble gases

inertial seismographs
Term introduced in 1981.
BT seismographs

inertinite
Includes use on level 3 under sedimentary rocks(1) organic residues(2). See list I.
BA macerals
coal
organic residues
BT sedimentary rocks
NA fusinite
micrinite
sclerotinite

infiltration
Includes use on level 3 under hydrology(1).
SA ground water
hydrology
percolation
recharge
soils
water regimes
waterlogging

infiltration galleries
Term introduced in 1981.
SA ground water
underground channels

infinite models
Term introduced in 1981.
BA models
SA mathematical models
semi-infinite models

influence
Term introduced in 1981.
SA effects

information systems
SA automatic data processing
data bases
data handling
data retrieval
data storage
systems

Infracambrian
BA upper Precambrian
Precambrian
SA Eocambrian
Vendian

infrared
A valid term through 1977. After 1977, use infrared spectroscopy.

infrared methods
Includes use on level 2 under geophysical methods(1). See spectroscopy(1) methods(2).
BA geophysical methods
SA infrared surveys
methods
remote sensing
thermal emission

infrared spectra
Includes use on level 3 under chemical analysis(1) or spectroscopy(1). Before 1978, also search infrared.
SA chemical analysis
infrared spectroscopy
spectra
spectroscopy

infrared spectroscopy
Not valid term through 1977. After 1977, includes use on level 3 under spectroscopy(1) and under chemical element(1). Before 1978, also search spectroscopy AND infrared.
BT spectroscopy
SA analysis
infrared spectra

infrared surveys
Includes use on level 3 under area terms(1) geophysical surveys(2).
BA geophysical surveys
BT surveys
SA infrared methods
remote sensing

Ingaladhal
use Ingaldhal

Ingaldhal
Ore bearing region in Chitradurga District in central Mysore.
UF Ingaladhal
BT Mysore
India

Ingham County
S central Lower Peninsula. Also search Ingham.
BT Michigan
United States

Ingleborough
Mountain in W Yorkshire in N.
BT England
Great Britain
United Kingdom

Ingleton
Village and parish in W Yorkshire in N.
BT England
Great Britain
United Kingdom

Inglewood
City in Los Angeles County in S.
BT California
United States

Ingul River
S Ukraine. Also search Ingul.
BT Ukraine
USSR

Ingulets River
S Ukraine. Also search Ingulets.
BT Ukraine
USSR

Ingur (River)
use Inguri River

Inguri River
Flows into Black Sea. Also search Ingur; Inguri.
UF Ingur (River)
BT Georgian Republic
USSR

inhomogeneity
As of 1981, no longer a valid term for GeoRef. Use heterogeneity.

injection
Includes use on level 3 under waste disposal(1).
SA engineering geology
environmental geology
fluid injection
waste disposal

Inland Sea
Extends E-W with W Honshu on the N, Shikoku on the S, and Kyushu on the W. Connected with Pacific Ocean by 4 channels.
UF Seto Inland Sea
Seto Sea
Seto-chi-umi
Seto-Naikai
Setouchi
BT Japan
Asia

inlets
SA bays
coastal environment
shore features
shorelines
tidal inlets

Inn Valley
River valley. Index Bavaria and countries as applicable.
BT Europe
SA Austria
Bavaria
Switzerland
West Germany

innelite
BA orthosilicates
silicates
BT minerals

inner core
BA core
SA outer core

Inner Hebrides
Islands immediately off W coast of Scotland. Separated from Outer Hebrides by the Straits of The Minch, the Little Minch, and the Sea of Hebrides.
BT Scotland
Great Britain
United Kingdom
NT Mull Island
SA Hebrides
Islay
Isle of Skye
Outer Hebrides
Raasay
Rhum

Inner Mongolia
Autonomous region S, SE, and E of Mongolia.
BT China
NT Khingan Range
SA Amur Basin
Gobi Desert
Kerulen River

inner shelf
BA continental shelf
SA outer shelf

inner slope
Term introduced in 1978.

BA continental slope
SA outer slope

inner transition elements
use rare earths

Innsbruck
City in central.
BT Tyrol
Austria

Inocerami
Term introduced in 1981. Includes use on level 2 under Mollusca(1). See list F.
BA Pteriina
Cyrtodontida
Bivalvia
Mollusca
BT Invertebrata

Inoceramidae
Includes use on level 3 under Mollusca(1). See list F.
BA Bivalvia
Mollusca
BT Invertebrata

Inoceramus
Genus. Includes use on level 3 under Mollusca(1). See list F.
BA Bivalvia
Mollusca
BT Invertebrata

inorganic
A valid term through 1977. After 1977, use inorganic materials.

inorganic acids
Term introduced in 1981.
SA acids
organic acids

inorganic materials
Before 1978, also search inorganic AND materials.
SA materials
organic materials

inosilicates
use chain silicates

Inowroclaw
City in NW central part of country.
BT Bydgoszcz
Poland

INQUA
Acronym. Includes use on level 3 under associations(1).
UF International Association for Quaternary Research
SA associations

Insecta
Class. Includes use on level 1 and 2 as a fossil term (list F).
BT Mandibulata
Arthropoda
Invertebrata
NA Archostemata
Blattopteroida
Coleopteroida
Dermapteroida
Diptera
Ectotropha
Entotropha
Ephemeropteroida
Hemipteroida
Heteroptera
Hymenopteroida
Lepidopteroida
Mecopteroida
Neuropteroida
Odonatopteroida
Orthopteroida
Palaeodictyopteroida
Plecoptera
Plectoptera
Protorthoptera
Psocopteroida
Raphidiodea
Siphonapteroida
Thysanopteroida

Insectivora
Includes use on level 2 under Mammalia(1). See list F.
BA Mammalia
BT Tetrapoda
Vertebrata

inselbergs
SA erosion features
geomorphology

insolation
SA mechanical weathering
weathering

insoluble residues
Includes use on level 3 under sedimentary rocks(1). Also search IR.
UF IR
residues, insoluble
SA sedimentary rocks
siliceous composition
solubility

instabilities, plasma
use plasma instabilities

installations, marine
use marine installations

installations, submarine
use submarine installations

installations, underground
use underground installations

institutions
Term introduced in 1981.
SA associations

instruments
Includes use as a level 2 or 3 term appropriate to a large number of topics, e.g. on level 2 under meteorology(1), geochemistry(1), heat flow(1), and hydrology(1). See list G.
SA accelerometers
design
extensometers
gravimeters
magnetometers
ocean bottom seismographs
Seislog
seismographs
techniques
tiltmeters

intensity
General term as of 1981. For intensity of earthquakes, use seismic intensity. Includes use on level 3 under magnetosphere(1) cosmic rays(2) and under interplanetary space(1) cosmic rays(2). Also search intensities. Before 1981, also used on level 2 under earthquakes(1).
SA aurora
bow shock waves
cosmic rays
disturbances
interplanetary space
magnetic intensity
magnitude
particle precipitation
particles
seismic intensity
solar cycles
sudden commencements

interactions, boundary
use boundary interactions

interface, air-sea
use air-sea interface

interface, oil-gas
use oil-gas interface

interface, oil-water
use oil-water interface

interface, sediment-water
use sediment-water interface

interfaces
Term introduced in 1981.
SA air-sea interface
oil-gas interface
oil-water interface
sediment-water interface

interglacial environment
Term introduced in 1978. Before 1978, search interglacial.
SA environment
glacial environment
glacial geology
paleoclimatology

intergrowths
Includes use on level 3 under crystal growth(1).
SA crystal growth
crystallization
crystallography
crystals

interior
As of 1981, restricted to the Earth. Includes use on level 2 under Earth(1) and seismology(1). Before 1981, also included use on level 2 under Moon(1) and individual planets.
SA core
Earth
geodynamics
lunar interior
Moon
planetary interiors
seismology

intermediate earthquake
use intermediate-focus earthquakes

intermediate-focus earthquakes
Term introduced in 1978. Includes use on level 3 under earthquakes(1). Before 1978, search intermediate-focus.
UF intermediate earthquake
SA deep-focus earthquakes
earthquakes
focus
shallow-focus earthquakes

intermontane basins
SA basins

internal waves
Includes use on level 2 under ocean waves(1).
SA ocean waves
waves

International Association for Quaternary Research
use INQUA

international cooperation
Includes use on level 3 under associations(1).
UF cooperation, international
SA aid projects
ASEAN
associations
negotiations
policy
technical cooperation

International Decade of Ocean Exploration
SA oceanography

International Geological Congress
Includes use on level 3 under associations(1).
UF IGC
SA associations

International Monetary Fund
Term introduced in 1981.
SA economics

International Phase of Ocean Drilling
use IPOD

International Union of Geological Sciences
use IUGS

interplanetary comparison
Term introduced in 1981.
SA extraterrestrial geology
planetology
planets

interplanetary dust
Includes use on level 2 under interplanetary space(1).
SA cosmic dust
dust
interplanetary space
meteorites
particles

interplanetary space
Usually out-of-scope for GeoRef. Includes use on level 1. Also search space. Term set options are:
comets
subtopic
cosmic rays
albedo
cut-off rigidities
intensity
ionization
subtopic (flares, measuring platform, particles, solar cycles)
electromagnetic radiation
subtopic
instruments
name of instrument or platform or subtopic
interplanetary dust
subtopic
shock waves
subtopic
solar wind
subtopic
UF outer space
space
SA aeronomy
albedo
asteroids
astrophysics and solar physics
comets
cosmic dust
cosmic rays
cutoff rigidities
electromagnetic radiation
extraterrestrial geology
intensity
interplanetary dust
ionization
magnetosphere
meteoroids
meteors
particles
planetology
plasma
shock waves
solar cycles
solar flares
solar system
solar wind

interpretation
Includes use as a level 2 or 3 term appropriate to a large number of topics, e.g. on level 2 under foliation(1) and heat flow(1). See list G.

interstitial water
use pore water

intertidal environment
Term introduced in 1978. As of 1981, used for littoral environment. Before 1981, also search littoral environment; tidal environment. Before 1978, search intertidal; littoral.
SA ecology
environment
intertidal sedimentation
sedimentation
wave-cut platforms

intertidal sedimentation
Term introduced in 1981.
SA intertidal environment
sedimentation

Intertrappean Beds
Upper Cretaceous to lower Eocene. Sedimentary deposits of fossiliferous cherts freshwater limestone or of bole intercalated between the flows of Deccan Trap Basalt. Also search Deccan Intertrappean; Deccan Intertrappean Beds; Deccan Intertrappean Series; Intertrappean.
UF Deccan Intertrappean
Deccan Intertrappean Beds
Deccan Intertrappean Series
SA Cretaceous
Deccan Traps
Eocene
India
lower Eocene
Upper Cretaceous

intraclastic texture
use intraclasts

intraclasts
UF intraclastic texture
SA carbonate rocks
textures

intracontinental belts
Term introduced in 1981.
SA orogenic belts
orogeny

intracratonic basins
Term introduced in 1981.
SA basins
cratons
tectonics

intrafolial folds
Term introduced in 1978. Before 1978, search folds AND intrafolial.
BA folds

intramagmatic deposits
Term introduced in 1981.
SA mineral deposits

intraplate
As of 1981, no longer a valid term for GeoRef. Use intraplate tectonics. Included use on level 3 under plate tectonics(1).

intraplate tectonics
Term introduced in 1981. Includes use on level 3 under plate tectonics(1).
SA plate tectonics

Intrazonal soils
See list M.
BA soils
SA Bog soils
Brown forest soils
Hydromorphic soils
Meadow soils
Planosols
Solonchak soils
Solonetz soils

intrusion (ground water)
use salt-water intrusion

intrusions
Includes use on level 1 (list A). Used for the igneous rocks mass formed by emplacement of magma in pre-existing rock. It is also used for the process of emplacement. Term set options are:
kind of intrusion [batholiths, diatremes, dikes, laccoliths, layered intrusions, lopoliths, pipes, plugs, plutons, ring complexes, sills, stocks]
topic
topic [age, classification, composition, contact metamorphism, distribution, emplacement, evolution, genesis, geochemistry, mechanism, occurrence, petrology]
subtopic
UF invasion (intrusion)
irruption (intrusion)
NA batholiths
diatremes
dikes
laccoliths
layered intrusions
lopoliths
pipes
plugs
plutons
ring complexes
sills
stocks
SA assimilation
aureoles
breccia pipes
contamination
country rocks
crystallization
diapirism
differentiation
dike swarms
domes
emplacement
foliation
heat flow
hybridization
igneous rocks
intrusive rocks
lineation
magma chambers
magmas
metamorphic rocks
metamorphism
metasomatism
petrology
ring structures
swarms
veins
volcanology
zoning

intrusive
A valid term through 1977. After 1977, use intrusive rocks.

intrusive mountain
use batholiths

intrusive rocks
Term introduced in 1978. Includes use on level 3 under igneous rocks(1). Before 1978, also search igneous rocks AND intrusive.
SA hypabyssal rocks
igneous rocks
intrusions
rocks

Inuvik
Village in NW District of Mackenzie.
BT Northwest Territories
Canada

invasion (intrusion)
use intrusions

inventory
Includes use on level 2 and 3 under commodity terms (list C); on level 3 under environmental geology and under engineering geology(1).
SA engineering geology
environmental geology

Inverness
City in Inverness-shire in N.
CO N564000N580000
W0033000W0073000
BT Scotland
Great Britain
United Kingdom

Inverness-shire
County in NW.
BT Scotland
Great Britain
United Kingdom

inverse magnetization
Term introduced in 1981.
BA magnetization

inverse problem
Includes use on level 3 under geophysics(1), geophysical methods(1), and seismology(1). Also search inversion.
UF inversion (seismology)
SA direct problem
geophysical methods
geophysical surveys
geophysics
seismology

inverse thermoremanent magnetization
Term introduced in 1981.
BA remanent magnetization
SA partial thermoremanent magnetization
thermoremanent magnetization

inversion (seismology)
use inverse problem

Invertebrata
Includes use on level 1 and 2 as a fossil term (list F); on level 3 under soils(1) biota(2). Term is to be used on level 1 only when more specific terms do not apply or are too numerous to be recorded.
NT Archaeocyatha
Arthropoda
Brachiopoda
Bryozoa
NX Camaroidea
Cephalodiscida
NT Coelenterata
NX Dendroidea
Didymograptina
Didymograptus
Diplograptina
NT Echinodermata
NX Enteropneusta
Glossograptina
Graptolithina
Graptoloidea
Isograptus
NT Mollusca
NX Monograptina
Monograptus
Monograptus uniformis
NT Porifera
Protista
NX Pterobranchia
Rhabdopleurida
Stolonoidea
Tuboidea
NT worms
SA biota
ichnofossils
invertebrates
shells
tests

invertebrates
Term introduced in 1981 as the common name for Invertebrata. See list F. To be used for Invertebrata when they are used as tools in stratigraphy. Includes use on level 1 when more specific common names do not apply or are too numerous to be recorded. Before 1981, search Invertebrata AND biostratigraphy. If 1, term set options are:
biostratigraphy
age [List E]
SA biostratigraphy
Invertebrata

investment
Term introduced in 1981.
SA economics
investment legislation

investment legislation
Term introduced in 1981.
SA economics
investment
legislation

Inyo County
E California.
CO N354500N373000
W1154000W1184500
BT California
United States
NT Coso Hot Springs KGRA

Inyo Mountains
Range in W central Inyo County in E California. Also search White-Inyo Mountains and White-Inyo Range.
BT California
United States

Io
use ionium

Io/Th
Includes use on level 3 under absolute age(1) methods(2).
SA absolute age
ionium
thorium

Io/U
Includes use on level 3 under absolute age(1) methods(2).
SA absolute age
ionium
uranium

iodates
Includes use on level 2 under minerals(1). See list L.
BT minerals

iodides
Term introduced in 1981. Includes use on level 2 under minerals(1). See list L.
BA halides
BT minerals

iodine
Includes use on level 1 and 2 as a chemical element (list D). As of 1981, use iodine deposits for iodine as a commodity. Before 1981, included use on level 1 as a commodity term.
UF I
SA brines
elements
I-129
I-131
iodine deposits

iodine deposits
Term introduced in 1981. Includes use on level 1 and 2 as a commodity term (list C). Before 1981, search iodine AND deposits.
SA brines
iodine
salt

Iola Field
Term introduced in 1978. Gas field in Clay County in S central Illinois. Before 1978, search Iola; Iola gas field.
UF Iola gas field
BT Illinois
United States

Iola gas field
use Iola Field

ion densities and temperatures
Includes use on level 2 under ionosphere(1). Also search ion densities.
SA densities and temperatures
density
electron content
ionosphere
ions
temperature

ion exchange
As of 1978, term is used on level 2 under crystal chemistry(1). Also

search cation exchange.
UF base exchange
SA anions
cation exchange capacity
clay mineralogy
crystal chemistry
crystal structure
diffusion
electrolytes
geochemistry
ions
sea water
soils

ion probe
Includes use on level 3 under chemical analysis(1) methods(2) and under spectroscopy(1) methods(2). For methodology, not data.
UF probe, ion
SA analysis
chemical analysis
electron probe
ion probe data
ions
spectroscopy

ion probe data
Term introduced in 1978. Before 1978, search ion probe AND data.
SA chemical analysis
data
ion probe
spectroscopy

Ionian Islands
BT Mediterranean region
SA Greek Ionian Islands

Ionian Sea
Between SE coast of Italy and W Greece. Includes use on level 1 as an area term (list O). For term set options see list B.
CO N362000N403000
E0222000E0150000
BA East Mediterranean
Mediterranean Sea
SA DSDP Site 374

Ionian Zone
S, SW, and W.
BT Albania
SA Cretaceous
Jurassic
Mesozoic
Tertiary

ionium
Radioactive isotope of radon of uranium-radium series. Also search Io.
UF Io
SA Io/Th
Io/U
isotopes
radon

ionization
Includes use on level 2 under aeronomy(1); on level 3 under interplanetary space(1) cosmic rays(2).
SA aeronomy
cosmic rays
interplanetary space
ionosphere
ions
particle radiation
solar wind

ionosphere
Usually out-of-scope for GeoRef. Includes use on level 1 for special bibliographies. Term set options are:
albedo
[magnetic albedo, neutron albedo, also albedo of electromagnetic waves]
aurora
subtopic
currents
topic [e.g. field-aligned, ring, etc.]
disturbances
subtopic
D-region
subtopic
electrical field
subtopic
electrojet
equatorial regions
polar regions
subtopic
electron content
topic [e.g. density, region of atmosphere, temperature]
E-region
topic [e.g. density, sporadic E, temperature, variations]
F-region
topic [density, spread-F, temperature]
instruments
name of instrument or phenomena or platform
ion densities and temperatures
name of ion or phenomenon or platform
name of phenomena or platform
name of region
subtopic
particle precipitation
name of particle or phenomenon or platform
phenomenon or platform
scintillations
subtopic
propagation
name of phenomenon or subtopic
type of wave or platform
SA aeronomy
albedo
atmosphere
aurora
currents
D-region
disturbances
E-region
electrical field
electrojet
electron content
F-region
ion densities and temperatures
ionization
magnetosphere
meteorology
particle precipitation
particle radiation
polar regions
scintillations
solar cycles
stratosphere

ions
Includes use on level 2 under meteorology(1); on level 3 under geochemistry(1). Also search cations; anions.
SA anions
carbonate ion
cation exchange capacity
cations
chloride ion
electrolytes
fixation
fluoride ion
geochemistry
ion densities and temperatures
ion exchange
ion probe
ionization
meteorology
particle precipitation
particles
sulfate ion
trapped particles

Iowa
Includes use on level 1 as an area term (list O). For term set options see list B.
CO N402500N433000
W0901000W0963500
BA Midwest
United States
NX Adams County
Benton County
Boone County
Buchanan County
Butler County
Calhoun County
Carroll County
Cass County
Cherokee County
Clarke County
Clay County
Clinton County
Crawford County
Dallas County
NT Davenport
NX Delaware County
Dickinson County
Fayette County
Floyd County
Franklin County
Fremont County
Greene County
Grundy County
Hamilton County
Hancock County
Hardin County
Harrison County
Howard County
Humboldt County
Iowa County
Jackson County
Jasper County
Jefferson County
Johnson County
Jones County
Lee County
Lucas County
Lyon County
Madison County
NT Mahaska County
NX Marion County
Marshall County
Mitchell County
Monroe County
Montgomery County
Page County
Pocahontas County
Polk County
Scott County
Shelby County
Sioux County
Taylor County
Union County
Warren County
Washington County
Wayne County
Webster County
Winnebago County
SA Beil Limestone Member
Cedar Valley Formation
Cherokee Group
Colorado Group
Desmoinesian
Douglas Group
Ervine Creek Limestone
Forest City Basin
Kansas City Group
Keokuk Limestone
Labette Shale
Lawrence Formation
Lecompton Limestone
Lime Creek Formation
Maquoketa Formation
Marmaton Group
Midcontinent
Mississippi River
Mississippi Valley
Missouri River
Missouri River basin
Missouri River valley
Niagara Escarpment
Oread Limestone
Peoria Loess
Platteville Formation
Plattsburg Limestone
Plattsmouth Limestone Member
Pleasanton Group
Saint Louis Limestone
Saint Peter Sandstone
Sainte Genevieve Limestone
Salem Limestone
Shakopee Formation
Sioux Quartzite
Stanton Formation
Upper Mississippi Valley
Wabaunsee Group
Warsaw Formation
Wyandotte Limestone

Iowa County
Index states as applicable.
BX Iowa
Wisconsin
BT United States

iozite
use wustite

IPOD
Term introduced in 1981. International Phase of Ocean Drilling. Phase of Deep Sea Drilling Project beginning with Leg 45. Before 1981, also search International Phase of Ocean Drilling.
UF International Phase of Ocean Drilling
SA Deep Sea Drilling Project
DSDP Site 395
DSDP Site 396
DSDP Site 417
DSDP Site 418
Leg 45
Leg 46
Leg 47
Leg 48
Leg 49
Leg 50
Leg 51
Leg 52
Leg 53
Leg 54
Leg 55
Leg 56
Leg 57
Leg 58
Leg 59
Leg 60
Leg 61
Leg 62
Leg 63
Leg 64
Leg 65
Leg 66

Ipswich
City in SE.
BT Queensland
Australia

Ipswich Coal Measures
SE Queensland.
BT Triassic
SA Australia
Queensland

Ipswichian
Europe. See list E.
BA upper Pleistocene
Pleistocene
Quaternary
BT Cenozoic

Ir
use iridium

IR
use insoluble residues

Iran
Includes use on level 1 as an area

term (list O). For term set options see list B.
CO N250000N393000
E0632000E0440000
UF Persia
BA Middle East
Asia
NT Anarak
Azerbaijan
Dasht-e-Bayaz
Djadjerud Valley
Elburz
Fars
Kerman
Lar
Lut Desert
Songhor
Talysh Mountains
Teheran
SA Bakhtiari Formation
Caspian Basin
Caspian Sea
Geirud Formation
Kopet-Dag Range
Kura Lowland
Zagros

Iraq
Includes use on level 1 as an area term (list O). For term set options see list B.
CO N290000N371500
E0483000E0390000
BA Middle East
Asia
NT Kirkuk
SA Bakhtiari Formation
Euphrates River
Mesopotamia
Near East
Shiranish Formation
Sinjar Formation
Zagros
Zubair Formation

Irati Formation
BT Permian
SA Brazil
Goias
Mato Grosso
Minas Gerais
Parana
Passa Dois Group
Sao Paulo

Ireland
The Republic of Ireland, also called Eire, which occupies the 26 counties in the S, central, and NW of the island of Ireland. Includes use on level 1 as an area term (list O). For term set options see list B.
CO N513000N552000
W0063000W0103000
UF Eire
BA Western Europe
Europe
NT Clare
Connemara
Cork
Dingle Peninsula
Donegal
Dublin
Galway
Kerry
Kildare
Leinster
Limerick
Mayo
Meath
Sligo
Tipperary
Tynagh
Wexford
Wicklow
SA Northern Ireland

Irgiz
Village on Irgiz River in Aktyubinsk Oblast in E.
BT Kazakhstan
USSR

Irian Barat
use Irian Jaya

Irian Jaya
W half of island of New Guinea. Also search West Irian.
CO S091500S001000
E1410000E1305000
UF Dutch New Guinea
Irian Barat
Netherlands New Guinea
West Irian
West New Guinea
BT Indonesia
NT False Cape
SA New Guinea

iridium
Includes use on level 1 and 2 as a chemical element (list D). Also search Ir.
UF Ir
SA elements

Irish Sea
Between England and Ireland. Includes use on level 1 as an area term (list O). For term set options see list B.
CO N530000N552000
W0030000W0063000
BA European Atlantic
North Atlantic
Atlantic Ocean
SA Cardigan Bay

Irkutsk
Oblast W and N of Lake Baikal.
CO N510000N640000
E1190000E0960000
BT Russian Republic
USSR
SA Baikal rift zone

Irkutsk Amphitheater
use Irkutsk Basin

Irkutsk Basin
Coal Basin extending 150 miles NW along Trans-Siberian RR from Lake Baikal. Also search Cheremkhovo Basin and Irkutsk Amphitheater.
UF Cheremkhovo Basin
Cheremkovo coal basin
Irkutsk Amphitheater
BT Russian Republic
USSR

iron
Includes use on level 1 and 2 as a chemical element (list D). As of 1981, use iron ores for iron as a commodity. Before 1981, included use on level 1 as a commodity term.
UF Fe
SA awaruite
elements
Fe-55
Fe-57
ferric iron
ferromanganese composition
ferrosilicon
ferrous iron
ferruginous composition
heavy metals
hematite
iron meteorites
iron ores
iron-rich composition
josephinite
kamacite
native elements

iron alum
use halotrichite

Iron County
Index states as applicable.
BX Michigan
Missouri
Utah
Wisconsin
BT United States

iron formations
Includes use on level 3 under sedimentary rocks(1) chemically precipitated rocks(2). See list I. Also search iron formation.
UF formations, iron
iron-formations
BA chemically precipitated rocks
BT sedimentary rocks
SA ferruginous composition
iron-rich composition
itabirite
jaspilite
taconite

iron meteorites
Includes use on level 3 under meteorites(1). Before 1974, also search octahedrites.
UF metallic meteorites
meteoric iron
siderite (meteorite)
BT meteorites
NA ataxite
hexahedrite
octahedrite
SA achondrites
chondrites
Coldwater Meteorites
iron
kamacite
stony irons

iron ores
Term introduced in 1981. Includes use on level 1 and 2 as a commodity term (list C). Before 1981, search iron AND ore deposits.
SA hematite
iron
limonite
magnetite
native elements

iron oxides
BA oxides
BT minerals

iron spinel
use hercynite

iron sulfides
BA sulfides
BT minerals

iron-formations
use Iron formations

iron-rich composition
Term introduced in 1978. Before 1978, search iron-rich rocks; iron-rich sediments.
SA chemically precipitated rocks
composition
ferruginous composition
geochemistry
iron
iron formations
sedimentary rocks
sediments

iron-rich rocks
A valid term through mid-1978. Use iron-rich composition or iron formations.

iron-rich sediments
A valid term through mid-1978. Use ferruginous composition under sediments(1).

iron-stony meteorite
use stony irons

ironstone
Includes use on level 3 under sedimentary rocks(1) chemically precipitated rocks(2). See list I.
BA chemically precipitated rocks
BT sedimentary rocks

Iroquois County
E Illinois.
BT Illinois
United States

irradiation
Term introduced in 1981.

irrigation
As of 1978, term is used on level 2 under waterways(1).
SA ponds
soils
surface water
treatment
utilization
water
water quality
water regimes
water supply
waterways

irrotational wave
use P-waves

irruption (intrusion)
use intrusions

Irtish River
use Irtysh River

Irtysh River
Largest tributary of the Ob River. Index Sinkiang Uighur and Soviet republics as applicable. Also search Irtysh.
UF Irtish River
BT Asia
SA Kazakhstan
Ob River
Ob-Irtysh Interfluve
Russian Republic
Sinkiang Uighur

Isachsen Formation
Conformably overlies Deer Bay Formation on Axel Heiberg and Ellef Ringnes Islands. On Isachsen Peninsula of Ellef Rignes Island in Franklin District.
BT Lower Cretaceous
Cretaceous
SA Northwest Territories

Isar Valley
River valley. Index Austrian and German states as applicable. Also search Isar River.
BT Europe
SA Bavaria
Tyrol

Ischia
Island in Tyrrhenian Sea outside Bay of Naples.
BT Campania
Italy

Ise Bay
Inlet of Pacific Ocean on S coast.
BT Honshu
Japan

Isera Mountains
use Izera Mountains

Isere
Department in SE.
CO N444500N455000
E0063000E0044500
BT France
NT Grenoble
Gresivaudan
SA Pelvoux Massif
Vercors

Isere Valley
River valley in SE. Also search Isere.
BT France

Isergebirge
Mountain range of the W Sudeten Mountains. Index countries as applicable.
BT Europe
SA Czechoslovakia

Poland
Sudeten Mountains

Ishikari Bay
Inlet of Sea of Japan on W Hokkaido. Also search Ishikari.
BT Hokkaido
Japan
SA Ishikari Plain

Ishikari Plain
S of Ishikari Bay in SW Hokkaido. Also search Ishikari.
BT Hokkaido
Japan
SA Ishikari Bay

Ishikawa
Prefecture on the Sea of Japan N of Nagoya in central.
BT Honshu
Japan

Ishim
River which flows into the Irtysh River. Index Soviet republics as applicable.
BT USSR
SA Kazakhstan
Russian Republic

Iskar River
use Isker River

Isker River
NW central Bulgaria. Also search Isker.
UF Iskar River
Iskur River
BT Bulgaria

Iskur River
use Isker River

island arcs
Includes use on level 2 under plate tectonics(1).
UF arcs, island
volcanic arcs
SA continental margin
continents
islands
plate tectonics
tectonophysics
trenches

islands
Includes use on level 3, e.g. under tectonophysics(1).
SA atolls
barrier islands
continents
island arcs

Islas Malvinas
use Falkland Islands

Islay
Most southerly island of Inner Hebrides off W coast.
CO N553500N555500
W0060000W0063000
BT Scotland
Great Britain
United Kingdom
SA Inner Hebrides

Isle of Man
Island in Irish Sea.
CO N541000N543000
W0041500W0044500
BT United Kingdom
Western Europe
Europe

Isle of Pines
Island in Caribbean S of W Cuba.
BT Cuba
West Indies

Isle of Skye
Island of Inner Hebrides off NW coast. Also search Skye.
CO N570000N574000
W0054000W0064500
UF Skye
BT Scotland
Great Britain
United Kingdom
SA Inner Hebrides

Isle of Wight
Island in English Channel just S of Southampton.
CO N504000N504500
W0010000W0013000
BT England
Great Britain
United Kingdom

Isle Royale National Park
In NW Lake Superior. Includes Isle Royale and neighboring islands.
BT Michigan
United States

isochrons
Includes use on level 3 under geochronology(1) and absolute age(1).
SA absolute age
geochronology
seismology

isoclinal
A valid term through 1977. After 1977, use isoclinal folds.

isoclinal folds
Term introduced in 1978. Includes use on level 3 under folds(1) style(2). Before 1978, also search folds AND isoclinal; isoclines.
UF isoclines
BA folds
SA deformation

isoclines
use isoclinal folds

isograd maps
Term introduced in 1978. Before 1978, search maps AND isograds.
BA maps
SA isograds

isograds
SA facies
grade
isograd maps
maps
metamorphism

Isograptus
Genus. As of 1981, includes use on level 3 under Graptolithina(1) Didymograptina(2). See list F. Before 1981, included use on level 3 under Graptolithina(1) Graptoloidea(2).
BA Didymograptina
Graptoloidea
Graptolithina
BX Hemichordata
Invertebrata

isolation
Term introduced in 1981.
SA radiation

isomorphism
Used as a general term.
SA crystal chemistry
paleontology

isopach maps
Before 1978, also search maps AND isopach.
BA maps
SA isopleth maps
stratigraphy

isopleth maps
BA maps
SA isopach maps

isoprenoids
BA organic materials

isoseismic maps
BA maps
SA earthquakes
seismology

isostasy
To be used for the condition of equilibrium of the crust above the mantle, and related features. Includes use on level 1 (list A). Term set options are:
topic [anomalies, causes, changes, compensation, concepts, detection, genesis, geomorphologic effects, interpretation, mechanism, observations]
subtopic (no area term)
SA asthenosphere
changes of level
compensation
continents
crust
density
Earth
epeirogeny
eustacy
geodesy
geomorphologic effects
geomorphology
geophysical methods
geophysical surveys
glacial geology
mantle
Mohorovicic discontinuity
neotectonics
tectonics
tectonophysics
vertical movements

isothermal magnetization
Term introduced in 1981.
BA magnetization

isothermal remanent magnetization
BA remanent magnetization
SA paleomagnetism

isotope ratios
A valid term through 1972. After 1972, use ratios. See isotopes, absolute age and name of chemical element(1) isotopes(2).

isotopes
Includes use on level 1 (list A); on level 2 under chemical elements (list D). Used for the application of the study of radioactive and stable isotopes, especially their abundances, to geology, excluding absolute age dating. Also search isotopic. If 1, term set options are:
element
specific isotope or ratio
material (first-level terms only)
type of material
topic [abundance, analysis, fractionation, methods, ratios, tracers]
subtopic
NA radioactive isotopes
stable isotopes
SA absolute age
activation analysis
Al-26
analysis
Ar-36
Ar-37
Ar-38
Ar-39
Ar-40
argon
Be-10
Be-7
C-12
C-13
C-13/C-12
C-14
carbon
chemical analysis
cosmogenic elements
Cs-137
D/H
deuterium
elements
fallout
Fe-55
Fe-57
fractionation
geochemistry
ground water
He-3
He-4
He-4/He-3
helium
hydrogen
I-129
I-131
inclusions
ionium
K-40
krypton
lead
Li-6
Li-7
metals
meteorites
mineral deposits
Mn-53
Mn-54
N-15/N-14
Na-22
Na-24
Ne-20
Ne-21
Ne-22
neon
noble gases
O-16
O-18
O-18/O-16
organic materials
oxygen
Pa-231
paleoclimatology
Pb-204
Pb-206
Pb-206/Pb-204
Pb-207
Pb-207/Pb-204
Pb-207/Pb-206
Pb-208
Pb-208/Pb-204
Pb-210
Po-210
potassium
Pu-239
Pu-244
Ra-226
radioactive decay
radioactive tracers
radioactivity
radium
radon
ratios
Rn-222
rubidium
S-32
S-34
S-34/S-32
sea water
Si-32
spallation
Sr-86
Sr-87
Sr-87/Sr-86
Sr-90
Sr/Rb
standard materials
strontium
sulfur
tektites
Th-228
Th-230
Th-232
Th-232/Th-230
Th-234
thorium
tracers
tritium
U-234
U-235

U-238
U-238/U-234
uranium
uranium disequilibrium
Xe-124
Xe-126
Xe-128
Xe-129
Xe-131
Xe-132
xenon
Zr-95

isotropic materials
Term introduced in 1978. Before 1978, search isotropic.
SA materials

isotropy
Term introduced in 1981 for use with minerals and crystals.
SA minerals
optical properties

Israel
Includes use on level 1 as an area term (list O). For term set options see list B.
CO N293000N333000
E0353000E0342000
BA Middle East
Asia
NT Elath
Galilee
Haifa
Makhtesh Ramon
Mount Carmel
Negev
Sea of Galilee
SA Dead Sea
Dead Sea Rift
Great Rift Valley
Jerusalem
Jordan River
Jordan Valley
Mediterranean region
Mishash Formation
Palestine
Red Sea Basin
Samaria
Sinai

Issyk-kul Lake
NE Kirghizia. Also search Issyk-Kul.
BT Kirghizia
USSR

Istanbul
Province on both sides of Bosporus. Also a city on the Bosporus in Europe.
BT Turkey
Middle East

Istranca Mountains
Range along Black Sea coast in SE Bulgaria, and Turkey in Europe. Index countries as applicable.
UF Strandzha Mountains
SA Bulgaria
Turkey

Istria
Peninsula in NW jutting into Adriatic Sea. Index republics as applicable.
BT Yugoslavia
SA Croatia
Slovenia

Itabira
City in E central Minas Gerais.
UF Presidente Vargas
BT Minas Gerais
Brazil

itabirite
As of 1981, includes use on level 2 under metamorphic rocks(1). See list J. Before 1981, included use on level 3 under sedimentary rocks(1) chemically precipitated rocks(2).
BT metamorphic rocks
SA iron formations

Itaborai
City in S central.
BT Rio de Janeiro
Brazil

Itaka
Town in central Chita Oblast in Transbaikalia.
BT Russian Republic
USSR

Italy
Includes use on level 1 as an area term (list O). For term set options see list B.
CO N363000N473000
E0190000E0063000
BA Southern Europe
Europe
NT Abruzzi
Acceglio
Acqui
Adamello Massif
Alessandria
Alto Adige
NX Apennine Front
NA Apennines
NT Apulia
Argentera
Argentera Massif
Asti
Basilicata
Biella
Calabria
Campania
Campobasso
Canavese Zone
Dolomites
Emilia-Romagna
Friuli-Venezia Giulia
Ivrea
Ivrea Zone
Ivrea-Verbano Zone
Langhe
Lanzo Massif
Latium
Lessini Mountains
Liguria
Lombardy
Lucania
Marches
Marmolada Glacier
Molise
Novara
Piedmont Alps
Piemonte
Po River
Po Valley
NA Sardinia
NT Sesia Valley
Sesia Zone
Sesia-Lanzo Zone
Sicily
Strona Valley
Tagliamento Valley
Tiber Valley
Trentino-Alto Adige
Turin
Tuscany
Umbria
Valle d'Aosta
Venetia
Veneto
Vercelli
Villafranca d'Asti
SA Adriatic region
Alps
Brianconnais Zone
Carnic Alps
Ceneri Zone
Central Alps
Cottian Alps
Julian Alps
Karst region
Lago Maggiore
Lepontine Alps
Maiolica Limestone
Maritime Alps
Mediterranean region
Monte Rosa
Otztal Alps
Pennine Alps
Prealps
Rhaetian Alps
Savoy
Scaglia Formation
Simplon region
Val Gardena Sandstone
Vigarano Meteorite
Western Alps
Zillertal Alps

Itasca County
N central.
BT Minnesota
United States

iterative methods
Term introduced in 1981. Also search iteration; iterative.
SA mathematical geology
mathematical methods
mathematical models

Ithaca
City in Tompkins County in S central; one of Ionian Islands in Greece.
SA Greek Ionian Islands
New York

Ito
City on E Izu Peninsula on Sagami Sea in central.
BT Honshu
Japan

Iturup Island
Largest of the Kuril Islands, SW of Kamchatka. Also search Iturup.
BT Russian Republic
USSR
SA Kuril Islands

IUGS
Acronym. Includes use on level 3 under associations(1).
UF International Union of Geological Sciences
SA associations

Ivano-Frankovsk
City and oblast in W Ukraine. Formerly part of Poland. Also search Stanislawow.
UF Stanislav
Stanislawow
BT Ukraine
USSR

Ivigtut
Settlement on SW coast.
BT Greenland
Arctic region

Iviza
use Ibiza

Ivory Coast
Includes use on level 1 as an area term (list O). For term set options see list B.
CO N042000N104000
W0023000W0084500
BA West Africa
Africa
NT Bandama River
Toumodi
SA Nimba Mountains
Trou Sans Fond
Volta Basin

Ivrea
City NE of Turin in central.
BT Italy
SA Piemonte

Ivrea Zone
Central.
BT Italy
SA Piemonte

Ivrea-Verbano Zone
Central and NW.
BT Italy
SA Piemonte

Iwaki
Mountain peak in Aomori Prefecture in N Honshu. Also a river and city.
BT Honshu
Japan

Iwate
Prefecture in NW Honshu. Also a dormant volcano.
BT Honshu
Japan

Iwo-Jima
Center island of volcano group located 660 nautical miles S of Tokyo. Returned to Japan in 1968.
BT Pacific Ocean

ixiolite
BZ niobates
tantalates
BA oxides
BT minerals

Iya River
SW Irkutsk Oblast. Also search Iya.
BT Russian Republic
USSR

Izera Mountains
Range of the Sudeten Mountains in N Bohemia and Lower Silesia. Also search Isera Mountains. Index countries as applicable.
UF Isera Mountains
BT Europe
SA Czechoslovakia
Poland
Sudeten Mountains

Izu Islands
use Izu-shichito

Izu Peninsula
Extends into Pacific Ocean between Suruga Bay and Sagami Bay in S central Honshu. Also search Izu.
BT Honshu
Japan

Izu-shichito
Group of volcanic islands in Pacific Ocean off Izu Peninsula. Also search Izu Islands.
UF Izu Islands
BT Honshu
Japan
SA Hachijo-jima
Miyake-Jima
O-shima

Izumi
City in Osaka Prefecture.
BT Honshu
Japan

Izumi Group
Kii Peninsula in S central Honshu, and Shikoku.
BT Upper Cretaceous
Cretaceous
SA Honshu
Japan
Senonian
Shikoku

J

Jabalpur
City and district in central Madhya Pradesh. Also search Jubbulpore.
UF Jubbulpore
BT Madhya Pradesh
India

Jabalpur Series
Consists of clays, shales, earthy sandstones, and thin coal seams. Near city of Jabalpur.
BT Upper Jurassic
Jurassic
SA India
Madhya Pradesh

Jachymov
Town in Erzgebirge in NW.
BT Bohemia
Czechoslovakia

Jackfork Group
Divided into Wildhorse Mountain Formation, Prairie Mountain Formation, Markham Mill Formation, Wesley Formation, Game Refuge Formation. SE and central S Oklahoma; and SW Arkansas. Also search Jackfork Sandstone.
BT Carboniferous
SA Arkansas
Oklahoma

Jackson County
BX Alabama
Arkansas
Colorado
Florida
Georgia
Illinois
Indiana
Iowa
Kansas
Kentucky
Louisiana
Michigan
Minnesota
Mississippi
Missouri
North Carolina
Ohio
Oklahoma
Oregon
South Dakota
Tennessee
Texas
West Virginia
Wisconsin
BT United States
NT Russell Cave

Jackson Group
Includes Caddell (Moodys Marl) Formation, Wellborn Formation, Manning Formation, and Whitsett Formation, Gosport Sand, Yazoo Clay, Ocata Limestone, Barnwell Formation, McElroy Formation, Mosley Hill Formation. Gulf Coastal Plain from Georgia to S Texas.
BT upper Eocene
Eocene
SA Alabama
Florida
Georgia
Louisiana
Mississippi
Texas
Yazoo Clay

Jackson Hole
Valley in Teton County in NW.
BT Wyoming
United States

jacobsite
BA oxides
BT minerals
SA spinel group

Jacobsville Sandstone
Precambrian or Cambrian. Underlies Munising Formation. N Michigan.
SA Cambrian
Michigan
Precambrian

jade
BA chain silicates
silicates
BT minerals
SA amphibole group
jadeite
nephrite
pyroxene group

Jade Bay
Inlet of North Sea on N coast of Oldenburg region.
BT Lower Saxony
West Germany

jadeite
BA clinopyroxene
pyroxene group
chain silicates
silicates
BT minerals
SA jade

Jaen
As of 1981, refers to only the city. Before 1981, also included the province. In S central Spain.
BT Jaen Province
Andalusia
Spain

Jaen Province
Term introduced in 1981. Before 1981, search Jaen. In S central Spain.
CO N372300N383200
W0022800W0041800
BT Andalusia
Spain
NT Jaen

Jaguaribe
Town and river in NE part of country.
BT Ceara
Brazil

Jaipur
City and former princely state in E central.
BT Rajasthan
India

Jaisalmer
Town and former princely state in W.
BT Rajasthan
India

Jalisco
State in W.
BT Mexico
NT Guadalajara City

Jamaica
Includes use on level 1 as an area term (list O). For term set options see list B.
CO N174000N183000
W0761000W0782500
BA Greater Antilles
West Indies
NT Discovery Bay
Maroon Town
SA Blue Mountains

James Bay
Southern extension of Hudson Bay between NE Ontario and W Quebec.
BT Canada

James River
E central.
BT Virginia
United States

James W. Ellsworth Land
use Ellsworth Land

Jameson Land
Central E coast.
BT Greenland
Arctic region

jamesonite
UF gray antimony
BA sulfantimonites
sulfosalts
BT minerals

Jamkhandi
Town and former princely state in S.
BT Maharashtra
India

Jammu
City and district in S Jammu and Kashmir.
UF Jummoo
BT Jammu and Kashmir
India

Jammu and Kashmir
State in N.
CO N321000N370000
E0802300E0722000
UF Kashmir and Jammu
BT India
NT Anantnag
Doda
Jammu
Kashmir
Kashmir Valley
Kishtwar
Naubug
Riasi
Srinagar
Tawi Valley
SA Indus River
Muth Quartzite
Pakistan

Jamtland
County in W.
BT Sweden

Jan Mayen
Norwegian volcanic island between Greenland and Norway. Includes use on level 1 as an area term (list O). For term set options see list B.
CO N700000N720000
W0073000W0090000
BA Arctic region
NT Beerenberg
SA Arctic Ocean
Atlantic Ocean
Greenland Sea
Norwegian Sea

Jan Mayen Ridge
Term introduced in 1981. Off the coast of Jan Mayen, to the S.
CO N673000N713000
W0070000W0100000
BT Norwegian Sea
Arctic Ocean

Japan
Includes use on level 1 as an area term (list O). For term set options see list B.
CO N300000N450000
E1470000E1290000
BA Far East
Asia
NT Green tuff region
Hokkaido
Honshu
Inland Sea
Kuroko
Kyushu
Nankaido
O-shima
Ryukyu Islands
Sakura-jima
Shikoku
Tsugaru Strait
SA Akiyoshi Limestone
Futaba Group
Green Tuff Formation
Izumi Group
Median Tectonic Line
Mikabu System
Motojuku Formation
Muro Group
Narita Formation
Osaka Group
Semata Formation
Shimanto Group
Shimosa Group
Taishu Group
Uonuma Group
Usuginu Conglomerate

Japan Current
use Kuroshio

Japan Sea
Between Japan in the E; and Korea and Primorye Kray in the Soviet Far East on the W. Includes use on level 1 as an area term (list O). For term set options see list B. Before 1974, also search Sea of Japan.
CO N340000N450000
E1420000E1271500
BA West Pacific
Pacific Ocean
SA Leg 31

Japan Trench
Submarine depression extending from Kuril Islands along E coast of Hokkaido and E coast of N and central Honshu.
BT Pacific Ocean

Japanese Alps
Term introduced in 1981. Before 1981, also search Japan AND Alps.
BT Honshu
Japan

Jaramillo Event
Geomagnetic event.
BA lower Pleistocene
Pleistocene
Quaternary
BT Cenozoic
SA paleomagnetism
reversals

Jarlsberg
use Vestfold

jarosite
UF utahite
BA sulfates
BT minerals
SA alunite

jasper
BA silica minerals
framework silicates
silicates
BT minerals
SA chalcedony
chert
jasperoid

Jasper County
Index states as applicable.
BX Georgia
Illinois
Indiana
Iowa
Mississippi
Missouri
South Carolina
Texas
BT United States

Jasper National Park
On British Columbia border in W Alberta. Also search Jasper.
BT Alberta
Canada

jasperoid
See list I.
BA chert
chemically precipitated rocks
BT sedimentary rocks
SA chalcedony
jasper
limestone
siliceous composition

jaspilite
Includes use on level 3 under sedimentary rocks(1) chemically precipitated rocks(2). See list I. Also search jaspillite.
BA chemically precipitated rocks
BT sedimentary rocks

SA iron formations

Java
Island in the Greater Sunda group. Most populous and economically most important.
CO S090000S054000
E1144000E1050500
BT Indonesia

Java Sea
N of Java and S of Borneo.
BA Indonesian Seas
West Pacific
BT Pacific Ocean
NT Sunda Shelf

jaws
Includes use on level 3 under fossil groups(1). See list F.
SA bones
fossils
paleontology
skulls
teeth
Vertebrata

Jaxartes River
use Syr Darya

Jefferson County
BX Alabama
Arkansas
Colorado
Florida
Georgia
Idaho
Illinois
Indiana
Iowa
Kansas
Kentucky
Louisiana
Mississippi
Missouri
Montana
Nebraska
New York
Ohio
Oklahoma
Oregon
Pennsylvania
Tennessee
Texas
Washington
West Virginia
Wisconsin
BT United States

Jefferson Parish
W and S of New Orleans.
BT Louisiana
United States

Jemez Mountains
use Valle Grande Mountains

Jerome
Town in central Arizona and city in S Idaho. Index states as applicable.
BX Arizona
Idaho
BT United States

Jerusalem
District in E central Israel. Also a city. Index countries as applicable.
BT Middle East
SA Israel
Jordan

jetties
BT marine installations
SA groins

Jhabua
Village, district and former princely state in W.
BT Madhya Pradesh
India

Jhansi
City and division in S.
BT Uttar Pradesh
India

Jharia
Town in SE Bihar.
UF Jherria
BT Bihar
India
SA Jharia coal field

Jharia coal field
In SE Bihar. Also search Jharia Coalfield; Jharia.
UF Jharia Coalfield
BT Bihar
India
SA Jharia

Jharia Coalfield
use Jharia coal field

Jherria
use Jharia

Jhunjhunu
Town and district in NE.
BT Rajasthan
India

Jiu River valley
Index regions as applicable. Also search Jiu; Jiu River.
BT Romania
SA Transylvania
Walachia

joaquinite
BA orthosilicates
silicates
BT minerals

Joban coal field
A major coal field on Pacific Ocean in central Honshu. Also search Joban; Joban Coalfield.
UF Joban Coalfield
BT Honshu
Japan

Joban Coalfield
use Joban coal field

Jodhpur
City and region in central Rajasthan. Former princely state.
BT Rajasthan
India

Johanna Island
use Anjouan Island

Johannesburg
City in S central.
BT Transvaal
South Africa

Johannsenite
BA clinopyroxene
pyroxene group
chain silicates
silicates
BT minerals

John Day Formation
Tentatively divided into three unnamed members. N central Oregon.
BT Tertiary
Cenozoic
SA lower Miocene
Miocene
Oligocene
Oregon
upper Oligocene

Johnnie Formation
Underlies Stirling Quartzite and overlies Noonday Dolomite. E California and SE Nevada.
BT Precambrian
SA California
Nevada

Johns Valley Formation
Scott County Arkansas, and Ouachita region of SE Oklahoma.
BT Carboniferous
SA Arkansas
Mississippian
Oklahoma
Pennsylvanian

Johnson County
BX Arkansas
Georgia
Illinois
Indiana
Iowa
Kansas
Kentucky
Missouri
Nebraska
Tennessee
Texas
Wyoming
BT United States

Johore
State just N of Singapore on the Malay Peninsula.
BT Malaysia
SA West Malaysia

joints
Includes use on level 3 under fractures(1) style(2) and under structural analysis(1) theoretical studies(2).
BA fractures
SA bedding joints
columnar joints
conjugate folds
cross fractures
fissures
fracture cleavage
oblique orientation
release fractures
strike
structural analysis

Joliba River
use Niger River

Joliba Valley
use Niger Valley

Jones County
Term introduced in 1981. Index states as applicable.
BX Georgia
Iowa
Mississippi
North Carolina
South Dakota
Texas
BT United States

Jones Mountains
On Eights Coast in Ellsworth Land on Pacific Ocean side.
BT Antarctica

Joplin
City in Jasper and Newton counties in SW.
BT Missouri
United States

Jordan
Includes use on level 1 as an area term (list O). For term set options see list B.
CO N290000N333000
E0390000E0343000
BA Middle East
Asia
SA Dead Sea
Dead Sea Rift
Great Rift Valley
Jerusalem
Jordan River
Jordan Valley
Near East
Palestine
Red Sea Basin
Samaria
Wadi Araba

Jordan Rift valley
use Jordan Valley

Jordan River
Rises in Syria and flows S through the Sea of Galilee to N end of the Dead Sea. Index countries as applicable.
BT Middle East
SA Israel
Jordan
Syria

Jordan River valley
use Jordan Valley

Jordan Valley
River valley between Sea of Galilee and Dead Sea which is part of the Great Rift Valley. Also search Jordan Rift valley; Jordan River valley; Tiberias-Dead Sea Rift valley. Index countries as applicable.
UF Jordan Rift valley
Jordan River valley
BT Middle East
SA Great Rift Valley
Israel
Jordan
Syria

jordanite
BA sulfarsenites
sulfosalts
BT minerals

Jordanow
Town in S part of country.
BT Krakow
Poland

jordisite
BA sulfides
BT minerals

Jos Plateau
Central Nigeria. Also search Bauchi Plateau.
UF Bauchi Plateau
BT Nigeria

joseite
As of 1981, includes use on level 3 under minerals(1) tellurides(2). See list L. Before 1981, included use on level 3 under minerals(1) sulfides(2).
BA tellurides
sulfides
BT minerals

Joseph Bonaparte Gulf
Inlet of Timor Sea. Index Northern Territory and/or Western Australia as applicable.
BT Australia
SA Bonaparte Gulf basin
Northern Territory
Western Australia

Josephine County
Term introduced in 1981. SW Oregon.
CO N420000N424700
W1231300W1240500
BT Oregon
United States

josephinite
As of 1981, includes use on level 3 under minerals(1) alloys(2). See list L. Before 1981, included use on level 3 under minerals(1) native elements and alloys(2).
BA alloys
BT minerals
SA iron
nickel

Jotnian
BA upper Precambrian
Precambrian

Jotunheim Mountains
Range in S central Norway. Also search Jotunheimen; Jotunheimen Mountains.
UF Jotunheimen Mountains
BT Norway

Jotunheimen Mountains
use Jotunheim Mountains

jouravskite
BA sulfates
BT minerals

Juab County
W central.
BT Utah
United States

Juan de Fuca Ridge
Off coast of Washington State. Also search Juan de Fuca Rise.
UF Juan de Fuca Rise
BT Pacific Ocean
SA mid-ocean ridges

Juan de Fuca Rise
use Juan de Fuca Ridge

Juan de Fuca Strait
Between Vancouver Island on the N and mainland of Washington on the S. Index British Columbia and/or Washington as applicable. Also search Strait of Juan de Fuca.
UF Strait of Juan de Fuca
SA British Columbia
Washington

Jubbulpore
use Jabalpur

Judith River Formation
In Montana Group. Comprises two members: Parkman Sandstone and an upper unnamed member. N, Central, S, and SE Montana.
BT Upper Cretaceous
Cretaceous
SA Montana
Montana Group

Juglandaceae
BA Dicotyledoneae
angiosperms
BT Spermatophyta
Plantae
SA Juglans

Juglans
Genus.
BA Dicotyledoneae
angiosperms
BT Spermatophyta
Plantae
SA Filicopsida
gymnosperms
Juglandaceae
pteridophytes

Jujuy
Province in extreme NW Argentina. Also a city.
CO S250000S220000
W0640000W0673000
BT Argentina
SA Yacoraite Formation

julgoldite
BA orthosilicates
silicates
BT minerals

Julian Alps
Division of Eastern Alps. Index countries as applicable.
BT Alps
Europe
SA Eastern Alps
Italy
Yugoslavia

Jumilla
City in SE part of country.
BT Murcia Province
Murcia region
Spain

Jummoo
use Jammu

Jumna River
N central India. Also search Yamuna River. Index states as applicable.
UF Yamuna River
BT India
SA Himachal Pradesh
Uttar Pradesh

Juneau
City in panhandle near border of Yukon Territory.
BT Alaska
United States

Juneau Glacier
use Juneau ice field

Juneau ice field
N of Juneau in panhandle. Also search Juneau Icefield; Juneau Glacier.
UF Juneau Glacier
Juneau Icefield
BT Alaska
United States

Juneau Icefield
use Juneau ice field

Juniata Formation
Includes East Waterford Red Sandstone Member, Plummer Hollow Red Mudstone, Run Gap Red Sandstone Member. S central and E Pennsylvania, W Maryland, E Tennessee, W Virginia, and E West Virginia.
BT Upper Ordovician
Ordovician
SA Maryland
Pennsylvania
Richmond Group
Tennessee
Virginia
West Virginia

Junin
Department in W central.
BT Peru
SA Pucara Group

junior high school
Term introduced in 1978. Includes use on level 3 under education(1). Before 1978, search junior high.
SA college-level education
education
elementary school
high school

Jupiter
Includes use on level 1 and 2. See term set options under Moon(1).
SA Galilean satellites
outer planets
Pioneer Program
Pioneer 10
planetology
planets
satellites
solar system

Jura
Department on the Swiss border.
CO N461500N472000
E0061500E0051500
BT France
NT Morez
SA Jura Mountains

Jura Mountains
Index countries as applicable. Also search Jura.
CO N455000N473500
E0081500E0044500
BT Europe
NZ Swiss Jura Mountains
SA France
Jura
Switzerland

Jurassic
Includes use on level 1 as an age term (list E); on level 2 under paleoterms, e.g. paleoecology, paleogeography, paleomagnetism. Above Triassic, below Cretaceous.
BT Mesozoic
NT Athgarh Sandstone
Bowser Formation
Ferrar Group
Lathi Formation
NA Lower Jurassic
NT Mansalay Formation
NA Middle Jurassic
NT Norphlet Formation
Posidonia Shale
Shaunavon Formation
Solnhofen Limestone
NA Upper Jurassic
NT Younger Granites
SA Beacon Supergroup
Brianconnais Zone
Franciscan Formation
Gondwana System
Great Valley Sequence
Ionian Zone
Karroo System
Khorat Group
Kootenay Formation
Maiolica Limestone
Navajo Sandstone
Novorayskoe Formation
Pucara Group
Serra Geral Formation
Tal Formation

Jutland
Peninsula including Schleswig in Germany and the Danish mainland. Politically it applies only to the mainland of Denmark. Index Denmark and/or Schleswig-Holstein as applicable. Also search Jylland.
UF Jylland
BT Europe
NT Djursland
SA Denmark
Schleswig-Holstein

juvenile water
Term introduced in 1981.
SA ground water
magmas
water

Jylland
use Jutland

K

K
use potassium

K-feldspar
Also search potassium feldspar; potash feldspar.
UF potash feldspar
potassium feldspar
BA feldspar group
framework silicates
silicates
BT minerals
SA adularia
alkali feldspar
microcline
orthoclase
sanidine

K-40
Includes use on level 3 under isotopes(1).
SA isotopes
potassium

K/Ar
Includes use on level 3 under absolute age(1) methods(2). Also search potassium-argon.
SA absolute age
argon
potassium

Kaap Valsch
use False Cape

Kabardia
N part of Kabardin-Balkar A.S.S.R. which is on N slopes of Caucasus.
BT Russian Republic
USSR

Kabardin-Balkar
Autonomous Soviet Socialist Republic. Also search Kabardinia-Balkar, and Kabardino-Balkar.
UF Kabardinia-Balkar
Kabarnino-Balkar
BT Russian Republic
USSR

Kabardinia-Balkar
use Kabardin-Balkar

Kabarnino-Balkar
use Kabardin-Balkar

Kabinda
use Cabinda

Kabul
Province in NE. Also a city.
BT Afghanistan

Kabylia
Mountainous coastal region in N.
BT Algeria
SA Great Kabylia

Kadah
Northernmost state in West Malaysia on the Malay Peninsula.
BT Malaysia

Kadzharan
Town in S.
BT Armenia
USSR

kaersutite
BA clinoamphibole
amphibole group
chain silicates
silicates
BT minerals
SA hornblende

Kafan
City in SE.
BT Armenia
USSR

Kafue
Township and river in N.
BT Zambia

Kaganovich Peak
use Darvaz

Kagawa
Prefecture on the Inland Sea in NE.
BT Shikoku
Japan

Kagoshima
City and prefecture in S.
BT Kyushu
Japan

Kaibab Formation
Divided vertically into three members. N Arizona, SE California, SE Nevada and W Utah.
BT Permian
SA Arizona
California
Nevada
Utah

Kaikoura
Mountain range in NE.
BT South Island
New Zealand

Kaimur Sandstone
Constitutes part of the Semri Group in the Gwalior and Mewar areas. Upper Kaimur Sandstone and lower Kaimur Sandstone are included in Kaimur Series.
SA Cambrian
India
Madhya Pradesh
Precambrian

Rajasthan
Uttar Pradesh

kainite
BZ chlorides
sulfates
BT minerals

Kaipara Harbor
Inlet of Pacific Ocean on W coast of N extension of North Island. Also search Kaipara.
BT North Island
New Zealand

Kaiparowits Plateau
W of Colorado River between Escalante and Paria rivers in S central.
BT Utah
United States

Kaiserstuhl
Mountain group in SW.
BT Baden-Wurttemberg
West Germany

Kakagi Lake
East of Lake of the Woods in W.
BT Ontario
Canada

Kakanui
Mountains in SE, north of Dunedin.
BT South Island
New Zealand

Kakhetia
Region in E.
BT Georgian Republic
USSR

Kakioka
Term introduced in 1981. Town in Ibaraki prefecture, central Honshu.
CO N361400N361400
E1401100E1401100
BT Honshu
Japan

kakortokite
As of 1981, includes use on level 3 under igneous rocks(1) syenites(2). See list H. Before 1981, included use on level 3 under igneous rocks(1) syenite family(2).
BA syenites
BT igneous rocks
SA nepheline syenite

Kakuto
Village in SE on Lake Victoria.
BT Uganda

Kaladgi
Village in W central.
BT Mysore
India

Kaladgi System
Precambrian? Divided into upper Kaladgi and lower Kaladgi series.
BT Precambrian
SA India
Maharashtra
Mysore

Kalahari Desert
Plateau and desert region. Index countries as applicable.
BT Africa
SA Botswana
Namibia
South Africa

Kalamazoo County
SW Lower Peninsula.
BT Michigan
United States

Kalba Range
W branch of Altai Mountains in E Kazakhstan. Also search Kalba.
CO N490000N492000
E0830000E0823000
BT Kazakhstan
USSR
SA Altai Mountains

Kalgoorlie
Municipality in S.
CO S293000S293000
E1220000E1220000
BT Western Australia
Australia

Kalgoorlie System
Includes "Older Greenstones", Black Flag Group, Yindarlgooda Group, Kundana Group, and "Younger Greenstones". S part of state.
BT Archean
SA Western Australia

Kali Gandaki Valley
River valley in central.
BT Nepal

kaliborite
Also search paternoite.
UF heintzite
hintzeite
paternoite
BA borates
BT minerals

Kalimantan
All of island of Borneo excepting Sarawak, Sabah and Brunei.
CO S042000N043000
E1190000E1085500
BT Indonesia
SA Borneo

Kalinin
City and oblast.
BT Russian Republic
USSR

Kaliningrad
City and oblast.
BT Russian Republic
USSR

kaliophilite
BA nepheline group
framework silicates
silicates
BT minerals
SA aluminosilicates

Kalmar
County in S.
BT Sweden
NT Oland
Vastervik
SA Smaland

Kalmyk
Autonomous Soviet Socialist Republic on NW shore of Caspian Sea.
BT Russian Republic
USSR

kalsilite
BA nepheline group
framework silicates
silicates
BT minerals
SA aluminosilicates

Kama
Village in Kivu, Zaire; village in Thayetmyo district, Upper Burma. Use Kama River for the river.
SA Burma
Kama River
Zaire

Kama River
Rises in central Urals and flows into Volga River. Also search Kama AND USSR.
CO N550000N610000
E0560000E0520000
BT Russian Republic
USSR
SA Kama

kamacite
Meteorite mineral. As of 1981, includes use on level 3 under minerals(1) alloys(2). See list L. Before 1981, included use on level 3 under minerals(1) native elements and alloys(2).
BA alloys
BT minerals
SA iron
iron meteorites
meteorites
nickel
plessite
taenite

Kambalda
Settlement S of Kalgoorlie in S.
BT Western Australia
Australia

Kamchatka
Use for the oblast. Includes Kamchatka Peninsula.
CO N510000N650000
E1750000E1550000
UF Kamchatka Oblast
BT Russian Republic
USSR
SA Okhotsk region

Kamchatka Oblast
use Kamchatka

Kamchatka Peninsula
Between Okhotsk Sea and Bering Sea. As of 1981, also includes the Commander Islands and Karaginskiy Island.
CO N510000N600000
E1680000E1550000
BA Soviet Pacific region
BT Russian Republic
USSR
NT Tolbachik
SA Soviet Far East

Kamen
City in the Ruhr in W.
BT North Rhine-Westphalia
West Germany

Kamenka River
Flows into the Angara River in S Krasnoyarsk Kray.
BT Russian Republic
USSR

kames
Includes use on level 3 under glacial geology(1) glacial features(2).
BT glaciofluvial features
glacial features
SA cirques
eskers
glacial geology
glaciers

Kamikawa
Sub-prefecture in central.
BT Hokkaido
Japan

Kaminak Lake
SW District of Keewatin.
BT Northwest Territories
Canada

Kamloops
City in S.
BT British Columbia
Canada

kammererite
BA chlorite group
sheet silicates
silicates
BT minerals

Kamoto
Mine five miles W of village of Kolwezi in S part of country.
BT Shaba
Zaire

Kamp Valley
River valley in NE part of country.
BT Lower Austria
Austria

Kampinos Forest
E central.
BT Poland

Kampuchea
use Cambodia

Kamyshin
City on the Volga River in Volgograd Oblast.
BT Russian Republic
USSR

Kanagawa
Prefecture whose capital is Yokohama.
BT Honshu
Japan

Kanara
Region. Former district of North Kanara in W.
BT Mysore
India

Kanawha County
W central.
BT West Virginia
United States

Kanazawa
City in W.
BT Honshu
Japan

Kane County
BX Illinois
Utah
BT United States

Kanem
Prefecture NE of Lake Chad in W.
BT Chad

Kaneohe Bay
Wide inlet on E coast of Oahu Island.
BT Hawaii
United States

Kanev
City in Kiev Oblast in N central.
BT Ukraine
USSR

Kangaroo Island
In Indian Ocean S of Yorke Peninsula.
BT South Australia
Australia

Kangra
Town in W.
BT Himachal Pradesh
India

Kangwon
Province in SE North Korea, and in NE South Korea. Index countries as applicable.
BT Asia
SA North Korea
South Korea

Kanhan Valley
River valley in central Madhya Pradesh. Also search Kanhan.
BT Madhya Pradesh
India

Kanin Peninsula
Projects into Barents Sea on N coast of Nenets National Okrug in Arkhangelsk Oblast.
BT Russian Republic
USSR

Kansan
North America. Pertains to second glacial stage of Pleistocene Epoch.
BA lower Pleistocene
Pleistocene
Quaternary
BT Cenozoic

Kansas
Includes use on level 1 as an area

term (list O). For term set options see list B.
CO N370000N383000
W0943500W0980000
BA Midwest
United States
NT Abilene
Admire
NX Allen County
Anderson County
NT Arkansas City
Atchison
Atchison County
NX Barton County
Brown County
Butler County
Chase County
Chautauqua County
Cherokee County
Cheyenne County
Clark County
Clay County
NT Cloud County
NX Comanche County
NT Cowley County
NX Crawford County
NT Decatur County
NX Dickinson County
NT Doniphan County
NX Douglas County
Edwards County
Ellis County
NT Ellsworth County
Emporia
NX Eureka
NT Finney County
NX Franklin County
NT Garnett
Geary County
Gove County
NX Graham County
Grant County
Greenwood County
Hamilton County
Harper County
NT Harvey County
Haviland
Hodgeman County
Hugoton
Hugoton Field
Hutchinson
NX Jackson County
Jefferson County
Johnson County
Kansas City
NT Kansas River
Kansas River valley
Kingman County
NX Kiowa County
NT Kirwin
Labette County
NX Lane County
NT Leavenworth County
NX Lincoln County
Logan County
Lyon County
Manhattan
Marion County
Marshall County
McPherson County
Meade County
Mitchell County
Montgomery County
Morton County
Nemaha County
NT Ness County
Norton County
NX Osage County
NT Osborne County
NX Ottawa County
Pawnee County
Phillips County
Pottawatomie County
NT Pratt County
Rawlins County
Reno County
Republic County
NX Rice County
NT Riley County
Rooks County
NX Rush County
Russell County
Saline County
Scott County
NT Sedgwick Basin
NX Sedgwick County
Seward County
NT Shawnee County
NX Sheridan County
Sherman County
Smith County
NT Smoky Hill River basin
NX Stafford County
Stanton County
Stevens County
Sumner County
NT Topeka
Trego County
Tuttle Creek Dam
Wabaunsee County
Wallace County
NX Washington County
NT Wichita
NX Wichita County
Wilson County
NT Woodson County
Wyandotte County
SA Americus Limestone Member
Anadarko Basin
Arkansas River
Arkansas River valley
Ash Hollow Formation
Bartlesville Sand
Beil Limestone Member
Benton Formation
Bevier Coal
Burbank Sand
Cabaniss Formation
Cambridge Arch
Carlile Shale
Chase Group
Cherokee Group
Cheyenne Sandstone
Coldwater Meteorites
Colorado Group
Cottonwood Limestone
Crooked Creek Formation
Dakota Formation
Desmoinesian
Dockum Group
Douglas Group
Ervine Creek Limestone
Eskridge Shale
Excello Shale
Flint Hills
Forest City Basin
Fort Hays Limestone Member
Fort Riley Limestone
Graneros Shale
Great Plains
Greenhorn Limestone
Haviland Meteorite
Hughes Creek Shale
Hugoton Embayment
Hutchinson Salt Member
Kansas City Group
Kingsdown Formation
Kiowa Formation
Labette Shale
Lansing Group
Lawrence Formation
Lecompton Limestone
Loup Fork Group
Marmaton Group
McPherson Formation
Mentor Beds
Midcontinent
Missouri River
Missouri River basin
Missouri River valley
Montana Group
Morrison Formation
Nemaha Ridge
Neosho River valley
Niobrara Formation
Norton County Meteorite
Ogallala Formation
Oread Limestone
Peoria Loess
Plattsburg Limestone
Plattsmouth Limestone Member
Pleasanton Group
Red Eagle Limestone
Rexroad Formation
Roca Formation
Rock Lake Shale Member
Saint Peter Sandstone
Shawnee Group
Smoky Hill Chalk Member
Stanton Formation
Stearns Shale
Stone Corral Formation
Tonganoxie Sandstone
Valentine Formation
Verdigris River valley
Wabaunsee Group
Wellington Formation
Wreford Limestone
Wyandotte Limestone

Kansas City
Term introduced in 1981. Index states as applicable.
BX Kansas
Missouri
BT United States

Kansas City Group
Subdivided into Bronson, Linn, and Zarah subgroups. Includes Hertha, Ladore, Swope, Galesburg, Dennis, Fontana, Sarpy, Drum, Chanute, Iola, Lane, Wyandotte and Bonner Springs formations.
BT Missourian
Upper Pennsylvanian
Pennsylvanian
SA Iowa
Kansas
Missouri
Nebraska
Wyandotte Limestone

Kansas River
Flows into Missouri River at Kansas City.
BT Kansas
United States

Kansas River basin
use Kansas River valley

Kansas River valley
NE central Kansas. Also search Kansas River basin.
UF Kansas River basin
BT Kansas
United States

kansite
use mackinawite

Kansk
City E of Krasnoyarsk on the Trans-Siberian R.R. in S Krasnoyarsk Kray.
BT Russian Republic
USSR

Kansk-Achinsk Basin
S and SW Krasnoyarsk Kray along Trans-Siberian R.R.
BT Russian Republic
USSR

Kanto
use Kwanto Plain

kaolin
As of 1981, use kaolin deposits for kaolin as a commodity. Before 1981, included use on level 1 as a commodity term. As of 1981, includes use on level 3 under sediments(1) clastic sediments(2). See list N.
BA clastic sediments
BT sediments
SA clay
clay mineralogy
clay minerals
clays
fuller's earth
kaolin deposits
kaolinite
kaolinization
metakaolin

kaolin deposits
Term introduced in 1981. Includes use on level 1 and 2 as a commodity term (list C). Before 1981, search kaolin AND deposits.
SA clays
fuller's earth
kaolin
kaolinization
refractory materials

kaolinisation
use kaolinization

kaolinite
BA clay minerals
sheet silicates
silicates
BT minerals
SA kaolin

kaolinitization
use kaolinization

kaolinization
Includes use on level 3 under metasomatism(1) processes(2). Also search kaolinitization.
UF kaolinisation
kaolinitization
kaolisation
BA metasomatism
SA alteration
clay minerals
kaolin
kaolin deposits
processes
weathering

kaolisation
use kaolinization

Kapoeta Meteorite
Term introduced in 1981. Impact near Kapoeta in S Sudan. Before 1981, also search Kapoeta AND meteorites.
BA howardite
achondrites
BT meteorites
SA Sudan

Kara Sea
Between Novaya Zemlya Island on W; and Yamal and Tamyr peninsulas on the E.
CO N680000N820000
E1050000E0553000
BT Arctic Ocean

Kara-Bogaz Gulf
Large shallow gulf of the Caspian Sea in NW Krasnovodsk Kray. Also search Kara-Bogaz-Gol Gulf; Kara-Bogaz-Gol region.
UF Kara-Bogaz-Gol Gulf
Kara-Bogaz-Gol region
BT Turkmenia
USSR

Kara-Bogaz-Gol
Town on the Caspian Sea at entrance to Kara-Bogaz Gulf.
BT Turkmenia
USSR
SA Caspian Sea

Kara-Bogaz-Gol Gulf
use Kara-Bogaz Gulf

Kara-Bogaz-Gol region
use Kara-Bogaz Gulf

Kara-Kum
use Karakum

Kara-Mazar
Village in Leninabad Oblast in N.
BT Tadzhikistan
USSR

Kara-Tau (Range)
use Karatau Range

Karabakh
Mountain area. Now Nagorno-Karabakh Autonomous Oblast in S.
BT Azerbaidzhan
USSR

Karafuto
use Sakhalin

Karaganda
City and oblast in NE central.
BT Kazakhstan
USSR

Karaganda Basin
Coal basin in Karaganda Oblast in NE central. Also search Karaganda.
BT Kazakhstan
USSR

Karaginski Island
use Karaginskiy Island

Karaginskiy Island
In Bering Sea just off coast of N Kamchatka Peninsula. Also search Karaginski Island.
UF Karaginski Island
BT Russian Republic
USSR

Karakoram
Mountain range in N Jammu and Kashmir. Index countries as applicable. Also search Karakorum.
UF Karakorum
BT Asia
SA India
Pakistan

Karakorum
use Karakoram

Karakum
Desert area S of the Aral Sea stretching from the Caspian Sea to the Amu Darya River. Also search Kara-Kum; Karakumy.
CO N350000N433000
E0660000E0540000
UF Kara-Kum
Karakumy
Qaraqum
BX Turkmenia
BZ Central Asia
BT USSR

Karakumy
use Karakum

Karamazar
Village in Leninabad Oblast in N.
BT Tadzhikistan
USSR

Karanpura
Village in central.
BT Bihar
India

Karanpura Basin
use Karanpura coal field

Karanpura coal field
Central Bihar. Also search Karanpura; Karanpura Basin; Karanpura Coalfield.
UF Karanpura Basin
Karanpura Coalfield
BT Bihar
India

Karanpura Coalfield
use Karanpura coal field

Karatau
use Karatau Range

Karatau Range
W branch of the Tien Shan in SE Kazakhstan. Also search Karatau; Kara-Tau.
CO N420000N443000
E0710000E0670000
UF Kara-Tau (Range)
Karatau
BT Kazakhstan
USSR

Karategin Range
SW Tadzhikistan.
BT Tadzhikistan
USSR

Karawanken
Range of Eastern Alps. A continuation of Carnic Alps. Index countries as applicable. Also search Karawanken Mountains; Karawanken Alps.
UF Karawanken Alps
Karawanken Mountains
BT Alps
Europe
SA Austria
Carnic Alps
Eastern Alps
Yugoslavia

Karawanken Alps
use Karawanken

Karawanken Mountains
use Karawanken

Karelia
Karelian Autonomous Soviet Socialist Republic in NW European USSR.
CO N600000N680000
E0400000E0300000
BT Russian Republic
USSR
SA Ladoga Series
Lake Ladoga region

Karelian
Includes use as an age term (list E).
BA middle Precambrian
Precambrian
SA Karelian Orogeny

Karelian Orogeny
Term introduced in 1978. Before 1978, also search Karelian AND orogeny.
BT Precambrian
SA Karelian
orogeny

Karharbari
Town in E central.
BT Bihar
India

Karharbari Stage
Basal member of the Damuda Series. Named after Karharbari in E central Bihar.
BT Upper Carboniferous
Carboniferous
SA Bihar
Damuda Series
India

Kariba
Town on Lake Kariba in NW.
BT Zimbabwe

Kariba Lake
use Lake Kariba

Karibib
Town in W central.
BT Namibia

Karkonosze
use Karkonosze Mountains

Karkonosze Mountains
Highest range of the Sudeten Mountains. Index Bohemia and/or Poland as applicable. Also search Karkonosze.
UF Karkonosze
BT Europe
SA Bohemia
Poland
Sudeten Mountains

Karl-Marx-Stadt
As of 1981, refers to only the city. Before 1981, also included the district.
BT Karl-Marx-Stadt Bezirk
East Germany

Karl-Marx-Stadt Bezirk
Term introduced in 1981. District in SE. Before 1981, search Karl-Marx-Stadt.
CO N501000N510800
E0133800E0115000
BT East Germany
NT Altenberg
Frankenberg
Freiberg
Karl-Marx-Stadt
Schneeberg
Schwarzenberg
Vogtland
Wildenfels
Zwickau
SA Saxony

Karlovy Vary
Town and health resort in W Bohemia. Used for Carlsbad when referring to Czechoslovakia. Before 1977, also search Carlsbad AND Czechoslavakia.
BT Bohemia
Czechoslovakia

Karlsruhe
City in NW central.
BT Baden-Wurttemberg
West Germany

Karnataka
Term introduced in 1981. State in SW India.
CO N113000N183000
E0784500E0741000
BT India

Karnes County
S central.
BT Texas
United States

Karnul
use Kurnool

Karoo
use Karroo Basin

Karpathos
Term introduced in 1981. Island in the Dodecanese. Before 1981, also search Karpathos Island.
BT Greek Aegean Islands
Greece

karpinskyite
BA framework silicates
silicates
BT minerals

karren
BT solution features
SA geomorphology
karst

Karroo Basalts
use Karroo System

Karroo Basin
An arid tableland region consisting of North Karroo along the Orange River in N, Great Karroo or Central Karroo in S central, and Southern Karroo or Little Karoo along S coast. Also search Karroo.
UF Karoo
BT Cape Province
South Africa

Karroo System
Carboniferous to Jurassic of Namibia and South Africa. Divided into Dwyka Series, Ecca Series, Beaufort Series, and Stormberg Series. Also search Karroo Basalts; Karroo.
UF Karroo Basalts
SA Africa
Botswana
Carboniferous
Dwyka Series
Jurassic
Namibia
Permian
South Africa
Triassic

Karshi
Town in Kashkadarya Oblast in S Uzbekistan.
UF Bek-Budi
BT Uzbekistan
USSR

Karshi Steppe
In Kashkadarya Oblast in S Uzbekistan. Also search Karshi.
BT Uzbekistan
USSR

karst
Includes use on level 3 under geomorphology(1) solution features(2). Before 1977, Karst was infrequently used to indicate the region. Now use Karst region.
UF carst
BT solution features
SA caves
dolines
geomorphology
karren
sinkholes
solution cavities
thermokarst

Karst region
Limestone plateau N of Trieste and E of Isonzo River. Index countries as applicable. Also search Karst regions. Before 1977, Karst was used to indicate the region.
BT Europe
SA Italy
Yugoslavia

Kartalinia
use Kartlia

Kartlia
Hilly central region.
UF Kartalinia
BT Georgian Republic
USSR

Karvina
City in NE Moravia.
UF Karvinna
BT Moravia
Czechoslovakia

Karvinna
use Karvina

Kasai
Term introduced in 1981.
CO S081000S014500
E0261500E0194000
BT Zaire

Kasai River
Flows into the Congo. Also search Kasai. Index countries as applicable.
BT Africa
SA Angola
Zaire

Kasauli Series
BT lower Miocene
Miocene
SA Himachal Pradesh
India

Kashiwazaki
City on Japan Sea in central.
BT Honshu
Japan

Kashmir
Part of Jammu and Kashmir.
BT Jammu and Kashmir
India

Kashmir and Jammu
use Jammu and Kashmir

Kashmir Valley
Intermontane valley in the Himalayas of W Jammu and Kashmir.
UF Vale of Kashmir
BT Jammu and Kashmir
India

Kasimovian
Term introduced in 1981. Includes use on level 3 under age terms(1). See list E.
BA Upper Carboniferous
Carboniferous
BT Paleozoic

Kaskawulsh Glacier
In the Saint Elias Mountains in SW.
BT Yukon Territory
Canada

kasolite
BA orthosilicates
silicates
BT minerals

Kassel
City in N.
BT Hesse
West Germany

katagenesis
use catagenesis

Katanga
use Shaba

Katangan Orogeny
Term introduced in 1978. Before 1978, search Katangan AND orogeny.
BT Precambrian
SA Africa
orogeny
Zaire

katathermal processes
use hypothermal processes

Katherine
River and settlement in N central.
BT Northern Territory
Australia

Kathiawar
Peninsula jutting into Arabian Sea.
BT Gujarat
India

Katmai National Monument
At N end of Alaska Peninsula in S.
BT Alaska
United States

Katowice
Province in S central. Also a city.
BT Poland
NT Bielsko
Bytom
Chorzow
Cieszyn
Czestochowa
Czestochowa-Zawierce Basin
Goczalkowice
Ruda
Rybnik
Sosnowiec
Upper Silesian coal basin
Zabrze
Zawiercie
SA Silesian coal basin

Kattegat
Strait between Sweden and Jutland, Denmark. Also search Kattegatt.
UF Kattegatt
BT North Sea

Kattegatt
use Kattegat

Katzenbuckel
Mountain. Highest point in the Odenwald in N.
BT Baden-Wurttemberg
West Germany

Kauai
One of the major Hawaiian Islands.
BT Hawaii
United States

Kavalerovo
Village in SE Primorye Kray in Soviet Far East.
BT Russian Republic
USSR

Kaveri Basin
use Cauvery Basin

Kayenta Formation
In Glen Canyon Group. Upper Triassic(?). NE Arizona, SW Colorado, and S and SE Utah.
BT Upper Triassic
Triassic
SA Arizona
Colorado
Utah

Kazakhstan
Kazakh Soviet Socialist Republic. In Soviet Central Asia.
CO N410000N550000
E0860000E0470000
BT USSR
NT Akbastau
Akchatau
Aksu
Aktyubinsk
Akzhal
Alakol
Alma-Ata
Atasu
Balkhash
Balkhash region
Barsakelmes
Bet-Pak-Dala
Biikzhal
Boshchekul
NX Central Kazakhstan
NT Char
Chingis-Tau
Chu
Chu-Ili Mountains
Dzhezkazgan
NX Eastern Kazakhstan
NT Ekibastuz
Emba
Emba Field
Emba River
Guryev
Ili
Irgiz
Kalba Range
Karaganda
Karaganda Basin
Karatau Range
Kokchetav
Kustanay
Lake Balkhash
Leninogorsk
Mangyshlak Peninsula
Mirgalimsay
NX Mugodzhar Hills
NT Muyunkum
Pavlodar
Sarysu
Sayak
Semipalatinsk
Stepnyak
Tekeli
Tengiz
Tokrau Synclinorium
Turgay
Ulu-Tau
Uspenskiy
Zaisan
Zaisan Basin
Zerenda
Zhetybay
Zyryanovsk
SA Aral region
Asiatic USSR
Caspian Basin
Caspian Depression
Central Asia
Chirchik River
Dzhungarian Alatau
Ili Basin
Irtysh River
Ishim
Kulunda Steppe
Kyzylkum
Ob-Irtysh Interfluve
Pskem Range
Rudny Altai
Siberia
Siberian Lowland
Southern Urals
Soviet Altai Mountains
Steppes
Syr Darya
Tarbagatay Range
Turan
Turanian Platform
Turgay Basin
Ural River
Urals
Ustyurt
Uzen
Volga-Ural region
West Siberia

Kazan
City on left bank of Volga in Tatar ASSR.
BT Russian Republic
USSR

Kazanian
Europe. Above Kungurian, below Tatarian. Includes use on level 3 under age terms(1). See list E.
BA Upper Permian
Permian
BT Paleozoic

keatite
BA silica minerals
framework silicates
silicates
BT minerals

Keban Mine
In E central. Also search Keban.
BT Turkey

Kebnekaise
Peak in Kjolen Mountains in NW Norrbotten. Highest in country.
BT Norrbotten
Sweden

Kedabek
Town in W.
BT Azerbaidzhan
USSR

Keel
Village on Achill Island in Atlantic Ocean.
BT Mayo
Ireland

Keewatin
Town in W. Ontario and village in NE Minnesota. Index Ontario or Minnesota as applicable. Before 1978, also used for District of Keewatin.
SA District of Keewatin
Minnesota
Ontario

Keewatin District
use District of Keewatin

Kefallinia
use Cephalonia

Keg River Formation
Subsurface. Includes the Rainbow (reef) Member. Eastward in Saskatchewan, the Keg River Formation is termed the Winnipegosis Formation. NW Alberta.
BT Devonian
SA Alberta
Canada

Kelantan
State on Thailand border on the Malay Peninsula in West Malaysia.
BT Malaysia
SA West Malaysia

Kelheim
Town in central.
BT Bavaria
West Germany

Kemerovo
City in the Kuznetsk Basin of Kemerovo Oblast in S central Siberia.
BT Russian Republic
USSR

Kenai Formation
use Kenai Group

Kenai Group
Eocene(?) and Oligocene, Central S Alaska. Also search Kenai Formation.
UF Kenai Formation
BT Paleogene
Tertiary
Cenozoic
SA Alaska

Kenai Peninsula
Between Cook Inlet on W and Prince William Sound on E in S.
BT Alaska
United States
SA Coast Ranges

Keno Hill
Village in central.
BT Yukon Territory
Canada

Kenora
City and district in W.
BT Ontario
Canada

Kenoran Orogeny
A time of plutonism, metamorphism, and deformation during the Precambrian of the Canadian Shield. Before 1978, also search Kenoran AND orogeny.
BT Precambrian
SA Canada
orogeny

Kent
Term introduced in 1981. County in SE England.
CO N505400N513100
E0013000E0010000
BT England
Great Britain
United Kingdom

Kent County
Term introduced in 1981 Index states as applicable.
BX Delaware
Maryland
Michigan
Rhode Island
Texas
BT United States

Kentland
Town in Newton County in NW.
BT Indiana
United States

Kentucky
Includes use on level 1 as an area term (list O). For term set options see list B.
CO N363000N391000
W0820000W0894000
BA Southern U.S.

United States
NX Allen County
Anderson County
NT Benham
NX Boone County
NT Breathitt County
NX Butler County
Caldwell County
Campbell County
Carroll County
Carter County
Christian County
Clark County
Clay County
Clinton County
Columbia
Crittenden County
Cumberland
Cumberland County
NT Edmonson County
Elliott County
NX Fayette County
Floyd County
Franklin County
Fulton County
Grant County
Hancock County
Hardin County
Harrison County
Jackson County
Jefferson County
Johnson County
NT Kentucky River
Kentucky River basin
Kentucky River valley
Knott County
NX Lawrence County
Lee County
NT Leslie County
Letcher County
NX Lewis County
Lincoln County
Livingston County
Logan County
NT Louisville
NX Lyon County
Madison County
NT Mammoth Cave
NX Marion County
Marshall County
Mason County
McLean County
Meade County
Mercer County
Monroe County
Montgomery County
Morgan County
Nelson County
NT Owsley County
NX Pendleton County
Perry County
Pike County
Powell County
Pulaski County
NT Rough Creek fault zone
NX Rowan County
Russell County
Scott County
Shelby County
Taylor County
Union County
Warren County
Washington County
Wayne County
Webster County
SA Allegheny Group
Allegheny Plateau
Ames Limestone
Appalachian Basin
Appalachian Plateau
Bedford Shale
Berea Sandstone
Borden Group
Brassfield Formation
Breathitt Formation
Carbondale Formation
Caseyville Formation
Chattanooga Shale
Chesterian
Cincinnati Arch
Cumberland Plateau
Eden Shale
Fairview Formation
Fort Payne Formation
Illinois Basin
Keokuk Limestone
Lee Formation
Lexington Limestone
Logan Formation
Midcontinent
Midway Group
Mississippi Embayment
Mississippi River
Mississippi Valley
Murray Meteorite
New Albany Shale
New Madrid region
Newman Limestone
Ohio River
Ohio River valley
Ohio Shale
Pennington Formation
Pottsville Group
Richmond Group
Ripley Formation
Saint Louis Limestone
Saint Peter Sandstone
Sainte Genevieve Limestone
Salem Limestone
Tennessee River
Tennessee Valley
Versailles
Waldron Shale
Warsaw Formation
Wilcox Group

Kentucky River
Formed by junction of North and Middle forks 4 miles ENE of Beattyville.
BT Kentucky
United States
SA Kentucky River valley

Kentucky River basin
Term introduced in 1981. Drainage basin.
CO N365000N384200
W0823500W0853000
BT Kentucky
United States

Kentucky River valley
NE central.
BT Kentucky
United States
SA Kentucky River

Kenya
Includes use on level 1 as an area term (list O). For term set options see list B.
CO S043000N043000
E0420000E0340000
BA East Africa
Africa
NT East Rudolf
Fort Ternan
Kenya Rift valley
NX Lake Baringo
Lake Magadi
NT Mount Kenya
Nairobi
Nakuru Basin
Turkana District
SA East African Rift
Gregory Rift
Lake Natron
Lake Turkana
Lake Victoria
Ngorora Formation
Umba

Kenya Rift
use Kenya Rift valley

Kenya Rift valley
That segment of the Great Rift Valley running N and S through W central Kenya. Also search Kenya Rift; Kenya Rift zone.
UF Kenya Rift
Kenya Rift zone
BT Kenya
SA Great Rift Valley
Lake Baringo

Kenya Rift zone
use Kenya Rift valley

kenyaite
BA scapolite group
framework silicates
silicates
BT minerals

Keokuk Limestone
Uppermost formation of Osagian Series. Iowa, Illinois, W Kentucky, and E Missouri.
BT Osagian
Lower Mississippian
Mississippian
SA Illinois
Iowa
Kentucky
Missouri

Keonjhar
District in N.
BT Orissa
India

kerabitumen
use kerogen

Kerala
State on Arabian Sea in SW.
CO N080000N130000
E0773000E0744000
BT India
SA Warkalli Formation

keratophyre
As of 1981, includes use on level 3 under igneous rocks(1) trachytes(2). See list H. Before 1981, included use on level 3 under igneous rocks(1) trachyte-phonolite family(2).
BA trachytes
volcanic rocks
BT igneous rocks
SA quartz keratophyre

Kerch
City on the Kerch Peninsula in E Crimea Oblast.
BT Ukraine
USSR

Kerch Peninsula
E section of the Crimea between the Azov Sea on the N and the Black Sea on the S. Also search Kerch.
BT Ukraine
USSR

Kerguelen Islands
French island group, 1,400 miles off Antarctic mainland.
CO S500000S480000
E0710000E0680000
UF Desolation Islands
BT Antarctica

Kermadec Islands
Volcanic island group 600 miles NNW of New Zealand. Annexed to New Zealand in 1887. Also search Kermadec.
BT Pacific Ocean
SA Raoul Island

Kerman
Province in SE central. Also a city.
BT Iran
NT Zarand Basin

Kermine
Town in Bukhara Oblast in central.
BT Uzbekistan
USSR

Kern County
S central.
BT California
United States

kernite
UF rasorite
BA borates
BT minerals

kerogen
Includes use on level 2 under organic materials(1).
UF kerabitumen
petrologen
BA organic materials
SA bitumens
oil shale
petroleum
sedimentary rocks

Kerry
County in SW.
BT Ireland

kersantite
As of 1981, includes use on level 3 under igneous rocks(1) lamprophyres(2). See list H. Before 1981, included use on level 3 under igneous rocks(1) lamprophyre and carbonatite family(2).
BA lamprophyres
BT igneous rocks

Kerulen
use Choybalsan

Kerulen River
A headstream of the Amur River. Index Inner Mongolia and/or Mongolia as applicable. Also search Kerulen.
BT Asia
SA Inner Mongolia
Mongolia

kesterite
BA sulfides
BT minerals

Ketchikan Quadrangle
Term introduced in 1981. SE Alaska.
CO N550000N560000
W1300000W1320000
BT Alaska
United States

Kettleman Hills
Along W side of San Joaquin Valley.
BT California
United States

Keuper
Europe. Above Muschelkalk, below Jurassic. Includes use on level 3 under age terms(1). See list E.
BA Upper Triassic
Triassic
BT Mesozoic

Keweenaw County
NW Upper Peninsula.
BT Michigan
United States

Keweenaw Peninsula
Juts into Lake Superior in NW Upper Peninsula.
BT Michigan
United States

Keweenawan
Provincial series, Michigan and Wisconsin. Includes use on level 3 under age terms(1). See list E.
BA Precambrian
NT Portage Lake Lava Series
SA Michigan
Wisconsin

key beds
Term introduced in 1978 for use in stratigraphy(1). Before 1978, search marker beds.

UF beds, key
index beds
SA marker beds
stratigraphy

Key Lake
Term introduced in 1981. Location of uranium deposit in NE Saskatchewan.
CO N570000N570000
W1040000W1040000
BT Saskatchewan
Canada

Key Largo
Island off extreme SE coast. Largest of the Florida Keys and the first island link of the Overseas Highway.
BT Florida
United States

Key Largo Limestone
Pleistocene deposit older than Pamlico Sand. S Florida.
BT Pleistocene
SA Florida

Khabarovsk region
Term introduced in 1981. Territory in SE Russian Republic. Also search Khabarovsk.
BT Russian Republic
USSR

Khamar-Daban Range
S of Lake Baikal in Buryat A.S.S.R. Also search Khamar-Daban.
BT Russian Republic
USSR

Khammam
Town and district in E central.
BT Andhra Pradesh
India

Khangai
use Hangay Mountains

Khangai Mountains
use Hangay Mountains

Khanka Lake
N of Vladivostok. Index China and/or Russian Republic as applicable. Also search Khanka.
BT Asia
SA China
Manchuria
Russian Republic

Khapcheranga
Town in SW Chita Oblast in Transbaikalia.
BT Russian Republic
USSR

Kharga
use Kharga Oasis

Kharga Oasis
Valley and oasis in S central Egypt. Also search El Kharga; Kharga.
UF El Kharga
Kharga
The Great Oasis
BT Egypt

Kharkov
City and oblast in NE.
BT Ukraine
USSR

Khaskovo
Province in S. Also a city.
CO N411000N421000
E0264000E0244000
BT Bulgaria

Khatanga
Town in Taymyr National Okrug about 150 miles from mouth of Khatanga River in N central Siberia.
BT Russian Republic
USSR

Khatanga Basin
River basin in E Taymyr National Okrug in N central Siberia. Also search Khatanga.
CO N713000N740000
E1080000E1013000
BT Russian Republic
USSR
SA Kotui
Yenisei-Khatanga basin

Khatanga River
Formed by the union of the Kheta and Kotui rivers in N central Siberia. Flows into Khatanga Gulf an inlet of the Laptev Sea. Also search Khatanga.
BT Russian Republic
USSR
SA Kheta River
Kotui

Khawr al Bazam
use Khor al Bazam

Kheta River
Chief tributary of the Khatanga River in Taymr National Okrug in N central Siberia.
BT Russian Republic
USSR
SA Khatanga River

Khetri
Town in E.
BT Rajasthan
India

Khetri copper belt
In N Aravalli Range in central. Also search Khetri.
BT Rajasthan
India

Khewra
Village in N.
BT Punjab Province
Pakistan

Khibiny Massif
use Khibiny Mountains

Khibiny Mountains
Central Kola Peninsula in extreme NW European USSR. Also search Khibiny; Khibiny Massif.
CO N670000N680000
E0340000E0330000
UF Khibiny Massif
BT Russian Republic
USSR

Khingan Range
At E end of Mongolian Plateau in N Inner Mongolia. Also search Khingan; Greater Khingan Mountains.
UF Great Khingan Range
Greater Khingan Mountains
Greater Khingan Range
BT Inner Mongolia
China

Khios
use Chios

Khiva
Town in oasis region on left bank of the lower Amu Darya in Khorezm Oblast in NW Uzbekistan. The former Khanate of Khiva is coextensive with current Khorezm Oblast.
BT Uzbekistan
USSR

Khmer Republic
use Cambodia

Khobdo
use Kobdo

khondalite
Includes use on level 3 under metamorphic rocks(1). See list J.
BA metasedimentary rocks
BT metamorphic rocks

Khor al Bazam
A strait off coast of central United Arab Emirates W of Abu Dhabi.
UF Khawr al Bazam
BT Persian Gulf

Khorat Group
Includes the Kamawkala Limestone. May also include beds of Permian as well as Cretaceous age or younger. Also search Korat series. Widespread throughout the country.
UF Korat Series
BT Mesozoic
SA Jurassic
Thailand
Triassic

Khorat Plateau
Tableland between the Mekong River on N and E and Cambodia on the S.
UF Korat Plateau
BT Thailand

Khovu-Aksy
Village in Tuva ASSR just N of W Mongolia.
BT Russian Republic
USSR

Khubsugul
Province in N Mongolia.
UF Hobsogol
Hubsugul
BT Mongolia

Kicking Horse River valley
In Banff National Park area. Index provinces as applicable. Also search Kicking Horse River.
BT Canada
SA Alberta
British Columbia

Kiel
City on the Baltic Sea in N.
BT Schleswig-Holstein
West Germany

Kiel Bay
Off NE coast of Schleswig-Holstein.
BT Baltic Sea
Atlantic Ocean

Kielce
Province in SE central. Also a city.
BT Poland
NT Nida Basin
Ostrowiec Swietokrzyski
Sandomierz

kieserite
BA sulfates
BT minerals

Kiev
City and oblast in N central.
BT Ukraine
USSR

Kiev Member
Also search Kiev Series.
BT lower Tertiary
Tertiary
SA Ukraine
USSR

Kiglapait
Cape in N.
BT Labrador
Canada

Kii Peninsula
Between Kii Channel on W and Kumano Sea on E in S Honshu. Also search Kii.
BT Honshu
Japan

Kilauea
Active crater on E side of Mauna Loa in Volcanoes National Park. Also search Kilauea Volcano.
CO N191500N192500
W1550500W1552000
UF Kilauea Volcano
BT Hawaii
United States
SA Alae Crater

Kilauea Volcano
use Kilauea

Kilbourne Hole
Term introduced in 1981. Exploratory drill hole in New Mexico.
CO N315700N315900
W1065700W1065900
BT Dona Ana County
New Mexico
United States

kilchoanite
BA orthosilicates
silicates
BT minerals

Kildare
County in E central. Also a town.
BT Ireland

Kilimanjaro
Highest mountain in Africa. Near Kenya border in NE.
BT Tanzania

Kilo
use Bambu

Kilo-Mines
use Bambu

Kimberley
City. World's diamond center with mines nearby. Near Orange Free State border in N.
BT Cape Province
South Africa

kimberlite
As of 1981, includes use on level 3 under igneous rocks(1) ultramafics(2). See list H. Before 1981, included use on level 3 under igneous rocks(1) ultramafic family(2).
BA ultramafics
BT igneous rocks
SA diamond
diamonds

Kimmerian
use Cimmerian

Kimmeridgian
Europe. Above Oxfordian, below Portlandian. Includes use on level 3 under age terms(1). See list E.
BA Upper Jurassic
Jurassic
BT Mesozoic
SA Malm

Kinderhookian
Provincial series, North America. Above Chautauquan (Devonian), below Osagian. Includes use on level 3 under age terms(1). See list E.
BA Lower Mississippian
Mississippian
Carboniferous
BT Paleozoic

kinematics
Branch of dynamics that deals with aspects of motion apart from considerations of mass and force.
UF cinematics
SA kinetics
velocity

kinetics
SA energy
geochemistry
geophysics
kinematics
velocity

King County
BX Texas
Washington
BT United States

King George Island
Largest of the South Shetland Islands in British Antarctic Territory N of the Antarctic Peninsula.
UF Waterloo Island
BA South Shetland Islands
Scotia Sea Islands
BT Antarctica

Kingfisher County
Central.
BT Oklahoma
United States

Kingman County
S central.
BT Kansas
United States

Kings Bay
Inlet on NW coast of Norwegian island of Spitsbergen.
BT Spitsbergen
Svalbard
Arctic region

Kings County
Term introduced in 1981. Index states as applicable.
BX California
New York
BT United States

Kings Mountain
Isolated ridge. Index states as applicable.
BT United States
SA North Carolina
South Carolina

Kingsdown Formation
In Sanborn Group. Overlies Crooked Creek Formation; underlies Vanhem Formation. SW Kansas.
BT Pleistocene
SA Kansas

Kingsport Formation
In Knox Group. E Tennessee and SW Virginia.
BT Lower Ordovician
Ordovician
SA Knox Group
Tennessee
Virginia

kink folds
Term introduced in 1978. Includes use on level 3 under folds(1) style(2). Before 1978, also search folds AND kink.
BA folds
SA chevron folds
kink-band structures

kink-band structures
Term introduced in 1978. Includes use on level 3 under deformation(1) experimental studies(2). Also search kink bands; kink-band.
SA deformation
folds
kink folds
structures

Kinki District
The Osaka, Kobe and Kyoto region in S central.
BT Honshu
Japan

Kinnashih
use Chinkuashih

Kinnaur
Term introduced in 1981. District in Himachal Pradesh.
CO N310500N320000
E0790000E0774500
BT Himachal Pradesh
India

Kinney County
Term introduced in 1981. SW Texas.
CO N290100N294000
W1000700W1005200
BT Texas
United States

Kinshasa
City and Federal District on left bank of the Congo River in W Zaire.
UF Leopoldville
BT Zaire

Kinta Valley
River valley in NW West Malaysia.
BT Perak
Malaysia

Kintyre
Term introduced in 1981. Peninsula in Strathclyde region, SW Scotland.
CO N551800N555200
W0051900W0054700
BT Scotland

Kinwashih Mine
use Chinkuashih Mine

Kinzers Formation
Overlies Vintage Limestone; underlies Ledger Dolomite. Maryland, SE Pennsylvania, and Virginia.
BT Lower Cambrian
Cambrian
SA Maryland
Pennsylvania
Virginia

kinzigite
Includes use on level 3 under metamorphic rocks(1) granulites(2). See list J.
BA granulites
BT metamorphic rocks

Kiowa County
Index states as applicable.
BX Colorado
Kansas
Oklahoma
BT United States
SA Haviland Meteorite

Kiowa Formation
Overlies Cheyenne Sandstone Member; underlies Dakota Sandstone. Central S Kansas, E Colorado, W Oklahoma. Also search Kiowa Shale.
UF Kiowa Shale
BT Lower Cretaceous
Cretaceous
SA Colorado
Kansas
Oklahoma

Kiowa Shale
use Kiowa Formation

Kipawa Lake
SW Quebec. Also search Kipawa.
BT Quebec
Canada

Kirghiz (Soviet Socialist Republic)
use Kirghizia

Kirghizia
Kirghiz Soviet Socialist Republic in Soviet Central Asia. Also search Kirghiz; Kirgizia.
CO N390000N420000
E0800000E0690000
UF Kirghiz (Soviet Socialist Republic)
Kirgizia
BT USSR
NT Alai Range
Chatkal Range
Frunze
Issyk-kul Lake
Naryn
Suzak
Talas Range
SA Asiatic USSR
Central Asia
Fergana Basin
Pskem Range
Syr Darya
Tien Shan

Kirgizia
use Kirghizia

Kirishima
Mountain in Kirishima Range in S Kyushu. Also search Kirishima-Yama.
UF Kirishima-Yama
BT Kyushu
Japan

Kirishima-Yama
use Kirishima

Kirkbyocopina
Term introduced in 1981. Includes use on level 2 under Ostracoda(1). See list F.
BA Ostracoda
BT Crustacea
Mandibulata
Arthropoda
Invertebrata

Kirkcudbrightshire
County in S.
CO N544500N551500
W0033000W0043000
BT Scotland
Great Britain
United Kingdom

Kirkland Lake
Town in E Ontario near Quebec border.
BT Ontario
Canada

Kirkuk
Town near oil fields in NE.
BT Iraq
Middle East

Kirkwood Formation
Arikareean and Hemingfordian. E New Jersey.
BT middle Miocene
Miocene
SA New Jersey

Kirov
City and oblast in E central European USSR.
UF Vyatka
BT Russian Republic
USSR

Kirovabad
City in W Azerbaidzhan.
UF Gandja
Gandzha
BT Azerbaidzhan
USSR

Kirovograd
City and oblast in central S.
BT Ukraine
USSR

Kiruna
City in N central.
BT Norrbotten
Sweden

Kirwin
Town in Phillips County in N.
BT Kansas
United States

Kishtwar
Town in SW.
BT Jammu and Kashmir
India

Kiso Mountains
In central. Also search Kiso.
BT Honshu
Japan

Kistna River
use Krishna

Kitakami Massif
use Kitakami Mountains

Kitakami Mountains
N Honshu. Also search Kitakami; Kitakami Massif.
UF Kitakami Massif
BT Honshu
Japan

Kitami
City in NE.
BT Hokkaido
Japan

Kitsap Peninsula
Bounded on W by Hood Canal and on E by Puget Sound. W of Seattle in NW.
BT Washington
United States

Kittitas County
Term introduced in 1981. Central Washington.
CO N464300N473600
W1195600W1213000
BT Washington
United States

Kivu
Province in E.
CO S050000N010000
E0300000E0244000
BT Zaire
NT Nyiragongo

Kiya River
In Kemerovo and Tomsk oblasts E and NE of Novosibirsk.
BT Russian Republic
USSR

Kizel coal basin
In Central Urals of European USSR in Perm Oblast. Also search Kizel.
BT Russian Republic
USSR

Kizil Kum
use Kyzylkum

Kizil-Dere
use Kizyl-Dere

Kizildere
use Kizyl-Dere

Kizyl-Dere
Gorge near Iranian border in S Turkmenia. Also search Kizil-Dere; Kizildere.
UF Kizil-Dere
Kizildere
BT Turkmenia
USSR

Kladno
City NW of Prague.
BT Bohemia
Czechoslovakia

Klamath Mountains
Part of the Coast Ranges. Index states as applicable.
CO N410000N430000
W1230000W1243000
BT United States
SA California
Coast Ranges
Oregon

Kletno
Village in central.
BT Byelorussia
USSR

Klichka
Village in S Chita Oblast near Inner Mongolian border.
BT Russian Republic
USSR

klippen
Also search klippe.
SA allochthons
erosion features
faults
nappes

overthrust faults
tectonics

Klodawa
Town in central part of country.
BT Poznan
Poland

Klondike
Region in Yukon River basin in SW central.
BT Yukon Territory
Canada

Kluane Lake
Along N slope of Saint Elias Mountains in SW.
BT Yukon Territory
Canada

Klyuchevskaya Sopka
Active volcano in Eastern Range of E central Kamchatka Peninsula. Also search Klyuchevskaya. Highest mountain in Siberia.
BT Russian Republic
USSR

Knee Lake
NE Manitoba.
BT Manitoba
Canada

Knife Lake Group
Comprises rocks formerly called Ogishke Conglomerate, Agawa Iron-formation, and Knife Lake Slates. Includes many members which cannot be separated into the three former divisions.
BT Precambrian
SA Minnesota

Knik Arm
Extension of Cook Inlet N of Anchorage. Also search Knik River.
UF Knik River
BT Alaska
United States

Knik River
use Knik Arm

Knott County
Term introduced in 1981. E Kentucky.
CO N371000N373200
W0824300W0830900
BT Kentucky
United States

Knox County
Term introduced in 1981. NW central Illinois.
CO N404400N411000
W0895900W0902800
BT Illinois
United States

Knox Group
Includes Copper Ridge Dolomite, Chepultepec Dolomite or Limestone, Longview or Nittany Dolomite, Kingsport Formation, Mascot Dolomite, Newala Formation, Conococheague Limestone, Jonesboro Limestone. NW Georgia, W North Carolina, E Tennessee, and SW Virginia.
BT Paleozoic
SA Cambrian
Georgia
Kingsport Formation
Lower Ordovician
Mascot Dolomite
North Carolina
Ordovician
Tennessee
Upper Cambrian
Virginia

Knysna
Town in S.
BT Cape Province
South Africa

Kobdo
Town and river in W Mongolia.
UF Hobdo
Khobdo
BT Mongolia

kobellite
BA sulfantimonites
sulfosalts
BT minerals

Kobenhavn
use Copenhagen

Kobystan
Region near Baku in E.
BT Azerbaidzhan
USSR

Kocaeli
Province in NW Anatolia.
BT Turkey
Middle East

Kochi
City and prefecture in S.
BT Shikoku
Japan

Kochkar
Village in central Chelyabinsk Oblast in W Siberia.
BT Russian Republic
USSR

Kodaikanal
City in S.
BT Madras
India

Kodar Mountains
use Kodar Range

Kodar Range
NE of Lake Baikal in N Chita Oblast. Also search Kodar; Kodar Mountains.
UF Kodar Mountains
BT Russian Republic
USSR

Kodiak Island
In Gulf of Alaska SE of Alaska Peninsula.
BT Alaska
United States
SA Coast Ranges

Kodor River basin
In Abkhaz ASSR in W.
BT Georgian Republic
USSR

Koenigsberger ratio
Term introduced in 1981.
SA magnetic field
magnetic susceptibility
paleomagnetism
remanent magnetization

Kokchetav
City and oblast in N.
BT Kazakhstan
USSR

Kola
Town and river in Murmansk Oblast on the Kola Peninsula in extreme NW European USSR.
BT Russian Republic
USSR

Kola Peninsula
Between White Sea and Barents Sea in extreme NW European USSR.
CO N660000N700000
E0420000E0320000
BT Russian Republic
USSR
SA Allarechenskiy

Kolar
Town and district in E.
BT Mysore
India

Kolar Gold Field
use Kolar Gold Fields

Kolar Gold Fields
City in E Mysore. Center of gold-mining industry. Also search Kolar; Kolar Gold Field; Kolar Goldfields.
UF Kolar Gold Field
Kolar Goldfields
BT Mysore
India

Kolar Goldfields
use Kolar Gold Fields

kolbeckine
use herzenbergite

Kolbeinsey Island
Off N coast in Norwegian Sea.
BT Iceland
SA Norwegian Sea

Kolima Mountains
use Kolyma Uplift

Kolima River
use Kolyma River

Kolima River basin
use Kolyma River basin

Kolyma Massif
use Kolyma Uplift

Kolyma River
Rises in SE Cherskiy Range and flows generally N into the East Siberian Sea. Also search Kolyma.
UF Kolima River
BT Russian Republic
USSR

Kolyma River basin
N Khabarovsk Kray and E Yakutia. Also search Kolyma.
UF Kolima River basin
BT Russian Republic
USSR

Kolyma Uplift
Mountain range N of the Okhotsk Sea in Magadan oblast and NE Yakutia.
CO N610000N673000
E1650000E1420000
UF Kolima Mountains
Kolyma Massif
BA Northeastern USSR
BT Russian Republic
USSR

Kom Ombo
Village in S part of country.
UF Kum Umbu
BT Aswan
Egypt

Koma-ga-take
As of 1981, no longer a valid index term for GeoRef. Use Koma-ga-take 1 or Koma-ga-take 2 as applicable.

Koma-ga-take 1
Term introduced in 1981 for the volcano on Hokkaido. Also search Koma-ga-take AND Hokkaido.
BT Hokkaido
Japan

Koma-ga-take 2
Term introduced in 1981 for the volcano on Honshu. Before 1981, also search Koma-ga-take AND Honshu.
BT Honshu
Japan

komatiite
As of 1981, includes use on level 3 under igneous rocks(1) ultramafics(2). See list H. Before 1981, included use on level 3 under igneous rocks(1) ultramafic family(2).
BA ultramafics
BT igneous rocks

Komi
Autonomous Soviet Socialist Republic in NE, west of the Northern Urals.
BT Russian Republic
USSR

Kommunar
Town in Khakass Autonomous Oblast in SW Krasnoyarsk Kray.
BT Russian Republic
USSR

Komsomolsk
City in S Khabarovsk Kray in Soviet Far East. Also search Komsomolsk-on-Amur.
UF Komsomolsk-na-Amure
Komsomolsk-on-Amure
BT Russian Republic
USSR

Komsomolsk-na-Amure
use Komsomolsk

Komsomolsk-on-Amure
use Komsomolsk

Kona
Divisions, both North and South, of Hawaii County on W and SW side of island of Hawaii.
BT Hawaii
United States

Kondapalli
Town in NE.
BT Tamil Nadu
India

Kongo River
use Congo River

Kongsberg
Town in SE.
BT Norway

Konia
use Konya

Konieh
use Konya

Konin
Town in central part of country.
BT Poznan
Poland

Konstantinovka
City in Donetsk Oblast in SE.
BT Ukraine
USSR

Konya
Province in SW central Anatolia. Also a city.
UF Konia
Konieh
BT Turkey
Middle East

Koolau Range
Extends along E side of Oahu Island.
BT Hawaii
United States

Kootenai County
In northern panhandle.
BT Idaho
United States

Kootenay Formation
Includes Moose Mountain Member, Adanac, Hillcrest, and Mutz members. Also search Kootenai Formation.
BT Mesozoic
SA British Columbia
Canada
Cretaceous
Jurassic

Kootenay Lake
SE British Columbia.
UF Kutenai Lake
BT British Columbia
Canada

Kopaonik
Central Serbia. Also search Kopaonik Mountains.

UF Kopaonik Mountains
BT Serbia
Yugoslavia

Kopaonik Mountains
use Kopaonik

Kopet Dag (Range)
use Kopet-Dag Range

Kopet Dagh (Range)
use Kopet-Dag Range

Kopet-Dag Range
In SW Turkmenia. Before 1981, term included extension in Iran.
CO N350000N393000
E0630000E0523000
UF Kopet Dag (Range)
Kopet Dagh (Range)
BX Turkmenia
BZ Central Asia
BT USSR
SA Iran

Koralpe Range
Small range of Noric Alps in S Austria. Also search Koralpe.
BT Austria

Koraput
Village and district in SW.
BT Orissa
India

Korat Plateau
use Khorat Plateau

Korat Series
use Khorat Group

Kordofan
Province in central.
BT Sudan

Korea
A peninsula on E coast of Asia, since 1948 partitioned into two republics. Includes use on level 1 as an area term (list O). For term set options see list B. Use North Korea and/or South Korea on level 3.
CO N330000N430000
E1300000E1241500
BA Far East
Asia
NT Cheju Island
North Korea
South Korea

Korenevskaya Formation
Dnieper-Donets region.
SA Permian
Triassic
Ukraine
USSR

kornerupine
UF prismatine
BA orthosilicates
silicates
BT minerals

Korosten
City in Zhitomir Oblast in NW central.
BT Ukraine
USSR

Korsun
Old name of city in Cherkassy Oblast in central Ukraine. Now called Korsun-Shevchenkovski.
BT Ukraine
USSR

Koryak Range
Extends from neck of Kamchatka Peninsula in Khabarovsk Kray into the Chukchi National Okrug. Also search Koryak.
CO N600000N650000
E1800000E1650000
BA Soviet Pacific region
BT Russian Republic
USSR

Korytnica
Health resort on E slope of the Tatra Mountains.
BT Slovakia
Czechoslovakia

Kos
One of the islands of the Dodeconese off SW Turkey. Also search Coo.
UF Coo
Cos
BT Greek Aegean Islands
Greece

Kosaka Mine
Near town of Kosaka in N.
BT Honshu
Japan

Kosi Basin
River basin. Index Bihar and/or Nepal as applicable. Also search Kosi River.
BT Asia
SA Bihar
Nepal

Kosov
Town in Ivano-Frankovsk Oblast in SE.
BT Ukraine
USSR

Kosovo-Metohija
Term introduced in 1981. Autonomous region in S part of country.
BT Serbia
Yugoslavia

Kosseir
E Egypt, port on Red Sea. Also search Quseir.
UF Al-Qusayr
Al-Quseir
Quseir
BT Egypt

Kostolac
Village and arm of the Danube River in E Serbia.
UF Kostolats
BT Serbia
Yugoslavia

Kostolats
use Kostolac

Kostroma
City on left Bank of the Volga River. Also an oblast.
BT Russian Republic
USSR

Koszalin
Province in NW. Also a city.
BT Poland

Kota
Town in E central.
BT Madhya Pradesh
India

Koto
River in E and S which flows into the Ubangi River.
UF Kotto
BT Central African Republic

kotoite
BA borates
BT minerals

Kotto
use Koto

Kotui
River in Evenk and Taymyr National okrugs in N central Siberia. Along with the Kheta River, it forms Khatanga River. Also search Kotuy; Kotuy River.
UF Kotuy River
BT Russian Republic
USSR
SA Khatanga Basin
Khatanga River

Kotuy River
use Kotui

Kouchibouguac Bay
On N North Umberland Strait in E.
BT New Brunswick
Canada

Koyna
River in W.
BT Maharashtra
India

Kozhim
Village and river in Komi A.S.S.R. W of the Northern Urals.
BT Russian Republic
USSR

Kr
use krypton

Kraishte
use Krajiste

Krajiste
Highland. Index countries as applicable. Also search Kraishte; Krayishte.
UF Kraishte
Krayishte
BT Europe
SA Bulgaria
Yugoslavia

Krakatao
use Krakatoa

Krakatau
use Krakatoa

Krakatoa
Island volcano in center of Sunda Strait between Sumatra and Java. Also search Krakatao; Krakatau.
UF Krakatao
Krakatau
BT Indonesia

Krakow
Province in S Poland. Also a City province and a city. Also search Cracow.
UF Cracow
BT Poland
NT Andrychow
Bochnia
Jordanow
Krzeszowice
Miechow
Nowy Sacz
Nowy Targ
Olkusz
Pieniny Klippen Belt
Pieniny Mountains
Pilica
Podhale
Tarnow
Wadowice
Wieliczka
Zakopane

Krakow-Czestochowa Jura
Mountain range in S between Krakow and Czestochowa.
BT Poland

Krappfeld
Region in S part of country.
BT Carinthia
Austria

Krasnodar
City in Krasnodar Kray in the Northern Caucasus.
BT Russian Republic
USSR

Krasnoyarsk
City in Krasnoyarsk Kray in S central Siberia.
BT Russian Republic
USSR

Krayishte
use Krajiste

KREEP
Acronym for a basaltic lunar rock type first found in Apollo 12 fines and breccias and characterized by unusually high contents of potassium (K), rare-earth elements (REE), phosphorus (P), and other trace elements in comparison to other lunar rock types.
UF nonmare basalt
SA Moon
phosphorus
potassium
rare earths

Krefeld
City in W.
BT North Rhine-Westphalia
West Germany

Kremasta
Village in NW. Central Greece and Euboea.
BT Greece

Kremenchug
City in Poltava Oblast in NE central.
BT Ukraine
USSR

Kremnica
Town in W central.
BT Slovakia
Czechoslovakia

Kremnica Mountains
In W central.
BT Slovakia
Czechoslovakia

Krems
City on the Danube River at the mouth of the Krems River.
UF Krems an dar Donau
BT Lower Austria
Austria

Krems an dar Donau
use Krems

Kreuth
Resort in Bavarian Alps in S.
BT Bavaria
West Germany

Krishna
River which flows into the Bay of Bengal. Index states as applicable. Also search Kistna River.
UF Kistna River
BT India
SA Andhra Pradesh
Maharashtra
Mysore

Kristianstad
County in S. Also a city.
BT Sweden
SA Skane

Kristiansund
Seaport city, W Norway.
UF Christiansund
BT Norway

Krivoi Rog
use Krivoy Rog

Krivoi Rog Basin
use Krivoy Rog Basin

Krivoi Rog region
use Krivoy Rog Basin

Krivoi Rog Series
use Krivoy Rog Series

Krivoy Rog
City in W Dnepropetrovsk Oblast in S central Ukraine. Also search Krivoi Rog.
UF Krivoi Rog
BT Ukraine
USSR

Krivoy Rog Basin
Basin of the Ingulets River in the Krivoy Rog region. An industrial basin in W Dnepropetrovsk Oblast in S central Ukraine. Also search Krivoi Rog Basin; Krivoy Rog region.
CO N473000N480000
E0353000E0323000
UF Krivoi Rog Basin
Krivoi Rog region
BT Ukraine
USSR

Krivoy Rog Series
Also search Krivoi Rog Series.
UF Krivoi Rog Series
BT Archean
SA Ukraine
USSR

Krosno Beds
Lattorfian or Rupelian of some authors. In the Polish Carpathians.
BT Oligocene
SA Poland
Rupelian
Tongrian

Krusne Hory
use Erzgebirge

Krusnehory Mountains
use Erzgebirge

Krusny Hory Mountains
use Erzgebirge

krypton
Includes use on level 1 and 2 as a chemical element (list D).
UF Kr
BT noble gases
SA argon
elements
helium
isotopes
neon
radon
xenon

Krzeszowice
Town in S part of country.
BT Krakow
Poland

Ksiaz
Town in W central part of country.
UF Ksiaz Wielkopolski
BT Poznan
Poland

Ksiaz Wielkopolski
use Ksiaz

Kuala Lumpur
City in SE West Malaysia on the Malay Peninsula.
BT Selangor
Malaysia

Kuban
Steppe region in the W Northern Caucasus. Formerly an oblast.
BT Russian Republic
USSR

Kuban River
Rises in Georgian Republic in the Caucasus and flows N and NW into the Sea of Azov. Index Soviet republics as applicable. Also search Kuban.
BT USSR
SA Georgian Republic
Russian Republic

Kuban Valley
River valley. Index Soviet republics as applicable. Also search Kuban.
BT USSR
SA Georgian Republic
Russian Republic

Kuchinoerabu-Jima
Small island off S coast.
BT Kyushu
Japan

Kufara
use Kufra Basin

Kufra Basin
Group of 5 oases in central Libyan Desert.
UF Al-Kufrah
Kufara
BT Libya

Kugitang (-Tau)
use Kugitang-Tau

Kugitang-Tau
Mountain range in E Chardzou Oblast in E Turkmenia. S spur of Baisun-Tau. Also search Kugitang.
UF Kugitang (-Tau)
BT Turkmenia
USSR

Kuhn Island
Off NE coast.
BT Greenland
Arctic region

Kuibyshev
City and oblast. Also search Kuybyshev.
UF Kuybyshev
BT Russian Republic
USSR

Kujawy
Region in N central.
BT Poland

Kuju
Mountain in Oita Prefecture in central.
BT Kyushu
Japan

Kular Range
N central Yakutia. Also search Kular.
BT Russian Republic
USSR

Kul'dzhuktau
Mountains in central.
BT Uzbekistan
USSR

Kulunda Steppe
Between Ob River in Novosibirsk Oblast and Altay Kray on the E, and E Pavlodar Oblast in Kazakhstan on the W. Index Soviet republics as applicable.
BT USSR
SA Kazakhstan
Russian Republic

Kulyab
Town in SW.
BT Tadzhikistan
USSR

Kum Umbu
use Kom Ombo

Kuma Basin
River basin in Northern Caucasus. Also search Kuma; Kuma River.
BT Russian Republic
USSR

Kumamoto
City and prefecture in W central.
BT Honshu
Japan

Kumaon Himalaya
use Kumaun Himalayas

Kumaon Himalayas
use Kumaun Himalayas

Kumaun Himalayas
W central subdivision of the Himalayas. Index Nepal, Tibet, and Indian states as applicable. Also search Kumaon Himalaya; Kumaon Himalayas; Kumuan Himalaya.
UF Kumaon Himalaya
Kumaon Himalayas
Kumuan Himalaya
BT Asia
SA Himachal Pradesh
Himalayas
Nepal
Tibet
Uttar Pradesh

Kumuan Himalaya
use Kumaun Himalayas

Kunashir Island
Second largest of the Kuril Islands which are administratively part of Sakhalin Oblast. Also search Kunashir.
BT Russian Republic
USSR
SA Goryachiy Plyazh
Kuril Islands

Kunavaram Series
BT India

Kundelungu Plateau
SE part of country.
BT Zaire

Kungurian
Europe. Above Artinskian, below Kazanian. Includes use on level 3 under age terms(1). See list E.
BA Permian
BT Paleozoic
SA Lower Permian
Middle Permian

Kuno
Danish island in the Faeroe Islands.
BT Faeroe Islands
Denmark

kunzite
Transparent gem variety of spodumene.
BA clinopyroxene
pyroxene group
chain silicates
silicates
BT minerals
SA spodumene

Kupferschiefer
Term introduced in 1981. Lithofacies in Europe. A copper-bearing bituminous shale/schist.
SA Permian
Zechstein

Kura Depression
use Kura Lowland

Kura Lowland
Extensive plain along SW shores of Caspian Sea. Index Azerbaidzhan and/or Iran as applicable. Also search Kura and Kura Depression.
UF Kura Depression
SA Azerbaidzhan
Iran

Kura River
Flows into Caspian Sea S of Baku. Index Turkey and Soviet republics as applicable. Also search Kura.
SA Azerbaidzhan
Georgian Republic
Turkey

Kurama Range
Branch of the Tien Shan. Index Soviet republics as applicable. Also search Kurama.
BT USSR
SA Tadzhikistan
Tien Shan
Uzbekistan

kurchatovite
BA borates
BT minerals

Kureika River (region)
use Kureyka River region

Kureyka River region
N Krasnoyarsk Kray in NW central Siberia. Also search Kureika River; Kureyka.
UF Kureika River (region)
BT Russian Republic
USSR

Kurgan
City and oblast E of the Southern Urals.
BT Russian Republic
USSR

Kuril Islands
Group of 56 islands extending from Kamchatka Peninsula to Hokkaido. Administratively part of Sakhalin Oblast. Also search Kurile Islands; Kurils.
CO N433000N510000
E1570000E1450000
UF Kurile Islands
Kurils
BA Soviet Pacific region
BT Russian Republic
USSR
SA Iturup Island
Kunashir Island
Kuril Trench
Okhotsk region
Paramushir
Soviet Far East
Urup

Kuril Trench
E of the Kuril Islands and along E coast of SE Kamchatka Peninsula. Also search Kuril-Kamchatka; Kuril-Kamchatka Trench.
UF Kuril-Kamchatka Trench
BT Pacific Ocean
SA Kuril Islands

Kuril-Kamchatka Trench
use Kuril Trench

Kurile Islands
use Kuril Islands

Kurils
use Kuril Islands

Kurland Spit
use Courland Spit

kurnakovite
BA borates
BT minerals

Kurnool
Town in W Andra Pradesh.
UF Karnul
BT Andhra Pradesh
India

Kurnool System
Included in the Purana Group. Subdivided into Kundair Stage, Paniam Stage, Jammalamadugu Stage, Banganapalli Stage. Named after town of Kurnool.
UF Kurnul System
SA Andhra Pradesh
Cambrian
India
Precambrian

Kurnul System
use Kurnool System

Kuroko
As of 1976, usage restricted to geographic place name. Before 1976, term was also used to indicate type of ore deposit.
BT Japan

kuroko-type
Metallic sulfide ore. Before 1977, also search kuroko.
SA mineral deposits

Kuroshio
A branch of the equatorial current flowing along E coast of Taiwan,

thence NE along E coast of Honshu, past the Aleutians, and S along North American coast. More commonly known as Japan Current. Also search Kuroshio Current.
UF Japan Current
Kuroshio Current

Kuroshio Current
use Kuroshio

Kursk
City and Oblast in SW European USSR.
CO N510000N521500
E0380000E0341500
BT Russian Republic
USSR

Kursk magnetic anomaly
SW European USSR. Extensive iron-ore region, 90 miles long and 10 miles wide, in central SE Kursk Oblast in SW European USSR. Also search Kursk.
BT Russian Republic
USSR

Kursk Series
Named after city in Kursk Oblast in SW Russian Republic.
BT Proterozoic
Precambrian
SA Russian Republic
USSR

Kurt
Village S of Aral Sea.
BT Turkmenia
USSR

kurtosis
BA statistical analysis
SA histograms
skewness

Kushiro coal field
SE Hokkaido. Also search Kushiro.
BT Hokkaido
Japan

Kushva
City in Ural Mountains in W Sverdlovsk Oblast in West Siberia.
BT Russian Republic
USSR

Kustanai
use Kustanay

Kustanay
City and oblast in NW Kazakhstan.
UF Kustanai
BT Kazakhstan
USSR

Kutch
use Cutch

Kutenai Lake
use Kootenay Lake

Kutna Hora
Town in central.
BT Bohemia
Czechoslovakia

Kuwait
Includes use on level 1 as an area term (list O). For term set options see list B.
CO N283000N301500
E0484500E0463000
BA Arabian Peninsula
Asia
SA Middle East
Near East

Kuybyshev
use Kuibyshev

Kuzbas
use Kuznetsk Basin

Kuznetsk Alatau
Mountain system along borders of Khakass Autonomous Oblast of Krasnoyarsk Kray, and Kemerovo Oblast in SSW Siberia. An outlier of the Altai Mountains.
CO N520000N561500
E0890000E0823000
BZ West Siberia
BX Russian Republic
BT USSR
SA Altai Mountains

Kuznetsk Basin
Coal basin, 150 miles long and 65 miles wide, in Kemerovo Oblast extending from Tomsk to Novokuznetsk in SSW Siberia.
CO N520000N561500
E0890000E0840000
UF Kuzbas
BT Russian Republic
USSR

Kwanto Plain
Largest and most densely populated plain in country. In central part of island and includes Tokyo-Yokohama industrial region. Also search Kwanto.
UF Kanto
BT Honshu
Japan

Kwanza Basin
use Cuanza Basin

Kworra River
use Niger River

Kworra Valley
use Niger Valley

kyanite
Includes use on level 3 under minerals(1) orthosilicates(2). See list L. As of 1981, use kyanite deposits for kyanite as a commodity. Before 1981, included use on level 3 as a commodity term.
BA orthosilicates
silicates
BT minerals
SA andalusite
heavy minerals
kyanite deposits
sillimanite

kyanite deposits
Term introduced in 1981. Includes use on level 3 as a commodity term (list C) under ceramic materials(1). Before 1981, search kyanite AND deposits.
SA ceramic materials
heavy mineral deposits
kyanite

Kyffhauser Range
In SW. Also search Kyffhauser.
BT East Germany

Kyongsang Basin
In North Kyongsang and South Kyongsang provinces in SE.
BT South Korea

Kyoto
City and prefecture in central.
BT Honshu
Japan

Kyushu
Southernmost of the four main islands.
CO N305500N340000
E1320500E1292500
BT Japan
NT Aira Caldera
Ariake Bay
Aso
Aso Caldera
Beppu
Ebino
Fukuoka
Goto Islands
Iki Island
Kagoshima
Kirishima
Kuchinoerabu-Jima
Kuju
Miyazaki
Nagasaki
Oita
Ontake
Osumi Peninsula
Saga
Sasebo
Satsuma Peninsula
Tanega-shima
Tsushima
Unzen
Usuki
SA Green Tuff Formation
Sambagawa Belt
Shimanto Group
Taishu Group

Kyushu-Palau Ridge
Term introduced in 1981. NW Philippine Sea.
CO N110000N200000
E1370000E1340000
BT Philippine Sea
Pacific Ocean

Kyzyl Kum
use Kyzylkum

Kyzyl Kum Desert
use Kyzylkum

Kyzyl-Kum
use Kyzylkum

Kyzylkum
Desert SE of Aral Sea between the Amu Darya and the Syr Darya. Also search Kyzyl Kum; Kyzyl-Kum; Kyzyl Kum Desert. Index Soviet republics as applicable.
CO N400000N480000
E0733000E0580000
UF Kizil Kum
Kyzyl Kum
Kyzyl Kum Desert
Kyzyl-Kum
Qizil Qum
BA Central Asia
BT USSR
SA Kazakhstan
Uzbekistan

L

L chondrites
Term introduced in 1981. Includes use on level 3 under meteorites(1).
UF L-group chondrites
BA chondrites
BT meteorites
NA Barwell Meteorite
Bjurbole Meteorite
Rio Negro Meteorite

L waves
use surface waves

L-group chondrites
use L chondrites

La
use lanthanum

La Caridad
Settlement in NW central part of country.
BT Asturias
Spain

La Coruna
As of 1981, refers to only the city. Before 1981, also included the province. In extreme NW Spain on the Atlantic Ocean.
BT La Coruna Province
Galicia
Spain

La Coruna Province
Term introduced in 1981. Before 1981, search La Coruna. In extreme NW Spain.
CO N423200N434800
W0073800W0092000
BT Galicia
Spain
NT Cabo Ortegal
La Coruna
Santiago de Compostela

La Grande Soufriere
Term introduced in 1981. Volcano. Before 1981, also search Soufriere AND Guadeloupe.
BT Guadeloupe
SA Soufriere

La Habana
Term introduced in 1981. Province in W Cuba. Before 1981, also search Havana.
BT Cuba
NT Havana

La Jolla
NW section of San Diego in San Diego County.
CO N324800N325200
W1171500W1171800
BT California
United States

La Ligua
Town and river in central part of country.
BT Aconcagua Province
Chile

La Pampa
Province in central.
BT Argentina
SA Salado Basin

La Paz
Department in W Bolivia. Also a city.
CO S180000S115000
W0670000W0693000
BT Bolivia

La Rioja
Province in NW.
CO S330000S270000
W0650000W0700000
BT Argentina
NT Paganzo
SA Chanares Formation
Salado Basin
Sierras Pampeanas

La Ronge
Village and lake in N central.
BT Saskatchewan
Canada

La Salle County
Term introduced in 1981. Index states as applicable.
BX Illinois
Texas
BT United States

La Salle Limestone
In McLeansboro Group. NE Illinois.
BT Pennsylvanian
SA Illinois
Middle Pennsylvanian
Upper Pennsylvanian

La Ventana Sandstone
Of Mesaverde Formation. NW New Mexico.
BT Upper Cretaceous
Cretaceous
SA Mesaverde Group
New Mexico

Laba Basin
River basin in Krasnodar Kray of the Northern Caucasus. Also search Laba.

BT Russian Republic
USSR

Labette County
SE Kansas.
BT Kansas
United States

Labette Shale
In Marmaton Group.
BT Desmoinesian
Middle Pennsylvanian
Pennsylvanian
SA Iowa
Kansas
Marmaton Group
Missouri
Oklahoma

laboratory studies
As of 1981, for engineering application only. Prior to 1981, applicable to a large number of topics.
UF studies, laboratory
SA experimental studies
theoretical studies

Labrador
Includes use on level 1 as an area term (list O). For term set options see list B.
CO N513000N610000
W0553000W0673000
BA Eastern Canada
Canada
NT Churchill Falls
Kiglapait
Mistastin Lake
Nain
Nain Massif
Wabush
SA Churchill Province
Grenville Province
Labrador Trough
Newfoundland
Sokoman Formation
Superior Province
Ungava

Labrador Basin
E of Labrador and S of Greenland.
BT Atlantic Ocean

Labrador Sea
Between Labrador and SW Greenland.
CO N480000N600000
W0440000W0640000
BA North American Atlantic
North Atlantic
BT Atlantic Ocean

Labrador Trough
Term introduced in 1981. In Quebec and Labrador.
BT Canada
SA Labrador
Quebec

labradorite
As of 1981, includes use on level 3 under minerals(1) framework silicates, plagioclase(2). See list L. Before 1981, included use on level 3 under minerals(1) framework silicates, feldspar group(2).
BA plagioclase
feldspar group
framework silicates
silicates
BT minerals
SA igneous rocks

labradoritite
As of 1981, includes use on level 3 under igneous rocks(1) gabbros(2). See list H. Before 1981, included use on level 3 under igneous rocks(1) gabbro family(2).
BA gabbros
BT igneous rocks

labuntsovite
BA orthosilicates
silicates
BT minerals

Labyrinthodontia
Includes use on level 2 under Amphibia(1). See list F.
BA Amphibia
BT Tetrapoda
Vertebrata

Lac La Ronge
Lake in N central.
BT Saskatchewan
Canada

Laccadive Islands
Term introduced in 1981. Off SW coast of India.
CO N100000N122000
E0740000E0720000
BT India
SA Indian Ocean Islands

laccoliths
Includes use on level 2 under intrusions(1).
UF cistern rock
BA intrusions
SA batholiths
dikes
domes
lopoliths

Lacertilia
As of 1981, includes use on level 3 under Reptilia(1) Squamata(2). See list F. Before 1981, included use on level 3 under Reptilia(1) Lepidosauria(2).
BA Squamata
Lepidosauria
Reptilia
BT Tetrapoda
Vertebrata

Lachlan fold belt
Term introduced in 1981. Trends NW-SE across New South Wales and extends into NE Victoria and NE Tasmania.
CO S420000S300000
E1501500E1430000
BT Australia
SA New South Wales
Tasmania
Victoria

LaCoste-Romberg gravimeters
Term introduced in 1981.
BT gravimeters

lacustrine
A valid term through 1977. After 1977, use lacustrine features or (as of 1981) lacustrine environment.

lacustrine environment
Term introduced in 1981.
SA ecology
environment
lacustrine features
lacustrine sedimentation
lakes
paleoecology
sedimentation

lacustrine features
Includes use on level 2 under geomorphology(1). Also search lacustrine.
UF features, lacustrine
SA beaches
extinct lakes
geomorphology
glacial lakes
glaciolacustrine environment
lacustrine environment
lakes
limnology
shoals
terraces

lacustrine sedimentation
Term introduced in 1981.
SA lacustrine environment
sedimentation

Ladak
use Ladakh

Ladakh
Region in Jammu and Kashmir. S Ladakh divided between India and Pakistan. Index countries as applicable.
UF Ladak
BT Asia
SA India
Pakistan

Ladinian
Europe: upper Middle Triassic. Above Anisian, below Carnian. Includes use on level 3 under age terms(1). See list E.
BA Middle Triassic
Triassic
BT Mesozoic
SA Hallstatt Limestone

Ladoga Series
Term introduced in 1981. Used for the Precambrian metamorphic unit in the Lake Ladoga region of the USSR and Finland. Not to be confused with Quaternary deposits or Cambrian-Silurian sedimentary rocks of the same name in the same region. (These units will be assigned unique names as needed according to frequency.) Before 1981, also search Ladoga Formation; Ladoga/IT. Age may be upper Archean or lower Proterozoic.
BT Precambrian
SA Baltic Shield
Finland
Karelia
Russian Republic
USSR

Lafayette County
BX Arkansas
Florida
Louisiana
Mississippi
Missouri
Wisconsin
BT United States

Lagenidae
Includes use on level 3 under foraminifera(1) Rotaliina(2).
BA foraminifera
BT Protista
Invertebrata

Laghouat
Town and oasis in N central.
BT Algeria

Lago Maggiore
Lake. Index countries as applicable.
BT Europe
SA Italy
Switzerland

Lagomorpha
Includes use on level 2 under Mammalia(1). See list F.
BA Mammalia
BT Tetrapoda
Vertebrata

lagoonal
A valid term through 1977. After 1977, use lagoons or (as of 1981) lagoonal environment.

lagoonal environment
Term introduced in 1981.
SA ecology
environment
lagoonal sedimentation
lagoons
paleoecology
sedimentation

lagoonal sedimentation
Term introduced in 1981.
SA lagoonal environment
sedimentation

lagoons
Includes use on level 3 under geomorphology(1). As of 1981, use lagoonal environment for type of environment. Before 1981, included use on level 3 under ecology(1), paleoecology(1), and sedimentation(1) for type of environment.
SA atolls
barrier reefs
beaches
coastal environment
ecology
geomorphology
lagoonal environment
lakes
paleoecology
reefs
sedimentation

Lagos Lagoon
Between Lagos Island just off the Gulf of Guinea and the mainland. City of Lagos at W end of lagoon. Also search Lagos.
BT Nigeria

lahars
SA lava flows
mass movements
mudflows
volcanism

Lahn River
Flows S and SW into the Rhine River SE of Koblenz. Index states as applicable. Also search Lahn.
BT West Germany
SA Hesse
Rhineland-Palatinate

Lahn River valley
Index states as applicable. Also search Lahn.
BT West Germany
SA Hesse
Rhineland-Palatinate

Laisvall
Village in N part of country.
BT Norrbotten
Sweden

Lake Agassiz
Glacial lake which existed in Pleistocene Epoch. Covered parts of N central U. S. and S central Canada. Index Manitoba and states as applicable.
BT North America
SA Manitoba
Minnesota
North Dakota

Lake Albert
Index Uganda and/or Zaire as applicable.
BA East African Lakes
BT Africa
SA East African Rift
Uganda
Zaire

Lake Baikal
Largest freshwater lake in Eurasia. Bounded on N, E, and S by the Buryat ASSR, and on the W by Irkutsk Oblast. Also search Baikal AND lake or lakes.
CO N520000N560000
E1100000E1040000
UF Baikal (Lake)
BT Russian Republic
USSR
SA Baikal region
Baikal rift zone

Lake Balaton
W Hungary. Also search Balaton.

UF Balaton (Lake)
BT Hungary
SA Balaton region

Lake Balkhash
E Kazakhstan. Also search Balkhash.
CO N450000N470000
E0790000E0730000
UF Balkhash (Lake)
BT Kazakhstan
USSR
SA Balkhash region

Lake Baringo
Term introduced in 1981. W central Kenya, in Great Rift Valley.
CO N003000N004500
E0360700E0360000
UF Baringo Lake
BZ East African Lakes
BX Kenya
SA Great Rift Valley
Kenya Rift valley

lake biscuits
use algal biscuits

Lake Biwa
W central Honshu. Also search Biwa; Biwa Lake.
UF Biwa Lake
Omi (Lake)
BT Honshu
Japan

Lake Bonneville
Large prehistoric body of water in present Great Salt Lake area. Index states as applicable.
BT United States
SA Idaho
Nevada
Utah

Lake Bonney
Southeast.
BT South Australia
Australia

Lake Chad
Index countries as applicable.
CO N123000N143000
E0160000E0123000
BT Africa
SA Cameroon
Chad
Chad Basin
Niger
Nigeria

Lake Champlain
Between New York and Vermont extending into Quebec. Index Quebec and states as applicable.
SA Champlain Valley
New York
Quebec
Vermont

Lake Chatuge
Index states as applicable.
BT United States
SA Georgia
North Carolina

Lake Constance
Index countries as applicable. Also search Bodensee; Constance Lake.
UF Bodensee
Constance Lake
BT Europe
SA Austria
Germany
Rhine Basin
Rhine River
Switzerland
West Germany

Lake County
BX California
Colorado
Florida
Illinois
Indiana
Michigan
Minnesota
Montana
Ohio
Oregon
South Dakota
Tennessee
BT United States

Lake District
Mountain and lake region in NW.
BT England
Great Britain
United Kingdom

Lake Edward
S of Lake Albert. Index countries as applicable.
BA East African Lakes
BT Africa
SA East African Rift
Uganda
Zaire

Lake Erie
Index Ontario and states as applicable.
CO N412230N425500
W0785100W0832800
BA Great Lakes
BT North America
SA Michigan
New York
Ohio
Ontario
Pennsylvania

Lake Eyre
Northeast.
BT South Australia
Australia

Lake Geneva
Index countries as applicable. Also search Lake Leman; Leman Lake.
UF Lake Leman
Leman Lake
BT Europe
SA France
Geneva
Rhone River
Rhone Valley
Switzerland

Lake George
NE New York.
BT New York
United States

Lake Gosiute
Term introduced in 1981. Eocene lake in SW Wyoming.
CO N403000N424500
W1080000W1104000
BT Sweetwater County
Wyoming
United States
SA Green River Formation

Lake Huron
Index Michigan and/or Ontario.
BA Great Lakes
BT North America
SA Michigan
Ontario

Lake Jocassee
Term introduced in 1981. Reservoir on Keowee River, NW South Carolina, near Seneca.
CO N345710N350230
W0825330W0825945
BT South Carolina
United States
SA Oconee County
Pickens County

Lake Kariba
Formed by dam in Kariba Gorge of Zambezi River. Index countries as applicable. Also search Kariba; Kariba Lake.
UF Kariba Lake
BA East African Lakes
BT Africa
SA Zambia
Zimbabwe

Lake Kinneret
use Sea of Galilee

Lake Kivu
Index countries as applicable.
BA East African Lakes
BT Africa
SA East African Rift
Rwanda
Zaire

Lake Ladoga
E and NE of Leningrad.
CO N595000N615000
E0330000E0295000
BT Russian Republic
USSR

Lake Ladoga region
Term introduced in 1981. Region surrounding Lake Ladoga, extending into Finland. Index countries as applicable.
CO N595000N620000
E0333000E0283000
BT Europe
SA Baltic Shield
Finland
Karelia
Russian Republic
USSR

Lake Lappajarvi
SW central. Vaasa.
BT Finland

Lake Leman
use Lake Geneva

Lake Louise
Village and small lake in Banff National Park in SW.
BT Alberta
Canada

Lake Magadi
Soda lake in S central.
BX Kenya
BZ East African Lakes
SA East African Rift

Lake Malawi
Index countries as applicable. Also search Lake Nyasa.
UF Lake Nyasa
BA East African Lakes
BT Africa
SA East African Rift
Malawi
Mozambique
Tanzania

Lake Maracaibo
NW part of country.
BT Zulia
Venezuela
SA Maracaibo Basin

Lake Mead
Within Lake Mead National Recreation Area. Index states as applicable.
BT United States
SA Arizona
Nevada

Lake Mendota
In Dane County in S central.
BT Wisconsin
United States

Lake Michigan
Index states as applicable.
CO N414500N460000
W0844500W0875000
BA Great Lakes
BT North America
SA Illinois
Indiana
Michigan
United States
Wedron Formation
Wisconsin

Lake Natron
Index countries as applicable.
BA East African Lakes
BT Africa
SA East African Rift
Kenya
Tanzania

Lake Nyasa
use Lake Malawi

Lake of Bienne
use Biel Lake

Lake of Zurich
N Switzerland.
BT Switzerland

Lake Okeechobee
S central.
BT Florida
United States

Lake Onega
In Karelian ASSR. Also search Onega.
BT Russian Republic
USSR
SA Onega

Lake Ontario
Index New York and/or Ontario as applicable.
BA Great Lakes
BT North America
SA New York
Ontario

Lake Pontchartrain
S Louisiana.
BT Louisiana
United States

Lake Powell
BT United States
SA Arizona
Utah

Lake Rudolf
use Lake Turkana

Lake Rudolph
use Lake Turkana

lake sediments
Term introduced in 1981. Before 1981, also search sediments AND lakes.
SA lakes
sediments

Lake Sevan
N Armenia.
BT Armenia
USSR
SA Sevan

Lake Superior
Index Ontario and states as applicable.
CO N462500N490000
W0842000W0921500
BA Great Lakes
BT North America
SA Michigan
Minnesota
Ontario
Wisconsin

Lake Suwa
W central.
BT Honshu
Japan

Lake Tahoe
Index states as applicable. Also search Lake Tahoe Basin.
UF Lake Tahoe Basin
BT United States
SA California
Nevada

Lake Tahoe Basin
use Lake Tahoe

Lake Tanganyika
Index countries as applicable.
BA East African Lakes
BT Africa
SA Burundi
East African Rift
Tanzania
Zaire
Zambia

Lake Tiberias
use Sea of Galilee

Lake Timiskaming
Index provinces as applicable.
BT Canada
SA Ontario
Quebec

Lake Titicaca
Index countries as applicable.
BT South America
SA Bolivia
Peru

Lake Torrens
E central.
BT South Australia
Australia

Lake Turkana
New name for Lake Rudolf. Primarily in Kenya. Index countries as applicable. Also search Lake Rudolf; Lake Rudolph; Rudolf.
UF Lake Rudolf
Lake Rudolph
Rudolf (Lake)
BA East African Lakes
BT Africa
SA East African Rift
Ethiopia
Kenya
Sudan

Lake Valley Formation
Includes Alamogordo, Arcente, Dona Ana, and Nunn members.
BT Lower Mississippian
Mississippian
SA New Mexico

Lake Vanda
Term introduced in 1981. Saline lake in dry-valleys region of S Victoria Land, near McMurdo Sound.
CO S773500S773500
E1613900E1613900
BT Victoria Land
Antarctica

Lake Victoria
Index countries as applicable.
BA East African Lakes
BT Africa
SA Kenya
Tanzania
Uganda

Lake Winnipeg
S central.
BT Manitoba
Canada

lakes
Includes use on level 3 under geomorphology(1) lacustrine features(2). Before 1981, also included use on level 3 under ecology(1), paleoecology(1), and sedimentation(1) for type of environment. As of 1981, use lacustrine environment for type of environment.
SA drainage
ecology
eutrophication
extinct lakes
glacial features
glacial lakes
glaciolacustrine environment
hydrogen sulfide
hydrology
hydrosphere
lacustrine environment
lacustrine features
lagoons
lake sediments
lava lakes
limnology
paleoecology
periglacial features
ponds
salt lakes
sedimentation
seiches
surface water
water balance
water supply
watersheds

Lambton County
Term introduced in 1981. S Ontario.
CO N423000N432000
W0814500W0824000
BT Ontario
Canada

lamellae
Restricted to use with minerals. Not used in biological sense.
SA deformation
minerals

laminations
Includes use on level 3 under sedimentary structures(1) planar bedding structures(2); on level 3 under foliation(1) style(2) and under structural anaylsis(1) theoretical studies(2). See list K. Before 1976, also search lamination.
BA planar bedding structures
sedimentary structures
SA bedding
cross-bedding
cross-laminations
foliation
ripple drift-cross laminations
stratification

Lampang
Term introduced in 1981. Town and province in NW.
BT Thailand

lamproite
As of 1981, includes use on level 3 under igneous rocks(1) lamprophyres(2). See list H. Before 1981, included use on level 3 under igneous rocks(1) trachyte-phonolite family(2).
BA lamprophyres
BT igneous rocks

lamprophyre
Includes use on level 3. As of 1981, not to be used in the igneous rocks set. See list H. Before 1981, included use on level 3 under igneous rocks(1) lamprophyre and carbonatite family(2).
BA lamprophyres
BT igneous rocks

lamprophyre and carbonatite family
As of 1981, no longer a valid term for GeoRef. Use lamprophyres or carbonatites as applicable. See list H. Before 1981, included use on level 2 under igneous rocks(1).

lamprophyres
Term introduced in 1981. Includes use on level 2 under igneous rocks(1). See list H. Before 1981, search lamprophyre and carbonatite family.
BT igneous rocks
NA camptonite
kersantite
lamproite
lamprophyre
minette
monchiquite
spessartite

Lancara Formation
Lower Acadian. Overlies The Herreira Sandstone. NW part of country.
BT Acadian
Middle Cambrian
Cambrian
SA Spain

Lancashire
County in NW.
CO N532000N543000
W0020000W0031500
BT England
Great Britain
United Kingdom

Lancaster County
Term introduced in 1981. Index states as applicable.
BX Nebraska
Pennsylvania
South Carolina
Virginia
BT United States

Lancaster Sound
Channel between Devon Island and N Baffin Island in E District of Franklin.
BT Northwest Territories
Canada

Lance Formation
Includes Colgate Sandstone Member, Cannonball Marine Member, Ludlow Lignitic Member, Tullock Member, Hell Creek Member, Torrington Member, and Ilo Ridge Member. Montana, W North Dakota, W South Dakota, and Wyoming.
BT Upper Cretaceous
Cretaceous
SA Colorado
Montana
North Dakota
South Dakota
Wyoming

land canyons
use canyons

land forms
use landforms

land leases
Term introduced in 1978.
SA conservation
reclamation

land subsidence
A level 1 term as of 1978. Includes use on level 2 under foundations(1) and geologic hazards(1). Used for geotechnical studies on land subsidence. If 1, term set options are:
causes
subtopic
controls
subtopic
foundations
subtopic
mines
subtopic
site exploration
subtopic
settlement
subtopic
stability
subtopic
solution features
subtopic
SA collapse structures
dams
engineering geology
explosions
foundations
geologic hazards
highways
mechanics
mines
oil and gas fields
rock mechanics
settlement
site exploration
slope stability
soil mechanics
solution features
stability
subsidence
tunnels
underground installations

land use
A level 1 term as of 1978. Includes use on level 2 under impact statements(1). If 1, term set options are:
agriculture
subtopic
arid environment
subtopic
changes
subtopic
classification
subtopic
conservation
subtopic
controls
subtopic
effects
subtopic
human ecology
subtopic
impact statements
subtopic
inventory
subtopic
legislation
subtopic
management
subtopic
maps
cartography
mines
subtopic
natural resources
subtopic
planning
subtopic
preservation
subtopic
reclamation
subtopic
recreation
subtopic
regional planning
subtopic
remote sensing
subtopic
soils
subtopic
underground space
subtopic
urban planning
subtopic
SA agriculture
airfields
arid environment
conservation
engineering geology
environmental geology
geologic hazards
human ecology
impact statements
legislation
management
maps
mine reactivation
mines
natural resources
planning
pollution
preservation
RARE II regions
reclamation
recreation
regional planning
remote sensing

soils
strip mining
underground space
urban planning
urbanization
waste disposal

Landenian
Belgium. Includes use on level 3 under age terms(1). See list E.
BA upper Paleocene
Paleocene
Paleogene
Tertiary
BT Cenozoic

Lander County
N central.
CO N391000N410000
W1163500W1174500
BT Nevada
United States

Landes
Department in SW.
CO N433000N443000
E0000800W0013000
BT France
NT Capbreton

landfills
Includes use on level 3 under environmental geology(1) waste disposal(2); on level 3 under engineering geology(1).
NT sanitary landfills
SA engineering geology
environmental geology
waste disposal

landform description
Includes use on level 2 under geomorphology(1). Before 1975, also search surficial geology.
SA aerial photography
description
geomorphology
landform evolution
landforms
landscapes
morphometry
mountains
relief

landform evolution
Includes use on level 2 under geomorphology(1).
SA beaches
erosion
etching
evolution
geomorphology
landform description
paleorelief
relief
relief inversion

landforms
Also search landforms; relief features.
UF land forms
relief features
SA geomorphology
hills
landform description
landscapes
mountains
plains
plateaus
relief exposure
valleys

landing sites
UF sites, landing
SA Apollo Program
Mars
Moon
Venus
Viking Program

Land's End
Cape. SW end of Cornwall. Use for Lands End.
BT England
Great Britain
United Kingdom

Landsat
Term introduced in 1978. Includes use on level 3 under remote sensing(1). Before 1978, also search ERTS; ERTS-1.
SA geophysical methods
geophysical surveys
imagery
remote sensing
satellite methods

landscapes
SA geomorphology
landform description
landforms

landslides
As of 1978, term is used on level 2 under geologic hazards(1), and slope stability(1).
SA avalanches
creep
engineering geology
erosion
geologic hazards
geomorphology
mass movements
mudflows
rockslides
slope stability
slopes
slumping
soils
talus slopes

Lane County
Index states as applicable.
BX Kansas
Oregon
BT United States

Lanersbach
Village SE of Innsbruck.
BT Tyrol
Austria

Laneuville-devant-Nancy Boring
Boring at or near outer SE suburb of Nancy.
BT Meurthe-et-Moselle
France

Laney Shale
use Laney Shale Member

Laney Shale Member
Of Green River Formation. SW Wyoming. Also search Laney Shale.
UF Laney Shale
BT middle Eocene
Eocene
SA Colorado
Green River Formation
Wyoming

langbeinite
BA sulfates
BT minerals

Langebaanweg
Village in SW.
BT Cape Province
South Africa

Langesund Fjord
Inlet of the Skaggerak.
BT Telemark
Norway

Langhe
Region NW of Genoa.
BT Italy
SA Piemonte

Langhian
Europe. See list E.
BA Miocene
Neogene
Tertiary
BT Cenozoic

Langkawi Islands
Group in Andaman Sea near entrance to Strait of Malacca. Administrated by Kedah state in West Malaysia. Also search Langkawi.
BT Malaysia

Languedoc
Historical region of S.
BT France

Lansing
City in Clinton, Eaton, and Ingham counties in S central Lower Peninsula.
BT Michigan
United States

Lansing Group
Includes Lane Shale, Plattsburg Limestone, Vilas Shale, Stanton Limestone.
BT Missourian
Upper Pennsylvanian
Pennsylvanian
SA Kansas
Plattsburg Limestone
Stanton Formation

lanthanide series
use rare earths

lanthanides
use rare earths

lanthanoans
use rare earths

lanthanum
Includes use on level 1 and 2 as a chemical element (list D).
UF La
SA cerium
elements
rare earths

Lanzarote
Easternmost of the Spanish controlled island group.
BT Canary Islands

Lanzo Massif
NW of Turin in Piemonte.
BT Italy
SA Piemonte

Laos
Includes use on level 1 as an area term (list O). For term set options see list B.
CO N140000N224000
E1074000E1000000
BA Far East
Asia
SA Indochina

lapilli
SA ejecta
fragments
pyroclastics

Laplace azimuths
use Laplace transformations

Laplace points
use Laplace transformations

Laplace tidal equation
use Laplace transformations

Laplace transformations
Term introduced in 1978. Search Laplace azimuths; Laplace points; Laplace tidal equation; Laplace transforms.
UF Laplace azimuths
Laplace points
Laplace tidal equation
Laplace transforms
transformations, Laplace
SA geodesy

Laplace transforms
use Laplace transformations

Lapland
Region N of the Arctic Circle including the Kola Peninsula of the Russian Republic. Index countries as applicable.
BT Europe
SA Finland
Norway
Sweden
USSR

Laptev Sea
Along N coast of Siberia between Taymyr Peninsula and New Siberian Islands.
CO N710000N813000
E1420000E0963000
BT Arctic Ocean

Lar
River valley in N, and city in Laristan region in S.
BT Iran

Lara
State in NW.
BT Venezuela

Laramian Orogenic Phase
use Laramide Orogeny

Laramian Orogeny
use Laramide Orogeny

Laramic Orogeny
use Laramide Orogeny

Laramide Orogeny
A time of deformation typically developed in the eastern Rocky Mountains of the United States, whose several phases extended from Upper Cretaceous until the end of the Paleocene. Term introduced in 1978. Before 1978, search orogeny AND Laramide; Laramian.
UF Laramian Orogenic Phase
Laramian Orogeny
Laramic Orogeny
Laramide Revolution
SA Cordilleran Orogeny
Cretaceous
orogeny
Tertiary

Laramide Revolution
use Laramide Orogeny

Laramie Basin
High plateau in SE between Laramie Mountains on E and Medicine Bow Mountains on the W.
BT Wyoming
United States

Laramie County
Extreme SE.
BT Wyoming
United States

Laramie Formation
Overlies Fox Hills Formation; unconformably underlies Denver Formation. Denver Basin region of E Colorado.
BT Upper Cretaceous
Cretaceous
SA Colorado

Laramie Mountains
A range of the Rocky Mountains. Index states as applicable. Also search Laramie Range.
UF Laramie Range
BT United States
SA Colorado
Rocky Mountains
Wyoming

Laramie Range
use Laramie Mountains

Lardeau Group
Includes Triune and Sharon Creek Formations.
UF Lardeau Series
BT Cambrian
SA British Columbia

Canada
Precambrian

Lardeau Series
use Lardeau Group

Larderello
Village in central part of country.
BT Tuscany
Italy

Largentiere
Village in S part of country.
BT Ardeche
France

Larimer County
On the Wyoming border in N.
CO N401500N410000
W1045000W1062000
BT Colorado
United States

larnite
BA orthosilicates
silicates
BT minerals

Larvik
Town at head of Larvik Fjord near entrance to Oslofjord.
BT Vestfold
Norway

Las Animas County
SE central.
CO N370000N375000
W1030500W1051500
BT Colorado
United States

Las Vegas
City in Clark County in extreme S.
BT Nevada
United States

Las Vegas Valley
Term introduced in 1981. S Nevada.
CO N360000N363000
W1150000W1153000
BT Clarke County
Nevada
United States

Las Villas
Province in W central Cuba.
BT Cuba
West Indies

LASA
Array of seismographs. Includes use on level 3 under seismology(1).
SA arrays
Montana

Lassen Peak
Active volcano in Lassen Volcanic National Park in NE. Also search Mount Lassen.
UF Mount Lassen
BT California
United States

Lassen Volcanic National Park
NE California.
BT California
United States

Latah County
On the Washington border in northern panhandle.
BT Idaho
United States

Latdorfian
use Lattorfian

late diagenesis
Term introduced in 1978.
BA diagenesis
SA early diagenesis

Latemar Massif
In Dolomites.
BT Trentino-Alto Adige
Italy
SA Dolomites

Laterite soils
use laterites

laterites
Includes use on level 3 under soils(1) soil group(2). See list M. Also search laterite; laterite soils; lateritic soils.
UF Laterite soils
BA soils
SA bauxite
bauxitization
biorhexistasy
laterization
Latosols

laterization
SA bauxitization
laterites
soils

Lathi Beds
use Lathi Formation

Lathi Formation
UF Lathi Beds
BT Jurassic
SA India
Rajasthan

Latin America
Term introduced in 1981.
SA America

latite
As of 1981, includes use on level 3 under igneous rocks(1) trachyandesites(2). See list H. Before 1981, included use on level 3 under igneous rocks(1) andesite-rhyolite family(2).
BA trachyandesites
volcanic rocks
BT igneous rocks
SA quartz latite
trachybasalt

latitude, paleo-
use paleolatitude

Latium
Autonomous region in W central.
CO N411400N425000
E0140000E0113000
BT Italy
NT Alban Hills
Frosinone
Rome
Sabatini Mountains
Tolfa Hills
University of Rome
Viterbo
SA Tiber Valley

Latosols
See list M.
BA soils
SA laterites

lattice
Includes use on level 3 under crystal structure(1). Also search Bravais lattice; crystal lattice; space lattice.
UF Bravais lattice
crystal lattice
space lattice
translation lattice
SA cell dimensions
crystal structure
habit
lattice
lattice parameters
substitution
twinning
unit cell

lattice constants
use lattice parameters

lattice parameters
Also search lattice constants.
UF lattice constants
SA crystal chemistry
crystal growth
crystal structure
crystallography
lattice
unit cell

Lattorfian
Europe. See list E.
UF Latdorfian
BA upper Eocene
Eocene
Paleogene
Tertiary
BT Cenozoic
SA Tongrian

Latvia
Latvian Soviet Socialist Republic.
CO N560000N580000
E0280000E0210000
BA Baltic Plain
Russian Platform
BT USSR
NT Riga
SA Baltic region
Baltic Shield
Dvina River
European USSR
Russian Plain

Lau Basin
West of the Tonga Islands and S of the Lau group of the Fiji Islands
BT Pacific Ocean

laumonite
use laumontite

laumontite
UF laumonite
lomonite
lomontite
BA zeolite group
framework silicates
silicates
BT minerals
SA leonhardite

Lauraceae
Family.
BA Dicotyledoneae
angiosperms
BT Spermatophyta
Plantae

Laurasia
The protocontinent of the Northern Hemisphere which is a combination of Laurentia and Eurasia. It included most of North America, Greenland, and much of Eurasia excluding India. Includes use on level 2 under continental drift(1).
SA continental drift
Eurasia
Gondwana
Northern Hemisphere
Pangaea

Laurentian Highlands
use Canadian Shield

Laurentian Plateau
use Canadian Shield

Laurentide ice sheet
SA Saint Lawrence Lowlands
Saint Lawrence Valley

Laurion
City SE of Athens. Central Greece and Euboea. Also search Laurium.
UF Laurium
Lavrion
BT Greece

Laurium
use Laurion

Lausitz
use Lusatia

lava
This set is used for all molten extrusives and the rocks that solidify from them. Includes use on level 1 (list A); on level 2 under isotopes(1). If 1, term set options are:
topic [age, alteration, analysis, classification, composition, distribution, flow mechanism, genesis, geochemistry, nomenclature, observations, occurrence, petrology, properties, temperature, viscosity]
subtopic [e.g. electrical properties, magnetic properties, pillow lava, pillow structure]
NT aa lava
pahoehoe
SA basalt
eruptions
extrusive rocks
flow mechanism
igneous rocks
inclusions
lava channels
lava fields
lava flows
lava lakes
lava tubes
magmas
mud volcanoes
paleomagnetism
perlite
petrology
pillow lava
pillow structure
plugs
viscosity
volcanic features
volcanic rocks
volcanism
volcanoes
volcanology

lava channels
BT volcanic features
SA channels
lava

lava domes
use shield volcanoes

lava fields
UF fields, lava
BT volcanic features
SA lava

lava flows
Includes use on level 3 under lava(1) and under volcanology(1).
UF flow (volcanic)
nappe (volcanic)
SA flow mechanism
flows
lahars
lava
lava tubes
volcanic features
volcanology

lava lakes
BT volcanic features
SA lakes
lava

lava tubes
Includes use on level 3 under lava(1). Before 1981, also search lava tunnels.
UF tubes, lava
SA lava
lava flows
volcanic features

lava tunnels
As of 1981, no longer a valid term for GeoRef. Use lava tubes.

Laval Basin
BT France

Lavrion
use Laurion

Lawrence County
BX Alabama
Arkansas
Illinois
Indiana
Kentucky
Mississippi

Missouri
Ohio
Pennsylvania
South Dakota
Tennessee
BT United States

Lawrence Formation
Includes two unnamed shale members separated by Amazonia Limestone Member, Ireland Limestone Member. SW Iowa, E Kansas, NW Missouri, and SE Nebraska.
UF Lawrence Shale
BT Virgilian
Upper Pennsylvanian
Pennsylvanian
SA Douglas Group
Iowa
Kansas
Missouri
Nebraska

Lawrence Shale
use Lawrence Formation

lawsonite
BA orthosilicates
silicates
BT minerals
SA aluminosilicates

Laxfordian
Europe. See list E.
BA upper Precambrian
Precambrian

layer, active
use active layer

layer, basaltic
use basaltic layer

layer, boundary
use boundary layer

layer, granitic
use granitic layer

layer, nepheloid
use nepheloid layer

layered
A valid term on level 2 through 1977. After 1977, use layered intrusions on level 2. For engineering geology, use layered materials.

layered intrusions
Not a valid term from 1975 through 1977. After 1977, includes use on level 2. Before 1978, also search intrusions AND layered.
BA intrusions
SA igneous rocks

layered materials
Term introduced in 1981. Includes use on level 3 under engineering geology(1). Before 1981, also search layered media; layered; layers.
SA engineering geology
materials

layered media
As of 1981, no longer a valid term for GeoRef. Use layered materials.

layers
A valid term through 1978. See layered materials.

lazurite
BZ sodalite group
sulfates
sulfides
BT minerals

Le Havre
City on the English Channel.
BT Seine-Maritime
France

Le Mans
City in NE central part of country.
BT Sarthe
France

Le Mont-Dore
use Mont-Dore

Le Trou Sans Fond
use Trou Sans Fond

leaching
Includes use on level 3 under soils(1) genesis(2); on level 3 under geochemistry(1) and under weathering(1).
UF lixiviation
SA alteration
degradation
desalinization
geochemistry
mining geology
percolation
soils
solution
solution mining
weathering

lead
Includes use on level 1 and 2 as a chemical element (list D). As of 1981, use lead ores for lead as a commodity. Before 1981, included use on level 1 as a commodity term.
UF Pb
SA elements
heavy metals
isotopes
lead ores
lead-zinc deposits
Pb-204
Pb-206
Pb-206/Pb-204
Pb-207
Pb-207/Pb-204
Pb-207/Pb-206
Pb-208
Pb-208/Pb-204
Pb-210
Pb/Pb
Pb/Th
U-238/Pb-206
U/Pb
U/Th/Pb

lead glance
use galena

lead ores
Term introduced in 1981. Includes use on level 1 and 2 as a commodity term (list C). Before 1981, search lead AND ore deposits.
SA lead
lead-zinc deposits
mississippi valley-type

lead-zinc deposits
As of 1978, term includes use on level 1 and 2 as a commodity (list C). Before 1978, also search lead-zinc.
SA lead
lead ores
mississippi valley-type
zinc
zinc ores

Leadville
City in Lake County in central.
BT Colorado
United States

Leadville Formation
Includes Gilman Sandstone Member.
UF Leadville Limestone
BT Mississippian
SA Colorado
Lower Mississippian
Upper Mississippian
Yule Marble

Leadville Limestone
use Leadville Formation

leaf
use leaves

leakage anomalies
Term introduced in 1981.
SA anomalies
dispersion patterns
geochemistry
hydrothermal alteration
mineral deposits

least-squares analysis
Also search least-squares method.
UF least-squares method
BA statistical analysis
SA analysis
mathematical methods
regression analysis

least-squares method
use least-squares analysis

Leavenworth County
NE Kansas.
BT Kansas
United States

leaves
Also search leaf.
UF leaf
SA humus
paleobotany
Plantae
vegetation

Leba
Town and river in N part of country.
BT Gdansk
Poland

Lebanon
Includes use on level 1 as an area term (list O). For term set options see list B.
CO N330000N344500
E0364500E0350000
BA Middle East
Asia
NT Beirut
SA Mediterranean region

Lebanon County
SE central.
BT Pennsylvania
United States

Lebedin
City in Sumy Oblast in NE central.
BT Ukraine
USSR

lebensspuren
Includes use on level 3 under sedimentary structures(1) biogenic structures(2). See list K. Before 1981, also search ichnofossils.
BA biogenic structures
sedimentary structures
SA borings
burrows
ichnofossils
tracks
trails

Lebombo Mountains
Index Mozambique and/or Transvaal as applicable. Also search Lebombo.
BT Africa
SA Mozambique
Transvaal

Lecce
City and province in heel of Italian boot.
BT Apulia
Italy

lechatelierite
BA silica minerals
framework silicates
silicates
BT minerals

Lechtal Alps
W Tyrol.
BT Tyrol
Austria

Lecompton Formation
use Lecompton Limestone

Lecompton Limestone
In Shawnee Group. Comprises Spring Branch Limestone, Doniphan Shale, Big Springs Limestone, Queen Hill Shale, Beil Limestone, King Hill Shale, Avoca Limestone members. E Kansas, NW Missouri, SE Nebraska, SW Iowa, central N Oklahoma. Also search Lecompton Formation.
UF Lecompton Formation
BT Virgilian
Upper Pennsylvanian
Pennsylvanian
SA Beil Limestone Member
Iowa
Kansas
Missouri
Nebraska
Oklahoma
Shawnee Group

Leczyca
Town in central.
BT Lodz
Poland

Leda Clay
BT Cenozoic
SA Canada
Ontario
Ottawa

Leduc
Town S of Edmonton in central.
BT Alberta
Canada

Lee County
BX Alabama
Arkansas
Florida
Georgia
Illinois
Iowa
Kentucky
Mississippi
North Carolina
South Carolina
Texas
Virginia
BT United States

Lee Formation
Includes Rockcastle Sandstone. Defined as all of lower Pennsylvanian rocks from Mississippian-Pennsylvanian contact to top of Zachariah Coal Bed. E Kentucky, E Tennessee, and SW Virginia.
BT Lower Pennsylvanian
Pennsylvanian
SA Kentucky
Tennessee
Virginia

Leeward Islands
The northern chain of islands in the Lesser Antilles extending from the Virgin Islands in N to Dominica in S.
BT Lesser Antilles
NA Guadeloupe
NT Montserrat
NA Virgin Islands
SA Netherlands Antilles

Leg 1
Term introduced in 1981. Gulf of Mexico and NW Atlantic Ocean, off Florida and SW of Bermuda. Sites 1-7.
CO N230000N310000
W0670000W0930000
SA Atlantic Ocean
Deep Sea Drilling Project
Gulf of Mexico

Leg 10
Term introduced in 1981. Gulf of Mexico. Sites 85-97.

CO N200000N260000
W0840000W0960000
SA Deep Sea Drilling Project
Gulf of Mexico

Leg 11
Term introduced in 1981. NW Atlantic Ocean. Sites 98-108, from N of the Bahamas to NW of Bermuda near the continental margin of the United States.
CO N230000N404500
W0690000W0801000
SA Atlantic Ocean
Deep Sea Drilling Project
DSDP Site 105

Leg 12
Term introduced in 1981. N Atlantic Ocean. Sites 111-119, from N of the Grand Banks to the Labrador Sea, to SW of Iceland, to W of Scotland, to N of Spain.
CO N450000N600000
W0070000W0490000
SA Atlantic Ocean
Deep Sea Drilling Project

Leg 13
Term introduced in 1981. Mediterranean Sea. Sites 120-134. (Site 120 is nearby in the Atlantic Ocean.)
CO N320000N410000
E0290000W0120000
SA Deep Sea Drilling Project
DSDP Site 125
DSDP Site 132
Mediterranean Sea

Leg 14
Term introduced in 1981. N Atlantic Ocean off the coasts of Africa and South America. Sites 135-144.
CO N030000N360000
W0100000W0550000
SA Atlantic Ocean
Deep Sea Drilling Project

Leg 15
Term introduced in 1981. Caribbean Sea. Sites 145-154.
CO N100000N170000
W0630000W0810000
SA Caribbean Sea
Deep Sea Drilling Project

Leg 16
Term introduced in 1981. E equatorial Pacific Ocean. Sites 155-163.
CO S020000N150000
W0800000W1510000
SA Deep Sea Drilling Project
DSDP Site 157
DSDP Site 158
DSDP Site 163
Pacific Ocean

Leg 17
Term introduced in 1981. N central and equatorial Pacific Ocean. Sites 164-171.
CO N020000N200000
W1610000E1730000
SA Deep Sea Drilling Project
DSDP Site 167
Pacific Ocean

Leg 18
NE Pacific Ocean. Sites 172-182, from off California and Oregon to the Gulf of Alaska.
CO N290000N590000
W1220000W1350000
SA Deep Sea Drilling Project
Pacific Ocean

Leg 19
Term introduced in 1981. Bering Sea and N Pacific Ocean. Sites 183-193, roughly following Aleutian Trench to E of the Kuril Trench.
CO N450000N580000
W1610000E1550000
SA Deep Sea Drilling Project
Pacific Ocean

Leg 2
Term introduced in 1981. N Atlantic Ocean. Sites 8-12, from NW of Bermuda to NW of Cape Verde Islands.
CO N190000N360000
W0253000W0680000
SA Atlantic Ocean
Deep Sea Drilling Project

Leg 20
Term introduced in 1981. NW Pacific Ocean. Sites 194-202, from E of the Japan and Bonin trenches, to E of the Mariana Trench, and W of the Marshall Islands.
CO N120000N340000
E1570000E1460000
SA Deep Sea Drilling Project
Pacific Ocean

Leg 21
Term introduced in 1981. SW Pacific Ocean. Sites 203-210, primarily W of the Tonga and Kermadec trenches and New Zealand, and in the Coral Sea.
CO S370000S130000
W1740000E1520000
SA Coral Sea
Deep Sea Drilling Project
Pacific Ocean

Leg 22
Term introduced in 1981. E Indian Ocean. Sites 211-218.
CO S200000N090000
E1030000E0860000
SA Deep Sea Drilling Project
DSDP Site 214
Indian Ocean

Leg 23
Term introduced in 1981. Arabian Sea and Red Sea. Sites 219-230.
CO N060000N220000
E0730000E0380000
SA Arabian Sea
Deep Sea Drilling Project
DSDP Site 225
DSDP Site 227
DSDP Site 228
Indian Ocean
Red Sea

Leg 24
Term introduced in 1981. W Indian Ocean and Gulf of Aden. Sites 231-238.
CO S120000N150000
E0710000E0480000
SA Arabian Sea
Deep Sea Drilling Project
Gulf of Aden
Indian Ocean

Leg 25
W Indian Ocean. Sites 239-249.
CO S340000S020000
E0530000E0360000
SA Deep Sea Drilling Project
Indian Ocean

Leg 26
Term introduced in 1981. S Indian Ocean. Sites 250-258.
CO S380000S230000
E1130000E0390000
SA Deep Sea Drilling Project
DSDP Site 253
DSDP Site 254
Indian Ocean

Leg 27
E Indian Ocean. Sites 259-263.
CO S300000S100000
E1240000E1100000
SA Deep Sea Drilling Project
DSDP Site 259
DSDP Site 260
DSDP Site 261
DSDP Site 262
DSDP Site 263
Indian Ocean

Leg 28
Term introduced in 1981. E Indian Ocean, and Antarctic Ocean S of Australia and New Zealand. Sites 264-274.
CO S780000S340000
E1790000E1040000
SA Antarctic Ocean
Deep Sea Drilling Project
DSDP Site 264
DSDP Site 270
DSDP Site 274
Indian Ocean

Leg 29
Term introduced in 1981. Pacific and Antarctic oceans, S of Australia and New Zealand. Sites 275-284.
CO S570000S400000
E1770000E1430000
SA Antarctic Ocean
Deep Sea Drilling Project
DSDP Site 284

Leg 3
Term introduced in 1981. S Atlantic Ocean. Sites 13-22, primarily spanning the Mid-Atlantic Ridge at approximately S30°, W to the Rio Grande Rise.
CO S310000N100000
W0050000W0360000
SA Atlantic Ocean
Deep Sea Drilling Project

Leg 30
Term introduced in 1981. SW Pacific Ocean. Sites 285-289 in the South Fiji Basin, New Hebrides, Coral Sea, and Ontong Java Plateau.
CO S270000S000000
E1760000E1530000
SA Deep Sea Drilling Project
Pacific Ocean

Leg 31
Term introduced in 1981. Japan Sea and W Philippine Sea. Sites 290-302.
CO N120000N420000
E1380000E1240000
SA Deep Sea Drilling Project
DSDP Site 292
DSDP Site 296
DSDP Site 297
Japan Sea
Pacific Ocean
Philippine Sea

Leg 32
Term introduced in 1981. NW Pacific Ocean. Sites 303-313, from E of the Kuril and Bonin trenches to the Emperor Seamounts and the Necker Ridge.
CO N200000N410000
E1800000E1540000
SA Deep Sea Drilling Project
Pacific Ocean

Leg 33
Term introduced in 1981. Central equatorial Pacific Ocean. Sites 314-316.
CO S150000N160000
W1460000W1690000
SA Deep Sea Drilling Project
DSDP Site 315
Pacific Ocean

Leg 34
Term introduced in 1981. SE Pacific Ocean. Sites 319-321.
CO S140000S090000
W0810000W1020000
SA Deep Sea Drilling Project
DSDP Site 319
DSDP Site 320
DSDP Site 321
Pacific Ocean

Leg 35
Term introduced in 1981. Antarctic Ocean, S of the SE Pacific Ocean. Sites 322-325.
CO S700000S600000
W0730000W0990000
SA Antarctic Ocean
Deep Sea Drilling Project
DSDP Site 322
DSDP Site 323
DSDP Site 325

Leg 36
Term introduced in 1981. Drake Passage, Scotia Sea, and SW Atlantic Ocean. Sites 326-331.
CO S570000S370000
W0360000W0660000
SA Antarctic Ocean
Atlantic Ocean
Deep Sea Drilling Project
Drake Passage
DSDP Site 327
DSDP Site 329
DSDP Site 330
Scotia Sea

Leg 37
Term introduced in 1981. In vicinity of the Mid-Atlantic Ridge, W of the Azores, N Atlantic Ocean. Sites 332-335.
CO N360000N380000
W0330000W0360000
SA Atlantic Ocean
Deep Sea Drilling Project
DSDP Site 332
DSDP Site 333
DSDP Site 334
DSDP Site 335

Leg 38
Term introduced in 1981. Norwegian and Greenland seas. Sites 336-352.
CO N630000N770000
E0080000W0130000
SA Arctic Ocean
Atlantic Ocean
Deep Sea Drilling Project

Leg 39
Term introduced in 1981. W and SE Atlantic Ocean. Sites 353-359.
CO S380000N110000
W0040000W0450000
SA Atlantic Ocean
Deep Sea Drilling Project
DSDP Site 356
DSDP Site 357

Leg 4
Term introduced in 1981. W Atlantic Ocean, NE of Brazil to the Puerto Rico Trench area and the Caribbean Sea. Sites 23-31.
CO S070000N210000
W0300000W0730000
SA Atlantic Ocean
Caribbean Sea
Deep Sea Drilling Project

Leg 40
Term introduced in 1981. SE Atlantic Ocean. Sites 360-365.
CO S360000S110000
E0190000E0090000
SA Atlantic Ocean
Deep Sea Drilling Project
DSDP Site 361
DSDP Site 362
DSDP Site 363
DSDP Site 364

Leg 41
Term introduced in 1981. NE Atlantic Ocean, off West Africa. Sites

366-370.
CO N050000N330000
W0100000W0220000
SA Atlantic Ocean
Deep Sea Drilling Project
DSDP Site 367
DSDP Site 368
DSDP Site 369
DSDP Site 370

Leg 42A
Term introduced in 1981. Mediterranean Sea. Sites 371-378.
CO N340000N410000
E0320000E0040000
SA Deep Sea Drilling Project
DSDP Site 372
DSDP Site 374
Mediterranean Sea

Leg 42B
Term introduced in 1981. Black Sea. Sites 379-381.
CO N230000N440000
E0370000E0290000
SA Black Sea
Deep Sea Drilling Project
DSDP Site 379
DSDP Site 380
DSDP Site 381

Leg 43
Term introduced in 1981. Central NW Atlantic Ocean. Sites 382-387.
CO N310000N410000
W0510000W0680000
SA Atlantic Ocean
Deep Sea Drilling Project

Leg 44
Term introduced in 1981. Central NW Atlantic Ocean. Sites 388-392.
CO N280000N360000
W0690000W0770000
SA Atlantic Ocean
Deep Sea Drilling Project
DSDP Site 391

Leg 45
First cruise of the International Phase of Ocean Drilling (IPOD) of the Deep Sea Drilling Project. N central Atlantic Ocean. Sites 395-396.
CO N220000N230000
W0430000W0470000
SA Atlantic Ocean
Deep Sea Drilling Project
DSDP Site 395
DSDP Site 396
IPOD

Leg 46
N central Atlantic Ocean. Site 396B, near the Mid-Atlantic Ridge axis.
CO N225200N230300
W0442700W0443700
SA Atlantic Ocean
Deep Sea Drilling Project
IPOD

Leg 47
Term introduced in 1981. NE Atlantic Ocean. Sites 397-398, off NW Africa near the Canary Islands and off SW Europe near Portugal.
CO N260000N410000
W0100000W0160000
SA Atlantic Ocean
Deep Sea Drilling Project
IPOD

Leg 48
Bay of Biscay and Rockall Plateau, NE Atlantic Ocean. Sites 399-406.
CO N470000N570000
W0080000W0240000
SA Atlantic Ocean
Deep Sea Drilling Project
IPOD

Leg 49
Term introduced in 1981. N Atlantic Ocean near Iceland and along the Mid-Atlantic Ridge. Sites 407-414.
CO N320000N640000
W0250000W0340000
SA Atlantic Ocean
Deep Sea Drilling Project
IPOD

Leg 5
Term introduced in 1981. NE Pacific Ocean. Sites 32-43, off California and across the Mendocino, Murray, and Clarion fracture zones at approximately W140°.
CO N130000N410000
W1260000W1520000
SA Deep Sea Drilling Project
Pacific Ocean

Leg 50
Term introduced in 1981. NE Atlantic Ocean, off Morocco. Sites 415-416.
CO N310000N330000
W0100000W0120000
SA Atlantic Ocean
Deep Sea Drilling Project
IPOD

Leg 51
Term introduced in 1981. NW Atlantic Ocean, E of Florida, N of Puerto Rico. Site 417.
CO N240000N260000
W0670000W0690000
SA Atlantic Ocean
Deep Sea Drilling Project
DSDP Site 417
IPOD

Leg 52
Term introduced in 1981. NW Atlantic Ocean, E of Florida, N of Puerto Rico. Sites 417-418.
CO N240000N260000
W0670000W0690000
SA Atlantic Ocean
Deep Sea Drilling Project
DSDP Site 417
DSDP Site 418
IPOD

Leg 53
Term introduced in 1981. NW Atlantic Ocean. Site 418.
CO N240000N260000
W0670000W0690000
SA Atlantic Ocean
Deep Sea Drilling Project
DSDP Site 418
IPOD

Leg 54
Term introduced in 1981. E equatorial Pacific. Sites 419-429, on the East Pacific Rise and Galapagos Rift.
CO N000000N100000
W0860000W1070000
SA Deep Sea Drilling Project
IPOD
Pacific Ocean

Leg 55
Term introduced in 1981. NW Pacific Ocean. Sites 430-433, in the Emperor Seamounts.
CO N370000N450000
E1710000E1700000
SA Deep Sea Drilling Project
IPOD
Pacific Ocean

Leg 56
Term introduced in 1981. NW Pacific Ocean. Sites 434-437, near the Japan Trench.
CO N390000N400000
E1460000E1430000
SA Deep Sea Drilling Project
IPOD
Pacific Ocean

Leg 57
Term introduced in 1981. NW Pacific Ocean. Sites 438-441, near the Japan Trench.
CO N390000N410000
E1450000E1430000
SA Deep Sea Drilling Project
IPOD
Pacific Ocean

Leg 58
Term introduced in 1981. N Philippine Sea. Sites 442-446.
CO N240000N300000
E1380000E1320000
SA Deep Sea Drilling Project
IPOD
Pacific Ocean
Philippine Sea

Leg 59
Term introduced in 1981. Philippine Sea. Sites 447-451.
CO N150000N190000
E1450000E1320000
SA Deep Sea Drilling Project
IPOD
Pacific Ocean
Philippine Sea

Leg 6
Term introduced in 1981. NW Pacific Ocean and Philippine Sea. Sites 44-60 from N of the Mid-Pacific Mountains to the Shatsky Rise, the E Philippine Sea, and the Mariana Trench.
CO N080000N340000
W1680000E1400000
SA Deep Sea Drilling Project
Pacific Ocean
Philippine Sea

Leg 60
Term introduced in 1981. SE Philippine Sea. Sites 452-461 from the Mariana Trough to E of the Mariana Trench.
CO N170000N190000
E1490000E1430000
SA Deep Sea Drilling Project
IPOD
Pacific Ocean
Philippine Sea

Leg 61
Term introduced in 1981. W central Pacific Ocean. Site 462 in the Nauru Basin.
CO N070000N080000
E1660000E1650000
SA Deep Sea Drilling Project
IPOD
Pacific Ocean

Leg 62
Term introduced in 1981. N central Pacific Ocean. Sites 463-466 in the Mid-Pacific Mountains and on Hess Rise.
CO N200000N400000
W1780000W1730000
SA Deep Sea Drilling Project
IPOD
Pacific Ocean

Leg 63
Term introduced in 1981. NE Pacific Ocean. Sites 467-473 off S California and Baja California.
SA Deep Sea Drilling Project
IPOD
Pacific Ocean

Leg 64
Term introduced in 1981. Gulf of California. Sites 474-481.
SA Deep Sea Drilling Project
Gulf of California
IPOD
Pacific Ocean

Leg 65
Term introduced in 1981. At mouth of the Gulf of California. Sites 482-485.
SA Deep Sea Drilling Project
Gulf of California
IPOD
Pacific Ocean

Leg 66
Term introduced in 1981. NE Pacific Ocean off S Mexico. Sites 486-493.
SA Deep Sea Drilling Project
IPOD
Pacific Ocean

Leg 7
Term introduced in 1981. W equatorial and N central Pacific Ocean. Sites 61-67.
CO S020000N250000
W1570000E1410000
SA Deep Sea Drilling Project
Pacific Ocean

Leg 8
Term introduced in 1981. E equatorial Pacific Ocean. Sites 68-75.
CO S130000N170000
W1340000W1650000
SA Deep Sea Drilling Project
Pacific Ocean

Leg 9
Term introduced in 1981. E equatorial Pacific Ocean. Sites 76-84.
CO S150000N080000
W0820000W1460000
SA Deep Sea Drilling Project
Pacific Ocean

legends
Term introduced in 1981. Before 1981, search legend.
SA cartography
maps

legislation
As of 1978, term is used on level 2 under land use(1).
SA antitrust legislation
decision-making
economic geology
engineering geology
environmental geology
investment legislation
land use
mining legislation
policy

legrandite
BA arsenates
BT minerals

Leguminosae
Family.
BA Dicotyledoneae
angiosperms
BT Spermatophyta
Plantae

lehm
use loess

Leicestershire
County in central.
CO N523000N530000
W0004000W0014000
BT England
Great Britain
United Kingdom
SA Barwell Meteorite

leifite
BZ fluorides
framework silicates
BT minerals
SA aluminosilicates

Leine Valley
River valley in SE and central Lower Saxony. Also search Leine.

BT Lower Saxony
Germany

Leinster
Province in E comprising 12 counties.
BT Ireland

Leipzig
As of 1981, refers to only the city. Before 1981, also included the district.
BT Leipzig Bezirk
East Germany

Leipzig Bezirk
Term introduced in 1981. District in S central. Before 1981, search Leipzig.
CO N505000N514000
E0132000E0120800
BT East Germany
NT Leipzig
SA Saxony

Leiria
District in W. Also a city.
BT Portugal

Leman Lake
use Lake Geneva

Lemhi County
E central.
BT Idaho
United States

Lemhi Range
E central Idaho. Primarily in Lemhi and Butte counties.
BT Idaho
United States

Lena Basin
River basin. E central and N Siberia including areas in Irkutsk Oblast, and Yakutia. Also search Lena; Lena region; Lena River basin.
UF Lena region
Lena River basin
BT Russian Republic
USSR
SA Angara-Lena Basin

Lena region
use Lena Basin

Lena River
Rises in the Baikal Mountains W of Lake Baikal in Irkutsk Oblast and flows NE into a great bend in Yakutia, and then NW through a large delta into the Laptev Sea. Also search Lena.
BT Russian Republic
USSR

Lena River basin
use Lena Basin

length of day
Term introduced in 1981.
SA climate

Leninabad
City in NW.
BT Tadzhikistan
USSR

Leningrad
City on the Gulf of Finland. Also an oblast.
BT Russian Republic
USSR

Leningrad Mining Institute
In Leningrad.
BT Russian Republic
USSR

Leninogorsk
City in Vostochno Kazakhstan Oblast in extreme E Kazakhstan.
UF Ridder
BT Kazakhstan
USSR

Lenkoran
City on Caspian Sea near Iranian border.
BT Azerbaidzhan
USSR

lenses
Any lense-shaped deposit.
SA mineral deposits
sedimentary rocks

Lenticulina
Genus. Includes use on level 3 under foraminifera(1) Nodosariacea(2).
BA Nodosariidae
Nodosariacea
Rotaliina
foraminifera
BT Protista
Invertebrata

Leon
As of 1981, refers to only the city in Spain. Before 1981, referred to the city, the province, and the region in Spain; the city and department in Nicaragua; and the city in Mexico. Before 1981, search Leon AND Spain.
BT Leon Province
Leon region
Spain
SA Mexico
Nicaragua

Leon City
Term introduced in 1981 for the city in central Mexico. Before 1981, search Leon AND Mexico.
BT Guanajuato
Mexico

Leon Department
Term introduced in 1981 for the department in Nicaragua. Before 1981, search Leon AND Nicaragua. In W part of country.
BT Nicaragua

Leon Province
Term introduced in 1981. Before 1981, search Leon AND Spain. In NW part of country.
CO N420000N431500
W0044500W0070500
BT Leon region
Spain
NT Leon

Leon region
Term introduced in 1981. Before 1981, search Leon AND Spain. In NW part of country.
CO N401500N431500
W0044500W0070600
BT Spain
NT Leon Province
Salamanca Province
Zamora Province

Leonardian
Provincial series, North America. Above Wolfcampian, below Guadalupian. Includes use on level 3 under age terms(1). See list E.
BA Lower Permian
Permian
BT Paleozoic
SA San Andres Formation

leonardite
use leonhardite

leonhardite
Also search leonardite.
UF leonardite
BA zeolite group
framework silicates
silicates
BT minerals
SA laumontite

leopoldite
use sylvite

Leopoldville
use Kinshasa

Leoville
Village in W central.
BT Saskatchewan
Canada

Leperditicopida
Includes use on level 2 under Ostracoda(1). See list F.
BA Ostracoda
BT Crustacea
Mandibulata
Arthropoda
Invertebrata

lepidocrocite
BA oxides
BT minerals

Lepidocyclina
Includes use on level 3 under foraminifera(1) Orbitoididae(2).
BA Orbitoididae
Orbitoidacea
Rotaliina
foraminifera
BT Protista
Invertebrata

Lepidocystoidea
Includes use on level 2 under Echinodermata(1). See list F.
BA Crinozoa
Echinodermata
BT Invertebrata

lepidolite
BZ fluorides
mica group
BT minerals
SA cookeite

lepidomelane
BA mica group
sheet silicates
silicates
BT minerals

Lepidoptera
As of 1981, includes use on level 3 under Insecta(1) Lepidopteroida(2). See list F. Before 1981, included use on level 2 under Insecta(1).
BA Lepidopteroida
Insecta
BT Mandibulata
Arthropoda
Invertebrata
SA Raphidiodea

Lepidopteroida
Term introduced in 1981. Includes use on level 2 under Insecta(1). See list F.
BA Insecta
BT Mandibulata
Arthropoda
Invertebrata
NA Lepidoptera

Lepidosauria
Includes use on level 2 under Reptilia(1). See list F.
BA Reptilia
BT Tetrapoda
Vertebrata
NA Eosuchia
Rhynchocephalia
Squamata

Lepontine Alps
Division of Central Alps. Index countries as applicable.
CO N460000N464500
E0091500E0081500
BT Alps
Europe
SA Central Alps
Gotthard Massif
Italy
Piedmont Alps
Switzerland

Lepospondyli
Includes use on level 2 under Amphibia(1). See list F.
BA Amphibia
BT Tetrapoda
Vertebrata

leptite
Includes use on level 3 under metamorphic rocks(1) granulites(2). See list J. Before 1978, also search leptynite.
UF leptynite
BA granulites
BT metamorphic rocks

Leptocythere
Genus. Includes use on level 3 under Ostracoda(1). See list F.
BA Cytheracea
Podocopida
Ostracoda
BT Crustacea
Mandibulata
Arthropoda
Invertebrata

leptynite
use leptite

Lerida
Province in Catalonia in NE.
CO N425200N411500
E0015100E0002000
BT Catalonia
Spain
NT Montsech
Pobla de Segur
Segre Valley
Tremp

Lesbos
Island in E Aegean Sea off NW Anatolia.
BT Greek Aegean Islands
Greece

Leslie County
Term introduced in 1981. SE Kentucky.
CO N365200N372000
W0831000W0833500
BT Kentucky
United States

Lesna
Town in Lower Silesia in SW part of the country.
BT Wroclaw
Poland

Lesotho
Former British colony of Basutoland. Includes use on level 1 as an area term (list O). For term set options see list B. Also search Basutoland.
CO S304000S283500
E0293000E0270000
UF Basutoland
BA Southern Africa
Africa
SA Orange River

Lesser Antilles
The smaller islands of the Antilles extending in an arc from Puerto Rico to the islands N of Venezuela. Index Barbados, Trinidad and island groups as applicable. As of 1977, includes use as a level 1 area term (list O). See list B for term set options.
CO N100000N190000
W0590000W0700200
BA West Indies
NA Barbados
NT Dominica
Leeward Islands
NA Netherlands Antilles
Trinidad and Tobago
NT Windward Islands
SA Antilles
Caribbean region

Lesser Caucasus
Mountain system formed by the N frontal ranges of the Armenian Highland and separated from the Greater Caucasus by the Colchis and Kura lowlands. Index Soviet republics as applicable.
CO N380000N420000
E0490000E0413000
BA Caucasus
BT USSR
SA Armenia
Azerbaidzhan
Georgian Republic
Shakhdag Range
Trialet Range

Lesser Himalayas
The central range of the Himalayas running parallel to the Greater Himalayas to the N. Index countries as applicable.
BT Asia
SA Bhutan
Himalayas
India
Nepal
Sikkim

Lesser Khingan Mountains
E Heilungkiang. Separates Amur River from Sungari Valley. Also search Lesser Khingan Range.
UF Lesser Khingan Range
Little Khingan Mountains
BT Heilungkiang
China

Lesser Khingan Range
use Lesser Khingan Mountains

Lesser Sunda Islands
Chain of islands E from Bali to and including Alor and Timor, but not Wetar.
CO S110000S080000
E1273000E1154000
BT Indonesia
NT Bali
Mount Agung
Pantar
NX Timor

Lesser Walachia
use Oltenia

Lessini Mountains
In SW Dolomites in NE.
BT Italy
SA Dolomites

Letcher County
Term introduced in 1981. SE Kentucky.
CO N365700N372100
W0823400W0831000
BT Kentucky
United States

letter
A valid term through mid-1978; used in title annotation when document is correspondence. No longer a valid index term.

leucite
UF Vesuvian garnet
BA framework silicates
silicates
BT minerals

leucitite
As of 1981, includes use on level 3 under igneous rocks(1) alkali gabbros(2). See list H. Before 1981, included use on level 3 under igneous rocks(1) alkali basalt family(2) and ultramafic family(2).
BA alkali gabbros
gabbros
BT igneous rocks
SA ultramafics

leucogranite
As of 1981, includes use on level 3 under igneous rocks(1) granites(2). See list H. Before 1981, included use on level 3 under igneous rocks(1) granite-granodiorite family(2).
BA granites
BT igneous rocks
SA granite

leucosphenite
BA sheet silicates
silicates
BT minerals

Leuven
use Louvain

Levant
As of 1981, no longer a valid term for GeoRef. Before 1981, included use on level 3, referring to countries along the E shore of the Mediterranean. Use Middle East.

Levantinian
Austria. Includes use on level 3 under age terms(1). See list E.
BA upper Pliocene
Pliocene
Neogene
Tertiary
BT Cenozoic

levees
SA channels
embankments
marine geology
streams

level, changes of
use changes of level

leveling
UF levelling
SA cartography
geodesy

levelling
use leveling

levels
Includes use on level 2 under ground water(1).
SA drawdown
ground water
potentiometric surface
reservoirs
water table

Levis Shale
BT Devonian
SA Quebec

Lewis and Clark County
W central.
BT Montana
United States

Lewis County
Term introduced in 1981. Index states as applicable.
BX Idaho
Kentucky
Missouri
New York
Tennessee
Washington
West Virginia
BT United States

Lewis Shale
Lewis and Meeteetse formations throughout N central Wyoming considered stratigraphic equivalents with Lewis being applied generally in areas where marine strata are present and Meeteetse in areas where rocks are nonmarine in character. W Colorado, NW New Mexico, and S and central Wyoming.
BT Upper Cretaceous
Cretaceous
SA Colorado
New Mexico
Wyoming

Lewisian
Europe. See list E.
BA Precambrian

lexicons
Includes use on level 1 (list A) for systematic arrangement of words in a particular language, or of a considerable number of them, and their definition. Term set options are: topic (List B or extraterrestrial geology or geophysics)
language (if not English)
SA bibliography
catalogs
dictionaries
glossaries
mineralogy
paleontology
petrology
sedimentary petrology
stratigraphy
structural geology

Lexington Limestone
Comprises Curdsville Limestone, Logana Formation, Jessamine Limestone, Benson Limestone, Brannon Limestone with Woodburn Phosphatic Member. Central Kentucky.
BT Middle Ordovician
Ordovician
SA Kentucky

Lherz
Lake in the Pyrenees in S.
BT France

lherzolite
As of 1981, includes use on level 3 under igneous rocks(1) peridotites(2). See list H. Before 1981, included use on level 3 under igneous rocks(1) ultramafic family(2).
BA peridotites
ultramafics
BT igneous rocks
SA garnet lherzolite
peridotite
spinel lherzolite

Li
use lithium

Li-6
Includes use on level 3 under isotopes(1).
SA isotopes
lithium

Li-7
Includes use on level 3 under isotopes(1).
SA isotopes
lithium

Liaoning
Province in S.
BT Manchuria
China

Liard River
Index British Columbia and territories as applicable.
BT Canada
SA British Columbia
Northwest Territories
Yukon Territory

Lias
use Lower Jurassic

Liassic
As of 1981, no longer a valid term for GeoRef. Use Lower Jurassic. Before 1981, included use on level 3 under age terms(1).

Liberia
Includes use on level 1 as an area term (list O). For term set options see list B.
CO N040000N080000
W0070000W0120000
BA West Africa
Africa
NT Bong Range
Liberian Shield
SA Nimba Mountains

Liberian Shield
Inland plateau.
BT Liberia

Libya
Includes use on level 1 as an area term (list O). For term set options see list B.
CO N193000N330000
E0250000E0093000
BA North Africa
Africa
NT Cyrenaica
Fezzan
Kufra Basin
Murzuk Basin
Sirte Basin
Tibesti
Tripoli
Tripolitania
Zelten
SA Libyan Desert
Mediterranean region
Near East
Sahara

Libyan Desert
Desert area of the E Sahara, W of the Nile. Index countries as applicable.
BT Africa
SA Egypt
Libya
Sudan
Western Desert

lichenometry
Includes use on level 2 under geochronology(1).
SA geochronology
lichens

lichens
Includes use on level 1 and 2 as a fossil term (list F).
BT thallophytes
Plantae
SA algae
fungi
lichenometry

Lichida
Includes use on level 2 under Trilobita(1). See list F.
BA Trilobita
BT Trilobitomorpha
Arthropoda
Invertebrata

Liechtenstein
Independent principality between NE Switzerland and W Austria.
BA Central Europe
BT Europe
SA Rhine Basin
Rhine River

Liege
Province in E Belgium. Also a city.
CO N501400N504800
E0063000E0045800
BT Belgium
NT Hohe Venn
Stavelot
Vesdre Valley

life
A valid level 2 term through 1977 used in combination with origin. After 1977, use life origin on level 2 under paleontology(1).

life assemblage
use biocenoses

life origin
Term introduced in 1978. Includes use on level 2 under paleontology(1). Before 1978, also search origin AND life.

SA exobiology
extraterrestrial geology
paleontology

light minerals
BT minerals
SA calcite
dolomite
heavy minerals
muscovite
opaque minerals
quartz

lightning
Includes use on level 3 under meteorology(1) electrical phenomena(2).
SA currents
electrical phenomena
meteorology
thunderstorms
whistlers

lightweight aggregate
use aggregate

lignite
Includes use on level 1 and 2 as a commodity term (list C); on level 3 under sedimentary rocks(1) organic residues(2). See list I. Before 1981, also search brown coal.
BA coal
organic residues
BT sedimentary rocks
SA anthracite
coal
organic materials
peat

Liguria
Autonomous region in NW.
CO N434500N444000
E0100500E0074500
BT Italy
NT Genoa
Ligurian Alps
Ligurian Apennines
Monte Antola
Savona

Ligurian Alps
E extention of Maritime Alps along coast of Ligurian Sea.
BT Liguria
Italy
SA Alps
Maritime Alps

Ligurian Apennines
Extend from near Savona SE to just N of La Spezia along coast of Ligurian Sea.
BT Liguria
Italy
SA Apennines
Monte Antola

Ligurian Sea
Enclosed by Italian autonomous regions of Liguria and Tuscany on the N and E, and by Corsica on the S.
BT Mediterranean Sea

Likhvin
use Chekalin

Liliidae
Term introduced in 1981. Includes use on level 2 under angiosperms(1). See list F.
BA Monocotyledoneae
angiosperms
BT Spermatophyta
Plantae

Lille
City near Belgian frontier in N part of country.
BT Nord Department
France

lillianite
BA sulfobismuthites
sulfosalts

BT minerals

Lilydale Limestone
Upper Yarra District in S central Victoria.
BT Devonian
SA Australia
Victoria

Lima
Department in W. Also a city.
BT Peru
NT Casapalca
Santa Eulalia
SA Pucara Group

Limagne
Fertile lowland in central part of country. Index departments as applicable.
BT France
SA Allier
Puy-de-Dome

Limburg
Name of Town and province in Belgium. Also province in SE Netherlands, and a region in W Europe. Index countries as applicable.
BT Europe
SA Belgium
Maastricht
Netherlands

limburgite
As of 1981, includes use on level 3 under igneous rocks(1) ultramafics(2). See list H. Before 1981, included use on level 3 under igneous rocks(1) ultramafic family(2).
BA ultramafics
BT igneous rocks

lime
Used as a level 3 commodity term (list C).
SA calcium carbonate
limestone deposits

Lime Creek Formation
Includes Juniper Hill, Cerro Gordo, and Owen members. Central N Iowa.
BT Upper Devonian
Devonian
SA Iowa

Limerick
County in W central.
BT Ireland

limestone
Includes use on level 3 under sedimentary rocks(1) carbonate rocks(2). See list I. As of 1981, use limestone deposits for limestone as a commodity. Before 1981, included use on level 1 as a commodity term.
BA carbonate rocks
BT sedimentary rocks
NA algal limestone
dolomitic limestone
oolitic limestone
SA algal mounds
biosparite
calcarenite
calcilutite
calcite
carbonate sediments
chalk
dolomitization
jasperoid
limestone deposits
metalimestone
micrite
oolite
oolitic texture
ophicalcite
pisolites
travertine

Limestone Alps
BT Austria
SA Alps

Limestone County
Index states as applicable.
BX Alabama
Texas
BT United States

limestone deposits
Term introduced in 1981. Includes use on level 1 and 2 as a commodity term (list C). Before 1981, search limestone AND deposits.
SA building stone
cement materials
chalk deposits
construction materials
dimension stone
dolostone deposits
lime
limestone
marble deposits

limit, elastic
use elastic limit

limits, Atterberg
use Atterberg limits

Limnocardium
Genus. Includes use on level 3 under Mollusca(1). See list F.
BA Bivalvia
Mollusca
BT Invertebrata

limnology
Includes use on level 2 under hydrology(1); on level 3 under geomorphology(1) lacustrine features(2).
SA geomorphology
glacial lakes
hydrology
lacustrine features
lakes
paleolimnology
ponds
seiches

limonite
As of 1981, includes use on level 3 as a commodity term (list C) under iron ores(1). Before 1981, included use as a commodity term under iron(1).
BA oxides
BT minerals
SA hydrogoethite
iron ores

Limousin
Historical region in S central.
CO N450000N461000
E0030000E0003000
BT France

Limpopo Basin
River basin. Also search Limpopo. Index Transvaal and countries as applicable.
UF Crocodile River basin
BT Africa
SA Botswana
Mozambique
Transvaal
Zimbabwe

Limpopo Belt
Term introduced in 1981. Orogenic belt at the N margin of the Kaapvaal Craton in S Africa. Consists of Precambrian metamorphic rocks (the Messina Formation in South Africa). Index countries as applicable.
BT Africa
SA Botswana
Mozambique
Precambrian
South Africa
Zimbabwe

Limski Channel
On Istrian Peninsula SW of Rijeka.
BT Croatia
Yugoslavia

Lincoln County
BX Arkansas
Colorado
Georgia
Idaho
Kansas
Kentucky
Louisiana
Maine
Minnesota
Mississippi
Missouri
Montana
Nebraska
Nevada
New Mexico
North Carolina
Oklahoma
Oregon
Oklahoma
Oregon
South Dakota
Tennessee
Washington
West Virginia
Wisconsin
Wyoming
BT United States

Lincoln Creek Formation
Vaguely defined. Also search Lincoln Formation. SW Washington.
BT Oligocene
SA Washington

Lincolnshire
County on the North Sea in E.
CO N524000N534500
E0002000W0010000
BT England
Great Britain
United Kingdom

Lincolnshire Limestone
Includes Edison Cherty Limestone Member and Hogskin Member. SW Virginia and NE Tennessee.
BT Middle Ordovician
Ordovician
SA Tennessee
Virginia

Lindlar
Village in S.
BT North Rhine-Westphalia
West Germany

line defects
use crystal dislocations

Line Islands
Group S of Hawaiian Islands, N and S of Equator. Some belong to U.S. and some to U.K.
BT Polynesia
NT Fanning Island
Vostok

line of strike
use strike

lineaments
Includes use on level 3 under tectonics(1).
SA fault zones
faults
remote sensing
tectonics

linear distortion
Term introduced in 1981.
SA distortion
elastic waves
seismology

linear materials
Term introduced in 1981.
SA engineering geology
materials

linear orientation
Term introduced in 1978. Before 1978, search linear.
SA orientation

lineation
Small-scale features. Includes use on level 1 (list A); on level 2 and 3 under structural analysis(1). Also search lineations. If level 1, term set options are:
experimental studies
subtopic
genesis
subtopic
interpretation
subtopic
style
subtopic [e.g. boudinage, elongate minerals, flow lines, mullions, slickensides]
SA bedding
boudinage
cleavage
deformation
elongate minerals
fabric
flow lines
folds
foliation
fractures
igneous rocks
intrusions
metamorphic rocks
metamorphism
mullions
orientation
parting lineation
preferred orientation
slickensides
structural analysis
structural geology
style
tectonics

lines, flow
use flow lines

Lingula
Genus. Includes use on level 3 under Brachiopoda(1) Inarticulata(2).
BA Inarticulata
Brachiopoda
BT Invertebrata

linnaeite
Also search linneite.
UF cobalt pyrites
linneite
BA sulfides
BT minerals
SA carrollite
polydymite

linneite
use linnaeite

Linz
City in NW part of country.
BT Upper Austria
Austria

Lipari Island
BT Sicily
Italy
SA Lipari Islands

Lipari Islands
Group of small volcanic islands in SE Tyrrhenian Sea off N coast of Sicily. Also search Lipari.
CO N382000N385000
E0152000E0142000
UF Aeolian Islands
BT Sicily
Italy
SA Lipari Island
Vulcano

liparite
As of 1981, includes use on level 3 under igneous rocks(1) rhyolites(2). See list H. Before 1981, included use on level 3 under igneous rocks(1) andesite-rhyolite family(2).
BA rhyolites
volcanic rocks
BT igneous rocks

liparite porphyry
As of 1981, includes use on level 3 under igneous rocks(1) rhyolites(2). See list H. Before 1981, included use on level 3 under igneous rocks(1) andesite-rhyolite family(2).
BA rhyolites
volcanic rocks
BT igneous rocks
SA porphyry

lipids
BA organic materials

Lippe
Former German state in NW part of country. Also a river.
BT North Rhine-Westphalia
West Germany

liquefaction
Not to be used under phase equilibria. As of 1981, includes use on level 2 under slope stability(1).
SA geologic hazards
liquefaction potential
mass movements
sediments
slope stability
slumping
soil mechanics

liquefaction potential
Term introduced in 1981.
SA engineering geology
geologic hazards
liquefaction
slope stability

liquid
use liquid phase

liquid phase
Term introduced in 1978. Before 1978, also search liquid.
UF liquid
SA fluid phase
gaseous phase
melts
solid phase

liquid waste
As of 1978, term is used on level 2 under waste disposal(1).
UF waste, liquid
SA detergents
effluents
engineering geology
environmental geology
ground water
industrial waste
radioactive waste
solid waste
storage
waste disposal
waste water

Lisburne Group
Includes Wachsmuth Limestone, Alapah Limestone, Kayak Shale, Kanayut Conglomerate. N Alaska.
BT Paleozoic
SA Alaska
Lower Mississippian
Mississippian
Pennsylvanian
Permian
Upper Mississippian

Lissamphibia
Includes use on level 2 under Amphibia(1). See list F.
BA Amphibia
BT Tetrapoda
Vertebrata
NA Anura

list, faunal
use faunal list

listvenite
BA mica group
sheet silicates
silicates
BT minerals

listwanite
Compositional term. Includes use on level 3 under metamorphic rocks(1). See list J.
BA schists
BT metamorphic rocks

Litchfield County
NW Connecticut.
BT Connecticut
United States

lithic texture
As of 1981, no longer a valid term for GeoRef. See textures. Before 1981, included use on level 3 under sediments(1) and sedimentary rocks(1).

lithification
Includes use on level 3 under sedimentation(1) or diagenesis(1).
SA cementation
coal
compaction
compression
consolidation
crystallization
diagenesis
recrystallization
sedimentation
sediments

lithiophilite
BA phosphates
BT minerals

lithiophorite
BA oxides
BT minerals

lithium
Includes use on level 1 and 2 as a chemical element (list D). As of 1981, use lithium ores for lithium as a commodity. Before 1981, included use on level 1 as a commodity term.
UF Li
SA alkali metals
amblygonite
elements
Li-6
Li-7
lithium ores

lithium ores
Term introduced in 1981. Includes use on level 1 and 2 as a commodity term (list C). Before 1981, search lithium AND ore deposits.
SA lithium

lithofacies
Includes use on level 2 under sedimentary rocks(1), sediments(1), and reefs(1).
SA facies
reefs
sedimentary rocks
sediments
stratigraphic wedges

lithologic controls
Includes use on level 3 under mineral deposits, genesis(1) controls(2). Also search lithologic.
SA controls
mineral deposits
stratigraphic controls

lithologic maps
Term introduced in 1978. Before 1978, search maps AND lithologic.
BA maps

lithologic traps
use stratigraphic traps

lithology
As of 1981, no longer a valid term for GeoRef. Before 1981, included use on level 3.

lithophile elements
Term introduced in 1978. Before 1978, also search lithophile.
SA elements

Lithophyllum
As of 1981, includes use on level 3 under algae(1) Corallinaceae(2). Before 1981, included use on level 3 under algae(1) Rhodophyta(2). See list F.
BA Corallinaceae
Rhodophyta
algae
BT thallophytes
Plantae

lithosiderite
use stony irons

lithosphere
Includes use on level 3 under tectonophysics(1) and under plate tectonics(1).
SA asthenosphere
Benioff zone
biosphere
core
crust
Earth
lower crust
mantle
Mohorovicic discontinuity
plate tectonics
tectonophysics
tectonosphere

lithostratigraphy
Includes use on level 2 under sedimentary rocks(1), metamorphic rocks(1) and sediments(1).
UF petrostratigraphy
rock-stratigraphy
SA metamorphic rocks
sedimentary rocks
sediments
stratigraphic wedges
stratigraphy

Lithothamnion
use Lithothamnium

Lithothamnium
Genus. As of 1981, includes use on level 3 under algae(1) Corallinaceae(2). Before 1981, included use on level 3 under algae(1) Rhodophyta(2). Also search Lithothamnion. See list F.
UF Lithothamnion
BA Corallinaceae
Rhodophyta
algae
BT thallophytes
Plantae

Lithuania
Lithuanian Soviet Socialist Republic.
CO N530000N560000
E0270000E0210000
BA Russian Platform
BT USSR
NT Sventoji River
Vilna
SA Baltic region
Courland Spit
European USSR
Neman River basin
Russian Plain

Litopterna
Includes use on level 2 under Mammalia(1). See list F.
BA Mammalia
BT Tetrapoda

L

Vertebrata

Little Belt Mountains
Range of the Rocky Mountains in central.
BT Montana
United States
SA Rocky Mountains

Little Falls Dolomite
use Little Falls Formation

Little Falls Formation
Represents an offshore, more carbonate phase of Potsdam Sandstone, Galway Formation, and perhaps even part of Hoyt Limestone. E central and E New York.
UF Little Falls Dolomite
BT Upper Cambrian
Cambrian
SA New York

Little Khingan Mountains
use Lesser Khingan Mountains

Little Missouri River basin
Index states as applicable. Also search Little Missouri River.
BT United States
SA Montana
North Dakota
South Dakota
Wyoming

Little Walachia
use Oltenia

Littleton Formation
Includes Gove Member, Hubbard Hill Member, May Pond Member, Dakin Hill Member, Pittsfield Member, Jenness Pond Member, Durgin Brook Member. Central and S New Hampshire, N central Massachusetts, and SE Vermont.
BT Lower Devonian
Devonian
SA Massachusetts
New Hampshire
Vermont

littoral
A valid term through 1977. After 1977, use littoral environment; as of 1981, use intertidal environment.

littoral drift
Also search longshore drift.
UF longshore drift
shore drift
SA longshore currents
progradation
sediments

littoral environment
As of 1981, no longer a valid term for GeoRef. Use intertidal environment. Term was introduced in 1978.

Lituolacea
Includes use on level 2 under foraminifera(1). See list F.
BA Textulariina
foraminifera
BT Protista
Invertebrata
NA Ammobaculites
Ataxophragmiidae
Lituolidae
Orbitolinidae

Lituolidae
Family. Includes use on level 3 under foraminifera(1) Lituolacea(2).
BA Lituolacea
Textulariina
foraminifera
BT Protista
Invertebrata
NA Haplophragmoides

Liverpool Bay
Inlet of Irish Sea off Liverpool. Also search Liverpool.
BT England
Great Britain
United Kingdom

living culture
As of 1981, no longer a valid term for GeoRef. Before 1981, included use on level 3. See living taxa.

living fossils
Term introduced in 1981.
SA fossils
living taxa

living materials
As of 1981, no longer a valid term for GeoRef. Use living taxa. Term was introduced in 1978 to distinguish from fossiliferous materials. Before 1978, search living.

living taxa
Term introduced in 1981. Before 1981, also search living materials; living culture. Used when organisms are studied while they are still living.
SA biota
ecology
faunal studies
floral studies
fossils
living fossils

Livingston County
BX Illinois
Kentucky
Louisiana
Michigan
Missouri
New York
BT United States

Livingston Island
One of South Shetland Islands in British Antarctic Territory off the Antarctic Peninsula.
BA South Shetland Islands
Scotia Sea Islands
BT Antarctica

Livorno
Province in NW Tuscany. Also a city on the Ligurian Sea now known as Leghorn.
BT Tuscany
Italy

lixiviation
use leaching

lizardite
BA serpentine group
sheet silicates
silicates
BT minerals

Ljubljana
City in NW part of country.
UF Lyublyana
BT Slovenia
Yugoslavia

LL chondrites
Term introduced in 1981. Includes use on level 3 under meteorites(1).
UF LL-group chondrites
BA chondrites
BT meteorites
NA Chainpur Meteorite

LL-group chondrites
use LL chondrites

Llandeilian
Europe. Above Llanvirnian, below lower Caradocian. Includes use on level 3 under age terms(1). See list E.
BA Middle Ordovician
Ordovician
BT Paleozoic
SA Borrowdale Volcanic Series

Llandoverian
Valentian. Europe. Above Ashgillian (Ordovician), below Tarannon. Includes use on level 3 under age terms(1). See list E.
BA Lower Silurian
Silurian
BT Paleozoic

Llandovery
Town in S.
BT Wales
Great Britain
United Kingdom

Llano
Town in Llano County in central.
BT Texas
United States

Llano County
Central.
BT Texas
United States

Llano Estacado
Vast plateau. Index states as applicable.
UF Staked Plain
BT United States
SA New Mexico
Oklahoma
Texas

Llano Uplift
BT Texas

Llanos
Vast plains drained by the Orinoco River. Index countries as applicable.
BT South America
SA Colombia
Venezuela

Llanvirnian
Europe. Above Arenigian, below Llandeilian. Includes use on level 3 under age terms(1). See list E.
BA Middle Ordovician
Ordovician
BT Paleozoic
SA Borrowdale Volcanic Series

Llewn Promontory
use Lleyn Peninsula

Lleyn Peninsula
Headland extending SW into Saint Georges channel from NW Wales.
UF Llewn Promontory
BT Wales
Great Britain
United Kingdom

Llogrebat River basin
In Catalonia in NE part of country. Also search Llogrebat River.
BT Barcelona Province
Catalonia
Spain

load casts
Includes use on level 3 under sedimentary structures(1) soft sediment deformation(2) , turbidity current structures(2), and bedding plane irregularities(2). See list K.
UF casts, load
BA turbidity current structures
sedimentary structures
SA bedding plane irregularities
flame structures
flute casts
soft sediment deformation
sole marks

load pressure
Term introduced in 1978.
SA engineering geology
pressure

loading
Includes use on level 3 under soil mechanics(1).
SA cyclic loading
soil mechanics

loam
Includes use on level 3 under soils(1). See list M.
BA soils
SA clay
loess
sand
silt

Loanda
use Luanda

loans
Term introduced in 1981.
SA economics

localization
Used for concentration of minerals, e.g. under sedimentation(1), mineral deposits, genesis(1), or metasomatism(1).
SA concentration
mineral deposits
mineralization
minerals
sedimentation

Loch Lomond
S central.
BT Scotland
Great Britain
United Kingdom

Lochkovian
Europe. See list E.
BA Lower Devonian
Devonian
BT Paleozoic

Lockport Formation
Includes Gasport Dolomite, Suspension Bridge Dolomite, Eramosa Dolomite members, and DeCew Waterlime, Gasport Limestone, Goat Island, Oak Orchard members, Devils Hole Dolomite, and Oakfield Limestone.
BT Niagaran
Middle Silurian
Silurian
SA Michigan
New York
Ontario

locomotion
Includes use on level 3 under appropriate fossil group (list F).
SA fossils

Lodeve
Town in S part of country.
BT Herault
France

Lodeve Basin
S part of country. Also search Lodeve.
BT Herault
France

Lodgepole Formation
In Madison Group. Includes Paine member, Woodhurst Member, Little Chief Canyon Member. SW Montana, NE Utah, and E Wyoming. Also search Lodgepole Limestone.
UF Lodgepole Limestone
BT Lower Mississippian
Mississippian
SA Madison Group
Montana
Utah
Wyoming

Lodgepole Limestone
use Lodgepole Formation

Lodz
Province in central part of country. Also a city.
BT Poland
NT Belchatow
Leczyca
Lowicz
Tomaszow Mazowiecki

Widawka Basin
Wielun

loellingite
use lollingite

loess
Includes use on level 3 under sediments(1) clastic sediments(2); on level 3 under geomorphology(1) eolian features(2). See list N.
UF bluff formation
lehm
BA clastic sediments
BT sediments
SA drift
eolian features
geomorphology
loam
marl
silt
soil mechanics
terrigenous materials

Lofoten Islands
Island group in Norwegian Sea off NW mainland. Also search Lofoten.
BT Nordland
Norway

log, road
use road log

Logan County
Term introduced in 1981. Index states as applicable.
BX Arkansas
Colorado
Illinois
Kansas
Kentucky
Nebraska
North Dakota
Ohio
Oklahoma
West Virginia
BT United States

Logan Formation
Comprises Beyer, Allensville, and Vinton members. NE Kentucky, and Ohio.
BT Mississippian
SA Kentucky
Ohio

Logar
Province in E.
BT Afghanistan

logging, acoustical
use acoustical logging

logging, caliper
use caliper logging

logging, dipmeter
use dipmeter logging

logging, electrical
use electrical logging

logging, electromagnetic
use electromagnetic logging

Logone River
Flows into the Shari River at Fort Lamy in Chad S of Lake Chad. Index countries as applicable. Also search Logone.
BT Africa
SA Cameroon
Chad

Logrono
As of 1981, refers to only the city. Before 1981, also included the province. In N central Spain.
BT Logrono Province
Old Castile
Spain

Logrono Province
Term introduced in 1981. Before 1981, search Logrono. In N central Spain.
CO N415600N424000
W0014100W0030800
BT Old Castile
Spain
NT Logrono

Logudoro
Region in NW.
BT Sardinia
Italy

Loir-et-Cher
Department in W central.
CO N471200N481000
E0021500E0003200
BT France
SA Beauce

Loire
Department in S central.
CO N452000N462000
E0050000E0034000
BT France
NT Saint-Etienne coal basin
SA Monts du Lyonnais

Loire River
Longest river in France. Rises in Ardeche Department in SE and flows N and NE into Loiret Department where it turns E emptying into the Bay of Biscay at Saint-Nazaire.
BT France

Loire Valley
River valley in S central, central, and E central part of country. Also search Loire.
BT France

Loire-Atlantique
Department in W France. Also search Loire-Inferieure.
CO N470000N475000
W0005000W0023000
BT France
NT Ancenis
Nantes

Loire-Inferieure
A valid term through 1976. After 1976, use Loire-Atlantique.

Loiret
Term introduced in 1981. Department in N central.
CO N473000N481500
E0031000E0013000
BT France

lollingite
As of 1981, includes use on level 3 under minerals(1) arsenides(2). See list L. Before 1981, included use on level 3 under minerals(1) sulfides(2). Also search loellingite.
UF loellingite
BA arsenides
sulfides
BT minerals
SA arsenopyrite

Lom Depression
In vicinity of Lom River and city of Lom in NW part of country.
BT Vidin
Bulgaria

Lombardy
Autonomous region in N.
CO N444000N464000
E0112500E0083000
BT Italy
NT Bergamo
Brescia
Como
Milan
Pavia
Sondrio
Trompia Valley
Valtellina
Varese
SA Po River
Po Valley
Sesia-Lanzo Zone

Lome Convention
Term introduced in 1981.
SA economics

lomonite
use laumontite

Lomonosov Range
use Lomonosov Ridge

Lomonosov Ridge
N of New Siberian Islands across top of world to N of Greenland.
UF Lomonosov Range
BT Arctic Ocean

lomontite
use laumontite

Lompoc
City in Santa Barbara County in SW.
BT California
United States

Lonar Crater
Term introduced in 1981. Impact crater in the basalt lava flows of the Deccan Plateau.
CO N195845N195845
E0763400E0763400
BX Deccan Plateau
Maharashtra
BT India

London
City on the Thames River in SE.
BT England
Great Britain
United Kingdom

London Basin
SE England.
CO N503500N544000
E0014500W0031000
BT England
Great Britain
United Kingdom

London Clay
Rests on Reading, Woolwich, Blackheath or Oldhaven Beds except in Norfolk to the west of Great Yarmouth. Overlain by the Claygate or Bagshot Beds, except in East Anglia. Occupies greater part of a large triangular area in London Basin bounded roughly by line from Herne Bay in Kent to Croyden, Farnham, and Newbury and from there NE to Aldeburgh in Suffolk.
BT Ypresian
lower Eocene
Eocene
SA England
United Kingdom

London Metal Exchange
Term introduced in 1981.
SA economics

Londonderry
County in NW Northern Ireland. Also a city.
BT Northern Ireland
United Kingdom

Long Beach
City in Los Angeles County in S.
BT California
United States

Long Island
Lying between Long Island Sound on N and Atlantic Ocean on S.
CO N403000N411500
W0715000W0741500
BT New York
United States
NX Queens County

Long Island Sound
Body of water between S shore of Connecticut and N shore of Long Island, New York.
BT Atlantic Ocean

long period
use long-period waves

long-period seismographs
Term introduced in 1981.
BT seismographs

long-period waves
Includes use on level 3 under seismology(1). Before 1978, search long period or long-period AND waves.
UF long period
BA elastic waves
SA seismology
traveltime
waves

longitudinal
A valid term through 1977 used in combination with folds, fractures, or faults. See longitudinal orientation.

longitudinal faults
use strike faults

longitudinal orientation
As of 1981, no longer a valid term for GeoRef. Term was introduced in 1978 and included use in relation to folds. Before 1978, also search folds or fractures AND longitudinal.

longitudinal wave
use P-waves

longshore bars
Term introduced in 1978.
UF barrier bars
BT bars
SA barrier islands

longshore currents
BT currents
SA littoral drift
ocean circulation
ocean waves

longshore drift
use littoral drift

lonsdaleite
As of 1981, includes use on level 3 under minerals(1) native elements(2). See list L. Before 1981, included use on level 3 under minerals(1) native elements and alloys(2).
BA native elements
BT minerals
SA carbon
chaoite
diamond
graphite
meteorites

loparite
As of 1981, includes use on level 3 under minerals(1) niobates(2). See list L. Before 1981, included use on level 3 under minerals(1) oxides(2).
BA niobates
oxides
BT minerals

lopoliths
Includes use on level 2 under intrusions(1).
BA intrusions
SA batholiths
igneous rocks
laccoliths

Lord Howe Island
Volcanic island 435 miles NE of Sydney. Dependency of New South Wales.
BT Pacific Ocean

Lord Howe Ridge
use Lord Howe Rise

Lord Howe Rise
Extends from SW of New Caledonia to W of New Zealand. Also search Lord Howe Ridge.
UF Lord Howe Ridge

Lord Howe-Chesterfield Ridge
Lord Howe-New Zealand Ridge
Lord Howe-New Zealand Rise
BT Pacific Ocean

Lord Howe-Chesterfield Ridge
use Lord Howe Rise

Lord Howe-New Zealand Ridge
use Lord Howe Rise

Lord Howe-New Zealand Rise
use Lord Howe Rise

Lorrain Formation
Huronian Supergroup of the Cobalt Group. Overlies The Gowganda conformably and gradationally.
UF Lorrain Series
BT Proterozoic
Precambrian
SA Ontario
Quebec

Lorrain Series
use Lorrain Formation

Lorraine
Region and former province on Belgium and Luxembourg borders in NE.
BT France

Los Alamos Scientific Laboratory
Los Alamos County, 35 miles NW of Santa Fe, in N central.
BT New Mexico
United States
SA University of California

Los Angeles
City in Los Angeles County in S.
BT California
United States

Los Angeles Basin
Greater Los Angeles area.
BT California
United States

Los Angeles County
S California.
CO N334500N344500
W1174000W1185000
BT California
United States
NT Palmdale Bulge

Los Islands
Group of small islands in Atlantic Ocean off Conakry.
UF Iles de Loos
BT Guinea

Los Pedroches
Region N of city of Cordoba in SW central part of country.
BT Cordoba Province
Andalusia
Spain

Los Testigos
Small group of islands SW of Grenada and N of Sucre, Venezuela.
BT Caribbean Sea

Lost Burro Formation
Includes Lippincott Member, Quartz Spring Sandstone Member. S California.
BT Devonian
SA California
Middle Devonian
Upper Devonian

Lost City
Ancient Indian city now covered by Lake Mead. Relics now housed in Overton in Clark County 5 miles N of original location in S.
BT Nevada
United States

Lot
Department in S central.
CO N441000N450500
E0021100E0005900
BT France
SA Quercy

Lot-et-Garonne
Department in SW.
CO N435700N444500
E0010000W0001000
BT France
NT Durance

Lotharingian
Europe. Above Sinemurian, below Pliensbachian. Includes use as level 3 term under age terms(1). See list E.
BA Lower Jurassic
Jurassic
BT Mesozoic

Louisiade Archipelago
Island group in Solomon Sea SE of E tip of New Guinea.
BT Papua New Guinea
Australasia

Louisiana
Includes use on level 1 as an area term (list O). For term set options see list B.
CO N290000N330000
W0890000W0940500
BA Southern U.S.
United States
NT Atchafalaya Bay
Barataria Bay
Baton Rouge
Belle Isle
Calcasieu Parish
Cameron Parish
Claiborne Parish
NX Columbia
NT East Bay
Evangeline Parish
Grand Isle
NX Jackson County
Jefferson County
NT Jefferson Parish
NX Lafayette County
NT Lake Pontchartrain
NX Lincoln County
Livingston County
NT Mississippi Delta
New Orleans
Rapides Parish
NX Richland County
NT Saint Mary Parish
South Pass
Vermilion Parish
Winnfield salt dome
SA Amite River
Bossier Formation
Buckner Formation
Citronelle Formation
Claiborne Group
Cook Mountain Formation
Cotton Valley Group
Fleming Formation
Frio Formation
Grand River
Gulf Coastal Plain
Jackson Group
Mississippi River
Mississippi Valley
Norphlet Formation
Queen City Formation
Red River
Red River valley
Sabine Lake
Schuler Formation
Smackover Formation
Trinity Group
Vicksburg Group
Washita Group
Wilcox Group
Woodbine Formation
Yazoo Clay

Louisville
City on the Ohio River in Jefferson County.
BT Kentucky
United States

Loup Fork Group
Miocene, Pliocene, and Pleistocene(?). Also search Loup Fork Formation.
BT Cenozoic
SA Colorado
Kansas
Miocene
Nebraska
New Mexico
Pleistocene
Pliocene
South Dakota
Texas
Wyoming

Lourdes
Town in SW part of country.
BT Hautes-Pyrenees
France

Louvain
City E of Brussels in central part of country.
UF Leuven
BT Brabant
Belgium

Love waves
Includes use on level 3 under seismology(1).
UF Q waves
BA surface waves
elastic waves
SA Q
seismology
waves

Loveland
City in Larimer County in N.
BT Colorado
United States

Lovelock
City in Pershing County in NW.
BT Nevada
United States

Lovozero
Village in Murmansk Oblast on Kola Peninsula in NW European USSR.
CO N670000N680000
E0350000E0340000
BT Russian Republic
USSR

Lovozero Massif
Murmansk Oblast on the Kola Peninsula in NW European USSR.
BT Russian Republic
USSR

Low Archipelago
use Tuamotu Islands

Low Jesenik Mountains
use Nizky Jezenik Mountains

low pressure
Includes use on level 3 under phase equilibria(1).
BT pressure
SA high pressure
metamorphism
phase equilibria

Low Tatra Mountains
Section of the Carpathians parallel to and S of the Tatra Mountains in N central Slovakia. Also search Low Tatra; Low Tatras.
UF Low Tatras
BT Slovakia
Czechoslovakia
SA Carpathians
Tatra Mountains

Low Tatras
use Low Tatra Mountains

low temperature
Includes use on level 3 under phase equilibria(1).
BT temperature
SA metamorphism
phase equilibria

low velocity zone
use low-velocity zones

low water
Term introduced in 1981.
SA tides
water

low-angle faults
Term introduced in 1978. Before 1978, search faults AND low-angle.
BA faults

low-energy environment
Term introduced in 1978. Before 1978, search low-energy.
SA environment
high-energy environment
sedimentation

low-grade
A valid term through 1977. After 1977, use low-grade metamorphism.

low-grade metamorphism
Term introduced in 1978. Includes use on level 3 under metamorphism(1). Before 1978, search low-grade AND metamorphism.
BA metamorphism
SA grade

low-velocity layer
As of 1981, no longer a valid term for GeoRef. Use low-velocity zones. Included use on level 2 under mantle(1).

low-velocity layers
use low-velocity zones

low-velocity zones
Term introduced in 1978 as a general term, i.e. applicable to crust, mantle, etc. As of 1981, includes use on level 2 under mantle(1). Also search low velocity zone; low velocity zones; low-velocity layer, low-velocity layers.
UF low velocity zone
low-velocity layers
zones, low-velocity
SA crust
mantle
seismology

Lower Austria
State in NE.
CO N472500N490200
E0170500E0142500
BT Austria
NT Bad Deutsch Altenburg
Eichkogel
Kamp Valley
Krems
Schwechat Valley
Spitz
Vienna
Wiener Wald
SA Enns Valley
Semmering
Wechsel

lower boundary
A valid general term through 1976. For stratigraphic meaning use boundary.

Lower Cambrian
See list E.
BA Cambrian
BT Paleozoic
NA Aldanian
NT Antietam Formation
Chilhowee Group
NA Harlech Stage
NT Kinzers Formation
Murphy Marble
Rome Formation
Shady Dolomite

Usa Series
Yudoma Series
SA Middle Cambrian
Tintic Quartzite
Upper Cambrian
Wood Canyon Formation

Lower Carboniferous
As of 1981, no longer a valid term for GeoRef. Use Dinantian. Before 1981, included use on level 3 under age terms(1). See list E.

lower Cenozoic
See list E.
BA Cenozoic

Lower Cretaceous
See list E.
BA Cretaceous
BT Mesozoic
NA Albian
NT Antlers Sands
NA Aptian
NT Areado Formation
Baquero Formation
NA Barremian
Berriasian
NT Cedar Mountain Formation
Cheyenne Sandstone
Christopher Formation
Fall River Formation
NA Gargasian
Gault
NT Glen Rose Formation
NA Hauterivian
NT Isachsen Formation
Kiowa Formation
Mannville Formation
Muddy Sandstone
NA Neocomian
NT Pasayten Group
NA Purbeckian
NT Speeton Clay
NA Urgonian
Valanginian
Wealden
NT Zubair Formation
SA Benton Formation
Blairmore Group
Colorado Group
Comanchean
Dakota Formation
Edwards Formation
Fredericksburg Group
Gondwana System
Graneros Shale
Mancos Shale
Mentor Beds
Middle Cretaceous
Paluxy Formation
Potomac Group
Upper Cretaceous
Vraconian
Washita Group

lower crust
BA crust
SA asthenosphere
basaltic layer
granitic layer
lithosphere
upper crust
upper mantle

Lower Devonian
See list E.
BA Devonian
BT Paleozoic
NT Coeymans Formation
NA Dittonian
Emsian
Gedinnian
NT Helderberg Group
NA Helderbergian
NT Littleton Formation
NA Lochkovian
NT Manlius Formation
Matagamon Sandstone
Oriskany Sandstone
NA Siegenian
SA Downtonian
Hidden Valley Dolomite
Middle Devonian
Read Bay Formation
Rondout Formation
Upper Devonian

lower Eocene
See list E.
BA Eocene
Paleogene
Tertiary
BT Cenozoic
NT Ager Formation
Aquia Formation
NA Cuisian
NT San Jose Formation
NA Sparnacian
NT Wilcox Group
Willwood Formation
Wind River Formation
NA Ypresian
SA Deccan Traps
Green River Formation
Intertrappean Beds
middle Eocene
Nanjemoy Formation
Rose Canyon Formation
Sinjar Formation
Umpqua Formation
upper Eocene

Lower Greensand
Aptian and lower Albian.
BT Cretaceous
SA England
United Kingdom

lower Holocene
See list E.
BA Holocene
Quaternary
BT Cenozoic

Lower Jurassic
Also search Liassic; Lias.
UF Lias
BA Jurassic
BT Mesozoic
NA Carixian
Lotharingian
lower Liassic
NT Luxembourg Sandstone
NA middle Liassic
NT Sunrise Formation
NA upper Liassic
SA Aalenian
Middle Jurassic
Novorayskoe Formation
Upper Jurassic

lower Liassic
Term introduced in 1981. Includes use on level 3 under age terms(1). See list E.
BA Lower Jurassic
Jurassic
BT Mesozoic
NA Hettangian
Sinemurian

Lower Lusatia
That part of Lusatia between the Neisse and the Bober rivers in SW.
BT Poland
SA Lusatia
Upper Lusatia

lower mantle
Includes use on level 3 under mantle(1).
BA mantle
SA upper mantle

lower Mesozoic
See list E.
BA Mesozoic

lower Miocene
See list E.
BA Miocene
Neogene
Tertiary
BT Cenozoic
NA Altonian
Aquitanian
Awamoan
Burdigalian
NT Chipola Formation
NA Eggenburgian
Hemingfordian
NT Kasauli Series
NA Saucesian
NT Waitemata Group
SA Arikareean
Badenian
Hawthorn Formation
John Day Formation
middle Miocene
upper Miocene

Lower Mississippi Valley
Term introduced in 1978.
BA Mississippi Valley
BT United States

Lower Mississippian
See list E.
BA Mississippian
Carboniferous
BT Paleozoic
NT Fort Payne Formation
NA Kinderhookian
NT Lake Valley Formation
Lodgepole Formation
NA Osagian
NT Pocono Formation
SA Borden Group
Catskill Formation
Leadville Formation
Lisburne Group
Madison Group
Monte Cristo Limestone
Rampart Group
Upper Mississippian
Valmeyeran

lower Oligocene
See list E.
BA Oligocene
Paleogene
Tertiary
BT Cenozoic
NT Chadron Formation
NA Chadronian
Sannoisian
Stampian
SA middle Oligocene
upper Oligocene

Lower Ordovician
See list E.
BA Ordovician
BT Paleozoic
NA Arenigian
Canadian Series
NT Fillmore Formation
Kingsport Formation
Manitou Formation
Mascot Dolomite
Shakopee Formation
Smithville Formation
NA Tremadocian
NT White Limestone
SA Antelope Valley Limestone
Arbuckle Group
Deadwood Formation
Knox Group
Middle Ordovician
Pogonip Group
Upper Ordovician

lower Paleocene
See list E.
BA Paleocene
Paleogene
Tertiary
BT Cenozoic
NA Danian
SA Difunta Group
upper Paleocene

lower Paleozoic
See list E.
BA Paleozoic
NT Ashe Formation
Glenarm Series
SA Wissahickon Formation

Lower Pennsylvanian
See list E.
BA Pennsylvanian
Carboniferous
BT Paleozoic
NT Caseyville Formation
Haymond Formation
Lee Formation
NA Morrowan
NT Pocahontas Formation
SA Mansfield Formation
Marble Falls Group
Middle Pennsylvanian
Pottsville Group
Upper Pennsylvanian

Lower Permian
See list E.
BA Permian
BT Paleozoic
NA Artinskian
Autunian
NT Barakar Stage
NA Leonardian
Sakmarian
Wolfcampian
SA Bird Spring Formation
Guadalupian
Kungurian
Middle Permian
Rotliegendes
San Andres Formation
Tensleep Sandstone
Upper Permian

lower Pleistocene
See list E.
BA Pleistocene
Quaternary
BT Cenozoic
NA Aftonian
Calabrian
Danube Stage
Jaramillo Event
Kansan
Nebraskan
Olduvai Event
Villafranchian
NT Waccamaw Formation
SA Matuyama Epoch
Uonuma Group
upper Pleistocene

lower Pliocene
See list E.
BA Pliocene
Neogene
Tertiary
BT Cenozoic
NA Gilbert Epoch
Plaisancian
Tabianian
SA Capistrano Formation
Meotian
middle Pliocene
upper Pliocene
Yakima Basalt

lower Precambrian
As of 1981, no longer a valid term for GeoRef. Use Archean. Before 1981, included use on level 3.

lower Proterozoic
See list E.
BA Proterozoic
Precambrian
NA Aphebian

lower Quaternary
See list E.
BA Quaternary
BT Cenozoic

Lower Rhine Basin
That section of the Rhine Basin between Bonn and the North Sea; in the Netherlands it includes the basins of the Lower Rhine, Lek, and Waal rivers. Index countries as applicable. Also search Lower Rhine.
BT Europe
SA Netherlands
Rhine Basin
Rhine Valley
West Germany

Lower Rhine Graben
BT Rhine Graben
Europe

Lower Saxony
State in N.
CO N511500N535000
E0113000E0070000
BT West Germany
NT Bad Pyrmont
Bramsche
East Frisian Islands
Eilenriede
Emsland
Gehrden
Gottingen
Helmstedt
Hildesheim
Jade Bay
Leine Valley
Osnabruck
Rammelsberg
Weser-Ems
SA Ems River
Teutoburg Forest
Weser River
Westphalia

Lower Silesia
Most of former German Silesia which became part of Poland following World War II. It lies on both sides of the upper Oder River centering on Wroclaw with Czechoslovakia and the Neisse River constituting the SE and E borders respectively.
BT Poland
SA Silesia
Silesian coal basin
Upper Silesia

Lower Silurian
See list E.
BA Silurian
BT Paleozoic
NA Alexandrian
NT Brassfield Formation
NA Llandoverian
NT Medina Formation
NA Tarannon
NT Tuscarora Formation
SA Middle Silurian
Rockwood Formation
Upper Silurian

lower Tertiary
See list E.
BA Tertiary
BT Cenozoic
NT Kiev Member
Taishu Group
SA Bauru Formation
Ewekoro Formation
middle Tertiary
Paleogene
upper Tertiary

Lower Triassic
See list E.
BA Triassic
BT Mesozoic
NA Bunter
NT Panchet Series
NA Scythian
NT Serebryanka Formation
Thaynes Formation
NA Werfenian
SA Hercynian Orogeny
Middle Triassic
Moenkopi Formation
Sadlerochit Formation
Upper Triassic

Lower Tunguska River
Rises in N central Irkutsk Oblast and flows N crossing into Evenk National Okrug then W into the Yenisei River at Turukhansk.
BT Russian Republic
USSR
SA Tunguska
Tunguska River

Lower Zaire
Term introduced in 1981.
CO S060000S040000
E0163000E0121000
BT Zaire

Lowicz
Town in central part of country.
BT Lodz
Poland

Loxoconcha
Includes use on level 3 under Ostracoda(1). See list F.
BA Podocopida
Ostracoda
BT Crustacea
Mandibulata
Arthropoda
Invertebrata

Loyalty Islands
Island group 60 miles E of New Caledonia belonging to France.
BT Pacific Ocean

Lozere
Department in S France.
CO N440800N450000
E0040000E0030000
BT France

Lu
use lutetium

Luanda
District in NW part of country. Also a city.
UF Loanda
BT Angola

Lubbock
City in Lubbock County in NW.
BT Texas
United States

Lubbock County
Term introduced in 1981. NW Texas.
CO N332000N335200
W1013000W1020400
BT Texas
United States

Luben
use Lubin

Lubin
City in SW part of country.
UF Luben
BT Wroclaw
Poland

Lubin Legnicki
Region in and around cities of Lubin and Legnica in SW part of country.
BT Wroclaw
Poland

Lublin
Province in SE part of country. Also a city.
BT Poland
NT Lublin Upland
Lukow
Tomaszow Lubelski

Lublin Upland
SW Lublin. Also search Lublin.
BT Lublin
Poland

Lucania
Ancient district in S including Basilicata autonomous region and part of Salerno Province of Campania. Index autonomous regions as applicable.
BT Italy
SA Basilicata
Campania

Lucas County
Term introduced in 1981. Index states as applicable.
BX Iowa
Ohio
BT United States

Lucca
City and province in central part of country.
BT Tuscany
Italy

Lucerne
Canton in N central Switzerland. Also a city.
UF Luzern
BT Switzerland

Lucinidae
Family.
BA Bivalvia
Mollusca
BT Invertebrata

Luderitz
Town in SW Namibia. Formerly Angra Pequena.
BT Namibia

Ludian
Europe. Above Bartonian, below Tongrian (Oligocene). Includes use on level 3 under age terms(1). See list E.
BA upper Eocene
Eocene
Paleogene
Tertiary
BT Cenozoic
SA Priabonian

Ludlovian
Europe. Above Wenlockian, below Gedinnian (Devonian). Includes use on level 3 under age terms(1). See list E.
BA Upper Silurian
Silurian
BT Paleozoic

ludwigite
BA borates
BT minerals

Lugo
Term introduced in 1981. Province in NW.
CO N421800N434600
W0065000W0075800
BT Galicia
Spain

lujavrite
As of 1981, includes use on level 3 under igneous rocks(1) syenites(2). See list H. Before 1981, included use on level 3 under igneous rocks(1) syenite family(2).
BA syenites
BT igneous rocks
SA nepheline syenite

Lukov
use Lukow

Lukow
Town in E part of country.
UF Lukov
BT Lublin
Poland

Lukuga River valley
Eastern.
BT Zaire

luminescence
Includes use on level 3 under discipline or material, e.g. under minerals(1), Moon(1), and under geochemistry(1).
SA cathodoluminescence
electromagnetic radiation
fluorescence
geochemistry
minerals
Moon
thermoluminescence
X-ray fluorescence

Luna County
SW New Mexico.
BT New Mexico
United States

Luna Program
Term introduced in 1981. Includes use on level 3 under Moon(1).
SA extraterrestrial geology
Moon
planetology
remote sensing

Luna 16
Includes use on level 3 under Moon(1).
SA Moon
remote sensing

Luna 20
Includes use on level 3 under Moon(1).
SA Moon
remote sensing

Luna 24
Term introduced in 1981. Includes use on level 3 under Moon(1).
SA Moon
remote sensing

lunar breccia
Term introduced in 1981. Before 1981, also search Moon AND breccia.
SA breccia
Moon

lunar craters
Term introduced in 1981. Before 1981, search Moon AND craters.
SA craters
Moon

lunar crust
Term introduced in 1981. Also search Moon AND crust. Includes use on level 2 under Moon(1).
SA crust
Moon

lunar interior
Term introduced in 1981. Before 1981, also search Moon AND interior. Includes use on level 2 under Moon(1).
SA interior
Moon
planetary interiors

lunar mantle
Term introduced in 1981. Also search Moon AND mantle.
SA mantle
Moon

lunar materials
As of 1981, no longer a valid term for GeoRef. Use Moon and type of material. Use lunar samples for actual samples.

lunar samples
Also search lunar rocks. Before 1981, also search lunar materials.
SA lunar soils
Moon
samples

lunar soils
Term reintroduced in 1981. A valid term through mid-1978. Before 1981, also search Moon AND soils.
SA lunar samples
Moon

Lund
City in extreme S part of country.
BT Malmohus
Sweden

Lusaka
City in central.
BT Zambia

Lusatia
Region in SE East Germany, and in SW Poland. It comprises both Lower and Upper Lusatia. Also search Lausitz. Index countries as applicable.
UF Lausitz
BT Europe
SA East Germany
Lower Lusatia
Poland
Upper Lusatia

Lushs Bight Group
Also search Lush's Bight Group.
BT Ordovician
SA Newfoundland

Lusitanian
Europe. Above Oxfordian, below Kimmeridgian. Includes use on level 3 under age terms(1). See list E.
BA Upper Jurassic
Jurassic
BT Mesozoic
SA Malm
Rauracian

Lut Desert
Great sandy and stony desert in E central Iran. Also search Dash-e-Lut Basin; Dash-i-Lut.
UF Dasht-e-Lut Basin
Dasht-i-Lut
BT Iran

Lutetian
Europe. Above Cuisian, below Auversian. Includes use on level 3 under age terms (1). See list E.
BA Eocene
Paleogene
Tertiary
BT Cenozoic

lutetium
Includes use on level 1 and 2 as a chemical element (list D).
UF Lu
SA elements
rare earths

Lutzow-Holm Bay
Between Prince Olav Coast and Prince Harald Coast in Queen Maud Land of the Norwegian Sector. Also search Lutzow-Holmbukta.
UF Lutzow-Holmbukta
BT Antarctica

Lutzow-Holmbukta
use Lutzow-Holm Bay

Luvisols
BA soils

Luxembourg
Includes use on level 1 as an area term (list O). For term set options see list B.
CO N493000N501500
E0063000E0054500
BA Western Europe
Europe
SA Ardennes
Luxembourg Sandstone
Moselle River
Moselle Valley

Luxembourg Province
Term introduced in 1981.
CO N493000N502800
E0060200E0045100
BT Belgium

Luxembourg Sandstone
Hettangian and Sinemurian. Also called Hettange Sandstone or Orval Sandstone.
BT Lower Jurassic
SA Belgium
Luxembourg

Luzern
use Lucerne

Luzerne County
NE Pennsylvania.
BT Pennsylvania
United States

Luzon
Northernmost and most important island.
BT Philippine Islands
NT Bondoc Peninsula
Camarines Norte
Taal
Wawa

luzonite
BA sulfosalts
BT minerals
SA famatinite

Lvov
City and oblast in SW European USSR. Also search L'vov.
UF Lwow
BT Ukraine
USSR

Lvov Basin
In Lvov Oblast on the Polish border. Also search Lvov.
BT Ukraine
USSR

L'vov Volyn Basin
use Lvov-Volyn Basin

Lvov-Volyn Basin
Includes parts of Lvov Oblast and adjoining Volyn Oblast along the Polish border. Also search L'vov Volyn Basin.
UF L'vov Volyn Basin
BT Ukraine
USSR

Lwow
use Lvov

Lyangar
Town in Samarkand Oblast in S central.
BT Uzbekistan
USSR

Lycian Taurus
Mountains in ancient district of SW Anatolia called Lycia.
BT Turkey
Middle East

Lycopoda
use Lycopsida

Lycopodiales
use Lycopsida

Lycopsida
Includes use on level 2 under pteridophytes(1). See list F. Before 1981, also search lycopods; Lycopoda. As of 1981, search lycopods only for biostratigraphy.
UF Lycopoda
Lycopodiales
BA pteridophytes
BT Plantae
NA Sigillaria
Stigmaria

Lyngen Peninsula
Along Lyngen Fjord 30 miles E of Tromso in N.
BT Troms
Norway

Lynn Lake
Town in NW.
BT Manitoba
Canada

Lyon
use Lyons

Lyon County
Term introduced in 1981. Index states as applicable.
BX Iowa
Kansas
Kentucky
Minnesota
Nevada
BT United States

Lyons
City in SE central part of country. Also search Lyon.
UF Lyon
BT Rhone
France

lysoclines
SA carbonate compensation depth
solution

Lystrosaurus
As of 1981, includes use on level 3 under Reptilia(1) Therapsida(2). See list F. Before 1981, included use on level 3 under Reptilia(1) Synapsida(2).
BA Therapsida
Synapsida
Reptilia
BT Tetrapoda
Vertebrata

Lytoceratida
Term introduced in 1981. Includes use on level 2 under Mollusca(1). See list F.
BA Ammonoidea
Tetrabranchiata
Cephalopoda
Mollusca
BT Invertebrata

Lyublyana
use Ljubljana

M

M-discontinuity
use Mohorovicic discontinuity

maars
BT volcanic features
SA craters
volcanism

Maas River
use Meuse River

Maas Valley
use Meuse Valley

Maastricht
City near the German border in extreme SE part of country.
UF Maestricht
BT Netherlands
SA Limburg

Maastrichtian
use Maestrichtian

Macacu River
Central Rio de Janeiro.
BT Rio de Janeiro
Brazil

macaluba
use mud volcanoes

Macanao Peninsula
On Margarita Island 15 miles off NE coast of Venezuela.
BT Nueva Esparta
Venezuela

Macedonia
Region in Bulgaria, Greece, and Yugoslavia.
BT Europe
NZ Greek Macedonia
Yugoslav Macedonia
SA Bulgaria
Greece
Salonika
Serbo-Macedonian Massif
Skopje
Vourinos
Yugoslavia

macerals
Includes use on level 3 under sedimentary rocks(1) organic residues(2). See list I.
UF micropetrological unit
BA coal
organic residues
BT sedimentary rocks
NA durain
exinite
inertinite
vitrain
vitrinite
SA petrography

Machaeridia
Includes use on level 2 under Echinodermata(1). See list F.
BA Homalozoa
Echinodermata
BT Invertebrata

Mackay
Town in Custer County in S central.
BT Idaho
United States

Mackenzie Delta
On Beaufort Sea in District of Mackenzie. Also search Mackenzie; Mackenzie River delta.
UF Mackenzie River delta
BT Northwest Territories
Canada

Mackenzie District
use District of Mackenzie

Mackenzie Mountains
Index territories as applicable. Also search Mackenzie.
BT Canada
SA Northwest Territories
Rocky Mountains
Yukon Territory

Mackenzie River delta
use Mackenzie Delta

Mackenzie River valley
W District of Mackenzie. Also search Mackenzie; Mackenzie River; Mackenzie Valley.
UF Mackenzie Valley
BT Northwest Territories
Canada

Mackenzie Valley
use Mackenzie River valley

mackinawite
UF kansite
BA sulfides
BT minerals

Macleay River
E New South Wales.
BT New South Wales
Australia

Macomb County
On Lake St. Clair just N of Detroit.
BT Michigan
United States

M

Macon
City in Bibb County in central.
BT Georgia
United States

Macon County
Index states as applicable.
BX Alabama
Georgia
Illinois
Missouri
North Carolina
Tennessee
BT United States

Macquarie Island
850 miles SE of Tasmania by whom it is administered.
BT Pacific Ocean

Macquarie Ridge
Extends between Macquarie Island on S and South Island of New Zealand on N.
BT Pacific Ocean

Macropodidae
Family. Includes use on level 3 under Mammalia(1) Marsupialia(2).
BA Marsupialia
Mammalia
BT Tetrapoda
Vertebrata

Macroscelida
Term introduced in 1981. Includes use on level 2 under Mammalia(1). See list F.
BA Mammalia
BT Tetrapoda
Vertebrata

Mactra
Genus. Includes use on level 3 under Mollusca(1). See list F.
BA Bivalvia
Mollusca
BT Invertebrata

Madagascar
A valid term through 1974. After 1974, use Malagasy Republic.

Madagascar Basin
SE of Madagascar. Also search Madagascar.
BT Indian Ocean

Madan
Village in SE Rhodope Mountains in S part of country.
BT Plovdiv
Bulgaria

Madeira
Island group belonging to Portugal 600 miles SW of Lisbon and 400 miles W of Morocco. Includes use on level 1 as an area term (list O). For term set options see list B.
CO N323000N330700
W0161300W0173000
UF Madeira Archipelago
Madeira Island
Madeira Islands
Madeiras
BA Atlantic Ocean Islands
NT Porto Santo Island

Madeira Archipelago
use Madeira

Madeira Island
use Madeira

Madeira Islands
use Madeira

Madeiras
use Madeira

Madera County
Central California.
BT California
United States

Madhya Pradesh
State in central.
CO N180000N270000
E0840000E0740000
BT India
NT Balaghat
Bastar
Chhindwara
Chhindwara District
Jabalpur
Jhabua
Kanhan Valley
Kota
Pauni
Pench Valley
Raipur
Rewa
Satna
Sidhi
Umaria
Umrer
SA Bagh Beds
Bhander Group
Bijawar System
Bundelkhand
Damuda Series
Deccan Plateau
Jabalpur Series
Kaimur Sandstone
Mahanadi Valley
Narmada River
Narmada Valley
Satpura Range
Semri Series
Son Valley
Trans-Aravalli Vindhyan Basin

Madison
City in Dane County in S central.
BT Wisconsin
United States

Madison County
Index states as applicable.
BX Alabama
Arkansas
Florida
Georgia
Idaho
Illinois
Indiana
Iowa
Kentucky
Mississippi
Missouri
Montana
Nebraska
New York
North Carolina
Tennessee
Texas
Virginia
BT United States

Madison Group
Lower and upper Mississippian. Consists of Lodgepole Limestone, Mission Canyon Limestone, and Charles Formation. Also search Madison Formation.
BT Mississippian
SA Colorado
Idaho
Lodgepole Formation
Lower Mississippian
Montana
Upper Mississippian
Utah
Wyoming

Madoc
Village in SE.
BT Ontario
Canada

Madonie Mountains
NW central.
BT Sicily
Italy

Madras
City on the Bay of Bengal in SE part of country.
BT Tamil Nadu
India
NT Kodaikanal

Madreporaria
use Scleractinia

Madrid
As of 1981, refers to only the city. Before 1981, also included the province. In central Spain.
BT Madrid Province
New Castile
Spain

Madrid Province
Term introduced in 1981. Before 1981, search Madrid. In central Spain.
CO N395700N411000
W0030500W0043600
BT New Castile
Spain
NT Madrid

Maestrazgo
Mountainous district in E Spain. Index provinces as applicable.
BT Spain
SA Castellon de la Plana Province
Teruel

Maestricht
use Maastricht

Maestrichtian
Europe. Above Campanian, below Danian (Tertiary). Includes use on level 3 under age terms(1). See list E.
UF Maastrichtian
BA Senonian
Upper Cretaceous
Cretaceous
BT Mesozoic
SA Rosario Formation

mafic
A valid term through 1977. After 1977, use mafic composition on level 2 under igneous rocks(1).

mafic composition
Term introduced in 1978. Includes use on level 2 under igneous rocks(1). Before 1978, also search igneous rocks AND mafic. Also search basic rocks.
SA composition
igneous rocks
mafic magmas
ultramafic composition

mafic magmas
Term introduced in 1981.
SA mafic composition
magmas

Magadan
City on N shore of Okhotsk Sea. Also an oblast.
BT Russian Republic
USSR
SA Okhotsk region

magadiite
BA scapolite group
framework silicates
silicates
BT minerals

Magallanes
Southernmost province.
BT Chile

Magdalena
Department in N Colombia. Village in W central New Mexico. Towns in Argentina, Bolivia and Peru. Index countries as applicable.
SA Argentina
Bolivia
Colombia
Mexico
New Mexico
Peru

Magdalena Basin
use Magdalena Valley

Magdalena Delta
At the mouth of the Magdalena River on the Caribbean Sea near Barranquilla. Also search Magdalena.
BT Bolivar
Colombia

Magdalena Mountains
In Socorro County in W central.
BT New Mexico
United States

Magdalena River
Rises on E slope of Andes in S Colombia and flows N into the Caribbean Sea near Barranquilla. Also search Magdalena AND appropriate area.
BT Colombia

Magdalena Valley
River valley extending from NW to SW. Also search Magdalena Basin.
UF Magdalena Basin
BT Colombia

Magdalenian
Archaeologic classification. Europe, North Africa. Upper Paleolithic.
BA Paleolithic
BT Cenozoic
SA Pleistocene
Quaternary

Magdeburg
As of 1981, refers to only the city. Before 1981, also included the district.
BT Magdeburg Bezirk
East Germany

Magdeburg Bezirk
Term introduced in 1981. District in W central. Before 1981, search Magdeburg.
CO N513500N530300
E0122500E0103500
BT East Germany
NT Elbingerode
Halberstadt
Magdeburg
Stassfurt
SA Saxony-Anhalt

maghemite
UF oxymagnite
BA oxides
BT minerals
SA spinel group
titanomaghemite

magma chambers
Includes use on level 3 under magmas(1) and under intrusions(1).
UF chambers, magma
magma reservoir
SA intrusions
magmas

magma reservoir
use magma chambers

magmas
Includes use on level 1 (list A); on level 2 under phase equilibria(1). Used for naturally occurring mobile rock material, generated within the Earth and capable of intrusion and extrusion. Also search magma. If 1, term set options are:
topic [age, classification, composition, contamination, differentiation, evolution, genesis, geochemistry, properties, temperature, viscosity]
subtopic
SA acidic magmas
anatexis

assimilation
asthenosphere
contamination
country rocks
crust
crystallites
crystallization
cumulates
differentiation
emplacement
fractional crystallization
gases
hot spots
hybridization
igneous activity
igneous processes
igneous rocks
immiscibility
inclusions
intrusions
juvenile water
lava
mafic magmas
magma chambers
magmatic associations
melts
palingenesis
partial melting
petrology
phase equilibria
pneumatolysis
segregation
viscosity
volcanology

magmatic associations
Term introduced in 1981.
SA magmas

magmatism
A valid term through 1974. Use igneous activity or magmas.

magnesian calcite
Also search magnesium calcite.
UF magnesium calcite
BA carbonates
BT minerals
SA calcite

magnesian limestone
As of 1981, no longer a valid term for GeoRef. Use dolomitic limestone.

magnesian spar
use dolomite

magnesioferrite
UF magnoferrite
BA oxides
BT minerals
SA spinel group

magnesioriebeckite
BA clinoamphibole
amphibole group
chain silicates
silicates
BT minerals

magnesite
As of 1981, use magnesite deposits for magnesite as a commodity. Before 1981, included use on level 1 as a commodity term.
BA carbonates
BT minerals
SA magnesite deposits

magnesite deposits
Term introduced in 1981. Includes use on level 1 and 2 as a commodity term (list C). Before 1981, search magnesite AND deposits.
SA magnesite
refractory materials

magnesium
Includes use on level 1 and 2 as a chemical element (list D). As of 1981, use magnesium ores for magnesium as a commodity. Before 1981, included use on level 1 as a commodity term.
UF Mg
SA elements
magnesium ores

magnesium calcite
use magnesian calcite

magnesium ores
Term introduced in 1981. Includes use on level 1 and 2 as a commodity term (list C). Before 1981, search magnesium AND ore deposits.
SA magnesium

magnetic anomalies
Includes use on level 3 under geophysical methods(1).
SA anomalies
geophysical methods
geophysical surveys
gravity anomalies
magnetic field
magnetic methods
magnetic surveys
paleomagnetism
sea-floor spreading

magnetic declination
Term introduced in 1981. Also search declination AND magnetic field.
SA magnetic field
paleomagnetism

magnetic domains
Level 2 term introduced in 1978 under paleomagnetism(1).
NT multi domains
pseudo-single domains
single domains
SA domains
magnetic properties
paleomagnetism

magnetic field
Includes use on level 2 under astrophysics and solar physics(1), aurora(1), Earth(1), Moon(1), and any of the planets. Before 1974, also search magnetic fields.
UF field, magnetic
SA astrophysics and solar physics
aurora
coastal effects
coercivity
core
cosmic rays
currents
declination
dipole moment
diurnal variations
Earth
Earth-current methods
electrical field
electrojet
electromagnetic field
electromagnetic induction
electromagnetic radiation
electromagnetic surveys
electromagnetic waves
geophysical methods
geophysical surveys
induction
Koenigsberger ratio
magnetic anomalies
magnetic declination
magnetic inclination
magnetic intensity
magnetic methods
magnetic storms
magnetic surveys
magnetic susceptibility
magnetic tail
magnetization
magnetohydrodynamics
magnetometers
magnetotosphere
magnetotelluric methods
magnetotelluric surveys
micropulsations
Moon
natural remanent magnetization
paleomagnetism
plasma instabilities
pole positions
pulsations
relief effects
remanent magnetization
reversals
seasonal variations
secular variations
solar wind
spherical harmonic analysis
thermoremanent magnetization
trapped particles
variations
whistlers

magnetic hysteresis
Term introduced in 1981. Before 1981, also search hysteresis AND paleomagnetism; hysteresis AND magnetization; hysteresis AND demagnetization; hysteresis AND magnetic.
SA demagnetization
magnetization
paleomagnetism
properties

magnetic inclination
Term introduced in 1981. Also search inclination AND magnetic field.
SA magnetic field
paleomagnetism

magnetic intensity
Term introduced in 1981.
SA intensity
magnetic field
magnetic storms
magnetosphere
paleomagnetism

magnetic iron ore
use magnetite

magnetic methods
Includes use on level 2 under geophysical methods(1).
BA geophysical methods
SA airborne methods
deep magnetic sounding
magnetic anomalies
magnetic field
magnetic properties
magnetic surveys
magnetometers
methods

magnetic minerals
BT minerals
SA ilmenite
magnetic properties
magnetite
titanomagnetite

magnetic polarization
use magnetization

magnetic properties
Includes use on level 3 under appropriate material name, e.g. under lava(1).
SA Curie point
magnetic domains
magnetic methods
magnetic minerals
magnetic susceptibility
properties
remanent magnetization

magnetic storms
Includes use on level 2 under magnetosphere(1); on level 3 under aurora(1) magnetic field(2).
SA aurora
disturbances
magnetic field
magnetic intensity
magnetosphere
main phase
plasmasphere
polar regions
solar cycles
solar flares
solar wind
storms
substorms
sudden commencements
trapped particles

magnetic survey maps
Term introduced in 1978. Before 1978, also search maps AND magnetic surveys.
BA maps
SA geophysical surveys
magnetic surveys

magnetic surveys
Includes use on level 3 under area terms(1) geophysical surveys(2). Before 1974, also search aeromagnetic surveys; aeromagnetic.
BA geophysical surveys
BT surveys
SA airborne methods
deep magnetic sounding
magnetic anomalies
magnetic field
magnetic methods
magnetic survey maps
magnetometers

magnetic susceptibility
Includes use on level 3 under paleomagnetism(1).
UF magnetic susceptibility anisotropy
volume susceptibility (magnetic)
SA anisotropy
Curie point
geophysical surveys
Koenigsberger ratio
magnetic field
magnetic properties
paleomagnetism

magnetic susceptibility anisotropy
use magnetic susceptibility

magnetic tail
Includes use on level 2 under magnetosphere(1).
SA magnetic field
magnetopause
magnetosphere
solar wind

magnetism, paleo-
use paleomagnetism

magnetite
Includes use on level 3 under minerals(1). See list L. As of 1981, includes use on level 3 as a commodity term (list C) under iron ores(1). Before 1981, included use on level 3 as a commodity term under iron(1).
UF magnetic iron ore
octahedral iron ore
BA oxides
BT minerals
SA heavy minerals
iron ores
magnetic minerals
spinel group
titanomagnetite

magnetization
The magnetic moment per unit volume.
UF magnetic polarization
NA direct magnetization
induced magnetization
inverse magnetization
isothermal magnetization

M

piezoremanent magnetization
primary magnetization
saturation magnetization
secondary magnetization
spontaneous magnetization
SA adiabatic demagnetization
chemical demagnetization
chemical remanent magnetization
demagnetization
magnetic field
magnetic hysteresis
natural remanent magnetization
paleomagnetism
remagnetization
remanent magnetization
thermal demagnetization
thermoremanent magnetization

magnetohydrodynamics
SA core
magnetic field

magnetometers
SA instruments
magnetic field
magnetic methods
magnetic surveys

magnetopause
Includes use on level 2 under magnetosphere(1).
SA bow shock waves
magnetic tail
magnetosheath
magnetosphere
solar wind

magnetosheath
Includes use on level 2 under magnetosphere(1).
SA bow shock waves
magnetopause
magnetosphere

magnetosphere
Includes use on level 1 for special indexes. Term set options are:
bow shock waves
topic
coordinate systems
topic
cosmic rays
albedo
cutoff rigidities
intensity
electrical field
topic or platform
general
subtopic
instruments
name of instrument
magnetic tail
topic or platform
magnetic storms
topic [e.g. initial phase, main phase, plasmasphere, recovery phase, substorms, sudden-commencements, trapped particles]
magnetopause
topic or platform
magnetosheath
topic or platform
plasma instabilities
topic or platform
plasma motion
topic [circulation, convection]
plasmapause
topic or platform
solar wind
subtopic
techniques
name of technique
type of phenomena
trapped particles
type of particle
type of phenomena
variations
topic [e.g. micropulsations, solar eclipses]
propagation
subtopic
type of wave
whistlers
(platform if satellite)
topic
SA aeronomy
albedo
astrophysics and solar physics
atmosphere
aurora
bow shock waves
coordinate systems
cosmic rays
cutoff rigidities
disturbances
Earth
electrical field
geophysics
interplanetary space
ionosphere
magnetic field
magnetic intensity
magnetic storms
magnetic tail
magnetopause
magnetosheath
main phase
meteorology
plasma instabilities
plasma motion
plasmapause
plasmasphere
pulsations
solar wind
substorms
sudden commencements
trapped particles
variations
whistlers

magnetostratigraphy
Term introduced in 1978.
BT stratigraphy

magnetotelluric methods
Includes use on level 2 under geophysical methods(1).
BA geophysical methods
SA Earth-current methods
electrical field
electromagnetic field
electromagnetic methods
magnetic field
magnetotelluric surveys
methods

magnetotelluric surveys
Includes use on level 3 under area terms(1) geophysical surveys(2).
BA geophysical surveys
BT surveys
SA Earth-current surveys
electromagnetic methods
electromagnetic surveys
magnetic field
magnetotelluric methods

Magnitogorsk
City just E of the South Urals in Chelyasinsk Oblast.
BT Russian Republic
USSR

magnitude
Includes use on level 2 under earthquakes(1). As of 1981, restricted to magnitude of earthquakes.
UF earthquake magnitude
SA earthquakes
intensity
Richter Scale
seismic intensity

magnitude-frequency ratio
Term introduced in 1981.
UF b
SA earthquakes
seismology

magnoferrite
use magnesioferrite

Magnoliidae
Term introduced in 1981. Includes use on level 2 under angiosperms(1). See list F.
BA Dicotyledoneae
angiosperms
BT Spermatophyta
Plantae

Magothy Formation
Consists essentially of light-gray crossbedded coarse sand containing small amounts of glauconite and pyrite; particles of carbonaceous matter or lignite common throughout.
BT Upper Cretaceous
Cretaceous
SA Delaware
Maryland
New Jersey
New York

Magura
Village W of Dacca.
BT Bangladesh

Mahanadi Valley
River valley. Index states as applicable. Also search Mahanadi River.
BT India
SA Madhya Pradesh
Orissa

Mahantango Formation
Central Pennsylvania. In Hamilton Group. Consists principally of greenish-gray, thin - to medium-bedded slightly carbonaceous shale. Sporadic sandstone zones appear throughout area and thicken eastward. In central Pennsylvania.
BT Middle Devonian
Devonian
SA Hamilton Group
Pennsylvania

Maharashtra
State in W central.
CO N153000N220000
E0804000E0723000
BT India
NT Bhandara
Bombay
Jamkhandi
Koyna
NX Lonar Crater
NT Nagpur
Poona
Wardha River valley
SA Deccan Plateau
Godavari River
Godavari Valley
Kaladgi System
Krishna
Pranhita-Godavari Valley
Satpura Range

Mahaska County
S central.
BT Iowa
United States

Mahasu
District in NW part of country.
BT Himachal Pradesh
India

Mahe Island
Most important of Seychelles, a British Colony, 1100 miles E of Kenya and 700 miles N of the Malagasy Republic. Also search Mahe.
BT Seychelles

Mahoning County
NE Ohio.
BT Ohio
United States

Maikop
City in Adygey Autonomous Oblast in the Northern Caucusus. Also search Maykop.
UF Maykop
BT Russian Republic
USSR

Maikop Series
Oligocene-Miocene. The following horizons have been established: Zuramakent, Riki, Mutsidakal, Miatly, and Khadum. Northern Caucusus and Crimea.
BT Tertiary
SA Russian Republic
Ukraine
USSR

Maimecha
use Maymecha

Maimecha-Kotui
use Maymecha-Kotuy

main phase
Includes use on level 3 under magnetosphere(1) magnetic storms(2).
SA magnetic storms
magnetosphere

Main River
Also search Main. Index states as applicable.
BT West Germany
SA Baden-Wurttemberg
Bavaria
Hesse

Maine
Includes use on level 1 as an area term (list O). For term set options see list B.
CO N430000N473000
W0670000W0710500
BA New England
Eastern U.S.
United States
NT Aroostook County
Augusta
NX Cumberland County
Franklin County
Hancock County
Lincoln County
NT Oxford County
Penobscot Bay
NX Presque Isle
Somerset County
Washington County
York County
SA Matagamon Sandstone
Merrimack Synclinorium
Rangeley Lakes
Saco River
Waterville Formation

Maine-et-Loire
Department in NW central.
CO N470000N474500
E0001500W0012000
BT France

maintenance
As of 1978, term is used on level 2 under reservoirs(1).
SA reservoirs

Mainz
City on the Rhine River in W central part of country.
BT Rhineland-Palatinate
West Germany
SA Mainz Basin

Mainz Basin
Area around confluence of the Main and Rhine Rivers. Index states as applicable.
BT West Germany
SA Hesse
Mainz
Rhineland-Palatinate

Maiolica Limestone
Jurassic-Cretaceous. White, marly limestone. In Lombardy, Italy. Also search Maiolica Formation.
BT Mesozoic
SA Cretaceous
Italy
Jurassic

Maizuru
City in Kyoto Prefecture in central.
BT Honshu
Japan

Majdan Pek
use Majdanpek

Majdanpek
City in E Serbia.
UF Majdan Pek
BT Serbia
Yugoslavia

Major County
NW central.
BT Oklahoma
United States

major elements
Includes use on level 3 under soils(1) nutrients(2).
SA elements
major-element analyses
minor elements
nutrients
trace elements

major-element analyses
For analytical methods. Includes use on level 3 under chemical analysis(1) and spectroscopy(1). For data, see material name.
SA analysis
chemical analysis
elements
major elements
minor-element analyses
spectroscopy
trace-element analyses

Majorca
Largest of Spain's Balearic Islands off E coast of Spanish mainland.
CO N391000N400000
E0033000E0021500
UF Mallorca
BT Balearic Islands
Spain

Majunga Basin
use Betsiboka Basin

Makhachkala
City in Dagestan Autonomous Soviet Socialist Republic on Caspian Sea.
UF Petrovsk
BT Russian Republic
USSR

Makhtesh Ramon
Canyon in the Negev.
BT Israel

Malacca Strait
use Strait of Malacca

Malacca Straits
use Strait of Malacca

malachite
BA carbonates
BT minerals

Malacostraca
Includes use on level 2 under Arthropoda(1). See list F. As of 1981, used for Decapoda. Before 1981, also search Decapoda; decapods.
BA Crustacea
Mandibulata
Arthropoda
BT Invertebrata
NA Brachyura
Callianassa
Ophiomorpha

Malaga
As of 1981, refers to only the city. Before 1981, also included the province. In S Spain on the Mediterranean Sea.
BT Malaga Province
Andalusia
Spain

Malaga Province
Term introduced in 1981. Before 1981, search Malaga. In S central Spain.
CO N361800N371800
W0034500W0053000
BT Andalusia
Spain
NT Malaga
SA Serrania de Ronda

Malagasy Republic
Formerly Madagascar. Became Malagasy Republic in 1958. Includes use on level 1 as an area term (list O). For term set options see list B. Before 1975, also search Madagascar.
CO S254000S115200
E0503000E0430000
BZ Indian Ocean Islands
Africa
NT Ambre Mountain
Andriamena
Betsiboka Basin
Sakoa Basin
Tananarive
Tulear Basin

Malaita
Long, narrow island in SE Solomon Islands.
BT Solomon Islands

Malakal
Town in S part of country.
BT Sudan

Malatya
Province in E central Anatolia. Also a city.
BT Turkey
Middle East

Malawi
Formerly Nyasaland. At one time it was part of British Central African Protectorate and later part of the Federation of Rhodesia and Nyasaland. Includes use on level 1 as an area term (list O). For term set options see list B.
CO S173000S090000
E0363000E0330000
UF Nyasaland
BA East Africa
Africa
SA East African Rift
Lake Malawi

Malay Archipelago
Largest island group in world off SE coast of Asia between Pacific and Indian oceans. Index countries as applicable. Includes use on level 1 or 2 as an area term (list O). If 1, see term set options under list B.
NX Borneo
NA New Guinea
NX Timor
SA Celebes Sea
Far East
Indonesia
Malaysia
Oceania
Papua New Guinea

Malay Peninsula
Comprises West Malaysia and SW part of Thailand. Index countries as applicable.
BT Far East
SA Indochina
Malaysia
Pahang
Thailand
West Malaysia

Malaya
A valid index term through 1976. After 1976, see Malaysia. Former Federation of Malaya which was a federation of 9 Malay states of the Malay Peninsula, plus 2 of the Straits settlements, Malacca and Penang. It constituted what is now West Malaysia of the present Federation of Malaysia.

malayaite
BA ring silicates
silicates
BT minerals

Malaysia
Or officially known as Federation of Malaysia. Independent federation, SE Asia, consisting of eleven states (West Malaysia) on the Malay Peninsula and two states (East Malaysia) on the island of Borneo. Includes use on level 1 as an area term (list O). For term set options see list B.
CO N013000N073000
E1190000E1000000
BA Far East
Asia
NT Darvel Bay
East Malaysia
Johore
Kadah
Kelantan
Langkawi Islands
Pahang
Perak
Sabah
Sandakan
Sarawak
Selangor
West Malaysia
SA Borneo
Indochina
Malay Archipelago
Malay Peninsula

Maldive Islands
Group of 19 atolls in the Indian Ocean 300 miles SW of southern tip of India. Includes use on level 1 as an area term (list O). For term set options see list B.
CO S010000N072000
E0740000E0720000
BA Indian Ocean Islands

Malekula
An island of the New Hebrides which lies E of Queensland, Australia, between the Solomon Islands and New Caledonia.
BT New Hebrides

Malgobek
City in North Ossetian Autonomous Soviet Socialist Republic in the Northern Caucasus. A petroleum producing center.
BT Russian Republic
USSR

Malheur County
Extreme E and SE.
CO N420000N442500
W1170500W1181500
BT Oregon
United States

Mali
Includes use on level 1 as an area term (list O). For term set options see list B.
CO N100500N250000
E0040000W0130000
BA West Africa
Africa
NT Taoudenni
SA Mali-Niger Syneclise
Niger River
Niger Valley
Sahara
Sahel
Senegal Basin
Senegal River
Tanezrouft
West African Shield

Mali-Niger Syneclise
Index countries as applicable.
UF Mali-Nigeria Syneclise
BT Africa
SA Benin
Mali
Niger
Nigeria

Mali-Nigeria Syneclise
use Mali-Niger Syneclise

Malines
City N of Brussels.
UF Mechelen
Mechlin
BT Antwerp
Belgium

Mallorca
use Majorca

Malm
Middle Europe. Above Dogger, below Neocomian (Cretaceous). Includes use on level 3 under age terms (1). See list E.
BA Upper Jurassic
Jurassic
BT Mesozoic
SA Kimmeridgian
Lusitanian
Oxfordian
Portlandian
Volgian

Malmohus
County in S.
BT Sweden
NT Halsingborg
Lund
University of Lund
SA Skane

Malta
Comprises 3 islands. Includes use on level 1 as an area term (list O). For term set options see list B.
CO N354500N361000
E0144000E0141000
BA Southern Europe
Europe
SA Mediterranean region

Malvern
City in Worcestershire in W central.
BT England
Great Britain
United Kingdom

Malvern Hills
Between Worcestershire and Herefordshire in W England.
BT England
Great Britain
United Kingdom

Mama
Town on the Vitim River at the mouth of the Mama River in NE Irkutsk Oblast.
BT Russian Republic
USSR

Mama River
Rises NE of Lake Baikal and flows NE into Vitim River at Mama. Also search Mama.
BT Russian Republic
USSR

Mamainse Point
Village on Lake Superior N of Sault Ste. Marie.

BT Ontario
Canada

Mammalia
Includes use on level 1 and 2 as a fossil term (list F).
BT Tetrapoda
Vertebrata
NA Acreodi
Amblypoda
Anagalida
Arctocyonia
Artiodactyla
Astrapotheria
Carnivora
Cetacea
Chiroptera
Condylarthra
Creodonta
Dermoptera
Desmostylia
Dinocerata
Diprotodonta
Docodonta
Edentata
Embrithopoda
Hyracoidea
Insectivora
Lagomorpha
Litopterna
Macroscelida
Marsupialia
Monotremata
Multituberculata
Nomarthra
Notoungulata
Pantodonta
Pantotheria
Paucituberculata
Perissodactyla
Pholidota
Polyprotodontia
Primates
Proboscidea
Proteutheria
Pyrotheria
Rodentia
Scandentia
Sirenia
Symmetrodonta
Taeniodonta
Theria
Tillodontia
Triconodonta
Tubulidentata
Ungulata
Xenarthra
Xenungulata
SA fossil man
mammals

mammals
Term introduced in 1981 as the common name for Mammalia. See list F. To be used for Mammalia when they are used as tools in stratigraphy. Includes use on level 1. Before 1981, search Mammalia AND biostratigraphy. If 1, term set options are:
biostratigraphy
age [List E]
SA biostratigraphy
Mammalia

Mammoth Cave
Cave and national park in Edmonson County in SW central.
BT Kentucky
United States

Mammut
use Mastodon

Mammuthus
Genus. As of 1981, includes use on level 3 under Mammalia(1) Elephantoidea(2). See list F. Before 1981, included use on level 3 under Mammalia(1) Proboscidea(2). Before 1981, also search mammoths. As of 1981, search mammoths only for biostratigraphy.
BA Elephantidae
Elephantoidea
Proboscidea
Mammalia
BT Tetrapoda
Vertebrata
NA Mammuthus primigenius

Mammuthus primigenius
BA Mammuthus
Elephantidae
Elephantoidea
Proboscidea
Mammalia
BT Tetrapoda
Vertebrata

Mammutidae
use Mastodontidae

man
As of 1981, no longer a valid term for GeoRef. See fossil man; Mammalia. Before 1981, included use on level 3. Before 1978, included use in combination with fossil (i.e. man, fossil) on level 1 and 2.

management
Used as a general term. As of 1978, term is used on level 2 under land use(1) and shorelines(1).
SA land use
planning
shorelines
soil management

Managua
Department in W part of country. Also a city.
BT Nicaragua

Manawatu River valley
SW North Island. Also search Manawatu.
BT North Island
New Zealand

Manche
Department on the English Channel in NW.
CO N483000N494500
W0004500W0020000
BT France
NT Cotentin Peninsula
SA Normandy

Manchester
City in Lancashire in NW England, and city in Hillsborough County in S New Hampshire. Index countries as applicable.
SA New Hampshire
United Kingdom
United States

Manchuria
Region in NE.
BT China
NT Liaoning
SA Amur Basin
Amur River
Argun River
Khanka Lake

Mancos Shale
Lower and upper Cretaceous. In Colorado, it is restricted to thick succession of shale, sandy shale, and thin-bedded sandstone overlying Niobrara Formation and underlying Mesaverde Group. NE Arizona, W Colorado, NW New Mexico, E Utah, and S and central Wyoming.
BT Cretaceous
SA Arizona
Colorado
Ferron Sandstone Member
Lower Cretaceous
New Mexico
Upper Cretaceous
Utah
Wyoming

Mandi
City in Mandi District in NW part of country.
BT Himachal Pradesh
India

Mandibulata
Term introduced in 1981. Includes use on level 2 under Arthropoda(1). See list F.
BA Arthropoda
BT Invertebrata
NA Crustacea
NT Insecta
NA Myriapoda

Mangalia
Town on the Black Sea in Constanta County.
BT Dobruja
Romania

Mangalore
City in SW.
BT Mysore
India

manganblende
use alabandite

manganese
Includes use on level 1 and 2 as a chemical element (list D). As of 1981, use manganese ores for manganese as a commodity. Before 1981, included use on level 1 as a commodity term.
UF Mn
SA elements
ferromanganese composition
heavy metals
manganese ores
manganese oxides
Mn-53
Mn-54

manganese ores
Term introduced in 1981. Includes use on level 1 and 2 as a commodity term (list C). Before 1981, search manganese AND ore deposits.
SA manganese

manganese oxides
Also search manganese oxide.
BA oxides
BT minerals
SA manganese
psilomelane

manganite
BA oxides
BT minerals

manganosite
BA oxides
BT minerals

mangerite
As of 1981, includes use on level 3 under igneous rocks(1) diorites(2). See list H. Before 1981, included use on level 3 under igneous rocks(1) diorite family(2).
BA diorites
BT igneous rocks
SA plutonic rocks

Mangishlak Peninsula
use Mangyshlak Peninsula

Mangla Dam
SE of Rawalpindi in N part of country. Also search Mangla.
BT Punjab Province
Pakistan

mangrove swamps
Term introduced in 1978.
SA ecology
swamps

Mangyshlak Peninsula
On E coast of N Caspian Sea. Also search Mangyshlak.
CO N423000N450000
E0543000E0500000
UF Mangishlak Peninsula
BT Kazakhstan
USSR

Manhattan
Island and borough of New York City at N end of New York Bay; and city in Riley County in NE central Kansas. Index states as applicable.
BX Kansas
New York
BT United States

Manhattan Formation
Dominantly a garnetiferous quartz-biotite-plagioclase gneiss characterized by sillimanite and locally much muscovite. SE New York, and W Connecticut. Also search Manhattan Schist.
BT Paleozoic
SA Connecticut
New York
New York City Group

Manicouagan Crater
Term introduced in 1981. Shock-metamorphic structure in E central Quebec. Before 1981, also search Manicouagan cryptoexplosion structure.
BT Quebec
Canada
SA cryptoexplosion features
Manicouagan Lake

Manicouagan Lake
Term introduced in 1981. Lake and reservoir on the Manicouagan River.
BT Quebec
Canada
SA Manicouagan Crater

Manicouagan River
Term introduced in 1981. E central Quebec.
BT Quebec
Canada

Manihiki Plateau
Undersea feature in the N Cook Islands area E of the Samoa Islands.
BT Pacific Ocean

Manila Trench
Just W of Luzon.
BT South China Sea
Pacific Ocean

Manildra
Village in E central.
BT New South Wales
Australia

Manistee County
On Lake Michigan in the Lower Peninsula.
BT Michigan
United States

Manitoba
Includes use on level 1 as an area term (list O). For term set options see list B.
CO N490000N600000
W0893000W1020000
BA Western Canada
Canada
NT Beresford Lake
Bernic Lake
Bird River
Churchill
Flin Flon
Fort Churchill
Knee Lake
Lake Winnipeg
Lynn Lake
Nelson River

Nelson River basin
Riding Mountain National Park
Setting Lake
Snow Lake
Steinbach
Tanco Pegmatite
Winnipeg
SA Amisk Group
Assiniboine River
Canadian Shield
Churchill Province
Elk Point Basin
Elk Point Group
English River Belt
Hudson Bay Lowlands
Lake Agassiz
Missi Group
Prairie Evaporite
Red River Formation
Reindeer Lake
Rice Lake
Rice Lake Group
Saskatchewan River
Superior Province
Williston Basin
Winnipeg Formation
Winnipegosis Formation

Manitou Formation
Consists of finely to coarsely crystalline limestone and minor amounts of dolomite. E Colorado.
BT Lower Ordovician
Ordovician
SA Colorado

Manitoulin Island
Largest lake island in world. In northern Lake Huron at NW end of Georgian Bay. Also search Manitoulin.
UF Grand Manitoulin
BT Ontario
Canada
SA Niagara Escarpment

Manitouwadge
Town N of Lake Superior.
BT Ontario
Canada

Manlius Formation
According to the latest published lexique, the age of Manlius Limestone (in Helderberg Group) is Lower Devonian.
BT Lower Devonian
Devonian
SA New York

Manning Park
Provincial park in Cascade Mountains E of Vancouver.
BT British Columbia
Canada

Mannville Formation
In the lower Mannville Formation, basal well-sorted quartzose sandstones were deposited as shoreline sediments or as channel-fills. In the upper Mannville, stratigraphic traps formed in part by intertonguing of sandstones and shales. Also search Manville Group.
UF Manville Formation
BT Lower Cretaceous
Cretaceous
SA Alberta
Canada

manpower
Term introduced in 1981.
SA economics

Mansalay Formation
Composed of hard calcareous mudstones, and calcareous and siliceous carbonaceous sandstone. S Mindoro Island.
BT Jurassic
SA Mindoro
Philippine Islands

Mansehra
Village in N part of country.
BT Punjab Province
Pakistan

Mansfeld Syncline
Also search Mansfeld.
BT Halle Bezirk
East Germany

Mansfield Formation
Lower and Middle Pennsylvanian. Contains large amounts of shale, thin beds of coal under clay, and limestone, and is only locally predominantly sandstone.
BT Pennsylvanian
SA Indiana
Lower Pennsylvanian
Middle Pennsylvanian

mantle
Includes use on level 1 (list A); on level 2 under seismology(1). Used for the zone of Earth below the crust and above the core (to a depth of 3480 km). If 1, term set options are: topic [age, composition, earthquakes, elastic waves, evolution, genesis, geochemistry, interpretation, low-velocity zones, processes, properties, temperature]
subtopic (no area term)
NA lower mantle
upper mantle
SA asthenosphere
continental drift
convection
convection currents
core
crust
degassing
discontinuities
Earth
earthquakes
elastic waves
geosynclines
geothermal gradient
heat flow
heat sources
horizontal discontinuities
hot spots
isostasy
lithosphere
low-velocity zones
lunar mantle
Mohorovicic discontinuity
partial melting
phase transitions
plate tectonics
plates
sea-floor spreading
seismology
tangential discontinuities
tectonophysics
transition zones
velocity structure
vertical discontinuities

manuals
Includes use on level 3 as a term appropriate to a large number of topics, e.g. under education(1). See list G. Also search manual.
SA guidebook
textbooks

Manville Formation
use Mannville Formation

Mapimi
Town in NE.
BT Durango
Mexico

mapping
A valid term through 1977. From 1978 on use cartography. To search, see note under cartography.

maps
Includes use on level 1 (list A); on level 2 under geomorphology(1), land use(1), and environmental geology(1); on level 3 under area(1) discipline from list B(2). Also search map. If 1, term set options are: cartography [for method, instrument, program]
subtopic
topic (for global maps; List B except for areal geology)
subtopic (no area term) e.g. type of map [e.g. economic geology, environmental geology, geologic maps, geomorphologic maps, hydrogeologic maps, photogeologic maps, soils maps, surficial geology, tectonic maps]
NA bathymetric maps
cave maps
contour maps
economic geology maps
geochemical maps
geologic maps
geomorphologic maps
geophysical maps
gravity survey maps
hydraulic maps
hydrogeologic maps
hydrologic maps
index maps
isograd maps
isopach maps
isopleth maps
isoseismic maps
lithologic maps
magnetic survey maps
paleogeographic maps
photogeologic maps
prediction maps
seismotectonic maps
shaded relief maps
soils maps
stratigraphic maps
structural maps
structure contour maps
tectonic maps
topographic maps
SA aerial photography
areal geology
atlas
automatic cartography
cartography
catalogs
coordinates
engineering geology
environmental geology
explanatory text
geodesy
geomorphology
geophysical methods
geophysical surveys
isograds
land use
legends
mineral exploration
Moon
photogeology
road log
structural geology
surficial geology
surveys

Maputo
New name for city of Lourenco Marques located on Delagoa Bay in S part of country.
BT Mozambique

Maquoketa Formation
In Missouri, it is typically thin laminated shale interbedded with shaly limestone members. W Illinois, E Iowa, S Minnesota, E Missouri, and SW Wisconsin.
BT Upper Ordovician
Ordovician
SA Illinois
Iowa
Minnesota
Missouri
Wisconsin

Maracaibo Basin
NW part of country. Index Lake Maracaibo and states as applicable.
BT Venezuela
SA Lake Maracaibo
Merida
Trujillo
Zulia

Maracaibo Falcon Plain
Term introduced in 1981.
CO N081000N121500
W0693000W0730000
BT Venezuela

Marajo
Largest island in Amazon Delta.
BT Para
Brazil

Maramures
County in N.
BT Transylvania
Romania

Maranhao
State in NE.
CO S103000S010000
W0415000W0490000
BT Brazil
NT Barreirinhas Basin
SA Brazilian Shield
Parnaiba Basin
Tocantins River region

Maranhao Basin
River basin in central.
BT Goias
Brazil

Marathon Basin
North of Big Bend National Park in W.
BT Texas
United States

Marathon County
Central Wisconsin.
BT Wisconsin
United States

Marathon Geosyncline
Term introduced in 1981. W Texas.
CO N294000N320000
W1020000W1043000
BT Texas
United States
SA Trans-Pecos

marble
As of 1981, no longer a valid term for GeoRef. Use marbles or marble deposits as applicable. Before 1981, included use on level 3 under metamorphic rocks(1) marbles(2) and on level 1 as a commodity term.

Marble Canyon
Gorge along the Colorado River. Often considered the upper part of Grand Canyon. Established as National Monument in 1969.
BT Arizona
United States

marble deposits
Term introduced in 1981. Includes use on level 1 and 2 as a commodity term (list C). Before 1981, search marble AND deposits.
SA building stone
calcite deposits
construction materials
dimension stone
dolostone deposits
limestone deposits
marbles

Marble Falls Group
Lower and middle Pennsylvanian. Comprises Sloan Formation below and Big Saline Formation above. Central Texas. Also search Marble Falls Formation.
BT Pennsylvanian
SA Lower Pennsylvanian
Middle Pennsylvanian
Texas

marbles
Includes use on level 2 under metamorphic rocks(1). See list J. Also search marble; crystalline limestone. As of 1981, use marble deposits for marble as a commodity.
BT metamorphic rocks
NA brucite marble
calciphyre
ophicalcite
ophite
SA dolomite
marble deposits

Marburg
City on Lahn River N of Frankfurt.
UF Marburg an der Lahn
BT Hesse
West Germany

Marburg an der Lahn
use Marburg

Marca Shale Member
Of Moreno Formation. Underlies Dos Palos Shale Member; overlies Tierra Loma Shale Member. S California.
BT Upper Cretaceous
Cretaceous
SA California
Moreno Formation

marcasite
BA sulfides
BT minerals

Marche
As of 1981, no longer a valid term for GeoRef. Referred to the historical region in central France.

Marches
Autonomous region in E central Italy. Also search The Marches.
CO N424200N435800
E0135700E0120800
UF The Marches
BT Italy
NT Ancona
Ascoli Piceno
Pesaro

Marcy Massif
Mt. Marcy. Highest peak in the Adirondack Mountains.
BT New York
United States

Mardin Formation
Cretaceous-Paleocene. Composed of clayey schist and marly sandstone. In Mardin in SE Anatolia.
BT Mesozoic
SA Cretaceous
Paleocene
Turkey

mare
use maria

Mare Crisium
use Sea of Crises

Mare Fecunditatis
use Sea of Fertility

Mare Imbrium
use Sea of Rains

Mare Nectaris
use Sea of Nectar

Mare Orientale
use Eastern Sea

Mare Serenitatis
use Sea of Serenity

Mare Spumans
use Foaming Sea

Mare Tranquillitatis
use Sea of Tranquillity

Mare Undarum
use Sea of Waves

Margarita Island
In the Caribbean Sea 15 miles off NE coast.
BT Nueva Esparta
Venezuela

margarite
BA mica group
sheet silicates
silicates
BT minerals

margin, continental
use continental margin

marginal basins
Includes use on level 3 under plate tectonics(1) evolution(2) and subduction(2).
SA basins
bottom features
continental slope
marginal seas
ocean basins
ocean floors
plate tectonics
subduction

marginal seas
Includes use on level 3 under plate tectonics(1) evolution(2) and subduction(2).
UF seas, marginal
SA continents
marginal basins
plate tectonics
subduction

marginal trench
use trenches

margins
No longer a valid term for GeoRef as of 1981. Use plate boundaries.

Marguerite Bay
Inlet on W coast of the Antarctic Peninsula between Adelaide and Alexander I Islands.
BT Antarctica

maria
Plural of mare. Includes use on level 3 under Moon(1). Also search mare.
UF mare
SA highlands
mascons
Moon

Mariana Islands
Group of islands 1500 miles E of Philippine Islands. The islands, not including Guam, formerly were part of the U.S. Trust Territory of the Pacific Islands. Commonwealth status within the U.S. was achieved on March 24, 1976. Includes use on level 1 as an area term (list O). For term set options see list B.
CO N120000N210000
E1470000E1440000
UF Marianas
Marianas Islands
BA Micronesia
NT Guam
Saipan
SA Pacific Ocean

Mariana Trench
Term introduced in 1981. E of the Mariana Islands.
CO N100000N223000
E1490000E1440000
BT Pacific Ocean

Mariana Trough
Term introduced in 1981. Basin W of the Mariana Islands.
CO N120000N223000
E1453000E1430000
BT Pacific Ocean

Marianas
use Mariana Islands

Marianas Islands
use Mariana Islands

Marianna Limestone
In Vicksburg Group. Consists of soft chalky limestone, locally called chimney rock; some local, hard limestone; tough to hard ledges in the chimney rock; sandy glauconitic limestone; marl; calcareous sand; and lignitic clay. S Alabama, W Florida, and S Mississippi.
BT middle Oligocene
Oligocene
SA Alabama
Florida
Mississippi
Vicksburg Group

Maricopa County
SW central Arizona.
BT Arizona
United States

Marie Byrd Land
Large section E of Ross Ice Shelf and Ross Sea and extending E to Ellsworth Land. Claimed for U.S. by Richard E. Byrd in 1924.
BT Antarctica

Mariinsk
City in Kemerovo Oblast on Trans-Siberian R.R.
BT Russian Republic
USSR

Marilia
City in W central.
BT Sao Paulo
Brazil

Marin County
Just N of San Francisco.
BT California
United States

Marinduque
Island and province just S of Luzon.
BT Philippine Islands

marine
A valid index term through 1977. After 1977, use marine environment for type of environment or marine methods for geophysical method, or use marine platforms for type of platform.

marine environment
Term introduced in 1978. Includes use on level 3 under ecology(1) and under sedimentation(1). Before 1978, also search marine.
SA ecology
environment
glaciomarine environment
ice shelves
marine sedimentation
paleoecology
sedimentation
spits
submarine environment
tidal flats

marine features
A valid term through 1975. Use shore features or ocean floors, or use the specific feature.

marine geology
Includes use on level 1 (list A). Used for that aspect of the study of the ocean which deals specifically with the ocean floor and the ocean-continent border. Term set options are: topic [applications, bibliography, bottom features, catalogs, concepts, education, experimental studies, geochemistry, history, instruments, methods, nomenclature, objectives, observations, practice, principles, textbooks, theoretical studies]
subtopic (no area term)
UF geological oceanography
SA bathymetry
benthonic taxa
bottom features
continental rise
continental shelf
continental slope
Deep Sea Drilling Project
environmental geology
geology
geophysical methods
levees
nodules
ocean basins
ocean circulation
ocean floors
ocean waves
oceanography
reefs
sea water
sea-floor spreading
sedimentary petrology
sedimentation
sedimentology
sediments
submarine canyons

marine gravimeters
Term introduced in 1981.
BT gravimeters

marine installations
Term introduced in 1976 and used on level 1 beginning in 1978. Before 1978, included use on level 2 under engineering geology(1). Term set options are:
construction
subtopic
design
subtopic
experimental studies
subtopic
feasibility studies
subtopic
foundations
subtopic
gravity platforms
subtopic
instruments
subtopic
marine platforms
subtopic
methods
subtopic
pipelines
subtopic
site exploration
subtopic
submarine installations
subtopic
theoretical studies
subtopic
UF installations, marine
NT breakwaters
gravity platforms
groins
jetties
marine platforms
piers
submarine installations
SA construction
design
engineering geology
explosions
feasibility studies
foundations
geologic hazards
nuclear facilities

pipelines
rock mechanics
shorelines
site exploration
soil mechanics
waste disposal
waterways

marine methods
Term introduced in 1978. Includes use on level 3 under geophysical methods(1). Before 1978, also search marine.
BA geophysical methods
SA geophysical surveys
methods
sonobuoys

marine platforms
Term introduced in 1978. Before 1978, search platforms AND marine.
BT marine installations
SA gravity platforms
platforms

marine sedimentation
Term introduced in 1981.
SA marine environment
marine sediments
sedimentation

marine sediments
Term introduced in 1981. Includes use on level 2 under sediments(1). See list N.
BT sediments
SA marine sedimentation

marine transport
Includes use on level 3 under sedimentation(1).
SA sedimentation
transport

Mariner Program
Term introduced in 1981. Includes use on level 3 under Mars(1).
SA extraterrestrial geology
Mars
planetology
remote sensing

Mariner 10
Includes use on level 3 under Mars(1).
SA Mars
remote sensing

Mariner 6
Includes use on level 3 under Mars(1).
SA Mars
remote sensing

Mariner 7
Includes use on level 3 under Mars(1).
SA Mars
remote sensing

Mariner 9
Includes use on level 3 under Mars(1).
SA Mars
remote sensing

Marion County
Index states as applicable.
BX Alabama
Arkansas
Florida
Georgia
Illinois
Indiana
Iowa
Kansas
Kentucky
Mississippi
Missouri
Ohio
Oregon
South Carolina
Tennessee
Texas
West Virginia
BT United States

Marion Island
Subantarctic island 1200 miles SE of Capetown and just SW of Prince Edward Islands. Annexed by South Africa.
BT Indian Ocean

Mariposa County
Central California.
BT California
United States

Maritime Alps
S division of Western Alps. Index countries as applicable.
BT Alps
Europe
SA France
Italy
Ligurian Alps
Piedmont Alps
Western Alps

Maritime Kray
use Primorye

Maritime Provinces
Often called the Maritimes. Index provinces as applicable. Introduced as level 1 area term in 1978.
UF Maritimes
BA Eastern Canada
Canada
NA New Brunswick
Nova Scotia
Prince Edward Island

Maritime Territory
use Primorye

Maritimes
use Maritime Provinces

Maritsa River
Rises in Bulgaria and flows into Aegean Sea. Also search Maritsa. Index countries as applicable.
UF Evros River
Meric River
BT Europe
SA Bulgaria
Greece
Turkey

Mariupol
use Zhdanov

marker beds
As of 1978, term is restricted for use in seismology(1). Before 1978, term also included use for key beds (stratigraphy).
UF beds, marker
SA key beds
seismology

market stabilization
Term introduced in 1981.
SA economics

markings, current
use current markings

Markov chain analysis
Term introduced in 1981.
BA statistical analysis

marl
Includes use on level 3 under sedimentary rocks(1) clastic rocks(2). See list I. Before 1981, also included use on level 3 under sedimentary rocks(1) carbonate rocks(2).
UF calcareous clay
marlstone
BA clastic rocks
BT sedimentary rocks
SA argillaceous texture
carbonate rocks
clay
loess
terrigenous materials

marlstone
use marl

Marmaton Group
Includes (ascending) Fort Scott, Labette, Pawnee, Bandera, Altamont, Nowata, and Lenapan formations. SW Iowa, E Kansas, W Missouri, and NE Oklahoma.
BT Desmoinesian
Middle Pennsylvanian
Pennsylvanian
SA Iowa
Kansas
Labette Shale
Missouri
Oklahoma
Wewoka Formation

Marmolada Glacier
At Marmolada Peak, the highest in the Dolomites. Index Autonomous regions as applicable.
BT Italy
SA Trentino-Alto Adige
Veneto

Marne
Department in NE.
CO N483000N493000
E0051000E0031500
BT France

Marne Valley
River valley in NE central France. Also search Marne River.
BT France

Maroon Town
Town in NW.
BT Jamaica
West Indies

Marquesas Islands
Term introduced in 1981. Group of ten islands in French Polynesia.
CO S110000S080000
W1380000W1420000
BT French Polynesia
Polynesia

Marquette
City on Lake Superior in Marquette County, Upper Peninsula.
BT Michigan
United States

Marquette County
Term introduced in 1981. Index states as applicable.
BX Michigan
Wisconsin
BT United States

Marquette Iron Range
use Marquette Range

Marquette Range
Iron range in Marquette County in Upper Peninsula.
UF Marquette Iron Range
BT Michigan
United States

Mars
Includes use on level 1 and 2. See entry under Moon(1) for term set options.
NT Polar Cap
Tharsis
SA Deimos
landing sites
Mariner Program
Mariner 10
Mariner 6
Mariner 7
Mariner 9
marsquakes
Moon
Phobos
planetology
planets
satellites
solar system
terrestrial planets
Viking Program

Marseilles
City on Gulf of Lion.
BT Bouches-du-Rhone
France

Marseilles Basin
Term introduced in 1981. Structural basin in SE France.
CO N431000N432500
E0054000E0050000
BT France

Marshall County
Index states as applicable.
BX Alabama
Illinois
Indiana
Iowa
Kansas
Kentucky
Minnesota
Mississippi
Oklahoma
South Dakota
Tennessee
West Virginia
BT United States

Marshall Islands
Group of 34 atolls and coral islands, including Eniwetok and Kwajalein, SW of Hawaii and E of Guam. Part of the U.S. Trust Territory of the Pacific Islands. Includes use on level 1 as an area term (list O). For term set options see list B.
CO N043000N150000
E1730000E1610000
BA Micronesia
NT Eniwetok Atoll
SA Pacific Ocean

Marshalltown Formation
In Matawan Group. Overlies Englishtown Formation; underlies Wenonah Sand.
BT Upper Cretaceous
Cretaceous
SA Delaware
New Jersey

marshes
Includes use on level 3 under geomorphology(1). As of 1981, use paludal environment for type of environment.
SA bogs
geomorphology
paludal environment
salt marshes
swamps

Marsica
Mountain range in E central part of country.
BT Abruzzi
Italy

marsquakes
Term introduced in 1981. Includes use on level 3 under Mars(1).
SA earthquakes
Mars

Marsupialia
Includes use on level 2 under Mammalia(1). See list F.
BA Mammalia
BT Tetrapoda
Vertebrata
NA Macropodidae

Martha's Vineyard
Island in Atlantic Ocean off SW coast of Cape Cod.
BT Massachusetts
United States

Martin Formation
Composed of three members (ascending): conglomerate sandstone

and dolomite limestone member; sandstone and limestone member; and calcareous sandstone, sandy limestone and shale member. Central and E Arizona.
BT Devonian
SA Arizona
Middle Devonian
Upper Devonian

Martin Lake
Formed by dam across Tallapoosa River in E central.
BT Alabama
United States

Martin River Glacier
NE of Cordova in S.
BT Alaska
United States

Martinique
Island and department of France in Windward Islands.
BA Windward Islands
BT Lesser Antilles

Martinsburg Formation
According to the latest published Lexique, the age of Martinsburg Shale is middle and upper Ordovician. In Pennsylvania, comprises (ascending) Cocalio Shale, unnamed and undifferentiated dark shales, Jonestown Beds (new), and Fairview Shochary Sandstones. Maryland, New Jersey, SE Pennsylvania, Tennessee, W Virginia, and West Virginia. Also search Martinsburg Shale.
BT Ordovician
SA Maryland
Middle Ordovician
New Jersey
Pennsylvania
Upper Ordovician
Virginia
West Virginia

martite
BA oxides
BT minerals
SA hematite

Maryborough Basin
SE Queensland. Also search Maryborough.
BT Queensland
Australia

Maryland
Includes use on level 1 as an area term (list O). For term set options see list B.
CO N375500N394300
W0750500W0793000
BA Eastern U.S.
United States
NX Allegany County
NT Anne Arundel County
Baltimore
Baltimore County
Calvert County
NX Carroll County
NT Charles County
NX Cumberland
NT Harford County
NX Howard County
Kent County
Montgomery County
NT Patuxent River
Prince Georges County
Saint Marys County
Savage River
NX Somerset County
Washington County
Worcester County
SA Allegheny Front
Allegheny Group
Allegheny Mountains
Ames Limestone
Anacostia River basin
Antietam Formation
Appalachian Basin
Appalachian Plateau
Aquia Formation
Atlantic Coastal Plain
Baltimore Gneiss
Bloomsburg Formation
Blue Ridge Province
Calvert Formation
Carolina Bays
Catoctin Formation
Catskill Delta
Catskill Formation
Chesapeake Bay
Chilhowee Group
Choptank Formation
Choptank River
Clinton Group
Coeymans Formation
Conemaugh Group
Delmarva Peninsula
Dunkard Group
Genesee Group
Glenarm Series
Greenland Gap Group
Hamilton Group
Helderberg Group
Juniata Formation
Kinzers Formation
Magothy Formation
Martinsburg Formation
McKenzie Formation
Monmouth Group
Monongahela Group
Nanjemoy Formation
Newark Group
Newark-Gettysburg Basin
Onondaga Limestone
Oriskany Sandstone
Piedmont
Pocono Formation
Potomac Group
Potomac River
Potomac River basin
Pottsville Group
Raritan Formation
Rochester Formation
Saint Marys Formation
Sharon Conglomerate
South Mountain
Susquehanna River
Susquehanna River basin
Tonoloway Limestone
Tuscarora Formation
Valley and Ridge Province
Wicomico Formation
Williamsport Sandstone
Wissahickon Formation
Yorktown Formation

Marysvale
City in Piute County in S central.
BT Utah
United States

Mascarene Basin
E of N Malagasy Republic.
BT Indian Ocean

Mascarene Islands
Group between 400 and 500 miles E of the Malagasy Republic. Index islands as applicable.
BT Indian Ocean
NT Rodriguez Island
SA Mauritius
Reunion

mascons
Includes use on level 3 under Moon(1).
SA maria
Moon

Mascot Dolomite
In Knox Group. Consists of light- and dark-gray dolomite and limestone; moderately cherty; base marked by chert matrix sandstone. E Tennessee and Virginia. Index states as applicable.
BT Lower Ordovician
Ordovician
SA Knox Group
Tennessee
Virginia

maskelynite
BT minerals
SA aluminosilicates
glasses
plagioclase

Mason County
Index states as applicable.
BX Illinois
Kentucky
Michigan
Texas
Washington
West Virginia
BT United States

mass
Includes use on level 3 under Earth(1).
SA mass balance
mass spectra
mass spectroscopy

mass balance
Includes use on level 3 under glacial geology(1) glaciers(2). Also search balance.
UF balance, mass
mass budget
SA ablation
glacial geology
glaciers
mass

mass budget
use mass balance

mass movements
Includes use on level 2 under geomorphology(1). Before 1981, also search mass wasting.
UF movements, mass
SA creep
flows
geomorphology
lahars
landslides
liquefaction
mudflows
rockslides
slope stability
slumping

mass spectra
Term introduced in 1981.
SA mass
mass spectroscopy
spectra

mass spectrometry
A valid term through 1973. After 1973, see methods(2) under chemical analysis(1), and spectroscopy(1). After 1977, use mass spectroscopy.

mass spectroscopy
Not a valid term from 1972 through 1977. After 1977, includes use on level 3 under spectroscopy(1) methods(2) and under chemical element (1). Before 1978, also search mass spectrometry; spectroscopy AND mass.
BT spectroscopy
SA mass
mass spectra

mass transfer
As of 1981, use only in the chemical sense. Before 1981, included use as a general term.
UF transfer, mass
SA geochemistry

mass wasting
As of 1981, no longer a valid term for GeoRef. Use mass movements.

Massachusetts
Includes use on level 1 as an area term (list O). For term set options see list B.
CO N411500N425500
W0695500W0733000
BA New England
Eastern U.S.
United States
NT Barnstable County
Berkshire County
Berkshire Hills
Boston
Buzzards Bay
NX Cambridge
NT Cape Cod
NX Essex County
Franklin County
NT Hampden County
NX Hampshire County
NT Martha's Vineyard
NX Middlesex County
NT Monomoy Island
Nantucket Island
NX Suffolk County
NT Wachusett-Marlborough Tunnel
Ware
Woods Hole
Woods Hole Oceanographic Institution
Worcester
NX Worcester County
SA Atlantic Coastal Plain
Bronson Hill Anticlinorium
Connecticut River
Connecticut Valley
Green Mountains
Littleton Formation
Merrimack River valley
Merrimack Synclinorium
Normanskill Formation
Springfield

Massiac
Village in S central part of country.
BT Cantal
France

Massif Central
use Central Massif

massifs
As of 1981, no longer a valid term for GeoRef. Use complexes.

massive
A valid index term through 1977. After 1977, use massive bedding for the sedimentary structure, and use massive deposits for ore deposits.

massive bedding
Term introduced in 1978. Includes use on level 3 under sedimentary structures(1) planar bedding structures(2). Before 1978, also search massive AND bedding.
BA planar bedding structures
sedimentary structures
SA bedding

massive deposits
Term introduced in 1978. Includes use on level 3 for ore deposits. Before 1978, also search massive AND ore deposits; massive AND deposits.
SA mineral deposits

massive hydraulic fracturing
Term introduced in 1981.
SA hydraulic fracturing

Mastodon
Genus. As of 1981, includes use on level 3 under Mammalia(1) Mastodontoidea(2). See list F. Before 1981, included use on level 3 under Mammalia(1) Proboscidea(2). Also search Mammut.

UF Mammut
BA Mastodontidae
Mastodontoidea
Proboscidea
Mammalia
BT Tetrapoda
Vertebrata

Mastodontidae
Family. As of 1981, includes use on level 3 under Mammalia(1) Mastodontoidea(2). See list F. Before 1981, included use on level 3 under Mammalia(1) Proboscidea(2). Also search Mammutidae.
UF Mammutidae
BA Mastodontoidea
Proboscidea
Mammalia
BT Tetrapoda
Vertebrata
NA Mastodon

Mastodontoidea
Term introduced in 1981. Includes use on level 2 under Mammalia(1). See list F.
BA Proboscidea
Mammalia
BT Tetrapoda
Vertebrata
NA Mastodontidae

Matadi
Port city on Congo River in W part of county. Bas-Zaire.
BT Zaire

Matagami
Lake SE of James Bay.
UF Mattagami
BT Quebec
Canada

Matagamon Sandstone
BT Lower Devonian
Devonian
SA Maine

Matagorda Bay
Inlet of the Gulf of Mexico in SE.
BT Texas
United States

Matanuska Valley
River valley NE of Anchorage in S Alaska. Also search Matanuska River.
BT Alaska
United States

Matera
City and province in S part of country.
BT Basilicata
Italy

materials
Includes use on level 2 under diagenesis(1) and in combination with properties (i.e. materials, properties) on level 2 under engineering geology(1), rock mechanics(1), and soil mechanics(1).
SA absorbent materials
amorphous materials
anelastic materials
anisotropic materials
banded materials
cement materials
ceramic materials
cohesionless materials
cohesive materials
colloidal materials
competent materials
conductive materials
consistent materials
construction materials
continuous materials
diagenesis
discontinuous materials
dissolved materials
education
elastic materials
elastoplastic materials
engineering geology
expansive materials
fossiliferous materials
granular materials
heterogeneous materials
inorganic materials
isotropic materials
layered materials
linear materials
natural materials
non-linear materials
organic materials
overconsolidated materials
parent materials
particulate materials
plastic materials
porous materials
properties
refractory materials
rock mechanics
soil mechanics
standard materials
suspended materials
synthetic materials
terrigenous materials
toxic materials
unconsolidated materials
viscoelastic materials
viscoplastic materials
viscous materials

mathematical geology
For mathematics as applied to geology. Includes use on level 1 (list A). Term set options are:
topic [bibliography, concepts, education, history, interpretation, methods, nomenclature, objectives, philosophy, principles, symposia, textbooks, theoretical studies]
subtopic
SA algorithms
analog simulation
automatic data processing
biometry
canonical analysis
cluster analysis
discriminant analysis
equations
factor analysis
finite difference analysis
finite element analysis
Fortran IV
Fourier analysis
functions
geology
geometry
geophysics
iterative methods
mathematical methods
mathematical models
multivariate analysis
regression analysis
simulation
statistical analysis
trend-surface analysis

mathematical methods
Used as a general term. Before 1981, also search quantitative methods. Before 1978, also search quantitative; quantitative AND methods.
SA autocorrelation
eigenvalues
finite difference analysis
functions
iterative methods
least-squares analysis
mathematical geology
mathematical transformations
methods
numerical analysis
probability
statistical analysis

mathematical models
BA models
SA automatic data processing
infinite models
iterative methods
mathematical geology
semi-infinite models

mathematical transformations
Term introduced in 1981.
SA mathematical methods
seismology

matildite
Also search schapbachite.
UF plenargyrite
schapbacite
BA sulfobismuthites
sulfosalts
BT minerals

Mato Grosso
State in SW central Brazil.
CO S240000S072000
W0501000W0613000
UF Matto Grosso
BT Brazil
SA Botucatu Formation
Brazilian Shield
Irati Formation

Matra Mountains
S spur of the Carpathians. Also search Matra.
BT Hungary
SA Carpathians

matrix
The fine-grained interstitial material of an igneous rock. Also search ground mass; groundmass.
UF ground mass
groundmass
SA igneous rocks
megacrysts
phenocrysts

Matsukawa
Town in N central.
BT Honshu
Japan

Matsushiro
Town in central.
BT Honshu
Japan

Matsuyama
City in W near Inland Sea.
BT Shikoku
Japan

Mattagami
use Matagami

Matto Grosso
use Mato Grosso

Mattoon Formation
McLeansboro Group. Predominantly shale and sandstone.
BT Pennsylvanian
SA Illinois

maturity
Used as a general term.

Matuyama Epoch
Geomagnetic epoch.
UF Matuyama Reversed
BA Cenozoic
SA lower Pleistocene
Neogene
paleomagnetism
Quaternary
reversals
upper Pliocene
upper Tertiary

Matuyama Reversed
use Matuyama Epoch

maucherite
As of 1981, includes use on level 3 under minerals(1) arsenides(2). See list L. Before 1981, included use on level 3 under minerals(1) sulfides(2).
BA arsenides
sulfides
BT minerals

Maui
Island in the Hawaiian Islands NW of the island of Hawaii.
BT Hawaii
United States

Maumee River valley
NE Indiana and NW Ohio. Index states as applicable. Also search Maumee River.
BT United States
SA Indiana
Ohio

Mauna Kea
Extinct volcano on N central island of Hawaii.
CO N193000N194000
W1552000W1553500
UF Mauna Kea Volcano
BT Hawaii
United States

Mauna Kea Volcano
use Mauna Kea

Mauna Loa
Volcano on S central island of Hawaii.
CO N192000N193500
W1552000W1554000
UF Mauna Loa Volcano
BT Hawaii
United States

Mauna Loa Volcano
use Mauna Loa

Maures Massif
On the Mediterranean coast at W end of the Riviera. Also search Maures; Monts des Maures.
UF Monts des Maures
BT Var
France

Mauritania
Includes use on level 1 as an area term (list O). For term set options see list B.
CO N145500N273000
W0045500W0173000
BA West Africa
Africa
NT Adrar
Akjoujt
Aoulloul
Fort Gouraud
Nouakchott
Richat Mountain
SA Cap Blanc
Reguibat Ridge
Sahara
Sahel
Senegal Basin
Senegal River

Mauritius
An independent state in the Mascarene Islands about 450 miles E of the Malagasy Republic. Includes use on level 1 as an area term (list O). For term set options see list B.
CO S204000S195000
E0580000E0570000
BA Indian Ocean Islands
SA Mascarene Islands

Maury Channel
Sea channel E of central Reykjanes Ridge and S of Iceland.
BT Atlantic Ocean

Maury County
Term introduced in 1981. Central Tennessee.
CO N352300N355000
W0864700W0872300
BT Tennessee

United States

Maverick Basin
Term introduced in 1981. Structural basin in SW Texas and NE Mexico.
CO N280000N293000
W0992500W1013000
BT North America
SA Mexico
Texas

Maverick County
On the Rio Grande in SW.
BT Texas
United States

mawsonite
BA sulfides
BT minerals

maximum entropy analysis
Term introduced in 1981.
SA seismology

May-sur-Orne
Village on the Orne River in NW part of country.
BT Calvados
France

Maya Mountains
Range in S Belize. Also search Maya AND appropriate area term.
BT Belize
Central America

Maya River basin
Chiefly in central Khabarovsk Kray but also in Yakutia. Also search Maya River.
BT Russian Republic
USSR

Mayaguez Bay
Off city of Mayaguez in W.
BT Puerto Rico
West Indies

Mayenne
Department in NW.
CO N474500N483500
W0000200W0011000
BT France

Maykop
use Maikop

Maymecha
River. Rises in Evenk National Okrug and flows N into the Kheta River in the Taymyr National Okrug. Also search Maimecha.
UF Maimecha
BT Russian Republic
United States

Maymecha Kotuy
use Maymecha-Kotuy

Maymecha-Kotui
use Maymecha-Kotuy

Maymecha-Kotuy
Region in valleys of the Maymecha, Kotuy, and Kheta rivers in the Taymyr and Evenki national okrugs. Also search Maimecha-Kotui.
CO N700000N711500
E1033000E0993000
UF Maimecha-Kotui
Maymecha Kotuy
Maymecha-Kotui
BT Russian Republic
USSR

Mayo
County in NW.
CO N533000N542000
W0084000W0101500
BT Ireland
NT Keel

Mayurbhanj
District in E part of country.
BT Orissa
India

Mazama Ash
At site of Mount Mazama, a prehistoric volcanic mountain in the Cascade Mountains of Klamath County in S Oregon. Caldera now occupied by Crater Lake.
BT Oregon
United States

Mazatlan
City on the Pacific Ocean.
BT Sinaloa
Mexico

Mazon Creek
NE Illinois.
BT Illinois
United States

McHenry County
Index states as applicable.
BX Illinois
North Dakota
BT United States

McIntosh County
Index states as applicable.
BX Georgia
North Dakota
Oklahoma
BT United States

McKenzie County
Term introduced in 1981. W North Dakota.
CO N471900N480800
W1023400W1040300
BT North Dakota
United States

McKenzie Formation
In Maryland, it includes Rabble Run Sandstone Member. Underlies Bloomsburg Formation and overlies Rochester Formation. W Maryland, central Pennsylvania, N Virginia, and NE West Virginia.
BT Middle Silurian
Silurian
SA Maryland
Pennsylvania
Virginia
West Virginia

McKinley County
NW New Mexico.
BT New Mexico
United States
NT Ambrosia Lake District
Fort Wingate
Grants District

McLean County
Index states as applicable.
BX Illinois
Kentucky
North Dakota
BT United States

McMurdo Ice Shelf
Part of Ross Ice Shelf in McMurdo Sound area in Ross Dependency on the Pacific Ocean side.
BT Antarctica
SA Ross Ice Shelf

McMurdo Sound
Inlet of SW Ross Sea at the edge of the Ross Ice Shelf beween Ross Island and coast of Victoria Land. Site of major U.S. research and exploration base in Ross Dependency on the Pacific Ocean side.
BT Antarctic Ocean

McPherson County
Index states as applicable.
BX Kansas
Nebraska
South Dakota
BT United States

McPherson Formation
Consists of early Pleistocene stream deposits, later and coarser Pleistocene stream channel deposits, and still later Pleistocene silt, clay, and fine sand. Central Kansas.
BT Pleistocene
SA Kansas

Meade Basin
River basin S of Point Barrow in N. Coal mining in area.
BT Alaska
United States

Meade County
Index states as applicable.
BX Kansas
Kentucky
South Dakota
BT United States

Meadow soils
See list M.
BA soils
SA Intrazonal soils

Meagher County
W central Montana.
BT Montana
United States

meanders
Includes use on level 3 under geomorphology(1) fluvial features(2).
BT fluvial features
SA floodplains
geomorphology
rivers
rivers and streams
sinuosity
streams

measurement
Includes use on level 2 under heat flow(1).
SA conductivity
geodesy
geothermal gradient
heat flow
satellite measurements
thermal conductivity

Meath
County on the Irish Sea in NE.
BT Ireland

mechanical controls
Includes use on level 3 under mineral deposits, genesis(1) controls(2).
SA controls
mineral deposits

mechanical erosion
use abrasion

mechanical properties
SA anisotropy
creep
deformation
elastic limit
elastic properties
elasticity
engineering geology
friction
hardness
physical properties
plasticity
Poisson's ratio
properties
rigidity
shear strength
soil mechanics
strain
strength
tensile strength
weathering
yield strength

mechanical weathering
Term introduced in 1978. Before 1978, also search weathering AND mechanical.
BT weathering
SA insolation
physical weathering

mechanics
Includes use on level 2 under faults(1) and folds(1); on level 3 under plate tectonics(1).
SA decollement
faults
flexural-slip
folds
land subsidence
plate tectonics
rock mechanics
shear
soil mechanics
stick-slip

mechanism
Includes use as a level 2 or 3 term appropriate to a large number of topics, e.g. on level 2 under orogeny(1), under salt tectonics(1), and under plate tectonics(1). See list G.
SA flow mechanism
focal mechanism

Mechelen
use Malines

Mechlin
use Malines

Mecklenburg
Region on the Baltic Sea in N. A former German state.
BT East Germany

Mecklenburg County
Term introduced in 1981. Index states as applicable.
BX North Carolina
Virginia
BT United States

Mecopteroida
Term introduced in 1981. Includes use on level 2 under Insecta(1). See list F.
BA Insecta
BT Mandibulata
Arthropoda
Invertebrata

Mecsek Mountains
Near Yugoslav border in S part of country. In Baranya. Also search Mecsek.
UF Baranya Mountains
BT Hungary

media
As of 1981, no longer a valid term for GeoRef. Use materials.

median masses
As of 1981, no longer a valid term for GeoRef. See tectonics; orogenic belts.

Median Tectonic Line
Southwest Japan.
SA Japan
plate tectonics
tectonics

median valley
UF valley, median
SA Mid-Atlantic Ridge
mid-ocean ridges
rift zones
rifting

medical geology
Also search health.
UF health
SA environmental geology
geology
human ecology
pollution

medicinal waters
Term introduced in 1981.
SA mineral waters
thermal waters
water

Medicine Bow Mountains
Range of the Rocky Mountains. In-

dex states as applicable.
BT United States
SA Colorado
Rocky Mountains
Wyoming

Medicine Hat
City in SE.
BT Alberta
Canada

Medicine Lake
In Siskiyou County, N California.
BT Siskiyou County
California
United States

Medina Formation
Also search Medina Group.
UF Medina Group
BT Lower Silurian
Silurian
SA Michigan
New York
Ontario

Medina Group
use Medina Formation

Mediterranean (region)
use Mediterranean region

Mediterranean region
The Mediterranean Sea and its islands, the Adriatic Sea, and parts of those countries along their shores in S Europe, N Africa, and the Middle East. Index Adriatic Sea, Mediterranean Sea, countries, and islands as applicable. Also search Mediterranean. Includes use on level 1 or 2 as an area term (list O). If 1, see term set options under list B.
CO N300000N473000
E0380000W0050000
UF Mediterranean (region)
NT Aegean Islands
Ionian Islands
SA Adriatic Sea
Albania
Algeria
Corsica
Cyprus
Egypt
France
Greece
Israel
Italy
Lebanon
Libya
Malta
Mediterranean Sea
Mesogaea
Morocco
Sardinia
Sicily
Spain
Syria
Tunisia
Turkey
Yugoslavia

Mediterranean Ridge
Between Crete and the coast of E Libya.
BT Mediterranean Sea

Mediterranean Sea
Enclosed by Europe on the W and N, Asia on the E, and Africa on the S. Includes use on level 1 or 2 as an area term (list O). If 1, see term set options under list B. As of 1981, includes the Black Sea.
CO N301000N464000
E0420000W0053000
NT Balearic Basin
East Mediterranean
El Alboran Island
Florence Rise
Gulf of Pozzuoli
Ligurian Sea
Mediterranean Ridge
Minorca Rise
Strait of Sicily
Tyrrhenian Basin
West Mediterranean
SA DSDP Site 125
DSDP Site 132
DSDP Site 372
DSDP Site 374
Leg 13
Leg 42A
Mediterranean region
Sardinia
Turkey

Medo
use Tokyo

Medoc
District between the Gironde River and the Bay of Biscay in NW.
BT Gironde
France

meerschaum
use sepiolite

Meerut
City, district, and division in N part of country.
BT Uttar Pradesh
India

meetings
A valid term through 1975. After 1975, use associations or symposia. Also search meeting.

megacrysts
SA igneous rocks
matrix
metamorphic rocks
textures

megacyclothems
See list K.
BA cyclothems
planar bedding structures
sedimentary structures

Megalopolis Basin
In Arcadia Department in central Peloponnesus. Also search Megalopolis.
BT Peloponnesus
Greece

megaripples
Includes use on level 3 under sedimentary structures(1) bedding plane irregularities(2). See list K.
BA bedding plane irregularities
sedimentary structures
SA sand waves

megaspores
Includes use on level 2 under palynomorphs(1). See list F.
BA palynomorphs
SA miospores
spores

Megion Field
Oil field in Khanty-Mansino National Okrug in West Siberian Plain. Also search Megion.
UF Megion Oil Field
BT Russian Republic
USSR

Megion Oil Field
use Megion Field

Megri
Town in S near Iranian border.
BT Armenia
USSR

Megrinskiy Pluton
In Megrinskiy Mountains in S.
BT Armenia
USSR

Meguma Group
Composed of Goldenville and Halifax formations. In Nova Scotia.
BT Ordovician
SA Canada
Nova Scotia

Mehadia
Village at W edge of the Transylvanian Alps.
BT Banat
Romania

Mehedinti Plateau
SW part of country. Index regions as applicable.
BT Romania
SA Banat
Walachia

Meighen Island
W of Axel Heiberg Island in District of Franklin.
BT Northwest Territories
Canada

meimechite
As of 1981, includes use on level 3 under igneous rocks(1) ultramafics(2). See list H. Before 1981, included use on level 3 under igneous rocks(1) ultramafic family(2). Before 1975, also search meymechite.
UF meymechite
BA ultramafics
BT igneous rocks

meionite
BZ carbonates
scapolite group
BT minerals
SA chlorides
sulfates

Mekong Delta
On the South China Sea S of Saigon.
BT Vietnam

Melanesia
Collective name for islands NE of Australia. Index island groups as applicable. Includes use on level 1 and 2 as an area term (list O). For term set options see list B.
CO S230000S050000
W1710000E1550000
BT Oceania
NA Fiji
New Caledonia
New Hebrides
Solomon Islands
SA Bismarck Archipelago
Pacific Ocean

melange
Includes use on level 3 under deformation(1), and under structural analysis(1); on level 3 under plate tectonics(1).
UF block clay
SA boudinage
deformation
plate tectonics
structural analysis

melanite
BA garnet group
nesosilicates
orthosilicates
silicates
BT minerals

melanocerite
BA orthosilicates
silicates
BT minerals

melanophlogite
BA silica minerals
framework silicates
silicates
BT minerals

melanterite
BA sulfates
BT minerals

melaphyre
As of 1981, includes use on level 3 under igneous rocks(1) basalts(2). See list H. Before 1981, included use on level 3 under igneous rocks(1) basalt family(2).
BA basalts
volcanic rocks
BT igneous rocks
SA basalt
porphyry

Melbourne
City on the Tasman Sea in SE part of country.
BT Victoria
Australia

Melbourne region
Term introduced in 1981. Also search Melbourne.
CO S381000S373000
E1452000E1443000
BT Victoria
Australia

Meldon Aplite
Nothumberland County in extreme NE.
BT England
Great Britain
United Kingdom

melilite
UF mellilite
BA melilite group
sorosilicates
orthosilicates
silicates
BT minerals

melilite group
Includes use in combination with orthosilicates (i.e. orthosilicates, melilite group) on level 2 under minerals(1). See list L.
BA sorosilicates
orthosilicates
silicates
BT minerals
NA akermanite
gehlenite
melilite

melilitite
As of 1981, includes use on level 3 under igneous rocks(1) volcanic rocks(2). See list H. Before 1981, included use on level 3 under igneous rocks(1) alkali basalt family(2) and ultramafic family(2).
BA volcanic rocks
BT igneous rocks
SA ultramafics

Melilla
Term introduced in 1981.
BT Morocco

mellilite
use melilite

melnikovite
use greigite

Melones Fault
Term introduced in 1981. Major structural feature in the Sierra Nevada of California.
CO N372500N400700
W1195000W1211500
BT California
United States
SA Sierra Nevada

Melosira
BA diatoms
algae
BT thallophytes
Plantae

melt water
use meltwater

melteigite
As of 1981, includes use on level 3 under igneous rocks(1) alkali gabbros(2). See list H. Before 1981, included use on level 3 under igneous rocks(1) alkali gabbro family(2).
BA alkali gabbros
gabbros
BT igneous rocks
SA ijolite
urtite

melting
Includes use on level 2 under phase equilibria(1).
SA fusion
partial melting
phase equilibria

melting relations
A valid term through mid-1978. Use melting or phase equilibria.

melts
Includes use on level 3 under phase equilibria(1).
SA liquid phase
magmas
partial melting
petrology
phase equilibria

meltwater
Includes use on level 3 under glacial geology(1) glaciers(2). Also search melt water.
UF melt water
SA glacial geology
glaciers
hydrology
ice
snow
water

Melville Peninsula
N projection of Canadian mainland in District of Franklin between Committee Bay on the W and Foxe Basin on the E.
BT Northwest Territories
Canada

Memel River basin
use Neman River basin

menaccanite
use ilmenite

Mendeleyev Ridge
N of Alaska and E of Siberia.
BT Arctic Ocean

Mendeleyev Volcano
On Kunashir, southernmost of the Kuril Islands. Also search Medeleyeva Volcano.
UF Mendeleyeva Volcano
BT Russian Republic
USSR

Mendeleyeva Volcano
use Mendeleyev Volcano

Mendip Hills
SW England. Also search Mendip.
BT England
Great Britain
United Kingdom

Mendocino (Cape)
use Cape Mendocino

Mendocino County
Along Pacific Ocean in NW. Also search Mendocino.
BT California
United States

Mendocino fracture zone
Running in an E-W direction approximately 1250 miles W of N California.
BT Pacific Ocean

Mendoza
Province in W.
CO S380000S330000
W0660000W0710000
BT Argentina
NT Payun Matru
SA Salado Basin

Menefee Formation
In Mesaverde Group. Includes Cleary Coal Member and above that a bed formerly called Allison Barren Member. SW Colorado, and NW New Mexico.
BT Upper Cretaceous
Cretaceous
SA Colorado
Mesaverde Group
New Mexico

meneghinite
BA sulfantimonites
sulfosalts
BT minerals

Menorca
use Minorca

Menorca Rise
use Minorca Rise

Mentor Beds
According to the latest published Lexique, the age of Mentor Sandstone Member (or Belvedere Formation) is Lower Cretaceous. In central Kansas.
BT Comanchean
Cretaceous
SA Kansas
Lower Cretaceous

Meotian
European stage, Black Sea area. Above Sarmatian, below Cimmerian. Includes use on level 3 under age terms(1). See list E.
BA upper Miocene
Miocene
Neogene
Tertiary
BT Cenozoic
SA lower Pliocene
Pliocene

Meramecian
Provincial series, North America. Above Osagian, below Chesterian. Includes use on level 3 under age terms(1). See list E.
BA Upper Mississippian
Mississippian
Carboniferous
BT Paleozoic
NT Saint Louis Limestone
Sainte Genevieve Limestone
Salem Limestone
Warsaw Formation
SA Chesterian
Valmeyeran

Merano
City in Bolzano Province in NE part of country.
BT Trentino-Alto Adige
Italy

Mercalli scale, modified
use modified Mercalli scale

Merced County
W center.
BT California
United States

Mercenaria
Genus. Includes use on level 3 under Mollusca(1). See list F.
BA Bivalvia
Mollusca
BT Invertebrata

Mercer County
Index states as applicable.
BX Illinois
Kentucky
Missouri
New Jersey
North Dakota
Ohio
Pennsylvania
West Virginia
BT United States

mercury
Includes use on level 1 and 2 as a chemical element (list D). As of 1981, use mercury ores for mercury as a commodity. Use Mercury Planet for the planet. Also search quicksilver. Before 1981, included use on level 1 as a commodity term.
UF Hg
quicksilver
SA cinnabar
elements
mercury ores
Mercury Planet
native elements

mercury ores
Term introduced in 1981. Includes use on level 1 and 2 as a commodity term (list C). Before 1981, search mercury AND ore deposits.
SA mercury
Mercury Planet
native elements

Mercury Planet
Includes use on level 1 and 2. See entry under Moon(1) for term set options. Before 1972, also search Mercury in combination with the category (extraterrestrial geology).
SA mercury
mercury ores
planetology
planets
solar system
terrestrial planets

Meric River
use Maritsa River

Merida
State in W Venezuela, and city in Yucatan in N part of Yucatan Peninsula in Mexico. Index countries as applicable.
SA Maracaibo Basin
Mexico
Venezuela

Merionethshire
County in W.
CO N523000N530000
W0031500W0041000
BT Wales
Great Britain
United Kingdom

Merostomata
Includes use on level 2 under Arthropoda(1). See list F.
BA Chelicerata
Arthropoda
BT Invertebrata
NA Eurypterida
Xiphosura

Merrimack River valley
Index states as applicable.
BT United States
SA Massachusetts
New Hampshire

Merrimack Synclinorium
Index states as applicable.
BT United States
SA Maine
Massachusetts
New Hampshire

Merseburg
City in SW central.
BT Halle Bezirk
East Germany

Mersey River
Name of rivers in N Tasmania and in NW England. Index England or Australia as applicable.
SA Australia
England
Mersey Valley

Mersey Valley
River valley in NW England and in Tasmania.
SA Australia
England
Mersey River

merwinite
BA orthosilicates
silicates
BT minerals

merzlota
use frozen ground

Mesa County
W central Colorado.
BT Colorado
United States

Mesabi Range
Iron-ore bearing range of low hills in NE.
BT Minnesota
United States

Mesaverde Group
Throughout San Juan Basin of SW Colorado and NW New Mexico, composed of Menefee Formation, Point Lookout Sandstone, Cliff House Sandstone, Gallup Sandstone, and Crevasse Canyon Formation. Arizona, W Colorado, NW New Mexico, E Utah, and Wyoming. Also search Mesaverde Formation.
BT Upper Cretaceous
Cretaceous
SA Almond Formation
Arizona
Blackhawk Formation
Colorado
Gallup Sandstone
La Ventana Sandstone
Menefee Formation
New Mexico
Utah
Wyoming

Meseta
Geographic term for the entire interior of Spain covering almost 3/4 of country and consisting of an immense plateau with Madrid at its center.
BT Spain

Mesogaea
Term introduced in 1978. Includes use on level 3 under paleogeography(1).
SA Mediterranean region
paleogeography
Tethys

Mesogastropoda
Term introduced in 1981. Includes use on level 2 under Mollusca(1). See list F.
BA Gastropoda
Mollusca
BT Invertebrata

mesolite
BA zeolite group
framework silicates
silicates
BT minerals

Mesolithic
Archaeologic classification.
UF Middle Stone Age
BA Holocene
Quaternary
BT Cenozoic

Mesopotamia
Region in SW Asia between Tigris and Euphrates rivers. Primarily in

Iraq but partly in NE Syria. Index countries as applicable.
BT Middle East
SA Iraq
Syria

Mesosauria
Term introduced in 1981. Includes use on level 2 under Reptilia(1). See list F.
BA Synapsida
Reptilia
BT Tetrapoda
Vertebrata
NA Proganosauria

mesosiderites
UF grahamite
BA stony irons
BT meteorites

mesothermal processes
Term introduced in 1981. Also search mesothermal.
SA mineral deposits
processes

Mesozoic
Includes use on level 1 as an age term (list E); on level 2 under paleoterms, e.g. paleoecology, paleogeography, paleomagnetism. From the end of the Paleozoic to the beginning of the Cenozoic.
BT Phanerozoic
NT Caracas Group
Cretaceous
Franciscan Formation
Great Valley Sequence
Jurassic
Khorat Group
Kootenay Formation
NA lower Mesozoic
NT Maiolica Limestone
Mardin Formation
NA middle Mesozoic
NT Navajo Sandstone
Pucara Group
Serra Geral Formation
Shimanto Group
Triassic
NA upper Mesozoic
SA Alpine Orogeny
Beacon Supergroup
Botucatu Formation
Brianconnais Zone
Ferrar Group
Himalayan Orogeny
Ionian Zone
Shuswap Complex

Messel
Village near Darmstadt in S.
BT Hesse
West Germany

Messina
City and province in NE.
BT Sicily
Italy

Messinian
BA upper Miocene
Miocene
Neogene
Tertiary
BT Cenozoic

meta-andesite
Also search metaandesite. See list J.
UF metaandesite
BA metaigneous rocks
BT metamorphic rocks
SA andesite

meta-arkose
Also search metaarkose. See list J.
UF metaarkose
BA metasedimentary rocks
BT metamorphic rocks
SA arkose

meta-limestone
use metalimestone

metaandesite
use meta-andesite

metaarkose
use meta-arkose

metabasalt
See list J.
BA metaigneous rocks
BT metamorphic rocks
SA basalt

metabasite
See list J.
BA metaigneous rocks
BT metamorphic rocks

metabentonite
Includes use as a rock and a mineral. See list J.
BX metamorphic rocks
minerals
SA bentonite
clay minerals
metasedimentary rocks

metabolism
Includes use under fossil groups(1) or paleontology(1).
SA nutrition

metacinnabar
Also search metacinnabarite.
UF metacinnabarite
BA sulfides
BT minerals
SA onofrite

metacinnabarite
use metacinnabar

metaconglomerate
See list J.
BA metasedimentary rocks
BT metamorphic rocks
SA conglomerate

metadiabase
See list J. Before 1981, also search metadolerite.
BA metaigneous rocks
BT metamorphic rocks
SA diabase

metadiorite
See list J.
BA metaigneous rocks
BT metamorphic rocks
SA diorite

metadolerite
As of 1981, no longer a valid term for GeoRef. Use metadiabase.

metagabbro
See list J.
BA metaigneous rocks
BT metamorphic rocks
SA gabbro

metagranite
See list J.
BA metaigneous rocks
BT metamorphic rocks
SA granite

metagraywacke
Also search metagreywacke. See list J.
UF metagreywacke
BA metasedimentary rocks
BT metamorphic rocks
SA graywacke

metagreywacke
use metagraywacke

metahalloysite
use halloysite

metaigneous
A valid index term on level 2 through 1977. After 1977, use metaigneous rocks.

metaigneous rocks
Includes use on level 2 under metamorphic rocks(1). Before 1978, also search metaigneous. See list J.
BT metamorphic rocks
NA meta-andesite
metabasalt
metabasite
metadiabase
metadiorite
metagabbro
metagranite
metaperidotite
metapyroxenite
metarhyolite
metatuff
NZ serpentinite
SA igneous rocks
rocks

metakaolin
Also search metakaolinite.
UF metakaolinite
SA clay minerals
kaolin

metakaolinite
use metakaolin

metal ores
Term introduced in 1981. Includes use on level 1 and 2 as a commodity term (list C). Before 1981, search metals AND ore deposits.
SA base metals
metallurgy
metals
native elements
nonmetal deposits
ore minerals
placers
polymetallic ores
rare earth deposits

metalimestone
Also search meta-limestone. See list J.
UF meta-limestone
BA metasedimentary rocks
BT metamorphic rocks
SA limestone

Metaline Falls
Town in Pend Oreille County in extreme NE.
BT Washington
United States

metallic meteorites
use iron meteorites

metallogenic provinces
Includes use on level 3 under metallic commodities (list C) or economic geology(1).
UF provinces, metallogenic
SA economic geology
mineral deposits
mineralization

metallogeny
Includes use on level 3 under geosynclines(1) processes (2) and under economic geology (1).
SA economic geology
geosynclines
mineral deposits

metallography
Not a valid index term for GeoRef. See mineralogy(1) methods(2) and petrology(1) methods(2).

metallurgy
SA alloys
hydrometallurgy
metal ores
metals

metals
Includes use on level 1 and 2 as a chemical element (list D). As of 1981, use metal ores for metals as a commodity. Before 1981, included use on level 1 as a commodity term.
NT alkali metals
heavy metals
rare metals
SA alloys
base metals
complexing
economic geology
elements
geochemistry
isotopes
metal ores
metallurgy
native elements
nonmetals
ore minerals
pollution
rare earths
trace metals

metamict
No longer a valid term for GeoRef as of 1981. Use metamict minerals.

metamict minerals
Term introduced in 1981. Includes use on level 3 under minerals(1). Before 1981, search metamict.
SA crystal structure
metamictization
minerals
radiation damage

metamictization
Includes use on level 3 under minerals(1) and mineralogy(1).
SA crystal structure
metamict minerals
minerals
radiation damage

metamorphic aureoles
use aureoles

metamorphic belts
Includes use on level 3 under plate tectonics(1).
UF belts, metamorphic
SA plate tectonics

metamorphic differentiation
Not a valid index term for GeoRef. See metamorphism(1) migration of elements(2) and metamorphism(1) differentiation(2).

metamorphic grade
use grade

metamorphic rocks
Includes use on level 1 (list A); on level 2 under paragenesis(1), phase equilibria(1), and weathering(1). If 1, term set options are:
rock group [amphibolites, cataclasites, eclogite, garnetite, gneisses, granulites, hornfels, itabirite, marbles, metaigneous rocks, metaplutonic rocks, metasedimentary rocks, metavolcanic rocks, migmatites, mylonites, phyllites, phyllonites, quartzites, schists, slates]
rock name
topic [Used for general studies where several rock types are discussed or where rock is not assignable to one of the groups above; age, classification, composition, correlation, distribution, evolution, experimental studies, facies, genesis, geochemistry, lithostratigraphy, mineral assemblages, occurrence, petrography, petrology, properties, textures]
subtopic
NT amphibolites
cataclasites
eclogite
garnetite
gneisses
NA gondite
NT granulites

hornfels
itabirite
marbles
NX metabentonite
NT metaigneous rocks
metaplutonic rocks
metasedimentary rocks
metavolcanic rocks
migmatites
mylonites
phyllites
phyllonites
quartzites
schists
slates
SA actinolite facies
amphibolite facies
banded structures
basement
blueschist facies
complexes
crystalline rocks
deformation
epidote-amphibolite facies
fabric
facies
folds
foliation
geochemistry
geochronology
gneissic texture
granulite facies
greenschist facies
impactite
intrusions
lineation
lithostratigraphy
megacrysts
metamorphism
metasomatic rocks
metasomatism
mineral assemblages
mullions
paragenesis
pelitic texture
petrography
petrology
phase equilibria
porphyroblastic texture
prehnite-pumpellyite facies
protoliths
rocks
schistosity
skarn
suevite
zeolite facies

metamorphic zone
use aureoles

metamorphism
Includes use on level 1 (list A). Term set options are:
kind of metamorphism [anchimetamorphism, autometamorphism, burial metamorphism, contact metamorphism, dynamic metamorphism, grade, polymetamorphism, prograde metamorphism, regional metamorphism, retrograde metamorphism, shock metamorphism, thermal metamorphism, ultrametamorphism]
topic (high-grade metamorphism, low-grade metamorphism)
topic [For more than one kind or for a kind not assignable to one of the categories above; age, causes, classification, concepts, distribution, environment, evolution, genesis, mechanism, migration of elements, processes, P-T conditions, rates, rheomorphism, temperature, theoretical studies, zoning]
subtopic
NA autometamorphism
burial metamorphism
contact metamorphism
dynamic metamorphism
NT glauconitization
NA granitization
high-grade metamorphism
low-grade metamorphism
migmatization
phyllonitization
polymetamorphism
prograde metamorphism
regional metamorphism
retrograde metamorphism
shock metamorphism
thermal metamorphism
ultrametamorphism
SA anatexis
aureoles
cataclasis
complexes
deformation
diagenesis
feldspathization
folds
foliation
geochemistry
geochronology
grade
high pressure
high temperature
igneous rocks
intrusions
isograds
lineation
low pressure
low temperature
metamorphic rocks
metaplutonic rocks
metasomatism
meteor craters
migration of elements
orogeny
P-T conditions
paragenesis
petrology
phase equilibria
protoliths
recrystallization
rheomorphism
shatter cones
shock waves
slaty cleavage
uralitization
zoning

metapelite
See list J.
BA metasedimentary rocks
BT metamorphic rocks

metaperidotite
See list J.
BA metaigneous rocks
BT metamorphic rocks
SA peridotite

metaplutonic rocks
Term introduced in 1978. Includes use on level 2 under metamorphic rocks(1). Before 1978, also search metamorphic rocks AND metaplutonic. See list J.
BT metamorphic rocks
SA metamorphism
plutonic rocks
rocks

metapyroxenite
See list J.
BA metaigneous rocks
BT metamorphic rocks
SA pyroxenite

metarhyolite
See list J.
BA metaigneous rocks
BT metamorphic rocks
SA rhyolite

metasandstone
See list J.
BA metasedimentary rocks
BT metamorphic rocks
SA sandstone

metasedimentary
A valid index term on level 2 through 1977. After 1977, use metasedimentary rocks.

metasedimentary rocks
Includes use on level 2 under metamorphic rocks(1). Before 1978, also search metasedimentary. See list J.
BT metamorphic rocks
NA khondalite
meta-arkose
metaconglomerate
metagraywacke
metalimestone
metapelite
metasandstone
metasiltstone
NZ paragneiss
SA metabentonite
rocks
sedimentary rocks

metasediments
A valid term through 1974. Use metasedimentary rocks.

metasiltstone
See list J.
BA metasedimentary rocks
BT metamorphic rocks
SA siltstone

metasomatic rocks
As of 1978, term is used on level 1 for any rock produced by replacement processes at constant volume with little disturbance of textural or structural features. Includes use on level 3 under petrology(1) and under metasomatism(1). Also search metasomatite and metasomatites. Term set options are:
rock group [greisen, propylite, skarn]
rock name or process or subtopic
topic [age, classification, composition, distribution, evolution, facies, genesis, geochemistry, mineral assemblages, occurrence, petrography, petrology, textures]
subtopic
UF metasomatite
metasomatites
NA fenite
greisen
propylite
NZ serpentinite
NA skarn
tactite
talc rock
SA alteration
metamorphic rocks
metasomatism
rocks

metasomatism
Includes use on level 1 (list A). For documents before 1976, also search replacement. Term set options are:
materials
name of material [e.g. carbonate rocks, granite, igneous rocks, sedimentary rocks]
processes
name of process [e.g. albitization, alunitization, analcimization, argillization, dolomitization, granitization, greisenization, hydrothermal alteration, kaolinization, laumonitization, microclinization, muscovitization, palagonitization, propylitization, pyrometasomatism, scapolitization, serpentinization, silicification, zeolitization]
topic [age, causes, environment, experimental studies, geochemistry, interpretation, mechanism, rates, theoretical studies]
subtopic
NA albitization
chloritization
feldspathization
fenitization
greisenization
hydrothermal alteration
kaolinization
propylitization
pyritization
pyrometasomatism
sericitization
serpentinization
spilitization
uralitization
wallrock alteration
zeolitization
SA alteration
diagenesis
dolomitization
enrichment
fenite
geochemistry
granitization
hydrothermal conditions
igneous rocks
intrusions
metamorphic rocks
metamorphism
metasomatic rocks
mineral deposits
P-T conditions
palingenesis
paragenesis
petrology
phase equilibria
plutons
pneumatolysis
silicification
zoning

metasomatite
use metasomatic rocks

metasomatites
use metasomatic rocks

metastibnite
BA sulfides
BT minerals
SA stibnite

metatuff
See list J.
BA metaigneous rocks
BT metamorphic rocks
SA tuff

metavolcanic
A valid term through 1977. After 1977, use metavolcanic rocks on level 2 under metamorphic rocks(1).

metavolcanic rocks
Term introduced in 1978. Includes use on level 2 under metamorphic rocks(1). Before 1978, also search metamorphic rocks AND metavolcanic. See list J.
BT metamorphic rocks
SA rocks
volcanic rocks

Metazoa
Used as a general fossil term. Also search individual phyla and groups.
SA Protista

meteor craters
Includes use on level 1 (list A); on level 3 under area(1) geomorphology(2) and under geomorphology(1) impact features(2). Used for impact craters formed by the falling of a large meteorite onto a surface. Also search meteor crater. If 1, term set options are:
topic [age, causes, classification,

concepts, distribution, evolution, genesis, interpretation, observations, occurrence, patterns]
subtopic
UF meteorite crater
SA astroblemes
breccia
craters
cryptoexplosion features
deformation
geochronology
geomorphology
impact features
impacts
metamorphism
meteorites
shatter cones
shock metamorphism
tectonics

meteoric iron
use iron meteorites

meteorite crater
use meteor craters

meteorite flux
Term introduced in 1981. Includes use on level 2 under meteorites(1). Before 1981, search meteorites AND flux.
SA meteorites

meteorites
Includes use on level 1 (list A) and 2. Used for any meteoroid that has fallen to the Earth's surface in one piece or in fragments without being completely vaporized. Term set options are:
topic [age, bibliography, catalogs, collections, composition, detection, distribution, experimental studies, meteorite flux, genesis, geochemistry, interpretation, isotopes, organic materials, phase equilibria, properties, radioactivity, textures]
name of meteorite or type [e.g. achondrites, chondrites, howardite, iron meteorites, stony irons] or subtopic
UF cosmolites
skystones
NT achondrites
Allan Hills Meteorites
chondrites
Coldwater Meteorites
NX eucrite
NT iron motoorito
micrometeorites
Odessa Meteorite
stony irons
Yamato Meteorites
SA asteroids
australite
chaoite
chondrules
cohenite
comets
extraterrestrial geology
geochemistry
impact features
impactite
impacts
interplanetary dust
isotopes
kamacite
lonsdaleite
meteor craters
meteorite flux
meteoroids
meteors
moissanite
Moon
petrology
planetology
plessite
shatter cones
solar system
suevite
taenite
tektites
troilite

meteoroids
SA asteroids
cosmic dust
interplanetary space
meteorites
meteors
planets

meteorology
Primarily used for special indexes. Includes use on level 1. Term set options are:
aerosols
topic [composition, convection, particles, radioactive tracers, turbidity]
boundary interactions
topic
circulation
topic [meridional circulation, zonal circulation, etc.]
composition
elements or compound or region of atmosphere
convection
topic
diffusion
models
property diffused
type of diffusion
electromagnetic waves
topic [currents, lightning, raindrops, space charge, thunderstorms]
experimental studies
topic
instruments
name of instrument
type of phenomenon
ions
name of ion
particles
name of particle or topic or subtopic
models
storms
models
type of storm
techniques
type of phenomenon
technique
temperature
topic
theoretical studies
topic
turbulence
area
subtopic
topic
water
topic [atmospheric precipitation, clouds, droplets, humidity, isotopes, raindrops, name of isotope]
waves
type of wave
winds
type of wind
SA aeronomy
aerosols
atmosphere
atmospheric precipitation
aurora
boundary interactions
circulation
climate
clouds
convection
currents
diffusion
droplets
electrical phenomena
electromagnetic waves
geophysics
humidity
hurricanes
hydrologic cycle
hydrology
ionosphere
ions
lightning
magnetosphere
particles
raindrops
storms
thunderstorms
turbidity
turbulence
water
waves
winds

meteors
Includes use on level 3 under meteorites(1) and under planetology(1).
UF shooting star
SA comets
interplanetary space
meteorites
meteoroids
planetology

methane
Includes use on level 3 under organic materials(1) hydrocarbons(2).
BA organic materials
SA ethane
hydrocarbons
natural gas

method, transmission
use transmission method

methods
This term applies to theoretical and experimental studies, while the term techniques is used for discussion of samples. Includes use as a level 2 or 3 term appropriate to a large number of topics, e.g. on level 2 under absolute age(1), geochemistry(1), geophysical methods(1) organic materials(1), and soils(1) and mineral exploration(1). See list G.
SA acoustical methods
airborne methods
analytical methods
Ar/Ar
autoradiography
biogeochemical methods
common-depth-point method
deep-tow methods
Earth-current methods
electrical methods
electromagnetic methods
electron paramagnetic resonance
electron probe
fission-track dating
gamma-gamma methods
gamma-ray methods
geobotanical methods
geochemical methods
geological methods
geomorphological methods
geophysical methods
graphic methods
gravity methods
ground methods
helicopter methods
high-resolution methods
hydrological methods
infrared methods
magnetic methods
magnetotelluric methods
marine methods
mathematical methods
microscope methods
microwave methods
new methods
photogeologic methods
physical methods
quantitative analysis
radar methods
radio-frequency spectroscopy
radio-wave methods
radioactivity methods
satellite methods
scanning method
seismic methods
self-potential methods
single-crystal method
sonar methods
statistical analysis
techniques
U-238/Pb-206
U/He
U/Pb
U/Th/Pb
wet methods

Methow River valley
Okanogan County on British Columbia border. Also search Methow River.
BT Washington
United States

Metohija
District in SW section of country forming part of autonomous province of Kosovo-Metohija.
BT Serbia
Yugoslavia

Meurthe-et-Moselle
Department in NE.
CO N483000N493000
E0071500E0053500
BT France
NT Laneuville-devant-Nancy Boring

Meuse
Department in NE.
CO N482500N493500
E0055200E0045000
BT France
NT Verdun

Meuse River
Rises in NE France, flows N across E Belgium and, as the Maas River, flows through the Netherlands centering the North Sea through its estuary, the Hollandsch Diep. Index countries as applicable. Also search Maas River; Meuse.
UF Maas River
BT Europe
SA Belgium
France
Netherlands

Meuse Valley
River valley. Known as the Maas Valley in the Netherlands. Index countries as applicable.
UF Maas Valley
BT Europe
SA Belgium
France
Netherlands

Mexicali
City on California border.
BT Baja California
Mexico

Mexico
Includes use on level 1 or 2 as an area term (list O). If 1, see term set options under list B.
CO N143000N324300
W0864500W1170000
BT North America
NT Baja California
Campeche
Chiapas
Chihuahua
Coahuila
Colorado River delta
Durango
NX Federal District
NT Guanajuato
Guerrero

M

Hidalgo
Jalisco
Mexico state
Michoacan
Nuevo Leon
Oaxaca
Puebla
Quintana Roo
San Luis Potosi
Sierra Madre
Sierra Madre del Sur
Sierra Madre Occidental
Sierra Madre Oriental
Sinaloa
Sonora
Tabasco
Tamaulipas
Valley of Mexico
Veracruz
Yucatan
Yucatan Peninsula
Zacatecas
Zaragoza
SA Allende Meteorite
Colorado River basin
Difunta Group
Leon
Magdalena
Maverick Basin
Merida
North American Cordillera
Rosario Formation
San Felipe Formation
Toluca Meteorite
Zuloaga Limestone

Mexico Basin
use Sigsbee Deep

Mexico City
City.
UF Mexico D.F.
BT Federal District
Mexico

Mexico D.F.
use Mexico City

Mexico state
State N, E and W of Federal District.
BT Mexico
SA Valley of Mexico

meymechite
use meimechite

Mezen
Town on the Mezen River in Arkhangelsk Oblast in NW European USSR.
BT Russian Republic
USSR

Mezen River basin
N central Arkhangelsk Oblast in NW European USSR. Also search Mezen River.
BT Russian Republic
USSR

Mezhrechye
Village in W.
BT Byelorussia
USSR

Mg
use magnesium

Miami
City in Dade County in SE.
BT Florida
United States

Miami Limestone
According to the latest published Lexique, the age of Miami Oolite is Pleistocene. It is a soft white oolitic limestone, containing streaks of thin irregular layers of calcite separating less crystalline streaks. In S Florida. Also search Miami Oolite.
BT Pleistocene
SA Florida

Mianwali
Town and district in NE part of country.
BT Punjab Province
Pakistan

Miaoli
Town in NW.
BT Taiwan

miargyrite
BA sulfantimonites
sulfosalts
BT minerals

miaskite
As of 1981, includes use on level 3 under igneous rocks(1) syenites(2). See list H. Before 1981, included use on level 3 under igneous rocks(1) syenite family(2).
BA syenites
BT igneous rocks

mica
As of 1981, no longer a valid term for GeoRef. See mica group. Use mica deposits for mica as a commodity. Included use on level 1 as a commodity term and on level 3 under minerals(1) sheet silicates, mica group(2).

mica deposits
Term introduced in 1981. Includes use on level 1 and 2 as a commodity term (list C). Before 1981, search mica AND deposits.
SA glauconite deposits
mica group
vermiculite deposits

mica group
Includes use in combination with sheet silicates (i.e. sheet silicates, mica group) on level 2 under minerals(1). See list L.
BA sheet silicates
silicates
BT minerals
NA biotite
celadonite
clintonite
fluor-phlogopite
glauconite
NZ lepidolite
NA lepidomelane
listvenite
margarite
muscovite
paragonite
phengite
phlogopite
polylithionite
sericite
NZ zinnwaldite
SA hydrobiotite
hydromica
hydromuscovite
mica deposits
vermiculite

mica schist
As of 1981, no longer a valid term for GeoRef. Use terms that indicate specific mica (i.e. biotite schist or muscovite schist).

Michelle Formation
Composed of black calcareous shale and aphanitic silty limestone. In W Northwest Territories and Yukon Territory.
BT Middle Devonian
Devonian
SA Canada
Northwest Territories
Yukon Territory

michenerite
BA bismuthides
sulfides
BT minerals

Michigan
Includes use on level 1 as an area term (list O). For term set options see list B.
CO N414500N473000
W0823000W0901500
BA Midwest
United States
NT Alpena County
Ann Arbor
NX Berrien County
Calhoun County
Cass County
NT Charlevoix County
NX Clinton County
Crawford County
Dickinson County
Genesee County
NT Glennie
Grand Rapids
Grand Traverse County
Gratiot County
Houghton County
Ingham County
NX Iron County
NT Isle Royale National Park
NX Jackson County
NT Kalamazoo County
NX Kent County
NT Keweenaw County
Keweenaw Peninsula
NX Lake County
NT Lansing
NX Livingston County
NT Macomb County
Manistee County
Marquette
NX Marquette County
NT Marquette Range
NX Mason County
NT Michigan Basin
Michigan Lower Peninsula
Michigan Upper Peninsula
NX Midland County
Monroe County
NT Negaunee Iron Formation
Ontonagon County
NX Ottawa County
NT Porcupine Mountains
NX Presque Isle
NT Roscommon County
NX Saint Clair County
NT University of Michigan
NX Wayne County
NT White Pine
White Pine Mine
SA Animikie Group
Berea Sandstone
Canadian Shield
Charlevoix
Clinton Group
Copper Harbor Conglomerate
Detroit River Group
Grand River
Great Lakes region
Green Bay
Jacobsville Sandstone
Keweenawan
Lake Erie
Lake Huron
Lake Michigan
Lake Superior
Lockport Formation
Medina Formation
Midcontinent
Niagara Escarpment
Nonesuch Shale
Penokean Orogeny
Portage Lake Lava Series
Saginaw Formation
Saint Clair River
Saint Clair River delta
Saint Peter Sandstone
Salina Group
Sault Sainte Marie
Southern Province
Sylvania Formation
Traverse Group
Trenton Group
Utica Shale

Michigan Basin
Lower Peninsula.
BT Michigan
United States

Michigan Lower Peninsula
Term introduced in 1981. S part of state S of Straits of Mackinack. Before 1981, search Michigan AND Lower Peninsula; Southern Peninsula.
UF Southern Peninsula
BT Michigan
United States
SA Michigan Upper Peninsula

Michigan Upper Peninsula
Term introduced in 1981. N part of state between Lake Michigan and Lake Superior. Before 1981, search Michigan AND Upper Peninsula.
BT Michigan
United States
SA Michigan Lower Peninsula

Michipicoten Island
NE Lake Superior.
BT Ontario
Canada

Michoacan
State in SW..
BT Mexico
NT Paricutin

micrinite
Includes use on level 3 under sedimentary rocks(1) organic residues(2). See list I.
BA inertinite
macerals
coal
organic residues
BT sedimentary rocks

micrite
Includes use on level 3 under sedimentary rocks(1) carbonate rocks(2). See list I.
BA carbonate rocks
BT sedimentary rocks
SA biomicrite
limestone

microbreccia
See list I.
BA clastic rocks
BT sedimentary rocks
SA breccia
sandstone
terrigenous materials

microcline
BA feldspar group
framework silicates
silicates
BT minerals
SA amazonite
K-feldspar

Microcodium
Includes use on level 3 under algae(1).
BA algae
BT thallophytes
Plantae

microcomputers
Term introduced in 1978. Before 1981, also search microprocessors.
SA automatic data processing
computers
hand calculators
minicomputers

microcracks
Before 1981, also search microfractures.
BT cracks
SA deformation
engineering geology

fractures
rock mechanics

microcraters
Includes use on level 2 under Moon(1).
SA craters
Moon

microdiorite
As of 1981, includes use on level 3 under igneous rocks(1) diorites(2). See list H. Before 1981, included use on level 3 under igneous rocks(1) diorite family(2).
BA diorites
BT igneous rocks
SA diorite

microearthquakes
Includes use on level 2 under seismology(1).
SA earthquakes
seismology

microelement
use trace elements

microfacies
Includes use on level 3 under micropaleontology(1).
SA facies
micropaleontology
petrography
thin sections

microfauna
As of 1981, no longer a valid term for GeoRef. Use microfossils.

microflora
use microfossils

microfossils
Includes use on level 3 under fossil groups (list F). Before 1981, also search microflora; microfauna.
UF microflora
NZ problematic microfossils
SA fossils
index fossils
micropaleontology
otoliths
pollen

microfractures
As of 1981, no longer a valid term for GeoRef. Use microcracks.

microgabbro
As of 1981, includes use on level 3 under igneous rocks(1) gabbros(2). See list H. Before 1981, included use on level 3 under igneous rocks(1) gabbro family(2).
BA gabbros
BT igneous rocks
SA gabbro

microgranite
As of 1981, includes use on level 3 under igneous rocks(1) granites(2). See list H. Before 1981, included use on level 3 under igneous rocks(1) granite-granodiorite family(2).
BA granites
BT igneous rocks
SA granite

microhardness
Includes use on level 3 under minerals(1).
SA hardness
minerals
physical properties

microlite
BZ niobates
fluorides
tantalates
BT minerals
SA pyrochlore

micrometeorites
Very small meteorite or meteoritic particle with diameter generally less than a millimeter.
BT meteorites

micromorphology
Includes use on level 3 under soils(1) morphology(2).
SA morphology
soils

Micronesia
Collective name for islands E of the Philippine Islands and S of Japan. Index island groups as applicable. Includes use on level 1 or 2 as an area term (list O). If 1, see term set options under list B.
CO S040000N210000
E1770000E1440000
BT Oceania
NT Caroline Islands
Gilbert Islands
NA Mariana Islands
Marshall Islands
NT Nauru Island

microorganisms
For several types. Includes use on level 3 under soils(1) biota(2); on level 3 under geochemistry(1).
SA algae
bacteria
biota
faunal studies
floral studies
fungi
pollen
Protista
soils

micropaleontology
Used for the discipline as a whole, not for specific applications. Includes use on level 1 (list A). See list F (microfossil groups). Term set options are:
topic [applications, bibliography, catalogs, concepts, experimental studies, history, instruments, methods, photography, practice, techniques, theoretical studies]
subtopic [e.g. sample preparation]
SA foraminifera
fossils
microfacies
microfossils
paleobotany
paleontology
palynology
sample preparation
stratigraphy

micropegmatite
As of 1981, includes use on level 3 under igneous rocks(1) granites(2). See list H. Before 1981, included use on level 3 under igneous rocks(1) granite-granodiorite family(2).
BA granites
BT igneous rocks
SA graphic granite
pegmatite

microperthite
BA feldspar group
framework silicates
silicates
BT minerals
SA perthite

micropetrological unit
use macerals

microplankton
use plankton

microplates
Includes use on level 3 under plate tectonics(1).
SA plate tectonics
plates

microprocessors
As of 1981, no longer a valid term for GeoRef. Use microcomputers.

micropulsations
Includes use on level 3 under aurora(1) magnetic field(2).
SA aurora
magnetic field
pulsations

microscope methods
Includes use on level 3 under petrology(1) and mineralogy(1).
SA methods
mineralogy
optical mineralogy
ore microscopy
petrography
petrology
polished sections
reflectance
reflectivity
thin sections
undulatory extinction
universal stage

microscopy, electron
use electron microscopy

microscopy, ore
use ore microscopy

microseisms
Includes use on level 2 under seismology(1); on level 3 under earthquakes(1).
SA earthquakes
noise
seismology

microstructure
A valid term through 1978. After 1978, use ultrastructure.

microstylolites
See list K.
BA secondary structures
sedimentary structures
SA stylolites

microsyenite
As of 1981, includes use on level 3 under igneous rocks(1) syenites(2). See list H. Before 1981, included use on level 3 under igneous rocks(1) syenite family(2).
BA syenites
BT igneous rocks
SA syenite

microtectonics
A valid term through 1975. After 1975, use structural analysis.

microtektites
BA tektites

microwave methods
Term introduced in 1978. Before 1978, also search microwave. Includes use on level 3 under remote sensing(1) and well-logging(1).
BA geophysical methods
SA methods
radar methods
remote sensing
well-logging

microwave spectroscopy
Term introduced in 1978. Includes use on level 3 under spectroscopy(1), and under chemical element(1) analysis(2). Before 1978, also search spectroscopy AND microwave.
BT spectroscopy
SA analysis

Mid-Antarctic Ridge
Term introduced in 1981.
BT Antarctic Ocean

Mid-Arctic Ocean Ridge
Term introduced in 1981.
BT Arctic Ocean
SA mid-ocean ridges

Mid-Atlantic Ridge
Extends parallel to the continental margins in mid ocean in both the North and South Atlantic Ocean. Rises 6000 ft. above the ocean floor and surfaces as the Azores, Ascension, St. Helena, and Tristan de Cunha islands.
CO S570000N600000
E0030000W0620000
BT Atlantic Ocean
NZ North Atlantic Ridge
South Atlantic Ridge
SA Atlantis fracture zone
FAMOUS
median valley
mid-ocean ridges

Mid-Cayman Rise
Term introduced in 1981. NW Caribbean Sea, S of Cuba.
CO N190000N200000
W0740000W0820000
BT Caribbean Sea
Atlantic Ocean

Mid-Continent
use Midcontinent

Mid-Indian Ocean Ridge
use Mid-Indian Ridge

Mid-Indian Ridge
Extends from S of the Maldives, S and SE into Antarctic waters. Also search Indian Ocean Ridge; Indian Ridge; Mid-Indian Ocean Ridge.
UF Indian Ocean Ridge
Indian Ridge
Mid-Indian Ocean Ridge
BT Indian Ocean
NA Central Indian Ridge
Southeast Indian Ridge
SA mid-ocean ridges

mid-ocean ridges
Includes use on level 2 under ocean floors(1). Also search mid-oceanic ridges; mid-oceanic ridge; midoceanic ridges.
UF mid-oceanic ridge
mid-oceanic ridges
midoceanic ridges
SA aseismic ridges
bottom features
Carlsberg Ridge
Central Indian Ridge
Chile Ridge
East Pacific Rise
fracture zones
Galapagos Rift
Juan de Fuca Ridge
median valley
Mid-Arctic Ocean Ridge
Mid-Atlantic Ridge
Mid-Indian Ridge
ocean floors
Pacific-Antarctic Ridge
sea-floor spreading
Southeast Indian Ridge
Southwest Indian Ridge
transform faults

mid-oceanic ridge
use mid-ocean ridges

mid-oceanic ridges
use mid-ocean ridges

Mid-Pacific Rise
Term introduced in 1978.
SA Pacific Ocean

Midcontinent
Mid-America between the Appalachians and the Rocky Mountains. Index states as applicable. Also search Mid-continent.
UF Mid-Continent
BT United States
SA Arkansas

Colorado
Illinois
Indiana
Iowa
Kansas
Kentucky
Michigan
Minnesota
Missouri
Montana
Nebraska
New Mexico
North Dakota
Ohio
Oklahoma
South Dakota
Tennessee
Texas
Wisconsin
Wyoming

Middle America Trench
Off W coast of Mexico and Central America.
BT Pacific Ocean

Middle Atlas
Range of the Atlas Mountains N and E of the High Atlas Mountains in North Central.
BT Morocco
SA Atlas Mountains
Moroccan Atlas Mountains

Middle Cambrian
See list E.
BA Cambrian
BT Paleozoic
NA Acadian
Barrandian
NT Burgess Shale
Flathead Sandstone
Pagoda Formation
SA Lower Cambrian
Tintic Quartzite
Upper Cambrian

Middle Carboniferous
See list E.
BA Carboniferous
BT Paleozoic

Middle Cretaceous
See list E.
BA Cretaceous
BT Mesozoic
NT Bhuj Series
SA Albian
Aptian
Cenomanian
Lower Cretaceous
Turonian
Upper Cretaceous

Middle Devonian
See list E.
BA Devonian
BT Paleozoic
NT Cedar Valley Formation
Columbus Limestone
NA Couvinian
NT Delaware Limestone
Detroit River Group
NA Eifelian
NT Elk Point Group
NA Givetian
NT Hamilton Group
Mahantango Formation
Michelle Formation
Nahanni Formation
Onondaga Limestone
Prairie Evaporite
Saint Laurent Limestone
Sylvania Formation
Tioga Bentonite
Tully Limestone
Winnipegosis Formation
SA Catskill Formation
Genesee Group
Lost Burro Formation
Lower Devonian
Martin Formation
Muth Quartzite
Traverse Group
Upper Devonian

Middle East
An indefinite and unofficial term comprising a region including Cyprus and the countries of SW Asia. Index countries as applicable. Includes use on level 1 or 2 as an area term (list O). If 1, see term set options under list B.
CO N250000N420000
E0632000E0253000
NA Cyprus
NT Dead Sea
Dead Sea Rift
NA Iran
Iraq
Israel
NT Jerusalem
NA Jordan
NT Jordan River
Jordan Valley
NA Lebanon
NT Mesopotamia
Samaria
NA Syria
Turkey
NT Wadi Araba
SA Arabian Peninsula
Egypt
Kuwait
Oman
Qatar
Red Sea Basin
Sinjar Formation
Southern Yemen
United Arab Emirates
Yemen
Zubair Formation

middle Eocene
See list E.
BA Eocene
Paleogene
Tertiary
BT Cenozoic
NT Claiborne Group
Cook Mountain Formation
Laney Shale Member
Queen City Formation
Tyee Formation
NA Ulatisian
SA Biarritzian
Castle Hayne Limestone
Green River Formation
lower Eocene
Nanjemoy Formation
Narizian
Rose Canyon Formation
Umpqua Formation
upper Eocene

Middle Franconia
Part of old historical region of Franconia. A hilly region in Franconian Jura in N central.
BT Bavaria
West Germany

middle Holocene
Term introduced in 1978. See list E.
BA Holocene
Quaternary
BT Cenozoic

Middle Jurassic
See list E.
BA Jurassic
BT Mesozoic
NA Aalenian
Bajocian
Bathonian
Callovian
Dogger
SA Bowser Formation
Lower Jurassic
Upper Jurassic

middle Liassic
Term introduced in 1981. Includes use on level 3 under age terms(1).
See list E.
BA Lower Jurassic
Jurassic
BT Mesozoic
NA Domerian
Pliensbachian

middle Mesozoic
See list E.
BA Mesozoic
SA Caracas Group
upper Mesozoic

middle Miocene
See list E.
BA Miocene
Neogene
Tertiary
BT Cenozoic
NT Calvert Formation
Choptank Formation
NA Helvetian
NT Kirkwood Formation
San Onofre Breccia
NA Serravallian
SA Badenian
Hawthorn Formation
lower Miocene
Monterey Formation
Saint Marys Formation
Siwalik System
upper Miocene

Middle Mississippian
See list E.
BA Mississippian
Carboniferous
BT Paleozoic

Middle Nile Valley
Term introduced in 1981. Used for the Nile Valley in Egypt.
CO N220000N301000
E0330000E0303000
BZ Nile Valley
BX Egypt

middle Oligocene
See list E.
BA Oligocene
Paleogene
Tertiary
BT Cenozoic
NT Marianna Limestone
Vicksburg Group
SA Brule Formation
lower Oligocene
upper Oligocene

Middle Ordovician
See list E.
BA Ordovician
BT Paleozoic
NT Bigby-Cannon Limestone
Black River Group
NA Blackriverian
Champlainian
Chazyan
NT Cloridorme Formation
Everton Formation
Lexington Limestone
Lincolnshire Limestone
NA Llandeilian
Llanvirnian
NT Normanskill Formation
Platteville Formation
Saint Peter Sandstone
Schenectady Formation
Simpson Group
Swan Peak Formation
Trenton Group
NA Trentonian
NT Winnipeg Formation
SA Antelope Valley Limestone
Bala
Chickamauga Group
Lower Ordovician
Martinsburg Formation
Pogonip Group
Upper Ordovician
Viola Limestone

middle Paleocene
See list E.
BA Paleocene
Paleogene
Tertiary
BT Cenozoic

middle Paleozoic
See list E.
BA Paleozoic

Middle Pennsylvanian
See list E.
BA Pennsylvanian
Carboniferous
BT Paleozoic
NT Allegheny Group
NA Atokan
NT Breathitt Formation
Carbondale Formation
NA Desmoinesian
NT Paradox Member
Staunton Formation
Strawn Series
SA La Salle Limestone
Lower Pennsylvanian
Mansfield Formation
Marble Falls Group
Pottsville Group
Upper Pennsylvanian

Middle Permian
See list E.
BA Permian
BT Paleozoic
SA Kungurian
Lower Permian
Rotliegendes
Upper Permian

middle Pleistocene
See list E.
BA Pleistocene
Quaternary
BT Cenozoic

middle Pliocene
See list E.
BA Pliocene
Neogene
Tertiary
BT Cenozoic
NT Bone Valley Formation
SA lower Pliocene
upper Pliocene

middle Precambrian
See list E.
BA Precambrian
NA Carpentarian
Karelian
Svecofennian
SA Aphebian

middle Proterozoic
See list E.
BA Proterozoic
Precambrian

middle Quaternary
See list E.
BA Quaternary
BT Cenozoic

Middle Silurian
See list E.
BA Silurian
BT Paleozoic
NT Clinton Group
McKenzie Formation
NA Niagaran
NT Roberts Mountains Formation
Rochester Formation
Waldron Shale
SA Lower Silurian
Read Bay Formation
Rockwood Formation
Upper Silurian

Middle Stone Age
use Mesolithic

middle Tertiary
See list E.
BA Tertiary
BT Cenozoic
SA lower Tertiary
San Juan Formation
upper Tertiary

Middle Triassic
See list E.
BA Triassic
BT Mesozoic
NA Anisian
Ladinian
Muschelkalk
SA Lower Triassic
Moenkopi Formation
Serebryanka Formation
Upper Triassic

Middle Tunguska River
use Stony Tunguska River

Middle West (United States)
use Midwest

Middlesex County
Index states as applicable.
BX Connecticut
Massachusetts
New Jersey
Virginia
BT United States

Midland
City in Midland County in W.
BT Texas
United States

Midland Basin
W Texas. Also search Midland.
BX Permian Basin
Texas

Midland County
Index states as applicable.
BX Michigan
Texas
BT United States

Midland Valley
Term introduced in 1981.
CO N543500N565500
W0020000W0051500
BT Scotland
Great Britain
United Kingdom

Midlands
Region including the highly industrialized central counties.
CO N513000N533000
W0013000W0023000
BT England
Great Britain
United Kingdom

Midlothian
Formerly Edinburghshire, county. SE Scotland on S shore of Firth of Forth.
CO N554000N560000
W0024500W0034500
BT Scotland
Great Britain
United Kingdom

midoceanic ridges
use mid-ocean ridges

Midway
Two small islands (Eastern and Sand), parts of a low coral atoll under U.S. Navy control 1300 miles WNW of Honolulu. Also search Midway Atoll.
UF Midway Atoll
BT Pacific Ocean

Midway Atoll
use Midway

Midway Group
Term introduced in 1981. In SE United States from W Georgia to S Texas, Arkansas, SW Illinois, Kentucky, SE Missouri, and W Tennessee.
BT Paleocene
SA Alabama
Arkansas
Florida
Georgia
Illinois
Kentucky
Mississippi
Missouri
Tennessee
Texas

Midwest
A level 1 area term as of 1978. N part of central U.S. Region comprising states N of the Ohio and Missouri Rivers plus the E edge of the Great Plains including Kansas and Nebraska. Also search United States AND north-central or north central.
CO N360000N490000
W0803000W1040000
UF Middle West (United States)
North Central (United States)
North-Central (United States)
BA United States
NA Illinois
Indiana
Iowa
Kansas
Michigan
Minnesota
Missouri
Nebraska
North Dakota
Ohio
South Dakota
Wisconsin

Mie
Prefecture in S central Honshu.
UF Miye
BT Honshu
Japan

Miechow
Town in S part of country.
BT Krakow
Poland

Miette Complex
Devonian reef complex. Distinguish from Miette Group.
BT Devonian
SA Canada
Miette Group

Miette Group
Composed of a basal unit of argillite and argillaceous sandstone, succeeded by 2,000 feet of sandstone, grit, conglomerate, and argillite, which is overlain by 3,000 feet of argillite, and at the top by a carbonate. Distinguish from the Devonian Miette Complex.
BT Proterozoic
Precambrian
SA Alberta
British Columbia
Canada
Miette Complex

migmatite
As of 1981, no longer a valid term for GeoRef. Use migmatites. Before 1981, included use on level 3 under metamorphic rocks(1) migmatites(2).

migmatites
Includes use on level 2 under metamorphic rocks(1). See list J. Before 1981, also search migmatite.
BT metamorphic rocks
NA agmatite
anatexite
ectinite
embrechite
SA migmatization
ptygmatic folds

migmatitization
use migmatization

migmatization
Includes use on level 3 under metamorphism(1).
UF migmatitization
BA metamorphism
SA migmatites

migration
Includes use on level 3 under soils(1) genesis(2) and under petroleum(1) genesis(2).
SA petroleum
soils

migration of elements
Includes use on level 2 under metamorphism(1); on level 3 under geochemistry(1) and under weathering(1).
SA elements
enrichment
geochemistry
metamorphism
mobilization
thermal waters
weathering

Mihara (Mount)
use Mount Mihara

Mihara Volcano
use Mount Mihara

Mihara-Yama
use Mount Mihara

Miho Bay
Inlet of Sea of Japan in S.
BT Honshu
Japan

Mikabu System
According to the Japan Lexique, the age of Mikabu Series is Paleozoic. Consists of amphibole, pyroxene, semi-schists, and phyllites with limestones and quartzites.
BT Paleozoic
SA Honshu
Japan

Mikawa
Former province in central Honshu. Now part of Aichi Prefecture.
BT Honshu
Japan

Mikhailovka
City in Volgograd Oblast 110 miles NW of Volgograd in European USSR.
BT Russian Republic
USSR

Milam County
NE of Austin in central.
BT Texas
United States

Milan
City and province in NW part of country.
UF Milano
BT Lombardy
Italy

Milano
use Milan

milarite
BA ring silicates
silicates
BT minerals
SA aluminosilicates

Mildura
Town on the Murray River 300 miles NW of Melbourne in NW.
BT Victoria
Australia

Miliolacea
Includes use on level 2 under foraminifera(1). See list F.
BA Miliolina
foraminifera
BT Protista
Invertebrata
NA Miliolidae
Ophthalmidium
Quinqueloculina

Miliolidae
Family. Includes use on level 3 under foraminifera(1) Miliolacea(2).
BA Miliolacea
Miliolina
foraminifera
BT Protista
Invertebrata

Miliolina
Includes use on level 2 under foraminifera(1). See list F.
BA foraminifera
BT Protista
Invertebrata
NA Alveolinellidae
Miliolacea

military geology
Term introduced in 1981.
SA geology

Milk River Formation
Composed of well-sorted sandstones. In S Alberta.
BT Upper Cretaceous
Cretaceous
SA Alberta
Canada

Mill Creek
Rises in Lassen Volcanic National Park and flows into Sacramento River.
BT California
United States

Millard County
W central Utah.
BT Utah
United States
NT Black Rock

millerite
UF nickel pyrites
BA sulfides
BT minerals

Millevaches Plateau
Term introduced in 1981. Tableland of the Central Massif, in Correze and Creuse departments, central France.
CO N451500N460000
E0023000E0014500
BT Central Massif
France
SA Correze
Creuse

Millstone Grit
Provincial series, Europe.
BA Upper Carboniferous
Carboniferous
BT Paleozoic

Milne Land
An island in Scoresby Sound which is a deep inlet of Greenland Sea on E coast.
BT Greenland
Arctic region

Milwaukee
City on Lake Michigan in Milwaukee County in SE.
BT Wisconsin
United States

mimetite
BZ arsenates
chlorides
BT minerals

Minas Basin
NE extension of Bay of Fundy in central Nova Scotia. Also search Minas AND appropriate area term.
BT Nova Scotia
Canada
SA Bay of Fundy
Cobequid Bay

Minas Gerais
State in E central.
CO S233000S140000
W0394500W0510000
BT Brazil
NT Araxa
Itabira
Ouro Preto
Pocos de Caldas
SA Areado Formation
Bambui Group
Bauru Formation
Brazilian Shield
Irati Formation
Sao Francisco Basin

Minas Series
BT Precambrian
SA Brazil

Mindanao
Second largest and southernmost major island.
BT Philippine Islands

Mindel
Europe. Above Gunz, below Riss. Includes use on level 3 under age terms(1). See list E.
BA Pleistocene
Quaternary
BT Cenozoic

Mindel-Riss
use Mindel/Riss Interglacial

Mindel-Riss Interglacial
use Mindel/Riss Interglacial

Mindel/Riss Interglacial
Europe. Term applied in the Alps to the second interglacial stage of the Pleistocene Epoch, after the Mindel glacial stage and before the Riss. Also search Mindel-Riss; Mindel-Riss Interglacial.
UF Mindel-Riss
Mindel-Riss Interglacial
BA Pleistocene
Quaternary
BT Cenozoic

Mindoro
Island in central Philippines SW of Luzon.
BT Philippine Islands
SA Mansalay Formation

mine drainage
Term introduced in 1981.
SA acid mine drainage
drainage
environmental geology
mines

mine reactivation
Term introduced in 1981.
SA economics
land use
mines

mineragraphy
A valid term through 1976. After 1976, use ore microscopy.

mineral assemblages
Includes use on level 2 under metamorphic rocks(1).
SA autometamorphism
metamorphic rocks
paragenesis

mineral associations
A valid term through 1977. Use mineral assemblages.

mineral cleavage
Term introduced in 1981. Before 1981, search cleavage AND minerals.
SA cleavage
minerals

mineral collecting
As of 1981, no longer a valid term for GeoRef. Use collecting.

mineral composition
Includes use on level 3 under igneous rocks(1), sedimentary rocks(1), metamorphic rocks(1), and under soils(1) composition(2).
SA composition

Mineral County
Index states as applicable.
BX Colorado
Montana
Nevada
West Virginia
BT United States

mineral data
Includes use on level 2 under clay mineralogy(1); on level 3 under minerals(1).
SA areal studies
clay mineralogy
data
mineralogy
minerals

mineral deposits
Do not include water resources or fuels. Used for substantial discussions of the genesis of ore deposits. Includes use on level 1 (list A) in combination with genesis (i.e. mineral deposits, genesis). Term set options are:
commodity (List C)
topic
controls
type of control [geochemical controls, hydrogeological controls, lithologic controls, mechanical controls, paleogeographic controls, stratigraphic controls, structural controls]
processes
type of process [endogene processes, epigene processes, exhalative processes, exogene processes, hydrothermal processes, igneous processes, metamorphism, plate tectonics, sedimentary processes, supergene processes, syngenesis, volcanism, weathering]
topic [age, causes, concepts, environment, experimental studies, interpretation, mechanism, patterns, theoretical studies]
subtopic
SA alpine-type
bauxitization
blind deposits
cementation deposits
controls
detrital deposits
discoveries
disseminated deposits
economic evaluation
economic geology
endogene processes
enrichment
epigene processes
epithermal processes
exhalative processes
exogene processes
fluid inclusions
gangue
genesis
geochemical controls
geochemistry
haloes
host rocks
hydrogeological controls
hydrothermal processes
hypabyssal processes
hypothermal processes
igneous processes
impregnated deposits
inclusions
intramagmatic deposits
isotopes
kuroko-type
leakage anomalies
lenses
lithologic controls
localization
massive deposits
mechanical controls
mesothermal processes
metallogenic provinces
metallogeny
metasomatism
mineral exploration
mineral resources
mineralization
mining
mining geology
mississippi valley-type
ore bodies
ore transport
ore-forming fluids
oxidation zone
paleogeographic controls
paragenesis
phase equilibria
placers
pneumatolysis
porphyry copper
reflectivity
roll-type deposits
sedimentary processes
stockwork deposits
stratabound deposits
stratiform deposits
stratigraphic controls
structural controls
sublimation
supergene processes
syngenesis
telethermal processes
vein systems
veins
wallrock alteration
weathering

mineral economics
SA economic geology
economics
mineral resources

mineral exploration
Used for substantial discussions of exploration for "commodities" (list C) excluding fuel deposits and water resources. Used primarily for the application of the methods listed below. Includes use on level 1 (list A). Term set options are:
methods [biogeochemical methods, geobotanical methods, geochemical methods, geological methods, geomorphological methods, geophysical methods, hydrological methods, methods, photogeologic methods, remote sensing]
topic
topic [applications, bibliography, concepts, history, instruments, objectives, ore guides, programs, techniques]
subtopic
SA biogeochemical methods
dispersion patterns
economic geology
electromagnetic methods
exploration
geobotanical methods
geochemical methods
geochemistry
geological methods
geomorphological methods
geophysical methods
geophysical surveys
haloes
hydrological methods
maps
mineral deposits
mineral resources
minerals
networks
ore guides
photogeologic methods
pit sections
placers
primary dispersion
programs
remote sensing
secondary dispersion
stream sediments

mineral facies
A valid term through 1976. Use facies.

mineral inclusions
Includes use on level 2 under inclusions(1).
BT inclusions
SA fluid inclusions
host rocks
xenoliths

mineral localities
Term introduced in 1981. Use only for collecting.
SA collecting
popular geology

mineral resources
For very general treatments. Includes use on level 1 as a commodity term (list C); on level 2 under ocean floors(1).
SA depletion
economic geology
environmental geology
exploration
mineral deposits
mineral economics
mineral exploration
minerals
ocean floors
resources

mineral sequence
use paragenesis

mineral soap
use bentonite

mineral synthesis
A valid term through 1970. After 1970, see synthesis under crystal growth(1) and under phase equilibria(1).

mineral waters
Includes use on level 2 and 3 under springs(1).
SA brines
ground water
medicinal waters
springs
thermal waters
water

mineral zoning
A valid term through 1975. After 1975, use zoning.

mineralization
Restricted to ore deposits. Not used for paleontology.
SA localization
metallogenic provinces
mineral deposits
ore-forming fluids

mineralogy
Used for the discipline as a whole.

Includes use on level 1 (list A); on level 2 under areas (list B), under symposia(1), and under education(1). If 1, term set options are:
topic [applications, bibliography, catalogs, classification, concepts, education, experimental studies, history, instruments, methods, nomenclature, objectives, philosophy, practice, principles, symposia, textbooks, theoretical studies]
subtopic
SA atomic packing
bibliography
catalogs
clay mineralogy
collecting
crystal chemistry
crystal growth
crystal structure
crystallography
decrepitation
lexicons
microscope methods
mineral data
minerals
optical mineralogy
optical properties
ore microscopy
petrology
phase equilibria
reflectance
reflectivity

minerals
Includes use on level 1 (list A); on level 2 under phase equilibria(1) and weathering(1). Used for descriptions of mineral occurrence or of minerals themselves. Also search mineral. If 1, term set options are:
mineral group (List L)
mineral species
topic [chemical analysis, crystal chemistry, crystal structure, mineral data, occurrence, optical properties, spectra] (more than one single mineral)
subtopic
NT accessory minerals
alloys
antimonates
antimonites
arsenates
arsenites
authigenic minerals
borates
carbonates
chromates
coexisting minerals
elongate minerals
halides
heavy minerals
industrial minerals
iodates
light minerals
magnetic minerals
maskelynite
NX metabentonite
NT miscellaneous minerals
mixed-layer minerals
molybdates
native elements
new minerals
nitrates
opaque minerals
ore minerals
organic compounds
oxalates
oxides
phosphates
secondary minerals
selenates
selenites
silicates
sulfates
sulfides
sulfosalts
tellurates
tellurites
tungstates
uranium minerals
vanadates
NX wehrlite
SA authigenesis
birefringence
cell dimensions
chemical analysis
clay mineralogy
color centers
crystal chemistry
crystal dislocations
crystal field
crystal growth
crystal structure
crystallography
crystals
dehydration
fluid inclusions
formula
gems
habit
hardness
inclusions
isotropy
lamellae
localization
luminescence
metamict minerals
metamictization
microhardness
mineral cleavage
mineral data
mineral exploration
mineral resources
mineralogy
optical dispersion
optical extinction
optical properties
overgrowths
paragenesis
phenocrysts
pleochroism
polymorphism
polytypism
pseudomorphism
reflectance
reflectivity
refractive index
single-crystal method
spectra
standard materials
synthesis
synthetic materials
thermodynamic properties
transformations
typomorphism
undulatory extinction
unit cell
water of crystallization
water of dehydration

minerals in meteorites
Not a valid index term. Place minerals in appropriate group. See list L.

mines
As of 1978, term is used on level 2 under land subsidence(1), land use(1), and underground installations(1).
SA acid mine drainage
land subsidence
land use
mine drainage
mine reactivation
mining
mining geology
open-pit mining
quarries
solution mining
underground installations

minette
As of 1981, includes use on level 3 under igneous rocks(1) lamprophyres(2). See list H. Before 1981, included use on level 3 under igneous rocks(1) lamprophyre and carbonatite family(2).
BA lamprophyres
BT igneous rocks

minicomputers
SA automatic data processing
computers
hand calculators
microcomputers

mining
NT solution mining
surface mining
SA fines
mineral deposits
mines
mining geology
mining legislation
recovery
spoils
tailings

mining geology
Geology applied to mining operations. Includes use on level 1 (list A). Term set options are:
topic [applications, bibliography, catalogs, classification, concepts, evaluation, history, instruments, methods, nomenclature, objectives, practice, production control, symposia, technology, textbooks]
subtopic
SA dredging
economic geology
engineering geology
environmental geology
evaluation
explosions
gases
geology
geophysical methods
geophysical surveys
hydraulic fracturing
leaching
mineral deposits
mines
mining
open-pit mining
production control
quarries
reclamation
recovery
solution mining
strip mining
technology

mining legislation
Term introduced in 1981.
SA legislation
mining

mining, open-pit
use open-pit mining

Minneapolis
City in Hennepin County in SE central.
BT Minnesota
United States

Minnelusa Formation
Pennsylvanian and Permian. Composed of sandstones and evaporite deposits. W South Dakota, and NE Wyoming.
BT Paleozoic
SA Pennsylvanian
Permian
South Dakota
Wyoming

Minnesota
Includes use on level 1 as an area term (list O). For term set options see list B.
CO N433000N490000
W0894500W0971000
BA Midwest
United States
NX Benton County
Cass County
Clay County
Clearwater County
Cook County
Douglas County
NT Duluth
Duluth Complex
NX Ely
Grant County
NT Itasca County
NX Jackson County
Lake County
Lincoln County
Lyon County
Marshall County
NT Mesabi Range
Minneapolis
Minnesota River valley
NX Murray County
NT New Ulm
NX Pennington County
Polk County
Pope County
NT Red Lake
NX Rice County
Saint Louis County
NT Saint Paul
NX Scott County
Stevens County
NT Two Harbors
Vermilion Range
NX Washington County
SA Animikie Group
Benton Formation
Biwabik Iron Formation
Cedar Valley Formation
Dakota Formation
Franconia Formation
Great Lakes region
Gunflint Iron Formation
Keewatin
Knife Lake Group
Lake Agassiz
Lake Superior
Maquoketa Formation
Midcontinent
Mississippi River
Mississippi Valley
Niobrara Formation
North Shore Volcanics
Penokean Orogeny
Pierre Shale
Platteville Formation
Rainy Lake
Rainy River
Saganaga Lake
Saint Peter Sandstone
Shakopee Formation
Sioux Quartzite
Southern Province
Superior Province
Thomson Slate
Upper Mississippi Valley

Minnesota River valley
S Minnesota.
BT Minnesota
United States

minor elements
Includes use on level 3. See list D.
SA elements
major elements
nutrients
trace elements

minor planets
use asteroids

minor-element analyses
Includes use on level 3 under chemical analysis(1) methods (2). Before 1974, also search minor element analyses.
SA analysis
chemical analysis
elements
major-element analyses
trace-element analyses

Minorca
Second largest of Spain's Balearic Islands off E coast of Spanish mainland.
UF Menorca
BT Balearic Islands
Spain

Minorca Rise
Term introduced in 1981. Off Minorca in the Mediterranean Sea. Before 1981, also search Menorca Rise.
CO N393500N403000
E0050000E0034500
UF Menorca Rise
BT Mediterranean Sea

Minsk
City and oblast in central.
BT Byelorussia
USSR

Minturn Formation
Consists chiefly of grayish to greenish sandstones, conglomerates, and shales or siltstones. Central Colorado.
BT Pennsylvanian
SA Colorado

Minusinsk
Town in SW Krasnoyarsk Kray on the Yenisei River in SSW Siberia.
BT Russian Republic
USSR

Minusinsk Basin
Coal basin in extreme S Krasnoyarsk Kray in SSW Siberia. A good agricultural region. Also search Minusinsk.
CO N520000N553000
E0950000E0870000
BX Russian Republic
BZ West Siberia
BT USSR

minyulite
BZ fluorides
phosphates
BT minerals

Miocene
World. Above Oligocene, below Pliocene. Includes use on level 1 as an age term (list E); on level 2 under paleo- terms, e.g. paleoecology, paleogeography, paleomagnetism; on level 3 under age terms(1).
BA Neogene
Tertiary
BT Cenozoic
NT Arikaree Group
NA Badenian
Barstovian
NT Fleming Formation
Green Tuff Formation
Hawthorn Formation
NA Langhian
lower Miocene
middle Miocene
NT Monterey Formation
Motojuku Formation
Ngorora Formation
Oktemberyan Series
NA Ottnangian
NT Pirabas Formation
Saint Marys Formation
NA upper Miocene
Vallesian
Vindobonian
NT Wood Mountain Formation
SA Columbia River Basalt
Cuddalore Series
Hemphillian
John Day Formation
Loup Fork Group
Neyveli Lignite
Pannonian
Poltava Series
Siwalik System
Snoqualmie Batholith
Warkalli Formation

miogeosynclines
Includes use on level 3 under geosynclines(1) or tectonics(1).
UF miomagmatic zone
SA eugeosynclines
geosynclines
tectonics

Miogypsina
Genus. Includes use on level 3 under foraminifera(1) Rotaliacea(2).
BA Miogypsinidae
Rotaliacea
Rotaliina
foraminifera
BT Protista
Invertebrata

Miogypsinidae
Family. Includes use on level 3 under foraminifera(1) Rotaliacea(2).
BA Rotaliacea
Rotaliina
foraminifera
BT Protista
Invertebrata
NA Miogypsina

miomagmatic zone
use miogeosynclines

miospores
Includes use on level 2 under palynomorphs(1). See list F.
BA palynomorphs
NA Cicatricosisporites
Cladophlebis
Classopollis
Dryas
Florinites
SA Artemisia
exine
megaspores
monolete taxa
spores
sporopollenin

mirabilite
BA sulfates
BT minerals

Mirador Formation
BT Eocene
SA Venezuela

Miramichi Bay
Inlet of Gulf of Saint Lawrence. Also search Miramichi.
BT New Brunswick
Canada

Miranda
State on the Caribbean Sea.
BT Venezuela

Mirgalimsay
Village in S near Syr Darya River.
BT Kazakhstan
USSR

Mirny
use Mirnyy Station

Mirnyy
City in Yakutia.
BT Yakutia
Russian Republic
USSR

Mirnyy Station
Term introduced in 1981. USSR IGY station on Davis Sea near Shackleton Ice Shelf on the Indian Ocean side. Before 1981, also search Mirnyy AND Antarctica; Mirny AND Antarctica.
UF Mirny
BT Antarctica

mirrorstone
use muscovite

Mirsk
Town near Czechoslovak border.
BT Wroclaw
Poland

Mirzapur
City in SE.
BT Uttar Pradesh
India

miscellanea
Includes use on level 2 under fossil group(1) if fossil is not yet classified. See list F.

miscellaneous minerals
Includes use on level 2 under minerals(1) for the case of several mineral groups when the topic is not emphasized or for a mineral of unknown affinity. See list L.
BT minerals
SA uranium minerals

miscibility
use immiscibility

miscibility gap
SA phase equilibria

Mishash Formation
BT Cretaceous
SA Israel

Missi Group
Its basal unit is normally composed of a conglomerate consisting of angular to subrounded boulders or pebbles along with minor amounts of chert, jasper and quartz. The unit grades upwards to graywacke and subgraywacke with thin beds of conglomerate. In the Flin Flon area of W Manitoba and E Saskatchewan.
BT Precambrian
SA Canada
Manitoba
Saskatchewan

Mississippi
Includes use on level 1 as an area term (list O). For term set options see list B.
CO N301500N350000
W0880500W0914000
BA Southern U.S.
United States
NX Adams County
Benton County
Calhoun County
Carroll County
Clarke County
Clay County
De Soto County
Franklin County
Greene County
Hancock County
Harrison County
NT Hattiesburg
NX Jackson County
Jasper County
Jefferson County
Jones County
Lafayette County
Lawrence County
Lee County
Lincoln County
Madison County
Marion County
Marshall County
Monroe County
Montgomery County
Newton County
NT Pascagoula River basin
NX Perry County
Pike County
Pontotoc County
NT Rankin County
NX Scott County
Smith County
NT Tishomingo
NX Union County
Warren County
Washington County
Wayne County
Webster County
Wilkinson County
SA Amite River
Arkansas River
Black Warrior Basin
Buckner Formation
Chattanooga Shale
Chickasawhay Formation
Citronelle Formation
Claiborne Group
Clayton Formation
Cotton Valley Group
Fort Payne Formation
Gulf Coastal Plain
Jackson Group
Marianna Limestone
Midway Group
Mississippi Embayment
Mississippi River
Mississippi Valley
New Madrid region
Pottsville Group
Ripley Formation
Smackover Formation
Talladega Group
Tuscaloosa Formation
Vicksburg Group
Warsaw Formation
Wilcox Group
Yazoo Clay

Mississippi Delta
On Gulf of Mexico 100 miles SE of New Orleans. Also search Mississippi River delta.
UF Mississippi River delta
BT Louisiana
United States
SA Mississippi River

Mississippi Embayment
A geosynclinal area including a section of the U S Gulf Coast and extending northward up the Mississippi Valley. Index states as applicable.
BT United States
SA Alabama
Arkansas
Illinois
Kentucky
Mississippi
Mississippi Valley
Missouri
New Madrid region
Tennessee

Mississippi River
Rises in Lake Itasca in NW Minnesota and flows SE and S into the Gulf of Mexico. Index states as applicable.
BT United States
SA Arkansas
Illinois
Iowa
Kentucky
Louisiana
Minnesota
Mississippi
Mississippi Delta
Mississippi River basin
Mississippi Valley
Missouri
Tennessee
Wisconsin

Mississippi River basin
Drainage basin including the Ohio and Missouri river systems comprises 2/5 of the total U.S. area covering all or parts of 31 states plus S Alberta and Saskatchewan. It drains much of the interior lowlands, Great Plains, central Gulf coastal plain, a portion of the Appalachians, and the Rocky Mountains W to the Continental Divide. Index countries as ap-

plicable.
SA Canada
Mississippi River
United States

Mississippi River delta
use Mississippi Delta

Mississippi River valley
use Mississippi Valley

Mississippi Valley
A level 1 area term as of 1978. River valley. Also search Mississippi River valley. Index states as applicable.
UF Mississippi River valley
BA United States
North America
NA Lower Mississippi Valley
Upper Mississippi Valley
SA Arkansas
Illinois
Iowa
Kentucky
Louisiana
Minnesota
Mississippi
Mississippi Embayment
Mississippi River
Missouri
Tennessee
Wisconsin

mississippi valley type
use mississippi valley-type

mississippi valley-type
Also search mississippi valley type.
UF mississippi valley type
SA lead ores
lead-zinc deposits
mineral deposits
zinc ores

Mississippian
Includes use on level 1 as an age term (list E); on level 2 under paleoterms, e.g. paleoecology, paleogeography, paleomagnetism. After the Devonian and before the Pennsylvanian. Approximate equivalent of Lower Carboniferous of European usage.
BA Carboniferous
BT Paleozoic
NT Bear Gulch Limestone Member
Borden Group
Leadville Formation
Logan Formation
NA Lower Mississippian
NT Madison Group
NA Middle Mississippian
NT Monte Cristo Limestone
Newman Limestone
Price Formation
Rampart Group
Redwall Limestone
Stanley Group
NA Upper Mississippian
Valmeyeran
NT Windsor Group
Yule Marble
SA Antler Orogeny
Bartlesville Sand
Bedford Shale
Berea Sandstone
Bird Spring Formation
Catskill Formation
Chattanooga Shale
Dinantian
Johns Valley Formation
Lisburne Group
Pennsylvanian
Springer Formation
Woodford Shale

Missoula Group
Includes numerous named formations: among these are Marsh Shale in Helena region; Striped Peak and Libby formations in NW Montana, five formations near Missoula, and others in and S of Glacier National Park. In N Idaho and NW Montana.
BT Precambrian
SA Belt Supergroup
Bonner Formation
Idaho
Montana

Missouri
Includes use on level 1 as an area term (list O). For term set options see list B.
CO N360000N403500
W0890500W0954500
BA Midwest
United States
NX Barton County
Benton County
Boone County
Buchanan County
Butler County
Caldwell County
Carroll County
Carter County
Cass County
Christian County
Clark County
Clay County
Clinton County
Crawford County
Dade County
Dallas County
NT Decaturville
Devils Icebox
NX Douglas County
Franklin County
Greene County
Grundy County
Harrison County
Howard County
Iron County
Jackson County
Jasper County
Jefferson County
Johnson County
NT Joplin
NX Kansas City
Lafayette County
Lawrence County
Lewis County
Lincoln County
Livingston County
Macon County
Madison County
Marion County
Mercer County
Monroe County
Montgomery County
Morgan County
NT New Madrid
New Madrid County
NX Newton County
Osage County
Perry County
Phelps County
Pike County
Platte County
Polk County
Pulaski County
Putnam County
Randolph County
NT Rolla Quadrangle
Saint Charles County
NX Saint Clair County
NT Saint Francois Mountains
Saint Louis
NX Saint Louis County
Saline County
Scott County
Shannon County
Shelby County
Sullivan County
Texas County
Warren County
Washington County
Wayne County
Webster County
SA Bainbridge Formation
Bonneterre Formation
Cabaniss Formation
Cedar City Formation
Chattanooga Shale
Cherokee Group
Chesterian
Clayton Formation
Desmoinesian
Douglas Group
Ervine Creek Limestone
Everton Formation
Excello Shale
Fayetteville Formation
Fernvale Formation
Forest City Basin
Grand River
Kansas City Group
Keokuk Limestone
Labette Shale
Lawrence Formation
Lecompton Limestone
Maquoketa Formation
Marmaton Group
Midcontinent
Midway Group
Mississippi Embayment
Mississippi River
Mississippi Valley
Missouri River
Missouri River basin
Missouri River valley
New Madrid region
Oread Limestone
Ozark Mountains
Plattsburg Limestone
Plattsmouth Limestone Member
Pleasanton Group
Red Eagle Limestone
Richmond Group
Ripley Formation
Rock Lake Shale Member
Saint Laurent Limestone
Saint Louis Limestone
Saint Peter Sandstone
Sainte Genevieve Limestone
Salem Limestone
Smithville Formation
Springfield
Stanton Formation
Tonganoxie Sandstone
Upper Mississippi Valley
Versailles
Wabaunsee Group
Warsaw Formation
White River
Wilcox Group
Wyandotte Limestone

Missouri River
Rises in S Montana and flows E and SE to join the Mississippi River N of St. Louis. Index states as applicable.
BT United States
SA Iowa
Kansas
Missouri
Missouri River basin
Missouri River valley
Montana
Nebraska
North Dakota
South Dakota

Missouri River basin
Drainage basin. Index states and provinces as applicable.
BT North America
SA Alberta
Canada
Colorado
Iowa
Kansas
Missouri
Missouri River
Montana
Nebraska
North Dakota
Saskatchewan
South Dakota
United States
Wyoming

Missouri River valley
Index states as applicable.
BT United States
SA Iowa
Kansas
Missouri
Missouri River
Montana
Nebraska
North Dakota
South Dakota

Missourian
Provincial series, North America: lower Upper Pennsylvanian, above Desmoinesian, below Virgilian. Includes use on level 3 under age terms(1). See list E.
UF Missourian Series
BA Upper Pennsylvanian
Pennsylvanian
Carboniferous
BT Paleozoic
NT Kansas City Group
Lansing Group
Plattsburg Limestone
Pleasanton Group
Rock Lake Shale Member
Seminole Formation
Stanton Formation
Wann Formation
Wyandotte Limestone

Missourian Series
use Missourian

Mistastin Lake
N Labrador.
BT Labrador
Canada

Mitchell County
Index states as applicable.
BX Georgia
Iowa
Kansas
North Carolina
Texas
BT United States

Mitsuishi
Town in SW Honshu. Sometimes spelled Mituisi.
UF Mituisi
BT Honshu
Japan

Mitterberg
Village in W central part of country.
BT Salzburg State
Austria

Mituisi
use Mitsuishi

Miura Peninsula
SE Honshu. Extends into Sagami Sea S of Yokohama and Tokyo Bay.
BT Honshu
Japan

mixed crystals
use solid solution

mixed-layer minerals
Term introduced in 1978. Before 1978, search clay minerals AND mixed-layer.
BT minerals
SA clay minerals

mixing
Used as a general term.
SA processes
separation

mixite
BA arsenites
BT minerals

SA mixtite

mixtite
See list I.
BA clastic rocks
BT sedimentary rocks
SA diamictite
mixite

Miyagi
Prefecture in N.
BT Honshu
Japan

Miyake-Jima
Island of Izu-shichito group in Greater Tokyo, S of O-shima Island and Sagami Sea. Active volcano on island.
UF Miyaki Island
Miyaki-Jima Island
BT Honshu
Japan
SA Izu-shichito

Miyaki Island
use Miyake-Jima

Miyaki-Jima Island
use Miyake-Jima

Miyazaki
City and Prefecture in SE.
BT Kyushu
Japan

Miye
use Mie

Mizunami
Town in central.
BT Honshu
Japan

Mizusawa
Town in N.
BT Honshu
Japan

MM scale
use modified Mercalli scale

Mn
use manganese

Mn-53
Includes use on level 3 under isotopes(1).
SA isotopes
manganese

Mn-54
Includes use on level 3 under isotopes(1).
SA isotopes
manganese

Mo
use molybdenum

Moab
City in Grand County in E.
BT Utah
United States

Moapa Valley
NE of Las Vegas in Clark County in S.
BT Nevada
United States

Mobile
City on Mobile Bay in Mobile County in S.
BT Alabama
United States

mobile belts
Includes use on level 3 under geosynclines(1).
UF belts, mobile
SA crust
fold belts
geosynclines
orogenic belts
tectonics

Mobile County
On Mobile Bay in S.
BT Alabama
United States

mobility
Used as a general term.
SA erosion
tectonics

mobilization
Restricted to geochemical meaning.
SA deformation
geochemistry
migration of elements

Mocamedes
use Mossamedes

Moctezuma
Town in E central.
BT Sonora
Mexico

modal analysis
Includes use on level 3 under mineralogy(1) methods(2), under petrology(1) methods(2), and under sedimentary petrology(1) methods(2).
SA analysis
geochemistry
petrology

models
Includes use on level 2 under ground water(1).
NA infinite models
mathematical models
one-dimensional models
physical models
semi-infinite models
spherical models
three-dimensional models
two-dimensional models
two-layer models
two-phase models
SA automatic data processing
representative basins
simulation

Modena
City and province in central.
BT Emilia-Romagna
Italy

modern
Used for present-day studies in conjunction with Holocene.
SA Holocene

modified Mercalli scale
Term introduced in 1978. Includes use on level 3 under earthquakes(1).
UF Mercalli scale, modified
MM scale
SA earthquakes
seismic intensity
seismology

Modoc County
In extreme NE. Also search Modoc.
BT California
United States

Modoc Plateau
High, semiarid, and volcanic plateau in NE California. Also search Modoc.
BT California
United States

modulus, bulk
use bulk modulus

modulus of compression
use compressibility

modulus of elasticity
use elastic constants

modulus of incompressibility
use bulk modulus

modulus of rigidity
use shear modulus

modulus, shear
use shear modulus

Moenkopi Formation
Lower and middle (?) Triassic or Triassic (?). In SW Utah, subdivided into (ascending) Timpoweap, Lower Red, Virgin Limestone, Middle Red, Shnabkaib, and Upper Red members. Arizona, California, Colorado, Nevada, and S Utah.
BT Triassic
SA Arizona
California
Colorado
Lower Triassic
Middle Triassic
Nevada
Utah

Moerintherioidea
Term introduced in 1981. Includes use on level 2 under Mammalia(1). See list F.
BA Proboscidea
Mammalia
BT Tetrapoda
Vertebrata

Moesia
Ancient region S of the lower Danube. Includes Bulgaria, SE Romania, and Serbia. Index Serbia and countries as applicable.
BT Europe
SA Bulgaria
Romania
Serbia

Moesian Platform
Index countries as applicable.
BT Europe
SA Bulgaria
Romania

Mogilev
City on the Dnieper River. Also an oblast.
UF Mogilev on the Dnieper
BT Byelorussia
USSR

Mogilev on the Dnieper
use Mogilev

Mogilno
Town in NW central.
BT Bydgoszcz
Poland

Mogollon Mountains
Term introduced in 1981. SW New Mexico in Catron County.
CO N330000N334000
W1083000W1085000
BT Catron County
New Mexico
United States
SA Datil-Mogollon volcanic field
Mogollon Plateau
Mogollon Rim

Mogollon Plateau
Tableland S of Winslow in E central.
BT Arizona
United States
NT Mogollon Rim
SA Mogollon Mountains

Mogollon Rim
Term introduced in 1981. S escarpment of Mogollon Plateau.
CO N335000N344000
W1090000W1113500
BT Mogollon Plateau
Arizona
United States
SA Mogollon Mountains

Mohave County
S of Lake Mead in NW.
BT Arizona
United States

Mohave Desert
use Mojave Desert

Mohawk River Valley
use Mohawk Valley

Mohawk Valley
River valley in E central New York.
UF Mohawk River Valley
BT New York
United States

Mohns Ridge
Term introduced in 1981. NE of Jan Mayen in the Arctic Ocean.
CO N700000N740000
E0100000W0080000
BT Arctic Ocean

Moho
use Mohorovicic discontinuity

Mohorovicic discontinuity
Includes use on level 1 (list A) for discussions which stress the crust-mantle transition. Term set options are:
topic [causes, composition, depth, detection, geometry, identification, interpretation, Mohole project, observations, patterns, properties, temperature]
subtopic (no area term)
UF M-discontinuity
Moho
SA asthenosphere
basement
continental drift
core
crust
depth
discontinuities
Earth
earthquakes
elastic waves
isostasy
lithosphere
mantle
plate tectonics
sea-floor spreading
seismology
tectonophysics

Moinian
Europe. See list E.
BA upper Precambrian
Precambrian

moissanite
Meteorite mineral. As of 1981, includes use on level 3 under minerals(1) carbides(2). See list L. Before 1981, included use on level 3 under minerals(1) native elements and alloys(2).
BA carbides
alloys
BT minerals
SA carbon
meteorites
silicides
silicon

moisture
Includes use on level 3 under soils(1). Also search moisture content.
UF moisture content
SA atmospheric precipitation
desiccation
humidity
soils
water
water vapor

moisture content
use moisture

Mojave Desert
Arid basin in S California. Also search Mojave.
UF Mohave Desert
BT California
United States
NT Antelope Valley

Mokoia
Village in SW.
BT North Island
New Zealand

molasse
Includes use on level 3 under sedimentary rocks(1) clastic rocks(2). See list I.
BA clastic rocks
BT sedimentary rocks
SA terrigenous materials

Molasse Basin
A geologic-geographic term which refers to the feature in Europe. Index countries as applicable. Also search molasse in combination with country name.
BT Europe
NZ Swiss Molasse Basin
SA Austria
Germany
Switzerland

Moldanubian
Also search Moldanubian Complex; Moldanubian Zone; Moldanubicum; Moldanubicum Complex; Moldanubikum.
UF Moldanubian Complex
Moldanubian Zone
Moldanubicum
Moldanubicum Complex
Moldanubikum
BT Precambrian
SA Archean
Czechoslovakia

Moldanubian Complex
use Moldanubian

Moldanubian Zone
use Moldanubian

Moldanubicum
use Moldanubian

Moldanubicum Complex
use Moldanubian

Moldanubikum
use Moldanubian

Moldava Valley
River valley. Also search Moldava River and Moldova Valley.
UF Moldova Valley
BT Moldavia
Romania

Moldavia
Region E of Transylvania and N of E Walachia in Romania, and the Moldavian Soviet Socialist Republic in the USSR. Index countries as applicable.
BT Europe
NT Bacau
Bessarabian
Bicaz Valley
Bistrita Valley
Gutii Mountains
Iacobeni
Moldava Valley
Moldavian Platform
Piatra-Neamt
Rarau Massif
Suceava
Teleajen Valley
Tiraspol
Trotus Valley
SA Dniester River
Dniester-Prut Interfluve
European USSR
Prut River
Romania
Romanian Plain
Russian Plain
Russian Platform
Siret River
USSR

Moldavian Plateau
use Moldavian Platform

Moldavian Platform
Also search Moldavian Plateau.
UF Moldavian Plateau
BT Moldavia
Romania

moldavite
BA tektites

Moldova Noua
Town in E near the Serbian border. Also search Moldova-Noua.
UF Moldova-Noua
BT Banat
Romania

Moldova Valley
use Moldava Valley

Moldova-Noua
use Moldova Noua

molecular structure
SA crystal structure

Molise
Term introduced in 1981. Region on the Adriatic Sea.
CO N412300N420500
E0151000E0140000
BT Italy

mollisol
use active layer

Mollisols
Includes use on level 3 under soils(1). See list M.
BA soils

Mollusca
Includes use on level 1 and 2 as a fossil term (list F). Before 1981, also search mollusks. As of 1981, search mollusks only for biostratigraphy.
BT Invertebrata
NA Aplacophora
Bivalvia
Cephalopoda
Gastropoda
Hyolithes
Monoplacophora
Polyplacophora
Rostroconchia
Scaphopoda
Tentaculitida
SA Dacryoconarida
mollusks
pearls
reef builders

mollusks
As of 1981, restricted to use as the common name for Mollusca. See list F. To be used for Mollusca when they are used as tools in stratigraphy. Includes use on level 1. Before 1981, search (mollusks OR Mollusca) AND biostratigraphy. If 1, term set options are:
biostratigraphy
age [List E]
SA biostratigraphy
Mollusca

Molodezhnaya Station
USSR station on Alasheyev Bight in Enderby Land between Lutzow-Holm Bay and Cape Ann on the Indian Ocean side. Also search Molodezhnaya.
BT Antarctica

Molokai
Island in the Hawaiian Islands between Oahu and Maui.
BT Hawaii
United States

Molotov
City W of the Urals in Perm Oblast.
BT Russian Republic
USSR
SA Perm

Molteno
Town in E.
BT Cape Province
South Africa

Moluccas
Group of islands between Celebes and New Guinea.
CO S071000N030000
E1350000E1240000
UF Spice Islands
BT Indonesia

Mcluya
use Moulouya River

molybdates
Includes use on level 2 under minerals(1). See list L.
BT minerals
NA ferrimolybdite
powellite
umohoite
wulfenite
SA tungstates

molybdenite
BA sulfides
BT minerals
SA molybdenum

molybdenum
Includes use on level 1 and 2 as a chemical element (list D). As of 1981, use molybdenum ores for molybdenum as a commodity. Before 1981, included use on level 1 as a commodity term.
UF Mo
SA elements
molybdenite
molybdenum ores

molybdenum ores
Term introduced in 1981. Includes use on level 1 and 2 as a commodity term (list C). Before 1981, search molybdenum AND ore deposits.
SA molybdenum

Monaco
On the Mediterranean Sea near the French-Italian border. Includes use on level 1 as an area term (list O). For term set options see list B.
BA Western Europe
Europe
NT Monte Carlo

monazite
As of 1981, use monazite deposits for monazite as a commodity. Before 1981, included use on level 1 as a commodity term.
BA phosphates
BT minerals
SA heavy minerals
monazite deposits

monazite deposits
Term introduced in 1981. Includes use on level 1 and 2 as a commodity term (list C). Before 1981, search monazite AND deposits.
SA heavy mineral deposits
monazite
placers
rare earth deposits
thorium ores

Monchegorsk
City in Murmansk Oblast about 70 miles S of Murmansk in extreme NW European USSR.
BT Russian Republic
USSR

monchiquite
As of 1981, includes use on level 3 under igneous rocks(1) lamprophyres(2). See list H. Before 1981, included use on level 3 under igneous rocks(1) lamprophyre and carbonatite family(2).
BA lamprophyres
BT igneous rocks

Moneron Island
Soviet island in Japan Sea 30 miles off SW Sakhalin.
BT Russian Republic
USSR

monetite
BA phosphates
BT minerals

Monghyr
City in NE central.
BT Bihar
India

Mongolia
Includes use on level 1 as an area term (list O). For term set options see list B.
CO N420000N520000
E1200000E0870000
UF Outer Mongolia
BA Far East
Asia
NT Altan Teeli
Bayan-Nuurin-khotnor Basin
Bayn Dzak
Choybalsan
Hangay Mountains
Ikh-Khayrkhan
Khubsugul
Kobdo
Mongolian Altai
Selenga
Ulan Bator
SA Altai Mountains
Altai-Sayan region
Amur Basin
Djadokhta Formation
Gobi Desert
Kerulen River
Selenga River valley
Tannu-Ola Range
Yenisei Basin
Yenisei-Khatanga basin

Mongolian Altai
Southeastern extension of the Altai Mountains.
BT Mongolia
SA Altai Mountains

monitoring
SA geophysical surveys

Monmouth County
On the Atlantic Ocean in E.
BT New Jersey
United States

Monmouth Group
Includes (ascending) Mount Laurel, Navesink, Red Bank, and Tinton formations. Delaware, NE Maryland, and New Jersey.
BT Upper Cretaceous
Cretaceous
SA Delaware
Maryland
Navesink Formation
New Jersey

Monmouthshire
County in SE.
CO N513000N520000
W0024000W0032000
BT Wales
Great Britain
United Kingdom

Mono Basin
Mono County in E.
BT California
United States

Mono County
Along the Nevada border in E.
CO N373000N384000
W1175000W1193000
BT California

United States

Mono Craters
Range of about 20 geologically recent volcanic cones just S of Mono Lake, Mono County in E.
BT California
United States

Mono Lake
In central Mono County in E.
BT California
United States

monoclines
Not valid term from 1975 through 1977. After 1977, includes use on level 3 under folds(1) style(2). Before 1978, also search folds AND monoclinal.
BA folds
SA flexure folds

monoclinic system
Term introduced in 1978. Before 1978, search monoclinic.
SA crystallography

Monocotyledoneae
Includes use on level 2 under angiosperms(1). See list F.
BA angiosperms
BT Spermatophyta
Plantae
NA Alismidae
Arecidae
Commelinidae
Gramineae
Liliidae
Palmae
SA Dicotyledoneae

Monograptina
Includes use on level 2 under Graptolithina(1). See list F.
BA Graptoloidea
Graptolithina
BX Hemichordata
Invertebrata
NA Monograptus

Monograptus
Genus. Includes use on level 3 under Graptolithina(1) Monograptina(2).
BA Monograptina
Graptoloidea
Graptolithina
BX Hemichordata
Invertebrata
NA Monograptus uniformis

Monograptus uniformis
Includes use on level 3 under Graptolithina(1) Monograptina(2).
BA Monograptus
Monograptina
Graptoloidea
Graptolithina
BX Hemichordata
Invertebrata

monohydrocalcite
BA carbonates
BT minerals

monolete taxa
Term introduced in 1978. Before 1978, also search monoletes.
UF monoletes
SA miospores
palynomorphs

monoletes
use monolete taxa

Monomoy Island
A 10 mile sandspit S of Chantham on Cape Cod. Sometimes it is connected to land and is not an island.
BT Massachusetts
United States

Monongahela Group
In Pennsylvania, contains Pittsburgh Coal Seam at base; top of Waynesburg Coal marks upper boundary. W Maryland, E Ohio, W Pennsylvania, W Virginia, and West Virginia.
BT Pennsylvanian
SA Maryland
Ohio
Pennsylvania
Virginia
West Virginia

Monoplacophora
Includes use on level 2 under Mollusca(1). See list F.
BA Mollusca
BT Invertebrata

Monotis
Genus. Includes use on level 3 under Mollusca(1). See list F.
BA Bivalvia
Mollusca
BT Invertebrata

Monotremata
Includes use on level 2 under Mammalia(1). See list F.
BA Mammalia
BT Tetrapoda
Vertebrata

Monroe County
Index states as applicable.
BX Alabama
Arkansas
Florida
Georgia
Illinois
Indiana
Iowa
Kentucky
Michigan
Mississippi
Missouri
New York
Ohio
Pennsylvania
Tennessee
West Virginia
Wisconsin
BT United States

Mons Basin
Coal mining area near the French border. Also search Mons.
BT Hainaut
Belgium

monsoons
SA hurricanes
storms

Mont Blanc
In Savoy Alps near Italian border. Highest mountains in Alps.
BT Haute-Savoie
France
SA Savoy Alps

Mont Dore
use Mont-Dore

Mont-Dore
Town and thermal station in S central part of country. Also search Mont Dore.
UF Le Mont-Dore
Mont Dore
Mont-Dore-les-Bains
BT Puy-de-Dome
France

Mont-Dore-les-Bains
use Mont-Dore

Mont-Saint-Michel Bay
use Bay of Saint-Michel

Montagne Noire
Southernmost range of Central Massif. Index departments as applicable.
BT France
SA Aude
Central Massif
Tarn

Montague Island
At entrance of Prince William Sound, SE of Anchorage in S.
BT Alaska
United States

Montana
Includes use on level 1 as an area term (list O). For term set options see list B.
CO N443000N490000
W1040200W1160200
BA Western U.S.
United States
NT Beaverhead County
Bell Creek Field
Belt Basin
Big Belt Mountains
NX Big Horn County
NT Big Snowy Mountains
Billings
NX Blaine County
NT Boulder Batholith
Broadwater County
Butte
Butte District
NX Carbon County
Carter County
NT Cascade County
Chouteau County
NX Custer County
NT Fergus County
Flathead County
Flathead Lake
Flint Creek Range
Gallatin County
NX Garfield County
NT Glacier County
Glacier National Park
Granite County
Hebgen Lake
Helena
NX Jefferson County
Lake County
NT Lewis and Clark County
NX Lincoln County
NT Little Belt Mountains
NX Madison County
NT Meagher County
NX Mineral County
NT Mount Morgan
NX Park County
Phillips County
NT Powder River County
NX Powell County
NT Rattlesnake Mountain
Ravalli County
NX Richland County
Roosevelt County
NT Rosebud County
Ruby Range
NX Sheridan County
NT Shonkin Sag Laccolith
Silver Bow County
NX Teton County
NT Three Forks
Tobacco Root Mountains
NX Valley County
NT Vulture Mountain
SA Absaroka Range
Altyn Limestone
Arikaree Group
Battle Mountain High
Bear Gulch Limestone Member
Bearpaw Formation
Beartooth Mountains
Beaverhead Formation
Beaverhead Mountains
Belt Supergroup
Benton Formation
Bighorn Basin
Bighorn Mountains
Bitterroot Range
Bonner Formation
Carlile Shale
Colorado Group
Columbia River basin
Dakota Formation
Deadwood Formation
Elk Point Group
Flathead Sandstone
Fort Union Formation
Fox Hills Formation
Frontier Formation
Gallatin Range
Graneros Shale
Great Plains
Greenhorn Limestone
Hell Creek Formation
Idaho Batholith
Judith River Formation
Lance Formation
LASA
Little Missouri River basin
Lodgepole Formation
Madison Group
Midcontinent
Missoula Group
Missouri River
Missouri River basin
Missouri River valley
Montana Group
Morrison Formation
Mowry Shale
Muddy Sandstone
Niobrara Formation
Pagoda Formation
Parkman Sandstone
Phosphoria Formation
Pierre Shale
Pilgrim Formation
Pinedale Glaciation
Powder River basin
Prichard Formation
Purcell System
Ravalli Group
Red River Formation
Rocky Mountain Trench
Sentinel Butte Formation
Stillwater Complex
Sundance Formation
Sweetgrass Arch
Tensleep Sandstone
Thaynes Formation
Tongue River Member
U. S. Rocky Mountains
Wallace Formation
Wasatch Formation
Western Interior
Western Overthrust Belt
White River Group
Williston Basin
Winnipeg Formation
Winnipegosis Formation
Yellowstone National Park

Montana Group
Includes Bearpaw Shale, Claggett Shale, Cody Shale (upper part), Eagle Sandstone, Fox Hills Sandstone, Horsethief Sandstone, Judith River Formation, Lennep Sandstone, Parkman Sandstone, Pierre Shale, Trinidad Sandstone, Two Medicine Formation, and Virgelle Sandstone.
BT Upper Cretaceous
Cretaceous
SA Bearpaw Formation
Cody Shale
Colorado
Judith River Formation
Kansas
Montana
New Mexico
North Dakota
Parkman Sandstone
Pierre Shale
South Dakota
Utah
Wyoming

Montastrea
Genus. Includes use on level 3 under Coelenterata(1) Scleractinia(2).
BA Scleractinia
Anthozoa

Coelenterata
BT Invertebrata

Montauban Meteorite
use Orgueil Meteorite

Monte Amiata
Extinct volcano in the Apennines.
BT Tuscany
Italy

Monte Antola
Peak in Ligurian Apennines in NW part of country.
BT Liguria
Italy
SA Ligurian Apennines

Monte Baldo
Mountain range between Lake Garda and Adige River in NE part of country.
BT Veneto
Italy

Monte Capanne
W end of island of Elba in Mediterranean Sea off Tuscany.
BT Tuscany
Italy

Monte Carlo
Resort on the Ligurian Sea in the Riviera.
BT Monaco

Monte Cristo Limestone
Lower and Upper Mississippian. Consists of (ascending) Dawn Limestone, Anchor Limestone, Bullion Limestone, and Yellowpine Limestone members. SE California and SE Nevada. Index states as applicable.
BT Mississippian
SA California
Lower Mississippian
Nevada
Upper Mississippian

Monte Gargano
use Gargano

Monte Pisano
Mountain group near Pisa in NW.
BT Tuscany
Italy

Monte Rosa
Highest mountain group in the Pennine Alps. Index countries as applicable.
BT Europe
SA Italy
Pennine Alps
Switzerland

Monte Somma
Semicircular ridge on N and E sides of Vesuvius in E.
BT Campania
Italy

Monteagle Limestone
BT Upper Mississippian
Mississippian
SA Alabama
Tennessee

montebrasite
BA phosphates
BT minerals
SA amblygonite

Montenegro
Constituent republic in SW.
CO N415000N433800
E0202000E0183000
BT Yugoslavia
NT Niksic

Monterey
City in Monterey County on the Pacific Ocean.
BT California
United States

Monterey Bay
Inlet of Pacific Ocean in Santa Cruz and Monterey Counties.
BT California
United States

Monterey Canyon
Submarine canyon just off coast of central California.
UF Monterey Deep-Sea Channel
Monterey Gorge
Monterey Seavalley
Monterey Submarine Canyon
Monterey Trough
BT Pacific Ocean

Monterey County
On Pacific Ocean in W.
BT California
United States

Monterey Deep-Sea Channel
use Monterey Canyon

Monterey Deep-Sea Fan
use Monterey Fan

Monterey Fan
Off coast of central California just to the W of the Monterey Canyon.
UF Monterey Deep-Sea Fan
BT Pacific Ocean

Monterey Formation
Middle and upper Miocene. Comprises (ascending) Gould Shale, Devilwater Silt, McDonald Shale, Antelope Shale, and Chico-Martinez Chert members. W California.
BT Miocene
SA California
middle Miocene
upper Miocene

Monterey Gorge
use Monterey Canyon

Monterey Seavalley
use Monterey Canyon

Monterey Submarine Canyon
use Monterey Canyon

Monterey Trough
use Monterey Canyon

Montes de Toledo
Mountain range in S central part of country. Index provinces as applicable.
BT Spain
SA Ciudad Real
Toledo Province

Montesano Formation
Largely massive coarse-grained light-brown sandstones, with many intercalated lenses of conglomerate and grit; shales subordinate in lower part but common in upper part. SW and NW Washington.
BT upper Miocene
Miocene
SA Washington

Montevideo
Department in S Uruguay. Also a city on the Rio de la Plata.
BT Uruguay

Montezuma County
Term introduced in 1981. SW Colorado.
CO N370000N373800
W1075800W1090230
BT Colorado
United States

Montgomery County
Index states as applicable.
BX Alabama
Arkansas
Georgia
Illinois
Indiana
Iowa
Kansas
Kentucky
Maryland
Mississippi
Missouri
New York
North Carolina
Ohio
Pennsylvania
Tennessee
Texas
Virginia
BT United States

Monti Berici
Range of volcanic hills SW of Vicenza in NE part of country.
BT Veneto
Italy

Monti Caronie
use Nebrodi Mountains

Monti Lessini
Mountain group separated from Monte Baldo group by Adige River in NE part of country.
BT Veneto
Italy

Montian
Europe. Above Danian, below Thanetian. Includes use on level 3 under age terms(1). See list E.
BA Paleocene
Paleogene
Tertiary
BT Cenozoic

monticellite
BA olivine group
nesosilicates
orthosilicates
silicates
BT minerals

Monticello
City in San Juan County in SW.
BT Utah
United States

montmorillonite
BA clay minerals
sheet silicates
silicates
BT minerals
SA beidellite
nontronite
smectite

Montpellier
City in S part of country.
BT Herault
France

Montreal
City on an island in the Saint Lawrence River.
BT Quebec
Canada

Monts des Maures
use Maures Massif

Monts Dome
As of 1981, use Chaine des Puys.

Monts du Lyonnais
Index departments as applicable.
BT France
SA Loire
Rhone

Montsech
Foothills of the Pyrenees.
BT Lerida
Catalonia
Spain

Montserrat
Volcanic island in the Leeward Islands belonging to the United Kingdom.
BT Leeward Islands
Lesser Antilles

monzodiorite
As of 1981, includes use on level 3 under igneous rocks(1) diorites(2). See list H. Before 1981, included use on level 3 under igneous rocks(1) diorite family(2).
BA diorites
BT igneous rocks
SA plutonic rocks

monzonite
Includes use on level 3. As of 1981, not to be used in the igneous rocks set. See list H. Before 1981, included use on level 3 under igneous rocks(1) syenite family(2).
BA monzonites
syenites
BT igneous rocks

monzonites
Term introduced in 1981. Includes use on level 2 under igneous rocks(1). See list H. Before 1981, search monzonite.
BA syenites
BT igneous rocks
NA monzonite

Moon
See also other planets, i.e. Mars, Venus, etc. Includes use on level 1. Term set options are:
name of discipline [List B except for oceanography, paleobotany and paleontology]
topic (similar to terms called for on 3rd level in List B)
theoretical studies
subtopic
topic [age, anomalies, atmosphere, bibliography, catalogs, composition, concepts, environment, evolution, exploration, genesis, gravity field, lunar crust, lunar interior, magnetic field, microcraters, motions, observations, paleomagnetism, planetary interiors, satellites, surface properties]
subtopic
NX Apennine Front
NT Eastern Sea
Foaming Sea
Fra Mauro
Hadley Rille
North Ray Crater
Ocean of Storms
Sculptured Hills
Sea of Crises
Sea of Fertility
Sea of Nectar
Sea of Rains
Sea of Serenity
Sea of Tranquillity
Sea of Waves
Shorty Crater
South Ray Crater
Taurus-Littrow
SA accretion
agglutinates
albedo
Apollo Program
Apollo 11
Apollo 12
Apollo 14
Apollo 15
Apollo 16
Apollo 17
Apollo 9
asteroids
atmosphere
breccia
condensation
coordinates
cosmic rays
craters
Earth
Eratosthenian
exploration

Explorer Program
Explorer 35
farside
gravity field
highlands
interior
KREEP
landing sites
luminescence
Luna Program
Luna 16
Luna 20
Luna 24
lunar breccia
lunar craters
lunar crust
lunar interior
lunar mantle
lunar samples
lunar soils
magnetic field
maps
maria
Mars
mascons
meteorites
microcraters
moonquakes
motions
obliquity of the ecliptic
orange material
particles
planetology
planets
rilles
satellites
selenodesy
shaded relief maps
solar flares
solar system
solar wind
Sun
surface features
surface properties
Surveyor Program
Surveyor 3

moon rocks
A valid term through mid-1978. Use lunar samples.

moonquakes
Includes use on level 3 under Moon(1).
SA earthquakes
Moon

moonstone
BA feldspar group
framework silicates
silicates
BT minerals
SA gems

Moore County
Index states as applicable.
BX North Carolina
Tennessee
Texas
BT United States

Moore County Meteorite
Term introduced in 1981. Before 1981, also search Moore County AND meteorites.
BA eucrite
achondrites
BT meteorites
SA North Carolina

Moose River basin
SW of lower James Bay.
BT Ontario
Canada

Mor
See list M.
UF raw humus
BA soils
SA humus
Mull

Mora County
Term introduced in 1981. NE New Mexico.
CO N354500N361600
W1042000W1054500
BT New Mexico
United States

morainal plains
use outwash plains

moraines
Includes use on level 3 under glacial geology(1) glacial features(2).
SA drift
drumlins
glacial features
glacial geology
glaciers
outwash plains
till

Morava
use Moravia

Morava River valley
River valley in NE central Serbia; and a river valley in NE Austria and central Czechoslovakia. Index Austria, Czechoslovakian regions and Serbia. Also search Morava; Morava River.
BT Europe
SA Austria
Moravia
Serbia
Slovakia

Moravia
Region between Bohemia and Slovakia.
CO N483500N502800
E0185000E0151500
UF Morava
BT Czechoslovakia
NT Brno
Karvina
Moravian Karst
Nizky Jezenik Mountains
Ostrava
Ostrava-Karvina
Stramberk
Svratka
SA Morava River valley
Sudeten
Sudeten Mountains
Sudetic Basin

Moravian Karst
BT Moravia
Czechoslovakia

Moravska Ostrava
use Ostrava

Morbihan
Department in NW.
CO N471500N481500
W0020300W0035000
BT France
NT Groix
Vilaine Bay

mordenite
Also search ashtonite.
UF arduinite
ashtonite
flokite
ptilolite
BA zeolite group
framework silicates
silicates
BT minerals
SA aluminosilicates

Morecambe Bay
Inlet of the Irish Sea in NW.
BT England
Great Britain
United Kingdom

morencite
use nontronite

Moreno Formation
In Chico Group. Upper Cretaceous and Paleocene (?). Includes four members (ascending): Dosados Sand and Shale, Tierra Loma Shale (including Mercy Sandstone lens), Marca Shale, and Dos Palos Shale (includes Cima Sandstone lens). S California.
BT Upper Cretaceous
Cretaceous
SA California
Marca Shale Member
Paleocene

Moreton
Village on Cape York Peninsula in N.
BT Queensland
Australia

Moreton Bay
In SE just N of Brisbane.
BT Queensland
Australia

Morez
Town in in E part of country.
BT Jura
France

Morgan County
Index states as applicable.
BX Alabama
Colorado
Georgia
Illinois
Indiana
Kentucky
Missouri
Ohio
Tennessee
Utah
West Virginia
BT United States

Morin Anorthosite
use Morin Complex

Morin Complex
Also search Morin Anorthosite.
CO N454500N463000
W0733000W0743000
UF Morin Anorthosite
Morin Plutonic Complex
BT Quebec
Canada

Morin Plutonic Complex
use Morin Complex

Morlaix
Town on inlet of English Channel.
BT Finistere
France

Mormoiron
Village in SE part of country.
BT Vaucluse
France

Mormon Mountains
SE Nevada.
BT Nevada
United States

Moroccan Atlas Mountains
Term introduced in 1981 for the Atlas Mountains in Morocco.
CO N303000N351000
W0010000W0100000
BZ Atlas Mountains
BX Morocco
SA Anti-Atlas
High Atlas
Middle Atlas

Morocco
Includes use on level 1 as an area term (list O). For term set options see list B.
CO N274000N360000
W0010000W0131500
BA North Africa
Africa
NT Anti-Atlas
Beni Bouchera
Bou Azzer
Cape Bojador
Casablanca
Ceuta
High Atlas
Melilla
Middle Atlas
NX Moroccan Atlas Mountains
NT Moulouya River
Rabat
Rehamna
Rif
Tafilalt
Tangier
Tarfaya
Taza
NX Western Sahara
SA Atlas Mountains
Mediterranean region
Sahara
Taourirt

Moroto Mountain
Just E of town of Moroto near Kenya border.
BT Uganda

morphology
Includes use on level 2 under fossil terms (list F), and under aurora(1) and soils(1).
NT functional morphology
SA aurora
coiling
description
extremities
horizon differentiation
horizons
ichnofossils
micromorphology
nervous system
physiology
septa
soils
spinal column
sutures
valves

morphometry
General term used under geomorphology(1).
SA geomorphology
landform description

morphostructures
SA geomorphology

Morrison Formation
In Plata Mountains of Colorado, comprises (ascending) Pony Express Limestone, Bilk Creek Sandstone, Wanakah Marl (restricted), and Junction Creek Sandstone.
BT Upper Jurassic
Jurassic
SA Arizona
Colorado
Kansas
Montana
New Mexico
Oklahoma
South Dakota
Utah
Wyoming

Morrow Formation
Composed of shales with some thin limestones and occasional sandstones. SE Arkansas, and central E and NE Oklahoma.
BT Pennsylvanian
SA Arkansas
Morrowan
Oklahoma

Morrow Series
use Morrowan

Morrowan
Provincial series, North America. Above Chesterian (Mississippian),

below Atokan. Includes use on level 3 under age terms(1). See list E.
UF Morrow Series
Morrowan Series
BA Lower Pennsylvanian
Pennsylvanian
Carboniferous
BT Paleozoic
NT Bloyd Formation
SA Chesterian
Morrow Formation
Springer Formation

Morrowan Series
use Morrowan

Mortagne
Town in NW Orne.
UF Mortagne-au-Perchey
BT Orne
France

Mortagne-au-Perchey
use Mortagne

Morton County
Index states as applicable.
BX Kansas
North Dakota
BT United States

Morvan
Northernmost spur of the Central Massif. Index departments as applicable.
UF Morvan Massif
Morvan Mountains
BT France
SA Central Massif
Cote-d'Or
Nievre
Saone-et-Loire
Yonne

Morvan Massif
use Morvan

Morvan Mountains
use Morvan

Mosabani Mine
use Mosaboni copper mines

Mosaboni copper mines
SE Bihar, India. Also search Mosaboni.
UF Mosabani Mine
Mosaboni Mine
Mosaboni Mines
Musabani copper mines
Mushabani copper mines
BT Bihar
India

Mosaboni Mine
use Mosaboni copper mines

Mosaboni Mines
use Mosaboni copper mines

Mosasauridae
As of 1981, includes use on level 3 under Reptilia(1) Squamata(2). See list F. Before 1981, included use on level 3 under Reptilia(1) Lepidosauria(2). Before 1981, also search mosasaurs. As of 1981, search mosasaurs only for biostratigraphy. Also search Mosasaurus.
BA Squamata
Lepidosauria
Reptilia
BT Tetrapoda
Vertebrata

Mosbas
use Moscow Basin

Moscovian
Russia. Includes use on level 3 under age terms(1). See list E.
BA Upper Carboniferous
Carboniferous
BT Paleozoic

moscovite
use muscovite

Moscow
City and oblast.
UF Moskva
BT Russian Republic
USSR

Moscow Basin
Lignite basin extending over 600 miles in an arc from Borovichi in Leningrad Oblast to Skopin, SE of Moscow, in Ryazan Oblast. Also search Moscow.
UF Mosbas
Moscow coal basin
BT Russian Republic
USSR

Moscow coal basin
use Moscow Basin

Moscow River
use Moskva River

Moscow State University
use Moscow University

Moscow Syneclise
Central European USSR. Also search Moscow.
CO N540000N573000
E0400000E0350000
BT Russian Republic
USSR
SA Moscow-Pechora Syneclise

Moscow University
In Moscow. Also search Moscow State University.
UF Moscow State University
BT Russian Republic
USSR

Moscow-Pechora Syneclise
Term introduced in 1981.
CO N503000N693000
E0620000E0300000
BA Russian Platform
BT USSR
SA Moscow Syneclise
Pechora Basin
Pechora River

Mosel Valley
use Moselle Valley

Moselle
Department in NE.
CO N483000N493200
E0074000E0055300
BT France
SA Saar Basin

Moselle River
Rises in NE France and flows N and NE entering the Rhine River at Koblenz. Index countries as applicable. Also search Moselle.
BT Europe
SA France
Luxembourg
West Germany

Moselle Valley
River valley. Index countries as applicable.
UF Mosel Valley
BT Europe
SA France
Germany
Luxembourg

Moskva
use Moscow

Moskva River
Flows through Moscow and then SE into the Oka River.
UF Moscow River
BT Russian Republic
USSR

Mosquito Range
S part of Park Range near Leadville.
BT Colorado
United States

Mossamedes
District and desert basin in extreme SW part of country. Also a city on the Atlantic Ocean.
UF Mocamedes
BT Angola

Mossbauer
A valid index term through 1977. After 1977, use Mossbauer spectroscopy.

Mossbauer spectra
Term introduced in 1978. Before 1978, also search Mossbauer.
UF Mossbauer spectrum
SA Mossbauer spectroscopy
spectra

Mossbauer spectroscopy
Term introduced in 1978. Includes use on level 3 under spectroscopy(1) and under chemical element(1). Before 1978, also search Mossbauer AND spectroscopy.
BT spectroscopy
SA analysis
Mossbauer spectra

Mossbauer spectrum
use Mossbauer spectra

Motagua Fault
Term introduced in 1981. SE Guatemala.
CO N144500N153000
W0884000W0905000
BT Guatemala

motion, ground
use ground motion

motion, plasma
use plasma motion

motion, strong
use strong motion

motions
Includes use on level 2 under Earth(1) and Moon(1); on level 3 under aurora(1) morphology(2).
SA aurora
Chandler wobble
Earth
Moon
oscillations

Motojuku Formation
BT Miocene
Neogene
Tertiary
SA Honshu
Japan

Moulouya River
Rises in central part of country and flows NE into Mediterranean Sea. Also search Moulouya.
UF Moluya
Mulwiya
BT Morocco

mounds
Includes use on level 3 under sedimentary structures(1) bedding plane irregularities(2). See list K.
BA bedding plane irregularities
sedimentary structures

mounds, algal
use algal mounds

Mount Agoeng
use Mount Agung

Mount Agung
Highest volcanic peak on Bali.
UF Mount Agoeng
BT Lesser Sunda Islands
Indonesia

Mount Asama
Active volcano in central.
BT Honshu
Japan

Mount Carmel
Mountain in N near Mediterranean coast.
BT Israel
Middle East

Mount Elbrus
use Elbrus

Mount Etna
Active volcano in NE Sicily. Also search Etna and Mt. Etna.
CO N374500N374500
E0150100E0150100
UF Etna
Mt. Etna
BT Sicily
Italy

Mount Fuji
use Fujiyama

Mount Girnar
Village in S central Kathiawar Peninsula. Also search Mount Girnar Massif.
UF Mount Girnar Massif
BT Gujarat
India

Mount Girnar Massif
use Mount Girnar

Mount Hakone
Extinct volcano in central Honshu. Also search Hakone; Hakone Volcano.
UF Hakone Volcano
BT Honshu
Japan
SA Hakone

Mount Hood
Peak in Cascade Range in Clackamas and Hood River counties. Also search Mt. Hood.
CO N452500N452500
W1214000W1214000
UF Mt. Hood
BT Oregon
United States

Mount Isa
Town in W.
CO S204500S204500
E1393000E1393000
BT Queensland
Australia

Mount Kenya
Extinct volcano in central.
BT Kenya

Mount Lassen
use Lassen Peak

Mount Lofty Ranges
SE South Australia.
UF Mt. Lofty Ranges
BT South Australia
Australia

Mount Lyell
Peak in Sierra Nevada near junction of Madera, Mariposa, and Tuolumne counties in E central.
UF Mt. Lyell
BT California
United States

Mount Mihara
Active volcano on central O-shima of Izu-shichito Island Group S of Sagami Bay. Also search Mihara.
UF Mihara (Mount)
Mihara Volcano
Mihara-Yama
Mt. Mihara
BT Honshu
Japan

Mount Morgan
Peak in Glacier National Park in NW.
BT Montana

United States

Mount Rainier
Volcanic peak of Cascade Range in Mount Rainier National Park in W central Washington. Also search Mt. Rainier.
UF Mt. Rainier
BT Washington
United States
SA Cascade Range

Mount Rainier National Park
W central part of state. Mt. Rainier occupies one fourth of park area in W central Washington.
UF Mt. Rainier National Park
BT Washington
United States

Mount Saint Helens
Peak in NW Skamania County in S Washington.
UF Mount St. Helens
Mt. St. Helens
BT Washington
United States

Mount St. Helens
use Mount Saint Helens

Mount Wood
Peak of St. Elias Mountains in SW.
BT Yukon Territory
Canada

mountain building
A valid term through 1973. After 1973, use orogeny.

mountain crystal
use quartz crystal

mountains
Includes use on level 3 under geomorphology(1) landform description(2).
SA geomorphology
highlands
hills
landform description
landforms

Mousterian
Archeologic classification.
BA Paleolithic
BT Cenozoic
SA Pleistocene
Quaternary

Mouthoumet Massif
S part of country. Also search Mouthoumet.
BT Aude
France

Mouydir
Mountains in S central.
BT Algeria

movement
Includes use on level 2 under ground water(1) and plate tectonics(1); on level 3 under soils(1) water regimes(2).
SA ground water
ice movement
plate tectonics
soils
water regimes

movements, mass
use mass movements

movements, vertical
use vertical movements

Mowry Shale
In Colorado Group. Consists of hard lighter-gray shales and thin bedded sandstones that weather light gray and form ridges. Montana, W South Dakota, and Wyoming. Index states as applicable.
BT Upper Cretaceous
Cretaceous
SA Colorado Group
Montana
South Dakota
Wyoming

Mozambique
Includes use on level 1 as an area term (list O). For term set options see list B.
CO S270000S100000
E0410000E0300000
BA East Africa
Africa
NT Maputo
Mozambique Belt
Tete
Zambezia
SA East African Rift
Lake Malawi
Lebombo Mountains
Limpopo Basin
Limpopo Belt
Zambezi Valley

Mozambique Belt
Orogenic belt. Mountains in N extend to the rim of the Great Rift Valley.
BT Mozambique

Mozambique Channel
Strait between the Malagasy Republic and Mozambique.
CO S261500S103000
E0491000E0323000
BT Indian Ocean

mroseite
BA tellurites
BT minerals

Mt. Elbrus
use Elbrus

Mt. Etna
use Mount Etna

Mt. Fuji
use Fujiyama

Mt. Fuji volcano
use Fujiyama

Mt. Hood
use Mount Hood

Mt. Lofty Ranges
use Mount Lofty Ranges

Mt. Lyell
use Mount Lyell

Mt. Mihara
use Mount Mihara

Mt. Rainier
use Mount Rainier

Mt. Rainier National Park
use Mount Rainier National Park

Mt. St. Helens
use Mount Saint Helens

Mubarek
Village SW of Samarkand.
BT Uzbekistan
USSR

mud
Includes use on level 3 under sediments(1) clastic sediments(2). See list N. Before 1981, also included use on level 3 under sediments(1) carbonate sediments(2).
BA clastic sediments
BT sediments
SA clay
mud banks
mud lumps
mud volcanoes
ooze
silt
terrigenous materials

mud balls, armored
use armored mud balls

mud banks
Also search mudbanks.
UF mudbanks
SA fluvial features
mud
shore features

mud flats
UF mudflats
BT shore features
SA tidal flats

mud flows
use mudflows

mud lumps
Includes use on level 3 under sedimentary structures(1) bedding plane irregularities(2); on level 3 under geomorphology(1) shore features(2) and fluvial features(2). See list K.
UF mudlumps
BA bedding plane irregularities
sedimentary structures
SA clay
mud

mud volcanoes
Includes use on level 1 (list A). Used for an accumulation, usually conical, of mud and rock ejected by volcanic gases; also for a similar accumulation formed by escaping petroliferous gases. Term set options are:
topic [age, causes, classification, detection, distribution, evolution, genesis, interpretation, mechanism, occurrence, temperature]
subtopic (no area term)
UF hervidero
macaluba
SA ejecta
geomorphology
lava
mud
volcanoes
volcanology

mudbanks
use mud banks

mudcracks
Includes use on level 3 under sedimentary structures(1) bedding plane irregularities(2). See list K. Before 1981, also search shrinkage cracks.
BA bedding plane irregularities
sedimentary structures
SA gilgai

Muddy Sandstone
Of Thermopolis Shale. Overlies a lower black shale member and underlies a black shale member. Surface and subsurface in central N Wyoming and subsurface in central S Montana. Also search Muddy Formation.
UF Muddy Sandstone Member
BT Lower Cretaceous
Cretaceous
SA Montana
Wyoming

Muddy Sandstone Member
use Muddy Sandstone

mudflats
use mud flats

mudflows
As of 1978, term is used on level 2 under geologic hazards(1), and slope stability(1). Also search mud flows.
UF mud flows
SA erosion
geologic hazards
lahars
landslides
mass movements
mudslides
slope stability

mudlumps
use mud lumps

mudslides
SA earthflows
mudflows

mudstone
Includes use on level 3 under sedimentary rocks(1) clastic rocks(2). See list I.
BA clastic rocks
BT sedimentary rocks
SA claystone
siltstone
terrigenous materials

Mudurnu
Village in NW part of Anatolia, SW of Bolu.
BT Turkey

Muenster
use Munster

Muenster in Westfalen
use Munster

Muensterland
use Munsterland

Mugan Steppe
Part of Kura Lowland S of Aras and Kura rivers in SE Azerbaidzhan. Also search Mugan.
BT Azerbaidzhan
USSR

mugearite
As of 1981, includes use on level 3 under igneous rocks(1) alkali basalts(2). See list H. Before 1981, included use on level 3 under igneous rocks(1) basalt family(2).
BA alkali basalts
basalts
volcanic rocks
BT igneous rocks

Mugodzhar Hills
Southernmost extension of the Urals in Aktyubinsk Oblast. Also search Mugodzhar; Mugodzhar Mountains; Mugodzhars; Mugodzhary.
CO N480000N520000
E0610000E0550000
UF Mugodzhar Mountains
Mugodzhar Range
Mugodzhars
Mugodzhary
BX Kazakhstan
BZ Urals
BT USSR

Mugodzhar Mountains
use Mugodzhar Hills

Mugodzhar Range
use Mugodzhar Hills

Mugodzhars
use Mugodzhar Hills

Mugodzhary
use Mugodzhar Hills

Mukhor-Tala
Village E of Lake Baikal in Buryat Autonomous Soviet Socialist Republic.
BT Russian Republic
USSR

Mule Ear Diatreme
In Mule Ear Peaks, a mountain in S Brewster County, in W. Also search Mule Ear.
BT Texas
United States

Mull
See list M.
BA soils
SA humus
Mor
Mull Island

Mull Island
Term introduced in 1978. Island of

the Inner Hebrides, Argyll, Scotland. Before 1978, also search Mull. After 1978, Mull is restricted for use as the soil.
CO N561500N564000
W0054000W0063000
BT Inner Hebrides
Scotland
SA Mull

mullions
Includes use on level 3 under lineation(1) style(2) and under structural analysis(1).
SA lineation
metamorphic rocks
structural analysis

mullite
BA orthosilicates
silicates
BT minerals

multi domains
Term introduced in 1981.
BT magnetic domains

multichannel filters
Term introduced in 1981.
BT filters

multiphase flow
Term introduced in 1981.
SA ground water
hydraulics
hydrology
petroleum
surface water

multispectral analysis
Term introduced in 1978. Before 1978, search multispectral AND analysis.
SA analysis
remote sensing

Multituberculata
Includes use on level 2 under Mammalia(1). See list F.
BA Mammalia
BT Tetrapoda
Vertebrata

multivariate analysis
BA statistical analysis
SA analysis
mathematical geology

Multnomah County
Term introduced in 1981. NW Oregon.
CO N452700N454500
W1214900W1225500
BT Oregon
United States

Mulwiya
use Moulouya River

Munchberg Gneiss Massif
In Fichtelgebirge area in NE Bavaria. Also search Munchberg Gneiss.
UF Munchberg Massif
BT Bavaria
West Germany

Munchberg Massif
use Munchberg Gneiss Massif

Munchen
use Munich

Mundrabilla
Village in SE.
BT Western Australia
Australia

Munich
City in S Bavaria.
UF Munchen
BT Bavaria
West Germany

Munster
City in N North Rhine-Westphalia.
UF Muenster
Muenster in Westfalen
BT North Rhine-Westphalia
West Germany

Munsterland
Area around Munster in N North Rhine-Westphalia. Also search Munsterland.
UF Muensterland
BT North Rhine-Westphalia
West Germany

Muntenia
Region and former province in E.
BT Walachia
Romania

Muntii Metalici
Metal Mountains. S part of Apuseni Mountains in W central part of country.
BT Transylvania
Romania
SA Apuseni Mountains

muong nong type
Also search muong nong.
UF muong nong-type
BA tektites

muong nong-type
use muong nong type

Murchison Meteorite
Term introduced in 1981. Impact at Murchison, Victoria. Before 1981, also search Murchison AND meteorites.
BA carbonaceous chondrites
chondrites
BT meteorites
SA Australia
Victoria

Murcia
As of 1981, no longer a valid term for GeoRef. Use Murcia Province or Murcia region as applicable. Before 1981, used for the province.

Murcia Province
Term introduced in 1981. SE Spain. Before 1981, also search Murcia.
CO N372200N384500
W0004000W0022000
BT Murcia region
Spain
NT Caravaca
Jumilla

Murcia region
Term introduced in 1981. In SE Spain.
CO N372200N392800
W0004000W0025500
BT Spain
NT Albacete
Murcia Province

Mures
County in N central part of country.
UF Mures Province
BT Transylvania
Romania

Mures Province
use Mures

Murgab Basin
River basin. Index Afghanistan and/or Turkmenia as applicable. Also search Murgab.
UF Murghab (Basin)
BT Asia
SA Afghanistan
Turkmenia

Murge
Plateau region SW of Bari near the heel of the Italian boot.
BT Apulia
Italy

Murghab (Basin)
use Murgab Basin

Muricacea
Includes use on level 3 under Mollusca(1). See list F.
BA Gastropoda
Mollusca
BT Invertebrata

Muridae
Family. As of 1981, includes use on level 3 under Mammalia(1) Myomorpha(2). Before 1981, included use on level 3 under Mammalia(1) Rodentia(2).
BA Myomorpha
Rodentia
Mammalia
BT Tetrapoda
Vertebrata

Murmansk
City on the Arctic Ocean in extreme NW European USSR. Also an oblast.
BT Russian Republic
USSR

Muro Group
SA Cretaceous
Japan
Tertiary

Murphy Marble
Sequence (ascending) is Valleytown Formation, Murphy Marble, Andrews Schist, and Nottely Quartzite. N Georgia, W North Carolina and E Tennessee.
BT Lower Cambrian
Cambrian
SA Georgia
North Carolina
Tennessee

Murray Basin
River basin. Index states as applicable. Also search Murray AND basin.
CO S360000S320000
E1450000E1400000
BT Australia
SA New South Wales
South Australia
Victoria

Murray County
Index states as applicable.
BX Georgia
Minnesota
Oklahoma
BT United States

Murray fracture zone
NE of Hawaii. Also search Murray AND fracture zone.
UF Murray Seascarp
BT Pacific Ocean

Murray Meteorite
Term introduced in 1981. Impact in Calloway County, Kentucky. Before 1981, also search Murray AND meteorites.
BA carbonaceous chondrites
chondrites
BT meteorites
SA Kentucky

Murray River
The major river of Australia. It rises in E Victoria and flows NW and then S into Encounter Bay on the Indian Ocean. Index states as applicable. Also search Murray.
BT Australia
SA New South Wales
South Australia
Victoria

Murray Seascarp
use Murray fracture zone

Murrumbidgee River
Flows W from Great Dividing Range near Canberra to join Murray River in S.
BT New South Wales
Australia

Murrumbidgee Valley
River valley in S.
BT New South Wales
Australia

Mururoa Atoll
In Tuamotu Archipelago.
BT Tuamotu Islands
French Polynesia
Polynesia

Murzuk Basin
Dune region in SW Libya. Also search Murzuk.
UF Murzuq
BT Libya
Africa

Murzuq
use Murzuk Basin

Musabani copper mines
use Mosaboni copper mines

Musashino
City in Tokyo Prefecture.
BT Honshu
Japan

Muschelkalk
Europe. Includes use on level 3 under age terms(1). See list E.
BA Middle Triassic
Triassic
BT Mesozoic

Musci
Includes use on level 2 under bryophytes(1). See list F.
BA bryophytes
BT Plantae

muscovite
Also search white mica.
UF common mica
mirrorstone
moscovite
potash mica
white mica
BA mica group
sheet silicates
silicates
BT minerals
SA hydromuscovite
illite
light minerals
phengite
sericite

muscovite granite
As of 1981, includes use on level 3 under igneous rocks(1) granites(2). See list H. Before 1981, included use on level 3 under igneous rocks(1) granite-granodiorite family(2).
BA granites
BT igneous rocks
SA granite

muscovite schist
Term introduced in 1981. Includes use on level 3 under metamorphic rocks(1) schists(2). See list J.
BA schists
BT metamorphic rocks

museums
Includes use on level 1 (list A). Used for papers discussing the functioning of a geology museum and/or its history. Term set options are:
topic (list B)
name of museum (in original language)
SA associations
catalogs
collections
education
geology
organization
survey organizations

Musgrave Range
use Musgrave Ranges

Musgrave Ranges
Index Northern Territory and/or South Australia as applicable. Also search Musgrave.
UF Musgrave Range
BT Australia
SA Northern Territory
South Australia

Mushabani copper mines
use Mosaboni copper mines

Muskogee County
E Oklahoma.
BT Oklahoma
United States

Muskox Intrusion
In District of Mackenzie.
BT Northwest Territories
Canada

Mussoorie
Hill station and sanitarium in N.
BT Uttar Pradesh
India

Mustang Island
In Neuces County between Corpus Christi Bay and the Gulf of Mexico.
BT Texas
United States

Mustelidae
Includes use on level 3 under Mammalia(1). See list F. Before 1981, also search mustelids. As of 1981, search mustelids only for biostratigraphy.
BA Carnivora
Mammalia
BT Tetrapoda
Vertebrata

Muth Quartzite
Middle to Upper Devonian. Includes a thick succession of snow-white to greenish quartzites.
BT Devonian
SA India
Jammu and Kashmir
Middle Devonian
Upper Devonian
Uttar Pradesh

Muyun-Kum
use Muyunkum

Muyunkum
Sandy desert in Dzhambul Oblast, S of the Chu River near Kirghizia border.
UF Muyun-Kum
BT Kazakhstan
USSR

mylonite
As of 1981, no longer a valid term for GeoRef. Use mylonites. Before 1981, included use on level 3 under metamorphic rocks(1) mylonites(2).

mylonites
Includes use on level 2 under metamorphic rocks(1). See list J. Also search mylonite.
BT metamorphic rocks
NA blastomylonite
pseudotachylite

mylonitization
SA deformation
faults
phyllonitization

Myodocopida
As of 1981, includes use on level 3 under Ostracoda(1). See list F. Before 1981, included use on level 2 under Ostracoda(1).
BA Ostracoda
BT Crustacea
Mandibulata
Arthropoda
Invertebrata

Myodocopina
Term introduced in 1981. Includes use on level 2 under Ostracoda(1). See list F.
BA Ostracoda
BT Crustacea
Mandibulata
Arthropoda
Invertebrata

Myoko Mountain
W central Honshu. Also search Myoko Volcano.
UF Myoko Volcano
Myoko-san
Myoko-zan
BT Honshu
Japan

Myoko Volcano
use Myoko Mountain

Myoko-san
use Myoko Mountain

Myoko-zan
use Myoko Mountain

Myomorpha
Term introduced in 1981. Includes use on level 2 under Mammalia(1). See list F.
BA Rodentia
Mammalia
BT Tetrapoda
Vertebrata
NA Arvicolidae
Cricetidae
Gliridae
Muridae

Myophoria
Genus. Includes use on level 3 under Mollusca(1). See list F.
BA Bivalvia
Mollusca
BT Invertebrata

Myriapoda
Includes Archipolypoda, Chilopoda, Diploda, Pauropoda, and Symphyla. Includes use on level 2 under Arthropoda(1).
BA Mandibulata
Arthropoda
BT Invertebrata

myrmekite
BA plagioclase
feldspar group
framework silicates
silicates
BT minerals

Mysore
State on the Arabian Sea in SW.
CO N135000N183000
E0783500E0740500
BT India
NT Bagalkot
Bangalore
Belgaum
Bellary
Bijapur
Chickmagalur
Chitaldrug schist belt
Chitradurga
Dharwar
Gulbarga
Hassan
Ingaldhal
Kaladgi
Kanara
Kolar
Kolar Gold Fields
Mangalore
Sagar
Sandur
SA Badami Series
Cauvery Basin
Deccan Plateau
Kaladgi System
Krishna

Mytilus
Genus. Includes use on level 3 under Mollusca(1). See list F.
BA Bivalvia
Mollusca
BT Invertebrata
NA Mytilus edulis

Mytilus edulis
Includes use on level 3 under Mollusca(1). See list F.
BA Mytilus
Bivalvia
Mollusca
BT Invertebrata

Myxomycetes
Term introduced in 1981. Includes use on level 2 under fungi(1). See list F.
BA fungi
BT thallophytes
Plantae

Myzostomia
As of 1981, includes use on level 2 under worms(1). See list F. Before 1981, included use on level 2 under Annelida(1).
BA worms
BT Invertebrata

N

N
use nitrogen

N-15/N-14
SA isotopes
nitrogen

Na
use sodium

Na-22
Includes use on level 3 under isotopes(1).
SA isotopes
sodium

Na-24
Includes use on level 3 under isotopes(1).
SA isotopes
sodium

Nabburg
Town in E.
BT Bavaria
West Germany

Nacimiento Mountains
Range in NW New Mexico. Also search Nacimiento Uplift.
UF Nacimiento Range
Nacimiento Uplift
BT New Mexico
United States
SA Fenton Hill

Nacimiento Range
use Nacimiento Mountains

Nacimiento Uplift
use Nacimiento Mountains

Nacozari de Garcia
Town in NE Sonora. Also search Nacozari.
BT Sonora
Mexico

nacrite
BA clay minerals
sheet silicates
silicates
BT minerals

Nagano
City in central.
BT Honshu
Japan

Nagano Prefecture
Central Honshu.
BT Honshu
Japan

Nagasaki
City and prefecture in W.
BT Kyushu
Japan

Nagpur
City, district, and division in NE.
BT Maharashtra
India

Naha test well
In Navajo Indian Reservation in NE.
BT Arizona
United States

Nahanni Formation
British Columbia, Mackenzie District of Northwest Territories, and Yukon Territory.
BT Middle Devonian
Devonian
SA British Columbia
Canada
Northwest Territories
Yukon Territory

nahcolite
BA carbonates
BT minerals

Nahe
River which flows into Rhine at Bingen. Index states as applicable.
BT West Germany
SA Rhineland-Palatinate
Saar-Nahe Basin
Saarland

Naiba River basin
use Nayba River basin

Nain
Village on central coast.
BT Labrador
Canada

Nain Anorthosite
use Nain Massif

Nain Anorthosite Massif
use Nain Massif

Nain Complex
use Nain Massif

Nain Massif
Anorthosite massif. Also search Nain Anorthosite; Nain Anorthosite Massif; Nain Complex.
UF Nain Anorthosite
Nain Anorthosite Massif
Nain Complex
BT Labrador
Canada

Naini Tal
Town, hill station and district in NE Uttar Pradesh.
CO N292000N292500
E0793000E0792200
UF Naini-Tal
Nainital
BT Uttar Pradesh
India

Naini-Tal
use Naini Tal

Nainital
use Naini Tal

Nairobi
City in S part of country.
BT Kenya

Naka-no-umi
Inlet of Sea of Japan in SW Honshu. Search Nakanoumi Lake.
UF Nakano umi
Nakanoumi Lake
BT Honshu
Japan

Nakano umi
use Naka-no-umi

Nakanoumi Lake
use Naka-no-umi

Nakhichevan
Town and Autonomous Soviet Socialist Republic on the Iranian border separated from Azerbaidzhan proper by a narrow strip of Armenia.
BT Azerbaidzhan
USSR

Nakhla
Village.
BT Qatar
Arabian Peninsula

nakhlite
Term introduced in 1981. Includes use on level 3 under meteorites(1).
BA achondrites
BT meteorites

Nakuru Basin
In Great Rift Valley in Lake Nakuru area of W central Rift Valley Province. Also search Nakuru.
BT Kenya

Nama Beds
use Nama System

Nama Group
use Nama System

Nama Series
use Nama System

Nama Supergroup
use Nama System

Nama System
In SW and S Africa. Also search Nama Group; Nama Beds; Nama Series; Nama Supergroup.
UF Nama Beds
Nama Group
Nama Series
Nama Supergroup
Nama Tillite
BT Cambrian
SA Namibia
South Africa

Nama Tillite
use Nama System

Namaland
use Namaqualand

Namaqualand
Coastal region of sandy plains and bare hills. Index Cape Province and/or Namibia as applicable.
UF Namaland
BT Africa
SA Cape Province
Namibia

Namib Desert
Arid region extending along entire coast.
BT Namibia

Namibia
Term introduced in 1981 to replace South-West Africa. Includes use on level 1 as an area term (list O). For term set options see list B. Also search South-West Africa; German Southwest Africa.
CO S290000S164000
E0251500E0113000
BA Southern Africa
Africa
NT Karibib
Luderitz
Namib Desert
Otavi
Tsumeb
Walvis Bay
Windhoek
SA Damara System
Dwyka Series
Gibeon Meteorite
Kalahari Desert
Karroo System
Nama System
Namaqualand
Orange River
Zambezi Valley

Namur
Province in S.
BT Belgium
NT Dinant
Dinant Basin

Namurian
Europe. Includes use on level 3 under age terms(1). See list E.
BA Upper Carboniferous
Carboniferous
BT Paleozoic

Nanao
City on W coast on E side of Noto Peninsula.
BT Honshu
Japan

Nandewar Mountains
Range in NE.
BT New South Wales
Australia

Nanjemoy Formation
In Pamunkey Group. Includes Potapaco Clay Member, Woodstock Greensand, Marlboro Clay Member. E Maryland and E Virginia.
BT Eocene
SA lower Eocene
Maryland
middle Eocene
Virginia

Nankaido
Former division including Awaji Islands, S central part of S coast of Honshu, and all of Shikoku. Index major islands as applicable.
BT Japan
SA Honshu
Shikoku

nannoconids
As of 1981, to be used for Nannoconus when they are used as tools in stratigraphy. See list F. Before 1981, included use on level 3 under algae(1) nannofossils(2). Before 1981, search (nannoconids OR Nannoconus) AND biostratigraphy.
SA biostratigraphy
Nannoconus

Nannoconus
Genus. Includes use on level 3 under algae(1) nannofossils(2). Before 1981, also search nannoconids. As of 1981, search nannoconids only for biostratigraphy.
BA nannofossils
algae
BT thallophytes
Plantae
SA nannoconids

nannofossils
Includes discoasters and Nannoconus. Includes use on level 2 under algae(1).
UF calcareous nannofossils
BA algae
BT thallophytes
Plantae
NA Braarudosphaeridae
discoasters
Helicopontosphaera
Nannoconus
Sphenolithus
SA Coccolithophoraceae
fossils
index fossils
problematic fossils

Nantes
City on the Loire River.
BT Loire-Atlantique
France

Nantucket (Island)
use Nantucket Island

Nantucket Island
In Atlantic Ocean, S of Cape Cod. Also search Nantucket.
UF Nantucket (Island)
BT Massachusetts
United States

Napa County
W central.
BT California
United States

Naples
City on the Tyrrhenian Sea.
UF Napoli
BT Campania
Italy

Napoli
use Naples

nappe (volcanic)
use lava flows

nappes
Includes use on level 3 under folds(1) orientation(2), under faults(1) displacements(2), and under tectonics(1) gravity sliding(2).
SA allochthons
displacements
faults
folds
klippen
orientation
overthrust faults
recumbent folds
tectonics

Nara
City and prefecture E of Osaka.
BT Honshu
Japan

Narbada River
use Narmada River

Narbada Valley
use Narmada Valley

Narbonne
City near the Mediterranean Sea.
BT Aude
France

Nari Series
Underlies the Gaj and overlies the Khirthar Series.
BT Oligocene
SA Gujarat
India
Pakistan
Sind

Narita Formation
Tokyo Bay area.
BT Pliocene
SA Honshu
Japan

Narizian
North America. Above Ulatisian, below Fresnian. Includes use on level 3 under age terms(1). See list E.
BA Eocene
Paleogene
Tertiary
BT Cenozoic
SA middle Eocene
upper Eocene

Narmada River
Rises in the Maikala Range in Madhya Pradesh and flows W into Gulf of Cambay. Index states as applicable. Also search Narmada.
UF Narbada River
Nerbudda River
BT India
SA Gujarat
Madhya Pradesh

Narmada Valley
River valley. Index states as applicable.
UF Narbada Valley
Nerbuda Valley
BT India
SA Gujarat
Madhya Pradesh

Narnaul
Town 80 miles WSW of Delhi.
BT Haryana
India

Narpaign Fjord
use Narpaing Fjord

Narpaing Fiord
use Narpaing Fjord

Narpaing Fjord
Narrow inlet of Davis Strait on Baffin Island, District of Keewatin.
UF Narpaign Fjord
Narpaing Fiord
BT Northwest Territories
Canada

Narrabeen Group
Includes the Gosford Formation and Clifton Sub-Group. N of Sydney.
BT Triassic
SA Australia
New South Wales

Narragansett Bay
Inlet of Atlantic Ocean in SE.
BT Rhode Island
United States

Narrows
Strait between W end of Long Island and Staten Island separating Upper New York Bay from Lower New York Bay.
BT New York
United States

Narugo
Town in N.
BT Honshu
Japan

Narvik
City on Ofot Fjord in N part of country.
BT Nordland
Norway

Narym
Village on Ob River in central Tomsk Oblast in W central Siberia.
BT Russian Republic
USSR

Naryn
Town on Naryn River in Naryn Oblast in SE.
BT Kirghizia
USSR

Nasca Plate
use Nazca Plate

Nashville
Term introduced in 1981. City in central Tennessee in Davidson County.
CO N361000N361000
W0865000W0865000
BT Davidson County
Tennessee
United States

Nassau County
Index states as applicable.
BX Florida
New York
BT United States

Nassellaria
use Nassellina

Nassellina
Suborder. Includes use on level 2 under Radiolaria(1). See list F. Also search Nassellaria.
UF Nassellaria
BA Osculosida
Radiolaria
BT Protista
Invertebrata

nasturan
use pitchblende

Nasu (dake)
use Nasudake

Nasu Mountain
use Nasudake

Nasudake
Volcanic peak NE of Nikko in N central Honshu. Also search Nasu Mountain.
UF Nasu (dake)
Nasu Mountain
BT Honshu
Japan

Natal
South African province on the Indian Ocean.
CO S311000S264500
E0325500E0285000
BT South Africa
NT Durban
Tugela Basin
Zululand

native elements
Term introduced in 1981. Includes use on level 2 under minerals(1). See list L. Before 1981, search native elements and alloys.
BT minerals
NA carbonado
chaoite
diamond
graphite
lonsdaleite
SA alloys
antimony
antimony ores
arsenic
arsenic ores
bismuth
bismuth ores
copper
copper ores
gold
gold ores
iron
iron ores
mercury
mercury ores
metal ores
metals
nonmetal deposits
nonmetals
silver
silver ores
sulfur
sulfur deposits

native elements and alloys
As of 1981, no longer a valid term for GeoRef. Use native elements or alloys as applicable. See list L. Before 1981, included use on level 2 under minerals(1).

natroborocalcite
use ulexite

natrolite
BA zeolite group
framework silicates
silicates
BT minerals

Natrona County
Central Wyoming.
BT Wyoming
United States

natural
A valid level 1 term through 1977 used in combination with gas (i.e. gas, natural). After 1977, use natural gas as level 1 term.

natural coke
use coke coal

natural curiosities
Term introduced in 1981.
SA popular geology

natural gas
Not a valid term through 1977. After 1977, includes use on level 1 as a commodity term. Before 1978, also search gas AND natural; petroleum-gas; petroleum-natural gas.
SA condensates
energy sources
fuel resources
gas storage
gases
gasification
giant fields
hydrocarbons
hydrogen sulfide
methane
oil and gas fields
oil-gas interface
petroleum
pipelines
reservoir properties
reservoir rocks
source rocks
stratigraphic traps
structural traps
traps

natural materials
Term introduced in 1968. To distinguish from synthetic materials.
SA materials
synthetic materials

natural remanence
use natural remanent magnetization

natural remanent magnetism
use natural remanent magnetization

natural remanent magnetization
Includes use on level 3 under paleomagnetism(1).
UF natural remanence
natural remanent magnetism
NRM
BA remanent magnetization
SA magnetic field
magnetization
paleomagnetism
remagnetization

natural resources
As of 1978, term is used on level 2 under conservation(1), land use(1), and reclamation(1).
SA conservation
depletion
environmental geology
land use
reclamation
resources

natural selection
UF selection, natural
SA evolution

Naubug
Village SE of Srinager.
BT Jammu and Kashmir
India

naujaite
As of 1981, includes use on level 3 under igneous rocks(1) syenites(2). See list H. Before 1981, included use on level 3 under igneous rocks(1) syenite family(2).
BA syenites
BT igneous rocks

Nauru Island
Term introduced in 1981.
UF Pleasant Island
BT Micronesia

Nautiloidea
As of 1981, includes use on level 2 under Mollusca(1). See list F. Before 1981, included use on level 3 under Mollusca(1) Cephalopoda(2).
BA Tetrabranchiata
Cephalopoda
Mollusca
BT Invertebrata
NA Aturia
Hercoglossa danica
Nautilus

Nautilus
Genus. As of 1981, includes use on level 3 under Mollusca(1) Nautiloidea(2). See list F. Before 1981, included use on level 3 under Mollusca(1) Cephalopoda(2).
BA Nautiloidea
Tetrabranchiata
Cephalopoda
Mollusca
BT Invertebrata

Navajo County
NE Arizona.
BT Arizona
United States

Navajo Indian Reservation
NE Arizona, NW New Mexico, and SE Utah. Index states as applicable.
BT United States
SA Arizona
New Mexico
Utah

Navajo Sandstone
In Glen Canyon Group. Upper Triassic(?)and Jurassic. N Arizona, W Colorado, NW New Mexico, and SE Utah.
BT Mesozoic
SA Arizona
Colorado
Jurassic
New Mexico
Upper Triassic
Utah

Navarra
Term introduced in 1981 for the former kingdom and province in NE Spain. Before 1981, also search Navarre AND Spain.
CO N415500N431800
W0004000W0023000
BT Spain
NT Asturreta
Pamplona

Navarre
As of 1981, refers to only the region in SW France. Before 1981, also referred to the former kingdom and province of N Spain. Before 1981, search Navarre AND France.
BT France
SA Spain

Navesink Formation
In Monmouth Group.
BT Upper Cretaceous
Cretaceous
SA Delaware
Monmouth Group
New Jersey

Navicula
BA diatoms
algae
BT thallophytes
Plantae

Naxos
Largest island of the Cyclades in the Aegean Sea.
BT Cyclades
Greek Aegean Islands
Greece

Nayba River basin
S Sakhalin Island. Also search Nayba.
UF Naiba River basin
BT Russian Republic
USSR

Nazca Plate
S of the Galapagos Islands. Also search Nasca Plate.
UF Nasca Plate
BT Pacific Ocean
SA Galapagos Rift
plate tectonics
plates

Nazca Ridge
Term introduced in 1981. SW of Peru.
CO S253000S140000
W0760000W0873000
BA South American Pacific
BT Pacific Ocean

Nb
use niobium

Nchanga
Mining township in N near Zaire border.
BT Zambia

Nd
use neodymium

Ne
use neon

Ne-20
Includes use on level 3 under isotopes(1).
SA isotopes
neon

Ne-21
Includes use on level 3 under isotopes(1).
SA isotopes
neon

Ne-22
Includes use on level 3 under isotopes(1).
SA isotopes
neon

Neanderthal
As of 1981, includes use on level 2 under Mammalia(1). See list F. Before 1981, included use on level 3 under Mammalia(1) Primates(2). Also search Neanderthal man; Homo neanderthalensis.
UF Homo neanderthalensis
Neanderthal man
BA Hominidae
Primates
Mammalia
BT Tetrapoda
Vertebrata
SA fossil man

Neanderthal man
use Neanderthal

Near East
An indefinite and unofficial term including the countries of the Middle East plus Libya, Sudan, and occasionally Greece. Index countries as applicable.
SA Cyprus
Egypt

Greece
Iraq
Jordan
Kuwait
Libya
Oman
Qatar
Southern Yemen
Sudan
Syria
Turkey
United Arab Emirates
Yemen

nearshore
A valid term through 1977. After 1977, use nearshore environment.

nearshore environment
Term introduced in 1978. Before 1978, search nearshore.
SA environment
nearshore sedimentation

nearshore sedimentation
Term introduced in 1981.
SA nearshore environment
sedimentation

Nebraska
Includes use on level 1 as an area term (list O). For term set options see list B.
CO N400000N430000
W0952000W1040500
BA Midwest
United States
NX Adams County
NT Beaver Lake
NX Blaine County
Boone County
Butler County
Cass County
Chase County
Cheyenne County
Clay County
Colfax County
Custer County
Douglas County
Franklin County
Garfield County
Grant County
Hamilton County
Howard County
Jefferson County
Johnson County
Lancaster County
Lincoln County
Logan County
Madison County
McPherson County
Nemaha County
Pawnee County
Phelps County
Pierce County
Platte County
NT Platte River
NX Polk County
Saline County
Seward County
Sheridan County
Sherman County
Sioux County
Stanton County
Valley County
Washington County
Wayne County
Webster County
Wheeler County
York County
SA Americus Limestone Member
Ash Hollow Formation
Beil Limestone Member
Benton Formation
Brule Formation
Cambridge Arch
Carlile Shale
Chadron Arch
Chadron Formation
Chase Group
Cherokee Group
Colorado Group
Cottonwood Limestone
Dakota Formation
Denver Basin
Desmoinesian
Douglas Group
Ervine Creek Limestone
Eskridge Shale
Forest City Basin
Graneros Shale
Great Plains
Greenhorn Limestone
Hughes Creek Shale
Kansas City Group
Lawrence Formation
Lecompton Limestone
Loup Fork Group
Midcontinent
Missouri River
Missouri River basin
Missouri River valley
Nemaha Ridge
Niobrara Formation
Ogallala Formation
Oread Limestone
Peoria Loess
Pierre Shale
Plattsburg Limestone
Plattsmouth Limestone Member
Red Eagle Limestone
Roca Formation
Rock Lake Shale Member
Sand Hills
Shawnee Group
Sioux Quartzite
South Platte River valley
Spearfish Formation
Stanton Formation
Stearns Shale
Sundance Formation
Valentine Formation
Wabaunsee Group
White River Group
Wreford Limestone
Wyandotte Limestone

Nebraskan
Pertaining to first glacial stage of Pleistocene Epoch in North America. See list E.
BA lower Pleistocene
Pleistocene
Quaternary
BT Cenozoic

Nebrodi Mountains
N Sicily.
UF Monti Caronie
BT Sicily
Italy

nebula, solar
use solar nebula

Neckar River
Rises in the Black Forest and flows N past Heidelberg into the Rhine River. Also search Neckar.
BT Baden-Wurttemberg
West Germany

Needles
Town on the Colorado River in San Bernardino County in SE.
BT California
United States

Neftechala
Town in SE.
BT Azerbaidzhan
USSR

Neftyanyye Kamni
Oil well in Caspian Sea off Apsheron Peninsula.
BT Azerbaidzhan
USSR
SA Caspian Sea

Negaunee Formation
use Negaunee Iron Formation

Negaunee Iron Formation
Upper Peninsula.
UF Negaunee Formation
BT Michigan
United States

Negev
Desert region in S Israel. Also search Negev Desert.
UF Negev Desert
BT Israel
Middle East

Negev Desert
use Negev

negotiations
Term introduced in 1981.
SA economics
international cooperation

neighborite
As of 1981, includes use on level 3 under minerals(1) fluorides(2). See list L. Before 1981, included use on level 3 under minerals(1) halides(2).
BA fluorides
halides
BT minerals

Neill Island
Small island E of South Andaman Island in the Andaman Islands in E Bay of Bengal. Also search Neill AND appropriate area term.
BT India

Neiveli
use Neyveli

Nellore
City in S.
BT Andhra Pradesh
India

Nellore mica belt
In Nellore area. Also search Nellore.
BT Andhra Pradesh
India

Nelson County
Index states as applicable.
BX Kentucky
North Dakota
Virginia
BT United States

Nelson River
Flows NE out of N Lake Winnipeg into Hudson Bay at Fort Nelson.
BT Manitoba
Canada

Nelson River basin
N central.
BT Manitoba
Canada

Nemaha County
Index states as applicable.
BX Kansas
Nebraska
BT United States

Nemaha Ridge
Index states as applicable.
BT United States
SA Kansas
Nebraska
Oklahoma

Neman River basin
Index Soviet republics as applicable. Also search Neman AND appropriate area term.
UF Memel River basin
Niemen River basin
Nyeman River basin
BT USSR
SA Byelorussia
Lithuania

Nematoida
Includes use on level 2 under worms(1). See list F.
BA worms
BT Invertebrata

Nematomorpha
Includes use on level 2 under worms(1). See list F.
BA worms
BT Invertebrata

Nemerta
Includes use on level 2 under worms(1). See list F.
BA worms
BT Invertebrata

Nemuro Peninsula
E Hokkaido. Also search Nemuro.
BT Hokkaido
Japan

nenadkevichite
BZ niobates
orthosilicates
BT minerals

Nenets
National okrug of NE Arkhangelsk Oblast in N European USSR.
BT Russian Republic
USSR

Neocomian
Europe. Includes use on level 3 under age terms(1). See list E.
BA Lower Cretaceous
Cretaceous
BT Mesozoic
SA Berriasian
Hauterivian
Valanginian

neodymium
Includes use on level 1 and 2 as a chemical element (list D).
UF Nd
SA elements
rare earths

Neogastropoda
Term introduced in 1981. Includes use on level 2 under Mollusca(1). See list F.
BA Gastropoda
Mollusca
BT Invertebrata

Neogene
As of 1978, term includes use on level 1 and 2 (list E). An interval of geologic time incorporating the Miocene and Pliocene of the Tertiary period; the upper Tertiary.
BA Tertiary
BT Cenozoic
NT Capistrano Formation
NA Hemphillian
Miocene
Pannonian
Pliocene
SA Arikareean
Matuyama Epoch
Siwalik System
upper Tertiary

Neolithic
Archaeologic classification.
BA Holocene
Quaternary
BT Cenozoic

neon
Includes use on level 1 and 2 as a chemical element (list D). Also search Ne.
UF Ne
BT noble gases
SA argon
elements
helium
isotopes
krypton
Ne-20
Ne-21

Ne-22
radon
xenon

Neornithes
Includes use on level 2 under Aves(1). See list F.
BA Aves
BT Tetrapoda
Vertebrata
NA Odontornithes
SA Archaeornithes

Neosho River valley
Index states as applicable. The lower course of the Neosho is called Grand River in Oklahoma. Also search Neosho River.
SA Kansas
Oklahoma

neotectonics
A level 1 term as of 1978. Used for the study of the last structures and structural history of the Earth's crust, after the Miocene and during the later Tertiary and the Quaternary. Before 1976, also search paleotectonics. Term set options are:
concepts
subtopic
effects
topic [e.g. changes of level, geomorphologic effects, seismicity]
measurement
subtopic
observations
subtopic
rates
subtopic [e.g. absolute age, geodetic coordinates, remote sensing, satellite measurements]
subsidence
subtopic
uplifts
subtopic
SA changes of level
crust
epeirogeny
faults
folds
foliation
fractures
geodesy
geomorphology
horizontal movements
isostasy
reactivation
seismology
subsidence
tectonics
tectonophysics
triangulation
uplifts
vertical movements

neotypes
SA fossils
type specimens

Nepal
Includes use on level 1 as an area term (list O). For term set options see list B.
CO N270000N301500
E0880000E0800000
BA Indian Peninsula
Asia
NT Kali Gandaki Valley
SA Himalayas
Kosi Basin
Kumaun Himalayas
Lesser Himalayas
Siwalik Range

nepheline
BA nepheline group
framework silicates
silicates
BT minerals

nepheline basalt
As of 1981, includes use on level 3 under igneous rocks(1) alkali basalts(2). See list H. Before 1981, included use on level 3 under igneous rocks(1) alkali basalt family(2).
BA alkali basalts
basalts
volcanic rocks
BT igneous rocks
SA ankaratrite
basalt
olivine nephelinite

nepheline group
Includes use in combination with framework silicates (i.e. framework silicates, nepheline group) on level 2 under minerals(1). See list L.
BA framework silicates
silicates
BT minerals
NA kaliophilite
kalsilite
nepheline
petalite

nepheline syenite
As of 1981, includes use on level 3 under igneous rocks(1) syenites(2). See list H. Before 1981, included use on level 3 under igneous rocks(1) syenite family(2).
BA syenites
BT igneous rocks
SA foyaite
kakortokite
lujavrite
syenite

nephelinite
As of 1981, includes use on level 3 under igneous rocks(1) alkali gabbros(2). See list H. Before 1981, included use on level 3 under igneous rocks(1) alkali basalt family(2) and ultramafic family(2).
BA alkali gabbros
gabbros
BT igneous rocks
SA olivine nephelinite
ultramafics

nepheloid layer
Also search nepheloid.
UF layer, nepheloid
nepheloid zone
SA continental rise
sea water

nepheloid zone
use nepheloid layer

nephrite
BA clinoamphibole
amphibole group
chain silicates
silicates
BT minerals
SA greenstone
jade

Nephrolepidina
Includes use on level 3 under foraminifera(1) Orbitoidacea(2).
BA Orbitoidacea
Rotaliina
foraminifera
BT Protista
Invertebrata

nepouite
use garnierite

Neptune
Includes use on level 1 and 2. See entry under Moon(1) for term set options.
SA outer planets
planetology
planets
solar system

neptunite
BA chain silicates
silicates
BT minerals

neptunium
Includes use on level 1 and 2 as chemical element (list D).
UF Np
SA elements

Nerbuda Valley
use Narmada Valley

Nerbudda River
use Narmada River

Nereites
Genus.
BA ichnofossils

Neretva Valley
River valley in Bosnia and Herzegovina.
BT Yugoslavia

neritic environment
use sublittoral environment

nervous system
Term introduced in 1981.
SA morphology
Vertebrata

nesosilicates
Term introduced in 1981. Includes use on level 2 under minerals(1). See list L.
BA orthosilicates
silicates
BT minerals
NA garnet group
humite group
olivine group

nesquehonite
BA carbonates
BT minerals

Ness County
W central Kansas.
BT Kansas
United States

Netherlands
Also search Holland. Includes use on level 1 as an area term (list O). For term set options see list B.
CO N504500N533000
E0071500E0031500
UF Holland
BA Western Europe
Europe
NT Amsterdam
Delta Area
Groningen
Maastricht
Overijssel
Utrecht
Wadden Zee
Zeeland
SA Campine
European Platform
Limburg
Lower Rhine Basin
Meuse River
Meuse Valley
North Sea Coast
Rhine Basin
Rhine River
Rhine Valley
Scheldt River

Netherlands Antilles
Formerly known as Curacao territory. Islands of Aruba, Bonaire, and Curacao in the Caribbean Sea off the coast of Venezuela, plus the Dutch section of St. Martin at N end of Leeward Islands. Includes use on level 1 as an area term (list O). For term set options see list B.
BA Lesser Antilles
West Indies
NT Bonaire
Curacao
SA Leeward Islands

Netherlands Guiana
use Surinam

Netherlands New Guinea
use Irian Jaya

network deposits
use stockwork deposits

networks
Includes use on level 3 under mineral exploration(1) and geophysical methods(1). As of 1981, for use in seismology, use arrays.
SA arrays
geodesy
geophysical methods
geophysical surveys
mineral exploration
triangulation

Neubrandenburg Bezirk
Term introduced in 1981. District in NE.
CO N530000N540500
E0142500E0121500
BT East Germany

Neuburg
City in W central Bavaria. Also Neuburg an der Donau.
UF Neuburg au der Donau
BT Bavaria
West Germany

Neuburg au der Donau
use Neuburg

Neuchatel
Canton in W Switzerland. Also a city on Lake Neuchatel.
BT Switzerland

Neuquen
Province in W central.
CO S410000S360000
W0670000W0730000
BT Argentina
SA Neuquen Basin

Neuquen Basin
River basin in W central part of country. Index provinces as applicable.
BT Argentina
SA Neuquen
Rio Negro Province

Neuropteris
BA Pteridospermae
gymnosperms
BT Spermatophyta
Plantae
SA pteridophytes

Neuropteroida
Term introduced in 1981. Includes use on level 2 under Insecta(1). See list F.
BA Insecta
BT Mandibulata
Arthropoda
Invertebrata

Neuse River
Rises in the Piedmont and flows SE into Pamlico Sound.
BT North Carolina
United States

neutral stress
use pore pressure

neutron
A valid term through mid-1978 used as a modifier for activation analysis. Use neutron activation analysis.

neutron activation analysis
Includes use on level 3 under chemical analysis(1) methods(2). Before 1978, also search activation anal

ysis AND neutron.
SA activation analysis
analysis
chemical analysis
quantitative analysis

neutron diffraction analysis
Includes use on level 3 under chemical element(1) analysis(2). See list D.
SA analysis
electron diffraction analysis
X-ray diffraction analysis

neutrons
Used as a general term.
SA electrons
particles

Nevada
Includes use on level 1 as an area term (list O). For term set options see list B.
CO N350000N420000
W1140500W1200000
BA Western U.S.
United States
NT Arrow Canyon Range
Beatty
Black Rock Desert
Carlin
Carlin Mine
Churchill County
NX Clark County
NT Copper Canyon
Cortez Mountains
Dixie Valley
NX Douglas County
NT Elko County
NX Ely
NT Esmeralda County
NX Eureka
NT Eureka County
Fairview Peak
Getchell Mine
Goldfield
Grant Range
NX Humboldt County
NT Humboldt River valley
Lander County
Las Vegas
NX Lincoln County
NT Lost City
Lovelock
NX Lyon County
Mineral County
NT Moapa Valley
Mormon Mountains
Nevada Test Site
Nye County
Pahute Mesa
Pequop Mountains
Pershing County
Pinon Range
Pyramid Lake
Reno
Roberts Mountains
Ruby Mountains
Santa Rosa Range
Shoshone Mountains
Silver Peak Mountains
Snake Range
Spring Mountains
Tonopah
Walker Lake
Walker River
Washoe County
White Pine County
Winnemucca
Yerington
Yucca Flat
SA Amargosa Desert
Antelope Valley Limestone
Antler Orogeny
Basin and Range Province
Battle Mountain High
Bird Spring Formation
Chinle Formation
Coconino Sandstone
Colorado River
Colorado River basin
Columbia River basin
Fillmore Formation
Great Basin
Johnnie Formation
Kaibab Formation
Lake Bonneville
Lake Mead
Lake Tahoe
Moenkopi Formation
Monte Cristo Limestone
Pogonip Group
Roberts Mountains Formation
Shadow Mountains
Snake River basin
Stirling Quartzite
Sunrise Formation
Supai Formation
Tor Formation
Toroweap Formation
Truckee River
Virgin River valley
White Mountains
Wood Canyon Formation

Nevada County
Index states as applicable.
BX Arkansas
California
BT United States

Nevada Test Site
NW of Las Vegas in S.
CO N363500N375200
W1152000W1170500
BT Nevada
United States
SA Rainier Mesa

New Albany Shale
Includes Devonian Blocher and Blackiston formations, and Mississippian Sanderson, Underwood, and Henryville formations. Indiana and N central Kentucky.
BT Upper Devonian
Devonian
SA Indiana
Kentucky

New Britain
Largest island in the Bismarck Archipelago NE of New Guinea.
BT Papua New Guinea
Australasia
SA Bismarck Archipelago

New Brunswick
Includes use on level 1 as an area term (list O). For term set options see list B.
CO N450000N480500
W0634500W0690000
BA Maritime Provinces
Eastern Canada
Canada
NT Bathurst
NX Charlotte County
NT Fredericton
Kouchibouguac Bay
Miramichi Bay
Saint John
Woodstock
SA Chaleur Bay
Restigouche Estuary
Windsor Group

New Caledonia
French overseas territory E of Queensland, Australia. Also its main island. Includes use on level 1 as an area term (list O). For term set options see list B.
CO S223000S200000
E1670000E1640000
BA Melanesia
NT Noumea
Ouegoa
SA Pacific Ocean

New Caledonia Basin
Between central Lord Howe Rise and Norfolk Island in SW.
BT Pacific Ocean

New Castile
Term introduced in 1981. Region in central Spain.
CO N382100N412000
E0010800W0052500
UF Castilla la Nueva
BA Castile
BT Spain
NT Ciudad Real
Cuenca Province
Guadalajara Province
Madrid Province
Toledo Province

new data
Used as a general term.
SA data

New Delhi
Capital city SSW of Delhi.
BT Delhi
India

new energy sources
Term introduced in 1981.
SA biomass
economic geology
energy
energy sources
solar energy
tidal energy
wind energy

New England
Index states as applicable. Includes use on level 1 as an area term (list O). For term set options see list B.
CO N420000N473000
W0670000W0733000
BA Eastern U.S.
United States
NA Connecticut
Maine
Massachusetts
New Hampshire
Rhode Island
Vermont
SA New England Range

New England Batholith
BT Australia
SA New England Range
New South Wales
Queensland

New England Plateau
use New England Range

New England Range
Term introduced in 1981. NE New South Wales. Part of the Great Dividing Range. Before 1981, also search New England AND Australia.
UF New England Plateau
BT New South Wales
Australia
SA New England
New England Batholith

new genera
A valid term through 1973. Use new taxa.

new genus
A valid term through 1973. Use new taxa.

New Guinea
Refers to whole island. Index Irian Jaya and/or Papua New Guinea. Includes use on level 1 as an area term (list O). For term set options see list B.
CO S101500S023000
E1510000E1410000
BA Malay Archipelago
SA Indian Plate
Irian Jaya
Pacific mobile belt
Papua New Guinea

New Hampshire
Includes use on level 1 as an area term (list O). For term set options see list B.
CO N424500N452000
W0704500W0723500
BA New England
Eastern U.S.
United States
NX Carroll County
NT Grafton County
NX Great Bay
Hillsborough County
NT Ossipee Mountains
NX Portland
Rockingham County
Sullivan County
SA Bronson Hill Anticlinorium
Connecticut River
Connecticut Valley
Littleton Formation
Manchester
Merrimack River valley
Merrimack Synclinorium
Rangeley Lakes
Saco River
White Mountains

New Hanover County
Extreme SE.
BT North Carolina
United States

New Hebrides
Group of islands NE of New Caledonia and W of Fiji. Under joint British and French administration. Includes use on level 1 as an area term (list O). For term set options see list B.
CO S203000S130000
E1690000E1660000
BA Melanesia
NT Malekula
SA Pacific Ocean

New Ireland
Island in the Bismarck Archipelago NNE of New Britain in SW Pacific.
BT Papua New Guinea
Australasia
SA Bismarck Archipelago

New Jersey
Includes use on level 1 as an area term (list O). For term set options see list B.
CO N385500N412000
W0735500W0753500
BA Eastern U.S.
United States
NT Atlantic County
Beemerville
NX Cumberland County
Essex County
Great Bay
Mercer County
Middlesex County
NT Monmouth County
Newark
Ocean County
Pine Barrens
Sandy Hook
NX Somerset County
Sussex County
NT Trenton
NX Union County
Warren County
Woodbury
SA Atlantic Coastal Plain
Carolina Bays
Delaware Bay
Delaware River
Delaware River basin
Glenarm Series
Hudson River
Hudson Valley
Kirkwood Formation

Magothy Formation
Marshalltown Formation
Martinsburg Formation
Monmouth Group
Navesink Formation
Newark Basin
Newark Group
Newark-Gettysburg Basin
Piedmont
Raritan Formation
Valley and Ridge Province
Woodbury Clay

New Liskeard
Town in SE.
BT Ontario
Canada

New Madrid
Town on Mississippi River in New Madrid County.
BT Missouri
United States

New Madrid County
Term introduced in 1981. SE Missouri.
CO N362200N365300
W0891900W0895800
BT Missouri
United States

New Madrid region
Term introduced in 1981. Region of earthquakes in the Mississippi Embayment. Index states as applicable.
UF New Madrid Seismic Zone
BT United States
SA Arkansas
Illinois
Indiana
Kentucky
Mississippi
Mississippi Embayment
Missouri
Tennessee

New Madrid Seismic Zone
use New Madrid region

new methods
Used as a general term, e.g. on level 3 under absolute age(1).
SA absolute age
methods

New Mexico
Includes use on level 1 as an area term (list O). For term set options see list B.
CO N313000N370000
W1030000W1090500
BA Southwestern U.S.
United States
NT Albuquerque
Ambrosia Lake
Animas Mountains
Bernalillo County
Black Range
Carlsbad
Carlsbad Caverns
Catron County
Chama Basin
Chaves County
Clovis
NX Colfax County
Curry County
NT Datil-Mogollon volcanic field
Dona Ana County
NX Eddy County
NT Florida Mountains
Ghost Ranch
NX Grant County
NT Grants
Grants mineral belt
NX Guadalupe County
Harding County
Hidalgo County
NT Hillsboro Quadrangle
NX Lincoln County
NT Los Alamos Scientific Laboratory
Luna County
Magdalena Mountains
McKinley County
Mora County
Nacimiento Mountains
NX Otero County
NT Picuris Range
Questa Mine
Rio Arriba County
NX Roosevelt County
NT Sacramento Mountains
NX San Juan County
NT San Lorenzo Quadrangle
San Mateo Mountains
Sandia Mountains
Sandoval County
Santa Fe County
Santa Rita
Sierra Blanca
NX Sierra County
NT Silver City
Socorro
Socorro County
Taos County
Tatum Basin
Tierra Amarilla
Torrance County
Tucumcari
Tusas Mountains
NX Union County
NT Ute Creek
Valencia County
Valle Grande Mountains
Valles Caldera
White Sands
Zuni Mountains
SA Basin and Range Province
Bell Canyon Formation
Benton Formation
Bidahochi Formation
Canadian River
Capitan Formation
Carlile Shale
Castile Formation
Chihuahua tectonic belt
Chinle Formation
Colorado Group
Colorado Plateau
Colorado River basin
Cutler Formation
Dakota Formation
Delaware Basin
Dockum Group
Fort Hays Limestone Member
Four Corners
Gallup Sandstone
Gila River
Graneros Shale
Great Plains
Greenhorn Limestone
Guadalupe Mountains
La Ventana Sandstone
Lake Valley Formation
Lewis Shale
Llano Estacado
Loup Fork Group
Magdalena
Mancos Shale
Menefee Formation
Mesaverde Group
Midcontinent
Montana Group
Morrison Formation
Navajo Indian Reservation
Navajo Sandstone
Niobrara Formation
Ogallala Formation
Ojo Alamo Sandstone
Paradox Basin
Pasamonte Meteorite
Pecos River valley
Pedregosa Basin
Permian Basin
Raton Basin
Red River
Red River valley
Redonda Formation
Rio Grande
Rio Grande Rift
Rio Grande Valley
San Andres Formation
San Francisco Mountains
San Jose Formation
San Juan Basin
San Juan River
San Luis Valley
Sangre de Cristo Mountains
Santa Fe
Smoky Hill Chalk Member
Supai Formation
Todilto Formation
Trans-Pecos
U. S. Rocky Mountains
Wasatch Formation
Yates Formation

new minerals
Used to indicate new mineral names.
BT minerals

new names
Used as a general term.
SA nomenclature
taxonomy

New Orleans
City on the Mississippi River in Orleans Parish.
BT Louisiana
United States

New Quebec
use Ungava

New Red Sandstone
England. Permian and Triassic. Includes use on level 3 under age terms(1). See list E.
SA Permian
Triassic

New River
Rises in NW North Carolina and joins with Gauley River in West Virginia to form the Kanawha River. Index states as applicable.
BT United States
SA North Carolina
Virginia
West Virginia

New Siberian Islands
Term introduced in 1981. Island group in Arctic Ocean between Laptev Sea and East Siberian Sea. Part of Yakutia.
CO N730000N770000
E1510000E1350000
BA Northeastern USSR
BT Russian Republic
USSR

New South Wales
Includes use on level 1 as an area term (list O). For term set options see list B.
CO S373000S281500
E1533000E1410000
BA Australia
Australasia
NT Armidale
Australian Capital Territory
Broken Bay
Broken Hill
Bungonia Caves
Cooma
Gosford
Hunter Valley
Macleay River
Manildra
Murrumbidgee River
Murrumbidgee Valley
Nandewar Mountains
New England Range
Round Mountain
Sydney
Sydney Basin
Tamworth
Wagga Wagga
Yass
Yeoval
SA Bulli Seam
Great Artesian Basin
Hawkesbury Sandstone
Illawarra Coal Measures
Lachlan fold belt
Murray Basin
Murray River
Narrabeen Group
New England Batholith
Singleton Coal Measures
Snowy Mountains
Tasman Geosyncline
Willyama Complex

new species
A valid term through 1973. Use new taxa.

new taxa
Includes use on level 3 under fossil group(1) whenever a new taxonomic group is discussed. See list F.
UF taxa, new
SA revision
taxonomy

New Ulm
City in Brown County in S central.
BT Minnesota
United States

New World Island
In Notre Dame Bay off NE.
BT Newfoundland
Canada

New York
Includes use on level 1 as an area term (list O). For term set options see list B.
CO N403000N450000
W0715500W0794500
BA Eastern U.S.
United States
NT Adirondack Mountains
Albany
NX Albany County
Allegany County
NT Bay Park
Becraft Mountain
Benson Mines
Binghamton
Blue Mountain Lake
Catskill Mountains
Cattaraugus County
NX Chautauqua County
NT Chenango River
NX Clinton County
Columbia County
Delaware County
NT Dutchess County
NX Erie County
Essex County
NT Finger Lakes
Fire Island
NX Franklin County
Fulton County
Genesee County
NT Gouverneur
Great South Bay
NX Greene County
Hamilton County
NT Herkimer County
Hudson Highlands
NX Jefferson County
Kings County
NT Lake George
NX Lewis County
Livingston County
NT Long Island
NX Madison County
Manhattan
NT Marcy Massif
Mohawk Valley
NX Monroe County
Montgomery County

NT Narrows
NX Nassau County
NT New York Bight
New York City
Ogdensburg
NX Oneida County
NT Oneida Lake
Onondaga County
NX Orange County
NT Oswego County
Peekskill
Port Jervis
NX Putnam County
NT Rensselaer County
Rochester
Saint Lawrence County
Schoharie County
NX Seneca County
NT Seneca Lake
Shawangunk Mountains
Springwater
Staten Island
NX Suffolk County
Sullivan County
NT Ticonderoga
Ulster County
NX Warren County
Washington County
Wayne County
NT Westchester County
SA Adirondack Anorthosite
Allegheny Plateau
Appalachian Basin
Appalachian Plateau
Atlantic Coastal Plain
Black River Group
Canadian Shield
Catskill Delta
Catskill Formation
Champlain Valley
Clinton Group
Coeymans Formation
Delaware River
Delaware River basin
Genesee Group
Great Lakes region
Grenville Province
Hamilton Group
Helderberg Group
Hudson River
Hudson Valley
Ithaca
Lake Champlain
Lake Erie
Lake Ontario
Little Falls Formation
Lockport Formation
Magothy Formation
Manhattan Formation
Manlius Formation
Medina Formation
New York City Group
Newark Basin
Newark Group
Newark-Gettysburg Basin
Niagara Escarpment
Niagara Falls
Niagara River
Normanskill Formation
Onondaga Limestone
Oriskany Sandstone
Piedmont
Potsdam Sandstone
Rochester Formation
Rondout Formation
Saint Lawrence Lowlands
Saint Lawrence River
Saint Lawrence Valley
Schenectady Formation
Susquehanna River
Susquehanna River basin
Syracuse
Tioga Bentonite
Trenton Group
Tully Limestone
Utica Shale
Valley and Ridge Province

New York Bight
CO N390000N410000
W0720000W0750000
BT New York
United States
SA Atlantic Coastal Plain

New York City
City on W tip of Long Island, Manhattan Island, Staten Island, and S tip of mainland.
BT New York
United States
NX Queens County

New York City Group
Precambrian or Paleozoic (pre-Upper Devonian). Includes Fordham Gneiss, Inwood Marble, Manhattan Formation. W Connecticut and SE New York.
SA Connecticut
Manhattan Formation
New York
Paleozoic
Precambrian

New Zealand
Includes use on level 1 as an area term (list O). For term set options see list B.
CO S473000S343000
E1783000E1663000
BA Australasia
NT Alpine Fault
Broadlands
Foveaux Strait
North Island
South Island
Tasman orogenic zone
SA Indian Plate
Oceania
Pacific mobile belt
Torlesse Supergroup
Waitemata Group

Newark
City on Newark Bay just W of Jersey City and New York City.
BT New Jersey
United States

Newark Basin
Index states as applicable.
BT United States
SA New Jersey
New York
Pennsylvania

Newark Group
Includes New Oxford Formation, Gettysburg Shale, Stockton Formation, Lockatong Formation, Brunswick Formation, New Haven Arkose, Meriden Formation, Portland Arkose, Pekin Formation, Cumnock Formation, Sanford Formation, Talcott Basalt, Shuttle Meadow Formation, Holyoke Basalt, East Berlin Formation, Hampden Basalt, Otterdale Sandstone.
BT Upper Triassic
Triassic
SA Connecticut
Delaware
Maryland
New Jersey
New York
North Carolina
Virginia

Newark-Gettysburg Basin
Newark Basin plus southwestward extension into SE Pennsylvania and N central Maryland. Index states as applicable.
BT United States
SA Maryland
New Jersey
New York
Pennsylvania

newberyite
BA phosphates
BT minerals

Newcastle
City on the Tyne River in Northumberland County in N England. Also search Newcastle upon Tyne.
UF Newcastle upon Tyne
BT England
Great Britain
United Kingdom

Newcastle upon Tyne
use Newcastle

Newfoundland
As of 1981, refers to the island of Newfoundland. Before 1981, referred to the province, which includes the island of Newfoundland plus Labrador. Includes use on level 1 as an area term (list O). For term set options see List B.
CO N463000N604500
W0523000W0673000
BA Eastern Canada
Canada
NT Avalon Peninsula
Bay of Islands
Bell Island
Betts Cove
Burlington Peninsula
Daniel's Harbour
New World Island
Notre Dame Bay
Port au Port Peninsula
Saint John's
White Bay
SA Davidsville Group
Gander Lake Group
Grenville Province
Harbour Main Group
Labrador
Lushs Bight Group
Signal Hill Formation

Newman Limestone
Overlies Grainger Formation; underlies Pennington Formation. E Kentucky, E Tennessee, and SW Virginia.
BT Mississippian
SA Kentucky
Tennessee
Virginia

Newmarkt
use Nowy Targ

Newport
City at mouth of Narragansett Bay in S Rhode Island, and city on the Pacific Ocean in W Oregon. Index states as applicable.
BX Oregon
Rhode Island
BT United States

Newport Bay
Dredged harbor in Orange County in S.
BT California
United States

Newport-Inglewood Fault
In Los Angeles area in S California. Also search Newport-Inglewood fault zone.
UF Newport-Inglewood fault zone
BT California
United States

Newport-Inglewood fault zone
use Newport-Inglewood Fault

Newton County
Index states as applicable.
BX Arkansas
Georgia
Indiana
Mississippi
Missouri
Texas
BT United States

Neyveli
Town in E Tamil Nadu.
UF Neiveli
BT Tamil Nadu
India

Neyveli Lignite
Associated with sands, conglomerates and clays of Cuddalore Sandstone Formation. E and NE of Tamil Nadu. Also search Neyveli.
BT Tertiary
SA India
Miocene
Pliocene
Tamil Nadu
upper Miocene

Ngorora Formation
BT Miocene
Neogene
Tertiary
SA Kenya

Ni
use nickel

Niagara Escarpment
Index states and province as applicable. Cuesta of Paleozoic rocks sloping away from the Canadian Shield, extending from central New York to N of Lake Huron and Lake Michigan, continuing southwestward into Wisconsin, Illinois, and Iowa.
BT North America
SA Bruce Peninsula
Great Lakes region
Illinois
Iowa
Manitoulin Island
Michigan
New York
Niagara Falls
Ontario
Wisconsin

Niagara Falls
Great falls of the Niagara River. Index state and province as applicable. Also search Niagara.
CO N430600N430600
W0790300W0790300
BT North America
SA New York
Niagara Escarpment
Ontario

Niagara River
Flows between Lake Erie and Lake Ontario going over the Niagara Escarpment. Index province and state as applicable. Also search Niagara.
BT North America
SA New York
Ontario

Niagaran
Provincial series, North America. Above Alexandrian, below Cayugan. Includes use on level 3 under age terms(1). See list E.
BA Middle Silurian
Silurian
BT Paleozoic
NT Bainbridge Formation
Lockport Formation

Nicaragua
Includes use on level 1 as an area term (list O). For term set options see list B.
CO N104000N150000
W0830500W0874000
BA Central America
NT Leon Department
Managua
SA Leon

Nicaragua Rise
Between Jamaica and Nicaragua.
UF Nicaraguan Rise
BT Caribbean Sea

Nicaraguan Rise
use Nicaragua Rise

niccolite
As of 1981, includes use on level 3 under minerals(1) arsenides(2). See list L. Before 1981, included use on level 3 under minerals(1) sulfides(2).
UF nickeline
nicolite
BA arsenides
sulfides
BT minerals
SA nickel

Nice
City on the Mediterranean Sea on the Riviera.
BT Alpes-Maritimes
France

nickel
Includes use on level 1 and 2 as a chemical element (list D). As of 1981, use nickel ores for nickel as a commodity. Before 1981, included use on level 1 as a commodity term.
UF Ni
SA awaruite
elements
heavy metals
josephinite
kamacite
niccolite
nickel ores
pentlandite

nickel glance
use gersdorffite

nickel ores
Term introduced in 1981. Includes use on level 1 and 2 as a commodity term (list C). Before 1981, search nickel AND ore deposits.
SA nickel

nickel pyrites
use millerite

nickel-antimony glance
use ullmannite

nickeline
use niccolite

Nicobar Island
use Nicobar Islands

Nicobar Islands
Island group in Bay of Bengal NW of Sumatra, forming S part of Andaman and Nicobar Islands. Also search Nicobar Island.
UF Nicobar Island
Nicobars
BA Bengal Islands
BT India

Nicobars
use Nicobar Islands

Nicola Group
S British Columbia.
BT Triassic
SA British Columbia
Canada

Nicolaus
Village N of Sacramento in Sutter County in N central.
BT California
United States

nicolite
use niccolite

nicopyrite
use pentlandite

Nida Basin
River basin in S part of country. Also search Nida; Nida River basin.
UF Nida River basin
BT Kielce
Poland

Nida River basin
use Nida Basin

Niemen River basin
use Neman River basin

Nievre
Department in central part of country.
CO N463000N473000
E0041000E0024500
BT France
NT Nivernais Plateau
SA Morvan
Yonne Valley

Niger
Includes use on level 1 as an area term (list O). For term set options see list B.
CO N120000N233000
E0160000E0001500
BA West Africa
Africa
NT Agades
SA Chad Basin
Lake Chad
Mali-Niger Syneclise
Niger River
Niger Valley
Sahara
Sahel
West African Shield
Younger Granites

Niger Delta
On the Gulf of Guinea.
BT Nigeria

Niger River
Rises in Guinea and flows NE into a great curve and then SE into the Gulf of Guinea. Index countries as applicable.
UF Joliba River
Kworra River
BT Africa
SA Benin
Guinea
Mali
Niger
Nigeria

Niger Valley
River valley. Index countries as applicable. Also search Niger.
UF Joliba Valley
Kworra Valley
BT Africa
SA Benin
Guinea
Mali
Niger
Nigeria

Nigeria
Includes use on level 1 as an area term (list O). For term set options see list B.
CO N040000N140000
E0143000E0023000
BA West Africa
Africa
NT Bida
East-Central Nigeria
Ewekoro
Ibadan
Jos Plateau
Lagos Lagoon
Niger Delta
Northern Nigeria
Sokoto Basin
Southeastern Nigeria
West-Central Nigeria
Western Nigeria
Zaria
SA Benue Valley
Chad Basin
Ewekoro Formation
Lake Chad
Mali-Niger Syneclise
Niger River
Niger Valley
Younger Granites

nigerite
BA oxides
BT minerals

niggliite
As of 1981, includes use on level 3 under minerals(1) alloys(2). See list L. Before 1981, included use on level 3 under minerals(1) native elements and alloys(2).
BA alloys
BT minerals
SA platinum
tin

Niigata
City and prefecture in NW.
BT Honshu
Japan

Nikitovka
City in central Stalino Oblast in Donbas.
BT Ukraine
USSR

Nikopol
City in Dnepropetrovsk Oblast in E central.
BT Ukraine
USSR

Niksic
Town in SW part of country.
BT Montenegro
Yugoslavia

Nile Delta
On Mediterranean Sea E of Alexandria.
CO N301000N314000
E0322000E0295000
UF Nile River delta
BT Egypt
SA Nile River

Nile River
The longest river in the world. Its remotest headstream is the Luvironza River in Burundi. The flow is through Lake Victoria and Lake Albert, and then, as the White Nile, into Sudan where it unites with the Blue Nile from Ethiopia as the Nile proper which flows into the Mediterranean Sea. Index countries as applicable.
CO N220000N310000
E0370000E0310000
BT Africa
SA Blue Nile
Burundi
Egypt
Ethiopia
Nile Delta
Sudan
Uganda

Nile River delta
use Nile Delta

Nile Valley
River valley of the Nile River proper. Index countries as applicable.
BT Africa
NZ Middle Nile Valley
SA Egypt
Sudan

Nilgiri
Former Indian state near coast of NW Bay of Bengal.
BT Orissa
India

Nilssoniales
Includes use on level 2 under gymnosperms(1). See list F.
BA gymnosperms
BT Spermatophyta
Plantae

Nimba Mountains
West African bulge. Index countries as applicable. Also search Nimba.
BT Africa
SA Guinea
Ivory Coast
Liberia

Ninetyeast Ridge
Undersea ridge extending N and S along line of 90° E longitude from just S of Nicobar Islands to approximately 30° S latitude.
CO S350000N080000
E0920000E0850000
BT Indian Ocean

niningerite
Meteorite mineral.
BA sulfides
BT minerals

Niniyur Group
S part of country.
BT Danian
lower Paleocene
Paleocene
Paleogene
Tertiary
SA Cretaceous
India
Tamil Nadu
Upper Cretaceous

niobates
As of 1981, includes use on level 2 under minerals(1). See list L. Before 1981, included use on level 3 under minerals(1) oxides(2).
UF columbates
BA oxides
BT minerals
NA aeschynite
NZ betafite
columbite
euxenite
fergusonite
ixiolite
NA loparite
NZ microlite
nenadkevichite
pyrochlore
samarskite
stibiotantalite
tantalite
tapiolite
wodginite
NA zirconolite
SA niobotantalates

niobium
Includes use on level 1 and 2 as a chemical element (list D). As of 1981, use niobium ores for niobium as a commodity. Before 1981, included use on level 1 as a commodity term.
UF columbium
Nb
SA elements
niobium ores
tantalum

niobium ores
Term introduced in 1981. Includes use on level 1 and 2 as a commodity term (list C). Before 1981, search niobium AND ore deposits.
SA niobium
tantalum ores

niobotantalates
Term introduced in 1981. Includes use on level 2 under minerals(1). See list L.
BA oxides
BT minerals
SA niobates

tantalates

Niobrara Chalk
use Niobrara Formation

Niobrara Formation
In Colorado Group. Includes Fort Hays Limestone, Smoky Hill Chalk members. Also search Niobrara Chalk.
UF Niobrara Chalk
BT Upper Cretaceous
Cretaceous
SA Colorado
Colorado Group
Fort Hays Limestone Member
Kansas
Minnesota
Montana
Nebraska
New Mexico
North Dakota
Smoky Hill Chalk Member
South Dakota
Wyoming

Nipissing
Term introduced in 1981. District in SE central.
CO N452000N472000
W0773000W0802500
BT Ontario
Canada

Nipissing Diabase
BT Precambrian
SA Canada
Ontario
Quebec

Nisqually Glacier
On S slope of Mt Rainier in Mt. Rainier National Park in W central.
BT Washington
United States

nitrate deposits
Term introduced in 1981. Includes use on level 1 and 2 as a commodity term (list C). Before 1981, search nitrates AND deposits.
SA nitrates

nitrates
Includes use on level 2 under minerals(1). See list L. As of 1981, use nitrate deposits for nitrate as a commodity. Before 1981, included use on level 1 as a commodity term.
BT minerals
SA nitrate deposits

nitrides
As of 1981, includes use on level 2 under minerals(1). See list L. Before 1981, included use on level 3 under minerals(1) native elements and alloys(2).
BA alloys
BT minerals

nitrogen
Includes use on level 1 and 2 as a chemical element (list D); on level 3 under soils(1). Also search N.
UF N
SA elements
fertilizers
N-15/N-14
soils

Nittany Valley
In Centre and Clinton counties in central.
BT Pennsylvania
United States

Nitzschia
Genus.
BA diatoms
algae
BT thallophytes
Plantae

nivation
UF snow-patch erosion
BT erosion
SA glaciers

Nivernais Hills
use Nivernais Plateau

Nivernais Plateau
In central part of country.
UF Nivernais Hills
BT Nievre
France

Nizhni Novgorod
use Gorki

Nizhniy Ufaley
use Ufaley

Nizke Jesenik Mountains
use Nizky Jezenik Mountains

Nizky Jezenik Mountains
In NW near the Polish border. Also search Nizky Jezenik.
UF Low Jesenik Mountains
Nizke Jesenik Mountains
BT Moravia
Czechoslovakia

noble gases
Includes use on level 1 and 2. See list D for term set options. Also see list C (commodities).
UF inert gases
rare gases
NT argon
helium
krypton
neon
radon
xenon
SA gases
isotopes

Nodosaria
Genus. Includes use on level 3 under foraminifera(1) Nodosariacea(2).
BA Nodosariidae
Nodosariacea
Rotaliina
foraminifera
BT Protista
Invertebrata

Nodosariacea
Includes use on level 2 under foraminifera(1). See list F.
BA Rotaliina
foraminifera
BT Protista
Invertebrata
NA Nodosariidae

Nodosariidae
Family. Includes use on level 3 under foraminifera(1) Nodosariacea(2).
BA Nodosariacea
Rotaliina
foraminifera
BT Protista
Invertebrata
NA Lenticulina
Nodosaria

nodules
Includes use on level 1 (list A). As of 1981, restricted to nodules on the ocean floor. For nodules in kimberlite, use xenoliths. For nodules within sedimentary formations such as chert nodules, use concretions. If level 1, term set options are:
kind of nodule [ferromanganese composition, manganese composition]
topic
topic [age, classification, composition, distribution, genesis, observations, properties]
subtopic

SA concretions
ferromanganese composition
marine geology
ocean floors
oceanography
sedimentation

Noeggerathiales
Includes use on level 2 under pteridophytes(1). See list F.
BA pteridophytes
BT Plantae

Nohi Rhyolite
Complex in central Japan.
BT Honshu
Japan

noise
Includes use on level 3 under seismology(1) or geophysical methods(1). Before 1978, also search seismic noise.
SA filters
geophysical methods
microseisms
seismology
signal-to-noise ratio
signals

Nojiri Lake
Central Honshu.
BT Honshu
Japan

Nomarthra
Term introduced in 1981. Includes use on level 2 under Mammalia(1). See list F.
BA Mammalia
BT Tetrapoda
Vertebrata

Nome
City on S side of Seward Peninsula in W.
BT Alaska
United States

nomenclature
Includes use as a level 2 or 3 term appropriate to a large number of topics, e.g. on level 2 under sedimentary petrology(1). See list G.
SA classification
definition
homonymy
identification
new names

nomograms
UF nomographs
SA automatic data processing

nomographs
use nomograms

non-glacial
A valid index term through 1976. See non-glacial ice. Before 1976 also search ice AND non-glacial.

non-glacial ice
As of 1981, no longer a valid term for GeoRef. Use ice. Included use on level 3 under glacial geology(1) periglacial features(2).

non-linear distortion
Term introduced in 1981.
SA distortion
elastic waves
seismology

non-linear materials
Term introduced in 1981.
SA engineering geology
materials

Noncalcarea
use Silicispongiae

Nonesuch Shale
In Oronto Group. N Michigan and NE Wisconsin.
BT Precambrian
SA Michigan
Wisconsin

Nong-Son
Village in Quang Nam Province in N South Vietnam.
UF Nongson
BT Vietnam
Asia

Nongson
use Nong-Son

nonmare basalt
use KREEP

nonmetal deposits
Term introduced in 1981. Includes use on level 1 and 2 as a commodity term (list C). Before 1981, search nonmetals AND deposits.
SA industrial minerals
metal ores
native elements
nonmetals

nonmetals
Includes use on level 1 and 2 as a chemical element (list D). As of 1981, use nonmetal deposits for nonmetals as a commodity. Before 1981, included use on level 1 as a commodity term.
SA economic geology
elements
metals
native elements
nonmetal deposits

nontectonics
use atectonic processes

nonterrigenous
A valid term through 1977. Included use in combination with clastic rocks or clastic sediments.

nonterrigenous materials
As of 1981, no longer a valid term for GeoRef. Term was introduced in 1978. Included use on level 3 under sediments(1) and sedimentary rocks(1). Before 1978, also search nonterrigenous AND clastic rocks or clastic sediments.

nontronite
Also search chloropal.
UF chloropal
gramenite
morencite
pinguite
BA clay minerals
sheet silicates
silicates
BT minerals
SA montmorillonite

Noonday Dolomite
Underlies Johnnie Formation; overlies Kingston Peale Formation with disconformity or slight angular unconformity.
BT Precambrian
SA California

Noranda
Mining city in SW.
BT Quebec
Canada

norbergite
BZ fluorides
humite group
BT minerals

Nord Department
Department in extreme N.
CO N495800N510500
E0041500E0020800
BT France
NT Dunkirk
Lille
Valenciennes
SA Nord-Pas-de-Calais Basin

N

Nord Fjord
Inlet of Norwegian Sea.
BT Norway

Nord-Pas-de-Calais Basin
In extreme N part of country. Index departments as applicable.
BT France
SA Nord Department
Pas-de-Calais

Nord-Trondelag
County N of Trondheim Fjord.
BT Norway
SA Trondelag

Nordaustlandet
Norwegian for North East Land. One of the islands of the Spitsbergen group in Barents Sea NE of island of Spitsbergen.
BT Spitsbergen
Svalbard
Arctic region

Nordland
County in W.
BT Norway
NT Bleikvassli
Lofoten Islands
Narvik
Ofoten
Rana Fjord
SA Vesteralen

Nordlingen
use Ries Crater

Nordlinger Ries
use Ries Crater

Nordlinger Ries Crater
use Ries Crater

nordstrandite
BA oxides
BT minerals

Norfolk
As of 1981, refers to the county in England.
CO N522500N530000
E0014500E0001000
BT England
Great Britain
United Kingdom

Norfolk City
Term introduced in 1981 for the city in SE Virginia. Before 1981, also search Norfolk AND Virginia.
BT Virginia
United States

Norfolk Island
Midway between New Caledonia and N New Zealand. An external territory of Australia.
BT Pacific Ocean

Norfolk Ridge
Between New Caledonia and Norfolk Island in SW.
BT Pacific Ocean

Norian
Europe. Above Carnian, below Rhaetian. Includes use on level 3 under age terms(1). See list E.
BA Upper Triassic
Triassic
BT Mesozoic

Norilsk
Town in Taymyr National Okrug in N Siberia. Also search Noril'sk.
CO N692000N692200
E0880300E0880100
BT Russian Republic
USSR

Norilsk region
A mining area in the Taymyr National Okrug connected to the town of Dudinka on the Yenisei River by a 60 mile narrow gage rail line. Also search Norilsk.
CO N690000N694500
E0890000E0870000
BT Russian Republic
USSR

norite
As of 1981, includes use on level 3 under igneous rocks(1) gabbros(2). See list H. Before 1981, included use on level 3 under igneous rocks(1) gabbro family(2).
BA gabbros
BT igneous rocks

normal
A valid term through 1977. After 1977, use normal faults or normal folds.

normal earthquake
use shallow-focus earthquakes

normal faults
Term introduced in 1978. Includes use on level 3 under faults(1) displacements(2). Before 1978, also search faults AND normal; slump faults.
UF gravity faults
normal slip faults
BA faults
SA block structures

normal folds
Term introduced in 1978. Before 1978 search folds AND normal.
BA folds
SA symmetric folds

normal slip faults
use normal faults

Norman Wells
Village on the Mackenzie River in W District of Mackenzie.
BT Northwest Territories
Canada

Normandie
use Normandy

Normandy
Historical region of NW part of country. Index departments as applicable.
UF Normandie
BT France
SA Calvados
Eure
Manche
Orne
Seine-Maritime

Normanskill Chalk
use Normanskill Formation

Normanskill Formation
Comprises Mount Merino Chert and Shale, Austin Glen Grit and Shale. NW Massachusetts, E New York, and SW Vermont.
UF Normanskill Chalk
BT Middle Ordovician
Ordovician
SA Massachusetts
New York
Vermont

Norphlet Formation
Underlies Smackover Formation; overlies Louann Salt. Subsurface.
BT Jurassic
SA Arkansas
Louisiana
Texas

Norrbotten
Northernmost county.
BT Sweden
NT Gallivare
Kebnekaise
Kiruna
Laisvall
SA Skellefte

Norseman
Town in S central.
BT Western Australia
Australia

norsethite
BA carbonates
BT minerals

North Africa
A region. Index countries as applicable.
CO N190000N373000
E0360000W0153000
BA Africa
NA Algeria
Egypt
Libya
Morocco
Tunisia
SA Central Africa
East Africa
Southern Africa
West Africa

North America
Includes use on level 1 or 2 as an area term (list O). If 1, see term set options under list B. To retrieve all documents, individual countries and physiographic regions should also be searched (see list O). Includes Canada, United States, Mexico, and Saint Pierre and Miquelon.
CO N080000N840000
W0100000W1730000
NA Appalachians
NT Canada
NA Canadian Shield
NT Coast Mountains
Coast Ranges
Colorado River basin
Denali Fault
Great Appalachian Valley
NA Great Lakes
Great Lakes region
Great Plains
Gulf Coastal Plain
NT Lake Agassiz
Maverick Basin
Mexico
Missouri River basin
Niagara Escarpment
Niagara Falls
Niagara River
North American Cordillera
Okanagan Valley
Rio Grande Rift
Rocky Mountain Trench
NA Rocky Mountains
NT Rocky Mountains foreland
Saint Clair River
Saint Clair River delta
Saint Elias Mountains
Saint Lawrence Lowlands
Saint Lawrence River
Saint Pierre and Miquelon
United States
NA Western Interior
NT White River
Williston Basin
SA America
Beringia
North American Plate
Pacific mobile belt

North American Atlantic
Term introduced in 1981 to indicate a region of the Atlantic Ocean.
BA North Atlantic
BT Atlantic Ocean
NA Caribbean Sea
Gulf of Mexico
Gulf of Saint Lawrence
Labrador Sea

North American Cordillera
Term introduced in 1981. Before 1981, search Cordillera AND North America.
BT North America
NX Canadian Cordillera
SA Alaska Range
Canada
Canadian Rocky Mountains
Coast Mountains
Coast Ranges
Mexico
Rocky Mountains
Sierra Madre
Sierra Madre del Sur
Sierra Madre Occidental
Sierra Madre Oriental
Sierra Nevada
U. S. Rocky Mountains
United States

North American Pacific
Term introduced in 1981 to indicate a region of the Pacific Ocean.
BA Pacific Ocean
NA Gulf of Alaska
Gulf of California

North American Plate
Includes North America N of central America as well as all the North Atlantic Ocean W of the Mid-Ocean Ridge.
SA North America
plate tectonics
plates

North Arcot
District in E central.
BT Tamil Nadu
India

North Atlantic
Term introduced in 1978. Before 1978, search Atlantic Ocean AND north.
BA Atlantic Ocean
NA Azores-Gibraltar Ridge
Cape Verde Atlantic
European Atlantic
Gulf of Guinea
Guyanese Atlantic
North American Atlantic
NZ North Atlantic Ridge
SA Northeast Atlantic
Northwest Atlantic
South Atlantic
Southeast Atlantic
Southwest Atlantic
West Atlantic

North Atlantic Ridge
Term introduced in 1981.
BZ Mid-Atlantic Ridge
North Atlantic
BT Atlantic Ocean

North Australian Seas
Term introduced in 1981. Includes seas of the Pacific Ocean just N of Australia.
BA West Pacific
BT Pacific Ocean
NA Arafura Sea
Timor Sea

North Austrian Alps
Term introduced in 1981.
CO N470000N482200
E0162000E0093300
BZ Alps
BX Austria

North Austrian Crystallines
Term introduced in 1981.
CO N480700N490200
E0160800E0132500
BT Austria

North Austrian Molasse
Term introduced in 1981.
CO N475500N482700
E0161700E0124500
BT Austria

North Bay
City on Lake Nipissing in SE.

BT Ontario
Canada

North Borneo
use Sabah

North Bulgarian Hills
Term introduced in 1981.
CO N431200N441300
E0274500E0223000
BT Bulgaria

North Carolina
Includes use on level 1 as an area term (list O). For term set options see list B.
CO N335000N363500
W0753000W0841500
BA Eastern U.S.
United States
NX Alleghany County
NT Beaufort
Beaufort County
NX Brunswick County
NT Buncombe County
NX Burke County
Caldwell County
NT Cape Fear Arch
Cape Hatteras
Cape Lookout
Carteret County
Castle Hayne
Charlotte
NX Chatham County
Cherokee County
Clay County
Cumberland County
NT Dare County
Davidson
NX Davidson County
NT Durham County
NX Franklin County
Graham County
NT Grandfather Mountain
NX Greene County
Jackson County
Jones County
Lee County
Lincoln County
Macon County
Madison County
Mecklenburg County
Mitchell County
Montgomery County
Moore County
NT Neuse River
New Hanover County
Onslow Bay
NX Orange County
NT Outer Banks
Pamlico River
Pamlico Sound
NX Polk County
NT Pungo River
NX Randolph County
Rockingham County
Rowan County
NT Stokes County
NX Surry County
Union County
Warren County
Washington County
Wayne County
Wilson County
SA Appalachian Basin
Appalachian Plateau
Ashe Formation
Atlantic Coastal Plain
Beaufort Formation
Blue Ridge Mountains
Blue Ridge Province
Brevard Zone
Carolina Bays
Carolina slate belt
Castle Hayne Limestone
Chilhowee Group
Dan River basin
Duplin Formation
Great Smoky Mountains
Kings Mountain
Knox Group
Lake Chatuge
Moore County Meteorite
Murphy Marble
New River
Newark Group
Peedee Formation
Piedmont
Roan Supergroup
Rome Formation
Sand Hills
Shady Dolomite
Talladega Group
Trent Valley
Tuscaloosa Formation
Valley and Ridge Province
Waccamaw Formation
Wicomico Formation
Wilmington
Yorktown Formation

North Caucasus
use Northern Caucasus

North Central (United States)
use Midwest

North Dakota
Includes use on level 1 as an area term (list O). For term set options see list B.
CO N455500N490000
W0963500W1040500
BA Midwest
United States
NX Adams County
NT Billings County
NX Burke County
Cass County
Dunn County
Eddy County
NT Grand Forks
NX Grant County
Logan County
McHenry County
McIntosh County
NT McKenzie County
NX McLean County
Mercer County
Morton County
Nelson County
NT Oliver County
NX Pierce County
Richland County
Sheridan County
Sioux County
NT Slope County
NX Stark County
NT Stutsman County
Walsh County
NX Ward County
SA Colorado Group
Dakota Formation
Elk Point Group
Fort Union Formation
Fox Hills Formation
Golden Valley Formation
Great Plains
Hell Creek Formation
Lake Agassiz
Lance Formation
Little Missouri River basin
Midcontinent
Missouri River
Missouri River basin
Missouri River valley
Montana Group
Niobrara Formation
Pierre Shale
Prairie Evaporite
Red River Formation
Sentinel Butte Formation
Souris River basin
Tongue River Member
Wasatch Formation
White River Group
Williston Basin
Winnipeg Formation
Winnipegosis Formation

North Devon Island
use Devon Island

North German Plain
Roughly the northern one-third of West Germany and the northern half of East Germany. Index countries as applicable.
BT Europe
SA East Germany
West Germany

North Island
The northernmost of 3 main islands of New Zealand.
CO S413000S343000
E1783000E1720000
BT New Zealand
NT Auckland
Coromandel Peninsula
Gisborne
Hawke's Bay
Kaipara Harbor
Manawatu River valley
Mokoia
Raukumara Peninsula
Ruapehu
Taranaki
Tarawera volcanic complex
Taupo
Taupo volcanic zone
Te Aroha
Waikato Basin
Wairakei
Wairarapa
Wanganui
Wanganui Valley
Wellington
White Island
SA Tasman orogenic zone
Waitemata Group

North Korea
Officially Democratic People's Republic of Korea. Republic, on E coast of Asia, bounded on N by China, on NE by USSR, on E by Japan Sea, on S by South Korea, and on W by the Yellow Sea and Korea Bay.
CO N374500N430000
E1303000E1241500
BT Korea
SA Kangwon
South Korea

North Ossetia
North Ossetian Autonomous Soviet Socialist Republic. On N slopes of central Caucasus Mountains.
BT Russian Republic
USSR

North Pacific
Term introduced in 1978. Before 1978, search Pacific Ocean AND north.
BA Pacific Ocean
SA East Pacific
Equatorial Pacific
Northeast Pacific
Northwest Pacific
Pacific Basin
South Pacific
Southeast Pacific
Southwest Pacific
West Pacific

North Polar Sea
use Arctic Ocean

North Pole
The N extremity of the Earth's axis at 90° N latitude and the point from which all directions are S.
SA Polar regions

North Pole deposit
Term introduced in 1981. Barite deposit in Western Australia. Also search North Pole AND Australia.
BT Western Australia
Australia

North Ray Crater
BT Moon

North Rhine
North portion of former Prussian Rhine Province which included Aachen, Cologne and Dusseldorf.
BT North Rhine-Westphalia
West Germany

North Rhine-Westphalia
State.
CO N502000N523000
E0093000E0060000
BT West Germany
NT Aachen
Bergisch Gladbach
Bochum
Cologne
Ibbenburen
Kamen
Krefeld
Lindlar
Lippe
Munster
Munsterland
North Rhine
Paderborn
Plettenberg
Ruhr
Sauerland
Siebengebirge
University of Bonn
Wiehen Mountains
SA Ems River
Essen Beds
Rhenish Schiefergebirge
Rhineland
Steinheim
Teutoburg Forest
Weser River
Westphalia

North Saskatchewan River
Rises in Columbia Ice Field at foot of Mt. Saskatchewan in Alberta. Index provinces as applicable.
BT Canada
SA Alberta
Saskatchewan
Saskatchewan River
South Saskatchewan River

North Sea
Between the European continent on the S and E and Great Britain on the W. Also search North Sea Basin. Includes use on level 1 as an area term (list O). For term set options see list B.
CO N510000N611000
E0110000W0040000
UF North Sea Basin
BA European Atlantic
North Atlantic
Atlantic Ocean
NT Dogger Bank
Kattegat
Norwegian Channel
Skagerrak
SA Forties Field
North Sea region

North Sea Basin
use North Sea

North Sea Coast
Index England, Scotland and European countries as applicable.
BT Europe
SA Belgium
Denmark
England
France
Germany
Netherlands
Norway
Scotland

Sweden

North Sea region
SA North Sea

North Shore Volcanic Group
use North Shore Volcanics

North Shore Volcanics
In Keweenawan Group. NE Minnesota. Also search North Shore Volcanic Group.
UF North Shore Volcanic Group
BT Precambrian
SA Minnesota

North Siberian Plain
Term introduced in 1981.
CO N700000N750000
E1300000E0810000
BA Siberian Platform
BT Russian Republic
USSR

North Slope
Arctic plains N of Brooks Range to Arctic Ocean.
BT Alaska
United States
SA Prudhoe Bay

North Sudetic Basin
North part of the Sudetic Basin in the Sudeten Mountains. Index countries as applicable.
BT Europe
SA Czechoslovakia
Poland
Sudeten Mountains
Sudetic Basin

North-Central (United States)
use Midwest

North-West Frontier Province
W Pakistan.
BT Pakistan
NT Chitral
Hazara
Peshawar
Swat
SA Sulaiman Range

Northamptonshire
County in central.
CO N520000N524000
W0002000W0012000
BT England
Great Britain
United Kingdom

Northeast Atlantic
Term introduced in 1978. Before 1978, search Atlantic Ocean AND northeast.
BA Atlantic Ocean
SA East Atlantic
North Atlantic
Northwest Atlantic
South Atlantic
Southeast Atlantic
Southwest Atlantic
West Atlantic

Northeast Pacific
Term introduced in 1978. Before 1978, search Pacific Ocean AND northeast.
BA Pacific Ocean
SA East Pacific
Equatorial Pacific
North Pacific
Northwest Pacific
South Pacific
Southeast Pacific
Southwest Pacific
West Pacific

Northeast Providence Channel
Strait NE of Nassau between Great Abaco and Eleuthera islands.
BT Bahamas
West Indies

Northeastern German Plain
Term introduced in 1981.
CO N514000N544000
E0144500E0104000
BT Germany
SA East Germany

Northeastern Hungarian Hills
Term introduced in 1981.
CO N474500N483500
E0214200E0184500
BT Hungary

Northeastern India
Term introduced in 1981.
CO N214000N293500
E0972000E0880000
BT India
NA Assam
Tripura

Northeastern Polish Plain
Term introduced in 1981.
CO N505500N542700
E0240500E0181000
BT Poland

Northeastern USSR
Term introduced in 1978. Before 1978, search USSR AND northeast. As of 1981, term refers to the region W of the Soviet Pacific region.
CO N423000N770000
W1700000E1140000
BT Russian Republic
USSR
NA Amur region
Chukchi Peninsula
Dzhugdzhur region
Kolyma Uplift
New Siberian Islands
Sikhote-Alin Range
Stanovoy Range
Verkhoyansk region

Northern Andes
Term introduced in 1978. Before 1978, search Andes AND north or northern.
BA Andes
BT South America
SA Central Andes
Southern Andes

Northern Apennines
Term introduced in 1981. Before 1981, also search Apennines AND North. Extend SE to Scheggia Pass.
CO N432000N450000
E0124500E0073000
BT Apennines
Italy

Northern Appalachians
Term introduced in 1978. Before 1978, search Appalachians AND north or northern.
BA Appalachians
BT North America
SA Central Appalachians
Southern Appalachians

Northern California
Term introduced in 1981. Before 1981, also search California AND North.
CO N390000N420000
W1200000W1242500
BT California
United States

Northern Caucasia
use Northern Caucasus

Northern Caucasus
N part of the Caucasus comprising Chechen-Ingush ASSR, S half of Krasnodar Kray, and the Daghestan, Kabardin-Balkar, and North Ossetian ASSR. Also search Ciscaucasia.
CO N440000N460000
E0450000E0380000
UF Ciscaucasia
North Caucasus
Northern Caucasia
BZ Caucasus
BX Russian Republic
BT USSR
SA Caucasus Foreland
Transcaucasia

Northern Dvina River
Chief river of the White Sea Basin in NW European USSR.
BT Russian Republic
USSR

Northern German Hills
Term introduced in 1981.
CO N511700N522800
E0120700E0073500
BT Germany

Northern Great Plains
Term introduced in 1978. Before 1978, search Great Plains AND north or northern.
BA Great Plains
BT North America
SA Southern Great Plains

Northern Hemisphere
Used when discussing many large areas too numerous to mention. Includes use on level 1 and 2 as an area term (list O). If 1, see term set options under list B.
SA Eastern Hemisphere
Laurasia
Pangaea
Southern Hemisphere
Western Hemisphere

Northern Highlands
Term introduced in 1981.
CO N563000N584000
W0030000W0062000
BT Scotland
SA Grampian Highlands

Northern Ireland
Six counties comprising the NE part of island of Ireland. Also search Ulster. Includes use on level 1 as an area term (list O). For term set options see list B.
CO N540000N553000
W0053000W0080000
UF Ulster
BA United Kingdom
Western Europe
Europe
NT Antrim
Belfast
Fermanagh
Giant's Causeway
Londonderry
Tyrone
SA Ireland

Northern Light Lake
In SW just N of Minnesota border.
BT Ontario
Canada

Northern Limestone Alps
The northern part of the Dinaric Alps which are a great belt of limestone ranges and plateaus along the Dalmatian coast of the Adriatic Sea.
BT Croatia
Yugoslavia
SA Alps
Dinaric Alps

Northern Nigeria
Term introduced in 1981.
CO N062500N135000
E0144000E0024500
BT Nigeria

Northern Norway
Term introduced in 1981. S boundary is Trondjeim.
CO N633000N711000
E0310500E0093000
BT Norway

Northern Range
N Trinidad.
BT Trinidad and Tobago
West Indies

Northern Rhodesia
use Zambia

Northern Rocky Mountains
Term introduced in 1978. Before 1978, also search Rocky Mountains AND north or northern.
BA Rocky Mountains
BT North America
SA Canadian Rocky Mountains
Central Rocky Mountains
Southern Rocky Mountains
U. S. Rocky Mountains

Northern Swiss Alps
Term introduced in 1981.
CO N460700N472200
E0093800E0064600
BA Swiss Alps
BT Switzerland

Northern Territory
Includes use on level 1 as an area term (list O). For term set options see list B.
CO S260000S110000
E1380000E1290000
BA Australia
Australasia
NT Alice Springs
Darwin
Gosses Bluff
Katherine
Tennant Creek
Victoria Valley
SA Amadeus Basin
Arunta Complex
Bitter Springs Formation
Carpentaria Basin
Georgina Basin
Great Artesian Basin
Henbury Meteorite
Joseph Bonaparte Gulf
Musgrave Ranges
Simpson Desert

Northern Urals
Term introduced in 1978. Before 1978, search Urals AND north or northern.
CO N600000N650000
E0620000E0560000
BX Russian Republic
BZ Urals
BT USSR
SA Polar Urals

Northumberland
County in N on border of Scotland.
CO N544500N554500
W0013000W0023000
BT England
Great Britain
United Kingdom

Northumberland Strait
Channel of the Gulf of Saint Lawrence between Prince Edward Island on the N, and New Brunswick and Nova Scotia on the SE and E respectively.
BT Canada

Northwest Atlantic
Term introduced in 1978. Before 1978, search Atlantic Ocean AND northwest.
BA Atlantic Ocean
SA East Atlantic
North Atlantic
Northeast Atlantic
South Atlantic
Southeast Atlantic
Southwest Atlantic
West Atlantic

Northwest Pacific
Term introduced in 1978. Before 1978, search Pacific Ocean AND northwest.
BA Pacific Ocean
SA East Pacific
Equatorial Pacific
North Pacific
Northeast Pacific
South Pacific
Southeast Pacific
Southwest Pacific
West Pacific

Northwest Territories
Index districts as applicable. Includes use on level 1 as an area term (list O). For term set options see list B.
CO N600000N840000
W0600000W1360000
BA Canada
NT Agricola Lake
Amund Ringnes Island
Axel Heiberg Island
Baffin Island
Banks Island
Barnes ice cap
Bathurst Island
Bear-Slave Operation
Belcher Islands
Boothia Peninsula
Cape Dyer
Coppermine River
Cornwallis Island
Cumberland Peninsula
Devon Island
District of Franklin
District of Keewatin
District of Mackenzie
Ellef Ringnes Island
Ellesmere Island
Fort Good Hope
Fort Ross
Great Bear Lake
Great Slave Lake
Inuvik
Kaminak Lake
Lancaster Sound
Mackenzie Delta
Mackenzie River valley
Meighen Island
Melville Peninsula
Muskox Intrusion
Narpaing Fjord
Norman Wells
Otto Fiord
Queen Elizabeth Islands
Radstock Bay
Richards Island
Somerset Island
South Nahanni River
Sverdrup Basin
Tanquary Fiord
Tuktoyaktuk Peninsula
Wager
Yellowknife
SA Baffin Bay
Bear Province
Canadian Shield
Carswell Structure
Christopher Formation
Churchill Province
Edmonton Formation
Eureka Sound Formation
Franklin Mountains
Horn Plateau Formation
Hudson Bay
Hudson Bay Lowlands
Hurwitz Group
Isachsen Formation
Liard River
Mackenzie Mountains
Michelle Formation
Nahanni Formation
Peel River
Peel Sound Formation
Prince of Wales Island
Ramparts Formation
Richardson Mountains
Selwyn Mountains
Slave Province
Yellowknife Group

Northwestern German Plain
Term introduced in 1981.
CO N521700N550500
E0113500E0064000
BT Germany
SA West Germany

Northwestern Polish Plain
Term introduced in 1981.
CO N495500N545000
E0220000E0141000
BT Poland

Northwestern Transdanubia
Term introduced in 1981.
CO N465500N480200
E0182000E0161500
BZ Transdanubia
BX Hungary

Northwestern USSR
Term introduced in 1978. Before 1978, search USSR AND northwest.
BA USSR

Norton County
NW Kansas.
BT Kansas
United States

Norton County Meteorite
Term introduced in 1978 to distinguish from place name, Norton County. Before 1978, also search Norton County for the meteorite.
BA chladnite
achondrites
BT meteorites
SA Kansas

Norton Sound
Term introduced in 1981. Arm of the Bering Sea, S of Seward Peninsula, W Alaska.
CO N631600N645500
W1604500W1660000
BT Bering Sea

Norway
Includes use on level 1 as an area term (list O). For term set options see list B.
CO N580000N710000
E0310000E0040000
BA Scandinavia
Western Europe
Europe
NT Arendal
Bergen
Bjerkrem-Sogndal Massif
Drammen
Finnmark
Folldal
Gudbrandsdalen
Hardangervidda
Jotunheim Mountains
Kongsberg
Kristiansund
Nord Fjord
Nord-Trondelag
Nordland
Northern Norway
Oslo
Ringerike
Rogaland
Sogn
Solund Islands
Sor-Trondelag
Soroy
Southern Norway
Telemark
Troms
Trondelag
Tunsbergdalsbreen
Valdres
Vest-Agder
Vesteralen
Vestfold
SA Arctic region
Baltic Shield
Fennoscandia
Lapland
North Sea Coast
Porsanger Dolomite Formation
Ringerike Sandstone
Sparagmite Group
Sulitjelma

Norwegian Channel
Undersea feature just off SW Norway.
UF Norwegian Deep
Norwegian Trough
BT North Sea
Atlantic Ocean

Norwegian Deep
use Norwegian Channel

Norwegian Sea
Off coast of Norway, opening N on Greenland Sea, NE on Barents Sea, S on North Sea, and SW on the open Atlantic.
BT Arctic Ocean
NT Jan Mayen Ridge
Voring Plateau
SA Jan Mayen
Kolbeinsey Island

Norwegian Trough
use Norwegian Channel

Norwich
City in E.
BT England
Great Britain
United Kingdom

nosean
Also search noselite.
UF noselite
BA sodalite group
framework silicates
silicates
BT minerals

noselite
use nosean

Nothosauria
Term introduced in 1981. Includes use on level 2 under Reptilia(1). See list F.
BA Sauropterygia
Euryapsida
Reptilia
BT Tetrapoda
Vertebrata

Noto Peninsula
Large headland projecting N into Japan Sea in central.
BT Honshu
Japan

Notoungulata
Order. Includes use on level 2 under Mammalia(1). See list F.
UF Notungulata
BA Mammalia
BT Tetrapoda
Vertebrata

Notre Dame Bay
Inlet of Atlantic Ocean on N coast of island of Newfoundland.
BT Newfoundland
Canada

Nottingham
City in N central.
BT England
Great Britain
United Kingdom

Nottinghamshire
County in N central.
CO N524500N533000
W0004000W0013000
BT England
Great Britain
United Kingdom

Notungulata
use Notoungulata

Nouakchott
City in SW near coast.
BT Mauritania

Noumea
Town on SW coast of French island of New Caledonia E of Queensland, Australia.
BT New Caledonia
Pacific Ocean

Nova Scotia
Includes use on level 1 as an area term (list O). For term set options see list B.
CO N433000N470000
W0594500W0661500
BA Maritime Provinces
Eastern Canada
Canada
NT Annapolis Valley
Antigonish County
Cape Breton Island
Cape Sable
Chedabucto Bay
Cobequid Bay
Halifax County
Minas Basin
Pictou County
Sable Island
Yarmouth
SA Meguma Group
Windsor Group

Nova Scotia Shelf
use Scotian Shelf

Nova Scotian Shelf
use Scotian Shelf

novaculite
Includes use on level 3 under sedimentary rocks(1) clastic rocks(2). See list I.
UF razor stone
BA clastic rocks
BT sedimentary rocks
SA chemically precipitated rocks

Novaky
Village in W central.
BT Slovakia
Czechoslovakia

Novara
City and province in NW part of country.
BT Italy
SA Piemonte

Novaya Zemlya
Two large islands of Arkhangelsk Oblast between Barents and Kara seas.
CO N700000N770000
E0700000E0500000
BX Russian Republic
BZ Urals
BT USSR

Novgorod
City S of Leningrad.
BT Russian Republic
USSR

Novoraskoe Series
use Novorayskoe Formation

Novorayskoe Formation
In Donbas in E Ukraine.
UF Novoraskoe Series
SA Donets Basin
Jurassic
Lower Jurassic
Triassic
Ukraine
Upper Triassic
USSR

Novosibirsk
City and oblast N of NE Kazakhstan in W Siberia.
BT Russian Republic
USSR

Nowa Ruda
Town near the border of Czechoslovakia in SW part of country.
BT Wroclaw
Poland

Nowa Sol
City on the Odor River in W part of country.
BT Zielona Gora
Poland

Nowy Sacz
City in N foothills of Carpathians in S part of country.
BT Krakow
Poland

Nowy Targ
City at foot of Tatra Mountains in S part of country.
UF Newmarkt
BT Krakow
Poland

Np
use neptunium

NRM
use natural remanent magnetization

nsutite
BA oxides
BT minerals

Nubia
Region in Nile Valley extending from about 16° N to include Aswan and First Cataract. Index countries as applicable.
BT Africa
SA Egypt
Sudan

Nubian Sandstone
Cretaceous?
BT Cretaceous
SA Africa
Egypt
Sudan

Nubian Shield
Term introduced in 1981. E Egypt, E Sudan, and N Ethiopia.
CO N140000N273000
E0410000E0300000
BT Africa
SA Egypt
Ethiopia
Sudan

nuclear energy
Also search nuclear AND energy.
SA energy
energy sources
nuclear facilities

nuclear explosions
Includes use on level 3 under seismology(1) explosions(2); on level 2 under explosions(1).
BT explosions
SA engineering geology
seismology
shatter cones

nuclear facilities
A level 1 term as of 1978. Before 1978, included use on level 2 under engineering geology(1). Also search nuclear AND facilities. If 1, term set options are:
design
subtopic
earthquakes
subtopic
faults
subtopic
feasibility studies
subtopic
foundations
subtopic
geologic hazards
subtopic
impact statements
subtopic
pollution
subtopic
rock mechanics
subtopic
seepage
subtopic
site exploration
subtopic
soil mechanics
subtopic
UF facilities, nuclear
SA design
earthquakes
engineering geology
faults
feasibility studies
foundations
geologic hazards
impact statements
marine installations
nuclear energy
power plants
rock mechanics
seepage
seismic risk
site exploration
soil mechanics

nuclear fission
use fission

nuclear magnetic resonance
Includes use on level 3 under spectroscopy(1) methods(2); on level 3 under chemical element(1) analysis(2). See list D.
SA analysis
electron paramagnetic resonance
resonance
spectroscopy

nucleation
Restricted to use under crystal growth.
SA crystal growth

Nuculanidae
Includes use on level 3 under Mollusca(1). See list F.
BA Bivalvia
Mollusca
BT Invertebrata
SA Nuculidae

Nuculidae
Includes use on level 3 under Mollusca(1). See list F.
BA Bivalvia
Mollusca
BT Invertebrata
SA Nuculanidae

Nueces River
Flows into Nueces Bay at head of Corpus Christi Bay in S.
BT Texas
United States

Nueva Esparta
State comprising an island group in the Caribbean Sea off N coast.
BT Venezuela
NT Macanao Peninsula
Margarita Island

Nuevo Leon
State in NE.
BT Mexico
SA Rio Grande
Rio Grande Valley
Sierra Madre Oriental

Nugssuak
use Nugssuaq

Nugssuaq
Peninsula in central part of W coast. Also search Nugssuak.
UF Nugssuak
BT Greenland
Arctic region

Nullarbor Plain
Extends inland along entire Great Australian Bight in S part of country. Index states as applicable.
BT Australia
SA South Australia
Western Australia

numerical analysis
BA statistical analysis
SA analysis
mathematical methods

numerical filters
Term introduced in 1981.
BT filters

numerical seismograms
Term introduced in 1981.
BT seismograms

Nummulites
Genus. Includes use on level 3 under foraminifera(1) Nummulitidae(2).
BA Nummulitidae
Rotaliacea
Rotaliina
foraminifera
BT Protista
Invertebrata
NA Operculina

Nummulitidae
Includes use on level 2 under foraminifera(1). See list F.
BA Rotaliacea
Rotaliina
foraminifera
BT Protista
Invertebrata
NA Nummulites
SA Heterostegina

Nunivak Island
Second largest island in Bering Sea. Off W coast.
BT Alaska
United States

Nura-Tau
Range in N Samarkand Oblast. Also search Nuratau.
CO N380000N400000
E0660000E0640000
UF Nuratau
BT Uzbekistan
USSR

Nuratau
use Nura-Tau

Nurek
Village in NE Dushanbe Oblast in W.
BT Tadzhikistan
USSR

Nuremberg
City in N central Bavaria.
UF Nurnberg
BT Bavaria
West Germany

Nurnberg
use Nuremberg

nutrients
Includes use on level 2 under soils(1). See list M.
SA fertilization
fertilizers
major elements
minor elements
nutrition
soils
trace elements

nutrition
Includes use on level 3 under fossil groups (list F) or paleontology(1).
SA metabolism
nutrients
paleontology

Nuwuk
use Point Barrow

N.Y. Commodity Exchange
Term introduced in 1981.
SA economics

Ny Friesland
Region on NE island of Spitsbergen.
BT Spitsbergen
Svalbard
Arctic region

Nyasaland
use Malawi

Nye County
Central and S.
BT Nevada
United States
NT Railroad Valley
Rainier Mesa
Round Mountain Quadrangle
Trap Spring Field

Nyeman River basin
use Neman River basin

Nyiragongo
Volcano at N end of Lake Kivu in E part of country.
UF Nyiragongo Volcano
BT Kivu
Zaire

Nyiragongo Volcano
use Nyiragongo

Nykoeping
use Nykoping

Nykoping
City on Baltic Sea.
UF Nykoeping
BT Sweden

O

O
use oxygen

O-shima
Island or groups of islands in Japan. Index as applicable.
UF Vries Island
BT Japan
SA Amami-O-shima
Izu-shichito

O-16
Includes use on level 3 under isotopes(1).
SA isotopes
O-18/O-16
oxygen

O-16/O-18
use O-18/O-16

O-18
Includes use on level 3 under isotopes(1).
SA isotopes
O-18/O-16
oxygen

O-18/O-16
Includes use on level 3 under isotopes(1). Also search O-16/O-18; O-16 AND O-18.
UF O-16/O-18
SA geologic thermometry
isotopes
O-16
O-18
oxygen
paleoclimatology

stable isotopes

Oahu
Third largest of the Hawaiian islands. Island on which Honolulu is located.
CO N211500N214300 W1573900W1581800
BT Hawaii
United States

Oamaru
Town on Pacific Ocean in SE.
BT South Island
New Zealand

Oasis region
Term introduced in 1981.
BT Algeria

Oaxaca
State in S Mexico. Also a city.
BT Mexico
SA Sierra Madre del Sur

Ob River
Rises in Altai Mountains and flows NE and N through West Siberian Plain into Gulf of Ob on Kara Sea. Also search Ob.
BT Russian Republic
USSR
SA Irtysh River
Ob-Irtysh Interfluve

Ob-Irtysh Interfluve
Extends from source of both rivers in Altai Mountains to their conjunction in West Siberian Plain. Index Soviet Republics as applicable.
BT USSR
SA Irtysh River
Kazakhstan
Ob River
Russian Republic

obduction
Includes use on level 2 under plate tectonics(1).
SA plate tectonics
subduction

Oberalp Pass
Term introduced in 1981. In the Alps, on border of Uri and Graubunden cantons, S central Switzerland.
CO N463800N464100 E0083800E0084200
BT Switzerland
SA Alps
Graubunden
Swiss Alps
Uri

Oberhalbstein
Valley NW of Saint Moritz in E part of country.
BT Graubunden
Switzerland

objectives
Includes use as a level 2 or 3 term appropriate to a large number of topics, e.g. on level 2 under geology(1). See list G.

oblique orientation
Term introduced in 1978. Includes use as orientation of folds relative to spatially associated macroscopic structures such as large folds, fold systems and orogenic zones. Includes use on level 3 under folds(1) orientation(2); on level 3 under faults(1) orientation(2) and fractures(1) style(2). Before 1978, also search folds or faults AND oblique.
SA faults
folds
fractures
joints
orientation

obliquity
As of 1981, no longer a valid term for GeoRef. See oblique orientation; obliquity of the ecliptic. Included use as a general term and for planet obliquity.

obliquity of the ecliptic
Term introduced in 1981. Includes use on level 3 under Earth(1) motions(2), and under individual planets(1). Before 1981, also search planet name AND tilt.
SA climate
Earth
Moon
paleoclimatology

observations
Includes use as a level 2 or 3 term appropriate to a large number of topics, e.g. on level 2 under seismology(1). See list G.
SA orbital observations

observatories
Includes use on level 2 under geophysics(1) and seismology(1).
SA geophysics
seismology

obsidian
As of 1981, includes use on level 3 under igneous rocks(1) glasses(2). See list H. Before 1981, included use on level 3 under igneous rocks(1) pyroclastics and glasses(2).
BA glasses
volcanic rocks
BT igneous rocks
SA volcanic glass

Ocala Group
Includes Tivola Tongue, Inglis Formation, Williston Formation, Crystal River Formation.
BT upper Eocene
Eocene
SA Alabama
Crystal River Formation
Florida
Floridan Aquifer
Georgia

occurrence
Includes use as a level 2 or 3 term appropriate to a large number of topics, e.g. on level 2 under metamorphic rocks(1). See list G. Also search occurrences.

ocean basins
Used for studies on the major ocean basins, their origin, evolution and present configuration. For basins found within the oceans and for sedimentation studies, see ocean floors. Includes use on level 1 (list A). Term set options are:
topic [age, evolution, genesis, patterns]
subtopic (no area term)
SA abyssal plains
basins
continental drift
crust
Deep Sea Drilling Project
geosynclines
marginal basins
marine geology
ocean floors
oceanography
paleo-oceanography
paleogeography
plate tectonics
sea-floor spreading
tectonophysics

ocean bottom seismographs
Term introduced in 1978.
UF ocean-bottom seismometers
BT seismographs
SA instruments
seismic methods
seismology

ocean circulation
Level 1 term introduced in 1978. Primarily used for special indexes. Before 1978, also search circulation AND oceans. Term set options are:
currents
(name of current)
topic [anomalies, biocirculation, boundary layer, causes, climate-induced circulation, Coriolis force, detection, diffusion, distribution, Ekman spiral, genesis, patterns, thermal circulation, thermohaline circulation, tides, turbulence]
subtopic
SA biocirculation
bottom currents
boundary layer
circulation
climate-induced circulation
continental shelf
continental slope
Coriolis force
current directions
currents
diffusion
Ekman spiral
longshore currents
marine geology
ocean floors
ocean waves
oceanography
paleocirculation
sedimentation
shorelines
thermal circulation
thermohaline circulation
tides
turbulence
upwelling
waterways

Ocean County
Term introduced in 1981. E New Jersey, on the Atlantic Ocean.
CO N393000N401100 W0740200W0743400
BT New Jersey
United States

ocean floors
Used for discussions of processes taking place on the ocean floor as well as features thereof. For tectonics, see ocean basins. Includes use on level 1 (list A). Term set options are:
topic [anomalies, bottom features, exploration, mid-ocean ridges, mineral resources, seamounts, sedimentation, submarine canyons, trenches, troughs]
subtopic (no area term)
SA abyssal plains
bathymetric maps
bathymetry
bottom features
continental margin
continental rise
continental shelf
continental slope
Deep Sea Drilling Project
dredging
fracture zones
marginal basins
marine geology
mid-ocean ridges
mineral resources
nodules
ocean basins
ocean circulation
oceanography
paleo-oceanography
paleobathymetry
relief
sea-floor spreading
seamounts
sedimentation
sediments
submarine canyons
submarine fans
topography
trenches
troughs

Ocean of Storms
Also search Oceanus Procellarum.
UF Oceanus Procellarum
BT Moon

ocean waves
Includes use on level 1 (list A) for special indexes or for papers relating physical oceanography to marine geology. Term set options are:
topic [anomalies, breaking waves, catastrophic waves, causes, effects, genesis, ideal waves, internal waves, shoaling, transformations]
subtopic
NT breaking waves
catastrophic waves
ideal waves
SA circulation
diffraction
internal waves
longshore currents
marine geology
ocean circulation
oceanography
propagation
seiches
shoaling
shorelines
surges
transformations
tsunamis
waves

ocean, world
use world ocean

ocean-bottom seismometers
use ocean bottom seismographs

ocean-floor spreading
use sea-floor spreading

Oceania
Collective name for the islands and island groups of the central and south Pacific Ocean. Index island divisions as applicable: Melanesia, Micronesia, and Polynesia. Includes use on level 1 or 2 as an area term (list O). If 1, see term set options under list B.
UF Oceanica
NT Melanesia
Micronesia
Polynesia
SA Australasia
Australia
Malay Archipelago
New Zealand
Pacific Ocean

oceanic
A valid term through 1977. See oceanic type; oceanic crust.

oceanic crust
Term introduced in 1981. Also search oceanic type AND crust; oceanic AND crust.
SA crust
plate tectonics

oceanic ridges
A valid term through 1971. After 1971, use mid-ocean ridges.

oceanic trench
use trenches

oceanic type
As of 1981, no longer a valid term for GeoRef. Use oceanic crust.

Oceanica
use Oceania

oceanite
As of 1981, includes use on level 3 under igneous rocks(1) basalts(2). See list H. Before 1981, included use on level 3 under igneous rocks(1) basalt family(2).
BA basalts
volcanic rocks
BT igneous rocks

oceanography
Includes use on level 1 (list A) for special indexes and for treating the discipline as a whole; on level 2 under area terms (list B), bibliography(1), education(1), continental shelf(1), and continental slope(1). In January 1976, the following terms were deleted from level 2: boundary layer, diffusion, distribution, and turbulence. These terms are now used on level 2 under ocean circulation(1). When oceanography is a level 1 term, set options are:
experimental studies
topic
instruments
name of instrument or platform
phenomenon
sea ice
topic
techniques
topic
theoretical studies
topic
topic [applications, bibliography, catalogs, classification, concepts, education, general, history, methods, nomenclature, objectives, practice, principles, research, symposia, textbooks]
subtopic
UF oceanology
SA abyssal plains
air-sea interface
bathymetry
bibliography
bottom water
boundary layer
catalogs
continental shelf
continental slope
convection
Deep Sea Drilling Project
diffusion
echo sounding
education
expeditions
FAMOUS
GEOSECS
International Decade of Ocean Exploration
marine geology
nodules
ocean basins
ocean circulation
ocean floors
ocean waves
paleo-oceanography
pelagic environment
sea ice
sea water
sedimentary petrology
sedimentation
shoals
submersibles
turbulence
upwelling

oceanology
use oceanography

oceans
A valid level 1 term through 1977 used in combination with circulation (i.e. oceans, circulation). After 1977, use ocean circulation on level 1.

Oceanus Procellarum
use Ocean of Storms

Ocoee Series
Provincial series, Virginia, Tennessee, North Carolina, and Georgia. Includes use on level 3 under age terms(1). See list E.
BA Precambrian

Oconee County
Index states as applicable.
BX Georgia
South Carolina
BT United States
SA Lake Jocassee

octahedral iron ore
use magnetite

octahedrite
Term introduced in 1981.
BA iron meteorites
BT meteorites
NA Canyon Diablo Meteorites
Gibeon Meteorite
Henbury Meteorite
Toluca Meteorite
Weekeroo Station Meteorite
Wolf Creek Meteorite
SA Coldwater Meteorites
Odessa Meteorite

octahedrite (mineral)
use anatase

octahedrites
A valid term through 1975. After 1975, use iron meteorites. See octahedrite.

Octocorallia
Includes use on level 2 under Coelenterata(1). See list F.
BA Anthozoa
Coelenterata
BT Invertebrata

Odate
City in N.
BT Honshu
Japan

Odenwald
Mountainous region in S central part of country. Index states as applicable.
UF Odenwald Massif
Odenwald Mountains
BT Germany
SA Baden-Wurttemberg
Bavaria
Hesse

Odenwald Massif
use Odenwald

Odenwald Mountains
use Odenwald

Oder Valley
River valley. Index countries as applicable. Also search Odra River valley.
UF Odra River valley
BT Europe
SA Czechoslovakia
East Germany
Poland

Odessa Meteorite
Term introduced in 1981. Refers to both the meteorite found in Texas (octahedrite) and to the meteorite found in the Ukraine (chondrite). Index Texas or Ukraine, octahedrite or chondrites as applicable. Before 1981, also search Odessa AND meteorites.
BT meteorites
SA chondrites
octahedrite
Texas
Ukraine

Odessa Oblast
Term introduced in 1981. Oblast on the Black Sea. Before 1981, also search Odessa AND USSR.
BT Ukraine
USSR

Odonata
As of 1981, includes use on level 3 under Insecta(1) Odonatopteroida(2). See list F. Before 1981, included use on level 2 under Insecta(1).
BA Odonatopteroida
Insecta
BT Mandibulata
Arthropoda
Invertebrata

Odonatopteroida
Term introduced in 1981. Includes use on level 2 under Insecta(1). See list F.
BA Insecta
BT Mandibulata
Arthropoda
Insecta
NA Odonata

Odontopleurida
Includes use on level 2 under Trilobita(1). See list F.
BA Trilobita
BT Trilobitomorpha
Arthropoda
Invertebrata

Odontornithes
Subclass. Includes use on level 3 under Aves(1) Neornithes(2).
BA Neornithes
Aves
BT Tetrapoda
Vertebrata

Odra River valley
use Oder Valley

Oerebro
use Orebro

Oetztal Alps
use Otztal Alps

Oetztal Massif
use Otztal Alps

Oetztaler
use Otztal Alps

off-lying
use offshore

Officer Basin
Sedimentary basin.
CO S300000S240000
E1280000E1240000
BT Australia
SA South Australia
Western Australia

offretite
BA zeolite group
framework silicates
silicates
BT minerals
SA aluminosilicates

offshore
Used to indicate area situated off or at a distance from the shore. Not used as an adjective.
UF off-lying
SA continental shelf
continental slope
shorelines

Oficina
Oil field near town of El Tigre in NE central part of country.
BT Anzoategui
Venezuela

Ofot Fjord
use Ofoten

Ofoten
Fjord. NE extension of Vest Fjord on which Narvik is located.
UF Ofot Fjord
BT Nordland
Norway

Oga Peninsula
Extends into Japan Sea in N.
BT Honshu
Japan

Ogallala Formation
Includes Burge Sands, Valentine Beds, Ash Hollow Member, Kimball Member. NE Colorado, W and central Kansas, W Nebraska, E New Mexico, W Oklahoma, NW Texas and SE Wyoming. Also search Ogallala Aquifer.
BT Pliocene
SA Ash Hollow Formation
Colorado
Kansas
Nebraska
New Mexico
Oklahoma
Texas
Valentine Formation
Wyoming

Ogasawara
Town in Yamanashi Prefecture in central.
BT Honshu
Japan

Ogasawara Islands
use Bonin Islands

Ogdensburg
City on the Saint Lawrence River in Saint Lawrence County.
BT New York
United States

Ogilvie Mountains
Range in central.
BT Yukon Territory
Canada

Oguni
Village in N central.
BT Honshu
Japan

Ohio
Includes use on level 1 as an area term (list O). For term set options see list B.
CO N382500N420000
W0803030W0845000
BA Midwest
United States
NX Adams County
NT Akron
NX Allen County
Butler County
Carroll County
NT Cincinnati
NX Clark County
NT Cleveland
NX Clinton County
Columbus
Crawford County
Delaware County
Erie County
Fayette County
Franklin County
Fulton County
NT Geauga County
NX Greene County
Hamilton County
Hancock County
Hardin County
Harrison County
Highland County
Jackson County
Jefferson County
Lake County
Lawrence County
Logan County

Lucas County
NT Mahoning County
NX Marion County
Mercer County
Monroe County
Montgomery County
Morgan County
NT Ohio State University
NX Ottawa County
Perry County
Pike County
Portage County
Putnam County
Richland County
NT Sandusky County
Sandusky River basin
Scioto River basin
NX Seneca County
Shelby County
Stark County
Summit County
NT Toledo
Tuscarawas County
NX Union County
Urbana
Warren County
Washington County
Wayne County
Wood County
NT Wyandot County
SA Allegheny Group
Allegheny Plateau
Appalachian Basin
Appalachian Plateau
Athens
Bedford Shale
Berea Sandstone
Brassfield Formation
Cincinnati Arch
Columbus Limestone
Conemaugh Group
Delaware Limestone
Detroit River Group
Dunkard Basin
Dunkard Group
Eden Shale
Fairview Formation
Great Lakes region
Lake Erie
Logan Formation
Maumee River valley
Midcontinent
Monongahela Group
Ohio River
Ohio River valley
Ohio Shale
Olentangy Shale
Pottsville Group
Richmond Group
Saint Peter Sandstone
Salina Group
Sharon Conglomerate
Springfield
Sylvania Formation
Tioga Bentonite
Trenton Group
Whitewater River valley

Ohio Range
N of the Horlick Mountains and S of Marie Byrd Land.
BT Antarctica

Ohio River
Formed by confluence of the Allegheny and Monongahela rivers at Pittsburgh. It flows W and SW into the Mississippi River at Cairo, Illinois. Index states as applicable.
BT United States
SA Illinois
Indiana
Kentucky
Ohio
Pennsylvania
West Virginia

Ohio River valley
Index states as applicable. Also search Ohio Valley.
UF Ohio Valley
BT United States
SA Illinois
Indiana
Kentucky
Ohio
Pennsylvania
West Virginia

Ohio Shale
Includes Cleveland Member, Huron Member, Chagrin Member. Ohio, N central Kentucky, and West Virginia.
BT Upper Devonian
Devonian
SA Kentucky
Ohio
West Virginia

Ohio State University
In Columbus, Franklin County.
BT Ohio
United States

Ohio Valley
use Ohio River valley

oil (petroleum)
use petroleum

oil and gas fields
For detailed descriptions of individual fields or for discussions of the origin of several fields. Includes use on level 1 (list A). See also list C (commodities). Term set options are: topic [classification, distribution, genesis]
subtopic (no area term)
UF fields, oil and gas
gas fields, oil and
NT giant fields
SA Bell Creek Field
economic geology
Forties Field
geothermal fields
Hassi Messaoud Field
Hugoton Field
land subsidence
natural gas
petroleum
petroleum engineering
reservoir properties
reservoir rocks
stratigraphic traps
traps

oil sands
Includes use on level 1 as a commodity term (list C). As of 1981, includes use on level 3 under sedimentary rocks(1) organic residues(2). See list I.
UF tar sands
BA organic residues
BT sedimentary rocks
SA fuel resources
oil shale
organic materials
petroleum

oil seeps
Also search petroleum seepage.
UF petroleum seepage
seeps, oil
SA bitumens
petroleum
seepage

oil shale
Includes use on level 1 as a commodity term (list C).
UF shale oil
BA organic residues
BT sedimentary rocks
SA anthraxolite
fuel resources
kerogen
oil sands
organic materials
petroleum
shale
torbanite

oil spills
As of 1978, term is used on level 2 under pollution(1).
UF spills, oil
SA petroleum
pollution

oil-gas interface
Term introduced in 1978. Before 1978, search interface AND oil-gas.
UF interface, oil-gas
SA gases
interfaces
natural gas
oil-water interface
petroleum

oil-water contact
use oil-water interface

oil-water interface
Term introduced in 1978. Before 1978, search oil-water AND interface; oil-water contact.
UF interface, oil-water
oil-water contact
SA interfaces
oil-gas interface
petroleum
water

Oise
Department in N.
CO N490000N494500
E0031000E0014000
BT France
NT Clermont
Creil
SA Oise River valley

Oise River valley
Index departments as applicable. Also search Oise Valley.
UF Oise Valley
BT France
SA Aisne
Oise
Val-d'Oise

Oise Valley
use Oise River valley

Oita
City and prefecture in NE.
BT Kyushu
Japan

Ojika Peninsula
Between Ishinomaki Bay and the Pacific Ocean in NE.
BT Honshu
Japan

Ojo Alamo Sandstone
NW New Mexico.
BT Upper Cretaceous
Cretaceous
SA New Mexico

Ojo de Liebre Lagoon
Inlet of Sebastian Vizcaino Bay on NW coast.
BT Baja California
Mexico

Oka Complex
Term introduced in 1981. Carbonatite complex mined for pyrochlore. In S Quebec.
BT Quebec
Canada

Oka River
River W of Lake Baikal in central Irkutsk Oblast which flows into the Angara River, and a river in central European USSR which flows into the Volga at Gorkiy. Also search Oka AND USSR.
BT Russian Republic
USSR

Okanagan Valley
River valley. Index British Columbia and/or Washington as applicable. Called Okanogan Valley in the U.S.
UF Okanagen Valley
Okanogan Valley
BT North America
SA British Columbia
Washington

Okanagen Valley
use Okanagan Valley

Okanogan County
N Washington.
BT Washington
United States

Okanogan Valley
use Okanagan Valley

Okayama
City and prefecture on N side of Inland Sea W of Kobe.
BT Honshu
Japan

Okefenokee Swamp
Mostly in SE Georgia. Index states as applicable.
BT United States
SA Florida
Georgia

Okehampton
Town in Devonshire in SW.
BT England
Great Britain
United Kingdom

Okhotsk
Town on NW coast of Okhotsk Sea in Khabarovsk Kray in E Siberia.
CO N580000N600000
E1450000E1400000
BT Russian Republic
USSR

Okhotsk Massif
Just N of Magadan in S Magadan Oblast in E Siberia.
UF Okhotskoye Ploskogor'ye
BT Russian Republic
USSR

Okhotsk region
Term introduced in 1981. The Okhotsk Sea and the land areas surrounding it.
CO N433000N630000
E1660000E1340000
BT Russian Republic
USSR
SA Kamchatka
Kuril Islands
Magadan
Sakhalin

Okhotsk Sea
West of Kamchatka Peninsula and the Kuril Islands. Includes use on level 1 as an area term (list O). For term set options see list B. Before 1974, also search Sea of Okhotsk.
CO N440000N630000
E1650000E1350000
BA West Pacific
Pacific Ocean

Okhotsk-Chukchi
Large region extending NE of Okhotsk Sea, including area of Chukchi Peninsula.
BT Russian Republic
USSR
SA Chukchi Peninsula
Soviet Far East

Okhotsk-Chukchi volcanic belt
Includes numerous ranges extending S from the Chukchi Peninsula around both sides of Okhotsk Sea. Many volcanoes are still active in Kamchatka Peninsula. Also search Okhotsk-Chukchi.

CO N580000N700000
W1700000E1400000
UF Okhotsk-Chukot volcanic belt
BT Russian Republic
USSR

Okhotsk-Chukot volcanic belt
use Okhotsk-Chukchi volcanic belt

Okhotskoye Ploskogor'ye
use Okhotsk Massif

Oki Archipelago
use Oki Islands

Oki Islands
In SE Japan Sea 44 miles off W coast.
CO N360000N370000
E1340000E1323000
UF Oki Archipelago
Oki Retto
BT Honshu
Japan
SA Dogo

Oki Retto
use Oki Islands

Okinawa
Major island in Okinawa Island group midway between the main islands of Japan and Taiwan.
BT Ryukyu Islands
Japan

Oklahoma
Includes use on level 1 as an area term (list O). For term set options see list B.
CO N333500N370000
W0942500W1030000
BA Southwestern U.S.
United States
NT Arbuckle Mountains
NX Beaver County
Blaine County
NT Caddo County
NX Carter County
Cherokee County
NT Cimmaron County
Coal County
NX Comanche County
NT Cotton
NX Custer County
Delaware County
Dewey County
Ellis County
Garfield County
Grant County
Harper County
NT Hughes County
NX Jackson County
Jefferson County
NT Kingfisher County
NX Kiowa County
Lincoln County
Logan County
NT Major County
NX Marshall County
McIntosh County
Murray County
NT Muskogee County
NX Osage County
Ottawa County
NT Pawhuska Rock Plain
NX Pawnee County
NT Payne County
Picher
NX Pontotoc County
Pottawatomie County
NT Stillwater
NX Texas County
NT Tulsa
Tulsa County
NX Washington County
NT Wichita Mountains
Woods County
Woodward
SA Americus Limestone Member
Anadarko Basin
Antlers Sands
Arbuckle Group
Arkansas River
Arkansas River valley
Arkoma Basin
Atoka Formation
Bartlesville Sand
Blaine Formation
Bloyd Formation
Burbank Sand
Cabaniss Formation
Canadian River
Chase Group
Chattanooga Shale
Cheyenne Sandstone
Cottonwood Limestone
Crooked Creek Formation
Dakota Formation
Desmoinesian
Dockum Group
Eskridge Shale
Excello Shale
Fayetteville Formation
Fernvale Formation
Flint Hills
Fort Riley Limestone
Fredericksburg Group
Grand River
Great Plains
Hennessey Formation
Hugoton Embayment
Hunton Group
Jackfork Group
Johns Valley Formation
Kiowa Formation
Labette Shale
Lecompton Limestone
Llano Estacado
Marmaton Group
Midcontinent
Morrison Formation
Morrow Formation
Nemaha Ridge
Neosho River valley
Ogallala Formation
Oread Limestone
Ouachita Mountains
Ozark Mountains
Paluxy Formation
Panhandle
Pitkin Limestone
Pleasanton Group
Reagan Sandstone
Red Eagle Limestone
Red Fork Sandstone
Red River
Red River valley
Rexroad Formation
Roca Formation
Saint Peter Sandstone
Savanna Formation
Seminole Formation
Simpson Group
Springer Formation
Stanley Group
Stanton Formation
Trinity Group
Verdigris River valley
Viola Limestone
Wabaunsee Group
Wann Formation
Wapanucka Limestone
Washita Group
Washita River valley
Wellington Formation
Wewoka Formation
Woodbine Formation
Woodford Shale
Wreford Limestone

Oklo
BT Gabon

Oktemberyan Series
BT Miocene
Neogene
Tertiary
SA Armenia
USSR

Oktyabr
Town in SE Uzbekistan.
BT Uzbekistan
USSR

Okujiri
use Okushiri Island

Okushiri Island
In Japan Sea off SW coast.
UF Okujiri
Okushiri-shima
BT Hokkaido
Japan

Okushiri-shima
use Okushiri Island

Oland
Term introduced in 1978 to refer to island and province in Baltic Sea off SE coast of Sweden. Before 1978, also search Oland Island.
UF Oland Island
BT Kalmar
Sweden

Oland Island
use Oland

Old Castile
Term introduced in 1981. Region in N central Spain.
CO N400500N433000
W0015000W0054500
UF Castilla la Vieja
BA Castile
BT Spain
NT Avila Province
Burgos Province
Logrono Province
Palencia Province
Santander Province
Segovia
Soria Province
Valladolid Province

Old Crow
Village in N.
BT Yukon Territory
Canada

Old Faithful Geyser
In Yellowstone National Park.
BT Wyoming
United States

Old Red Sandstone
Great Britain. Includes use on level 3 under age terms(1). See list E.
BT Devonian
NA Downtonian
SA Dittonian

Old Stone Age
use Paleolithic

Older Dryas
Term used primarily in Europe for an interval of late glacial time following the Bolling and preceding the Allerod. Before 1978, search Dryas. See list E.
BA Weichselian
upper Pleistocene
Pleistocene
Quaternary
BT Cenozoic
SA Dryas

Oldest Dryas
Term introduced in 1978. Used primarily in Europe for an interval of late-glacial time preceding the Bolling. Before 1978, search Dryas. See list E.
BA Weichselian
upper Pleistocene
Pleistocene
Quaternary
BT Cenozoic
SA Dryas

Oldman Formation
Of the Belly River Group.
BT Upper Cretaceous
Cretaceous
SA Alberta

Oldoinyo Lengai
Village in N part of country.
BT Tanzania

Olduvai Event
Term introduced in 1981. Geomagnetic event.
BA lower Pleistocene
Pleistocene
Quaternary
BT Cenozoic
SA paleomagnetism
reversals

Olduvai Gorge
Site of rich fossil beds in N part of country. Also search Olduvai.
CO S031500S031500
E0353000E0353000
BT Tanzania

O'Leary Peak
Coconino County in N central.
BT Arizona
United States

Olekma
River in N Chita Oblast and S Yakutia in SE central Siberia.
BT Russian Republic
USSR

Olekma-Vitim Highlands
In NE Buryat ASSR and NW Chita Oblast in N Transbaikalia.
UF Olekma-Vitim mountain country
Olekma-Vitim Mountains
BT Russian Republic
USSR

Olekma-Vitim mountain country
use Olekma-Vitim Highlands

Olekma-Vitim Mountains
use Olekma-Vitim Highlands

Olenek River
Rises in Central Siberian Plateau and flows NE through Yakutia into Laptev Sea. Also search Olenek.
BT Russian Republic
USSR

Olenidae
Family. Includes use on level 3 under Trilobita(1) Ptychopariida(2).
BA Ptychopariida
Trilobita
BT Trilobitomorpha
Arthropoda
Invertebrata

Olentangy Shale
Plum Brook Shale has been correlated with Olentangy, and so-called Prout Limestone has been regarded as member of Olentangy Shale. Central Ohio.
BT Upper Devonian
Devonian
SA Ohio

Oligocene
World. Above Eocene, below Miocene. Includes use on level 1 as an age term (list E); on level 2 under paleo- terms, e.g. paleoecology, paleogeography, paleomagnetism.
BA Paleogene
Tertiary
BT Cenozoic
NT Bembridge Marls
Brule Formation
Cypress Hills Formation
Frio Formation
Grimmertingen
Krosno Beds
Lincoln Creek Formation

NA lower Oligocene
middle Oligocene
NT Nari Series
NA Rupelian
upper Oligocene
NT White River Group
SA Duchesne River Formation
John Day Formation
Poltava Series
Refugian
San Lorenzo Formation
Saucesian
Sespe Formation
Twin River Formation

Oligochaetia
As of 1981, includes use on level 2 under worms(1). See list F. Before 1981, included use on level 2 under Annelida(1).
BA worms
BT Invertebrata

oligoclase
As of 1981, includes use on level 3 under minerals(1) framework silicates, plagioclase(2). See list L. Before 1981, included use on level 3 under minerals(1) framework silicates, feldspar group(2).
BA plagioclase
feldspar group
framework silicates
silicates
BT minerals

olistoliths
Includes use on level 3 under sedimentary structures(1) turbidity current structures(2) and soft sediment deformation(2). See list K.
BA soft sediment deformation
sedimentary structures
SA olistostromes
turbidity current structures

olistostromes
Includes use on level 3 under sedimentary structures(1) turbidity current structures(2) and soft sediment deformation(2). See list K.
BA soft sediment deformation
sedimentary structures
SA olistoliths
turbidity current structures

Oliver County
W central North Dakota.
BT North Dakota
United States

olivine
Also search individual minerals such as fayalite, forsterite, and tephroite. See olivine group.
BA olivine group
nesosilicates
orthosilicates
silicates
BT minerals

olivine basalt
As of 1981, includes use on level 3 under igneous rocks(1) basalts(2). See list H. Before 1981, included use on level 3 under igneous rocks(1) basalt family(2).
BA basalts
volcanic rocks
BT igneous rocks
SA alkali olivine basalt

olivine diabase
As of 1981, includes use on level 3 under igneous rocks(1) diabase(2). See list H. Before 1981, included use on level 3 under igneous rocks(1) basalt family(2). Before 1981, also search olivine dolerite.
BA diabase
BT igneous rocks

olivine dolerite
As of 1981, no longer a valid term for GeoRef. Use olivine diabase. Before 1981, included use on level 3 under igneous rocks(1) basalt family(2).

olivine gabbro
As of 1981, includes use on level 3 under igneous rocks(1) gabbros(2). See list H. Before 1981, included use on level 3 under igneous rocks(1) gabbro family(2).
BA gabbros
BT igneous rocks
SA gabbro

olivine group
Includes use in combination with orthosilicates (i.e. orthosilicates, olivine group) on level 2 under minerals(1). See list L.
BA nesosilicates
orthosilicates
silicates
BT minerals
NA fayalite
forsterite
monticellite
olivine
peridot
tephroite

olivine nephelinite
As of 1981, includes use on level 3 under igneous rocks(1) alkali gabbros(2). See list H. Before 1981, included use on level 3 under igneous rocks(1) alkali basalt family(2).
BA alkali gabbros
gabbros
BT igneous rocks
SA ankaratrite
nepheline basalt
nephelinite

olivine tholeiite
As of 1981, includes use on level 3 under igneous rocks(1) basalts(2). See list H. Before 1981, included use on level 3 under igneous rocks(1) basalt family(2).
BA basalts
volcanic rocks
BT igneous rocks
SA tholeiite

olivinite
As of 1981, includes use on level 3 under igneous rocks(1) ultramafics(2). See list H. Before 1981, included use on level 3 under igneous rocks(1) ultramafic family(2).
BA ultramafics
BT igneous rocks

Olkusz
Town in S part of country.
BT Krakow
Poland

Olsztyn
Province in N part of country. Also city SE of Gdansk.
BT Poland

Olt River
Central and S Romania, in Transylvania and Walachia. Also search Olt and Olt River valley.
UF Oltul
BT Romania
SA Transylvania
Walachia

Olt River valley
Index regions as applicable. Also search Olt, Olt River, and Olt Valley.
UF Olt Valley
Oltul Valley
BT Romania
SA Transylvania
Walachia

Olt Valley
use Olt River valley

Oltenia
A sub-region. W part of Walachia.
UF Lesser Walachia
Little Walachia
BT Walachia
Romania

Oltul
use Olt River

Oltul Valley
use Olt River valley

Olympic Mountains
Part of the Coast Ranges in Olympic National Park on the Olympic Peninsula in NW.
BT Washington
United States
SA Coast Ranges

Olympic Peninsula
Between the Pacific Ocean and Puget Sound in NW.
BT Washington
United States

Olympus
Highest mountain in Greece in NE part of country.
BT Thessaly
Greece

Oman
Formerly Muscat and Oman. Includes use on level 1 as an area term (list O). For term set options see list B.
CO N170000N270000
E0600000E0523000
BA Arabian Peninsula
Asia
NT Oman Mountains
SA Middle East
Near East

Oman Mountains
BT Oman
Arabian Peninsula

Omi (Lake)
use Lake Biwa

Omine Mine
Near town of Omine in SW.
BT Honshu
Japan

Omineca Mountains
NW British Columbia.
BT British Columbia
Canada

Omo
Village in SW part of country.
BT Ethiopia

Omo Basin
use Omo Valley

Omo River
Flows through SW Ethiopia into Lake Turkana (Lake Rudolph). Also search Omo.
BT Ethiopia

Omo Valley
River valley in SW Ethiopia. Also search Omo; Omo Basin.
UF Omo Basin
BT Ethiopia

Omolon
River in N Khabarovsk Kray, W Chukchi National Okrug, and NE Yakutia which flows into the Kolyma River in NE Siberia.
BT Russian Republic
USSR

Omolon Block
NE Siberia in the Omolon River basin. Also search Omolon.
BT Russian Republic
USSR

omphacite
BA clinopyroxene
pyroxene group
chain silicates
silicates
BT minerals
SA augite

Omsk
City and oblast in SW Siberia.
BT Russian Republic
USSR

Omulevka
River in N Magadan Oblast in NE Siberia.
BT Russian Republic
USSR

Onaping Formation
Of the Whitewater Group.
BT Huronian
Proterozoic
Precambrian
SA Canada
Ontario

oncolites
Includes use on level 3 under sedimentary structures(1) biogenic structures(2). See list K.
BA biogenic structures
sedimentary structures
SA stromatolites

one-dimensional models
Term introduced in 1978. Before 1978, search one dimensional; one-dimensional AND models.
BA models
SA three-dimensional models
two-dimensional models

Onega
City on Onega Bay at mouth of Onega River in NW Arkhangelsk Oblast in NW European USSR.
BT Russian Republic
USSR
SA Lake Onega

O

Oneida County
Index states as applicable.
BX Idaho
New York
Wisconsin
BT United States

Oneida Lake
SE of Lake Ontario in central.
BT New York
United States

Onikobe
Region.
BT Honshu
Japan

onofrite
BZ selenites
sulfides
BT minerals
SA metacinnabar
selenium

Onondaga County
W central New York.
BT New York
United States

Onondaga Limestone
Includes Springfield Center Member, Babcock Hill Member, Edgecliff Member, Nedrow Member, Moorehouse Member, Seneca Member, Needmore Shale, Selinsgrove Limestone. W Maryland, New York, Pennsylvania, Western Virginia, and N West Virginia.
BT Middle Devonian
Devonian
SA Maryland

New York
Pennsylvania
Virginia
West Virginia

onshore
As of 1981, no longer a valid term for GeoRef. Used to distinguish from offshore.

Onslow Bay
Between Cape Lookout and Cape Fear.
BT North Carolina
United States

Ontake
On a peninsula in Kagoshima Bay in S Kyushu. Also search Ontake Volcano.
UF Ontake Volcano
BT Kyushu
Japan

Ontake Volcano
use Ontake

Ontario
Includes use on level 1 as an area term (list O). For term set options see list B.
CO N420000N570000
W0740000W0950000
BA Eastern Canada
Canada
NT Algoma
Bancroft
Batchawana Bay
Blind River
Brent Crater
Bruce Peninsula
Callander Bay
Cobalt
Cochrane
District of Kenora
Echo Bay
Elliot Lake
Flack Lake
NX Frontenac County
NT Georgian Bay
Gowganda
Guelph
Haliburton County
Kakagi Lake
Kenora
Kirkland Lake
Lambton County
Madoc
Mamainse Point
Manitoulin Island
Manitouwadge
Michipicoten Island
Moose River basin
New Liskeard
Nipissing
North Bay
Northern Light Lake
Ottawa
Parry Sound
Quirke Lake
Renfrew County
Saint Catharines
Sudbury
Sudbury Basin
Sudbury Irruptive
Temagami Mine
Thunder Bay
Timiskaming District
Timmins
Toronto
Uchi Lake
Welland Canal
Whetstone Lake
Windsor
SA Abitibi Belt
Animikie Group
Canadian Shield
Churchill Province
Detroit River Group
English River Belt
Espanola Formation
Flinton Group
Gowganda Formation
Grand River
Great Lakes region
Grenville
Grenville Front
Grenville Province
Hudson Bay Lowlands
Keewatin
Lake Erie
Lake Huron
Lake Ontario
Lake Superior
Lake Timiskaming
Leda Clay
Lockport Formation
Lorrain Formation
Medina Formation
Niagara Escarpment
Niagara Falls
Niagara River
Nipissing Diabase
Onaping Formation
Osler Series
Ottawa Valley
Rainy Lake
Rainy River
Rice Lake
Rochester Formation
Saganaga Lake
Saint Clair River
Saint Clair River delta
Saint Lawrence Lowlands
Saint Lawrence River
Saint Lawrence Valley
Sault Sainte Marie
Southern Province
Superior Province
Sylvania Formation
Timiskaming
Trent Valley
Utica Shale

ontogeny
Includes use on level 2 and 3 under fossil group(1). See list F.
SA fossils
paleontology
phylogeny

Ontonagon County
In Upper Peninsula.
BT Michigan
United States

Ontong Java Plateau
Between Solomon Islands on the W and Lord Howe Islands (Ontong Java Islands) on the E.
UF Antong Java Rise
Ontong Java Rise
BT Pacific Ocean

Ontong Java Rise
use Ontong Java Plateau

onvarovite
use uvarovite

Onverwacht Group
BT Precambrian
SA South Africa

onyx marble
use alabaster

oolite
Includes use on level 3 under sedimentary rocks(1) carbonate rocks(2); on level 3 under sediments(1) carbonate sediments(2). See list I (sed. rocks), list N (sediments). Also search oolites.
UF eggstone
ooliths
roestone
SA carbonate rocks
carbonate sediments
limestone
oolitic limestone
oolitic texture
pellets
sedimentary rocks

oolites
A valid term through 1976. After 1976, use oolite.

ooliths
use oolite

oolitic limestone
BA limestone
carbonate rocks
BT sedimentary rocks
SA oolite

oolitic texture
Term introduced in 1978. Includes use on level 3 under sediments(1) and sedimentary rocks(1). Before 1978, also search oolitic.
SA limestone
oolite
sedimentary rocks
sediments
textures

ooze
Includes use on level 3 under sediments(1) clastic sediments(2).
BA clastic sediments
BT sediments
SA calcareous composition
mud
sapropel

opal
Also search opaline.
UF opaline
BA silica minerals
framework silicates
silicates
BT minerals
SA hyalite

opaline
use opal

opaque minerals
Includes use on level 3 under minerals(1).
BT minerals
SA light minerals

open fractures
Term introduced in 1978. Includes use on level 3 under fractures(1) style(2). Before 1978, also search fractures AND open.
BA fractures
SA cracks

open pit mines
use open-pit mining

open pit mining
use open-pit mining

open systems
Used as a general term.
SA closed systems
systems

open-pit mining
Includes use on level 3 under mining geology(1) and land use(1). Also search open pit mines; open pit mining.
UF mining, open-pit
open pit mines
open pit mining
opencast mining
opencut mining
BT surface mining
mining
SA mines
mining geology
strip mining

opencast mining
use open-pit mining

opencut mining
use open-pit mining

Operculina
Includes use on level 3 under foraminifera(1) Nummulitidae(2).
BA Nummulites
Nummulitidae
Rotaliacea
Rotaliina
foraminifera
BT Protista
Invertebrata

ophicalcite
See list J.
BA marbles
BT metamorphic rocks
SA limestone

Ophiocistioidea
Includes use on level 2 under Echinodermata(1). See list F.
BA Echinozoa
Echinodermata
BT Invertebrata

ophiolite
Includes use on level 3 under plate tectonics(1) and as of 1981, includes use on level 3 under igneous rocks(1) ultramafics(2). See list H. Before 1981, included use on level 3 under igneous rocks(1) ultramafic family(2).
BA ultramafics
BT igneous rocks
SA geosynclines
plate tectonics

Ophiomorpha
Includes use on level 3 under Arthropoda(1) Malacostraca(2).
BA Malacostraca
Crustacea
Mandibulata
Arthropoda
BT Invertebrata
NA Ophiomorpha nodosa
SA ichnofossils

Ophiomorpha nodosa
Includes use on level 3 under Arthropoda(1) Malacostraca(2).
BA Ophiomorpha
Malacostraca
Crustacea
Mandibulata
Arthropoda
BT Invertebrata
SA ichnofossils

ophite
Includes use on level 3 under metamorphic rocks(1) marbles(2). See list J.
BA marbles
BT metamorphic rocks

Ophiuroidea
As of 1981, includes use on level 2 under Echinodermata(1). Before 1981, included use on level 3 under Echinodermata(1) Stelleroidea(2). See list F.
BA Stelleroidea
Asterozoa
Echinodermata
BT Invertebrata

Ophthalmidium
Genus. Includes use on level 3 under foraminifera(1) Miliolacea(2).
BA Miliolacea
Miliolina
foraminifera
BT Protista
Invertebrata

Opole
Province in SW part of country. Also a city.
BT Poland
NT Raciborz

optical
A valid term through 1977. After 1977, use optical spectroscopy.

optical dispersion
Term introduced in 1981.
SA minerals
optical properties
refractive index

optical extinction
Term introduced in 1981. Includes use on level 3 under minerals(1) optical properties(2). Before 1981, also search extinction AND minerals.
SA extinction
minerals
optical properties
undulatory extinction

optical methods
As of 1981, no longer a valid term for GeoRef. See microscope methods.

optical mineralogy
Includes use on level 2 under geochronology(1); on level 3 under minerals(1) methods(2).
SA geochronology
microscope methods
mineralogy
ore microscopy

optical properties
As of 1978, term is used on level 2 under minerals(1).
SA birefringence
color centers
fluorescence
isotropy
mineralogy
minerals
optical dispersion
optical extinction
pleochroism
polarization
properties
refractive index
scattering

optical spectra
Used for data, to distinguish from methodology. Before 1978, search optical AND spectroscopy.
SA spectra
spectroscopy

optical spectroscopy
Term introduced in 1978. Includes use on level 3 under spectroscopy(1) methods (2) and under chemical element(1) analysis(2). Before 1978, also search spectroscopy AND optical.
BT spectroscopy
SA analysis

optimal filters
Term introduced in 1981.
BT filters

optimization
Used as a general term.

Oquirrh Mountains
S of Great Salt Lake.
UF Oquirrh Range
BT Utah
United States

Oquirrh Range
use Oquirrh Mountains

Oran
No longer a valid term for GeoRef as of 1981. Use Oran region.

Oran region
Term introduced in 1981.
CO N320500N362200
E0023000W0021300
BT Algeria
NT Descartes

Orange County
Index states as applicable.
BX California
Florida
Indiana
New York
North Carolina
Texas
Vermont
Virginia
BT United States

Orange Free State
Province in E central.
CO S304500S264000
E0295000E0241500
BT South Africa
NT Swartkrans
Vredefort Dome
SA Orange River
Vaal River

orange material
SA Moon

Orange River
Rises in the Drakensberg Range in Lesotho and flows W into the Pacific Ocean. Index Namibia, Lesotho and provinces of South Africa.
BT Africa
SA Cape Province
Lesotho
Namibia
Orange Free State

Orava Valley
River valley in N Slovakia. Also search Orava.
BT Slovakia
Czechoslovakia

orbicular texture
Term introduced in 1978. Before 1978, search orbicular AND texture.
SA spherulites
textures

orbital observations
Term introduced in 1978. Before 1978, search observations AND orbital.
SA observations

Orbitoidacea
Includes use on level 2 under foraminifera(1). See list F.
BA Rotaliina
foraminifera
BT Protista
Invertebrata
NA Amphistegina
Asterocyclina
Cibicides
Discocyclina
Nephrolepidina
Orbitoididae

Orbitoidae
use Orbitoididae

Orbitoididae
Includes use on level 2 under foraminifera(1). See list F. Also search Orbitoidae.
UF Orbitoidae
BA Orbitoidacea
Rotaliina
foraminifera
BT Protista
Invertebrata
NA Lepidocyclina

Orbitolina
Genus. Includes use on level 3 under foraminifera(1) Lituolacea(2).
BA Orbitolinidae
Lituolacea
Textulariina
foraminifera
BT Protista
Invertebrata

Orbitolinidae
Family. Includes use on level 3 under foraminifera(1) Lituolacea(2).
BA Lituolacea
Textulariina
foraminifera
BT Protista
Invertebrata
NA Orbitolina

Orbulina
Genus. Includes use on level 3 under foraminifera(1) Globigerinacea(2).
BA Globigerinidae
Globigerinacea
Rotaliina
foraminifera
BT Protista
Invertebrata
NA Orbulina universa

Orbulina universa
Includes use on level 3 under foraminifera(1) Globigerinacea(2).
BA Orbulina
Globigerinidae
Globigerinacea
Rotaliina
foraminifera
BT Protista
Invertebrata

Orca Basin
Term introduced in 1981. Depression on the continental slope of the N central Gulf of Mexico, containing brines.
CO N265200N270300
W0911300W0912700
BT Gulf of Mexico

order-disorder
As of 1978, term is used on level 2 under crystal chemistry(1).
SA bonding
crystal chemistry
substitution

Ordovician
Includes use on level 1 as an age term (list E); on level 2 under paleoterms, e.g. paleoecology, paleogeography, paleomagnetism. After the Cambrian, before the Silurian.
BT Paleozoic
NT Antelope Valley Limestone
NA Bala
NT Borrowdale Volcanic Series
Chickamauga Group
Davidsville Group
Gander Lake Group
NA Lower Ordovician
NT Lushs Bight Group
Martinsburg Formation
Meguma Group
NA Middle Ordovician
NT Pogonip Group
Shap Granite
Skiddaw Slates
NA Upper Ordovician
NT Viola Limestone
NA Viruan
SA Arbuckle Group
Knox Group
Taconic Orogeny
Talladega Group

Ordubad
City in Nakhichevan Autonomous Soviet Socialist Republic (W enclave of Azerbaidzhan).
BT Azerbaidzhan
USSR

ore bodies
UF bodies, ore
SA mineral deposits

ore deposits
As of 1981, no longer a valid term for GeoRef. See list C. Before 1981, included use for all papers on metal commodities.

ore finding
Not a valid index term for GeoRef. See mineral exploration.

ore guides
Includes use on level 2 under mineral exploration(1).
UF guides, ore
SA mineral exploration

ore microscopy
Includes use on level 3 under mineralogy(1) methods(2) and under petrology(1) methods(2). See list C (commodities). Before 1976, also search mineragraphy.
UF microscopy, ore
SA microscope methods
mineralogy
optical mineralogy
universal stage

ore minerals
Includes use on level 3 under economic geology(1).
BT minerals
SA economic geology
metal ores
metals
ore-forming fluids

ore of sedimentation
use placers

ore rolls
use roll-type deposits

ore sources
As of 1981, no longer a valid term for GeoRef. See mineral deposits.

ore transport
Includes use on level 3 under mineral deposits, genesis(1). Before 1978, search ore deposits AND transport.
SA mineral deposits
transport

ore-forming fluids
Includes use on level 3 under mineral deposits, genesis(1). Before 1981, also search mineralizers; hydrothermal solutions; hydrothermal fluids.
UF fluids, ore-forming
SA hydrothermal alteration
hydrothermal processes
mineral deposits
mineralization
ore minerals

Oread Limestone
In Shawnee Group. Includes Toronto Limestone, Snyderville Shale, Leavenworth Limestone, Heebner Shale, Plattsmouth Limestone, Heumader Shale, Kereford Limestone members. SW Iowa, E Kansas, NW Missouri, SE Nebraska, and N Oklahoma.
BT Virgilian
Upper Pennsylvanian
Pennsylvanian
SA Iowa
Kansas
Missouri
Nebraska
Oklahoma
Plattsmouth Limestone Member
Shawnee Group

Orebro
County in S central part of country. Also a city. Search Oerebro.
UF Oerebro
BT Sweden

Oregon
Includes use on level 1 as an area term (list O). For term set options see list B.
CO N420000N462000

W1163500W1243500
BA Pacific Coast
Western U.S.
United States
NX Baker County
Benton County
NT Clarno
Clatsop County
NX Columbia County
NT Coos Bay
Coos County
Crater Lake
NX Crook County
Curry County
Douglas County
Grant County
NT Harney County
NX Jackson County
Jefferson County
NT Josephine County
NX Lake County
Lane County
Lincoln County
NT Malheur County
NX Marion County
NT Mazama Ash
Mount Hood
Multnomah County
NX Newport
Polk County
Portland
NT Riddle
Rogue River
NX Sherman County
NT Sixes River
Steens Mountain
Tillamook County
NX Union County
NT Wallowa Mountains
NX Washington County
Wheeler County
NT Willamette Valley
Willow Creek
Yaquina Bay
SA Bald Mountain
Basin and Range Province
Battle Mountain High
Blue Mountains
Cascade Range
Clarno Formation
Coast Ranges
Columbia Plateau
Columbia River
Columbia River Basalt
Columbia River basin
Columbia River estuary
Great Basin
John Day Formation
Klamath Mountains
Snake River
Snake River basin
Snake River canyon
Tyee Formation
Umpqua Formation
Yamhill Formation

Orenburg
City on Ural River in the Southern Urals in European USSR. Also an oblast.
UF Chkalov
BT Russian Republic
USSR

Orense
As of 1981, refers to only the city. Before 1981, also included the province. In NW Spain.
BT Orense Province
Galicia
Spain

Orense Province
Term introduced in 1981. Before 1981, search Orense. In NW Spain.
CO N414700N423500
W0064500W0082000
BT Galicia
Spain
NT Orense

ores in sedimentary rocks
A valid term through 1973. Use sedimentary rocks or sedimentation in conjunction with the individual ore.

ores, polymetallic
use polymetallic ores

Oresund
Strait connecting the Kattegat with the Baltic Sea. Index countries as applicable.
BT Europe
SA Denmark
Sweden

organic
A valid term through 1977. After 1977, use organic materials.

organic acids
Term introduced in 1981.
SA acids
inorganic acids

organic carbon
Includes use on level 3 under organic materials(1).
BA organic materials
SA carbon

organic compounds
Includes use on level 2 under minerals(1). See list L.
BT minerals
NA amber
ozocerite
SA compounds
porphyrins

organic materials
Includes use on level 1 (list A); on level 2 under soils(1). Used for discussions of mostly very small concentrations of organic materials in rocks. Also search organic matter. If 1, term set options are:
kind of material [amino acids, bitumens, carbohydrates, fatty acids, humates, humic acids, hydrocarbons, kerogen, phenols]
specific kind of material or topic
topic [abundance, age, alteration, analysis, classification, composition, detection, distribution, experimental studies, genesis, geochemistry, identification, nomenclature, observations, occurrence, properties, varieties]
subtopic
NA amino acids
aromatic hydrocarbons
bitumens
carbohydrates
carotenoids
chlorophyll
collagen
conchiolin
enzymes
ethane
fatty acids
fulvic acids
humates
humic acids
humus
hydrocarbons
isoprenoids
kerogen
lipids
methane
organic carbon
phenols
pigments
porphyrins
proteins
purines
resins
sapropel
sporopollenin
sugars
SA abundance
anthracite
anthraxolite
carbon
carbonaceous composition
caustobiolith
chemical analysis
coal
complexing
geochemistry
inorganic materials
isotopes
lignite
materials
oil sands
oil shale
organic residues
organo-metallics
peat
petroleum
polymerization
reflectance
sedimentary rocks
sedimentation
sediments
soils

organic mound
use bioherms

organic residues
As of 1981, includes use on level 2 under sedimentary rocks(1) and sediments(1). See lists I and N. Before 1981, included use on level 2 under sedimentary rocks(1). Before 1976, included use on level 2 under sedimentary rocks(1) and sediments(1). Before 1981, also search organic sediments.
UF residues, organic
BX sedimentary rocks
sediments
NA caustobiolith
coal
guano
gyttja
oil sands
oil shale
peat
torbanite
SA organic materials

organic sediments
As of 1981, no longer a valid term for GeoRef. Use organic residues. Before 1981, included use on level 2 under sediments(1).

organisms
As of 1981, no longer a valid term for GeoRef. Before 1981, included use on level 3 under soils(1) genesis(2) as type of factor. See biota.

organization
As of 1981, includes use on level 2 under survey organizations(1). Before 1981, included use on level 2 under surveys(1).
SA associations
museums
survey organizations
symposia

organo-metallic complexes
use organo-metallics

organo-metallic compounds
use organo-metallics

organo-metallics
Term introduced in 1981. Includes use on level 3 under organic materials(1) and geochemistry(1).
UF organo-metallic complexes
organo-metallic compounds
SA geochemistry
organic materials
porphyrins

Orgueil Meteorite
Term introduced in 1981. Before 1981, also search Montauban AND meteorites; Orgueil AND meteorites.
UF Montauban Meteorite
BA carbonaceous chondrites
chondrites
BT meteorites
SA France
Tarn-et-Garonne

orientation
Includes use as attitude of fold elements with respect to external coordinates. Includes use on level 2 under faults(1) and folds(1); on level 3 to index orientation of grains under structural analysis(1).
SA dip
discordant folds
drag folds
fabric
faults
folds
horizontal orientation
inclined folds
linear orientation
lineation
nappes
oblique orientation
overturned folds
petrofabrics
plunging folds
preferred orientation
recumbent folds
schistosity
strike
structural analysis
style
superposed folds
transverse faults
vertical orientation

Oriente
Province in E.
BT Cuba
West Indies

origin
A valid level 2 term through 1977 used in combination with life. After 1977, use life origin on level 2 under paleontology(1).

Orinoco River
Rises in Serra Parima Mountains in S Venezuela, turns E in central Venezuela and empties through a wide delta into the Atlantic Ocean. Index countries as applicable. Also search Orinoco.
BT South America
SA Colombia
Venezuela

Oriskany Sandstone
Comprises Shriver Chert, and Ridgeley Sandstone Member. W Maryland, New York, Pennsylvania, Western Virginia and E West Virginia.
BT Lower Devonian
Devonian
SA Maryland
New York
Pennsylvania
Virginia
West Virginia

Orissa
State on the Bay of Bengal in E.
CO N174000N223000
E0873000E0812000
BT India
NT Balasore
Bolangir
Cuttack
Dhenkanal
Keonjhar
Koraput
Mayurbhanj

Nilgiri
Rampur coal field
Sambalpur
Talchir
Talchir coal field
SA Deccan Plateau
Gangpur Series
Mahanadi Valley
Singhbhum Granite
Singhbhum shear zone
Talchir Series

Orkney Islands
Archipelago off NE coast.
CO N584500N593000
W0061000W0080000
BT Scotland

Orleans
City in N central part of country.
BT France

Orlov
Old name for city of Khalturin just W of Kirov in Kirov Oblast of European USSR.
BT Russian Republic
USSR

ornamentation
SA paleontology
shells

Orne
Department in NW.
CO N481000N490000
E0010000W0005000
BT France
NT Mortagne
SA Eure Valley
Normandy

Ornithischia
As of 1981, includes use on level 2 under Reptilia(1). See list F. Before 1981, included use on level 3 under Reptilia(1) Archosauria(2).
BA Archosauria
Reptilia
BT Tetrapoda
Vertebrata
NA Hadrosauridae

orogenesis
A valid term through 1974. After 1974, use orogeny.

orogenic belts
Term introduced in 1981.
SA fold belts
geosynclines
intracontinental belts
mobile belts
orogeny
tectonics

orogeny
Includes use on level 1 (list A). Used for discussions of either individual orogenies or detailed general treatments on several orogenies. Before 1974, also search mountain building; before 1975, also search orogenesis. Term set options are:
topic [absolute age, causes, evolution, mechanism, periodicity]
name of orogeny
subtopic
SA absolute age
Acadian Phase
Allegheny Orogeny
Alpine Orogeny
Andean Orogeny
Antler Orogeny
Appalachian Phase
Assyntic Orogeny
Asturian Orogeny
Avalonian Orogeny
Baikalian Phase
Cadomian Orogeny
Caledonian Orogeny
Cordilleran Orogeny
deformation
epeirogeny
faults
folds
geosynclines
Grenvillian Orogeny
Himalayan Orogeny
Hudsonian Orogeny
igneous activity
intracontinental belts
Karelian Orogeny
Katangan Orogeny
Kenoran Orogeny
Laramide Orogeny
metamorphism
orogenic belts
Pan-African Orogeny
Penokean Orogeny
periodicity
plate tectonics
Pyrenean Orogeny
Saalian Phase
structural geology
Taconic Orogeny
taphrogeny
tectonics
volcanology

Orosei
Village and gulf of Tyrrhenian Sea.
BT Sardinia
Italy

Oroville
Term introduced in 1981. City in Butte County, N California. Site of August, 1975 earthquake.
CO N393100N393100
W1213400W1213400
BT Butte County
California
United States

Oroville Dam
Earth fill dam on Feather River which forms Oroville Reservoir in Butte County of N central California. Also search Oroville Reservoir; reservoirs AND Oroville; surface reservoirs AND Oroville.
BT California
United States
SA dams
reservoirs

orpiment
UF yellow arsenic
BA sulfides
BT minerals

Orsha
City in Vitebsk Oblast in N.
BT Byelorussia
USSR

Orsk
City in Orenburg Oblast in Southern Urals near the Kazakh border.
BT Russian Republic
USSR

Orthida
Includes use on level 2 under Brachiopoda(1). See list F.
BA Articulata
Brachiopoda
BT Invertebrata
NA Enteletacea

orthite
use allanite

ortho-amphibolite
use orthoamphibolite

orthoamphibole
Term introduced in 1981. Includes use in combination with chain silicates (i.e. chain silicates, orthoamphibole) on level 2 under minerals(1). See list L.
BA amphibole group
chain silicates
silicates
BT minerals
NA anthophyllite
gedrite
holmquistite
SA amosite

orthoamphibolite
See list J.
UF ortho-amphibolite
BA amphibolites
BT metamorphic rocks

orthoclase
BA feldspar group
framework silicates
silicates
BT minerals
SA adularia
anorthoclase
K-feldspar
sanidine

orthoenstatite
BA orthopyroxene
pyroxene group
chain silicates
silicates
BT minerals

orthogneiss
Term introduced in 1981. Includes use on level 3 under metamorphic rocks(1) gneisses(2). See list J.
BA gneisses
BT metamorphic rocks

Orthopteroida
Term introduced in 1981. Includes use on level 2 under Insecta(1). See list F.
BA Insecta
BT Mandibulata
Arthropoda
Invertebrata

orthopyroxene
As of 1981, includes use in combination with chain silicates (i.e. chain silicates, orthopyroxene) on level 2 under minerals(1). See list L. Before 1981, included use on level 3 under minerals(1) chain silicates, pyroxene group(2).
BA pyroxene group
chain silicates
silicates
BT minerals
NA bronzite
enstatite
oulito
hypersthene
orthoenstatite

orthoquartzite
Includes use on level 3 under sedimentary rocks(1) clastic rocks(2). See list I. Also search sedimentary quartzite.
UF orthoquartzitic sandstone
sedimentary quartzite
BA clastic rocks
BT sedimentary rocks
SA ferruginous quartzite
quartzites
sandstone
terrigenous materials

orthoquartzitic sandstone
use orthoquartzite

orthosilicates
Sorosilicates and nesosilicates. Includes use on level 2 under minerals(1); in combination with epidote group, garnet group, humite group, melilite group, and olivine group (i.e. orthosilicates, epidote group) to form terms on level 2 under minerals(1). See list L.
BA silicates
BT minerals
NA andalusite
braunite
chevkinite
chloritoid
coffinite
cyrtolite
datolite
dumortierite
euclase
eucryptite
gadolinite
garrelsite
grandidierite
harkerite
hemimorphite
ilvaite
innelite
joaquinite
julgoldite
kasolite
kilchoanite
kornerupine
kyanite
labuntsovite
larnite
lawsonite
melanocerite
merwinite
mullite
NZ nenadkevichite
NA nesosilicates
perrierite
phenakite
pumpellyite
roggianite
sapphirine
sillimanite
sorosilicates
NZ spurrite
NA staurolite
stillwellite
thalenite
NZ thaumasite
NA thorite
thortveitite
titanite
NZ topaz
NA uranophane
vesuvianite
viridine
willemite
yttrialite
zircon
NZ zunyite

Orulgan Mountains
Extend N and S, E of Lena River in N central Russian Republic. Also search Orulgan; Orulgan Range
UF Orulgan Range
BT Russian Republic
USSR

Orulgan Range
use Orulgan Mountains

Oruro
Department in W part of the country on the Chilean border. Also a city.
BT Bolivia

os
use eskers

Os
use osmium

Osage County
Index states as applicable.
BX Kansas
Missouri
Oklahoma
BT United States

Osagean
use Osagian

Osagian
Provincial series, North America. Above Kinderhookian, below Meramecian. Includes use on level 3 under age terms(1). See list E. Before 1976, also search Osagean.
UF Osagean

BA Lower Mississippian
Mississippian
Carboniferous
BT Paleozoic
NT Keokuk Limestone
SA Valmeyeran

Osaka
City and prefecture in S.
BT Honshu
Japan

Osaka Group
BT Cenozoic
SA Honshu
Japan
Pleistocene
Pliocene

Osborne County
N Kansas.
BT Kansas
United States

oscillations
As of 1978 term is restricted to refer to motions under Earth(1).
SA Earth
free oscillations
high-level mode
motions
resonance
spheroidal mode
toroidal mode
vibration

Osculosida
Order. Includes use on level 3 under Radiolaria(1).
BA Radiolaria
BT Protista
Invertebrata
NA Nassellina

Osetia
use Ossetia

Oshima
Fishing town in Kumamoto Prefecture W Kyushu, Japan.
UF Amami
BT Ryukyu Islands
Japan

Oshima Peninsula
N of Hakodate in Oshima sub-prefecture in SW.
BT Hokkaido
Japan

Osler Series
SW Ontario.
BT Precambrian
SA Canada
Ontario

Oslo
County in SE Norway. Also a city on Oslo Fjord an inlet of the Skagerrak.
BT Norway

osmium
Includes use on level 1 and 2 as a chemical element (list D).
UF Os
SA elements

Osnabruck
City in NW part of country. Also search Osnabrueck.
UF Osnabrueck
BT Lower Saxony
West Germany

Osnabrueck
use Osnabruck

Osogovo Mountains
Index countries as applicable. Also search Osogov; Osogovo.
BT Europe
SA Bulgaria
Yugoslavia

Osor
Village in NE part of country.
BT Gerona Province
Catalonia
Spain

Ossetia
Region of the central Caucasus. Divided into North Ossetian ASSR, and the South Ossetian Autonomous Oblast. Index Soviet Republics as applicable.
UF Osetia
BT USSR
SA Georgian Republic
Russian Republic

Ossipee Mountains
In Carroll County in E.
BT New Hampshire
United States

Ostashkovichi
Village in S.
BT Byelorussia
USSR

Osteichthyes
Includes use on level 2 under Pisces(1). See list F.
BA Pisces
BT Vertebrata
NA Acanthodii
Actinistia
Actinopterygii
Brachiopterygii
Crossopterygii
Dipnoi
Rhipidistia

osteology
SA bones
paleontology

Ostracoda
Includes use on level 1 and 2 as a fossil term (list F).
BT Crustacea
Mandibulata
Arthropoda
Invertebrata
NA Archeocopida
Bairdiomorpha
Beyrichicopina
Cladocopina
Cypridocopina
Cytherocopina
Entomozocopina
Eridostraca
Kirkbyocopina
Leperditicopida
Myodocopida
Myodocopina
Paleocopida
Platycopida
Podocopida
SA ostracods
plankton

Ostracodermi
Term introduced in 1981. Includes use on level 3 under Agnatha(1). See list F. Before 1981, also search ostracoderms. As of 1981, search ostracoderms only for biostratigraphy.
BA Agnatha
BT Vertebrata

ostracods
Term introduced in 1981 as the common name for Ostracoda. See list F. To be used for Ostracoda when they are used as tools in stratigraphy. Includes use on level 1. Before 1981, search Ostracoda AND biostratigraphy. If 1, term set options are:
biostratigraphy
age [List E]
SA biostratigraphy
Ostracoda

Ostrava
City in N Moravia.
UF Moravska Ostrava
BT Moravia
Czechoslovakia

Ostrava-Karvina
Coal mining and steel producing region near the Polish border in N.
BT Moravia
Czechoslovakia

Ostrea
Genus. Includes use on level 3 under Mollusca(1). See list F.
BA Ostreidae
Bivalvia
Mollusca
BT Invertebrata

Ostreacea
Term introduced in 1981. Includes use on level 2 under Mollusca(1). See list F.
BA Cyrtodontida
Bivalvia
Mollusca
BT Invertebrata

Ostreidae
Includes use on level 3 under Mollusca(1). See list F.
BA Bivalvia
Mollusca
BT Invertebrata
NA Ostrea

Ostrowiec
use Ostrowiec Swietokrzyski

Ostrowiec Swietokrzyski
City in SW central part of country. Also search Ostrowiec.
UF Ostrowiec
BT Kielce
Poland

Osumi Peninsula
Between Kagoshima Bay and Osumi Strait in S.
BT Kyushu
Japan

osumilite
BA ring silicates
silicates
BT minerals
SA aluminosilicates

Oswego County
On SE Lake Ontario.
BT New York
United States

Otago
Provincial district in S.
BT South Island
New Zealand

Otago Peninsula
On E side of Otago Harbor in Dunedin area on SE.
BT South Island
New Zealand

Otake
Town on Hiroshima Bay in SW.
BT Honshu
Japan

Otavi
Town in N.
BT Namibia

Otero County
Index states as applicable.
BX Colorado
New Mexico
BT United States

Othris
use Othrys

Othrys
Mountain range in central part of country. Index administrative regions as applicable.
UF Othris
Othrys Massif
BT Greece
SA Thessaly

Othrys Massif
use Othrys

otoliths
Includes use on level 3 under Pisces(1).
SA fossils
microfossils
Pisces

Otranto
Town on Strait of Otranto in heel of Italian boot.
BT Apulia
Italy

Ottawa
City in SE.
BT Ontario
Canada
SA Leda Clay

Ottawa County
Index states as applicable.
BX Kansas
Michigan
Ohio
Oklahoma
BT United States

Ottawa Valley
River valley. Index provinces as applicable.
BT Canada
SA Ontario
Quebec

Ottnangian
Europe. See list E.
BA Miocene
Neogene
Tertiary
BT Cenozoic

Otto Fiord
use Otto Fjord

Otto Fjord
NW Ellesmere Island, District of Franklin. Also search Otto.
UF Otto Fiord
BT Northwest Territories
Canada

Otway Basin
Primarily in W and SW Victoria. Index states as applicable.
BT Australia
SA South Australia
Victoria

Otztal Alps
Mountain range of the Eastern Alps. Index countries as applicable. Also search Oetztal Alps.
UF Oetztal Alps
Oetztal Massif
Oetztaler
Otztaler Alps
BT Alps
Europe
SA Austria
Eastern Alps
Italy
Stubai Alps

Otztaler Alps
use Otztal Alps

Ouachita Mountains
Index states as applicable.
CO N340000N344000
W0943000W0960000
BT United States
SA Arkansas
Oklahoma

Ouegoa
Village.
BT New Caledonia

Ougarta
Village W of the Great Western Erg

in W central.
BT Algeria

Ouray County
SW Colorado.
BT Colorado
United States

Ouro Preto
City in S.
BT Minas Gerais
Brazil

outcropping
use outcrops

outcrops
Used as a general term. Also search outcropping.
UF outcropping
SA bedrock

Outer Banks
Chain of sandy barrier islands stretching length of coast.
UF The Banks
BT North Carolina
United States

outer core
BA core
SA inner core

Outer Hebrides
Outer group of the Hebrides, W of Little Minch. Sometimes called Long Island.
CO N564500N583000
W0061000W0080000
BT Scotland
Great Britain
United Kingdom
SA Hebrides
Inner Hebrides

outer mantle
use upper mantle

Outer Mongolia
use Mongolia

outer planets
SA Jupiter
Neptune
planetology
planets
Pluto
Saturn
terrestrial planets
Uranus

outer shelf
BA continental shelf
SA inner shelf

outer slope
Term introduced in 1978.
UF slope, outer
BA continental slope
SA inner slope

outer space
use interplanetary space

Outokumpu
Village in SE part of country.
BT Finland

outwash
Includes use on level 3 under sediments(1) clastic sediments(2); on level 3 under glacial geology(1) glacial features(2). See list N.
UF glacial outwash
outwash drift
BA clastic sediments
BT sediments
SA glacial features
glacial geology
glaciofluvial features
glaciofluvial sedimentation
terrigenous materials

outwash aprons
use outwash plains

outwash drift
use outwash

outwash plains
Also search sandurs.
UF morainal plains
outwash aprons
overwash plain
sandurs
BT glaciofluvial features
glacial features
SA moraines
plains

overburden
SA regolith
sedimentary cover

overconsolidated materials
Term introduced in 1978. Includes use on level 3 under engineering geology(1). Before 1978, search overconsolidated AND materials.
SA consolidation
engineering geology
foundations
materials
soil mechanics

overgrowths
SA crystal growth
minerals

Overijssel
Province in E.
BT Netherlands
NT Twente

overthrust
A valid term through 1977. After 1977, use overthrust faults.

Overthrust Belt
As of 1981, no longer a valid term for GeoRef. See Western Overthrust Belt.

overthrust faults
Term introduced in 1978. Includes use on level 3 under faults(1) displacements(2). Before 1978, also search faults AND overthrust.
BA faults
SA allochthons
autochthons
displacements
klippen
nappes
thrust faults
windows

Overton County
N Tennessee.
BT Tennessee
United States

overturned folds
Term introduced in 1978. Includes use as folds defined on the basis of orientation in relation to the geographic horizontal plane. Includes use on level 3 under folds(1) orientation(2). Before 1978, also search folds AND overturned.
BA folds
SA orientation

overwash
As of 1981, no longer a valid term for GeoRef. Term was introduced in 1978 and used on level 3.

overwash plain
use outwash plains

Oviedo
use Asturias

Ovruch Series
Divided into two subseries.
BT Proterozoic
Precambrian
SA Ukraine
USSR

Owens Valley
River valley between Sierra Nevada on the W, and White and Inyo Mountains on E.
BT California
United States

Owsley County
Term introduced in 1981. E central Kentucky.
CO N371500N373200
W0833000W0835300
BT Kentucky
United States

Owyhee County
Extreme SW Idaho.
BT Idaho
United States

oxalates
Term introduced in 1981. Includes use on level 2 under minerals(1). See list L.
BT minerals

Oxford
City 52 miles WNW of London.
BT England
Great Britain
United Kingdom

Oxford Clay
BT Oxfordian
Jurassic
SA England
United Kingdom

Oxford County
W Maine.
BT Maine
United States

Oxfordian
Europe. Above Callovian, below Kimmeridgian. Includes use on level 3 under age terms(1). See list E.
BA Upper Jurassic
Jurassic
BT Mesozoic
NT Oxford Clay
SA Malm

Oxfordshire
County in central.
CO N513000N521000
W0005000W0014500
BT England
Great Britain
United Kingdom

oxidation
Includes use on level 3 under geochemistry(1) processes(2).
SA Eh
geochemistry
hydrolysis
oxidation zone
oxygen
processes
reduction

oxidation zone
UF zone, oxidation
SA Eh
mineral deposits
oxidation

oxidation-reduction potential
use Eh

oxides
Includes niobates and tantalates. Includes use on level 2 under minerals(1). See list L.
BT minerals
NA akaganeite
alexandrite
anatase
armalcolite
baddeleyite
bayerite
birnessite
bixbyite
boehmite
brannerite
bromellite
brookite
brucite
cassiterite
chrome spinel
chromite
chrysoberyl
coronadite
corundum
cryptomelane
cuprite
delafossite
diaspore
ferropseudobrookite
franklinite
gahnite
geikielite
germanates
gibbsite
goethite
groutite
hausmannite
hematite
hercynite
heterogenite
hollandite
hydrogoethite
hydroxides
ilmenite
iron oxides
jacobsite
lepidocrocite
limonite
lithiophorite
maghemite
magnesioferrite
magnetite
manganese oxides
manganite
manganosite
martite
nigerite
niobates
niobotantalates
nordstrandite
nsutite
periclase
perovskite
pitchblende
pseudobrookite
psilomelane
pyrolusite
ramsdellite
rancieite
rutile
sapphire
senarmontite
specularite
spinel
spinel group
taaffeite
tantalates
thorianite
titanomaghemite
titanomagnetite
todorokite
trevorite
ulvospinel
uraninite
valentinite
wustite
zincite
SA alumina

Oxisols
Includes use on level 3 under soils(1). See list M.
BA soils

Oxnard
City in Ventura County in S.
BT California
United States

Oxus
use Amu Darya

oxygen
Includes use on level 1 and 2 as a chemical element (list D). Also search O.
UF O
SA elements

isotopes
O-16
O-18
O-18/O-16
oxidation
ozone
stable isotopes

oxymagnite
use maghemite

oxysulfides
As of 1981, includes use on level 2 under minerals(1). See list L. Before 1981, included use on level 3 under minerals(1) sulfides(2).
BA sulfides
BT minerals

Ozark Highlands
use Ozark Mountains

Ozark Mountains
Eroded tableland. Also search Ozarks; Ozark uplift. Index states as applicable.
UF Ozark Highlands
Ozark Uplift
Ozarks
BT United States
SA Arkansas
Missouri
Oklahoma

Ozark Uplift
use Ozark Mountains

Ozarkodina
Genus.
BA Conodonta

Ozarks
use Ozark Mountains

Ozernoye
Village in SE Kamchatka on Okhotsk Sea just N of Ozernovskiy in Khabarovsk Kray.
BT Russian Republic
USSR

ozocerite
BA organic compounds
BT minerals

ozone
SA atmosphere
oxygen

Ozun
use Uzon

P

P
use phosphorus

P-T conditions
Includes use on level 2 under metamorphism(1). As of 1981, includes use on level 2 under inclusions(1) and fluid inclusions(1).
UF conditions, P-T
SA fluid inclusions
geologic barometry
geologic thermometry
grade
inclusions
metamorphism
metasomatism
phase equilibria
pressure
temperature

P-waves
Includes use on level 3 under seismology(1). Also search compressional wave; compressional waves; longitudinal wave.
UF compressional wave
dilatational wave
irrotational wave
longitudinal wave
pressure wave
primary wave
push wave
push-pull wave
BA body waves
elastic waves
SA PcP-waves
PKiP-waves
PKP-waves
PKS-waves
PPP-waves
PPS-waves
S-waves
seismology
SKP-waves
SKS-waves
waves

Pa
use protactinium

Pa-231
Includes use on level 3 under isotopes(1).
SA isotopes
protactinium

Pacaya
Volcano in S central Guatemala. Also search Pacaya Volcano.
UF Pacaya Volcano
BT Guatemala

Pacaya Volcano
use Pacaya

Pachelma
Town in Penza Oblast W of Kuibyshev in S central European USSR.
BT Russian Republic
USSR

Pachuca
City NE of Mexico City.
UF Pachuca de Soto
BT Hidalgo
Mexico

Pachuca de Soto
use Pachuca

Pachypteris
BA Caytoniales
gymnosperms
BT Spermatophyta
Plantae
SA Cycadales

Pacific Basin
Generally the entire ocean floor of both the North and South Pacific up to the continental shelf.
BT Pacific Ocean
SA North Pacific
Pacific Plate
South Pacific

Pacific Coast
Region comprising those states fronting on the Pacific Ocean. Index states as applicable. Also search West Coast. Introduced as level 1 area term in 1978.
UF West Coast
BA Western U.S.
United States
NA California
Oregon
Washington
SA Pacific region

Pacific mobile belt
Crustal regions of tectonic activity along the coastal fringes of continents and some entire islands of the Pacific. Index continents and islands as applicable.
BT Pacific region
SA Asia
New Guinea
New Zealand
North America
South America

Pacific Ocean
Includes use on level 1 or 2 as an area term (list O). If 1, see term set options under list B.
NT Aleutian Ridge
Aleutian Trench
Astoria Canyon
Auckland Islands
Bauer Deep
Bay of Plenty
Bismarck Sea
Blanco fracture zone
Bonin Islands
Bowie Seamount
Campbell Island
Cascadia Basin
Cascadia Channel
NA Central Pacific
NT Chatham Rise
China Sea
Circum-Pacific region
Cobb Seamount
Cocos Plate
NA East Pacific
NT East Pacific Rise
Emperor Seamounts
NA Equatorial Pacific
NT Erimo Seamount
Fiji Plateau
Formosa Strait
Galapagos Rift
Gorda Rise
Guadalupe Island
Guatemala Basin
Gulf of Panama
Hachijo-jima
NZ Hawaii
NT Hawaiian Ridge
Iwo-Jima
Japan Trench
Juan de Fuca Ridge
Kermadec Islands
Kuril Trench
Lau Basin
Lord Howe Island
Lord Howe Rise
Loyalty Islands
Macquarie Island
Macquarie Ridge
Manihiki Plateau
Mariana Trench
Mariana Trough
Mendocino fracture zone
Middle America Trench
Midway
Monterey Canyon
Monterey Fan
Murray fracture zone
Nazca Plate
New Caledonia Basin
Norfolk Island
Norfolk Ridge
NA North American Pacific
North Pacific
Northeast Pacific
Northwest Pacific
NT Ontong Java Plateau
Pacific Basin
Panama Basin
Peru-Chile Trench
Raoul Island
Redondo Canyon
San Diego Trough
Santa Barbara Basin
Shatsky Rise
Shikoku Basin
Solomon Sea
NA South American Pacific
South Pacific
South Polynesian Pacific
Southeast Pacific
Southwest Pacific
NT Sulu Sea
Tasman Basin
Tonga Trench
Torres Strait
Wake
NA West Pacific
SA Antarctic Ocean
Caroline Islands
Cook Islands
DSDP Site 157
DSDP Site 158
DSDP Site 163
DSDP Site 167
DSDP Site 284
DSDP Site 292
DSDP Site 296
DSDP Site 297
DSDP Site 315
DSDP Site 319
DSDP Site 320
DSDP Site 321
East Pacific Ocean Islands
Fiji
French Polynesia
Gilbert Islands
Indian Plate
Leg 16
Leg 17
Leg 18
Leg 19
Leg 20
Leg 21
Leg 30
Leg 31
Leg 32
Leg 33
Leg 34
Leg 5
Leg 54
Leg 55
Leg 56
Leg 57
Leg 58
Leg 59
Leg 6
Leg 60
Leg 61
Leg 62
Leg 63
Leg 64
Leg 65
Leg 66
Leg 7
Leg 8
Leg 9
Mariana Islands
Marshall Islands
Melanesia
Mid-Pacific Rise
New Caledonia
New Hebrides
Oceania
Pacific Plate
Pacific region
Pacific-Antarctic Ridge
Polynesia
Ross Sea
Solomon Islands
Tonga
Tuvalu

Pacific Plate
Includes most of the Pacific Basin with the Eurasian Plate on the W and N; the American, Cocos, and Nazca Plates on the E; the Indian Plate on the SW; and the Antarctic Plate on the S.
SA Pacific Basin
Pacific Ocean
plate tectonics
plates

Pacific region
Includes use on level 1 or 2 as an area term (list O). If 1, see term set options under list B.
NT Pacific mobile belt
SA Circum-Pacific region

Far East
Pacific Coast
Pacific Ocean

Pacific-Antarctic Ridge
Mid-ocean ridge extending from the Balleny fracture zone S of Macquarie Island to the East Pacific Rise.
SA Antarctic Ocean
East Pacific Rise
mid-ocean ridges
Pacific Ocean

packing
Includes use on level 3 under sediments(1) and sedimentation(1). Before 1978, also used for atomic packing under crystal structure.
SA atomic packing
sedimentary rocks
sedimentation
textures

packstone
Includes use on level 3 under sedimentary rocks(1) carbonate rocks(2). See list I.
BA carbonate rocks
BT sedimentary rocks

Pacoima Dam
On Pacoima River in Los Angeles County NW of Los Angeles. Also search Pacoima.
BT California
United States

Paderborn
City in E.
BT North Rhine-Westphalia
West Germany

Padre Island
Narrow barrier island along Gulf of Mexico from SE of Corpus Christi to near the Mexican border.
BT Texas
United States

Padstow
Town in SW near N coast of Cornwall.
BT England
Great Britain
United Kingdom

Padurea Craiului Mountains
Range in SW. Also search Padurea Craiului.
BT Transylvania
Romania

Paganzo
Village in central.
BT La Rioja
Argentina

Paganzo Group
BT Paleozoic
SA Argentina
Carboniferous
Permian

Page County
Term introduced in 1981. Index states as applicable.
BX Iowa
Virginia
BT United States

Pageland
Town in Chesterfield County in N.
BT South Carolina
United States

Pagoda Formation
Underlies Pentagon Shale; overlies Dearborn Limestone. NW Montana.
UF Pagoda Limestone
BT Middle Cambrian
Cambrian
SA Montana

Pagoda Limestone
use Pagoda Formation

Pahang
State on the Malay Peninsula in West Malaysia.
BT Malaysia
SA Malay Peninsula
West Malaysia

pahoehoe
Includes use on level 3 under lava(1).
BT lava
SA aa lava
volcanism

Pahute Mesa
Tableland in Nye County in S.
UF Paiute Mesa
BT Nevada
United States

Pai Khoi
use Pai-Khoi

Pai-Khoi
Mountain range on Yugor Peninsula between Barents and Kara seas. N extension of Northern Urals.
CO N683000N691500
E0650000E0604500
UF Pai Khoi
BT Russian Republic
USSR

Paiute Mesa
use Pahute Mesa

Pajaro Valley
River valley S of San Jose in W California. Also search Pajaro River.
BT California
United States

Pakistan
Formerly consisted of East Pakistan and West Pakistan which were separated by about 1,000 miles of Indian territory. After East Pakistan became the independent state of Bangladesh in 1971, West Pakistan and Pakistan became coextensive. Also search West Pakistan. Includes use on level 1 as an area term (list O). For term set options see list B.
CO N233500N373000
E0751500E0601500
BA Indian Peninsula
Asia
NT Baluchistan
North-West Frontier Province
Punjab Province
Sind
Sulaiman Range
SA Hindu Kush
Hyderabad
Indian Plate
Indus River
Jammu and Kashmir
Karakoram
Ladakh
Nari Series
Ramnagar
Siwalik Range
Siwalik System

Paktia
Province in E.
BT Afghanistan

Palaeodictyopteroida
Term introduced in 1981. Includes use on level 2 under Insecta(1). See list F.
BA Insecta
BT Mandibulata
Arthropoda
Invertebrata

Palaeofavosites
Genus. Includes use on level 3 under Coelenterata(1) Tabulata(2).
BA Favositidae
Tabulata
Anthozoa
Coelenterata
BT Invertebrata
SA Favosites

Palaeoloxodon naumanni
As of 1981, includes use on level 3 under Mammalia(1) Elephantoidea(2). See list F. Before 1981, included use on level 3 under Mammalia(1) Proboscidea(2).
BA Elephantidae
Elephantoidea
Proboscidea
Mammalia
BT Tetrapoda
Vertebrata

palagonite
As of 1981, includes use on level 3 under igneous rocks(1) glasses(2). See list H. Before 1981, included use on level 3 under igneous rocks(1) pyroclastics and glasses(2).
BA glasses
volcanic rocks
BT igneous rocks

Palamau
District in coal mining area in W.
BT Bihar
India

Palatinate
Historical region in two parts. Lower or Rhine Palatinate on both sides of Rhine River S of the Main River; and the Upper Palatinate in E Bavaria. Index states as applicable.
BT Germany
SA Baden-Wurttemberg
Bavaria
Hesse
Rhineland-Palatinate
Upper Palatinate

Palawan
Long, narrow island between N Borneo and W Philippines.
UF Paragua
BT Philippine Islands

Palencia
As of 1981, refers to only the city. Before 1981, also included the province. In N Spain.
BT Palencia Province
Old Castile
Spain

Palencia Province
Term introduced in 1981. Before 1981, search Palencia. In N Spain.
CO N414600N430300
W0035200W0050000
BT Old Castile
Spain
NT Palencia

paleo-oceanography
SA bathymetry
ocean basins
ocean floors
oceanography
paleocirculation
sea-floor spreading

paleobathymetry
SA bathymetry
ocean floors
paleocirculation
sea-floor spreading

paleobiology
SA biochemistry
biology
fossils
paleontology

paleobotany
For general discussion of fossil plants. See names of major floral groups (list F). Includes use on level 1 (list A); on level 2 under area terms (list B). If level 1, term set options are:
topic [applications, bibliography, catalogs, classification, concepts, education, evolution, history, instruments, methods, nomenclature, objectives, practice, principles, symposia, textbooks]
subtopic
UF botany, paleo-
SA bacteria
bibliography
biogeography
bryophytes
catalogs
floral studies
fossilization
fossils
fungi
holotypes
leaves
micropaleontology
paleoecology
paleontology
palynology
photosynthesis
Plantae
Protista
seeds
sporangia
staining
vegetation
wood

Paleocene
World. Lower Tertiary, above Gulfian (Cretaceous), below Eocene. Includes use on level 1 as an age term (list E); on level 2 under paleoterms, e.g. paleoecology, paleogeography; on level 3 under age terms(1).
BA Paleogene
Tertiary
BT Cenozoic
NT Beaufort Formation
Clayton Formation
NA lower Paleocene
middle Paleocene
NT Midway Group
NA Montian
NT Pinyon Conglomerate
Ravenscrag Formation
Sentinel Butte Formation
Tongue River Member
NA upper Paleocene
SA Beaverhead Formation
Brianconnais Zone
Difunta Group
Fort Union Formation
Mardin Formation
Moreno Formation
Paskapoo Formation
Scaglia Formation
Sinjar Formation
Sparnacian
Wasatch Formation

paleochannels
Not a valid index term for GeoRef. See channels under sedimentary structures(1) planar bedding structures(2).

paleocirculation
SA circulation
ocean circulation
paleo-oceanography
paleobathymetry

paleoclimatology
Includes use on level 1 (list A). Used for treatments of the climate of a given period of time in the geologic past. If 1, term set options are:
age (single term from list E)
(area)
topic [applications, changes, concepts, cycles, evolution, indicators, interpretation, methods, patterns, temperature]

subtopic
UF climatology, paleo-
SA arctic environment
arid environment
atmosphere
Boreal
boreal environment
C-13/C-12
climate
cycles
fluctuations
fossils
glacial geology
glaciation
humid environment
indicators
interglacial environment
isotopes
O-18/O-16
obliquity of the ecliptic
paleoecology
paleogeography
S-34/S-32
subtropical environment
taiga environment
temperate environment
tropical environment

Paleocopida
As of 1981, includes use on level 3 under Ostracoda(1). See list F. Before 1981, included use on level 2 under Ostracoda(1).
BA Ostracoda
BT Crustacea
Mandibulata
Arthropoda
Invertebrata

paleocurrents
Includes use on level 3 under sedimentation(1) provenance(2) or paleogeography(1).
SA currents
paleogeography
provenance
sedimentation

paleoecology
Includes use on level 1 (list A); on level 2 under fossil terms (list F). Used for the study of the relationships between organisms and their environments, the death of organisms, and their burial and postburial history in the geologic past based on fossil fauna and flora and their stratigraphic position. If 1, term set options are:
age [List E]
area
fossil group [List F]
age [List E]
topic [analysis, changes, indicators, interpretation, sedimentation]
subtopic [e.g. name of environment: deltaic environment, estuarine environment, fluvial environment, lacustrine environment, lagoonal environment, marine environment, paludal environment, reefs, terrestrial environment]
SA adaptation
aerobic environment
anaerobic environment
archaeology
arctic environment
beaches
benthonic taxa
biogeography
biostratigraphy
biotypes
boreal environment
changes of level
communities
deltaic environment
deltas
ecology
environment
estuaries
estuarine environment
fluvial environment
fresh water
fresh-water environment
indicators
lacustrine environment
lagoonal environment
lagoons
lakes
marine environment
paleobotany
paleoclimatology
paleoenvironment
paleogeography
paleontology
paludal environment
peat bogs
ponds
reefs
salinity
salt marshes
sedimentation
shelf environment
taphonomy
terrestrial environment

paleoenvironment
Distinguish from paleoecology. Includes use on level 3 under age terms(1) sedimentary petrology(2).
SA environment
environmental analysis
historical geology
paleoecology
paleogeography

Paleogene
As of 1978, includes use on level 1 and 2 (list E). An interval of geologic time incorporating the Oligocene, Eocene, and Paleocene of the Tertiary; the lower Tertiary.
BA Tertiary
BT Cenozoic
NT Duchesne River Formation
NA Eocene
Ilerdian
NT Kenai Group
NA Oligocene
Paleocene
NT Sinjar Formation
Wasatch Formation
SA Arikareean
lower Tertiary

paleogeographic controls
Includes use on level 3 under mineral deposits, genesis(1) controls(2).
SA controls
mineral deposits
paleogeography

paleogeographic maps
Term introduced in 1978. Before 1978, also search maps AND paleogeographic.
BA maps

paleogeography
Includes use on level 1 (list A). Used for the geography of ancient times, specifically the study and description of the physical geography of the geologic past. Term set options are:
age [List E; a single term]
(area)
topic [applications, changes, concepts, interpretation, maps, methods, patterns, principles]
subtopic (no area term)
SA alluvial plains
basins
biogeography
buried channels
buried valleys
changes of level
continental drift
geography
geomorphology
geosynclines
Mesogaea
ocean basins
paleoclimatology
paleocurrents
paleoecology
paleoenvironment
paleogeographic controls
reconstruction
sebkha environment
sedimentary basins
sedimentary structures
sedimentation
shorelines
stratigraphy
syntectonic processes
transgression

paleolatitude
UF latitude, paleo-
SA paleomagnetism
polar wandering

paleolimnology
SA glacial lakes
limnology

Paleolithic
Archaeologic classification.
UF Old Stone Age
BA Cenozoic
NA Acheulian
Magdalenian
Mousterian
SA Pleistocene
Quaternary
Tertiary

paleomagnetism
Used for the study of natural remanent magnetization. Includes use on level 1 (list A); on level 2 under Moon(1) and geochronology(1); on level 3 under plate tectonics(1). If 1, term set options are:
age [List E]
(area)
topic [applications, causes, changes, concepts, experimental studies, geochemistry, interpretation, methods, patterns, polar wandering, pole positions, reversals, stability]
subtopic
UF magnetism, paleo-
SA adiabatic demagnetization
anhysteretic remanent magnetization
Brunhes Epoch
chemical demagnetization
chemical remanent magnetization
coercivity
continental drift
crust
Curie point
declination
demagnetization
depositional remanent magnetization
dipole moment
Earth
Gauss Epoch
geochemistry
geochronology
geophysical methods
geophysics
Gilbert Epoch
high-resolution methods
induction
isothermal remanent magnetization
Jaramillo Event
Koenigsberger ratio
lava
magnetic anomalies
magnetic declination
magnetic domains
magnetic field
magnetic hysteresis
magnetic inclination
magnetic intensity
magnetic susceptibility
magnetization
Matuyama Epoch
natural remanent magnetization
Olduvai Event
paleolatitude
plate tectonics
polar wandering
pole positions
remagnetization
remanent magnetization
reversals
sea-floor spreading
stability
stratigraphy
tectonophysics
thermal demagnetization
thermoremanent magnetization
viscous remanent magnetization

paleontology
Used for discipline as a whole. Includes use on level 1 (list A); on level 2 under area terms (list B), age terms (list E), continental drift(1), symposia(1), bibliography(1), and automatic data processing(1). See also names of fossil groups (list F). If 1, term set options are:
topic [applications, bibliography, catalogs, classification, concepts, education, evolution (as a concept), fossilization, history, instruments, life origin, nomenclature, practice, principles, symposia, taxonomy (principles of), textbooks]
subtopic
SA adaptation
bibliography
biocenoses
biogeography
biometry
biostratigraphy
bones
catalogs
collections
continental drift
faunal provinces
fossilization
fossils
geochronology
glossaries
historical geology
holotypes
isomorphism
jaws
lexicons
life origin
micropaleontology
nutrition
ontogeny
ornamentation
osteology
paleobiology
paleobotany
paleoecology
palynology
phylogeny
problematic fossils
radiation
reproduction
sample preparation
shells
skeletons
skulls
species diversity
staining
stratigraphy
synonymy
taxonomy
thanatocenoses
ultrastructure
wood

paleorelief
SA landform evolution
relief

paleosalinity
As of 1978, term is used on level 2 under fluid inclusions(1).
SA brackish water
fluid inclusions
geochemistry
salinity
salt
sea water

Paleosols
UF fossil soils
BA soils

paleotectonics
A valid term through 1975. After 1975, see tectonics(1) and paleogeography(1). See also neotectonics.

paleotemperature
A valid term through 1973. After 1973, use temperature under paleoclimatology(1) and see geologic thermometry.

Paleozoic
Includes use on level 1 as an age term (list E); on level 2 under paleoterms, e.g. paleoecology, paleogeography, paleomagnetism.
BT Phanerozoic
NT Antler Orogeny
Arbuckle Group
Baralaba Coal Measures
Bedford Shale
Berea Sandstone
Bird Spring Formation
Broken River Formation
Caledonian Orogeny
Cambrian
Carboniferous
Casper Formation
Catskill Formation
Chattanooga Shale
Deadwood Formation
Devonian
Dunkard Group
Geirud Formation
Hercynian Orogeny
Hidden Valley Dolomite
Honaker Trail Formation
Hunton Group
Knox Group
Lisburne Group
NA lower Paleozoic
NT Manhattan Formation
NA middle Paleozoic
NT Mikabu System
Minnelusa Formation
Ordovician
Paganzo Group
Peel Sound Formation
Permian
Petersburg Granite
Read Bay Formation
Ringerike Sandstone
Rondout Formation
Sambagawa Belt
Silurian
Silvretta Group
Supai Formation
Taconic Orogeny
Talchir Formation
Talladega Group
Tensleep Sandstone
Tulghes Series
NA upper Paleozoic
NT Weber Sandstone
Wissahickon Formation
Woodford Shale
SA Beacon Supergroup
Botucatu Formation
Ceneri Zone
New York City Group
Shuswap Complex
Sparagmite Group

Palermo
City on Bay of Palermo in NW.
BT Sicily
Italy

Palestine
Approximately cooextensive with Israel and the W bank of the Jordan River. Index countries as applicable.
SA Israel
Jordan

palingenesis
As of 1978, term is restricted to meaning in petrology. Term is not used in reference to paleontology.
SA anatexis
assimilation
contamination
magmas
metasomatism
petrology

palladium
Includes use on level 1 and 2 as a chemical element (list D).
UF Pd
SA elements

pallasites
Used as igneous rock until 1977. Now used as meteorite.
BA stony irons
BT meteorites

Palliser Formation
BT Upper Devonian
Devonian
SA Alberta
Canada

Palm Beach County
SE Florida.
BT Florida
United States

Palm Springs
City in Riverside County in S.
BT California
United States

Palma
City on Majorca.
UF Palma de Majorca
BT Balearic Islands
Spain

Palma de Majorca
use Palma

Palmae
Family coextensive with order Palmales. Also search Palmales.
BA Monocotyledoneae
angiosperms
BT Spermatophyta
Plantae

Palmatolepis
Genus.
BA Conodonta

Palmdale Bulge
Term introduced in 1981. Uplift in vicinity of the San Andreas Fault, S California.
BT Los Angeles County
California
United States

Palmer Peninsula
use Antarctic Peninsula

Palo Alto
City S of San Francisco in Santa Clara County.
BT California
United States

Palos Verdes Hills
Occupy peninsula between Santa Monica Bay and San Pedro Bay S of Los Angeles.
UF San Pedro Hills
BT California
United States

paludal environment
Term introduced in 1981.
SA ecology
environment
marshes
paleoecology
paludal sedimentation
sedimentation
swamps

paludal sedimentation
Term introduced in 1981.
SA paludal environment
sedimentation

Paluxy Formation
In Trinity Group. SW Arkansas, SE Oklahoma, and E Texas.
BT Comanchean
Cretaceous
SA Arkansas
Lower Cretaceous
Oklahoma
Texas
Trinity Group

palygorskite
UF attapulgite
BA clay minerals
sheet silicates
silicates
BT minerals

palynology
Includes use on level 1 (list A). See list F. Used for the study of pollen of seed plants and spores of other embryophytic plants whether living or fossil, including their dispersal and applications. Term set options are: topic [applications, bibliography, catalogs, classification, concepts, education, fossilization, history, instruments, methods, nomenclature, practice, principles, symposia, techniques, textbooks]
subtopic
SA biogeography
geochronology
micropaleontology
paleobotany
paleontology
palynomorphs
pollen
pollen analysis
pollen diagrams
sample preparation
stratigraphy

palynomorphs
Includes use on level 1 and 2 as a fossil term (list F).
NA acritarchs
Chitinozoa
Dinoflagellata
megaspores
miospores
problematic palynomorphs
SA Artemisia
biostratigraphy
exine
floral studies
monolete taxa
palynology
plankton
Plantae
pollen
Pyrrhophyta
spores
Tasmanites

Pambak
NE section of city of Kirovakan in N central Armenia.
UF Bambak
BT Armenia
USSR

Pamir (Range)
use Pamirs

Pamir Range
use Pamirs

Pamirs
High altitude region in the USSR, mostly in Tadzhikistan. Before 1981, also included small parts of China, India, and Afghanistan. Also search Pamir; Pamir Range.
CO N370000N393000
E0750000E0650000
UF Pamir (Range)
Pamir Range
BA Central Asia
BT USSR
SA Tadzhikistan
Uzbekistan

Pamlico River
Bisects Beaufort County and flows into Pamlico Sound. Actually it is the estuary of the Tar River.
BT North Carolina
United States

Pamlico Sound
Between E North Carolina mainland and narrow islands off the coast.
BT North Carolina
United States

Pampa
use Pampas

Pampas
Vast, treeless, fertile plain extending N and S approximately 1000 miles from lower Parana River to Patagonia, and from the Atlantic to the Andean piedmont. The humid Pampas are in the E and the dry Pampas are in the W. Also search Pampa.
CO S395000S215500
W0533000W0694000
UF Pampa
BT Argentina
South America

Pampean Mountains
In the province of Cordoba.
BT Argentina

Pamplona
City in N part of country.
BT Navarra
Spain

Pan-African Orogeny
Term introduced in 1978. Includes use on level 3 under orogeny(1). Before 1978, also search orogeny AND Pan-African or Pan African.
SA orogeny

Panagyurishte
Plovdiv district, W Central Bulgaria in central Sredna Gora on Luda Yana River.
BT Bulgaria
SA Sredna Gora

Panama
Includes use on level 1 as an area term (list O). For term set options see list B.
CO N071000N094000
W0771000W0830200
BA Central America
NT Darien
Sansan

Panama Basin
SW of Panama and W of Colombia.
BT Pacific Ocean

Panama Canal Zone
10 mile strip of territory across Panama in which canal is located.
BT Central America

Panamint Range
W of Death Valley in Inyo County in E.
BT California
United States

Panasqueira
Village in S part of country.
BT Portugal

Panay Island
One of the Visayan Islands midway between Luzon and Mindanao. Also search Panay.
BT Philippine Islands

Panch Mahals
District in W part of country. Also search Panchmahal; Panchmahals.
UF Panchmahal
Panchmahals
BT Gujarat
India

Panchet Series
Series of rocks overlying the Damuda Series in the Raniganj coal field.
BT Lower Triassic
Triassic
SA Bihar
India

Panchmahal
use Panch Mahals

Panchmahals
use Panch Mahals

Panderodus
BA Conodonta

Pangaea
Name proposed for the supercontinent comprising all the landmasses or earth which existed about 300 million years ago prior to continental drift. Includes use on level 2 under continental drift(1).
UF Pangea
SA continental drift
Gondwana
Laurasia
Northern Hemisphere

Pangea
use Pangaea

Panhandle
An area or projection of land like the handle of a pan which occurs in a number of states. Index states as applicable.
BT United States
SA Florida
Oklahoma
Texas
West Virginia

Pannonia
Ancient Roman province S and W of the Danube. Index countries as applicable.
BT Europe
SA Austria
Hungary
Yugoslavia

Pannonian
Europe. Includes use on level 3 under age terms(1). See list E.
BA Neogene
Tertiary
BT Cenozoic
SA Miocene
Pliocene

Pannonian Basin
SE Austria, Hungary W of the Danube, N central Yugoslavia, and S central Czechoslovakia. Index countries as applicable.
BT Europe
NZ Slovakian Pannonian Basin
SA Austria
Czechoslovakia
Hungary
Yugoslavia

Pantar
Island of the Alor group 60 miles NW of Timor.
BT Lesser Sunda Islands
Indonesia

Pantellaria
use Pantelleria

Pantelleria
Volcanic Italian island in the Mediterranean Sea between Sicily and Tunisia.
UF Pantellaria
BT Sicily
Italy

pantellerite
As of 1981, includes use on level 3 under igneous rocks(1) rhyolites(2). See list H. Before 1981, included use on level 3 under igneous rocks(1) andesite-rhyolite family(2).
BA rhyolites
volcanic rocks
BT igneous rocks
SA comendite

Pantodonta
Term introduced in 1981. Includes use on level 2 under Mammalia(1). See list F.
BA Mammalia
BT Tetrapoda
Vertebrata

Pantotheria
Includes use on level 2 under Mammalia(1). See list F.
BA Mammalia
BT Tetrapoda
Vertebrata

Papaloapan River basin
Term introduced in 1981. In state of Veracruz, SE Mexico.
CO N180500N184500
W0953000W0961000
BT Veracruz
Mexico

Papua
Former Australian territory. Now the southern half of new state of Papua New Guinea on the island of New Guinea. Includes use on level 3. Included use as a first order area term until 1976. Now use Papua New Guinea on level 1.
SA Papua New Guinea

Papua New Guinea
Eastern half of the island of New Guinea. Comprises Papua and the former Australian U.N. trusteeship of The Territory of New Guinea plus the Bismarck Archipelago, and Bougainville, Buka and Green islands of the W Solomon Islands. Became an independent state on September 16, 1976. Includes use on level 1 as an area term (list O). Also search Papua-New Guinea.
CO S110000S003000
E1563000E1410000
UF Papua-New Guinea
BA Australasia
NT Bismarck Archipelago
Bougainville
Huon Peninsula
Louisiade Archipelago
New Britain
New Ireland
Port Moresby
Talasea
Vanimo region
SA Malay Archipelago
New Guinea
Papua
Solomon Islands

Papua-New Guinea
use Papua New Guinea

Para
State in N.
CO S095000N024000
W0460000W0592000
BT Brazil
NT Marajo
SA Barreiras Formation
Brazilian Shield
Tocantins River region

para-amphibolite
See list J.
BA amphibolites
BT metamorphic rocks
SA amphibolite

Parablastoidea
Includes use on level 2 under Echinodermata(1). See list F.
BA Crinozoa
Echinodermata
BT Invertebrata

paracelsian
As of 1981, includes use on level 3 under minerals(1) framework silicates, barium feldspar(2). See list L. Before 1981, included use on level 3 under minerals(1) framework silicates, feldspar group(2).
BA barium feldspar
feldspar group
framework silicates
silicates
BT minerals
SA celsian

Paracrinoidea
Includes use on level 2 under Echinodermata(1). See list F.
BA Crinozoa
Echinodermata
BT Invertebrata

Paradox Basin
In the "Four Corners" area of the SW where the boundaries of 4 states intersect. Index states as applicable.
BT United States
SA Arizona
Colorado
New Mexico
Utah

Paradox Member
Of Hermosa Formation. W Colorado and SE Utah.
BT Middle Pennsylvanian
Pennsylvanian
SA Colorado
Utah

paragenesis
For detailed treatment of mineral sequences in metamorphosed or altered rocks and mineral deposits. Includes use on level 1 (list A). Term set options are:
kind of mineral deposit (first level term only)
area
kind of rock (first level term only)
area
topic [changes, evolution, interpretation, observations, patterns, processes, rates]
subtopic
UF mineral sequence
paragenetic sequence
SA clay mineralogy
crystallography
economic geology
genesis
igneous rocks
metamorphic rocks
metamorphism
metasomatism
mineral assemblages
mineral deposits
minerals
phase equilibria
rocks
sedimentary rocks

paragenetic sequence
use paragenesis

paragneiss
See list J.
BZ gneisses
metasedimentary rocks
BT metamorphic rocks

paragonite
BA mica group
sheet silicates
silicates
BT minerals

Paragua
use Palawan

Paraguana Peninsula
Between the Caribbean Sea and the Gulf of Venezuela in NW part of country.
BT Falcon
Venezuela

Paraguay
Includes use on level 1 as an area term (list O). For term set options see list B.
CO S273000S191000
W0543000W0624500
BA South America
SA Botucatu Formation
Chaco
Parana Basin
Parana River
Serra Geral Formation

Parahiba
use Paraiba

Parahyba
use Paraiba

Paraiba
State in NE Brazil.
CO S082000S060000
W0343000W0384500
UF Parahiba
Parahyba
BT Brazil
SA Borborema

parallel faults
Term introduced in 1978. Includes use on level 3 under faults(1) patterns(2). Before 1978, also search faults AND parallel.
BA faults

parallel folds
use concentric folds

parameters
A valid general term through 1978. See factors.

Paramushir
Large island in the N Kuril Islands near S tip of Kamchatka Peninsula.
BT Russian Republic
USSR
SA Kuril Islands

Parana
State in S.
CO S265000S222500
W0481000W0543000
BT Brazil
SA Acungui Group
Brazilian Shield
Estrada Nova Formation
Irati Formation
Serra do Mar

Parana Basin
River basin. Index countries as applicable.
UF Parana River basin
BT South America
SA Argentina
Brazil
Paraguay

Parana River
Formed in S central Brazil by the confluence of the Rio Grande and the Paranaiba rivers. It flows SW and S into the Rio de la Plata. Index countries as applicable. Also search Parana.
BT South America
SA Argentina
Brazil
Paraguay

Parana River basin
use Parana Basin

Paranthropus
Genus. As of 1981, includes use on level 3 under Mammalia(1) Hominidae(2). See list F. Before 1981, included use on level 3 under Mammalia(1) Primates(2).
BA Hominidae
Primates
Mammalia
BT Tetrapoda
Vertebrata

pararammelsbergite
As of 1981, includes use on level 3 under minerals(1) arsenides(2). See list L. Before 1981, included use on level 3 under minerals(1) sulfides(2).
BA arsenides
sulfides
BT minerals
SA rammelsbergite

parasites
SA bacteria
epibiotism
fungi

paratacamite
As of 1981, includes use on level 3 under minerals(1) chlorides(2). See list L. Before 1981, included use on level 3 under minerals(1) halides(2).
BA chlorides
halides
BT minerals
SA atacamite

Paratethys
Includes use on level 3 under continental drift(1).
SA continental drift
Tethys

Paratunka
Village just W of Petropavlosk Kamchatskiy in S Kamchatka Oblast on Kamchatka Peninsula.
BT Russian Republic
USSR

Pareiasauria
Term introduced in 1981. Includes use on level 2 under Reptilia(1). See list F.
BA Anapsida
Reptilia
BT Tetrapoda
Vertebrata

parent materials
Includes use on level 3 under soils(1) genesis(2) as type of factor; on level 3 under weathering(1).
SA factors
horizons
materials
soils
weathering

parent rocks
use protoliths

pargasite
BA clinoamphibole
amphibole group
chain silicates
silicates
BT minerals

Paria Peninsula
NE part of country. Along with island of Trinidad nearly encloses Gulf of Paria.
BT Venezuela
SA Sucre

Paricutin
Volcano on site of former village of Paricutin in SW part of country.
BT Michoacan
Mexico

Paring Mountains
BT Romania

Paris
Department in N central. Includes the city of Paris.
CO N485000N490000
E0023000E0021000
UF Paris region
BT France

Paris Basin
Chief depression of N and N central France. Bounded by English Channel on NW, Amorican Massif on W, Massif Central on S, and plateaus of Langres and Lorraine on E. Also search Paris AND basin.
CO N480000N500000
E0050000E0010000
BT France

Paris region
use Paris

Park City
City in Summir County in NE.
BT Utah
United States

Park County
Index states as applicable.
BX Colorado
Montana
Wyoming
BT United States

Parker County
W of Fort Worth in N.
BT Texas
United States

Parkfield
Village in Monterey County near Fremont County line in W. Location of earthquake.
BT California
United States

Parkman Sandstone
In Montana Group. N Wyoming and S Montana.
BT Upper Cretaceous
Cretaceous
SA Montana
Montana Group
Wyoming

Parma
City in N central part of country.
BT Emilia-Romagna
Italy

Parnaiba Basin
River basin. Index states as applicable.
BT Brazil
SA Maranhao
Piaui

Parnalee
Located 16 miles S of Madurai where meteorite impacted.
BT Tamil Nadu
India

Parnassos
use Parnassus

Parnassus
One of highest massifs in central Greece. N of Gulf of Corinth.
UF Parnassos
BT Greece

Parras Basin
S Coahuila in NE central part of country. Also search Parras.
UF Parras de la Fuente
BT Coahuila
Mexico

Parras de la Fuente
use Parras Basin

Parry Sound
Town in SE on E shore of Georgian Bay.
BT Ontario
Canada

partial melting
Includes use on level 3 under magmas(1).
SA magmas
mantle
melting
melts
phase equilibria
plate tectonics

partial pressure
BT pressure
SA fugacity

partial thermoremanent magnetization
Term introduced in 1981.
BA remanent magnetization
SA inverse thermoremanent magnetization
thermoremanent magnetization

particle precipitation
Includes use on level 2 under ionosphere(1).
SA intensity
ionosphere
ions
particle radiation
particles
precipitation

particle radiation
Includes use on level 2 under astrophysics and solar physics(1).
SA astrophysics and solar physics
ionization
ionosphere
particle precipitation
particles
radiation
solar wind

particle-track dating
Cosmic-ray tracks. Includes use on level 2 under geochronology(1). Also search particle track.
UF dating, particle-track
SA cosmic rays
geochronology

particles
General term. Includes use on level 2 and 3 under meteorology(1); on level 3 under interplanetary space(1) cosmic rays(2).
SA activation energy
aerosols
agglutinates
cosmic dust
cosmic rays
electromagnetic radiation
electrons
fines
fission
fragments
grains
granulometry
intensity
interplanetary dust
interplanetary space
ions
meteorology
Moon
neutrons
particle precipitation
particle radiation
precipitation
rounding
roundness
scintillations
soils
solar wind
sorting
sphericity
trapped particles

particulate materials
Term introduced in 1981.
SA clastic sediments
granular materials
materials
unconsolidated materials

parting lineation
Before 1978, also search current lineations.
UF current lineations
current partings
BA bedding plane irregularities
sedimentary structures
SA lineation

partition coefficients
UF coefficients, partition
SA crystal chemistry
partitioning

partitioning
As of 1978, term is used on level 2 under crystal chemistry(1).
SA crystal chemistry
geochemistry
partition coefficients
phase equilibria

partridgeite
use bixbyite

Pas-de-Calais
Department in extreme N.
CO N500000N511500
E0031000E0014500
BT France
NT Boulogne
Boulonnais
SA Nord-Pas-de-Calais Basin

Pasadena
City in Los Angeles County.
BT California
United States

Pasamonte Meteorite
Term introduced in 1981. Impact near village of Pasamonte in Union County, New Mexico. Before 1981, also search Pasamonte AND meteorites.
BA eucrite
achondrites
BT meteorites
SA New Mexico
Union County

Pasayten Group
Stratigraphically above Dewdney Creek Formation. S British Columbia and Central N Washington.
BT Lower Cretaceous
Cretaceous
SA British Columbia
Washington

Pascagoula River basin
SE Mississippi.
BT Mississippi
United States

Pasco Basin
S central Washington.
BT Washington
United States

Paskapoo Formation
SA Alberta
Canada
Cretaceous
Eocene

Paleocene
Saskatchewan
Upper Cretaceous

Passa Dois Group
BT Permian
SA Brazil
Irati Formation

Passau
City on the Danube River at the Austrian border.
BT Bavaria
West Germany

passband filters
Term introduced in 1981.
BT filters

passive margins
Term introduced in 1981. Also search Atlantic-type.
UF Atlantic-type margins
SA aseismic margins
continental margin
plate tectonics

Patagonia
Region. A barren tableland S of the Pampas, between the Atlantic Ocean and the Andes. Index provinces as applicable.
CO S550000S370000
W0621000W0733000
BT Argentina
SA Chile
Chubut
Rio Negro Province
Santa Cruz

Patagonia Cordillera
use Patagonian Andes

Patagonian Andes
Mountain range extending the length of Patagonia. Index countries as applicable.
UF Patagonia Cordillera
BT South America
SA Argentina
Chile

patch reefs
Includes use in level 3 under reefs(1).
BT reefs

paternoite
use kaliborite

Patom Plateau
Between Vitim and Lena rivers in NE Irkutsk Oblast N of Lake Baikal. Also search Patom.
CO N540000N590000
E1180000E1060000
BT Russian Republic
USSR

patterned ground
Includes use on level 3 under glacial geology(1) periglacial features(2) and under geomorphology(1).
UF ground, patterned
SA frost action
geomorphology
glacial geology
periglacial features
polygons

patterns
Includes use as a level 2 or 3 term appropriate to a large number of topics, e.g. on level 2 under faults(1), fractures(1), and under soils(1). See list G.
SA dispersion patterns
drainage patterns
Ekman spiral
regional patterns

Patuxent River
Rises in central Maryland and flows S and SE into Chesapeake Bay.
BT Maryland
United States

Pau
City in extreme SW part of country.
BT Pyrenees-Atlantiques
France

Paucituberculata
Term introduced in 1981. Includes use on level 2 under Mammalia(1). See list F.
BA Mammalia
BT Tetrapoda
Vertebrata

Pauni
Town in central.
BT Madhya Pradesh
India

Pavia
City and province S of Milan.
BT Lombardy
Italy

Pavlodar
Town and oblast in NE.
BT Kazakhstan
USSR

Pavlograd
City in Dnepropetrovsk Oblast in E central.
BT Ukraine
USSR

Pawhuska Rock Plain
Osage county in N.
BT Oklahoma
United States

Pawnee County
Index states as applicable.
BX Kansas
Nebraska
Oklahoma
BT United States

Payne County
N central Oklahoma.
BT Oklahoma
United States

Payson
City in Utah County in N central.
BT Utah
United States

Payun Matru
Volcano on Payun Plateau in SW Mendoza.
UF Payun Matru Volcano
Payun-Matru
Payun-Matru Volcano
BT Mendoza
Argentina

Payun Matru Volcano
use Payun Matru

Payun-Matru
use Payun Matru

Payun-Matru Volcano
use Payun Matru

Pb
use lead

Pb-204
Includes use on level 3 under isotopes(1).
SA isotopes
lead
Pb-206/Pb-204
Pb-207/Pb-204
Pb-208/Pb-204

Pb-204/Pb-206
use Pb-206/Pb-204

Pb-204/Pb-208
use Pb-208/Pb-204

Pb-206
Includes use on level 3 under isotopes(1).
SA isotopes
lead
Pb-206/Pb-204
Pb-207/Pb-206

Pb-206 Pb-207
use Pb-207/Pb-206

Pb-206/Pb-204
Includes use on level 3 under isotopes(1). Also search Pb-204/Pb-206; Pb-204 AND Pb-206.
UF Pb-204/Pb-206
SA isotopes
lead
Pb-204
Pb-206

Pb-206/Pb-207
use Pb-207/Pb-206

Pb-207
Includes use on level 3 under isotopes(1).
SA isotopes
lead
Pb-207/Pb-204
Pb-207/Pb-206

Pb-207/Pb-204
Includes use on level 3 under isotopes(1). Also Search Pb-207 AND Pb-204.
SA isotopes
lead
Pb-204
Pb-207

Pb-207/Pb-206
Includes use on level 3 under isotopes(1). Also search Pb-206/P-207; Pb-207 AND Pb-206.
UF Pb-206 Pb-207
Pb-206/Pb-207
SA isotopes
lead
Pb-206
Pb-207

Pb-208
Includes use on level 3 under isotopes(1).
SA isotopes
lead
Pb-208/Pb-204

Pb-208/Pb-204
Includes use on level 3 under isotopes(1). Also search Pb-204/Pb-208; Pb-204 AND Pb-208.
UF Pb-204/Pb-208
SA isotopes
lead
Pb-204
Pb-208

Pb-210
Includes use on level 3 under absolute age(1) methods(2).
SA absolute age
isotopes
lead
Pb/Pb
Pb/Th
U/Pb

Pb/Pb
Includes use on level 3 under absolute age(1) methods(2). Also search lead-lead.
SA absolute age
lead
Pb-210
Pb/Th
U/Pb

Pb/Th
Includes use on level 3 under absolute age(1) methods(2). Also search Pb AND Th.
SA absolute age
lead
Pb-210
Pb/Pb
thorium
U/Pb

PcP-waves
Term introduced in 1981. Also search PcP waves.
BA body waves
elastic waves
SA P-waves
seismology
waves

Pd
use palladium

Peace River
Flows into Snake River just N of Lake Athabasca. Index provinces as applicable.
BT Canada
SA Alberta
British Columbia

Peak District
Plateau region in N central.
BT England
Great Britain
United Kingdom

pearceite
BA sulfarsenites
sulfosalts
BT minerals

Pearl Harbor
Inlet on S coast of Oahu 6 miles W of Honolulu.
BT Hawaii
United States

Pearlette Ash
use Pearlette Volcanic Ash

Pearlette Volcanic Ash
SW Kansas.
UF Pearlette Ash
BT Pleistocene
Quaternary

pearlite
use perlite

pearls
SA aragonite
gems
Mollusca

Peary Land
Region in extreme N on Arctic Ocean.
BT Greenland
Arctic region

peat
Includes use on level 1 as a commodity term (list C). As of 1981, includes use on level 3 under sediments(1) organic residues(2). See list N. Before 1981, included use on level 3 under sediments(1) organic sediments(2).
BA organic residues
BT sediments
SA biogenic structures
Bog soils
coal
humus
lignite
organic materials
peat bogs

peat bed
use peat bogs

peat bogs
Also search peat bed.
UF peat bed
peat moor
BT bogs
SA ecology
paleoecology
peat
swamps

peat moor
use peat bogs

pebbles
Includes use on level 3 under sedi-

ments(1) clastic sediments(2); on level 3 under geomorphology(1) and glacial geology(1). See list N.
BA clastic sediments
BT sediments
SA cobbles
gravel
rounding
shingle
terrigenous materials

Pechenga
Territory, formerly in N Finland, which was called Petsamo. Now it is part of Murmansk Oblast.
BT Russian Republic
USSR

Pechora
City in Komi Autonomous Soviet Socialist Republic on left bank of Pechora River in NE European USSR.
BT Russian Republic
USSR

Pechora Basin
Coal basin mainly in basin of Usa River in NE Komi Autonomous S.S.R. in NE European USSR. Also search Pechora.
CO N650000N680000
E0600000E0500000
BT Russian Republic
USSR
SA Moscow-Pechora Syneclise

Pechora River
Rises in the Central Urals of N Perm Oblast and flows N through the Komi A.S.S.R., and the Nenets National Okrug into Pechora Bay. Also search Pechora.
CO N620000N670000
E0593000E0520000
BT USSR
SA Moscow-Pechora Syneclise

Pecopteris
Genus. Includes use on level 3 under pteridophytes(1) Filicopsida(2).
BA Filicopsida
pteridophytes
BT Plantae
SA Glossopteris
gymnosperms
Pteridospermae

Pecos County
W Texas.
BT Texas
United States

Pecos River valley
Index states as applicable. Also search Pecos River.
BT United States
SA New Mexico
Texas

Pectinacea
As of 1981, includes use on level 2 under Mollusca(1). See list F. Before 1981, included use on level 3 under Mollusca(1) Bivalvia(2).
BA Bivalvia
Mollusca
BT Invertebrata

Pectinidae
Includes use on level 3 under Mollusca(1). See list F.
BA Bivalvia
Mollusca
BT Invertebrata

pectolite
BA chain silicates
silicates
BT minerals

pediments
Includes use on level 3 under geomorphology(1) erosion features(2).
SA erosion features
erosion surfaces
geomorphology

pediplains
UF desert pediplains
desert plains
pediplanes
SA erosion surfaces
peneplains

pediplanes
use pediplains

Pedregosa Basin
Index Mexican and U.S. states as applicable.
SA Arizona
Chihuahua
New Mexico
Sonora

Peedee Formation
Variable gray to green argillaceous sands and impure limestones. E North Carolina and E South Carolina.
BT Upper Cretaceous
Cretaceous
SA North Carolina
South Carolina

Peekskill
City in Westchester County just N of New York City.
BT New York
United States

Peel River
Flows into Mackenzie River. Index territories as applicable.
BT Canada
SA Northwest Territories
Yukon Territory

Peel Sound Formation
Franklin District.
BT Paleozoic
SA Canada
Devonian
Northwest Territories
Silurian

pegmatite
As of 1981, includes use on level 3 under igneous rocks(1) granites(2). See list H. Before 1981, included use on level 3 under igneous rocks(1) granite-granodiorite family(2). Also includes use on level 1 as a commodity term (list C). Also search pegmatites.
BA granites
BT igneous rocks
SA graphic granite
micropegmatite

Peichang
use Peikang

Peikang
Town in W central.
UF Peichang
BT Taiwan

pelagic environment
Term introduced in 1978. Before 1978, search pelagic.
SA environment
oceanography
pelagic sedimentation

pelagic sedimentation
Term introduced in 1981.
SA pelagic environment
sedimentation

Pelecypoda
A valid index term through 1975. After 1975, use Bivalvia.

pelite
use shale

pelitic gneiss
A valid metamorphic rock term through 1977. Use gneisses and pelitic texture.

pelitic rocks
use pelitic texture

pelitic schist
See list J.
BA schists
BT metamorphic rocks

pelitic texture
Term introduced in 1978. Before 1978, search pelitic AND texture; pelitic rocks.
UF pelitic rocks
SA metamorphic rocks
sedimentary rocks
textures

pelletization
Term introduced in 1981.
SA production

pellets
SA oolite
pisoliths

Pelmatozoa
BA Echinodermata
BT Invertebrata

Pelomedusidae
As of 1981, includes use on level 3 under Reptilia(1) Chelonia(2). Before 1981, included use on level 3 under Reptilia(1) Anapsida(2).
BA Chelonia
Anapsida
Reptilia
BT Tetrapoda
Vertebrata

Pelona Schist
Precambrian(?). S California.
BT Precambrian
SA California

Peloponnese
use Peloponnesus

Peloponnesos
use Peloponnesus

Peloponnesus
Administrative region and peninsula forming S part of the mainland.
CO N361500N382000
E0233000E0210000
UF Peloponnese
Peloponnesos
BT Greece
NT Argolis
Corinth
Gavrovo Zone
Megalopolis Basin
Tripolis
SA Hellenides

Peloritani Mountains
NE Sicily.
BT Sicily
Italy

Pelotas Basin
River basin. Index states as applicable.
BT Brazil
SA Rio Grande do Sul
Santa Catarina

Pelvoux Massif
Mountain group which contains Barre das Ecrins, the highest peak in the Dauphine Alps. Index departments as applicable. Also search Pelvoux, and Ecrins-Pelvoux Massif.
UF Ecrins-Pelvoux Massif
BT France
SA Dauphine Alps
Hautes-Alpes
Isere

Pelycosauria
Term introduced in 1981. Includes use on level 2 under Reptilia(1). See list F.
BA Synapsida
Reptilia
BT Tetrapoda
Vertebrata

Pembroke
Town in Pembrokeshire in SW.
BT Wales
Great Britain
United Kingdom

Pembroke County
use Pembrokeshire

Pembrokeshire
County in SW.
CO N513500N520500
W0043000W0051500
UF Pembroke County
BT Wales
Great Britain
United Kingdom

Pench Valley
River valley in central Madhya Pradesh. Also search Pench.
BT Madhya Pradesh
India

Pend Oreille County
Extreme NE.
BT Washington
United States

Pendleton County
Index states as applicable.
BX Kentucky
West Virginia
BT United States

pendulum gravimeters
Term introduced in 1981.
BT gravimeters

peneplains
Includes use on level 3 under geomorphology(1) erosion features(2) and landform evolution(2).
UF base-level peneplain
SA erosion features
erosion surfaces
geomorphology
pediplains

penetration
Not to be used for crystal growth.
SA engineering geology
rock mechanics
soil mechanics

penetration tests
Term introduced in 1981.
SA engineering geology
soil mechanics

Penghu Islands
Group of about 48 islands in Formosa Strait between Taiwan and the mainland of China.
UF Hoko Gunto
Hoko Shoto
Pescadores
BT Taiwan

Peninsular Ranges
Part of the Coast Ranges S of Los Angeles.
BT California
United States
SA Coast Ranges

Peninsular Shield
use Indian Shield

Pennine Alps
SW division of Central Alps extending from Great Saint Bernard Pass to Simplon Pass. Index countries as applicable. Also search Pennine.
CO N455000N461000
E0081500E0070000
BT Alps
Europe
SA Central Alps
Italy
Monte Rosa
Piedmont Alps
Switzerland

Pennine Chain
use Pennines

Pennine Hills
use Pennines

Pennine Range
use Pennines

Pennines
Long hill range extending from the Cheviot Hills in the N to the S Midlands.
CO N520800N554000
W0004000W0034000
UF Pennine Chain
Pennine Hills
Pennine Range
BT England

Pennington County
Index states as applicable.
BX Minnesota
South Dakota
BT United States

Pennington Formation
Comprises Stony Gap Sandstone, Avis Limestone, Falls Mills Sandstone members. N Alabama, NW Georgia, E Kentucky, E Tennessee, and SW Virginia.
BT Upper Mississippian
Mississippian
SA Alabama
Georgia
Kentucky
Tennessee
Virginia

penninite
BA sheet silicates
silicates
BT minerals

Pennsylvania
Includes use on level 1 as an area term (list O). For term set options see list B.
CO N394500N423500
W0744500W0803500
BA Eastern U.S.
United States
NX Adams County
NT Allegheny County
Altoona
NX Beaver County
Bedford County
NT Berks County
NX Butler County
Carbon County
NT Centre County
NX Chester County
Clinton County
Columbia County
NT Conemaugh
NX Crawford County
Cumberland County
Delaware County
Erie County
Fayette County
Franklin County
Fulton County
Greene County
NT Harrisburg
NX Jefferson County
Lancaster County
Lawrence County
NT Lebanon County
Luzerne County
NX Mercer County
Monroe County
Montgomery County
NT Nittany Valley
Pennsylvania State University
NX Perry County
NT Philadelphia
NX Pike County
NT Pittsburgh
Pittsburgh coal basin
NX Presque Isle
NT Reading
Schuylkill County
NX Somerset County
Sullivan County
Union County
NT University of Pennsylvania
NX Warren County
Washington County
Wayne County
Waynesboro
Westmoreland County
Woodbury
York County
SA Allegheny Front
Allegheny Group
Allegheny Mountains
Allegheny Plateau
Ames Limestone
Antietam Formation
Appalachian Basin
Appalachian Plateau
Baltimore Gneiss
Bedford Shale
Berea Sandstone
Black River Group
Bloomsburg Formation
Blue Ridge Province
Catskill Delta
Catskill Formation
Clinton Group
Coeymans Formation
Conemaugh Group
Delaware River
Delaware River basin
Dunkard Basin
Dunkard Group
Genesee Group
Glenarm Series
Great Lakes region
Hamilton Group
Helderberg Group
Juniata Formation
Kinzers Formation
Lake Erie
Mahantango Formation
Martinsburg Formation
McKenzie Formation
Monongahela Group
Newark Basin
Newark-Gettysburg Basin
Ohio River
Ohio River valley
Onondaga Limestone
Oriskany Sandstone
Perry Formation
Piedmont
Pocono Formation
Pottsville Group
Rochester Formation
Rondout Formation
Sharon Conglomerate
South Mountain
Susquehanna River
Susquehanna River basin
Tioga Bentonite
Tonoloway Limestone
Trenton Group
Tully Limestone
Tuscarora Formation
Valley and Ridge Province
Wissahickon Formation
York

Pennsylvania State University
University Park, Centre County.
BT Pennsylvania
United States

Pennsylvanian
Includes use on level 1 as an age term (list E); on level 2 under paleoterms, e.g. paleoecology, paleogeography, paleomagnetism.
BA Carboniferous
BT Paleozoic
NT Bevier Coal
Conemaugh Group
Excello Shale
Francis Creek Shale
La Salle Limestone
NA Lower Pennsylvanian
NT Mansfield Formation
Marble Falls Group
Mattoon Formation
NA Middle Pennsylvanian
NT Minturn Formation
Monongahela Group
Morrow Formation
Pottsville Group
Red Fork Sandstone
Saginaw Formation
Sharon Conglomerate
Springfield Coal Member
Tennessee Sandstone
NA Upper Pennsylvanian
NT Wabash Formation
Wapanucka Limestone
Westerly Granite
SA Allegheny Orogeny
Baralaba Coal Measures
Bartlesville Sand
Bird Spring Formation
Casper Formation
Dunkard Group
Honaker Trail Formation
Johns Valley Formation
Lisburne Group
Minnelusa Formation
Mississippian
Springer Formation
Stephanian
Supai Formation
Talladega Group
Tensleep Sandstone
Upper Carboniferous
Weber Sandstone

Penobscot Bay
Large inlet of Atlantic Ocean in S.
BT Maine
United States

Penokean Orogeny
A time of deformation and granite emplacement during the Precambrian. Index states as applicable. Before 1978, also search Penokean AND orogeny.
BT Precambrian
SA Michigan
Minnesota
orogeny

Penrhyn Slate
BT Precambrian
SA United Kingdom
Wales

Pensacola
City on Pensacola Bay in Escambia County in extreme W of the Florida panhandle.
BT Florida
United States

Pensacola Mountains
In the Horlick Mountains area SE of Ross Ice Shelf on the Pacific Ocean side.
BT Antarctica

Pentameracea
Superfamily. Includes use on level 3 under Brachiopoda(1) Pentamerida(2).
BA Pentamerida
Articulata
Brachiopoda
BT Invertebrata

Pentamerida
Includes use on level 2 under Brachiopoda(1). See list F.
BA Articulata
Brachiopoda
BT Invertebrata
NA Pentameracea

pentlandite
UF nicopyrite
BA sulfides
BT minerals
SA nickel

Pentoxylales
Term introduced in 1981. Includes use on level 2 under gymnosperms(1). See list F.
BA gymnosperms
BT Spermatophyta
Plantae

Penzhina Bay
NE extension of the Shelikhov Gulf of the Okhotsk Sea in NE Siberia. Also search Penzhina.
UF Penzhinskaya Bay
BT Russian Republic
USSR

Penzhinskaya Bay
use Penzhina Bay

Peoples Democratic Republic of Yemen
use Southern Yemen

People's Republic of the Congo
use Congo

Peoria Loess
Underlies Bignell Formation.
UF Peorian Loess
BT Pleistocene
SA Illinois
Iowa
Kansas
Nebraska

Peorian Loess
use Peoria Loess

Pequop Mountains
In Elko County in NE.
BT Nevada
United States

Perak
State on W coast of Malay Peninsula in West Malaysia.
BT Malaysia
NT Kinta Valley
SA West Malaysia

percolation
SA ground water
hydrology
infiltration
leaching
seepage
water

Peredovoy Range
In Kabardino-Balkarian ASSR in Northern Caucasus. Also search Peredovoy.
BT Russian Republic
USSR

Peremyshl
use Przemsyl

Peribaltic Syneclise
A depressed structure of the continental platform in the Baltic Sea area produced by slow crustal downwarp. Also search Baltic Syneclise.
UF Baltic Syneclise
BT Europe
SA Baltic region

periclase
BA oxides
BT minerals

Peridinium
BA Dinoflagellata
palynomorphs

peridot
Also search peridote.
UF peridote
peridote (olivine)
BA olivine group
nesosilicates
orthosilicates
silicates

BT minerals
SA gems
tourmaline

peridote
use peridot

peridote (olivine)
use peridot

peridotite
Includes use on level 3. As of 1981, not to be used in the igneous rocks set. See list H. Before 1981, included use on level 3 under igneous rocks(1) ultramafic family(2).
BA peridotites
ultramafics
BT igneous rocks
SA dunite
garnet peridotite
harzburgite
lherzolite
metaperidotite
plutonic rocks
wehrlite

peridotites
Term introduced in 1981. Includes use on level 2 under igneous rocks(1). See list H. Before 1981, search peridotite.
BA ultramafics
BT igneous rocks
NA dunite
garnet lherzolite
garnet peridotite
harzburgite
lherzolite
peridotite
spinel lherzolite
SA wehrlite

periglacial environment
Term introduced in 1978. Before 1978, search periglacial.
SA environment
frost action
glacial environment
glacial geology
glaciers
glaciolacustrine environment

periglacial features
Includes use on level 2 under glacial geology(1). Also search periglacial.
UF features, periglacial
SA congelifraction
cryopedology
cryoturbation
extinct lakes
firn
fossil ice wedges
frozen ground
glacial features
glacial geology
glaciers
hummocks
ice
ice wedges
lakes
patterned ground
permafrost
pingos
polygons
solifluction

periglacial phenomena
A valid term through 1976. Use periglacial features.

periodicity
Includes use on level 2 under orogeny(1); on level 3 under earthquakes(1) and under volcanology(1).
SA earthquakes
frequency
orogeny
sedimentation
seismology
volcanology

peripheral faults
Term introduced in 1978. Includes use on level 3 under faults(1) patterns(2). Before 1978, also search faults AND peripheral.
BA faults

Perisphinctida
Term introduced in 1981. Includes use on level 2 under Mollusca(1). See list F.
BA Ammonoidea
Tetrabranchiata
Cephalopoda
Mollusca
BT Invertebrata
NA Perisphinctidae

Perisphinctidae
As of 1981, includes use on level 3 under Mollusca(1) Perisphinctida(2). See list F. Before 1981, included use on level 3 under Mollusca(1) Cephalopoda(2).
BA Perisphinctida
Ammonoidea
Tetrabranchiata
Cephalopoda
Mollusca
BT Invertebrata

Perissodactyla
Includes use on level 2 under Mammalia(1). See list F.
BA Mammalia
BT Tetrapoda
Vertebrata
NA Ceratomorpha
Coelodonta antiquitatis
Equidae
Hippomorpha
Rhinocerotidae

peristerite
BA feldspar group
framework silicates
silicates
BT minerals

perlite
As of 1981, includes use on level 3 under igneous rocks(1) pyroclastics(2). See list H. Before 1981, included use on level 3 under igneous rocks(1) pyroclastics and glasses(2). Also includes use on level 3 as a commodity term (list C) under construction materials(1).
UF pearlite
BA pyroclastics
volcanic rocks
BT igneous rocks
SA construction materials
lava
volcanic glass

Perm
City and oblast W of the Urals.
BT Russian Republic
USSR
SA Molotov

permafrost
A level 1 term as of 1978. Used for geotechnical studies of permafrost; geomorphological studies should be under geomorphology. Term set options are:
active layer
subtopic
classification
subtopic
creep
subtopic
engineering properties
subtopic
experimental studies
subtopic
frost action
subtopic
frost heaving
subtopic
site exploration
subtopic
solifluction
subtopic
theoretical studies
subtopic
SA active layer
creep
cryopedology
engineering geology
engineering properties
freezing
frost action
frost heaving
frozen ground
geomorphology
glacial geology
ground ice
hummocks
ice
ice wedges
periglacial features
pingos
site exploration
soils
solifluction
thawing
thermal properties
thermokarst
tundra
Tundra soils
underground installations

permeability
Includes use on level 3 under engineering geology(1).
SA circulation
Darcy's law
diffusion
engineering geology
ground water
hydraulic conductivity
porosity
rock mechanics
roughness
sediments
thermal waters
transmissivity

permeability coefficient
use hydraulic conductivity

Permian
Includes use on level 1 as an age term (list E); on level 2 under paleoterms, e.g. paleoecology, paleogeography, paleomagnetism. Above Carboniferous, below Triassic (Mesozoic).
BT Paleozoic
NT Akiyoshi Limestone
Americus Limestone Member
Appalachian Phase
NA Asselian
NT Blaine Formation
Bulli Seam
Castile Formation
Chase Group
Coconino Sandstone
Cottonwood Limestone
Cutler Formation
Damuda Series
Ecca Series
Eskridge Shale
Fort Riley Limestone
NA Guadalupian
NT Hennessey Formation
Hughes Creek Shale
Hutchinson Salt Member
Illawarra Coal Measures
Irati Formation
Kaibab Formation
NA Kungurian
Lower Permian
Middle Permian
NT Passa Dois Group
Phosphoria Formation
Red Eagle Limestone
Rio Bonito Formation
NA Rotliegendes
NT Saalian Phase
San Andres Formation
NA Saxonian
NT Stearns Shale
Stone Corral Formation
NA Thuringian
NT Toroweap Formation
NA Upper Permian
NT Val Gardena Sandstone
Wellington Formation
Wreford Limestone
Yates Formation
SA Allegheny Orogeny
Baralaba Coal Measures
Baskunchak Series
Beacon Supergroup
Bird Spring Formation
Botucatu Formation
Brianconnais Zone
Casper Formation
Dunkard Group
Dwyka Series
Estrada Nova Formation
Geirud Formation
Gondwana System
Hercynian Orogeny
Honaker Trail Formation
Karroo System
Korenevskaya Formation
Kupferschiefer
Lisburne Group
Minnelusa Formation
New Red Sandstone
Paganzo Group
Sadlerochit Formation
Singleton Coal Measures
Slovakian Karst
Spearfish Formation
Supai Formation
Talchir Formation
Tensleep Sandstone
Weber Sandstone

Permian Basin
Term introduced in 1981. Structural basin in W Texas and SE New Mexico. Includes Midland, Delaware, and Marfa basins.
CO N283000N333000
W1003000W1062000
BT United States
NT Delaware Basin
NX Midland Basin
SA New Mexico
Texas

Pernambuco
State in NE Brazil.
CO S092000S071500
W0344500W0412500
BT Brazil
SA Barreiras Formation
Sao Francisco Basin

Pernik coal basin
15 miles SW of Sofia. Also search Pernik.
UF Dimitrovo coal basin
BT Bulgaria

perofskite
use perovskite

perovskite
UF perofskite
BA oxides
BT minerals
SA perovskite structure

perovskite structure
SA crystal structure
perovskite

perrierite
BA orthosilicates
silicates
BT minerals

Perris
City in Riverside County in S.

BT California
United States

Perry County
Index states as applicable.
BX Alabama
Arkansas
Illinois
Indiana
Kentucky
Mississippi
Missouri
Ohio
Pennsylvania
Tennessee
BT United States

Perry Formation
Consists of thick sequence of coarse clastic sediments containing interbedded basalt flows. Central Pennsylvania.
BT Upper Devonian
Devonian
SA Pennsylvania

Persani Mountains
SE Transylvania.
BT Transylvania
Romania

Pershing County
NW Nevada.
BT Nevada
United States

Persia
use Iran

Persian Gulf
Between Arabian Peninsula on the W and S, and Iran on E. Includes use on level 1 as an area term (list O). For term set options see list B. Also search Arabian Gulf.
CO N240000N310000
E0570000E0480000
UF Arabian Gulf
BA Arabian Sea
Indian Ocean
NT Khor al Bazam

Perth abyssal plain
Off SW Australia.
BT Indian Ocean

Perth Basin
Along the Indian Ocean from the Perth area on the S to the Carnarvon Basin in the N.
BT Western Australia
Australia

Perth Stade
Europe. Term introduced in 1978. Before 1978, also search Perth. See list E.
BA Weichselian
upper Pleistocene
Pleistocene
Quaternary
BT Cenozoic

perthite
BA feldspar group
framework silicates
silicates
BT minerals
SA antiperthite
cryptoperthite
microperthite

Perthshire
County in central.
BT Scotland
Great Britain
United Kingdom

Peru
Includes use on level 1 as an area term (list O). For term set options see list B.
CO S181500N000000
W0700000W0811000
BA South America
NT Arequipa
Ayacucho
Cerro de Pasco
Junin
Lima
SA Amazon Basin
Amazon River
Andes
Eastern Cordillera
Lake Titicaca
Magdalena
Pucara Group

Peru-Chile Trench
Just off Peruvian and Chilean coasts.
BT Pacific Ocean

Perugia
City and province in central part of country.
BT Umbria
Italy
NT Gubbio

Pervomaisk
City in Nikolayev Oblast in S.
BT Ukraine
USSR

Pesaro
City on Adriatic Sea.
BT Marches
Italy

Pescadores
use Penghu Islands

Peschany Island
use Peschanyy

Peschanyy
Island in Caspian Sea off S shore of Apsheron Peninsula.
UF Peschany Island
BT Azerbaidzhan
USSR

Peshawar
City, district and division near the Khyber Pass.
BT North-West Frontier Province
Pakistan

pesticides
Includes use on level 3 under environmental geology(1) waste disposal(2).
SA environmental geology
waste disposal

petalite
BA nepheline group
framework silicates
silicates
BT minerals
SA aluminosilicates

Petersburg
City S of Richmond. Within Dinwiddie County but independent of it.
BT Virginia
United States

Petersburg Granite
Late Paleozoic. Underlies Triassic sediments. E Virginia.
BT Paleozoic
SA Virginia

Petrified Forest National Park
In Painted Desert area in E.
BT Arizona
United States

petrified moss
use tufa

petrified wood
use fossil wood

petrofabric analysis
use structural analysis

petrofabrics
Includes use on level 3 under structural analysis(1). See fabric under igneous rocks(1), metamorphic rocks(1), and sedimentary rocks(1).
SA deformation
fabric
orientation
preferred orientation
structural analysis

petrogenesis
A valid term through 1973. Use igneous rocks and genesis; metamorphic rocks and genesis.

petrogeometry
use structural analysis

petrography
Includes use on level 2 under sedimentary rocks(1) for microscopic studies only; on level 2 under metamorphic rocks(1) and under igneous rocks(1).
SA igneous rocks
macerals
metamorphic rocks
microfacies
microscope methods
petrology
sedimentary rocks
sediments
thin sections
ultrastructure

petroleum
Includes use on level 1 and 2 as a commodity term (list C). Also search petroleum-gas; petroleum-natural gas.
UF oil (petroleum)
SA bitumens
condensates
Darcy's law
depletion
drilling
energy sources
fuel resources
geopressure
giant fields
heavy oil
hydrocarbons
kerogen
migration
multiphase flow
natural gas
oil and gas fields
oil sands
oil seeps
oil shale
oil spills
oil-gas interface
oil-water interface
organic materials
pipelines
reservoir properties
reservoir rocks
secondary recovery
seepage
source rocks
stratigraphic traps
structural traps
tertiary recovery
traps

petroleum engineering
Term introduced in 1978 on level 2 under engineering geology(1).
UF engineering, petroleum
SA engineering geology
hydraulic fracturing
oil and gas fields
sand equivalent tests
secondary recovery

petroleum seepage
use oil seeps

petrologen
use kerogen

petrology
Treated as a whole. For studies on igneous or metamorphic rocks. Includes use on level 1 (list A); on level 2 under area terms(1), bibliography(1), continental shelf(1), education(1), intrusions(1), lava(1), and sediments(1). See list B. If level 1, term set options are:
topic [applications, bibliography, catalogs, classification, concepts, education, experimental studies, general, history, instruments, methods, nomenclature, objectives, philosophy, practice, principles, symposia, textbooks, theoretical studies]
subtopic
SA banded structures
bibliography
catalogs
crystallography
education
fluid inclusions
homogenization
igneous rocks
inclusions
intrusions
lava
lexicons
magmas
melts
metamorphic rocks
metamorphism
metasomatism
meteorites
microscope methods
mineralogy
modal analysis
palingenesis
petrography
sedimentary petrology
sediments
silicate rocks
staining
standard materials
tektites
volcanology

petromorphology
use structural analysis

petrophysics
SA physical properties
reservoir rocks

Petrosani Basin
Coal basin in Hunedoara County in the Transylvanian Alps. Also search Petrosani.
UF Petroseni Basin
BT Transylvania
Romania

Petroseni Basin
use Petrosani Basin

petrostratigraphy
use lithostratigraphy

Petrovsk
use Makhachkala

pH
Includes use on level 3 under sea water(1). Also search acidity.
UF acidity
SA acidic composition
acids
alkalinity
geochemistry
physical properties
sea water

Phacopida
Includes use on level 2 under Trilobita(1). See list F.
BA Trilobita
BT Trilobitomorpha
Arthropoda
Invertebrata
NA Phacopina
Phacops

Phacopina
Suborder or genus. Includes use on level 3 under Trilobita(1) Phacopida(2).

BA Phacopida
Trilobita
BT Trilobitomorpha
Arthropoda
Invertebrata
SA Phacops

Phacops
Genus. Includes use on level 3 under Trilobita (1) Phacopida(2).
BA Phacopida
Trilobita
BT Trilobitomorpha
Arthropoda
Invertebrata
SA Phacopina

Phaeodarina
Includes use on level 2 under Radiolaria(1). See list F.
BA Radiolaria
BT Protista
Invertebrata

Phaeophyta
Including brown algae and seaweed. Includes use on level 2 under algae(1). See list F.
UF brown algae
seaweed
BA algae
BT thallophytes
Plantae

Phanerozoic
Includes use on level 1 and 2 as an age term (list E).
NT Cenozoic
Mesozoic
Paleozoic

pharmacolite
BA arsenates
BT minerals

pharmacosiderite
BA arsenates
BT minerals

phase diagrams
As of 1981, no longer a valid term for GeoRef. See phase equilibria.

phase distortion
Term introduced in 1981.
SA distortion
elastic waves
seismology

phase equilibria
For laboratory studies only. Includes use on level 1 (list A) and on level 2 under crystal chemistry(1) and crystal growth(1). If 1, term set options are:
material [e.g. igneous rocks, silicates; only 1st and 2nd level terms]
(system in chemical symbols) or topic
topic [anomalies, concepts, experimental studies, interpretation, melting, theoretical studies]
(system) or subtopic
SA carbon dioxide
crystal chemistry
crystal growth
crystallography
entropy
equations of state
equilibrium
free energy
gases
geochemistry
hydrothermal conditions
igneous rocks
immiscibility
inclusions
low pressure
low temperature
magmas
melting
melts
metamorphic rocks
metamorphism
metasomatism
mineral deposits
mineralogy
miscibility gap
P-T conditions
paragenesis
partial melting
partitioning
phase rule
phase transitions
sedimentation
segregation
solid solution
transformations

phase, gaseous
use gaseous phase

phase rule
Includes use on level 3 under phase equilibria(1).
UF Gibbs phase rule
SA phase equilibria

phase, solid
use solid phase

phase transitions
Term introduced in 1981. For phase changes within the Earth. For laboratory studies, use phase equilibria. Includes use on level 3 under Earth(1), crust(1), core(1), mantle(1), and seismology(1).
SA core
crust
Earth
mantle
phase equilibria
seismology

phase velocity
Term introduced in 1981.
SA elastic waves
seismology

phases
A valid general term through 1978. After 1978, see specific terms, e.g. liquid phase, solid phase, gaseous phase, etc.

Phelps County
Term introduced in 1981. Index states as applicable.
BX Missouri
Nebraska
BT United States

phenacite
use phenakite

phenakite
Also search phenacite.
UF phenacite
BA orthosilicates
silicates
BT minerals

phengite
BA mica group
sheet silicates
silicates
BT minerals
SA muscovite

phenocrysts
SA crystals
igneous rocks
matrix
minerals
porphyritic texture

phenols
Includes use on level 2 under organic materials(1).
BA organic materials

phenomena, electrical
use electrical phenomena

phenomena, surface
use surface phenomena

phi grade scale
use phi scale

phi scale
Includes use on level 3 under sedimentary rocks(1), sediments(1), and sedimentary petrology(1).
UF phi grade scale
BA statistical analysis
SA sedimentary rocks
sediments

Philadelphia
City on the Delaware River in SE Pennsylvania. Coextensive with Philadelphia County.
BT Pennsylvania
United States

Philip Island
In Indian Ocean E of Mornington Peninsula S of Melbourne.
BT Victoria
Australia

Philip Smith Mountains Quadrangle
Term introduced in 1981. NE Alaska.
CO N680000N690000
W1470000W1500000
BT Alaska
United States

Philippine Islands
Includes use on level 1 as an area term (list O). For term set options see list B.
CO N050000N190000
E1263000E1170000
UF Philippines
BA Far East
Asia
NT Cebu
Iloilo Basin
Luzon
Marinduque
Mindanao
Mindoro
Palawan
Panay Island
Samar
SA Mansalay Formation

Philippine Sea
Includes use on level 1 as an area term (list O). That part of the W Pacific Ocean with the Philippines Islands on the W, Taiwan and the Ryukyus on the NW, and the U.S. Trust Territory of the Pacific Islands on the E and SE.
CO N000000N350000
E1500000E1200000
BA West Pacific
Pacific Ocean
NT Daito Ridge
Kyushu-Palau Ridge
SA DSDP Site 292
DSDP Site 296
DSDP Site 297
Leg 31
Leg 58
Leg 59
Leg 6
Leg 60
Philippine Sea Plate

Philippine Sea Plate
A tectonic plate roughly covering the same area as the Philippine Sea and virtually surrounded by the SE edge of the Eurasian Plate and the Pacific Plate.
SA Philippine Sea
plate tectonics
plates

Philippines
use Philippine Islands

philippinite
BA tektites

Phillips County
Index states as applicable.
BX Arkansas
Colorado
Kansas
Montana
BT United States

phillipsite
BA zeolite group
framework silicates
silicates
BT minerals

philosophy
Use term under discipline. Includes use as a level 2 or 3 term appropriate to a large number of topics, e.g. on level 2 under geology(1). See list G.

Phlegraean Fields
Volcanic region W of Naples.
BT Campania
Italy

phlogopite
UF amber mica
brown mica
BA mica group
sheet silicates
silicates
BT minerals
SA fluor-phlogopite

Phobos
SA Deimos
Mars
satellites

Phoenix
City in Maricopa County in S central.
BT Arizona
United States

Pholadomyida
Term introduced in 1981. Includes use on level 2 under Mollusca(1). See list F.
BA Bivalvia
Mollusca
BT Invertebrata

Pholidota
Includes use on level 2 under Mammalia(1). See list F.
BA Mammalia
BT Tetrapoda
Vertebrata

phonolite
Includes use on level 3. As of 1981, not to be used in the igneous rocks set. See list H. Before 1981, included use on level 3 under igneous rocks(1) trachyte-phonolite family(2).
BA phonolites
volcanic rocks
BT igneous rocks

phonolites
Term introduced in 1981. Includes use on level 2 under igneous rocks(1). See list H. Before 1981, search trachyte-phonolite family.
BA volcanic rocks
BT igneous rocks
NA phonolite
tinguaite

Phoronida
Term introduced in 1981. Includes use on level 2 under worms(1). See list F.
BA worms
BT Invertebrata

phosgenite
BZ carbonates
chlorides
BT minerals

phosphate
As of 1981, no longer a valid term

for GeoRef. Use phosphate deposits for the commodity, phosphates for the mineral group, or phosphate rocks. Before 1981, included use on level 1 as a commodity term. See also phosphate composition.

phosphate composition
Term introduced in 1978. Included use on level 2 under nodules(1) until 1981. Before 1978, also search nodules AND phosphate.
SA composition

phosphate deposits
Term introduced in 1981. Includes use on level 1 and 2 as a commodity term (list C). Before 1981, search phosphate AND deposits. Also search apatite ores.
UF apatite ores

phosphate rocks
Refers to any rock containing phosphate. Includes use on level 3 under sedimentary rocks(1) chemically precipitated rocks(2). See list I. Before 1981, also search phosphorite.
BA chemically precipitated rocks
BT sedimentary rocks
NA bone beds

phosphates
Includes use on level 2 under minerals(1). See list L.
BT minerals
NA amblygonite
apatite
autunite
barbosalite
beraunite
NZ britholite
NA brushite
cacoxenite
carbonate apatite
chlorapatite
crandallite
NZ dahllite
NA eosphorite
florencite
fluorapatite
francolite
graftonite
herderite
NZ hinsdalite
NA hureaulite
hydroxylapatite
lithiophilite
NZ minyulite
NA monazite
monetite
montebrasite
newberyite
plumbogummite
NZ pyromorphite
NA rhabdophane
rockbridgeite
sarcopside
scholzite
strengite
struvite
triphylite
NZ triplite
NA turquoise
variscite
vivianite
wavellite
whitlockite
xenotime

phosphatization
SA diagenesis

phosphides
As of 1981, includes use on level 2 under minerals(1). See list L. Before 1981, included use on level 3 under minerals(1) native elements and alloys(2).
BA alloys
BT minerals
NZ schreibersite

Phosphoria Formation
Includes Sybille tongue (new), Forelle tongue, and Ervay tongue (new).
BT Permian
Paleozoic
SA Idaho
Montana
Utah
Wyoming

phosphorite
As of 1981, no longer a valid term for GeoRef. Use phosphate rocks. Before 1981, included use on level 3 under sedimentary rocks(1) chemically precipitated rocks(2).

phosphorus
Includes use on level 1 and 2 as a chemical element (list D); on level 3 under soils(1). Also search P.
UF P
SA elements
fertilizers
KREEP
soils

photogeologic maps
Term introduced in 1978. Before 1978, also search maps AND photogeologic.
BA maps

photogeologic methods
Includes use on level 2 under mineral exploration(1).
SA methods
mineral exploration
photogeology
photography
remote sensing

photogeology
Includes use on level 3 under engineering geology(1) and maps(1). See also mineral exploration(1) photogeologic methods(2) and photogrammetry(3) under engineering geology(1).
SA aerial photography
cartography
engineering geology
exploration
maps
photogeologic methods
photogrammetry
photography

photogrammetry
Includes use on level 3 under engineering geology(1) methods(2). See also photogrammetric studies.
SA aerial photography
cartography
definition
engineering geology
photogeology
photography
remote sensing

photographs
A valid term through 1978. After 1978, use photography.

photography
Includes use on level 2 under micropaleontology(1). Before 1978, also search photographs.
SA aerial photography
photogeologic methods
photogeology
photogrammetry

photometric
A valid term through 1977. After 1977, see photometry.

photometry
Includes use on level 3 under chemical analysis(1).
SA chemical analysis
flame photometry
spectroscopy

photosynthesis
Includes use on level 3 under geochemistry(1) processes(2).
SA geochemistry
paleobotany
processes
respiration

phreatic zone
use saturated zone

Phuket Group
May be the equivalent of the Megui Series, at least in part, of Burma. W side of peninsular Thailand, including island of Phuket, and lower Burma.
UF Phuket Series
BT Cambrian
SA Asia
Burma
Thailand

Phuket Series
use Phuket Group

phyllite
As of 1981, no longer a valid term for GeoRef. Use phyllites. Before 1981, included use on level 3 under metamorphic rocks(1) phyllites(2).

phyllites
Includes use on level 2 under metamorphic rocks(1). See list J. Before 1981, also search phyllite.
BT metamorphic rocks

Phylloceratida
Term introduced in 1981. Includes use on level 2 under Mollusca(1). See list F.
BA Ammonoidea
Tetrabranchiata
Cephalopoda
Mollusca
BT Invertebrata

phyllonite
As of 1981, no longer a valid term for GeoRef. Use phyllonites. Before 1981, included use on level 3 under metamorphic rocks(1) phyllonites(2).

phyllonites
Includes use on level 2 under metamorphic rocks(1). Before 1981, also search phyllonite.
BT metamorphic rocks
SA phyllonitization

phyllonitization
Term introduced in 1978.
BA metamorphism
SA mylonitization
phyllonites
processes
recrystallization

phyllosilicates
use sheet silicates

phylogeny
Includes use on level 3 under fossil group(1) and under paleontology(1). See list F.
SA ontogeny
paleontology

physical
A valid term through 1977. After 1977, use physical properties.

physical geography
use geography

physical geology
Includes use on level 3.
SA geology
historical geology

physical methods
Includes use on level 3 under soils(1) analysis(2).
SA methods
soils

physical models
BA models

physical properties
Includes use on level 3 under igneous rocks(1) properties(2) and under seismology(1). See also appropriate entries under material name. See under Earth(1), crust(1), core(1), Mohorovicic discontinuity(1), and mantle(1).
SA acoustical properties
bearing capacity
compressibility
dilatancy
friction
hardness
mechanical properties
microhardness
petrophysics
pH
porosity
properties
soils
specific heat
thermal properties

physical remanent magnetization
Term introduced in 1981.
BA remanent magnetization

physical weathering
Term introduced in 1978. Before 1978, search weathering AND physical.
BT weathering
SA mechanical weathering

physiographic provinces
UF provinces, physiographic
SA geomorphology
topography

physiography
A valid term through 1975. After 1975, see geomorphology.

physiology
Includes use under appropriate fossil term (list F).
SA morphology

phytogeography
use biogeography

phytoplankton
BT plankton
SA algae
diatoms

Piacenza
City and province in NW.
BT Emilia-Romagna
Italy

Piatra Neamt
use Piatra-Neamt

Piatra-Neamt
City in Neamt County in N.
UF Piatra Neamt
BT Moldavia
Romania

Piaui
State in NE.
CO S105000S024500
W0402500W0460000
BT Brazil
SA Parnaiba Basin

Piave Valley
River valley in N and NE Veneto. Also search Piave River.
BT Veneto
Italy

Picacho
Term introduced in 1981. Town in Pinal County, S central Arizona.

CO N324200N324200
W1113000W1113000
BT Pinal County
Arizona
United States

Picardy
Historical region in extreme N France. Abuts on English Channel between Calais in N and Treport on S, and extends E along Somme river and upper Oise River to Belgian border.
BT France

Picea
Term introduced in 1981. Genus.
BA Coniferales
gymnosperms
BT Spermatophyta
Plantae

Piceance Basin
use Piceance Creek basin

Piceance Creek basin
Rio Blanco County in NW Colorado. Also search Piceance Basin.
CO N393000N401500
W1075500W1083000
UF Piceance Basin
BT Colorado
United States

Picher
City in Ottawa County in extreme NE.
BT Oklahoma
United States

Pickens County
Term introduced in 1981. Index states as applicable.
BX Alabama
Georgia
South Carolina
BT United States
SA Lake Jocassee

pickeringite
BA sulfates
BT minerals

Pico Formation
Consists of three members: lower Pico marine, middle Pliocene; upper Pico marine, upper Pliocene; and continental upper Pliocene Sunshine Ranch member.
BT Pliocene
Neogene
Tertiary
SA California

picotite
use chrome spinel

picrite
Includes use on level 3 under igneous rocks(1). See list H. Before 1981, included use on level 3 under igneous rocks(1) basalt family(2) and ultramafic family(2).
BT igneous rocks
SA basalts
ultramafics

picrite porphyry
Includes use on level 3 under igneous rocks(1). See list H. Before 1981, included use on level 3 under igneous rocks(1) basalt family(2).
BT igneous rocks
SA basalts
porphyry
ultramafics

Pictou County
Term introduced in 1981. N Nova Scotia on Northumberland Strait.
CO N451700N455000
W0620500W0631000
BT Nova Scotia
Canada

Picuris Mountains
use Picuris Range

Picuris Range
N New Mexico.
UF Picuris Mountains
BT New Mexico
United States

Piedmont
An upland belt in the U. S. lying E of the Blue Ridge and Appalachian mountains and W of the coastal plain. For region in Italy, use Piemonte. Before 1981, search Piedmont AND United States.
BT United States
SA Alabama
Georgia
Maryland
New Jersey
New York
North Carolina
Pennsylvania
South Carolina
Virginia

Piedmont Alps
Those alpine ranges such as the Maritime Alps, Cottian Alps, Graian Alps, Pennine Alps, and Lepontine Alps, which lie within or on the borders of Piemonte.
BT Italy
SA Alps
Cottian Alps
Lepontine Alps
Maritime Alps
Pennine Alps
Piemonte

piedmontite
use piemontite

piedmonts
Term introduced in 1981. Generic term.
SA geomorphology

Piemonte
Term introduced in 1981. Region in NW Italy. Before 1981, search Piedmont AND Italy.
CO N440500N462900
E0091100E0063800
BT Italy
SA Acceglio
Acqui
Alessandria
Argentera
Argentera Massif
Asti
Biella
Canavese Zone
Ivrea
Ivrea Zone
Ivrea-Verbano Zone
Langhe
Lanzo Massif
Novara
Piedmont Alps
Po River
Po Valley
Sesia Valley
Sesia Zone
Sesia-Lanzo Zone
Simplon region
Strona Valley
Turin
Vercelli
Villafranca d'Asti

piemontite
Also search piedmontite.
UF piedmontite
BA epidote group
sorosilicates
orthosilicates
silicates
BT minerals

Pieniny Klippen Belt
In the Pieniny Mountains. Also search Pieniny.
BT Krakow
Poland
SA Pieniny Mountains

Pieniny Mountains
Range of the Beskids in S part of the country. Also search Pieniny.
UF Pieniny Range
BT Krakow
Poland
SA Pieniny Klippen Belt

Pieniny Range
use Pieniny Mountains

Pierce County
Term introduced in 1981. Index states as applicable.
BX Georgia
Nebraska
North Dakota
Washington
Wisconsin
BT United States

piercement
use diapirs

piercing fold
use diapirs

Pierre
City in Hughes County in central.
BT South Dakota
United States

Pierre Formation
use Pierre Shale

Pierre Shale
In Montana Group. E Colorado, W Minnesota, E Montana, Nebraska, North Dakota, South Dakota, and E Wyoming. Also search Pierre Formation.
UF Pierre Formation
BT Upper Cretaceous
Cretaceous
SA Colorado
Minnesota
Montana
Montana Group
Nebraska
North Dakota
South Dakota
Wyoming

piers
BT marine installations

piezometric surface
use potentiometric surface

piezoremanent magnetization
Term introduced in 1981.
BA magnetization

pigeonite
BA clinopyroxene
pyroxene group
chain silicates
silicates
BT minerals

pigments
BA organic materials
SA carotenoids

Pike County
Index states as applicable.
BX Alabama
Arkansas
Georgia
Illinois
Indiana
Kentucky
Mississippi
Missouri
Ohio
Pennsylvania
BT United States

Pikes Peak
Mountain W of Colorado Springs in El Paso County in central Colorado. Before 1978, also included use for Pikes Peak Batholith.
BT Colorado
United States
SA Pikes Peak Batholith

Pikes Peak Batholith
Term introduced in 1978. Before 1978, also search Pikes Peak for the batholith.
BT Colorado
United States
SA Pikes Peak

Pilbara gold field
NW Western Australia. Also search Pilbara.
UF Pilbara Goldfield
BT Western Australia
Australia

Pilbara Goldfield
use Pilbara gold field

piles
As of 1978, term is used on level 2 under foundations(1).
SA foundations

Pilgrim Formation
Threefold subdivision. Underlies Maywood or Red Lion Formation.
UF Pilgrim Limestone
BT Upper Cambrian
Cambrian
SA Montana

Pilgrim Limestone
use Pilgrim Formation

Pilica
Village in N.
BT Krakow
Poland

Pilica River basin
Term introduced in 1981. E central Poland.
CO N502000N515200
E0212000E0193500
BT Poland

pillars
Includes use as a general term.

pillow lava
Includes use on level 3 under lava(1).
SA basalt
lava
pillow structure
spilite

pillow structure
Includes use on level 3 under lava(1). Included use under sedimentary structures(1) soft sediment deformation(2) until 1976; now use ball-and-pillow.
SA ball-and-pillow
lava
pillow lava

pilot plants
Term introduced in 1981.
SA economics

Pilsen Basin
Valley in W in which city of Pilsen is located. Also search Pilsen.
UF Plzen Basin
BT Bohemia
Czechoslovakia

Piltdown
Locality in East Sussex in SE.
BT England
Great Britain
United Kingdom

Pima County
On the Mexican border in S.
BT Arizona

United States

Pinaceae
Family.
BA Coniferales
gymnosperms
BT Spermatophyta
Plantae
NA Pinus
SA Taxodiaceae

Pinal County
S central Arizona.
BT Arizona
United States
NT Picacho

Pinar del Rio
Westernmost province. Also a city.
BT Cuba
West Indies
NT Sierra de los Organos

Pinchi Lake
Central British Columbia.
BT British Columbia
Canada

Pindus Mountains
W central and NW Greece. Index administrative regions as applicable. Also search Pindus.
BT Greece
SA Epirus
Hellenides
Thessaly

Pine Barrens
Region of coastal plain of S and SE.
BT New Jersey
United States

Pine Valley Mountains
Washington County in SW.
BT Utah
United States

Pinedale Glaciation
Term introduced in 1981. Before 1981, also search Pinedale.
BT upper Quaternary
Quaternary
SA Colorado
Montana
Wyoming

Pinellas County
On Gulf of Mexico W of Tampa.
BT Florida
United States

pingos
Includes use on level 3 under glacial geology(1) periglacial features(2).
SA glacial geology
periglacial features
permafrost

pinnacle reefs
Includes use on level 3 under reefs(1).
UF coral pinnacle
reef pinnacle
BT reefs

Pinnacles National Monument
San Benito County in W.
BT California
United States

Pinnipedia
Term introduced in 1981. Includes use on level 2 under Mammalia(1). See list F.
BA Carnivora
Mammalia
BT Tetrapoda
Vertebrata

Pinon Range
Also search Pinon.
BT Nevada
United States

pinquite
use nontronite

Pinus
Genus. Includes use on level 3 under gymnosperms(1) Coniferales(2).
BA Pinaceae
Coniferales
gymnosperms
BT Spermatophyta
Plantae

Pinyon Conglomerate
Unconformably overlies Harebell Formation and unconformably underlies Colter Formation. Yellowstone National Park in NW Wyoming.
BT Paleocene
SA Wyoming

Pinzgau
Term introduced in 1981. Valley of the upper Salzach River in state of Salzburg. W central Austria.
CO N471500N472000
E0130400E0121000
BT Salzburg State
Austria
SA Salzach River

Pioneer Mountains
Custer and Blaine Counties in S central.
BT Idaho
United States

Pioneer Program
Term introduced in 1981. Includes use on level 3 under Jupiter(1).
SA extraterrestrial geology
Jupiter
planetology
remote sensing

Pioneer 10
SA Jupiter
remote sensing

pipelines
As of 1978, term is used on level 2 under marine installations(1).
SA marine installations
natural gas
petroleum
transportation

pipes
Includes use on level 2 under intrusions(1).
UF tuffsite
BA intrusions
NT breccia pipes
SA diatremes
igneous rocks
plugs
volcanoes

Pirabas Formation
BT Miocene
Neogene
Tertiary
SA Brazil

piracy
use stream capture

Pisa
City and province in NW.
BT Tuscany
Italy

Pisces
Includes use on level 1 and 2 as a fossil term (list F).
BT Vertebrata
NA Chondrichthyes
Osteichthyes
Placodermi
SA Agnatha
fish
gastroliths
otoliths

Pisgah Crater
BT California
United States

Pishpek
use Frunze

pisolites
SA limestone
pisoliths
pisolitic texture
sedimentary rocks

pisoliths
SA pellets
pisolites
pisolitic texture
sedimentary rocks
sediments

pisolitic texture
Term introduced in 1978. Includes use on level 3 under sediments(1) and sedimentary rocks(1). Before 1978, also search pisolitic AND specific sediment type or sedimentary rocks.
SA pisolites
pisoliths
sedimentary rocks
sediments
textures

pistacite
use epidote

pit sections
Term introduced in 1981.
SA mineral exploration
sections
soils

pitchblende
UF nasturan
BA oxides
BT minerals
SA uraninite

pitching folds
use plunging folds

pitchstone
As of 1981, includes use on level 3 under igneous rocks(1) glasses(2). See list H. Before 1981, included use on level 3 under igneous rocks(1) pyroclastics and glasses(2).
BA glasses
volcanic rocks
BT igneous rocks
SA volcanic glass

Pitesti
City in Arges County in central.
BT Walachia
Romania

Pithecanthropus
Genus. As of 1981, includes use on level 3 under Mammalia(1) Hominidae(2). See list F. Before 1981, included use on level 3 under Mammalia(1) Primates(2).
BA Hominidae
Primates
Mammalia
BT Tetrapoda
Vertebrata

Pitkin County
W central Colorado.
BT Colorado
United States

Pitkin Limestone
In Arkansas, underlies Cane Hill Member of Hale Formation. N Arkansas, and E Oklahoma.
BT Upper Mississippian
Mississippian
SA Arkansas
Oklahoma

Piton de la Fournaise
Volcanic peak on Reunion Island.
BT Reunion

Piton des Neiges
Highest peak on Reunion Island.
BT Reunion

Pittsburgh
City in Allegheny County in SW.
BT Pennsylvania
United States

Pittsburgh coal basin
Bituminous coal and steel producing region in the Pittsburgh area.
BT Pennsylvania
United States

Piute County
SW central.
BT Utah
United States

PKiP-waves
Term introduced in 1981.
BA body waves
elastic waves
SA P-waves
seismology
waves

PKP-waves
Term introduced in 1981.
BA body waves
elastic waves
SA P-waves
seismology
waves

PKS-waves
Term introduced in 1981.
BA body waves
elastic waves
SA P-waves
S-waves
seismology
waves

Placer County
W of Lake Tahoe in E.
BT California
United States

placers
Includes use on level 1 (list A). Used for a surficial mineral deposit formed by mechanical concentration of mineral particles from weathered debris. See also list C for type of commodity.
Term set options are:
kind of placer [e.g. diamonds, gold ores, heavy mineral deposits, platinum ores, tin ores; first level terms only]
area
topic [detection, distribution, exploration, genesis, identification, patterns, sampling]
subtopic
UF ore of sedimentation
SA diamonds
discoveries
gold ores
heavy mineral deposits
metal ores
mineral deposits
mineral exploration
monazite deposits
platinum ores
sampling
tin ores
weathering

Placodermi
Includes use on level 2 under Pisces(1). See list F.
BA Pisces
BT Vertebrata
NA Arthrodira

Placodontia
Term introduced in 1981. Includes use on level 2 under Reptilia(1). See list F.
BA Synapsida
Reptilia
BT Tetrapoda
Vertebrata

plagioclase
As of 1981, includes use in combination with framework silicates (i.e. framework silicates, plagioclase) on level 2 under minerals(1). See list L. Before 1981, included use on level 3 under minerals(1) framework silicates, feldspar group(2).
BA feldspar group
framework silicates
silicates
BT minerals
NA albite
andesine
anorthite
bytownite
labradorite
myrmekite
oligoclase
SA maskelynite

plagiogranite
As of 1981, includes use on level 3 under igneous rocks(1) diorites(2). See list H. Before 1981, included use on level 3 under igneous rocks(1) plutonic rocks(2).
BA diorites
BT igneous rocks
SA quartz diorite
trondhjemite

plagionite
BA sulfantimonites
sulfosalts
BT minerals

plains
SA abyssal plains
alluvial plains
coastal plains
floodplains
geomorphology
landforms
outwash plains
plateaus
playas
steppes
topography
tundra

Plaisancian
Europe. Above Pontian (Miocene), below Astian. Includes use on level 3 under age terms(1). See list E.
UF Plaisanzian
BA lower Pliocene
Pliocene
Neogene
Tertiary
BT Cenozoic

Plaisanzian
use Plaisancian

Plan de la Tour
Village in SE part of country.
BT Var
France

Plana
Town in W.
BT Bohemia
Czechoslovakia

planar bedding structures
Includes use on level 2 under sedimentary structures(1). See list K.
UF bedding structures, planar
BA sedimentary structures
NA bedding
cross-bedding
cross-laminations
cross-stratification
cut and fill
cyclothems
flaser bedding
imbrication
laminations
massive bedding
rhythmic bedding
ripple drift-cross laminations
sand bodies
stratification
varves
SA bars
buried channels
channels
graded bedding

planation
SA abrasion
erosion
erosion surfaces

planation surfaces
use erosion surfaces

plancheite
BA chain silicates
silicates
BT minerals

planes, fault
use fault planes

planetary interiors
Term introduced in 1981. Includes use on level 2 under individual planets. Before 1981, search planet name AND interior.
SA interior
lunar interior
planetology
planets

planetology
For studies of more than one planet, the relationship between planets, or the solar system as a whole. See also names of planets. Includes use on level 1 (list A). Term set options are: topic [atmosphere, bibliography, concepts, cosmic dust, cosmic rays, cosmochemistry, methods, nomenclature, principles, techniques, theoretical studies]
subtopic
SA accretion
Apollo Program
Apollo-Soyuz Program
asteroids
astronomy
astrophysics and solar physics
atmosphere
condensation
cosmic dust
cosmochemistry
Earth
exobiology
Explorer Program
extraterrestrial geology
impacts
interplanetary comparison
interplanetary space
Jupiter
Luna Program
Mariner Program
Mars
Mercury Planet
meteorites
meteors
Moon
Neptune
outer planets
Pioneer Program
planetary interiors
planets
Pluto
Saturn
Skylab
solar system
Sun
surface features
Surveyor Program
tektites
terrestrial planets
Uranus
Venus
Viking Program

planets
Includes use on level 3 under planetology(1).
SA asteroids
condensation
Earth
extraterrestrial geology
interplanetary comparison
Jupiter
Mars
Mercury Planet
meteoroids
Moon
Neptune
outer planets
planetary interiors
planetology
Pluto
Saturn
solar system
Sun
surface features
terrestrial planets
Uranus
Venus

plankton
Includes use on level 3 under various fossil groups(1). See list F. Also search microplankton.
UF microplankton
NT phytoplankton
zooplankton
SA algae
diatoms
Dinoflagellata
foraminifera
Ostracoda
palynomorphs
planktonic taxa
Protista
Radiolaria

planktonic
A valid term through 1977. After 1977, use planktonic taxa.

planktonic taxa
Term introduced in 1978. Includes use on level 3 under fossil group(1). Before 1978, also search planktonic.
UF taxa, planktonic
SA plankton

planning
As of 1978 term is used on level 2 under highways(1) and land use(1).
SA associations
highways
land use
management
programs
regional planning
urban planning

Planolites
Genus.
BA ichnofossils

Planorbis
Includes use on level 3 under Mollusca(1). See list F.
BA Gastropoda
Mollusca
BT Invertebrata

Planosols
Includes use on level 3 under soils(1). See list M.
BA soils
SA Intrazonal soils
soil group

Plantae
Includes use on level 1 and 2 as a fossil term (list F). In indexing, used in a general sense when specific plants are unknown.
NT bryophytes
pteridophytes
Pteropsida
Spermatophyta
thallophytes
SA floral studies
fossil wood
leaves
paleobotany
palynomorphs
vegetation

plasma
Not to be used for soil plasma.
SA interplanetary space

plasma instabilities
Includes use on level 2 under magnetosphere(1).
UF instabilities, plasma
SA corona
magnetic field
magnetosphere
plasma motion
plasmapause
solar wind

plasma motion
Includes use on level 2 under magnetosphere(1).
UF motion, plasma
SA magnetosphere
plasma instabilities
solar wind

plasmapause
Includes use on level 2 under magnetosphere(1).
SA magnetosphere
plasma instabilities

plasmasphere
Includes use on level 3 under magnetosphere(1) magnetic storms(2).
SA magnetic storms
magnetosphere

plaster of paris
use gypsum

plaster stone
use gypsum

plastic flow
Includes use on level 3 under deformation(1) field studies(2). Also search plastic deformation; plastic.
UF plastic strain
SA deformation
flow lines
plasticity

plastic materials
Term introduced in 1978. Before 1978, also search plastic AND materials.
SA elastic materials
engineering geology
materials
plasticity
rock mechanics
soil mechanics

plastic strain
use plastic flow

plasticity
As of 1978, term is used on level 2 under rock mechanics(1).
SA Atterberg limits
compressibility
deformation
elasticity
mechanical properties
plastic flow
plastic materials
rock mechanics
viscosity

plate boundaries
Term introduced in 1981. Before 1981, also search margins; plate margins; plate boundary. Includes use on level 3 under plate tectonics(1).
UF plate margins
SA plate tectonics
plates

plate collision
Term introduced in 1981. Before 1981, also search collision; collisions. Includes use on level 3 under plate tectonics(1).
SA plate tectonics
plates

plate convergence
Term introduced in 1981. Before 1981, search convergence. Includes use on level 3 under plate tectonics(1).
SA plate tectonics
plates

plate geometry
Term introduced in 1981. Before 1981, also search geometry AND plate tectonics. Includes use on level 2 under plate tectonics(1).
SA geometry
plate tectonics
plates

plate margins
use plate boundaries

plate tectonics
Includes use on level 1, as of 1976. Used for global tectonics based on an Earth model characterized by a small number of large, broad, thick plates. Before 1976, articles discussing plate tectonics were included under tectonophysics(1). See list B. Term set options are:
age
age term (List E)
topic [concepts, effects, evolution, indicators, island arcs, mechanism, movement, obduction, plate geometry, processes, rifting, subduction]
subtopic
SA African Plate
Antarctic Plate
Arabian Plate
aseismic margins
asthenosphere
Benioff zone
Caribbean Plate
Cocos Plate
continental drift
continental margin
continental type
convection
convection currents
crust
crustal shortening
deformation
Earth
earthquakes
Eurasian Plate
fault zones
faults
fracture zones
fracturing
Galapagos Rift
geodynamics
geosynclines
heat flow
hot spots
Indian Plate
indicators
intraplate tectonics
island arcs
lithosphere
mantle
marginal basins
marginal seas
mechanics
Median Tectonic Line
melange
metamorphic belts
microplates
Mohorovicic discontinuity
movement
Nazca Plate
North American Plate
obduction
ocean basins
oceanic crust
ophiolite
orogeny
Pacific Plate
paleomagnetism
partial melting
passive margins
Philippine Sea Plate
plate boundaries
plate collision
plate convergence
plate geometry
plates
plumes
polar wandering
rift zones
rifting
rotation
sea-floor spreading
seismology
seismotectonics
slabs
South American Plate
spherical harmonic analysis
spreading centers
subduction
subduction zones
taphrogeny
tectonics
tectonophysics
tectonosphere
transform faults
transition zones
trenches
triple junctions
volcanic belts
volcanology
West Siberian Plate

plate-bearing tests
Term introduced in 1981.
SA soil mechanics

plateaus
SA geomorphology
highlands
landforms
plains
topography

plates
Includes use on level 3 under plate tectonics(1).
SA African Plate
Antarctic Plate
Arabian Plate
Benioff zone
Caribbean Plate
Cocos Plate
crust
Eurasian Plate
Indian Plate
mantle
microplates
Nazca Plate
North American Plate
Pacific Plate
Philippine Sea Plate
plate boundaries
plate collision
plate convergence
plate geometry
plate tectonics
South American Plate
subduction zones
West Siberian Plate

platforms
Includes use on level 3 under engineering geology(1).
SA gravity platforms
marine platforms
tectonic platforms

platinum
Includes use on level 1 and 2 as a chemical element (list D). As of 1981, use platinum ores for platinum as a commodity. Before 1981, included use on level 1 as a commodity term.
UF Pt
SA elements
heavy metals
niggliite
platinum ores

platinum ores
Term introduced in 1981. Includes use on level 1 and 2 as a commodity term (list C). Before 1981, search platinum AND ore deposits.
SA placers
platinum

Platte County
Term introduced in 1981. Index states as applicable.
BX Missouri
Nebraska
Wyoming
BT United States

Platte River
Formed by confluence of the North Platte and South Platte rivers in SW central Nebraska and flows E into the Missouri River below Omaha.
BT Nebraska
United States

Platteville Formation
Includes Mifflin Limestone Member, Quimbys Mill Member, Pecatonica Member, McGregor Member, and Carimona Member, Hidden Falls Member, Magnolia Member, Glenwood Shale Member. E Iowa, NW Illinois, S Minnesota, and SW Wisconsin.
BT Middle Ordovician
Ordovician
SA Illinois
Iowa
Minnesota
Wisconsin

Plattsburg Limestone
In Lansing Group. Includes Merriam Limestone, Hickory Creek Shale and Spring Hill Limestone members. SW Iowa, E Kansas, NW Missouri, and SE Nebraska.
BT Missourian
Upper Pennsylvanian
Pennsylvanian
SA Iowa
Kansas
Lansing Group
Missouri
Nebraska

Plattsmouth Limestone
use Plattsmouth Limestone Member

Plattsmouth Limestone Member
Of Oread Limestone. SW Iowa, E Kansas, NW Missouri, and SE Nebraska.
UF Plattsmouth Limestone
BT Virgilian
Upper Pennsylvanian
Pennsylvanian
SA Iowa
Kansas
Missouri
Nebraska
Oread Limestone

Platycopida
Term introduced in 1981. Includes use on level 2 under Ostracoda(1). See list F.
BA Ostracoda
BT Crustacea
Mandibulata
Arthropoda
Invertebrata

Platyrrhina
Term introduced in 1981. Includes use on level 2 under Mammalia(1). See list F.
BA simians
Primates
Mammalia
BT Tetrapoda
Vertebrata

playas
SA basins
coastal environment
deserts
geomorphology
plains
sebkha environment

Pleasant Island
use Nauru Island

Pleasanton Group
Comprises Warrensburg Channel Sandstone, Sni Mills Limestone, Dawson Coal Horizon, Wayside Sandstone, Exline Limestone, Knobtown Sandstone, Hepler Sandstone, Checkerboard Limestone, and Ovid Coal. SW Iowa, E Kansas, NW Missouri, and NE Oklahoma.
BT Missourian
Upper Pennsylvanian
Pennsylvanian
SA Iowa
Kansas
Missouri
Oklahoma

Plecoptera
As of 1981, includes use on level 3 under Insecta(1). See list F. Before 1981, included use on level 2 under Insecta(1).
BA Insecta
BT Mandibulata
Arthropoda
Invertebrata

Plectoptera
As of 1981, includes use on level 3 under Insecta(1). See list F. Before 1981, included use on level 2 under Insecta(1).
BA Insecta
BT Mandibulata
Arthropoda
Invertebrata

Pleistocene
Glacial epoch. World. Above Pliocene (Tertiary), below Holocene. Includes use on level 1 as an age term (list E); on level 2 under paleoterms, e.g. paleoecology, paleogeography; on level 3 under age terms(1).
BA Quaternary
BT Cenozoic
NT Baltic Glaciation
Bishop Tuff
Crooked Creek Formation
NA Gunz
NT Key Largo Limestone
Kingsdown Formation
NA lower Pleistocene
NT McPherson Formation
Miami Limestone
NA middle Pleistocene
Mindel
Mindel/Riss Interglacial
NT Pearlette Volcanic Ash
Peoria Loess
NA Riss
Riss/Wurm Interglacial
NT Semata Formation
Shimosa Group
NA upper Pleistocene
NT Wedron Formation
Wicomico Formation
NA Woodfordian
SA Blancan
Brunhes Epoch
Loup Fork Group

Magdalenian
Mousterian
Osaka Group
Paleolithic
San Juan Formation
Santa Maria Formation
Siwalik System
Tulare Formation
Uonuma Group
Verde Formation

plenargyrite
use matildite

pleochroism
Includes use on level 3 under minerals(1) optical properties(2).
UF polychroism
SA crystallography
minerals
optical properties

Plesiosauria
As of 1981, includes use on level 2 under Reptilia(1). See list F. Before 1981, included use on level 3 under Reptilia(1) Euryapsida(2).
BA Sauropterygia
Euryapsida
Reptilia
BT Tetrapoda
Vertebrata

plessite
As of 1981, includes use on level 3 under minerals(1) alloys(2). See list L. Before 1981, included use on level 3 under minerals(1) native elements and alloys(2).
BA alloys
BT minerals
SA kamacite
meteorites
taenite

Plettenberg
City NE of Cologne in W.
BT North Rhine-Westphalia
West Germany

Pleven
Province in N Bulgaria. Also a city.
BT Bulgaria
NT Teteven

Pliensbachian
Europe. Includes use on level 3 under age terms(1). See list E.
BA middle Liassic
Lower Jurassic
Jurassic
BT Mesozoic

Pliocene
World. Above Miocene, below Pleistocene (Quaternary). Includes use on level 1 as an age term (list E); on level 2 under paleo- terms, e.g. paleoecology, paleogeography; on level 3 under age terms(1).
BA Neogene
Tertiary
BT Cenozoic
NT Ash Hollow Formation
Bakhtiari Formation
Bidahochi Formation
NA Cimmerian
NT Citronelle Formation
NA Dacian
Gauss Epoch
lower Pliocene
middle Pliocene
NT Narita Formation
Ogallala Formation
Pico Formation
Saint George Formation
San Diego Formation
NA upper Pliocene
NT Uvalde Gravel
Valentine Formation
SA Blancan
Columbia River Basalt
Cuddalore Series
Hemphillian
Loup Fork Group
Meotian
Neyveli Lignite
Osaka Group
Pannonian
Santa Maria Formation
Siwalik System
Snoqualmie Batholith
Tulare Formation
Uonuma Group
Verde Formation
Warkalli Formation

pliomagmatic zone
use eugeosynclines

Pliomys
As of 1981, includes use on level 3 under Mammalia(1) Myomorpha(2). See list F. Before 1981, included use on level 3 under Mammalia(1) Rodentia(2).
BA Cricetidae
Myomorpha
Rodentia
Mammalia
BT Tetrapoda
Vertebrata

Plock
City in central part of country.
UF Plokz
BT Warsaw
Poland

Ploesti
City N of Bucharest in Prahova County.
UF Ploiesti
BT Walachia
Romania

Ploiesti
use Ploesti

Plokz
use Plock

Plovdiv
Province in S central part of country. Also a city.
BT Bulgaria
NT Madan

plugs
Includes use on level 2 under intrusions(1).
BA intrusions
SA breccia pipes
lava
pipes
volcanoes

Plumas County
NE California.
BT California
United States

plumbogummite
BA phosphates
BT minerals

plumes
Includes use on level 3 under plate tectonics(1).
SA hot spots
plate tectonics

plunging folds
Term introduced in 1978. Folds defined on the basis of orientation in relation to the geographic horizontal plane. Includes use on level 3 under folds(1) orientation(2). Before 1978, also search folds AND plunging.
UF pitching folds
BA folds
SA orientation

Pluto
Includes use on level 1 and 2. See entry under Moon for term set options.
SA outer planets
planetology
planets
solar system

plutonic
A valid term through 1977. After 1977, use plutonic rocks on level 2 under igneous rocks(1).

plutonic rocks
Term introduced in 1978. Includes use on level 2 and under igneous rocks(1). Before 1978, also search igneous rocks AND plutonic. See list H.
BT igneous rocks
NA quartzolite
SA granodiorite
hypabyssal rocks
igneous processes
ijolite
mangerite
metaplutonic rocks
monzodiorite
peridotite
quartz syenite
rocks

plutonium
Includes use on level 1 and 2 as a chemical element (list D).
UF Pu
SA elements
Pu-239
Pu-244

plutons
Includes use on level 2 under intrusions(1).
BA intrusions
SA batholiths
emplacement
igneous rocks
metasomatism

Plzen Basin
use Pilsen Basin

Pm
use promethium

pneumatolysis
Term introduced in 1981. Also search pneumatolytic.
UF pneumatolytic alteration
SA contact metamorphism
gases
magmas
metasomatism
mineral deposits

pneumatolytic alteration
use pneumatolysis

Po
use polonium

Po Delta
On Adriatic Sea S of Venice.
UF Po River delta
BT Veneto
Italy

Po River
Rises in the Cottian Alps of W Piemonte and flows E into the Adriatic Sea. Index autonomous regions as applicable.
BT Italy
SA Emilia-Romagna
Lombardy
Piemonte
Veneto

Po River delta
use Po Delta

Po Valley
River valley. Index autonomous regions as applicable.
BT Italy
SA Emilia-Romagna
Lombardy
Piemonte
Veneto

Po-210
Includes use on level 3 under isotopes(1).
SA isotopes
polonium

Pobla de Segur
Town in NE part of country.
BT Lerida
Catalonia
Spain

Pocahontas County
Index states as applicable.
BX Iowa
West Virginia
BT United States

Pocahontas Formation
In Pottsville Group. SW Virginia and S West Virginia.
BT Lower Pennsylvanian
Pennsylvanian
SA Pottsville Group
Virginia
West Virginia

Pocatello Formation
Divided into two series: lower tillite series and upper varved slate series. SE Idaho.
BT Precambrian
SA Idaho

pocket calculators
use hand calculators

Pocono Formation
Includes Benezette Limestone, Patton Shale Member, Peters Mountain Sandstone, Second Mountain Member, Cove Mountain Member, Burgoon Sandstone Member. W Maryland, E Ohio, Pennsylvania, W Virginia, N West Virginia.
BT Lower Mississippian
Mississippian
SA Maryland
Pennsylvania
Virginia
West Virginia

Pocos de Caldas
Resort city in SW.
BT Minas Gerais
Brazil

Podhale
Highland basin in the Carpathians.
BT Krakow
Poland

Podocopida
As of 1981, includes use on level 3 under Ostracoda(1). See list F. Before 1981, included use on level 2 under Ostracoda(1).
BA Ostracoda
BT Crustacea
Mandibulata
Arthropoda
Invertebrata
NA Bairdiacea
Cavellinidae
Cyprididae
Cytheracea
Cytherellidae
Darwinula
Healdiidae
Loxoconcha

Podolia
Region between the Dniester and the Southern Bug rivers nearly coextensive with Khmelnitskiy Oblast in W.
BT Ukraine
USSR

Podsols
use Podzols

Podzols
Includes use on level 3 under soils(1). See list M.
UF Podsols
BA soils
SA Zonal soils

Pogonip Group
Comprises House Limestone, Fillmore Limestone, Wahwah Limestone, Juab Limestone, Kanosh Shale, Lehman Formation, Mazourka Formation, Goodwin Limestone, Ninemile Formation, Antelope Valley Limestone. SE California, E and S Nevada, and W Utah.
BT Ordovician
SA Antelope Valley Limestone
California
Lower Ordovician
Middle Ordovician
Nevada
Utah

Pogonophora
Term introduced in 1981. Includes use on level 2 under worms(1). See list F.
BA worms
BT Invertebrata

Pohang
City NNE of Pusan in SE South Korea. North Kyongsang.
BT South Korea

Poiana Rusca (Mountains)
use Poiana-Rusca Mountains

Poiana Rusca Massif
use Poiana-Rusca Mountains

Poiana-Rusca Massif
use Poiana-Rusca Mountains

Poiana-Rusca Mountains
Hunedoara County in SW Transylvania. Also search Poiana Rusca, Poiana Rusca Massif, and Poiana-Rusca Massif.
UF Poiana Rusca (Mountains)
Poiana Rusca Massif
Poiana-Rusca Massif
BT Transylvania
Romania

Point Arena
A promontory on the Pacific Ocean in Mendocino County in NW. Also a town.
BT California
United States

Point Barrow
Eskimo village, N Alaska, on Point Barrow headland.
UF Nuwuk
BT Alaska
United States

Point Loma
Peninsula sheltering San Diego Bay.
CO N324000N324400
W1171200W1171600
BT California
United States

Point Mugu
On the Pacific Ocean in Ventura County W of Los Angeles. Also site of Point Mugu Naval Missile center.
BT California
United States

Point Reyes
On the Pacific Ocean N of San Francisco in Marin County. Part of Point Reyes National Seashore.
BT California
United States

Point Sal
Term introduced in 1981. On central California coast.
CO N345200N345800
W1203700W1204400
BT Santa Barbara County
California
United States

Poisson's ratio
BA elastic constants
SA elasticity
engineering geology
mechanical properties
strain
stress
tunnels

Poitiers
City in W central.
BT Vienne
France

Poitou
Historical region of W central. Index departments as applicable.
BT France
SA Deux-Sevres
Vendee
Vienne

Pokutye
Upland region between E Beskids on the S and the Dniester River on the N in SW.
BT Ukraine
USSR

Poland
Includes use on level 1 as an area term (list O). For term set options see list B.
CO N490000N544500
E0241500E0141500
BA Central Europe
Europe
NT Bialowieza
Bialystok
Borzeta
Boza Wola
Bydgoszcz
Central Polish Glaciation
Foresudetic Monocline
Galezice
Gdansk
Grzybow
Kampinos Forest
Katowice
Kielce
Koszalin
Krakow
Krakow-Czestochowa Jura
Kujawy
Lodz
Lower Lusatia
Lower Silesia
Lublin
Northeastern Polish Plain
Northwestern Polish Plain
Olsztyn
Opole
Pilica River basin
NX Polish Carpathians
NT Polish Lowland
NX Polish Sudeten Mountains
NT Poznan
Rzeszow
Silesian coal basin
Southeastern Polish Hills
Sowie Mountains
Swiety Krzyz Mountains
Szczecin
Upper Silesia
Vistula River
Vistula River valley
Warsaw
Warta
Wroclaw
Zielona Gora
SA Beshchady Mountains
Brest Basin
Bug region
Bug River
Carpathian Foredeep
Carpathian Foreland
Carpathians
European Platform
Galitsiya
Isergebirge
Izera Mountains
Karkonosze Mountains
Krosno Beds
Lusatia
North Sudetic Basin
Oder Valley
Pomerania
Poznan Clays
Roztocze
Silesia
Snieznik
South Polish Glaciation
Spis
Subcarpathians
Sudeten Mountains
Sudetic Basin
Tatra Mountains
Western Carpathians

Polar Cap
As of 1981, refers only to Mars. Before 1981, search Polar Cap AND Mars. Before 1981, included use for the ice sheet centered at the Earth's South Pole.
BT Mars

Polar Continental Shelf
The continental shelf surrounding the Antarctic Continent. (Might apply also to continental shelves of Europe, Asia, and North America facing the Arctic Ocean and lying within the Arctic Circle).
BT Antarctica
SA continental shelf

polar migration
use polar wandering

polar regions
Term introduced in 1978. Includes use on level 3 under ionosphere(1) electrojet(2). Before 1978, also search polar.
SA electrojet
ionosphere
magnetic storms

Polar regions
Region surrounding both geographic poles. Also search Polar; Polar region. Index poles as applicable.
NA Antarctica
Arctic region
SA North Pole
South Pole

Polar Urals
Section of the Urals extending from approximately N65° to the Arctic Ocean, including Vaygach Island.
CO N650000N710000
E0680000E0573000
BX Russian Republic
BZ Urals
BT USSR
SA Northern Urals

polar wandering
As of 1978, term is used on level 2 under paleomagnetism(1).
UF polar migration
wandering, polar
SA continental drift
paleolatitude
paleomagnetism
plate tectonics
pole positions
rotation

polarization
A general term used mostly for optical properties. Not used for magnetization.
SA birefringence
induced polarization
optical properties

polarographic analysis
use polarography

polarography
Includes use on level 3 under chemical element(1) analysis(2). See list D. For methods, see under chemical analysis(1). For data, see appropriate material.
UF polarographic analysis
SA analysis
chemical analysis
diffusion
electrolysis
quantitative analysis

pole positions
Includes use on level 2 under paleomagnetism(1).
UF positions, pole
SA magnetic field
paleomagnetism
polar wandering
reversals

Polesie
use Polesye

Polesye
Lowland area, formerly in Poland, comprising the Pripet Marshes of the Pripet Basin. Also search Polesie. Index Soviet republics as applicable.
CO N513000N530000
E0320000E0250000
UF Polesie
BT USSR
SA Byelorussia
Pripet Basin
Ukraine

polianite
use pyrolusite

policy
Used as a general term. Also search policies.
SA decision-making
international cooperation
legislation
programs

Polish Carpathians
Term introduced in 1981.
CO N490000N502200
E0230200E0183200
BZ Carpathians
BX Poland

Polish Lowland
Eastern extension of the North German Plain in NE and N part of country. Also search Polish Lowlands.
CO N513000N544500
E0200000E0141500
BT Poland
Europe

Polish Sudeten Mountains
Term introduced in 1981.
CO N500500N511800
E0171000E0145000
BZ Sudeten Mountains
BX Poland

polished sections
SA microscope methods
sections
thin sections

polished surface
use slickensides

Polk County
Index states as applicable.
BX Arkansas
Florida
Georgia
Iowa
Minnesota
Missouri
Nebraska

North Carolina
Oregon
Tennessee
Texas
Wisconsin
BT United States

Polkowice
Town in Lower Silesia in SW part of country.
BT Zielona Gora
Poland

pollen
SA exine
microfossils
microorganisms
palynology
palynomorphs
pollen analysis
pollen diagrams
spores
sporopollenin

pollen analysis
Includes use on level 3 under palynology(1).
SA analysis
palynology
pollen
pollen diagrams

pollen diagrams
Includes use on level 3 under palynology(1).
SA palynology
pollen
pollen analysis

pollucite
BA framework silicates
silicates
BT minerals

pollutants
As of 1978, term is used on level 2 under pollution(1).
SA effluents
pollution
waste disposal

polluted water
Term introduced in 1981.
SA pollution
water

pollution
A level 1 term as of 1978. Used for geological studies on pollution of the environment. Includes use on level 2 under impact statements(1) and nuclear facilities(1). If 1, term set options are:
air
subtopic
case studies
subtopic
causes
subtopic
concepts
subtopic
controls
subtopic
detection
subtopic
effects
subtopic
experimental studies
subtopic
field studies
subtopic
ground water
subtopic
human ecology
subtopic
impact statements
subtopic
metals
subtopic
oil spills
subtopic
pollutants
subtopic
surface water
subtopic
waste disposal
subtopic
water
subtopic
SA acid mine drainage
air
atmosphere
case studies
conservation
contamination
effluents
environmental geology
fallout
geochemistry
geologic hazards
ground water
human ecology
hydrology
impact statements
impurities
industrial waste
land use
medical geology
metals
oil spills
pollutants
polluted water
purification
radioactive waste
radioactivity
reclamation
sewage
sulfur dioxide
sulfuric acid
surface water
thermal pollution
toxic materials
trace metals
waste disposal
water
water supply

polonium
Includes use on level 1 and 2 as a chemical element (list D).
UF Po
SA elements
Po-210

Polousnyy
A ridge in N Yakutia W of the Indigirka River.
BT Russian Republic
USSR

Poltava
City and Oblast in NE central.
UF Pultova
Pultowa
BT Ukraine
USSR

Poltava Series
In Donets Ridge, and W and NW Ukraine.
BT Tertiary
SA Miocene
Oligocene
Russian Republic
Ukraine
USSR

polybasite
BA sulfantimonites
sulfosalts
BT minerals

Polychaetia
As of 1981, includes use on level 2 under worms(1). See list F. Before 1981, included use on level 2 under Annelida(1).
BA worms
BT Invertebrata
NA Serpulidae
SA Chaetopoda

polychroism
use pleochroism

polydymite
BA sulfides
BT minerals
SA linnaeite

Polygnathus
Genus.
BA Conodonta

polygonal fractures
Term introduced in 1977. Includes use on level 3 under fractures(1) style(2). Before 1978, also search fractures AND polygonal.
BA fractures

polygons
Includes use on level 3 under glacial geology(1) periglacial features(2).
SA frost action
patterned ground
periglacial features

polyhalite
BA sulfates
BT minerals

polyhedra
NT tetrahedra
SA crystal structure

polylithionite
BA mica group
sheet silicates
silicates
BT minerals

polymerization
SA geochemistry
organic materials

polymetallic ores
Includes use on level 1 and 2 as a commodity term (list C).
UF ores, polymetallic
SA metal ores

polymetamorphism
Includes use on level 2 under metamorphism(1).
UF superimposed metamorphism
BA metamorphism
SA prograde metamorphism
retrograde metamorphism

polymorphism
Includes use on level 3 under fossil sets (list F) or under crystal structure(1) or minerals(1). Also search polymorphs. Before 1981, also search dimorphism.
UF polymorphs
SA crystal structure
minerals
polytypism

polymorphs
use polymorphism

Polynesia
Collective name for islands of the central and SE Pacific Ocean. Index island and island groups as applicable. Includes use on level 1 or 2 as an area term (list O). If 1, see term set options under list B.
CO S280000N290000
W1310000W1790000
BT Oceania
NT Cook Islands
French Polynesia
NZ Hawaii
NT Line Islands
NA Samoa
NT Tokelau
NA Tonga
NT Tuvalu
SA Pacific Ocean

polyphase processes
Also search polyphase.
SA processes

Polyplacophora
Includes use on level 2 under Mollusca(1). See list F.
BA Mollusca
BT Invertebrata

Polyprotodontia
Term introduced in 1981. Includes use on level 2 under Mammalia(1). See list F.
BA Mammalia
BT Tetrapoda
Vertebrata

polytypes
use polytypism

polytypism
Includes use on level 3 under crystal structure(1) or minerals(1). Also search polytypes.
UF polytypes
SA crystal structure
minerals
polymorphism

Polyzoa
use Bryozoa

Pomerania
Historical region on Baltic Sea extending from Stralsund, W of the Oder River, to the Vistula River on the E. Now primarily in Poland. Index countries as applicable.
BT Europe
SA East Germany
Poland

Ponce
City in S on Caribbean Sea.
BT Puerto Rico
West Indies

Pondicherry
A centrally administered territory known as an Union Territory composed of the four scattered former French settlements of Karikal, Pondicherry, and Yanaon on the Bay of Bengal and Mahe on the Indian Ocean. The city of Pondicherry is the capital.
BT India

ponds
Includes use on level 3 under geomorphology(1).
SA ecology
irrigation
lakes
limnology
paleoecology
reservoirs
surface water
water resources
water supply

Pongidae
Family. As of 1981, includes use on level 2 under Mammalia(1). See list F. Before 1981, included use on level 3 under Mammalia(1) Primates(2).
BA Primates
Mammalia
BT Tetrapoda
Vertebrata

Pontevedra
As of 1981, refers to only the city. Before 1981, also included the province. In NW Spain.
BT Pontevedra Province
Galicia
Spain

Pontevedra Province
Term introduced in 1981. Before 1981, search Pontevedra. In NW Spain.
CO N415100N425300
W0075300W0085500
BT Galicia

Spain
NT Pontevedra

Pontian
Europe. Above Sarmatian, below Plaisancian (Pliocene). Includes use on level 3 under age terms(1). See list E.
BA upper Miocene
Miocene
Neogene
Tertiary
BT Cenozoic

Pontic Mountains
In N part of country.
CO N395000N420500
E0423500E0300000
BT Turkey

Pontotoc County
Index states as applicable.
BX Mississippi
Oklahoma
BT United States

Poona
City, district, and division in W Maharashtra.
UF Pune
BT Maharashtra
India

Pope County
Index states as applicable.
BX Arkansas
Illinois
Minnesota
BT United States

Popigay
Village on Popigay River of E Taymyr National Okrug in NW Siberia.
BT Russian Republic
USSR

popular and elementary geology
A valid term through 1978. After 1978, use terms separately, i.e. elementary geology.

popular geology
Term introduced in 1978. Before 1978, search popular and elementary geology.
SA Atlantis
collecting
education
elementary geology
fossil localities
geology
mineral localities
natural curiosities

populations
As of 1981, restricted to statistics. Before 1981, used as a general term.
BA statistical analysis

porcelanite
use porcellanite

porcellanite
Includes use on level 3 under sedimentary rocks(1) clastic rocks(2). See list I.
UF porcelanite
BA clastic rocks
BT sedimentary rocks

Porcupine Mountains
Range in Gogebic and Ontonagon Counties in NW extremity of Upper Peninsula.
BT Michigan
United States

pore pressure
Includes use on level 3 under soil mechanics(1) or engineering geology(1).
UF neutral stress
BT pressure
SA engineering geology
high pressure
porous materials
soil mechanics
stress

pore water
Includes use on level 2 under sediments(1). Also search interstitial water.
UF interstitial water
SA connate waters
desiccation
fossil waters
porous materials
sediments
water

Porifera
Includes use on level 1 and 2 as a fossil term (list F).
UF Spongiae
BT Invertebrata
NA Calcispongea
Demospongea
Hyalospongea
Silicispongiae
SA Receptaculitaceae
reef builders
spicules

porosity
Includes use on level 3 under rock name, material, e.g. under sedimentary rocks(1), ground water(1), soils(1) and sediments(1).
SA compaction
Darcy's law
ground water
permeability
physical properties
porous materials
sedimentary rocks
sediments

porosity traps
use stratigraphic traps

porous materials
Term introduced in 1981. Before 1981, also search porous media.
SA materials
pore pressure
pore water
porosity

porous media
As of 1981, no longer a valid term for GeoRef. Use porous materials.

porphyrins
Includes use on level 3 under organic materials(1).
BA organic materials
SA organic compounds
organo-metallics

porphyrite
use porphyry

porphyritic texture
Term introduced in 1978. Before 1978, search porphyritic AND texture.
SA igneous rocks
phenocrysts
porphyry
textures
vitrophyre

porphyroblastic texture
Term introduced in 1978. Includes use on level 3 under metamorphic rocks(1). Before 1978, search porphyroblasts; porphyroblastic AND texture.
UF porphyroblasts
SA metamorphic rocks
textures

porphyroblasts
use porphyroblastic texture

porphyry
Also search porphyrite.
UF porphyrite
BA igneous rocks
SA andesite porphyry
dacite porphyry
diabase porphyry
diorite porphyry
granite porphyry
granodiorite porphyry
liparite porphyry
melaphyre
picrite porphyry
porphyritic texture
quartz porphyry
rhyolite porphyry
syenite porphyry

porphyry copper
Includes use on level 3 under mineral deposits, genesis(1).
SA copper
copper ores
mineral deposits

Porsang Fjord
Inlet of Arctic Ocean on N coast.
UF Porsangen Fjord
Porsangerfjord
BT Finnmark
Norway

Porsangen Fjord
use Porsang Fjord

Porsanger Dolomite Formation
Uppermost Precambrian or lowermost Cambrian. Porsanger Dolomite found in upper part of Porsanger Series.
SA Cambrian
Finnmark
Norway
Precambrian

Porsangerfjord
use Porsang Fjord

Port au Port Peninsula
On the Gulf of Saint Lawrence between Port au Port Bay and Saint George Bay.
BT Newfoundland
Canada

Port Campbell
Town on Indian Ocean in SW.
BT Victoria
Australia

Port Elizabeth
City on Indian Ocean in SE.
BT Cape Province
South Africa

Port Jervis
City and summer resort at meeting of borders with New Jersey and Pennsylvania in S.
BT New York
United States

Port Moresby
City on Coral Sea in SE.
BT Papua New Guinea
Australasia

Port Phillip Bay
Harbor of Melbourne.
BT Victoria
Australia

Port Royal Sound
Inlet of Atlantic Ocean between islands of St. Helena and Hilton Head in Beaufort County.
BT South Carolina
United States

Port Valdez
use Valdez

Portage County
Index states as applicable.
BX Ohio
Wisconsin
BT United States

Portage Lake Lava Series
Tops of conglomerate beds used for stratigraphic reference include: St. Louis Conglomerate, The Old Colony and Wolverine sandstones, and the Kingston, Calumet and Hecla, Houghton, Allouez, Pewabic West, and Hancock conglomerates. Upper Peninsula of Michigan. Also search Portage Lake.
BT Keweenawan
Precambrian
SA Michigan

Portel
Village in S part of country.
BT Aude
France

Portland
City in Rockingham County in SE New Hampshire, and city on the Willamette River in NW Oregon. Index states as applicable.
BX New Hampshire
Oregon
BT United States

Portlandian
Europe. Above Kimmeridgian, below Berriasian (Cretaceous). Includes use on level 3 under age terms(1). See list E.
BA Upper Jurassic
Jurassic
BT Mesozoic
NA Tithonian
SA Malm

Porto Santo Island
One of two inhabited islands of Portuguese island group. Also search Porto Santo.
BT Madeira

Portola Valley
Town in San Mateo County S of San Francisco.
BT California
United States

Portugal
Includes use on level 1 as an area term (list O). For term set options see list B.
CO N370000N421000
W0061000W0093000
BA Iberian Peninsula
Southern Europe
Europe
NT Alentejo
Algarve
Alto Alentejo
Baixo Alentejo
Beira Baixa
Braganca
Coimbra
Elvas
Estremadura
Evora
Leiria
Panasqueira
Setubal
Tomar
Tras-os-Montes
Vila Real
Viseu
SA Douro River
Tagus Basin
Tagus River
Timor

Portuguese Guinea
Not a valid index term for GeoRef since 1974. Use Guinea-Bissau.

Portuguese Timor
A valid index term through mid-1978. Included use on level 1 as an area term until 1977. After mid-

1978, use Timor.

Portuguese West Africa
use Angola

Porulosida
Order. Includes use on level 3 under Radiolaria(1).
BA Radiolaria
BT Protista
Invertebrata
NA Acantharina

Posen
use Poznan

Posidonia Shale
BT Jurassic
SA Germany
West Germany

positions, pole
use pole positions

posnjakite
BA sulfates
BT minerals

Possagno
Village in central.
BT Veneto
Italy

possibilities
Includes use on level 2 and 3. As of 1981, restricted to economic geology, especially referring to the economic potential of a mineral deposit. Before 1981, included use as a general term.
SA prediction maps

Postglacial
Not a valid term for GeoRef. Use Holocene.

postglacial environment
Term introduced in 1978. Before 1978, search postglacial.
SA environment
glacial environment
glacial geology
glaciers

Postmasburg
Village in N central.
BT Cape Province
South Africa

potash
Includes use on level 1 as a commodity term (list C). As of 1981, includes use on level 3 under sedimentary rocks(1) chemically precipitated rocks(2). See list I.
BA evaporites
chemically precipitated rocks
BT sedimentary rocks

potash alum
use alum

potash feldspar
use K-feldspar

potash mica
use muscovite

potassic composition
Term introduced in 1978. Before 1978, search potassic.
SA composition
potassium

potassium
Includes use on level 1 and 2 as a chemical element (list D). Also search K.
UF K
SA alkali metals
elements
isotopes
K-40
K/Ar
KREEP
potassic composition

potassium alum
use alum

potassium feldspar
use K-feldspar

potentiometric surface
Term introduced in 1981. Also search piezometric surface; piezometry AND ground water; piezometers AND ground water.
UF piezometric surface
SA ground water
hydrogeologic maps
levels
water table
wells

Potenza
City and province in S part of country.
BT Basilicata
Italy

potholes
Used as a general term under glacial geology(1) or geomorphology(1).
BT erosion features
SA erosion
sinkholes

Potomac Group
Consists of Patuxent, Arundel, and Patapsco formations.
BT Cretaceous
SA Delaware
Lower Cretaceous
Maryland
Upper Cretaceous
Virginia

Potomac River
Formed by the confluence of the North and South branches of the Potomac River SE of Cumberland, Maryland. It flows SE into Chesapeake Bay. Index District of Columbia and states as applicable.
BT United States
SA District of Columbia
Maryland
Virginia
West Virginia

Potomac River basin
Index states and District of Columbia as applicable.
BT United States
SA District of Columbia
Maryland
Virginia
West Virginia

Potsdam
As of 1981, refers to only the city. Before 1981, also included the district.
BT Potsdam Bezirk
East Germany

Potsdam Bezirk
Term introduced in 1981. District in central. Before 1981, search Potsdam AND East Germany.
CO N515000N532200
E0135500E0115200
BT East Germany
NT Potsdam

Potsdam Sandstone
Underlies Ticonderoga Formation. Central and E New York, and Vermont.
BT Upper Cambrian
Cambrian
SA New York
Vermont

Pottawatomie County
Index states as applicable.
BX Kansas
Oklahoma
BT United States

Pottsville Group
Includes Lee Formation, Norton Formation, Gladeville Sandstone, Wise Formation, Harlan Sandstone, Lookout Sandstone, Walden Sandstone, Tumbling Run Formation, Schuylkill Formation, Sharp Mountain Formation. Alabama, S Ohio, S Illinois, S Indiana, Kentucky, Maryland, NE Mississippi, Pennsylvania, Tennessee, Virginia, West Virginia.
BT Pennsylvanian
SA Alabama
Breathitt Formation
Illinois
Kentucky
Lower Pennsylvanian
Maryland
Middle Pennsylvanian
Mississippi
Ohio
Pennsylvania
Pocahontas Formation
Sharon Conglomerate
Tennessee
Virginia
West Virginia

Potwar Plateau
Lies between Indus and Jhelum rivers in N. Also search Potwar.
BT Punjab Province
Pakistan

Pouilly-en-Auxois
Village in E central part of country.
BT Cote-d'Or
France

Poway Conglomerate
Nearly horizontal, and in most places lies unconformably on Cretaceous Peninsular Range Batholith and accompanying metamorphic rocks. S California.
UF Poway Conglomerates
BT upper Eocene
Eocene
SA California

Poway Conglomerates
use Poway Conglomerate

Powder River basin
Index states as applicable.
CO N430000N452000
W1050000W1071500
BT United States
SA Montana
Wyoming

Powder River County
Term introduced in 1981. SE Montana.
CO N450000N454800
W1045700W1062000
BT Montana
United States

Powell
Town in Park County in NW.
BT Wyoming
United States

Powell County
Index states as applicable.
BX Kentucky
Montana
BT United States

powellite
BA molybdates
BT minerals

power plants
SA energy sources
nuclear facilities

Poza Rica
City in N Veracruz.
UF Poza Rica de Hidalgo
BT Veracruz
Mexico

Poza Rica de Hidalgo
use Poza Rica

Poznan
Province in W central Poland. Also a city province and a city.
UF Posen
BT Poland
NT Adamow Mine
Klodawa
Konin
Ksiaz

Poznan Clays
East Germany and Poland.
BT upper Tertiary
Tertiary
SA Germany
Poland

Pozzuoli
City on inlet of Gulf of Naples.
BT Campania
Italy

PPP-waves
Term introduced in 1981.
BA body waves
elastic waves
SA P-waves
seismology
waves

PPS-waves
Term introduced in 1981.
BA body waves
elastic waves
SA P-waves
S-waves
seismology
waves

Pr
use praseodymium

practice
To be used for geology as a profession after 1975. Includes use as a level 2 or 3 term appropriate to a large number of topics, e.g. on level 2 under geochemistry(1), geology(1), and economic geology(1). See list G.

Praecardiida
Term introduced in 1981. Includes use on level 2 under Mollusca(1). See list F.
BA Bivalvia
Mollusca
BT Invertebrata

Prague
City in central Bohemia.
UF Praha
BT Bohemia
Czechoslovakia

Praha
use Prague

Prahova
County in NE.
BT Walachia
Romania

Prairie Evaporite
Salt and anhydrite beds that form upper unit of Elk Point Group. Subsurface. Also search Prairie Evaporite Formation.
UF Prairie Evaporite Formation
BT Middle Devonian
Devonian
SA Elk Point Group
Manitoba
North Dakota
Saskatchewan

Prairie Evaporite Formation
use Prairie Evaporite

Pranhita-Godavari Valley
Combined valleys of the Pranhita and Godavari river systems. Index states as applicable.

BT India
SA Andhra Pradesh
Godavari Valley
Maharashtra

praseodymium
Includes use on level 1 and 2 as a chemical element (list D.)
UF Pr
SA elements
rare earths

prasinite
use greenschist

Pratt County
S Kansas.
BT Kansas
United States

Prbram
use Pribram

pre-Neanderthal
Term introduced in 1981. Includes use on level 2 under Mammalia(1). See list F.
BA Hominidae
Primates
Mammalia
BT Tetrapoda
Vertebrata

Prealps
Mostly in Italy. Also search Alpine Foreland.
UF Alpine Foreland
SA Alps
Italy

Prebetic Zone
A geographic term with stratigraphic-tectonic connotations.
BT Spain
SA Betic Cordillera
Subbetic Zone

Preboreal
Europe. An interval of post-glacial time following the Younger Dryas of the late-glacial Arctic interval and preceding the Boreal. Also search Subarctic.
UF Subarctic
BA Holocene
Quaternary
BT Cenozoic
SA Boreal

Precambrian
Includes use on level 1 as an age term (list E); on level 2 under paleoterms, e.g. paleoecology, paleogeography, paleomagnetism.
NT Acungui Group
NA Adelaidean
NT Adirondack Anorthosite
NA Algonkian
NT Altyn Limestone
Animikie Group
NA Archean
NT Assyntic Orogeny
Avalonian Orogeny
Baltimore Gneiss
Belt Supergroup
Bijawar System
Biwabik Iron Formation
Bonner Formation
Brockman Iron Formation
Catoctin Formation
Chuar Group
Copper Harbor Conglomerate
Cuddapah System
Delhi System
NA Eocambrian
NT Espanola Formation
Fig Tree Series
Flinton Group
Grenvillian Orogeny
Gunflint Iron Formation
Hamersley Group
Harbour Main Group
Hecla Hoek Formation
Hector Formation
NA Helikian
NT Highland Series
Hudsonian Orogeny
Johnnie Formation
Kaladgi System
Karelian Orogeny
Katangan Orogeny
Kenoran Orogeny
NA Keweenawan
NT Knife Lake Group
Ladoga Series
NA Lewisian
middle Precambrian
NT Minas Series
Missi Group
Missoula Group
Moldanubian
Nipissing Diabase
Nonesuch Shale
Noonday Dolomite
North Shore Volcanics
NA Ocoee Series
NT Onverwacht Group
Osler Series
Pelona Schist
Penokean Orogeny
Penrhyn Slate
Pocatello Formation
Prichard Formation
Prince Albert Group
NA Proterozoic
NT Purcell System
Ravalli Group
Roan Supergroup
Sioux Quartzite
Sokoman Formation
Stillwater Complex
Stirling Quartzite
Swaziland Sequence
Swaziland System
Thomson Slate
Transvaal Supergroup
Twilight Gneiss
Unkar Group
NA upper Precambrian
NT Vindhyan
Wallace Formation
Waterberg System
Wilcox Formation
Witwatersrand System
Wyman Formation
SA Baikalian Phase
Bhander Group
Cadomian Orogeny
Dalradian
Dore Lake Complex
Duluth Complex
Jacobsville Sandstone
Kaimur Sandstone
Kurnool System
Lardeau Group
Limpopo Belt
New York City Group
Porsanger Dolomite Formation
Shuswap Complex
Sparagmite Group
Tal Formation
Talladega Group
Wood Canyon Formation

Precambrian Shield
use Canadian Shield

precession
Used as a general term.
SA processes

precipitation
As of 1978, term refers to geochemistry. For meteorology, use atmospheric precipitation. Includes use on level 2 under sedimentation(1).
SA atmospheric precipitation
crystallization
drainage basins
geochemistry
hydrology
particle precipitation
particles
rivers and streams
runoff
sedimentation
storms
water

Precordillera
BT Argentina

precursors
Includes use on level 3 under seismology(1). Before 1981, also search premonitory phenomena.
SA seismology

Predazzo
City and resort in Dolomites in S.
BT Trentino-Alto Adige
Italy

prediction
Includes use on level 2 under earthquakes(1).
SA earthquakes

prediction maps
Term introduced in 1981.
BA maps
SA economics
possibilities
reserves

preferred orientation
Includes use on level 2 under structural analysis(1).
SA fabric
lineation
orientation
petrofabrics
structural analysis
tectonite

prehnite
BA sheet silicates
silicates
BT minerals
SA aluminosilicates

prehnite-pumpellyite facies
BT facies
SA metamorphic rocks

premonitory phenomena
As of 1981, no longer a valid term for GeoRef. Use precursors.

preobrazhenskite
BA borates
BT minerals

preparation
Includes use as a level 2 or 3 term appropriate to a large number of topics, e.g. on level 2 under coal(1). See list G. Also search preparations.
SA sample preparation

preservation
As of 1978, term is used on level 2 under land use(1).
SA land use

Presidente Vargas
use Itabira

Presidio County
W Texas.
BT Texas
United States

Presque Isle
City in Aroostook County in N Maine; small peninsula in Lake Erie in NW Pennsylvania; county in NE Michigan.
BX Maine
Michigan
Pennsylvania
BT United States

pressure
Includes use on level 3 under deformation(1); on level 3 under experimental studies(2) under appropriate headings in which pressure is a factor, e.g. faults(1), folds(1), fractures(1), foliation(1), metasomatism(1), magmas(1), and metamorphism(1).
NT atmospheric pressure
capillary pressure
high pressure
hydraulic pressure
hydrostatic pressure
low pressure
partial pressure
pore pressure
water pressure
SA compression
deformation
earth pressure
fluid pressure
geopressure
load pressure
P-T conditions
stress
tension

pressure solution
SA diagenesis
solution

pressure wave
use P-waves

Preston County
BT West Virginia
United States

Preston Peak
In Siskiyou Mountains in Siskiyou County in extreme N.
BT California
United States

Pretoria
City in S.
BT Transvaal
South Africa

prevention
A valid general term through mid-1978.

preventive measures
Term introduced in 1981.
SA environmental geology

Priabonian
Europe. Includes use on level 3 under age terms(1). See list E.
BA upper Eocene
Eocene
Paleogene
Tertiary
BT Cenozoic
SA Auversian
Bartonian
Ludian

Priapulida
Term introduced in 1981. Includes use on level 2 under worms(1). See list F.
BA worms
BT Invertebrata

Pribram
City SW of Prague in central Bohemia. Also search Prbram.
UF Prbram
BT Bohemia
Czechoslovakia

price
Term introduced in 1981.
SA economics

Price Formation
Includes Cloyd Conglomerate Member. SW Virginia.
BT Mississippian
SA Virginia

Prichard Formation
Belt Series. NE Idaho and NW Montana.
BT Precambrian
SA Idaho
Montana

Pridolian
Europe.
BA Upper Silurian
Silurian
BT Paleozoic

Prikumsk
City in Stavropol Kray in the Northern Caucasus.
UF Budennovsk
BT Russian Republic
USSR

primary dispersion
Term introduced in 1978. Includes use on level 3 under mineral exploration(1). Before 1978, search primary AND dispersion.
SA dispersion patterns
mineral exploration
secondary dispersion

primary magnetization
Term introduced in 1981.
BA magnetization

primary structures
Includes use on level 2 under sedimentary structures(1). See list K.
BA sedimentary structures
SA ball-and-pillow
bedding
ripple marks
secondary structures
structures

primary wave
use P-waves

Primates
Includes use on level 2 under Mammalia(1). See list F.
BA Mammalia
BT Tetrapoda
Vertebrata
NA Australopithecinae
Hominidae
Pongidae
Prosimii
simians
SA fossil man

Primorski Krai
use Primorye

Primorye
Primorye Kray (territory). New name for Primorski Krai which is the SE section of the Soviet Far East between Manchuria and the Japan Sea. Also search Maritime Territory.
CO N420000N500000
E1400000E1300000
UF Maritime Kray
Maritime Territory
Primorski Krai
BT Russian Republic
USSR

Prince Albert Group
BT Precambrian
SA Canada
Saskatchewan

Prince Charles Mountains
In Mac-Robertson Land near the Amery Ice Shelf on Indian Ocean side.
BT Antarctica

Prince Edward Island
Island in the Gulf of Saint Lawrence constituting a province. Includes use on level 1 as an area term (list O). For term set options see list B.
CO N455500N470500
W0620000W0643000
BA Maritime Provinces
Eastern Canada
Canada

Prince Georges County
NE, E, and SE of District of Columbia.
BT Maryland
United States

Prince of Wales Island
Largest island of Alexander Archipelago in SE Alaska, W of Ketchikan; and an island between Victoria and Somerset Islands in District of Franklin, Northwest Territories. Index Alaska and/or Northwest Territories as applicable.
SA Alaska
Northwest Territories

Prince Rupert Quadrangle
Term introduced in 1981. SE Alaska.
CO N540000N550000
W1300000W1320000
BT Alaska
United States

Prince William Sound
Inlet of Gulf of Alaska E of Kenai Peninsula.
BT Alaska
United States

principal components analysis
Term introduced in 1981. Also search principal component analysis.
BA statistical analysis

principles
Includes use as a level 2 or 3 term appropriate to a large number of topics, e.g. on level 2 under geology(1). See list G.

Pripet Basin
River basin most of which comprises the Pripet or Pinsk Marshes which is the largest tract of swamp in Europe. Largely coextensive with Polesye Lowland. Also search Pripet; Pripet Depression; Pripyat Basin. Index Soviet republics as applicable.
CO N510000N540000
E0330000E0240000
UF Pripet Depression
Pripyat Basin
BT USSR
SA Byelorussia
Polesye
Ukraine

Pripet Depression
use Pripet Basin

Pripyat Basin
use Pripet Basin

prismatine
use kornerupine

Privas
Town in SE central part of country.
BT Ardeche
France

probability
BA statistical analysis
SA mathematical methods
reliability
sampling
stochastic processes

probe, electron
use electron probe

probe, ion
use ion probe

problematic
A valid term through 1977 used on level 1 in combination with fossils (i.e. fossils, problematic). After 1977, use problematic fossils on level 1.

problematic fossils
Term introduced in 1978. Includes use on level 1 and 2. Before 1978, also search fossils AND problematic.
NA Calcisphaerulidae
Dacryoconarida
NZ problematic microfossils
SA chemical fossils
fossilization
fossils
Hyolithes
nannofossils
paleontology
Receptaculitaceae
Stromatoporoidea
Tentaculites
Tentaculitida
Tentaculitidae

problematic microfossils
Term introduced in 1981. Includes use on level 2 under problematic fossils(1). See list F.
BZ microfossils
problematic fossils

problematic palynomorphs
As of 1981, includes use on level 3 under palynomorphs(1). See list F. Before 1981, included use on level 2 under palynomorphs(1).
BA palynomorphs

problems
A valid general term through mid-1978.

Proboscidea
Includes use on level 2 under Mammalia(1). See list F.
BA Mammalia
BT Tetrapoda
Vertebrata
NA Barytherioidea
Deinotherioidea
Elephantoidea
Mastodontoidea
Moeritherioidea

processes
Includes use as a level 2 or 3 term appropriate to a large number of topics, e.g. on level 2 under geochemistry(1), sedimentation(1), and under mineral deposits, genesis(1). See list G. Also search process.
SA adsorption
aggradation
albitization
alteration
atectonic processes
autometamorphism
bauxitization
brecciation
calcification
calcitization
chloritization
coalification
combustion
compaction
complexing
contraction
cyclic processes
dedolomitization
diagenesis
diffusion
dissociation
dolomitization
electrolysis
endogene processes
epigene processes
epithermal processes
erosion
exhalative processes
exogene processes
filtration
flocculation
fractionation
fragmentation
granitization
hydration
hydrolysis
hydrothermal processes
hypabyssal processes
hypothermal processes
igneous processes
kaolinization
mesothermal processes
mixing
oxidation
photosynthesis
phyllonitization
polyphase processes
precession
propylitization
reduction
respiration
sedimentary processes
sedimentation
separation
serpentinization
silicification
stochastic processes
supergene processes
syntectonic processes
telethermal processes

production
Includes use on level 2 and 3 under commodity terms, e.g. on level 2 under energy sources(1). See list C. When referring to mining geology before May 1978, also search extraction.
SA exploitation
pelletization
production control
production value

production control
Includes use on level 2 under mining geology(1).
UF control, production
SA mining geology
production
technology

production value
Term introduced in 1981.
SA economics
production

productive capacity
Term introduced in 1981.
SA economics

productivity
Used as a general term.
SA economics

products
A valid general term through mid-1978.

Proetidae
Family. Includes use on level 3 under Trilobita(1).
BA Trilobita
BT Trilobitomorpha
Arthropoda
Invertebrata

profiles
As of 1981, no longer a valid term for GeoRef. See sections; geophysical profiles; soil profiles; geochemical profiles. Before 1981, included use on level 3 under soils(1) morphology(2).

profitability
Term introduced in 1981.
SA economics

Proganosauria
Term introduced in 1981. Includes use on level 2 under Reptilia(1). See list F.
BA Mesosauria
Synapsida
Reptilia
BT Tetrapoda
Vertebrata

progradation
SA littoral drift
sedimentation
shore features

prograde metamorphism
Term introduced in 1978. Includes use on level 2 under metamorphism(1). Before 1978, also search

metamorphism AND prograde.
BA metamorphism
SA polymetamorphism
retrograde metamorphism

programs
Includes use on level 2 under mineral exploration(1); on level 3 under automatic data processing(1) and education(1). Also search projects.
UF projects
SA associations
computer programs
computers
education
mineral exploration
planning
policy
reclamation
research

progress report
Includes use on level 3 under associations(1) and symposia(1).
BT report
SA annual report
associations
current research
research
survey organizations
symposia

projects
use programs

Prokopyevsk
City at S end of Kuznetsk Basin in Kemerovo Oblast in S Siberia.
BT Russian Republic
USSR

Prolecanitida
Term introduced in 1981. Includes use on level 2 under Mollusca(1). See list F.
BA Ammonoidea
Tetrabranchiata
Cephalopoda
Mollusca
BT Invertebrata

proluvium
See list N.
BA clastic sediments
BT sediments

prometheum
use promethium

promethium
Includes use on level 1 and 2 as a chemical element (list D).
UF Pm
prometheum
SA elements
rare earths

propagation
Includes use on level 3 under seismology(1) elastic waves(2). As of 1981, includes use on level 2 under ionosphere(1) and magnetosphere(1). Before 1981, also search wave propagation.
SA elastic waves
electromagnetic waves
ocean waves
seismology
shock waves
waves

properties
Includes use in combination with materials (i.e. materials, properties) on level 2 under engineering geology(1); as a level 2 or 3 term appropriate to a large number of topics; for physical or chemical properties on level 2 or 3 under commodity terms(1). See list C (commodities) and list G (general terms).
SA acoustical properties
chemical properties
dielectric properties
dynamic properties
elastic properties
electrical properties
electrochemical properties
engineering properties
magnetic hysteresis
magnetic properties
materials
mechanical properties
optical properties
physical properties
reservoir properties
solubility
surface properties
thermal properties
thermochemical properties
thermodynamic properties
tortuosity

Propria Geosyncline
In Propria region of the lower Sao Francisco River near the Atlantic Ocean.
BT Sergipe
Brazil
SA geosynclines

propylite
As of 1978, term is used on level 2 under metasomatic rocks(1). Before 1978, included use on level 3 under igneous rocks(1) andesite-rhyolite family(2).
BA metasomatic rocks
SA andesite
andesites
propylitization

propylitization
Includes use on level 3 under metasomatism(1) processes(2).
BA metasomatism
SA hydrothermal alteration
processes
propylite

Prosimii
Term introduced in 1981. Includes use on level 2 under Mammalia(1). See list F.
BA Primates
Mammalia
BT Tetrapoda
Vertebrata

Prosobranchia
Includes use on level 3 under Mollusca(1). See list F.
BA Gastropoda
Mollusca
BT Invertebrata

prosopite
As of 1981, includes use on level 3 under minerals(1) fluorides(2). See list L. Before 1981, included use on level 3 under minerals(1) halides(2).
BA fluorides
halides
BT minerals

Prospect Mountain Quartzite
use Stirling Quartzite

prospecting
As of 1981, no longer a valid term for GeoRef. Use exploration.

protactinium
Includes use on level 1 and 2 as a chemical element (list D).
UF Pa
SA elements
Pa-231

proteins
Includes use on level 3 under organic materials(1).
BA organic materials
SA amino acids
collagen
conchiolin

Proterozoic
As of 1978, includes use on level 1 and 2 (list E).
UF Agnotozoic
BA Precambrian
NT Athabasca Formation
Badami Series
Bambui Group
Bitter Springs Formation
Damara System
NA Huronian
NT Kursk Series
Lorrain Formation
NA lower Proterozoic
middle Proterozoic
NT Miette Group
Ovruch Series
Udokan Series
NA upper Proterozoic
NT Windermere System
Wolkberg Group
SA Algonkian
Helikian
Shuswap Complex
upper Precambrian

Proteutheria
Term introduced in 1981. Includes use on level 2 under Mammalia(1). See list F.
BA Mammalia
BT Tetrapoda
Vertebrata

Protista
Includes use on level 1 and 2 as a fossil term (list F). Before 1976, also search Protozoa.
BT Invertebrata
NA ebridians
NT foraminifera
Radiolaria
NA Silicoflagellata
Thecamoeba
Tintinnidae
SA algae
bacteria
Metazoa
microorganisms
paleobotany
plankton
zooxanthellae

Protochordata
Term introduced in 1981.
BT Chordata

protodolomite
BA carbonates
BT minerals

protoenstatite
Artificial mineral.
BA pyroxene group
chain silicates
silicates
BT minerals

protoliths
Term introduced in 1978. Also search parent rocks.
UF parent rocks
SA igneous rocks
metamorphic rocks
metamorphism

protons
Used as a general term.
SA cosmic rays
electrons

Protopivskaya Formation
Central Ukraine and NW Donets Basin.
BT Upper Triassic
Triassic
SA Ukraine
USSR

Protorthoptera
As of 1981, includes use on level 3 under Insecta(1). See list F. Before 1981, included use on level 2 under Insecta(1).
BA Insecta
BT Mandibulata
Arthropoda
Invertebrata

Protozoa
A valid index term through 1976. After 1976, use Protista.

proustite
BA sulfarsenites
sulfosalts
BT minerals

provenance
Includes use on level 2 under sedimentation(1) and under sediments(1).
UF source area
sourceland
SA paleocurrents
sedimentary rocks
sedimentation
sediments

Provence
Historical region in SE. Index departments as applicable.
CO N430000N444500
E0074500E0043000
BT France
SA Alpes-de-Haute Provence
Alpes-Maritimes
Bouches-du-Rhone
Var
Vaucluse

Provence Alps
In SE Provence. Index departments as applicable.
BT France
SA Alpes-de-Haute Provence
Alpes-Maritimes
Alps
Var

provinces
A valid general term through 1976. See specific term, e.g. metallogenic provinces, physiographic provinces, etc.

provinces, faunal
use faunal provinces

provinces, floral
use floral provinces

provinces, metallogenic
use metallogenic provinces

provinces, physiographic
use physiographic provinces

provincialism
use provinciality

provinciality
Used mostly for fossils. See list F.
UF provincialism
SA fossils

Provo
City S of Salt Lake City in Utah County.
BT Utah
United States

Prudhoe Bay
Inlet of the Arctic Ocean on the North Slope.
BT Alaska
United States
SA North Slope

Prut River
Rises in W central Carpathians and flows into Danube River below Galati. Index Romania and Soviet republics as applicable. Also search Prut.
UF Pruth River
BT Europe
SA Moldavia
Romania
Ukraine

Pruth River
use Prut River

Przemsyl
City near Ukrainian border in SE part of country.
UF Peremyshl
BT Rzeszow
Poland

psammite
use sandstone

psammitic texture
use arenaceous texture

pseudo-single domains
Term introduced in 1981.
BT magnetic domains

pseudobrookite
BA oxides
BT minerals
SA armalcolite
ferropseudobrookite

pseudogalena
use sphalerite

Pseudogleys
See list M.
BA soils
SA Gleys

pseudoleucite
BA framework silicates
silicates
BT minerals

pseudomorphism
Includes use on level 3 under mineralogy(1). Before 1976, also search replacement.
SA minerals

pseudotachylite
Includes use on level 3 under metamorphic rocks(1) mylonites(2). See list J.
BA mylonites
BT metamorphic rocks

Psiloceratida
Term introduced in 1981. Includes use on level 2 under Mollusca(1). See list F.
BA Ammonoidea
Tetrabranchiata
Cephalopoda
Mollusca
BT Invertebrata

psilomelane
BA oxides
BT minerals
SA manganese oxides

Psilophytales
use Psilopsida

Psilopsida
Including Psilophytales. Includes use on level 2 under pteridophytes(1). See list F.
UF Psilophytales
BA pteridophytes
BT Plantae

Pskem Range
Outlier of Kirghiz Range. Index Soviet republics as applicable.
BT USSR
SA Kazakhstan
Kirghizia

Pskov
City 155 miles SW of Leningrad. Also an oblast.
BT Russian Republic
USSR

Psocopteroida
Term introduced in 1981. Includes use on level 2 under Insecta(1). See list F.
BA Insecta
BT Mandibulata
Arthropoda
Invertebrata

Pt
use platinum

Pteranodon
As of 1981, includes use on level 3 under Reptilia(1) Pterosauria(2). See list F. Before 1981, included use on level 3 under Reptilia(1) Archosauria(2).
BA Pterosauria
Archosauria
Reptilia
BT Tetrapoda
Vertebrata

Pteridophyllen
Includes use on level 2 under pteridophytes(1). See list F.
BA pteridophytes
BT Plantae

pteridophytes
Includes use on level 1 and 2 as a fossil term (list F).
BT Plantae
NA Filicopsida
Lycopsida
Noeggerathiales
Psilopsida
Pteridophyllen
Sphenopsida
SA ferns
Juglans
Neuropteris
Pteropsida
sporangia
spores

Pteridospermae
As of 1981, includes use on level 3 under gymnosperms(1). See list F. Before 1981, included use on level 2 under gymnosperms(1).
BA gymnosperms
BT Spermatophyta
Plantae
NA Callipteris
Neuropteris
SA Dicroidium
Pecopteris
Ptilophyllum
Sphenopteris
Taeniopteris

Pteriina
Term introduced in 1981. Includes use on level 2 under Mollusca(1). See list F.
BA Cyrtodontida
Bivalvia
Mollusca
BT Invertebrata
NA Inocerami

Pterobranchia
Includes use on level 1 and 2 as a fossil term (list F).
BX Hemichordata
Invertebrata
NA Cephalodiscida
Rhapdopleurida

Pteropoda
As of 1981, includes use on level 2 under Mollusca(1). See list F. Before 1981, included use on level 3 under Mollusca(1) Gastropoda(2).
BA Gastropoda
Mollusca
BT Invertebrata

Pteropsida
Used as a general term.
BT Plantae
SA angiosperms
gymnosperms
pteridophytes
thallophytes

Pterosauria
As of 1981, includes use on level 2 under Reptilia(1). See list F. Before 1981, included use on level 3 under Reptilia(1) Archosauria(2).
BA Archosauria
Reptilia
BT Tetrapoda
Vertebrata
NA Pteranodon

ptilolite
use mordenite

Ptilophyllum
Includes use on level 3 under gymnosperms(1) Cycadales(2).
BA Cycadales
gymnosperms
BT Spermatophyta
Plantae
SA Bennettitales
Pteridospermae

Ptychopariida
Includes use on level 2 under Trilobita(1). See list F.
BA Trilobita
BT Trilobitomorpha
Arthropoda
Invertebrata
NA Asaphidae
Cryptolithus
Olenidae
Scutelluidae

ptygma
use ptygmatic folds

ptygmatic folds
Term introduced in 1978. Includes use on level 3 under folds(1) style(2). Before 1978, also search folds AND ptygmatic.
UF ptygma
BA folds
SA migmatites

Pu
use plutonium

Pu-239
Includes use on level 3 under isotopes(1).
SA isotopes
plutonium

Pu-244
Includes use on level 3 under isotopes(1).
SA isotopes
plutonium

publications
Used as a general term.
SA associations
survey organizations

Pucara Group
Includes Paria Formation. W part of Country. Upper Triassic and Jurassic.
BT Mesozoic
SA Junin
Jurassic
Lima
Peru
Triassic
Upper Triassic

Puck Bay
Inlet of the Gulf of Danzig on the Baltic Sea. Also search Puck.
BT Gdansk
Poland

Puebla
State E of Mexico City. Also a city.
BT Mexico
SA Sierra Madre Oriental

Puebla de Guzman
Town in SW part of country.
BT Huelva Province
Andalusia
Spain

Pueblo County
SE central.
CO N374500N383000
W1040500W1050500
BT Colorado
United States

Puente Formation
Includes Papel Blanco Shale, Blanco Sandstone, Cubierto Shale, Hunter Sandstone and Conglomerate, Peculiar Shale, Mahala Sandstone and Conglomerate, Sycamore Canyon Member, La Vida Member, Soquel Member, Yorba Member. S California.
BT upper Miocene
Miocene
SA California

Puente Hills
S part of Los Angeles County.
BT California
United States

Puerto Cabello
City W of Caracas on the Caribbean Sea.
BT Carabobo
Venezuela

Puerto Deseado
Town and Bay in S part of country.
BT Argentina
SA Santa Cruz

Puerto Rico
Includes use on level 1 as an area term (list O). For term set options see list B.
CO N175000N183000
W0654000W0671500
BA Greater Antilles
West Indies
NT Anasco Bay
Arecibo
Cabo Rojo
Mayaguez Bay
Ponce
San Juan
SA Robles Formation
San Juan Formation

Puerto Rico Trench
NNE of Puerto Rico.
UF Puerto Rico Trough
BT Atlantic Ocean

Puerto Rico Trough
use Puerto Rico Trench

Puget Lowland
Area surrounding Puget Sound and the broad trough extending S.
BT Washington
United States

Puget Sound
Arm of Pacific Ocean extending S from E end of Juan de Fuca Strait.
CO N471000N483000
W1221000W1231500
BT Washington
United States

Puglia
use Apulia

Pulaski County
Index states as applicable.
BX Arkansas
Georgia
Illinois
Indiana
Kentucky
Missouri
Virginia
BT United States

Pulaski Fault
SW Virginia.
BT Virginia
United States

pulaskite
As of 1981, includes use on level 3 under igneous rocks(1) syenites(2). See list H. Before 1981, included use on level 3 under igneous rocks(1) syenite family(2).
BA syenites
BT igneous rocks

pull apart structures
use boudinage

pulsating spring
use geysers

pulsations
Includes use on level 3 under Earth(1).
SA Earth
magnetic field
magnetosphere
micropulsations
vibration

Pultova
use Poltava

Pultowa
use Poltava

Pultusk
City in E central part of country.
BT Warsaw
Poland

pumice
As of 1981, includes use on level 3 under igneous rocks(1) pyroclastics(2). See list H. Before 1981, included use on level 3 under igneous rocks(1) pyroclastics and glasses(2) and on level 1 as a commodity term. As of 1981, use pumice deposits for pumice as a commodity.
BA pyroclastics
volcanic rocks
BT igneous rocks
SA pumice deposits
scoria

pumice deposits
Term introduced in 1981. Includes use on level 1 and 2 as a commodity term (list C). Before 1981, search pumice AND deposits.
SA abrasives
aggregate
pumice

pump tests
Term introduced in 1978. Before 1978, also search pumping; pumping tests.
UF pumping tests
SA ground water

pumpellyite
BA orthosilicates
silicates
BT minerals
SA aluminosilicates

pumping tests
use pump tests

punch cards
Also search punched cards.
UF punched cards
SA automatic data processing

Punchbowl Formation
Described in Valyermo Quadrangle as consisting of two facies. S California.
BT upper Miocene
Miocene
SA California

punched cards
use punch cards

Pune
use Poona

Pungo River
Tidal estuary near the Pamlico River off Pamlico Sound.
BT North Carolina
United States

Punjab
No longer a valid term for GeoRef as of 1981. Use Punjab State or Punjab Province as applicable. Before 1981, referred to both the state in India and the province in Pakistan.

Punjab Province
Term introduced in 1981. In E central part of country.
BT Pakistan
NT Attock
Khewra
Mangla Dam
Mansehra
Mianwali
Potwar Plateau
Salt Range

Punjab State
Term introduced in 1981. In NW part of country.
CO N295000N323000
E0770000E0735000
BT India
NT Chandigarh
Spiti

Punta Mosquito Formation
BT Eocene
Paleogene
Tertiary
SA Venezuela

Pupillidae
Includes use on level 3 under Mollusca(1). See list F.
BA Gastropoda
Mollusca
BT Invertebrata

Purbeckian
Europe. Includes use on level 3 under age terms(1). See list E.
BA Lower Cretaceous
Cretaceous
BT Mesozoic

Purcell System
Also search Purcell Series.
BT Precambrian
SA British Columbia
Idaho
Montana

purchases
Term introduced in 1981.
SA economics

pure coal
use vitrain

purification
Includes use on level 3 under ground water(1) and water resources(1).
SA beneficiation
ground water
hydrology
impurities
pollution
water
water quality
water resources

purines
BA organic materials

purple copper ore
use bornite

Purulia
City in E part of country.
BT West Bengal
India

push wave
use P-waves

push-pull wave
use P-waves

Putnam County
Index states as applicable.
BX Florida
Georgia
Illinois
Indiana
Missouri
New York
Ohio
Tennessee
West Virginia
BT United States

Putrid Sea
use Sivash

Puy-de-Dome
Department in S central.
CO N451500N461500
E0040000E0023000
BT France
NT Chaine des Puys
Clermont-Ferrand
Mont-Dore
SA Limagne

Pyramid Lake
Washoe County in NW.
BT Nevada
United States

pyrargyrite
BA sulfantimonites
sulfosalts
BT minerals

Pyrenean Orogenic Phase
use Pyrenean Orogeny

Pyrenean Orogeny
One of the 30 or more short-lived orogenies during Phanerozoic time identified by Stille, in this case during the late Eocene, between the Bartonian and Ludian stages. Before 1978, search Pyrenean AND orogeny.
UF Pyrenean Orogenic Phase
BT Eocene
SA orogeny

Pyrenees
Mountain range extending from the Bay of Biscay to the SW coast of the Gulf of Lion. Index countries as applicable. Includes use on level 1 as an area term (list O). For term set options see list B.
CO N420000N430000
E0031500W0020000
BA Europe
NZ French Pyrenees
Spanish Pyrenees
SA Andorra
France
Spain

Pyrenees-Atlantiques
Department in SW France. Also search Basses-Pyrenees.
UF Basses-Pyrenees
BT France
NT Biarritz
Pau

Pyrenees-Orientales
Department in E Pyrenees.
CO N421500N430000
E0031500E0013000
BT France
NT Agly Massif

pyrite
Includes use on level 3 under minerals(1) sulfides(2). See list L. As of 1981, use pyrite ores for pyrite as a commodity. Before 1981, included use on level 1 as a commodity term.
BA sulfides
BT minerals
SA bravoite
framboidal texture
pyrite ores

pyrite ores
Term introduced in 1981. Includes use on level 1 and 2 as a commodity term (list C). Before 1981, search pyrite AND ore deposits.
SA pyrite
sulfur deposits

pyritization
BA metasomatism
SA diagenesis
hydrothermal alteration

pyroaurite
BA carbonates
BT minerals

pyrochlore
BZ fluorides
niobates
tantalates
BT minerals
SA betafite
microlite

pyroclastics
Includes use on level 3 under sedimentary rocks(1) clastic rocks(2), under sediments(1) clastic sediments(2), and as of 1981, on level 2 under igneous rocks(1). See lists H, I, and N. Before 1981, included use on level 3 under igneous rocks(1) pyroclastics and glasses(2). Also search tephra. Before 1981, also search pyroclastics and glasses.
UF tephra
BX igneous rocks
sedimentary rocks
sediments
NA andesite tuff
ash flows
ash-flow tuff
green tuff
ignimbrite
perlite
pumice
rhyolite tuff
scoria
tuff
tuffite
volcanic ash
welded tuff
SA agglomerate
ash falls
bentonite
clastic rocks
clastic sediments
ejecta
eruptions
glasses
lapilli
terrigenous materials
volcanic breccia
volcanic rocks

pyroclastics and glasses
As of 1981, no longer a valid term for GeoRef. Use pyroclastics or glasses as applicable. See list H. Before 1981, included use on level 2 under igneous rocks(1).

pyrolusite
Also search polianite.
UF polianite
BA oxides
BT minerals

pyrolysis
Includes use on level 3 under geochemistry(1) processes(2).
SA geochemistry

pyrometasomatism
BA metasomatism
SA hydrothermal alteration

pyromorphite
BZ chlorides
phosphates
BT minerals
SA apatite

pyrope
BA garnet group
nesosilicates
orthosilicates
silicates
BT minerals

pyrophyllite
BA sheet silicates
silicates
BT minerals

Pyrotheria
Term introduced in 1981. Includes use on level 2 under Mammalia(1). See list F.
BA Mammalia
BT Tetrapoda
Vertebrata

pyroxene
As of 1981, no longer a valid term for GeoRef. See pyroxene group. Before 1981, included use on level 3 under minerals(1) chain silicates, pyroxene group(2).

pyroxene andesite
As of 1981, includes use on level 3 under igneous rocks(1) andesites(2). See list H. Before 1981, included use on level 3 under igneous rocks(1) andesite-rhyolite family(2).
BA andesites
volcanic rocks
BT igneous rocks
SA andesite

pyroxene granulite
See list J.
BA granulites
BT metamorphic rocks

pyroxene group
Includes use in combination with chain silicates (i.e. chain silicates,pyroxene group) on level 2 under minerals(1). See list L. Also search pyroxenes.
UF pyroxenes
BA chain silicates
silicates
BT minerals
NA alkalic pyroxene
clinopyroxene
ferrosilite
orthopyroxene
protoenstatite
pyroxferroite
SA jade

pyroxenes
use pyroxene group

pyroxenite
As of 1981, includes use on level 3 under igneous rocks(1) ultramafics(2). See list H. Before 1981, included use on level 3 under igneous rocks(1) ultramafic family(2).
BA ultramafics
BT igneous rocks
SA bronzitite
clinopyroxenite
garnet pyroxenite
metapyroxenite
websterite

pyroxferroite
BA pyroxene group
chain silicates
silicates
BT minerals

pyroxmangite
BA chain silicates
silicates
BT minerals

Pyrrhophyta
As of 1981, includes use on level 2 under algae(1). See list F. Before 1981, included use on level 3.
BA algae
BT thallophytes
Plantae
SA Dinoflagellata
palynomorphs

pyrrhotine
use pyrrhotite

pyrrhotite
Also search pyrrhotine.
UF dipyrite
pyrrhotine
BA sulfides
BT minerals
SA troilite

Q

Q
Ratio of the peak seismic energy in a cycle to the energy dissipated.
SA earthquakes
elastic waves
Love waves
seismology

Q waves
use Love waves

Qaraqum
use Karakum

Qatar
Includes use on level 1 as an area term (list O). For term set options see list B.
CO N243000N262000
E0514000E0504000
BA Arabian Peninsula
Asia
NT Nakhla
SA Middle East
Near East

Qizil Qum
use Kyzylkum

qualitative analysis
SA analysis
chemical analysis
quantitative analysis

quality
A valid general term through 1978. After 1978 use specific term, e.g. water quality.

quantitative analysis
Includes use on level 3 under chemical analysis(1) methods(2).
SA analysis
chemical analysis
flame photometry
methods
neutron activation analysis
polarography
qualitative analysis

quantitative geomorphology
Includes use on level 3 under geomorphology(1).
SA geometry
geomorphology

quantitative methods
As of 1981, no longer a valid term for GeoRef. See mathematical methods. Term was introduced in 1978.

quarries
Includes use on level 3 under commodities (list C).
SA mines
mining geology
strip mining

quartz
BA silica minerals
framework silicates
silicates
BT minerals
SA agate
amethyst
chalcedony
cristobalite
light minerals
quartz crystal

quartz crystal
Includes use on level 1 as a commodity term (list C).
UF berg crystal
crystal, quartz
mountain crystal
rock crystal
SA quartz

quartz diabase
As of 1981, includes use on level 3 under igneous rocks(1) diabase(2). See list H. Before 1981, included use on level 3 under igneous rocks(1) basalt family(2). Before 1981, also search quartz dolerite.
BA diabase
BT igneous rocks

quartz diorite
Includes use on level 3. As of 1981, not to be used in the igneous rocks set. See list H. Before 1981, included use on level 3 under igneous rocks(1) diorite family(2).
BA quartz diorites
diorites
BT igneous rocks
SA diorite
plagiogranite
tonalite

quartz diorites
Term introduced in 1981. Includes use on level 2 under igneous rocks(1). See list H. Before 1981, search quartz diorite.
BA diorites
BT igneous rocks
NA quartz diorite

quartz dolerite
As of 1981, no longer a valid term for GeoRef. Use quartz diabase. Before 1981, included use on level 3 under igneous rocks(1) basalt family(2).

quartz keratophyre
As of 1981, includes use on level 3 under igneous rocks(1) trachytes(2). See list H. Before 1981, included use on level 3 under igneous rocks(1) trachyte-phonolite family(2).
BA trachytes
volcanic rocks
BT igneous rocks
SA keratophyre

quartz latite
As of 1981, includes use on level 3 under igneous rocks(1) rhyodacites(2). See list H. Before 1981, included use on level 3 under igneous rocks(1) andesite-rhyolite family(2).
BA rhyodacites
volcanic rocks
BT igneous rocks
SA latite
rhyodacite

quartz monzonite
As of 1981, includes use on level 3 under igneous rocks(1) granites(2). See list H. Before 1981, included use on level 3 under igneous rocks(1) granite-granodiorite family(2).
BA granites
BT igneous rocks

quartz porphyry
As of 1981, includes use on level 3 under igneous rocks(1) rhyolites(2). See list H. Before 1981, included use on level 3 under igneous rocks(1) andesite-rhyolite family(2).
BA rhyolites
volcanic rocks
BT igneous rocks
SA beresite
porphyry

quartz sand
See list N.
BA clastic sediments
BT sediments
SA sand

quartz syenite
As of 1981, includes use on level 3 under igneous rocks(1) syenites(2). See list H. Before 1981, included use on level 3 under igneous rocks(1) syenite family(2).
BA syenites
BT igneous rocks
SA plutonic rocks
syenite

quartz-fiber gravimeters
Term introduced in 1981.
BT gravimeters

quartzite
As of 1981, no longer a valid term for GeoRef. Use quartzites for metamorphic rocks and orthoquartzite for sedimentary rocks. Before 1981, included use on level 3 under metamorphic rocks(1) and sedimentary rocks(1).

quartzites
Term introduced in 1981. Includes use on level 2 under metamorphic rocks(1). See list J. Before 1981, search quartzite AND metamorphic rocks.
BT metamorphic rocks
SA ferruginous quartzite
orthoquartzite

quartzolite
Term introduced in 1981. Includes use on level 3 under igneous rocks(1) plutonic rocks(2). See list H. Before 1981, search silexite.
BA plutonic rocks
BT igneous rocks

Quaternary
Includes use on level 1 as an age term (list E); on level 2 under paleo-terms, e.g. paleoecology, paleogeography, paleomagnetism. Consists of Pleistocene and Holocene.
BT Cenozoic
NA Brunhes Epoch
Holocene
lower Quaternary
middle Quaternary
Pleistocene
upper Quaternary
SA Blancan
Magdalenian
Matuyama Epoch
Mousterian
Paleolithic
Tertiary
Tulare Formation
Uonuma Group
upper Cenozoic
Verde Formation

Quebec
Includes use on level 1 as an area term (list O). For term set options see list B.
CO N450000N630000
W0570000W0790000
BA Eastern Canada
Canada
NT Abitibi
Anticosti Island

Chibougamau
Dore Lake Complex
NX Frontenac County
NT Gaspe
Gaspe Peninsula
Gatineau Valley
Horne Mine
Hull
Kipawa Lake
Manicouagan Crater
Manicouagan Lake
Manicouagan River
Matagami
Montreal
Morin Complex
Noranda
Oka Complex
Quebec City
Rouyn
Saguenay Valley
Saint Lawrence Estuary
Schefferville
Sept-Iles
Thetford Mines
Val d'Or
SA Abitibi Belt
Canadian Shield
Chaleur Bay
Champlain Valley
Charlevoix
Churchill Province
Cloridorme Formation
Green Mountains
Grenville Front
Grenville Province
Hudson Bay Lowlands
Labrador Trough
Lake Champlain
Lake Timiskaming
Levis Shale
Lorrain Formation
Nipissing Diabase
Ottawa Valley
Restigouche Estuary
Saint Lawrence Lowlands
Saint Lawrence River
Saint Lawrence Valley
Sokoman Formation
Superior Province
Timiskaming
Ungava

Quebec City
City on the Saint Lawrence River near where the river widens into its estuary.
BT Quebec
Canada

Queen Alexandra Range
W of the S Ross Ice Shelf and NW of Queen Maud Range on the Pacific Ocean side.
BT Antarctica

Queen Charlotte Islands
Group of islands in Pacific Ocean separated from mainland by Hecate Strait.
BT British Columbia
Canada
SA Coast Ranges

Queen City Formation
In Claiborne Group. Includes Arp Member, Omen Glauconitic Sandstone Member, and unnamed upper sand member. NW Louisiana and E Texas.
BT middle Eocene
Eocene
SA Claiborne Group
Louisiana
Texas

Queen Elizabeth Islands
Large group in Arctic Ocean N of Parry Channel in District of Franklin.
BT Northwest Territories
Canada
SA Amund Ringnes Island

Queen Maud Mountains
use Queen Maude Range

Queen Maude Land
Large region between Coats Land and Enderby Land in Norwegian Sector on the Atlantic Ocean side. Also search Dronning Maud Land.
UF Dronning Maud Land
BT Antarctica

Queen Maude Range
SSW of S tip of Ross Ice Shelf on the Pacific Ocean side.
UF Queen Maud Mountains
BT Antarctica
SA Beardmore Glacier

Queens County
Term introduced in 1981. SE New York on Long Island.
CO N403300N404800
W0734100W0735900
BX Long Island
New York City
BT New York
United States

Queensland
Includes use on level 1 as an area term (list O). For term set options see list B.
CO S290000S100000
E1530000E1380000
BA Australia
Australasia
NT Adavale Basin
Bowen
Bowen Basin
Brisbane
Burdekin Delta
Burdekin River
Cape York Peninsula
Charters Towers
Darling Downs
Eromanga Basin
Galilee Basin
Heron Island
Ipswich
Maryborough Basin
Moreton
Moreton Bay
Mount Isa
Rockhampton
Selwyn Range
Springsure
Surat Basin
Townsville
Vale of Glamorgan
Weipa
Yarrol Basin
SA Baralaba Coal Measures
Broken River Formation
Carpentaria Basin
Cooper Basin
Georgina Basin
Great Artesian Basin
Ipswich Coal Measures
New England Batholith
Simpson Desert
Tasman Geosyncline

queenstownite
use Darwin glass

Quercus
Genus.
BA Dicotyledoneae
angiosperms
BT Spermatophyta
Plantae

Quercy
Region in SW France. Index departments as applicable.
BT France
SA Lot
Tarn-et-Garonne

Querigut Massif
In the eastern Pyrenees. Also search Querigut.
UF Querigut Mountains
BT Ariege
France

Querigut Mountains
use Querigut Massif

Quesnel Lake
SE British Columbia.
BT British Columbia
Canada

Questa Mine
Taos County in N New Mexico. Also search Questa.
BT New Mexico
United States

quick clay
Also search quickclays; quick clays.
UF quickclays
SA clay
clays
engineering geology

quickclays
use quick clay

quicksilver
use mercury

Quillan
Town in S part of country.
BT Aude
France

Quinqueloculina
Genus. Includes use on level 3 under foraminifera(1) Miliolacea(2).
BA Miliolacea
Miliolina
foraminifera
BT Protista
Invertebrata

Quintana Roo
Territory on the Caribbean Sea on E Yucatan Peninsula.
BT Mexico
SA Gulf Coastal Plain
Yucatan Peninsula

Quirke Lake
N of Lake Huron.
BT Ontario
Canada

Quseir
use Kosseir

R

R waves
use Rayleigh waves

Ra
use radium

Ra-226
Includes use on level 3 under isotopes(1).
SA isotopes
radium

Raasay
Island of the Inner Hebrides NE of Skye Island off NW.
BT Scotland
Great Britain
United Kingdom
SA Inner Hebrides

Rabat
City on the Atlantic Ocean in NW.
BT Morocco

Rabaul Caldera
On New Britain Island.
BT Bismarck Archipelago
Papua New Guinea

Rabbit Lake
Term introduced in 1981. Location of uranium deposit in central Saskatchewan.
CO N531000N531000
W1074600W1074600
BT Saskatchewan
Canada

Raca
use Racha

racemization
Includes use on level 2 under geochronology(1).
SA amino acids
geochronology

Racha
Village in central Serbia.
UF Raca
BT Serbia
Yugoslavia

Raciborz
City on Oder River in SW part of country.
UF Ratibor
BT Opole
Poland

radar
A valid term through 1977. After 1977, use radar methods.

radar methods
Term introduced in 1978. Includes use on level 3 under geophysical methods(1). Before 1978, search radar.
BA geophysical methods
SA geophysical surveys
methods
microwave methods
remote sensing
sonar methods

radial faults
Term introduced in 1978. Includes use on level 3 under faults(1) patterns(2). Before 1978, also search faults AND radial.
BA faults

radiation
As of 1978, term is restricted for use on level 3 under paleontology(1) for evolution. Before 1978, term could also have been used for radioactivity.
SA electromagnetic radiation
isolation
paleontology
particle radiation
radioactivity

radiation damage
Includes use on level 2 under geochronology(1).
UF damage, radiation
SA fission tracks
fission-track dating
geochronology
metamict minerals
metamictization
radioactivity

radio-frequency spectroscopy
Term introduced in 1978. Includes use on level 3 under spectroscopy(1) methods(2). Before 1978, also search spectroscopy AND radio frequency. See list D.
BT spectroscopy
SA analysis
methods

radio-wave methods
Term introduced in 1978. Before 1978, search radio waves.
SA electromagnetic radiation
electromagnetic waves
methods

radioactivation analysis
use activation analysis

radioactive decay
Term introduced in 1978. Includes use on level 3 under isotopes(1) or absolute age(1). Before 1978, search decay AND radioactive.
UF decay, radioactive
SA absolute age
isotopes

radioactive elements
use radioactive isotopes

radioactive isotopes
Includes use on level 3 under isotopes(1). Also search radioactive elements; radionuclides.
UF radioactive elements
BA isotopes
SA radioactivity

radioactive tracers
Includes use on level 3 under meteorology(1) aerosols(2); and under isotopes(1).
SA aerosols
isotopes
radioactivity
tracers

radioactive waste
As of 1978, term is used on level 2 under waste disposal(1).
UF waste, radioactive
SA engineering geology
environmental geology
industrial waste
liquid waste
pollution
radioactivity
solid waste
storage
waste disposal

radioactivity
For natural or induced radioactivity. Includes use on level 2 under well-logging(1) and on level 3 under soils(1), heat flow(1), and pollution(1). Before 1978, also search radiation.
SA alpha-ray spectroscopy
autoradiography
electromagnetic radiation
engineering geology
fallout
fission tracks
gamma rays
gamma-gamma methods
gamma-ray methods
gamma-ray spectroscopy
heat flow
isotopes
pollution
radiation
radiation damage
radioactive isotopes
radioactive tracers
radioactive waste
radioactivity methods
radioactivity surveys
scintillations
tracers
waste disposal
well-logging

radioactivity methods
Includes use on level 2 under geophysical methods(1).
BA geophysical methods
SA methods
radioactivity
radioactivity surveys

radioactivity surveys
Includes use on level 3 under area terms(1) geophysical surveys(2).
BA geophysical surveys
BT surveys
SA radioactivity
radioactivity methods

radiography
use autoradiography

radiography, X-ray
use X-ray radiography

Radiolaria
Includes use on level 1 and 2 as a fossil term (list F).
BT Protista
Invertebrata
NA Osculosida
Phaeodarina
Porulosida
Spumellina
SA plankton
radiolarians

radiolarians
Term introduced in 1981 as the common name for Radiolaria. See list F. To be used for Radiolaria when they are used as tools in stratigraphy. Includes use on level 1. Before 1981, search Radiolaria AND biostratigraphy. If 1, term set options are:
biostratigraphy
age [List E]
SA biostratigraphy
Radiolaria

radiolarite
Includes use on level 3 under sedimentary rocks(1) clastic rocks(2). See list I.
BA clastic rocks
BT sedimentary rocks

radiometric properties
As of 1981, no longer a valid term for GeoRef. Before 1981, included use on level 3 under Moon(1) surface properties(2). Term was introduced in 1978. Before 1978, also search radiometric; radiometry.

radionuclides
A valid term through 1978. After 1978 use radioactive isotopes.

radium
Includes use on level 1 and 2 as a chemical element (list D).
UF Ra
SA elements
isotopes
Ra-226
uranium

Radnorshire
County in E.
BT Wales
Great Britain
United Kingdom

radon
Includes use on level 1 and 2 as a chemical element (list D).
UF Rn
BT noble gases
SA argon
elements
ionium
isotopes
krypton
neon
Rn-222

radon-222
use Rn-222

Radstock Bay
SW coast of Devon Island on Lancaster Sound in District of Franklin.
BT Northwest Territories
Canada

Raft River
In N Utah and S Idaho. Index states as applicable.
BT United States
SA Idaho
Raft River basin
Utah

Raft River basin
Index states as applicable. Also search Raft River.
BT United States
SA Idaho
Raft River
Utah

rafting, ice
use ice rafting

Ragavapuram (Shales)
use Raghavapuram Shales

Raghavapuram Shales
Also search Raghavapuram.
UF Ragavapuram (Shales)
BT Upper Jurassic
Jurassic
SA Andhra Pradesh
India

Railroad Valley
Term introduced in 1981. SE central Nevada.
CO N380000N384000
W1153500W1160300
BT Nye County
Nevada
United States

rain
use rainfall

Rainbow Mountain
Peak in Salmon River Mountains in Valley County in W central.
BT Idaho
United States

raindrops
Includes use on level 3 under meteorology(1) electromagnetic waves(2) and water(2).
SA atmospheric precipitation
clouds
droplets
electrical phenomena
humidity
meteorology
rainfall
water

rainfall
Includes use on level 3 under hydrology(1). Also search rain.
UF rain
SA atmospheric precipitation
hydrology
raindrops
watersheds

Rainier Mesa
Term introduced in 1981. Outlier of the Belted Range, in NW portion of the Nevada Test Site.
CO N371000N371245
W1161215W1161430
BT Nye County
Nevada
United States
SA Nevada Test Site

Rainy Lake
Index Minnesota and/or Ontario.
SA Minnesota
Ontario

Rainy River
Flows from Rainy Lake to Lake of the Woods. Index Minnesota and/or Ontario as applicable.
SA Minnesota
Ontario

Raipur
City in SE.
BT Madhya Pradesh
India

Rajahmundry
City in NE.
BT Andhra Pradesh
India

Rajasthan
State in NW.
CO N230000N301000
E0775000E0693000
BT India
NT Ajmer
Alwar
Aravalli Range
Barmer
Bhilwara
Bikaner
Jaipur
Jaisalmer
Jhunjhunu
Jodhpur
Khetri
Khetri copper belt
Sirohi
Udaipur
SA Aravalli System
Kaimur Sandstone
Lathi Formation
Trans-Aravalli Vindhyan Basin

Rajgir
Village in N central.
BT Bihar
India

Rajmahal
Town in E.
BT Bihar
India

Rajmahal Hills
Low range of hills S and W of the Ganges River in E.
BT Bihar
India

Rakhov
Town in Transcarpathian Oblast in W.
BT Ukraine
USSR

Rakhov Massif
In central Carpathian Mountains of Transcarpathian Oblast. Also search Rakhov.
BT Ukraine
USSR

Raman spectroscopy
Term introduced in 1978. Includes use on level 3 under spectroscopy(1) methods(2) and under chemical element(1). Before 1978, also search spectroscopy AND Raman.
BT spectroscopy
SA analysis

Ramapithecus
Genus. As of 1981, includes use on level 3 under Mammalia(1) Hominidae(2). See list F. Before 1981, included use on level 3 under Mammalia(1) Primates(2).
BA Hominidae
Primates
Mammalia
BT Tetrapoda
Vertebrata

Ramgarh coal field
S Bihar, India. Also search Ramgarh.
UF Ramgarh Coalfield
BT Bihar
India

Ramgarh Coalfield
use Ramgarh coal field

Rammelsberg
Mountain in the Harz Mountains.
BT Lower Saxony
West Germany

rammelsbergite
As of 1981, includes use on level 3 under minerals(1) arsenides(2). See list L. Before 1981, included use on

level 3 under minerals(1) sulfides(2).
BA arsenides
sulfides
BT minerals
SA pararammelsbergite

Ramnagar
Town in SE Uttar Pradesh, India, and village NW of Lahore in Punjab, Pakistan. Index countries as applicable.
BT Asia
SA India
Pakistan

Rampart Group
Probably Lower Mississippian. Yukon-Tanana region in E central and E Alaska.
BT Mississippian
SA Alaska
Lower Mississippian

Ramparts Formation
Mackenzie District. Also search Ramparts.
BT Devonian
SA Canada
Northwest Territories

Rampur coal field
NW Orissa. Also search Rampur.
BT Orissa
India

ramsdellite
BA oxides
BT minerals

Ran
use Rana Fjord

Rana Fjord
On the Norwegian Sea in central part of country. Also search Rana.
UF Ran
Ranen Fjord
BT Nordland
Norway

Ranchi
City and District in S.
BT Bihar
India

rancieite
BA oxides
BT minerals

Rand
use Witwatersrand

Rand Mountains
Range in the Mohave Desert. Also search Rand.
BT California
United States

Randolph County
Index states as applicable.
BX Alabama
Arkansas
Georgia
Illinois
Indiana
Missouri
North Carolina
West Virginia
BT United States

random processes
use stochastic processes

Ranen Fjord
use Rana Fjord

range
Usage is restricted to stratigraphy. Also search stratigraphic range.
UF stratigraphic range
SA biogeography
biostratigraphy
fossils
stratigraphy
taxonomy

range, basin
use basin range structure

Rangeley Lakes
Index states as applicable. Also search Rangeley.
BT United States
SA Maine
New Hampshire

Rangely
Town in Rio Blanco County in NW.
BT Colorado
United States

Rangely Anticline
In NW near Utah border. Also search Rangely.
BT Colorado
United States

Rangely Field
Term introduced in 1978. In NW near Utah border. Before 1978, also search Rangely; Rangely oil field.
UF Rangely oil field
BT Colorado
United States

Rangely oil field
use Rangely Field

Ranger
Town in Eastland County in central.
BT Texas
United States

Rangitata River
E central South Island. Also search Rangitata.
BT South Island
New Zealand

Raniganj
City on N bank of Damodar River.
BT West Bengal
India

Raniganj (Stage)
use Raniganj Stage

Raniganj coal field
Also search Raniganj and Raniganj Coalfield.
CO N233500N233900
E0870300E0870900
UF Raniganj Coalfield
BT West Bengal
India

Raniganj Coalfield
use Raniganj coal field

Raniganj Stage
The highest division of The Damuda Series. Also search Raniganj.
UF Raniganj (Stage)
BT Upper Permian
Permian
SA Damuda Series
India
West Bengal

rank
To be used only for classification of coal.
SA coal

Rankin County
S central.
BT Mississippi
United States

Raoul Island
Largest of the volcanic Kermadec Islands which are a dependency of New Zealand 500 miles NE of Auckland.
BT Pacific Ocean
SA Kermadec Islands

rapakivi
As of 1981, includes use on level 3 under igneous rocks(1) granites(2). See list H. Before 1981, included use on level 3 under igneous rocks(1) granite-granodiorite family(2).
UF wiborgite
BA granites
BT igneous rocks

Raphidiodea
As of 1981, includes use on level 3 under Insecta(1). See list F. Before 1981, included use on level 2 under Insecta(1).
BA Insecta
BT Mandibulata
Arthropoda
Invertebrata
SA Diptera
Hymenoptera
Lepidoptera

Rapid City
City in Pennington County in SW.
BT South Dakota
United States

Rapides Parish
Central.
BT Louisiana
United States

Rappahannock River
Rises in the Blue Ridge Mountains and flows into Chesapeake Bay.
BT Virginia
United States

Rarau Massif
Suceava County in N Moldavia. Also search Rarau.
UF Rarau Mountains
BT Moldavia
Romania

Rarau Mountains
use Rarau Massif

rare earth deposits
Term introduced in 1981. Includes use on level 1 and 2 as a commodity term (list C). Before 1981, search rare earths AND deposits.
SA metal ores
monazite deposits
rare earths

rare earths
Includes use on level 1 and 2 as a chemical element (list D). As of 1981, use rare earth deposits for rare earths as a commodity. Before 1981, included use on level 1 as a commodity term. Also search lanthanides.
UF inner transition elements
lanthanide series
lanthanides
lanthanoans
SA cerium
dysprosium
elements
erbium
europium
gadolinium
holmium
KREEP
lanthanum
lutetium
metals
neodymium
praseodymium
promethium
rare earth deposits
samarium
scandium
thulium
ytterbium
yttrium

rare gases
use noble gases

RARE II regions
Term introduced in 1981. Tracts of land belonging to the U. S. Forest Service that are being evaluated for their potential as wilderness areas. Full name of program: Roadless Area Review & Evaluation, Phase II.
BT United States
SA conservation
land use

rare metals
BT metals

Raritan Formation
Comprises Amboy Stonewane Clay, Old Bridge Sand Member, South Amboy Fire Clay, Sayreville Sand Member, Woodbridge Clay, Farrington Sand Member, and Raritan Fire Clay.
BT Upper Cretaceous
Cretaceous
SA Delaware
Maryland
New Jersey

Ras al-Khaimah
Sheikdom.
BT United Arab Emirates

rasorite
use kernite

Rat Island
Small island in center of Rat Islands in W Aleutian Islands.
BT Alaska
United States
SA Aleutian Islands

Rat Islands
Group in W Aleutian Islands.
BT Alaska
United States
SA Aleutian Islands

rates
Includes use as a level 2 or 3 term appropriate to a large number of topics, e.g. on level 2 under sedimentation(1). See list G.

Ratibor
use Raciborz

ratios
Includes use on level 2 under isotopes(1); on level 3 under name of element(1) isotopes(2) and under absolute age(1). See list D. Before 1973, also search isotope ratios.
SA isotopes

Raton Basin
SE Colorado and NE New Mexico. Index states as applicable. Also search Raton.
BT United States
SA Colorado
New Mexico

Rattlesnake Mountain
BT Montana
United States

Raukumara Peninsula
NE North Island.
BT North Island
New Zealand

Rauracian
Europe. Above Argovian, below Sequanian. Includes use on level 3 under age terms(1). See list E.
BA Upper Jurassic
Jurassic
BT Mesozoic
SA Lusitanian

Ravalli County
Term introduced in 1981. W Montana.
CO N452800N463800
W1133200W1143300
BT Montana
United States

Ravalli Group
Belt Series. Includes Altyn Formation, Appekunny Formation, Grinnell Formation. NE Idaho, and NW

Montana.
BT Precambrian
SA Altyn Limestone
Belt Supergroup
Idaho
Montana

Ravenscrag Formation
BT Paleocene
SA Canada
Saskatchewan

raw humus
use Mor

Rawlins County
NW Kansas.
BT Kansas
United States

ray paths
use raypaths

Rayleigh waves
Includes use on level 3 under seismology(1).
UF R waves
BA surface waves
elastic waves
SA seismology
Stoneley waves
waves

raypaths
Term introduced in 1981. Also search ray paths.
UF ray paths
SA geophysical methods
geophysics
seismology

rays, alpha
use alpha rays

rays, cosmic
use cosmic rays

rays, gamma
use gamma rays

Razdan
River which serves as an outlet of Lake Sevan.
UF Zanga
BT Armenia
USSR

razor stone
use novaculite

Rb
use rubidium

Rb/Sr
use Sr/Rb

Re
use rhenium

reactions
Used only for chemical reactions. Also search chemical reactions.
UF chemical reactions
SA endothermic reactions
geochemistry

reactivation
Includes use on level 3 under faults(1) displacements(2).
SA displacements
faults
neotectonics

Read Bay Formation
Middle Silurian to lower Devonian. Cornwallis Island in Franklin District. Also search Read; Read Bay.
BT Paleozoic
SA Canada
Devonian
Lower Devonian
Middle Silurian
Silurian
Upper Silurian

Reading
City in Berks County in SE.
BT Pennsylvania
United States

Reagan Sandstone
In Timbered Hills Group. Central S Oklahoma.
BT Upper Cambrian
Cambrian
SA Oklahoma

reagents
SA geochemistry
solutions

realgar
UF red arsenic
red orpiment
sandarac
BA sulfides
BT minerals

Reasi
use Riasi

Recent
Not a valid index term for GeoRef. See Holocene.

Receptaculitaceae
BA Chlorophyceae
Chlorophyta
algae
BT thallophytes
Plantae
SA Coelenterata
Echinodermata
foraminifera
Porifera
problematic fossils

recharge
Includes use on level 2 under ground water(1).
UF ground-water increment
ground-water recharge
ground-water replenishment
increment
SA aquifers
artificial recharge
discharge
ground water
infiltration

Rechitsa
City in Gomel Oblast in S.
BT Byelorussia
USSR

Recife-Joao Pessoa
Two city region in neighboring states in NE. Recife is in Pernambuco and Joao Pessoa is in Paraiba.
BT Brazil

Recita
use Resita

reclamation
A level 1 term as of 1978. Used for geological studies on the reclamation of the natural environment. Includes use on level 2 under impact statements(1) and land use(1). Also search land reclamation. If 1, term set options are:
environment
beaches
drainage basins
mines
open-pit mining
strip mining
natural resources
type (floods, land, soils, waste water, etc.)
topic (experimental studies, methods, practice, programs)
subtopic
SA conservation
deforestation
depletion
environment
environmental geology
impact statements
land leases
land use
mining geology
natural resources
pollution
programs
recovery
soils
strip mining
tailings
treatment
utilization
waste disposal

reclined
A valid term through 1976. After 1976, use recumbent folds.

reclined folds
use recumbent folds

Reconcavo Basin
Fertile coastal lowland surrounding Todos os Santos Bay W of Salvador. Also search Reconcavo.
BT Bahia
Brazil

reconstruction
Includes use on level 3 under paleogeography(1) or continental drift(1).
SA continental drift
paleogeography

recording
A valid general term through 1976.

recovery
A general term used mostly for mining geology(1) or commodity sets (list C). Use secondary or tertiary recovery for petroleum if applicable.
SA mining
mining geology
reclamation
secondary recovery
tertiary recovery
water recovery

recreation
As of 1978, term is used on level 2 under land use(1).
SA land use
soils

recrystallization
Includes use on level 3 under deformation(1) field studies(2).
SA authigenesis
crystallization
deformation
grains
lithification
metamorphism
phyllonitization

rectorite
Also search allevardite.
UF allevardite
BA clay minerals
sheet silicates
silicates
BT minerals

recumbent
A valid term through 1977. After 1977, use recumbent folds.

recumbent folds
Term introduced in 1978. Includes use on level 3 under folds(1) orientation(2). Before 1978, search folds AND recumbent; folds AND reclined.
UF reclined folds
BA folds
SA nappes
orientation

recursive filters
Term introduced in 1981.
BT filters

recycling
Term introduced in 1981.
SA economic geology
economics
environmental geology

red algae
use Rhodophyta

red arsenic
use realgar

red beds
Includes use on level 3 under sedimentary rocks(1) clastic rocks(2). See list I. Also search redbeds.
UF red rock
redbeds
BA clastic rocks
BT sedimentary rocks
SA ferruginous composition
sandstone
terrigenous materials

Red Deer River
Rises in Banff National Park and flows into South Saskatchewan River near the Saskatchewan border.
BT Alberta
Canada

Red Eagle Limestone
In Council Grove Group. Includes Glenrock Limestone Member, Bennett Shale Member, Howe Limestone Member.
BT Permian
SA Kansas
Missouri
Nebraska
Oklahoma

Red Fork Sandstone
Said to lie higher than Glenn Sand, lower than Skinner Sand. NE Oklahoma.
BT Pennsylvanian
SA Oklahoma

Red Hill
Mountain in Haleakala National Park on Maui Island.
BT Hawaii
United States

Red Lake
Beltrami County in N. Includes both Upper Red Lake and Lower Red Lake.
BT Minnesota
United States

Red Mountain
Ridge primarily in Jefferson County in N central.
BT Alabama
United States

Red Mountain Formation
Overlies Chickamauga Limestone; underlies Frog Mountain Sandstone. N Alabama, and NW Georgia.
BT Silurian
SA Alabama
Georgia

red orpiment
use realgar

Red Peak Formation
In Chugwater Group. NW Wyoming.
BT Triassic
SA Wyoming

Red River
Rises in high plains of E New Mexico and flows E and SE into the Mississippi River. Index states as applicable.
BT United States
SA Arkansas
Louisiana
New Mexico
Oklahoma
Texas

Red River Formation
In Bighorn Group.
BT Upper Ordovician
Ordovician

SA Manitoba
Montana
North Dakota
South Dakota
Wyoming

Red River valley
Index states as applicable.
BT United States
SA Arkansas
Louisiana
New Mexico
Oklahoma
Texas

red rock
use red beds

Red Sea
Between NE Africa and the Arabian Peninsula connecting the Mediterranean Sea with the Indian Ocean. Includes use on level 1 or 2 as an area term (list O). If 1, see term set options under list B.
CO N100000N300000
E0430000E0340000
BA Indian Ocean
NT Atlantis II Deep
Chain Deep
Discovery Deep
Gulf of Aqaba
Gulf of Suez
SA DSDP Site 225
DSDP Site 227
DSDP Site 228
Great Rift Valley
Leg 23
Red Sea Basin
Red Sea region

Red Sea Basin
Index Red Sea, Djibouti and countries as applicable.
SA Djibouti
Egypt
Ethiopia
Israel
Jordan
Middle East
Red Sea
Red Sea region
Saudi Arabia
Southern Yemen
Sudan
Yemen

Red Sea region
Introduced as a level 1 and 2 area term (list O) in 1976. This is an artificial term used to indicate the coastal region immediately adjacent to the Red Sea and the immediate littoral zone. For term set options see list B.
SA Red Sea
Red Sea Basin

Red soils
See list M.
BA soils
SA Zonal soils

red zinc ore
use zincite

redbeds
use red beds

Redlichiida
Includes use on level 2 under Trilobita(1). See list F.
BA Trilobita
BT Trilobitomorpha
Arthropoda
Invertebrata

Redon
Town in E Brittany.
BT France

Redonda Formation
Unconformably overlies Chinle Formation; unconformably underlies Entrada Sandstone. NE New Mexico.
BT Upper Triassic
Triassic
SA New Mexico

Redondo Canyon
Off S California.
UF Redondo submerged valley
BT Pacific Ocean

Redondo submerged valley
use Redondo Canyon

Redonian
North America. See list E.
BA upper Pliocene
Pliocene
Neogene
Tertiary
BT Cenozoic

redox
use Eh

redox potential
use Eh

reduction
Includes use on level 3 under geochemistry(1) processes(2).
SA Eh
geochemistry
oxidation
processes

Redwall Limestone
Comprises four unnamed members. N Arizona.
BT Mississippian
SA Arizona

reedmergnerite
BA framework silicates
silicates
BT minerals

reef builders
Term introduced in 1981.
SA algae
Brachiopoda
Bryozoa
Coelenterata
Echinodermata
Mollusca
Porifera
reefs
stromatolites

reef pinnacle
use pinnacle reefs

reefs
Includes use on level 1 (list A); on level 3 for type of environment under sedimentation(1). Before 1974, also search coral reefs. If used on level 1, term set options are:
topic [age, distribution, ecology, evolution, lithofacies, paleoecology]
subtopic (no area term)
NT atolls
barrier reefs
fringing reefs
patch reefs
pinnacle reefs
SA biogenic structures
bioherms
biostromes
changes of level
Coelenterata
continental shelf
ecology
geomorphology
hermatypic taxa
lagoons
lithofacies
marine geology
paleoecology
reef builders
sedimentary rocks
sedimentary structures
sedimentation
shoals
spits

Reefton
Village in NW.
BT South Island
New Zealand

Reeves County
W Texas.
BT Texas
United States

refinement
As of 1978, term is used on level 2 under crystal structure(1).
SA crystal structure

reflectance
Applies to specific samples of coal, organic materials, and sometimes minerals. For ore deposits, use reflectivity. When discussing methods, use microscope methods. Includes use on level 3 under methods(2) or properties(2) under appropriate material or discipline, e.g. under minerals(1), mineralogy(1), and crystallography(1).
SA coal
microscope methods
mineralogy
minerals
organic materials
reflectivity

reflection
Includes use on level 3 under seismology(1). Used also on level 2 under ocean waves(1) through 1977. As of 1981, restricted to seismic meaning only.
SA diffraction
geophysical surveys
reflectograms
refraction
seismic methods
seismic surveys
seismology

reflection methods
Term introduced in 1978. Before 1978, search reflection AND seismic methods or geophysical methods.
BA seismic methods
BT geophysical methods
SA reflectograms

reflection waves
As of 1981, no longer a valid index term for GeoRef. From 1978-1980, used on level 2 under ocean waves(1). Before 1978, also search ocean waves AND reflection.

reflectivity
For ore deposits. Does not include reflectivity of a bed of rocks to seismic waves. Includes use on level 3 under methods(2) or properties(2) under appropriate material or discipline, e.g. under minerals(1), and crystallography(1). When discussing methods, use microscope methods.
SA albedo
crystallography
microscope methods
mineral deposits
mineralogy
minerals
reflectance

reflectograms
Term introduced in 1978.
SA reflection
reflection methods

refraction
Includes use on level 3 under seismology(1). Used also on level 2 under ocean waves(1) through 1977. As of 1981, restricted to seismic meaning only. For minerals, use refractive index.
SA diffraction
geophysical surveys
reflection
refractive index
seismic methods
seismic surveys
seismology

refraction methods
Term introduced in 1978. Before 1978, search refraction AND seismic methods or geophysical methods.
BA seismic methods
BT geophysical methods

refraction waves
As of 1981, no longer a valid index term for GeoRef. From 1978-1980, used on level 2 under ocean waves(1). Before 1978, also search ocean waves AND refraction.

refractive index
Includes use on level 3 under minerals(1). Also search index of refraction; refractivity.
UF index of refraction
refractivity
SA birefringence
minerals
optical dispersion
optical properties
refraction

refractivity
use refractive index

refractory clay
use fireclay

refractory materials
Includes use as level 3 commodity term under ceramic materials(1). See list C.
SA ceramic materials
clays
construction materials
fireclay
flint clay
graphite deposits
industrial minerals
kaolin deposits
magnesite deposits
materials

Refugian
North America. Above Fresnian, below Zemorrian. Includes use on level 3 under age terms(1). See list E.
BA Eocene
Paleogene
Tertiary
BT Cenozoic
SA Oligocene

Regensburg
City on Danube River in E central.
BT Bavaria
West Germany

Reggio
use Reggio di Calabria

Reggio Calabria
use Reggio di Calabria

Reggio di Calabria
City and province on Strait of Messina.
UF Reggio
Reggio Calabria
BT Calabria
Italy

regimes, water
use water regimes

Regina
City in S.
BT Saskatchewan
Canada

regional
After 1977, used only as a general term for geography. A valid level 2 term through 1977 used in combi-

nation with metamorphism(1); after 1977, regional metamorphism is used on level 2.

regional anomalies
Term introduced in 1981.
SA anomalies

regional metamorphism
Term introduced in 1978. Includes use on level 2 under metamorphism(1). Before 1978, also search metamorphism AND regional.
BA metamorphism
SA burial metamorphism
dynamic metamorphism

regional patterns
Includes use on level 2 under heat flow(1).
SA geothermal gradient
heat flow
heat sources
patterns

regional planning
As of 1978, term is used on level 2 under land use(1).
SA environmental geology
land use
planning
urban planning

regolith
Includes use on level 3 under planet sets(1).
SA debris
fragments
overburden
sedimentary cover
soils

regression
Term is used in stratigraphy for Pre-Quaternary sedimentation. For Quaternary, use changes of level. For statistical meaning, use regression analysis.
SA changes of level
sedimentation
stratigraphy
transgression

regression analysis
BA statistical analysis
SA analysis
least-squares analysis
mathematical geology

Reguibat Ridge
In NW Mauritania and Western Sahara.
BT Africa
SA Mauritania
Western Sahara

Rehamma
use Rehamna

Rehamna
Tribal area N of Marrakech in W central Morocco. Also search Rehamma.
UF Rehamma
BT Morocco

Reichenstein
use Zloty Stok

Reindeer Lake
Term introduced in 1981. NE Saskatchewan and NW Manitoba.
CO N562000N575500
W1013000W1032000
BT Canada
SA Manitoba
Saskatchewan

relation
A valid general term through 1977.

relative age
Includes use on level 3 under geochronology(1).
SA absolute age
age
geochronology

release fractures
Term introduced in 1978. Includes use on level 3 under fractures(1) genesis(2). Before 1978, also search fractures AND release.
BA fractures
SA joints
stress

reliability
Used as a general term.
SA accuracy
errors
probability
sampling
statistical analysis

relief
Includes use on level 3 under geomorphology(1) or ocean floors(1).
SA geomorphology
landform description
landform evolution
ocean floors
paleorelief
relief exposure
relief inversion
terrains
topography

relief effects
Term introduced in 1981 for use in relation to the Earth's magnetic field.
SA effects
magnetic field

relief exposure
Term introduced in 1981.
SA geomorphology
landforms
relief

relief features
use landforms

relief inversion
Term introduced in 1981.
SA geomorphology
landform evolution
relief

remagnetization
SA demagnetization
magnetization
natural remanent magnetization
paleomagnetism
secondary magnetization

remanent magnetism
use remanent magnetization

remanent magnetization
Includes use on level 3 under paleomagnetism(1).
UF remanent magnetism
NA anhysteretic remanent magnetization
chemical remanent magnetization
depositional remanent magnetization
inverse thermoremanent magnetization
isothermal remanent magnetization
natural remanent magnetization
partial thermoremanent magnetization
physical remanent magnetization
thermochemical remanent magnetization
thermoremanent magnetization
viscous remanent magnetization
SA adiabatic demagnetization
chemical demagnetization
coercivity
demagnetization
Koenigsberger ratio
magnetic field
magnetic properties
magnetization
paleomagnetism
thermal demagnetization

remobilization
As of 1981, no longer a valid term for GeoRef. Included use in petrology.

remote sensing
A level 1 term as of 1978. Used for both methods and applications. Includes use on level 2 under mineral exploration(1) and land use(1), and on level 3 in area sets under geophysical surveys(2). If 1, term set options are:
aerial photography
subtopic
applications
subtopic
automatic data processing
subtopic
imagery
subtopic
instruments
subtopic
interpretation
subtopic
methods
subtopic
photogeologic methods
subtopic
UF sensing, remote
SA aerial photography
airborne methods
albedo
Apollo Program
Apollo 11
Apollo 12
Apollo 13
Apollo 14
Apollo 15
Apollo 16
Apollo 17
Apollo 9
Apollo-Soyuz Program
engineering geology
environmental geology
Explorer Program
Explorer 35
geophysical methods
geophysical surveys
ground truth
helicopter methods
high-resolution methods
imagery
infrared methods
infrared surveys
land use
Landsat
lineaments
Luna Program
Luna 16
Luna 20
Luna 24
Mariner Program
Mariner 10
Mariner 6
Mariner 7
Mariner 9
microwave methods
mineral exploration
multispectral analysis
photogeologic methods
photogrammetry
Pioneer Program
Pioneer 10
radar methods
satellite methods
Skylab
sonar methods
Surveyor Program
Surveyor 3
thermal emission
Viking Program

Rendzinas
Includes use on level 3 under soils(1). See list M.
BA soils

Renfrew County
Term introduced in 1981. SE Ontario.
CO N450300N462000
W0762000W0782500
BT Ontario
Canada

renierite
BA sulfogermanates
sulfosalts
BT minerals

Renmark
Town on Murray River in SE.
BT South Australia
Australia

Rennes
City in E Brittany.
BT France

Reno
City near the California border in Washoe County in W.
BT Nevada
United States

Reno County
S central.
BT Kansas
United States

Rensselaer County
E New York.
BT New York
United States

replacement
A valid term through 1975. After 1975, use metasomatism or pseudomorphism.

report
Includes use on level 3 under symposia(1). Also search brief report.
NT progress report
SA annual report
associations
atlas
current research
symposia

representative basins
Term introduced in 1981.
SA drainage basins
ground water
hydrogeology
hydrology
models
water resources

reproduction
A general term used on level 3 under paleontology(1) or fossil sets (list F).
SA paleontology
sexual dimorphism

reptiles
Term introduced in 1981 as the common name for Reptilia. See list F. To be used for Reptilia when they are used as tools in stratigraphy. Includes use on level 1. Before 1981, search Reptilia AND biostratigraphy. If 1, term set options are:
biostratigraphy
age [List E]
SA biostratigraphy
Reptilia

Reptilia
Includes use on level 1 and 2 as a fossil term (list F).
BT Tetrapoda
Vertebrata
NA Anapsida
Archosauria

R

dinosaurs
Euryapsida
Ichthyopterygia
Lepidosauria
Synapsida
SA eggs
gastroliths
reptiles

Republic County
On the Nebraska border in N.
BT Kansas
United States

research
Includes use on level 2 under geology(1).
SA associations
bibliography
current research
geology
programs
progress report
survey organizations

reserves
Includes use on level 2 and 3 under commodity terms(1), e.g. on level 2 under coal(1). See list C.
SA prediction maps
resources

reservoir properties
Not to be used in relation to surface reservoirs.
SA Darcy's law
natural gas
oil and gas fields
petroleum
properties

reservoir rocks
Includes use on level 3 under engineering geology(1) petroleum engineering(2).
SA geopressure
geothermal energy
natural gas
oil and gas fields
petroleum
petrophysics
rocks
roughness
sedimentation
source rocks

reservoirs
A level 1 term as of 1978. Before 1978, included use on level 2 under engineering geology(1). Used for geological studies on surface reservoirs only. For subsurface reservoirs, use petroleum engineering under engineering geology.
construction
subtopic
dams
subtopic
design
subtopic
earthquakes
subtopic
experimental studies
subtopic
feasibility studies
subtopic
field studies
subtopic
maintenance
subtopic
seepage
subtopic
site exploration
subtopic
storage
subtopic
SA construction
dams
deposition
design
drawdown
earthquakes
embankments
engineering geology
feasibility studies
floods
gas storage
geologic hazards
hydrogeology
hydrology
levels
maintenance
Oroville Dam
ponds
rock mechanics
seepage
site exploration
slope stability
soil mechanics
storage
waste disposal
water balance
water storage
water supply
waterways

residual anomalies
Term introduced in 1981.
SA anomalies

residual clays
Term introduced in 1981. Includes use on level 3 under sediments(1) clastic sediments(2). See list N.
BA clastic sediments
BT sediments
SA clay
clays

residues, insoluble
use insoluble residues

residues, organic
use organic residues

residuum
Includes use on level 3 under sediments(1) clastic sediments(2). See list N.
BA clastic sediments
BT sediments
SA terrigenous materials
weathering

resinite
Includes use on level 3 under sedimentary rocks(1) organic residues(2). See list I.
BA exinite
macerals
coal
organic residues
BT sedimentary rocks

resins
Also search fossil resins; mineral resins.
BA organic materials
SA amber

resistivity
Includes use on level 3 under geophysical methods(1) electrical methods(2). Also search electrical resistivity.
UF electrical resistivity
SA conductivity
electrical conductivity
electrical logging
electrical methods
electrical surveys
geophysical methods
geophysical surveys
well-logging

Resita
City in Caras-Severin County in W part of country.
UF Recita
BT Banat
Romania

resonance
SA electron paramagnetic resonance
nuclear magnetic resonance
oscillations
vibration

resources
Includes use on level 2 and 3 under commodity terms(1), e.g. on level 2 under energy sources(1). See list C.
SA conservation
energy sources
mineral resources
natural resources
reserves
water resources

respiration
Term introduced in 1981.
SA geochemistry
photosynthesis
processes

response
As of 1981, no longer a valid term for GeoRef. Used as a general term.

Restigouche Estuary
W extremity of Chaleur Bay. Index provinces as applicable.
BT Canada
SA New Brunswick
Quebec

retention
The amount of water from precipitation that has not escaped as runoff or through evapotranspiration.
SA atmospheric precipitation
hydrology
runoff
water

Retezat Mountains
Group in W Transylvanian Alps in SW.
BT Transylvania
Romania
SA Transylvanian Alps

retrieval, data
use data retrieval

retrograde
A valid term through 1977. After 1977, use retrograde metamorphism on level 2 under metamorphism(1).

retrograde metamorphism
Term introduced in 1978. Includes use on level 2 under metamorphism(1). Before 1978, also search metamorphism AND retrograde; diaphthoresis.
UF diaphthoresis
retrogressive metamorphism
BA metamorphism
SA dynamic metamorphism
polymetamorphism
prograde metamorphism

retrogressive metamorphism
use retrograde metamorphism

Reunion
One of the Mascarene Islands and a French overseas territory 425 miles E of the Malagasy Republic. Includes use on level 1 as an area term (list O). For term set options see list B.
CO S205000S212500
E0560000E0550000
BA Indian Ocean Islands
NT Piton de la Fournaise
Piton des Neiges
Saint-Louis
SA Mascarene Islands

Revelstoke
City in SE.
BT British Columbia
Canada

reversals
Includes use on level 2 under paleomagnetism(1).
SA Brunhes Epoch
Earth
Gauss Epoch
Gilbert Epoch
Jaramillo Event
magnetic field
Matuyama Epoch
Olduvai Event
paleomagnetism
pole positions

reverse
A valid term through 1977. After 1977, use reverse faults.

reverse faults
Term introduced in 1978. Includes use on level 3 under faults(1) displacements(2). Before 1978, also search faults AND reverse.
BA faults
SA displacements
thrust faults

reverse slip faults
use thrust faults

review
Used as a general term.
SA bibliography
book reviews
critical review

revision
Used for revision of classification and taxonomy, especially in fossils (list F).
SA fossils
new taxa
taxonomy

Rewa
City and former state in N.
BT Madhya Pradesh
India

reworking
Used in relation to fossils.
SA deposition
fossils
sedimentation
sediments

Rexroad Formation
Underlies Angell Member of Ballard Formation.
BT upper Pliocene
Pliocene
SA Kansas
Oklahoma

Reykjanes Peninsula
WSW of Reykjavik. Also search Reykjanes.
CO N633000N643000
W0220000W0240000
BT Iceland

Reykjanes Ridge
SW of Iceland.
BT Atlantic Ocean

Reynolds Creek
Flows into Snake River in extreme SW.
BT Idaho
United States

Rh
use rhodium

rhabdite
use schreibersite

Rhabdomesidae
Family. Includes use on level 3 under Bryozoa(1) Cryptostomata(2).
BA Cryptostomata
Bryozoa
BT Invertebrata

rhabdophane
BA phosphates
BT minerals

Rhaetian
Europe. Above Norian, below Hettangian (Jurassic). Includes use on level 3 under age terms(1). See list E.
BA Upper Triassic
Triassic
BT Mesozoic

Rhaetian Alps
Division of Central Alps principally in Graubunden Canton, Switzerland. Index countries as applicable.
BT Alps
SA Austria
Central Alps
Italy
Switzerland

Rhapdopleurida
Includes use on level 2 under Pterobranchia(1). See list F.
BA Pterobranchia
BX Hemichordata
Invertebrata

Rhein River
use Rhine River

Rhein Valley
use Rhine Valley

Rheingau
Region on S slope of Rheingau Mountains on right bank of Rhine E of Rudesheim in extreme W.
BT Hesse
West Germany

Rheinland
use Rhineland

Rhenish Hesse
Administrative division along the left bank of the Rhine River with Worms and Mainz the major cities.
BT Rhineland-Palatinate
West Germany

Rhenish Schiefergebirge
Rhenish Slate Mountains. Extensive plateau with Belgium and Luxembourg on the W, the Lahn River on the E, the Bonn area in the N, and including the Hunsruck mountain region on the S. Also search Rhenish Slate Mountains and Schiefergebirge. Index states as applicable.
CO N495700N513200
E0091200E0064500
UF Rhenish Slate Mountains
Schiefergebirge
BT Germany
SA Hesse
North Rhine-Westphalia
Rhineland-Palatinate
West Germany
Westerwald

Rhenish Slate Mountains
use Rhenish Schiefergebirge

rhenium
Includes use on level 1 and 2 as a chemical element (list D).
UF Re
SA elements

rheology
Includes use on level 3 under deformation(1).
SA deformation
flow lines
flow mechanism
viscosity

rheomorphism
Includes use on level 2 under metamorphism(1).
SA metamorphism

Rhin River
use Rhine River

Rhin Valley
use Rhine Valley

Rhine Basin
Index Lake Constance and countries as applicable. Also search Rhine, and Rhine region.
UF Rhine region
BT Europe
SA Austria
France
Lake Constance
Liechtenstein
Lower Rhine Basin
Netherlands
Switzerland
West Germany

Rhine Graben
Rift valley extending N from Basel to Mainz. Also search Rhine Trough.
UF Rhine Trough
BT Europe
NT Lower Rhine Graben
NZ Swiss Rhine Graben
NT Upper Rhine Graben
SA Germany
Switzerland

Rhine region
use Rhine Basin

Rhine River
Formed by confluence of the Hinterrhein and the Vorderrhein rivers in SE Switzerland; it flows through Lake Constance and then W, N, and NW to the North Sea. Index Lake Constance and countries as applicable. Also search Rhine.
UF Rhein River
Rhin River
Rijn River
BT Europe
SA Austria
France
Lake Constance
Liechtenstein
Netherlands
Switzerland
West Germany

Rhine Trough
use Rhine Graben

Rhine Valley
Commonly recognized as that part of the river valley extending from Basel, Switzerland, to the North Sea. Index countries as applicable. Also search Rhine.
UF Rhein Valley
Rhin Valley
Rijn Valley
BT Europe
SA France
Lower Rhine Basin
Netherlands
Switzerland
Upper Rhine Valley
West Germany

Rhine Westphalian Basin
Term introduced in 1981.
CO N493000N522000
E0085700E0060000
BT Germany
SA West Germany

Rhineland
Region. The part of West Germany W of the Rhine River (left bank). Index states as applicable.
UF Rheinland
BT West Germany
SA North Rhine-Westphalia
Rhineland-Palatinate

Rhineland-Palatinate
State in W.
CO N490000N510000
E0083000E0061000
BT West Germany
NT Eifel
Hunsruck
Mainz
Rhenish Hesse
SA Hunsruck Shale
Lahn River
Lahn River valley
Mainz Basin
Nahe
Palatinate
Rhenish Schiefergebirge
Rhineland
Saar Basin
Saar-Nahe Basin
Westerwald

Rhinocerotidae
Includes use on level 3 under Mammalia(1) Perissodactyla(2). Also search Rhinocerotoidea. Before 1981, also search rhinoceros. As of 1981, search rhinoceros only for biostratigraphy.
BA Perissodactyla
Mammalia
BT Tetrapoda
Vertebrata

Rhipidistia
As of 1981, includes use on level 2 under Pisces(1). See list F. Before 1981, included use on level 3 under Pisces(1) Osteichthyes(2).
BA Osteichthyes
Pisces
BT Vertebrata

Rhode Island
Includes use on level 1 as an area term (list O). For term set options see list B.
CO N411000N420200
W0710700W0715500
BA New England
Eastern U.S.
United States
NX Cumberland
Kent County
NT Narragansett Bay
NX Newport
Washington County
SA Atlantic Coastal Plain
Westerly Granite

Rhodes
Largest island of the Dodecanese off SW Turkey.
BT Dodecanese
Greek Aegean Islands
Greece

Rhodesia
A valid term through 1980. After 1980, use Zimbabwe.

Rhodesian Plateau
Part of great South African plateau.
BT Zimbabwe

rhodium
Includes use on level 1 and 2 as a chemical element (list D).
UF Rh
SA elements

rhodochrosite
Also search dialogite.
UF dialogite
BA carbonates
BT minerals

rhodonite
BA chain silicates
silicates
BT minerals

Rhodope
Department in central.
BT Greece

Rhodope Massif
use Rhodope Mountains

Rhodope Mountains
Major mountain system primarily in S Bulgaria. Index countries as applicable. Also search Rhodope and Rhodope Massif.
CO N411000N421000
E0253000E0233000
UF Rhodope Massif
BT Europe
NZ Bulgarian Rhodope Mountains
SA Bulgaria
Greece

Rhodophyta
Including red algae. Includes use on level 2 under algae(1). See list F.
UF red algae
BA algae
BT thallophytes
Plantae
NA Corallinaceae
Gymnocodiaceae

Rhoen Mountains
use Rhon Mountains

Rhon Mountains
Index Suhl District and West German states as applicable. Also search Rhon; Rhoen Mountains.
UF Rhoen Mountains
BT West Germany
SA Bavaria
Hesse

Rhone
Department in E central.
CO N453000N462000
E0050300E0041500
BT France
NT Lyons
SA Monts du Lyonnais

Rhone Basin
use Rhone Valley

Rhone Delta
On Gulf of Lion W of Marseilles.
BT Bouches-du-Rhone
France

Rhone River
Rises in Rhone Glacier in S central Switzerland; it flows through Lake Geneva and then W and S into the Gulf of Lion. Index Lake Geneva and countries as applicable.
BT Europe
SA France
Lake Geneva
Switzerland

Rhone River valley
use Rhone Valley

Rhone Valley
Index Lake Geneva and countries as applicable. Also search Rhone Basin; Rhone River valley.
UF Rhone Basin
Rhone River valley
BT Europe
SA France
Lake Geneva
Saone-Rhone Basin
Switzerland

rhonite
BA chain silicates
silicates
BT minerals
SA aenigmatite

Rhum
Isle of Rhum. One of the Inner Hebrides S and SE of the Isle of Skye. Also search Rhum Island; Rum.
UF Rhum Island
Rum
BT Scotland
Great Britain
United Kingdom
SA Inner Hebrides

Rhum Island
use Rhum

rhyacolite
use sanidine

R

Rhynchocephalia
Term introduced in 1981. Includes use on level 2 under Reptilia(1). See list F.
BA Lepidosauria
Reptilia
BT Tetrapoda
Vertebrata

Rhynchonellida
Includes use on level 2 under Brachiopoda(1). See list F.
BA Articulata
Brachiopoda
BT Invertebrata
NA Rhynchonellidae

Rhynchonellidae
Family. Includes use on level 3 under Brachiopoda(1) Rhynchonellida(2).
BA Rhynchonellida
Articulata
Brachiopoda
BT Invertebrata

rhyodacite
Includes use on level 3. As of 1981, not to be used in the igneous rocks set. See list H. Before 1981, included use on level 3 under igneous rocks(1) andesite-rhyolite family(2).
BA rhyodacites
volcanic rocks
BT igneous rocks
SA dellenite
quartz latite

rhyodacites
Term introduced in 1981. Includes use on level 2 under igneous rocks(1). See list H. Before 1981, search rhyodacite.
BA volcanic rocks
BT igneous rocks
NA dellenite
quartz latite
rhyodacite

rhyolite
Includes use on level 3. As of 1981, not to be used in the igneous rocks set. See list H. Before 1981, included use on level 3 under igneous rocks(1) andesite-rhyolite family(2).
BA rhyolites
volcanic rocks
BT igneous rocks
SA metarhyolite
rhyolitic composition
volcanic rocks

rhyolite porphyry
As of 1981, includes use on level 3 under igneous rocks(1) rhyolites(2). See list H. Before 1981, included use on level 3 under igneous rocks(1) andesite-rhyolite family(2).
BA rhyolites
volcanic rocks
BT igneous rocks
SA porphyry

rhyolite tuff
As of 1981, includes use on level 3 under igneous rocks(1) pyroclastics(2). See list H. Before 1981, included use on level 3 under igneous rocks(1) pyroclastics and glasses(2).
BA pyroclastics
volcanic rocks
BT igneous rocks
SA tuff

rhyolites
Term introduced in 1981. Includes use on level 2 under igneous rocks(1). See list H. Before 1981, search andesite-rhyolite family.
BA volcanic rocks
BT igneous rocks
NA comendite
liparite
liparite porphyry
pantellerite
quartz porphyry
rhyolite
rhyolite porphyry

rhyolitic composition
Term introduced in 1978. Before 1978, search rhyolitic; rhyolitic AND composition.
SA composition
rhyolite

rhythmic bedding
Includes use on level 3 under sedimentary structures(1) planar bedding structures(2). See list K.
BA planar bedding structures
sedimentary structures
SA bedding
rhythmite

rhythmite
SA cyclothems
rhythmic bedding
sedimentation

Rians
Village in SE part of country.
BT Var
France

Riasi
Town and district in SW Jammu and Kashmir.
UF Reasi
BT Jammu and Kashmir
India

Ribeira de Iguape River
Flows into the Atlantic Ocean near Iguape in S.
BT Sao Paulo
Brazil

Rice County
Index states as applicable.
BX Kansas
Minnesota
BT United States

Rice Lake
Lake in SE Ontario, and lake in Manitoba. Index provinces as applicable.
BT Canada
SA Manitoba
Ontario

Rice Lake Group
BT Archean
Precambrian
SA Canada
Manitoba

Rich County
Term introduced in 1981. N Utah.
CO N411900N420000
W1110400W1113200
BT Utah
United States

Richards Island
Large island in Beaufort Sea at mouth of Mackenzie River, District of Mackenzie.
BT Northwest Territories
Canada

Richardson Mountains
Index territories as applicable.
BT Canada
SA Northwest Territories
Yukon Territory

Richat Mountain
E of S border of Spanish Sahara. Also search Richat.
BT Mauritania

Richland
City on the Columbia River in Benton County in S.
BT Washington
United States

Richland County
Term introduced in 1981. Index states as applicable.
BX Illinois
Louisiana
Montana
North Dakota
Ohio
South Carolina
Wisconsin
BT United States

Richmond
City on the James River in Henrico County in E central.
BT Virginia
United States

Richmond Group
Includes Queenston Shale, Arnheim Formation, Waynesville Formation, Liberty Formation, Whitewater Formation, Sequatchie Formation, Fernvale Limestone, Marnie Shale, Oswego Sandstone, Juniata Formation, Saluda Formation, Elkhorn Formation.
BT Upper Ordovician
Ordovician
SA Fernvale Formation
Illinois
Indiana
Juniata Formation
Kentucky
Missouri
Ohio

Richter Scale
Term introduced in 1981. Numerical scale of earthquake magnitude.
SA earthquakes
magnitude
seismology

richterite
BA clinoamphibole
amphibole group
chain silicates
silicates
BT minerals

Ridder
use Leninogorsk

Riddle
Town in Douglas County in SW.
BT Oregon
United States

ridges
As of 1981, no longer a valid term for GeoRef. See geomorphology.

Riding Mountain National Park
SW Manitoba. Also search Riding Mountain.
BT Manitoba
Canada

riebeckite
BA clinoamphibole
amphibole group
chain silicates
silicates
BT minerals
SA crocidolite

Ries Basin
use Ries Crater

Ries Crater
Term introduced in 1978 for the meteor crater. Before 1978, the terms Ries Basin and Nordlingen were used. Also search Ries Basin; Nordlingen; Nordlinger Ries; Nordlinger Ries Crater; Ries.
UF Nordlingen
Nordlinger Ries
Nordlinger Ries Crater
Ries Basin
BT Bavaria
West Germany

Rif
Hilly region constituting part of former Spanish Morocco extending along Mediterranean Sea from E of Melilla on E to Ceuta on W.
CO N340000N360000
W0020500W0063500
UF Er Rif
Er Riff
Riff
BT Morocco

Riff
use Rif

Rifle
Town on Colorado River in Garfield County in W.
BT Colorado
United States

Rift Valley
use Great Rift Valley

rift valleys
As of 1981, no longer a valid term for GeoRef. Use rift zones.

rift zones
Includes use on level 3 under plate tectonics(1). Before 1981, also search rift valleys.
UF zones, rift
SA fault zones
median valley
plate tectonics
tectonics
tectonophysics

rifting
Includes use on level 2 under plate tectonics(1).
SA median valley
plate tectonics
sea-floor spreading
taphrogeny

Riga
City on Gulf of Riga.
BT Latvia
USSR

right-lateral faults
Term introduced in 1978. Before 1978, search faults AND right-lateral; faults AND right-slip.
UF dextral faults
right-lateral slip faults
right-slip faults
BA faults

right-lateral slip faults
use right-lateral faults

right-slip faults
use right-lateral faults

rigidities, cutoff
use cutoff rigidities

rigidity
SA elasticity
mechanical properties
rock mechanics
stress

rigidity modulus
use shear modulus

Rijn River
use Rhine River

Rijn Valley
use Rhine Valley

Rila
Village in W part of country.
BT Sofia
Bulgaria

Rila Mountains
Range in SW at W end of the Rhodope Mountains.
BT Bulgaria

Riley County
NE central.
BT Kansas

United States

rilles
Includes use on level 3 under Moon(1).
SA Moon
valleys

Rincon Formation
Bonaire.
BT Upper Cretaceous
Cretaceous
SA West Indies

ring
A valid term through 1976. After 1976, use ring dikes.

ring complexes
As of 1978, term is used on level 2 under intrusions(1).
BA intrusions
SA complexes
dikes
ring dikes
ring structures

ring dikes
Includes use on level 3 under intrusions(1) dikes(2). Before 1976, also search ring.
UF ring-fracture intrusion
BT dikes
intrusions
SA dike swarms
ring complexes
ring structures
swarms

ring silicates
Includes use on level 2 under minerals(1). See list L. Also search cyclosilicates.
UF cyclosilicates
BA silicates
BT minerals
NA axinite
barylite
bertrandite
beryl
buergerite
cordierite
dravite
ekanite
elbaite
emerald
NZ eudialyte
NA malayaite
milarite
osumilite
roedderite
tourmaline

ring structures
Includes use on level 3 under intrusions(1). Also search ring structure.
SA dikes
intrusions
ring complexes
ring dikes
structures

ring-fracture intrusion
use ring dikes

Ringerike
Region NW of Oslo.
BT Norway

Ringerike Sandstone
Included in Ringerike Series. S Norway.
BT Paleozoic
SA Devonian
Norway
Silurian

rings, tree
use tree rings

rinneite
As of 1981, includes use on level 3 under minerals(1) chlorides(2). See list L. Before 1981, included use on level 3 under minerals(1) halides(2).
BA chlorides
halides
BT minerals

Rio Arriba County
On the Colorado border in NW.
BT New Mexico
United States

Rio Blanco Basin
River basin in NW part of country. Also search Rio Blanco.
BT Salta
Argentina

Rio Blanco County
On the Utah border in NW.
CO N394000N402000
W1070000W1090500
BT Colorado
United States

Rio Bonito Formation
BT Permian
Paleozoic
SA Brazil
upper Paleozoic

Rio Bravo
use Rio Grande

Rio Bravo del Norte
use Rio Grande

Rio Claro
City in E central.
BT Sao Paulo
Brazil

Rio de Janeiro
State in SE Brazil. Before 1981, also referred to the city.
CO S233000S205000
W0410000W0450000
BT Brazil
NT Angra dos Reis
Guanabara
Itaborai
Macacu River
Rio de Janeiro City

Rio de Janeiro City
Term introduced in 1981. Before 1981, also search Rio de Janeiro.
BT Rio de Janeiro
Brazil

Rio de la Plata
Estuary of the combined Parana and Uruguay rivers. Index countries as applicable.
UF Rio de la Plata Estuary
River Plate
BT South America
SA Argentina
Uruguay

Rio de la Plata Estuary
use Rio de la Plata

Rio Grande
River which rises in SW Colorado and flows S and then SE into the Gulf of Mexico. Index Mexican and U.S. states as applicable.
UF Rio Bravo
Rio Bravo del Norte
Rio Grande River
SA Chihuahua
Coahuila
Colorado
New Mexico
Nuevo Leon
Tamaulipas
Texas

Rio Grande Depression
Lower Rio Grande Valley. Index Mexican and U. S. states as applicable.
SA Tamaulipas
Texas

Rio Grande do Norte
State on the Atlantic Ocean in extreme NE.
CO S070000S045000
W0345500W0384000
BT Brazil
SA Borborema

Rio Grande do Sul
State in extreme S.
CO S335000S270000
W0493000W0574000
BT Brazil
NT Encruzilhada do Sul
Sao Gabriel
SA Pelotas Basin

Rio Grande Rift
Index Mexican and U.S. states as applicable. Also search Rio Grande Rift Zone.
CO N313000N370000
W1053000W1073000
UF Rio Grande Rift Zone
BT North America
SA Chihuahua
Chihuahua tectonic belt
Coahuila
New Mexico
Texas

Rio Grande Rift Zone
use Rio Grande Rift

Rio Grande River
use Rio Grande

Rio Grande Valley
Index Mexican and U. S. states as applicable.
SA Chihuahua
Coahuila
Colorado
New Mexico
Nuevo Leon
Tamaulipas
Texas

Rio Negro
River in NW South America. Flows into the Amazon River.
BT South America
SA Brazil
Colombia
Venezuela

Rio Negro Department
Term introduced in 1981. In W Uruguay. Also search Rio Negro AND Uruguay.
CO S332500S322000
W0563500W0582500
BT Uruguay

Rio Negro Meteorite
Term introduced in 1981. Impact in Brazil. Before 1981, also search Rio Negro AND meteorites.
BA L chondrites
chondrites
BT meteorites
SA Brazil

Rio Negro Province
Term introduced in 1981. In central Argentina. Also search Rio Negro AND Argentina.
CO S420000S373000
W0630000W0720000
BT Argentina
SA Neuquen Basin
Patagonia

Rio Tinto
Town in copper mining area of W Sierra Moreno in SW part of country.
UF Riotinto
BT Huelva Province
Andalusia
Spain

Rioni Basin
River basin in W Georgian Republic. Also search Rioni; Rioni River.
BT Georgian Republic
USSR

Riotinto
use Rio Tinto

Riphean
Europe. See list E.
BA upper Precambrian
Precambrian
SA Eocambrian
Sparagmite Group

Ripley Formation
In Selma Group. Includes Coon Creek Tongue, Chiwapa Sandstone Member, McNairy Sand Member, Keownville Limestone Member. Alabama, Georgia, S Illinois, W Kentucky, SE Missouri, N and central Mississippi, and W Tennessee.
BT Upper Cretaceous
Cretaceous
SA Alabama
Georgia
Illinois
Kentucky
Mississippi
Missouri
Tennessee

ripple drift-cross laminations
Includes use on level 3 under sedimentary structures(1) planar bedding structures(2). See list K.
UF ripple-cross-laminations
BA planar bedding structures
sedimentary structures
SA cross-laminations
laminations

ripple marks
Asymmetrical, interference, etc. Includes use on level 3 under sedimentary structures(1) bedding plane irregularities(2). See list K.
BA bedding plane irregularities
sedimentary structures
SA primary structures
sand waves

ripple-cross-laminations
use ripple drift-cross laminations

rise, continental
use continental rise

risk, seismic
use seismic risk

Riss
Europe. Above Mindel, below Wurm. Includes use on level 3 under age terms (1). See list E.
BA Pleistocene
Quaternary
BT Cenozoic

Riss-Wurm
use Riss/Wurm Interglacial

Riss-Wurm Interglacial
use Riss/Wurm Interglacial

Riss/Wurm Interglacial
Europe. A term applied in the Alps to the third interglacial stage of the Pleistocene Epoch, after the Riss glacial stage and before the Wurm. Also search Riss-Wurm Interglacial, and Riss-Wurm. See list E.
UF Riss-Wurm
Riss-Wurm Interglacial
BA Pleistocene
Quaternary
BT Cenozoic
SA Sangamonian

Rita Blanca Lake
Formed by Rita Blanca Dam on Rita Blanca Creek in Texas panhandle.
BT Texas
United States

Riukiu Islands
use Ryukyu Islands

river capture
use stream capture

river piracy
use stream capture

river plain
use alluvial plains

River Plate
use Rio de la Plata

rivers
Includes use on level 3 under geomorphology(1) fluvial features(2).
SA channel geometry
channels
drainage basins
drainage patterns
estuaries
floodplains
fluvial environment
fluvial features
geomorphology
hydrogeology
hydrology
hydrosphere
meanders
rivers and streams
runoff
stream capture
stream transport
streams
surface water
water supply
watersheds
waterways

rivers and streams
Includes use on level 2 under hydrology(1) and under waterways(1).
SA channel geometry
drainage basins
estuaries
floods
fluvial environment
fluvial features
hydrology
hydrosphere
meanders
precipitation
rivers
runoff
stream capture
stream gradient
stream transport
streams
valleys
waterways

Riverside
City SE of Los Angeles in Riverside County.
BT California
United States

Riverside County
Southern California.
BT California
United States

Rn
use radon

Rn-222
Includes use on level 3 under isotopes(1). Also search thoron.
UF radon-222
thoron
SA isotopes
radon

road log
UF log, road
SA field trips
guidebook
maps

road tests
Term introduced in 1981.
SA engineering geology
highways

Roan Supergroup
Interlayered with Carolina Gneiss in Spruce Pine District. N Georgia, W North Carolina, NW South Carolina, and E Tennessee.
BT Precambrian
SA Georgia
North Carolina
South Carolina
Tennessee

Roane County
Index states as applicable.
BX Tennessee
West Virginia
BT United States

Roanoke
City in Roanoke County in SW.
BT Virginia
United States

Roaring Brook Valley
Valley in Hartford County and another in Tolland County.
BT Connecticut
United States

robbery
use stream capture

Robertinacea
Superfamily. Includes use on level 2 under foraminifera(1).
BA Rotaliina
foraminifera
BT Protista
Invertebrata

Roberts Mountains
Eureka County in central.
BT Nevada
United States

Roberts Mountains Formation
Underlies Rabbit Hill Formation. E Nevada, and W Utah.
BT Middle Silurian
Silurian
SA Nevada
Utah

Robles Formation
BT Cretaceous
SA Puerto Rico
West Indies

Roca Formation
Underlies Sallyards Limestone Member of Grenola Limestone; overlies Howe Limestone Member of Red Eagle Limestone. NE Kansas, SE Nebraska, and N Oklahoma.
BT Wolfcampian
Lower Permian
Permian
SA Kansas
Nebraska
Oklahoma

Roccamonfina
Village in extinct volcanic crater in NW.
BT Campania
Italy

Rochechouart
Town in W central part of country.
BT Haute-Vienne
France

Rochester
City in Monroe County on S Lake Ontario.
BT New York
United States

Rochester Formation
Includes Keefer Sandstone Member, Decew Waterlime Bed.
BT Middle Silurian
Silurian
SA Maryland
New York
Ontario
Pennsylvania
West Virginia

rock bursts
As of 1978, term is used on level 2 under geologic hazards(1).
UF bursts, rock
SA geologic hazards

Rock Creek Park
Also search Rock Creek.
BT District of Columbia
United States

rock crystal
use quartz crystal

rock failure
use failures

rock glaciers
SA boulders
glaciers

Rock Island
City in Rock Island County in NE.
BT Illinois
United States

Rock Lake Shale
use Rock Lake Shale Member

Rock Lake Shale Member
Of Stanton Limestone. E Kansas, NW Missouri, and SE Nebraska.
UF Rock Lake Shale
BT Missourian
Upper Pennsylvanian
Pennsylvanian
SA Kansas
Missouri
Nebraska
Stanton Formation

rock mechanics
A level 1 term as of 1978. Includes use on level 2 under nuclear facilities(1), tunnels(1), and underground installations(1). Used for geotechnical studies. If 1, term set options are:
applications
subtopic
case studies
subtopic
concepts
subtopic
deformation
subtopic
elasticity
subtopic
excavations
subtopic
experimental studies
subtopic
field studies
subtopic
frost action
subtopic
failures
subtopic
materials, properties
subtopic
methods
subtopic
plasticity
subtopic
site exploration
subtopic
techniques
subtopic
theoretical studies
subtopic
SA Atterberg limits
case studies
cohesive materials
competent materials
compressive strength
congelifraction
cracks
cryoturbation
dams
deformation
ductility
earthquakes
elastic constants
elastic materials
elasticity
engineering geology
excavations
explosions
extensometers
failures
finite strain analysis
foundations
frost action
geologic hazards
highways
hydraulic fracturing
land subsidence
marine installations
materials
mechanics
microcracks
nuclear facilities
penetration
permeability
plastic materials
plasticity
reservoirs
rigidity
shear strength
site exploration
slope stability
soil mechanics
strain
strength
stress
structural analysis
testing
triaxial tests
tunnels
underground installations
vibration

rock salt
use halite

rock slides
use rockslides

rock slip
use rockslides

Rock Springs
Town in Sweetwater County in SW.
BT Wyoming
United States

rock-stratigraphy
use lithostratigraphy

Rockall Bank
About 225 miles W of the Hebrides.
BT Atlantic Ocean

Rockall Plateau
Rise or submarine plateau just N of Rockall Bank W of the Hebrides.
UF Rockall Rise
BT Atlantic Ocean

Rockall Rise
use Rockall Plateau

Rockall Trench
use Rockall Trough

Rockall Trough
SE of Rockall Bank W of the Hebrides.
UF Rockall Trench
BT Atlantic Ocean

rockbridgeite
BA phosphates
BT minerals

rockfalls
As of 1978, term is used on level 2 under slope stability(1).
SA slope stability

Rockhampton
City in E.
BT Queensland
Australia

Rockingham County
Index states as applicable.
BX New Hampshire
North Carolina
Virginia
BT United States

rocks
Includes use on level 3 under engineering geology(1) materials, properties(2); on level 3 under inclusions(1).
SA aggregate
bedrock
cap rocks
carbonate rocks
chemically precipitated rocks
country rocks
crystalline rocks
extrusive rocks
fluid inclusions
host rocks
hypabyssal rocks
inclusions
intrusive rocks
metaigneous rocks
metamorphic rocks
metaplutonic rocks
metasedimentary rocks
metasomatic rocks
metavolcanic rocks
paragenesis
plutonic rocks
reservoir rocks
sedimentary rocks
silicate rocks
source rocks
standard rocks
volcanic rocks

rockslides
Also search rock slides.
UF rock slides
rock slip
SA landslides
mass movements

Rockwood Formation
Underlies Hancock Limestone; overlies Sequatchie Formation. E Tennessee and NW Georgia.
BT Silurian
SA Georgia
Lower Silurian
Middle Silurian
Tennessee

Rocky Mountain Trench
A trough in the Northern Rockies, E of the interior plateau of British Columbia, between a series of parallel ranges extending WNW from Montana to the headwaters of the Yukon River. Index British Columbia, Washington, and Yukon Territory as applicable.
BT North America
SA British Columbia
Montana
Rocky Mountains
Yukon Territory

Rocky Mountains
Mountain system in W North America extending from N Alaska to the Mexican frontier. Index Alaska, and countries as applicable. Includes use on level 1 as an area term (list O). For term set options see list B.
BA North America
NX Canadian Rocky Mountains
NA Central Rocky Mountains
Northern Rocky Mountains
Southern Rocky Mountains
NX U. S. Rocky Mountains
SA Absaroka Range
Alaska
Big Belt Mountains
Bighorn Mountains
Bitterroot Range
Brooks Range
Canada
Canadian Cordillera
Cariboo Mountains
Elk Mountains
Flint Creek Range
Franklin Mountains
Front Range
Laramie Mountains
Little Belt Mountains
Mackenzie Mountains
Medicine Bow Mountains
North American Cordillera
Rocky Mountain Trench
Rocky Mountains foreland
San Juan Mountains
Sangre de Cristo Mountains
Sawatch Range
Selkirk Mountains
Tobacco Root Mountains
Uinta Mountains
United States
Wasatch Range
Wet Mountains
Wind River Range

Rocky Mountains foreland
Term introduced in 1981. Stable region E of the Rocky Mountains.
BT North America
SA Rocky Mountains

Rocroi
Town in N part of country.
UF Rocroy
BT Ardennes
France

Rocroy
use Rocroi

Rodentia
Includes use on level 2 under Mammalia(1). See list F. Before 1981, also search rodents. As of 1981, search rodents only for biostratigraphy.
BA Mammalia
BT Tetrapoda
Vertebrata
NA Hystricomorpha
Myomorpha
Sciuromorpha
SA rodents

rodents
Term introduced in 1981 as the common name for Rodentia. See list F. To be used for Rodentia when they are used as tools in stratigraphy. Before 1981, search (rodents OR Rodentia) AND biostratigraphy.
SA biostratigraphy
Rodentia

Rodez Pass
use Rodez Trough

Rodez Trough
Between Segala Plateau in S and Causse du Comtal in N. Index departments as applicable. Also search Rodez; Rodez Pass.
UF Rodez Pass
BT France
SA Aveyron
Haute-Garonne
Tarn

rodingite
As of 1981, includes use on level 3 under igneous rocks(1) gabbros(2). See list H. Before 1981, included use on level 3 under igneous rocks(1) gabbro family(2).
BA gabbros
BT igneous rocks

Rodna
Village in NE Cluj County in NE.
BT Transylvania
Romania

Rodna Mountains
Range of the Carpathians along border between Maramures and Cluj Counties in N.
UF Rodnei Mountains
BT Transylvania
Romania
SA Carpathians

Rodnei Mountains
use Rodna Mountains

Rodrigues Island
use Rodriguez Island

Rodriguez Island
A dependency of Mauritius in the Mascarene Islands about 500 miles E of the Malagasy Republic.
UF Rodrigues Island
BT Mascarene Islands
Indian Ocean

roedderite
Meteorite mineral.
BA ring silicates
silicates
BT minerals

roemerite
BA sulfates
BT minerals

roestone
use oolite

Rogaland
County in SW.
BT Norway
NT Sandnes
Sogndal
Stavanger

roggianite
BA orthosilicates
silicates
BT minerals

Rogue River
SW Oregon. Also search Rogue.
BT Oregon
United States

Rokko Mountains
NE of Kobe in S central.
BT Honshu
Japan

roll orebodies
use roll-type deposits

roll-type deposits
Term introduced in 1978. Before 1978, search roll-type. Also search ore rolls.
UF ore rolls
roll orebodies
SA mineral deposits
uranium ores

Rolla Quadrangle
Term introduced in 1981. SE Missouri.
CO N370000N380000
W0900000W0920000
BT Missouri
United States

Romagna
Historical region in N central part of country on the Adriatic Sea.
BT Emilia-Romagna
Italy

Romanche Deep
use Romanche Trench

Romanche fracture zone
Cuts across the Mid-Atlantic Ridge between the bulge of South America and Liberia.
BT Atlantic Ocean

Romanche Gap
use Romanche Trench

Romanche Trench
Extends in a NW-SE direction to the SW of the Romanche fracture zone between bulge of South America and Liberia.
UF Romanche Deep
Romanche Gap
BT Atlantic Ocean

Romania
Includes use on level 1 as an area term (list O). For term set options see list B.
CO N434000N481000
E0294500E0201500
UF Roumania
Rumania
BA Southern Europe
Europe
NT Apuseni Mountains
Arges River
Bistrita Mountains
Bucegi Mountains
Dacian Basin
Dobruja
Fagaras Mountains
Getic Nappe
Haghimas Syncline
Jiu River valley
Mehedinti Plateau
Olt River
Olt River valley
Paring Mountains
Romanian Plain
Transylvania
Transylvanian Alps
Vrancea
Vulcan Mountains
Walachia
SA Balkan Peninsula
Banat
Black Sea region
Bukovina
Carpathian Foredeep
Carpathian Foreland
Carpathians
Danube Delta
Danube Plain
Danube River
Danube Valley
Dobruja Basin
Eastern Carpathians
Moesia
Moesian Platform
Moldavia
Prut River
Somes Basin
Subcarpathians
Tisza River
Tulghes Series

Romania Plain
use Romanian Plain

Romanian
Provincial series, Europe.
BA upper Pliocene
Pliocene
Neogene
Tertiary
BT Cenozoic

Romanian Plain
Area between the Transylvanian Alps on the N, and the Danube River on the S and E. Index regions as applicable. Also search Romanian.
UF Romania Plain
BT Romania
SA Moldavia
Walachia

Romashkino Field
Term introduced in 1978. Orenburg Oblast in Southern Urals. Before 1978, also search Romashkino; Romashkino oil field.
UF Romashkino oil field
BT Russian Republic
USSR

Romashkino oil field
use Romashkino Field

Rome
City on the Tiber River. Also search Roma.
BT Latium
Italy

Rome Formation
Named for exposures south of Rome, Georgia.
BT Lower Cambrian
Cambrian
SA Alabama
Georgia
North Carolina
Tennessee
Virginia

Ronda Massif
use Serrania de Ronda

Ronda, Serrania de
use Serrania de Ronda

Ronda Sierra
use Serrania de Ronda

Rondonia
Territory. In W part of country on Bolivian border.
CO S135000S075500
W0600000W0664500
UF Guapore
BT Brazil

Rondout Formation
Includes Fuyk Sandstone Member.
BT Paleozoic
SA Devonian
Lower Devonian
New York
Pennsylvania
Silurian
Upper Silurian

Rooks County
NW central.
BT Kansas
United States

Roorkee
Town in N Uttar Pradesh.
UF Rurki
BT Uttar Pradesh
India

Roosevelt County
Index states as applicable.
BX Montana
New Mexico
BT United States

Roosevelt Hot Springs
use Roosevelt Hot Springs KGRA

Roosevelt Hot Springs KGRA
Term introduced in 1981. Known Geothermal Resources Area. Before 1981, also search Roosevelt Hot Springs.
CO N381500N385000
W1122000W1133000
UF Roosevelt Hot Springs
BT Utah
United States

root clay
use underclay

roquesite
BA sulfides
BT minerals

Roraima
Term introduced in 1981. Territory in N.
CO S013000N051000
W0585000W0644500
BT Brazil

Roraima Formation
Tentatively dated from lower Proterozoic to as young as Miocene. In the Pacaraima Mountains.
SA Brazil
Guyana
South America
Venezuela

Ros River valley
W central Ukraine. Also search Ros; Ros River.
BT Ukraine
USSR

Rosario Formation
BT Upper Cretaceous
Cretaceous
SA Baja California
Campanian
Maestrichtian
Mexico

Roscommon County
N central Lower Peninsula.
BT Michigan
United States

Rose Canyon Formation
Overlies Torrey Sand; underlies Poway Conglomerate. San Diego County.
UF Rose Canyon Shale
BT Eocene
SA California
lower Eocene
middle Eocene

Rose Canyon Shale
use Rose Canyon Formation

Rosebery
Village in W.
BT Tasmania
Australia

Rosebud County
Term introduced in 1981. E central Montana.
CO N451000N465200
W1060700W1075500
BT Montana
United States

Rosetown
Town in SW.
BT Saskatchewan
Canada

Rosia
Village 25 miles SE of Oradea in Bihor County in W.
BT Transylvania
Romania

Rosidae
Term introduced in 1981. Includes use on level 2 under angiosperms(1). See list F.
BA Dicotyledoneae
angiosperms
BT Spermatophyta
Plantae

Ross Barrier
use Ross Ice Shelf

Ross Dependency
Section lying S of 60° S and between 160° E and 150° W extending to the South Pole. Placed under jurisdiction of New Zealand in 1923 by act of British Parliament. Also search Ross.
BT Antarctica

Ross Ice Shelf
Covers S part of Ross Sea between Marie Byrd Land and Victoria Land with its S end at the foot of Queen Maude Range on the edge of the Antarctic continent. Also search Ross.
UF Ross Barrier
BT Antarctica
SA Beardmore Glacier
McMurdo Ice Shelf
Shackleton Glacier

Ross Island
In Ross Sea at W end of Ross Ice Shelf separated from Victoria Land by McMurdo Sound. Also search Ross.
BT Antarctica

Ross Sea
Arm of S Pacific Ocean just N of Ross Ice Shelf between Victoria Land and Edward VII Peninsula. Also search Ross.
CO S783000S713000
W1580000E1620000
BT Antarctic Ocean
SA DSDP Site 270
DSDP Site 274
Pacific Ocean

Ross-shire
A separate county until the 17th century but now part of Ross and Cromarty County in N.
CO N573000N580000
W0043000W0053000
BT Scotland
Great Britain
United Kingdom

Rosses Granite
BT Donegal
Ireland

Rossland
City in SE.
BT British Columbia
Canada

Rostock Bezirk
Term introduced in 1981. District in N.
CO N534500N544000
E0141300E0104500
BT East Germany

Rostov
Oblast.
BT Russian Republic
USSR
NT Rostov-na-Donu

Rostov-na-Donu
Term introduced in 1981. City in Rostov (oblast).
BT Rostov
Russian Republic
USSR

Rostroconchia
Includes use on level 2 under Mollusca(1). See list F.
BA Mollusca
BT Invertebrata

Rotalia
Genus. Includes use on level 3 under foraminifera(1) Rotaliacea(2).
BA Rotaliacea
Rotaliina
foraminifera
BT Protista
Invertebrata

Rotaliacea
Includes use on level 2 under foraminifera(1). See list F.
BA Rotaliina
foraminifera
BT Protista
Invertebrata
NA Ammonia
Elphidium
Heterostegina
Miogypsinidae
Nummulitidae
Rotalia

Rotaliina
Includes use on level 2 under foraminifera(1). See list F.
BA foraminifera
BT Protista
Invertebrata
NA Buliminacea
Carterinacea
Cassidulinacea
Discorbacea
Globigerinacea
Nodosariacea
Orbitoidacea
Robertinacea
Rotaliacea
Spirillinacea

Rotalipora
Genus. Includes use on level 3 under foraminifera(1) Globigerinacea(2).
BA Globigerinacea
Rotaliina
foraminifera
BT Protista
Invertebrata

rotation
Includes use on level 3 under plate tectonics(1) and Earth(1).
SA Earth
plate tectonics
polar wandering

rotational wave
use S-waves

Rotliegendes
Europe. Below Zechstein. Includes use on level 3 under age terms(1). See list E.
BA Permian
BT Paleozoic
SA Autunian
Lower Permian
Middle Permian
Saxonian

Rottleberode
Village in SW central part of country.
BT Halle Bezirk
East Germany

Rouen
City on the Seine River in N part of country.
BT Seine-Maritime
France

Rouergue
Ancient province. Primarily in Aveyron Department in S central. Index departments as applicable.
BT France
SA Aveyron
Tarn-et-Garonne

Rough Creek fault zone
W Kentucky.
BT Kentucky
United States

roughness
SA hydraulics
permeability
reservoir rocks

Roumania
use Romania

Round Mountain
On E spur of Great Dividing Range in NE.
BT New South Wales
Australia

Round Mountain Quadrangle
Term introduced in 1981. S central Nevada.
CO N383000N390000
W1170000W1173000
BT Nye County
Nevada
United States

rounding
Includes use on level 3 under sedimentary rocks(1).
SA particles
pebbles
roundness
sand
sedimentary rocks
sorting

roundness
SA particles
rounding
sediments
shape analysis
sphericity

Roussillon
Region and former province roughly coextensive with Pyrenees-Orientales Department on the Spanish border.
BT France

Rouyn
Mining city in SW near N Ontario border.
BT Quebec
Canada

Rowan County
Index states as applicable.
BX Kentucky
North Carolina
BT United States

Rowne
use Zloty Stok

Royal Creek
BT Yukon Territory
Canada

Rozdol
Town in Lvov Oblast in E.
BT Ukraine
USSR

rozenite
BA sulfates
BT minerals

Roztocze
Mountain range. Index Poland and/or Ukraine as applicable.
UF Tomaszow-Lvov Ridge
BT Europe
SA Poland
Ukraine

Ru
use ruthenium

Ruanda
use Rwanda

Ruapehu
Volcano in Tongariro National Park in S central North Island.
UF Ruapehu Volcano
BT North Island
New Zealand

Ruapehu Volcano
use Ruapehu

rubblerock
use breccia

rubidium
Includes use on level 1 and 2 as a chemical element (list D). Also search Rb.
UF Rb
SA alkali metals
elements
isotopes
Sr/Rb

Ruby Mountains
Range in Elko and White Pine counties in NE Nevada. Also search Ruby Range.
BT Nevada
United States
SA Ruby Range

Ruby Range
N extension of Snowcrest Mountains in SW Montana, lies just W of Ruby River.
BT Montana
United States
SA Ruby Mountains

Ruda
Town in coal mining region in S part of country.
BT Katowice
Poland

Rudistae
As of 1981, includes use on level 2 under Mollusca(1). See list F. Before 1981, included use on level 3 under Mollusca(1) Bivalvia(2). Before 1981, also search rudists. As of 1981, search rudists only for biostratigraphy.
BA Bivalvia
Mollusca
BT Invertebrata
SA rudists

rudists
As of 1981, to be used for Rudistae when they are used as tools in stratigraphy. See list F. Before 1981, included use on level 3 under Mollusca(1) Bivalvia(2). Before 1981, search (rudists OR Rudistae) AND biostratigraphy.
SA biostratigraphy
Rudistae

Rudki
Town in Lvov Oblast in E.
BT Ukraine
USSR

Rudnichny
Town in W Chelyabinsk Oblast in Southern Urals of W Siberia.
UF Rudnichnyy
BT Russian Republic
USSR

Rudnichnyy
use Rudnichny

Rudny Altai
Region in the Altai Mountains in Vostochno Kazakhstan Oblast and in Altai Kray. Index Soviet republics as applicable.
CO N480000N520000
E0873000E0823000
UF Rudnyy Altai
Rudnyy Altay
BT USSR
SA Kazakhstan
Russian Republic

Rudnyy Altai
use Rudny Altai

Rudnyy Altay
use Rudny Altai

Rudolf (Lake)
use Lake Turkana

Ruegen (Island)
use Rugen Island

Rugen Island
In the Baltic Sea just off Rostock District. Largest island of East Germany. Also search Ruegen; Rugen.
UF Ruegen (Island)
BT East Germany

Rugosa
Includes use on level 2 under Coelenterata(1). See list F.
BA Anthozoa
Coelenterata
BT Invertebrata
SA Scleractinia
Tabulata
Tetracorallia

Ruhla
Town in Thuringian Forest in SW part of country.
BT Erfurt Bezirk
East Germany

Ruhr
Major coal-mining and industrial region. Includes the Ruhr River valley and the Dusseldorf area to the S along the Rhine River.
UF Ruhr Basin
Ruhr Valley
BT North Rhine-Westphalia
West Germany

Ruhr Basin
use Ruhr

Ruhr Valley
use Ruhr

Rum
use Rhum

Rumania
use Romania

Rumilly
Town in E part of country.
UF Rumilly-Albanais
BT Haute-Savoie
France

Rumilly-Albanais
use Rumilly

Ruminantia
As of 1981, includes use on level 2 under Mammalia(1). See list F. Before 1981, included use on level 3 under Mammalia(1) Artiodactyla(2).
BA Artiodactyla
Mammalia
BT Tetrapoda
Vertebrata
NA Bovidae
Camelidae
Cervidae

Runnels County
W central.
BT Texas
United States

runoff
Includes use on level 3 under hydrology(1) rivers and streams(2).
SA discharge
hydrology
precipitation
retention
rivers
rivers and streams
sediment yield
streams
surface water
water balance
water yield
watersheds

runout
use water yield

Rupelian
Europe. Above Tongrian, below Chattian. Includes use on level 3 under age terms(1). See list E.
BA Oligocene
Paleogene
Tertiary
BT Cenozoic
SA Krosno Beds
Stampian

Rurki
use Roorkee

Rusca Montana
Village in the Poiana-Rusca Mountains in Caras-Severin County in E part of country.
UF Rusca-Montana
BT Banat
Romania

Rusca-Montana
use Rusca Montana

Ruse
Province in N.
BT Bulgaria

Rush County
Index states as applicable.
BX Indiana
Kansas
BT United States

Rusk County
Term introduced in 1981. Index states as applicable.
BX Texas
Wisconsin
BT United States

Rusophycus
BA ichnofossils
SA Trilobita

Russell Cave
Term introduced in 1981. NE Alabama, in Jackson County.
CO N345600N345900
W0854700W0855000
BT Jackson County
Alabama
United States

Russell County
Index states as applicable.
BX Alabama
Kansas
Kentucky
Virginia
BT United States

Russia
use USSR

Russian
Used to indicate language of a document.

Russian Basin
use Russian Plain

Russian Plain
Comprises most of European USSR W of the Urals and is bordered on the S by the Carpathians, the Crimean Mountains and the Caucasus. Index Soviet republics as applicable. Also search Russian AND plain; Russian Basin.
UF East European Plain
Russian Basin
BT USSR
SA Byelorussia
Estonia
Latvia
Lithuania
Moldavia
Russian Platform
Russian Republic
Ukraine

Russian Platform
Ancient platform of Precambrian crystalline rocks overlain by sedimentary deposits of the Russian or East European Plain. Also search East European Platform; Russian AND platform. Index Soviet republics as applicable.
CO N523000N840000
E0660000E0190000
BT USSR
NA Baltic Plain
Byelorussia
NZ Crimea
NA Dnieper-Donets Basin
NZ Franz Josef Land
NA Lithuania
Moscow-Pechora Syneclise
NZ Timan Ridge
Transcarpathia
Ukrainian Shield
NA Ukrainian Syneclise
Voronezh-Volga Anteclise
SA Moldavia
Russian Plain
Russian Republic
Soviet Carpathians
Soviet Fennoscandia
Ukraine

Russian Republic
Russian Soviet Federated Socialist Republic (RSFSR) in both European and Asiatic USSR.
CO N370000N830000
W1700000E0280000
BT USSR
NT Abakan
Aldan
Aldan Plateau

R

Aldan River
Allarechenskiy
Altai
Anabar Bay
Anabar River
Anadyr Basin
Anadyr Range
Anapa
Angara River
Arbarastakh
Argun
Arkhangelsk
Arlan
Astrakhan
Avacha
Azov
Baikal Mountains
NX Baikal region
NT Baikal rift zone
Bakal
Balei
Bashkiria
Batenev Ridge
Belaya Gora
Belgorod
Belomorsk
Belozero
Biryusa
Blyava
Bodaibo
Bolshezemelskaya Tundra
Bryansk
Bureya
Buryat
Caucasus Foreland
NX Central Urals
NT Chadobets Uplift
Chaya Massif
Chechen-Ingush
Chekalin
Chelyabinsk
Chita
Chuya
Chuya Alps
Chuya Basin
Dagestan
Darasun
Don Basin
Don River
Dzhagdy Range
Dzhida River
Elbrus
NX Franz Josef Land
NT Gay
Golovanevsk
Gorki
Gornaya Shoriya
Gorny Altai
Goryachiy Plyazh
Grozny
Igarka
Ilmen Mountains
Imandra
Indigirka River
Irkutsk
Irkutsk Basin
Itaka
Iturup Island
Iya River
Kabardia
Kabardin-Balkar
Kalinin
Kaliningrad
Kalmyk
Kama River
Kamchatka
Kamenka River
Kamyshin
Kanin Peninsula
Kansk
Kansk-Achinsk Basin
Karaginskiy Island
Karelia
Kavalerovo
Kazan
Kemerovo
Khabarovsk region
Khamar-Daban Range
Khapcheranga
Khatanga
Khatanga Basin
Khatanga River
Kheta River
Khibiny Mountains
Khovu-Aksy
Kirov
Kiya River
Kizel coal basin
Klichka
Klyuchevskaya Sopka
Kochkar
Kodar Range
Kola
Kola Peninsula
Kolyma River
Kolyma River basin
Komi
Kommunar
Komsomolsk
Kostroma
Kotui
Kozhim
Krasnodar
Krasnoyarsk
Kuban
Kuibyshev
Kular Range
Kuma Basin
Kunashir Island
Kureyka River region
Kurgan
Kursk
Kursk magnetic anomaly
Kushva
NX Kuznetsk Alatau
NT Kuznetsk Basin
Laba Basin
Lake Baikal
Lake Ladoga
Lake Onega
Lena Basin
Lena River
Leningrad
Leningrad Mining Institute
Lovozero
Lovozero Massif
Lower Tunguska River
Magadan
Magnitogorsk
Maikop
Makhachkala
Malgobek
Mama
Mama River
Mariinsk
Maya River basin
Maymecha
Maymecha-Kotuy
Megion Field
Mendeleyev Volcano
Mezen
Mezen River basin
Mikhailovka
Minusinsk
NX Minusinsk Basin
NT Molotov
Monchegorsk
Moneron Island
Moscow
Moscow Basin
Moscow Syneclise
Moscow University
Moskva River
Mukhor-Tala
Murmansk
Narym
Nayba River basin
Nenets
Norilsk
Norilsk region
North Ossetia
Northeastern USSR
NX Northern Caucasus
NT Northern Dvina River
NX Northern Urals
Novaya Zemlya
NT Novgorod
Novosibirsk
Ob River
Oka River
Okhotsk
Okhotsk Massif
Okhotsk region
Okhotsk-Chukchi
Okhotsk-Chukchi volcanic belt
Olekma
Olekma-Vitim Highlands
Olenek River
Omolon
Omolon Block
Omsk
Omulevka
Onega
Orenburg
Orlov
Orsk
Orulgan Mountains
Ozernoye
Pachelma
Pai-Khoi
Paramushir
Paratunka
Patom Plateau
Pechenga
Pechora
Pechora Basin
Penzhina Bay
Peredovoy Range
Perm
NX Polar Urals
NT Polousnyy
Popigay
Prikumsk
Primorye
Prokopyevsk
Pskov
Romashkino Field
Rostov
Rudnichny
Ryazan
Sadon
Sakmara
Salair
Salair Ridge
Salym
Samara Bend
Sangilen Mountains
Saratov
Sarbay
Segozero
Selennyakh
Serov
Sette-Daban Range
NX Severnaya Zemlya
NT Sheveluch
Shilka Valley
Shoriya Mountains
Sibay
Siberian Platform
Slyudyanka
Smolensk
Sochi
Solikamsk
Soviet Far East
Soviet Pacific region
Sredniy Vasyugan
Stavropol region
Stony Tunguska River
Suchan Basin
Sunzha
Surgut
Sutam River
Sverdlovsk
Sysert
Tagil Basin
Taiga
Taman
Taman Peninsula
Tambov
Tareya
Tas-Khayakhtakh Range
Tatar
Tatar Arch
Tatar Strait
Tataria
Taygonos Peninsula
NX Taymyr
NT Taymyr Peninsula
Taz Basin
Tetyukhe
NX Timan Ridge
NT Timan-Pechora region
Tomsk
Transbaikalia
Tuapse
Tula
Tunguska Basin
Tunguska River
Tunguska Syneclise
Turukhan
Turukhansk
NX Tuva
NT Tuymazy
Tyrny-Auz
Tyumen
Uchaly
Uchur River basin
Uda River
Udachnaya
Udmurtia
Udokan Mountains
Ufa
Ufaley
Ulkan
Ulyanovsk
Unda
Ural region
Ural-Tau
Uralian Foreland
Urup
USSR Academy of Sciences
Ust-Kut
Ust-Yenisei Basin
Uzon
Valdai
Vaygach Island
Verkhoyansk
Verkhoyansk Range
Vetrenyy Ridge
Vilyuy River
Vilyuy River basin
Vitim
Vitim Plateau
Vladimir
Vladivostok
Volga region
Volga River
Volga-Don region
Volga-Urals
Volgograd
Vologda
Vorkuta
Voronezh
Voronezh Anteclise
Vuoriyarvi
Vyatka River
Vyatka-Kama Interfluve
Vychegda River
NX Western Transbaikalia
NT Wrangel Island
Yakutia
Yakutsk
Yamal
Yamal-Nenets
Yana
Yana-Indigirka Lowland
Yaroslavl
Yenisei River
Yessey
Yudoma
Yuryuzan
Zeya
Zeya-Bureya Basin
Zmeinogorsk
Zyryanka
SA Amur Basin
Amur River
Argun River

Asiatic USSR
Azov region
Baikalian Phase
Baltic region
Baltic Shield
Barabash Suite
Baskunchak Series
Caspian Basin
Caspian Depression
Caucasus
Courland Spit
Dnieper Basin
Dnieper River
Dnieper-Donets Basin
Donets Basin
Dvina River
European USSR
Greater Caucasus
Irtysh River
Ishim
Khanka Lake
Kuban River
Kuban Valley
Kulunda Steppe
Kursk Series
Ladoga Series
Lake Ladoga region
Maikop Series
Ob-Irtysh Interfluve
Ossetia
Poltava Series
Rudny Altai
Russian Plain
Russian Platform
Scythian Platform
Serebryanka Formation
Siberia
Siberian Lowland
Southern Urals
Soviet Altai Mountains
Steppes
Tannu-Ola Range
Terek River
Tunguska Series
Tura Formation
Udokan Series
Ural River
Urals
Usa Series
Uzen
Vetluga Series
Vodino
Volga-Ural region
Vyshkovo
West Siberia
Yenisei Basin
Yenisei-Khatanga basin
Yudoma Series

ruthenium
Includes use on level 1 and 2 as a chemical element (list D).
UF Ru
SA elements

rutile
BA oxides
BT minerals
SA anatase
brookite
heavy minerals
rutile structure

rutile structure
Term introduced in 1978.
SA rutile

Rutland
City in Rutland County in W.
CO N523500N524500
W0003000W0004500
BT Vermont
United States

Rwanda
Formerly Ruanda which was part of the Belgium trust territory of Ruanda-Urundi. Includes use on level 1 as an area term (list O). For term set options see list B.
CO S025000S010500
E0305000E0285000
UF Ruanda
BA Central Africa
Africa
SA East African Rift
Lake Kivu

Ryazan
City and oblast. City is located 120 miles SE of Moscow.
BT Russian Republic
USSR

Rybnik
City in S part of country.
BT Katowice
Poland

Ryoke
Region midway between Yokohama and Nagoya.
BT Honshu
Japan

Ryoke Belt
UF Ryoke Metamorphic Belt
BT Honshu
Japan

Ryoke Metamorphic Belt
use Ryoke Belt

Ryujima
Island in Japan Sea off the Noto Peninsula.
BT Honshu
Japan

Ryukyu Islands
600 mile chain of islands in W Pacific Ocean between Taiwan and Kyushu. Returned to Japan from U.S. control in 1972.
CO N240000N301000
E1301000E1225000
UF Riukiu Islands
BT Japan
NT Amami-O-shima
Okinawa
Oshima

Rzeszow
Province in NW part of country.
BT Poland
NT Dukla
Przemsyl
Tarnobrzeg

S

S
use sulfur

S-waves
Includes use on level 3 under seismology(1).
UF rotational wave
secondary wave
shake wave
shear wave
tangential wave
transverse wave
BA body waves
elastic waves
NA SH-waves
SV-waves
SA P-waves
PKS-waves
PPS-waves
ScS-waves
seismology
SKP-waves
SKS-waves
waves

S-32
Includes use on level 3 under isotopes(1).
SA isotopes
S-34/S-32
sulfur

S-32/S-34
use S-34/S-32

S-34
Includes use on level 3 under isotopes(1).
SA isotopes
S-34/S-32
sulfur

S-34/S-32
Includes use on level 3 under isotopes(1). Also search S-32/S-34; S-32 AND S-34.
UF S-32/S-34
SA geologic thermometry
isotopes
paleoclimatology
S-32
S-34
stable isotopes
sulfur

Saale glacial stage
use Saalian

Saale Glaciation
use Saalian

Saale River
Rises in the Fichtelgebirge Range in NE Bavaria and flows N into the Elbe River SE of Magdeburg. Index countries as applicable. Also search Saale.
BT Europe
SA East Germany
West Germany

Saalian
Term applied in northern Europe to the middle glacial stage of the Pleistocene Epoch, after the Elster glacial stage and before the Warthe; equivalent to the Riss and Illinoian glacial stages.
UF Saale glacial stage
Saale Glaciation
BA upper Pleistocene
Pleistocene
Quaternary
BT Cenozoic
SA Central Polish Glaciation

Saalian Orogenic Phase
use Saalian Phase

Saalian Phase
Orogenic phase. Before 1977 also search Saalian AND orogeny.
UF Saalian Orogenic Phase
BT Permian
SA orogeny

Saanich Inlet
Arm of Strait of Georgia off SE Vancouver Island, NW of Victoria.
BT British Columbia
Canada

Saar
use Saarland

Saar Basin
River basin. Index French departments and West German states as applicable. Also search Saar.
BT Europe
SA Bas-Rhin
Moselle
Rhineland-Palatinate
Saarland

Saar Nahe (Basin)
use Saar-Nahe Basin

Saar-Nahe Basin
As of 1981, refers to a region in Saarland and Rhineland-Palatinate, corresponding roughly to the Saar and Nahe river basins in West Germany. Before 1981, part of France was included.
CO N490800N500700
E0081000E0061100
UF Saar Nahe (Basin)
BT Germany
SA Nahe
Rhineland-Palatinate
Saarland
West Germany

Saarbrucken Anticline
BT Saarland
West Germany

Saarland
State constituting an industrial region, formerly known as the Saar. Achieved statehood within West Germany in 1957. Also search Saar.
CO N490700N493800
E0072100E0062000
UF Saar
BT West Germany
NT Saarbrucken Anticline
SA Nahe
Saar Basin
Saar-Nahe Basin

Sabah
State of East Malaysia on NE Borneo. Formerly British North Borneo. Also called North Borneo.
CO N040000N074000
E1193000E1150000
UF British North Borneo
North Borneo
BT Malaysia
SA Borneo
East Malaysia

Sabana de Bogota
Plateau 55 miles long and 25 miles wide with Bogota at its center.
BT Colombia

Sabatini Mountains
N of Lake Bracciano NW of Rome. Also search Sabatini.
BT Latium
Italy

Sabine Lake
Formed by expansion of Sabine River which flows through the lake and Sabine Pass to Gulf of Mexico. Index states as applicable.
BT United States
SA Louisiana
Texas

Sabkah
use Sabkha

Sabkha
Village on right bank of Euphrates River in N central Syria.
UF Sabkah
BT Syria
Middle East

sabkha environment
use sebkha environment

Sable Island
Low, sandy island in North Atlantic 115 miles off mainland.
BT Nova Scotia
Canada

Sable Island Bank
Marine bank in vicinity of Sable Island approximately 115 miles off mainland of Nova Scotia.
BT Atlantic Ocean

sabulous texture
use arenaceous texture

Saco River
Flows into Atlantic Ocean from S Maine. Index states as applicable.
BT United States
SA Maine
New Hampshire

Sacoglossa
Term introduced in 1981. Includes use on level 2 under Mollusca(1). See list F.
BA Gastropoda
Mollusca
BT Invertebrata

Sacramento
City in Sacramento County in N central.
BT California
United States

Sacramento County
N central.
BT California
United States

Sacramento Mountains
Range in Otero County NE of El Paso in S.
BT New Mexico
United States

Sacramento Valley
River valley. Northern half of Central Valley. Extends from Lake Shasta to San Joaquin Valley. Also search Sacramento River.
BT California
United States
SA Central Valley

Sadlerochit Formation
Permian and lower Triassic: N Alaska.
SA Alaska
Lower Triassic
Permian
Triassic

Sado Island
Mountainous island in E Japan Sea off NW coast of Honshu.
BT Honshu
Japan

Sadon
Town in lead-zinc-silver mining area in North Ossetian A.S.S.R. in Northern Caucasus.
BT Russian Republic
USSR

Safaga
City on N Red Sea.
BT Egypt

safflorite
As of 1981, includes use on level 3 under minerals(1) arsenides(2). See list L. Before 1981, included use on level 3 under minerals(1) sulfides(2).
BA arsenides
sulfides
BT minerals

Saga
Prefecture in NW.
BT Kyushu
Japan

Sagami Bay
Inlet of the Pacific Ocean SW of Yokohama.
UF Sagami Sea
Sagami-nada
BT Honshu
Japan

Sagami Sea
use Sagami Bay

Sagami-nada
use Sagami Bay

Saganaga Lake
Chain of lakes near Lake Superior. Index Minnesota and/or Ontario as applicable.
SA Minnesota
Ontario

Sagar
Town in NW.
BT Mysore
India

Saghalin
use Sakhalin

Saginaw Formation
Includes Verne Limestone Member, Eaton, Ionia, and Woodville Sandstone members. Lower Peninsula.
BT Pennsylvanian
SA Michigan

Saguache County
S Colorado.
CO N374000N383000
W1053000W1070000
BT Colorado
United States

Saguenay Valley
River valley between Lake Saint Jean and Saint Lawrence River NE of Quebec City. Also search Saguenay; Saguenay River.
BT Quebec
Canada

Sahara
Vast arid region extending across North Africa from the Atlantic Ocean to the Red Sea. Index countries as applicable. Includes use on level 1 as an area term (list O). For term set options see list B. From 1977-80, documents on the Spanish Sahara were indexed under Sahara as the level-1 term. As of 1981, use Western Sahara for the Spanish Sahara.
CO N140000N350000
E0380000E0172000
UF Sahara Desert
BA Africa
NZ Western Sahara
SA Ahaggar
Algeria
Chad
Egypt
Libya
Mali
Mauritania
Morocco
Niger
Sudan
Tanezrouft
Tunisia

Sahara Desert
use Sahara

Sahel
Region. A transitional steppe belt just S of the Sahara. Index countries as applicable.
BT Africa
SA Chad
Mali
Mauritania
Niger
Senegal
Upper Volta

Sahul Shelf
SW of W Timor Island.
BT Timor Sea
Pacific Ocean

Saihun River
use Syr Darya

Saint Catharines
City NW of Niagara Falls on Welland Canal in S Ontario. Also search St. Catharines.
UF St. Catherines
BT Ontario
Canada

Saint Charles County
Term introduced in 1981. E Missouri.
CO N383200N385800
W0900700W0905800
BT Missouri
United States

Saint Clair County
Index states as applicable. Also search St. Clair County.
UF St. Clair County
BX Alabama
Illinois
Michigan
Missouri
BT United States

Saint Clair River
Connects Lake Huron with Lake Saint Claire. Index Michigan and/or Ontario as applicable. Also search St. Claire River.
UF St. Clair River
BT North America
SA Michigan
Ontario

Saint Clair River delta
Formed at river mouth in Lake Saint Clair. Index Michigan and/or Ontario as applicable. Also search St. Clair River delta.
UF St. Clair River delta
BT North America
SA Michigan
Ontario

Saint Croix
Term introduced in 1978. Largest and most populous of the U.S. Virgin Islands which are E of Puerto Rico. Before 1978, also search Saint Croix Island; St. Croix.
CO N173500N175000
W0643500W0645500
UF Saint Croix Island
St. Croix
St. Croix Island
BA U. S. Virgin Islands
Virgin Islands
Leeward Islands
BT Lesser Antilles

Saint Croix Island
use Saint Croix

Saint Elias Mountains
Range near the Pacific Ocean. Index Alaska and/or Yukon Territory as applicable. Also search St. Elias Mountains.
UF St. Elias Mountains
BT North America
SA Alaska
Coast Ranges
Yukon Territory

Saint Francois Mountains
WNW of Cape Girardeau in SE Missouri. Also search St. Francois Mountains.
CO N372000N380500
W0901000W0904000
UF St. Francois Mountains
BT Missouri
United States

Saint Gall
Canton in NE Switzerland.
UF Saint Gallen
St. Gallen
BT Switzerland

Saint Gallen
use Saint Gall

Saint George Formation
Underlies Battery Formation; overlies Jurassic rocks. NW California. Also search St. George Formation.
UF St. George Formation
BT Pliocene
SA California

Saint Gotthard
use Gotthard Massif

Saint Helena
British island in S Atlantic Ocean about 1200 miles from W coast of Africa. Also search St. Helena.
CO S170000S150000
W0050000W0070000
UF Saint Helena Island
St. Helena
St. Helena Island
BT Atlantic Ocean Islands

Saint Helena Island
use Saint Helena

Saint John
City on Bay of Fundy. Also search St. John.
UF St. John
BT New Brunswick
Canada

Saint John's
City on the Atlantic Ocean in S Newfoundland. Also search St. John's.
UF St. John's
BT Newfoundland
Canada

Saint Johns River basin
NE Florida. Also search St. Johns River; St. Johns River basin.
UF St. Johns River basin
BT Florida
United States

Saint Laurent Limestone
Regarded as lower Hamilton (Cazenovia Stage) and below Lingle Limestone. S Illinois and E Missouri. Also search St. Laurent Limestone.
UF St. Laurent Limestone
BT Middle Devonian
Devonian
SA Illinois
Missouri

Saint Lawrence County
On the Saint Lawrence River. Also search St. Lawrence County.
UF St. Lawrence County
BT New York
United States
NT Balmat-Edwards District

Saint Lawrence Estuary
The lower, wide part of the river from below Quebec City to its mouth on the Gulf of Saint Lawrence. Also search St. Lawrence Estuary.
UF St. Lawrence Estuary
BT Quebec
Canada
SA Saint Lawrence Valley

Saint Lawrence Lowlands
Along both sides of river from Lake Ontario to Quebec. Index New York and Canadian provinces as applicable. Also search St. Lawrence Lowlands.
UF St. Lawrence Lowland
St. Lawrence Lowlands
BT North America
SA Laurentide ice sheet
New York
Ontario
Quebec
Saint Lawrence Valley

Saint Lawrence River
Flows NE out of Lake Ontario into the Gulf of Saint Lawrence. Index New York and Canadian provinces as applicable. Also search Saint Lawrence AND river; St. Lawrence River.
UF St. Lawrence River
BT North America
SA Canada
New York
Ontario
Quebec

Saint Lawrence Valley
River valley extending to the Gulf of Saint Lawrence. Index New York and

Canadian provinces as applicable. Also search Saint Lawrence AND valley; St. Lawrence Valley; St. Lawrence River valley.
UF St. Lawrence River valley
St. Lawrence Valley
SA Laurentide ice sheet
New York
Ontario
Quebec
Saint Lawrence Estuary
Saint Lawrence Lowlands

Saint Louis
City on the Mississippi River in E Missouri. Also search St. Louis.
UF St. Louis
BT Missouri
United States

Saint Louis County
Index states as applicable. Also search St. Louis County.
UF St. Louis County
BX Minnesota
Missouri
BT United States

Saint Louis Limestone
Comprises Croton and Verdi members. Also search St. Louis Limestone.
UF St. Louis Limestone
BT Meramecian
Mississippian
SA Alabama
Georgia
Illinois
Indiana
Iowa
Kentucky
Missouri
Tennessee
Upper Mississippian
Virginia

Saint Lucia
Island. A self-governing state in association with United Kingdom in Windward Islands of the Lesser Antilles. Also search St. Lucia.
UF St. Lucia
BA Windward Islands
BT Lesser Antilles

Saint Mary Parish
On Gulf of Mexico. Also search St. Mary Parish.
UF St. Mary Parish
BT Louisiana
United States

Saint Marys County
Term introduced in 1981. S Maryland.
CO N380200N383130
W0761900W0765300
BT Maryland
United States

Saint Marys Formation
In Chesapeake Group. Delaware, E Maryland and E Virginia. Also search St. Marys Formation.
UF St. Marys Formation
BT Miocene
SA Delaware
Maryland
middle Miocene
upper Miocene
Virginia

Saint Paul
City on the Mississippi River E of Minneapolis in Ramsey County in E. One of the "Twin Cities". Also search St. Paul.
UF St. Paul
BT Minnesota
United States

Saint Paul Island
Most northerly of Pribilof Islands.
BT Bering Sea
Pacific Ocean

Saint Paul Rocks
Group of uninhabited volcanic, rocky islets belonging to Brazil about 600 miles NE of Natal.
UF Saint Paul's Rocks
St. Paul's Rock
St. Paul's Rocks
BT Atlantic Ocean

Saint Paul's Rocks
use Saint Paul Rocks

Saint Peter Sandstone
Overlies Prairie du Chien Group and underlies Glenwood Shale Member of Platteville Formation. Also search St. Peter Sandstone.
UF St. Peter Sandstone
BT Middle Ordovician
Ordovician
SA Arkansas
Illinois
Indiana
Iowa
Kansas
Kentucky
Michigan
Minnesota
Missouri
Ohio
Oklahoma
Wisconsin

Saint Pierre and Miquelon
Term introduced in 1981. Consists of two islands in the Atlantic Ocean just S of Newfoundland. Includes use on level 1 as an area term (list O). For term set options see list B.
CO N464500N471000
W0560800W0563000
BT North America

Saint Severin
use Saint-Severin

Saint Thomas
Island in the U.S. Virgin Islands E of Puerto Rico. Also search St. Thomas.
UF St. Thomas
BA U. S. Virgin Islands
Virgin Islands
Leeward Islands
BT Lesser Antilles

Saint Vincent
Self governing British state in the Windward Islands comprising Saint Vincent Island and the northern Grenadines. Also search St. Vincent.
UF St. Vincent
BA Windward Islands
BT Lesser Antilles
NT Soufriere

Saint Vincent Bay
use Saint Vincent Gulf

Saint Vincent Gulf
Inlet of Indian Ocean between Yorke Peninsula and mainland. Also search Saint Vincent; St. Vincent.
UF Saint Vincent Bay
BT South Australia
Australia

Saint-Chinian
Village in S part of country.
BT Herault
France

Saint-Etienne Basin
use Saint-Etienne coal basin

Saint-Etienne coal basin
In E central part of country. Also search Saint-Etienne.
UF Saint-Etienne Basin
BT Loire
France

Saint-Girons
Town at foot of central Pyrenees.
BT Ariege
France

Saint-Louis
Town on Reunion Island E of the Malagasy Republic.
BT Reunion

Saint-Severin
Village in E part of country. Also search St. Severin.
UF Saint Severin
St. Severin
BT Charente
France

Saint-Sylvestre Massif
In the Monts d'Ambazac in E central part of country.
BT Haute-Vienne
France

Saint-Vallier
Village in Provence Alps in extreme SE part of country.
UF Saint-Vallier-de-Thiey
BT Alpes-Maritimes
France

Saint-Vallier-de-Thiey
use Saint-Vallier

Sainte Genevieve Limestone
Includes Fredonia Limestone, Rosiclare Sandstone, and Levias Limestone members. N Alabama, Georgia, S Illinois, Indiana, Iowa, Kentucky, E Missouri, Tennessee. Also search Ste. Genevieve Formation; Ste. Genevieve Limestone.
UF Ste. Genevieve Formation
Ste. Genevieve Limestone
BT Meramecian
Upper Mississippian
Mississippian
SA Alabama
Georgia
Illinois
Indiana
Iowa
Kentucky
Missouri
Tennessee

Sainte-Baume Massif
In lower Provence Alps in SE part of country. Also search Sainte-Baume.
BT Var
France

Sainte-Marie-aux-Mines
Town near crest of the Vosges Mountains in NE part of country.
BT Haut-Rhin
France

Sainte-Victoire Massif
use Sainte-Victoire Mountain

Sainte-Victoire Mountain
In S part of country.
UF Sainte-Victoire Massif
BT Bouches-du-Rhone
France

Saipan
United States island N of Guam.
BT Mariana Islands
Micronesia

Saitama
Prefecture N of Tokyo.
BT Honshu
Japan

Sakar Mountains
Between Maritsa and Tundzha rivers in SE Bulgaria. Also search Sakar; Sakar Mountain.
BT Bulgaria

sakhaite
BZ borates
carbonates
BT minerals

Sakhalin
Island N of Hokkaido in W Okhotsk Sea. Also search Sakhalin Island.
CO N460000N550000
E1450000E1413000
UF Karafuto
Saghalin
Sakhalin Island
BA Soviet Pacific region
BT Russian Republic
USSR
SA Okhotsk region
Soviet Far East

Sakhalin Island
use Sakhalin

Sakmara
Village in central Orenburg Oblast in Southern Urals in European USSR.
CO N520000N523000
E0552000E0522000
BT Russian Republic
USSR

Sakmarian
Europe. Above Stephanian (Carboniferous), below Artinskian. Includes use on level 3 under age terms(1). See list E.
BA Lower Permian
Permian
BT Paleozoic

Sakoa Basin
Main source of coal in SW Malagasy Republic. Also search Sakoa.
BT Malagasy Republic

Saksagan River
In Dnepropetrovsk Oblast in E central Ukraine. Also search Saksagan.
BT Ukraine
USSR

Sakura-jima
Peninsula. NW projection of Osumi Peninsula in S Kyushu. An island until 1914. Also search Sakurajima.
UF Sakurajima
BT Japan

Sakurajima
use Sakura-jima

Salado Basin
600 mile-long basin of the combined Desaguadero and Salado river systems. Index provinces as applicable. Also search Salado.
BT Argentina
SA La Pampa
La Rioja
Mendoza
San Luis

Salair
City on Salair Ridge in W Kemerovo Oblast in S Siberia.
BT Russian Republic
USSR

Salair Ridge
Along borders of Altai Kray and Kemerovo Oblast in S Siberia. Also search Salair.
BT Russian Republic
USSR

Salaj
County in NW.
BT Transylvania
Romania

Salamanca
As of 1981, refers to only the city. Before 1981, also included the province. In W Spain.
BT Salamanca Province
Leon region
Spain

Salamanca Province
Term introduced in 1981. Before

S

1981, search Salamanca. In W Spain on the Portuguese border.
CO N401300N411800
W0050800W0065200
BT Leon region
Spain
NT Salamanca

Salat Valley
River valley in S part of country. Index departments as applicable. Also search Salat River.
BT France
SA Ariege
Haute-Garonne

Salem Limestone
Includes Kidd, Fults, Chalfin, and Rocher members. S Illinois , S Indiana, SE Iowa, W and central Kentucky, and E Missouri. Also search Salem.
BT Meramecian
Upper Mississippian
Mississippian
SA Illinois
Indiana
Iowa
Kentucky
Missouri

Salentina Peninsula
SE Apulia. The heel of the Italian boot. Also search Salentina and Salentine Peninsula.
BT Apulia
Italy

Salerno
City on Gulf of Salerno SE of Naples.
BT Campania
Italy

sales
Term introduced in 1981.
SA economics

Salida
City in Chaffee County in central.
BT Colorado
United States

Salina Group
Includes Vernon Shale, Camillus Shale, Bertie Formation, Akron Dolomite, Syracuse Formation. Michigan and N Ohio.
BT Upper Silurian
Silurian
SA Michigan
Ohio

Salinas
City in Monterey County in W.
BT California
United States

Salinas River valley
use Salinas Valley

Salinas Valley
River valley in San Luis Obispo and Monterey counties in W California. Also search Salinas River; Salinas River valley.
UF Salinas River valley
BT California
United States

saline composition
Term introduced in 1978. Before 1978, search saline.
SA composition
ground water
salinity

Saline County
Term introduced in 1981. Index states as applicable.
BX Arkansas
Illinois
Kansas
Missouri
Nebraska
BT United States

saline water
use salt water

Salinian Block
Term introduced in 1981. Block structure in the California Coast Ranges.
CO N343000N390000
W1190000W1240000
BX California
Coast Ranges

salinity
Includes use on level 2 under soils(1) and under paleoecology(1).
SA brackish water
brackish-water environment
desalinization
paleoecology
paleosalinity
saline composition
salt
salt marshes
sea water
sedimentation
soils
Solonchak soils
springs
treatment

Salisbury
City in NE.
BT Zimbabwe

Salmo
Village in S.
BT British Columbia
Canada

Salmon
City in Lemhi County in E.
BT Idaho
United States

Salmon River
Rises in central part of state and flows N then W, and again N to empty into Snake River in W Idaho. Also search Salmon.
BT Idaho
United States
SA Salmon River breaks

Salmon River breaks
Large gorge or canyon in lower course of Salmon River.
BT Idaho
United States
SA Salmon River

Salonika
City on Gulf of Salonika in N part of country.
UF Saloniki
Thessaloniki
BT Greece
SA Macedonia

Saloniki
use Salonika

Salsigne Mine
N of Carcassonne in S part of country. Also search Salsigne.
BT Aude
France

salt
Includes use on level 1 and 2 as a commodity term (list C); on level 2 under mineral deposits, genesis(1). As of 1981, includes use on level 3 under sedimentary rocks(1) chemically precipitated rocks(2). See list I.
BA evaporites
chemically precipitated rocks
BT sedimentary rocks
SA brines
bromine
bromine deposits
desalinization
diapirs
evaporite deposits
fractional crystallization
halite
iodine deposits
paleosalinity
salinity
salt domes
salt tectonics
sea water
sodium chloride

Salt Creek
Central.
BT Wyoming
United States

salt domes
Includes use on level 3 under salt tectonics(1), salt(1) and under folds(1) style(2).
SA cap rocks
diapirism
diapirs
domes
folds
salt
salt tectonics

Salt Lake City
City in Salt Lake County in N.
BT Utah
United States

Salt Lake County
N Utah.
BT Utah
United States

salt lakes
SA brackish water
brines
lakes
salt water

salt marshes
SA brackish-water environment
marshes
paleoecology
salinity
sebkha environment
sedimentation

Salt Range
Between the Indus and the Jhelum rivers.
CO N330000N340000
E0730000E0720000
BT Punjab Province
Pakistan

Salt River
Rises in E part of state and flows W into Gila River W of Phoenix.
BT Arizona
United States

salt tectonics
Includes use on level 1 (list A). Used for the study of the structure and mechanism of emplacement of salt domes. Term set options are:
topic [causes, evolution, interpretation, mechanism, processes]
subtopic (no area term)
UF halokinesis
SA cap rocks
deformation
diapirism
diapirs
emplacement
faults
folds
salt
salt domes
tectonics
tectonophysics

salt water
Includes use on level 3 under ground water(1) when comparing with fresh water. Also search saline water.
UF saline water
SA brackish water
brines
dispersion
fresh water
ground water
salt lakes
salt-water intrusion
sea water
water

salt-water contamination
Not a valid term for GeoRef. Use salt-water intrusion.

salt-water intrusion
Includes use on level 2 under ground water(1).
UF encroachment (ground water)
intrusion (ground water)
sea-water encroachment
sea-water intrusion
SA brines
dispersion
ground water
salt water

Salta
Province in NNW.
UF Salta Province
BT Argentina
NT Rio Blanco Basin
SA Sierras Pampeanas
Yacoraite Formation

Salta Province
use Salta

saltation
SA sedimentation
transport

Salton Sea
Shallow saline lake just N of Imperial Valley in S.
CO N331000N333000
W1153500W1160500
BT California
United States

Salton Trough
Depression including from NW to SE, the Coachella Valley, the Salton Sea, and the Imperial Valley in S.
CO N310000N340000
W1140000W1170000
BT California
United States

Saltville Fault
SW Virginia. Also search Saltville.
BT Virginia
United States

Salvador
City on the Atlantic Ocean in E part of country. Formerly Sao Salvador or Bahia.
UF Sao Salvador
BT Bahia
Brazil

Salym
Stream in NW Novosibirsk Oblast in S Siberia.
BT Russian Republic
USSR

Salzach River
Primarily in Salzburg State. Index Austrian and German states as applicable. Also search Salzach.
BT Europe
SA Bavaria
Pinzgau
Salzburg State
Upper Austria

Salzburg
As of 1981, refers to only the city. Before 1981, referred to both the state and the city in central Austria.
BT Salzburg State
Austria

Salzburg State
Term introduced in 1981. Before 1981, search Salzburg.

CO N465500N480300
E0140000E0120400
BT Austria
NT Bad Gastein
Mitterberg
Pinzgau
Salzburg
SA Salzach River
Salzkammergut
Venediger Group

Salzkammergut
Lake and mountain region of Eastern Alps. Index states as applicable.
BT Austria
SA Eastern Alps
Salzburg State
Styria
Upper Austria

Samar
Island on E side of central part of archipelago.
BT Philippine Islands

Samara Bend
Region in Kuibyshev Oblast within oxbow of the middle Volga River where it reaches its easternmost point. Also search Samara.
BT Russian Republic
USSR

Samaria
Region extending from the Mediterranean Sea to the Jordan River S of Galilee and N of Judaea. Includes most of the Israeli occupied West bank of the Jordan River. Index countries as applicable.
BT Middle East
SA Israel
Jordan

samarium
Includes use on level 1 and 2 as a chemical element (list D).
UF Sm
SA elements
rare earths

Samarkand
City and oblast in central Uzbekistan.
UF Samarqand
BT Uzbekistan
USSR

Samarqand
use Samarkand

samarskite
BZ niobates
tantalates
BA oxides
BT minerals

Sambagawa Belt
Innermost and oldest of three belts of different metamorphic grades of schist in a Paleozoic Group. SW Honshu, Kyushu, and Shikoku. Also search Sambagawa; Sambagawa Schist; Sanbagawa Belt.
UF Sanbagawa Belt
BT Paleozoic
SA Honshu
Kyushu
Shikoku

Sambalpur
Town in N.
BT Orissa
India

Samoa
Group of volcanic islands in SW central Pacific Ocean. American Samoa is in E part of group, and independent Western Samoa comprises the W part. Includes use on level 1 as an area term (list O). For term set options see list B.
BA Polynesia

Samos
Island in the Aegean Sea off W coast of Turkey.
BT Greek Aegean Islands
Greece

sample preparation
Includes use on level 3 under chemical analysis(1) techniques(2) and under micropaleontology(1); on level 3 under soils(1) analysis(2) under palynology(1), and under paleontology(1).
SA chemical analysis
decrepitation
differential thermal analysis
micropaleontology
paleontology
palynology
preparation
samples
sampling
soil sampling
soils
staining
techniques
thermal analysis

samples
Used for physical samples, not for statistical meaning.
SA lunar samples
sample preparation
sampling
thin sections

sampling
Includes use on level 2 under placers(1).
SA analysis
field studies
placers
probability
reliability
sample preparation
samples
soil sampling
techniques

San Andreas Fault
A fault system or zone extending for more than 600 miles from the Pacific Ocean at Point Arena through the San Francisco Peninsula and on into S California. Also search San Andreas; San Andreas fault system; San Andreas fault zone; San Andreas System.
CO N340000N400000
W1173000W1243000
UF San Andreas fault system
San Andreas fault zone
San Andreas System
BT California
United States
SA San Gregorio Fault

San Andreas fault system
use San Andreas Fault

San Andreas fault zone
use San Andreas Fault

San Andreas System
use San Andreas Fault

San Andres Formation
In Manzano Group. Central and SE New Mexico. Also search San Andres.
BT Permian
SA Guadalupian
Leonardian
Lower Permian
New Mexico
Upper Permian

San Antonio
City in Bexar County in S central.
BT Texas
United States

San Benito County
W California.
BT California
United States

San Bernardino
City in extreme SW San Bernardino County in S.
BT California
United States

San Bernardino County
S California.
CO N340000N354500
W1141500W1174000
BT California
United States

San Bernardino Mountains
In San Bernardino and Riverside counties in S.
BT California
United States
SA Coast Ranges
San Giorgio Mountain
Transverse Ranges

San Buenaventura
use Ventura

San Carlos Indian Reservation
SE central Arizona. Also search San Carlos.
BT Arizona
United States

San Clemente Island
One of the Santa Barbara Islands in the Pacific Ocean W of San Diego. Also search San Clemente.
BT California
United States

San Diego
City on the Pacific Ocean in San Diego County in S.
CO N324500N324500
W1171000W1171000
BT California
United States

San Diego County
S California.
CO N323000N333000
W1161000W1174000
BT California
United States

San Diego Formation
Rests with angular unconformity upon Rose Canyon shale member of the La Jolla, or on Poway Formation, or overlaps them and rests with marked unconformity upon Black Mountain volcanics; unconformably underlies Sweitzer Formation. S California.
BT Pliocene
SA California

San Diego Trough
Just off San Diego.
BT Pacific Ocean

San Felipe Formation
Also search San Felipe.
BT Cretaceous
SA Mexico

San Fernando
City in Los Angeles County. An enclave of Los Angeles.
BT California
United States

San Fernando Valley
Fertile basin about 20 miles NW of downtown Los Angeles.
BT California
United States

San Francisco
City on the Pacific Ocean in W. Coextensive with San Francisco County.
BT California
United States

San Francisco Bay
Inlet connected to the Pacific Ocean via the Golden Gate in W.
BT California
United States
SA San Francisco Bay region

San Francisco Bay region
Term Introduced in 1978.
SA San Francisco Bay

San Francisco County
Coextensive with city of San Francisco.
BT California
United States

San Francisco de la Selva
use Copiapo

San Francisco Mountain
One of the three peaks N of Flagstaff in Coconino County known as San Francisco Peaks.
UF Humphreys Peak
BT Arizona
United States
SA San Francisco Peaks

San Francisco Mountains
Range primarily in Catron County, New Mexico. Index states as applicable.
BT United States
SA Arizona
New Mexico

San Francisco Peaks
An eroded volcano 10 miles N of Flagstaff in Coconino County with Agassiz, Fremont and San Francisco (Humphreys) peaks on its rim. Also search San Francisco Volcanic Field.
UF San Francisco Volcanic Field
BT Arizona
United States
SA San Francisco Mountain

San Francisco Peninsula
Extends S from San Francisco the length of San Francisco Bay.
BT California
United States

San Francisco Volcanic Field
use San Francisco Peaks

San Gabriel Mountains
Range SW of the Mojave Desert primarily in Los Angeles County in S. Also search San Gabriel.
BT California
United States
SA Coast Ranges
Transverse Ranges

San Giorgio Mountain
Highest peak of San Bernardino Mountains in San Bernardino County in S.
BT California
United States
SA San Bernardino Mountains

San Gregorio Fault
Term introduced in 1981. W central California. Component of San Andreas fault system.
CO N361000N375500
W1214500W1225000
BT California
United States
SA San Andreas Fault

San Jacinto Fault
Riverside County in S California. Also search San Jacinto.
BT California
United States

San Joaquin County
In N San Joaquin Valley in central. Also search San Joaquin.
BT California

United States

San Joaquin Valley
Southern half of Central Valley. Extends from Buena Vista Lake in S to Sacramento Valley in N. Also search San Joaquin.
CO N354000N380500
W1190000W1214000
BT California
United States
SA Central Valley

San Jose
City in Santa Clara County in W.
BT California
United States

San Jose Formation
Lithology of formation is highly variable, both vertically and horizontally. N New Mexico, and S Colorado.
BT lower Eocene
Eocene
SA Colorado
New Mexico

San Juan
As of 1981, refers to the city on the Atlantic Ocean in N Puerto Rico. Before 1981, search San Juan AND Puerto Rico. Before 1981, also referred to province in Argentina.
BT Puerto Rico

San Juan Basin
Index states as applicable.
BT United States
SA Arizona
Colorado
New Mexico
San Juan County
San Juan District
San Juan Mountains
San Juan River
San Juan volcanic field
Utah

San Juan Bautista
Town in San Benito County in W.
BT California
United States

San Juan County
Index states as applicable.
BX Colorado
New Mexico
Utah
Washington
BT United States
SA San Juan Basin
San Juan District
San Juan Mountains
San Juan River
San Juan volcanic field

San Juan District
Term introduced in 1981. Mining region in SW Colorado. Before 1981, also search San Juan AND Colorado.
BT Colorado
United States
SA San Juan Basin
San Juan County
San Juan Mountains
San Juan River
San Juan volcanic field

San Juan Formation
Middle and late Tertiary in SW Colorado; Pleistocene in Puerto Rico.
BT Cenozoic
SA Colorado
middle Tertiary
Pleistocene
Puerto Rico
upper Tertiary

San Juan Islands
Group of islands off NW Washington and E of Vancouver Island.
BT Washington
United States

San Juan Mountains
Range of the Rocky Mountains in SW.
CO N370000N374500
W1063000W1073000
BT Colorado
United States
SA Rocky Mountains
San Juan Basin
San Juan County
San Juan District
San Juan River
San Juan volcanic field

San Juan Province
Term introduced in 1981. In W Argentina. Before 1981, also search San Juan AND Argentina.
CO S320000S283000
W0670000W0703000
BT Argentina
SA Barreal
Sierras Pampeanas
Talacasto

San Juan River
Rises in S Colorado and flows SW, bends W, then NW emptying into the Colorado River in SE Utah. Index states as applicable. Also search San Juan AND (Colorado OR New Mexico OR Utah).
BT United States
SA Colorado
New Mexico
San Juan Basin
San Juan County
San Juan District
San Juan Mountains
San Juan volcanic field
Utah

San Juan volcanic field
In San Juan Mountains of SW Colorado. Also search San Juan Volcanics.
BT Colorado
United States
SA San Juan Basin
San Juan County
San Juan District
San Juan Mountains
San Juan River

San Lorenzo Formation
Includes Twobar Shale and Rices Mudstone. In Santa Cruz Mountain region of S California. Also search San Lorenzo.
BT Tertiary
SA California
Eocene
Oligocene

San Lorenzo Quadrangle
Term introduced in 1981. SW New Mexico.
CO N324500N330000
W1074500W1080000
BT New Mexico
United States
SA Grant County
Sierra County

San Luis
Province in W central.
BT Argentina
SA Salado Basin
Sierras Pampeanas

San Luis Obispo
City in San Luis Obispo County in SW.
BT California
United States

San Luis Obispo County
SW California.
BT California
United States

San Luis Potosi
State in E central Mexico. Also city.
BT Mexico
NT Valles
SA Sierra Madre Oriental

San Luis Valley
Once bottom of extensive lake. In Saguache, Alamosa, and Conejos counties in S Colorado and in Taos County in New Mexico. Index states as applicable.
BT United States
SA Colorado
New Mexico

San Manuel
Village in Pinal County in S central.
BT Arizona
United States

San Marcos Arch
S central Texas. Also search San Marcos.
BT Texas
United States

San Marino
Term introduced in 1981 as a level 1 area term (list O). Republic in the Apennines, N part of Italian peninsula near the Adriatic coast.
BA Southern Europe
Europe

San Mateo County
Just S of San Francisco.
BT California
United States

San Mateo Mountains
SW Socorro County in SW central.
BT New Mexico
United States

San Miguel Island
Northernmost of Santa Barbara Islands in Pacific Ocean SW of Santa Barbara. Also search San Miguel AND appropriate area.
BT California
United States

San Nicolas Island
In central Santa Barbara Islands in the Pacific Ocean off SW California. Also search San Nicolas.
BT California
United States

San Onofre Breccia
Underlies Monterey Formation; overlies Cozy Dell. San Diego County in S California.
BT middle Miocene
Miocene
SA California

San Pedro
Harbor in Los Angeles County. Former city which was annexed to Los Angeles in 1909.
BT California
United States

San Pedro Hills
use Palos Verdes Hills

San Pedro Valley
In San Mateo County in W California. Also search San Pedro AND California.
BT California
United States

San Rafael Swell
E central Utah. Also search San Rafael.
BT Utah
United States

San Saba County
Central.
BT Texas
United States

San Salvador
Island in E central Bahamas, and a city in El Salvador. Index Bahamas and/or El Salvador as applicable.
SA Bahamas
El Salvador

San San
use Sansan

San Sebastian
City on Bay of Biscay in Basque country.
BT Guipuzcoa
Basque Provinces
Spain

Sanbagawa Belt
use Sambagawa Belt

sand
Includes use on level 3 under sediments(1) clastic sediments(2). See list N. As of 1981, use sands or gravel deposits for sand as a commodity. See list C. Before 1981, included use on level 1 as a commodity term. Also includes use as a material in engineering geology.
BA clastic sediments
BT sediments
SA aggregate
alluvium
arenaceous texture
carbonate sediments
construction materials
grains
gravel
greensand
loam
quartz sand
rounding
sands
sandstone
silica
silt
soils
terrigenous materials

sand bodies
Includes use on level 3 under sedimentary structures(1) planar bedding structures(2). See list K.
BA planar bedding structures
sedimentary structures
SA bedding plane irregularities

sand equivalent tests
Term introduced in 1981.
SA drilling
petroleum engineering

Sand Hills
Belt, 20 to 40 miles wide, of low, sandy hills extending along inner border of coastal plain from central North Carolina to central Georgia. Also large area of stable dunes covered with vegetation in NW central Nebraska. Index states as applicable.
BT United States
SA Georgia
Nebraska
North Carolina
South Carolina

Sand Wash Basin
Term introduced in 1981. NW Colorado.
CO N403000N410000
W1071500W1090500
BX Colorado
Green River basin
BT United States

sand waves
Includes use on level 3 under sedimentary structures(1) bedding plane irregularities(2). See list K.
BA bedding plane irregularities
sedimentary structures
SA antidunes

dunes
megaripples
ripple marks

Sandakan
City on Sandakan Harbor on Sulu Sea in Sabah in East Malaysia.
BT Malaysia

sandarac
use realgar

Sandelzhausen
Village N of Munich.
BT Bavaria
West Germany

Sandia Mountains
NE of Albuquerque in N central.
BT New Mexico
United States

Sandnes
Town S of Stavanger in SW part of country.
BT Rogaland
Norway

Sandomierz
Town on Vistula River in SE central part of country.
UF Sandomir
BT Kielce
Poland

Sandomir
use Sandomierz

Sandoval County
NW central.
BT New Mexico
United States
NT Fenton Hill

sands
As of 1981, includes use on level 1 and 2 as a commodity term (list C) for glass, ceramic, chemical use, etc. Before 1981, search sand AND deposits. For sand used as a construction material, use gravel deposits. Before 1981, included use as a type of material in engineering geology; as of 1981, use sand.
SA aggregate
construction materials
gravel deposits
sand
silica

sandstone
Includes use on level 3 under sedimentary rocks(1) clastic rocks(2). See list I. As of 1981, use sandstone deposits for sandstone as a commodity. Before 1981, included use on level 1 as a commodity term.
UF psammite
BA clastic rocks
BT sedimentary rocks
SA arenaceous texture
arkose
gaize
graywacke
greensand
metasandstone
microbreccia
orthoquartzite
red beds
sand
sandstone deposits
sandstone dikes
subgraywacke
terrigenous materials

sandstone deposits
Term introduced in 1981. Includes use on level 1 and 2 as a commodity term (list C). Before 1981, search sandstone AND deposits.
SA building stone
construction materials
dimension stone
sandstone

sandstone dikes
Includes use on level 3 under sedimentary structures(1) soft sediment deformation(2). See list K.
BA soft sediment deformation
sedimentary structures
SA clastic dikes
sandstone

Sandur
Town in W central.
BT Mysore
India

sandurs
use outwash plains

Sandusky County
Term introduced in 1981. N Ohio.
CO N411500N413000
W0824900W0832500
BT Ohio
United States
SA Sandusky River basin

Sandusky River basin
Term introduced in 1981. N Ohio.
CO N404000N413000
W0824200W0832500
BT Ohio
United States
SA Crawford County
Sandusky County
Seneca County
Wyandot County

Sandy Hook
Peninsula in NE Monmouth County constituting S side of entrance to lower New York Bay.
CO N402300N402900
W0735500W0770500
BT New Jersey
United States

sandy texture
use arenaceous texture

Sangamon
River which flows into the Illinois River in central.
BT Illinois
United States

Sangamonian
Pertaining to third interglacial stage of Pleistocene Epoch in North America. See list E.
BA upper Pleistocene
Pleistocene
Quaternary
BT Cenozoic
SA Riss/Wurm Interglacial

Sangilen Mountains
S Tuva ASSR just N of Mongolian border. Also search Sangilen.
BT Russian Republic
USSR

Sangre de Cristo Mountains
Range of the Rocky Mountains. Index states as applicable. Also search Sangre de Cristo Range.
UF Sangre de Cristo Range
Sangre de Cristo Uplift
BT United States
SA Colorado
New Mexico
Rocky Mountains

Sangre de Cristo Range
use Sangre de Cristo Mountains

Sangre de Cristo Uplift
use Sangre de Cristo Mountains

Sangun Zone
Term introduced in 1981. Region of metamorphic rocks, Honshu, Japan.
BT Honshu
Japan

sanidine
UF glassy feldspar
rhyacolite
BA feldspar group
framework silicates
silicates
BT minerals
SA K-feldspar
orthoclase

sanitary landfills
BT landfills
SA waste disposal

Sannoisian
Europe. Lower Oligocene in France overlain by Stampian; same as Lattorfian. Includes use on level 3 under age terms(1). See list E.
BA lower Oligocene
Oligocene
Paleogene
Tertiary
BT Cenozoic
SA Tongrian

Sanpete County
Central Utah. Also search Sanpete.
BT Utah
United States

Sansan
Village in NW Panama.
UF San San
BT Panama

Santa Ana
City in Orange County at base of Santa Ana Mountains in S.
BT California
United States

Santa Ana Mountains
Range along border between Orange and Riverside counties in S California. Also search Santa Ana AND California.
BT California
United States
SA Coast Ranges

Santa Barbara
City on Santa Barbara Channel in Santa Barbara County in SW.
BT California
United States

Santa Barbara Basin
Undersea feature between Santa Barbara County and the N Santa Barbara Islands.
BT Pacific Ocean

Santa Barbara Channel
Between Santa Barbara County and N Santa Barbara Islands.
BT California
United States

Santa Barbara County
SW California.
CO N342500N351000
W1192500W1204500
BT California
United States
NT Point Sal

Santa Barbara Islands
use Channel Islands

Santa Catalina Island
Island in central Santa Barbara Islands in Pacific Ocean off Long Beach. Also search Santa Catalina AND California.
BT California
United States

Santa Catalina Mountains
Small range in NE Pima County in S Arizona. Also search Santa Catalina AND Arizona.
BT Arizona
United States

Santa Catarina
State in S.
CO S292000S255000
W0483000W0535000
BT Brazil
NT Santa Catarina Island
SA Brazilian Shield
Estrada Nova Formation
Pelotas Basin
Serra do Mar

Santa Catarina Island
Island in Atlantic Ocean just off central Santa Catarina. Also search Santa Catarina.
BT Santa Catarina
Brazil

Santa Clara
City in Santa Clara County in W.
BT California
United States

Santa Clara County
S of San Francisco Bay.
BT California
United States

Santa Clara Valley
River valley in Los Angeles and Ventura counties. Also the S extension of the San Francisco Bay depression in Santa Clara and San Benito Counties is called the Santa Clara Valley.
BT California
United States

Santa Cruz
Province in S Argentina and department in E Bolivia. Index countries as applicable.
BT South America
SA Argentina
Baquero Formation
Bolivia
Patagonia
Puerto Deseado

Santa Cruz County
Index states as applicable. Also search Santa Cruz AND appropriate area.
BX Arizona
California
BT United States

Santa Cruz Island
Easternmost of northern islands in Santa Barbara Islands off Ventura County. Also search Santa Cruz AND California. For the "Santa Cruz Island" in the Galapagos Islands, use Chaves Island.
CO N335000N341000
W1193000W1195500
BT California
United States
SA Chaves Island

Santa Cruz Mountains
One of the Coast Ranges extending along the W side of Santa Clara Valley just S of San Francisco Bay. Also search Santa Cruz AND California.
BT California
United States
SA Coast Ranges

Santa Eulalia
Town on Santa Eulalia River with water reservoir and hydroelectric plant servicing Lima.
BT Lima
Peru

Santa Fe
Province in NE central Argentina, and a city in N central New Mexico. Index Argentina and/or New Mexico as applicable.
SA Argentina
New Mexico

Santa Fe County
N central.
BT New Mexico
United States

Santa Isabel
Volcanic island of the British Solomon Islands in E central.
BT Solomon Islands

Santa Lucia Range
One of the Coast Ranges in Monterey and San Luis Obispo counties in W.
BT California
United States
SA Coast Ranges

Santa Margarita Formation
Includes Quatal Red Clay Member. S California. Also search Santa Margarita.
BT upper Miocene
Miocene
SA California

Santa Maria
City in Santa Barbara County in SW.
BT California
United States

Santa Maria Formation
S California.
BT Cenozoic
SA California
Pleistocene
Pliocene

Santa Marta
City on Caribbean Sea.
BT Colombia

Santa Monica Mountains
E-W range paralleling N shore of Santa Monica Bay in S. Also search Santa Monica AND California.
BT California
United States
SA Coast Ranges

Santa Rita
Village in Grant County in SW.
BT New Mexico
United States

Santa Rita Mountains
Santa Cruz County in SE. Also search Santa Rita.
BT Arizona
United States

Santa Rosa
City in Sonoma County N of San Francisco.
BT California
United States

Santa Rosa Mountains
SE California. Range along W side of Coachella Valley.
BT California
United States

Santa Rosa Range
In NE Humboldt County in N.
BT Nevada
United States

Santa Ynez Mountains
E-W coastal range bordering Santa Barbara Channel in SW. Also search Santa Ynez.
BT California
United States
SA Coast Ranges

Santana Formation
BT Cretaceous
SA Brazil

Santander
As of 1981, no longer a valid index term for GeoRef. Use Santander Province or Santander Department as applicable. Before 1981, "Santander" referred to both the department in Colombia and the province in Spain.

Santander Department
Term introduced in 1981. In E Colombia. Before 1981, search Santander AND Colombia. When referring to Spain, use Santander Province.
BT Colombia
NT California City

Santander Massif
In the E central Cantabrian Mountains in N part of country.
BT Santander Province
Old Castile
Spain

Santander Province
Term introduced in 1981. In N Spain on the Bay of Biscay. Before 1981, search Santander AND Spain. When referring to Colombia, use Santander Department.
CO N424500N433000
W0031000W0045300
BT Old Castile
Spain
NT Santander Massif
SA Cantabrian Basin

Santee River
Flows into Atlantic Ocean. SE central South Carolina. Also search Santee.
BT South Carolina
United States

Santiago
Province in central Chile. Also a city.
BT Chile

Santiago de Compostela
City in NW part of country.
BT La Coruna Province
Galicia
Spain

Santiago del Estero
Province in N.
BT Argentina

Santiaguito
Volcano.
BT Guatemala

Santo Domingo
City on the Caribbean Sea.
UF Ciudad Trujillo
BT Dominican Republic

Santonian
Europe. Above Coniacian, below Campanian. Includes use on level 3 under age terms(1). See list E.
BA Senonian
Upper Cretaceous
Cretaceous
BT Mesozoic

Santorin
Volcano. A portion of the crater forms the island of Thera, southernmost of the Cyclades, in the Aegean Sea. Also search Santorin Volcano.
UF Santorin Volcano
Santorini Volcano
BT Greek Aegean Islands
Greece
SA Thera

Santorin Volcano
use Santorin

Santorini
use Thera

Santorini Volcano
use Santorin

Sao Francisco Basin
Drainage basin of Sao Francisco River. Index states as applicable. Also search Sao Francisco; Sao Francisco River.
BT Brazil
SA Alagoas
Bahia
Minas Gerais
Pernambuco
Sergipe

Sao Gabriel
City in extreme S part of country.
BT Rio Grande do Sul
Brazil

Sao Miguel Island
Largest island in group. In Portugal's Ponta Delgado District. Also search Sao Miguel.
BT Azores

Sao Paulo
State in S Brazil.
CO S252000S194000
W0442000W0530000
BT Brazil
NT Amparo
Marilia
Ribeira de Iguape River
Rio Claro
SA Bauru Formation
Botucatu Formation
Brazilian Shield
Estrada Nova Formation
Irati Formation
Serra do Mar
Serra Geral Formation
Tubarao Group

Sao Salvador
use Salvador

Sao Tome
Term introduced in 1981. Portuguese island in the Gulf of Guinea.
CO S003000N004000
E0070000E0060000
BT Atlantic Ocean Islands

Saone Valley
River valley in E central part of country extending from the Vosges Mountains to the Rhone river at Lyon. Also search Saone.
BT France
SA Saone-Rhone Basin

Saone-et-Loire
Department in E central.
CO N461000N471000
E0053000E0033500
BT France
NT Autun
SA Morvan

Saone-Rhone Basin
Term introduced in 1981.
BT France
SA Rhone Valley
Saone Valley

Saoura
No longer a valid term for GeoRef as of 1981. Use Saoura region. Referred to department on Moroccan border in Algeria.

Saoura region
Term introduced in 1981. Before 1981, search Saoura.
CO N204000N333700
E0022000W0084200
BT Algeria
NT Bechar

Sapelo Island
In Atlantic Ocean off McIntosh County.
BT Georgia
United States

saponite
BA clay minerals
sheet silicates
silicates
BT minerals
SA smectite

Sappada
Village on Piave River in NE part of country.
BT Veneto
Italy

sapphire
BA oxides
BT minerals
SA corundum
gems

sapphirine
BA orthosilicates
silicates
BT minerals
SA aluminosilicates

saprolite
Includes use on level 3 under sedimentary rocks(1) clastic rocks(2). See list I.
BA clastic rocks
BT sedimentary rocks

sapropel
Includes use on level 3 under organic materials(1).
BA organic materials
SA anaerobic environment
gyttja
hydrocarbons
ooze

sapropelic coal
use sapropelite

sapropelite
See list I.
UF sapropelic coal
BA coal
organic residues
BT sedimentary rocks

Saragossa
As of 1981, refers to only the city. Before 1981, also included the province. In NE Spain. When referring to Mexico, use Zaragoza.
BT Saragossa Province
Aragon
Spain
SA Zaragoza

Saragossa Province
Term introduced in 1981. Before 1981, search Saragossa. In NE Spain.
CO N405500N424500
E0002300W0021000
BT Aragon
Spain
NT Saragossa
SA Calatayud-Teruel Basin
Zaragoza

Sarajevo
City in E Bosnia and Herzegovina.
UF Serajevo
BT Yugoslavia

Saratoga Chalk
Includes a lower chalk member and an upper argillaceous-arenaceous unit. SW Arkansas.
BT Upper Cretaceous
Cretaceous
SA Arkansas

Saratov
City on W bank of the Volga River. Also an oblast.
BT Russian Republic
USSR

Sarawak
State of East Malaysia in NW Borneo. Former British protectorate governed by the Brooke family. Joined Malaysia in 1963. Included use as a first order area term until 1976. Now includes use on level 3. See under Malaysia(1).
CO N005000N050000
E1155000E1093000
BT Malaysia
SA Borneo
East Malaysia

Sarbay
River in Kuybyshev Oblast in the Middle Volga region.
BT Russian Republic
USSR

sarcopside
BA phosphates
BT minerals

Sardinia
Island and autonomous region in the Mediterranean Sea. Includes use on level 1 as an area term (list O). For term set options see list B.
CO N385000N413000
E0094500E0081500
BA Italy
Southern Europe
Europe
NT Bosano
Gerrei
Iglesiente
Logudoro
Orosei
Sarrabus
Sulcis
SA Mediterranean region
Mediterranean Sea

Sargasso Sea
Large tract of relatively still water between the West Indies and Bermuda. Named after Sargasso weed which floats there.
BT Atlantic Ocean

Sargent
Village in S Santa Clara County in W.
BT California
United States

Sarmatian
Europe. Includes use on level 3 under age terms(1). See list E.
BA upper Miocene
Miocene
Neogene
Tertiary
BT Cenozoic

Sarrabus
Region in SE.
BT Sardinia
Italy

Sarthe
Department in NW.
CO N473000N483000
E0005000W0003000
BT France
NT Le Mans

Sary-Su
use Sarysu

Sary-Su River
use Sarysu

Sarysu
River in central Kazakhstan. Flows S into desert but becomes dry before reaching Syr Darya. Also search Sary-Su; Sary-Su River.
UF Sary-Su
Sary-Su River
BT Kazakhstan
USSR

Sasca-Montana
Village in SW.
BT Banat
Romania

Sasebo
City on inlet of east China Sea in NW.
BT Kyushu
Japan

Saskatchewan
Includes use on level 1 as an area term (list O). For term set options see list B.
CO N490000N600000
W1012000W1100000
BA Western Canada
Canada
NT Amisk Lake
Athabasca District
Beaverlodge
Coronation Mine
Cree Lake
Delmas
Eldorado
Esterhazy
Estevan
Hanson Lake
Key Lake
La Ronge
Lac La Ronge
Leoville
Rabbit Lake
Regina
Rosetown
Saskatoon
Tazin Lake
Wapawekka Lake
Wollaston Lake Belt
SA Assiniboine River
Athabasca Formation
Belly River Formation
Canadian Shield
Carswell Structure
Churchill Province
Cold Lake
Cypress Hills Formation
Edmonton Formation
Elk Point Basin
Elk Point Group
Great Plains
Missi Group
Missouri River basin
North Saskatchewan River
Paskapoo Formation
Prairie Evaporite
Prince Albert Group
Ravenscrag Formation
Reindeer Lake
Saskatchewan River
Shaunavon Formation
Souris River basin
South Saskatchewan River
Whitemud Formation
Williston Basin
Winnipeg Formation
Wood Mountain Formation
Wynyard

Saskatchewan River
Formed by confluence of the North and South Saskatchewan rivers in central Saskatchewan. Empties into N Lake Winnipeg. Index provinces as applicable.
BT Canada
SA Manitoba
North Saskatchewan River
Saskatchewan
South Saskatchewan River

Saskatoon
City in S central.
BT Saskatchewan
Canada

satellite
A valid term through 1977. After 1977, use satellite methods.

satellite measurements
Includes use on level 2 under geodesy(1).
SA geodesy
geodetic coordinates
measurement

satellite methods
Term introduced in 1978. Used for artificial satellites. For natural satellites, see satellites. Includes use on level 3 under geophysical methods(1). Before 1978, search satellite AND methods.
BA geophysical methods
SA geophysical surveys
Landsat
methods
remote sensing
satellites

satellites
Used for natural satellites other than the Earth's Moon. As of 1981, includes use on level 2 under specific planets, e.g. Jupiter(1). Before 1981, included use on level 3 in planet sets. For artificial satellites, use remote sensing.
SA Deimos
Galilean satellites
Jupiter
Mars
Moon
Phobos
satellite methods
Saturn
Tethys Satellite

Satna
Town in NE Madhya Pradesh.
UF Sutna
BT Madhya Pradesh
India

Satpura Range
Line of hills forming N limit of Deccan Plateau. Primarily in Madhya Pradesh. Index states as applicable. Also search Satpura.
BT India
SA Madhya Pradesh
Maharashtra
Uttar Pradesh

Satsuma Peninsula
Between East China Sea and Kagoshima in SW Kyushu. Also search Satsuma.
BT Kyushu
Japan

Satu Mare
County in NW.
BT Transylvania
Romania

saturated materials
As of 1981, no longer a valid term for GeoRef. See saturated zone. Term was introduced in 1978. Before 1978, search saturated.

saturated zone
Term introduced in 1981.
UF phreatic zone
SA ground water
soils
unsaturated zone
water table

saturation
Used as a general term, e.g. under sea water(1) or springs(1).
SA concentration
geochemistry
sea water
soils
solubility
springs
water

saturation magnetization
Term introduced in 1981.
BA magnetization

Saturn
Includes use on level 1 and 2. See entry under Moon(1) for term set options.
SA Iapetus Satellite
outer planets
planetology
planets
satellites
solar system
Tethys Satellite

Sau Alpe
use Sau Alps

Sau Alps
Mountains in NW Carinthia. Also search Saualpe.
UF Sau Alpe
Saualpe
BT Carinthia
Austria
SA Alps

Saualpe
use Sau Alps

Sauce Grande River valley
S Buenos Aires Province. Also search Sauce Grande; Sauce Grande River.
BT Buenos Aires Province
Argentina

Saucesian
North America. Above Zemorrian, below Relizian. Includes use on level 3 under age terms(1). See list E.
BA lower Miocene
Miocene
Neogene
Tertiary
BT Cenozoic
SA Oligocene

Saudi Arabia
Includes use on level 1 as an area term (list O). For term set options see list B.
CO N170000N323000
E0570000E0344500
BA Arabian Peninsula
Asia
NT Hail
SA Arabian Plate
Arabian Shield
Red Sea Basin

Sauerland
Region S and E of the Ruhr River.
BT North Rhine-Westphalia
West Germany

Saugeen Peninsula
use Bruce Peninsula

Sault Sainte Marie
Twin cities on Saint Mary's River between Lake Huron and Lake Superior. Near Sault Saint Marie Canals. Index Michigan and/or Ontario. Also search Sault Ste. Marie.
UF Sault Ste. Marie
SA Michigan
Ontario

Sault Ste. Marie
use Sault Sainte Marie

Saurashtra
Region and former state comprising greater part of Kathiawar Peninsula in W part of country.
BT Gujarat
India

Saurischia
As of 1981, includes use on level 2 under Reptilia(1). See list F. Before 1981, included use on level 3 under Reptilia(1) Archosauria(2).
BA Archosauria
Reptilia
BT Tetrapoda
Vertebrata
NA Sauropoda
Theropoda

Sauropoda
As of 1981, includes use on level 3 under Reptilia(1) Saurischia(2). See list F. Before 1981, included use on level 3 under Reptilia(1) Archosauria(2).
BA Saurischia
Archosauria
Reptilia

BT Tetrapoda
Vertebrata

Sauropterygia
Term introduced in 1981. Includes use on level 2 under Reptilia(1). See list F.
BA Euryapsida
Reptilia
BT Tetrapoda
Vertebrata
NA Nothosauria
Plesiosauria

sausage structure
use boudinage

Sausar Group
use Sausar Series

Sausar Series
Indian Shield region of central S and S India. Also search Sausar Group.
UF Sausar Group
BT Archean
SA India

Savage River
In Garrett County in extreme W.
BT Maryland
United States

Savanna Formation
In Krebs Group. Includes Spaniard Limestone Member, Spiro Sandstone, Sam Creek Limestone, Doneley Limestone. W Arkansas and E and S Oklahoma. Also search Savanna.
BT Desmoinesian
Middle Pennsylvanian
Pennsylvanian
SA Arkansas
Oklahoma

Savannah River
Serves as boundary between Georgia and South Carolina, and flows into Atlantic Ocean. Index states as applicable.
BT United States
SA Georgia
South Carolina

Savannah River Plant
U.S. Atomic Energy Commission plant in Aiken and Barnwell counties on Savannah River.
BT South Carolina
United States

Savoia
use Savoy

Savoie
Department in SE.
CO N450000N455000
E0070000E0053000
BT France
NT Ambin Massif
Tarentaise
Vanoise

Savona
City and province near Genoa in NW part of country.
BT Liguria
Italy

Savoy
Historical region of SE France and NW Italy, now chiefly in French departments of Haute-Savoie and Savoie. Index countries as applicable.
UF Savoia
BT Europe
SA France
Italy

Savoy Alps
NW offshoots of Graian Alps S of Lake Geneva. Also search Savoy.
BT Haute-Savoie
France
SA Alps
Mont Blanc
Vanoise

Sawatch Mountains
use Sawatch Range

Sawatch Range
Range of the Rocky Mountains in central.
UF Sawatch Mountains
BT Colorado
United States
SA Rocky Mountains

Sawtooth Range
Large group of mountain ranges in S central Idaho.
BT Idaho
United States

Saxonian
Europe. Above Autunian, below Thuringian. Includes use on level 3 under age terms(1). See list E.
BA Permian
BT Paleozoic
SA Rotliegendes

Saxonian Massif
Term introduced in 1981.
CO N502200N523000
E0150400E0110000
BT Germany
SA East Germany

Saxony
Former German state now part of East Germany. Index districts as applicable.
BT East Germany
SA Cottbus Bezirk
Dresden Bezirk
Karl-Marx-Stadt Bezirk
Leipzig Bezirk
Saxony-Thuringia

Saxony-Anhalt
Former East German state. Index districts as applicable.
BT East Germany
SA Halle Bezirk
Magdeburg Bezirk

Saxony-Thuringia
Region in S and SE East Germany including the former German states of Saxony and Thuringia.
BT East Germany
SA Saxony
Thuringia

Sayak
Village N of W Lake Balkash in W central.
BT Kazakhstan
USSR

Sayan
E-W range just N of Mongolia extending from S Krasnoyarsk Kray on W across Tuva ASSR into W Irkutsk Oblast on the E. Sayan Mountains used from 1976-1978. Also search Sayan Mountains; Sayan Range.
UF Sayan Mountains
Sayan Range
BT USSR
NA Eastern Sayan
Western Sayan

Sayan Mountains
use Sayan

Sayan Range
use Sayan

Sb
use antimony

Sc
use scandium

Scaglia Formation
Cretaceous to Eocene. In the Apennines and NE Italy.
SA Cretaceous
Eocene
Italy
Paleocene

scales, time
use time scales

Scandentia
Term introduced in 1981. Includes use on level 2 under Mammalia(1). See list F.
BA Mammalia
BT Tetrapoda
Vertebrata

Scandinavia
Region comprising Denmark, Norway, Sweden, and Finland. Iceland sometimes included. Index countries as applicable. Includes use on level 1 as an area term (list O). For term set options see list B.
CO N543000N710000
E0320000E0050000
BA Western Europe
Europe
NA Denmark
Finland
Norway
Sweden
SA Caledonides

scandium
Includes use on level 1 and 2 as a chemical element (list D).
UF Sc
SA elements
rare earths

Scania
use Skane

scanning
A valid term through 1978 on level 3 under electron microscopy(1). Use scanning method.

scanning electromicroscopy data
use SEM data

scanning method
Term introduced in 1978. Includes use on level 3 under electron microscopy(1). Before 1978, also search scanning.
SA electron microscopy
methods
SEM data
TEM data

Scaphites
Genus. Includes use on level 3 under Mollusca(1). See list F.
BA Ammonoidea
Tetrabranchiata
Cephalopoda
Mollusca
BT Invertebrata

Scaphopoda
Includes use on level 2 under Mollusca(1). See list F.
BA Mollusca
BT Invertebrata
NA Dentalium

scapolite
BA scapolite group
framework silicates
silicates
BT minerals

scapolite group
Includes use in combination with framework silicates (i.e. framework silicates, scapolite group) on level 2 under minerals(1). See list L.
BA framework silicates
silicates
BT minerals
NA kenyaite
magadiite
NZ meionite
NA scapolite

scarps
Also search escarpments.
UF escarpments
SA cliffs
fault scarps
faults
geomorphology
slopes

scattering
A general term used for properties of waves.
SA absorption and scattering
elastic waves
electron microscopy
optical properties
spectroscopy
wave dispersion

scawtite
BA chain silicates
silicates
BT minerals

Schaffhausen
Northernmost canton. Also a city.
BT Switzerland

schapbacite
use matildite

scheelite
BA tungstates
BT minerals

Schefferville
Town in NE near Labrador border.
BT Quebec
Canada

Schela
Village in SW.
BT Walachia
Romania

Schelde River
use Scheldt River

Scheldt River
Rises in N France and empties into North Sea through two estuaries, the East and West Scheldt. Index countries as applicable.
UF Schelde River
BT Europe
SA Belgium
France
Netherlands

Schenectady Formation
Underlies Indian Ladder Beds; overlies Snake Hill Formation. E central New York. Also search Schenectady.
BT Middle Ordovician
Ordovician
SA New York

Schiefergebirge
use Rhenish Schiefergebirge

schist
As of 1981, no longer a valid term for GeoRef. Use schists. Before 1981, included use on level 3 under metamorphic rocks(1) schists(2).

schistosity
Includes use on level 3 under foliation(1) style(2) and under structural analysis(1).
BA foliation
SA axial-plane structures
cleavage
flow cleavage
metamorphic rocks
orientation
schists
slaty cleavage
slip cleavage
structural analysis

schists
Includes use on level 2 under metamorphic rocks(1). See list J. Before 1981, also search schist.
BT metamorphic rocks

NA biotite schist
blueschist
calc-schist
chlorite schist
epidiorite
glaucophane schist
greenschist
greenstone
listwanite
muscovite schist
pelitic schist
SA schistosity

schizomycetes
use bacteria

Schizophoria
Genus. Includes use on level 3 under Brachiopoda(1) Orthida(2).
BA Enteletacea
Orthida
Articulata
Brachiopoda
BT Invertebrata

Schizophyta
use Cyanophyta

Schleswig-Holstein
State in extreme N.
CO N531500N550000
E0112500E0081000
BT West Germany
NT Helgoland
Holstein
Kiel
Sylt
SA Jutland

schlieren
SA igneous rocks

Schlotheim
Town in E.
BT Halle Bezirk
East Germany

Schneeberg
Town in Erzgebirge.
BT Karl-Marx-Stadt Bezirk
East Germany

Schoharie County
E central.
BT New York
United States

scholzite
BA phosphates
BT minerals

schorlomite
BA garnet group
nesosilicates
orthosilicates
silicates
BT minerals

schreibersite
As of 1981, includes use on level 3 under minerals(1) phosphides(2). See list L. Before 1981, included use on level 3 under minerals(1) native elements and alloys(2).
UF rhabdite
BZ phosphides
sulfides
BT minerals

Schuler Formation
In Cotton Valley Group. Includes Morgan Sands Zone, Jones Sand, Dorcheat and Shongaloo members. Subsurface.
BT Upper Jurassic
Jurassic
SA Arkansas
Cotton Valley Group
Louisiana
Texas

schuppen structure
use tectonic wedges

Schuylkill County
E Central.
BT Pennsylvania
United States

Schwagerina
Genus. Includes use on level 3 under foraminifera(1) Fusulinidae(2).
BA Fusulinidae
Fusulinina
foraminifera
BT Protista
Invertebrata

Schwartzburg Anticline
use Schwarzburg Anticlinorium

Schwarzburg Anticlinorium
In Thuringia.
UF Schwartzburg Anticline
BT Gera Bezirk
East Germany

Schwarzenberg
City in W.
BT Karl-Marx-Stadt Bezirk
East Germany

Schwechat Valley
River valley SE of Vienna in E.
BT Lower Austria
Austria

Schwerin
As of 1981, refers to only the city. Before 1981, also included the district.
BT Schwerin Bezirk
East Germany

Schwerin Bezirk
Term introduced in 1981. District in NW. Before 1981, search Schwerin.
CO N525200N540000
E0123000E0103500
BT East Germany
NT Schwerin

scintillations
Includes use on level 2 under ionosphere(1).
SA gamma rays
ionosphere
particles
radioactivity

Scioto drainage basin
use Scioto River basin

Scioto River basin
Central and S central.
UF Scioto drainage basin
BT Ohio
United States

Sciuromorpha
Term introduced in 1981. Includes use on level 2 under Mammalia(1). See list F.
BA Rodentia
Mammalia
BT Tetrapoda
Vertebrata
NA Castoridae

Scleractinia
Includes use on level 2 under Coelenterata(1). See list F. Also search Madreporaria; madrepores.
UF Hexacorallia
Madreporaria
BA Anthozoa
Coelenterata
BT Invertebrata
NA Montastrea
SA Rugosa
Tabulata

sclerites
SA Echinodermata
fossils
Holothuroidea

sclerotinite
Term introduced in 1981.
BA inertinite
macerals
coal
organic residues
BT sedimentary rocks

scolecite
BA zeolite group
framework silicates
silicates
BT minerals

scolecodonts
Includes use on level 2 under worms(1). See list F.
BA worms
BT Invertebrata
SA Annelida

Scoresby Land
Region in E on Greenland Sea between King Oscar Fjord on N and Scoresby Sound on the S.
BT Greenland
Arctic region

Scoresby Sound
Deep inlet and fjord system of Greenland Sea on central E coast. Also search Scoresby Sund.
UF Scoresby Sund
BT Greenland
Arctic region

Scoresby Sund
use Scoresby Sound

scoria
As of 1981, includes use on level 3 under igneous rocks(1) pyroclastics(2). See list H. Before 1981, included use on level 3 under igneous rocks(1) volcanic rocks(2) and pyroclastics and glasses(2).
BA pyroclastics
volcanic rocks
BT igneous rocks
SA pumice

scorodite
BA arsenates
BT minerals

Scotia Arc
use Scotia Ridge

Scotia Ridge
Extends in an arc in the South Atlantic Ocean from S of the Falkland Islands eastward to the South Sandwich Islands and then westward to S of the South Orkney Islands. Also search Scotia Arc.
UF Scotia Arc
BT Atlantic Ocean

Scotia Sea
Part of the South Atlantic Ocean within the arc of the Scotia Ridge and E of Drake Passage.
CO S620000S530000
W0270000W0650000
BT Antarctic Ocean
SA Atlantic Ocean
Leg 36

Scotia Sea Islands
Term introduced in 1981 for islands off Antarctica in the Scotia Sea.
CO S640000S530000
W0250000W0650000
BT Antarctica
NA South Georgia
South Orkney Islands
South Sandwich Islands
South Shetland Islands

Scotian Shelf
Off Nova Scotia. Also search Nova Scotia Shelf.
UF Nova Scotia Shelf
Nova Scotian Shelf
BT Atlantic Ocean

Scotland
Includes use on level 1 as an area term (list O). For term set options see list B.
CO N544000N610000
W0004500W0083000
BA Great Britain
United Kingdom
Western Europe
Europe
NT Aberdeen
Aberdeenshire
Angus
Ardnamurchan
Argyllshire
Arran
Ayrshire
Banffshire
Berwickshire
Cairngorm Mountains
Caithness
Dundee
East Lothian
Edinburgh
Fair Isle
Fife
Firth of Clyde
Forth Valley
Gairloch
Galloway
Girvan
Glasgow
Glen Coe
Grampian Highlands
Great Glen Fault
Harris
Hebrides
Huntly
Inner Hebrides
Inverness
Inverness-shire
Islay
Isle of Skye
Kintyre
Kirkcudbrightshire
Loch Lomond
Midland Valley
Midlothian
Northern Highlands
Orkney Islands
Outer Hebrides
Perthshire
Raasay
Rhum
Ross-shire
Scottish Lowlands
Scourie
NZ Shetland Islands
NT Southern Uplands
Sutherland
Tay Estuary
Unst
SA Cheviot Hills
Esk Trough
North Sea Coast

Scott County
Index states as applicable.
BX Arkansas
Illinois
Indiana
Iowa
Kansas
Kentucky
Minnesota
Mississippi
Missouri
Tennessee
Virginia
BT United States

Scottish Lowlands
Term introduced in 1981. Region S of Dumbarton-Stonehaven line to the Southern Uplands. Before 1981, search Lowlands AND Scotland.
BT Scotland
Great Britain
United Kingdom

scour casts
Includes use on level 3 under sedi-

mentary structures(1) bedding plane irregularities(2). See list K.
BA bedding plane irregularities
sedimentary structures
SA flute casts

scour marks
Includes use on level 3 under sedimentary structures(1) bedding plane irregularities(2). See list K.
BA bedding plane irregularities
sedimentary structures
SA current markings
flute casts

Scourie
Village on North Minch in NW.
BT Scotland
Great Britain
United Kingdom

Scripps Institution of Oceanography
In La Jolla, a NW section of San Diego, in San Diego County.
BT California
United States

ScS-waves
Term introduced in 1981.
BA body waves
elastic waves
SA S-waves
seismology
waves

Sculptured Hills
BT Moon

Scutelluidae
Includes use on level 3 under Trilobita(1) Ptychopariida(2).
BA Ptychopariida
Trilobita
BT Trilobitomorpha
Arthropoda
Invertebrata

Scyphozoa
Includes use on level 2 under Coelenterata(1). See list F.
BA Coelenterata
BT Invertebrata
NA Conularida

Scythian
Europe. Above Tatarian (Permian), below Anisian. Includes use on level 3 under age terms(1). See list E.
UF Skythian
BA Lower Triassic
Triassic
BT Mesozoic
SA Werfenian

Scythian Platform
Broad, sedimentary plain N and NE of the Black Sea. Index Soviet republics as applicable.
BT USSR
SA Russian Republic
Ukraine

Se
use selenium

sea fan
use submarine fans

sea floor spreading
use sea-floor spreading

sea ice
Includes use on level 2 under oceanography(1).
SA ice
oceanography
sea water

sea level
Not a valid term for GeoRef. Use changes of level for recent sea level. Use regression or transgression when referring to older sea levels.

sea mounts
use seamounts

Sea of Azof
use Azov Sea

Sea of Azov
use Azov Sea

Sea of Crises
Also search Crisium.
UF Mare Crisium
BT Moon

Sea of Fertility
UF Mare Fecunditatis
BT Moon

Sea of Galilee
Freshwater lake lying in N part of the Great Rift Valley. Also search Lake Tiberias; Lake Kinneret.
UF Lake Kinneret
Lake Tiberias
BT Israel
SA Great Rift Valley

Sea of Japan
Not a valid index term for GeoRef since 1973. Use Japan Sea. Before 1974, also search Sea of Japan.

Sea of Marmara region
Term introduced in 1981.
CO N390500N420500
E0310000E0254000
BT Turkey

Sea of Nectar
Also search Mare Nectaris.
UF Mare Nectaris
BT Moon

Sea of Okhotsk
Not a valid index term for GeoRef since 1973. Use Okhotsk Sea. Before 1974, also search Sea of Okhotsk.

Sea of Rains
Also search Mare Imbrium; Imbrium Basin.
UF Imbrium Basin
Mare Imbrium
BT Moon

Sea of Serenity
Also search Mare Serenitatis.
UF Mare Serenitatis
BT Moon

Sea of Tranquillity
Also search Mare Tranquillitatis.
UF Mare Tranquillitatis
BT Moon

Sea of Waves
UF Mare Undarum
BT Moon

sea water
Includes use on level 1 (list A); on level 2 under isotopes(1). Used for studies on the composition and properties of the water of the oceans. For ancient water, see sets such as sedimentation or geochemistry. Term set options are:
composition
subtopic
evolution
subtopic
experimental studies
subtopic
genesis
subtopic
geochemistry
subtopic
theoretical studies
subtopic
SA atmosphere
brackish water
brines
density
desalinization
dispersion
geochemistry
halmyrolysis
ion exchange
isotopes
marine geology
nepheloid layer
oceanography
paleosalinity
pH
salinity
salt
salt water
saturation
sea ice
solubility
suspended materials
upwelling

sea-floor spreading
Includes use on level 1 (list A); on level 2 under tectonophysics(1). Used for topics related to the hypothesis that the oceanic crust is increasing by convective upwelling of magma along the mid-ocean ridges or world rift system. If 1, term set options are:
topic [causes, concepts, evolution, mechanism, rates]
subtopic (no area term)
UF ocean-floor spreading
sea floor spreading
spreading concept
spreading-floor hypothesis
SA Benioff zone
continental drift
crust
fracture zones
geosynclines
heat flow
magnetic anomalies
mantle
marine geology
mid-ocean ridges
Mohorovicic discontinuity
ocean basins
ocean floors
paleo-oceanography
paleobathymetry
paleomagnetism
plate tectonics
rifting
spreading centers
tectonophysics

sea-floor trench
use trenches

sea-water encroachment
use salt-water intrusion

sea-water intrusion
use salt-water intrusion

sealing
Term introduced in 1981.
SA engineering geology

seamounts
Includes use on level 2 under ocean floors(1). Also search guyots.
UF guyots
sea mounts
SA ocean floors

seams, coal
use coal seams

Searles Lake
Dry lake in NW San Bernardino County in S.
BT California
United States

seas, marginal
use marginal seas

seasonal
A valid term through 1977. After 1977, use seasonal variations.

seasonal variations
Term introduced in 1978. Before 1978, search seasonal AND variations.
SA magnetic field
secular variations
variations

seat clay
use underclay

seat earth
use underclay

Seattle
City on Puget Sound in King County in NW.
BT Washington
United States

seawalls
Term introduced in 1978 on level 2 under shorelines(1).
SA shorelines

seaweed
use Phaeophyta

Sebes
Town in Alba County in SW.
BT Transylvania
Romania

Sebes Mountains
SW Transylvania.
BT Transylvania
Romania

Sebkha el Melah
BT Tunisia

sebkha environment
Term introduced in 1978. Before 1978, search sebkha.
UF sabkha environment
SA environment
paleogeography
playas
salt marshes
sedimentation
supratidal environment

Sebkra de Tindouf
use Tindouf Basin

secondary dispersion
Term introduced in 1978 on level 3 under mineral exploration(1).
SA dispersion patterns
mineral exploration
primary dispersion

secondary magnetization
Term introduced in 1981.
BA magnetization
SA remagnetization

secondary minerals
BT minerals

secondary recovery
SA petroleum
petroleum engineering
recovery
tertiary recovery

secondary structures
Includes use on level 2 under sedimentary structures(1). See list K.
BA sedimentary structures
NA armored mud balls
concretions
cone-in-cone
geodes
microstylolites
septaria
stylolites
SA primary structures
structures

secondary wave
use S-waves

sections
Not to be used for thin sections. Used for cross-sections in stratigraphy.
SA borehole sections
pit sections
polished sections

stratigraphy
thin sections
type sections

secular variations
Includes use on level 3 under Earth(1).
UF geomagnetic secular variation
SA magnetic field
seasonal variations
variations

Sedan
City near Belgian border.
BT Ardennes
France

Sedgwick Basin
S central.
BT Kansas
United States

Sedgwick County
Index states as applicable.
BX Colorado
Kansas
BT United States

sediment sampling
As of 1981, no longer a valid term for GeoRef. Use sampling and sediments. Included use on level 3 under mineral exploration(1). Term was introduced in 1978.

sediment transport
A valid term through 1976. Use transport under sedimentation(1).

sediment yield
SA erosion
runoff
sediments
streams
yields

sediment-water interface
Term introduced in 1978. Before 1978, search interface AND sediment-water.
UF interface, sediment-water
SA interfaces
sedimentation
sediments
water

sedimentary basins
Includes use on level 3 under paleogeography(1). Term should be used in paleogeographic papers, not for recent basins.
SA basins
paleogeography

sedimentary bodies
Term introduced in 1981.
SA sedimentary rocks
sedimentation
sediments

sedimentary cover
UF cover, sedimentary
SA basement
overburden
regolith
tectonics

sedimentary fault
use growth faults

sedimentary petrology
Treated as a whole. Includes use on level 1 (list A); on level 2 under area terms(1), bibliography(1), education(1), and symposia(1). If level 1, term set options are:
topic [applications, bibliography, catalogs, classification, concepts, education, history, instruments, methods, nomenclature, philosophy, practice, principles, symposia, techniques, textbooks]
subtopic
SA bibliography
catalogs
clay mineralogy
diagenesis
domains
glossaries
lexicons
marine geology
oceanography
petrology
sedimentary rocks
sedimentary structures
sedimentation
sedimentology
sediments
shape analysis

sedimentary processes
Includes use on level 3 under mineral deposits, genesis(1) processes(2).
SA deposition
mineral deposits
processes

sedimentary quartzite
use orthoquartzite

sedimentary rocks
See list I. Includes use on level 1 (list A); on level 2 under paragenesis(1) and weathering(1). If level 1, term set options are:
rock group [carbonate rocks, chemically precipitated rocks, clastic rocks, organic residues]
rock name
topic [classification, composition, diagenesis, environmental analysis, geochemistry, lithofacies, lithostratigraphy, petrography, petrology, pore water, properties, provenance, textures]
subtopic [e.g. fabric, grains, grain size, rounding, sorting, surface textures, etc.]
NT carbonate rocks
chemically precipitated rocks
clastic rocks
NA grapestone
NX organic residues
pyroclastics
SA agglomerate
anhydrite
arenaceous texture
argillaceous texture
arkosic composition
authigenesis
bauxite
boulders
calcareous composition
catagenesis
cement
clasts
clay mineralogy
concretions
diagenesis
dolomitic composition
dolomitization
environmental analysis
fabric
ferruginous composition
fossiliferous materials
glauconitic composition
grains
greensand
gypsum
insoluble residues
iron-rich composition
kerogen
lenses
lithofacies
lithostratigraphy
metasedimentary rocks
oolite
oolitic texture
organic materials
packing
paragenesis
pelitic texture
petrography
phi scale
pisolites
pisoliths
pisolitic texture
porosity
provenance
reefs
rocks
rounding
sedimentary bodies
sedimentary petrology
sedimentary structures
sedimentation
sedimentology
sediments
shape analysis
shells
siliceous composition
size distribution
sorting
terrigenous materials
wildflysch

sedimentary structures
See list K for types and names of structures. Includes use on level 1 (list A). Used for a structure in a sedimentary rock, formed either contemporaneously with deposition (primary) or subsequently to deposition (secondary). Term set options are:
topic [classification, environmental analysis, genesis, interpretation, nomenclature, patterns]
subtopic
type of structure [bedding plane irregularities, biogenic structures, cylindrical structures, planar bedding structures, primary structures, secondary structures, soft sediment deformation, turbidity current structures]
specific type (List K)
NA bedding plane irregularities
biogenic structures
cylindrical structures
planar bedding structures
primary structures
secondary structures
soft sediment deformation
turbidity current structures
SA bars
channels
deposition
diagenesis
dunes
environmental analysis
fluvial features
glacial geology
ice wedges
paleogeography
reefs
sedimentary petrology
sedimentary rocks
sedimentation
sediments
solution cavities
spherules
stromatolites
structures

sedimentation
Includes use on level 1 (list A); on level 2 under paleoecology(1). Used for the act or process of forming or accumulating sediment in layers, including all processes from transport through diagenesis. Before 1976, also search aggregation. If level 1, term set options are:
environment
type of environment [coastal environment, deltaic environment, eolian environment, estuarine environment, intertidal environment, lacustrine environment, lagoonal environment, marine environment, paludal environment, reefs, terrestrial environment]
topic [controls, cyclic processes, deposition, diagenesis, precipitation, processes, provenance, rates]
subtopic
transport
type of transport [glacial transport, ice-rafting, marine transport, stream transport, turbidity currents, wind transport]
SA ablation
accretion
atectonic processes
base surges
basins
beaches
biochemical sedimentation
bioclastic sedimentation
biofacies
bioturbation
braided streams
cementation
changes of level
channels
chemical sedimentation
coastal environment
continental shelf
continental slope
controls
current directions
cyclic processes
cyclothems
deep-sea sedimentation
deltaic environment
deltaic sedimentation
deltas
deposition
detrital sedimentation
diagenesis
dolomitization
engineering geology
environment
environmental analysis
eolian environment
eolian sedimentation
estuaries
estuarine environment
estuarine sedimentation
flow structures
fluvial sedimentation
fresh water
fresh-water sedimentation
geochemical controls
geopressure
glacial geology
glacial sedimentation
glacial transport
glaciofluvial sedimentation
glaciomarine sedimentation
granulometry
heavy minerals
high-energy environment
hydrology
ice rafting
intertidal environment
intertidal sedimentation
lacustrine environment
lacustrine sedimentation
lagoonal environment
lagoonal sedimentation
lagoons
lakes
lithification
localization
low-energy environment
marine environment
marine geology
marine sedimentation
marine transport
nearshore sedimentation
nodules
ocean circulation
ocean floors
oceanography
organic materials
packing

paleocurrents
paleoecology
paleogeography
paludal environment
paludal sedimentation
pelagic sedimentation
periodicity
phase equilibria
precipitation
processes
progradation
provenance
reefs
regression
reservoir rocks
reworking
rhythmite
salinity
salt marshes
saltation
sebkha environment
sediment-water interface
sedimentary bodies
sedimentary petrology
sedimentary rocks
sedimentary structures
sedimentology
sediments
shallow-water environment
shelf environment
silicification
slope environment
slumping
sorting
stream transport
structural controls
swamps
syntectonic processes
taiga environment
tectonic controls
terrestrial environment
terrestrial sedimentation
tidal flats
transport
turbidity currents
uniformitarianism
wind transport

sedimentology
For the discipline.
SA geomorphology
marine geology
sedimentary petrology
sedimentary rocks
sedimentation
sediments

sediments
See list N. Includes use on level 1 (list A); on level 2 under weathering(1). Used for unconsolidated solid fragmental material, or a mass of such material, that originates from weathering of rocks. If 1, term set options are:
sediment group [carbonate sediments, clastic sediments, marine sediments, organic residues]
sediment name
topic [classification, composition, diagenesis, distribution, environmental analysis, genesis, geochemistry, lithofacies, lithostratigraphy, petrography, petrology, pore water, properties, provenance, textures]
subtopic [e.g. grain size]
NT carbonate sediments
clastic sediments
marine sediments
NX organic residues
pyroclastics
SA anhydrite
arenaceous texture
argillaceous texture
arkosic composition
biofacies
breccia
calcareous composition
cement
clasts
clay mineralogy
concretions
dehydration
deposition
diagenesis
dolomitic composition
dolomitization
engineering properties
entropy
environmental analysis
evaporites
fabric
ferruginous composition
filtration
fines
fossiliferous materials
glacial geology
glauconitic composition
grain size
grains
granulometry
gypsum
halite
heavy minerals
hydrology
iron-rich composition
lake sediments
liquefaction
lithification
lithofacies
lithostratigraphy
littoral drift
marine geology
ocean floors
oolitic texture
organic materials
permeability
petrography
petrology
phi scale
pisoliths
pisolitic texture
pore water
porosity
provenance
reworking
roundness
sediment yield
sediment-water interface
sedimentary bodies
sedimentary petrology
sedimentary rocks
sedimentary structures
sedimentation
sedimentology
shape analysis
shells
siliceous composition
size distribution
soils
sorting
sphericity
spherules
stream sediments
suspension
terrigenous materials
unconsolidated materials
volcanic ash
weathering
wildflysch

seeds
SA paleobotany

Seeland
use Zealand

seepage
Includes use on level 2 under engineering geology(1).
SA drainage
engineering geology
foundations
ground water
nuclear facilities
oil seeps
percolation
petroleum
reservoirs
soil mechanics
soils
tunnels
underground installations
waste disposal
water
waterways

seeps, oil
use oil seeps

Segovia
Province in central.
CO N404000N413600
W0031500W0044200
BT Old Castile
Spain

Segozero
Lake in S central Karelian ASSR in NW European USSR.
BT Russian Republic
USSR

Segre Valley
River valley in Catalonia. Also search Segre; Segre River.
BT Lerida
Catalonia
Spain

segregation
Term introduced in 1981.
SA fractional crystallization
fractionation
geochemistry
magmas
phase equilibria

seiches
SA lakes
limnology
ocean waves
waves

Seiland
Island in Arctic Ocean off NW.
BT Finnmark
Norway

Seine
Former department which constituted Paris proper and a ring of industrial and residential suburbs. Department officially abolished January 1, 1960.
BT France

Seine Estuary
The tidal mouth of the Seine River emptying into the Bay of the Seine at Le Havre. Index departments as applicable.
BT France
SA Calvados
Eure
Seine-Maritime

Seine River
Rises in Cote-d'Or Department in E and flows NW through Paris and into the English Channel near Le Havre.
BT France

Seine Valley
River valley in N central and N.
BT France

Seine-et-Marne
Department in N central.
CO N481000N491000
E0033000E0021500
BT France
NT Fontainebleau
SA Yonne Valley

Seine-et-Oise
Former department outside Paris suburbs which completely surrounded former Seine Department. Officially abolished January 1, 1968.
BT France

Seine-Maritime
Department in NW.
CO N491500N500500
E0015000E0000000
BT France
NT Caux
Dieppe
Le Havre
Rouen
SA Normandy
Seine Estuary

Seine-Saint-Denis
Term introduced in 1981. Department in N.
CO N485000N490000
E0023500E0021500
BT France

Seislog
Term introduced in 1978.
SA instruments
seismic methods
seismology

seismic data
A valid term through 1977. Use data and earthquakes.

seismic effects
A valid term through 1977. Use effects and earthquakes.

seismic energy
Term introduced in 1981 for energy produced by earthquakes.
SA earthquakes
energy
seismology

seismic focus
use focus

seismic intensity
Term introduced in 1981. Includes use on level 2 under earthquakes(1).
SA amplitude
earthquakes
intensity
magnitude
modified Mercalli scale

seismic methods
Includes use on level 2 under geophysical methods(1).
BA geophysical methods
NA reflection methods
refraction methods
SA common-depth-point method
deep seismic sounding
elastic waves
filters
methods
ocean bottom seismographs
reflection
refraction
Seislog
seismic surveys
signal-to-noise ratio
Vibroseis

seismic moment
Term introduced in 1981.
SA displacements
earthquakes
faults
seismology
shear modulus

seismic noise
A valid term through 1978. After 1978, use noise.

seismic response
Term introduced in 1981.
SA earthquakes
engineering geology
seismology

seismic risk
Includes use on level 3 under nuclear facilities(1) or geologic hazards(1).
UF risk, seismic
SA engineering geology

geologic hazards
nuclear facilities
seismic zoning

seismic sea waves
use tsunamis

seismic sounding
use deep seismic sounding

seismic sources
Includes use on level 2 under seismology(1).
UF sources, seismic
SA earthquakes
elastic waves
explosions
seismology

seismic studies
A valid term through 1976. Use seismic surveys and theoretical studies or experimental studies.

seismic surge
use tsunamis

seismic surveys
Includes use on level 3 under area terms(1) geophysical surveys(2).
BA geophysical surveys
BT surveys
SA crust
explosions
reflection
refraction
seismic methods
seismology
Vibroseis

seismic waves
A valid term through 1975. After 1975, use elastic waves.

seismic zoning
Term introduced in 1981.
SA earthquakes
engineering geology
environmental geology
seismic risk
seismology

seismicity
Includes use on level 2 under seismology(1).
SA seismology
seismotectonics

seismograms
Includes use on level 3 under earthquakes(1) or seismology(1).
UF earthquake record
NT numerical seismograms
synthetic seismograms
theoretical seismograms
SA coda waves
earthquakes
seismographs
seismology
signal-to-noise ratio
spectral analysis

seismographs
Includes use on level 3 under seismology(1). Before 1981, also search seismometers.
NT electromagnetic seismographs
horizontal-component seismographs
inertial seismographs
long-period seismographs
ocean bottom seismographs
short-period seismographs
strain seismographs
three-component seismographs
vertical-component seismographs
SA earthquakes
high-level mode
instruments
seismograms
seismology
spheroidal mode
toroidal mode

seismology
For general treatments of the subject. See earthquakes for specific treatments. Includes use on level 1 (list A); on level 2 under area terms(1), bibliography(1) and symposia(1). See list B. If level 1, term set options are:
topic [catalogs, core, crust, earthquakes, elastic waves, experimental studies, explosions, interior, mantle, methods, microearthquakes, microseisms, observatories, properties, seismicity, seismic sources, theoretical studies, tsunamis, volcanology]
subtopic [e.g. elastic properties, physical properties, P-waves, S-waves]
SA accelerograms
accelerometers
acoustical properties
acoustical waves
aftershocks
Airy waves
amplitude
amplitude distortion
anelastic materials
anelasticity
anisotropy
arrays
arrival time
attenuation
Bacukus-Gilbert analysis
Benioff zone
body waves
coda waves
Conrad discontinuity
core
crust
curved seismic interface
deconvolution
deep seismic sounding
deformation
diffraction
dilatancy
direct problem
discontinuities
distortion
Earth
earthquakes
elastic properties
elastic waves
energy relaxation
epicenters
explosions
faults
filters
focus
geodesy
geophysical methods
geophysics
ground motion
group velocity
Gruneisen parameters
half-space
heterogeneity
high-level mode
high-resolution methods
homogeneity
horizontal discontinuities
inclined seismic interface
induced earthquakes
interior
inverse problem
isochrons
isoseismic maps
linear distortion
long-period waves
Love waves
low-velocity zones
magnitude-frequency ratio
mantle
marker beds
mathematical transformations
maximum entropy analysis
microearthquakes
microseisms
modified Mercalli scale
Mohorovicic discontinuity
neotectonics
noise
non-linear distortion
nuclear explosions
observatories
ocean bottom seismographs
P-waves
PcP-waves
periodicity
phase distortion
phase transitions
phase velocity
PKiP-waves
PKP-waves
PKS-waves
plate tectonics
PPP-waves
PPS-waves
precursors
propagation
Q
Rayleigh waves
raypaths
reflection
refraction
Richter Scale
S-waves
ScS-waves
Seislog
seismic energy
seismic moment
seismic response
seismic sources
seismic surveys
seismic zoning
seismicity
seismograms
seismographs
seismotectonics
SH-waves
shock waves
short-period waves
signal distortion
signal-to-noise ratio
signals
SKP-waves
SKS-waves
sounding
spectral analysis
spherical models
spheroidal mode
Stoneley waves
strain
strain relaxation
stress
strong motion
surface waves
SV-waves
swarms
tangential discontinuities
tectonophysics
teleseismic signals
Thomson-Haskell analysis
three-dimensional models
tilt
tiltmeters
toroidal mode
traveltime
traveltime curves
traveltime residuals
tsunamis
two-dimensional models
velocity structure
vertical discontinuities
vibration
Vibroseis
volcanology
wave dispersion
Wiener-Hopf analysis

seismometers
As of 1981, no longer a valid term for GeoRef. Use seismographs.

seismotectonic maps
Term introduced in 1981. Before 1981, search seismotectonics AND maps.
BA maps

seismotectonics
SA plate tectonics
seismicity
seismology
tectonics

Selachii
As of 1981, includes use on level 3 under Pisces(1) Elasmobranchii(2). See list F. Before 1981, included use on level 3 under Pisces(1) Chondrichthyes(2). Before 1981, also search selachians. As of 1981, search selachians only for biostratigraphy.
BA Elasmobranchii
Chondrichthyes
Pisces
BT Vertebrata

Selangor
State on W coast of Malay Peninsula in West Malaysia.
BT Malaysia
NT Kuala Lumpur
SA West Malaysia

selection, natural
use natural selection

selenates
Term introduced in 1981. Includes use on level 2 under minerals(1). See list L. Before 1981, search selenates and selenites.
BT minerals

selenates and selenites
As of 1981, no longer a valid term for GeoRef. Use selenates or selenites as applicable. Before 1981, included use on level 2 under minerals(1).

Selenga
Province on the Buryat ASSR border in N Mongolia.
UF Selenge
BT Mongolia

Selenga River valley
Extends to SE Lake Baikal in Russian Republic. Also search Selenga. Index countries as applicable.
UF Selenge River valley
BT Asia
SA Mongolia
USSR

Selenge
use Selenga

Selenge River valley
use Selenga River valley

selenides
As of 1981, includes use on level 2 under minerals(1). See list L. Before 1981, included use on level 3 under minerals(1) sulfides(2).
BA sulfides
BT minerals
NA eskebornite
umangite

selenite
Used for the variety of gypsum. Not related to the group name, selenites.
BA sulfates
BT minerals
SA gypsum

selenites
Term introduced in 1981. Includes use on level 2 under minerals(1). See list L. Before 1981, search selenates and selenites.
BT minerals
NZ onofrite

S

selenium
Includes use on level 1 and 2 as chemical element (list D).
UF Se
SA elements
onofrite

Selennyakh
River which rises in the N Cherskiy Range and flows into the Indigirka River in NE Yakutia.
UF Selenyak
BT Russian Republic
USSR

selenodesy
SA geodesy
Moon

Selenyak
use Selennyakh

self-potential methods
Term introduced in 1978. Includes use on level 3 under geophysical methods(1) electrical methods(2). Before 1978, also search self-potential.
UF spontaneous polarization
SA electrical methods
electrical surveys
geophysical surveys
methods

Seligdar
Term introduced in 1981. In SE Yakutia.
CO N584300N584300
E1253000E1253000
BT Yakutia
Russian Republic
USSR

seligmannite
BA sulfarsenites
sulfosalts
BT minerals

Selkirk Mountains
Range in the Rocky Mountains in SE.
UF Selkirks Mountains
BT British Columbia
Canada
SA Esplanade Range
Rocky Mountains

Selkirks Mountains
use Selkirk Mountains

Selukwe
Town in central.
BT Zimbabwe

Selva
use Copiapo

Selwyn Mountains
Term introduced in 1981. On border of Yukon Territory and Northwest Territories, W of the Mackenzie Mountains. Index provinces as applicable. Before 1981, also search Selwyn Range AND Canada.
BT Canada
SA Northwest Territories
Selwyn Range
Yukon Territory

Selwyn Range
W central.
BT Queensland
Australia
SA Selwyn Mountains

SEM data
Term introduced in 1978. For the method, see scanning method.
UF scanning electromicroscopy data
SA data
electron microscopy
scanning method
TEM data
transmission method

Semata Formation
BT Pleistocene
Quaternary
SA Japan

Semenic Mountains
Central Banat. Also search Semenic.
BT Banat
Romania

semi-arid environment
Term introduced in 1978. Before 1978, also search semi-arid; semiarid; semiarid regions.
UF semiarid regions
SA arid environment
climate
ecology
environment
geomorphology

semi-infinite models
Term introduced in 1981.
BA models
SA infinite models
mathematical models

semiarid regions
use semi-arid environment

Seminole Formation
In Skiatook Group. Includes Lenapah Limestone, Nowata Shale, Memorial Shale. NE, central, and central S Oklahoma.
BT Missourian
Upper Pennsylvanian
Pennsylvanian
SA Oklahoma

Semipalatinsk
City and oblast in NE.
BT Kazakhstan
USSR

semitropical environment
use subtropical environment

Semmering
Resort area in Eastern Alps.
BT Austria
SA Lower Austria
Styria

Semri Series
BT Cambrian
SA Bihar
India
Madhya Pradesh

semseyite
BA sulfantimonites
sulfosalts
BT minerals

senarmontite
BA oxides
BT minerals

Sendai
City near E coast in N.
BT Honshu
Japan

Seneca County
Term introduced in 1981. Index states as applicable.
BX New York
Ohio
BT United States
SA Sandusky River basin

Seneca Lake
One of the Finger Lakes in Yates and Seneca counties. Also search Seneca.
CO N422000N424500
W0764500W0770000
BT New York
United States

Senegal
Formerly a republic in French Community. Achieved independence in 1960. Includes use on level 1 as an area term (list O). For term set options see list B.
CO N123000N164000
W0112000W0173000
BA West Africa
Africa
NT Dakar
SA Sahel
Senegal Basin
Senegal River

Senegal Basin
River basin. Index countries as applicable. Also search Senegal.
UF Senegal River basin
BT Africa
SA Guinea
Mali
Mauritania
Senegal

Senegal River
Rises in the highlands of Guinea and flows N and then NW into the Atlantic Ocean at Saint-Louis in Senegal. Index countries as applicable. Also search Senegal.
BT Africa
SA Guinea
Mali
Mauritania
Senegal

Senegal River basin
use Senegal Basin

Senonian
Europe. Above Turonian, below Danian. Includes use on level 3 under age terms(1). See list E.
BA Upper Cretaceous
Cretaceous
BT Mesozoic
NA Campanian
Coniacian
NT Futaba Group
NA Maestrichtian
Santonian
SA Izumi Group

sensing, remote
use remote sensing

sensitive clays
Term introduced in 1981. Also search sensitive clay.
SA clay
clays

Sentinel Butte Formation
Overlies Tongue River Member. NE Montana, and SW North Dakota.
UF Sentinel Butte Shale Member
BT Paleocene
SA Montana
North Dakota

Sentinel Butte Shale Member
use Sentinel Butte Formation

separation
Used as a general term, e.g. on level 3 under micropaleontology(1).
SA mixing
processes

sepiolite
Also search meerschaum.
UF meerschaum
BA clay minerals
sheet silicates
silicates
BT minerals

Sept-Iles
City on the Saint Lawrence River in SE.
BT Quebec
Canada

septa
UF septae
SA fossils
morphology

septae
use septa

septaria
Includes use on level 3 under sedimentary structures(1) secondary structures(2). See list K.
BA secondary structures
sedimentary structures
SA concretions
cone-in-cone

Septibranchia
Term introduced in 1981. Includes use on level 2 under Mollusca(1). See list F.
BA Bivalvia
Mollusca
BT Invertebrata

Sequanian
Europe. See list E.
BA Upper Jurassic
Jurassic
BT Mesozoic

Sequatchie River valley
use Sequatchie Valley

Sequatchie Valley
River valley in SE central Tennessee. Also search Sequatchie River valley.
UF Sequatchie River valley
BT Tennessee
United States

Sequoia
Genus. Includes use on level 3 under gymnosperms(1) Coniferales(2).
BA Taxodiaceae
Coniferales
gymnosperms
BT Spermatophyta
Plantae

Serajevo
use Sarajevo

Serbia
Constituent republic in E and NE Yugoslavia. Formerly Servia.
CO N420500N461000
E0230000E0185500
UF Servia
BT Yugoslavia
NT Aleksinac
Belgrade
Bor
Kopaonik
Kosovo-Metohija
Kostolac
Majdanpek
Metohija
Racha
Soko Banja
Timok Basin
Trepca Mine
Vojvodina
Zapadna Morava
Zlatibor Mountains
SA Moesia
Morava River valley
Serbo-Macedonian Massif

Serbo-Macedonian Massif
Macedonia and S Serbia. Index republics as applicable.
BT Yugoslavia
SA Macedonia
Serbia

Serebryanka Formation
Lower and middle Triassic (?). SE and E Ukraine and adjoining area in Russian Republic.
BT Lower Triassic
Triassic
SA Middle Triassic
Russian Republic
Ukraine

Sereth River
use Siret River

Sergipe
State in NE.
CO S113000S092000
W0362000W0382000
BT Brazil
NT Aracaju
Propria Geosyncline
SA Sao Francisco Basin
Sergipe-Alagoas Basin

Sergipe Basin
use Sergipe-Alagoas Basin

Sergipe-Alagoas Basin
In NE part of country. Index states as applicable. Also search Sergipe Basin.
UF Sergipe Basin
BT Brazil
SA Alagoas
Sergipe

sericite
BA mica group
sheet silicates
silicates
BT minerals
SA muscovite

sericitization
BA metasomatism
SA hydrothermal alteration

Serov
City in Sverdlovsk Oblast E of Urals.
BT Russian Republic
USSR

Serozem
use Sierozems

serpentine
BA serpentine group
sheet silicates
silicates
BT minerals

serpentine group
Includes use in combination with sheet silicates (i.e. sheet silicates, serpentine group) on level 2 under minerals(1). See list L.
BA sheet silicates
silicates
BT minerals
NA antigorite
cerolite
chrysotile
garnierite
lizardite
serpentine

serpentine rock
use serpentinite

serpentinite
Compositional term.
UF serpentine rock
BZ metaigneous rocks
metasomatic rocks
SA serpentinization

serpentinization
Includes use on level 3 under metasomatism(1) processes(2).
BA metasomatism
SA alteration
hydrothermal alteration
processes
serpentinite

Serpulidae
Family. As of 1981, includes use on level 3 under worms(1) Polychaetia(2). See list F. Before 1981, included use on level 3 under Annelida(1) Polychaetia(2). Before 1981, also search serpulids. As of 1981, search serpulids only for biostratigraphy.
BA Polychaetia
worms
BT Invertebrata

Serra do Mar
Coastal mountain range in S and SE. Index states as applicable.
BT Brazil
SA Parana
Santa Catarina
Sao Paulo

Serra do Navio
Town in central.
BT Amapa
Brazil

Serra Geral Formation
BT Mesozoic
SA Brazil
Cretaceous
Jurassic
Paraguay
Sao Paulo

Serrania de Cuenca
Mountain range in E central. Index provinces as applicable.
BT Spain
SA Cuenca Province
Guadalajara Province

Serrania de Ronda
Spur of the Cordillera Penibetica in Andalusia. Index provinces as applicable.
UF Ronda Massif
Ronda, Serrania de
Ronda Sierra
Sierra de Ronda
BT Spain
SA Cadiz Province
Malaga Province

Serravallian
Europe. See list E.
BA middle Miocene
Miocene
Neogene
Tertiary
BT Cenozoic

Servia
use Serbia

Sesia Valley
River valley E of Lake Maggiore in N.
BT Italy
SA Piemonte

Sesia Zone
N Piemonte.
BT Italy
SA Piemonte

Sesia-Lanzo Zone
N Piemonte and N Lombardy. Index autonomous regions as applicable.
BT Italy
SA Alps
Lombardy
Piemonte

Sespe Formation
Overlying Coldwater Sandstone Member of Tejon Formation and underlying Vaqueros Formation. In S California.
BT Tertiary
SA California
Eocene
Oligocene
upper Eocene

Sete
City on Gulf of Lion in S Herault.
UF Cette
BT Herault
France

Seto Inland Sea
use Inland Sea

Seto Sea
use Inland Sea

Seto-chi-umi
use Inland Sea

Seto-Naikai
use Inland Sea

Setouchi
use Inland Sea

Sette Daban (Range)
use Sette-Daban Range

Sette-Daban Range
NW Khabarovsk Kray and SE Yakutia. Also search Sette Daban; Sette-Daban.
UF Sette Daban (Range)
BT Russian Republic
USSR

Setting Lake
Central.
BT Manitoba
Canada

settlement
As of 1978, term is used on level 2 under foundations(1), land subsidence(1), and soil mechanics(1).
SA engineering geology
foundations
land subsidence
soil mechanics

Setubal
District in SW.
BT Portugal

Sevan
Town on shore of Lake Sevan.
BT Armenia
USSR
SA Lake Sevan

Seven Devils Mountains
Range in Adams and Idaho counties in W.
BT Idaho
United States

Severin
Former province whose capital was Caransebes. Much of the province is now Caras-Severin County.
BT Banat
Romania

Severn Valley
River valley. Index political divisions as applicable. Also search Severn River.
BT United Kingdom
SA England
Wales

Severnaya Zemlya
Term introduced in 1981.
CO N780000N820000
E1060000E0900000
BZ West Siberia
BX Russian Republic
BT USSR

Sevier County
Index states as applicable. Also search Sevier AND appropriate state.
BX Arkansas
Tennessee
Utah
BT United States

Sevier orogenic belt
The Sevier Plateau area in Garfield, Piute, and Sevier counties in S central Utah. Also search Sevier AND Utah.
UF Sevier Plateau
BT Utah
United States

Sevier Plateau
use Sevier orogenic belt

Sevilla
use Seville

Seville
As of 1981, refers to only the city. Before 1981, also included the province. In SW Spain.
UF Sevilla
BT Seville Province
Andalusia
Spain

Seville Province
Term introduced in 1981. Before 1981, search Seville. In SW Spain.
CO N365000N381300
W0044000W0063300
BT Andalusia
Spain
NT Carmona
Seville

sewage
Includes use on level 3 under waste disposal(1) or pollution(1).
SA pollution
waste disposal
water quality

Seward County
Index states as applicable.
BX Kansas
Nebraska
BT United States

Seward Peninsula
On Bering Strait between Kotzebue and Norton sounds in W Alaska. Also search Seward AND Alaska.
BT Alaska
United States

Seward Quadrangle
Term introduced in 1981. S Alaska.
CO N600000N610000
W1470000W1500000
BT Alaska
United States

sexual dimorphism
Includes use on level 3 under fossil groups (list F).
SA reproduction

Seychelles
Level 1 area term as of 1978. Officially designated as the Republic of Seychelles. Island group about 700 miles NE of Malagasy Republic.
CO S050000S040000
E0560000E0550000
BA Indian Ocean Islands
NT Aldabra Island
Mahe Island

SH waves
use SH-waves

SH-waves
Also search SH waves.
UF SH waves
BA S-waves
body waves
elastic waves
SA seismology
waves

Shaba
Province in S Zaire.
CO S130000S050000
E0310000E0220000
UF Katanga
BT Zaire
NT Kamoto

Shackleton Glacier
One of glaciers feeding Ross Ice Shelf from Queen Maud Range. Also Shackleton Ice Shelf off Queen Mary Coast extending into Indian Ocean is sometimes called Shackleton Glacier.
BT Antarctica
SA Ross Ice Shelf

Shackleton Range
SE of the Filchner Ice Shelf which is on the Weddell Sea on the Atlantic Ocean side.
BT Antarctica

shaded relief maps
Term introduced in 1981.
BA maps
SA Moon

Shadow Mountains
Index states as applicable.
BT United States
SA California
Nevada

Shady Dolomite
Equivalent to Tomstown Dolomite. N Alabama, NW Georgia, W North Carolina, E Tennessee, and SW Virginia.
BT Lower Cambrian
Cambrian
SA Alabama
Chilhowee Group
Georgia
North Carolina
Tennessee
Virginia

shake wave
use S-waves

Shakh-Dag Range
use Shakhdag Range

Shakhdag Range
In the Lesser Caucasus on N shore of Lake Sevan. Index Soviet republics as applicable.
UF Shakh-Dag Range
BT USSR
SA Armenia
Azerbaidzhan
Lesser Caucasus

Shakopee Dolomite Member
use Shakopee Formation

Shakopee Formation
In Prairie du Chien Group. Overlies and interbedded with Root Valley Sandstone. N Illinois, Iowa, S Minnesota, and S Wisconsin.
UF Shakopee Dolomite Member
BT Lower Ordovician
Ordovician
SA Illinois
Iowa
Minnesota
Wisconsin

shale
Does not include oil shale. Use clays for shale as brick clay. Includes use on level 3 as a commodity term (list C) under construction materials(1) and on level 3 under sedimentary rocks(1) clastic rocks(2). See list I. Before 1976, also search pelite.
UF bloating shale
pelite
BA clastic rocks
BT sedimentary rocks
SA argillaceous texture
black shale
clays
construction materials
oil shale
slate deposits
slates
terrigenous materials
torbanite

shale oil
use oil shale

shallow
A valid term through 1977. After 1977, use shallow-water environment.

shallow earthquake
use shallow-focus earthquakes

shallow-focus earthquakes
Term introduced in 1978. Includes use on level 3 under earthquakes(1). Before 1978, search earthquakes AND shallow focus.
UF normal earthquake
shallow earthquake
SA deep-focus earthquakes
earthquakes
focus
intermediate-focus earthquakes

shallow-water environment
Term introduced in 1978. Before 1978, search shallow-water or shallow.
SA ecology
environment
sedimentation

Shamlug
Town in NE.
BT Armenia
USSR

Shan State
E Burma.
UF Federated Shan States
Shan States
BT Burma

Shan States
use Shan State

Shan-tung
use Shandong

Shan-tung Province
use Shandong

Shandong
Term introduced in 1981. Province in E central on the Yellow Sea. Also search Shantung, Shan-tung.
CO N343000N380000
E1223500E1150000
UF Shan-tung
Shan-tung Province
Shantung
Shantung Province
BT China

Shannon County
Index states as applicable.
BX Missouri
South Dakota
BT United States

Shansi
Province in N central.
BT China

Shantung
use Shandong

Shantung Province
use Shandong

Shap Granite
The Shap Granite and associated igneous and metamorphic rocks have been called Shap Andesites and Shap Rhyolites. English Lake District in NW England.
UF Shap Rhyolites
BT Ordovician
SA England
United Kingdom

Shap Rhyolites
use Shap Granite

shape
As of 1981, no longer a valid term for GeoRef. See shape analysis for grain shape. See figure of Earth and geoid for shape of the Earth. Before 1981, included use on level 2 under Earth(1) and core(1), and on level 3 under sedimentary rocks(1) and sediments(1).

shape analysis
SA analysis
grains
granulometry
roundness
sedimentary petrology
sedimentary rocks
sediments
sphericity
textures

Sharjah
Emirate. One of federation of 7 states at S end of Persian Gulf.
BT United Arab Emirates
Arabian Peninsula

Shark Bay
Large inlet of Indian Ocean. Also search Sharks Bay.
UF Sharks Bay
BT Western Australia
Australia

Sharks Bay
use Shark Bay

Sharon Conglomerate
In Pottsville Group. Maryland, E Ohio, W Pennsylvania, and N West Virginia.
BT Pennsylvanian
SA Maryland
Ohio
Pennsylvania
Pottsville Group
West Virginia

Sharps Meteorite
Term introduced in 1981. Impact near Sharps, in Richmond County, Virginia. Before 1981, also search Sharps AND meteorites.
BA chondrites
BT meteorites
SA Virginia

Shasta County
N central.
BT California
United States

Shatskiy Rise
use Shatsky Rise

Shatsky Rise
In mid Northwest Pacific Basin E of Tokyo.
UF Shatskiy Rise
BT Pacific Ocean

shatter cones
Includes use on level 3 under metamorphism(1) shock metamorphism(2); under deformation(1) field studies(2); and under geomorphology(1) cryptoexplosion features(2).
UF cones, shatter
SA cryptoexplosion features
deformation
explosions
impact features
metamorphism
meteor craters
meteorites
nuclear explosions
shock waves

shattuckite
BA chain silicates
silicates
BT minerals

Shaunavon Formation
Subsurface.
BT Jurassic
SA Canada
Saskatchewan

Shawangunk Mountains
Part of the Kittatinny Mountains in Ulster County in S.
BT New York
United States

Shawnee County
NE Kansas.
BT Kansas
United States

Shawnee Group
Comprises Oread Limestone, Kanwaka Shale, Lecompton Limestone, Tecumseh Shale, Deer Creek Limestone, Calhoun Shale, Topeka Limestone, Severy Shale, Howard Limestone, White Cloud Shale, Happy Hollow Limestone, Cedar Vale Shale, Rulo Limestone, Silver Lake-Auburn Shale Interval, Reading Limestone, Harveyville Shale, Elmont or Preston limestone, and Williard Shale.
BT Virgilian
Upper Pennsylvanian
Pennsylvanian
SA Kansas
Lecompton Limestone
Nebraska
Oread Limestone

shear
As of 1978, used only to indicate the mechanics of deformation. Before 1978, used to indicate style of structure. Now use similar folds for style of structure.
UF shear strain
SA deformation
mechanics
shear zones
similar folds
strain
stress

shear cleavage
use slip cleavage

shear folds
use similar folds

shear modulus
UF Coulombs' modulus
modulus of rigidity
modulus, shear
rigidity modulus
torsion modulus
BA elastic constants
SA bulk modulus
elasticity
seismic moment

shear strain
use shear

shear strength
Includes use on level 3 under soil mechanics(1), rock mechanics(1), or engineering geology(1). Also search cohesion.
UF cohesion
SA mechanical properties
rock mechanics
soil mechanics
strain
strength
stress
tensile strength

shear stress
Includes use on level 3 under engineering geology(1) or deformation(1).
UF tangential stress
SA deformation
engineering geology
stress

shear wave
use S-waves

shear zones
Includes use on level 3 under faults(1) effects(2) and under tectonics(1).
UF zones, shear
SA breccia
faults
fractures
shear
strain
structural analysis
tectonics

shear-cleavage folds
use cleavage folds

shearing
A valid term through 1975. After 1975, see shear. See also fractures, foliation, and structural analysis.

sheet erosion
Term introduced in 1981.
SA erosion
water erosion

sheet silicates
Includes use on level 2 under minerals(1); in combination with chlorite group, clay minerals, mica group, and serpentine group (i.e. sheet silicates, chlorite group) to form terms on level 2 under minerals(1). See list L. Also search phyllosilicates.
UF phyllosilicates
BA silicates
BT minerals
NA apophyllite
astrophyllite
bavenite
berthierine
chlorite group
chrysocolla
clay minerals
cymrite
gillespite
gyrolite
hisingerite
hydrobiotite
hydromica
hydromuscovite
leucosphenite
mica group
penninite
prehnite
pyrophyllite
serpentine group
stilpnomelane
talc
zussmanite
SA asbestos
bentonite
clay mineralogy

sheets
As of 1981, no longer a valid term for GeoRef. See intrusions. Before 1981, included use on level 2 under intrusions(1).

Sheffield
City in Yorkshire in N.
BT England
Great Britain
United Kingdom

Shelby County
Term introduced in 1981. Index states as applicable.
BX Alabama
Illinois
Indiana
Iowa
Kentucky
Missouri
Ohio
Tennessee
Texas
BT United States

shelf
A valid term through 1978 used to distinguish older sedimentation from more recent. After 1978, use shelf environment.

shelf, continental
use continental shelf

shelf environment
Term introduced in 1978. Use this term or slope environment for older sedimentation. For recent sedimentation, use continental shelf or continental slope. Before 1978, search shelf.
SA continental shelf
environment
paleoecology
sedimentation
slope environment

shells
Includes use on level 3 under fossil group(1) for type of material. See list F. Before 1975, also search shell.
SA fossils
Invertebrata
ornamentation
paleontology
sedimentary rocks
sediments
tests

shelves, ice
use ice shelves

Shemakha
City on E Caucasus.
BT Azerbaidzhan
USSR

Shemya Island
One of the Semichi Islands at W end of Aleutian Islands. Also search Shemya.
BT Alaska
United States

Shenandoah Valley
River valley between the Allegheny and Blue Ridge mountains. Primarily in Virginia. Index states as applicable.
BT United States
SA Virginia
West Virginia

Shensi
Province in N central.
BT China
SA Han River basin

Sheridan County
Index states as applicable.
BX Kansas
Montana
Nebraska
North Dakota
Wyoming
BT United States

Sherman County
Index states as applicable.
BX Kansas
Nebraska
Oregon
Texas
BT United States

Shetland Islands
Archipelago off N Scotland 50 miles NE of Orkney Islands. Also search Shetland. Includes use on level 1 as an area term (list O). For term set options see list B.
CO N595000N610000
W0004500W0014500
UF Shetlands
Zetland
BZ Scotland
Atlantic Ocean
SA Fair Isle
South Shetland Islands
Unst

Shetlands
use Shetland Islands

Sheveluch
Volcano on E central Kamchatka Peninsula. Also search Sheveluch Volcano.
CO N563801N563801
E1611900E1611900
UF Sheveluch Volcano
Shiveluch Volcano
BT Russian Republic
USSR

Sheveluch Volcano
use Sheveluch

shield volcanoes
Also search lava domes.
UF basaltic domes
lava domes
BA volcanoes
SA stratovolcanoes
volcanism

shields
Usage restricted to tectonics.
SA basement
cratons
crust
tectonics

Shiga
Prefecture in S.
BT Honshu
Japan

Shikoku
Smallest of the 4 major Japanese islands in S.
CO N323500N342500
E1344500E1320000
BT Japan
NT Ehime
Kagawa
Kochi
Matsuyama
Tokushima
Tosa Bay
Yoshino River
SA Izumi Group
Nankaido
Sambagawa Belt
Shimanto Group

Shikoku Basin
Undersea feature between Shikoku on the NW and the Bonin and Volcano islands to the SE.
BT Pacific Ocean

Shikotsu
Lake in SW.
BT Hokkaido
Japan

Shilka Valley
River valley in SW central Chita Oblast in Transbaikalia. Also search Shilka; Shilka River.
BT Russian Republic
USSR
SA Argun River

Shillong
City in W central.
BT Assam
India

Shillong Plateau
Undulating tableland in W.
BT Assam
India

Shimane
Prefecture in SW.
BT Honshu
Japan

Shimanto Group
Includes Terazoma Series, Nishigawa Series, and Higashigawa Series. Trias-Cretaceous, presumably. Kyushu and S Shikoku. Also search Shimanto.
BT Mesozoic
SA Cretaceous
Japan
Kyushu
Shikoku
Triassic

Shimokita Peninsula
In Mutsu Bay area in extreme N.
BT Honshu
Japan

Shimosa Group
BT Pleistocene
Quaternary
SA Japan

Shinano River
W central Honshu. Flows N into Japan Sea. Longest river in Japan.
BT Honshu
Japan

shingle
Includes use on level 3 under sediments(1) clastic sediments(2). See list N.
BA clastic sediments
BT sediments
SA cobbles
gravel
pebbles

Shinji Lake
SW Honshu.
BT Honshu
Japan

Shiobara
Town in central.
BT Honshu
Japan

Shiraki Steppe
Semidesert plain in E Georgian Republic. Also search Shiraki.
BT Georgian Republic
USSR

Shiranish Formation
BT Cretaceous
SA Iraq

Shirasu
Stream NW of Tokyo in central.
BT Honshu
Japan

Shiraz
Term introduced in 1981. City in Fars (province), SW central Iran.
CO N293800N293800
E0523400E0523400
BT Fars
Iran

Shirley Basin
NW Carbon County in S central.
BT Wyoming
United States

Shiveluch Volcano
use Sheveluch

Shizukuishi
Town in N.
BT Honshu
Japan

Shizuoka
Prefecture. Also a city 55 miles SW of Tokyo.
BT Honshu
Japan

shoaling
Includes use on level 2 under ocean waves(1).
SA ocean waves
waves

shoals
SA fluvial features
geomorphology
lacustrine features
oceanography
reefs
shore features
spits

shock
A valid term through 1977. After 1977, use shock metamorphism on level 2 under metamorphism(1).

shock metamorphism
Term introduced in 1978. Includes use on level 2 under metamorphism(1). Before 1978, also search metamorphism AND shock.
BA metamorphism
SA cryptoexplosion features
meteor craters

shock waves
Includes use on level 2 under interplanetary space(1); on level 3 under deformation(1) experimental studies(2), and under seismology(1).
SA bow shock waves
deformation
foreshocks
interplanetary space
metamorphism
propagation
seismology
shatter cones
waves

Shonai River
In the Nagoya area of S central.
BT Honshu
Japan

Shonkin Sag Laccolith
In Highwood Mountains area of N central.
BT Montana
United States

shonkinite
As of 1981, includes use on level 3 under igneous rocks(1) syenites(2). See list H. Before 1981, included use on level 3 under igneous rocks(1) syenite family(2) and alkali gabbro family(2).
BA syenites
BT igneous rocks
SA alkali gabbros

shooting star
use meteors

shore drift
use littoral drift

shore features
Includes use on level 2 under geomorphology(1).
UF features, shore
NT mud flats
SA bars
bays
beach ridges
beaches
benches
cheniers
cliffs
deltas
estuaries
fjords
geomorphology
inlets
mud banks
progradation
shoals
shorelines
spits
terraces
wave-cut platforms

shore reef
use fringing reefs

shorelines
A level 1 term as of 1978. Used for geological studies on the engineering aspects of shorelines. Before 1978, included use on level 2 under engineering geology(1). Term set options are:
barrier islands
subtopic
beaches
subtopic
changes
subtopic
construction
subtopic
design
subtopic
dynamics
subtopic
erosion
subtopic
hydraulics
subtopic
management
subtopic
seawalls
subtopic
stabilization
subtopic
UF coastlines
SA barrier islands
bays
beach ridges
beaches
breakwaters
changes of level
cheniers
coastal environment
construction
design
dynamics
engineering geology
erosion
geomorphology
hydraulics
inlets
management
marine installations
ocean circulation
ocean waves
offshore
paleogeography
seawalls
shore features
slope stability
stabilization
tidal inlets
waterways

Shoriya Mountains
At SW end of Kuznetsk Alatau just S of Kuznetsk Basin. Also search Shoriya.
UF Shoriya Range
BT Russian Republic
USSR

Shoriya Range
use Shoriya Mountains

short-period seismographs
Term introduced in 1981.
BT seismographs

short-period waves
Term introduced in 1981.
BA elastic waves
SA seismology
traveltime
waves

shortening, crustal
use crustal shortening

shortite
BA carbonates
BT minerals

Shorty Crater
BT Moon

Shoshone County
In N Idaho.
BT Idaho
United States

Shoshone Mountains
W of Toiyabe Range and Reese River in central Nevada. Also search Shoshone Range.
UF Shoshone Range
BT Nevada
United States

Shoshone Range
use Shoshone Mountains

shoshonite
As of 1981, includes use on level 3 under igneous rocks(1) basalts(2). See list H. Before 1981, included use on level 3 under igneous rocks(1) basalt family(2).
BA basalts
volcanic rocks
BT igneous rocks

shrinkage cracks
As of 1981, no longer a valid term for GeoRef. Use mudcracks. Included use on level 3 under sedimentary structures(1) bedding plane irregularities(2).

Shropshire
County on Welsh border.
CO N522000N530000
W0021500W0031500
BT England
Great Britain
United Kingdom

Shugnan
Village in Gorno Badakhshan Autonomous Oblast in Pamirs in E.
BT Tadzhikistan
USSR

Shuswap Complex
Underlies the southern culmination of the Omineca Geanticline. Comprises rocks ranging in age from Proterozoic to early Mesozoic. S British Columbia. Also search Shuswap metamorphic complex.
UF Shuswap metamorphic complex
BT British Columbia
Canada
SA Mesozoic
Paleozoic
Precambrian
Proterozoic

Shuswap metamorphic complex
use Shuswap Complex

Shventoyi River
use Sventoji River

Si
use silicon

Si-32
Includes use on level 3 under isotopes(1).
SA isotopes
silicon

sial
use granitic layer

Siam
use Thailand

Sibay
Town in the Southern Urals in Bashkir A.S.S.R. in SE European USSR.
BT Russian Republic
USSR

Siberia
Region in Asia extending from the Urals to the Pacific Ocean, and from the Arctic Ocean to the Chinese and Mongolian borders and including N Kazakhstan. Index Soviet republics as applicable.
CO N500000N800000
W1700000E0600000
BT USSR
SA Beringia
Kazakhstan
Russian Republic

Siberia-Soviet Far East
use Soviet Far East

Siberian fold belt
Term introduced in 1981.
CO N490000N610000
E1210000E0820000
BT USSR
NZ Baikal region
Tuva
Western Transbaikalia
SA Soviet Altai Mountains

Siberian Lowland
Region comprising the West Siberian Plain which extends from the Urals in the W to the Yenisei River in the E, and from the Kara Sea in the N to the Kazakh Hills in the S. Also search Siberian, West Siberian Basin, West Siberian Lowland and West Siberian Plain.
CO N500000N733000
E0950000E0620000
UF West Siberian Basin
West Siberian Lowland
West Siberian Plain
BA West Siberia
BT USSR
SA Kazakhstan
Russian Republic

Siberian Platform
Region comprising the Central Siberian Plateau which lies between the Yenisei River on the W, and the Lena River on the E. It is bounded on the N by the Arctic Ocean and on the S by the Upper Tunguska and the upper Lena Rivers. Also search Siberian, and West Siberian Platform.
CO N523000N750000
E1390000E0820000
UF Central Siberian Plateau
West Siberian Platform
BT Russian Republic
USSR
NA Aldan Shield
Anabar Shield
Angara-Lena Basin
North Siberian Plain
Tunguska
Vilyuy Syneclise
Yakutia region
Yenisei Ridge
SA West Siberian Plate

Sibiu
City and county in S.
BT Transylvania
Romania

Sicilia
use Sicily

Sicilian
Europe. See list E.
BA upper Pleistocene
Pleistocene
Quaternary
BT Cenozoic

Sicily
Island in Mediterranean Sea and autonomous region.
CO N364500N381500
E0150000E0120000
UF Sicilia
BT Italy
NT Caltanissetta
Ciminna Basin
Enna
Lipari Island
Lipari Islands
Madonie Mountains
Messina
Mount Etna
Nebrodi Mountains
Palermo
Pantelleria
Peloritani Mountains
Stromboli
Trapani
Ustica Island
Vulcano
SA Mediterranean region
Syracuse

side-scanning methods
Term introduced in 1981.
SA acoustical methods
geophysical methods
sonar methods

sideraerolite
use stony irons

siderite
Includes use on level 3 under minerals (1). Also search chalybite.
BA carbonates
BT minerals
SA chalybite

siderite (meteorite)
use iron meteorites

siderolite
use stony irons

siderophile elements
SA elements

Sidhi
Village and district in E.
BT Madhya Pradesh
India

Sidobre Massif
In S Central Massif NE of Castres. Also search Sidobre.
BT Tarn
France
SA Central Massif

Siebengebirge
Hills in the Westerwald on right bank of Rhine River SSE of Bonn.
BT North Rhine-Westphalia
West Germany

Siedlce
City in E part of country.
UF Syedlets
BT Warsaw
Poland

Siegenian
Europe. Above Gedinnian, below Emsian. Includes use on level 3 under age terms(1). See list E.
BA Lower Devonian
Devonian
BT Paleozoic

Siena
City and province in S central.
BT Tuscany
Italy

Sierozems
Includes use on level 3 under soils(1). See list M.
UF Cerozem
Gray desert soil
Gray earth
Serozem
BA soils
SA Zonal soils

Sierra Ancha
Ridge in central Gila County in E central.
BT Arizona
United States

Sierra Blanca
Range in Otero and Lincoln counties in S central.
BT New Mexico
United States

Sierra County
Index states as applicable.
BX California
New Mexico
BT United States
SA Hillsboro Quadrangle
San Lorenzo Quadrangle

Sierra de Gador
Range in SE part of country.
BT Almeria Province
Andalusia
Spain

Sierra de Gredos
Range W of Madrid in W central.
BT Spain

Sierra de Guadarrama
Range N of Madrid separating Old Castile from New Castile.
BT Spain

Sierra de la Demanda
Range of the Cordillera Iberica in Old Castile in N.
BT Spain

Sierra de los Filabres
Range in W central.
BT Almeria Province
Andalusia
Spain

Sierra de los Organos
Range in W.
BT Pinar del Rio
Cuba

Sierra de Ronda
use Serrania de Ronda

Sierra Gorda
E range of S Sierra Madre Occidental in central part of country.
BT Guanajuato
Mexico

Sierra Leone
Former British colony and protectorate. Became independent in 1961. Includes use on level 1 as an area term (list O). For term set options see list B.
CO N070000N100000
W0110000W0133000
BA West Africa
Africa

Sierra Leone Rise
Term introduced in 1981. SW of West Africa in the Atlantic Ocean.
CO N023000N093000
W0180000W0260000
BT Atlantic Ocean

Sierra Madre
Chief mountain system extending 1500 miles SE from U.S. border including 3 separate ranges which enclose the great central plateau. Index Sierras as applicable. Before 1978, this term also included Sierra Madre Range.
BT Mexico
SA North American Cordillera
Sierra Madre del Sur
Sierra Madre Occidental
Sierra Madre Oriental
Sierra Madre Range

Sierra Madre del Sur
Coastal range in SW Mexico. Index states as applicable.
BT Mexico
SA Guerrero
North American Cordillera
Oaxaca
Sierra Madre

Sierra Madre Occidental
Range running for 700 miles parallel to the Pacific Ocean. Index states as applicable.
BT Mexico
SA Chihuahua
Durango
North American Cordillera
Sierra Madre
Sonora
Zacatecas

Sierra Madre Oriental
Range running parallel to the Gulf of Mexico. Index states as applicable.
BT Mexico
SA Coahuila
Hidalgo
North American Cordillera
Nuevo Leon
Puebla
San Luis Potosi
Sierra Madre
Tamaulipas
Veracruz

Sierra Madre Range
Term introduced in 1978 for range in Continental Divide, S Wyoming, just W of North Platte River. Before 1978, also search Sierra Madre AND Wyoming.
BT Wyoming
United States
SA Sierra Madre

Sierra Morena
A valid term through mid-1978. See Betic Cordillera.

Sierra Nevada
Mountain range in E California. Before 1981, search Sierra Nevada AND California. See Spanish Sierra Nevada.
CO N353000N400000
W1174500W1210000
BT California
United States
SA Cascade Range
Melones Fault
North American Cordillera
Spanish Sierra Nevada

Sierra Nevada Batholith
Term introduced in 1981. E central California.
CO N350000N402000
W1180000W1210000
BT California
United States

Sierra Nevada de Santa Marta
Mountain range in N on Caribbean coast.
BT Colombia

Sierras Pampeanas
Term introduced in 1981. Mountainous region in NW Argentina, E of the Andes. Bounded on the N by the Puna and the Andean foothills in Salta, on the E by the Chaco, on the S by the dry Pampas, and on the W by the Andean foothills.
CO S340000S250000
W0640000W0693000
BT Argentina
SA Catamarca
Cordoba Department
La Rioja
Salta
San Juan Province
San Luis
Tucuman

Sierrita Mountains
BT Arizona
United States

Sifton Basin
N British Columbia. Also search Sifton.
BT British Columbia
Canada

Sigillaria
Genus.
BA Lycopsida
pteridophytes
BT Plantae

signal distortion
Term introduced in 1981.
SA distortion
elastic waves
seismology

Signal Hill Formation
Unconformably underlies Random Formation and overlies Avalonian Formation.
BT Huronian
Proterozoic
Precambrian
SA Canada
Newfoundland

signal-to-noise ratio
Term introduced in 1981. Also search signal to noise ratios.
SA geophysical methods
geophysical surveys
noise
seismic methods
seismograms
seismology
signals

signals
Includes use on level 3 under seismology(1).
SA filters
geophysical methods
noise
seismology
signal-to-noise ratio
teleseismic signals

Signy Island
One of the South Orkney Islands SE of the Falkland Islands.
BA South Orkney Islands
Scotia Sea Islands
BT Antarctica

Sigsbee Deep
In SW central part of Gulf of Mexico.
UF Mexico Basin
BT Gulf of Mexico

Sikhote Alin
A valid term through 1976. After 1976, use Sikhote-Alin Range.

Sikhote-Alin Range
Mountain range in Primorye and Khabarovsk krays along the Japan Sea and the Tatar Strait. Also search Sikhote-Alin.
CO N423000N543000
E1420000E1310000
BA Northeastern USSR
BT Russian Republic
USSR
SA Soviet Far East

Sikkim
Former British and Indian protectorate between Bhutan and Nepal. Became an associated Indian state on Sept. 7, 1974. Included use on level 1 as an area term until 1977. Now use term on level 3; use India on level 1.
BT India
Asia
SA Himalayas
Lesser Himalayas
Siwalik Range
Siwalik System

Sila Massif
In the Southern Apennines forming central part of toe of the Italian boot.
BT Calabria
Italy

silcrete
As of 1981, includes use on level 3 under sedimentary rocks(1) chemically precipitated rocks(2). See list I. Before 1981, included use on level 3 under sediments(1) chemically precipitated sediments(2).
BA chemically precipitated rocks
BT sedimentary rocks
SA terrigenous materials

Silesia
Region in E central Europe lying mostly in SW Poland. It comprises both Upper and Lower Silesia plus small areas in SE East Germany and N Moravia. Index countries as applicable.
BT Europe
SA Czechoslovakia
East Germany
Lower Silesia
Poland
Upper Silesia

Silesian
See list E.
BA Carboniferous
BT Paleozoic

Silesian Basin
use Silesian coal basin

Silesian coal basin
In Lower Silesia in Wroclaw Province, and in Upper Silesia in Katowice Province. Index provinces as applicable.
UF Silesian Basin
BT Poland
SA Katowice
Lower Silesia
Upper Silesia
Upper Silesian coal basin
Wroclaw

silexite
As of 1981, no longer a valid term for GeoRef. Use quartzolite (for the igneous rock) or chert as applicable. Before 1981, included use on level 3 under igneous rocks(1) granite-granodiorite family(2).

silica
As of 1981, includes use on level 3 as a commodity term under sands(1). See list C. Before 1981, included use as a commodity term on level 3 under sand(1).
SA abrasives
sand
sands
silicates
siliceous composition
silicon

silica group
use silica minerals

silica minerals
Includes use in combination with framework silicates (i.e. framework silicates, silica minerals) on level 2 under minerals(1). See list L. Also search silica group.
UF silica group
BA framework silicates
silicates
BT minerals
NA agate
amethyst
chalcedony
coesite
cristobalite
hyalite
jasper
keatite
lechatelierite
melanophlogite
opal
quartz
stishovite
tridymite

silicate minerals
use silicates

silicate rocks
SA petrology
rocks
siliceous composition
standard materials

silicates
In indexing, use this term only for broad treatments of the entire class of minerals; otherwise, use a narrower term, e.g. orthosilicates, ring silicates, chain silicates, etc. Includes use on level 2 under minerals(1). See list L. As of 1978, the term silicates is autoposted to all individual silicate minerals. Before 1978 also search individual silicate groups. Also search the term silicate minerals.
UF silicate minerals
BT minerals
NA aluminosilicates
asbestos
borosilicates
NZ britholite
NA chain silicates
chlorophaeite
framework silicates
NZ helvite
NA orthosilicates
ring silicates
sheet silicates
SA clay mineralogy
silica
silicon

siliceous composition
Term introduced in 1978. Includes use on level 3 under sediments(1). Before 1978 also search siliceous in combination with specific sediment type. Also search siliceous rocks.
SA clastic rocks
composition
insoluble residues
jasperoid
sedimentary rocks
sediments
silica
silicate rocks
siliceous sinter

siliceous rocks
A valid term through 1977. After 1977, use siliceous composition.

siliceous sinter
As of 1981, includes use on level 3 under sedimentary rocks(1) chemically precipitated rocks(2). See list I. Before 1981, included use on level 3 under sediments(1) chemically precipitated sediments(2). Also search sinter AND siliceous.
BA chemically precipitated rocks
BT sedimentary rocks
SA siliceous composition

silicides
As of 1981, includes use on level 2 under minerals(1). See list L. Before 1981, included use on level 3 under minerals(1) native elements and alloys(2).
BA alloys
BT minerals
NA ferrosilicon
SA moissanite

silicification
Includes use on level 3 under diagenesis(1), under metasomatism(1) processes(2), or under sedimentation(1).
UF silification
SA chertification
diagenesis
fossilization
metasomatism
processes
sedimentation

silicified wood
use fossil wood

Silicispongiae
UF Noncalcarea
BA Porifera
BT Invertebrata

Silicoflagellata
Includes use on level 2 under Protista(1). See list F.
BA Protista
BT Invertebrata
NA Dictyocha
Distephanus

silicon
Includes use on level 1 and 2 as a chemical element (list D). Also search Si.
UF Si
SA elements
ferrosilicon
moissanite
Si-32
silica
silicates

silification
use silicification

Siljan
Lake in central part of country.
BT Sweden

sillimanite
Includes use on level 3 under minerals(1) orthosilicates(2). See list L. As of 1981, use sillimanite deposits for sillimanite as a commodity. Before 1981, included use on level 3 as a commodity term under ceramic materials(1). Also search fibrolite.
UF fibrolite
BA orthosilicates
silicates
BT minerals
SA andalusite
kyanite
sillimanite deposits

sillimanite deposits
Term introduced in 1981. Includes use on level 3 as a commodity term (list C) under ceramic materials(1). Before 1981, search sillimanite AND deposits.
SA ceramic materials
sillimanite

sillimanite gneiss
See list J.
BA gneisses
BT metamorphic rocks

sills
Includes use on level 2 under intrusions(1).
BA intrusions
SA dikes

silt
Includes use on level 3 under sediments(1) clastic sediments(2). See list N. Also search silts.
BA clastic sediments
BT sediments
SA alluvium
clay
loam
loess
mud
sand
siltstone
terrigenous materials

siltite
use siltstone

siltstone
Includes use on level 3 under sedimentary rocks(1) clastic rocks(2). See list I. Also search siltite.
UF siltite
BA clastic rocks
BT sedimentary rocks
SA metasiltstone
mudstone
silt
terrigenous materials

Silurian
Includes use on level 1 as an age term (list E); on level 2 under paleoterms, e.g. paleoecology; paleogeography, paleomagnetism. Above Ordovician, below Devonian. Also search Gotlandian.
UF Gothlandian
Gotlandian
BT Paleozoic
NT Aberystwyth Grits
NA Lower Silurian
Middle Silurian
NT Red Mountain Formation
Rockwood Formation
NA Upper Silurian
NT Waterville Formation
SA Broken River Formation
Hidden Valley Dolomite
Hunton Group
Peel Sound Formation
Read Bay Formation
Ringerike Sandstone
Rondout Formation
Taconic Orogeny
Talladega Group

silvanite
use sylvanite

silver
Includes use on level 1 and 2 as a chemical element (list D). As of 1981, use silver ores for silver as a commodity. Before 1981, included use on level 1 as a commodity term.
UF Ag
SA argentite
elements
heavy metals
native elements
silver ores

Silver Bow County
SW Montana.
BT Montana
United States

Silver City
Town in Grant County in SW.
BT New Mexico
United States

silver ores
Term introduced in 1981. Includes use on level 1 and 2 as a commodity term (list C). Before 1981, search silver AND ore deposits.
SA native elements
silver

Silver Peak Mountains
Small range in W Esmeralda County in SW Nevada. Also search Silver Peak.
UF Silver Peak Range
BT Nevada
United States

Silver Peak Range
use Silver Peak Mountains

Silverton Caldera
Near town of Silverton in San Juan County in San Juan Mountains of SW. Also search Silverton.
BT Colorado
United States

Silvretta Group
Mountain group. Also search Silvretta.
UF Silvretta Massif
BT Paleozoic
SA Switzerland
Tyrol

Silvretta Massif
use Silvretta Group

sima
use basaltic layer

Simbirsk
use Ulyanovsk

Simferopol
City in the S central Crimea.
BT Ukraine
USSR

Simi Hills
In Ventura County NW of Los Angeles.
BT California
United States

simians
Term introduced in 1981. Includes

use on level 2 under Mammalia(1). See list F.
BA Primates
Mammalia
BT Tetrapoda
Vertebrata
NA Cynomorpha
Platyrrhina

similar folds
Term introduced in 1978. Includes use on level 3 under folds(1) style(2). Before 1978, search similar AND folds.
UF shear folds
BA folds
SA concentric folds
shear

Simla
Town and hill resort in S.
BT Himachal Pradesh
India

Simla Hills
Hill and mountain area of the outer Kumaun Himalayas around Simla. Also search Simla.
BT Himachal Pradesh
India

Simleu Basin
In Bihor Mountains near town of Simleu Silvaniei in NW. Also search Simleu.
BT Transylvania
Romania

Simplon region
Includes Simplon Pass, Simplon Road, and Simplon Tunnel area. Index Italy and/or Switzerland. Also search Simplon.
BT Europe
SA Italy
Piemonte
Switzerland

Simpson Desert
Primarily in SE Northern Territory. Index Northern Territory and states as applicable.
BT Australia
SA Northern Territory
Queensland
South Australia

Simpson Group
Comprises Joins, Oil Creek, McLish, Tulip Creek, Bromide, and Corbin Ranch formations. Central and S Oklahoma.
BT Middle Ordovician
Ordovician
SA Oklahoma

simulation
SA analog simulation
automatic data processing
digital simulation
mathematical geology
models

Sinai
Peninsula extending from the Mediterranean Sea to the Red Sea. It is bounded on the W by the Suez Canal and the Gulf of Suez, and on the E by Israel and the Gulf of Aqaba. Also search Sinai Peninsula.
CO N274200N313000
E0345500E0322000
UF Sinai Peninsula
BT Egypt
SA Israel

Sinai Peninsula
use Sinai

Sinaia
Town and health resort at SE foot of Bucegi Mountains of the Transylvanian Alps in Prahova County.
BT Walachia
Romania

Sinaloa
State in W.
BT Mexico
NT Mazatlan

Sind
Province in SE.
BT Pakistan
NT Chor
SA Nari Series

Sinemurian
Europe. Above Hettangian, below Pliensbachian. Includes use on level 3 under age terms(1). See list E.
BA lower Liassic
Lower Jurassic
Jurassic
BT Mesozoic

Singapore
Island republic and city off the southern tip of the Malay Peninsula. Includes use on level 1 as an area term (list O). For term set options see list B.
CO N010000N014500
E1042000E1032000
BA Far East
Asia

Singhbhum
District, Chota Nagpur division, S Bihar.
BT Bihar
India
SA Singhbhum Granite

Singhbhum Granite
Also search Singhbhum.
BT Archean
Precambrian
SA Bihar
India
Orissa
Singhbhum

Singhbhum shear zone
Index states as applicable. Also search Singhbhum.
BT India
SA Bihar
Orissa

single domains
Term introduced in 1981.
BT magnetic domains

single-crystal method
Term introduced in 1978. Before 1978, search single crystal.
SA crystal growth
crystal structure
methods
minerals

Singleton Coal Measures
E New South Wales.
SA Australia
Carboniferous
New South Wales
Permian
Triassic

Sini
Village in Singhbhum District. Site of limestone quarries for Jamshedpur iron and steel works.
BT Bihar
India

Sinjar Formation
Paleocene-lower Eocene: In the Jabel Sinjar Mountains.
UF Sinjar Limestone Formation
BT Paleogene
Tertiary
Cenozoic
SA Eocene
Iraq
lower Eocene
Middle East
Paleocene

Sinjar Limestone Formation
use Sinjar Formation

sinkholes
Includes use on level 3 under geomorphology(1) solution features(2).
SA dolines
geomorphology
karst
potholes
solution features

Sinkiang
use Sinkiang Uighur

Sinkiang Province
use Sinkiang Uighur

Sinkiang Uighur
Autonomous region in NW China. Also search Sinkiang.
UF Sinkiang
Sinkiang Province
BT China
NT Dzhungaria
SA Dzhungarian Alatau
Ili Basin
Irtysh River
Tarbagatay Range
Turkestan

sinking
Use land subsidence or subsidence as applicable.

sinks
SA geochemistry

sinnerite
BA sulfides
BT minerals

sinter
A valid term through 1977. After 1977, use siliceous sinter.

sinuosity
SA channels
fluvial features
meanders
streams

Sioux County
Index states as applicable.
BX Iowa
Nebraska
North Dakota
BT United States

Sioux Quartzite
Has been correlated with Baraboo Quartzite in Wisconsin. NW Iowa, SW Minnesota, NE Nebraska, and SE South Dakota.
BT Precambrian
SA Iowa
Minnesota
Nebraska
South Dakota

Siphonapteroida
Term introduced in 1981. Includes use on level 2 under Insecta(1). See list F.
BA Insecta
BT Mandibulata
Arthropoda
Invertebrata

Sipunculoida
Includes use on level 2 under worms(1). See list F.
BA worms
BT Invertebrata

Sir Darya
use Syr Darya

Sirenia
Includes use on level 2 under Mammalia(1). See list F.
BA Mammalia
BT Tetrapoda
Vertebrata

Siret River
Rises on the E slopes of the Carpathians in the Ukraine and flows SE into the Danube River just N of Galati. Index Moldavia and/or Ukraine. Also search Siret.
UF Sereth River
Siretul River
BT Europe
SA Moldavia
Ukraine

Siretul River
use Siret River

Sirohi
Former Indian state now part of Rajasthan. Also a town in SW.
BT Rajasthan
India

Sirte Basin
On the Gulf of Sidra in N Libya. Also search Sirte.
UF Syrte
BT Libya

Sisian
Village in S.
BT Armenia
USSR

Siskiyou County
On Oregon border.
CO N410000N420000
W1213000W1214500
BT California
United States
NT Medicine Lake

site exploration
Includes use on level 2 under engineering geology(1). Before 1978, also search site selection.
UF site selection
SA dams
engineering geology
exploration
explosions
foundations
geologic hazards
highways
land subsidence
marine installations
nuclear facilities
permafrost
reservoirs
rock mechanics
slope stability
soil mechanics
tunnels
underground installations
waste disposal

site selection
use site exploration

sites, archaeological
use archaeological sites

sites, landing
use landing sites

Sitka Sound
On W side of Baranof Island in SE Alaska. Entrance to Sitka from Gulf of Alaska. Also search Sitka.
BT Alaska
United States

Sivamalai
A hill in the Coimbatore District in E.
BT Tamil Nadu
India

Sivash
Salt lagoons and marshes in N and NE Crimea.
UF Putrid Sea
BT Ukraine
USSR

Siwalik Formation
use Siwalik System

Siwalik Group
use Siwalik System

Siwalik Hills
use Siwalik Range

Siwalik Range
Range of foothills parallel with the main Himalayan system and extending 1000 miles SE from N Punjab in Pakistan to Sikkim. Also search Siwalik and Siwaliks. Index Sikkim and countries as applicable.
UF Siwalik Hills
Siwaliks
BT Asia
SA India
Nepal
Pakistan
Sikkim

Siwalik Sandstone
use Siwalik System

Siwalik Series
use Siwalik System

Siwalik System
Middle Miocene-Pleistocene: Extends beyond limits of Siwalik Range to include parts in Baluchistan and Northwest Frontier Province in Pakistan, Tipam Series in Assam, and Irrawaddy Series in Burma. Includes Lower (Kamlial and Chinji), Middle (Nagri and Dhokpathan) and Upper (Tatrot, Pinjor and Boulder conglomerates) Siwalik Sub-groups. Also search Siwalik; Siwalik Series; Siwalik Group; Siwalik Formation; Siwalik Sandstone.
UF Siwalik Formation
Siwalik Group
Siwalik Sandstone
Siwalik Series
BT Cenozoic
SA Burma
India
middle Miocene
Miocene
Neogene
Pakistan
Pleistocene
Pliocene
Sikkim
Tertiary
upper Miocene
upper Tertiary

Siwaliks
use Siwalik Range

Sixes River
N Curry County in SW Oregon. Flows into the Pacific Ocean.
BT Oregon
United States

size
As of 1981, no longer a valid term for GeoRef. See grain size.

size analysis
A valid term through mid-1978. Use granulometry.

size distribution
SA distribution
grain size
granulometry
sedimentary rocks
sediments
textures

Sjaeland
use Zealand

sjogrenite
BA carbonates
BT minerals

Skaergaard Intrusion
BT Greenland

Skagerrak
Arm of the North Sea between Norway and Denmark.
BT North Sea
Atlantic Ocean

Skagit County
Term introduced in 1981. NW Washington.
CO N481800N484000
W1204000W1224500
BT Washington
United States

Skagit Valley
River valley. Primarily in NW Washington. Index British Columbia and/or Washington as applicable. Also search Skagit; Skagit River.
SA British Columbia
Washington

Skamania County
On the Oregon border in SW.
BT Washington
United States

Skane
Region in S Sweden. Index counties as applicable. Also search Scania.
UF Scania
BT Sweden
SA Kristianstad
Malmohus

skarn
Compositional term. As of 1978, includes use on level 2 under metasomatic rocks(1). Before 1978, included use on level 3 under metamorphic rocks(1).
BA metasomatic rocks
SA metamorphic rocks

Skeena Mountains
NW central British Columbia. E of Coast Mountains.
BT British Columbia
Canada

Skeidararjokull
Glacier.
BT Iceland

skeletons
Includes use on level 3 under fossil group(1) for type of material; on level 3 under paleontology(1). See list F.
SA bones
fossils
paleontology
skulls
teeth
Vertebrata

Skellefte
River in N Sweden. Flows SE into Gulf of Bothnia. Index counties as applicable.
BT Sweden
SA Norrbotten
Vasterbotten

skewness
BA statistical analysis
SA kurtosis

Skiddaw Group
use Skiddaw Slates

Skiddaw Slates
Arenig and Llanvirn Series. Cumberland County in NW England. Also search Skiddaw Group.
UF Skiddaw Group
BT Ordovician
SA England
United Kingdom

Skolithos
BA ichnofossils

Skomer Island
In Saint Georges Channel off Pembrokeshire.
BT Wales
Great Britain
United Kingdom

Skopje
City on Vardar River in N Macedonia.
UF Skoplje
BT Yugoslavia
SA Macedonia

Skoplje
use Skopje

SKP-waves
Term introduced in 1981.
BA body waves
elastic waves
SA P-waves
S-waves
seismology
waves

SKS-waves
Term introduced in 1981.
BA body waves
elastic waves
SA P-waves
S-waves
seismology
waves

skulls
Includes use on level 3 under fossil group(1) for type of material. See list F.
SA bones
fossils
jaws
paleontology
skeletons
teeth
Vertebrata

skutterudite
As of 1981, includes use on level 3 under minerals(1) arsenides(2). See list L. Before 1981, included use on level 3 under minerals(1) sulfides(2).
BA arsenides
sulfides
BT minerals

Skye
use Isle of Skye

Skylab
SA extraterrestrial geology
planetology
remote sensing

skystones
use meteorites

Skythian
use Scythian

slabs
Includes use on level 3 under plate tectonics(1).
SA plate tectonics

Slak Dolny
use Wroclaw

Slanic
Town in Prahova County in N central.
BT Walachia
Romania

Slany
Town in W central.
BT Bohemia
Czechoslovakia

slate
As of 1981, no longer a valid term for GeoRef. Use slate deposits or slates as applicable. Before 1981, included use on level 1 as a commodity term and on level 3 under metamorphic rocks(1) slates(2).

slate deposits
Term introduced in 1981. Includes use on level 1 and 2 as a commodity term (list C). Before 1981, search slate AND deposits.
SA shale

slates
Includes use on level 2 under metamorphic rocks(1). See list J.
BT metamorphic rocks
SA shale
slaty cleavage

Slatina
City in Olt County in SW.
BT Walachia
Romania

slaty cleavage
Includes use on level 3 under foliation(1) style(2).
BA foliation
SA axial-plane structures
cleavage
flow cleavage
metamorphism
schistosity
slates
slip cleavage

Slave Province
Structural province of the Canadian Shield.
CO N620000N690000
W1050000W1170000
BT Canadian Shield
SA Northwest Territories

slickensides
Includes use on level 3 under faults(1) effects(2) and under lineation(1) style(2).
UF polished surface
SA breccia
faults
lineation

sliding, gravity
use gravity sliding

Sligo
County in NW. Also a town.
BT Ireland

slip cleavage
Term introduced in 1978. Before 1978, search slip AND cleavage; shear cleavage; crenulation cleavage.
UF crenulation cleavage
shear cleavage
strain-slip cleavage
BA foliation
SA cleavage
schistosity
slaty cleavage

Slocan mining camp
SE British Columbia. Also search Slocan.
BT British Columbia
Canada

slope, continental
use continental slope

Slope County
Term introduced in 1981. SW North Dakota.
CO N461600N463900
W1025400W1040300
BT North Dakota
United States

slope environment
Term introduced in 1978. Use this term or shelf environment for older sedimentation. For recent sedimentation, use continental slope or continental shelf. Before 1978, search slope.
SA continental shelf
continental slope
environment
sedimentation
shelf environment

slope, outer
use outer slope

slope stability
A level 1 term as of 1978. Used for geological studies on the engineering aspects of mass movements. For other aspects, see geomorphology. Before 1978, included use on level 2 under engineering geology(1). Term set options are:
creep
subtopic
debris flows
subtopic
earthflows
subtopic
embankments
subtopic
erosion
subtopic
excavations
subtopic
experimental studies
subtopic
failures
subtopic
field studies
subtopic
landslides
subtopic
liquefaction
subtopic
mass movements
subtopic
mudflows
subtopic
rockfalls
subtopic
site exploration
subtopic
stabilization
subtopic
talus slopes
subtopic
theoretical studies
subtopic
SA avalanches
controls
creep
dams
debris flows
earthflows
earthquakes
embankments
engineering geology
erosion
excavations
explosions
failures
foundations
geologic hazards
geomorphology
highways
land subsidence
landslides
liquefaction
liquefaction potential
mass movements
mudflows
reservoirs
rock mechanics
rockfalls
shorelines
site exploration
slopes
slumping
soil mechanics
solifluction
stability
stabilization
talus slopes
tunnels
underground installations

slopes
Includes use on level 3 under geomorphology(1).
SA cliffs
continental slope
embankments
fault scarps
geomorphology
glacis
landslides
scarps
slope stability
talus slopes

Slovakia
Region in E.
CO N474500N495500
E0223000E0160000
UF Slovensko
BT Czechoslovakia
NT Banska Stiavnica
Bojnice
Bratislava
Brezno
Gemer
Hodrusa
Korytnica
Kremnica
Kremnica Mountains
Low Tatra Mountains
Novaky
Orava Valley
Slovakian Karst
Spis-Gemer
Sturovo
Vah Valley
SA Morava River valley
Spis
Tokaj-Eperjes Mountains

Slovakian Carpathians
Term introduced in 1981.
CO N480500N495500
E0223000E0160000
BZ Carpathians
BX Czechoslovakia

Slovakian Karst
Near the Hungarian border in SE.
BT Slovakia
Czechoslovakia
SA Permian
Triassic

Slovakian Pannonian Basin
Term introduced in 1981.
BZ Pannonian Basin
BX Czechoslovakia

Slovenia
Constituent republic in NE Yugoslavia.
CO N451500N465000
E0170000E0133000
UF Slovenija
BT Yugoslavia
NT Celje
Ljubljana
SA Istria

Slovenija
use Slovenia

Slovenske Rudohorie
use Spis-Gemer

Slovensko
use Slovakia

sludging
use solifluction

slump faults
Not a valid term for GeoRef. See normal faults or growth faults.

slump structures
Includes use on level 3 under sedimentary structures(1) soft sediment deformation(2). See list K. Also search slump structure.
BA soft sediment deformation
sedimentary structures
SA collapse structures
convoluted beds
slumping
structures

slumping
Includes use on level 3 under sedimentation(1). Also search slump; slumps.
UF slumps
SA continental shelf
continental slope
creep
erosion
geomorphology
landslides
liquefaction
mass movements
sedimentation
slope stability
slump structures

slumps
use slumping

Slyudyanka
City at SW end of Lake Baikal in S Irkutsk Oblast.
BT Russian Republic
USSR

Sm
use samarium

Smackover Formation
Overlies Eagle Mills Formation; underlies Buckner Formation. Subsurface.
BT Upper Jurassic
Jurassic
SA Alabama
Arkansas
Louisiana
Mississippi
Texas

Smaland
Plateau region S of Lake Vattern in S Sweden. Index counties as applicable.
BT Sweden
SA Kalmar

smaragd
use emerald

smectite
BA clay minerals
sheet silicates
silicates
BT minerals
SA fuller's earth
montmorillonite
saponite

Smith County
Index states as applicable.
BX Kansas
Mississippi
Tennessee
Texas
BT United States

Smithers
Village in W central.
BT British Columbia
Canada

Smithsonian Institution
BT District of Columbia
United States

smithsonite
UF zinc spar
BA carbonates
BT minerals
SA azurite

Smithville Formation
Above Powell Formation and below Black Rock Limestone in N Arkansas, and below Everton Formation in SE Missouri. N Arkansas and SE Missouri.
BT Lower Ordovician
Ordovician
SA Arkansas
Missouri

Smokies
use Great Smoky Mountains

Smoky Hill Chalk Member
Of Niobrara Formation. E Colorado, W Kansas, NE New Mexico, and SE South Dakota.
BT Upper Cretaceous
Cretaceous
SA Colorado
Kansas
New Mexico
Niobrara Formation
South Dakota

Smoky Hill River basin
W central and central Kansas. Also search Smoky Hill River.
BT Kansas
United States

Smoky River
Rises in Jasper National Park and flows NNE into Peace River.
BT Alberta
Canada

Smolensk
City on left bank of upper Dnieper River SW of Moscow. Also an oblast.
BT Russian Republic
USSR

smythite
BA sulfides
BT minerals

Sn
use tin

Snaefellsnes Peninsula
Juts into Denmark Strait between Breidi Fjord and Faxa Bay in W Iceland. Also search Snaefellsnes.
BT Iceland

Snake Range
Mountain range in E White Pine County in E.
BT Nevada
United States

Snake River
Rises in Yellowstone National Park and flows S, then W, N, and again W into the Columbia River in SE Washington. Index states as applicable.
BT United States
SA Idaho
Oregon
Washington
Wyoming

Snake River basin
Index states as applicable.
BT United States
SA Idaho
Nevada
Oregon
Utah
Washington
Wyoming

Snake River canyon
Grand Canyon of the Snake River. Extends N-S between Wallowa Mountains, Oregon, and Seven Devils Mountains, Idaho. Index states as applicable.
UF Hell's Canyon
BT United States
SA Idaho
Oregon

Snake River plain
Crescent-shaped lava tableland across S central.
CO N421500N441500
W1113000W1160000
BT Idaho
United States

Snieznik
Highest peak in Kralicky Sneznik Mountains in NE Bohemia and Lower Silesia. Index countries as applicable.
BT Europe
SA Czechoslovakia
Poland

Snohomish County
NNE of Seattle.
BT Washington
United States

Snoqualmie Batholith
Probably intermediate in age between lower part of Keechelus Andesitic Series and Fifes Peak Andesite. W central Washington.
BT Tertiary
SA Miocene
Pliocene
Washington

snow
Includes use on level 2 under hydrology(1); as level 3 under glacial geology(1) glaciers(2).
SA atmospheric precipitation
firn
glacial geology
glaciers
hydrology
hydrosphere
ice
meltwater
thawing
water

Snow Lake
Village in gold-mining region 70 miles E of Flin Flon in W.
BT Manitoba
Canada

snow-patch erosion
use nivation

Snowdon
use Snowdonia

Snowdonia
Highest mountain in Wales, in Caenarvonshire 10 miles SE of Caernarvon; consists of 5 peaks, separated by passes. Also search Snowdon.
UF Snowdon
BT Wales
Great Britain
United Kingdom

Snowy Mountains
Range of the Australian Alps. Index states as applicable.
BT Australia
SA New South Wales
Victoria

soap clay
use bentonite

soapstone
As of 1981, includes use on level 3 as a commodity term (list C) under talc deposits(1). Before 1981, included use on level 3 as a commodity term under talc(1).
SA talc
talc deposits

Sob River basin
Near border of Moldavia, USSR, in SW Ukraine. Also search Sob.
BT Ukraine
USSR

Sobotka
Town in N.
BT Bohemia
Czechoslovakia

Sochi
City in S Krasnodar Kray on Black Sea near Georgian border.
BT Russian Republic
USSR

Society Islands
Includes use on level 1 as an area term (list O). For term set options see list B.
CO S190000S150000
W1470000W1550000
BA French Polynesia
Polynesia
NA Tahiti

Socorro
City in Socorro County in W central.
BT New Mexico
United States

Socorro County
W central.
BT New Mexico
United States

soda ash
use sodium carbonate

sodalite
BA sodalite group
framework silicates
silicates
BT minerals

sodalite group
Includes use in combination with framework silicates (i.e. framework silicates,sodalite group) on level 2 under minerals(1). See list L.
BA framework silicates
silicates
BT minerals
NA hauyne
hydrosodalite
NZ lazurite
NA nosean
sodalite
tugtupite

sodium
Includes use on level 1 and 2 as a chemical element (list D). Also search Na.
UF Na
SA alkali metals
elements
Na-22
Na-24

sodium carbonate
Includes use on level 1 as a commodity term (list C). Also search soda ash.
UF soda ash
SA carbonates
trona

sodium chloride
Includes use on level 3 under geochemistry(1).
SA chlorides
evaporites
geochemistry
halides
salt

sodium sulfate
Includes use on level 1 as a commodity term (list C).
SA sulfates

Sofia
Province in W Bulgaria. Also a city.
UF Sofiya
Sophia
BT Bulgaria
NT Rila
Yetropole

Sofiya
use Sofia

soft clays
Term introduced in 1981. Also search soft clay.
SA clay
clays

soft coal
use bituminous coal

soft sediment deformation
Includes use on level 2 under sedimentary structures(1). See list K.
BA sedimentary structures
NA ball-and-pillow
boudinage
clastic dikes
convoluted beds
flame structures
flow structures
olistoliths
olistostromes
sandstone dikes
slump structures
SA flute casts
load casts
sole marks
tool marks

Sogn
Mountain region around Sogne Fjord in SW part of country.
BT Norway

Sogndal
Village on North Sea.
BT Rogaland
Norway

soil dynamics
Term introduced in 1981.
SA dynamics
soil mechanics
soils

soil flow
use solifluction

soil fluction
use solifluction

soil group
Includes use on level 2 under soils(1). See list M.
SA Bog soils
Brown forest soils
Chernozems
Chestnut soils
Desert soils
Planosols
soils
Solonchak soils
Solonetz soils

soil horizon
use horizons

soil management
Includes use on level 3 under soils(1) utilization(2).
SA fertilizers
management
soils
tillage
treatment
utilization

soil mechanics
A level 1 term as of 1978. Used for geotechnical studies. Includes use on level 2 under tunnels(1), underground installations(1), and nuclear facilities(1). If 1, term set options are:
applications
subtopic
case studies
subtopic
concepts
subtopic
deformation
subtopic
earth pressure
subtopic
elasticity
subtopic
experimental studies
subtopic
frost action
subtopic
liquefaction
subtopic
materials, properties
subtopic
methods
subtopic
settlement
subtopic
site exploration
subtopic
techniques
subtopic
theoretical studies
subtopic
SA alluvium
Atterberg limits
bearing capacity
case studies
clays
cohesive materials
compaction
competent materials
compressibility
compressive strength
congelifraction
consolidation
consolidometer tests
cryoturbation
dams
deformation
earth pressure
earthquakes
elastic constants
elasticity
engineering geology
engineering properties
explosions
finite strain analysis
foundations
frost action
frozen ground
geologic hazards
granular materials
highways
hydraulics
land subsidence
liquefaction
loading
loess
marine installations
materials
mechanical properties
mechanics
nuclear facilities
overconsolidated materials
penetration
penetration tests
plastic materials
plate-bearing tests
pore pressure
reservoirs
rock mechanics
seepage
settlement
shear strength
site exploration
slope stability
soil dynamics
soils
testing
triaxial tests
tunnels
unconsolidated materials
underground installations
water pressure

soil profiles
Term introduced in 1981. Before 1981, search soils AND profiles. Includes use on level 3 under soils(1).
SA soils

soil sampling
Term introduced in 1978. Includes use on level 3 under mineral exploration.
SA sample preparation
sampling
soils

soil ulmin
use humus

soil zone
use horizons

soils
Includes use on level 1 (list A); on level 2 under area terms(1), education(1), bibliography(1), and sym-

posia(1). Used for general pedology as well as specific topics. If level 1, term set options are:
analysis
biological methods
chemical analysis
physical methods
sample preparation
biota
bacteria
fungi
Invertebrata
microorganisms (for several types)
Protista
Vertebrata
geochemistry
ion exchange
nitrogen
phosphorus
(This heading of second rank picks up papers on cation exchange relations, the nitrogen status of soils, transformations of nitrogen, phosphorus fixation, potassium fixation and the like. Listing of nitrogen and phosphorus is illustrative; names of all other elements are allowed.
classification
soil group [List M; names of groups in different systems may be used; Paleosols may also be used]
composition
chemical composition
mineral composition
organic materials
conservation
erosion control
fertility maintenance
physical properties
(This covers maintenance of structure, pore space, good tilth and the like so that root penetration would be facilitated. Included should be papers on deterioration in the physical properties of soils and its prevention.)
correlation
subtopic
erosion
landslides
water erosion
wind erosion
fertilizers
farms
field studies
methods or type of methods (e.g. remote sensing, sampling, applications, programs, techniques)
genesis
factors or type of factor (e.g. biota, climate, geomorphology, parent materials, time factor)
horizon differentiation
leaching
migration
organic materials (gains, losses, etc.)
transformations
morphology
color
horizons
micromorphology
soil profiles
textures
nutrients
major elements
toxic substances
trace elements
organic materials
abundance
composition
distribution
patterns
subtopic
properties
subtopic
salinity
occurrence
treatment
soil group
name of group
surveys (only for detailed study of a particular area)
area
utilization
agriculture
fertilization
forestry
housing
reclamation
recreation
soil management
tillage
waste disposal
animal waste
human waste
water regimes
characterization
drainage
irrigation
movement (in soil)
storage (storage in soil)
yields
forests
fruits
vegetables

NA Alfisols
Alluvial soils
Andosols
Aridisols
Bog soils
Brown forest soils
Brown soils
Calcareous soils
Chernozems
Chestnut soils
Clay soils
Desert soils
Entisols
Gleys
Gray forest soils
Histosols
Hydromorphic soils
Inceptisols
Intrazonal soils
laterites
Latosols
loam
Luvisols
Meadow soils
Mollisols
Mor
Mull
Oxisols
Paleosols
Planosols
Podzols
Pseudogleys
Red soils
Rendzinas
Sierozems
Solonchak soils
Solonetz soils
Spodosols
Terra rossa
Tundra soils
Ultisols
Vertisols
Zonal soils

SA active layer
agriculture
alluvium
bacteria
bedrock
biorhexistasy
biota
calcification
calcrete
caliche
capillarity
capillary water
catenas
characterization
chemical composition
clay mineralogy
climate
colluvium
color
conservation
creep
degradation
desalinization
desiccation
drainage
duricrust
ecology
eluvium
engineering geology
engineering properties
environmental geology
erosion
erosion control
expansive materials
factors
fertilization
fertilizers
field capacity
field studies
forestry
forests
foundations
frozen ground
fruits
fuller's earth
geomorphology
gilgai
glacial geology
gyttja
highways
horizon differentiation
horizons
humus
infiltration
ion exchange
irrigation
land use
landslides
laterization
leaching
micromorphology
microorganisms
migration
moisture
morphology
movement
nitrogen
nutrients
organic materials
parent materials
particles
permafrost
phosphorus
physical methods
physical properties
pit sections
reclamation
recreation
regolith
salinity
sample preparation
sand
saturated zone
saturation
sediments
seepage
soil dynamics
soil group
soil management
soil mechanics
soil profiles
soil sampling
soils maps
solifluction
storage
surficial geology
surveys
tillage
topography
transformations
treatment
tundra
unconsolidated materials
unsaturated zone
utilization
vegetation
waste disposal
water erosion
water regimes
waterlogging
weathering
weathering crust
wind erosion
yields

soils maps
Term introduced in 1981.
BA maps
SA soils

Sokh
Village S of Kokand near Kirghiz border in E.
BT Uzbekistan
USSR

Soko Banja
Village near Nis in E Serbia.
UF Soko Banya
BT Serbia
Yugoslavia

Soko Banya
use Soko Banja

Sokolov
Town on Ohre River in W Bohemia.
UF Falkenau
Falknov
BT Bohemia
Czechoslovakia

Sokolov Basin
W Bohemia. Also search Sokolov.
BT Bohemia
Czechoslovakia

Sokoman Formation
In Knob Lake Group.
UF Sokoman Iron Formation
BT Precambrian
SA Canada
Labrador
Quebec

Sokoman Iron Formation
use Sokoman Formation

Sokoto Basin
River basin in N Nigeria. Also search Sokoto.
BT Nigeria

Solano County
Term introduced in 1981. Central California.
CO N380200N383200
W1213500W1222300
BT California
United States

solar activity
Term includes use under a variety of topics as a cause of whatever is being discussed in the document.
SA disturbances
solar flares
Sun

solar cycles
Includes use on level 3 under interplanetary space(1) cosmic rays(2).
SA astrophysics and solar physics
cosmic rays
cycles
disturbances
intensity
interplanetary space
ionosphere
magnetic storms
solar wind

Sun

solar energy
Includes use on level 3 under energy sources(1).
SA energy
energy sources
new energy sources
Sun

solar flares
Term introduced on level 2 under astrophysics and solar physics(1) in 1978. Before 1978, search flares.
SA astrophysics and solar physics
cosmic rays
disturbances
interplanetary space
magnetic storms
Moon
solar activity
solar wind
Sun

solar nebula
UF nebula, solar
SA Sun

solar physics, astrophysics and
use astrophysics and solar physics

solar system
Includes use on level 3 under planetology(1). See also names of planets.
SA asteroids
comets
Earth
extraterrestrial geology
interplanetary space
Jupiter
Mars
Mercury Planet
meteorites
Moon
Neptune
planetology
planets
Pluto
Saturn
Sun
Uranus
Venus

solar wind
Includes use on level 2 under interplanetary space(1) and under magnetosphere(1).
UF wind, solar
SA aurora
comets
corona
electromagnetic radiation
interplanetary space
ionization
magnetic field
magnetic storms
magnetic tail
magnetopause
magnetosphere
Moon
particle radiation
particles
plasma instabilities
plasma motion
solar cycles
solar flares
Sun

sole markings
A valid term through 1971. After 1971, use sole marks.

sole marks
Includes use on level 3 under sedimentary structures(1) turbidity current structures(2) and bedding plane irregularities(2). See list K. Before 1972, also search sole markings.
BA turbidity current structures
sedimentary structures
SA bedding plane irregularities
flute casts
load casts
soft sediment deformation

Solemyida
Term introduced in 1981. Includes use on level 2 under Mollusca(1). See list F.
BA Bivalvia
Mollusca
BT Invertebrata

Solenhofen
use Solnhofen

Solenhofen Limestone
use Solnhofen Limestone

solfataras
BT fumaroles
SA gases
sublimates
volcanism

solid phase
Term introduced in 1978. Includes use as a general term under geochemistry(1). Also search solids.
UF phase, solid
solids
SA fluid phase
gaseous phase
geochemistry
liquid phase
solid solution

solid solution
Includes use on level 3 under phase equilibria(1). Also search mixed crystals.
UF mixed crystals
SA phase equilibria
solid phase
solution
solutions

solid waste
As of 1978, term is used on level 2 under waste disposal(1).
UF waste, solid
SA engineering geology
environmental geology
industrial waste
liquid waste
radioactive waste
waste disposal

solidification
use consolidation

solids
use solid phase

solifluction
As of 1978, term is used on level 2 under permafrost(1).
UF sludging
soil flow
soil fluction
solifluxion
SA creep
engineering geology
glacial geology
periglacial features
permafrost
slope stability
soils

solifluxion
use solifluction

Solikamsk
City in Perm Oblast W of the Urals.
BT Russian Republic
USSR

solitary waves
As of 1981, no longer a valid term for GeoRef. Term was introduced in 1978. Included use on level 2 under ocean waves(1). Before 1978, search ocean waves AND solitary.

Solnhofen
Village in W central Bavaria. Also search Solenhofen.
UF Solenhofen
BT Bavaria
West Germany

Solnhofen Limestone
In Bavaria, West Germany. Also search Solenhofen; Solenhofen Limestone; Solnhofen.
UF Solenhofen Limestone
BT Jurassic
SA Bavaria
West Germany

Solomon Islands
Group of islands E of New Guinea in the SW Pacific Ocean. Bougainville, Buka, and Green Islands in the W are part of Papua New Guinea while the remaining 10 large islands and 4 groups of small islands remain a protectorate of the United Kingdom. Includes use on level 1 as an area term (list O). For term set options see list B.
CO S120000S063000
E1630000E1550000
BA Melanesia
NT Malaita
Santa Isabel
SA Australasia
Bougainville
Pacific Ocean
Papua New Guinea

Solomon Sea
Enclosed on the W by New Guinea, on the N by the Bismarck Archipelago, on the E by the Solomon Islands and on the S by the Coral Sea.
BT Pacific Ocean

Solonchak soils
Saline. Includes use on level 3 under soils(1). See list M. Also search saline soils.
BA soils
SA Intrazonal soils
salinity
soil group
Solonetz soils

Solonetz soils
Includes use on level 3 under soils(1). See list M.
BA soils
SA Intrazonal soils
soil group
Solonchak soils

solongoite
BA borates
BT minerals

solubility
Includes use on level 3 under geochemistry(1) properties(2).
SA chemical analysis
concentration
geochemistry
immiscibility
insoluble residues
properties
saturation
sea water
solution

Solund Islands
Island group in the North Sea at mouth of Sogne Fjord. Also search Solund.
BT Norway

solutes
SA geochemistry
solution

solution
Includes use on level 3 under geochemistry(1) processes(2). Before 1978, also search dissolution.
SA adsorption
aqueous solutions
caves
concentration
electrolysis
exsolution
geochemistry
leaching
lysoclines
pressure solution
solid solution
solubility
solutes
solution features
solutions

solution cavities
Includes use on level 3 under geomorphology(1) solution features(2).
UF cavities, solution
SA caves
geomorphology
karst
sedimentary structures
solution features

solution features
Includes use on level 2 under geomorphology(1) and land subsidence(1).
UF features, solution
NT karren
karst
SA caverns
caves
geomorphology
land subsidence
sinkholes
solution
solution cavities
speleology

solution mining
Includes use on level 3 under mining geology(1).
BT mining
SA leaching
mines
mining geology

solutions
Used for material or mixture. Singular form used to indicate process.
SA aqueous solutions
geochemistry
reagents
solid solution
solution

Somali Basin
E of the Somali Republic and N of the Seychelles.
BT Indian Ocean

Somali Republic
Comprises former British Somaliland and Trust Territory of Somalia (formerly Italian Somaliland). Also search Somalia. Includes use on level 1 as an area term (list O). For term set options see list B.
CO S014000N120000
E0513000E0410000
UF Somalia
BA East Africa
Africa

Somalia
use Somali Republic

Somasteroidea
As of 1981, includes use on level 2 under Echinodermata(1). See list F. Before 1981, included use on level 3 under Echinodermata(1) Stelleroidea(2).
BA Stelleroidea
Asterozoa
Echinodermata
BT Invertebrata

Somerset
County in SW England.
CO N505000N512000
W0021500W0035000

BT England
Great Britain
United Kingdom

Somerset County
Index states as applicable.
BX Maine
Maryland
New Jersey
Pennsylvania
BT United States

Somerset Island
In central District of Franklin N of Boothia Peninsula.
CO N720000N740000
W0900000W0960000
BT Northwest Territories
Canada

Somes Basin
River basin in NE Hungary and NW Romania. Index countries as applicable. Also search Somes; Somes River.
UF Somesul Basin
BT Europe
SA Hungary
Romania

Somesul Basin
use Somes Basin

Somme
Department in N.
CO N493000N503000
E0031000E0011500
BT France
SA Somme River valley

Somme River valley
Index departments as applicable. Also search Somme Valley.
UF Somme Valley
BT France
SA Aisne
Somme

Somme Valley
use Somme River valley

Son Valley
River valley. Index states as applicable. Also search Son River.
BT India
SA Bihar
Madhya Pradesh

sonar methods
Term introduced in 1978. Includes use on level 3 under geophysical methods(1). Before 1978, search sonar.
BA geophysical methods
SA acoustical methods
acoustical surveys
echo sounding
geophysical surveys
methods
radar methods
remote sensing
side-scanning methods
sounding

Sondre Strom Fjord
Inlet of Davis Strait on Arctic Circle in SW.
BT Greenland
Arctic region

Sondrio
City and province in N.
BT Lombardy
Italy

Songhor
Village in Kermanshahan, W Iran.
UF Sonqor
Sunqur
BT Iran

sonic logging
Not a valid term for GeoRef. See acoustical logging.

sonic waves
use acoustical waves

sonobuoys
SA geophysical methods
geophysical surveys
marine methods

Sonoma County
W California. Also search Sonoma.
BT California
United States

Sonora
State in NW.
CO N263000N323000
W1083000W1150500
BT Mexico
NT Cananea
Moctezuma
Nacozari de Garcia
SA Colorado River
Colorado River delta
Pedregosa Basin
Sierra Madre Occidental
Sonoran Desert

Sonoran Desert
Index Sonora and U.S. states as applicable.
SA Arizona
California
Sonora

Sonqor
use Songhor

Sonyea Formation
use Sonyea Group

Sonyea Group
Includes Middlesex Shale, Pultenay Shale, Rock Stream Siltstone, and Cashaqua Shale members. W New York. Also search Sonyea Formation.
UF Sonyea Formation
BT Upper Devonian
Devonian

Sophia
use Sofia

Sor Rondane Mountains
use Sor-Rondane Mountains

Sor-Rondane Mountains
In Queen Maud Land near Princess Ragnhild Coast in Norwegian Sector on Atlantic Ocean side. Also search Sor Rondane Mountains; Sor Rondane.
UF Sor Rondane Mountains
BT Antarctica

Sor-Trondelag
County S of Trondheim Fjord in central.
BT Norway
NT Trondheim
SA Trondelag

Sorachi
River in W central.
BT Hokkaido
Japan

Soria
As of 1981, refers to only the city. Before 1981, also included the province. In N central Spain.
BT Soria Province
Old Castile
Spain

Soria Province
Term introduced in 1981. Before 1981, search Soria. In N central Spain.
CO N410400N420900
W0014800W0033200
BT Old Castile
Spain
NT Soria

Soroka
use Belomorsk

sorosilicates
Term introduced in 1981. Includes use on level 2 under minerals(1). See list L.
BA orthosilicates
silicates
BT minerals
NA epidote group
melilite group

Soroy
Island in Norwegian Sea off NW. Finnmark County.
BT Norway

sorption
Includes use on level 3 under geochemistry(1) processes(2).
SA absorption
adsorption
geochemistry

Sorrento Peninsula
On S side of Bay of Naples. Also search Sorrento.
BT Campania
Italy

sorting
Includes use on level 3 under sedimentary rocks(1). See also textures under appropriate rock types.
SA grains
particles
rounding
sedimentary rocks
sedimentation
sediments

Sosnowice
use Sosnowiec

Sosnowiec
City in S part of country.
UF Sosnowice
BT Katowice
Poland

Soufriere
As of 1981, refers to volcano on Saint Vincent. Before 1981, also search Soufriere AND Saint Vincent.
UF Soufriere Volcano
BT Saint Vincent
Lesser Antilles
SA La Grande Soufriere

Soufriere Volcano
use Soufriere

Souk-el-Arba salt works
In Souk-el-Arba area in NW Tunisia.
UF Souk-el-Arba Works
BT Tunisia

Souk-el-Arba Works
use Souk-el-Arba salt works

sound waves
use acoustical waves

sounding
Includes use on level 3 under geophysical methods(1).
SA deep seismic sounding
deep sounding
echo sounding
geophysical methods
geophysical surveys
seismology
sonar methods

Sounds National Park
use Fiordland National Park

source area
use provenance

source rocks
SA natural gas
petroleum
reservoir rocks
rocks

sourceland
use provenance

sources
A valid general term through 1978.

sources, energy
use energy sources

sources, heat
use heat sources

sources, seismic
use seismic sources

Souris River basin
Index North Dakota and Canadian provinces as applicable. Also search Souris; Souris River.
SA Alberta
North Dakota
Saskatchewan

South Africa
Includes use on level 1 as an area term (list O). For term set options see list B.
CO S350000S220000
E0330000E0160000
BA Southern Africa
Africa
NT Cape Province
Natal
Orange Free State
Transvaal
Vaal River
SA Dwyka Series
Ecca Series
Fig Tree Series
Kalahari Desert
Karroo System
Limpopo Belt
Nama System
Onverwacht Group
Stormberg Series
Swaziland Sequence
Swaziland System
Transvaal Supergroup
Waterberg System
Witwatersrand System
Wolkberg Group

South America
Includes use on level 1 or 2 as an area term (list O). If 1, see term set options under list B. To retrieve all documents, individual countries and physiographic regions should also be searched (see list O).
CO S560000N130000
W0350000W0820000
NT Amazon Basin
Amazon River
NA Andes
Argentina
Bolivia
Brazil
NT Chaco
NA Chile
Colombia
Ecuador
NZ Falkland Islands
NA French Guiana
NT Guajira Peninsula
Guiana Basin
Guianas
NA Guyana
NT Guyana Shield
Lake Titicaca
Llanos
Orinoco River
NA Paraguay
NT Parana Basin
Parana River
Patagonian Andes
NA Peru
NT Rio de la Plata
Rio Negro
Santa Cruz
NA Surinam
NT Tierra del Fuego

NA Uruguay
Venezuela
SA America
Botucatu Formation
Galapagos Islands
Gondwana
Pacific mobile belt
Roraima Formation
South American Plate

South American Atlantic
Term introduced in 1981 to indicate a region of the Atlantic Ocean.
BA South Atlantic
BT Atlantic Ocean

South American Pacific
Term introduced in 1981 to indicate a region of the Pacific Ocean.
BA Pacific Ocean
NA Carnegie Ridge
Chile Ridge
Cocos Ridge
Nazca Ridge

South American Plate
Term introduced in 1981. Includes the South American continent (excluding a small portion in the Caribbean Plate) and part of the Atlantic Ocean W of the Mid-Atlantic Ridge. Extends northward to the Caribbean region and southward to the Scotia Ridge.
SA plate tectonics
plates
South America

South Arcot
Region on Coromandel Coast in E.
BT Tamil Nadu
India

South Atlantic
Term introduced in 1978. Before 1978, search Atlantic Ocean AND south.
BA Atlantic Ocean
NA South American Atlantic
NZ South Atlantic Ridge
NA Southeast Atlantic
SA East Atlantic
North Atlantic
Northeast Atlantic
Northwest Atlantic
Southeast Atlantic
Southwest Atlantic
West Atlantic

South Atlantic Ridge
Term introduced in 1981.
BZ Mid-Atlantic Ridge
South Atlantic
BT Atlantic Ocean

South Australia
Includes use on level 1 as an area term (list O). For term set options see list B.
CO S380000S260000
E1410000E1290000
BA Australia
Australasia
NT Adelaide
Adelaide Geosyncline
Balcanoona
Beltana
Coorong Lagoon
Encounter Bay
Eyre Peninsula
Fleurieu Peninsula
Flinders Ranges
Kangaroo Island
Lake Bonney
Lake Eyre
Lake Torrens
Mount Lofty Ranges
Renmark
Saint Vincent Gulf
Spilsby Island
Willunga
Yorke Peninsula
SA Cooper Basin
Eucla Basin
Gambier Embayment
Giles Complex
Great Artesian Basin
Murray Basin
Murray River
Musgrave Ranges
Nullarbor Plain
Officer Basin
Otway Basin
Simpson Desert
Weekeroo Station Meteorite

South Austrian Alps
Term introduced in 1981.
CO N462200N470200
E0145800E0122500
BZ Alps
BX Austria

South Austrian Molasse
Term introduced in 1981.
CO N463500N474000
E0164300E0151200
BT Austria

South Canadian River
use Canadian River

South Carolina
Includes use on level 1 as an area term (list O). For term set options see list B.
CO N320400N351200
W0783200W0831500
BA Eastern U.S.
United States
NT Aiken
Aiken County
NX Anderson County
NT Barnwell County
NX Calhoun County
Charleston
NT Charleston County
NX Cherokee County
Chester County
Columbia
NT Georgetown County
NX Greenwood County
NT Horry County
NX Jasper County
NT Lake Jocassee
NX Lancaster County
Lee County
Marion County
Oconee County
NT Pageland
NX Pickens County
NT Port Royal Sound
NX Richland County
NT Santee River
Savannah River Plant
NX Union County
York County
SA Atlantic Coastal Plain
Blue Ridge Province
Carolina Bays
Carolina slate belt
Castle Hayne Limestone
Duplin Formation
Georgetown
Hawthorn Formation
Kings Mountain
Peedee Formation
Piedmont
Roan Supergroup
Sand Hills
Savannah River
Tuscaloosa Formation
Waccamaw Formation
Wicomico Formation

South Carpathians
use Transylvanian Alps

South Cascade Glacier
BT Washington
United States

South China Sea
Bounded on N by China and Taiwan, on the E by the Philippine Islands, on the S by Malaysia and on the W by Vietnam. As of 1977, includes use as level 1 area term (list O). For term set options see list B.
CO N000000N250000
E1220000E0991000
BA West Pacific
Pacific Ocean
NT Gulf of Siam
Gulf of Tonkin
Manila Trench
SA China Sea

South Dakota
Includes use on level 1 as an area term (list O). For term set options see list B.
CO N423000N455500
W0962700W1040500
BA Midwest
United States
NT Badlands National Monument
NX Butte County
Campbell County
Clark County
Clay County
Custer County
Dewey County
Douglas County
NT Fall River County
NX Grant County
Harding County
NT Homestake Mine
NX Jackson County
Jones County
Lake County
Lawrence County
Lincoln County
Marshall County
McPherson County
Meade County
Pennington County
NT Pierre
Rapid City
NX Shannon County
Union County
SA Arikaree Group
Ash Hollow Formation
Bald Mountain
Benton Formation
Black Hills
Brule Formation
Carlile Shale
Chadron Arch
Chadron Formation
Colorado Group
Dakota Formation
Deadwood Formation
Elk Point Group
Fall River Formation
Fort Hays Limestone Member
Fort Union Formation
Fox Hills Formation
Grand River
Graneros Shale
Great Plains
Greenhorn Limestone
Hell Creek Formation
Lance Formation
Little Missouri River basin
Loup Fork Group
Midcontinent
Minnelusa Formation
Missouri River
Missouri River basin
Missouri River valley
Montana Group
Morrison Formation
Mowry Shale
Niobrara Formation
Pierre Shale
Red River Formation
Sioux Quartzite
Smoky Hill Chalk Member
Spearfish Formation
Sundance Formation
Tongue River Member
Valentine Formation
White River Group
Williston Basin
Winnipeg Formation
Winnipegosis Formation

South Georgia
Island 800 miles E of the Falkland Islands of which it is a dependency.
CO S550000S540000
W0350000W0380000
UF South Georgia Island
BA Scotia Sea Islands
BT Antarctica
SA Falkland Islands

South Georgia Island
use South Georgia

South Greenland
Term introduced in 1978. Before 1978, search Greenland AND south or southern.
BA Greenland
SA East Greenland
West Greenland

South Island
Largest Island of New Zealand S and SW of North Island.
CO S473000S403000
E1743000E1663000
BT New Zealand
NT Canterbury
Christchurch
Dunedin
Fiordland National Park
Fox Glacier
Haast River
Inangahua
Kaikoura
Kakanui
Oamaru
Otago
Otago Peninsula
Rangitata River
Reefton
Southern Alps
Southland
Westland
SA Tasman orogenic zone

South Korea
Officially Republic of Korea. Bounded on N by North Korea, on E by Japan Sea, on S by the Korea Strait, and on W by the Yellow Sea.
CO N330000N383000
E1293000E1260000
BT Korea
NT Kyongsang Basin
Pohang
Ulsan
Yongyang
SA Kangwon
North Korea

South Massif
use Causses

South Mountain
Ridge in W Maryland and S Pennsylvania. Index states as applicable.
BT United States
SA Maryland
Pennsylvania

South Nahanni River
SW District of Mackenzie.
BT Northwest Territories
Canada

South Orkney Islands
British islands 850 miles NE of Antarctic Peninsula.
CO S620000S600000
W0430000W0470000
UF South Orkneys
BA Scotia Sea Islands
BT Antarctica
NA Signy Island

SA Falkland Islands

South Orkneys
use South Orkney Islands

South Pacific
Term introduced in 1978. Before 1978, search Pacific Ocean AND south.
BA Pacific Ocean
SA East Pacific
Equatorial Pacific
North Pacific
Northeast Pacific
Northwest Pacific
Pacific Basin
Southeast Pacific
Southwest Pacific
West Pacific

South Pass
One of the channels at the mouth of the Mississippi River.
BT Louisiana
United States

South Platte River valley
Index states as applicable. Also search South Platte; South Platte River.
BT United States
SA Colorado
Nebraska

South Pole
The S extremity of the Earth's axis at 90° S latitude, and the point from which all directions are N.
BT Antarctica
SA Antarctic Polar Cap
Polar regions

South Polish Glaciation
Refers to glaciation in S Poland which was roughly the southern limit of Pleistocene glaciers in central Europe.
BT Cromerian
upper Pleistocene
Pleistocene
SA Poland

South Polynesian Pacific
Term introduced in 1981 to indicate a region of the Pacific Ocean.
BA Pacific Ocean

South Ray Crater
BT Moon

South Sandwich Islands
Group of small volcanic islands at E end of Scotia Sea about 1350 miles E of Cape Horn. Part of Falkland Island Dependencies.
CO S580000S540000
W0250000W0300000
BA Scotia Sea Islands
BT Antarctica
SA Atlantic Ocean
Falkland Islands

South Saskatchewan River
Rises in the Rocky Mountains of W Alberta. Index provinces as applicable.
BT Canada
SA Alberta
North Saskatchewan River
Saskatchewan
Saskatchewan River

South Shetland Islands
N of the Antarctic Peninsula and S of Drake Passage. Part of British Antarctic Territory.
CO S630000S610000
W0540000W0630000
UF South Shetlands
BA Scotia Sea Islands
BT Antarctica
NA Deception Island
King George Island
Livingston Island
SA Shetland Islands

South Shetlands
use South Shetland Islands

South Victoria Land
use Victoria Land

South Wales
Term introduced in 1981. Region S of narrowest part of country.
CO N512000N522500
W0024000W0053000
BT Wales

South Wales coal field
Concentrated industrial area N of Bristol Channel. Also search South Wales.
UF South Wales Coalfield
BT Wales
Great Britain
United Kingdom

South Wales Coalfield
use South Wales coal field

South-West Africa
A valid index term through 1980. After 1980, use Namibia.

South-West England
Term introduced in 1981.
CO N495600N513000
W0023500W0054500
BT England

Southeast Asia
Term introduced in 1981. Before 1981, also search Asia AND southeast.
BT Asia

Southeast Atlantic
Term introduced in 1978. Before 1978, search Atlantic Ocean AND southeast.
BA South Atlantic
Atlantic Ocean
SA East Atlantic
North Atlantic
Northeast Atlantic
Northwest Atlantic
South Atlantic
Southwest Atlantic
West Atlantic

Southeast Indian Ridge
Term introduced in 1981.
CO S600000S330000
E1560000E0783000
UF Indian-Antarctic Ridge
Indo-Antarctic Ridge
BA Mid-Indian Ridge
BT Indian Ocean
SA mid-ocean ridges

Southeast Pacific
Term introduced in 1978. Before 1978, search Pacific Ocean AND southeast.
BA Pacific Ocean
SA East Pacific
Equatorial Pacific
North Pacific
Northeast Pacific
Northwest Pacific
South Pacific
Southwest Pacific
West Pacific

Southeastern Nigeria
Term introduced in 1981.
CO N042500N065500
E0093500E0072000
BT Nigeria

Southeastern Polish Hills
Term introduced in 1981.
CO N494500N514000
E0241000E0174500
BT Poland

Southeastern Transdanubia
Term introduced in 1981.
CO N454500N462500
E0185000E0174000
BZ Transdanubia
BX Hungary

Southeastern Venezuela
Term introduced in 1981.
CO N004000N082500
W0602000W0675500
BT Venezuela

Southern Africa
Term introduced in 1978. Before 1978, search Africa AND southern.
CO S350000S153000
E0334000E0120000
BA Africa
NA Botswana
Lesotho
Namibia
South Africa
Swaziland
Zimbabwe
SA Central Africa
East Africa
North Africa
West Africa

Southern Alps
W central.
BT South Island
New Zealand
SA Alps

Southern Andes
Term introduced in 1978. Before 1978, also search Andes AND south or southern.
BA Andes
BT South America
SA Central Andes
Northern Andes

Southern Apennines
Term introduced in 1981. Before 1981, also search Apennines AND South. Northern extent is Sella di Vinchiaturo Pass.
CO N375500N413000
E0164500E0130000
BT Apennines
Italy

Southern Appalachians
Term introduced in 1978. Before 1978, also search Appalachians AND south or southern.
BA Appalachians
BT North America
SA Central Appalachians
Northern Appalachians

Southern Atlantic Coastal Plain
Term introduced in 1981. Before 1981, also search Atlantic Coastal Plain AND South.
CO N250000N340000
W0780000W0840000
BT Atlantic Coastal Plain

Southern California
Term introduced in 1978. Before 1978, search California AND south.
CO N390000N420000
W1200000W1242500
BT California
United States

Southern California Batholith
BT California
United States

Southern Carpathians
use Transylvanian Alps

Southern Cook Islands
use Cook Islands

Southern Europe
Term introduced in 1981.
BA Europe
NA Albania
Bulgaria
Greece
Iberian Peninsula
Italy
Malta
Romania
San Marino
Yugoslavia

Southern Great Plains
Term introduced in 1978. Before 1978, search Great Plains AND south or southern; also search Southern High Plains.
UF Southern High Plains
BA Great Plains
BT North America
SA Northern Great Plains

Southern Hemisphere
Used when discussing many large areas too numerous to mention. Includes use on level 1 and 2 as an area term (list O). If 1, see list B for term set options.
SA Eastern Hemisphere
Gondwana
Northern Hemisphere
Western Hemisphere

Southern High Plains
use Southern Great Plains

Southern Highlands
Highland region. Former province of Tanganyika bounded on SW by Zambia, S by Malawi and Lake Nyasa.
BT Tanzania

Southern Norway
Term introduced in 1981. N boundary is Trondjeim.
CO N580000N635000
E0125500E0044000
BT Norway

Southern Peninsula
use Michigan Lower Peninsula

Southern Province
Structural province of the Canadian Shield. Primarily in the Lake Superior region. Index Ontario or states as applicable.
CO N440000N491500
W0803000W0963000
BT Canadian Shield
SA Michigan
Minnesota
Ontario
Wisconsin

Southern Rhodesia
A valid index term through 1972. After 1972, use Rhodesia.

Southern Rocky Mountains
Term introduced in 1978. Before 1978, also search Rocky Mountains AND south or southern.
BA Rocky Mountains
BT North America
SA Central Rocky Mountains
Northern Rocky Mountains
U. S. Rocky Mountains

Southern Swiss Alps
Term introduced in 1981.
BA Swiss Alps
BT Switzerland

Southern Transdanubia
Term introduced in 1981.
CO N454500N472200
E0190200E0160500
BZ Transdanubia
BX Hungary

Southern Uplands
Between the Scottish lowlands to the N and the Cheviot Hills on the English border to the S.
BT Scotland
Great Britain
United Kingdom

Southern Urals
Section of the Urals N of the Mu-

godzhar Hills.
CO N510000N560000
E0610000E0550000
BA Urals
BT USSR
SA Kazakhstan
Russian Republic
Ural-Tau

Southern U.S.
Term introduced in 1981. Index states as applicable. Includes use on level 1 as an area term. For term set options see List B.
CO N290000N390000
W0814000W0943000
BA United States
NA Alabama
Arkansas
Kentucky
Louisiana
Mississippi
Tennessee

Southern Yemen
Peoples Democratic Republic of Yemen. Includes use on level 1 as an area term (list O). For term set options see list B.
CO N123000N183000
E0530000E0432000
UF Peoples Democratic Republic of Yemen
BA Arabian Peninsula
Asia
NT Aden
SA Arabian Shield
Middle East
Near East
Red Sea Basin
Yemen

Southland
A land district in SW.
BT South Island
New Zealand

Southwest Atlantic
Term introduced in 1978. Before 1978, search Atlantic Ocean AND southwest.
BA Atlantic Ocean
SA East Atlantic
North Atlantic
Northeast Atlantic
Northwest Atlantic
South Atlantic
Southeast Atlantic
West Atlantic

Southwest Indian Ridge
Term introduced in 1981.
CO S413000S223000
E0673000E0440000
BT Indian Ocean
SA mid-ocean ridges

Southwest Pacific
Term introduced in 1978. Before 1978, search Pacific Ocean AND southwest.
BA Pacific Ocean
SA East Pacific
Equatorial Pacific
North Pacific
Northeast Pacific
Northwest Pacific
South Pacific
Southeast Pacific
West Pacific

Southwestern German Hills
Term introduced in 1981.
CO N473200N504200
E0123700E0080200
BT Germany

Southwestern U.S.
Term introduced in 1978. Includes use on level 1 as an area term. Before 1978, also search United States AND southwest or southwestern.
CO N260000N370000
W0933000W1144500
BA United States
NA Arizona
New Mexico
Oklahoma
Texas

Southwestern USSR
Term introduced in 1978. Before 1978, search USSR AND southwest.
BT USSR

Soviet Altai Mountains
Term introduced in 1981 for the Altai Mountains in the USSR. Also index Siberian fold belt.
CO N473000N523000
E0890000E0810000
BZ Altai Mountains
BX USSR
SA Kazakhstan
Russian Republic
Siberian fold belt

Soviet Arctic
That area N of the Arctic Circle in European USSR and Siberia, and including Soviet islands in the Arctic Ocean. In Siberia, arctic climatic conditions prevail far S of the Arctic Circle because of its great continental expanse. Also search USSR AND Arctic region.
CO N663200N900000
W1700000E0300000
BT USSR
SA Arctic region

Soviet Carpathians
Term introduced in 1981 for the Carpathians in the USSR. Also index Russian Platform.
CO N480000N500000
E0260000E0230000
BZ Carpathians
BX USSR
SA Eastern Carpathians
Russian Platform

Soviet Central Asia
use Central Asia

Soviet Far East
Region occupying easternmost Siberia including Amur Oblast, Khabarovsk Kray, Primorye Kray, and Sakhalin Oblast. Also search USSR AND Far East.
CO N420000N750000
W1700000E1300000
UF Siberia-Soviet Far East
BT Russian Republic
USSR
SA Amur Basin
Far East
Kamchatka Peninsula
Kuril Islands
Okhotsk-Chukchi
Sakhalin
Sikhote-Alin Range

Soviet Fennoscandia
Term introduced in 1981. Also index Russian Platform.
CO N600000N700000
E0420000E0280000
BZ Fennoscandia
BX USSR
SA Russian Platform

Soviet Pacific region
Term introduced in 1981.
CO N433000N650000
E1800000E1413000
BT Russian Republic
USSR
NA Kamchatka Peninsula
Koryak Range
Kuril Islands
Sakhalin

sovite
As of 1981, includes use on level 3 under igneous rocks(1) carbonatites(2). See list H. Before 1981, included use on level 3 under igneous rocks(1) lamprophyre and carbonatite family(2).
BA carbonatites
BT igneous rocks

Sowerbyella
BA Strophomenida
Articulata
Brachiopoda
BT Invertebrata

Sowie Mountains
Range of the Sudeten Mountains in Lower Silesia in SW Poland.
UF Eulengebirge
BT Poland
SA Sudeten Mountains

SO2
use sulfur dioxide

space
use interplanetary space

space groups
Includes use on level 3 under crystal structure(1).
SA crystal structure

space lattice
use lattice

space, underground
use underground space

Spain
Includes use on level 1 as an area term (list O). For term set options see list B.
CO N360000N434500
E0043000W0093000
BA Iberian Peninsula
Southern Europe
Europe
NT Alpujarras
Andalusia
Aragon
Asturian Arc
Asturias
NA Balearic Islands
NT Basque Provinces
Betic Cordillera
Betic Zone
Calatayud-Teruel Basin
Cantabrian Basin
Cantabrian Mountains
Cantabrian region
Castile
Catalonia
Catalonian Coastal Ranges
Duero Basin
Ebro Basin
Ebro River
Extremadura
Galaico Massif
Galicia
Guadalquivir
Guadalquivir Basin
Hercinico Centro
Hercinico Sur
Iberian Mountains
Iberica
Leon region
Maestrazgo
Meseta
Montes de Toledo
Murcia region
Navarra
Prebetic Zone
Serrania de Cuenca
Serrania de Ronda
Sierra de Gredos
Sierra de Guadarrama
Sierra de la Demanda
NX Spanish Pyrenees
NT Spanish Sierra Nevada
Subbetic Zone
Valencia region
SA Ager Formation
Andalusian
Canary Islands
Douro River
ENADIMSA
Lancara Formation
Mediterranean region
Navarre
Pyrenees
Tagus Basin
Tagus River
Vitoria

spallation
SA cosmic rays
isotopes

Spanish Guinea
A valid term through 1975. After 1975, use Equatorial Guinea.

Spanish Peaks
Two mountains in Huerfano and Las Animas counties in S.
BT Colorado
United States

Spanish Pyrenees
Term introduced in 1981.
BX Spain
BZ Pyrenees

Spanish Sahara
A valid level 1 term through 1976. From 1977-80, documents on the Spanish Sahara were indexed using Sahara as the level 1 term and the term west. As of 1981, use Western Sahara.

Spanish Sierra Nevada
Term introduced in 1981. Mountain range in S Spain. Before 1981, also search Sierra Nevada AND Spain.
CO N364500N373500
W0020000W0043000
BT Spain
SA Sierra Nevada

sparagmite
Includes use on level 3 under sedimentary rocks(1) clastic rocks(2). See list I.
BA clastic rocks
BT sedimentary rocks
SA terrigenous materials

Sparagmite Division
use Sparagmite Group

Sparagmite Group
Late Precambrian-early Paleozoic.
UF Sparagmite Division
Sparagmite Series
SA Norway
Paleozoic
Precambrian
Riphean
Sweden
upper Precambrian

Sparagmite Series
use Sparagmite Group

Sparnacian
Europe. Above Thanetian, below Ypresian of France. Includes use on level 3 under age terms(1). See list E.
BA lower Eocene
Eocene
Paleogene
Tertiary
BT Cenozoic
SA Paleocene

Spathognathodus
Genus.
BA Conodonta

spatial distribution
Term introduced in 1981.
SA distribution

earthquakes
volcanoes

spatial frequency filters
Term introduced in 1981.
BT filters

spatial variations
Used as a general term.
SA variations

Spearfish Formation
Underlies Gypsum Spring Formation, geographically extended into Black Hills area. NW Nebraska, W South Dakota, and E Wyoming.
SA Nebraska
Permian
South Dakota
Triassic
Wyoming

species
As of 1981, no longer a valid term for GeoRef. Included use on level 3 under paleontology(1).

species diversity
SA diversity
evolution
fossils
paleontology

specific gravity
Includes use on level 3 under appropriate material name, e.g. under soils(1) and sediments(1).
SA density

specific heat
UF heat, specific
SA geochemistry
heat capacity
physical properties
temperature
thermal conductivity
thermal properties
thermodynamic properties

specimens, type
use type specimens

spectra
As of 1978, term is used on level 2 under minerals(1). Also search spectrum.
UF spectrum
SA EPR spectra
geochemistry
infrared spectra
mass spectra
minerals
Mossbauer spectra
optical spectra
spectroscopy
ultraviolet spectra
X-ray fluorescence spectra

spectral analysis
Includes use on level 3 under seismology(1). As of 1981, restricted to seismology.
SA analysis
seismograms
seismology

spectrography
use spectroscopy

spectrometry
A valid term through 1974. After 1974, use spectroscopy.

spectrophotometry
use spectroscopy

spectroscopy
Used for methodology. For data, see under appropriate material. Includes use on level I; on level 3 under chemical element(1) analysis(2). Before 1973, also search mass spectroscopy; before 1975, also search spectrometry. If used on level I, term set options are:
methods
name of method [alpha-ray spectroscopy, atomic absorption, electron probe, emission spectroscopy, flame photometry, gamma-ray spectroscopy, infrared spectroscopy, ion probe, laser probe, mass spectroscopy, microwave spectroscopy, Mossbauer spectroscopy, neutron spectroscopy, nuclear magnetic resonance, optical spectroscopy, radio-frequency spectroscopy, Raman spectroscopy, ultraviolet spectroscopy, X-ray spectroscopy, X-ray fluorescence]
techniques
topic [e.g. sample preparation]
UF spectrography
spectrophotometry
NT alpha-ray spectroscopy
emission spectroscopy
gamma-ray spectroscopy
infrared spectroscopy
mass spectroscopy
microwave spectroscopy
Mossbauer spectroscopy
optical spectroscopy
radio-frequency spectroscopy
Raman spectroscopy
ultraviolet spectroscopy
X-ray spectroscopy
SA absorption
alpha rays
analysis
atomic absorption
chemical analysis
chromatography
clay mineralogy
colorimetry
crystallography
differential thermal analysis
electron microscopy
electron paramagnetic resonance
electron probe
flame photometry
geochemistry
infrared spectra
ion probe
ion probe data
major-element analyses
nuclear magnetic resonance
optical spectra
photometry
scattering
spectra
standard materials
surface properties
thermal analysis
trace-element analyses
X-ray analysis
X-ray diffraction analysis
X-ray fluorescence

spectrum
use spectra

specularite
UF gray hematite
BA oxides
BT minerals
SA hematite

Speeton Clay
Cretaceous: Yorkshire.
BT Lower Cretaceous
Cretaceous
SA Barremian
England
United Kingdom

speleology
Includes use on level 3 under geomorphology(1) solution features(2).
SA caves
exploration
geomorphology
solution features
speleothems

speleothems
SA caves
speleology

Spermatophyta
Term introduced in 1981. Includes use on level 1 and 2 as a fossil term (list F).
BT Plantae
NT angiosperms
gymnosperms

sperrylite
As of 1981, includes use on level 3 under minerals(1) arsenides(2). See list L. Before 1981, included use on level 3 under minerals(1) sulfides(2).
BA arsenides
sulfides
BT minerals

Spessart
Low mountain range between the Odenwald and the Rhon Mountains in NW Bavaria. Also search The Spessart.
UF The Spessart
BT Bavaria
West Germany

spessartine
BA garnet group
nesosilicates
orthosilicates
silicates
BT minerals

spessartite
As of 1981, includes use on level 3 under igneous rocks(1) lamprophyres(2). See list H. Before 1981, included use on level 3 under igneous rocks(1) lamprophyre and carbonatite family(2).
BA lamprophyres
BT igneous rocks

Sphaeroidinella dehiscens
Includes use on level 3 under foraminifera(1) Globigerinacea(2).
BA Globigerinidae
Globigerinacea
Rotaliina
foraminifera
BT Protista
Invertebrata

sphaerolites
use spherulites

sphalerite
UF pseudogalena
zinc blende
BA sulfides
BT minerals

sphene
use titanite

Sphenolithus
BA nannofossils
algae
BT thallophytes
Plantae
SA discoasters

Sphenophyllum
Genus of Paleozoic fossil plants.
BA Sphenopsida
pteridophytes
BT Plantae

Sphenopsida
Including Articulatae. Includes use on level 2 under pteridophytes(1). See list F.
BA pteridophytes
BT Plantae
NA Articulatae
Equisetales
Sphenophyllum
NZ Sphenopteris

Sphenopteris
Genus. Includes use on level 3 under pteridophytes(1) Filicopsida(2) or Sphenopsida(2).
BZ Filicopsida
Sphenopsida
BA pteridophytes
BT Plantae
SA Articulatae
Pteridospermae

spherical harmonic analysis
Includes use on level 3 under plate tectonics(1).
SA analysis
Earth
magnetic field
plate tectonics

spherical models
Term introduced in 1981.
BA models
SA seismology

sphericity
SA particles
roundness
sediments
shape analysis

spheroidal mode
Term introduced in 1981.
SA oscillations
seismographs
seismology

spherules
Includes use on level 3 under tektites(1).
SA sedimentary structures
sediments
tektites

spherulites
UF sphaerolites
SA orbicular texture

Sphinctozoa
Includes use on level 3 under Porifera(1) Calcispongea(2).
BA Calcispongea
Porifera
BT Invertebrata

Spice Islands
use Moluccas

spicules
SA Porifera

spilite
As of 1981, includes use on level 3 under igneous rocks(1) alkali basalts(2). See list H. Before 1981, included use on level 3 under igneous rocks(1) alkali basalt family(2).
BA alkali basalts
basalts
volcanic rocks
BT igneous rocks
SA pillow lava
spilitization

spilitization
Includes use on level 3 under metasomatism(1).
BA metasomatism
SA albitization
spilite

spills, oil
use oil spills

Spilsby Island
Largest of Sir Joseph Banks Islands in Spencer Gulf about 5 miles off SE coast of Eyre Peninsula. Also search Spilsby.
BT South Australia
Australia

spinal column
Term introduced in 1981.
SA morphology
Vertebrata

spinel
BA oxides

BT minerals
SA chrome spinel
spinel group

spinel group
BA oxides
BT minerals
SA gahnite
hercynite
jacobsite
maghemite
magnesioferrite
magnetite
spinel
ulvospinel

spinel lherzolite
As of 1981, includes use on level 3 under igneous rocks(1) peridotites(2). See list H. Before 1981, included use on level 3 under igneous rocks(1) ultramafic family(2).
BA peridotites
ultramafics
BT igneous rocks
SA lherzolite

spinifex texture
Term introduced in 1981. Before 1981, also search spinifex.
SA igneous rocks
textures

spirals, growth
use growth spirals

Spiriferida
Includes use on level 2 under Brachiopoda(1). See list F.
BA Articulata
Brachiopoda
BT Invertebrata
NA Atrypidae
Cyrtospirifer
Spiriferidina

Spiriferidae
Family. Includes use on level 3 under Brachiopoda(1) Spiriferida(2).
BA Spiriferidina
Spiriferida
Articulata
Brachiopoda
BT Invertebrata

Spiriferidina
Suborder. Includes use on level 3 under Brachiopoda(1) Spiriferida(2).
BA Spiriferida
Articulata
Brachiopoda
BT Invertebrata
NA Spiriferidae
Spiriferina

Spiriferina
Genus. Includes use on level 3 under Brachiopoda(1) Spiriferida(2).
BA Spiriferidina
Spiriferida
Articulata
Brachiopoda
BT Invertebrata

Spirillinacea
Includes use on level 2 under foraminifera(1). See list F.
BA Rotaliina
foraminifera
BT Protista
Invertebrata

Spis
Region. An historic area in dispute before WWI between Hungary, Austria, and Russia. Index Poland and/or Slovakia as applicable.
BT Europe
SA Poland
Slovakia

Spis-Gemer
The Slovak Ore Mountains. A range of the Carpathians in S Slovakia. Also search Spis-Gemer Mountains.
UF Slovenske Rudohorie
Spis-Gemer Mountains
BT Slovakia
Czechoslovakia
SA Carpathians

Spis-Gemer Mountains
use Spis-Gemer

Spiti
Region in NE.
BT Punjab State
India

spits
SA beaches
coastal environment
geomorphology
marine environment
reefs
shoals
shore features

Spitsbergen
Norwegian archipelago, 360 miles N of Norway, including the main island of Spitsbergen plus North East Land, Edge Island, and Barents Island. Part of the Svalbard Island group. Includes use on level 1 as an area term (list O). For term set options see list B.
CO N760000N810000
E0300000E0100000
BA Svalbard
Arctic region
NT Brogger Peninsula
Hornsund
Kings Bay
Nordaustlandet
Ny Friesland
Vestspitsbergen
SA Hecla Hoek Formation

Spitz
Village on left bank of Danube River.
BT Lower Austria
Austria

Split
City on the central Dalmatian Coast on the Adriatic Sea.
BT Croatia
Yugoslavia

Spodosols
Includes use on level 3 under soils(1). See list M.
BA soils

spodumene
BA clinopyroxene
pyroxene group
chain silicates
silicates
BT minerals
SA aluminosilicates
kunzite

spoils
Term introduced in 1981.
SA environmental geology
mining
tailings
waste disposal

Spokane
City in Spokane County in E.
BT Washington
United States

Spokane County
BT Washington
United States

Spoleto
City in Perugia Province in central part of country.
BT Umbria
Italy

Spongiae
use Porifera

spongolite
Includes use on level 3 under sedimentary rocks(1) clastic rocks(2). See list I.
BA clastic rocks
BT sedimentary rocks

spontaneous fission-track dating
use fission-track dating

spontaneous magnetization
Term introduced in 1981.
BA magnetization

spontaneous polarization
use self-potential methods

Spor Mountain
BT Utah
United States

sporangia
Also search sporangium.
UF sporangium
SA angiosperms
gymnosperms
paleobotany
pteridophytes
spores

sporangium
use sporangia

spores
Includes use on level 3 under palynomorphs(1) miospores(2) or megaspores(2).
SA exine
megaspores
miospores
palynomorphs
pollen
pteridophytes
sporangia
sporopollenin

sporinite
Includes use on level 3 under sedimentary rocks(1) organic residues(2). See list I.
BA exinite
macerals
coal
organic residues
BT sedimentary rocks

sporopollenin
UF sporopollenine
BA organic materials
SA exine
miospores
pollen
spores

sporopollenine
use sporopollenin

spreading centers
Includes use on level 3 under plate tectonics(1).
UF centers, spreading
SA plate tectonics
sea-floor spreading

spreading concept
use sea-floor spreading

spreading-floor hypothesis
use sea-floor spreading

spring gravimeters
Term introduced in 1981.
BT gravimeters

Spring Mountains
In W Clark County near California line.
BT Nevada
United States

Springdale
Village in Washington County in SW Utah. Gateway to Zion National Park.
BT Utah
United States

Springer Formation
Chester and Morrow Series. Conformably overlies Goddard Shale Central S Oklahoma.
BT Carboniferous
SA Chesterian
Mississippian
Morrowan
Oklahoma
Pennsylvanian

Springfield
City. Index states as applicable.
BT United States
SA Illinois
Massachusetts
Missouri
Ohio

Springfield Coal Member
Of Carbondale Formation. W and N Illinois.
BT Pennsylvanian
SA Carbondale Formation
Illinois

springs
Includes use on level 1 (list A). Use for papers stressing spring hydrology. Term set options are:
topic [composition, genesis, geochemistry, hot springs, mineral waters, temperature]
subtopic (no area term)
NA hot springs
SA circulation
exsurgence
fumaroles
geysers
ground water
hydrochemistry
hydrogeology
hydrology
mineral waters
salinity
saturation
thermal waters
water
water resources

Springsure
Village 165 miles WSW of Rockhampton in E central.
BT Queensland
Australia

Springwater
Village in Livingston County in W central.
BT New York
United States

Spumellaria
use Spumellina

Spumellina
Includes use on level 2 under Radiolaria(1). See list F. Also search Spumellaria.
UF Spumellaria
BA Radiolaria
BT Protista
Invertebrata

spurrite
BZ orthosilicates
carbonates
BT minerals

Squamata
As of 1981, includes use on level 2 under Reptilia(1). See list F. Before 1981, included use on level 3 under Reptilia(1) Lepidosauria(2).
BA Lepidosauria
Reptilia
BT Tetrapoda
Vertebrata
NA Lacertilia
Mosasauridae

Sr
use strontium

Sr-86
Includes use on level 3 under isotopes(1).
SA isotopes
Sr-87/Sr-86
strontium

Sr-86/Sr-87
use Sr-87/Sr-86

Sr-87
Includes use on level 3 under isotopes(1).
SA isotopes
Sr-87/Sr-86
strontium

Sr-87/Sr-86
Includes use on level 3 under isotopes(1). Also search Sr-86/Sr-87; Sr-86 AND Sr-87.
UF Sr-86/Sr-87
SA isotopes
Sr-86
Sr-87
strontium

Sr-90
Includes use on level 3 under isotopes(1).
SA isotopes
strontium

Sr/Rb
Includes use on level 3 under absolute age(1) methods(2).
UF Rb/Sr
SA absolute age
isotopes
rubidium
strontium

Sredna Gora
Mountain range between the Balkan and Rhodope mountains in central Bulgaria.
CO N421500N424500
E0250000E0233000
UF Sredna Gora Mountains
BT Bulgaria
SA Panagyurishte

Sredna Gora Mountains
use Sredna Gora

Sredniy Vasyugan
Village on the Vasyugan River in NW Tomsk Oblast in S central Siberia.
BT Russian Republic
USSR

Sri Lanka
Formerly Ceylon. Includes use on level 1 as an area term (list O). For term set options see list B. Before 1974, also search Ceylon.
CO N060000N100000
E0823000E0790000
BA Indian Peninsula
Asia
SA Highland Series
Indian Shield

Srikakulam
City on Bay of Bengal in NE.
BT Andhra Pradesh
India

Srinagar
City and district in the Vale of Kashmir.
BT Jammu and Kashmir
India

St. Catherines
use Saint Catharines

St. Clair County
use Saint Clair County

St. Clair River
use Saint Clair River

St. Clair River delta
use Saint Clair River delta

St. Croix
use Saint Croix

St. Croix Island
use Saint Croix

St. Elias Mountains
use Saint Elias Mountains

St. Francois Mountains
use Saint Francois Mountains

St. Gallen
use Saint Gall

St. George Formation
use Saint George Formation

St. Gotthard
use Gotthard Massif

St. Helena
use Saint Helena

St. Helena Island
use Saint Helena

St. John
use Saint John

St. John's
use Saint John's

St. Johns River basin
use Saint Johns River basin

St. Laurent Limestone
use Saint Laurent Limestone

St. Lawrence County
use Saint Lawrence County

St. Lawrence Estuary
use Saint Lawrence Estuary

St. Lawrence Lowland
use Saint Lawrence Lowlands

St. Lawrence Lowlands
use Saint Lawrence Lowlands

St. Lawrence River
use Saint Lawrence River

St. Lawrence River valley
use Saint Lawrence Valley

St. Lawrence Valley
use Saint Lawrence Valley

St. Louis
use Saint Louis

St. Louis County
use Saint Louis County

St. Louis Limestone
use Saint Louis Limestone

St. Lucia
use Saint Lucia

St. Mary Parish
use Saint Mary Parish

St. Marys Formation
use Saint Marys Formation

St. Paul
use Saint Paul

St. Paul's Rock
use Saint Paul Rocks

St. Paul's Rocks
use Saint Paul Rocks

St. Peter Sandstone
use Saint Peter Sandstone

St. Severin
use Saint-Severin

St. Thomas
use Saint Thomas

St. Vincent
use Saint Vincent

stability
Includes use on level 2 under paleomagnetism(1), foundations(1), land subsidence(1), and tunnels(1).
SA engineering geology
foundations
geochemistry
land subsidence
paleomagnetism
slope stability
stabilization
tunnels
underground installations

stabilization
As of 1978, term is used on level 2 under shorelines(1) and slope stability(1).
SA shorelines
slope stability
stability

stable isotopes
BA isotopes
SA C-13/C-12
carbon
D/H
deuterium
hydrogen
O-18/O-16
oxygen
S-34/S-32
sulfur

Stablo
use Stavelot

Stafford County
Index states as applicable.
BX Kansas
Virginia
BT United States

Staffordshire
County in W central.
CO N523000N531500
W0013000W0023000
BT England
Great Britain
United Kingdom

stage, universal
use universal stage

stages
A valid general term through 1977. Use specific term, e.g. stratigraphic units.

stainierite
use heterogenite

staining
SA paleobotany
paleontology
petrology
sample preparation

Staked Plain
use Llano Estacado

Stalin
use Brasov

Stalinabad
use Dushanbe

Stalingrad
use Volgograd

Stampian
France. Includes use on level 3 under age terms(1). See list E.
BA lower Oligocene
Oligocene
Paleogene
Tertiary
BT Cenozoic
SA Rupelian

standard deviation
UF deviation, standard
BA statistical analysis

standard materials
Level 1 term introduced in 1976. Used for rocks or minerals or other materials that have been designated as standard by geological laboratories. Before 1976, search standard rocks. Includes use on level 1 (list A). Term set options are:
material name (1st level terms only)
type
topic [age, analysis, alteration, catalogs, classification, experimental studies, identification, nomenclature, observations, preparation, properties]
subtopic or name of laboratory
SA absolute age
chemical analysis
crystallography
geochemistry
igneous rocks
isotopes
materials
minerals
petrology
silicate rocks
spectroscopy
standard rocks
thermal analysis

standard rocks
Includes use on level 3 under standard materials (1). A first order term until 1976; now use standard materials on level 1.
SA rocks
standard materials

standardization
Used as a general term.

Stanislaus County
W central.
BT California
United States

Stanislav
use Ivano-Frankovsk

Stanislawow
use Ivano-Frankovsk

Stanley Group
Includes Hatton Tuff Lentil, Ten Mile Creek Formation, Moyers Formation, Chickasaw Creek Formation. W Arkansas, and central S and SE Oklahoma.
BT Mississippian
SA Arkansas
Oklahoma

stannite
UF tin pyrites
BA sulfostannates
sulfosalts
BT minerals

stannoidite
BA sulfides
BT minerals

Stanovoi Range
use Stanovoy Range

Stanovoy Range
Mountain range between Yakutia and Amur Oblast with Khabarovsk Kray on the E. Also search Stanovoi Range; Stanovoy.
CO N530000N573000
E1350000E1150000
UF Stanovoi Range
Stanovoy Upland
BA Northeastern USSR
BT Russian Republic
USSR

Stanovoy Upland
use Stanovoy Range

Stansbury Mountains
In Wasatch National Forest in NW.
BT Utah
United States

Stanton County
Index states as applicable.
BX Kansas
Nebraska
BT United States

Stanton Formation
In Lansing Group. Includes Captain Creek Limestone, Eudora Shale, Stoner Limestone, Rock Lake Shale, South Bend Limestone members. SW Iowa, E Kansas, NW Missouri, SE

Nebraska, and NE Oklahoma.
UF Stanton Limestone Member
BT Missourian
Pennsylvanian
SA Iowa
Kansas
Lansing Group
Missouri
Nebraska
Oklahoma
Rock Lake Shale Member

Stanton Limestone Member
use Stanton Formation

Stark County
Index states as applicable.
BX Illinois
North Dakota
Ohio
BT United States

Starobin
Town in S central.
BT Byelorussia
USSR

Stassfurt
Salt mining city in W part of country.
BT Magdeburg Bezirk
East Germany

Staten Island
S of Manhattan and across The Narrows from Long Island on the E. Constitutes Borough of Richmond in New York City.
BT New York
United States

statistical analysis
Used as a general term for studies that use statistics in analyzing data. As of 1981, includes statistical methods.
NA autocorrelation
canonical analysis
cluster analysis
correlation coefficient
correspondence analysis
dendrograms
discriminant analysis
factor analysis
finite difference analysis
finite element analysis
histograms
kurtosis
least-squares analysis
Markov chain analysis
multivariate analysis
numerical analysis
phi scale
populations
principal components analysis
probability
regression analysis
skewness
standard deviation
trend-surface analysis
variance analysis
SA analysis
autocorrelation
automatic data processing
biometry
equations
geometry
mathematical geology
mathematical methods
methods
reliability
statistical distribution

statistical distribution
Term introduced in 1981.
SA distribution
statistical analysis

statistical methods
As of 1981, no longer a valid term for GeoRef. Use statistical analysis. Before 1981, included use on level 2 under mineral exploration(1) and on level 3 under disciplines(1).

Staunton Formation
Overlies Brazil Formation. SW Indiana.
BT Middle Pennsylvanian
Pennsylvanian
SA Indiana

staurolite
BA orthosilicates
silicates
BT minerals

Stavanger
City on Stavanger Fjord on North Sea S of Bergen.
BT Rogaland
Norway

Stavelot
Town in the N Ardennes.
UF Stablo
BT Liege
Belgium

Stavelot-Venn Massif
Primarily in the N Ardennes of Belgium. Index countries as applicable.
BT Europe
SA Belgium
West Germany

Stavers Island
use Vostok

Stavropol region
In Stavropol Kray (formerly Ordzhonikidze Kray) in the Northern Caucasus. Also search Stavropol.
UF Voroshilovsk (region)
BT Russian Republic
USSR

Ste. Genevieve Formation
use Sainte Genevieve Limestone

Ste. Genevieve Limestone
use Sainte Genevieve Limestone

steam coal
Term introduced in 1981. Includes use on level 3 under sedimentary rocks(1) organic residues(2). See list I.
BA coal
organic residues
BT sedimentary rocks

Stearns Shale
In Council Grove Group. Kansas, and SE Nebraska.
BT Permian
SA Kansas
Nebraska

Stebnik
Town in Lvov Oblast in SW.
BT Ukraine
USSR

Steele Glacier
Emanates from Steele Mountain in Saint Elias Mountains near Alaska border in SW.
BT Yukon Territory
Canada

Steens Mountain
Mountain mass in SW Harney County in SE.
CO N420000N431000
W1181000W1185000
BT Oregon
United States

Stegodon
As of 1981, includes use on level 3 under Mammalia(1) Elephantoidea(2). See list F. Before 1981, included use on level 3 under Mammalia(1) Proboscidea(2).
BA Elephantidae
Elephantoidea
Proboscidea
Mammalia
BT Tetrapoda
Vertebrata

Steinach
Town in Thuringian Forest in SE Suhl.
BT East Germany

Steinbach
Town 39 miles SE of Winnipeg.
BT Manitoba
Canada

Steinheim
Town in two different states. Index states as applicable.
BT West Germany
SA Hesse
North Rhine-Westphalia

Steinheim Basin
BT West Germany

stellerite
BA zeolite group
framework silicates
silicates
BT minerals

Stelleroidea
As of 1981, includes use on level 3 under Echinodermata(1). See list F. Before 1981, included use on level 2 under Echinodermata(1).
BA Asterozoa
Echinodermata
BT Invertebrata
NA Asteroidea
Ophiuroidea
Somasteroidea

step faults
Term introduced in 1978. Includes use on level 3 under faults(1) systems(2). Before 1978, also search faults AND step.
BA faults
SA thrust faults

Stephanian
Europe. Above Westphalian, below Sakmarian of Permian. Includes use on level 3 under age terms(1). See list E.
BA Upper Carboniferous
Carboniferous
BT Paleozoic
SA Pennsylvanian

stephanite
BA sulfantimonites
sulfosalts
BT minerals

Stepnyak
City in Kokchetav Oblast in N.
BT Kazakhstan
USSR

Steppe
use Steppes

steppes
SA geomorphology
plains
taiga environment

Steppes
An extensive, treeless, semi-arid, grassland area of the mid-latitudes extending from the western border of the Soviet Union to the Altai Mountains in the E. Index Soviet republics as applicable. Also search Steppe.
UF Steppe
BT USSR
SA Kazakhstan
Russian Republic
Ukraine

Sterea Ellas
Term introduced in 1981.
CO N373500N392000
E0240500E0204000
BT Greece

stereochemistry
use crystal chemistry

Sterling County
W central Texas. Also search Sterling.
BT Texas
United States

sternbergite
BA sulfides
BT minerals

Stettin
use Szczecin

Stevens County
Index states as applicable.
BX Kansas
Minnesota
Washington
BT United States

stevensite
BA clay minerals
sheet silicates
silicates
BT minerals

Stewart Valley
River valley in W central Yukon Territory. Also search Stewart River.
BT Yukon Territory
Canada

stibiotantalite
BZ niobates
tantalates
BA oxides
BT minerals

stibnite
Also search antimonite.
UF antimonite
BA sulfides
BT minerals
SA metastibnite

stick-slip
Includes use on level 3 under faults(1) mechanics(2); on level 3 under earthquakes(1). Also search stick slip.
SA earthquakes
faults
mechanics

stiff clays
Term introduced in 1981. Also search stiff clay.
SA clay
clays

Stigmaria
Genus.
BA Lycopsida
pteridophytes
BT Plantae

stilbite
Also search desmine.
UF desmine
BA zeolite group
framework silicates
silicates
BT minerals

Stillwater
City in Payne County in N central.
BT Oklahoma
United States

Stillwater Complex
Consists of four zones: basal, ultramafic zone, banded zone and upper zone.
BT Precambrian
SA Montana

stillwellite
BA orthosilicates
silicates
BT minerals

stilpnomelane
BA sheet silicates
silicates

BT minerals
SA aluminosilicates

Stirling Quartzite
Considered synonym for Prospect Mountain Quartzite. E California and SE Nevada. Also search Prospect Mountain Quartzite.
UF Prospect Mountain Quartzite
BT Precambrian
SA California
Nevada

stishovite
BA silica minerals
framework silicates
silicates
BT minerals

stochastic methods
use stochastic processes

stochastic models
use stochastic processes

stochastic processes
Includes use on level 3 under mathematical geology(1). Also search random process; random processes; stochastic methods; stochastic models.
UF random processes
stochastic methods
stochastic models
SA probability
processes

Stockholm
County. Also a city on the Baltic Sea.
BT Sweden

stocks
Includes use on level 2 under intrusions(1).
BA intrusions
SA batholiths
igneous rocks

stockwork deposits
Term introduced in 1978. Includes use on level 3 under commodities (list C) or under mineral deposits, genesis(1). Before 1978, search ore deposits AND stockwork, or search deposits AND stockwork.
UF network deposits
SA mineral deposits

stoichiometry
SA geochemistry

Stokes County
On the Virginia border in NW North Carolina.
BT North Carolina
United States

Stolonoidea
Includes use on level 2 under Graptolithina(1). See list F.
BA Graptolithina
BX Hemichordata
Invertebrata

stolzite
BA tungstates
BT minerals

Stomachorda
use Hemichordata

Stomiosphaera
BA Tintinnidae
Protista
BT Invertebrata

stone, building
use building stone

Stone Corral Formation
In Sumner Group. E Kansas.
BT Permian
SA Kansas

stone, dimension
use dimension stone

Stone Mountain
Huge gray granite monadnock in De Kalb County near Atlanta.
BT Georgia
United States

Stonehenge
Prehistoric assemblage of stones on the Salisbury Plain 7 miles N of Salisbury in S.
BT England
Great Britain
United Kingdom

Stoneley waves
Term introduced in 1981.
BA surface waves
elastic waves
SA Rayleigh waves
seismology
waves

stony irons
Includes use on level 3 under meteorites(1).
UF iron-stony meteorite
lithosiderite
sideraerolite
siderolite
BT meteorites
NA mesosiderites
pallasites
SA achondrites
chondrites
iron meteorites

Stony Tunguska River
Rises in SE Evenk National Okrug and flows WNW into mid course of the Yenisei River.
UF Middle Tunguska River
BT Russian Republic
USSR
SA Tunguska
Tunguska River

storage
As of 1978, term is used on level 2 under reservoirs(1).
SA data storage
drainage basins
gas storage
ground water
liquid waste
radioactive waste
reservoirs
soils
underground installations
waste disposal
water regimes
water storage

Stormberg Series
BT Upper Triassic
Triassic
SA South Africa

storms
Includes use on level 2 under meteorology(1) and geologic hazards(1).
SA atmosphere
climate
dust storms
geologic hazards
hurricanes
magnetic storms
meteorology
monsoons
precipitation
substorms
thunderstorms
winds

strain
Includes use on level 3 under deformation(1) and under engineering geology(1).
SA deformation
elastic limit
elastic strain
elasticity
engineering geology
finite strain analysis
Hooke's law
mechanical properties
Poisson's ratio
rock mechanics
seismology
shear
shear strength
shear zones
strainmeters
stress
structural analysis
viscoelasticity

strain relaxation
Term introduced in 1981.
SA seismology

strain seismographs
Term introduced in 1981.
BT seismographs

strain-slip cleavage
use slip cleavage

strainmeters
SA deformation
extensometers
geodesy
strain

Strait of Georgia
Channel between Vancouver Island on the W and the mainland of British Columbia and Washington on the E. Index British Columbia and/or Washington as applicable.
SA British Columbia
Washington

Strait of Gibraltar
Passage connecting Mediterranean Sea and the Atlantic Ocean between Spain and Morocco.
BT Atlantic Ocean
SA Gibraltar

Strait of Juan de Fuca
use Juan de Fuca Strait

Strait of Malacca
Channel between the S Malay Peninsula and the island of Sumatra connecting the Indian Ocean with the South China Sea. Also search Malacca Strait; Malacca Straits.
UF Malacca Strait
Malacca Straits
Straits of Malacca
BT Asia

Strait of Sicily
Between Sicily and Tunisia.
BT Mediterranean Sea

Straits of Florida
Wide channel between Florida Keys and Cuba connecting the Atlantic Ocean with the Gulf of Mexico. Also search Florida Strait; Florida Straits.
UF Florida Strait
Florida Straits
BT Atlantic Ocean

Straits of Malacca
use Strait of Malacca

Stramberk
Town SSW of Ostrava in NE.
BT Moravia
Czechoslovakia

Strandzha Mountains
use Istranca Mountains

Strasbourg
City on the Rhine River in Alsace.
UF Strassburg
BT Bas-Rhin
France

Strassburg
use Strasbourg

strata
use stratigraphic units

strata-bound deposits
use stratabound deposits

stratabound deposits
Term introduced in 1978. Includes use on level 3 under commodities (list C) or mineral deposits, genesis(1). Before 1978, search ore deposits AND stratabound; deposits AND stratabound; strata-bound.
UF strata-bound deposits
SA mineral deposits
stratiform deposits

Strathcona Mine
Near Edmonton in central.
BT Alberta
Canada

stratification
Includes use on level 3 under sedimentary structures(1) planar bedding structures(2). See list K.
BA planar bedding structures
sedimentary structures
SA bedding
cross-stratification
deposition
laminations

stratified volcanic cone
use stratovolcanoes

stratified volcano
use stratovolcanoes

stratiform deposits
Term introduced in 1978. Includes use on level 3 under commodities (list C) or mineral deposits, genesis(1). Before 1978, search ore deposits AND stratiform; deposits AND stratiform.
UF stratiform ore deposits
SA mineral deposits
stratabound deposits

stratiform ore deposits
use stratiform deposits

stratigraphic
A valid term through 1977. After 1977, use stratigraphic maps.

stratigraphic controls
Includes use on level 3 under mineral deposits, genesis(1) controls(2).
SA controls
lithologic controls
mineral deposits
structural controls

stratigraphic gaps
Term introduced in 1981.
SA stratigraphy
unconformities

stratigraphic geology
use stratigraphy

stratigraphic maps
Term introduced in 1978. Before 1978, search maps AND stratigraphic.
BA maps
SA structural maps

stratigraphic range
use range

stratigraphic traps
Includes use on level 3 under oil and gas fields(1).
UF lithologic traps
porosity traps
BA traps
SA natural gas
oil and gas fields
petroleum
stratigraphic wedges
structural traps

stratigraphic units
Term introduced in 1978. Includes use on level 3 under stratigraphy(1). Also search strata.

UF strata
units, stratigraphic
SA stratigraphy

stratigraphic wedges
Term introduced in 1981. Used for beds that pinch out.
SA lithofacies
lithostratigraphy
stratigraphic traps
stratigraphy

stratigraphy
Used for the discipline as a whole. See under age terms (list E). Includes use on level 1 (list A); on level 2 under area terms(1), bibliography(1), education(1), continental shelf(1), and symposia(1). If level 1, term set options are:
topic [applications, bibliography, catalogs, classification, concepts, education, history, methods, nomenclature, objectives, philosophy, practice, principles, textbooks]
subtopic
UF stratigraphic geology
NT magnetostratigraphy
SA bibliography
biogeography
biostratigraphy
boundary
catalogs
changes of level
chronostratigraphy
continental drift
correlation
diachronism
education
geochronology
glacial geology
glossaries
historical geology
index fossils
isopach maps
key beds
lexicons
lithostratigraphy
micropaleontology
paleogeography
paleomagnetism
paleontology
palynology
range
regression
sections
stratigraphic gaps
stratigraphic units
stratigraphic wedges
stratotypes
transgression
type sections
unconformities
uniformitarianism

stratosphere
Includes use on level 3 under aeronomy(1) composition(2).
SA aeronomy
atmosphere
ionosphere
troposphere

stratotypes
SA boundary
stratigraphy
type sections

stratovolcanoes
UF bedded volcano
composite cone
composite volcano
stratified volcanic cone
stratified volcano
BA volcanoes
SA shield volcanoes

Strawn Series
Comprises Millsap Lake and Lone Camp groups. Central and N Texas.
BT Middle Pennsylvanian
Pennsylvanian
SA Texas

stream action
Not a valid index term for GeoRef. See geomorphology(1) and sedimentation(1).

stream capture
Includes use on level 3 under geomorphology(1). Also search stream piracy; capture.
UF capture
piracy
river capture
river piracy
robbery
stream piracy
stream robbery
SA rivers
rivers and streams
streams

stream gradient
Term introduced in 1978. Before 1978, search streams AND gradient.
SA rivers and streams
streams

stream order
Includes use on level 3 under geomorphology(1) fluvial features(2).
UF channel order
SA drainage patterns
fluvial features
geomorphology
streams

stream piracy
use stream capture

stream robbery
use stream capture

stream sediments
Includes use on level 3 under mineral exploration(1). Also search stream sampling.
SA fluvial environment
fluvial features
mineral exploration
sediments
streams

stream transport
Includes use on level 3 under sedimentation(1). Also search fluvial transport.
UF fluvial transport
SA bedload
rivers
rivers and streams
sedimentation
streams
transport

streams
Includes use on level 3 under geomorphology(1). As of 1981, use fluvial environment for type of environment. Before 1981, included use on level 3 under sedimentation(1) for type of environment. Also search alluvial.
NA braided streams
ephemeral streams
SA bedload
canals
cascades
channels
efficiency
exsurgence
fluvial environment
fluvial features
hydrosphere
levees
meanders
rivers
rivers and streams
runoff
sediment yield
sinuosity
stream capture
stream gradient
stream order
stream sediments
stream transport
surface water
watersheds

strengite
BA phosphates
BT minerals

strength
SA compressive strength
deformation
fracture strength
mechanical properties
rock mechanics
shear strength
stress
tensile strength
yield strength

stress
Includes use on level 3 under deformation(1).
SA bearing capacity
creep
deformation
elasticity
extensometers
failures
finite strain analysis
Hooke's law
Poisson's ratio
pore pressure
pressure
release fractures
rigidity
rock mechanics
seismology
shear
shear strength
shear stress
strain
strength
structural analysis
tension
torsion
viscoelasticity
yield strength

stretch modulus
use Young's modulus

striations
Includes use on level 3 under sedimentary structures(1) bedding plane irregularities(2). See list K.
BA bedding plane irregularities
sedimentary structures
SA glacial features
glacial geology
grooves

strike
Includes use on level 3 under faults(1) and fractures(1). Term used to indicate strike faults through 1977. After 1977, use strike faults when discussing type of faults.
UF line of strike
SA bedding
dip
faults
folds
fractures
joints
orientation
strike faults
strike-slip faults

strike faults
Term introduced in 1978. Includes use on level 3 under faults(1) orientation(2). Before 1978, also search faults AND longitudinal; faults AND strike.
UF longitudinal faults
BA faults
SA strike

strike-shift faults
use strike-slip faults

strike-slip faults
Term introduced in 1978. Includes use on level 3 under faults(1) displacements(2). Before 1978, also search strike-slip.
UF strike-shift faults
BA faults
SA strike
transcurrent faults
transform faults

strip mining
Includes use on level 3 under mining geology(1) or land use(1).
BT surface mining
mining
SA coal
environmental geology
land use
mining geology
open-pit mining
quarries
reclamation

stromatolites
Includes use on level 2 under algae(1); on level 3 under sedimentary structures(1) biogenic structures(2). See list F (fossils) and list K (sedimentary structures).
BZ biogenic structures
algae
NA Collenia
SA algal mats
calcareous algae
oncolites
reef builders
sedimentary structures

Stromatoporoidea
Includes use on level 2 under Coelenterata(1). See list F.
BA Coelenterata
BT Invertebrata
SA problematic fossils

Stromboli
Active volcano on Stromboli Island of Lipari group in Tyrrhenian Sea.
BT Sicily
Italy

stromeyerite
BA sulfides
BT minerals

Strona Valley
River valley W of Lake Maggiore in N.
BT Italy
SA Piemonte

strong motion
Includes use on level 3 under seismology(1).
UF motion, strong
SA ground motion
seismology

strontianite
BA carbonates
BT minerals

strontium
Includes use on level 1 and 2 as a chemical element (list D). Also search Sr.
UF Sr
SA elements
isotopes
Sr-86
Sr-87
Sr-87/Sr-86
Sr-90
Sr/Rb

Strophomena
Genus. Includes use on level 3 under Brachiopoda(1) Strophomenida(2).
BA Strophomenida

Articulata
Brachiopoda
BT Invertebrata

Strophomenida
Includes use on level 2 under Brachiopoda(1). See list F.
BA Articulata
Brachiopoda
BT Invertebrata
NA Sowerbyella
Strophomena

structural
A valid term through 1977. After 1977, use structural maps.

structural analysis
Used for the analysis of structural features on a relatively small scale (from thin section to outcrop). The analysis may lead to interpretation of larger-scale features such as folds, fractures, and faults, or tectonics. Includes use on level 1 (list A). Before 1976, also search fabric analysis; structural petrology; microtectonics. Term set options are:
concepts
subtopic
experimental studies
subtopic
faults
subtopic
folds
subtopic
foliation
subtopic
fractures
subtopic
interpretation
topic [e.g. axial-plane structures, boudinage, cleavage, elongate minerals, fold axes, folds, foliation, fractures, joints, laminations, layering, linear deformation, lineation, melange, mullions, petrofabrics, planar deformation, preferred orientation, schistosity, slickensides, etc...]
lineation
subtopic
methods
subtopic [e.g. electron microscopy, universal stage, X-ray analysis]
preferred orientation
subtopic
principles
subtopic
theoretical studies
subtopic
UF petrofabric analysis
petrogeometry
petromorphology
SA analysis
axial-plane structures
boudinage
breccia
cleavage
deformation
dip
electron microscopy
elongate minerals
engineering geology
fabric
faults
flow cleavage
fold axes
folds
foliation
fractures
geometry
geophysics
joints
lineation
melange
mullions
orientation
petrofabrics
preferred orientation
rock mechanics
schistosity
shear zones
strain
stress
structural geology
tectonics
tectonophysics
universal stage
X-ray analysis

structural basins
use basins

structural complexes
No longer a valid term for GeoRef as of 1981. Use complexes. Included use on level 3 under tectonics(1).

structural controls
Includes use on level 3 under mineral deposits, genesis(1) controls(2). As of 1981, restricted to economic geology.
SA controls
mineral deposits
sedimentation
stratigraphic controls
tectonic controls

structural features
A valid term through mid-1978, used under geomorphology, volcanics, structural geology, and igneous rocks.

structural geology
Used for the discipline as a whole. Includes use on level 1 (list A); on level 2 under area terms(1). If 1, term set options are:
topic [applications, bibliography, catalogs, classification, concepts, education, experimental studies, history, instruments, methods, nomenclature, philosophy, practice, principles, textbooks, theoretical studies]
subtopic
SA allochthons
bibliography
catalogs
faults
folds
foliation
fractures
geology
glossaries
lexicons
lineation
maps
orogeny
structural analysis
tectonics

structural maps
Term introduced in 1978. Before 1978, search maps AND structural.
UF structure maps
BA maps
SA stratigraphic maps
structure contour maps

structural petrology
A valid term through 1974. After 1974, use structural analysis.

structural traps
BA traps
SA natural gas
petroleum
stratigraphic traps

structure
As of 1981, no longer a valid term for GeoRef. Included use as a level 2 and 3 term appropriate to a large number of topics e.g. on level 2 under crust(1), tectonics(1), and tectonophysics(1).

structure contour maps
BA maps
SA contour maps
structural maps
tectonic maps

structure maps
use structural maps

structures
As of 1978, used on level 2 under foundations(1). Term is restricted to engineering geology.
SA axial-plane structures
banded structures
biogenic structures
buildings
collapse structures
cylindrical structures
deep-seated structures
flame structures
flow structures
foundations
kink-band structures
primary structures
ring structures
secondary structures
sedimentary structures
slump structures
turbidity current structures

Struma River valley
Index countries as applicable. Also search Struma; Struma River.
UF Struma Valley
BT Europe
SA Bulgaria
Greece

Struma Valley
use Struma River valley

Strunian
Term introduced in 1981. Includes use on level 3 under age terms(1). See list E.
BA Upper Devonian
Devonian
BT Paleozoic

struvite
BA phosphates
BT minerals

Strzegom
City in Lower Silesia in central.
BT Wroclaw
Poland

Strzegom-Sobotka Granitoid Massif
use Strzegom-Sobotka Massif

Strzegom-Sobotka Massif
Granitoid massif in Lower Silesia. Also search Strzegom-Sobotka.
UF Strzegom-Sobotka Granitoid Massif
BT Wroclaw
Poland

Strzelin
City of Lower Silesia in SE.
BT Wroclaw
Poland

Stubai Alps
A NE group of Otztal Alps in central Tyrol. Also search Stubai.
UF Stubai Massif
Stubai Mountains
BT Tyrol
Austria
SA Alps
Otztal Alps

Stubai Massif
use Stubai Alps

Stubai Mountains
use Stubai Alps

studies, areal
use areal studies

studies, case
use case studies

studies, experimental
use experimental studies

studies, faunal
use faunal studies

studies, feasibility
use feasibility studies

studies, field
use field studies

studies, floral
use floral studies

studies, laboratory
use laboratory studies

studies, theoretical
use theoretical studies

Stump Formation
Overlies Preuss Sandstone and underlies Bechler and Ephraim conglomerates undifferentiated. SE Idaho and W Wyoming. Also search Stump Sandstone.
UF Stump Sandstone
BT Upper Jurassic
Jurassic
SA Idaho
Wyoming

Stump Sandstone
use Stump Formation

Sturovo
Town on left bank of Danube River in SW.
BT Slovakia
Czechoslovakia

Stutsman County
E central.
BT North Dakota
United States

Stuttgart
City in Arkansas, and a city in Baden-Wurttemberg. Index Arkansas and/or West Germany as applicable.
SA Arkansas
West Germany

style
Includes use on level 2 under folds(1), foliation(1), fractures(1) and lineation(1).
SA folds
foliation
fractures
lineation
orientation

stylolites
Includes use on level 3 under sedimentary rocks(1) diagenesis(2); on level 3 under sedimentary structures(1) secondary structures(2); on level 3 under sediments(1) and sedimentation(1). See list K.
BA secondary structures
sedimentary structures
SA microstylolites

Stylommatophora
Term introduced in 1981. Includes use on level 2 under Mollusca(1). See list F.
BA Gastropoda
Mollusca
BT Invertebrata

Stylophora
Includes use on level 2 under Echinodermata(1). See list F.
BA Homalozoa
Echinodermata
BT Invertebrata

Styria
State in SE.
CO N463500N474500
E0161500E0133000
BT Austria

NT Graz
Hartberg
SA Salzkammergut
Semmering
Totes Gebirge
Wechsel

suanite
BA borates
BT minerals

subaerial environment
Term introduced in 1978. Before 1978 search subaerial.
SA environment

subantarctic regions
Term introduced in 1978. Pertaining or relating to the regions immediately outside of the Antarctic circle. Before 1978, also search subantarctic.
SA Antarctic Ocean
Antarctica

Subarctic
use Preboreal

subarctic regions
Term introduced in 1978. Pertaining or relating to the regions immediately outside of the Arctic circle or to areas that have characteristics such as climate, vegetation, and animals similar to these regions. Before 1978, also search subarctic.
SA Arctic region

subarkose
Includes use on level 3 under sedimentary rocks(1) clastic rocks(2). See list I.
BA clastic rocks
BT sedimentary rocks
SA arkose

Subbetic Zone
A geographic term with stratigraphic-tectonic connotations.
BT Spain
SA Betic Cordillera
Prebetic Zone

Subboreal
Europe. A term used primarily in Europe for an interval of postglacial time following the Atlantic and preceding the Subatlantic.
BA Holocene
Quaternary
BT Cenozoic

Subcarpathians
Sub ranges of the Carpathians. Index Ukraine and countries as applicable.
BT Europe
SA Carpathian Foreland
Carpathians
Czechoslovakia
Hungary
Poland
Romania
Ukraine

subduction
Includes use on level 2 under plate tectonics(1); on level 3 under tectonophysics(1).
SA Benioff zone
marginal basins
marginal seas
obduction
plate tectonics
subduction zones
tectonophysics
trenches

subduction zones
Includes use on level 3 under plate tectonics(1) and tectonophysics(1).
UF zones, subduction
SA Benioff zone
plate tectonics
plates
subduction
tectonophysics
trenches

subgraywacke
Includes use on level 3 under sedimentary rocks(1) clastic rocks(2). See list I.
BA clastic rocks
BT sedimentary rocks
SA graywacke
sandstone
terrigenous materials

Sublette County
W Wyoming.
BT Wyoming
United States

sublimates
SA fumaroles
gases
solfataras
sublimation
volcanism
volcanoes

sublimation
SA evaporation
geochemistry
mineral deposits
sublimates

sublittoral environment
Term introduced in 1978. Before 1978, search sublittoral. Before 1981, also search neritic; neritic environment.
UF neritic environment
SA environment

submarine
A valid term through 1977. After 1977, use submarine environment.

submarine canyons
Includes use on level 2 under ocean floors(1). Before 1976, also search canyons for submarine canyons.
UF submarine valleys
SA bottom features
canyons
continental margin
continental shelf
continental slope
marine geology
ocean floors
submarine fans
turbidity currents

submarine cone
use submarine fans

submarine delta
use submarine fans

submarine dunes
Term introduced in 1981. Includes use on level 3 under ocean floors(1).
BA dunes

submarine environment
Term introduced in 1978. Before 1978, search submarine.
SA environment
marine environment

submarine fans
Also search deep sea fans; sea fan.
UF abyssal cones
abyssal fans
deep-sea fans
sea fan
submarine cone
submarine delta
SA ocean floors
submarine canyons
turbidity currents

submarine geology
Not a valid index term for GeoRef. See marine geology, oceanography, ocean floors, and ocean basins.

submarine installations
As of 1978, term is used on level 2 under marine installations(1). Term used only on level 3 from 1976 through 1977.
UF installations, submarine
BT marine installations
SA engineering geology
tunnels

submarine valleys
use submarine canyons

submarine weathering
use halmyrolysis

submergence
SA changes of level

submersibles
SA FAMOUS
oceanography

subsequent folds
use superposed folds

subsidence
As of 1978, includes use on level 2 under neotectonics(1). Also includes use on level 3 under tectonics(1).
SA land subsidence
neotectonics
tectonics
uplifts
vertical movements

subsidies
Term introduced in 1981.
SA economics

substitution
Includes use on level 3 under crystal chemistry(1).
SA crystal chemistry
lattice
order-disorder

substorms
Includes use on level 3 under magnetosphere(1) magnetic storms(2).
SA disturbances
magnetic storms
magnetosphere
storms

substrates
Includes use on level 3 under appropriate fossil group (list F).

subsurface
A valid term through mid-1978 under engineering geology(1). See subsurface reservoirs.

subsurface reservoirs
As of 1981, no longer a valid term for GeoRef. See petroleum engineering; reservoir rocks. Included use on level 3 under engineering geology(1) petroleum engineering(2). Before 1978, search reservoirs AND subsurface.

subtidal environment
Term introduced in 1978. Before 1978, search subtidal.
SA environment

subtropical environment
Term introduced in 1978. Before 1978, search subtropical.
UF semitropical environment
SA climate
ecology
environment
paleoclimatology
tropical environment

subways
Term introduced in 1978 on level 2 under tunnels(1).
SA tunnels

Suceava
Town in Suceava County in N.
BT Moldavia
Romania

Suchan Basin
Coal basin in Primorye Kray 60 miles ENE of Vladivostok. Also search Suchan.
BT Russian Republic
USSR

Sucre
A department in N Colombia, a state in N Venezuela and a city in S central Bolivia. Index countries as applicable.
BX Bolivia
Colombia
Venezuela
BT South America
SA Araya Peninsula
Cumana
Paria Peninsula

Sudan
Includes use on level 1 as an area term (list O). For term set options see list B.
CO N030000N220000
E0383000E0220000
BA East Africa
Africa
NT Darfur
Gezira
Kordofan
Malakal
SA Blue Nile
Kapoeta Meteorite
Lake Turkana
Libyan Desert
Near East
Nile River
Nile Valley
Nubia
Nubian Sandstone
Nubian Shield
Red Sea Basin
Sahara

Sudbury
Mining city N of Georgian Bay in SE.
CO N463000N463000
W0810100W0810100
BT Ontario
Canada

Sudbury Basin
Mining basin N of Georgian Bay in SE Ontario. Also search Sudbury.
BT Ontario
Canada

Sudbury Irruptive
An intrusive region around Sudbury in SE Ontario.
UF Sudbury Nickel Irruptive
BT Ontario
Canada

Sudbury Nickel Irruptive
use Sudbury Irruptive

sudden commencements
Includes use on level 3 under magnetosphere(1) magnetic storms(2).
UF commencements, sudden
SA disturbances
intensity
magnetic storms
magnetosphere

Sudeten
Region. All the borderlands of Bohemia and Moravia formerly inhabited by German speaking people (Sudeten Germans). Index Bohemia and/or Moravia as applicable.
UF Sudetenland
BT Czechoslovakia
SA Bohemia
Moravia

Sudeten Mountains
Ranges between NW Czechoslovakia and SW Poland. Also search Sudeten and Sudetes. Index Poland and Czechoslovak regions as applicable.
CO N495000N510500

E0180000E0145000
UF Sudetes
Sudetes Mountains
Sudetic Mountains
Sudety Mountains
BT Europe
NZ Czech Sudeten Mountains
Polish Sudeten Mountains
SA Bohemia
Isergebirge
Izera Mountains
Karkonosze Mountains
Moravia
North Sudetic Basin
Poland
Sowie Mountains

Sudetenland
use Sudeten

Sudetes
use Sudeten Mountains

Sudetes Mountains
use Sudeten Mountains

Sudetic Basin
In the Sudeten Mountains. Index countries as applicable. Also search Sudetic.
UF Sudetic Depression
BT Europe
SA Czechoslovakia
Moravia
North Sudetic Basin
Poland

Sudetic Depression
use Sudetic Basin

Sudetic Mountains
use Sudeten Mountains

Sudety Mountains
use Sudeten Mountains

sudoite
BA chlorite group
sheet silicates
silicates
BT minerals
SA aluminosilicates

suevite
SA breccia
cryptoexplosion features
impact features
impactite
metamorphic rocks
meteorites

Suez (Canal)
use Suez Canal

Suez Canal
Sea level canal crossing the Isthmus of Suez between the E Mediterranean Sea and the Gulf of Suez. Also search Suez.
UF Suez (Canal)
BT Egypt
Africa

Suffield
Town in Hartford County in N.
BT Connecticut
United States

Suffolk
City in SE.
CO N515500N523500
E0014500W0003000
BT England
Great Britain
United Kingdom

Suffolk County
Index states as applicable.
BX Massachusetts
New York
BT United States

sugars
BA organic materials

Suhl Bezirk
Term introduced in 1981. District in SW.
CO N501100N505400
E0112000E0095200
BT East Germany

Suidae
As of 1981, includes use on level 3 under Mammalia(1) Suiformes(2). See list F. Before 1981, included use on level 3 under Mammalia(1) Artiodactyla(2).
BA Suiformes
Artiodactyla
Mammalia
BT Tetrapoda
Vertebrata

Suiformes
Term introduced in 1981. Includes use on level 2 under Mammalia(1). See list F.
BA Artiodactyla
Mammalia
BT Tetrapoda
Vertebrata
NA Suidae

Sulaiman Range
W of the Indus River. Index provinces as applicable.
BT Pakistan
SA Baluchistan
North-West Frontier Province

Sulawesi
use Celebes

Sulcis
Region in extreme SW.
BT Sardinia
Italy

sulfantimonates
BA sulfosalts
BT minerals
NA cylindrite
famatinite

sulfantimonites
BA sulfosalts
BT minerals
NA boulangerite
bournonite
chalcostibite
freibergite
freieslebenite
NZ geocronite
NA heteromorphite
jamesonite
kobellite
meneghinite
miargyrite
plagionite
polybasite
pyrargyrite
semseyite
stephanite
tetrahedrite
NZ vrbaite
NA zinckenite

sulfarsenates
Term introduced in 1981.
BA sulfosalts
BT minerals
NA enargite

sulfarsenites
BA sulfosalts
BT minerals
NA dufrenoysite
NZ geocronite
NA gratonite
jordanite
pearceite
proustite
seligmannite
tennantite
NZ vrbaite

sulfate ion
Term introduced in 1981.
SA geochemistry
ions
sulfates

sulfates
Includes use on level 2 under minerals(1). See list L.
BT minerals
NA alabaster
alum
alunite
anglesite
anhydrite
barite
bassanite
NZ beudantite
NA brochantite
celestite
chalcanthite
copiapite
coquimbite
epsomite
ettringite
glauberite
gypsum
halotrichite
hexahydrite
NZ hinsdalite
NA jarosite
jouravskite
NZ kainite
NA kieserite
langbeinite
NZ lazurite
NA melanterite
mirabilite
pickeringite
polyhalite
posnjakite
roemerite
rozenite
selenite
svanbergite
NZ thaumasite
NA thenardite
voltaite
NZ wenkite
SA hauyne
meionite
sodium sulfate
sulfate ion

sulfides
Includes antimonides, arsenides, bismuthides, oxysulfides, selenides, and tellurides. Includes use on level 2 under minerals(1). See list L.
UF sulphides
BT minerals
NA acanthite
aikinite
alabandite
alloclasite
anilite
antimonides
argentite
arsenides
arsenosulfides
berndtite
betekhtinite
bismuthides
bismuthinite
bohdanowiczite
bornite
bravoite
briartite
carrollite
cattierite
chalcocite
chalcopyrite
cinnabar
copper sulfides
covellite
cubanite
digenite
djurleite
galena
NZ genthelvite
NA gersdorffite
greenockite
greigite
gudmundite
heazlewoodite
NZ helvite
NA herzenbergite
idaite
iron sulfides
jordisite
kesterite
NZ lazurite
NA linnaeite
mackinawite
marcasite
mawsonite
metacinnabar
metastibnite
millerite
molybdenite
niningerite
NZ onofrite
NA orpiment
oxysulfides
pentlandite
polydymite
pyrite
pyrrhotite
realgar
roquesite
NZ schreibersite
NA selenides
sinnerite
smythite
sphalerite
stannoidite
sternbergite
stibnite
stromeyerite
talnakhite
tellurides
troilite
tungstenite
ullmannite
ultrabasite
vaesite
valleriite
violarite
willyamite
wurtzite
SA hydrogen sulfide

sulfobismuthites
Includes use on level 3 under minerals(1) sulfosalts(2). See list L.
BA sulfosalts
BT minerals
NA cosalite
emplectite
galenobismutite
hammarite
lillianite
matildite
wittichenite

sulfogermanates
BA sulfosalts
BT minerals
NZ canfieldite
NA germanite
renierite

sulfosalts
Includes sulfantimonates, sulfantimonites, sulfarsenates, sulfarsenites, sulfobismuthites, sulfogermanates, sulfostannates, sulfovanadates. Includes use on level 2 under minerals(1). See list L.
BT minerals
NA berryite
gustavite
luzonite
sulfantimonates
sulfantimonites
sulfarsenates
sulfarsenites
sulfobismuthites
sulfogermanates
sulfostannates

sulfovanadates

sulfostannates
BA sulfosalts
BT minerals
NZ canfieldite
NA stannite

sulfovanadates
BA sulfosalts
BT minerals
NA sulvanite

sulfur
Includes use on level 1 and 2 as a chemical element (list D). As of 1981, use sulfur deposits for sulfur as a commodity. Before 1981, included use on level 1 as a commodity term.
UF S
sulphur
SA elements
isotopes
native elements
S-32
S-34
S-34/S-32
stable isotopes
sulfur deposits

sulfur deposits
Term introduced in 1981. Includes use on level 1 and 2 as a commodity term (list C). Before 1981, search sulfur AND deposits.
SA native elements
pyrite ores
sulfur

sulfur dioxide
UF SO2
SA atmosphere
pollution

sulfuric acid
UF H2SO4
SA acid mine drainage
acids
pollution
waste disposal

Sulina
Town on the Black Sea in Tulcea County in NE.
BT Dobruja
Romania

Sulitjelma
Peak in Kjolen Mountains on Norwegian-Swedish border in N part of Scandinavian Peninsula. Index countries as applicable.
BT Europe
SA Norway
Sweden

Sullivan County
Index states as applicable.
BX Indiana
Missouri
New Hampshire
New York
Pennsylvania
Tennessee
BT United States

sulphides
use sulfides

sulphur
use sulfur

Sulu Sea
Large interisland sea between the Philippine Islands and NE Borneo with Palawan Island on the NW and the Sulu Archipelago on the SW. Also search Sulu.
BT Pacific Ocean

sulvanite
BA sulfovanadates
sulfosalts
BT minerals

Sumatra
One of the islands of the Malay Archipelago SE and S of the Malay Peninsula. The second largest Indonesian island.
CO S060000N054500
E1061000E0950500
BT Indonesia
NT Toba Lake

Summit County
Index states as applicable.
BX Colorado
Ohio
Utah
BT United States

Sumner County
Index states as applicable.
BX Kansas
Tennessee
BT United States

Sun
Includes use on level 1. See entry under Moon(1) for term set options.
SA astrophysics and solar physics
corona
Moon
planetology
planets
solar activity
solar cycles
solar energy
solar flares
solar nebula
solar system
solar wind

Sunda Shelf
The continental shelf between Borneo and Java.
UF Borneo-Java Shelf
BT Java Sea
Pacific Ocean

Sundance Formation
Includes Canyon Springs Sandstone, Stockade Beaver Shale, Hulett Sandstone, Lak, and Redwater Shale members. Central N Colorado, central S Montana, NW Nebraska, SW South Dakota, and SW Wyoming.
BT Upper Jurassic
Jurassic
SA Colorado
Montana
Nebraska
South Dakota
Wyoming

Sunlight
Peak in SW Colorado and mountain in NW Wyoming. Index states as applicable.
BT United States
SA Colorado
Wyoming

Sunqur
use Songhor

Sunrise Formation
Overlies Gabbs Formation; unconformably underlies Dunlap Formation. SW Nevada.
BT Lower Jurassic
Jurassic
SA Nevada

sunspots
As of 1978, term is used on level 2 under geologic hazards(1).
SA astrophysics and solar physics
geologic hazards

Sunzha
Mountain range. N outlier of the central Greater Caucasus in Northern Caucasus.
BT Russian Republic
USSR
SA Greater Caucasus

Supai Formation
In Aubrey Group. Includes Esplanade Sandstone Member, Fort Apache Limestone, Kinishba Beds, Amos Wash Member, Big "A" Member, Apache Member, Corduroy Member, Packard Member, Oak Creek Member. N Arizona, E California, E Nevada, W New Mexico, and S Utah.
BT Paleozoic
SA Arizona
California
Nevada
New Mexico
Pennsylvanian
Permian
Utah

supergene processes
Includes use on level 3 under mineral deposits, genesis(1) processes(2).
SA enrichment
mineral deposits
processes

superimposed folds
use superposed folds

superimposed metamorphism
use polymetamorphism

Superior Province
Index Minnesota and provinces as applicable.
CO N460000N614000
W0660000W0990000
BT Canadian Shield
NT Abitibi Belt
English River Belt
SA Labrador
Manitoba
Minnesota
Ontario
Quebec

superposed folds
Term introduced in 1978. Includes use as orientation of folds relative to spatially associated macroscopic structures such as large folds, fold systems, and orogenic zones. Includes use on level 3 under folds(1) orientation(2). Before 1978, also search folds AND superposed; folds AND cross.
UF cross folds
subsequent folds
superimposed folds
transverse folds
BA folds
SA orientation

Superstition Mountains
S central Arizona. Also search Superstition.
BT Arizona
United States

superstructure
SA crystal structure

supply
Term introduced in 1981.
SA economics

supply, water
use water supply

supralittoral environment
use supratidal environment

supratidal environment
Term introduced in 1978. Before 1978, search supratidal.
UF supralittoral environment
SA ecology
environment
sebkha environment
tidal flats

Sur fault zone
In the Point Sur area in W California. Also search Sur-Nacimiento fault zone.
UF Sur-Nacimiento fault zone
BT California
United States

Sur-Nacimiento fault zone
use Sur fault zone

Surakhany
Town on central Apsheron Peninsula in oil fields near Baku. Also search Surakhany oil field.
BT Azerbaidzhan
USSR

Surat Basin
Sedimentary basin in SE.
CO S290000S270000
E1510000E1470000
BT Queensland
Australia

surface
A valid term through 1977. See surface reservoirs. Use surficial geology for maps.

surface features
Also see under individual planets, e.g. Mars(1).
UF features, surface
SA geomorphology
Moon
planetology
planets
surface properties

surface mining
Term introduced in 1981. Includes use on level 3 under land use(1) and mining geology(1).
BT mining
NT open-pit mining
strip mining

surface phenomena
Includes use on level 2 under astrophysics and solar physics(1).
UF phenomena, surface
SA astrophysics and solar physics

surface properties
For photographic characteristics, electromagnetic responses, etc. Includes use on level 2 under Moon(1).
SA electrical properties
Moon
properties
spectroscopy
surface features

surface reservoirs
As of 1981, no longer a valid term for GeoRef. Use reservoirs. Included use on level 3 under engineering geology(1). Term was introduced in 1978. Before 1978, also search surface AND reservoirs.

surface water
Includes use on level 3 under environmental geology(1). As of 1981, includes use on level 2 under pollution(1).
SA environmental geology
floods
ground water
hydrochemistry
hydrology
hydrosphere
irrigation
lakes
multiphase flow
pollution
ponds
rivers
runoff
streams
swamps
water
water hardness
water resources
water supply

watersheds

surface waves
Includes use on level 3 under seismology(1).
UF L waves
BA elastic waves
NA coda waves
Love waves
Rayleigh waves
Stoneley waves
SA Airy waves
body waves
earthquakes
engineering geology
ground motion
seismology
waves

surfaces, erosion
use erosion surfaces

surficial geology
Not valid term through 1977. After 1977, includes use on level 3 under geomorphology(1) and maps(1). Before 1978, also search maps AND surficial; search surface AND maps.
SA geology
geomorphology
maps
soils

surges
As of 1978, term is restricted for use with ocean waves. Before 1978, term was also used for glacier surges.
SA base surges
glacier surges
ocean waves
waves

Surgut
Town on right bank of Ob River in S Khanty-Mansino National Okrug in W Siberia.
BT Russian Republic
USSR

Surinam
Includes use on level 1 as an area term (list O). For term set options see list B.
CO N015000N060000
W0535900W0580500
UF Dutch Guiana
Netherlands Guiana
DA South America
NT Wilhelmina Mountains
SA Guiana Basin
Guianas
Guyana Shield

Surkhan Darya
River which rises in the Hissar Range of Tadzhikistan and flows SSW into the Amu Darya near Termez. Index Soviet republics as applicable.
BT USSR
SA Tadzhikistan
Uzbekistan

Surkhan Darya basin
River basin, primarily in Surkhan Darya Oblast in S Uzbekistan. Index Soviet republics as applicable.
UF Surkhan-Darya basin
BT USSR
SA Tadzhikistan
Uzbekistan

Surkhan-Darya basin
use Surkhan Darya basin

Surrey
County SW of London.
CO N511000N513000
E0001000W0004500
BT England
Great Britain
United Kingdom

Surry County
Index states as applicable. Also search Surry.
BX North Carolina
Virginia
BT United States

Surtsey
Volcanic island off S.
BT Iceland

Suruga Bay
Inlet of Pacific Ocean SW of Yokohama on SE coast.
BT Honshu
Japan

survey organizations
Term introduced in 1981. Includes use on level 1 (list A). Used for the work of geological surveys, national or local. Before 1981, search *surveys. Term set options are:
topic [annual report, current research, history, organization, research]
name of survey [e.g. U. S. Geological Survey, etc.]
UF geologic surveys
SA annual report
associations
bibliography
catalogs
current research
ENADIMSA
Geological Survey of Canada
history
Illinois State Geological Survey
museums
organization
progress report
publications
research
surveys
U. S. Geological Survey

Surveyor Program
Term introduced in 1981. Includes use on level 3 under Moon(1).
SA extraterrestrial geology
Moon
planetology
remote sensing

Surveyor 3
Includes use on level 3 under Moon(1). Also search Surveyor III.
SA Moon
remote sensing

surveys
As of 1981, restricted to actual surveying and its results. Before 1981, also used on level 1 for the work of geological surveys. See survey organizations. Includes use on level 2 under soils(1), ground water(1), hydrology(1), geochemistry(1), and geodesy(1); and on level 3 in area sets under geophysical surveys(2).
NT geophysical surveys
SA exploration
geochemistry
geodesy
ground water
hydrology
maps
soils
survey organizations
triangulation

susceptibility
A valid term through 1977. Use magnetic susceptibility.

Susitna River basin
N of Anchorage in S Alaska. Also search Susitna; Susitna River.
BT Alaska
United States

suspended materials
Before 1978, search materials AND suspended.
SA dissolved materials
flocculation
materials
sea water

suspension
SA geochemistry
hydrology
sediments
transport

suspension current
use turbidity currents

Susquehanna River
Rises in Otsego Lake in central New York and flows S emptying into N Chesapeake Bay. Index states as applicable.
BT United States
SA Maryland
New York
Pennsylvania

Susquehanna River basin
Primarily in Pennsylvania. Index states as applicable. Also search Susquehanna Valley.
UF Susquehanna Valley
BT United States
SA Maryland
New York
Pennsylvania

Susquehanna Valley
use Susquehanna River basin

Sussex
County on English Channel. Divided into two administrative counties: East and West Sussex.
CO N504500N511500
E0004500W0010000
BT England
Great Britain
United Kingdom

Sussex County
Index states as applicable.
BX Delaware
New Jersey
Virginia
BT United States

Sustut Basin
River basin in N central.
BT British Columbia
Canada

Sutam River
S Yakutia near N border of Amur Oblast in SE Siberia. Also search Sutam.
BT Russian Republic
USSR

Sutherland
County in highlands of extreme N.
CO N575000N583000
W0033500W0052500
BT Scotland
Great Britain
United Kingdom

Sutna
use Satna

Sutton County
In broken uplands of Edwards Plateau in W Texas. Also search Sutton.
BT Texas
United States

sutures
SA fossils
morphology

Sutwik Island Quadrangle
Term introduced in 1981. S Alaska.
CO N560000N570000
W1560000W1580000
BT Alaska
United States

Suva
Town with one of best harbors in South Pacific on SE coast of Viti Levu Island.
BT Fiji

Suvalkai
use Suwalki

Suvalki
use Suwalki

Suwalki
Region E of the Masurian Lakes in extreme NE part of country. Also a city.
UF Suvalkai
Suvalki
BT Bialystok
Poland

Suwannee Limestone
Upper part of formation considered to be equivalent to lower Chickasawhay Marl of E Mississippi, and lower part possibly equivalent of Byram Formation and Marianna Limestone of Florida. S central Georgia, and E Florida.
BT upper Oligocene
Oligocene
SA Florida
Georgia

Suzak
Village in central.
BT Kirghizia
USSR

SV-waves
Term introduced in 1981.
BA S-waves
body waves
elastic waves
SA seismology
waves

Svalbard
Norwegian Island group including Spitsbergen group and Bear Island.
BT Arctic region
NT Bear Island
NA Spitsbergen
SA Arctic Ocean

svanbergite
BA sulfates
BT minerals

Svanetia
Region on S slopes of the Greater Caucasus in NW.
BT Georgian Republic
USSR

Svecofennian
Europe. See list E.
BA middle Precambrian
Precambrian

Sventoji River
Central Lithuania. Also search Sventoji.
UF Shventoyi River
BT Lithuania
USSR

Sverdlovsk
City and oblast in foothills of the Central Urals in W Siberia.
UF Ekaterinburg
Yekaterinburg
BT Russian Republic
USSR

Sverdrup Basin
In the Sverdrup Channel area of the Sverdrup Islands in N District of Franklin.
BT Northwest Territories
Canada

Svinita
Village in W part of country in S.
BT Banat
Romania

Svratka
Village in W central.
BT Moravia
Czechoslovakia

Swabia
Region. Duchy of medieval Germany. Nearly coextensive with Baden-Wurttemberg, Hesse, and W Bavaria. Index states as applicable.
BT West Germany
SA Baden-Wurttemberg
Bavaria
Hesse

Swabian Alb
Mountain range between the Neckar and the upper Danube rivers.
UF Swabian Jura
BT Baden-Wurttemberg
West Germany

Swabian Jura
use Swabian Alb

swamps
As of 1981, use paludal environment for type of environment.
SA bogs
brackish-water environment
ecology
geomorphology
mangrove swamps
marshes
paludal environment
peat bogs
sedimentation
surface water

Swan Hills
Hilly region S of Lesser Slave Lake in central.
BT Alberta
Canada

Swan Peak Formation
Includes Watson Ranch Tongue. SE Idaho and NE Utah.
BT Middle Ordovician
Ordovician
SA Idaho
Utah

Swansea
City on Bristol Channel in S.
BT Wales
Great Britain
United Kingdom

swarms
As of 1981, restricted to earthquakes. Includes use on level 3 under earthquakes(1).
SA dike swarms
earthquakes
intrusions
ring dikes
seismology

Swartkrans
Mountains near Lesotho border in SE.
BT Orange Free State
South Africa

Swat
District in the valley of the Swat River NNE of Peshawar. Former princely state.
BT North-West Frontier Province
Pakistan

Swauk Formation
Unconformably overlies metamorphic basement of unknown age. Central and N central Washington.
BT Eocene
SA Washington

Swaziland
Borders on Mozambique and South Africa. Administered by a British High Commissioner until independence in 1968. Includes use on level 1 as an area term (list O). For term set options see list B.
CO S273000S260000
E0321500E0310000
BA Southern Africa
Africa
SA Swaziland Sequence
Swaziland System

Swaziland Sequence
In South Africa, and Swaziland.
BT Precambrian
SA South Africa
Swaziland

Swaziland System
In South Africa and Swaziland.
BT Precambrian
SA South Africa
Swaziland

Sweden
Includes use on level 1 as an area term (list O). For term set options see list B.
CO N551500N691500
E0241500E0110000
BA Scandinavia
Western Europe
Europe
NT Alno
Blekinge
Dalarna
Gota Valley
Goteborg
Gotland
Halland
Jamtland
Kalmar
Kristianstad
Malmohus
Norrbotten
Nykoping
Orebro
Siljan
Skane
Skellefte
Smaland
Stockholm
Uppsala
Varmland
Vasterbotten
SA Arctic region
Baltic region
Baltic Shield
Fennoscandia
Lapland
North Sea Coast
Oresund
Sparagmite Group
Sulitjelma

Sweetgrass Arch
Index Alberta and/or Montana as applicable.
SA Alberta
Montana

Sweetwater County
SW Wyoming.
CO N410000N422000
W1073000W1101000
BT Wyoming
United States
NT Lake Gosiute

swelling soils
use expansive materials

Swietokrzyskie Mountains
use Swiety Krzyz Mountains

Swiety Krzyz Mountains
Mountain group between the Vistula and Pilica rivers in SE central part of country. Also search Holy Cross Mountains; Swiety Krzyz.
CO N503500N510000
E0213000E0202000
UF Holy Cross Mountains
Swietokrzyskie Mountains
BT Poland
SA Galezice

Swiss Alps
Those ranges of the Alps mountain system which lie within the country.
BX Switzerland
BZ Alps
NA Central Swiss Alps
Eastern Swiss Alps
Northern Swiss Alps
Southern Swiss Alps
SA Oberalp Pass
Tavetsch

Swiss Jura Mountains
Term introduced in 1981.
BZ Jura Mountains
BX Switzerland

Swiss Molasse Basin
Term introduced in 1981.
CO N460700N474800
E0094100E0055600
BZ Molasse Basin
BX Switzerland

Swiss Rhine Graben
Term introduced in 1981.
BZ Rhine Graben
BX Switzerland

Switzerland
Includes use on level 1 as an area term (list O). For term set options see list B.
CO N454500N474500
E0103000E0055000
BA Central Europe
Europe
NT Aar Massif
Aar Valley
Aargau
Basel
Bergell Massif
Bern
Bernese Alps
Geneva
Glarus
Glarus Alps
Gotthard Massif
Graubunden
Lake of Zurich
Lucerne
Neuchatel
Oberalp Pass
Saint Gall
Schaffhausen
NX Swiss Alps
Swiss Jura Mountains
Swiss Molasse Basin
Swiss Rhine Graben
NT Thur Valley
Ticino
Uri
Valais
Vaud
Zurich
SA Alps
Arve Valley
Ceneri Zone
Central Alps
Inn Valley
Jura Mountains
Lago Maggiore
Lake Constance
Lake Geneva
Lepontine Alps
Molasse Basin
Monte Rosa
Pennine Alps
Rhaetian Alps
Rhine Basin
Rhine Graben
Rhine River
Rhine Valley
Rhone River
Rhone Valley
Silvretta Group
Simplon region
Western Alps

sychnodymite
use carrollite

Sydney
City on the Pacific Ocean in E.
CO S335500N335500
E1511000E1511000
BT New South Wales
Australia

Sydney Basin
Including the Sydney area, it extends from NW of Newcastle on the N to W of Wollongong on the S.
BT New South Wales
Australia

Syedlets
use Siedlce

syenite
Includes use on level 3. As of 1981, not to be used in the igneous rocks set. See list H. Before 1981, included use on level 3 under igneous rocks(1) syenite family(2).
BA syenites
BT igneous rocks
SA microsyenite
nepheline syenite
quartz syenite

syenite family
As of 1981, no longer a valid term for GeoRef. Use syenites, alkali syenites, or monzonites as applicable. See list H. Before 1981, included use on level 2 under igneous rocks(1).

syenite porphyry
As of 1981, includes use on level 3 under igneous rocks(1) syenites(2). See list H. Before 1981, included use on level 3 under igneous rocks(1) syenite family(2).
BA syenites
BT igneous rocks
SA porphyry

syenites
Term introduced in 1981. Includes use on level 2 under igneous rocks(1). See list H. Before 1981, search syenite family.
BT igneous rocks
NA agpaite
albitite
albitophyre
alkali syenites
bostonite
foyaite
NZ granosyenite
NA kakortokite
lujavrite
miaskite
microsyenite
monzonites
naujaite
nepheline syenite
pulaskite
quartz syenite
shonkinite
syenite
syenite porphyry
SA appinite

syenodiorite
As of 1981, includes use on level 3 under igneous rocks(1) diorites(2). See list H. Before 1981, included use on level 3 under igneous rocks(1) diorite family(2).
BA diorites
BT igneous rocks

Sylmar Fault
Named after town in San Fernando Valley in Los Angeles County. Also search Sylmar.
BT California
United States

Sylt
Main island of the North Frisian group just off the mainland in the North Sea.
UF Sylt Island
BT Schleswig-Holstein
West Germany

Sylt Island
use Sylt

Sylvania Formation
Underlies Detroit River Formation; overlies Bois Blanc Formation. SE Michigan, NW Ohio, and Ontario.
BT Middle Devonian
Devonian
SA Michigan
Ohio
Ontario

sylvanite
As of 1981, includes use on level 3 under minerals(1) tellurides(2). See list L. Before 1981, included use on level 3 under minerals(1) sulfides(2).
UF silvanite
BA tellurides
sulfides
BT minerals

sylvine
use sylvite

sylvite
Also search sylvine. As of 1981, includes use on level 3 under minerals(1) chlorides(2). See list L. Before 1981, included use on level 3 under minerals(1) halides(2).
UF leopoldite
sylvine
BA chlorides
halides
BT minerals

symbiosis
Includes use on level 3 under appropriate fossil group (list F).
SA epibiotism

symmetric folds
Term introduced in 1978. Includes use on level 3 under folds(1) style(2). Before 1978, search folds AND symmetric.
BA folds
SA asymmetric folds
normal folds

Symmetrodonta
Order. Includes use on level 2 under Mammalia(1).
BA Mammalia
BT Tetrapoda
Vertebrata

symmetry
Used as a general term, e.g. on level 3 under crystal growth(1) or crystal structure(1) or under fossil groups (list F). Also search asymmetry.
UF asymmetry

symmicton
use diamicton

symposia
Includes use on level 1 (list A); as a level 2 or 3 term appropriate to a large number of topics, e.g. on level 2 under sedimentary petrology(1). See list G. Also search meetings; symposium. If level 1, term set options are:
topic [list B or extraterrestrial geology, general, geophysics]
subtopic
UF colloquia
symposium
SA annual report
associations
geology
organization
progress report
report

symposium
use symposia

Synapsida
Includes use on level 2 under Reptilia(1). See list F.
BA Reptilia
BT Tetrapoda
Vertebrata
NA Mesosauria
Pelycosauria
Placodontia
Therapsida

synchisite
UF synchysite
BZ carbonates
fluorides
BT minerals

synchysite
use synchisite

synclinal
A valid term through 1977. After 1977, use synclines.

synclines
Term introduced in 1978. Includes use on level 3 under folds(1) style(2). Before 1978, also search folds AND synclinal.
BA folds
SA anticlines
synform folds

synclinoria
Includes use on level 3 under folds(1) systems(2).
BA folds
SA anticlinoria
geanticlines
geosynclines
systems

synform folds
Term introduced in 1978. Includes use on level 3 under folds(1) style(2). Before 1978, search folds AND synform.
BA folds
SA antiform folds
synclines

syngenesis
Includes use on level 3 under mineral deposits, genesis(1) processes(2).
BA diagenesis
SA mineral deposits

synonymy
SA paleontology
taxonomy

syntectonic processes
Term introduced in 1978. Before 1978, search syntectonic.
SA paleogeography
processes
sedimentation
tectonics

synthesis
As of 1978, term is used on level 2 under crystal growth(1). Before 1971, also search mineral synthesis.
SA crystal growth
minerals
synthetic materials

synthetic
A valid term through 1977 used in combination with the name of the mineral on level 3 to index artificial minerals. After 1977, use synthetic materials.

synthetic materials
Term introduced in 1978 to index artificial minerals. Before 1978, also search synthetic AND name of mineral.
SA materials
minerals
natural materials
synthesis

synthetic seismograms
Term introduced in 1981.
BT seismograms

Syr Darya
River which rises in the Tien Shan Mountains and flows W and NW into the Aral Sea. Index Soviet republics as applicable.
UF Jaxartes River
Saihun River
Sir Darya
Syr Darya River
Syr-Darya
BT USSR
SA Kazakhstan
Kirghizia
Tadzhikistan
Uzbekistan

Syr Darya River
use Syr Darya

Syr-Darya
use Syr Darya

Syracuse
City in Onondaga County in New York, and city in SE Sicily. Index New York and/or Sicily as applicable.
SA New York
Sicily

Syria
Includes use on level 1 as an area term (list O). For term set options see list B.
CO N323000N371500
E0423000E0353000
BA Middle East
Asia
NT Damascus
Golan Heights
Sabkha
SA Euphrates River
Great Rift Valley
Jordan River
Jordan Valley
Mediterranean region
Mesopotamia
Near East

Syrte
use Sirte Basin

Sysert
City 25 miles SSE of Sverdlovsk in Sverdlovsk Oblast in W Siberia.
BT Russian Republic
USSR

systematics
A valid term through 1977. Use taxonomy.

systems
Includes use on level 2 under faults(1), folds(1), and fractures(1).
SA anticlinoria
block structures
closed systems
coordinate systems
en echelon faults
faults
folds
fractures
geothermal systems
grabens
horsts
information systems
open systems
synclinoria

systems analogs
Includes use on level 2 under ground water(1).
SA ground water

szaibelyite
Also search ascharite.
UF ascharite
BA borates
BT minerals

Szczecin
Province in NW Poland. Also a city.
UF Stettin
BT Poland
NT Wolin

Szechuan
use Szechwan

Szechwan
Province in S central China.
UF Szechuan
BT China

Szolnok
County in E central Hungary. Also a city.
BT Hungary

T

Ta
use tantalum

taaffeite
BA oxides
BT minerals

Taal
On Volcano Island in center of Lake Taal in Batangas Province in S Luzon. Also search Taal Volcano.
UF Taal Volcano
BT Luzon
Philippine Islands

Taal Volcano
use Taal

Tabasco
State in SE.
BT Mexico
SA Gulf Coastal Plain

Tabianian
Europe. See list E.
BA lower Pliocene
Pliocene
Neogene
Tertiary
BT Cenozoic

Tabulata
Includes use on level 2 under Coelenterata(1). See list F.
BA Anthozoa
Coelenterata
BT Invertebrata
NA Chaetetidae
Favositidae
Heliolitidae
SA Rugosa
Scleractinia

Tachira
State in W.
BT Venezuela

Taconian Orogeny
use Taconic Orogeny

Taconic Allochthon
In Northern Appalachians. Index countries as applicable. Also search Taconic.
SA Canada
United States

Taconic Orogeny
Term introduced in 1978. An orogeny in the latter part of the Ordovician period, named for the Taconic Range of eastern New York State and well developed through

most of the northern Appalachians in U.S. and Canada. Before 1978, search Taconic AND orogeny; Taconian; Taconian Orogeny.
UF Taconian Orogeny
BT Paleozoic
SA Canada
Ordovician
orogeny
Silurian
United States

taconite
Includes use on level 3 under sedimentary rocks(1) chemically precipitated rocks(2). See list I.
BA chemically precipitated rocks
BT sedimentary rocks
SA iron formations

tactite
Compositional term. Before 1978, included use on level 3 under metamorphic rocks(1).
BA metasomatic rocks

Tadzhik Basin
use Tadzhik Depression

Tadzhik Depression
North section of Afghan-Tadzhik Depression in S and SW Tadzhikistan. Also search Tadzhik; Tadzhik Basin; Tadzhikistan Depression.
UF Tadzhik Basin
Tadzhikistan Depression
BT Tadzhikistan
USSR
SA Afghan-Tadzhik Depression

Tadzhikhistan
use Tadzhikistan

Tadzhikistan
Tadzhik Soviet Socialist Republic in Soviet Central Asia. Also search Tadzhikhistan; Tadzhikstan.
CO N360000N410000
E0750000E0670000
UF Tadzhikhistan
Tadzhikstan
Tajikistan
BT USSR
NT Altyn-Topkan
Darvaz
Darvaza Range
Dushanbe
Garm
Hissar Range
Kara-Mazar
Karamazar
Karategin Range
Kulyab
Leninabad
Nurek
Shugnan
Tadzhik Depression
Tary-Ekam
Turkestan Range
Vakhsh
Vanchskiy Range
Zeravshan Range
Zeravshan-Hissar
SA Afghan-Tadzhik Depression
Amu Darya
Asiatic USSR
Badakhshan
Central Asia
Fergana Basin
Hindu Kush
Kurama Range
Pamirs
Surkhan Darya
Surkhan Darya basin
Syr Darya
Zeravshan River

Tadzhikistan Depression
use Tadzhik Depression

Tadzhikstan
use Tadzhikistan

taele
use frozen ground

Taeniodonta
Includes use on level 2 under Mammalia(1). See list F.
BA Mammalia
BT Tetrapoda
Vertebrata

Taeniopteris
Genus. Includes use on level 3 under pteridophytes(1) Filicopsida(2).
BA Filicopsida
pteridophytes
BT Plantae
SA Cycadales
Glossopteris
gymnosperms
Pteridospermae

taenite
As of 1981, includes use on level 3 under minerals(1) alloys(2). See list L. Before 1981, included use on level 3 under minerals(1) native elements and alloys(2).
BA alloys
BT minerals
SA kamacite
meteorites
plessite

Tafilalet
use Tafilalt

Tafilalt
Oasis in SE Morocco.
UF Tafilalet
Tafilelt
Tafilet
BT Morocco

Tafilelt
use Tafilalt

Tafilet
use Tafilalt

Tagil Basin
River basin in Sverdlovsk Oblast E of the Central Urals. Also search Tagil; Tagil River.
BT Russian Republic
USSR

Tagliamento Valley
River valley extending from Carnic Alps to the head of the Gulf of Venice. Index autonomous regions as applicable. Also search Tagliamento; Tagliamento River.
BT Italy
SA Friuli-Venezia Giulia
Veneto

Tagus Basin
River basin in central Spain and S central Portugal. Index countries as applicable. Also search Tagus.
UF Tagus River basin
BT Europe
SA Portugal
Spain

Tagus River
Longest river in the Iberian Peninsula. Rises in E central Spain and flows W and SW entering the Atlantic Ocean at Lisbon. Index countries as applicable.
BT Europe
SA Portugal
Spain

Tagus River basin
use Tagus Basin

Tahiti
Island of E group of Society Islands in French Polynesia. Includes use on level 1 as an area term (list O). For term set options see list B.
CO S180000S170000
W1490000W1500000
BA Society Islands
French Polynesia
Polynesia

Taiga
City in NW Kemerovo Oblast.
UF Tayga
BT Russian Republic
USSR

taiga environment
Term introduced in 1978. Before 1978, also search taiga.
SA climate
ecology
environment
geomorphology
paleoclimatology
sedimentation
steppes
tundra

Taigonos Peninsula
use Taygonos Peninsula

tailings
SA mining
reclamation
spoils
waste disposal

Taimyr
use Taymyr

Taimyr (Peninsula)
use Taymyr Peninsula

Taimyr Peninsula
use Taymyr Peninsula

Tainan
City on SW coast.
BT Taiwan

Taipei
City at N end of island.
BT Taiwan

Taishu Group
Chiefly composed of sandstone and shale in alternation, and intercalating arkose sandstone and red-coloured conglomerates. On Tsushima Island a part of Nagasaki Prefecture, Kyushu.
BT lower Tertiary
Tertiary
SA Japan
Kyushu

Taito
use Taitung

Taitung
City on SE coast.
UF Taito
BT Taiwan

Taiwan
Island off Fuklen Province of China. Seat of Chinese Nationalist government known as the Republic of China. Includes use on level 1 as an area term (list O). For term set options see list B. Before 1977, also search Formosa.
CO N220000N253000
E1230000E1200000
BA Far East
Asia
NT Chiayi-Hsinying area
Chinkuashih
Chinkuashih Mine
Hsinchu
Miaoli
Peikang
Penghu Islands
Tainan
Taipei
Taitung
Taoyuan
Tatun Shan
Tiehchenshan

Taiwan Strait
use Formosa Strait

Tajikistan
use Tadzhikistan

Takanuki
Village NE of Utsunomiya in E central.
BT Honshu
Japan

Tal Formation
Consists of 2 members: the quartzite member and the limestone member. The formation has been considered of Jurassic-Cretaceous age, but current literature indicates an unconformity between the 2 members so that the quartzite member is Precambrian and the limestone member is Cretaceous. Also search Tal Beds; Tal Limestone; Tal Series.
SA Cretaceous
India
Jurassic
Precambrian
Uttar Pradesh

Talacasto
Village in central.
BT Argentina
SA San Juan Province

Talas Ala-Tau
use Talas Range

Talas Range
Branch of the Tien Shan in NW Kirghizia. Also search Talas.
UF Talas Ala-Tau
BT Kirghizia
USSR
SA Tien Shan

Talasea
Settlement on E side of Willaumez Peninsula on N coast of New Britain, Bismarck Archipelago.
BT Papua New Guinea
Australasia

talc
As of 1981, use talc deposits for talc as a commodity. Before 1981, included use on level 1 as a commodity term.
BA sheet silicates
silicates
BT minerals
SA soapstone
talc deposits

talc deposits
Term introduced in 1981. Includes use on level 1 and 2 as a commodity term (list C). Before 1981, search talc AND deposits.
SA soapstone
talc

talc rock
Compositional term. Before 1978, included use on level 3 under metamorphic rocks(1).
BA metasomatic rocks

Talcher
use Talchir

Talcher coal field
use Talchir coal field

Talcher Coalfield
use Talchir coal field

Talchir
Region. Former princely state in Orissa States. Also search Talcher.
UF Talcher
BT Orissa
India

Talchir coal field
E central Orissa. Also search Talcher coal field, Talcher Coalfield, and Talchir.
UF Talcher coal field
Talcher Coalfield

BT Orissa
India

Talchir Formation
Upper Carboniferous to Permo-Carboniferous. Consists of boulder bed with boulders set in an ill-sorted matrix, often containing angular clastics, and is followed by greenish shales, silty shales, and sandstones. In peninsular India.
BT Paleozoic
SA Carboniferous
India
Permian
Upper Carboniferous

Talchir Series
Upper Carboniferous. At the base is a glacial boulder bed (Talchir Boulder Bed). Overlain by Damuda Series. In Orissa.
BT Upper Carboniferous
Carboniferous
SA India
Orissa

Talin
use Tallin

Talkeetna Mountains Quadrangle
Term introduced in 1981. S Alaska.
CO N620000N630000
W1470000W1500000
BT Alaska
United States

Talkeetna Quadrangle
Term introduced in 1981. S Alaska.
CO N620000N630000
W1500000W1530000
BT Alaska
United States

Talladega Front
E Talladega County in E central.
BT Alabama
United States

Talladega Group
According to the latest published Lexique, the age of Talladega Formation is Precambrian (?) to Carboniferous (?). In NW Georgia, the Talladega Series has been grouped into 10 formations (ascending): Pinelog Quartzite, Hiawassee Slate, Great Smoky Formation, Natahala Schist, Tusquitee Quartzite, Brasstown Schist, Valleytown Schist, Murphy Marble, Andrews Schist and Nottley Quartzite. In E Alabama, North Carolina, NW Georgia, and Tennessee.
BT Paleozoic
SA Alabama
Cambrian
Devonian
Georgia
Mississippi
North Carolina
Ordovician
Pennsylvanian
Precambrian
Silurian
Tennessee

Tallapoosa County
E central.
BT Alabama
United States

Tallin
City on the Gulf of Finland. Also search Tallinn.
UF Talin
Tallinn
BT Estonia
USSR

Tallinn
use Tallin

talnakhite
BA sulfides
BT minerals

talus
A valid term through 1978. Use talus slopes.

talus fan
use alluvial fans

talus slopes
Term introduced in 1978 on level 2 under slope stability(1). Before 1978, search talus; debris slopes.
UF debris slopes
SA cliffs
colluvium
erosion
erosion features
geomorphology
landslides
slope stability
slopes

Talysh Mountains
NW extremity of the Elburz in NW Iran. Also search Talysh.
BT Iran

Tamagawa
River flowing into Tokyo Bay between Tokyo and Yokohama.
BT Honshu
Japan

Taman
Village on N shore of Taman Peninsula near Kerch Strait in W Krasnodar Kray in Northern Caucasus.
BT Russian Republic
USSR

Taman Peninsula
Peninsula of W Krasnodar Kray jutting into Kerch Strait connecting Black Sea and Azov Sea. Also search Taman.
BT Russian Republic
USSR

Tamar Valley
River valley in Cornwall and Devon in SW England. Also search Tamar River.
BT England
Great Britain
United Kingdom

Tamaulipas
State in E.
BT Mexico
SA Gulf Coastal Plain
Rio Grande
Rio Grande Depression
Rio Grande Valley
Sierra Madre Oriental

Tamba Plateau
In the Kyoto area. Also search Tamba.
BT Honshu
Japan

Tambov
City 260 miles SE of Moscow. Also an oblast.
BT Russian Republic
USSR

Tamdy-Tau
use Tamdytau

Tamdytau
Mountains in the Kyzylkum SE of the Aral Sea.
UF Tamdy-Tau
Tamdytau Range
BT Uzbekistan
USSR

Tamdytau Range
use Tamdytau

Tamil Nadu
State in extreme S and SE. Formerly Madras.
CO N080000N135000
E0803000E0762000
BT India
NT Ariyalur
Coimbatore
Kondapalli
Madras
Neyveli
North Arcot
Parnalee
Sivamalai
South Arcot
Tiruchirapalli
Tirupati
Vridhachalam
SA Cauvery Basin
Cuddalore Series
Neyveli Lignite
Niniyur Group

Tampa Bay
Inlet of Gulf of Mexico on W coast of Hillsborough County in W.
BT Florida
United States

Tampere
City in SW part of country.
BT Finland

Tamworth
Town in E central.
BT New South Wales
Australia

Tana Fjord
Inlet of Arctic Ocean which receives the Tana River in NE.
BT Finnmark
Norway

Tananarive
City on the Indian Ocean in E.
BT Malagasy Republic

Tanco Pegmatite
BT Manitoba
Canada

Tanega-shima
Largest of the Osumi Islands off S Kyushu.
UF Tanegashima Island
BT Kyushu
Japan

Tanegashima Island
use Tanega-shima

Tanezrouft
Section of the Sahara primarily in Algeria. Index countries as applicable.
BT Africa
SA Algeria
Mali
Sahara

Tanganyika
Not a valid index term. Former British U.N. Trust Territory which became independent in 1961. See Tanzania.

tangential discontinuities
Term introduced in 1981. Includes use on level 3 under crust(1), mantle(1), and seismology(1). Used for a type of interface in seismology.
BT discontinuities
SA crust
mantle
seismology

tangential stress
use shear stress

tangential wave
use S-waves

Tanger
use Tangier

Tangier
City at W end of the Strait of Gibraltar.
UF Tanger
Tangiers
BT Morocco

Tangiers
use Tangier

Tannu Tuva
use Tuva

Tannu-Ola Range
On the Mongolian-Tuva A.S.S.R. border. Index Mongolia and/or Russian Republic as applicable. Also search Tannu-Ola.
BT Asia
SA Mongolia
Russian Republic

Tanquary Fiord
W central Ellesmere Island in District of Franklin.
BT Northwest Territories
Canada

tantalates
As of 1981, includes use on level 2 under minerals(1). See list L. Before 1981, included use on level 3 under minerals(1) oxides(2).
BA oxides
BT minerals
NZ betafite
columbite
euxenite
fergusonite
ixiolite
microlite
pyrochlore
samarskite
stibiotantalite
tantalite
tapiolite
wodginite
SA niobotantalates

tantalite
BZ niobates
tantalates
BA oxides
BT minerals

tantalum
Includes use on level 1 and 2 as a chemical element (list D). As of 1981, use tantalum ores for tantalum as a commodity. Before 1981, included use on level 1 as a commodity term.
UF Ta
SA elements
heavy metals
niobium
tantalum ores

tantalum ores
Term introduced in 1981. Includes use on level 1 and 2 as a commodity term (list C). Before 1981, search tantalum AND ore deposits.
SA niobium ores
tantalum

Tanzania
Former British U.N. Trust Territory of Tanganyika, which became independent in 1961. United with the island of Zanzibar to become Tanzania in 1964. Includes use on level 1 as an area term (list O). For term set options see list B.
CO S114000S010000
E0403000E0293000
BA East Africa
Africa
NT Kilimanjaro
Oldoinyo Lengai
Olduvai Gorge
Southern Highlands
SA East African Rift
Gregory Rift
Lake Malawi

Lake Natron
Lake Tanganyika
Lake Victoria
Umba
Zanzibar

tanzanite
BA epidote group
sorosilicates
orthosilicates
silicates
BT minerals
SA zoisite

Tanzawa Massif
use Tanzawa Mountains

Tanzawa Mountainland
use Tanzawa Mountains

Tanzawa Mountains
Between Fujiyama and Yokohama. Also search Tanzawa.
UF Tanzawa Massif
Tanzawa Mountainland
BT Honshu
Japan

Taos County
On the Colorado border in N New Mexico. Also search Taos.
BT New Mexico
United States

Taoudeni
use Taoudenni

Taoudeni Basin
use Taoudenni

Taoudenni
An oasis in the Sahara in NW Mali.
UF Taoudeni
Taoudeni Basin
Taoudenni Basin
BT Mali

Taoudenni Basin
use Taoudenni

Taourirt
Town in NE Morocco, and village in W central Algeria. Index countries as applicable.
BT Africa
SA Algeria
Morocco

Taoyuan
Town in N.
BT Taiwan

taphocenoses
use thanatocenoses

taphocoenosis
use thanatocenoses

taphonomy
Includes use on level 3 under fossil group(1). See list F.
SA fossilization
paleoecology
thanatocenoses

taphrogeny
Implies rifting. Includes use on level 3 under plate tectonics(1).
SA orogeny
plate tectonics
rifting
tectonics

tapiolite
BZ niobates
tantalates
BA oxides
BT minerals

tar sands
use oil sands

Taranaki
Provincial district in W.
BT North Island
New Zealand

Tarannon
Term introduced in 1981. Includes use on level 3 under age terms(1). See list E.
BA Lower Silurian
Silurian
BT Paleozoic

Taranto
City on N Gulf of Taranto on heel of Italian boot.
BT Apulia
Italy

Tarapaca
Northernmost province.
BT Chile

Tarawera Complex
use Tarawera volcanic complex

Tarawera volcanic complex
N central North Island.
UF Tarawera Complex
BT North Island
New Zealand

Tarbagatai Range
use Tarbagatay Range

Tarbagatay Mountains
use Tarbagatay Range

Tarbagatay Range
Northern outlier of the Tien Shan. Index Kazakhstan and/or Sinkiang Uighur as applicable. Also search Tarbagatai Range; Tarbagatay.
UF Tarbagatai Range
Tarbagatay Mountains
BT Asia
SA Kazakhstan
Sinkiang Uighur
Tien Shan

Tarentaise
Alpine valley of upper Isere River in Savoy Alps.
BT Savoie
France

Tareya
Village in NE Taymyr National Okrug in NW Siberia.
BT Russian Republic
USSR

Tarfaia
use Tarfaya

Tarfaya
Town in extreme SW Morocco. A former Spanish enclave. Also search Tarfaia.
UF Tarfaia
BT Morocco

Tarkhankut Peninsula
In extreme W Crimea jutting into Black Sea. Also search Tarkhankut.
BT Ukraine
USSR

Tarn
Department in S.
CO N432000N441200
E0025500E0013000
BT France
NT Sidobre Massif
SA Montagne Noire
Rodez Trough

Tarn-et-Garonne
Department in SW.
CO N434000N443000
E0020000E0004500
BT France
SA Orgueil Meteorite
Quercy
Rouergue

Tarnobrzeg
Town on the Vistula River in SE part of country.
BT Rzeszow
Poland

Tarnow
City in E.
BT Krakow
Poland

Taro Valley
River valley extending in a SW-NE direction from the N Apennines to the Po River. Also search Taro.
BT Emilia-Romagna
Italy

Tarragona
Province in Catalonia in NE.
CO N403000N413700
E0014000E0001000
BT Catalonia
Spain

Tarrant County
County in which Fort Worth is located.
BT Texas
United States

Tartary
use Tataria

Tartu
City on Ema River W of Lake Peipus.
UF Dorpat
BT Estonia
USSR

Tarumae
Active volcano. S of Sapporo in S Hokkaido. Also search Tarumae Volcano.
UF Tarumae Volcano
BT Hokkaido
Japan

Tarumae Volcano
use Tarumae

Tarvisio
Town near Austrian and Yugoslav borders.
BT Friuli-Venezia Giulia
Italy

Tary Ekan
use Tary-Ekam

Tary-Ekam
Village in Leninabad Oblast in N Tadzhikistan. Also search Tary-Ekan.
UF Tary Ekan
Tary-Ekan
BT Tadzhikistan
USSR

Tary-Ekan
use Tary-Ekam

Tas-Khayakhtakh Range
Northern section of Cherski Range N of Okhotsk Sea in NE Yakutia. Also search Tas-Khayakhtakh.
BT Russian Republic
USSR

Tashkent
City and oblast in NE.
BT Uzbekistan
USSR

Tasman Basin
Extends S of Tasman Sea proper between South Tasmanian Rise and Macquarie Ridge SW of New Zealand. Also search Tasman.
UF Tasman-Becken
BT Pacific Ocean

Tasman Geosyncline
Index states as applicable. Also search Tasman AND geosynclines.
BT Australia
SA geosynclines
New South Wales
Queensland
Tasmania
Victoria

Tasman orogenic zone
Index islands as applicable. Also search Tasman.
BT New Zealand
SA North Island
South Island

Tasman Sea
Between SE Australia and Tasmania on W and New Zealand on the E. Also search Tasman. As of 1977, includes use as a level 1 area term (list O). For set options see list B.
CO S510000S300000
E1743000E1465500
BA West Pacific
Pacific Ocean
NT Dampier Ridge
SA DSDP Site 284

Tasman-Becken
use Tasman Basin

Tasmania
Island and state S of Victoria. Includes use on level 1 as an area term (list O). For term set options see list B.
CO S434000S393000
E1483000E1435000
BA Australia
Australasia
NT Burnie
Coles Bay
Flinders Island
Great Lake
Hobart
Rosebery
Tasmania Basin
Zeehan
SA Heemskirk Granite
Lachlan fold belt
Tasman Geosyncline
Wynyard

Tasmania Basin
Term introduced in 1981. In Tasmania.
BT Tasmania
Australia

Tasmanites
Includes use on level 2 under algae(1). See list F.
BA Chlorophyta
algae
BT thallophytes
Plantae
SA palynomorphs

Tassili des Adjer
use Tassili n'Ajjer

Tassili n'Ajjer
An arid plateau region which is a NE extension of the Ahaggar Mountains in SE Algeria. Also search Tassili.
UF Tassili des Adjer
BT Algeria

Tatar
Tatar Autonomous Soviet Socialist Republic. In area around great bend of Volga River where it heads S at Kazan.
CO N540000N560000
E0540000E0480000
UF Tatarstan
BT Russian Republic
USSR

Tatar Arch
E central European USSR centering on Tatar A.S.S.R. Also search Tatar.
BT Russian Republic
USSR

Tatar Strait
Between E coast of Primorye Kray and W coast of Sakhalin Island at the N end of Japan Sea. Also search Tatar.
BT Russian Republic
USSR

Tataria
Originally applied to area of Mongol Empire at its height. Now, since dis-

placement of Crimean Tatars, it can be applied to the Tartar A.S.S.R and surrounding areas inhabited by Tatars.
UF Tartary
Tatary
BT Russian Republic
USSR

Tatarian
Europe. Above Kazanian, below Scythian (Triassic). Includes use on level 3 under age terms(1). See list E.
BA Upper Permian
Permian
BT Paleozoic

Tatarstan
use Tatar

Tatary
use Tataria

Tatra Mountains
Chief mountain group of central Carpathians. Also search High Tatra; High Tatra Mountains; High Tatras; Tatras. Index countries as applicable.
UF High Tatra
High Tatra Mountains
High Tatras
Tatras
BT Carpathians
Europe
SA Czechoslovakia
Low Tatra Mountains
Poland

Tatras
use Tatra Mountains

Tatum Basin
SE New Mexico.
BT New Mexico
United States

Tatun Shan
Mountain 9 miles N of Taipei. Also search Tatun.
BT Taiwan

Taubach
Term introduced in 1981. Town in SW.
CO N505600N505600
E0112400E0112400
BT East Germany

Tauern Tunnel
Railroad tunnel through the Hohe Tauern in SW Austria. Also search Tauern.
BT Austria
SA Hohe Tauern

Taunus
Mountain range E of the Rhine River and N of the lower Main River.
UF Taunus Mountains
BT Hesse
Germany

Taunus Mountains
use Taunus

Taupo
Town on Lake Taupo in central.
BT North Island
New Zealand

Taupo volcanic zone
Mountain region surrounding Lake Taupo in central North Island. Also search Taupo.
UF Taupo Zone
BT North Island
New Zealand

Taupo Zone
use Taupo volcanic zone

Taurus Littrow
use Taurus-Littrow

Taurus Mountains
Mountain chain on Mediterranean coast of Turkey.
CO N355000N383000
E0374000E0283000
BT Turkey

Taurus-Littrow
Includes use on level 3 under Moon(1). Also search Taurus Littrow; Taurus-Littrow Valley; Tranquillity Base.
UF Taurus Littrow
Taurus-Littrow Valley
BT Moon

Taurus-Littrow Valley
use Taurus-Littrow

Tavetsch
Term introduced in 1981. Valley of the upper Vorderrhein River.
CO N363500N364500
E0085200E0083900
BT Graubunden
Switzerland
SA Alps
Swiss Alps

tavistockite
use carbonate apatite

Tavua
Village on Viti Levu Island.
BT Fiji

Tawi River
use Tawi Valley

Tawi Valley
River valley in SW Jammu and Kashmir. Also search Tawi River.
UF Tawi River
BT Jammu and Kashmir
India

taxa, ahermatypic
use ahermatypic taxa

taxa, anaerobic
use anaerobic taxa

taxa, benthonic
use benthonic taxa

taxa, colonial
use colonial taxa

taxa, endemic
use endemic taxa

taxa, extinct
use extinct taxa

taxa, hermatypic
use hermatypic taxa

taxa, heteromorphic
use heteromorphic taxa

taxa, new
use new taxa

taxa, planktonic
use planktonic taxa

taxes
Term introduced in 1981.
SA economics

Taxodiaceae
Family. Includes use on level 3 under gymnosperms(1) Coniferales(2).
BA Coniferales
gymnosperms
BT Spermatophyta
Plantae
NA Sequoia
SA Pinaceae
Taxodium

Taxodium
BA Coniferales
gymnosperms
BT Spermatophyta
Plantae
SA Taxodiaceae

taxonomy
Includes use on level 2 under fossil group(1); on level 2 under paleontology(1). See list F.
SA holotypes
homonymy
new names
new taxa
paleontology
range
revision
synonymy

taxonomy review
A valid term through 1973. Use taxonomy.

Tay Estuary
N of the Firth of Forth on the North Sea. Also search Firth of Tay.
UF Firth of Tay
BT Scotland
Great Britain
United Kingdom

Tayga
use Taiga

Taygonos Peninsula
On N Okhotsk Sea between Gizhiga and Penzhina bays. Also search Taigonos; Taigonos Peninsula; Taygonos.
UF Taigonos Peninsula
BT Russian Republic
USSR

Taylor County
Index states as applicable.
BX Florida
Georgia
Iowa
Kentucky
Texas
West Virginia
Wisconsin
BT United States

Taylor Glacier
In the McMurdo Sound area in S Victoria Land just NW of Ross Ice Shelf. Also search Taylor AND appropriate area.
BT Antarctica

Taylor Valley
At McMurdo Sound in S Victoria Land in the Taylor Glacier area NW of Ross Ice Shelf. Also search Taylor.
BT Antarctica

Taymyr
Approximately the N and NW portion of the Taymyr Peninsula.
CO N720000N775000
E1140000E0800000
UF Taimyr
BX Russian Republic
BZ West Siberia
BT USSR

Taymyr Peninsula
Between the Yenisei and the Khatanga rivers in the Taymyr National Okrug in northernmost Siberia. Also search Taimyr; Taimyr Peninsula; Taymyr.
CO N720000N775000
E1140000E0800000
UF Taimyr (Peninsula)
Taimyr Peninsula
BT Russian Republic
USSR

Taz Basin
River basin in E and NE Yamal Nenets National Okrug in NW Siberia. Also search Taz; Taz River.
UF Taz River
BT Russian Republic
USSR

Taz River
use Taz Basin

Taza
Town E of Fez in N.
BT Morocco

Tazewell County
SW Virginia. Also search Tazewell.
BT Virginia
United States

Tazin Lake
Extreme NW.
BT Saskatchewan
Canada

Tb
use terbium

Tbilisi
City in SE Georgian Republic, USSR. Also search Tiflis.
UF Tiflis
BT Georgian Republic
USSR

Tc
use technetium

Tchornozem
use Chernozems

Te
use tellurium

Te Aroha
Town 70 miles SE of Auckland in N central.
BT North Island
New Zealand

Teapot Dome
Near Casper in central.
BT Wyoming
United States

technetium
Includes use on level 1 and 2 as chemical element (list D).
UF Tc
SA elements

technical cooperation
Term introduced in 1981.
SA international cooperation

techniques
This term deals with sampling. Includes use as a level 2 or 3 term appropriate to a large number of topics, e.g. on level 2 under meteorology(1), mineral exploration(1), and spectroscopy(1). See list G.
SA instruments
methods
sample preparation
sampling

technology
Includes use on level 2 under mining geology(1).
SA mining geology
production control
technology transfer

technology transfer
Term introduced in 1981.
SA economics
technology

tectite
use tektites

tectonic
A valid term through 1977. Used in reference to maps. See tectonic maps.

tectonic breccia
Term introduced in 1981.
BA breccia
clastic rocks
BT sedimentary rocks
SA tectonics

tectonic controls
A valid term through 1978. Term reintroduced in 1981. Includes use on level 3 under sedimentation(1) and geomorphology(1). Use structural controls when related to genesis of mineral deposits.
SA controls

geomorphology
sedimentation
structural controls

tectonic elements
Term introduced in 1981.
SA faults
folds
fractures
tectonics

tectonic maps
Term introduced in 1978. Includes use on level 3 under maps(1). Before 1978, also search maps AND tectonic.
BA maps
SA structure contour maps
tectonics

tectonic platforms
Term introduced in 1981. Before 1981, search platforms. Includes use on level 3 under tectonics(1) evolution(2).
SA continents
cratons
platforms
tectonics

tectonic wedges
Term introduced in 1981.
UF schuppen structure
SA faults
imbrication
tectonics

tectonics
Used for the structural makeup and structural evolution of regions. See also structural analysis. Includes use on level 1. Term set options are:
concepts
topic
evolution
(type of structure, e.g. Alpine Orogeny, Appalachian Phase, basement, basin range structure, Caledonian Orogeny, geanticlines, Hercynian Orogeny, lineaments, rift zones, shear zones, tectonic platforms, etc.)
gravity sliding
topic
vertical movements
topic (subsidence, uplifts)
UF geotectonics
SA allochthons
alpine-type
Andean Orogeny
Appalachian Phase
atectonic processes
aulacogens
autochthons
basement
basin range structure
basins
breccia
changes of level
continental drift
continental margin
Cordilleran Orogeny
cratons
crust
crustal shortening
decollement
deep-seated structures
deformation
epeirogeny
eugeosynclines
faults
fold belts
folds
foliation
fractures
geanticlines
geodynamics
geosynclines
glaciotectonics
gravity sliding
Himalayan Orogeny
horizontal movements
imbrication
intracratonic basins
isostasy
klippen
lineaments
lineation
Median Tectonic Line
meteor craters
miogeosynclines
mobile belts
mobility
nappes
neotectonics
orogenic belts
orogeny
plate tectonics
rift zones
salt tectonics
sedimentary cover
seismotectonics
shear zones
shields
structural analysis
structural geology
subsidence
syntectonic processes
taphrogeny
tectonic breccia
tectonic elements
tectonic maps
tectonic platforms
tectonic wedges
tectonophysics
tectonosphere
undation
uplifts
vertical movements
volcanology
windows

tectonite
SA deformation
fabric
preferred orientation

tectonophysics
Used for treatments of the application of physics in tectonics. Until 1976 this set included documents dealing with plate tectonics(1). Includes use on level 1; on level 2 under area terms(1), bibliography(1), and symposia(1). See list B. If level 1, term set options are:
topic [bibliography, concepts, convection, convection currents, experimental studies, methods, observations, practice, processes, symposia, theoretical studies]
subtopic
SA asthenosphere
bibliography
changes of level
continental drift
continental margin
convection
convection currents
core
crust
deformation
Earth
Earth tides
faults
folds
foliation
fractures
geophysics
geosynclines
heat flow
island arcs
isostasy
lithosphere
mantle
Mohorovicic discontinuity
neotectonics
ocean basins
paleomagnetism
plate tectonics
rift zones
salt tectonics
sea-floor spreading
seismology
structural analysis
subduction
subduction zones
tectonics

tectonosphere
SA biosphere
hydrosphere
lithosphere
plate tectonics
tectonics

tectosilicates
use framework silicates

teeth
Includes use on level 3 under fossil group(1) for type of material. See list F.
SA bones
jaws
skeletons
skulls
Vertebrata

Tehachapi Mountains
Range running E-W between S end of Sierra Nevada and the Coast Ranges in Kern County in S California. Also search Tehachapi.
BT California
United States

Tehama County
Term introduced in 1981. N California.
CO N394700N402500
W1212000W1230500
BT California
United States

Teheran
Province in NW central Iran. Also a city at S foot of the Elburz.
UF Tehran
BT Iran

Tehran
use Teheran

Tehri
use Tehri Garhwal

Tehri Garhwal
District in Himalayas in NW Uttar Pradesh. Also search Tehri.
UF Tehri
BT Uttar Pradesh
India

Teign Valley
River valley in Devonshire in SW England. Also search Teign River.
BT England
Great Britain
United Kingdom

Tejon Formation
Subdivided into 4 members (ascending): Uvas Conglomerate, Liveoak, Metralla Sandstone, and Reed Canyon Silt. Underlies Tecuya Formation; overlies basement complex. In W California.
BT upper Eocene
Eocene
SA California

Tekeli
Town in Taldy Kurgan Oblast near Sinkiang Uighur border in E.
BT Kazakhstan
USSR

tektites
Includes use on level 1 (list A). Used for the descriptions of small, rounded, pitted, black to green or yellow bodies of silicate glass of nonvolcanic origin. Term set options are:
topic [age, classification, composition, distribution, genesis, mineral composition, properties]
(type of tektite) or subtopic (no area term)
UF tectite
NA australite
Darwin glass
indochinite
microtektites
moldavite
muong nong type
philippinite
SA asteroids
comets
glasses
isotopes
meteorites
petrology
planetology
spherules

tele
use frozen ground

Teleajen River
use Teleajen Valley

Teleajen Valley
River valley in Prahova County in NE Moldavia. Also search Teleajen; Teleajen River.
UF Teleajen River
BT Moldavia
Romania

Telemark
County in S.
BT Norway
NT Bamble
Langesund Fjord

Teleosauridae
As of 1981, includes use on level 3 under Reptilia(1) Crocodilia(2). See list F. Before 1981, included use on level 3 under Reptilia(1) Archosauria(2).
BA Crocodilia
Archosauria
Reptilia
BT Tetrapoda
Vertebrata

Teleostei
As of 1981, includes use on level 2 under Pisces(1). See list F. Before 1981, included use on level 3 under Pisces(1) Osteichthyes(2).
BA Actinopterygii
Osteichthyes
Pisces
BT Vertebrata
NA Cyprinidae

teleseismic signals
SA earthquakes
seismology
signals

telethermal processes
Term introduced in 1981. Also search telethermal.
SA mineral deposits
processes

telinite
Term introduced in 1981.
BA vitrinite
macerals
coal
organic residues
BT sedimentary rocks

Tell
The Mediterranean coastal region of Algeria which is favored by a Mediterranean type climate.
BT Algeria

tellurates
Term introduced in 1981. Includes use on level 2 under minerals(1).

See list L. Before 1981, search tellurates and tellurites.
BT minerals

tellurates and tellurites
As of 1981, no longer a valid term for GeoRef. Use tellurates or tellurites as applicable. See list L. Before 1981, included use on level 2 under minerals(1).

tellurbismuth
use tellurobismuthite

telluric currents
Not a valid index term for GeoRef. See geophysical methods(1) Earth-current methods(2).

telluric methods
use Earth-current methods

telluric surveys
use Earth-current surveys

tellurides
As of 1981, includes use on level 2 under minerals(1). See list L. Before 1981, included use on level 3 under minerals(1) sulfides(2).
BA sulfides
BT minerals
NA altaite
calaverite
coloradoite
NZ hedleyite
NA hessite
joseite
sylvanite
tellurobismuthite
tetradymite
SA wehrlite

tellurites
Term introduced in 1981. Includes use on level 2 under minerals(1). See list L. Before 1981, search tellurates and tellurites.
BT minerals
NA mroseite

tellurium
Includes use on level 1 and 2 as a chemical element (list D).
UF Te
SA elements
heavy metals

tellurobismuthite
Also search tellurbismuth. As of 1981, includes use on level 3 under minerals(1) tellurides(2). See list L. Before 1981, included use on level 3 under minerals(1) sulfides(2).
UF tellurbismuth
BA tellurides
sulfides
BT minerals

TEM data
Term introduced in 1978. For the method, see transmission method.
UF transmission electromicroscopy data
SA data
electron microscopy
scanning method
SEM data
transmission method

Temagami Mine
Near village of Temagami on E end of Lake Temagami near Quebec border. Also search Temagami; Timagami.
UF Timagami Mine
BT Ontario
Canada

Temblor Range
Along SW San Joaquin Valley in W and SW Kern County.
BT California
United States

Temiscamingue
use Timiskaming

Temiskaming
use Timiskaming

temperate environment
Term introduced in 1978. Before 1978, search temperate.
SA climate
ecology
environment
paleoclimatology

temperature
For temperatures of mineral formations, see phase equilibria(1). Includes use on level 2 under core(1), Earth(1), heat flow(1), lava(1), mantle(1) and meteorology(1); on level 2 under paleoclimatology(1) for paleotemperatures. Before 1974, also search paleotemperatures. Also includes use as a level 2 or 3 term appropriate to a large number of topics. See list G. Also search temperatures.
NT high temperature
low temperature
SA climate
cooling
densities and temperatures
fluid inclusions
geologic thermometry
ion densities and temperatures
P-T conditions
specific heat
thermal properties

temporal distribution
Term introduced in 1981.
SA distribution
earthquakes
volcanoes

Tenerife
Largest island in the Spanish controlled Canary Islands off NW Africa.
UF Teneriffe
BT Canary Islands

Teneriffe
use Tenerife

Tengiz
Salt lake NW of Karaganda in N central Kazakhstan. Also search Teniz.
UF Teniz
BT Kazakhstan
USSR

Teniz
use Tengiz

Tennant Creek
Village 280 miles N of Alice Springs in central.
BT Northern Territory
Australia

tennantite
BA sulfarsenites
sulfosalts
BT minerals

Tennessee
Includes use on level 1 as an area term (list O). For term set options see list B.
CO N350000N364500
W0814000W0901500
BA Southern U.S.
United States
NX Anderson County
Bedford County
Benton County
Blount County
NT Buffalo River
NX Campbell County
NT Cannon County
NX Carroll County
Carter County
NT Central Basin
NX Chester County
Clay County
Coffee County
Columbia
Cumberland County
Davidson County
NT Ducktown
NX Fayette County
Franklin County
Giles County
Greene County
Grundy County
Hamilton County
Hancock County
Hardeman County
Hardin County
NT Henry County
NX Jackson County
Jefferson County
Johnson County
Lake County
Lawrence County
Lewis County
Lincoln County
Macon County
Madison County
Marion County
Marshall County
NT Maury County
NX Monroe County
Montgomery County
Moore County
Morgan County
NT Overton County
NX Perry County
Polk County
Putnam County
Roane County
Scott County
NT Sequatchie Valley
NX Sevier County
Shelby County
Smith County
Sullivan County
Sumner County
NT Trousdale County
NX Union County
Warren County
Washington County
Wayne County
White County
Williamson County
Wilson County
Woodbury
SA Appalachian Basin
Appalachian Plateau
Ashe Formation
Bangor Limestone
Bigby-Cannon Limestone
Brassfield Formation
Chattanooga Shale
Chesterian
Chickamauga Group
Chilhowee Group
Clayton Formation
Cumberland Plateau
Fernvale Formation
Fort Payne Formation
Great Smoky Mountains
Juniata Formation
Kingsport Formation
Knox Group
Lee Formation
Lincolnshire Limestone
Mascot Dolomite
Midcontinent
Midway Group
Mississippi Embayment
Mississippi River
Mississippi Valley
Monteagle Limestone
Murphy Marble
New Madrid region
Newman Limestone
Pennington Formation
Pottsville Group
Ripley Formation
Roan Supergroup
Rockwood Formation
Rome Formation
Saint Louis Limestone
Sainte Genevieve Limestone
Shady Dolomite
Talladega Group
Tennessee River
Tennessee Valley
Tuscaloosa Formation
Valley and Ridge Province
Waldron Shale
Warsaw Formation
Wilcox Group

Tennessee River
Formed near Knoxville in E Tennessee. It flows SW, W, and then N entering the Ohio River at Paducah in W Kentucky. Index states as applicable.
BT United States
SA Alabama
Kentucky
Tennessee

Tennessee Sandstone
Name has either been abandoned by author or rejected for use by U.S. Geological Survey. Typically exposed in Tennessee Ridge, Sebastian County, in W Arkansas.
BT Pennsylvanian
SA Arkansas

Tennessee Valley
River valley. Index states as applicable.
BT United States
SA Alabama
Kentucky
Tennessee

tensile strength
SA deformation
elastic properties
engineering geology
failures
mechanical properties
shear strength
strength
tension

tension
Includes use on level 3 under deformation(1) field studies(2) and under fractures(1) genesis(2).
SA compression
deformation
fractures
pressure
stress
tensile strength
torsion
yield strength

Tensleep Formation
use Tensleep Sandstone

Tensleep Sandstone
In Montchauve Group. Pennsylvanian and lower Permian. Overlies Amsden Formation. Also search Tensleep Formation.
UF Tensleep Formation
BT Paleozoic
SA Lower Permian
Montana
Pennsylvanian
Permian
Wyoming

Tentaculites
Genus. As of 1981, includes use on level 2 under Mollusca(1). See list F. Before 1981, included use on level 3 under problematic fossils(1).
BA Tentaculitidae
Tentaculitida
Mollusca
BT Invertebrata
SA problematic fossils

Tentaculitida
Order. Includes use on level 3 under Mollusca(1). See list F.
BA Mollusca
BT Invertebrata
NA Tentaculitidae
SA problematic fossils

Tentaculitidae
Family. Includes use on level 3 under Mollusca(1). See list F.
BA Tentaculitida
Mollusca
BT Invertebrata
NA Tentaculites
SA problematic fossils

tephra
use pyroclastics

tephrite
As of 1981, includes use on level 3 under igneous rocks(1) alkali basalts(2). See list H. Before 1981, included use on level 3 under igneous rocks(1) alkali basalt family(2).
BA alkali basalts
basalts
volcanic rocks
BT igneous rocks

tephrochronology
Includes use on level 2 under geochronology(1).
SA geochronology

tephroite
BA olivine group
nesosilicates
orthosilicates
silicates
BT minerals

Teplice
City in the Erzgebirge NNW of Prague.
UF Teplice-Sanov
BT Bohemia
Czechoslovakia

Teplice-Sanov
use Teplice

Ter River basin
In Catalonia. Index provinces as applicable. Also search Ter River.
BT Catalonia
Spain
SA Barcelona Province
Gerona Province

terbium
Includes use on level 1 and 2 as a chemical element (list D).
UF Tb
SA elements

Terceira Island
Central island of Portuguese controlled group. Site of U. S. Lajes Air Force Base. Also search Terceira.
BT Azores

Tereblya
Village in SW near the Romanian border.
BT Ukraine
USSR

Terebratulida
Includes use on level 2 under Brachiopoda(1). See list F.
BA Articulata
Brachiopoda
BT Invertebrata
NA Terebratulidae
Terebratulina

Terebratulidae
Family. Includes use on level 3 under Brachiopoda(1) Terebratulida(2).
BA Terebratulida
Articulata
Brachiopoda
BT Invertebrata

Terebratulina
Genus. Includes use on level 3 under Brachiopoda(1) Terebratulida(2).
BA Terebratulida
Articulata
Brachiopoda
BT Invertebrata

Terek River
Flows into the Caspian Sea N of Makhachkala in the Northern Caucasus. Index Soviet republics as applicable. Also search Terek.
BT USSR
SA Georgian Republic
Russian Republic

Terlingua
Village in Brewster County in W.
BT Texas
United States

terminology
A valid term through mid-1978. Use nomenclature.

Terni
City NE of Rome.
BT Umbria
Italy

Terra rosa
use Terra rossa

Terra rossa
See list M. Also search Terra rosa.
UF Terra rosa
BA soils

terraces
Includes use on level 3 under appropriate headings in geomorphology(1) such as erosion features(2), fluvial features(2), landform description(2), landform evolution(2), and shore features(2).
SA benches
changes of level
erosion features
floodplains
fluvial features
lacustrine features
shore features

terrain classification
Term introduced in 1978. Before 1978, search classification AND terrains.
SA classification
geomorphology
terrains

terrains
Also search terranes.
UF terranes
SA geomorphology
relief
terrain classification
topography

terranes
use terrains

terrestrial
A valid term through 1977. After 1977, use terrestrial environment for type of environment. See terrestrial materials.

terrestrial environment
Term introduced in 1978. Includes use on level 3 under ecology(1), paleoecology(1), and sedimentation(1). Before 1978, also search terrestrial. Use to distinguish from marine environment.
SA ecology
environment
paleoecology
sedimentation
terrestrial sedimentation

terrestrial materials
No longer a valid term for GeoRef as of 1981. Use interplanetary comparison if applicable. Term was introduced in 1978 to distinguish from lunar samples.

terrestrial planets
SA Earth
Mars
Mercury Planet
outer planets
planetology
planets
Venus

terrestrial sedimentation
Term introduced in 1981.
SA sedimentation
terrestrial environment

terrigenous
A valid level 2 term through 1977 used in combination with clastic rocks or clastic sediments. After 1977, use terrigenous materials only on level 3.

terrigenous materials
Term introduced in 1978. Includes use on level 3 under sediments(1) and sedimentary rocks(1). Before 1978, also search terrigenous AND clastic rocks or clastic sediments.
SA arenite
argillite
arkose
bentonite
black shale
boulders
breccia
cinerite
clastic rocks
clastic sediments
clay
claystone
cobbles
colluvium
conglomerate
contourite
coquina
diamictite
drift
dust
eluvium
eolianite
fanglomerate
flint clay
flysch
gravel
graywacke
loess
marl
materials
microbreccia
molasse
mud
mudstone
orthoquartzite
outwash
pebbles
pyroclastics
red beds
residuum
sand
sandstone
sedimentary rocks
sediments
shale
silcrete
silt
siltstone
sparagmite
subgraywacke
till
tillite
tilloid
tonstein
turbidite
volcanic ash

Tertiary
Includes use on level 1 as an age term (list E); on level 2 under paleoterms, e.g. paleoecology, paleogeography, paleomagnetism.
BT Cenozoic
NA Arikareean
NT Barreiras Formation
Columbia River Basalt
Cuddalore Series
John Day Formation
NA lower Tertiary
NT Maikop Series
NA middle Tertiary
Neogene
NT Neyveli Lignite
NA Paleogene
NT Poltava Series
San Lorenzo Formation
Sespe Formation
Snoqualmie Batholith
Twin River Formation
Twin Sisters Dunite
NA upper Tertiary
NT Warkalli Formation
Yakima Basalt
SA Bauru Formation
Blancan
Brianconnais Zone
Cretaceous
Ewekoro Formation
Ionian Zone
Laramide Orogeny
Muro Group
Paleolithic
Quaternary
Siwalik System
Tulare Formation
Uonuma Group
Verde Formation
Villafranchian

tertiary recovery
Term introduced in 1978 to be used in relation to petroleum.
SA petroleum
recovery
secondary recovery

Teruel
Province in NE central.
CO N395200N412200
E0001700W0015000
BT Aragon
Spain
SA Calatayud-Teruel Basin
Maestrazgo

teschenite
As of 1981, includes use on level 3 under igneous rocks(1) alkali gabbros(2). See list H. Before 1981, included use on level 3 under igneous rocks(1) alkali gabbro family(2).
BA alkali gabbros
gabbros
BT igneous rocks

testing
Term introduced in 1981.
SA engineering geology
rock mechanics
soil mechanics

tests
Includes use on level 3 under fossil group(1) for type of material. See list F. As of 1976, term is only to be used with fossils.
SA Invertebrata
shells

Tete
Town on the Zambezi River in NW.
BT Mozambique

Teterev River valley
NW Ukraine. Also search Teterev; Teterev River.
UF Teterev Valley
BT Ukraine

USSR

Teterev Valley
use Teterev River valley

Teteven
Town in N part of country.
BT Pleven
Bulgaria

Tethys
As of 1978, term is used on level 2 under continental drift(1). An elongated east-west sea, similar to the Mediterranean Sea, that separated Europe and Africa and extended across southern Asia in Pre-Tertiary time. Index continents as applicable.
UF Tethys Sea
SA Africa
Asia
continental drift
Europe
Mesogaea
Paratethys

Tethys Satellite
Term introduced in 1978.
SA satellites
Saturn

Tethys Sea
use Tethys

Teton County
Index states as applicable.
BX Idaho
Montana
Wyoming
BT United States

Teton National Forest
BT Wyoming
United States

Tetrabranchiata
Term introduced in 1981. Includes use on level 2 under Mollusca(1). See list F.
BA Cephalopoda
Mollusca
BT Invertebrata
NA Ammonoidea
Nautiloidea

Tetracorallia
Includes use on level 3 under Coelenterata(1) Anthozoa(2). See list F. Before 1981, also search tetracorals. As of 1981, search tetracorals only for biostratigraphy.
BA Anthozoa
Coelenterata
BT Invertebrata
SA Rugosa
Zoantharia

tetradymite
As of 1981, includes use on level 3 under minerals(1) tellurides(2). See list L. Before 1981, included use on level 3 under minerals(1) sulfides(2).
BA tellurides
sulfides
BT minerals

tetrahedra
BT polyhedra

tetrahedrite
BA sulfantimonites
sulfosalts
BT minerals
SA freibergite

Tetrapoda
As of 1981, includes use on level 1 and 2 as a fossil term (list F). Before 1981, included use on level 3 under Vertebrata(1). Before 1981, also search tetrapods. As of 1981, search tetrapods only for biostratigraphy.
BT Vertebrata
NT Amphibia
Aves
Mammalia
Reptilia

Tetyukhe
Town in E Primorye Kray 220 miles NE of Vladivostok.
BT Russian Republic
USSR

Teutoburg Forest
Range of hills S of Osnabruck in NW part of country. Index states as applicable.
UF Teutoburger Wald
BT West Germany
SA Lower Saxony
North Rhine-Westphalia

Teutoburger Wald
use Teutoburg Forest

Texas
Includes use on level 1 as an area term (list O). For term set options see list B.
CO N254500N363000
W0933000W1063000
BA Southwestern U.S.
United States
NX Anderson County
NT Atascosa County
Austin
Azalea Field
NX Baffin Bay
NT Baylor County
Beaumont
Big Bend National Park
Brazos River
Brenham
Brewster County
Burleson County
Burnet County
NX Caldwell County
Calhoun County
Cass County
Cherokee County
NT Christmas Mountains
NX Clay County
Comanche County
NT Corpus Christi Bay
Culberson County
Dallas
NX Dallas County
NT Davis Mountains
Del Rio
Diablo Platform
Dimmit County
NX Duval County
NT East Texas Field
Ector County
NX Edwards County
NT Edwards Plateau
El Paso
NX El Paso County
Ellis County
Fayette County
Floyd County
NT Fort Worth Basin
NX Franklin County
NT Frio County
Galveston
Galveston Bay
Galveston County
Galveston Island
Glass Mountains
Gonzales County
NX Guadalupe County
NT Guadalupe River
NX Hamilton County
Hardeman County
Hardin County
Harris County
Harrison County
Hidalgo County
NT Houston
NX Howard County
NT Hudspeth County
NX Jackson County
Jasper County
Jefferson County
Johnson County
Jones County
NT Karnes County
NX Kent County
King County
NT Kinney County
NX La Salle County
Lee County
Limestone County
NT Llano
Llano County
Llano Uplift
Lubbock
Lubbock County
NX Madison County
NT Marathon Basin
Marathon Geosyncline
NX Marion County
Mason County
NT Matagorda Bay
Maverick County
Midland
NX Midland Basin
Midland County
NT Milam County
NX Mitchell County
Montgomery County
Moore County
NT Mule Ear Diatreme
Mustang Island
NX Newton County
NT Nueces River
NX Orange County
NT Padre Island
Parker County
Pecos County
NX Polk County
NT Presidio County
Ranger
Reeves County
Rita Blanca Lake
Runnels County
NX Rusk County
NT San Antonio
San Marcos Arch
San Saba County
NX Shelby County
Sherman County
Smith County
NT Sterling County
Sutton County
Tarrant County
NX Taylor County
NT Terlingua
Travis County
NX Trinity County
NT Trinity River
Upton County
Uvalde County
Val Verde Basin
Val Verde County
Van Horn
Vinton Canyon
Waco
NX Ward County
Washington County
Wheeler County
Wichita County
Williamson County
Wilson County
NT Winkler County
NX Wise County
Wood County
NT Zavala County
SA Anadarko Basin
Antlers Sands
Ashmore Meteorite
Austin Group
Basin and Range Province
Bell Canyon Formation
Blaine Formation
Blancan
Bossier Formation
Buckner Formation
Canadian River
Canyon Group
Capitan Formation
Castile Formation
Chihuahua tectonic belt
Citronelle Formation
Claiborne Group
Cook Mountain Formation
Cotton Valley Group
Delaware Basin
Dockum Group
Edwards Formation
Escondido Formation
Fleming Formation
Franklin Mountains
Fredericksburg Group
Frio Formation
Glen Rose Formation
Great Plains
Guadalupe Mountains
Gulf Coastal Plain
Haymond Formation
Jackson Group
Llano Estacado
Loup Fork Group
Marble Falls Group
Maverick Basin
Midcontinent
Midway Group
Norphlet Formation
Odessa Meteorite
Ogallala Formation
Paluxy Formation
Panhandle
Pecos River valley
Permian Basin
Queen City Formation
Red River
Red River valley
Rio Grande
Rio Grande Depression
Rio Grande Rift
Rio Grande Valley
Sabine Lake
Schuler Formation
Smackover Formation
Strawn Series
Trans-Pecos
Trinity Group
Uvalde Gravel
Vicksburg Group
Washita Group
Washita River valley
Wilberns Formation
Wilcox Group
Woodbine Formation
Yates Formation

Texas County
Index states as applicable.
BX Missouri
Oklahoma
BT United States

text, explanatory
use explanatory text

textbooks
Use under disciplines. Includes use as a level 2 or 3 term appropriate to a large number of topics, e.g. on level 2 under economic geology(1), geology(1), and geochemistry(1). See list G.
SA book reviews
education
glossaries
manuals

Textulariina
Includes use on level 2 under foraminifera(1). See list F.
BA foraminifera
BT Protista
Invertebrata
NA Ammodiscacea
Lituolacea

textures
Includes use as a level 2 or 3 term appropriate to a large number of topics, e.g. on level 2 under meta-

morphic rocks(1). See list G.
SA aphanitic texture
arenaceous texture
argillaceous texture
framboidal texture
gneissic texture
graphic texture
intraclasts
megacrysts
oolitic texture
orbicular texture
packing
pelitic texture
pisolitic texture
porphyritic texture
porphyroblastic texture
shape analysis
size distribution
spinifex texture

TGA data
Term introduced in 1981.
SA thermogravimetric analysis

Th
use thorium

Th-228
Includes use on level 3 under isotopes(1).
SA isotopes
thorium

Th-230
Includes use on level 3 under isotopes(1).
SA isotopes
thorium

Th-230/Th-232
use Th-232/Th-230

Th-232
Includes use on level 3 under isotopes(1).
SA isotopes
thorium

Th-232/Th-230
Includes use on level 3 under isotopes(1). Also search Th-230/Th-232; Th-230 AND Th-232.
UF Th-230/Th-232
SA isotopes
thorium

Th-234
Includes use on level 3 under isotopes(1).
SA isotopes
thorium

Th/Th
Includes use on level 3 under absolute age(1) methods(2).
SA absolute age
Th/U
thorium

Th/U
Includes use on level 3 under absolute age(1) methods(2).
SA absolute age
Th/Th
thorium
U/Th/Pb
uranium

Thailand
Includes use on level 1 as an area term (list O). For term set options see list B.
CO N054500N203000
E1060000E0963000
UF Siam
BA Far East
Asia
NT Bangkok
Changwat Nakhon Phanom
Chiang Mai
Khorat Plateau
Lampang
SA Indochina
Khorat Group
Malay Peninsula
Phuket Group

Thalassinoides
BA ichnofossils

thalenite
BA orthosilicates
silicates
BT minerals

thallium
Includes use on level 1 and 2 as a chemical element (list D).
UF Tl
SA elements

thallophytes
Includes use on level 1 and 2 as a fossil term (list F).
BT Plantae
NT algae
bacteria
fungi
lichens
SA bryophytes
Pteropsida

Thames Estuary
The broad tidal mouth of the Thames River in SE England. Also search Thames.
BT England
Great Britain
United Kingdom

Thames River
Flows from the E slope of the Cotswold Hills E to the North Sea in SE England. Also search Thames.
UF The Thames
BT England
Great Britain
United Kingdom

thanatocenoses
Includes use on level 3 under fossil group(1) and under paleontology(1). See list F.
UF death assemblages
taphocenoses
taphocoenosis
thanatocoenosis
SA biocenoses
paleontology
taphonomy

thanatocoenosis
use thanatocenoses

Thanetian
Europe. Above Montian, below Ypresian (Eocene). Includes use on level 3 under age terms(1). See list E.
BA upper Paleocene
Paleocene
Paleogene
Tertiary
BT Cenozoic

Tharsis
As of 1978, term is restricted to morphologic features on Mars. Before 1978 term was also used for village in Spain.
BT Mars

thaumasite
BZ carbonates
orthosilicates
sulfates
BT minerals

thawing
Includes use on level 3 under glacial geology(1) or permafrost(1).
SA freezing
glacial geology
ice
permafrost
snow

Thaynes Formation
Consists of limestone, calcareous sandstone, sandstone, shale, and in the middle, red shale member. In SE Idaho, SW Montana, NE Utah, and SW Wyoming.
BT Lower Triassic
Triassic
SA Idaho
Montana
Utah
Wyoming

The Banks
use Outer Banks

The Coorong
use Coorong Lagoon

The Geysers
Eight miles NE of Geyserville in Sonoma County, W California.
UF Geysers, The
BT California
United States

The Great Oasis
use Kharga Oasis

The Himalaya
use Himalayas

The Marches
use Marches

The Rand
use Witwatersrand

The Spessart
use Spessart

The Thames
use Thames River

The Weald
Wooded region in Kent, Surrey, and Sussex counties of SE England. Also search Weald; Weald region.
CO N504500N513000
E0013000W0010000
UF Weald
Weald region
BT England
Great Britain
United Kingdom

Thecamoeba
Includes use on level 2 under Protista(1). See list F.
BA Protista
BT Invertebrata

Thecideidae
As of 1981, includes use on level 3 under Brachiopoda(1) Thecideidina(2). Before 1981, included use on level 3 under Brachiopoda(1) Articulata(2).
BA Thecideidina
Articulata
Brachiopoda
BT Invertebrata

Thecideidina
As of 1981, includes use on level 2 under Brachiopoda(1). Before 1981, included use on level 3 under Brachiopoda(1) Articulata(2).
BA Articulata
Brachiopoda
BT Invertebrata
NA Thecideidae

Thecodontia
Term introduced in 1981. Includes use on level 2 under Reptilia(1). See list F.
BA Archosauria
Reptilia
BT Tetrapoda
Vertebrata

Theiss River
use Tisza River

thenardite
BA sulfates
BT minerals

theoretical seismograms
Term introduced in 1981.
BT seismograms

theoretical studies
Includes use as a level 2 or 3 term appropriate to a large number of topics, e.g. on level 2 under faults(1), folds(1), fractures(1), heat flow(1), and meteorology(1). See list G.
UF studies, theoretical
SA experimental studies
laboratory studies
undation

theory
A valid term through mid-1978. Use theoretical studies.

Thera
Volcanic island. The southernmost of the Cyclades in the Aegean Sea. Also search Santorin.
UF Santorini
Thira
BT Cyclades
Greek Aegean Islands
Greece
SA Santorin

Therapsida
As of 1981, includes use on level 2 under Reptilia(1). See list F. Before 1981, included use on level 3 under Reptilia(1) Synapsida(2).
BA Synapsida
Reptilia
BT Tetrapoda
Vertebrata
NA Cynodontia
Lystrosaurus

Theria
Subclass. As of 1981, includes use on level 2 under Mammalia(1). See list F. Before 1981, included use on level 3 under Mammalia(1).
BA Mammalia
BT Tetrapoda
Vertebrata

thermal
A valid term through 1977. After 1977, use thermal metamorphism on level 2 under metamorphism(1), and use thermal circulation on level 2 under ocean circulation(1).

thermal alteration
Before 1978, also search thermal AND alteration.
SA alteration
hydrothermal alteration
thermal effects

thermal analysis
Level 1 term introduced in 1978. Used for methodology and not for data. For data, use thermal analysis data. Before 1978, thermogravimetric analysis and differential thermal analysis were used as level 1 terms. Term set options are:
differential thermal analysis
subtopic
methods
subtopic
techniques
subtopic
thermogravimetric analysis
subtopic
thermomagnetic analysis
subtopic
SA analysis
chemical analysis
clay mineralogy
differential thermal analysis
DTA data
electron microscopy
endothermic reactions
sample preparation

spectroscopy
standard materials
thermogravimetric analysis
thermomagnetic analysis
X-ray analysis

thermal aureole
use aureoles

thermal circulation
As of 1978 term is used on level 2 under ocean circulation(1). Before 1978, also search oceans AND circulation AND thermal.
SA circulation
ocean circulation

thermal conductivity
Includes use on level 2 under heat flow(1). Introduced as level 2 term in 1976. Before 1976, also search conductivity.
SA conductivity
electrical conductivity
geothermal gradient
heat flow
measurement
specific heat
thermal diffusivity

thermal demagnetization
Term introduced in 1981. Includes use on level 3 under paleomagnetism(1) and geophysics(1).
BT demagnetization
SA geophysics
magnetization
paleomagnetism
remanent magnetization

thermal diffusivity
UF diffusivity, thermal
SA heat flow
thermal conductivity

thermal discharge
use thermal pollution

thermal effects
Term introduced in 1981.
SA effects
thermal alteration

thermal emission
UF emission, thermal
SA infrared methods
remote sensing

thermal history
Includes use on level 3 under geosynclines(1) processes(2).
SA geosynclines
thermal regime

thermal metamorphism
Not a valid index term through 1977. After 1977, includes use on level 2 under metamorphism(1). Before 1978, also search metamorphism AND thermal. Also search thermometamorphism.
UF thermometamorphism
BA metamorphism

thermal pollution
Term introduced in 1978 on level 2 under pollution(1). Before 1978, also search pollution AND thermal; thermal discharge.
UF thermal discharge
SA pollution
waste disposal

thermal properties
SA engineering geology
heat capacity
permafrost
physical properties
properties
specific heat
temperature
thermodynamic properties

thermal regime
Term introduced in 1981.
SA engineering geology
geothermal systems
heat flow
hydrogeology
thermal history

thermal waters
Includes use on level 1 (list A). Term set options are:
topic [fumaroles, geochemistry, genesis, geysers, temperature]
subtopic
SA fractures
fumaroles
geothermal energy
geothermal gradient
geysers
ground water
heat flow
heat sources
hot springs
hydrochemistry
hydrogeology
hydrothermal alteration
medicinal waters
migration of elements
mineral waters
permeability
springs
trace elements
wallrock alteration
water
water resources

thermochemical properties
SA chemical properties
geochemistry
properties

thermochemical remanent magnetization
Term introduced in 1981.
BA remanent magnetization

thermodynamic properties
Includes use on level 3 under geochemistry(1) properties(2) and under appropriate material name.
SA enthalpy
entropy
free energy
fugacity
geochemistry
minerals
properties
specific heat
thermal properties

thermodynamics
As of 1981, no longer a valid term for GeoRef. See geochemistry.

thermography
use differential thermal analysis

thermogravimetric analysis
A valid level 1 term through 1977 used for methods. After 1977, use on level 2 under thermal analysis(1).
UF thermogravimetry
SA analysis
chemical analysis
differential thermal analysis
TGA data
thermal analysis

thermogravimetry
use thermogravimetric analysis

thermohaline circulation
Term introduced in 1978 on level 2 under ocean circulation(1). Before 1978, also search oceans AND circulation AND thermohaline.
SA circulation
ocean circulation

thermokarst
Also search cryokarst.
UF cryokarst
SA ground ice
karst
permafrost

thermoluminescence
Includes use on level 2 under geochronology(1); on level 3 under material name or discipline.
SA fluorescence
geochronology
luminescence

thermomagnetic analysis
As of 1978, term is used on level 2 under thermal analysis(1).
SA analysis
thermal analysis

thermometamorphism
use thermal metamorphism

thermometry
A valid term through mid-1978. Use geologic thermometry.

thermonatrite
BA carbonates
BT minerals

thermoremanence
use thermoremanent magnetization

thermoremanent magnetization
Includes use on level 3 under paleomagnetism(1). Also search thermoremanence; TRM.
UF thermoremanence
TRM
BA remanent magnetization
SA Curie point
inverse thermoremanent magnetization
magnetic field
magnetization
paleomagnetism
partial thermoremanent magnetization

Theropoda
As of 1981, includes use on level 3 under Reptilia(1) Saurischia(2). See list F. Before 1981, included use on level 3 under Reptilia(1) Archosauria(2).
BA Saurischia
Archosauria
Reptilia
BT Tetrapoda
Vertebrata

Thessaloniki
use Salonika

Thessaly
Administrative region in E central.
CO N390000N401200
E0231500E0210700
BT Greece
NT Olympus
SA Othrys
Pindus Mountains

Thetford Mines
City about 50 miles S of Quebec. Also search Thetford.
BT Quebec
Canada

thickness
Includes use on level 2 under crust(1).
SA crust

thin sections
Includes use on level 3 under electron microscopy(1).
SA electron microscopy
microfacies
microscope methods
petrography
polished sections
samples
sections

Thira
use Thera

Third World
Term introduced in 1981.
UF developing countries
SA economics

tholeiite
As of 1981, includes use on level 3 under igneous rocks(1) basalts(2). See list H. Before 1981, included use on level 3 under igneous rocks(1) basalt family(2).
BA basalts
volcanic rocks
BT igneous rocks
SA alkali basalt
olivine tholeiite
tholeiitic composition

tholeiitic basalt
As of 1981, includes use on level 3 under igneous rocks(1) basalts(2). See list H. Before 1981, included use on level 3 under igneous rocks(1) basalt family(2).
BA basalts
volcanic rocks
BT igneous rocks
SA basalt

tholeiitic composition
Term introduced in 1978. Before 1978, search tholeiitic.
SA basalt
composition
tholeiite

tholeiitic dolerite
As of 1981, includes use on level 3 under igneous rocks(1) diabase(2). See list H. Before 1981, included use on level 3 under igneous rocks(1) basalt family(2).
BA diabase
BT igneous rocks

Thomar
use Tomar

Thomas Range
Term introduced in 1981. W Utah.
CO N394000N400000
W1125000W1131500
BT Utah
United States

thomsenolite
As of 1981, includes use on level 3 under minerals(1) fluorides(2). See list L. Before 1981, included use on level 3 under minerals(1) halides(2).
BA fluorides
halides
BT minerals

Thomson Slate
Has been referred to in literature as Thomson, St. Louis, Cloquet, and Carlton slates. Has usually been considered equivalent of Animikie-Virginia Slate of Mesabi Range. In NE Minnesota and NW Wisconsin. Also search Thomson Formation.
BT Precambrian
SA Minnesota
Wisconsin

Thomson-Haskell analysis
Term introduced in 1981.
SA analysis
seismology

thomsonite
BA zeolite group
framework silicates
silicates
BT minerals

thorianite
BA oxides
BT minerals

thorite
BA orthosilicates
silicates
BT minerals

thorium
Includes use on level 1 and 2 as a chemical element (list D). As of 1981, use thorium ores for thorium as a commodity. Before 1981, included use on level 1 as a commodity term.
UF Th
SA elements
Io/Th
isotopes
Pb/Th
Th-228
Th-230
Th-232
Th-232/Th-230
Th-234
Th/Th
Th/U
thorium ores
U/Th/Pb

thorium ores
Term introduced in 1981. Includes use on level 1 and 2 as a commodity term (list C). Before 1981, search thorium AND ore deposits.
SA monazite deposits
thorium

Thorn
use Torun

thoron
use Rn-222

thortveitite
BA orthosilicates
silicates
BT minerals

Thrace
Administrative region in extreme NE Greece (Western Thrace), and all of Turkey in Europe (Eastern Thrace). Index countries as applicable.
BT Europe
NZ Greek Thrace
SA Greece
Turkey

Three Forks
Town in Gallatin County in SW.
BT Montana
United States

three-component seismographs
Term introduced in 1981.
BT seismographs

three-dimensional models
Term introduced in 1978. Includes use on level 3 under seismology(1). Before 1978, search three-dimensional AND models.
BA models
SA one-dimensional models
seismology
two-dimensional models

thrust
A valid term through 1977. After 1977, use thrust faults.

thrust faults
Term introduced in 1978. Includes use on level 3 under faults(1) displacements(2). Before 1978, also search faults AND thrust.
UF reverse slip faults
thrust slip faults
BA faults
SA crustal shortening
overthrust faults
reverse faults
step faults

thrust slip faults
use thrust faults

Thule
Settlement on coast of Hayes Peninsula N of Cape York in NW. U.S. Air Force base nearby.
BT Greenland
Arctic region

thulium
Includes use on level 1 and 2 as a chemical element (list D).
UF Tm
SA elements
rare earths

Thunder Bay
Inlet of NW Lake Superior. Also a city on the bay. Search Thunder.
BT Ontario
Canada

thunderstorms
Includes use on level 3 under meteorology(1) electrical phenomena(2).
SA electrical phenomena
lightning
meteorology
storms

Thur Valley
River valley SSW of Lake Constance extending to the Rhine River. Also search Thur River.
BT Switzerland

Thuringer Wald
use Thuringian Forest

Thuringia
A former German state in SE. Now a region including the Thuringian Forest.
CO N501200N510500
E0122000E0095200
BT East Germany
SA Erfurt Bezirk
Gera Bezirk
Saxony-Thuringia

Thuringia Basin
use Thuringian Basin

Thuringian
Term introduced in 1981. Includes use on level 3 under age terms(1). See list E.
BA Permian
BT Paleozoic

Thuringian Basin
Thuringian Forest region in Thuringia. Index districts as applicable.
UF Thuringia Basin
BT East Germany
SA Gera Bezirk

Thuringian Forest
Forested mountain range extending NW-SE in Thuringia.
UF Thuringer Wald
BT East Germany
SA Erfurt Bezirk
Gera Bezirk

Thuringian Hills
Term introduced in 1981 as a region in Germany.
CO N504000N513700
E0123000E0095500
BT Germany
SA East Germany

Thuringian Massif
Term introduced in 1981 as a region in Germany.
CO N500700N510200
E0130500E0101300
BT Germany
SA East Germany

Thysanopteroida
Term introduced in 1981. Includes use on level 2 under Insecta(1). See list F.
BA Insecta
BT Mandibulata
Arthropoda
Invertebrata

Ti
use titanium

Tiber River
use Tiber Valley

Tiber Valley
River valley. Index autonomous regions as applicable. Also search Tiber; Tiber River.
UF Tiber River
BT Italy
SA Latium
Tuscany
Umbria

Tibesti
Habitation in S central.
BT Libya

Tibesti Massif
Highest mountain group of the Sahara in NW Chad. Also search Tibesti Mountains.
UF Tibesti Mountains
BT Chad

Tibesti Mountains
use Tibesti Massif

Tibet
Autonomous region in extreme SW.
BT China
SA Brahmaputra River
Himalayas
Indian Plate
Indus River
Kumaun Himalayas

Ticino
Canton in Lepontine Alps in S.
BT Switzerland

Ticonderoga
Village on N outlet of Lake George in Essex County in NE.
BT New York
United States

tidal channels
SA channels
tidal flats
tidal inlets

tidal energy
Term introduced in 1981.
SA new energy sources

tidal environment
As of 1981, no longer a valid term for GeoRef. Use intertidal environment or estuarine environment as applicable.

tidal flats
Includes use on level 3 under geomorphology(1).
UF tide flat
SA brackish-water environment
coastal environment
marine environment
mud flats
sedimentation
supratidal environment
tidal channels
tides

tidal inlets
Includes use on level 3 under shorelines(1).
UF tidal outlets
SA inlets
shorelines
tidal channels

tidal outlets
use tidal inlets

tidal wave
use tsunamis

tide flat
use tidal flats

tides
Includes use on level 2 under aeronomy(1) and under ocean circulation(1).
SA aeronomy
Earth tides
low water
ocean circulation
tidal flats

Tieh-Chen Shan
use Tiehchenshan

Tiehchenshan
Hill region in NW Taiwan.
UF Tieh-Chen Shan
BT Taiwan

Tien Shan
Mountain region mostly in Kirghizia. Before 1981, also included part of China. Also search Tien-Shan.
CO N400000N440000
E0800000E0680000
UF Tien Shan Mountains
Tien Shan Range
Tien-Shan
Tien-Shan Range
BA Central Asia
BT USSR
SA Chatkal Range
Chu-Ili Mountains
Kirghizia
Kurama Range
Talas Range
Tarbagatay Range
Zeravshan Range

Tien Shan Mountains
use Tien Shan

Tien Shan Range
use Tien Shan

Tien-Shan
use Tien Shan

Tien-Shan Range
use Tien Shan

Tierra Amarilla
Village in Rio Arriba County in N.
BT New Mexico
United States

Tierra del Fuego
Archipelago comprising all islands S of Strait of Magellan to Drake Passage. Also name of main island. Index countries as applicable.
BT South America
SA Argentina
Chile

Tiflis
use Tbilisi

Tigre
Province in N.
BT Ethiopia

till
Includes use on level 3 under sediments(1) clastic sediments(2); on level 3 under glacial geology(1) glacial features(2). See list N.
BA clastic sediments
BT sediments
SA boulder clay
diamicton
drift
drumlins
glacial features
glacial geology
glacial transport
moraines
terrigenous materials
tillite
tilloid

tillage
Includes use on level 3 under soils(1) utilization(2).
SA soil management
soils
utilization
yields

Tillamook County
NW Oregon. Also search Tillamook.
BT Oregon
United States

tillite
Includes use on level 3 under sedimentary rocks(1) clastic rocks(2). See list I.
BA clastic rocks
BT sedimentary rocks
SA glacial features
terrigenous materials
till

Tillodontia
Order. Includes use on level 2 under Mammalia(1).
BA Mammalia
BT Tetrapoda
Vertebrata

tilloid
Includes use on level 3 under sedimentary rocks(1) clastic rocks(2). See list I.
BA clastic rocks
BT sedimentary rocks
SA terrigenous materials
till

tilt
Includes use on level 3 under seismology(1). As of 1981, for the tilt of a planet on its axis, use obliquity of the ecliptic.
SA geodesy
seismology
tiltmeters
volcanic earthquakes
volcanology

tiltmeters
Includes use on level 3 under seismology(1).
SA earthquakes
instruments
seismology
tilt
volcanology

Timagami Mine
use Temagami Mine

Timan Mountains
use Timan Ridge

Timan Range
use Timan Ridge

Timan Ridge
An eroded range of ancient sedimentary rocks in N European USSR dividing the eastern Pechora Basin from the plains to the W. Also search Timan.
CO N620000N673000
E0550000E0473000
UF Timan Mountains
Timan Range
BZ Russian Platform
BX Russian Republic
BT USSR

Timan-Pechora Basin
use Timan-Pechora region

Timan-Pechora Province
use Timan-Pechora region

Timan-Pechora region
Includes the Timan Ridge area; and the Pechora Basin in Komi ASSR, and Nenets National Okrug. In N and NE European USSR. Also search Timan-Pechora; Timan-Pechora Basin; Timan-Pechora Province.
UF Timan-Pechora Basin
Timan-Pechora Province
BT Russian Republic
USSR

time
As of 1981, no longer a valid term for GeoRef. Use time factor. Included use on level 3 under soils(1) genesis(2) as type of factor.

time domain analysis
Term introduced in 1981. Also search time domain.
SA analysis
geophysical methods

time factor
Term introduced in 1981. Includes use on level 3 under soils(1) genesis(2). Before 1981, search time.
SA factors

time scales
Includes use on level 2 and 3 under geochronology(1). See list B and list E (age terms).
UF scales, time
SA geochronology

time variations
Includes use on level 3 under aurora(1) morphology(2).
SA aurora
variations

Timiskaming
Lake in SW Quebec and SE Ontario. Index provinces as applicable.
UF Temiscamingue
Temiskaming
BT Canada
SA Ontario
Quebec
Timiskaming District

Timiskaming District
Term introduced in 1981. E central Ontario on Quebec border. Also search District of Timiskaming.
CO N470000N483000
W0793000W0820000
BT Ontario
Canada
SA Timiskaming

Timmins
Mining town 135 miles N of Sudbury between Lake Huron and James Bay.
BT Ontario
Canada

Timna
Ancient site in Asaylan area in E.
BT Arabian Peninsula

Timok Basin
River and coal basin near the Bulgarian and Romanian borders in E Serbia. Also search Timok; Timok River; Timok River basin.
UF Timok River basin
BT Serbia
Yugoslavia

Timok River basin
use Timok Basin

Timor
Island in the Lesser Sundas separated from Australia by the Timor Sea. W half of island belongs to Indonesia. The E half belongs to Portugal although it was occupied by Indonesia in November 1975. Index countries as applicable. Before 1978, also search Portuguese Timor.
BX Lesser Sunda Islands
Malay Archipelago
SA Indonesia
Portugal

Timor Sea
Between the island of Timor and NW Australia.
BA North Australian Seas
West Pacific
BT Pacific Ocean
NT Bonaparte Gulf basin
Sahul Shelf
Timor Trough

Timor Trench
use Timor Trough

Timor Trough
Undersea feature off the S coast of Timor.
UF Timor Trench
BT Timor Sea
Pacific Ocean

tin
Includes use on level 1 and 2 as a chemical element (list D). As of 1981, use tin ores for tin as a commodity. Before 1981, included use on level 1 as a commodity term.
UF Sn
SA elements
niggliite
tin ores

tin ores
Term introduced in 1981. Includes use on level 1 and 2 as a commodity term (list C). Before 1981, search tin AND ore deposits.
SA placers
tin

tin pyrites
use stannite

Tinaquillo
Town in NW part of country.
BT Cojedes
Venezuela

Tindouf
Village and oasis in extreme W.
BT Algeria

Tindouf Basin
Salt flats in extreme W Algeria. Also search Tindouf.
UF Sebkra de Tindouf
BT Algeria

tinguaite
As of 1981, includes use on level 3 under igneous rocks(1) phonolites(2). See list H. Before 1981, included use on level 3 under igneous rocks(1) trachyte-phonolite family(2).
BA phonolites
volcanic rocks
BT igneous rocks

Tintic District
Area around East Tintic Creek in Juab County in W central Utah. Also search Tintic AND Utah.
BT Utah
United States

Tintic Quartzite
Lower and middle Cambrian. Underlies Ophir Formation and unconformably overlies crystalline complex. Central and N Utah.
BT Cambrian
SA Lower Cambrian
Middle Cambrian
Utah

Tintina fault zone
Index Alaska and/or Yukon Territory as applicable.
SA Alaska
Yukon Territory

Tintinnidae
Family. Including Calpionellidae. Includes use on level 2 under Protista(1). See list F. Before 1981, also search tintinnids. As of 1981, search tintinnids only for biostratigraphy.
BA Protista
BT Invertebrata
NA Calpionella
Calpionellidae
Calpionellites
Stomiosphaera

Tioga Bentonite
According to latest published Lexique, Tioga Bentonite Bed is in Seneca Member of Onondaga Limestone. It is subsurface in Pennsylvania.
BT Middle Devonian
Devonian
SA New York
Ohio
Pennsylvania
Virginia
West Virginia

Tippecanoe County
W central.
BT Indiana
United States

Tipperary
County in S.
BT Ireland
NT Cashel

Tiraspol
City on the Dniester River in SE.
BT Moldavia
USSR

tirodite
BA clinoamphibole
amphibole group
chain silicates
silicates
BT minerals

Tirol
use Tyrol

Tiruchirapalli
City 200 miles SSW of Madras. Also search Trichinopoly.
UF Trichinopoly
BT Tamil Nadu
India

Tirupati
City in S part of country in central.
BT Tamil Nadu
India

Tisa River
use Tisza River

Tishomingo
Village and county in extreme NE.
BT Mississippi
United States

Tismana
Village in S foothills of Transylvanian Alps in NW.
BT Walachia
Romania

Tisza River
Rises in Carpathians in W Ukraine and flows W, SW and then S, into the Danube River N of Belgrade. Index countries as applicable. Also search Tisza.
UF Theiss River
Tisa River
BT Europe
SA Hungary
Romania
Ukraine

titanaugite
BA clinopyroxene
pyroxene group
chain silicates
silicates
BT minerals
SA augite

titanite
Before 1981, also search sphene.
UF sphene
BA orthosilicates
silicates
BT minerals
SA heavy minerals

titanium
Includes use on level 1 and 2 as a chemical element (list D). As of 1981, use titanium ores for titanium as a commodity. Also search Ti.

Before 1981, included use on level 1 as a commodity term.
UF Ti
SA elements
titanium ores

titanium ores
Term introduced in 1981. Includes use on level 1 and 2 as a commodity term (list C). Before 1981, search titanium AND ore deposits.
SA titanium

titanoclinohumite
BA humite group
nesosilicates
orthosilicates
silicates
BT minerals
SA clinohumite

titanomaghemite
BA oxides
BT minerals
SA maghemite

titanomagnetite
BA oxides
BT minerals
SA magnetic minerals
magnetite

Tithonian
Europe. Includes use on level 3 under age terms(1). See list E.
BA Portlandian
Upper Jurassic
Jurassic
BT Mesozoic

tjaele
use frozen ground

Tkibuli
City in W.
UF Tkvibuli
BT Georgian Republic
USSR

Tkvibuli
use Tkibuli

Tl
use thallium

Tm
use thulium

Toarcian
Europe. Includes use on level 3 under age terms(1). See list E.
BA upper Liassic
Lower Jurassic
Jurassic
BT Mesozoic

Toba Lake
In Barisan Mountains in N central Sumatra. Thought to occupy the crater of an extinct volcano. Also search Toba.
BT Sumatra
Indonesia

Tobacco Root Mountains
Range of Rocky Mountains in SW.
BT Montana
United States
SA Rocky Mountains

Tobago
Small island NE of Trinidad. A constituent part of Trinidad and Tobago.
BT Trinidad and Tobago
West Indies

tobermorite
BA chain silicates
silicates
BT minerals

Tocantins River region
N central and N part of country. Index states as applicable. Also search Tocantins.
BT Brazil
SA Goias
Maranhao
Para

Tochigi
Town near Utsonomiya in Tochigi Prefecture in E central.
BT Honshu
Japan

Todilto Formation
In San Rafael Group.
UF Todilto Limestone
BT Upper Jurassic
Jurassic
SA Arizona
Colorado
New Mexico

Todilto Limestone
use Todilto Formation

todorokite
BA oxides
BT minerals

Todos Santos Bay
Just S and SW of Ensenada. Also search Todos Santos.
BT Baja California
Mexico

Tofino Basin
SW Vancouver Island. Also search Tofino.
BT British Columbia
Canada

Tofua
Volcanic island. The largest of Haapai group in central.
BT Tonga
Polynesia

Togo
Formerly French Togo. Includes use on level 1 as an area term (list O). For term set options see list B.
CO N061000N111000
E0015000W0001000
BA West Africa
Africa
SA Volta Basin

Tohoku
Region, comprising N Honshu, which represents a transitional zone between cold Hokkaido and temperate and subtropical areas farther S.
BT Honshu
Japan

Tokachi
Volcanic peak in central Hokkaido. Also a sub prefecture.
BT Hokkaido
Japan

Tokachi Plain
Central.
BT Hokkaido
Japan

Tokai
use Tokaj

Tokaj
Town in NE part of country. Also search Tokai.
UF Tokai
Tokay
BT Hungary

Tokaj Mountains
In NE Hungary. Hungarian section of Tokaj-Eperjes Mountains. Also search Tokaj.
BT Hungary
SA Tokaj-Eperjes Mountains

Tokaj-Eperjes Mountains
Outlier of the Carpathians. Index Hungary and/or Slovakia as applicable.
BT Europe
SA Carpathians
Hungary
Slovakia
Tokaj Mountains

Tokay
use Tokaj

Tokelau
Term introduced in 1981. Island group N of Western Samoa.
UF Union Islands
BT Polynesia

Toki
City NE of Nagoya in central.
BT Honshu
Japan

Tokio
use Tokyo

Tokrau Synclinorium
N of Lake Balkhash. Also search Tokrau.
BT Kazakhstan
USSR

Tokushima
City on E coast.
BT Shikoku
Japan

Tokyo
City on Tokyo Bay in Tokyo Prefecture in SE Honshu.
UF Edo
Medo
Tokio
Yeddo
BT Honshu
Japan

Tokyo Bay
Term introduced in 1981. SE Honshu.
CO N344000N354200
E1401000E1393000
BT Honshu
Japan

Tokyo Imperial University
use University of Tokyo

Tolbachik
Term introduced in 1981. Volcano on Kamchatka Peninsula.
CO N554902N554902
E1602202E1602202
BT Kamchatka Peninsula
Kamchatka
Russian Republic
USSR

Toledo
As of 1981, refers to only the city in Ohio. Before 1981, also referred to the city and province in Spain. Search Toledo AND Ohio.
BT Ohio
United States

Toledo City
Term introduced in 1981 for the city in Spain. Before 1981, search Toledo AND Spain.
BT Toledo Province
New Castile
Spain

Toledo Province
Term introduced in 1981. Before 1981, search Toledo AND Spain. In central part of country.
CO N391600N401800
W0030000W0052500
BT New Castile
Spain
NT Toledo City
SA Montes de Toledo

Tolfa Hills
ENE of Civitavechia in NW Latium. Also search Tolfa.
BT Latium
Italy

Tolland County
Term introduced in 1981. NE Connecticut.
CO N413500N420300
W0720500W0723200
BT Connecticut
United States

Toluca Meteorite
Term introduced in 1981. Impact near Xiquipilco in Toluca area W of Mexico City. Before 1981, also search Toluca AND meteorites.
BA octahedrite
iron meteorites
BT meteorites
SA Mexico

Tomales Bay
Inlet of Pacific Ocean on NW coast of Marin County N of San Francisco.
BT California
United States

Tomar
City in W central part of country.
UF Thomar
BT Portugal

Tomaszow Lubelski
Town in SE part of country.
BT Lublin
Poland

Tomaszow Mazowiecki
Town SE of Lodz in central part of country.
BT Lodz
Poland

Tomaszow-Lvov Ridge
use Roztocze

Tombstone
City in Cochise County in SE.
BT Arizona
United States

Tomsk
City on right bank of Tom River near its junction with Ob River in S central Siberia. Also an oblast.
BT Russian Republic
USSR

tonalite
As of 1981, includes use on level 3 under igneous rocks(1) diorites(2). See list H. Before 1981, included use on level 3 under igneous rocks(1) diorite family(2).
BA diorites
BT igneous rocks
SA quartz diorite

tonalite gneiss
See list J.
BA gneisses
BT metamorphic rocks

Tonga
An archipelago of about 150 islands NE of New Zealand. Formerly a British protectorate which became independent in 1970. Also search Tonga Islands; Tonga island arc. Includes use on level 1 as an area term (list O). For term set options see list B.
CO S230000S150000
W1710000W1770000
UF Friendly Islands
Tonga Archipelago
Tonga Islands
BA Polynesia
NT Ata Caldera
Eua Island
Tofua
Tonga island arc
SA Pacific Ocean
Tonga Trench

Tonga Arc
use Tonga island arc

Tonga Archipelago
use Tonga

Tonga island arc
The islands of Tonga. Also search Tonga.
UF Tonga Arc
BT Tonga
Polynesia
SA Tonga Trench

Tonga Islands
use Tonga

Tonga Trench
E of the Tongtapu group of the S Tonga Islands.
BT Pacific Ocean
SA Tonga
Tonga island arc

Tonganoxie Sandstone
Of Stranger Formation. Underlies Westphalia Limestone Member. In E Kansas and NW Missouri. Also search Tonganoxie Sandstone Member.
UF Tunganoxie Sandstone Member
BT Virgilian
Upper Pennsylvanian
Pennsylvanian
SA Kansas
Missouri

Tongrian
Lattorfian. Europe. Above Ludian (Eocene), below Rupelian. Includes use on level 3 under age terms(1). See list E. Also search Lattorfian.
BA upper Eocene
Eocene
Paleogene
Tertiary
BT Cenozoic
SA Krosno Beds
Lattorfian
Sannoisian

Tongue of the Ocean
Strait between Andros Island on the W and New Providence Island and Exuma Cays on the E.
BT Bahamas
West Indies

Tongue River Member
In Fort Union Formation. In Montana and North Dakota, composed of light yellow, tan, and gray sandstones and shales; thin lenses of limestone; and numerous beds of lignite. In E Montana, SW North Dakota, South Dakota and NW Wyoming. Also search Tongue River Formation.
BT Paleocene
SA Montana
North Dakota
South Dakota
Wyoming

Tonkin Gulf
use Gulf of Tonkin

Tonnerre
Town in central part of country.
BT Yonne
France

Tono
Village in SW Washington, and village in N Honshu. Index Honshu and/or Washington as applicable.
SA Honshu
Washington

Tonoloway Limestone
In Cayuga Group. Underlies Keyser Limestone. Also search Tonoloway Formation.
BT Upper Silurian
Silurian
SA Maryland
Pennsylvania
Virginia
West Virginia

Tonopah
Village in Nye County in S central.
BT Nevada
United States

Tons Member
According to the India Lexique, the Tons Series is of Cambrian age. The Tons Series and Son Series are united to make up the Ken Subsystem, or lower main division of the Vindhyan System. In central India.
BT Cambrian
SA India

tonstein
Includes use on level 3 under sedimentary rocks(1) clastic rocks(2). See list I.
BA clastic rocks
BT sedimentary rocks
SA terrigenous materials

Tonto Basin
Valley in N Gila County in central.
BT Arizona
United States

Tooele County
NW Utah.
BT Utah
United States

tool marks
Includes use on level 3 under sedimentary structures(1) bedding plane irregularities(2). See list K.
BA bedding plane irregularities
sedimentary structures
SA current markings
soft sediment deformation

topaz
BZ fluorides
orthosilicates
BT minerals
SA gems

Topeka
City in Shawnee County in NE.
BT Kansas
United States

tophus
use tufa

Topley Intrusions
Central.
BT British Columbia
Canada

topographic maps
Term introduced in 1978. Includes use on level 3 under maps(1). Before 1978, also search maps AND topographic.
BA maps
SA contour maps

topography
Includes use on level 3 under geodesy(1).
SA cartography
geodesy
geomorphology
hills
ocean floors
physiographic provinces
plains
plateaus
relief
soils
terrains

Tor Formation
Overlies Antelope Valley Formation. In central Nevada. Also search Tor Limestone.
BT Devonian
SA Nevada

torbanite
See list I.
UF bitumenite
BA organic residues
BT sedimentary rocks
SA oil shale
shale

Torino
use Turin

Torlesse Supergroup
BT Upper Triassic
Triassic
SA New Zealand

toroidal mode
Term introduced in 1981.
SA oscillations
seismographs
seismology

Toronto
City at NW end of Lake Ontario.
CO N434000N434000
W0792800W0792800
BT Ontario
Canada

Toroweap Formation
In Aubrey Group. In Arizona can be subdivided into 3 main units (ascending): red to buff sandstone; calcareous sandstone and arenaceous limestone; and alternating red and buff sandstone, siltstone, and some shale. In NW Arizona, SE Nevada, and SW Utah.
BT Permian
SA Arizona
Nevada
Utah

Torquay
Resort city on English Channel in SW.
BT England
Great Britain
United Kingdom

Torrance County
Term introduced in 1981. Central New Mexico.
CO N341400N350300
W1051800W1063000
BT New Mexico
United States

Torres Strait
Between the island of New Guinea and the N tip of Cape York Peninsula, Queensland. Connects the Arafura Sea with the Coral Sea. Also search Torres.
BT Pacific Ocean

Torridonian
Europe. See list E.
BA upper Precambrian
Precambrian

tors
SA geomorphology
hills

torsion
Includes use on level 3 under deformation(1) field studies(2).
SA deformation
stress
tension

torsion faults
use wrench faults

torsion modulus
use shear modulus

Tortonian
Europe. Includes use on level 3 under age terms(1). See list E.
BA upper Miocene
Miocene
Neogene
Tertiary
BT Cenozoic
SA Helvetian

tortuosity
Property of materials.
SA geometry
properties

Torun
City in N central part of country.
UF Thorn
BT Bydgoszcz
Poland

Tosa Bay
Inlet of the Pacific Ocean on S coast. Also search Tosa.
BT Shikoku
Japan

tosudite
BA clay minerals
sheet silicates
silicates
BT minerals

total rock
use whole rock

Totes Gebirge
Mountain range in W central part of country. Index states as applicable.
BT Austria
SA Styria
Upper Austria

Tottori
City on Japan Sea NW of Kyoto.
BT Honshu
Japan

Toulon
City on the Mediterranean Sea.
BT Var
France

Toulouse
City on the Garonne River in S part of country.
BT Haute-Garonne
France

Toumodi
Village NW of Abidjan in S central.
BT Ivory Coast

Touraine
Historical region of NW central France. The area now comprises Indre-et-Loire Department and part of Indre Department. Index departments as applicable.
BT France
SA Indre
Indre-et-Loire

tourmaline
BA ring silicates
silicates
BT minerals
SA dravite
elbaite
heavy minerals
peridot

Tournai
City on Scheldt River near the French border.
UF Doornik
Tournay
BT Hainaut
Belgium

Tournaisian
Europe. Includes use on level 3 under age terms(1). See list E.
BA Dinantian
Carboniferous
BT Paleozoic

Tournay
use Tournai

Tours
City on the Loire River in NW central part of country.
BT Indre-et-Loire
France

Towada
In Towada-Hachimantai National Park in N Honshu. Also search Towada Volcano.
UF Towada Volcano
BT Honshu
Japan

Towada Volcano
use Towada

Towns County
On North Carolina border in N.
BT Georgia
United States

Townsville
City on Halifax Bay on the Coral Sea.
BT Queensland
Australia

toxic materials
Term introduced in 1978. Includes use on level 3 under environmental geology(1). Before 1978, also search toxic AND materials.
SA environmental geology
materials
pollution
toxicity
waste disposal

toxicity
Includes use on level 3 under environmental geology(1) waste disposal(2).
SA environmental geology
toxic materials
waste disposal

Toyama
City on S shore of Toyama Bay on Japan Sea.
BT Honshu
Japan

Toyoma
Town SE of Taira on the Pacific Ocean in central.
BT Honshu
Japan

trace elements
Includes use on level 2 under organic materials(1); on level 3 under soils(1) nutrients(2). Also search trace-elements.
UF guest element
microelement
trace-elements
SA elements
major elements
minor elements
nutrients
thermal waters
trace metals
trace-element analyses

trace fossils
A valid term through 1972. After 1972, use ichnofossils.

trace metals
Includes use on level 3 under commodities (list C).
SA environmental geology
geochemistry
metals
pollution
trace elements

trace-element analyses
For data, see appropriate material. For methods, includes use on level 3 under chemical analysis(1) methods(2) and under spectroscopy(1).
SA analysis
chemical analysis
elements
major-element analyses
minor-element analyses
spectroscopy
trace elements

trace-elements
use trace elements

tracer experiments
As of 1981, no longer a valid term for GeoRef. Use tracers.

tracers
Includes use on level 2 under isotopes(1). Before 1981, also search tracer experiments.
SA D/H
isotopes
radioactive tracers
radioactivity

trachyandesite
Includes use on level 3. As of 1981, not to be used in the igneous rocks set. See list H. Before 1981, included use on level 3 under igneous rocks(1) andesite-rhyolite family(2).
BA trachyandesites
volcanic rocks
BT igneous rocks

trachyandesites
Term introduced in 1981. Includes use on level 2 under igneous rocks(1). See list H. Before 1981, search trachyandesite.
BA volcanic rocks
BT igneous rocks
NA latite
trachyandesite

trachybasalt
Includes use on level 3. As of 1981, not to be used in the igneous rocks set. See list H. Before 1981, included use on level 3 under igneous rocks(1) alkali basalt family(2).
BA trachybasalts
alkali basalts
basalts
volcanic rocks
BT igneous rocks
SA latite
trachydolerite

trachybasalts
Term introduced in 1981. Includes use on level 2 under igneous rocks(1). See list H. Before 1981, search trachybasalt.
BA alkali basalts
basalts
volcanic rocks
BT igneous rocks
NA trachybasalt

trachydolerite
As of 1981, includes use on level 3 under igneous rocks(1) alkali basalts(2). See list H. Before 1981, included use on level 3 under igneous rocks(1) alkali basalt family(2).
BA alkali basalts
basalts
volcanic rocks
BT igneous rocks
SA trachybasalt

Trachyleberididae
Includes use on level 3 under Ostracoda(1). See list F. Also search Trachyleberidinae.
UF Trachyleberidinae
BA Cytheracea
Podocopida
Ostracoda
BT Crustacea
Mandibulata
Arthropoda
Invertebrata

Trachyleberidinae
use Trachyleberididae

trachyte
Includes use on level 3. As of 1981, not to be used in the igneous rocks set. See list H. Before 1981, included use on level 3 under igneous rocks(1) trachyte-phonolite family(2).
BA trachytes
volcanic rocks
BT igneous rocks

trachyte-phonolite family
As of 1981, no longer a valid term for GeoRef. Use trachytes or phonolites as applicable. See list H. Before 1981, included use on level 2 under igneous rocks(1).

trachytes
Term introduced in 1981. Includes use on level 2 under igneous rocks(1). See list H. Before 1981, search trachyte-phonolite family.
BA volcanic rocks
BT igneous rocks
NA keratophyre
quartz keratophyre
trachyte

tracks
Includes use on level 3 under sedimentary structures(1) and ichnofossils(1). Before 1977, also search tracks and trails.
BA biogenic structures
sedimentary structures
SA ichnofossils
lebensspuren
trails

tracks and trails
Not a valid index term for GeoRef. Use tracks or trails as separate terms.

trails
Includes use on level 3 under sedimentary structures(1) and ichnofossils(1). Before 1977, also search tracks and trails.
BA biogenic structures
sedimentary structures
SA ichnofossils
lebensspuren
tracks

Tranquillity Base
A valid term through 1977. Use Taurus-Littrow.

Trans-Alaska Pipeline
Pipeline constructed for the movement of crude oil from the North Slope at Prudhoe Bay on the Arctic Ocean in the NE to the port of Valdez off Prince William Sound in the SE.
BT Alaska
United States

Trans-Aravalli Vindhyan Basin
Index states as applicable.
BT India
SA Madhya Pradesh
Rajasthan

Trans-Pecos
Region W of the Pecos and E of the Rio Grande rivers in W Texas and S central New Mexico. Index states as applicable.
CO N290000N320000
W1010000W1063000
BT United States
SA Marathon Geosyncline
New Mexico
Texas

Transantarctic Mountains
A series of mountain ranges extending across the continent from the Filchner Ice Shelf on the Atlantic Ocean side through Victoria Land on the Pacific Ocean side separating East Antarctica from West Antarctica.
BT Antarctica

Transbaikal
use Transbaikalia

Transbaikalia
Region of SE Siberia E of Lake Baikal including Buryat ASSR; and Chita and Amur oblasts. Also search Transbaikal.
CO N490000N573000
E1250000E1023000
UF Transbaikal
BT Russian Republic
USSR

Transcarpathia
Region in extreme W which was, prior to World War II, that part of Czechoslovakia known as Ruthenia or the Carpatho-Ukraine. Now coextensive with the Transcarpathian Oblast.
CO N480000N500000
E0230000E0220000
BZ Russian Platform
BX Ukraine
BT USSR

Transcaucasia
Region S of the Caucasus. Index Soviet republics as applicable.
BT USSR
SA Armenia
Azerbaidzhan
Georgian Republic
Northern Caucasus

transcurrent faults
Term introduced in 1978. Includes use on level 3 under faults(1) displacements(2). Before 1978, also search faults AND transcurrent.
BA faults
SA displacements
strike-slip faults
transverse faults

Transdanubia
Fertile, hilly region between the Danube River and the Austrian border. Also search Dunantul.
UF Dunantul
BT Hungary
NZ Central Transdanubia
Northwestern Transdanubia
Southeastern Transdanubia
Southern Transdanubia

Transdanubian Central Mountains
use Bakony Mountains

transfer, heat
use heat transfer

transfer, mass
use mass transfer

transform
A valid term through 1977. After 1977, use transform faults.

transform faults
Term introduced in 1978. Includes use on level 3 under faults(1) displacements(2). Before 1978, also search faults AND transform.
BA faults
SA displacements
fracture zones
mid-ocean ridges
plate tectonics
strike-slip faults

transformations
As of 1978, term is used on level 2 under ocean waves(1). Includes use on level 3 under phase equilibria(1), minerals(1), crystal chemistry(1), or soils(1). Before 1978, transformation was used under ocean waves. Also search transformation.
SA changes
clay mineralogy
crystal chemistry

minerals
ocean waves
phase equilibria
soils

transformations, Laplace
use Laplace transformations

transgression
Term is used in stratigraphy for Pre-Quaternary sedimentation. For Quaternary, use changes of level.
SA changes of level
paleogeography
regression
stratigraphy

transient methods
Term introduced in 1981.
SA geophysical methods

transition
A valid general term through 1977.

transition zones
Includes use on level 3 under plate tectonics(1). Also search transition zone.
UF zones, transition
SA core
mantle
plate tectonics

translation lattice
use lattice

transmissibility coefficient
As of 1981, no longer a valid term for GeoRef. Use transmissivity.

transmission
A valid term through 1978. After 1978, use transmission method, e.g. under electron microscopy(1).

transmission electromicroscopy data
use TEM data

transmission method
Term introduced in 1978. Includes use on level 3 under electron microscopy(1). Before 1978 search transmission.
UF method, transmission
SA electron microscopy
SEM data
TEM data

transmissivity
Restricted to usage in relation to hydrology. Not to be used for remote sensing. Also search transmissibility.
UF coefficient of transmissibility
SA aquifers
ground water
hydraulic conductivity
hydrogeology
hydrology
permeability

transport
Includes use on level 2 under sedimentation(1).
SA degradation
glacial transport
ice rafting
marine transport
ore transport
saltation
sedimentation
stream transport
suspension
transportation
wind transport
winds

transportation
As of 1978, term is restricted to non-geological meaning.
SA pipelines
transport

Transvaal
Province in NE.
CO S281000S220500
E0320300E0244000
BT South Africa
NT Barberton
Barberton Mountain Land
Bushveld Complex
Johannesburg
Pretoria
Ventersdorp
Witwatersrand
SA Fig Tree Series
Lebombo Mountains
Limpopo Basin
Vaal River
Waterberg System
Witwatersrand System
Wolkberg Group

Transvaal Supergroup
BT Precambrian
SA South Africa

transverse faults
Term introduced in 1978. Includes use on level 3 under faults(1) orientation(2). Before 1978, also search faults AND transverse.
BA faults
SA orientation
transcurrent faults

transverse folds
use superposed folds

Transverse Ranges
In S California.
CO N333000N350000
W1153000W1193000
BT California
United States
SA San Bernardino Mountains
San Gabriel Mountains

transverse wave
use S-waves

Transylvania
Region bounded by the Carpathians on the E and the Transylvanian Alps on the S in NW and central part of country.
BT Romania
NT Alba-Iulia
Almas Valley basin
Aries Valley
Baia Mare
Baita
Baraolt
Baraolt Basin
Beius Basin
Bihor
Bihor Mountains
Bistrita
Brad
Brasov
Calimani Mountains
Cavnic
Chiuzbaia
Cluj
Covasna
Crai
Crisana-Maramures
Ditrau
Gilau
Gurghiu
Gurghiu Mountains
Harghita
Harghita Mountains
Hateg
Herja
Hoghiz
Maramures
Muntii Metalici
Mures
Padurea Craiului Mountains
Persani Mountains
Petrosani Basin
Poiana-Rusca Mountains
Retezat Mountains
Rodna
Rodna Mountains
Rosia
Salaj
Satu Mare
Sebes
Sebes Mountains
Sibiu
Simleu Basin
Transylvanian Basin
Vladeasa Mountain
SA Fagaras Mountains
Getic Nappe
Jiu River valley
Olt River
Olt River valley
Transylvanian Alps

Transylvania Basin
use Transylvanian Basin

Transylvanian Alps
Continuation of the Carpathians extending E-W in central part of country. Also search South Carpathians an Southern Carpathians. Index region as applicable.
UF South Carpathians
Southern Carpathians
BT Romania
SA Bucegi Mountains
Carpathians
Fagaras Mountains
Retezat Mountains
Transylvania
Vulcan Mountains
Walachia

Transylvanian Basin
Primarily a plateau. Also search Transylvania Basin; Transylvanian Depression; Transylvanian Plain.
UF Transylvania Basin
Transylvanian Depression
Transylvanian Plain
BT Transylvania
Romania

Transylvanian Depression
use Transylvanian Basin

Transylvanian Plain
use Transylvanian Basin

trap rock
No longer a valid term for GeoRef as of 1981. Use trap rocks. Before 1981, included use on level 3 as a commodity term under construction materials(1).

trap rocks
As of 1981, includes use on level 3 under igneous rocks(1) basalts(2). Before 1981, included use on level 3 under igneous rocks(1) basalt family(2). See list H. Before 1981, also search trap rock.
BA basalts
volcanic rocks
BT igneous rocks
SA basalt
construction materials
diabase

Trap Spring Field
Term introduced in 1981. Oil field in central Nevada.
CO N380000N385000
W1153000W1163000
BT Nye County
Nevada
United States

Trapani
Province in W.
BT Sicily
Italy

trapped particles
Includes use on level 2 under magnetosphere(1); on level 3 under magnetosphere(1) magnetic storms(2).
SA ions
magnetic field
magnetic storms
magnetosphere
particles

traps
As of 1978, term is restricted to use in relation to petroleum and natural gas, e.g. on level 3 under oil and gas fields(1).
NA stratigraphic traps
structural traps
SA natural gas
oil and gas fields
petroleum

Tras-os-Montes
Former province in NE Portugal. Now it comprises Braganca and Vila Real districts. Index districts as applicable.
BT Portugal
SA Braganca
Vila Real

travel time
use traveltime

traveltime
Includes use on level 3 under seismology(1). Also search travel time.
UF travel time
SA elastic waves
long-period waves
seismology
short-period waves
traveltime curves

traveltime curves
Also search hodograph(s).
UF curves, traveltime
hodographs
SA elastic waves
seismology
traveltime

traveltime residuals
Term introduced in 1981.
SA seismology

Traverse Group
Middle and upper Devonian. Comprises (ascending) Belle Shale, Rockport Quarry Limestone, Ferron Point Shale, Genshaw Formation, Koehler Limestone, Gravel Point Formation, and Beebe School Formation. In S Michigan.
BT Devonian
SA Michigan
Middle Devonian
Upper Devonian

travertine
Includes use on level 3 under sedimentary rocks(1) carbonate rocks(2). See list I. Before 1981, also included use on level 3 under sediments(1) chemically precipitated sediments(2) or carbonate sediments(2) and on level 3 under sedimentary rocks(1) chemically precipitated rocks(2).
UF calc-sinter
calcareous sinter
BA carbonate rocks
BT sedimentary rocks
SA calcite
carbonate sediments
chemically precipitated rocks
limestone
tufa

Travis County
Central.
BT Texas
United States

treatment
Includes use on level 3 under soils(1) salinity(2).
SA desalinization
fertilization
irrigation

reclamation
salinity
soil management
soils
utilization

tree rings
Includes use on level 2 under geochronology(1).
UF annual growth rings
rings, tree
SA geochronology

Trego County
NW central.
BT Kansas
United States

Tremadoc Bay
Inlet of Saint Georges Channel between Caernarvon and Merioneth counties in NW Wales. Also search Tremadoc.
BT Wales
Great Britain
United Kingdom

Tremadocian
Europe. Above Dolgellian (Cambrian), below Arenigian. Includes use on level 3 under age terms(1). See list E.
BA Lower Ordovician
Ordovician
BT Paleozoic

tremolite
BA clinoamphibole
amphibole group
chain silicates
silicates
BT minerals

Tremp
Town in Catalonia in NE part of country.
BT Lerida
Catalonia
Spain

Trempealeauan
North America. Upper Cambrian, above Franconian, below Lower Ordovician. See list E.
BA Upper Cambrian
Cambrian
BT Paleozoic

trenches
Includes use on level 2 under ocean floors(1); on level 3 under plate tectonics(1).
UF marginal trench
oceanic trench
sea-floor trench
SA Benioff zone
bottom features
island arcs
ocean floors
plate tectonics
subduction
subduction zones

trend surface analysis
use trend-surface analysis

trend-surface analysis
Also search trend surface analysis.
UF trend surface analysis
BA statistical analysis
SA analysis
automatic data processing
mathematical geology

Trent
use Trento

Trent Valley
River valley in central England, SE North Carolina, and SE Ontario. Index England, North Carolina, or Ontario as applicable. Also search Trent.
SA England
North Carolina
Ontario

Trentino
S part of Trentino-Alto Adige autonomous region.
BT Trentino-Alto Adige
Italy

Trentino-Alto Adige
Autonomous region in NE Italy. Formerly Venezia-Tridentina.
CO N454000N470800
E0123000E0102500
UF Venezia-Tridentina
BT Italy
NT Bolzano
Cima d'Asta
Latemar Massif
Merano
Predazzo
Trentino
Trento
Valsugana
SA Dolomites
Marmolada Glacier
Venetia

Trento
City in S Trentino-Alto Adige. Also search Trent.
UF Trent
BT Trentino-Alto Adige
Italy

Trenton
City on the Delaware River in Mercer County in W.
BT New Jersey
United States

Trenton Group
Standard section of group comprises (ascending) Rockland Limestone, Hull Formation, Sherman Fall Formation, Cobourg Formation, Collingwood Shale, and Gloucester Shale. In Georgia, Michigan, N Ohio, New York, Vermont, and W Virginia. Also search Trenton Limestone.
BT Middle Ordovician
Ordovician
SA Georgia
Michigan
New York
Ohio
Pennsylvania
Trentonian
Vermont
Virginia

Trentonian
North America. Upper Mohawkian, above Blackriverian. Includes use on level 3 under age terms(1). See list E.
BA Middle Ordovician
Ordovician
BT Paleozoic
SA Champlainian
Trenton Group

Trepca Mine
In the Kosovo region of S Serbia. Also search Trepca.
UF Trepcha Mine
BT Serbia
Yugoslavia

Trepcha Mine
use Trepca Mine

Trepostomata
Includes use on level 2 under Bryozoa(1). See list F.
BA Bryozoa
BT Invertebrata

trevorite
BA oxides
BT minerals

Trialet Mountains
use Trialet Range

Trialet Range
N range of the Lesser Caucasus in central Georgian Republic, USSR. Also search Trialet.
UF Trialet Mountains
BT Georgian Republic
USSR
SA Lesser Caucasus

triangulation
Includes use on level 3 under geodesy(1) or neotectonics(1).
SA cartography
geodesy
neotectonics
networks
surveys

Triassic
Includes use on level 1 as an age term (list E); on level 2 under paleoterms, e.g. paleoecology, paleogeography, paleomagnetism. Above Permian (of Paleozoic), below Jurassic.
BT Mesozoic
NA Andalusian
NT Chanares Formation
Fremouw Formation
Hallstatt Limestone
Hawkesbury Sandstone
Ipswich Coal Measures
NA Lower Triassic
Middle Triassic
NT Moenkopi Formation
Narrabeen Group
Nicola Group
Red Peak Formation
NA Upper Triassic
NT Wetterstein Limestone
SA Baskunchak Series
Beacon Supergroup
Botucatu Formation
Estrada Nova Formation
Gondwana System
Hercynian Orogeny
Karroo System
Khorat Group
Korenevskaya Formation
New Red Sandstone
Novorayskoe Formation
Pucara Group
Sadlerochit Formation
Shimanto Group
Singleton Coal Measures
Slovakian Karst
Spearfish Formation

triaxial tests
SA deformation
rock mechanics
soil mechanics
uniaxial tests

Trichinopoly
use Tiruchirapalli

Triconodonta
Includes use on level 2 under Mammalia(1). See list F.
BA Mammalia
BT Tetrapoda
Vertebrata

Tridacna
Genus. Includes use on level 3 under Mollusca(1). See list F.
BA Bivalvia
Mollusca
BT Invertebrata

tridymite
BA silica minerals
framework silicates
silicates
BT minerals
SA cristobalite

Trieste
City at head of the Adriatic Sea on Gulf of Trieste.
BT Friuli-Venezia Giulia
Italy

Trigonia
Genus. Includes use on level 3 under Mollusca(1). See list F.
BA Trigoniidae
Bivalvia
Mollusca
BT Invertebrata

Trigoniidae
Includes use on level 3 under Mollusca(1). See list F.
BA Bivalvia
Mollusca
BT Invertebrata
NA Trigonia

Trilobita
Class. Includes use on level 1 and 2 as a fossil term (list F).
BT Trilobitomorpha
Arthropoda
Invertebrata
NA Agnostida
Corynexochida
Dechenellidae
Lichida
Odontopleurida
Phacopida
Proetidae
Ptychopariida
Redlichiida
SA Crustacea
Cruziana
Rusophycus
trilobites
Trilobitoidea

trilobites
Term introduced in 1981 as the common name for Trilobita. See list F. To be used for Trilobita when they are used as tools in stratigraphy. Includes use on level 1. Before 1981, search Trilobita AND biostratigraphy. If 1, term set options are:
biostratigraphy
age [List E]
SA biostratigraphy
Trilobita

Trilobitoidea
Class. Includes use on level 3 under Arthropoda(1) Trilobitomorpha(2). See list F.
BA Trilobitomorpha
Arthropoda
BT Invertebrata
SA Trilobita

Trilobitomorpha
Includes use on level 2 under Arthropoda(1). See list F.
BA Arthropoda
BT Invertebrata
NT Trilobita
NA Trilobitoidea

Trilophosauria
Term introduced in 1981. Includes use on level 2 under Reptilia(1). See list F.
BA Araeoscelidia
Euryapsida
Reptilia
BT Tetrapoda
Vertebrata

Trinidad
Island off NE coast of Venezuela in the Lesser Antilles. Major part of independent state of Trinidad and Tobago. Included use on level 1 as an area term until 1977. Now index under Trinidad and Tobago(1).
CO N100200N105100
W0605300W0615600
BT Trinidad and Tobago
West Indies

Trinidad and Tobago
Comprises the islands of Trinidad

and Tobago, Atlantic Ocean, off NE coast of Venezuela. As of 1977, includes use as a level 1 area term (list O). For term set options see list B.
CO N100000N113000 W0601500W0620000
BA Lesser Antilles
West Indies
NT Northern Range
Tobago
Trinidad

Trinity County
Index states as applicable.
BX California
Texas
BT United States

Trinity Group
Lower Cretaceous. Subdivided into (ascending) Pearsall, Rodessa, Ferry Lake, and Rusk time-stratigraphic units. In SW Arkansas, NW Louisiana, central S and SE Oklahoma, and Texas.
BT Comanchean
Cretaceous
SA Antlers Sands
Arkansas
Glen Rose Formation
Louisiana
Oklahoma
Paluxy Formation
Texas

Trinity River
Flows into Trinity Bay off Galveston Bay on the Gulf of Mexico.
BT Texas
United States

triphylite
BA phosphates
BT minerals

triple junctions
Includes use on level 3 under plate tectonics(1) geometry(2).
SA plate tectonics

triplite
BZ fluorides
phosphates
BT minerals

Tripoli
City on the Mediterranean Sea in NW.
BT Libya

Tripolis
City in central.
BT Peloponnesus
Greece

Tripolitania
NW part of country which was a province until 1963. Under the Italians, it was a major province covering the entire W part of colony.
BT Libya
NT Garian

tripolite
use diatomaceous earth

Tripura
State E of Bangladesh in NE.
BA Northeastern India
BT India
SA Bengal

Tristan da Cunha
A group of 5 British volcanic islands in the central South Atlantic Ocean. Also the name of the main island of the group.
UF Tristan da Cunha Islands
BT Atlantic Ocean

Tristan da Cunha Islands
use Tristan da Cunha

Triticites
Genus. Includes use on level 3 under foraminifera(1) Fusulinidae(2).
BA Fusulinidae
Fusulinina
foraminifera
BT Protista
Invertebrata
NA Triticites ventricosus

Triticites ventricosus
Includes use on level 3 under foraminifera(1) Fusulinidae(2).
BA Triticites
Fusulinidae
Fusulinina
foraminifera
BT Protista
Invertebrata

tritium
Includes use on level 1 and 2. See list D (chemical elements). Also search H-3.
UF H-3
SA elements
hydrogen
isotopes

TRM
use thermoremanent magnetization

troctolite
As of 1981, includes use on level 3 under igneous rocks(1) gabbros(2). See list H. Before 1981, included use on level 3 under igneous rocks(1) gabbro family(2).
BA gabbros
BT igneous rocks

Troia
use Troy

troilite
BA sulfides
BT minerals
SA meteorites
pyrrhotite

Troja
use Troy

Trompia Valley
Valley of upper Mella River in E.
BT Lombardy
Italy

Troms
County in far N.
BT Norway
NT Lyngen Peninsula
SA Vesteralen

trona
Includes use as level 3 commodity term under sodium carbonate(1). See list C.
BA carbonates
BT minerals
SA sodium carbonate

Trona Village
Village in San Bernardino County in Mojave Desert in S. Before 1977, Trona was used to indicate the village.
BT California
United States

Trondelag
Region in central part of country between Norwegian Sea and Swedish border around Trondheim Fjord. Index counties as applicable.
BT Norway
SA Nord-Trondelag
Sor-Trondelag

Trondheim
City on Norwegian Sea.
BT Sor-Trondelag
Norway

trondhjemite
As of 1981, includes use on level 3 under igneous rocks(1) diorites(2). See list H. Before 1981, included use on level 3 under igneous rocks(1) diorite family(2).
BA diorites
BT igneous rocks
SA plagiogranite

Troodos Massif
Mountainous mass SW of Nicosia which includes Mt. Troodos, the highest peak on the island.
UF Troodos Mountain
Troodos Mountains
BT Cyprus
Middle East

Troodos Mountain
use Troodos Massif

Troodos Mountains
use Troodos Massif

tropical
A valid term through 1977. After 1977, use tropical environment.

tropical environment
Term introduced in 1978. Before 1978, search tropical.
SA climate
ecology
environment
paleoclimatology
subtropical environment

troposphere
Includes use on level 3 under aeronomy(1) composition(2).
SA aeronomy
atmosphere
stratosphere

Trotus Valley
River valley SSW of Bacau in W Moldavia. Also search Trotus; Trotus River.
BT Moldavia
Romania

Trou Sans Fond
Submarine canyon off SW Nigeria in the Bight of Benin.
UF Le Trou Sans Fond
Trou-Sans-Fond
BT Atlantic Ocean
SA Ivory Coast

Trou-Sans-Fond
use Trou Sans Fond

troughs
Includes use on level 2 under ocean floors(1).
SA aulacogens
bottom features
channels
ocean floors

Trousdale County
N Tennessee.
BT Tennessee
United States

Troy
Ancient ruined city. An archeological site, named Hissarlik, on the Menderes River S of the Dardanelles in NW Anatolia.
UF Ilion
Ilium
Troia
Troja
BT Turkey

Trucial Coast
A 350 mile section of S Persian Gulf littoral extending from Qatar to Ruus al Jubal. Often formerly used interchangeably with Trucial Oman and Trucial States. A valid term through 1974. After 1974, use United Arab Emirates.

Trucial Oman
Former name of 7 British protected states on the Trucial Coast now known as the United Arab Emirates. A valid term through 1972. After 1972, use United Arab Emirates.

Truckee River
Rises in Lake Tahoe in E California and flows NE and E into Pyramid Lake in NW Nevada. Index states as applicable. Also search Truckee.
BT United States
SA California
Nevada

Trujillo
State in NW.
BT Venezuela
SA Maracaibo Basin

Tsaritsyn
use Volgograd

tschermakite
BA clinoamphibole
amphibole group
chain silicates
silicates
BT minerals

Tschernosem
use Chernozems

Tschernosiom
use Chernozems

Tsugaru Strait
Channel between islands of Honshu and Hokkaido connecting the Pacific Ocean with the Japan Sea. Index islands as applicable. Also search Tsugaru.
BT Japan
SA Hokkaido
Honshu

Tsumeb
Mineral-collecting locality and mining center.
CO S191300S191300 E0174200E0174200
BT Namibia

tsunamis
Includes use on level 2 under seismology(1).
UF earthquake sea wave
seismic sea waves
seismic surge
tidal wave
SA catastrophic waves
earthquakes
geologic hazards
ocean waves
seismology
waves

Tsushima
Channel between Tsushima Island and NW Kyushu connecting the Japan Sea with the East China Sea.
BT Kyushu
Japan

Tuamotu Islands
Group of about 80 small islands in French Polynesia E of Society Islands and S of Marquesas Islands. Also search Tuamotu.
CO S230000S140000 W1340000W1490000
UF Dangerous Islands
Low Archipelago
BT French Polynesia
Polynesia
NT Mururoa Atoll

Tuapse
City on the Black Sea in S Krasnodar Kray in the Northern Caucasus.
BT Russian Republic
USSR

Tuar-Kyr
Village near Kara-Bogaz-Gol Gulf in W Turkmenia.
UF Tuarkyr
BT Turkmenia

USSR

Tuarkyr
use Tuar-Kyr

Tubarao Group
BT Carboniferous
Paleozoic
SA Brazil
Sao Paulo

tubes, lava
use lava tubes

Tuboidea
Includes use on level 2 under Graptolithina(1). See list F.
BA Graptolithina
BX Hemichordata
Invertebrata

Tubuai Islands
use Austral Islands

Tubulidentata
Includes use on level 2 under Mammalia(1). See list F.
BA Mammalia
BT Tetrapoda
Vertebrata

Tucano Basin
NE Bahia. Also search Tucano.
BT Bahia
Brazil

Tucson
City in Pima County in S.
CO N322000N322000
W1110000W1110000
BT Arizona
United States

Tucson Basin
S Arizona.
BT Arizona
United States

Tucson Mountains
W of Tucson in Pima County.
BT Arizona
United States

Tucuman
Province in NW.
BT Argentina
SA Sierras Pampeanas

Tucumcari
City in Quay County in NE.
BT New Mexico
United States

tufa
Includes use on level 3 under sedimentary rocks(1) chemically precipitated rocks(2) and carbonate rocks(2). Before 1981, also included use on level 3 under sediments(1) chemically precipitated sediments(2) and carbonate sediments(2). See list I (sed. rocks) and list N (sediments). Also search calcareous tufa.
UF calc-tufa
calcareous tufa
petrified moss
tophus
tuft
BA chemically precipitated rocks
BT sedimentary rocks
SA calcite
carbonate rocks
carbonate sediments
travertine

tuff
As of 1981, includes use on level 3 under igneous rocks(1) pyroclastics(2). See list H. Before 1981, included use on level 3 under igneous rocks(1) pyroclastics and glasses(2). Also includes use on level 3 under sediments(1) clastic sediments(2).
BA pyroclastics
volcanic rocks
BT igneous rocks
SA andesite tuff
ash
ash flows
ash-flow tuff
bentonite
green tuff
ignimbrite
metatuff
rhyolite tuff
tuffite
volcanic ash
welded tuff

tuff lava
use welded tuff

tuffite
As of 1981, includes use on level 3 under igneous rocks(1) pyroclastics(2). See list H. Before 1981, included use on level 3 under igneous rocks(1) pyroclastics and glasses(2).
BA pyroclastics
volcanic rocks
BT igneous rocks
SA tuff

tuffsite
use pipes

tuft
use tufa

Tugela Basin
River basin in central Natal. Also search Tugela; Tugela River.
BT Natal
South Africa

tugtupite
BA sodalite group
framework silicates
silicates
BT minerals

Tuimazy
use Tuymazy

Tuktoyaktuk Peninsula
Extends into the Beaufort Sea E of the Mackenzie River delta in District of Mackenzie. Also search Tuktoyaktuk.
CO N683000N703000
W1293000W1350000
BT Northwest Territories
Canada

Tula
City S of Moscow. Also an oblast.
BT Russian Republic
USSR

Tulameen coal area
S British Columbia.
BT British Columbia
Canada

Tulare County
S central.
BT California
United States

Tulare Formation
Pliocene and Pleistocene (?). Consists primarily of sandstone and conglomerate. In southern California.
BT Cenozoic
SA California
Pleistocene
Pliocene
Quaternary
Tertiary

Tulcea
City in the Danube Delta area in N.
BT Dobruja
Romania

Tulear Basin
Before 1978, also search Tulear.
BT Malagasy Republic

Tulghes Series
BT Paleozoic
SA Eastern Carpathians
Romania

Tully Limestone
In Susquehanna Group. Divided into several members in different areas: Apulia, Laurens, New Lisbon, Tinkers Falls, and West Brook.
BT Middle Devonian
Devonian
SA New York
Pennsylvania

Tulsa
City in Tulsa County in NE central.
BT Oklahoma
United States

Tulsa County
NE central.
CO N355000N362500
W0954500W0961500
BT Oklahoma
United States

tundra
SA ecology
geomorphology
permafrost
plains
soils
taiga environment

Tundra soils
Includes use on level 3 under soils(1). See list M.
BA soils
SA permafrost
Zonal soils

Tunganoxie Sandstone Member
use Tonganoxie Sandstone

tungstates
Includes use on level 2 under minerals(1). See list L.
BT minerals
NA ferberite
scheelite
stolzite
wolframite
SA molybdates

tungsten
Includes use on level 1 and 2 as a chemical element (list D). As of 1981, use tungsten ores for tungsten as a commodity. Before 1981, included use on level 1 as a commodity term.
UF W
SA tungsten ores

tungsten ores
Term introduced in 1981. Includes use on level 1 and 2 as a commodity term (list C). Before 1981, search tungsten AND ore deposits.
SA tungsten

tungstenite
BA sulfides
BT minerals

Tunguska
In GeoRef indexing, this term is used to indicate a general area. Name of three rivers flowing into the Yenisei River in central Siberia: the Lower Tunguska, Stony Tunguska, and the lower course of the Angara river which was called the Upper Tunguska. Index rivers as applicable.
CO N570000N710000
E1120000E0840000
BA Siberian Platform
BT Russian Republic
USSR
SA Angara River
Lower Tunguska River
Stony Tunguska River
Tunguska River

Tunguska Basin
Large coal basin in central Siberia between the Yenisei and Lena rivers in an area drained by the Lower Angara (Upper Tunguska), Stony Tunguska, and Lower Tunguska rivers. Also search Tunguska.
BT Russian Republic
USSR

Tunguska River
Tributary of the Amur River in the Soviet Far East near Khabarovsk. Not to be confused with the Lower Tunguska, Stony Tunguska, and Angara (Upper Tunguska) rivers that drain into the Yenesei River in central Siberia.
BT Russian Republic
USSR
SA Angara River
Lower Tunguska River
Stony Tunguska River
Tunguska

Tunguska Series
Composed of 2 sections (ascending): productive section with coal beds and a section of tufa containing intercalary beds of normal sedimentary rock. In Tunguska Basin of central Siberia.
BT upper Paleozoic
Paleozoic
SA Russian Republic
USSR

Tunguska Syneclise
The Tunguska Basin area in central Siberia. Also search Tunguska.
BT Russian Republic
USSR

Tunis
City on the Gulf of Tunis in N.
BT Tunisia

Tunisia
Includes use on level 1 as an area term (list O). For term set options see list B.
CO N303000N373000
E0120000E0073000
BA North Africa
Africa
NT Sebkha el Melah
Souk-el-Arba salt works
Tunis
Zarzis
SA Atlas Mountains
Mediterranean region
Sahara

tunnels
A valid level 1 term as of 1978. Used for geological studies on manmade tunnels. Before 1978, included use on level 2 under engineering geology(1). Term set options are:
construction
subtopic
design
subtopic
excavations
subtopic
experimental studies
subtopic
feasibility studies
subtopic
instruments
subtopic
rock mechanics
subtopic
seepage
subtopic
site exploration
subtopic
soil mechanics
subtopic
stability
subtopic

subways
  subtopic
theoretical studies
  subtopic
SA construction
  design
  engineering geology
  excavations
  feasibility studies
  foundations
  geologic hazards
  highways
  land subsidence
  Poisson's ratio
  rock mechanics
  seepage
  site exploration
  slope stability
  soil mechanics
  stability
  submarine installations
  subways
  underground installations

Tunsbergdalsbreen
Glacier emanating from larger Jostedalsbreen Glacier in W part of country.
BT Norway

Tuolumne County
E central.
BT California
  United States

Tura Formation
According to USSR Lexique, the age of Tura Series is upper Silurian. Central Urals and West Siberian Plain. Also search Tura.
BT Upper Silurian
  Silurian
SA Russian Republic

Turan
Desert lowland N, S, and SE of the Aral Sea. Index Soviet republics as applicable.
BT USSR
SA Kazakhstan
  Turkmenia
  Uzbekistan

Turanian Platform
Extends from the Caspian Sea on the W to the Tien Shan in the E. Index Soviet republics as applicable.
BT USSR
SA Kazakhstan
  Turkmenia
  Uzbekistan

turbidite
Includes use on level 3 under sedimentary rocks(1) clastic rocks(2). See list I. Also search turbidites.
BA clastic rocks
BT sedimentary rocks
SA terrigenous materials
  turbidity currents

turbidity
Includes use as level 3 under meteorology(1) aerosols(2).
SA aerosols
  meteorology

turbidity current structures
Includes use on level 2 under sedimentary structures(1). See list K.
BA sedimentary structures
NA graded bedding
  load casts
  sole marks
SA convoluted beds
  flow structures
  flute casts
  olistoliths
  olistostromes
  structures
  turbidity currents

turbidity currents
Includes use on level 3 under sedimentation(1) transport(2). For structures, see under sedimentary structures (list K).
UF suspension current
BT currents
SA density currents
  sedimentation
  submarine canyons
  submarine fans
  turbidite
  turbidity current structures

turbulence
Includes use on level 2 under aeronomy(1), meteorology(1) and under ocean circulation(1).
SA aeronomy
  boundary layer
  meteorology
  ocean circulation
  oceanography
  waterways

Turgai
use Turgay

Turgai Basin
use Turgay Basin

Turgai Depression
use Turgay Basin

Turgai Downwarp
use Turgay Basin

Turgai Gates
use Turgay Basin

Turgay
Oblast (region) NE of Aral Sea. Also a village. Also search Turgai.
UF Turgai
BT Kazakhstan
  USSR

Turgay Basin
An elongated depression joining the Turan Lowland and West Siberian Plain. Also search Turgai; Turgai Basin; Turgai Downwarp; Turgay.
CO N450000N570000
  E0680000E0573000
UF Turgai Basin
  Turgai Depression
  Turgai Downwarp
  Turgai Gates
  Turgay Depression
BA Central Asia
BT USSR
SA Kazakhstan

Turgay Depression
use Turgay Basin

Turin
City in W central Piemonte. Also search Torino.
UF Torino
BT Italy
SA Piemonte

Turkana District
Administrative district W of Lake Turkana (Rudolf) in NW.
BT Kenya

Turkestan
Region including S Kazakhstan and the rest of Soviet Central Asia, plus a small section of NE Afghanistan and W and SW Sinkiang Uighur. Index Afghanistan, Sinkiang Uighur, and Soviet Central Asia as applicable.
UF Turkistan
BT Asia
SA Afghanistan
  Sinkiang Uighur

Turkestan Range
W Tadzhikistan.
BT Tadzhikistan
  USSR

Turkey
Includes all of Turkey within its political boundaries. Includes use on level 1 as an area term (list O). For term set options see list B.
CO N355000N420000
  E0444500E0260000
BA Middle East
  Asia
NT Adana
  Alanya
  Amanos Mountains
  Amasra Basin
  Anatolia
  Ankara
  Antalya
  Ararat
  Bitlis
  Bosporus
  Elbistan
  Eskisehir
  Hatay
  Istanbul
  Keban Mine
  Kocaeli
  Konya
  Lycian Taurus
  Malatya
  Mudurnu
  Pontic Mountains
  Sea of Marmara region
  Taurus Mountains
  Troy
  Turkish Aegean region
  Tuz Golu
  Zonguldak
SA Balkan Peninsula
  Black Sea region
  Euphrates River
  Istranca Mountains
  Kura River
  Mardin Formation
  Maritsa River
  Mediterranean region
  Mediterranean Sea
  Near East
  Thrace

Turkish Aegean region
Term introduced in 1981.
CO N363000N400500
  E0313000E0260500
BT Turkey

Turkistan
use Turkestan

Turkmenia
Turkmen Soviet Socialist Republic. In Soviet Central Asia.
CO N350000N430000
  E0660000E0520000
UF Turkmenistan
BT USSR
NT Ashkhabad
  Badkhyz
  Balkhan
  Chardzhov
  Cheleken
  Cheleken Peninsula
  Darvaza
  Gaurdak
  Kara-Bogaz Gulf
  Kara-Bogaz-Gol
NX Karakum
NT Kizyl-Dere
NX Kopet-Dag Range
NT Kugitang-Tau
  Kurt
  Tuar-Kyr
SA Amu Darya
  Aral region
  Asiatic USSR
  Caspian Basin
  Central Asia
  Murgab Basin
  Turan
  Turanian Platform
  Ustyurt

Turkmenistan
use Turkmenia

Turku
Port city in SW part of country.
UF Abo
BT Finland

Turnyauz
use Tyrny-Auz

Turonian
Europe. Above Cenomanian, below Coniacian. Includes use on level 3 under age terms(1). See list E.
BA Upper Cretaceous
  Cretaceous
BT Mesozoic
SA Middle Cretaceous

turquoise
BA phosphates
BT minerals
SA gems

Turritella
Includes use on level 3 under Mollusca(1). See list F.
BA Turritellidae
  Gastropoda
  Mollusca
BT Invertebrata

Turritellidae
Includes use on level 3 under Mollusca(1). See list F.
BA Gastropoda
  Mollusca
BT Invertebrata
NA Turritella

Turukhan
River which flows into Yenisei River just N of Turukhansk in N Krasnoyarsk Kray in central Siberia.
BT Russian Republic
  USSR

Turukhansk
Town on Yenisei River at the mouth of the Lower Tunguska River in N Krasnoyarsk Kray in central Siberia.
BT Russian Republic
  USSR

Tusas Mountains
BT New Mexico
  United States

Tuscaloosa County
W central.
BT Alabama
  United States

Tuscaloosa Formation
Irregularly or obscurely bedded quartzitic and miracaceous sands interbedded with heterogeneous clays. Also lenticular pebble beds. Coastal plain from W Tennessee, NE Mississippi, and NW Alabama across Alabama, Georgia, South Carolina, and North Carolina.
BT Upper Cretaceous
  Cretaceous
SA Alabama
  Georgia
  Mississippi
  North Carolina
  South Carolina
  Tennessee

Tuscany
Autonomous region in W.
CO N422000N442000
  E0122500E0094000
BT Italy
NT Apuane Alps
  Arno River Basin
  Elba
  Florence
  Grosseto
  Larderello
  Livorno
  Lucca

T

Monte Amiata
Monte Capanne
Monte Pisano
Pisa
Siena
SA Tiber Valley

Tuscarawas County
E central.
BT Ohio
United States

Tuscarora Formation
In central Pennsylvania, it is gradational with overlying Rose Hill Formation and transitional with underlying Juniata Formation. In W Maryland, central S and E Pennsylvania, W Virginia, and E West Virginia. Also search Tuscarora Quartzite.
UF Tuscarora Quartzite
BT Lower Silurian
Silurian
SA Maryland
Pennsylvania
Virginia
West Virginia

Tuscarora Quartzite
use Tuscarora Formation

Tuttle Creek Dam
On Tuttle Creek forming reservoir in NE Kansas. Also search Tuttle Creek.
BT Kansas
United States

Tuva
Tuva Autonomous Soviet Socialist Republic. On the Mongolian border between the Altai Mountains on the W and the Sayan Mountains on the NE.
CO N493000N533000
E0990000E0890000
UF Tannu Tuva
BX Russian Republic
BZ Siberian fold belt
BT USSR

Tuvalu
Term introduced in 1981. Island group NE of Australia, N of Fiji Islands, and SE of Gilbert Islands.
CO S110000S050000
E1800000E1750000
UF Ellice Islands
BT Polynesia
SA Pacific Ocean

Tuymazy
Town W of Ufa in W Bashkiria in SE European USSR. Also search Tuimazy.
UF Tuimazy
BT Russian Republic
USSR

Tuz Golu
Salt lake in W central Anatolia.
UF Tuz Lake
BT Turkey
Middle East

Tuz Lake
use Tuz Golu

Twente
Region in SE Overijssel. Also search De Twente.
UF De Twente
BT Overijssel
Netherlands

Twiggs Clay
Consists of pale green hackly fuller's earth clay, green hackly clay, gray marl, and calcareous sand at base of member. In E Georgia.
UF Twiggs Clay Member
BT upper Eocene
Eocene
SA Georgia

Twiggs Clay Member
use Twiggs Clay

Twiggs County
Central.
BT Georgia
United States

Twilight Gneiss
According to latest published Lexique, the age of Twilight Granite is Precambrian. IN SE Colorado.
BT Precambrian
SA Colorado

Twin Lakes
N California and central Colorado. Index states as applicable.
BT United States
SA California
Colorado

Twin River Formation
Upper Eocene to upper Oligocene. Three mappable sequences recognized: lower member consisting of thin-bedded sandstone and siltstone; middle member of massive siltstone that grades westward into bedded siltstone and sandstone; and upper member composed chiefly of massive mudstone. In NW Washington.
BT Tertiary
SA Eocene
Oligocene
upper Eocene
upper Oligocene
Washington

Twin Sisters Dunite
An ultrabasic intrusive. In NW Washington.
BT Tertiary
SA Washington

twinning
As of 1978, term is used on level 2 under crystal growth(1).
SA crystal growth
epitaxy
lattice

Two Harbors
City on Lake Superior in Lake County in NE.
BT Minnesota
United States

two-dimensional models
Term introduced in 1978. Includes use on level 3 under seismology(1). Before 1978, search two-dimensional AND models.
BA models
SA one-dimensional models
seismology
three-dimensional models

two-layer models
Term introduced in 1981.
BA models

two-phase models
Term introduced in 1981.
BA models

Twocreekan
Term introduced in 1981. Substage of Wisconsinan stage. Includes use on level 3 under age terms(1). See list E. Before 1981, also search Twocreekian.
BA upper Pleistocene
Pleistocene
Quaternary
BT Cenozoic
SA Wisconsinan

Tyee Formation
Overlies Siletz River Volcanic Series. It consists of thick series of rhythmically bedded sandstone and intercalated siltstone. In W Oregon.
BT middle Eocene
Eocene
SA Oregon

Tylopoda
Term introduced in 1981. Includes use on level 2 under Mammalia(1). See list F.
BA Artiodactyla
Mammalia
BT Tetrapoda
Vertebrata

Tynagh
Village in S County Galway in W.
BT Ireland

Tyne River
Flows by Newcastle and enters the North Sea at Tynemouth in NE.
BT England
Great Britain
United Kingdom

type sections
SA sections
stratigraphy
stratotypes

type specimens
Used as a general term.
UF specimens, type
SA neotypes

types
A valid general term through 1977.

typomorphism
SA fossils
minerals

Tyrny-Auz
Town NW of Ordzhonikidze in the Northern Caucasus. Also search Turnyauz.
UF Turnyauz
BT Russian Republic
USSR

Tyrol
State between Germany and Italy in W.
CO N464500N474500
E0124300E0100400
UF Tirol
BT Austria
NT Baumkirchen
Innsbruck
Lanersbach
Lechtal Alps
Stubai Alps
Vernagt Glacier
SA Allgau Alps
Isar Valley
Silvretta Group
Venediger Group
Wetterstein Limestone

Tyrone
County in W.
BT Northern Ireland
United Kingdom

Tyrrhenian
Europe. Upper Pleistocene, Quaternary. See list E.
BA upper Pleistocene
Pleistocene
Quaternary
BT Cenozoic

Tyrrhenian Basin
Term introduced in 1981. Basin of the Tyrrhenian Sea.
CO N380000N430000
E0160000E0093000
BT Mediterranean Sea
SA Tyrrhenian Sea

Tyrrhenian Sea
Between Corsica and Sardinia on the W, the mainland of Italy on the E, and Sicily on the S. Also search Tyrrhenian. Includes use on level 1 as an area term (list O). For term set options see list B.
CO N380000N430000
E0160000E0093000
BA West Mediterranean
Mediterranean Sea
SA DSDP Site 132
Tyrrhenian Basin

Tyumen
City E of Sverdlovsk in W Siberia. Also an oblast.
BT Russian Republic
USSR

U

U
use uranium

U. S. Geological Survey
Includes use on level 3 under survey organizations(1). Also search United States Geological Survey; USGS; U.S.G.S.; U.S. Geological Survey.
UF United States Geological Survey
U.S. Geological Survey
U.S.G.S.
USGS
SA survey organizations

U. S. Rocky Mountains
Term introduced in 1981. Before 1981, search Rocky Mountains AND United States. Index states as applicable.
BX Rocky Mountains
United States
SA Central Rocky Mountains
Colorado
Idaho
Montana
New Mexico
North American Cordillera
Northern Rocky Mountains
Southern Rocky Mountains
Utah
Washington
Wyoming

U. S. Virgin Islands
Term introduced in 1981 for the Virgin Islands of the United States. Before 1981, search Virgin Islands.
BA Virgin Islands
Leeward Islands
BT Lesser Antilles
NA Saint Croix
Saint Thomas
SA United States

U-stage
use universal stage

U-234
Includes use on level 3 under isotopes(1).
SA isotopes
U-238/U-234
uranium

U-234/U-238
use U-238/U-234

U-235
Includes use on level 3 under isotopes(1).
SA isotopes
uranium

U-238
Includes use on level 3 under isotopes(1).
SA isotopes
U-238/U-234
uranium

U-238/Pb-206
Includes use on level 3 under absolute age(1) methods(2).
SA absolute age
lead
methods
U/He
U/Pb
U/Th/Pb
uranium

U-238/U-234
Includes use on level 3 under isotopes(1). Also search U-234/U-238; U-238 AND U-234.
UF U-234/U-238
SA isotopes
U-234
U-238
uranium

U/He
Includes use on level 3 under absolute age(1) methods(2).
SA absolute age
helium
methods
U-238/Pb-206
uranium

U/Pb
Includes use on level 3 under absolute age(1) methods(2).
SA absolute age
lead
methods
Pb-210
Pb/Pb
Pb/Th
U-238/Pb-206
uranium

U/Th/Pb
Includes use on level 3 under absolute age(1) methods(2).
SA absolute age
lead
methods
Th/U
thorium
U-238/Pb-206
uranium

Uaua
Village in NE.
BT Bahia
Brazil

Ube coal field
E of Shimonoseki in SW Honshu.
UF Ube Coalfield
BT Honshu
Japan

Ube Coalfield
use Ube coal field

Uchaly
Village in E Bashkiria in the Southern Urals in SE European USSR.
BT Russian Republic
USSR

Uchi Lake
Village on Uchi Lake in NW Ontario. Also search Uchi.
BT Ontario
Canada

Uchur River basin
Central W Khabarovsk Kray and SE Yakutia in E Siberia. Also search Uchur; Uchur River.
BT Russian Republic
USSR

Uda River
Rises in Irkutsk Oblast and flows into Angara River just above its junction with the Yenisei River in S central Siberia. Also search Uda.
UF Chuma River
BT Russian Republic
USSR

Udachnaya
Village just S of the Arctic Circle in W central Yakutia.
BT Russian Republic
USSR

Udaipur
City in S.
BT Rajasthan
India

Udmurt
use Udmurtia

Udmurtia
Udmurt Autonomous Soviet Sovialist Republic. W of the Urals and NE of Kazan. Also search Udmurt.
UF Udmurt
BT Russian Republic
USSR

Udokan Mountains
In area where Amur Oblast, Chita Oblast and Yakutia meet NE of Lake Baikal. Also search Udokan.
BT Russian Republic
USSR

Udokan Series
Lower Proterozoic. An aggregate of metamorphised sedimentary rocks including schists, phyllites, sandstones, and crystalline limestone. In NW Amur Oblast, N Chite Oblast, and S Yakutia NE of Lake Baikal.
BT Proterozoic
Precambrian
SA Russian Republic
USSR

Ufa
City in Bashkiria W of the Southern Urals.
BT Russian Republic
USSR

Ufaley
Town in NW Chelyabinsk Oblast in the Central Urals of W Siberia.
UF Nizhniy Ufaley
BT Russian Republic
USSR

Uganda
Former British protectorate which became independent in 1962. Includes use on level 1 as an area term (list O). For term set options see list B.
CO S020000N040000
E0350000E0290000
BA East Africa
Africa
NT Bukusu
Kakuto
Moroto Mountain
SA East African Rift
Lake Albert
Lake Edward
Lake Victoria
Nile River

Uinta Basin
NW Colorado and NE Utah. Index states as applicable.
UF Uintah Basin
BT United States
SA Colorado
Utah

Uinta County
Extreme SW.
BT Wyoming
United States

Uinta Mountains
Range chiefly in NE Utah. Index states as applicable.
BT United States
SA Rocky Mountains
Utah
Wyoming

Uintah Basin
use Uinta Basin

Uintah County
NE Utah.
BT Utah
United States

Ukraine
Ukrainian Soviet Socialist Republic.
CO N443000N523000
E0400000E0220000
BT USSR
NT Alma River
Artemovsk
Bakhchisarai
Bar
Belozerka
Beregovo
Bitkov
Boltyshka Depression
Borislav
Borkut Deposit
Borshchev
Cherkassy
Chernigov
NX Crimea
NT Crimean Mountains
Crimean Plain
Dnepropetrovsk
Dolina
Glinsk
Gorlovka Basin
Il'Intsa
Ingul River
Ingulets River
Ivano-Frankovsk
Kanev
Kerch
Kerch Peninsula
Kharkov
Kiev
Kirovograd
Konstantinovka
Korosten
Korsun
Kosov
Kremenchug
Krivoy Rog
Krivoy Rog Basin
Lebedin
Lvov
Lvov Basin
Lvov-Volyn Basin
Nikitovka
Nikopol
Odessa Oblast
Pavlograd
Pervomaisk
Podolia
Pokutye
Poltava
Rakhov
Rakhov Massif
Ros River valley
Rozdol
Rudki
Saksagan River
Simferopol
Sivash
Sob River basin
Stebnik
Tarkhankut Peninsula
Tereblya
Teterev River valley
NX Transcarpathia
NT Ukrainian Carpathians
NX Ukrainian Shield
NT Verkhovtsevo
Volhynia
Volyn
Volyn-Podolia
Yalta
Zaporozhe
Zhdanov
Zhitomir
SA Azov region
Beshchady Mountains
Brest Basin
Bug region
Bug River
Bukovina
Carpathian Foredeep
Carpathian Foreland
Carpathians
Danube Delta
Danube River
Danube Valley
Dnieper Basin
Dnieper River
Dnieper-Donets Basin
Dniester River
Dniester-Prut Interfluve
Donets Basin
Eastern Carpathians
European USSR
Galitsiya
Kiev Member
Korenevskaya Formation
Krivoy Rog Series
Maikop Series
Novorayskoe Formation
Odessa Meteorite
Ovruch Series
Polesye
Poltava Series
Pripet Basin
Protopivskaya Formation
Prut River
Roztocze
Russian Plain
Russian Platform
Scythian Platform
Serebryanka Formation
Siret River
Steppes
Subcarpathians
Tisza River
Ukrainian Syneclise
Vodino

Ukrainian Carpathians
E extension of the Carpathians in SW.
BT Ukraine
USSR
SA Carpathians

Ukrainian Shield
Extends from the S Polesye in the NW to the Azov Sea.
CO N450000N510000
E0390000E0280000
UF Volyno-Azov Massif
BZ Russian Platform
BX Ukraine
BT USSR

Ukrainian Syneclise
Term introduced in 1981.
CO N450000N530000
E0320000E0230000
BA Russian Platform
BT USSR
SA Ukraine

Ulan Bator
City in N central Mongolia.
UF Urga
BT Mongolia

Ulatisian
North America. Above Penutian, below Narizian. Includes use on level 3 under age terms(1). See list E.
BA middle Eocene
Eocene
Paleogene
Tertiary
BT Cenozoic

ulexite
UF boronatrocalcite
natroborocalcite
BA borates
BT minerals

Ulianovsk
use Ulyanovsk

Ulkan
Village NW of N Lake Baikal in Irkutsk Oblast.
BT Russian Republic
USSR

ullmannite
UF nickel-antimony glance
BA sulfides
BT minerals
SA willyamite

Ulmaceae
BA Dicotyledoneae
angiosperms
BT Spermatophyta
Plantae

Ulsan
Town in S part of country.
BT South Korea

Ulster
use Northern Ireland

Ulster County
S central.
BT New York
United States

Ultisols
Includes use on level 3 under soils(1). See list M.
BA soils

ultrabasic composition
As of 1981, no longer a valid term for GeoRef. Use ultramafic composition. Term was introduced in 1978.

ultrabasic rocks
Not a valid term for GeoRef. Use ultramafics or ultramafic composition.

ultrabasite
Also search diaphorite.
UF diaphorite
BA sulfides
BT minerals

ultramafic
A valid term through 1977. After 1977, use ultramafic composition.

ultramafic composition
Term introduced in 1978. Before 1978, search ultramafic; ultrabasic. Before 1981, also search ultrabasic composition.
SA composition
igneous rocks
mafic composition

ultramafic family
As of 1981, no longer a valid term for GeoRef. Use ultramafics or peridotites as applicable. See list H. Before 1981, included use on level 2 under igneous rocks(1).

ultramafic rocks
A valid term through 1976. Use ultramafics or ultramafic composition.

ultramafics
Term introduced in 1981. Includes use on level 2 under igneous rocks(1). See list H. Before 1981, search ultramafic family.
BT igneous rocks
NA ariegite
augitite
bronzitite
chromitite
clinopyroxenite
diallagite
eulysite
garnet pyroxenite
griquaite
hornblendite
kimberlite
komatiite
limburgite
meimechite
olivinite
ophiolite
peridotites
pyroxenite
websterite
SA ankaramite
leucitite
melilitite
nephelinite
picrite
picrite porphyry

ultrametamorphism
Includes use on level 2 under metamorphism(1).
BA metamorphism

ultrastructure
Includes use on level 3 under fossil groups (list F). Before 1978, also search microstructure.
SA paleontology
petrography

ultraviolet spectra
Term introduced in 1978. Before 1978, also search ultraviolet.
UF ultraviolet spectrum
SA spectra
ultraviolet spectroscopy

ultraviolet spectroscopy
Not a valid term from 1975 through 1977. After 1977, includes use on level 3 under spectroscopy(1) and under chemical element(1). Before 1978, also search spectroscopy AND ultraviolet.
BT spectroscopy
SA analysis
ultraviolet spectra

ultraviolet spectrum
use ultraviolet spectra

Ulu-Tau
Range W of the Kazakh Hills in Dzhezkazgan Oblast in central Kazakhstan. Also search Ulutau.
UF Ulutau
BT Kazakhstan
USSR

Ulutau
use Ulu-Tau

ulvospinel
BA oxides
BT minerals
SA spinel group

Ulyanovsk
City on the right bank of the Volga River S of Kazan in Ulyanovsk Oblast.
UF Simbirsk
Ulianovsk
BT Russian Republic
USSR

Umanak
Settlement on S shore of Umanak Fjord midway up W coast.
BT Greenland
Arctic region

umangite
As of 1981, includes use on level 3 under minerals(1) selenides(2). See list L. Before 1981, included use on level 3 under minerals(1) sulfides(2).
BA selenides
sulfides
BT minerals

Umaria
Town in NE.
BT Madhya Pradesh
India

Umba
River in Kenya which flows into Indian Ocean just N of Tanzanian boundary. Index countries as applicable.
BT Africa
SA Kenya
Tanzania

Umbria
Autonomous region in central.
CO N422000N434000
E0131500E0115000
BT Italy
NT Perugia
Spoleto
Terni
SA Tiber Valley

Umm al Qaiwain
use Umm al-Qaiwain

Umm al-Qaiwain
Emirate. One of federation of 7 states at S end of Persian Gulf.
UF Umm al Qaiwain
BT United Arab Emirates
Arabian Peninsula

umohoite
BA molybdates
BT minerals

Umpqua Formation
Lower to middle Eocene. Predominantly medium-grained sandstone with some shaly and conglomerate layers. In N California and SW Oregon.
BT Eocene
SA California
lower Eocene
middle Eocene
Oregon

Umrer
Town in central.
BT Madhya Pradesh
India

unconfined aquifers
BT aquifers
SA ground water
water table

unconformities
Includes use on level 3 under age terms(1). See list E. Also search unconformity.
SA stratigraphic gaps
stratigraphy

unconsolidated materials
Term introduced in 1978. Before 1978, search unconsolidated.
SA materials
particulate materials
sediments
soil mechanics
soils

UNCTAD
Term introduced in 1981.
SA United Nations

Unda
Village in S Chita Oblast in Transbaikalia.
BT Russian Republic
USSR

undation
SA crust
tectonics
theoretical studies

underclay
Also search seat earth.
UF coal clay
root clay
seat clay
seat earth
underearth
SA clay
coal seams
fireclay

underearth
use underclay

underground
A valid term through 1977. After 1977, use underground space.

underground channels
Term introduced in 1981.
SA channels
infiltration galleries
underground installations
waterways

underground installations
A valid level 1 term as of 1978. Used for geological studies on underground cavities (natural or otherwise), excluding tunnels. Before 1978, included use on level 2 under engineering geology(1). Term set options are:
construction
subtopic
design
subtopic
excavations
subtopic
experimental studies
subtopic
feasibility studies
subtopic
instruments
subtopic
mines
subtopic
rock mechanics
subtopic
seepage
subtopic
site exploration
subtopic
soil mechanics
subtopic
stability
subtopic
theoretical studies
subtopic
underground space
subtopic
waste disposal
subtopic
UF installations, underground
SA construction
design
engineering geology
excavations
feasibility studies
foundations
gas storage
geologic hazards
land subsidence
mines
permafrost
rock mechanics
seepage
site exploration
slope stability
soil mechanics
stability
storage
tunnels
underground channels
underground space
utilization
waste disposal

underground space
Term introduced in 1978 on level 2 under land use(1) and underground installations(1).
UF space, underground
SA land use
underground installations

underthrust faults
Term introduced in 1978. Before 1978, search underthrust AND faults.
BA faults

UNDP
Term introduced in 1981.
SA United Nations

undulatory extinction
Term introduced in 1981.

SA microscope methods
minerals
optical extinction

Ungava
Region including Quebec N of Eastmain River and SW Labrador. Also search New Quebec. Index Labrador and/or Quebec as applicable.
CO N580000N623000
W0690000W0780000
UF New Quebec
BT Canada
SA Labrador
Quebec

Ungulata
As of 1981, includes use on level 3 under Mammalia(1). See list F. Before 1981, included use on level 2 under Mammalia(1). Before 1981, also search ungulates. As of 1981, search ungulates only for biostratigraphy.
BA Mammalia
BT Tetrapoda
Vertebrata

uniaxial tests
Term introduced in 1978. Before 1978, search uniaxial.
SA engineering geology
triaxial tests

UNIDO
Term introduced in 1981.
SA United Nations

uniformitarianism
Includes use on level 3 under geology(1).
SA catastrophism
deposition
geology
history
sedimentation
stratigraphy

Union County
Index states as applicable.
BX Arkansas
Florida
Georgia
Illinois
Indiana
Iowa
Kentucky
Mississippi
New Jersey
New Mexico
North Carolina
Ohio
Oregon
Pennsylvania
South Carolina
South Dakota
Tennessee
BT United States
SA Pasamonte Meteorite

Union Islands
use Tokelau

Unionidae
Includes use on level 3 under Mollusca(1). See list F.
BA Bivalvia
Mollusca
BT Invertebrata

unit cell
UF cell, unit
SA crystal structure
lattice
lattice parameters
minerals

United Arab Emirates
Includes use on level 1 as an area term (list O). Federation of 7 states which was achieved in 1972. Formerly known as Trucial States, Trucial Oman, or Trucial Coast. Not strictly equivalent to Trucial Coast, but replaces Trucial Coast as first order term. For term set options see list B. Before 1973, also search Trucial Oman. Before 1975, also search Trucial Coast.
CO N224000N255000
E0563000E0510000
BA Arabian Peninsula
Asia
NT Abu Dhabi
Ajman
Dubai
Fujairah
Ras al-Khaimah
Sharjah
Umm al-Qaiwain
SA Middle East
Near East

United Kingdom
Index political divisions as applicable. Comprising Great Britain and Northern Ireland. Includes use on level 1 as an area term (list O). For term set options see list B.
CO N500000N610000
E0020000E0110000
BA Western Europe
Europe
NT Bristol Channel
NA English Channel Islands
NT Esk Trough
NA Great Britain
NT Isle of Man
NA Northern Ireland
NT Severn Valley
Wye Valley
SA Aberystwyth Grits
Barton Beds
Bembridge Marls
British Virgin Islands
London Clay
Lower Greensand
Manchester
Oxford Clay
Penrhyn Slate
Shap Granite
Skiddaw Slates
Speeton Clay
Upper Old Red Sandstone
Weald Clay
Wenlock Limestone

United Nations
SA associations
UNCTAD
UNDP
UNIDO

United States
Includes use on level 1 or 2 as an area term (list O). If 1, see term set options under list B.
BT North America
NT Absaroka Range
Allegheny Front
Allegheny Mountains
Allegheny Plateau
Amargosa Desert
Amite River
Anacostia River basin
Anadarko Basin
Appalachian Basin
Appalachian Plateau
Arkansas River
Arkansas River valley
Arkoma Basin
NA Atlantic Coastal Plain
NT Bald Mountain
NA Basin and Range Province
NT Battle Mountain High
Bear Lake
Bear River basin
Bear River Range
Beartooth Mountains
Beaverhead Mountains
Bighorn Basin
Bighorn Mountains
Bitterroot Range
Black Hills
Black Warrior Basin
Blue Ridge Mountains
Bronson Hill Anticlinorium
Cache Valley
Cambridge Arch
Canadian River
Carolina Bays
Cascade Range
Catskill Delta
Chadron Arch
Chattahoochee River
Choptank River
Cincinnati Arch
NA Colorado Plateau
Columbia Plateau
NT Columbia River estuary
Connecticut River
Connecticut Valley
Cumberland Plateau
Dan River basin
Delaware Bay
Delaware River
Delaware River basin
Delmarva Peninsula
Denver Basin
Dunkard Basin
NA Eastern U.S.
NT Flint Hills
Forest City Basin
Four Corners
Gallatin Range
Gila River
NA Great Basin
NT Great Smoky Mountains
Green Bay
Green River
Green River basin
Guadalupe Mountains
NZ Hawaii
NT Hudson River
Hudson Valley
Hugoton Embayment
Hurricane Ridge Syncline
Idaho Batholith
Illinois Basin
Kings Mountain
Klamath Mountains
Lake Bonneville
Lake Chatuge
Lake Mead
Lake Powell
Lake Tahoe
Laramie Mountains
Little Missouri River basin
Llano Estacado
Maumee River valley
Medicine Bow Mountains
Merrimack River valley
Merrimack Synclinorium
Midcontinent
NA Midwest
NT Mississippi Embayment
Mississippi River
NA Mississippi Valley
NT Missouri River
Missouri River valley
Navajo Indian Reservation
Nemaha Ridge
New Madrid region
New River
Newark Basin
Newark-Gettysburg Basin
Ohio River
Ohio River valley
Okefenokee Swamp
Ouachita Mountains
Ozark Mountains
Panhandle
Paradox Basin
Pecos River valley
Permian Basin
Piedmont
Potomac River
Potomac River basin
Powder River basin
Raft River
Raft River basin
Rangeley Lakes
RARE II regions
Raton Basin
Red River
Red River valley
Sabine Lake
Saco River
San Francisco Mountains
San Juan Basin
San Juan River
San Luis Valley
Sand Hills
Sangre de Cristo Mountains
Savannah River
Shadow Mountains
Shenandoah Valley
Snake River
Snake River basin
Snake River canyon
South Mountain
South Platte River valley
NA Southern U.S.
Southwestern U.S.
NT Springfield
Sunlight
Susquehanna River
Susquehanna River basin
Tennessee River
Tennessee Valley
Trans-Pecos
Truckee River
Twin Lakes
NX U. S. Rocky Mountains
NT Uinta Basin
Uinta Mountains
Valley and Ridge Province
Verdigris River valley
Virgin River valley
Wasatch Front
Wasatch Range
Washakie Basin
Washita River valley
Western Overthrust Belt
NA Western U.S.
NT White Mountains
Whitewater River valley
Wilmington
Yellowstone National Park
SA Appalachians
Canyon Group
conterminous regions
Great Lakes
Great Lakes region
Great Plains
Gulf Coastal Plain
Huronian
Lake Michigan
Manchester
Mississippi River basin
Missouri River basin
North American Cordillera
Rocky Mountains
Taconic Allochthon
Taconic Orogeny
U. S. Virgin Islands
Western Interior

United States Geological Survey
use U. S. Geological Survey

units, stratigraphic
use stratigraphic units

universal stage
Includes use on level 3 under structural analysis(1) methods(2).
UF Fedorov stage
stage, universal
U-stage
SA microscope methods
ore microscopy
structural analysis

University of Arizona
In Tucson.
BT Arizona
United States

U

University of Bonn
In Bonn.
BT North Rhine-Westphalia
West Germany

University of California
In Berkeley.
BT California
United States
SA Los Alamos Scientific Laboratory

University of Cambridge
use Cambridge University

University of Louvain
In Louvain E of Brussels.
BT Brabant
Belgium

University of Lund
In Lund in extreme S part of country.
BT Malmohus
Sweden

University of Michigan
In Ann Arbor W of Detroit.
BT Michigan
United States

University of Pennsylvania
In Philadelphia.
BT Pennsylvania
United States

University of Rome
In Rome.
BT Latium
Italy

University of Tokyo
In Tokyo.
UF Tokyo Imperial University
BT Honshu
Japan

University of Wisconsin
In Madison.
BT Wisconsin
United States

Unkar Group
Grand Canyon Series. Divided into (descending) Dox Sandstone, Shinumo Quartzite, Hakatai Shale, Bass Limestone, and Hotuata Conglomerate. In N Arizona.
BT Precambrian
SA Arizona

unsaturated zone
Term introduced in 1981.
UF vadose zone
zone of aeration
SA capillary water
ground water
saturated zone
soils
water regimes
water table

Unst
Northernmost large island of the Shetland Islands off N.
BT Scotland
Great Britain
United Kingdom
SA Shetland Islands

Unstrut River
Flows into the Saale River in SW East Germany. Also search Unstrut. Index districts as applicable.
BT East Germany
SA Erfurt Bezirk
Halle Bezirk

Unzen
Active volcano. In Unzen National Park on central Shimabar Peninsula E of Nagasaki. Also search Unzen Volcano.
UF Unzen Volcano
BT Kyushu
Japan

Unzen Volcano
use Unzen

Uonuma Group
Upper Pliocene to lower Pleistocene. Composed of Oguni and Tsukayama formations. In NW Honshu.
BT Cenozoic
SA Honshu
Japan
lower Pleistocene
Pleistocene
Pliocene
Quaternary
Tertiary
upper Pliocene

uplifts
As of 1978, includes use on level 2 under neotectonics(1). Also includes use on level 3 under tectonics(1).
SA domes
epeirogeny
neotectonics
subsidence
tectonics
vertical movements

upper atmosphere
BA atmosphere

Upper Austria
State in NW.
CO N472600N484700
E0150000E0124500
BT Austria
NT Gosau
Linz
Weyer
SA Enns Valley
Salzach River
Salzkammergut
Totes Gebirge

Upper Bavaria
An administrative division in the extreme S in which Munich, the Bavarian Alps and Salzburg Alps are located.
BT Bavaria
West Germany

upper boundary
A valid general term through 1976. Now use boundary for stratigraphic meaning.

Upper Cambrian
See list E.
BA Cambrian
BT Paleozoic
NT Bonneterre Formation
Franconia Formation
Little Falls Formation
Pilgrim Formation
Potsdam Sandstone
Reagan Sandstone
NA Trempealeauan
NT Wilberns Formation
SA Arbuckle Group
Deadwood Formation
Knox Group
Lower Cambrian
Middle Cambrian

Upper Carboniferous
See list E.
BA Carboniferous
BT Paleozoic
NA Bashkirian
Coal Measures
Gzhelian
NT Karharbari Stage
NA Kasimovian
Millstone Grit
Moscovian
Namurian
Stephanian
NT Talchir Series
NA Uralian
Westphalian
SA Dinantian
Gondwana System
Pennsylvanian
Talchir Formation

upper Cenozoic
See list E.
BA Cenozoic
SA Quaternary

Upper Cretaceous
See list E.
BA Cretaceous
BT Mesozoic
NT Almond Formation
Ariyalur Stage
Bagh Beds
Bearpaw Formation
Belly River Formation
Blackhawk Formation
Carlile Shale
NA Cenomanian
NT Cody Shale
Djadokhta Formation
Edmonton Formation
Ferron Sandstone Member
Fort Hays Limestone Member
Fox Hills Formation
Frontier Formation
Gallup Sandstone
Greenhorn Limestone
NA Gulfian
NT Harebell Formation
Hell Creek Formation
Izumi Group
Judith River Formation
La Ventana Sandstone
Lance Formation
Laramie Formation
Lewis Shale
Magothy Formation
Marca Shale Member
Marshalltown Formation
Menefee Formation
Mesaverde Group
Milk River Formation
Monmouth Group
Montana Group
Moreno Formation
Mowry Shale
Navesink Formation
Niobrara Formation
Ojo Alamo Sandstone
Oldman Formation
Parkman Sandstone
Peedee Formation
Pierre Shale
Raritan Formation
Rincon Formation
Ripley Formation
Rosario Formation
Saratoga Chalk
NA Senonian
NT Smoky Hill Chalk Member
NA Turonian
NT Tuscaloosa Formation
Woodbury Clay
SA Bauru Formation
Beaverhead Formation
Benton Formation
Blairmore Group
Colorado Group
Comanchean
Dakota Formation
Deccan Traps
Difunta Group
Ewekoro Formation
Fort Union Formation
Graneros Shale
Great Valley Sequence
Intertrappean Beds
Lower Cretaceous
Mancos Shale
Middle Cretaceous
Niniyur Group
Paskapoo Formation
Potomac Group
Vraconian
Washita Group

upper crust
BA crust
SA lower crust

Upper Devonian
See list E.
BA Devonian
BT Paleozoic
NA Famennian
Frasnian
NT Greenland Gap Group
Lime Creek Formation
New Albany Shale
Ohio Shale
Olentangy Shale
Palliser Formation
Perry Formation
Sonyea Group
NA Strunian
SA Catskill Formation
Chattanooga Shale
Genesee Group
Lost Burro Formation
Lower Devonian
Martin Formation
Middle Devonian
Muth Quartzite
Traverse Group

upper Eocene
See list E.
BA Eocene
Paleogene
Tertiary
BT Cenozoic
NA Auversian
NT Jackson Group
NA Lattorfian
Ludian
NT Ocala Group
Poway Conglomerate
NA Priabonian
NT Tejon Formation
NA Tongrian
NT Twiggs Clay
Yazoo Clay
SA Biarritzian
Castle Hayne Limestone
lower Eocene
middle Eocene
Narizian
Sespe Formation
Twin River Formation

Upper Franconia
Administration division in NE Bavaria. Part of old historical region of Franconia.
BT Bavaria
Germany
SA Franconia

upper Holocene
See list E.
BA Holocene
Quaternary
BT Cenozoic

Upper Jurassic
See list E.
BA Jurassic
BT Mesozoic
NA Argovian
NT Bossier Formation
Buckner Formation
Cotton Valley Group
Jabalpur Series
NA Kimmeridgian
Lusitanian
Malm
NT Morrison Formation
NA Oxfordian
Portlandian
NT Raghavapuram Shales
NA Rauracian
NT Schuler Formation
NA Sequanian
NT Smackover Formation
Stump Formation

Sundance Formation
Todilto Formation
NA Volgian
NT Zuloaga Limestone
SA Bowser Formation
Lower Jurassic
Middle Jurassic

upper Liassic
Term introduced in 1981. Includes use on level 3 under age terms(1).
See list E.
BA Lower Jurassic
Jurassic
BT Mesozoic
NA Toarcian

Upper Lusatia
That part of Lusatia in SE East Germany. Index districts as applicable.
BT East Germany
SA Cottbus Bezirk
Dresden Bezirk
Lower Lusatia
Lusatia

upper mantle
Includes use on level 3 under mantle(1).
UF outer mantle
BA mantle
SA asthenosphere
lower crust
lower mantle

upper Mesozoic
See list E.
BA Mesozoic
SA Caracas Group
middle Mesozoic

upper Miocene
See list E.
BA Miocene
Neogene
Tertiary
BT Cenozoic
NT Duplin Formation
NA Meotian
Messinian
NT Montesano Formation
NA Pontian
NT Puente Formation
Punchbowl Formation
Santa Margarita Formation
NA Sarmatian
Tortonian
NT Yorktown Formation
SA Capistrano Formation
Cuddalore Series
lower Miocene
middle Miocene
Monterey Formation
Neyveli Lignite
Saint Marys Formation
Siwalik System
Warkalli Formation

Upper Mississippi Valley
That part of the Mississippi Valley N of Cairo, Illinois. Index states as applicable.
BA Mississippi Valley
BT United States
SA Illinois
Iowa
Minnesota
Missouri
Wisconsin

Upper Mississippian
See list E.
BA Mississippian
Carboniferous
BT Paleozoic
NT Bangor Limestone
NA Chesterian
NT Fayetteville Formation
NA Meramecian
NT Monteagle Limestone
Pennington Formation
Pitkin Limestone
SA Bird Spring Formation
Borden Group
Leadville Formation
Lisburne Group
Lower Mississippian
Madison Group
Monte Cristo Limestone
Saint Louis Limestone
Valmeyeran
Warsaw Formation

Upper Old Red Sandstone
Represented by Nairn, Boghole, Alves-Scaat Craig, Rosebrae, Plateau, and Portishead beds; by Quartz Conglomerate; and by Farlow Sandstone. In Great Britain.
BT Devonian
SA Great Britain
United Kingdom

upper Oligocene
See list E.
BA Oligocene
Paleogene
Tertiary
BT Cenozoic
NA Chattian
NT Chickasawhay Formation
NA Egerian
NT Suwannee Limestone
SA Arikareean
Brule Formation
John Day Formation
lower Oligocene
middle Oligocene
Twin River Formation

Upper Ordovician
See list E.
BA Ordovician
BT Paleozoic
NA Ashgillian
Caradocian
Cincinnatian
Edenian
NT Fairview Formation
Fernvale Formation
Juniata Formation
Maquoketa Formation
Red River Formation
Richmond Group
Utica Shale
SA Bala
Chickamauga Group
Lower Ordovician
Martinsburg Formation
Middle Ordovician
Viola Limestone

Upper Palatinate
That part of the historical region of the Palatinate located in E.
BT Bavaria
West Germany
SA Palatinate

upper Paleocene
See list E.
BA Paleocene
Paleogene
Tertiary
BT Cenozoic
NA Landenian
Thanetian
SA lower Paleocene

upper Paleozoic
See list E.
BA Paleozoic
NT Dwyka Series
Tunguska Series
SA Rio Bonito Formation

Upper Pennsylvanian
See list E.
BA Pennsylvanian
Carboniferous
BT Paleozoic
NT Ames Limestone
Canyon Group
NA Missourian
Virgilian
SA La Salle Limestone
Lower Pennsylvanian
Middle Pennsylvanian

Upper Permian
See list E.
BA Permian
BT Paleozoic
NT Barabash Suite
NA Kazanian
NT Raniganj Stage
NA Tatarian
NT Usuginu Conglomerate
Vetluga Series
NA Zechstein
SA Estrada Nova Formation
Guadalupian
Lower Permian
Middle Permian
San Andres Formation

upper Pleistocene
See list E.
BA Pleistocene
Quaternary
BT Cenozoic
NA Cromerian
Devensian
Eemian
Holsteinian
Hoxnian
Illinoian
Ipswichian
Saalian
Sangamonian
Sicilian
Twocreekan
Tyrrhenian
Weichselian
Wisconsinan
Wurm
SA Brunhes Epoch
lower Pleistocene

upper Pliocene
See list E.
BA Pliocene
Neogene
Tertiary
BT Cenozoic
NA Akchagylian
Astian
Levantinian
Redonian
NT Rexroad Formation
NA Romanian
SA lower Pliocene
Matuyama Epoch
middle Pliocene
Uonuma Group
Yakima Basalt

upper Precambrian
See list E.
BA Precambrian
NA Hadrynian
Infracambrian
Jotnian
Laxfordian
Moinian
Riphean
Torridonian
SA Adelaidean
Algonkian
Proterozoic
Sparagmite Group
Vendian

upper Proterozoic
See list E.
BA Proterozoic
Precambrian
NA Brioverian
Vendian
SA Adelaidean

upper Quaternary
See list E.
BA Quaternary
BT Cenozoic
NT Pinedale Glaciation

Upper Rhine
use Upper Rhine Valley

Upper Rhine Graben
BT Rhine Graben
Europe

Upper Rhine Valley
As of 1981, restricted to that part of the Rhine Valley between Basel and Mainz in West Germany. Before 1981, included part of France.
CO N473000N502800
E0090700E0073200
UF Upper Rhine
BT Germany
SA Rhine Valley
West Germany

Upper Silesia
The eastern part of former German Silesia plus Polish Silesia lying SE of Lower Silesia centering on Katowice and Krakow.
UF Upper Silesian area
BT Poland
SA Lower Silesia
Silesia
Silesian coal basin

Upper Silesia coal basin
use Upper Silesian coal basin

Upper Silesian area
use Upper Silesia

Upper Silesian Basin
use Upper Silesian coal basin

Upper Silesian coal basin
N foot of the W Beskid Mountains in Upper Silesia.
UF Upper Silesia coal basin
Upper Silesian Basin
Upper Silesian coal field
BT Katowice
Poland
SA Silesian coal basin

Upper Silesian coal field
use Upper Silesian coal basin

Upper Silurian
See list E.
BA Silurian
BT Paleozoic
NT Bloomsburg Formation
NA Cayugan
Ludlovian
Pridolian
NT Salina Group
Tonoloway Limestone
Tura Formation
NA Wenlockian
SA Downtonian
Lower Silurian
Middle Silurian
Read Bay Formation
Rondout Formation
Williamsport Sandstone

upper Tertiary
See list E.
BA Tertiary
BT Cenozoic
NT Poznan Clays
SA Akchagylian
Gauss Epoch
Gilbert Epoch
lower Tertiary
Matuyama Epoch
middle Tertiary
Neogene
San Juan Formation
Siwalik System

Upper Triassic
See list E.
BA Triassic
BT Mesozoic
NA Carnian

NT Chinle Formation
Dockum Group
Kayenta Formation
NA Keuper
NT Newark Group
NA Norian
NT Protopivskaya Formation
Redonda Formation
NA Rhaetian
NT Stormberg Series
Torlesse Supergroup
SA Hallstatt Limestone
Lower Triassic
Middle Triassic
Navajo Sandstone
Novorayskoe Formation
Pucara Group

Upper Tunguska
use Angara River

Upper Volta
Former French protectorate which achieved independence in 1960. Includes use on level 1 as an area term (list O). For term set options see list B.
CO N092500N151000
E0022000W0053000
UF Voltaic Republic
BA West Africa
Africa
SA Sahel
Volta Basin

Upper Zaire
Term introduced in 1981.
CO S020000N053000
E0312000E0221000
BT Zaire

Uppsala
County in E Sweden. Also a city.
BT Sweden

Upton County
W Texas.
BT Texas
United States

upwelling
Includes use on level 3 under ocean circulation(1).
SA currents
ocean circulation
oceanography
sea water

Ural Mountains
use Urals

Ural Range
use Urals

Ural region
A mining and industrial region on both sides of the Central Urals comprising the Chelyabinsk, Sverdlovsk, Kurgan, Orenburg, and Perm oblasts; and Bashkiria and Udmurtia. Also search Ural.
UF Urals Region
BT Russian Republic
USSR

Ural River
Rises at S end of Urals and flows into the Caspian Sea. Index Soviet republics as applicable. Also search Ural.
BT USSR
SA Kazakhstan
Russian Republic

Ural-Tau
Low, water-divide mountain range in E Southern Urals.
BT Russian Republic
USSR
SA Southern Urals

Ural-Volga Interfluve
use Volga-Ural region

Ural-Volga region
use Volga-Ural region

Uralian
Russia. Above Gzhelian, below Sakmarian of Permian. Includes use on level 3 under age terms(1). See list E.
BA Upper Carboniferous
Carboniferous
BT Paleozoic

Uralian Foreland
The western slope and foothills of the Urals.
BT Russian Republic
USSR

uralitization
BA metasomatism
SA hydrothermal alteration
metamorphism

Urals
Mountain range extending from the Kara Sea in the N to W Kazakhstan border in the S separating Europe from Asia. Comprises the Northern, Central, and Southern Urals. Index ranges as applicable. Also search Ural Mountains.
CO N473000N773000
E0700000E0560000
UF Ural Mountains
Ural Range
BT USSR
NZ Central Urals
Mugodzhar Hills
Northern Urals
Novaya Zemlya
Polar Urals
NA Southern Urals
SA Kazakhstan
Russian Republic

Urals Region
use Ural region

uraninite
BA oxides
BT minerals
SA pitchblende

uranium
Includes use on level 1 and 2 as a chemical element (list D). As of 1981, use uranium ores for uranium as a commodity. Before 1981, included use on level 1 as a commodity term.
UF U
SA elements
Io/U
isotopes
radium
Th/U
U-234
U-235
U-238
U-238/Pb-206
U-238/U-234
U/He
U/Pb
U/Th/Pb
uranium disequilibrium
uranium minerals
uranium ores

uranium disequilibrium
SA absolute age
isotopes
uranium

uranium minerals
Includes use under commodities (list C).
BT minerals
SA miscellaneous minerals
uranium
uranium ores

uranium ores
Term introduced in 1981. Includes use on level 1 and 2 as a commodity term (list C). Before 1981, search uranium AND ore deposits.
SA roll-type deposits
uranium
uranium minerals

uranophane
BA orthosilicates
silicates
BT minerals

Uranus
Includes use on level 1 and 2. See entry under Moon(1) for term set options.
SA outer planets
planetology
planets
solar system

urban planning
As of 1978, term is used on level 2 under land use(1).
SA environmental geology
land use
planning
regional planning
urbanization

Urbana
City in E Illinois and in W central Ohio. Index states as applicable.
BX Illinois
Ohio
BT United States

urbanization
Includes use on level 3 under environmental geology(1).
SA conservation
environmental geology
land use
urban planning

ureilite
Term introduced in 1981. Includes use on level 3 under meteorites(1). Before 1981, also search ureilites.
BA achondrites
BT meteorites
NA Haveroe Meteorite

ureilites
As of 1981, no longer a valid term for GeoRef. Use ureilite.

Urga
use Ulan Bator

Urgonian
Europe. See list E.
BA Lower Cretaceous
Cretaceous
BT Mesozoic

Uri
Canton in central.
BT Switzerland
SA Oberalp Pass

Ursidae
Includes use on level 3 under Mammalia(1). See list F.
BA Carnivora
Mammalia
BT Tetrapoda
Vertebrata
NA Ursus

Ursus
Includes use on level 3 under Mammalia(1). See list F.
BA Ursidae
Carnivora
Mammalia
BT Tetrapoda
Vertebrata
NA Ursus spelaeus

Ursus spelaeus
Includes use on level 3 under Mammalia(1). See list F.
BA Ursus
Ursidae
Carnivora
Mammalia
BT Tetrapoda
Vertebrata

urtite
As of 1981, includes use on level 3 under igneous rocks(1) alkali gabbros(2). See list H. Before 1981, included use on level 3 under igneous rocks(1) alkali gabbro family(2).
BA alkali gabbros
gabbros
BT igneous rocks
SA ijolite
melteigite

Uruguay
Includes use on level 1 as an area term (list O). For term set options see list B.
CO S350000S300000
W0530000W0583000
BA South America
NT Montevideo
Rio Negro Department
SA Rio de la Plata

Urup
Volcanic island in S central Kuril Islands NE of Japan.
UF Urup Island
BT Russian Republic
USSR
SA Kuril Islands

Urup Island
use Urup

U.S. Geological Survey
use U. S. Geological Survey

Usa Series
Includes marble plus dolomites interspersed with beds of siliceous schist. In the Kuznetsk Alatau mountain region of Kemerovo Oblast and Khakass Autonomous Oblast E of Novosibirsk.
BT Lower Cambrian
Cambrian
SA Russian Republic
USSR

uses
A valid term through 1976. Use applications or utilization.

U.S.G.S.
use U. S. Geological Survey

USGS
use U. S. Geological Survey

Uspenski
use Uspenskiy

Uspenskiy
Town in Karaganda Oblast in E central Kazakhstan. Also search Uspenski.
UF Uspenski
BT Kazakhstan
USSR

USSR
Includes use on level 1 or 2 as an area term (list O). If 1, see term set options under list B.
CO N360000N820000
W1700000E0200000
UF Russia
NT Aral region
Aral Sea
Armenia
Asiatic USSR
Azerbaidzhan
Azov region
Azov Sea
Caucasus
Central Asia
Chirchik River
Courland Spit
Dnieper Basin
Dnieper River
Dniester River
Dniester-Prut Interfluve
Donets Basin

Dvina River
European USSR
Georgian Republic
Ishim
Kazakhstan
Kirghizia
Kuban River
Kuban Valley
Kulunda Steppe
Kurama Range
Neman River basin
NA Northwestern USSR
NT Ob-Irtysh Interfluve
Ossetia
Pechora River
Polesye
Pripet Basin
Pskem Range
Rudny Altai
Russian Plain
Russian Platform
Russian Republic
Sayan
Scythian Platform
Shakhdag Range
Siberia
Siberian fold belt
Southwestern USSR
NX Soviet Altai Mountains
NT Soviet Arctic
NX Soviet Carpathians
Soviet Fennoscandia
NT Steppes
Surkhan Darya
Surkhan Darya basin
Syr Darya
Tadzhikistan
Terek River
Transcaucasia
Turan
Turanian Platform
Turkmenia
Ukraine
Ural River
Urals
Uzbekistan
Uzen
Vodino
Volga-Ural region
Vyshkovo
West Siberia
Zangezur
Zeravshan River
SA Altai Mountains
Altai-Sayan region
Arctic region
Baikalian Phase
Barabash Suite
Baskunchak Series
Black Sea region
Caspian Sea
Eurasia
European Platform
Fennoscandia
Ilimaussaq
Kiev Member
Korenevskaya Formation
Krivoy Rog Series
Kursk Series
Ladoga Series
Lake Ladoga region
Lapland
Maikop Series
Moldavia
Novorayskoe Formation
Oktemberyan Series
Ovruch Series
Poltava Series
Protopivskaya Formation
Selenga River valley
Tunguska Series
Udokan Series
Usa Series
Vetluga Series
Yudoma Series

USSR Academy of Sciences
In Moscow.
BT Russian Republic
USSR

Ust-Kut
Town in N central Irkutsk Oblast NW of Lake Baikal.
BT Russian Republic
USSR

Ust-Urt
use Ustyurt

Ust-Urt Plateau
use Ustyurt

Ust-Yenisei Basin
That part of the lower Yenisei Basin from Ust-Port to Yenisei Bay which constitutes a drowned river mouth or estuary. Located in NW Taymyr-Nenets National Okrug.
BT Russian Republic
USSR

Ustica Island
Small island in the Tyrrhenian Sea NW of Sicily. Also search Ustica.
BT Sicily
Italy

Ustron
Village in central.
BT Byelorussia
USSR

Ustyurt
Desert plateau extending from the Caspian Sea to the Aral Sea. Also search Ust-Urt; Ust-Urt Plateau. Index Soviet republics as applicable.
CO N394500N450000
E0590000E0500000
UF Ust-Urt
Ust-Urt Plateau
BA Central Asia
BT USSR
SA Kazakhstan
Turkmenia
Uzbekistan

Usu
On NE Uchiura Bay in SW Hokkaido. Also search Usu Volcano.
UF Usu Volcano
BT Hokkaido
Japan

Usu Volcano
use Usu

Usuginu Conglomerate
Includes large boulders of limestone, compact slate, sandstone, granite, and porphyrite. In N Honshu.
BT Upper Permian
Permian
SA Honshu
Japan

Usuki
Town on Bungo Strait in NE.
BT Kyushu
Japan

Utah
Includes use on level 1 as an area term (list O). For term set options see list B.
CO N370000N420000
W1090500W1140500
BA Western U.S.
United States
NT Arches National Park
NX Beaver County
NT Bingham
Bonneville Salt Flats
Box Elder County
Cache County
NX Carbon County
NT Duchesne County
Emery
Emery County
NX Eureka
NT Fairfield
NX Garfield County
Grand County
NT Great Salt Lake
Henry Mountains
House Range
NX Iron County
NT Juab County
Kaiparowits Plateau
NX Kane County
NT Marysvale
Millard County
Moab
Monticello
NX Morgan County
NT Oquirrh Mountains
Park City
Payson
Pine Valley Mountains
Piute County
Provo
Rich County
Roosevelt Hot Springs KGRA
Salt Lake City
Salt Lake County
NX San Juan County
NT San Rafael Swell
Sanpete County
NX Sevier County
NT Sevier orogenic belt
Spor Mountain
Springdale
Stansbury Mountains
NX Summit County
NT Thomas Range
Tintic District
Tooele County
Uintah County
Utah County
Wasatch County
Wasatch Fault
Wasatch Plateau
NX Washington County
Wayne County
NT Weber County
SA Bald Mountain
Basin and Range Province
Bear Lake
Bear River basin
Bear River Range
Bird Spring Formation
Blackhawk Formation
Cache Valley
Cedar Mountain Formation
Chinle Formation
Coconino Sandstone
Colorado Plateau
Colorado River
Colorado River basin
Columbia River basin
Cutler Formation
Duchesne River Formation
Ferron Sandstone Member
Fillmore Formation
Four Corners
Frontier Formation
Great Basin
Green River
Green River basin
Green River Formation
Honaker Trail Formation
Kaibab Formation
Kayenta Formation
Lake Bonneville
Lake Powell
Lodgepole Formation
Madison Group
Mancos Shale
Mesaverde Group
Moenkopi Formation
Montana Group
Morrison Formation
Navajo Indian Reservation
Navajo Sandstone
Paradox Basin
Paradox Member
Phosphoria Formation
Pogonip Group
Raft River
Raft River basin
Roberts Mountains Formation
San Juan Basin
San Juan River
Snake River basin
Supai Formation
Swan Peak Formation
Thaynes Formation
Tintic Quartzite
Toroweap Formation
U. S. Rocky Mountains
Uinta Basin
Uinta Mountains
Virgin River valley
Wasatch Formation
Wasatch Front
Wasatch Range
Weber Sandstone
Western Interior
Western Overthrust Belt

Utah County
N central.
BT Utah
United States

utahite
use jarosite

Ute Creek
Flows into the Canadian River in NE.
BT New Mexico
United States

Utica Shale
Consists of Nowadaga, Loyal Creek, Hopkinson, Holland, and Patent members.
BT Upper Ordovician
Ordovician
SA Michigan
New York
Ontario

utilization
Includes use on level 2 under soils(1), and on level 3 under commodity terms (list C).
SA agriculture
fertilization
forestry
irrigation
reclamation
soil management
soils
tillage
treatment
underground installations

Utrecht
Province. Also a city S of Amsterdam.
BT Netherlands

Uttar Pradesh
State in N central.
CO N240000N320000
E0840000E0770000
BT India
NT Almora
Dehra Dun
Garhwal
Jhansi
Meerut
Mirzapur
Mussoorie
Naini Tal
Roorkee
Tehri Garhwal
SA Bundelkhand
Jumna River
Kaimur Sandstone
Kumaun Himalayas
Muth Quartzite
Satpura Range
Tal Formation

Uvalde County
Term introduced in 1981. SW Texas.
CO N290000N294000

W0992000W1001000
BT Texas
United States

Uvalde Gravel
Consists almost wholly of rounded flat cobbles, boulders, and occasional limestone pebbles. In Texas.
BT Pliocene
SA Texas

uvarovite
UF onvarovite
BA garnet group
nesosilicates
orthosilicates
silicates
BT minerals

Uvigerina
Genus. Includes use on level 3 under foraminifera(1) Buliminacea(2).
BA Buliminacea
Rotaliina
foraminifera
BT Protista
Invertebrata

Uvigerinidae
Family. Includes use on level 3 under foraminifera(1) Buliminacea(2).
BA Buliminacea
Rotaliina
foraminifera
BT Protista
Invertebrata

Uzbekistan
Uzbek Soviet Socialist Republic. In Soviet Central Asia.
CO N370000N450000
E0720000E0550000
BT USSR
NT Almalyk
Angren
Bukantau
Bukhara
Bukhara-Khiva
Chadak
Chirchik
Dengizkul
Karshi
Karshi Steppe
Kermine
Khiva
Kul'dzhuktau
Lyangar
Mubarek
Nura-Tau
Oktyabr
Samarkand
Sokh
Tamdytau
Tashkent
Ziaetdin
Zirabulak
SA Amu Darya
Aral region
Asiatic USSR
Central Asia
Chirchik River
Fergana Basin
Kurama Range
Kyzylkum
Pamirs
Surkhan Darya
Surkhan Darya basin
Syr Darya
Turan
Turanian Platform
Ustyurt
Zeravshan River

Uzen
Two steppe rivers, the Greater Uzen and Lesser Uzen, which flow parallel to one another in Saratov Oblast and NW Kazakhstan. Index Soviet republics as applicable.
BT USSR
SA Kazakhstan
Russian Republic

Uzon
Volcano. SE Kamchatka Peninsula.
CO N543007N543007
E1595408E1595408
UF Ozun
BT Russian Republic
USSR

V

V
use vanadium

Vaal River
Forms boundary between Transvaal and Orange Free State, and empties into Orange River in N Cape Province. Index provinces as applicable.
BT South Africa
SA Cape Province
Orange Free State
Transvaal

vacuum fusion analysis
Term introduced in 1978. Includes use on level 2 under chemical analysis(1).
SA analysis
chemical analysis

Vadia
use Wadia

vadose zone
use unsaturated zone

vaesite
BA sulfides
BT minerals

Vah River valley
use Vah Valley

Vah Valley
River valley in W Slovakia. Also search Vah; Vah River.
UF Vah River valley
BT Slovakia
Czechoslovakia

Vaigach Island
use Vaygach Island

Vail Pass
In Summit and Eagle counties W of Denver.
BT Colorado
United States

Vakhsh
River. A tributary of the Amu Darya in SW Tadzhikistan.
UF Vaksh
BT Tadzhikistan
USSR

Vaksh
use Vakhsh

Val d'Aosta
use Valle d'Aosta

Val d'Or
Town in SW.
BT Quebec
Canada

Val Gardena Sandstone
NE Italy.
BT Permian
SA Italy

Val Verde Basin
SW Texas.
BT Texas
United States

Val Verde County
SW Texas.
CO N291000N301500
W1005000W1014500
BT Texas
United States

Val-de-Marne
Term introduced in 1981. Department in N.
CO N484000N485500
E0023500E0021500
BT France

Val-d'Oise
Department just NNW of Paris.
CO N485200N491000
E0023000E0013000
BT France
NT Cormeilles-en-Parisis
SA Oise River valley

Valais
Canton in SW.
BT Switzerland
NT Binnental
Zermatt

Valanginian
Europe. Above Berriasian, below Hauterivian. Includes use on level 3 under age terms(1). See list E.
BA Lower Cretaceous
Cretaceous
BT Mesozoic
SA Neocomian

Valdai
Moraine region between Moscow and Leningrad. Also search Valdai Hills.
UF Valdai Hills
Valdai Uplands
BT Russian Republic
USSR

Valdai Hills
use Valdai

Valdai Uplands
use Valdai

Valdez
Town and port at head of an inlet off Prince William Sound in S Alaska. Terminal point of Trans-Alaska pipeline. Also search Port Valdez.
UF Port Valdez
BT Alaska
United States

Valdres
Mountainous region W and SW of Fagernes in S central part of country.
BT Norway

Vale of Glamorgan
In Brisbane area.
BT Queensland
Australia

Vale of Kashmir
use Kashmir Valley

Valence
City on left bank of Rhone River in SE part of country.
BT Drome
France

Valencia
As of 1981, refers to only the city. Before 1981, also included the province and the region. In SE Spain.
BT Valencia Province
Valencia region
Spain

Valencia County
W New Mexico.
BT New Mexico
United States

Valencia Province
Term introduced in 1981. Before 1981, search Valencia.
CO N394000N401300
W0000200W0013200
BT Valencia region
Spain
NT Bunol
Valencia

Valencia region
Term introduced in 1981. Before 1981, search Valencia.
CO N375100N404700
E0003100W0013300
BT Spain
NT Alicante Province
Castellon de la Plana Province
Valencia Province

Valenciennes
City near the Belgian frontier.
BT Nord Department
France

valency
Term introduced in 1978.
SA chemical properties
geochemistry

Valentine Formation
In Ogallala Group. Composed largely of unconsolidated layers of greenish silty marl sand and very few local lenses of diatomaceous marl and light bluish-gray volcanic ash. In W Kansas, NW Nebraska, and SW South Dakota.
BT Pliocene
SA Kansas
Nebraska
Ogallala Formation
South Dakota

valentinite
UF white antimony
BA oxides
BT minerals

Valladolid Province
Term introduced in 1981. Before 1981, search Valladolid. In N Spain.
CO N410500N421800
W0035800W0053300
BT Old Castile
Spain

Valle d'Aosta
Autonomous region in extreme NW Italy. Also search Aosta Valley; Val d'Aosta.
CO N452800N460000
E0075800E0064800
UF Aosta Valley
Val d'Aosta
BT Italy

Valle Grande Mountains
NW of Santa Fe in N New Mexico. Also search Jemez Mountains; Valle Grande.
UF Jemez Mountains
BT New Mexico
United States
SA Fenton Hill

valleriite
BA sulfides
BT minerals

Valles
City in E central part of country.
UF Ciudad de Valles
BT San Luis Potosi
Mexico

Valles Caldera
Also search Valles.
BT New Mexico
United States
SA Fenton Hill

Vallesian
Europe. See list E.
BA Miocene
Neogene
Tertiary
BT Cenozoic

Valley and Ridge Province
Folded mountain ridges and parallel valleys between the Appalachian Plateau on the W and the Older Appalachians on the E. Extends from along the Hudson River in the N into Alabama in the S. Index states as applicable. Also search Valley and Ridge.
BT United States
SA Alabama
Georgia
Great Valley Sequence
Maryland
New Jersey
New York
North Carolina
Pennsylvania
Tennessee
Virginia
West Virginia

Valley County
Index states as applicable.
BX Idaho
Montana
Nebraska
BT United States

valley, median
use median valley

Valley of Mexico
Large oval basin 50 miles by 40 miles which is a subdivision of the plateau of Anahuac in central part of country. Index Federal District and/or Mexico state as applicable.
BT Mexico
SA Federal District
Mexico state

Valley of Ten Thousand Smokes
Volcanic region in Katmai National Monument W of Mt. Katmai in SW.
BT Alaska
United States

valleys
Includes use on level 3 under geomorphology(1) erosion features(2), fluvial features(2), landform description(2), and landform evolution(2).
SA buried valleys
drainage basins
erosion
erosion features
fluvial features
geomorphology
landforms
rilles
rivers and streams

Valmeyeran
Provincial series, Illinois. Lower and Upper Mississippian. Equivalent to Osagian and Meramecian elsewhere. Includes use on level 3 under age terms(1). See list E.
BA Mississippian
Carboniferous
BT Paleozoic
SA Lower Mississippian
Meramecian
Osagian
Upper Mississippian

Valpacos
Town in N central part of country.
BT Vila Real
Portugal

Valparaiso
Province in central part of country. Also a city on the Pacific Ocean.
BT Chile

Valsequillo
Village in NW.
BT Cordoba Province
Andalusia
Spain

Valsugana
Valley of upper Brenta River in SE.
BT Trentino-Alto Adige
Italy

Valtellina
Valley of upper Adda River in N.
BT Lombardy
Italy

valves
Includes use on level 3 under appropriate fossil group (list F).
SA morphology

Van Horn
Village in Culberson County in W.
BT Texas
United States

vanadates
Includes use on level 2 under minerals(1). See list L.
BT minerals
NA carnotite
descloizite
NZ vanadinite
NA volborthite

vanadinite
BZ chlorides
vanadates
BT minerals

vanadium
Includes use on level 1 and 2 as a chemical element (list D). As of 1981, use vanadium ores for vanadium as a commodity. Before 1981, included use on level 1 as a commodity term.
UF V
SA elements
vanadium ores

vanadium garnet
BA garnet group
nesosilicates
orthosilicates
silicates
BT minerals
SA grossular

vanadium ores
Term introduced in 1981. Includes use on level 1 and 2 as a commodity term (list C). Before 1981, search vanadium AND ore deposits.
SA vanadium

Vanch Range
use Vanchskiy Range

Vanchskiy Mountains
use Vanchskiy Range

Vanchskiy Range
Near Afghan border in S central Tadzhikistan. Also search Vanch; Vanch Range.
UF Vanch Range
Vanchskiy Mountains
BT Tadzhikistan
USSR

Vancouver
City in SW British Columbia and a city on the Columbia River in SW Washington. Index state or province as applicable.
SA British Columbia
Washington

Vancouver Island
Large island off SW coast of mainland British Columbia.
CO N483000N510000
W1233000W1283000
BT British Columbia
Canada
SA Coast Ranges

Vanimo region
Term introduced in 1981. In NW.
CO S024000S024000
E1411700E1411700
BT Papua New Guinea

Vanoise
High mountain group of the Savoy Alps. Also search Vanoise Massif.
UF Vanoise Massif
BT Savoie
France
SA Alps
Savoy Alps

Vanoise Massif
use Vanoise

vapor, water
use water vapor

vaporization
use evaporation

Var
Department in SE.
CO N430000N434500
E0070000E0053000
BT France
NT Maures Massif
Plan de la Tour
Rians
Sainte-Baume Massif
Toulon
SA Esterel
Provence
Provence Alps

Varanger Fjord
Inlet of the Barents Sea S of the Varanger Peninsula in extreme NE part of country.
UF Varangerfjord
BT Finnmark
Norway

Varanger Halvoy
use Varanger Peninsula

Varanger Peninsula
Extreme NE part of country jutting into the Barents Sea.
UF Varanger Halvoy
Varangerhalvoeya
Varangerhalvoya
BT Finnmark
Norway

Varangerfjord
use Varanger Fjord

Varangerhalvoeya
use Varanger Peninsula

Varangerhalvoya
use Varanger Peninsula

Vardar River
Rises in S Yugoslavia and flows into Gulf of Salonika. Index countries as applicable.
BT Europe
SA Greece
Yugoslavia

Varese
City in NW.
BT Lombardy
Italy

variance analysis
BA statistical analysis
SA analysis

variations
Includes use on level 2 under magnetosphere(1); on level 3 under Earth(1). Also search variation.
SA disturbances
diurnal variations
magnetic field
magnetosphere
seasonal variations
secular variations
spatial variations
time variations

varieties
Includes use as level 2 or 3 term appropriate to a large number of topics. See list G.

Variscan
A valid index term through 1976. After 1976, use Hercynian Orogeny.

Variscan Orogeny
use Hercynian Orogeny

Variscides
Mountain system raised in the latter part of the Paleozoic era, particularly in central Europe; more or less equivalent to Hercynian.
BT Europe
SA Hercynian Orogeny

variscite
BA phosphates
BT minerals

Varmland
County in SW.
BT Sweden

Varna
Province along the Black Sea. Also a city.
BT Bulgaria

Varpalota
Town N of E Lake Balaton.
BT Veszprem
Hungary

Varta
use Warta

varved clay
A valid term through 1975. After 1975, use varves.

varves
As of 1978, includes use on level 2 under geochronology(1). See list K. Before 1976, also search varved clay.
BA planar bedding structures
sedimentary structures
SA geochronology
glacial features
glacial lakes
graded bedding

Vasterbotten
County in N.
BT Sweden
NT Boliden
SA Skellefte

Vastervik
City on Baltic Sea.
BT Kalmar
Sweden

vaterite
Artificial mineral.
BA carbonates
BT minerals

Vatnajokull
Glacier region and snow field in SE.
BT Iceland

Vatukoula
Town on N coast of Viti Levu island in S.
BT Fiji

Vaucluse
Department in SE.
CO N434000N442000
E0054500E0044500
BT France
NT Apt
Mormoiron
SA Provence

Vaud
Canton in W.
BT Switzerland

Vaygach Island
Between NE Nenets National Okrug and island of Novaya Zemlya off N European USSR. Also search Vaigach; Vaigach Island; Vaygach.
UF Vaigach Island
BT Russian Republic

USSR

vegetation
Includes use on level 3 under ecology(1).
SA ecology
floral studies
grasslands
leaves
paleobotany
Plantae
soils
wetlands

vein systems
Term introduced in 1981.
SA mineral deposits
veins

veins
General term not restricted to ore deposits.
SA fractures
host rocks
intrusions
mineral deposits
vein systems

Vejer de la Frontera
City near the Atlantic Ocean in SW.
BT Cadiz Province
Andalusia
Spain

Velay
Region of the Central Massif in W and central Haute-Loire. Also search Velay Massif; Velay Plateau.
BT Haute-Loire
France

Velebit Mountains
Range of Dinaric Alps along Adriatic Sea opposite Pag Island in W Croatia. Also search Velebit.
BT Croatia
Yugoslavia
SA Dinaric Alps

Velez Rubio
Town in Andalusia in N.
BT Almeria Province
Andalusia
Spain

velocity
Includes use as a general term, e.g. under seismology(1). Also search velocities.
SA Conrad discontinuity
flows
half-space
kinematics
kinetics
velocity structure
vibration

velocity structure
Term introduced in 1978.
SA crust
mantle
seismology
velocity

Vema fracture zone
Crosses Mid Atlantic Ridge near 11° N.
BT Atlantic Ocean

Vendee
Department on Bay of Biscay in W.
CO N462500N471000
W0004500W0020500
BT France
SA Poitou

Vendian
Europe. See list E.
BA upper Proterozoic
Proterozoic
Precambrian
SA Eocambrian
Infracambrian
upper Precambrian

Vendsyssel
Region of Jutland Peninsula N of Lim Fjord on the Skagerrak.
BT Denmark

Venediger Group
Mountain group in SE Tyrol and SW Salzburg State just S of W Hohe Tauern. Index states as applicable.
UF Venediger Gruppe
Venediger Massif
BT Austria
SA Salzburg State
Tyrol

Venediger Gruppe
use Venediger Group

Venediger Massif
use Venediger Group

Venerida
Term introduced in 1981. Includes use on level 2 under Mollusca(1). See list F.
BA Bivalvia
Mollusca
BT Invertebrata

Venetia
Region E of Lombardy. Also a former autonomous region now known as Veneto. Index autonomous regions as applicable.
BT Italy
SA Friuli-Venezia Giulia
Trentino-Alto Adige
Veneto

Veneto
Autonomous region in NE Italy. Formerly Venetia autonomous region.
CO N444800N464100
E0130500E0104000
UF Venezia Euganea
BT Italy
NT Alpago Basin
Belluno
Berici Hills
Cortina D'Ampezzo
Euganean Hills
Monte Baldo
Monti Berici
Monti Lessini
Piave Valley
Po Delta
Possagno
Sappada
Venice
Verona
Vicenza
SA Dolomites
Marmolada Glacier
Po River
Po Valley
Tagliamento Valley
Venetia

Venezia
use Venice

Venezia Euganea
use Veneto

Venezia-Tridentina
use Trentino-Alto Adige

Venezuela
Includes use on level 1 as an area term (list O). For term set options see list B.
CO N004500N121000
W0595500W0731500
BA South America
NT Anzoategui
Apure Guarico Plain
Aragua
Araya Peninsula
Barinas
Bolivar
Carabobo
Cojedes
Cordillera de la Costa
Cumana
Eastern Venezuela
Falcon
NX Federal District
NT Guarico
Lara
Maracaibo Basin
Maracaibo Falcon Plain
Miranda
Nueva Esparta
Paria Peninsula
Southeastern Venezuela
NX Sucre
NT Tachira
Trujillo
NX Venezuelan Andes
NT Venezuelan Islands
Yaracuy
Zulia
SA Amazon Basin
Andes
Caracas Group
Caribbean region
Guajira Peninsula
Llanos
Merida
Mirador Formation
Orinoco River
Punta Mosquito Formation
Rio Negro
Roraima Formation

Venezuela Basin
use Venezuelan Basin

Venezuelan Andes
Term introduced in 1981.
BZ Andes
BX Venezuela

Venezuelan Basin
Marine basin between island of Hispaniola on the N and Venezuela on S. Also search Venezuela Basin.
UF Venezuela Basin
BT Caribbean Sea

Venezuelan Islands
Term introduced in 1981.
CO N103000N120000
W0630000W0680000
BT Venezuela

Venice
City on Gulf of Venice off N Adriatic Sea.
UF Venezia
BT Veneto
Italy

Ventersdorp
Town in SW.
BT Transvaal
South Africa

vents
SA volcanology

Ventura
City WNW of Los Angeles in Ventura County.
UF San Buenaventura
BT California
United States

Ventura Basin
WNW of Los Angeles in Ventura and Santa Barbara counties.
BT California
United States

Ventura County
WNW of Los Angeles.
BT California
United States

Venus
Includes use on level 1 and 2. See entry under Moon(1) for term set options.
SA landing sites
planetology
planets
solar system
terrestrial planets

Veracruz
State on the Gulf of Mexico in E Mexico. Also a city.
CO N170000N221500
W0930000W0990000
BT Mexico
NT Papaloapan River basin
Poza Rica
SA Gulf Coastal Plain
Sierra Madre Oriental

Vercelli
City in E.
BT Italy
SA Piemonte

Vercors
Limestone massif of the Dauphine Alps SE part of country. Index departments as applicable.
UF Vercors Massif
BT France
SA Dauphine Alps
Drome
Isere

Vercors Massif
use Vercors

Verde Formation
Pliocene (?) or Pleistocene. Unconformably overlies Hickey Formation in Jerome area. Central Arizona.
BT Cenozoic
SA Arizona
Pleistocene
Pliocene
Quaternary
Tertiary

Verde Valley
River valley in central.
BT Arizona
United States

Verdigris River valley
SE Kansas, and NE Oklahoma. Index states as applicable. Also search Verdigris River.
BT United States
SA Kansas
Oklahoma

Verdon Valley
River valley in SE part of country. Also search Verdon.
BT Alpes-de-Haute Provence
France

Verdun
City on the Meuse River in NE part of country.
UF Verdun-sur-Meuse
BT Meuse
France

Verdun-sur-Meuse
use Verdun

Verkhovtsevo
Town in W central Dnepropetrovsk Oblast in S central.
BT Ukraine
USSR

Verkhoyansk
Town N of Arctic Circle on the Yana River in N central Yakutia in N central Siberia.
BT Russian Republic
USSR

Verkhoyansk Mountains
use Verkhoyansk Range

Verkhoyansk Range
Extends along the E banks of the Lena and lower Aldan rivers in N central Yakutia.
CO N630000N700000
E1400000E1240000
UF Verkhoyansk Mountains
Verkhoyanski Mountains

BT Russian Republic
USSR
SA Verkhoyansk region

Verkhoyansk region
Term introduced in 1981 for region of the Verkhoyansk and Cherski ranges.
CO N570000N730000
E1600000E1240000
BA Northeastern USSR
BT Russian Republic
USSR
SA Verkhoyansk Range

Verkhoyanski Mountains
use Verkhoyansk Range

vermiculite
Includes use on level 3 under minerals(1) sheet silicates, clay minerals(2). See list L. As of 1981, use vermiculite deposits for vermiculite as a commodity. Before 1981, included use on level 1 as a commodity term.
BA clay minerals
sheet silicates
silicates
BT minerals
SA clay mineralogy
mica group
vermiculite deposits

vermiculite deposits
Term introduced in 1981. Includes use on level 1 and 2 as a commodity term (list C). Before 1981, search vermiculite AND deposits.
SA clays
mica deposits
vermiculite

Vermilion Parish
On the Gulf of Mexico. Also search Vermilion AND Louisiana.
BT Louisiana
United States

Vermilion Range
Iron mining area in Saint Louis and Lake counties in NE. Also search Vermilion AND Minnesota.
BT Minnesota
United States

Vermont
Includes use on level 1 as an area term (list O). For term set options see list B.
CO N424500N450000
W0713000W0732500
BA New England
Eastern U.S.
United States
NT Addison County
Caledonia County
Castleton
Chittenden County
NX Essex County
Franklin County
Orange County
NT Rutland
NX Washington County
Windham County
SA Barre Granite
Champlain Valley
Connecticut River
Connecticut Valley
Green Mountains
Lake Champlain
Littleton Formation
Normanskill Formation
Potsdam Sandstone
Trenton Group
Wilcox Formation

Vernagt Glacier
In Otztal Alps S of the Wildspitze in S.
BT Tyrol
Austria

Vernyi
use Alma-Ata

Verona
City in Verona Province in W.
BT Veneto
Italy

Versailles
City in Yvelines Department of N France WSW of Paris; and a city in numerous U. S. states. Index France and states as applicable.
SA France
Illinois
Indiana
Kentucky
Missouri

Versilian
Europe. See list E.
BA Holocene
Quaternary
BT Cenozoic

Vertebrata
Includes use on level 1 and 2 as a fossil term (list F). Term is to be used only when more specific terms do not apply, or are too numerous to be recorded.
BT Chordata
NT Agnatha
Pisces
Tetrapoda
SA biota
bones
coprolites
extremities
fossil man
jaws
nervous system
skeletons
skulls
spinal column
teeth
vertebrates

vertebrates
Term introduced in 1981 as the common name for Vertebrata. See list F. To be used for Vertebrata when they are used as tools in stratigraphy. Includes use on level 1 when more specific common names do not apply or are too numerous to be recorded. Before 1981, search Vertebrata AND biostratigraphy. If 1, term set options are:
biostratigraphy
age [List E]
SA biostratigraphy
Vertebrata

Vertes
Town in E part of country.
BT Hungary

Vertes Mountains
N Central Hungary. Also search Vertes.
BT Hungary

vertical
A valid index term through 1977. After 1977, use vertical orientation for folds.

vertical discontinuities
Term introduced in 1981. Includes use on level 3 under crust(1), mantle(1), and seismology(1). Used for a type of interface in seismology.
BT discontinuities
SA crust
mantle
seismology

vertical movements
Includes use on level 3 under neotectonics(1) and as of 1981, on level 2 under tectonics(1). Before 1981, also search vertical tectonics.
UF movements, vertical
SA horizontal movements
isostasy
neotectonics
subsidence
tectonics
uplifts

vertical orientation
Term introduced in 1978. Reference to folds defined on the basis of orientation in relation to the geographic horizontal plane. Includes use on level 3 under folds(1) orientation(2). Before 1978, also search folds AND vertical.
SA folds
orientation

vertical tectonics
As of 1981, no longer a valid term for GeoRef. Use vertical movements. Included use on level 2 under tectonics(1).

vertical-component seismographs
Term introduced in 1981.
BT seismographs

Vertisols
Includes use on level 3 under soils(1). See list M.
BA soils

Vesdre Valley
River valley in E part of country.
BT Liege
Belgium

Vest Agder
use Vest-Agder

Vest-Agder
County in extreme S Norway.
UF Vest Agder
BT Norway
Europe

Vesteraalen
use Vesteralen

Vesteralen
Island group N of the Lofoten Islands in the Norwegian Sea off NW mainland. Index counties as applicable.
UF Vesteraalen
BT Norway
SA Nordland
Troms

Vestfold
County in SE.
UF Jarlsberg
BT Norway
NT Larvik

Vestman Islands
use Vestmannaeyjar

Vestmann Islands
use Vestmannaeyjar

Vestmannaeyjar
Small island group of volcanic origin S of Iceland.
UF Vestman Islands
Vestmann Islands
Westman Islands
BT Iceland

Vestspitsbergen
Old name for Spitsbergen, the main island, of the Spitsbergen Archipelago of the Svalbard Island group.
UF West Spitsbergen
BT Spitsbergen
Svalbard
Arctic region

Vesuvian garnet
use leucite

vesuvianite
Also search idocrase.
UF idocrase
BA orthosilicates
silicates
BT minerals

Vesuvio
use Vesuvius

Vesuvius
Active volcano on E side of the Bay of Naples. Also search Vesuvius Volcano.
UF Vesuvio
Vesuvius Volcano
BT Campania
Italy

Vesuvius Volcano
use Vesuvius

Veszprem
County in W Hungary. Also a city N of Lake Balaton.
BT Hungary
NT Varpalota

Vetluga Series
According to the USSR Lexique, the age of the Vetluga Horizon is upper Permian. It is composed of red spotted clay, and maroon and gray sandstone in an interwinning stratification not characterized paleontologically. NE of Moscow.
BT Upper Permian
Permian
SA Russian Republic
USSR

Vetrenyy Ridge
S of Onega Bay in W Arkhangelsk Oblast in NW European USSR. Also search Vetrenyy.
BT Russian Republic
USSR

Viatka River
use Vyatka River

vibrating-string gravimeters
Term introduced in 1981.
BT gravimeters

vibration
Includes use on level 3 under rock mechanics(1) or seismology(1). Also search vibrations.
SA earthquakes
oscillations
pulsations
resonance
rock mechanics
seismology
velocity

Vibroseis
SA elastic waves
seismic methods
seismic surveys
seismology

Vicenza
City W of Venice in Vicenza Province.
BT Veneto
Italy

Vich
City NE of Barcelona in NE part of country.
BT Barcelona Province
Catalonia
Spain

Vicksburg Group
Includes Mariana Limestone and Byram Formation. Gulf Coastal Plain. Vicksburg.
BT middle Oligocene
Oligocene
SA Alabama
Florida
Louisiana
Marianna Limestone
Mississippi
Texas

Victoria
Includes use on level 1 as an area term (list O). For term set options see list B.
CO S390000S340000
E1500000E1400000
BA Australia
Australasia
NT Gippsland
Gippsland Basin
Green Gully
Melbourne
Melbourne region
Mildura
Philip Island
Port Campbell
Port Phillip Bay
NX Western Port
SA Gambier Embayment
Lachlan fold belt
Lilydale Limestone
Murchison Meteorite
Murray Basin
Murray River
Otway Basin
Snowy Mountains
Tasman Geosyncline

Victoria Land
W of Ross Sea and W of NW Ross Ice Shelf, primarily in Ross Dependency on the Pacific Ocean side S of New Zealand.
UF South Victoria Land
BT Antarctica
NT Allan Hills
Lake Vanda
Yamato Mountains
SA Allan Hills Meteorites
Yamato Meteorites

Victoria Valley
River valley in NW.
BT Northern Territory
Australia

Vidda
use Hardangervidda

Vidin
Province in extreme NW Bulgaria. Also a city on the Danube River near the Yugoslav border.
BT Bulgaria
NT Lom Depression

Vienerwald
use Wiener Wald

Vienna
City on the Danube River in NE part of country.
UF Wien
BT Lower Austria
Austria

Vienna Basin
Index countries as applicable.
BT Europe
NZ Austrian Vienna Basin
SA Austria
Czechoslovakia

Vienne
Department in W central.
CO N460500N471000
E0011500W0000500
BT France
NT Poitiers
SA Poitou

Vietnam
North Vietnam (Democratic Republic of Vietnam) and South Vietnam (Republic of Vietnam) were combined into an unified nation on June 24, 1976. Apparently, the new country will be known as the Democratic Republic of Vietnam. Includes use on level 1 as an area term (list O). For term set options see list B. Before 1977, also search Vietnam AND north; search Vietnam AND south.
CO N083000N231500
E1092000E1020000
BA Far East
Asia
NT Baoha
Dalat
Mekong Delta
Nong-Son
Yenbay
SA Indochina

Vigarano Meteorite
Term introduced in 1981. Impact near village of Vigarano in Ferrara Province, NE Emilia-Romagna. Before 1981, also search Vigarano AND meteorites.
BA carbonaceous chondrites
chondrites
BT meteorites
SA Emilia-Romagna
Italy

Viking
As of 1981, no longer a valid term for GeoRef. Use Viking Program.

Viking Formation
Consists of sandstone which grades eastward into siltstone and shale and into the marine shales of the Ashville Formation. In SW Alberta, the sandstones grade into dark marine shales and sandstones of the Bow Island Formation. Oil reservoirs in beds equivalent to the lower Colorado are found in sandstone of the Viking Formation. NE and SW Alberta.
BT Cretaceous
SA Alberta

Viking Program
Term introduced in 1981. Includes use on level 3 under Mars(1).
SA extraterrestrial geology
landing sites
Mars
planetology
remote sensing

Vila Real
District in N.
BT Portugal
NT Valpacos
Vilarandelo
SA Tras-os-Montes

Vilaine Bay
At the mouth of the Vilaine River on the Bay of Biscay in SE Brittany. Also search Vilaine.
BT Morbihan
France

Vilarandelo
Village in N part of country.
BT Vila Real
Portugal

Villach
City W of Klagenfurt on the Drava River.
UF Beljak
BT Carinthia
Austria

Villafranca d'Asti
Village E of Turin in central.
BT Italy
SA Piemonte

Villafranchian
Europe. Includes use on level 3 under age terms(1). See list E.
BA lower Pleistocene
Pleistocene
Quaternary
BT Cenozoic
SA Calabrian
Tertiary

Villany Mountains
In S part of county. In Baranya.
BT Hungary

villiaumite
As of 1981, includes use on level 3 under minerals(1) fluorides(2). See list L. Before 1981, included use on level 3 under minerals(1) halides(2).
BA fluorides
halides
BT minerals

Vilna
City near the Byelorussian border in SE Lithuania.
UF Vilnius
Vilno
Vilnyus
Wilna
Wilno
BT Lithuania
USSR

Vilnius
use Vilna

Vilno
use Vilna

Vilnyus
use Vilna

Vilyui Basin
use Vilyuy River basin

Vilyui River
use Vilyuy River

Vilyui Syneclise
use Vilyuy Syneclise

Vilyuy Basin
use Vilyuy River basin

Vilyuy River
Rises on the Central Siberian Plateau of E Evenk National Okrug and flows through W and W central Yakutia into the Lena River. Also search Vilyui; Vilyuy.
UF Vilyui River
BT Russian Republic
USSR

Vilyuy River basin
In E Evenk National Okrug and W and W central Yakutia in central Siberia.
UF Vilyui Basin
Vilyuy Basin
BT Russian Republic
USSR

Vilyuy Sineclise
use Vilyuy Syneclise

Vilyuy Syneclise
The lower Vilyuy River basin in the area of the river's flow into Lena River in central Yakutia. Also search Vilyui Syneclise.
CO N593000N720000
E1280000E1110000
UF Vilyui Syneclise
Vilyuy Sineclise
BA Siberian Platform
BT Russian Republic
USSR

vimsite
BA borates
BT minerals

Vindhya Basin
use Vindhyan Basin

Vindhyan
Precambrian-Cambrian. Includes series of sandstones, shales, and limestones extensively exposed in the highlands of central India. Also search Vindhyan System.
UF Vindhyan System
BT Precambrian
SA Cambrian
India

Vindhyan Basin
Index countries as applicable.
UF Vindhya Basin
BT Asia
SA Bangladesh
India

Vindhyan System
use Vindhyan

Vindobonian
Europe. Includes use on level 3 under age terms(1). See list E.
BA Miocene
Neogene
Tertiary
BT Cenozoic

Vinton Canyon
BT Texas
United States

Viola Limestone
Middle and Upper Ordovician. In Patterson Ranch Group. Overlies Bromide Formation and underlies Fernvale. Central S and SW Oklahoma.
BT Ordovician
SA Middle Ordovician
Oklahoma
Upper Ordovician

violarite
BA sulfides
BT minerals

Virgilian
Provincial series, North America. Above Missourian, below Wolfcampian (Permian). Includes use on level 3 under age terms(1). See list E. Also search Virgilian Series.
UF Virgilian Series
BA Upper Pennsylvanian
Pennsylvanian
Carboniferous
BT Paleozoic
NT Beil Limestone Member
Douglas Group
Ervine Creek Limestone
Lawrence Formation
Lecompton Limestone
Oread Limestone
Plattsmouth Limestone Member
Shawnee Group
Tonganoxie Sandstone
Wabaunsee Group

Virgilian Series
use Virgilian

Virgin Islands
As of 1981, includes the British Virgin Islands and the U. S. Virgin Islands. From 1978-1980, term referred to only the U. S. Virgin Islands. Before 1978, included both British and U. S. Virgin Islands.
BA Leeward Islands
BT Lesser Antilles
NA British Virgin Islands
NA U. S. Virgin Islands

Virgin River valley
Index states as applicable. Also search Virgin River.
BT United States
SA Arizona
Nevada
Utah

Virginia
Includes use on level 1 as an area term (list O). For term set options see list B.
CO N363500N392800
W0751500W0833500
BA Eastern U.S.
United States
NT Accomack County
NX Alleghany County

NT Amelia
Augusta County
NX Bedford County
NT Blacksburg
NX Brunswick County
Buchanan County
NT Buckingham County
NX Campbell County
Carroll County
Charlotte County
Clarke County
Cumberland
Cumberland County
NT Danville
Dickenson County
NX Essex County
Floyd County
Franklin County
Giles County
Greene County
NT Hanover County
NX Highland County
NT James River
NX Lancaster County
Lee County
Madison County
Mecklenburg County
Middlesex County
Montgomery County
Nelson County
NT Norfolk City
NX Orange County
Page County
NT Petersburg
NX Pulaski County
NT Pulaski Fault
Rappahannock River
Richmond
Roanoke
NX Rockingham County
Russell County
NT Saltville Fault
NX Scott County
Stafford County
Surry County
Sussex County
NT Tazewell County
Virginia Beach
Wachapreague Inlet
NX Warren County
Washington County
Waynesboro
Westmoreland County
Wise County
NT Wythe County
NX York County
NT York River
SA Allegheny Front
Allegheny Group
Allegheny Mountains
Allegheny Plateau
Antietam Formation
Appalachian Basin
Appalachian Plateau
Aquia Formation
Ashe Formation
Atlantic Coastal Plain
Baltimore Gneiss
Bloomsburg Formation
Blue Ridge Mountains
Blue Ridge Province
Calvert Formation
Carolina Bays
Catoctin Formation
Catskill Formation
Chesapeake Bay
Chickamauga Group
Chilhowee Group
Choptank Formation
Clinton Group
Coeymans Formation
Conemaugh Group
Cumberland Plateau
Dan River basin
Delmarva Peninsula
Genesee Group
Glenarm Series
Greenland Gap Group
Helderberg Group
Hurricane Ridge Syncline
Juniata Formation
Kingsport Formation
Kinzers Formation
Knox Group
Lee Formation
Lincolnshire Limestone
Martinsburg Formation
Mascot Dolomite
McKenzie Formation
Monongahela Group
Nanjemoy Formation
New River
Newark Group
Newman Limestone
Onondaga Limestone
Oriskany Sandstone
Pennington Formation
Petersburg Granite
Piedmont
Pocahontas Formation
Pocono Formation
Potomac Group
Potomac River
Potomac River basin
Pottsville Group
Price Formation
Rome Formation
Saint Louis Limestone
Saint Marys Formation
Shady Dolomite
Sharps Meteorite
Shenandoah Valley
Tioga Bentonite
Tonoloway Limestone
Trenton Group
Tuscarora Formation
Valley and Ridge Province
Wicomico Formation
Williamsport Sandstone
Wissahickon Formation
Yorktown Formation

Virginia Beach
Independent city on the Atlantic Ocean 18 miles E of Norfolk in SE.
BT Virginia
United States

viridine
BA orthosilicates
silicates
BT minerals

Viruan
Europe. See list E.
BA Ordovician
BT Paleozoic

Viruddhachalam
use Vridhachalam

Visagapatnam
use Visakhapatnam

Visakhapatnam
City on the Coromandel Coast of the Bay of Bengal in NE Andhra Pradesh.
UF Visagapatnam
Vishakhapatnam
Vizagapatam
Vizagapatnam
BT Andhra Pradesh
India

viscoelastic materials
Term introduced in 1981.
SA engineering geology
materials

viscoelasticity
Includes use on level 3 under deformation(1) field studies(2).
SA deformation
elastic limit
elastic strain
elasticity
strain
stress

viscoplastic materials
Term introduced in 1981.
SA engineering geology
materials

viscosity
Includes use on level 2 under magmas(1) and lava(1); on level 3 under deformation(1) field studies(2).
SA deformation
flow mechanism
lava
magmas
plasticity
rheology
viscous materials

viscous materials
SA materials
viscosity

viscous remanent magnetization
BA remanent magnetization
SA paleomagnetism

Visean
Europe. Above Tournaisian, below Namurian. Includes use on level 3 under age terms(1). See list E.
BA Dinantian
Carboniferous
BT Paleozoic
NT Great Scar Limestone

Viseu
District in N central Portugal. Also a city NE of Coimbra.
UF Vizeu
BT Portugal

Vishakhapatnam
use Visakhapatnam

Visla River
use Vistula River

Vistula River
Rises on the N slope of the Carpathian Mountains and flows into the Baltic Sea at Gdansk. Also search Vistula.
UF Visla River
Weichsel River
Wisla River
BT Poland

Vistula River valley
S, E central, and N part of country. Also search Vistula and Vistula Valley.
UF Vistula Valley
BT Poland

Vistula Valley
use Vistula River valley

Viterbo
City and province.
BT Latium
Italy

Viti Levu
Largest island in Fiji group.
BT Fiji

Vitim
River which rises in central Buryatia and flows across NE Irkutsk Oblast into the Lena River on the SW border of Yakutia.
UF Vitim River
BT Russian Republic
USSR

Vitim Plateau
Gold mining area between the Vitim and Barguzin rivers E of Lake Baikal in NE Buryatia. Also search Vitim.
BT Russian Republic
USSR

Vitim River
use Vitim

Vitoria
City on the Atlantic Ocean 250 miles NE of Rio de Janeiro in Espirito Santo and city in Alava in N Spain. Index Brazil and/or Spain as applicable.
SA Brazil
Spain

vitrain
Includes use on level 3 under sedimentary rocks (1) organic residues(2). See list I.
UF pure coal
BA macerals
coal
organic residues
BT sedimentary rocks

vitrinite
Includes use on level 3 under sedimentary rocks(1) organic residues(2). See list I.
BA macerals
coal
organic residues
BT sedimentary rocks
NA collinite
telinite

vitrophyre
As of 1981, includes use on level 3 under igneous rocks(1) volcanic rocks(2). See list H. Before 1981, included use on level 3 under igneous rocks(1) andesite-rhyolite family(2).
BA volcanic rocks
BT igneous rocks
SA porphyritic texture

Vivarais
Ancient region now mostly in Ardeche department in SE part of country.
BT Ardeche
France

vivianite
BA phosphates
BT minerals

Viviparus
Includes use on level 3 under Mollusca(1). See list F.
BA Gastropoda
Mollusca
BT Invertebrata

Vizagapatam
use Visakhapatnam

Vizagapatnam
use Visakhapatnam

Vizcaya
One of the Basque Provinces in N.
CO N425500N432800
W0022200W0033100
BT Basque Provinces
Spain
NT Bilbao

Vizeu
use Viseu

Vladeasa Massif
use Vladeasa Mountain

Vladeasa Mountain
In the Bihor Mountains of W central Transylvania. Also search Vladeasa.
UF Vladeasa Massif
BT Transylvania
Romania
SA Bihor Mountains

Vladimir
City 110 miles of Moscow. Also an oblast, and a former principality of central Russia.
BT Russian Republic
USSR

Vladivostok
City on the southern tip of a peninsula extending into Peter the Great Bay on the Japan Sea in S Primorye Kray.

BT Russian Republic
USSR

VLF
As of 1981, no longer a valid term for GeoRef. See electromagnetic waves. Before 1981, included use on level 3 under ionosphere(1) wave propagation(2).

Vocontian Trough
BT France

Vodino
Village in 3 locations in the Ukraine; Also a village in Kuybyshev and Volgograd oblasts of the Russian Republic in the European USSR, and a village in Novosibirsk Oblast of the Russian Republic in Siberia. Index Soviet republics as applicable.
BT USSR
SA Russian Republic
Ukraine

Vogels Berg
use Vogelsberg

Vogelsberg
Mountain range in central Hesse.
UF Vogels Berg
BT Hesse
West Germany

Vogtland
Region NW of the W edge of the Erzgebirge around Plauen in S part of country.
BT Karl-Marx-Stadt Bezirk
East Germany

Voivodina
use Vojvodina

Vojvodina
Autonomous province on the Hungarian and Romanian borders in N Serbia.
UF Voivodina
Voyvodina
BT Serbia
Yugoslavia

volatile combustible
use volatiles

volatile elements
SA elements
volatiles

volatile matter
use volatiles

volatiles
Includes use as a general term on level 3, e.g. under geochemistry(1).
UF volatile combustible
volatile matter
SA coal
geochemistry
volatile elements

volborthite
BA vanadates
BT minerals

volcanic
A valid term through 1977. After 1977, use volcanic rocks on level 2 under igneous rocks(1).

volcanic arcs
use island arcs

volcanic ash
As of 1981, includes use on level 3 under igneous rocks(1) pyroclastics(2). See list H. Before 1981, included use on level 3 under igneous rocks(1) pyroclastics and glasses(2).
BA pyroclastics
volcanic rocks
BT igneous rocks
SA ash
ash falls
ash flows
bentonite
clastic sediments
dust
ejecta
sediments
terrigenous materials
tuff

volcanic belts
Includes use on level 3 under plate tectonics(1).
UF belts, volcanic
SA plate tectonics
volcanoes

volcanic breccia
Term introduced in 1981.
BA breccia
clastic rocks
BT sedimentary rocks
SA pyroclastics

volcanic clay
use bentonite

volcanic earthquakes
Term introduced in 1981.
SA earthquakes
eruptions
geologic hazards
tilt
volcanism
volcanoes
volcanology

volcanic features
Includes use on level 2 under geomorphology(1).
UF features, volcanic
NT calderas
cinder cones
lava channels
lava fields
lava lakes
maars
SA cauldrons
caves
diatremes
geomorphology
lava
lava flows
lava tubes
volcanism
volcanoes
volcanology

volcanic glass
Includes use on level 3. As of 1981, not to be used in the igneous rocks set. See list H. Before 1981, included use on level 3 under igneous rocks(1) pyroclastics and glasses(2). For basalt glass, use volcanic glass and basalt.
UF glass, volcanic
BA glasses
volcanic rocks
BT igneous rocks
SA crystallites
devitrification
ignimbrite
obsidian
perlite
pitchstone

volcanic rocks
Not a valid term from 1975 through 1977. After 1977, includes use on level 2 under igneous rocks(1). Before 1978, also search igneous rocks AND volcanic.
BT igneous rocks
NA andesites
basalts
dacites
glasses
granophyre
melilitite
phonolites
rhyodacites
rhyolites
trachyandesites
trachytes
vitrophyre
SA breccia pipes
lava
metavolcanic rocks
pyroclastics
rhyolite
rocks
volcanism
volcanoes
volcanology

volcanicity
use volcanism

volcanics
A valid term through 1976. Use vo'- canic rocks.

volcanism
Includes use on level 2 under volcanology(1); on level 3 under igneous rocks(1) volcanic rocks(2) and under lava(1); on level 3 under plate tectonics(1).
UF volcanicity
vulcanism
SA aa lava
ash falls
ash flows
breccia pipes
calderas
cauldrons
cinder cones
eruptions
explosive eruptions
fumaroles
hawaiian-type eruptions
heat sources
lahars
lava
maars
pahoehoe
shield volcanoes
solfataras
sublimates
volcanic earthquakes
volcanic features
volcanic rocks
volcanoes
volcanology

Volcano Island
use Vulcano

volcanoes
Used when discussing specific volcanoes. Includes use on level 2 under volcanology(1).
NA shield volcanoes
stratovolcanoes
SA ash falls
breccia pipes
calderas
cinder cones
eruptions
explosive eruptions
geologic hazards
lava
mud volcanoes
pipes
plugs
spatial distribution
sublimates
temporal distribution
volcanic belts
volcanic earthquakes
volcanic features
volcanic rocks
volcanism
volcanology

volcanology
Used for the discipline and specific treatments. Includes use on level 1 (list A); on level 2 under area terms(1), bibliography(1), education(1), and symposia(1). If level 1, term set options are:
topic [applications, bibliography, catalogs, classification, concepts, education, history, instruments, methods, nomenclature, practice, research, symposia, textbooks, theoretical studies]
subtopic (no area term)
volcanism [for ancient volcanoes and processes]
subtopic [causes, classification, concepts, eruptions, fumaroles, processes, volcanoes]
volcanoes [for recent activity only]
name of volcano or area
UF vulcanology
SA ash falls
atmosphere
bibliography
calderas
catalogs
cauldrons
continental drift
craters
crust
ejecta
eruptions
explosive eruptions
extrusive rocks
fumaroles
hawaiian-type eruptions
igneous activity
igneous rocks
intrusions
lava
lava flows
magmas
mud volcanoes
orogeny
periodicity
petrology
plate tectonics
seismology
tectonics
tilt
tiltmeters
vents
volcanic earthquakes
volcanic features
volcanic rocks
volcanism
volcanoes

Volga
A valid term through 1976. Use Volga River or Volga region.

Volga region
The middle and lower courses of the Volga River and its tributaries. It includes the Volga drainage basin with the exception of parts of the NE, NW, and SW. Also search Volga.
BT Russian Republic
USSR
SA Voronezh-Volga Anteclise

Volga River
Longest river in Europe. It rises in the Valdai Hills NE of Moscow and flows SE and then S into the N Caspian Sea. Also search Volga.
BT Russian Republic
USSR
SA Voronezh-Volga Anteclise

Volga-Don Interfluve
use Volga-Don region

Volga-Don region
Between the Don and Volga rivers in the Volga-Don Canal area.
UF Volga-Don Interfluve
BT Russian Republic
USSR

Volga-Ural Basin
use Volga-Ural region

Volga-Ural Interfluve
use Volga-Ural region

Volga-Ural region
Between the parallel courses of the lower Volga and lower Ural rivers.

Also search Ural-Volga; Ural-Volga Interfluve; Ural-Volga Region; Volga-Ural; Volga-Ural Basin. Index Soviet republics as applicable.
CO N560000N600000
E0580000E0400000
UF Ural-Volga Interfluve
Ural-Volga region
Volga-Ural Basin
Volga-Ural Interfluve
BT USSR
SA Kazakhstan
Russian Republic

Volga-Urals
Region between the middle course of the Volga River and the Urals.
BT Russian Republic
USSR

Volgian
Europe. See list E.
BA Upper Jurassic
Jurassic
BT Mesozoic
SA Malm

Volgograd
City in S Volgograd Oblast on the right bank of the Volga River about 280 miles from its mouth.
UF Stalingrad
Tsaritsyn
BT Russian Republic
USSR

Volhynia
Forested region with marshland and many lakes S and SW of the Pripet Marshes in NW Ukraine.
UF Volynia
BT Ukraine
USSR
SA Volyn

Vologda
City on the Vologda River in S Vologda Oblast 330 miles E of Leningrad.
BT Russian Republic
USSR

Volta Basin
River basin including Lake Volta. Index countries as applicable. Also search Volta.
BT Africa
SA Ghana
Ivory Coast
Togo
Upper Volta

Voltaic Republic
use Upper Volta

voltaite
BA sulfates
BT minerals

Voltzia Sandstone
BT Bunter
Triassic
SA France
Vosges

volume
Used as a general term.

volume elasticity
use bulk modulus

volume susceptibility (magnetic)
use magnetic susceptibility

Volyn
Oblast in NW Ukraine. A province of Poland prior to WW II.
UF Wolyn
BT Ukraine
USSR
SA Volhynia

Volyn-Podolia
An upland region between the Dniester and Dnieper rivers in W Ukraine. Also search Volyn-Podolian Platform; Volyn-Podolian Upland.
UF Volyn-Podolian Platform
Volyn-Podolian Upland
BT Ukraine
USSR

Volyn-Podolian Platform
use Volyn-Podolia

Volyn-Podolian Upland
use Volyn-Podolia

Volynia
use Volhynia

Volyno-Azov Massif
use Ukrainian Shield

vonsenite
BA borates
BT minerals

Vorarlberg
State in extreme W.
CO N464800N473500
E0101500E0093100
BT Austria

Voring Plateau
Underwater feature S of the Lofoten Basin off the central Norwegian coast.
BT Norwegian Sea
Arctic Ocean

Vorkuta
Town on the Vorkuta River in extreme NE Komi ASSR at N end of the Urals.
BT Russian Republic
USSR

Voronezh
City on the Voronezh River near its junction with the Don River in Voronezh Oblast N of E Ukraine.
CO N503000N530000
E0400000E0384500
BT Russian Republic
USSR

Voronezh Anteclise
Near the Ukrainian border in SW Voronezh Oblast. Also search Voronezh Massif.
UF Voronezh Anticline
Voronezh Anticlise
Voronezh Massif
BT Russian Republic
USSR
SA Voronezh-Volga Anteclise

Voronezh Anticline
use Voronezh Anteclise

Voronezh Anticlise
use Voronezh Anteclise

Voronezh Massif
use Voronezh Anteclise

Voronezh-Volga Anteclise
Term introduced in 1981.
CO N470000N563000
E0470000E0300000
BA Russian Platform
BT USSR
SA Volga region
Volga River
Voronezh Anteclise

Voroshilovsk (region)
use Stavropol region

Vosges
Department in NE.
CO N474500N483000
E0071500E0053000
BT France
SA Voltzia Sandstone
Vosges Mountains

Vosges Mountains
Range extending from Belfort Gap to West German border. Index departments as applicable. Also search Vosges.
CO N473000N493000
E0080000E0060000
BT France
SA Haut-Rhin
Vosges

Vostok
Small British uninhabited coral island in S.
UF Stavers Island
BT Line Islands
Polynesia

Vostok Station
Two soviet IGY stations. Vostok 1 Station is S of the Queen Mary Coast and Vostok 2 Station is midway between Queen Mary Coast and the South Pole near the South Geomagnetic Pole. Both stations are on the Indian Ocean side of the continent.
BT Antarctica

Vourinos
Mountain SW of Kozane in W Macedonia.
UF Burono
Vourinos Massif
BT Greece
SA Macedonia

Vourinos Massif
use Vourinos

Voyvodina
use Vojvodina

Vraca
use Vratsa

Vraconian
Europe. Includes use on level 3 under age terms(1). See list E.
BA Cretaceous
BT Mesozoic
SA Lower Cretaceous
Upper Cretaceous

Vrancea
Range of the Carpathians near their junction with the Transylvanian Alps.
UF Vrancei Mountains
BT Romania
SA Carpathians

Vrancei Mountains
use Vrancea

Vratsa
Province in NW Bulgaria. Also a city NNE of Sofia.
UF Vraca
Vrattsa
BT Bulgaria

Vrattsa
use Vratsa

vrbaite
BZ sulfantimonites
sulfarsenites
BA sulfosalts
BT minerals

Vredefort Dome
N Orange Free State. Also search Vredefort.
BT Orange Free State
South Africa

Vriddhachalam
use Vridhachalam

Vridhachalam
Town SW of Cuddalore in NE Tamil Nadu.
UF Viruddhachalam
Vriddhachalam
BT Tamil Nadu
India

Vries Island
use O-shima

Vulcan Mountains
Section of the Transylvanian Alps in W central Romania. Also search Vulcan.
BT Romania
SA Transylvanian Alps

vulcanism
use volcanism

Vulcano
Southernmost of the Lipari Islands in the Tyrrhenian Sea off NE Sicily.
UF Volcano Island
Vulcano Island
BT Sicily
Italy
SA Lipari Islands

Vulcano Island
use Vulcano

vulcanology
use volcanology

Vulture Mountain
Peak in Glacier National Park in NW.
BT Montana
United States

Vuori-Yarvi
use Vuoriyarvi

Vuoriyarvi
Village near the Finnish border in NW Karelian ASSR.
UF Vuori-Yarvi
BT Russian Republic
USSR

Vyatka
use Kirov

Vyatka River
Rises in W foothills of the Central Urals and flows into the Kama River in N Tatar ASSR. Also search Vyatka.
UF Viatka River
BT Russian Republic
USSR

Vyatka-Kama Interfluve
Region between the lower reaches of the Vyatka and Kama rivers in the Izhevsk area W of the Urals. Also search Vyatka-Kama.
CO N550000N600000
E0570000E0480000
BT Russian Republic
USSR

Vychegda River
Rises in the Komi ASSR and flows W into the Northern Dvina at Kotlas in SE Arkhangelsk Oblast in N European USSR. Also search Vychegda.
BT Russian Republic
USSR

Vyernyi
use Alma-Ata

Vyshkovo
Village N of Minsk in Byelorussia, and village in Kalinin Oblast NW of Moscow. Index Soviet republics as applicable.
BT USSR
SA Byelorussia
Russian Republic

W

W
use tungsten

Wabasca region
Term introduced in 1981. N Alberta.
CO N555800N555800
W1135600W1135600
BT Alberta
Canada

Wabash Formation
Comprises (ascending) Parker Lime-

stone, Buffkin Limestone, Graysville Limestone, and Livingston Limestone. In SW Indiana.
BT Pennsylvanian
SA Indiana

Wabaunsee County
NE central.
BT Kansas
United States

Wabaunsee Group
Includes (ascending) Severy Shale, Howard Limestone, Scranton Shale, Bern Limestone, Auburn Shale, Emporia Limestone, Willard Shale, Zeandale Limestone, Pillsbury Shale, Stotler Limestone, Root Shale, and Wood Siding Formation. E Kansas, SW Iowa, NW Missouri, SE Nebraska, and N Oklahoma.
BT Virgilian
Upper Pennsylvanian
Pennsylvanian
SA Iowa
Kansas
Missouri
Nebraska
Oklahoma

Wabush
Lake near the Quebec border in SW.
BT Labrador
Canada

Waccamaw Formation
Consists of soft limestones and loose gray to buff fine quartz sands in which occasional small quartz pebbles are present. S North Carolina and S and E South Carolina. Since second edition of Thesaurus, age has been redefined to be lower Pleistocene. Second edition stated lower Pliocene.
BT lower Pleistocene
Pleistocene
SA North Carolina
South Carolina

Wachapreague Inlet
Inlet connecting small bay off town of Wachapreague with Atlantic Ocean. In Accomack County on the eastern shore. Also search Wachapreague.
BT Virginia
United States

Wachusett-Marlborough Tunnel
Rock tunnel between the Wachusett Reservoir and Marlborough in the E central part of the state. Part of the tunnel system which provides water to Boston.
BT Massachusetts
United States

wackestone
BA carbonate rocks
BT sedimentary rocks

Waco
City on the Brazos River in McClennan County in central.
CO N313300N313300
W0971000W0971000
BT Texas
United States

Wadden Sea
use Wadden Zee

Wadden Zee
Outer section of the former Zuider Zee between the West Frisian Islands and the dike enclosing IJsselmeer.
UF Wadden Sea
Waddenzee
BT Netherlands

Waddenzee
use Wadden Zee

Wadi Araba
Depression along a valley extending from the Dead Sea to the Gulf of Aqaba. Part of the Great Rift Valley. Index countries as applicable.
UF Wadi Arabah
BT Middle East
SA Great Rift Valley
Jordan

Wadi Arabah
use Wadi Araba

Wadia
Former Western Kathiawar state of Western India States Agency; merged in 1948 with Saurashtra. Now part of Gujarat.
UF Vadia
BT Gujarat
India

Wadowice
Town on Skawa River in S part of country.
BT Krakow
Poland

Wager
Bay off Roes Welcome Sound of N Hudson Bay in NE District of Keewatin.
BT Northwest Territories
Canada

Wagga Wagga
Town on the Murrumbidgee River in S.
BT New South Wales
Australia

Waianae
Mountain range extending along SW side of Oahu Island. Also search Waianae Range.
UF Waianae Range
BT Hawaii
United States

Waianae Range
use Waianae

Waikato Basin
River basin in central.
BT North Island
New Zealand

Waimanalo
Village on SE coast of Oahu.
BT Hawaii
United States

Wairakei
Geothermal field N of Lake Taupo in central North Island. Also a village. Also search Wairakei geothermal field.
UF Wairakei geothermal field
BT North Island
New Zealand

Wairakei geothermal field
use Wairakei

wairakite
BA zeolite group
framework silicates
silicates
BT minerals

Wairarapa
Lake E of Wellington in S.
BT North Island
New Zealand

Waitemata Group
Comprises Manukau Breccias, Parnell Grit, Orakei Bay Greensand, Pakaurangi Beds, Albany Conglomerate, and Oneroa Beds all of which are known to be interbedded with the Waitemata Sandstone. In N North Island, New Zealand.
BT lower Miocene
Miocene
SA New Zealand
North Island

Wakayama
Prefecture in SW Honshu. Also a city on Kii Channel SSW of Osaka.
BT Honshu
Japan

Wake
Coral atoll and 3 islets (Wake, Peale, and Wilkes) between Hawaii and Guam in the N Pacific Ocean. Also search Wake Island.
UF Wake Island
BT Pacific Ocean

Wake Island
use Wake

Walachia
Region between the Danube River and the Transylvanian Alps.
UF Wallachia
BT Romania
NT Braila
Bucharest
Buzau
Buzau River
Muntenia
Oltenia
Pitesti
Ploesti
Prahova
Schela
Sinaia
Slanic
Slatina
Tismana
SA Fagaras Mountains
Jiu River valley
Mehedinti Plateau
Olt River
Olt River valley
Romanian Plain
Transylvanian Alps

Walbrzych
City on the Bobr River in SW part of country.
UF Waldenburg
Waldenburg in Schlesien
BT Wroclaw
Poland

Waldeck
Former German state in central part of country. Also a town in N.
BT Hesse
West Germany

Waldenburg
use Walbrzych

Waldenburg in Schlesien
use Walbrzych

Waldron Shale
A gray or greenish-gray calcareous shale with occasional thin beds of limestone or argillaceous shale. S Indiana, W central Kentucky, and central Tennessee.
BT Middle Silurian
Silurian
SA Indiana
Kentucky
Tennessee

Wales
Includes use on level 1 as an area term (list O). For term set options see list B.
CO N513000N533000
W0024000W0051500
BA Great Britain
United Kingdom
Western Europe
Europe
NT Aberystwyth
Anglesey
Arenig
Bala District
Brecknockshire
Caernarvonshire
Cardiff
Cardigan Bay
Cardiganshire
Carmarthenshire
Denbighshire
Glamorgan
Gower Peninsula
Harlech
Llandovery
Lleyn Peninsula
Merionethshire
Monmouthshire
Pembroke
Pembrokeshire
Radnorshire
Skomer Island
Snowdonia
South Wales
South Wales coal field
Swansea
Tremadoc Bay
SA Aberystwyth Grits
Bristol Channel
Penrhyn Slate
Severn Valley
Welsh Borderland
Wye Valley

Walfischbai
use Walvis Bay

Walfish Bay
use Walvis Bay

Walker Branch watershed
Term introduced in 1981. In Anderson County, NE Tennessee.
CO N355200N360000
W0841000W0842000
BT Anderson County
Tennessee
United States

Walker Lake
Term introduced in 1981. W Nevada in Mineral County. Fed by Walker River.
CO N383000N385500
W1183500W1185000
BT Nevada
United States

Walker River
Flows through Walker River Indian Reservation into Walker Lake in W central Nevada.
BT Nevada
United States

wall-rock alteration
use wallrock alteration

Wallace County
On the Colorado border in W.
BT Kansas
United States

Wallace Formation
In Piegan Group. Belt Series. Thin-bedded bluish and greenish more or less calcareous shales, underlain by rapidly alternating thin beds of argillite, calcareous sandstone, and impure limestone. NE Idaho, and W Montana.
BT Precambrian
SA Idaho
Montana

Wallachia
use Walachia

Wallowa Mountains
Range in NE.
BT Oregon
United States

wallrock alteration
Includes use on level 3 under metasomatism(1).
UF wall-rock alteration
BA metasomatism
SA alteration

country rocks
hydrothermal alteration
mineral deposits
thermal waters

Walsh County
On the Minnesota border in NE.
BT North Dakota
United States

Walton County
Index states as applicable.
BX Florida
Georgia
BT United States

Walvis Basin
use Cape Basin

Walvis Bay
Inlet of the Atlantic Ocean on the central W coast. Also a town which, along with the bay and immediate vicinity, constitutes an exclave of Cape Province, South Africa.
UF Walfischbai
Walfish Bay
BT Namibia

Walvis Ridge
Extends in a NE-SW direction off Namibia and South Africa.
UF Walvish Ridge
BT Atlantic Ocean

Walvish Ridge
use Walvis Ridge

wandering, polar
use polar wandering

Wanganui
City at mouth of the Wanganui River in SW.
BT North Island
New Zealand

Wanganui Valley
River valley in SW central North Island. Also search Wanganui; Wanganui River.
BT North Island
New Zealand

Wann Formation
In Ochelata Group. Includes both Clem Creek and Washington Irving sandstones. In NE Oklahoma.
BT Missourian
Upper Pennsylvanian
Pennsylvanian
SA Oklahoma

Wapanucka Limestone
Underlies Atoka Formation in McAlestar Basin and Ouachita Front. Overlies Bloyd Formation in McAlestar Basin and Caney Shale and Union Valley (sandstone) in Ouachita Front. In central S and SE Oklahoma.
BT Pennsylvanian
SA Oklahoma

Wapawekka Lake
At foot of the Wapawekka Hills in central Saskatchewan. Also search Wapawekka.
BT Saskatchewan
Canada

Warburton Meteorite
Term introduced in 1981. Impact in Warburton Range. Before 1981, also search Warburton AND meteorites.
BA ataxite
iron meteorites
BT meteorites
SA Australia
Western Australia

Ward County
Index states as applicable.
BX North Dakota
Texas
BT United States

Wardak
Province in E central.
BT Afghanistan

Wardha River valley
In central part of country in E Maharashtra. Also search Wardha.
BT Maharashtra
India

Ware
River in central.
BT Massachusetts
United States

Warkalli Formation
According to "Indian Stratigraphical Nomenclature", the age of Warkalli Beds ranges from upper Miocene to Pliocene. They consist of sandy clays and lignite seams. In Kerala in S India.
BT Tertiary
SA India
Kerala
Miocene
Pliocene
upper Miocene

Warren County
Index states as applicable.
BX Georgia
Illinois
Indiana
Iowa
Kentucky
Mississippi
Missouri
New Jersey
New York
North Carolina
Ohio
Pennsylvania
Tennessee
Virginia
BT United States

Warsaw
Province. Also a City province, and a city on both banks of the Vistula River in E central part of country.
UF Warschau
Warszawa
BT Poland
NT Plock
Pultusk
Siedlce

Warsaw Formation
In Osage Group. Lower division consists of massive fine-grained earthy geode-bearing limestone below, thin bed of locally brownish dolomite cherty limestone in middle and bluish-gray slightly calcareous geode-bearing shale above. N Alabama, Illinois, Indiana, Iowa, Kentucky, NE Mississippi, E Missouri, and Tennessee.
BT Meramecian
Upper Mississippian
Mississippian
SA Alabama
Illinois
Indiana
Iowa
Kentucky
Mississippi
Missouri
Tennessee
Upper Mississippian

Warschau
use Warsaw

Warszawa
use Warsaw

Warta
River which rises NW of Krakow and flows NW and W into the Oder River at Kostrzyn on the East German border.
UF Varta
Warthe
BT Poland

Warthe
use Warta

Warwick County
use Warwickshire

Warwickshire
County in central England.
CO N515500N524000
W0011500W0020000
UF Warwick County
BT England
Great Britain
United Kingdom

Wasatch County
N central.
BT Utah
United States

Wasatch Fault
N Utah.
BT Utah
United States

Wasatch Formation
Paleocene and Eocene. In Wasatch Plateau, consists of three members: lower member of sandstone, varicolored shale, conglomerate and small amounts of fresh-water limestone; and an upper member of varicolored shale and sandstone. W Colorado, central S and SE Montana, NE New Mexico, SW North Dakota, Utah, and W Wyoming.
BT Paleogene
Tertiary
SA Colorado
Eocene
Montana
New Mexico
North Dakota
Paleocene
Utah
Wyoming

Wasatch Front
Outer slope of the Wasatch Range. Index states as applicable.
BT United States
SA Idaho
Utah
Wasatch Range

Wasatch Mountains
use Wasatch Range

Wasatch Plateau
High tableland at S end of Wasatch Range in central.
BT Utah
United States

Wasatch Range
Range of Rocky Mountains extending from SE Idaho to central Utah. Index states as applicable. Also search Wasatch Mountains.
UF Wasatch Mountains
BT United States
SA Idaho
Rocky Mountains
Utah
Wasatch Front

Washakie Basin
Primarily in S Wyoming but also just within NW Colorado. Index states as applicable.
BT United States
SA Colorado
Wyoming

Washington
Includes use on level 1 as an area term (list O). For term set options see list B.
CO N453000N490000
W1165500W1244500
BA Pacific Coast
Western U.S.
United States
NX Adams County
Benton County
NT Blue Glacier
Chelan County
Clallam County
NX Clark County
Columbia County
NT Cowlitz County
NX Douglas County
Franklin County
Garfield County
Golden Horn Batholith
Grant County
NT Grays Harbor
Hanford Reservation
NX Jefferson County
King County
NT Kitsap Peninsula
Kittitas County
NX Lewis County
Lincoln County
Mason County
NT Metaline Falls
Methow River valley
Mount Rainier
Mount Rainier National Park
Mount Saint Helens
Nisqually Glacier
Okanogan County
Olympic Mountains
Olympic Peninsula
Pasco Basin
Pend Oreille County
NX Pierce County
NT Puget Lowland
Puget Sound
Richland
NX San Juan County
NT San Juan Islands
Seattle
Skagit County
Skamania County
Snohomish County
South Cascade Glacier
Spokane
Spokane County
NX Stevens County
NT Whatcom County
Yakima County
SA Belt Supergroup
Blue Mountains
Cascade Range
Chuckanut Formation
Coast Ranges
Columbia Plateau
Columbia River
Columbia River Basalt
Columbia River basin
Columbia River estuary
Juan de Fuca Strait
Lincoln Creek Formation
Montesano Formation
Okanagan Valley
Pasayten Group
Skagit Valley
Snake River
Snake River basin
Snoqualmie Batholith
Strait of Georgia
Swauk Formation
Tono
Twin River Formation
Twin Sisters Dunite
U. S. Rocky Mountains
Vancouver
Yakima Basalt

Washington County
Index states as applicable.
BX Alabama
Arkansas
Colorado
Florida
Georgia
Idaho
Illinois

Indiana
Iowa
Kansas
Kentucky
Maine
Maryland
Minnesota
Mississippi
Missouri
Nebraska
New York
North Carolina
Ohio
Oklahoma
Oregon
Pennsylvania
Rhode Island
Tennessee
Texas
Utah
Vermont
Virginia
Wisconsin
BT United States

Washita Group
Lower and Upper Cretaceous. In Oklahoma, includes (ascending) Duck Creek Formation, Fort Worth Limestone, Denton Clay, Weno Clay, Pawpaw Formation, Main Street Limestone, and Grayson Shale. SW Arkansas, NW Louisiana, S Oklahoma, and Texas.
BT Cretaceous
SA Arkansas
Louisiana
Lower Cretaceous
Oklahoma
Texas
Upper Cretaceous

Washita River valley
E Texas panhandle, and W and S central Oklahoma. Index states as applicable. Also search Washita River.
BT United States
SA Oklahoma
Texas

Washoe County
NW Nevada.
BT Nevada
United States

waste, agricultural
use agricultural waste

waste disposal
A level 1 term as of 1978. Includes use on level 2 under impact statements(1), pollution(1), and underground installations(1). Term set options are:
liquid waste
topic (effects, experimental studies, impact statements, methods, pollution, storage)
radioactive waste
subtopic
seepage
subtopic
site exploration
subtopic
solid waste
subtopic
thermal pollution
subtopic
topic
subtopic
UF disposal, waste
SA agricultural waste
conservation
controls
detergents
engineering geology
environmental geology
fluid injection
geologic hazards
ground water
human waste
impact statements
industrial waste
injection
land use
landfills
liquid waste
marine installations
pesticides
pollutants
pollution
radioactive waste
radioactivity
reclamation
reservoirs
sanitary landfills
seepage
sewage
site exploration
soils
solid waste
spoils
storage
sulfuric acid
tailings
thermal pollution
toxic materials
toxicity
underground installations
waste water

waste, human
use human waste

waste, industrial
use industrial waste

waste, liquid
use liquid waste

waste, radioactive
use radioactive waste

waste, solid
use solid waste

waste water
Includes use on level 3 under waste disposal(1) liquid waste(2).
UF water, waste
SA detergents
effluents
environmental geology
industrial waste
liquid waste
waste disposal
water

water
Includes use on level 2 under meteorology(1), under isotopes(1), and under pollution(1).
SA aqueous solutions
artesian waters
atmospheric precipitation
bottling
bottom water
brackish water
clouds
connate waters
drainage
droplets
environmental geology
fresh water
gauging
geochemistry
ground water
humidity
hydration
hydrologic cycle
hydrological methods
hydrology
hydrosphere
hydrothermal processes
ice
impurities
irrigation
juvenile water
low water
medicinal waters
meltwater
meteorology
mineral waters
moisture
oil-water interface
percolation
polluted water
pollution
pore water
precipitation
purification
raindrops
retention
salt water
saturation
sediment-water interface
seepage
snow
springs
surface water
thermal waters
waste water
water hardness
water pressure
water regimes
water storage
water supply
water vapor
water yield
waterways

water balance
Also search water budget.
UF balance, water
hydrologic budget
water budget
SA aquifers
drainage basins
hydrology
lakes
reservoirs
runoff

water biscuits
use algal biscuits

water budget
use water balance

water crop
use water yield

water cycle
use hydrologic cycle

water erosion
Includes use on level 3 under soils(1) erosion(2).
BT erosion
SA denudation
erosion control
erosion features
sheet erosion
soils
wind erosion

water hardness
Term introduced in 1981. Before 1981, also search hardness AND water.
SA ground water
hardness
surface water
water
water quality
water resources

water harnessing
Term introduced in 1981.
SA water resources

water of crystallization
Term introduced in 1981.
SA crystallization
geochemistry
minerals
water of dehydration

water of dehydration
Term introduced in 1981.
SA dehydration
geochemistry
minerals
water of crystallization

water opal
use hyalite

water pressure
BT pressure
SA soil mechanics
water

water quality
Includes use on level 3 under ground water(1), hydrology(1) and water resources(1).
SA fresh water
ground water
hydrology
irrigation
purification
sewage
water hardness
water resources
water supply

water recovery
Term introduced in 1978 for recovery in relation to ground water. Before 1978 search recovery AND ground water.
SA discharge
ground water
recovery
wells

water regimes
Includes use on level 2 under soils(1). Also search water regime.
UF regimes, water
SA characterization
drainage
ground water
infiltration
irrigation
movement
soils
storage
unsaturated zone
water

water resources
Includes use on level 1 as a commodity term (list C) for economically oriented papers.
SA aquifers
artesian waters
bottling
economic geology
energy sources
environmental geology
fresh water
ground water
hydrogeology
hydrology
ponds
purification
representative basins
resources
springs
surface water
thermal waters
water hardness
water harnessing
water quality
water storage
water supply
water yield
well screens

water storage
Includes use on level 3 under reservoirs(1).
SA dams
engineering geology
reservoirs
storage
water
water resources
water supply

water supply
UF supply, water
SA aquifers

hydrology
irrigation
lakes
pollution
ponds
reservoirs
rivers
surface water
water
water quality
water resources
water storage

water table
Term introduced in 1981.
SA capillary water
ground water
hydrogeology
hydrology
levels
potentiometric surface
saturated zone
unconfined aquifers
unsaturated zone

water vapor
UF vapor, water
SA evaporation
geochemistry
humidity
moisture
water

water, waste
use waste water

water yield
UF runout
water crop
SA drainage basins
runoff
water
water resources
yields

water-break
use breakwaters

Waterberg System
Composed primarily of sedimentary rocks. In Transvaal, South Africa.
BT Precambrian
SA South Africa
Transvaal

waterfalls
BT fluvial features
SA cascades

waterlogging
Term introduced in 1981.
SA hydrology
infiltration
soils

Waterloo Island
use King George Island

watersheds
Includes use on level 3 under hydrology(1).
SA drainage basins
floods
hydrology
lakes
rainfall
rivers
runoff
streams
surface water
waterways

Waterville Formation
According to the latest published Lexique, the age of Waterville Shale is Silurian. A series of shales, fine grained sandstones, and impure limestones; often pyritiferous. In S central Maine.
BT Silurian
SA Maine

waterways
A valid level 1 term as of 1978. Used for geological studies on man-made or man-modified water channels. Before 1978, included use on level 2 under engineering geology(1). Term set options are:
canals
subtopic
channels
subtopic
design
subtopic
erosion
subtopic
floods
subtopic
harbors
subtopic
hydraulics
subtopic
irrigation
subtopic
rivers and streams
subtopic
seepage
subtopic
SA canals
channels
controls
design
dredging
engineering geology
environmental geology
erosion
estuaries
floods
geometry
geomorphology
harbors
hydraulics
hydrology
irrigation
marine installations
ocean circulation
reservoirs
rivers
rivers and streams
seepage
shorelines
turbulence
underground channels
water
watersheds

Waterways Formation
BT Devonian
Paleozoic
SA Alberta

wave dispersion
Term introduced in 1981. Before 1981, also search waves AND dispersion; waves AND dispersal.
SA attenuation
distortion
elastic waves
scattering
seismology

wave propagation
As of 1981, no longer a valid term for GeoRef. Use propagation. Included use on level 2 under ionosphere(1) and magnetosphere(1).

wave-cut platforms
Term introduced in 1981. Includes use on level 3 under geomorphology(1) erosion features(2).
SA erosion features
erosion surfaces
geomorphology
intertidal environment
shore features

wavellite
BA phosphates
BT minerals

waves
Includes use on level 2 under aeronomy(1) and meteorology(1).
SA acoustical waves
aeronomy
Airy waves
body waves
bow shock waves
breaking waves
catastrophic waves
coda waves
elastic waves
electromagnetic waves
ideal waves
internal waves
long-period waves
Love waves
meteorology
ocean waves
P-waves
PcP-waves
PKiP-waves
PKP-waves
PKS-waves
PPP-waves
PPS-waves
propagation
Rayleigh waves
S-waves
ScS-waves
seiches
SH-waves
shoaling
shock waves
short-period waves
SKP-waves
SKS-waves
Stoneley waves
surface waves
surges
SV-waves
tsunamis

Wawa
Village NE of Manila in Rizal Province.
BT Luzon
Philippine Islands

Wayne County
Index states as applicable.
BX Georgia
Illinois
Indiana
Iowa
Kentucky
Michigan
Mississippi
Missouri
Nebraska
New York
North Carolina
Ohio
Pennsylvania
Tennessee
Utah
West Virginia
BT United States

Waynesboro
Borough in Franklin County, S Pennsylvania, and city in Augusta County in N central Virginia. Index states as applicable.
BX Pennsylvania
Virginia
BT United States

Weald
use The Weald

Weald Clay
Divided into 3 lithological groups: Group 1- buff grey, Group 2- variegated and Group 3- yellow. In Sussex in SE England.
BT Cretaceous
SA England
United Kingdom

Weald region
use The Weald

Wealden
Europe. Above Purbeckian, below Gault. Includes use on level 3 under age terms(1). See list E.
BA Lower Cretaceous
Cretaceous
BT Mesozoic

weathering
For treatments emphasizing the process. Includes use on level 1 (list A). See mineral deposits, genesis(1) processes(2). Term set options are:
topic [analysis, classification, environment, experimental studies, rates, textbooks]
subtopic
type of material [igneous rocks, metamorphic rocks, minerals, sedimentary rocks, sediments]
specific type (e.g. rock name)
NT chemical weathering
differential weathering
mechanical weathering
physical weathering
SA alteration
bauxitization
biorhexistasy
clay mineralogy
degradation
denudation
detritus
diagenesis
duricrust
ecology
engineering geology
enrichment
erosion
etching
exfoliation
geochemistry
geomorphology
hydrolysis
insolation
kaolinization
leaching
mechanical properties
migration of elements
mineral deposits
parent materials
placers
residuum
sediments
soils
weathering crust

weathering crust
As of 1981, includes use on level 3 under sedimentary rocks(1) chemically precipitated rocks(2). See list I. Before 1981, included use on level 3 under sediments(1) chemically precipitated sediments(2). Also includes use on level 3 under soils(1) and weathering(1).
UF crust, weathering
BA chemically precipitated rocks
BT sedimentary rocks
SA calcrete
caliche
duricrust
soils
weathering

Weber County
E of Great Salt Lake in N.
BT Utah
United States

Weber Sandstone
Pennsylvanian and Permian. In Duchesne River area of Utah, consists mainly of fine grained gray and white sandstone that weathers buff. W Colorado, and NE Utah.
BT Paleozoic
SA Colorado
Pennsylvanian
Permian
Utah

weberite
As of 1981, includes use on level 3 under minerals(1) fluorides(2). See list L. Before 1981, included use on level 3 under minerals(1) halides(2).
BA fluorides
halides
BT minerals

Webster County
Index states as applicable.
BX Georgia
Iowa
Kentucky
Mississippi
Missouri
Nebraska
West Virginia
BT United States

websterite
As of 1981, includes use on level 3 under igneous rocks(1) ultramafics(2). See list H. Before 1981, included use on level 3 under igneous rocks(1) ultramafic family(2).
BA ultramafics
BT igneous rocks
SA pyroxenite

Wechsel
An outlier of the Eastern Alps in E part of country. Index states as applicable.
BT Austria
SA Eastern Alps
Lower Austria
Styria

Weddell Sea
Arm of the S Atlantic Ocean between the Antarctic Peninsula on the W and Coats Land on the SE.
CO S780000S630000
W0180000W0600000
BT Antarctic Ocean

weddellite
BA carbonates
BT minerals
SA whewellite

wedges, fossil ice
use fossil ice wedges

wedges, ice
use ice wedges

Wedron Formation
Glacial till. In association with Wadsworth Till Member, Shorewood Till Member, Manitowoc Till Member, and Two Rivers Till Member. Indiana, Illinois, and under Lake Michigan.
BT Pleistocene
SA Illinois
Indiana
Lake Michigan

Weekeroo Station Meteorite
Term introduced in 1981. Impact at Weekeroo Station near Mannahill, Australia. Before 1981, also search Weekeroo Station AND meteorites.
BA octahedrite
iron meteorites
BT meteorites
SA Australia
South Australia

Wegener hypothesis
use continental drift

wehrlite
As of 1981, includes use on level 3 under igneous rocks(1) peridotites(2). See list H. Also a mineral. Before 1981, included use on level 3 under igneous rocks(1) ultramafic family(2).
BX igneous rocks
minerals
SA peridotite
peridotites
tellurides

Weichsel
use Weichselian

Weichsel River
use Vistula River

Weichselian
Term applied in Northern Europe to fourth and last glacial stage of the Pleistocene Epoch. See list E.
UF Weichsel
BA upper Pleistocene
Pleistocene
Quaternary
BT Cenozoic
NA Allerod
Brandenburg Stade
Brandon Stade
Frankfurt Stade
Older Dryas
Oldest Dryas
Perth Stade
Younger Dryas
SA Wisconsinan
Wurm

Weimar
City in SW part of the country.
BT Erfurt Bezirk
East Germany

weiner filters
Term introduced in 1981.
BT filters

Weinheim
City NE of Mannheim in NE.
BT Baden-Wurttemberg
West Germany

Weipa
Habitation in an aboriginal reserve on the W coast of Cape York Peninsula in N.
BT Queensland
Australia

Weisse Elster Basin
River basin in extreme W Bohemia and S central East Germany. Index Bohemia and/or East Germany as applicable. Also search Weisse Elster.
UF White Elster Basin
BT Europe
SA Bohemia
East Germany

Weissemburg in Bayern
use Weissenburg

Weissenberg
Town just N of Lobau near the Polish and Czechoslovak borders in Upper Lusatia.
BT East Germany

Weissenburg
Town about 30 miles S of Nuremberg in W central Bavaria.
UF Weissemburg in Bayern
Weissenburg-am-Sand
BT Bavaria
West Germany

Weissenburg-am-Sand
use Weissenburg

Weld County
N Colorado.
CO N400000N410000
W1033000W1051000
BT Colorado
United States

welded tuff
As of 1981, includes use on level 3 under igneous rocks(1) pyroclastics(2). See list H. Before 1981, included use on level 3 under igneous rocks(1) pyroclastics and glasses(2).
UF tuff lava
BA pyroclastics
volcanic rocks
BT igneous rocks
SA ignimbrite
tuff

well logging
use well-logging

well screens
Term introduced in 1981.
SA ground water
water resources
wells

well-logging
For treatments that stress methodology. Includes use on level 1 (list A). Also search well logging. Term set options are:
topic [applications, automatic data processing, instruments, interpretation, methods, techniques]
subtopic
type [acoustical logging, caliper logging, dipmeter logging, electrical logging, electromagnetic logging, radioactivity]
elaboration of type
UF well logging
NA acoustical logging
caliper logging
dipmeter logging
electrical logging
electromagnetic logging
SA automatic data processing
boreholes
cores
cuttings
drilling
gamma-gamma methods
gamma-ray methods
gamma-ray spectroscopy
geophysical methods
geophysical surveys
induced polarization
induction
microwave methods
radioactivity
resistivity
wells

Welland Canal
Ship canal connecting Lake Erie with Lake Ontario in SE Ontario. Also search Welland.
BT Ontario
Canada

Wellington
City on Port Nicholson an inlet of Cook Strait on S.
BT North Island
New Zealand

Wellington Formation
In Sumner Group. In Kansas, includes Milan Limestone Member at top; Hutchinson Salt Member in middle part but not exposed; Carlton Limestone Member below Hutchinson; Hollenburg Limestone Member near base. Central and S Kansas, and N Oklahoma.
BT Permian
SA Hutchinson Salt Member
Kansas
Oklahoma

wells
Includes use on level 3 under well-logging (1) and under engineering geology (1).
UF wells and drill holes
SA boreholes
cores
cuttings
drilling
engineering geology
potentiometric surface
water recovery
well screens
well-logging

wells and drill holes
use wells

weloganite
BA carbonates
BT minerals

Welsh Borderland
As of 1981, refers to region bordering on Wales that is entirely within England. Before 1981, part of Wales was included.
CO N513000N524500
W0021000W0031500
BT England
SA Wales

Welwitschiales
Term introduced in 1981. Includes use on level 2 under gymnosperms(1). See list F.
BA gymnosperms
BT Spermatophyta
Plantae

wenkite
BZ framework silicates
sulfates
BT minerals

Wenlock Edge
A limestone ridge in Shropshire in W.
BT England
Great Britain
United Kingdom

Wenlock Limestone
Upper, thin-bedded, lenticular and lower, more massive and crystalline limestones, underlying the "Upper Ludlow Rock" and above the "Lower Ludlow Rock". In Shropshire in W England.
BT Wenlockian
Upper Silurian
Silurian
SA England
United Kingdom

Wenlockian
Europe. Includes use on level 3 under age terms(1). See list E.
BA Upper Silurian
Silurian
BT Paleozoic
NT Wenlock Limestone

Werfenian
Triassic. Includes use on level 3 under age terms(1). See list E.
BA Lower Triassic
Triassic
BT Mesozoic
SA Scythian

Werillup Formation
BT Eocene
Paleogene
Tertiary
SA Western Australia

Werra
River which rises in Suhl District, East Germany, and unites with the Fulda River in Hesse, West Germany, to form the Weser River. Index countries as applicable.
BT Europe
SA East Germany
West Germany

Weser River
Formed by the confluence of the Fulda and Werra rivers. Flows primarily through Lower Saxony into the North Sea. Index states as applicable. Also search Weser.
BT West Germany
SA Hesse
Lower Saxony
North Rhine-Westphalia

Weser-Ems
Region between the Weser and Ems rivers in W and W central.
BT Lower Saxony
West Germany
SA Ems River

West Africa
Index countries as applicable.
CO N020000N250000
E0240000W0173000
BA Africa
NA Benin
Cameroon
Chad
Gambia
Ghana
Guinea
Guinea-Bissau
Ivory Coast
Liberia
Mali
Mauritania
Niger
Nigeria
Senegal
Sierra Leone
Togo
Upper Volta
SA Central Africa
East Africa
North Africa
Southern Africa

West African Craton
use West African Shield

West African Shield
Index countries as applicable. Also search West African Craton.
UF West African Craton
BT Africa
SA Mali
Niger

West Atlantic
Term introduced in 1978. Before 1978, also search Atlantic Ocean AND west or western.
BA Atlantic Ocean
SA East Atlantic
North Atlantic
Northeast Atlantic
Northwest Atlantic
South Atlantic
Southeast Atlantic
Southwest Atlantic

West Bengal
State bordering Bangladesh in E.
CO N213000N271000
E0900000E0854000
BT India
NT Bankura
Barakar
Burdwan
Calcutta
Cooch Behar
Darjeeling
Digha
Purulia
Raniganj
Raniganj coal field
SA Ajay River
Bengal
Damodar Valley
Raniganj Stage

West Bulgarian Hills
Term introduced in 1981.
CO N411900N425200
E0233000E0222000
BT Bulgaria

West Carpathians
use Western Carpathians

West Coast
use Pacific Coast

West Germany
Officially known as Federal Republic of Germany or Bundesrepublik Deutschland. In W central Europe, bounded on N by North Sea and Denmark, on E by East Germany and Czechoslovakia, on SE by Austria, on S by Austria and Switzerland, on SW by France, and on W by Luxembourg, Belgium and the Netherlands. Introduced as a level 1 area term in 1978.
CO N472000N550000
E0134500E0055500
BA Germany
Central Europe
Europe
NT Baden-Wurttemberg
Bavaria
Bergisches Land
Bergstrasse
Ems River
Franconia
Franconian Forest
Hamburg
Hanover
Hesse
Lahn River
Lahn River valley
Lower Saxony
Main River
Mainz Basin
Nahe
North Rhine-Westphalia
Rhineland
Rhineland-Palatinate
Rhon Mountains
Saarland
Schleswig-Holstein
Steinheim
Steinheim Basin
Swabia
Teutoburg Forest
Weser River
Westerwald
Westphalia
SA Alpenvorland
Bavarian Alps
Bavarian Massif
Berlin
Brunswick
Central Alps
Elbe River
Elbe Valley
Essen Beds
Frankfurt
Harz Foreland
Harz Mountains
Harz region
Inn Valley
Lake Constance
Lower Rhine Basin
Moselle River
North German Plain
Northwestern German Plain
Posidonia Shale
Rhenish Schiefergebirge
Rhine Basin
Rhine River
Rhine Valley
Rhine Westphalian Basin
Saale River
Saar-Nahe Basin
Solnhofen Limestone
Stavelot-Venn Massif
Stuttgart
Upper Rhine Valley
Werra
Zechstein

West Greenland
Term introduced in 1978. Before 1978, search Greenland AND west.
BA Greenland
SA South Greenland

West Indian Ocean
Term introduced in 1978. Before 1978, also search Indian Ocean AND west or western.
BA Indian Ocean

West Indies
Islands between SE North America and N South America enclosing the Caribbean Sea. Index island groups as applicable. Includes use on level 1 or 2 as an area term (list O). If 1, see term set options under list B.
CO N100000N253000
W0590000W0850000
NT Antilles
NA Bahamas
NT Caribbean Mountain Range
NA Cayman Islands
Greater Antilles
Lesser Antilles
SA America
Caribbean region
Grenville
Rincon Formation
Robles Formation

West Irian
use Irian Jaya

West Malaysia
That part of the Federation of Malaysia which is comprised of eleven states on the Malay Peninsula; Johore, Kedah, Kelantan, Malacca, Negri Sembilan, Pahang, Penang, Perak, Perlis, Selangor, and Trengganu.
BT Malaysia
SA Indochina
Johore
Kelantan
Malay Peninsula
Pahang
Perak
Selangor

West Mediterranean
Term introduced in 1978. Before 1978, search Mediterranean Sea AND west.
CO N351000N443000
E0161500W0053000
BT Mediterranean Sea
NA Alboran Sea
Gulf of Lion
Tyrrhenian Sea
SA East Mediterranean

West New Guinea
use Irian Jaya

West Pacific
Term introduced in 1978. Before 1078, search Pacific Ocean AND west.
BA Pacific Ocean
NA Bering Sea
Coral Sea
East China Sea
Indonesian Seas
Japan Sea
North Australian Seas
Okhotsk Sea
Philippine Sea
South China Sea
Tasman Sea
Yellow Sea
SA East Pacific
Equatorial Pacific
North Pacific
Northeast Pacific
Northwest Pacific
South Pacific
Southeast Pacific
Southwest Pacific

West Pakistan
The west wing of Pakistan prior to the independence of East Pakistan which became Bangladesh. A valid term through 1972. After 1972, use Pakistan. Also search Pakistan AND west.

West Siberia
Term introduced in 1981.
CO N500000N820000
E1140000E0620000
BT USSR
NZ Kuznetsk Alatau
Minusinsk Basin
Severnaya Zemlya
NA Siberian Lowland
NZ Taymyr
SA Kazakhstan
Russian Republic

West Siberian Basin
use Siberian Lowland

West Siberian Lowland
use Siberian Lowland

West Siberian Plain
use Siberian Lowland

West Siberian Plate
SA plate tectonics
plates
Siberian Platform

West Siberian Platform
use Siberian Platform

West Spitsbergen
use Vestspitsbergen

West Valley
Term introduced in 1981. Town in W New York in Cattaraugus County.
CO N422300N422300
W0783800W0783800
BT Cattaraugus County
New York
United States

West Virginia
Includes use on level 1 as an area term (list O). For term set options see list B.
CO N371500N404000
W0774500W0823000
BA Eastern U.S.
United States
NT Blackwater
NX Boone County
Calhoun County
Charleston
Clay County
NT Eskdale
NX Fayette County
Grant County
NT Greenbrier County
NX Hampshire County
Hancock County
Harrison County
Jackson County
Jefferson County
NT Kanawha County
NX Lewis County
Lincoln County
Logan County
Marion County
Marshall County
Mason County
Mercer County
Mineral County
Monroe County
Morgan County
Pendleton County
Pocahontas County
NT Preston County
NX Putnam County
Randolph County
Roane County
Taylor County
Wayne County
Webster County
NT Wheeling
NX Wood County
SA Allegheny Front
Allegheny Group
Allegheny Mountains
Allegheny Plateau
Ames Limestone
Antietam Formation
Appalachian Basin
Appalachian Plateau

Berea Sandstone
Bloomsburg Formation
Blue Ridge Mountains
Blue Ridge Province
Catoctin Formation
Catskill Delta
Clinton Group
Coeymans Formation
Conemaugh Group
Cumberland Plateau
Dunkard Basin
Dunkard Group
Genesee Group
Greenland Gap Group
Hamilton Group
Helderberg Group
Hurricane Ridge Syncline
Juniata Formation
Martinsburg Formation
McKenzie Formation
Monongahela Group
New River
Ohio River
Ohio River valley
Ohio Shale
Onondaga Limestone
Oriskany Sandstone
Panhandle
Pocahontas Formation
Pocono Formation
Potomac River
Potomac River basin
Pottsville Group
Rochester Formation
Sharon Conglomerate
Shenandoah Valley
Tioga Bentonite
Tonoloway Limestone
Tuscarora Formation
Valley and Ridge Province
Williamsport Sandstone

West-Central Nigeria
Term introduced in 1981.
CO N044000N073800
E0064500E0050000
BT Nigeria

Westchester County
Just N of New York City.
BT New York
United States

Westerly Granite
Pennsylvanian or younger. Finely crystalline gray rock, which shows minor variations in color and texture but which is petrographically the same. SE Connecticut and SW Rhode Island.
BT Pennsylvanian
SA Connecticut
Rhode Island

Western Alps
Ranges of the Alps in SE France, NW Italy, and SW Switzerland. Index countries as applicable.
BT Alps
Europe
SA Cottian Alps
France
Italy
Maritime Alps
Switzerland

Western Australia
Includes use on level 1 as an area term (list O). For term set options see list B.
CO S350000S140000
E1290000E1130000
BA Australia
Australasia
NT Barrow Island
Canning Basin
Carnarvon Basin
Cue
Dalgaranga
Darling Range
Eastern Goldfields
Fraser Range
Hamersley Basin
Hamersley Range
Kalgoorlie
Kambalda
Mundrabilla
Norseman
North Pole deposit
Perth Basin
Pilbara gold field
Shark Bay
Widgiemooltha
Wiluna
Wittenoom Gorge
Yilgarn
Yilgarn Block
SA Amadeus Basin
Brockman Iron Formation
Eucla Basin
Giles Complex
Hamersley Group
Joseph Bonaparte Gulf
Kalgoorlie System
Nullarbor Plain
Officer Basin
Warburton Meteorite
Werillup Formation
Wolf Creek Meteorite

Western Canada
Term introduced in 1981.
CO N480000N694000
W0890000W1410000
BA Canada
NA Alberta
British Columbia
Manitoba
Saskatchewan
Yukon Territory

Western Carpathians
Ranges of the Carpathians in S and SE Poland, and in Slovakia. Index countries as applicable. Also search West Carpathians.
UF West Carpathians
BT Carpathians
Europe
SA Czechoslovakia
Poland

Western Cordillera
Term introduced in 1981. Western range of the Andes. Index countries as applicable. Before 1981, also search Cordillera Occidental.
UF Cordillera Occidental
BT Andes
SA Chile
Colombia
Ecuador

Western Desert
Part of the Libyan Desert in W and W central.
CO N220000N314000
E0325000E0244000
BT Egypt
SA Libyan Desert

Western Europe
BA Europe
NA Andorra
Belgium
Corsica
France
Iceland
Ireland
Luxembourg
Monaco
Netherlands
Scandinavia
United Kingdom

Western Ghat Mountains
use Western Ghats

Western Ghats
Mountain range extending 800 miles along SW and W coast as far N as mouth of Tapti River on the Gulf of Cambay.
UF Western Ghat Mountains
BT India
SA Ghats

Western gravimeters
Term introduced in 1981.
BT gravimeters

Western Hemisphere
Used when discussing many large areas too numerous to mention. Includes use on level 1 and 2 as an area term (list O). If 1, see list B for term set options.
SA Eastern Hemisphere
Northern Hemisphere
Southern Hemisphere

Western Interior
Level 1 area term as of 1978. Tremendous region in North America including the Great Plains, the Rocky Mountains, the Basin and Range Province of the U. S., and the interior plateaus of Canada. Index countries as applicable.
BA North America
SA Canada
Idaho
Montana
United States
Utah
Wyoming

Western Islands
use Hebrides

Western Morava River
use Zapadna Morava

Western Nigeria
Term introduced in 1981.
CO N055000N091000
E0060000E0024000
BT Nigeria

Western Overthrust Belt
Term introduced in 1981. Before 1981, search Overthrust Belt; Thrust Belt.
BT United States
SA Idaho
Montana
Utah
Wyoming

Western Port
Term introduced in 1981. Inlet of Bass Strait, S Victoria, Australia.
CO S385000S381000
E1454000E1443000
BX Bass Strait
Victoria
BT Australia

Western Sahara
Term introduced in 1981. Includes use on level 1 as an area term (list O). For term set options see list B. Also search Sahara AND West; Spanish Sahara.
CO N212000N274000
W0084000W0171000
BZ Sahara
BX Morocco
SA Reguibat Ridge

Western Sayan
Term introduced in 1978. Before 1978, search Sayan AND west or western.
CO N500000N540000
E0950000E0860000
BA Sayan
Siberian fold belt
BT USSR

Western Transbaikalia
Term introduced in 1981.
CO N490000N560000
E1210000E1040000
BZ Siberian fold belt
BX Russian Republic
BT USSR

Western U.S.
Term introduced in 1978. Includes use as level 1 area term. Before 1978, also search United States AND west or western. As of 1981, term includes five more states: Colorado, Wyoming, Montana, Idaho, and Utah.
BA United States
NA Alaska
Colorado
Idaho
Montana
Nevada
Pacific Coast
Utah
Wyoming
SA Battle Mountain High

Western World
Term introduced in 1981. Used for the non-communist "industrialized world".
SA communist countries

Westerwald
Mountainous region extending NE from near Koblenz for about 70 miles between the Rhine, Sieg, and Lahn rivers. Geologically it is considered part of the Rhenish Schiefergebirge (Rhenish Slate Mountains). Index states as applicable.
BT West Germany
SA Hesse
Rhenish Schiefergebirge
Rhineland-Palatinate

Westland
Provincial district along the Tasman Sea in W.
CO S442000S420000
E1723000E1680000
BT South Island
New Zealand

Westman Islands
use Vestmannaeyjar

Westmoreland County
Index states as applicable. Also search Westmoreland AND appropriate area.
BX Pennsylvania
Virginia
BT United States

Westmorland
County in NW.
CO N541500N544500
W0021000W0031000
BT England
Great Britain
United Kingdom

Weston County
On the South Dakota border in NE Wyoming. Also search Weston AND Wyoming.
BT Wyoming
United States

Westphalia
Region and former Prussian province now in parts of 3 West German states. Index states as applicable.
BT West Germany
SA Hesse
Lower Saxony
North Rhine-Westphalia

Westphalian
Europe. Above Namurian, below Stephanian. Includes use on level 3 under age terms(1). See list E.
BA Upper Carboniferous
Carboniferous
BT Paleozoic
SA Essen Beds

wet
A valid term through 1977. After 1977, use wet methods.

wet methods
Term introduced in 1978. Includes use on level 3 under chemical analysis(1) methods(2). Before 1978, also search chemical analysis AND wet.
SA analysis
chemical analysis
methods

Wet Mountains
Range of Rocky Mountains in S central.
BT Colorado
United States
SA Rocky Mountains

wetlands
SA ecology
geomorphology
grasslands
vegetation

Wetterau
Region NNE of Frankfurt much of it along the Wetter River.
BT Hesse
West Germany

Wetterstein Limestone
According to the latest published Austrian Lexique, the age of Wetterstein Dolomite is Triassic. In Tyrol in W Austria.
BT Triassic
SA Austria
Tyrol

Wewoka Formation
In Marmaton Group. Conformably overlies Wetunka Formation and conformably underlies Holdenville Shale.
BT Desmoinesian
Middle Pennsylvanian
Pennsylvanian
SA Marmaton Group
Oklahoma

Wexford
County in SE Ireland. Also a city on Wexford Harbour off Saint George's Channel.
CO N521000N524500
W0061000W0070000
BT Ireland

Weyer
Village on the Enns River in SE.
BT Upper Austria
Austria

Wharton Basin
SSW of the Java Trench in E.
BT Indian Ocean

Whatcom County
On the British Columbia border E of the Strait of Georgia.
BT Washington
United States

wheel ore
use bournonite

Wheeler County
Index states as applicable.
BX Georgia
Nebraska
Oregon
Texas
BT United States

Wheeling
City on the Ohio River in the N panhandle in Ohio County.
BT West Virginia
United States

Whetstone Lake
BT Ontario
Canada

whewellite
BA carbonates
BT minerals
SA weddellite

whistlers
Includes use on level 2 under magnetosphere(1).
SA electromagnetic radiation
lightning
magnetic field
magnetosphere

white antimony
use valentinite

White Bay
Large inlet of the Atlantic Ocean in N Newfoundland.
BT Newfoundland
Canada

White County
Index states as applicable.
BX Arkansas
Georgia
Illinois
Indiana
Tennessee
BT United States

White Elster Basin
use Weisse Elster Basin

White Island
In Bay of Plenty off N central North Island. An active volcano is located on the island.
BT North Island
New Zealand

White Limestone
Name has either been abandoned by author or rejected for use by the U.S. Geological Survey. It was a color term applied in a titular sense to 160 feet of light-gray to dark-gray dolomitic limestone with shaly layers. Leadville District, Colorado.
BT Lower Ordovician
Ordovician
SA Colorado

white mica
use muscovite

White Mountain
Peak in the Sierra Nevada Mountains in E central California, a hill W of Prague, and a village on S Seward Peninsula in W Alaska. Index Bohemia and states as applicable.
SA Alaska
Bohemia
California

White Mountains
Range of the Appalachians in N central New Hampshire, range in E California and SW Nevada, and a range in Fort Apache Indian Reservation in E Arizona. Index states as applicable. Also search White-Inyo Mountains or search White-Inyo Range for California mountains.
BT United States
SA Arizona
California
Nevada
New Hampshire

White Pine
Village on the Iron River in Ontonogan County in W Upper Penninsula.
BT Michigan
United States

White Pine County
On the Utah border in E Nevada. Also search White Pine.
BT Nevada
United States

White Pine Mine
Copper mine in the Porcupine Mountains near Lake Superior in W Upper Peninsula. Also search White Pine.
BT Michigan
United States

white pyrites
use arsenopyrite

White River
River in Alaska and SW Yukon Territory; a river in Arkansas and S Missouri; and a river in SW Indiana. Index Yukon Territory and states as applicable.
BT North America
SA Alaska
Arkansas
Indiana
Missouri
Yukon Territory

White River Group
In Wyoming, includes Chadron, Brule, and Arikaree formations. NE Colorado, E Montana, Nebraska, North Dakota, South Dakota, Wyoming.
BT Oligocene
SA Brule Formation
Chadron Formation
Colorado
Montana
Nebraska
North Dakota
South Dakota
Wyoming

White River Plateau
NW central Colorado.
UF White River Uplift
BT Colorado
United States

White River Uplift
use White River Plateau

White Russia
use Byelorussia

White Sands
National Monument constituting a great expanse of white gypsum sand and dunes in Tularosa Basin in S New Mexico. Also a proving ground for testing rockets and guided missiles SW of White Sands National Monument.
BT New Mexico
United States

White Sea
Large inlet of the Barents Sea S of the Kola Peninsula in NW European USSR.
CO N640000N660000
E0440000E0350000
BT Arctic Ocean

Whitehorse
City on the Yukon River in S.
BT Yukon Territory
Canada

Whitemud Formation
Clay for china, ball, fire, and stoneware products are obtained from this formation. S and SW Saskatchewan.
BT Cretaceous
SA Saskatchewan

Whitewater River valley
Primarily E and SE Indiana, but also in extreme SW Ohio. Also search Whitewater River.
BT United States
SA Indiana
Ohio

whitlockite
BA phosphates
BT minerals

Whittier Fault
E of Los Angeles in S.
BT California
United States

whole rock
Includes use on level 3 under absolute age(1). Also search whole-rock.
UF total rock
whole-rock
SA absolute age

whole-rock
use whole rock

wiborgite
use rapakivi

Wichita
City on the Arkansas River in Sedgwick County in S central.
BT Kansas
United States

Wichita County
Index states as applicable.
BX Kansas
Texas
BT United States

Wichita Mountains
Range in SW.
CO N343000N350000
W0983000W0990000
BT Oklahoma
United States

Wicklow
County in E Ireland. Also a city on the Irish Sea, and a mountain range extending along E coast.
CO N524000N531500
W0060000W0064500
BT Ireland

Wicomico Formation
In Columbia Group. In Prince Georges County, Maryland, consists of coarse gravel bed at base and finer sand and silt above; color of silt ranges from yellow to drab to dirty white. Atlantic Coastal Plain from Delaware to Florida.
BT Pleistocene
SA Delaware
District of Columbia
Florida
Georgia
Maryland
North Carolina
South Carolina
Virginia

Widawka Basin
River basin in central part of country.
UF Widawka River basin
BT Lodz
Poland

Widawka River basin
use Widawka Basin

Widgiemooltha
Village in S central.
BT Western Australia
Australia

Wiehen Mountains
Low range of the Weser Mountains in NE.
BT North Rhine-Westphalia
West Germany

Wieliczka
Town SE of Krakow.
BT Krakow
Poland

Wielun
Town in central part of country.
BT Lodz
Poland

Wien
use Vienna

Wiener Wald
A spur of the Eastern Alps W and NW of Vienna and S of the Danube River.
CO N480000N482000
E0161500E0154000
UF Vienerwald
BT Lower Austria
Austria
SA Eastern Alps

Wiener-Hopf analysis
Term introduced in 1981.
SA analysis
seismology

Wiesbaden
City on the Rhine River 20 miles W of Frankfurt.
BT Hesse
West Germany

Wilberns Formation
Includes four members: Welge Sandstone, Morgan Creek Limestone, Point Peak Shale, and San Saba Limestone. In central Texas.
BT Upper Cambrian
Cambrian
SA Texas

Wilcox Formation
Predominantly green, white, and black schist enclosing thin dolomite beds near base of schist. Pegmatitic quartzose gneiss occurs near middle of the formation. Above the gneiss, dark schistose grits contrast with strictly argillaceous types below. In W central Vermont.
BT Precambrian
SA Vermont

Wilcox Group
In Louisiana, comprises (ascending) Converse, Lime Hill, "Hall Summit", Marthaville, Pendleton, Sabinetown, and Carrizo formations. Gulf Coastal Plain from Georgia to S Texas plus SW Illinois, W Kentucky, SE Missouri, and W Tennessee.
BT lower Eocene
Eocene
SA Florida
Georgia
Illinois
Kentucky
Louisiana
Mississippi
Missouri
Tennessee
Texas

Wildenfels
Town at the foot of the Erzgebirge in SW.
BT Karl-Marx-Stadt Bezirk
East Germany

wildflysch
SA clastic rocks
clastic sediments
flysch
sedimentary rocks
sediments

Wilhelmina Mountains
Range in central.
BT Surinam

Wilkes Land
Coastal region from the Shackleton Ice Shelf to the George V Coast on the Indian Ocean side.
BT Antarctica

Wilkinson Basin
Off the coast of Massachusetts.
BT Atlantic Ocean

Wilkinson County
Index states as applicable.
BX Georgia
Mississippi
BT United States

Will County
S and SW of Chicago.
BT Illinois
United States

Willamette River valley
use Willamette Valley

Willamette Valley
River valley in NW Oregon.
UF Willamette River valley
BT Oregon
United States

willemite
BA orthosilicates
silicates
BT minerals

Williamson County
Index states as applicable.
BX Illinois
Tennessee
Texas
BT United States

Williamsport Sandstone
In Maryland, includes Cedar Hill Limestone Member (reallocated). Abruptly overlies calcareous shales of Mackenzie Formation. W Maryland, W Virginia, and N West Virginia. Also search Williamsport Formation.
BT Cayugan
Upper Silurian
Silurian
SA Maryland
Upper Silurian
Virginia
West Virginia

Williston Basin
Index states and provinces as applicable.
BT North America
SA Manitoba
Montana
North Dakota
Saskatchewan
South Dakota

Willow Creek
Rises in Blue Mountains in NE and flows NW into Columbia River.
BT Oregon
United States

Willunga
Village S of Adelaide in SE.
BT South Australia
Australia

Willwood Formation
Variegated shales and hornblende-bearing sandstones conformably overlying Polecat Bench Formation near center of Bighorn Basin. In N Wyoming.
BT lower Eocene
Eocene
SA Wyoming

Willyama Complex
Great igneous and metamorphic complex comprised of slaty and schistose rocks, gneisses, amphibolites, and granites. In extreme W New South Wales.
BT Archean
SA New South Wales

willyamite
BA sulfides
BT minerals
SA ullmannite

Wilmington
City in New Castle County in N Delaware, a city on Los Angeles Harbor in Los Angeles County, California, and a city on Cape Fear River in New Hanover County, North Carolina. Index states as applicable.
BT United States
SA California
Delaware
North Carolina

Wilmington Canyon
Off the coast of Delaware. Also search Wilmington.
UF Wilmington submarine valley
BT Atlantic Ocean

Wilmington oil field
In the Wilmington area around Los Angeles Harbor in Los Angeles County. Also search Wilmington.
BT California
United States

Wilmington submarine valley
use Wilmington Canyon

Wilna
use Vilna

Wilno
use Vilna

Wilson County
Index states as applicable. Also search Wilson AND appropriate state.
BX Kansas
North Carolina
Tennessee
Texas
BT United States

Wilson Lake
A reservoir in Jerome County in S.
BT Idaho
United States

Wilts County
use Wiltshire

Wiltshire
County in S England.
CO N510000N514000
W0013000W0022000
UF Wilts County
BT England
Great Britain
United Kingdom

Wiluna
Town in a gold mining area in W central.
BT Western Australia
Australia

wind energy
Term introduced in 1981.
SA new energy sources
winds

wind erosion
Includes use on level 3 under soils(1) erosion(2).
BT erosion
SA ablation
eolian environment
eolian features
erosion control
soils
water erosion
wind transport
winds

Wind River
Rises in NW Fremont County and flows SE uniting with the Popo Agie River to form the Big Horn River in W central.
BT Wyoming
United States

Wind River basin
River basin along E slopes of Wind River Range in Fremont County in W central.
BT Wyoming
United States

Wind River Formation
Divides Lysite and Lost Cabin members, each of which consists of two facies. In W Wyoming.
BT lower Eocene
Eocene
SA Wyoming

Wind River Mountains
use Wind River Range

Wind River Range
Range of the Rocky Mountains in W central Wyoming. Also search Wind River Mountains.
CO N424500N433000
W1083000W1093000
UF Wind River Mountains
BT Wyoming
United States
SA Rocky Mountains

wind, solar
use solar wind

wind transport
Includes use on level 3 under sedimentation(1).
SA ablation
eolian environment
eolian features
sedimentation
transport
wind erosion
winds

Windermere System
Unconformably overlays the Purcell System in W and N Purcell Mountains of SE British Columbia.
BT Proterozoic
Precambrian
SA British Columbia

Windham County
Index states as applicable.
BX Connecticut
Vermont
BT United States

Windhoek
City in central.
BT Namibia

windows
Term introduced in 1981. Before 1981, also search fensters.
UF fensters
SA allochthons
autochthons
erosion features
overthrust faults
tectonics

winds
Includes use on level 2 under aeronomy(1) and under meteorology(1).
SA aeronomy
circulation
climate
Ekman spiral
eolian features
hurricanes
meteorology
storms
transport
wind energy
wind erosion
wind transport

Windsor
City on Detroit River across from Detroit.
BT Ontario
Canada

Windsor Group
Consists of thick members of massive to poorly bedded, red or red and greenish gray, mottled siltstone, shale, and sandstone intercalated with gypsum, salt, and thin tabular sequences of limestone and dolomite. New Brunswick and Nova Scotia.

BT Mississippian
SA Canada
New Brunswick
Nova Scotia

Windward Islands
The S chain of the Lesser Antilles extending from Martinique in the N to Grenada in the S.
BT Lesser Antilles
NA Carriacou
Grenada
Martinique
Saint Lucia
Saint Vincent

Winkler County
At the SE corner of New Mexico in W.
BT Texas
United States

Winnebago County
Index states as applicable.
BX Illinois
Iowa
Wisconsin
BT United States

Winnemucca
City on the Humboldt River in Humboldt County in NW.
BT Nevada
United States

Winnfield salt dome
In Winnfield area in Winn County in N central Louisiana.
UF Winnfield salt mine
BT Louisiana
United States

Winnfield salt mine
use Winnfield salt dome

Winnipeg
City S of Lake Winnipeg in SE.
BT Manitoba
Canada

Winnipeg Formation
Defined as the shale and sandstone section which underlies the limestone of Red River Formation and which rests upon Precambrian basement complex in Manitoba. Surface and subsurface in Manitoba, and subsurface in Saskatchewan; and subsurface in Montana, North Dakota, and South Dakota.
BT Middle Ordovician
Ordovician
SA Manitoba
Montana
North Dakota
Saskatchewan
South Dakota

Winnipegosis Formation
In Elk Point Group. In outcrop, overlies Elm Point Formation and underlies Dawson Bay Formation. Redefined in subsurface to include Elm Point Limestone. Surface and subsurface in Manitoba, and subsurface in NE Montana, W North Dakota, and NW South Dakota.
BT Middle Devonian
Devonian
SA Elk Point Group
Manitoba
Montana
North Dakota
South Dakota

Wisconsin
Includes use on level 1 as an area term (list O). For term set options see list B.
CO N423000N470000
W0864500W0925000
BA Midwest
United States
NX Adams County
NT Baraboo
NX Clark County
Columbia County
Crawford County
NT Dane County
NX Douglas County
Dunn County
Grant County
NT Green Lake
NX Iowa County
Iron County
Jackson County
Jefferson County
Lafayette County
NT Lake Mendota
NX Lincoln County
NT Madison
Marathon County
NX Marquette County
NT Milwaukee
NX Monroe County
Oneida County
Pierce County
Polk County
Portage County
Richland County
Rusk County
Taylor County
NT University of Wisconsin
NX Washington County
Winnebago County
NT Wolf River Batholith
NX Wood County
SA Animikie Group
Baraboo Quartzite
Canadian Shield
Franconia Formation
Great Lakes region
Green Bay
Keweenawan
Lake Michigan
Lake Superior
Maquoketa Formation
Midcontinent
Mississippi River
Mississippi Valley
Niagara Escarpment
Nonesuch Shale
Platteville Formation
Saint Peter Sandstone
Shakopee Formation
Southern Province
Thomson Slate
Upper Mississippi Valley

Wisconsin Range
N of the Horlick Mountains and S of Marie Byrd Land.
BT Antarctica

Wisconsinan
Stage. Includes use on level 3 under age terms(1). See list E.
BA upper Pleistocene
Pleistocene
Quaternary
BT Cenozoic
SA Twocreekan
Weichselian
Woodfordian

Wise County
Index states as applicable.
BX Texas
Virginia
BT United States

Wisla River
use Vistula River

Wissahickon Formation
Lower Paleozoic (?). Prevalent structure is schistose. Delaware, N Maryland, SE Pennsylvania, and Virginia.
BT Paleozoic
SA Delaware
Glenarm Series
lower Paleozoic
Maryland
Pennsylvania
Virginia

witherite
BA carbonates
BT minerals
SA aragonite

Wittenoom Gorge
Village N of the SE end of the Hamersley Range in NW.
BT Western Australia
Australia

wittichenite
BA sulfobismuthites
sulfosalts
BT minerals

Witwatersrand
Region on a ridge of auriferous rock about 62 miles long and 23 miles wide with Johannesburg nearly at its center in S Transvaal. Also search Witwatersrand Basin.
CO S270000S260000
E0300000E0260000
UF Rand
The Rand
Witwatersrand Basin
BT Transvaal
South Africa

Witwatersrand Basin
use Witwatersrand

Witwatersrand System
Contains detrital gold, uranium, and pyrite concentration. In Transvaal, South Africa.
BT Precambrian
SA South Africa
Transvaal

wodginite
BZ niobates
tantalates
BA oxides
BT minerals

Wolf Creek Meteorite
Term introduced in 1981. Impact S of Hall's Creek, in NE Western Australia. Before 1981, also search Wolf Creek AND meteorites.
BA octahedrite
iron meteorites
BT meteorites
SA Australia
Western Australia

Wolf River Batholith
BT Wisconsin
United States

Wolfcampian
Provincial series, North America. Above Virgilian (Pennsylvanian), below Leonardian. Includes use on level 3 under age terms(1). See list E.
BA Lower Permian
Permian
BT Paleozoic
NT Roca Formation

wolfram
use wolframite

wolframite
UF wolfram
BA tungstates
BT minerals

Wolfsberg
Town in S part of country.
BT Carinthia
Austria

Wolin
Island in the Baltic Sea just off the extreme NW part of country. Also a town on the SE shore of the island.
UF Wolin Island
Wollin
BT Szczecin
Poland

Wolin Island
use Wolin

Wolkberg Group
Composed of sedimentary rocks. In Transvaal, South Africa.
BT Proterozoic
Precambrian
SA South Africa
Transvaal

Wollaston Lake Belt
Fold Belt in NE. Also search Wollaston Lake.
BT Saskatchewan
Canada

wollastonite
BA chain silicates
silicates
BT minerals

Wollin
use Wolin

Wolverhampton
County borough NW of Birmingham in W central.
BT England
Great Britain
United Kingdom

Wolyn
use Volyn

wood
SA fossil wood
paleobotany
paleontology

Wood Canyon Formation
Overlies Stirling Quartzite; underlies Cadiz Formation.
SA California
Cambrian
Lower Cambrian
Nevada
Precambrian

Wood County
Index states as applicable.
BX Ohio
Texas
West Virginia
Wisconsin
BT United States

Wood Mountain Formation
Fluvial deposits of gravels. In S Saskatchewan.
BT Miocene
SA Saskatchewan

Wood River
City NE of St. Louis in Madison County in SW.
BT Illinois
United States

Woodbine Formation
In Oklahoma, lower member is principally crossbedded dark tuffaceous sand, red clay, and gravel lentils. Upper member mostly gray to brown crossbedded quartz and sandy gravel. SW Arkansas, W Louisiana, central S and SE Oklahoma, and Texas.
BT Gulfian
Upper Cretaceous
Cretaceous
SA Arkansas
Louisiana
Oklahoma
Texas

Woodbury
Town in central Tennessee. Borough in Bedford Co., S Pennsylvania. Town in W Connecticut. City in SW New Jersey. Index states as applicable.
BX Connecticut
New Jersey
Pennsylvania
Tennessee

BT United States

Woodbury Clay
In Matawan Group. Thick black clay which weathers to dove or light chocolate color. In New Jersey.
BT Upper Cretaceous
Cretaceous
SA New Jersey

Woodford Shale
Devonian and Mississippian. Upper part consists of alternating beds of black papery shale that weather light gray, and black chert in beds 1 to 4 inches thick. Central S and SE Oklahoma.
BT Paleozoic
SA Devonian
Mississippian
Oklahoma

Woodfordian
Substage of Wisconsinan stage.
BA Pleistocene
Quaternary
BT Cenozoic
SA Illinois
Wisconsinan

Woods County
On the Kansas border in N.
BT Oklahoma
United States

Woods Hole
Village in Falmouth town, Barnstable Co., SE Massachusetts. U. S. fish and wildlife station located here. Also site of important marine biological institute and of Oceanographic Institution.
BT Massachusetts
United States

Woods Hole Oceanographic Institution
Located in village of Woods Hole in town of Falmouth at SW tip of Cape Cod in Barnstable County.
BT Massachusetts
United States

Woodson County
SE Kansas.
BT Kansas
United States

Woodstock
Town on the Saint John River in W.
BT New Brunswick
Canada

Woodward
City SE of the panhandle in Woodward County.
BT Oklahoma
United States

Worcester
City in Worcester County in central.
BT Massachusetts
United States

Worcester County
Index states as applicable.
BX Maryland
Massachusetts
BT United States

Worcestershire
County in W central.
CO N515500N523000
W0014500W0023000
BT England
Great Britain
United Kingdom

Worden gravimeters
Term introduced in 1981.
BT gravimeters

world
use global

World Bank
Term introduced in 1981.
SA economics

world ocean
Used as a general term. Also see entries for specific oceans.
UF ocean, world
SA global

worm borings
use borings

worms
Includes use on level 1 and 2 as a fossil term (list F).
BT Invertebrata
NA Annelida
Chaetognatha
Chaetopoda
Echiurida
Myzostomia
Nematoida
Nematomorpha
Nemerta
Oligochaetia
Phoronida
Pogonophora
Polychaetia
Priapulida
scolecodonts
Sipunculoida

Wrangel Island
About 100 miles off N coast of Chukchi National Okrug in the Arctic Ocean NW of Alaska.
BT Russian Republic
USSR

Wrangell Mountains
Range near the Canadian border at the N end of the Alaskan panhandle.
BT Alaska
United States

Wreford Limestone
In Chase Group. In Kansas, includes (ascending) Threemile Limestone, Havensville Shale, and Schroyer Limestone members. E Kansas, SE Nebraska, and central Oklahoma. Also search Wreford Megacyclothem.
BT Permian
SA Chase Group
Kansas
Nebraska
Oklahoma

wrench faults
Term introduced in 1978. Includes use on level 3 under faults(1) displacements(2). Before 1978, also search faults AND wrench.
UF basculating faults
torsion faults
BA faults
SA displacements

Wright Valley
In the Wright Glacier area W of McMurdo Sound in Victoria Land in Ross Dependency on the Pacific Ocean side. Also search Wright AND appropriate area.
BT Antarctica

Wroclaw
Province in SW Poland. Also a city on the Oder River.
UF Breslau
Slak Dolny
BT Poland
NT Bardo Mountains
Boguszow
Boleslawiec
Bystrzyca
Lesna
Lubin
Lubin Legnicki
Mirsk
Nowa Ruda
Strzegom
Strzegom-Sobotka Massif
Strzelin
Walbrzych
Zloty Stok
SA Silesian coal basin

Wuerttemberg
use Wurttemberg

wulfenite
BA molybdates
BT minerals

Wurm
Europe, Above Riss, below Holocene. Includes use on level 3 under age terms(1). See list E.
BA upper Pleistocene
Pleistocene
Quaternary
BT Cenozoic
SA Weichselian

Wurttemberg
Former German state now part of Baden-Wurttemberg. Also search Wuerttemberg.
UF Wuerttemberg
BT Baden-Wurttemberg
West Germany

wurtzite
BA sulfides
BT minerals

wustite
Artificial mineral. Also search iozite.
UF iozite
BA oxides
BT minerals

Wutach River valley
use Wutach Valley

Wutach Valley
River valley in SW Baden-Wurttemberg. Also search Wutach; Wutach River.
UF Wutach River valley
BT Baden-Wurttemberg
West Germany

Wyandot County
Term introduced in 1981. N central Ohio.
CO N404000N410000
W0830700W0833200
BT Ohio
United States
SA Sandusky River basin

Wyandotte County
Just W of Kansas City, Missouri.
BT Kansas
United States

Wyandotte Limestone
In Kansas City Group. Includes (ascending) Frisbie Limestone, Quindaro Shale, Argentine Limestone, Island Creek Shale, and Farley Limestone members. SW Iowa, E Kansas, NW Missouri, and SE Nebraska. Also search Wyandotte Limestone.
BT Missourian
Upper Pennsylvanian
Pennsylvanian
SA Iowa
Kansas
Kansas City Group
Missouri
Nebraska

Wye Valley
River valley in E Wales, and W England. Index political divisions as applicable. Also search Wye River.
BT United Kingdom
SA England
Wales

Wyman Formation
Spotted schists and phyllites with a few interbedded dolomite. In S California.
BT Precambrian
SA California

Wynyard
Town in S central Saskatchewan, and a town on Bass Strait in NW Tasmania. Index province and/or state as applicable.
SA Saskatchewan
Tasmania

Wyoming
Includes use on level 1 as an area term (list O). For term set options see list B.
CO N410000N450000
W1040500W1110500
BA Western U.S.
United States
NX Albany County
NT Beaver Creek
NX Big Horn County
Campbell County
Carbon County
NT Casper
Converse County
NX Crook County
Fremont County
NT Gas Hills
Goshen County
Granite Mountains
Hanna Basin
Heart Mountain Fault
Jackson Hole
NX Johnson County
NT Laramie Basin
Laramie County
NX Lincoln County
NT Natrona County
Old Faithful Geyser
NX Park County
Platte County
NT Powell
Rock Springs
Salt Creek
NX Sheridan County
NT Shirley Basin
Sierra Madre Range
Sublette County
Sweetwater County
Teapot Dome
NX Teton County
NT Teton National Forest
Uinta County
Weston County
Wind River
Wind River basin
Wind River Range
SA Absaroka Range
Almond Formation
Arikaree Group
Ash Hollow Formation
Bald Mountain
Battle Mountain High
Bearpaw Formation
Beartooth Mountains
Benton Formation
Bighorn Basin
Bighorn Mountains
Black Hills
Brule Formation
Carlile Shale
Casper Formation
Chadron Formation
Cody Shale
Colorado Group
Colorado River basin
Columbia River basin
Dakota Formation
Deadwood Formation
Denver Basin
Fall River Formation
Flathead Sandstone
Fort Union Formation
Fox Hills Formation
Frontier Formation
Gallatin Range

Graneros Shale
Great Plains
Green River
Green River basin
Green River Formation
Greenhorn Limestone
Harebell Formation
Lance Formation
Laney Shale Member
Laramie Mountains
Lewis Shale
Little Missouri River basin
Lodgepole Formation
Loup Fork Group
Madison Group
Mancos Shale
Medicine Bow Mountains
Mesaverde Group
Midcontinent
Minnelusa Formation
Missouri River basin
Montana Group
Morrison Formation
Mowry Shale
Muddy Sandstone
Niobrara Formation
Ogallala Formation
Parkman Sandstone
Phosphoria Formation
Pierre Shale
Pinedale Glaciation
Pinyon Conglomerate
Powder River basin
Red Peak Formation
Red River Formation
Snake River
Snake River basin
Spearfish Formation
Stump Formation
Sundance Formation
Sunlight
Tensleep Sandstone
Thaynes Formation
Tongue River Member
U. S. Rocky Mountains
Uinta Mountains
Wasatch Formation
Washakie Basin
Western Interior
Western Overthrust Belt
White River Group
Willwood Formation
Wind River Formation
Yellowstone National Park

Wythe County
SW Virginia.
BT Virginia
United States

X

X-ray
A valid term through 1977. See X-ray analysis.

X-ray analysis
Level 1 term introduced in 1978. Used for methodology not data. For data, use X-ray data. X-ray diffraction analysis was used on level 1 through 1977. Before 1978, also search X-ray. Term set options are:
methods
topic or instruments or application
techniques
topic
X-ray diffraction analysis
subtopic
X-ray fluorescence
subtopic
X-ray radiography
subtopic
X-ray spectroscopy
subtopic
SA analysis
chemical analysis
clay mineralogy
crystal structure
crystallography
electron microscopy
spectroscopy
structural analysis
thermal analysis
X-ray data
X-ray diffraction analysis
X-ray fluorescence
X-ray radiography
X-ray spectroscopy

X-ray data
For methodology, use X-ray analysis.
SA data
X-ray analysis

X-ray diffraction analysis
A valid level 1 term through 1977 used for methodology. After 1977, used on level 2 under X-ray analysis(1).
SA analysis
chemical analysis
differential thermal analysis
electron diffraction analysis
neutron diffraction analysis
spectroscopy
X-ray analysis
X-ray fluorescence

X-ray fluorescence
As of 1978, term is used on level 2 under X-ray analysis(1) for technique or applications.
SA analysis
chemical analysis
fluorescence
luminescence
spectroscopy
X-ray analysis
X-ray diffraction analysis
X-ray fluorescence spectra

X-ray fluorescence spectra
Term introduced in 1978. Refers to data; to be distinguished from X-ray fluorescence as a method.
SA spectra
X-ray fluorescence

X-ray radiography
As of 1978, term is used on level 2 under X-ray analysis(1).
UF radiography, X-ray
SA X-ray analysis

X-ray spectroscopy
Term introduced in 1978. Includes use on level 2 under X-ray analysis(1) and on level 3 under chemical element(1) analysis(2). Before 1978, also search spectroscopy AND X-ray.
BT spectroscopy
SA analysis
X-ray analysis

X-rays
Use when referring to the rays themselves. Includes use on level 3 under aeronomy(1) ionization(2).
SA electromagnetic radiation
gamma rays

xanthochroite
use greenockite

xanthophyllite
use clintonite

xanthosiderite
use goethite

Xe
use xenon

Xe-124
Includes use on level 3 under isotopes(1).
SA isotopes
xenon

Xe-126
Includes use on level 3 under isotopes(1).
SA isotopes
xenon

Xe-128
Includes use on level 3 under isotopes(1).
SA isotopes
xenon

Xe-129
Includes use on level 3 under isotopes(1).
SA isotopes
xenon

Xe-131
Includes use on level 3 under isotopes(1).
SA isotopes
xenon

Xe-132
Includes use on level 3 under isotopes(1).
SA isotopes
xenon

Xenarthra
Term introduced in 1981. Includes use on level 2 under Mammalia(1). See list F.
BA Mammalia
BT Tetrapoda
Vertebrata

xenoliths
Includes use on level 2 under inclusions(1).
UF exogenous inclusions
BT inclusions
SA fluid inclusions
host rocks
mineral inclusions

xenon
Includes use on level 1 and 2 as a chemical element (list D).
UF Xe
BT noble gases
SA argon
elements
isotopes
krypton
neon
Xe-124
Xe-126
Xe-128
Xe-129
Xe-131
Xe-132

xenotime
BA phosphates
BT minerals

Xenungulata
Term introduced in 1981. Includes use on level 2 under Mammalia(1). See list F.
BA Mammalia
BT Tetrapoda
Vertebrata

Xiphosura
Subclass. Includes use on level 3 under Arthropoda(1) Merostomata(2).
BA Merostomata
Chelicerata
Arthropoda
BT Invertebrata

xonotlite
BA chain silicates
silicates
BT minerals

Y

Y
use yttrium

Yacoraite Formation
In Jujuy and Salta, Argentina.
BT Cretaceous
SA Argentina
Jujuy
Salta

Yahagi River
Flows into Chita Bay on the Pacific Ocean S of Nagoya in central Honshu. Also search Yahagi.
BT Honshu
Japan

Yake-Dake
Peak NE of Kure in W Honshu.
UF Yakedake
Yakeyama
BT Honshu
Japan

Yakedake
use Yake-Dake

Yakeyama
use Yake-Dake

Yakima Basalt
Upper Miocene and lower Pliocene. Typical plateau basalt. In Yakima East Quadrangle, it is exposed in four southeast-trending strips that coincide with crestal portions of four anticlinal axes. In E Washington.
BT Tertiary
SA lower Pliocene
upper Pliocene
Washington

Yakima County
Term introduced in 1981. S Washington.
CO N460200N470500
W1195200W1213200
BT Washington
United States

Yakut A.S.S.R.
use Yakutia

Yakutat Bay
Term introduced in 1981. SE Alaska, inlet of the Gulf of Alaska.
CO N592000N600000
W1391500W1410000
BT Alaska
United States
SA Gulf of Alaska

Yakutia
Yakut Autonomous Soviet Socialist Republic. Located in E central Siberia. Also search Yakut.
CO N550000N730000
E1620000E1080000
UF Yakut A.S.S.R.
BT Russian Republic
USSR
NT Mirnyy
Seligdar

Yakutia region
Term introduced in 1981. Region which surrounds and includes the city of Yakutsk, extending past the Aldan River on the E and N.
CO N573000N633000
E1350000E1270000
BA Siberian Platform
BT Russian Republic
USSR

Yakutsk
City on the bend of the Lena River in central Yakutia in E central Siberia.

BT Russian Republic
USSR

Yalta
City on the Black Sea in S Crimea.
BT Ukraine
USSR

Yamagata
City on the Mogami river W of Sendai in Yamagata Prefecture in N.
BT Honshu
Japan

Yamaguchi
City NE of Shimonoseki in Yamaguchi Prefecture in W.
BT Honshu
Japan

Yamal
Peninsula between the Kara Sea on the W and the Gulf of Ob on the E in the Yamal-Nenets National Okrug in NW Siberia.
UF Yamal Peninsula
BT Russian Republic
USSR

Yamal Peninsula
use Yamal

Yamal-Nenets
National Okrug in N Tyumen Oblast in NW Siberia.
UF Yamalo-Nenets
BT Russian Republic
USSR

Yamalo-Nenets
use Yamal-Nenets

Yamanashi
Prefecture W of Tokyo in central.
BT Honshu
Japan

Yamato Meteorites
Term introduced in 1981. Due to numbering of these meteorites, also truncate when searching. Before 1981, also search Yamato AND meteorites.
BT meteorites
SA Antarctica
Victoria Land

Yamato Mountains
In Victoria Land, Antarctica.
BT Victoria Land
Antarctica

Yamhill Formation
Comprises Mill Creek Beds, of predominantly dark-gray shale and siltstone with occasional beds of lime-cemented sandstone, and an overly of a sequence of massive sandstone beds which grades upward into more agrillaceous rock. In W Oregon.
BT Eocene
SA Oregon

Yamuna River
use Jumna River

Yana
River which rises in the Verkhoyansk Range of N central Yakutia and flows N into the Laptev Sea. Also search Yana River; Yana River region; Yana River basin.
UF Yana River
BT Russian Republic
USSR

Yana River
use Yana

Yana-Indigirka Depression
use Yana-Indigirka Lowland

Yana-Indigirka Lowland
Coastal lowland comprising the area around the lower courses and the lower interfluve of the Yana and the Indigirka rivers in N central Yakutia. Also search Yana-Indigirka Depression.
UF Yana-Indigirka Depression
BT Russian Republic
USSR

Yaquina Bay
Inlet of the Pacific Ocean at the mouth of the Yaquina River in W Oregon.
UF Yaquina Estuary
BT Oregon
United States

Yaquina Estuary
use Yaquina Bay

Yaracuy
State in NW.
BT Venezuela

Yarmouth
Town on the Atlantic Ocean in SW.
BT Nova Scotia
Canada

Yaroslavl
City NE of Moscow on the Volga River in Yaroslavl Oblast.
BT Russian Republic
USSR

Yarrol Basin
BT Queensland
Australia

Yass
River basin in SE New South Wales. Also search Yass Basin.
UF Yass Basin
BT New South Wales
Australia

Yass Basin
use Yass

Yates Formation
In Artesia Group. A 50-foot sandstone that is subsurface in Texas, and subsurface and surface in New Mexico.
BT Permian
SA New Mexico
Texas

Yatsudake Mountain
use Yatsugatake

Yatsugatake
Mountain peak NW of Tokyo in central Honshu.
UF Yatsudake Mountain
BT Honshu
Japan

Yavapai County
Central.
BT Arizona
United States

Yazoo Clay
In Jackson Group. Divided into (ascending) North Creek Clay, Cocoa Sand, Pachuta Clay, and Shubuta Clay members. SW Alabama, central N Louisiana and Mississippi. Also search Yazoo Formation.
BT upper Eocene
Eocene
SA Alabama
Jackson Group
Louisiana
Mississippi

Yb
use ytterbium

Yeddo
use Tokyo

Yekaterinburg
use Sverdlovsk

yellow arsenic
use orpiment

Yellow Sea
Between NE China and the Korean Peninsula. As of 1977, includes use as level 1 area term (list O). For term set options see list B.
CO N333000N411000
E1263000E1173000
BA West Pacific
Pacific Ocean

Yellowknife
Town on NW shore of Great Slave Lake at the mouth of the Yellowknife River in S District of Mackenzie.
BT Northwest Territories
Canada

Yellowknife Group
Composed of quartzite, conglomerate, graywacke, meta-basalt, meta-andesite, dacite, rhyolite, tuff, agglomerate; undifferentiated gabbro. In District of Mackenzie, Northwest Territories.
BT Archean
Precambrian
SA Canada
Northwest Territories

Yellowstone National Park
Largest and oldest of U. S. national parks primarily in extreme NW Wyoming. Index states as applicable. Also search Yellowstone.
CO N441000N450000
W1095000W1110500
UF Yellowstone Park
BT United States
SA Idaho
Montana
Wyoming

Yellowstone Park
use Yellowstone National Park

Yemen
Yemen Arab Republic. Includes use on level 1 as an area term (list O). For term set options see list B.
CO N124000N173000
E0471000E0423000
BA Arabian Peninsula
Asia
SA Arabian Shield
Middle East
Near East
Red Sea Basin
Southern Yemen

Yen Bai
use Yenbay

Yen Bay
use Yenbay

Yenbay
Town on the Red River NW of Hanoi.
UF Yen Bai
Yen Bay
BT Vietnam

Yenisei
Not a valid term for GeoRef. Use Yenisei Basin, Yenisei River, or Yenisei Ridge.

Yenisei Basin
River basin which extends into N Mongolia in the S and drains the mountain areas W of Lake Baikal and much of the Central Siberian Plateau. Also search Yenisei. Index Mongolia and/or Russian Republic as applicable.
BT Asia
SA Mongolia
Russian Republic
Yenisei-Khatanga basin

Yenisei Range
use Yenisei Ridge

Yenisei Ridge
Upland which extends N-S for 40[0] miles along the right bank of the Yenisei River between the Trans-Siberian R.R. and the Stony Tunguska River. A major gold-mining region. Also search Yenisei, Yenisei Range, and Yenisey Ridge.
CO N560000N630000
E0950000E0893000
UF Enisei Ridge
Yenisei Range
Yenisey Ridge
BA Siberian Platform
BT Russian Republic
USSR

Yenisei River
Rises in the E Sayan Mountains of [E] Tuva A.S.S.R.; it flows W for a short distance and then N across central Siberia into the Kara Sea via Yenisei Bay and Yenisei Gulf. Also search Yenisei.
UF Enisei River
Yenisey River
BT Russian Republic
USSR

Yenisei-Khatanga basin
Combined drainage basin of the Khatanga and the Yenisei rivers. The Khatanga basin drains much of the northern Central Siberian Plateau S of the Taymyr Peninsula. Index Mongolia and/or Russian Republic as applicable.
BT Asia
SA Khatanga Basin
Mongolia
Russian Republic
Yenisei Basin

Yenisey Ridge
use Yenisei Ridge

Yenisey River
use Yenisei River

Yeoval
Village NW of Sydney in E central.
BT New South Wales
Australia

Yerevan
City on the Razdan River in W Armenia.
UF Erevan
Erivan
BT Armenia
USSR

Yerington
City in Lyon County in W.
BT Nevada
United States

Yessei
use Yessey

Yessey
Village N of the Arctic Circle in N Evenk National Okrug in N central Siberia.
UF Essei
Yessei
BT Russian Republic
USSR

Yetropole
Town ENE of Sofia in W central part of country.
BT Sofia
Bulgaria

yield strength
Includes use on level 3 under deformation(1) field studies(2).
SA compression
deformation
elastic limit
elasticity
fracture strength
mechanical properties

strength
stress
tension

yields
Includes use on level 2 under soils(1). Before 1978, also search yield.
SA fertilizers
forests
fruits
sediment yield
soils
tillage
water yield

Yilgarn
Gold Field ENE of Perth in SW central Western Australia. Also search Yilgarn Goldfield.
CO S300000S300000
E1180000E1180000
UF Yilgarn goldfield
BT Western Australia
Australia

Yilgarn Block
UF Yilgarn Shield
BT Western Australia
Australia

Yilgarn goldfield
use Yilgarn

Yilgarn Shield
use Yilgarn Block

Ylojarvi
Village NW of Tampere in SW part of country. Copper and wolfram mines in area.
BT Finland

Yokohama
City on W shore of Tokyo Bay in Kanagawa Prefecture in SE.
BT Honshu
Japan

Yonezawa
City in Yamagata Prefecture in N.
BT Honshu
Japan

Yongyang
Village N of Taegu in E central part of country. North Kyongsang.
BT South Korea
Korea

Yonne
Department in NE central.
CO N472000N482700
E0042000E0025000
BT France
NT Auxerre
Avallon
Tonnerre
SA Morvan
Yonne Valley

Yonne River valley
use Yonne Valley

Yonne Valley
River valley in central part of country. Index departments as applicable.
UF Yonne River valley
BT France
SA Nievre
Seine-et-Marne
Yonne

York
City S of Harrisburg in York County in S Pennsylvania, and a city in Yorkshire in N England. Index England and/or Pennsylvania as applicable.
SA England
Pennsylvania

York County
Index states as applicable.
BX Maine
Nebraska
Pennsylvania
South Carolina
Virginia
BT United States

York River
An estuary receiving the Pamunkey and Mattaponi rivers at West Point and flowing SE to Chesapeake Bay.
BT Virginia
United States

Yorke Peninsula
Between Spencer Gulf on W and Gulf of Saint Vincent on E in SE.
BT South Australia
Australia

Yorkshire
County in N.
CO N532000N544000
E0001000W0023000
BT England
Great Britain
United Kingdom

Yorktown Formation
In Chesapeake Group. Divisible into 2 major zones. Zone 1 (at base) corresponds to beds exposed at Raysor Bridge, South Carolina; Zone 2 (upper) includes: Uppermost Yorktown beds at Suffolk, Virginia, the beds at Yorktown, Virginia, and the Chama-bearing bed which is correlated with the aluminous clay of Florida. E Maryland, E Virginia, and North Carolina.
BT upper Miocene
Miocene
SA Maryland
North Carolina
Virginia

Yosemite National Park
On W slope of the Sierra Madre in E central California. Also search Yosemite.
UF Yosemite Park
BT California
United States

Yosemite Park
use Yosemite National Park

Yoshino River
Rises in mountains of NW and flows E to Kii Channel at Tokushima. Also search Yoshino.
BT Shikoku
Japan

Younger Dryas
Term used primarily in Europe for an interval of late-glacial time following the Allerod and preceding the Preboreal. Before 1978, search Dryas.
BA Weichselian
upper Pleistocene
Pleistocene
Quaternary
BT Cenozoic
SA Dryas

Younger Granites
In S Niger and N Nigeria.
BT Jurassic
SA Niger
Nigeria

Young's modulus
UF stretch modulus
BA elastic constants
SA elasticity

Ypresian
Europe. Above Thanetian (Paleocene), below Cuisian. Includes use on level 3 under age terms(1). See list E.
BA lower Eocene
Eocene
Paleogene
Tertiary
BT Cenozoic
NT London Clay

ytterbium
Includes use on level 1 and 2 as a chemical element (list D).
UF Yb
SA elements
rare earths

yttrialite
BA orthosilicates
silicates
BT minerals

yttrium
Includes use on level 1 and 2 as a chemical element (list D).
UF Y
SA elements
rare earths

Yuba County
N central.
BT California
United States

Yucatan
State on N Yucatan Peninsula in SE.
BT Mexico
SA Gulf Coastal Plain
Yucatan Peninsula

Yucatan Basin
Undersea feature between the Yucatan Peninsula and Belize on the W and S Cuba on the NE and E. Also search Yucatan.
BT Caribbean Sea

Yucatan Channel
Strait between W tip of Cuba and NE Yucatan Peninsula connecting the Gulf of Mexico and the Caribbean Sea.
BT Atlantic Ocean

Yucatan Peninsula
Separates the Gulf of Mexico from the Caribbean Sea in SE Mexico. Index Quintana Roo and states as applicable. Also search Yucatan.
BT Mexico
SA Campeche
Caribbean region
Quintana Roo
Yucatan

Yucatan Shelf
N and W of the Yucatan Peninsula.
BT Gulf of Mexico
SA Campeche Bank

Yucca Flat
NNW of Frenchman Flat in the Nellis Air Force Range and AEC Nuclear Testing Site.
BT Nevada
United States

Yudoma
River which rises in extreme N Khabarovsk Kray in East Siberia and flows S and SW into the Maya River.
BT Russian Republic
USSR

Yudoma Series
Homogeneous bed of gray dolomite and of dolomite limestone. In N Khabarovsk Kray and S Yakutia in E Siberia.
BT Lower Cambrian
Cambrian
SA Russian Republic
USSR

yugawaralite
BA zeolite group
framework silicates
silicates
BT minerals

Yugoslav Macedonia
Term introduced in 1981. Region in S.
CO N405000N422200
E0230500E0203000
BZ Macedonia
BX Yugoslavia

Yugoslavia
Includes use on level 1 as an area term (list O). For term set options see list B.
CO N410000N470000
E0230000E0133000
BA Southern Europe
Europe
NT Banja Luka
Bosnia
Bosnia-Herzegovina
Croatia
Dalmatia
Herzegovina
Istria
Montenegro
Neretva Valley
Sarajevo
Serbia
Serbo-Macedonian Massif
Skopje
Slovenia
NX Yugoslav Macedonia
SA Adriatic region
Alps
Balkan Peninsula
Banat
Danube River
Danube Valley
Dinaric Alps
Eastern Alps
Julian Alps
Karawanken
Karst region
Krajiste
Macedonia
Mediterranean region
Osogovo Mountains
Pannonia
Pannonian Basin
Vardar River

Yukon River
Formed by confluence of Lewes and Pelly rivers in SW Yukon Territory. It flows NW into Alaska and then SE into Bering Sea S of Norton Sound. Index Alaska and/or Yukon Territory as applicable. Also search Yukon.
SA Alaska
Yukon Territory

Yukon Territory
Includes use on level 1 as an area term (list O). For term set options see list B.
CO N600000N700000
W1250000W1410000
BA Western Canada
Canada
NT Beaver River
Blow River
Dawson
Donjek Glacier
Donjek River
Hess Mountains
Kaskawulsh Glacier
Keno Hill
Klondike
Kluane Lake
Mount Wood
Ogilvie Mountains
Old Crow
Royal Creek
Steele Glacier
Stewart Valley
Whitehorse
SA Chitistone Pass
Coast Mountains
Denali Fault
Liard River
Mackenzie Mountains
Michelle Formation

Nahanni Formation
Peel River
Richardson Mountains
Rocky Mountain Trench
Saint Elias Mountains
Selwyn Mountains
Tintina fault zone
White River
Yukon River

Yukon-Tanana Upland
Mountainous area between the Yukon and Tanana rivers in SE central Alaska.
UF Yukon-Tanana Uplands
BT Alaska
United States

Yukon-Tanana Uplands
use Yukon-Tanana Upland

Yule Marble
Former trade name for marble quarried from Leadville Limestone in Gunnison County, Colorado. In 1928 the trade name became Yule Colorado Marble.
BT Mississippian
SA Colorado
Leadville Formation

Yuma
City on the Colorado River in Yuma County in extreme SW.
BT Arizona
United States

Yuma County
Index states as applicable.
BX Arizona
Colorado
BT United States

Yunnan
Province in SW.
BT China

Yuryuzan
City on the Yuryuzan River in W Chelyabinsk Oblast in the Southern Urals.
BT Russian Republic
USSR

Yvelines
Term introduced in 1981. Department in N.
CO N482500N491000
E0021200E0012500
BT France

Z

Zabrze
City in S part of country.
UF Hindenburg
Hindenburg in Oberschlesien
BT Katowice
Poland

Zacatecas
State in central.
BT Mexico
SA Sierra Madre Occidental
Zuloaga Limestone

Zagros
Mountain system primarily in W and SW Iran. Index Iran and/or Iraq as applicable. Also search Zagros Mountains.
UF Zagros Mountains
Zagros Range
BT Asia
SA Iran
Iraq

Zagros Mountains
use Zagros

Zagros Range
use Zagros

Zaire
Formerly Belgian Congo and now officially the Democratic Republic of the Congo. Includes use on level 1 as an area term (list O). For term set options see list B.
CO S130000N051500
E0320000E0120000
UF Belgian Congo
BA Central Africa
Africa
NT Bambu
Bandundu
Equatorial Zaire
Kasai
Kinshasa
Kivu
Kundelungu Plateau
Lower Zaire
Lukuga River valley
Matadi
Shaba
Upper Zaire
SA Congo
Congo Basin
Congo River
Kama
Kasai River
Katangan Orogeny
Lake Albert
Lake Edward
Lake Kivu
Lake Tanganyika

Zaire River
use Congo River

Zaisan
Lake primarily in Vostochno Kazakhstan Oblast in extreme E Kazakhstan. Also a town SE of the lake.
UF Zaisan Lake
Zaysan
BT Kazakhstan
USSR

Zaisan Basin
Includes Zaisan Lake and surrounding area between Narym Range of the Altai Mountains on the N and Tarbagatai Range on the S in Vostochno Oblast and Semipalatinsk Oblast in extreme E Kazakhstan. Also search Zaisan.
UF Zaisan Depression
Zaysan Basin
Zaysan Depression
BT Kazakhstan
USSR

Zaisan Depression
use Zaisan Basin

Zaisan Lake
use Zaisan

Zakopane
Town at the N foot of the Tatra Mountains in S part of country. The chief summer resort and winter sports center in Poland.
BT Krakow
Poland

Zambesi Valley
use Zambezi Valley

Zambeze Valley
use Zambezi Valley

Zambezi Valley
River valley in S central and SE Africa. Index Namibia and countries as applicable. Also search Zambezi; Zambezi River.
UF Zambesi Valley
Zambeze Valley
BT Africa
SA Angola
Botswana
Mozambique
Namibia
Zambia

Zambezia
District on the Mozambique Channel in E.
BT Mozambique

Zambia
Formerly Northern Rhodesia. Includes use on level 1 as an area term (list O). For term set options see list B.
CO S170000S080000
E0333000E0220000
UF Northern Rhodesia
BA East Africa
Africa
NT Copperbelt
Kafue
Lusaka
Nchanga
SA Congo Basin
Lake Kariba
Lake Tanganyika
Zambezi Valley

Zamora
As of 1981, refers to only the city. Before 1981, also included the province. In NW Spain.
BT Zamora Province
Leon region
Spain

Zamora Province
Term introduced in 1981. Before 1981, search Zamora. In NW Spain on Portuguese border.
CO N410700N421300
W0051800W0070300
BT Leon region
Spain
NT Zamora

Zanga
use Razdan

Zangezur
Mountain range in the Lesser Caucasus extending from E of Lake Sevan S to near the Iranian border. Contains copper and molybdenum ores. Index Soviet republics as applicable. Also search Zangezur Range.
UF Zangezur Range
Zangezurskiy Mountains
BT USSR
SA Armenia
Azerbaidzhan

Zangezur Range
use Zangezur

Zangezurskiy Mountains
use Zangezur

Zanzibar
Island in the Indian Ocean off NE coast of Tanzania. United with Tanganyika in 1964 to form Tanzania.
SA Tanzania

Zao
Volcanic mountain SW of Sendai in N central Honshu.
UF Zao Volcano
BT Honshu
Japan

Zao Volcano
use Zao

Zapadna Morava
River which rises in SW Serbia and flows N and then SW joining the Southern Morava River near Stalac to form the Morava River. Called the Moravica River in its upper course.
UF Western Morava River
BT Serbia
Yugoslavia

Zaporozhe
City on the left bank of the Dnieper River in Zaporozhe Oblast in S Ukraine.
UF Zaporozhye
BT Ukraine
USSR

Zaporozhye
use Zaporozhe

Zaragoza
Term introduced in 1978. Municipality in Mexico, 12 miles NNW of Mexico City. Use Saragossa Province for the province in Spain and Saragossa for the city in Spain.
BT Mexico
SA Saragossa
Saragossa Province

Zarand Basin
In Zarand area W of Dasht-i-Lut in N Kerman. Also search Zarand.
BT Kerman
Iran

Zaria
Town SW of Kano in N central part of country.
BT Nigeria

Zarzis
Town and oasis on the Mediterranean Sea in SE part of country.
BT Tunisia

Zavala County
Term introduced in 1981. SW Texas.
CO N284000N290800
W0992000W1001000
BT Texas
United States

Zawiercie
City in S part of country. Coal and iron mining nearby.
BT Katowice
Poland

Zaysan
use Zaisan

Zaysan Basin
use Zaisan Basin

Zaysan Depression
use Zaisan Basin

Zealand
Largest and easternmost of the major islands of Denmark. Island on which Copenhagen is located. Also search Sjaeland.
UF Seeland
Sjaeland
BT Denmark

Zechstein
Composed of evaporites (celestite and rock salt). In Denmark, East Germany, and West Germany. Also search Zechstein sedimentary rocks, and Zechstein evaporites. Includes use on level 3 under age terms(1). See list E.
BA Upper Permian
Permian
BT Paleozoic
SA Denmark
East Germany
Kupferschiefer
West Germany

Zeehan
Mining town in W.
BT Tasmania
Australia

Zeeland
Province consisting of several islands on the North Sea coast and part of the mainland S of the Western Scheldt Estuary in SW.
BT Netherlands

Zelten
Village W of Tripoli in extreme NW part of country.
BT Libya

zeolite
As of 1981, no longer a valid term for GeoRef. Use zeolite group or zeolites. Before 1981, included use on level 3 under minerals(1) framework silicates, zeolite group(2).

zeolite facies
BT facies
SA metamorphic rocks

zeolite group
Includes use in combination with framework silicates (i.e. framework silicates,zeolite group) on level 2 under minerals(1). See list L.
BA framework silicates
silicates
BT minerals
NA analcime
chabazite
clinoptilolite
edingtonite
epistilbite
erionite
faujasite
ferrierite
garronite
gmelinite
harmotome
heulandite
laumontite
leonhardite
mesolite
mordenite
natrolite
offretite
phillipsite
scolecite
stellerite
stilbite
thomsonite
wairakite
yugawaralite
zeolites

zeolites
As of 1981, not to be used in the minerals set. Before 1981, included use on level 3 under minerals(1) framework silicates, zeolite group(2).
BA zeolite group
framework silicates
silicates
BT minerals

zeolitization
BA metasomatism
SA hydrothermal alteration

Zeravshan Range
Branch of the Tien Shan extending from Alai Range W for about 200 miles in W Tadzhikistan. Sometimes considered part of the Pamir-Alai system. Also search Zeravshan.
UF Zeravshanskiy Range
BT Tadzhikistan
USSR
SA Tien Shan
Zeravshan-Hissar

Zeravshan River
Rises at W end of Alai Range and flows W disappearing in the desert near Bukhara. Index Soviet republics as applicable. Also search Zeravshan.
BT USSR
SA Tadzhikistan
Uzbekistan

Zeravshan-Hissar
Parallel ranges of the Tien Shan in W Tadzhikistan. The Zeravshan Range lies to the N of the Hissar Range.
BT Tadzhikistan
USSR
SA Hissar Range
Zeravshan Range

Zeravshanskiy Range
use Zeravshan Range

Zerenda
Village S of Kokchetav in Kokchetav Oblast in N Kazakhstan.
BT Kazakhstan
USSR

Zermatt
Resort village in Pennine Mountains in SW part of country.
BT Valais
Switzerland

Zetland
use Shetland Islands

Zeya
River in Transbaikalia which rises in the Stanovoi Mountains in E Chita Oblast and flows S and SE into the Amur River.
BT Russian Republic
USSR

Zeya-Bureya Basin
Fertile plain between the lower courses of the Zeya and Bureya rivers in SE Amur Oblast in Transbaikalia.
UF Zeya-Bureya Depression
Zeya-Bureya Plain
BT Russian Republic
USSR

Zeya-Bureya Depression
use Zeya-Bureya Basin

Zeya-Bureya Plain
use Zeya-Bureya Basin

Zhdanov
City in Donetsk Oblast on N shore of Sea of Azov.
UF Mariupol
BT Ukraine
USSR

Zhetybay
Village in Uralsk Oblast in extreme W.
BT Kazakhstan
USSR

Zhitkovichi
Town WNW of Mozyr in S
BT Byelorussia
USSR

Zhitomir
City W of Kiev on the Teterev River. Also an oblast.
BT Ukraine
USSR

Ziaetdin
Station on the Trans-Caspian RR SE of Navoi in S central.
UF Ziatdin
BT Uzbekistan
USSR

Ziatdin
use Ziaetdin

Zielona Gora
Province on the East German border in W Poland. Also a city.
BT Poland
NT Nowa Sol
Polkowice

Zillertal Alps
Range of Eastern Alps W of the Hohe Tauern in the Tyrol and in Italy's Trentino-Alto Adige. Index countries as applicable. Also search Zillertal.
UF Zillertaler Alps
BT Alps
Europe
SA Austria
Eastern Alps
Italy

Zillertaler Alps
use Zillertal Alps

Zimbabwe
Term introduced in 1981 to replace Rhodesia. Includes use on level 1 as an area term (list O). For term set options see list B. Before 1981, also search Rhodesia. Before 1973, also search Southern Rhodesia.
CO S223000S153000
E0330000E0251500
BA Southern Africa
Africa
NT Great Dyke
Kariba
Rhodesian Plateau
Salisbury
Selukwe
SA Bulawayan Group
Lake Kariba
Limpopo Basin
Limpopo Belt

zinc
Includes use on level 1 and 2 as a chemical element (list D). As of 1981, use zinc ores for zinc as a commodity. Before 1981, included use on level 1 as a commodity term.
UF Zn
SA elements
heavy metals
lead-zinc deposits
zinc ores

zinc blende
use sphalerite

zinc ores
Term introduced in 1981. Includes use on level 1 and 2 as a commodity term (list C). Before 1981, search zinc AND ore deposits.
SA lead-zinc deposits
mississippi valley-type
zinc

zinc spar
use smithsonite

zinc spinel
use gahnite

zincite
UF red zinc ore
BA oxides
BT minerals

zinckenite
Also search zinkenite.
UF zinkenite
BA sulfantimonites
sulfosalts
BT minerals

zinkenite
use zinckenite

zinnwaldite
See list L.
UF zinwaldite
BZ fluorides
mica group
BT minerals

zinwaldite
use zinnwaldite

Zirabulak
Station on the Trans-Caspian RR W of Kattakurgan in Samarkand Oblast in S central.
BT Uzbekistan
USSR

zircon
As of 1981, use zircon deposits for zircon as a commodity. Before 1981, included use on level 1 as a commodity term.
BA orthosilicates
silicates
BT minerals
SA cyrtolite
gems
heavy minerals
zircon deposits

zircon deposits
Term introduced in 1981. Includes use on level 1 and 2 as a commodity term (list C). Before 1981, search zircon AND deposits.
SA heavy mineral deposits
zircon

zirconium
Includes use on level 1 and 2 as a chemical element (list D). Also search Zr.
UF Zr
SA elements
Zr-95

zirconolite
As of 1981, used for the mineral zirkelite. Before 1981, also search zirkelite AND minerals.
BA niobates
oxides
BT minerals

zirkelite
As of 1981, restricted to use as the igneous rock; for the mineral, use zirconolite. Before 1981, included use on level 3 under igneous rocks(1) pyroclastics and glasses(2) and under minerals(1) oxides(2).
BA glasses
volcanic rocks
BT igneous rocks

Zlatibor Massif
use Zlatibor Mountains

Zlatibor Mountains
Range of the S Dinaric Alps in W Serbia.
UF Zlatibor Massif
BT Serbia
Yugoslavia
SA Dinaric Alps

Zloty Stok
Town E of Klodzko in Lower Silesia in SW part of country.
UF Reichenstein
Rowne
BT Wroclaw
Poland

Zmeinogorsk
Town SE of Rubtsovsk in S Altai Kray near the E Kazakhstan border.
BT Russian Republic
USSR

Zn
use zinc

Zoantharia
As of 1981, includes use on level 3 under Coelenterata(1) Anthozoa(2). Before 1981, included use on level 2 under Coelenterata(1). See list F.
BA Anthozoa
Coelenterata
BT Invertebrata
SA Tetracorallia

Zoanthiniaria
Includes use on level 2 under Coelenterata(1). See list F.
BA Anthozoa
Coelenterata
BT Invertebrata

Zod
Village in E.
BT Armenia
USSR

zodiacal dust
use cosmic dust

zoisite
BA epidote group
sorosilicates
orthosilicates
silicates
BT minerals
SA clinozoisite
tanzanite

Zonal soils
See list M.
BA soils
SA Brown soils
Chernozems
Chestnut soils
Desert soils
Podzols
Red soils
Sierozems
Tundra soils

zonation
A valid term through 1976. Use zoning.

zone, auroral
use auroral zone

zone, Benioff
use Benioff zone

zone of aeration
use unsaturated zone

zone of mobility
use asthenosphere

zone, oxidation
use oxidation zone

zones
A valid term through 1977. Use zoning.

zones, fault
use fault zones

zones, fracture
use fracture zones

zones, low-velocity
use low-velocity zones

zones, rift
use rift zones

zones, shear
use shear zones

zones, subduction
use subduction zones

zones, transition
use transition zones

Zonguldak
Province in NW Anatolia. Also a city on the Black Sea.
BT Turkey
Middle East
NT Amasra

zoning
Includes use on level 2 under metamorphism(1); on level 3 under metasomatism(1), under crystal growth(1) and under mineral name. Before 1975, also search mineral zoning. Also search zonation; zones.
SA crystal growth
intrusions
metamorphism
metasomatism

zoogeography
use biogeography

zooplankton
BT plankton

zooxanthellae
SA algae
Protista

Zr
use zirconium

Zr-95
Includes use on level 3 under isotopes(1).
SA isotopes
zirconium

Zubair Formation
Lower Aptian-Hauterivian. In N, central, and S Iraq. In S, comprises over 1200 feet of sandstones, siltstones, and shales.
BT Lower Cretaceous
Cretaceous
SA Aptian
Hauterivian
Iraq
Middle East

Zulia
State encircling almost all of Lake Maracaibo in NW.
BT Venezuela
NT Lake Maracaibo
SA Maracaibo Basin

Zuloaga Limestone
Coahuila and Zacatecas, Mexico. Also search Zuloaga Formation.
BT Upper Jurassic
Jurassic
SA Coahuila
Mexico
Zacatecas

Zululand
Region in NE.
BT Natal
South Africa

Zungaria
use Dzhungaria

Zuni Mountains
Range in McKinley and Valencia counties in W.
BT New Mexico
United States

zunyite
BZ halides
orthosilicates
BT minerals

Zurich
Canton in N part of country. Also a city at NW end of Lake of Zurich.
BT Switzerland

zussmanite
BA sheet silicates
silicates
BT minerals

Zwickau
City on the Mulde River S of Leipzig in S part of country.
BT Karl-Marx-Stadt Bezirk
East Germany

Zyryanka
Town on the Kolyma River at the mouth of the Zyryanka River in NE Yakutia. A coal mining center.
BT Russian Republic
USSR

Zyryanovsk
City SE of Ust Kamengorsk in Vostochno Kazakhstan Oblast in extreme E.
BT Kazakhstan
USSR

GUIDE TO INDEXING

This Guide is used by the GeoRef indexers for selecting index terms and arranging them into the three-level index entries found in the printed products made from the GeoRef database.

Each complete index entry (set) has three parts: first, second, and third-level terms. For example, the five entries below appeared in the 1981 Bibliography and Index of Geology:

(1) 1st level, 2nd level
3rd level

Illinois—areal geology
guidebook: Geological science field trip, Springfield area, Sangamon County, Illinois, New City, Springfield East, Springfield West, and Williamsville 7.5-minute quadrangles 25909

(2) 1st level, 2nd level
3rd level

Illinois—economic geology
coal: Roof strata of the Herrin (No. 61) Coal Member in mines of Illinois; their geology and stability; summary report 26933

(3) 1st level, 2nd level
3rd level

Illinois—engineering geology
land subsidence: Roof strata of the Herrin (No. 61) Coal Member in mines of Illinois; their geology and stability; summary report 26933

(4) 1st level, 2nd level
3rd level

Illinois—environmental geology
land use: Land use and land cover and associated maps for Aurora, Illinois 27017

(5) 1st level, 2nd level
3rd level

Illinois—sedimentary petrology
sedimentation: Sedimentology of the Vermilion River (Illinois and Indiana, U.S.A.) and its significance for lithofacies models of meandering alluvial streams 27733

All of the above documents were indexed with the term Illinois. Then in addition to Illinois document (5), for example, also was indexed with the terms sedimentary petrology (level-two term) and sedimentation (level-three term).

This Guide consists of a Main List which is found embedded in the Thesaurus, and several special lists of terms.

MAIN LIST

The Main List consists of sets of terms. Each set indicates, for each level-one term, specific terms and/or classes of terms which can be used as level-two and level-three terms.

This Main List is now embedded in the Thesaurus. The entries in the Main List are the terms which are valid on level one. These terms contain notes, often extensive, giving details of the construction of sets. In this manner, the breakdown of topics under each first-level term is regularized.

crust

topic (age, anomalies, composition, concepts, evolution, genesis, interpretation, observations, processes, properties, theoretical studies, thickness)

subtopic (no area term)

The above entry indicates that second-level terms under crust will be a topic such as age, composition, etc. For crust, the third-level terms under a topic term can be freely chosen, but should not include any area terms.

Sample entries under crust, according to this breakdown, taken from the 1981 Bibliography and Index of Geology are:

crust—processes
heat flow: Heat flow 26637
igneous processes: Magmatism and fundamental structural-tectonic regions 24938

crust—structure
continental type: Some problems of Pliocene-Quaternary movement of the continental crust 27805
deep seismic sounding: The structure of the Earth's crust and upper mantle from deep seismic sounding data 26631
interpretation: A method for studying the Earth's crust using magnetometer data 26439

SPECIAL LISTS

The special lists which follow are itemized on the contents page. They consist of categories of terms used as level-one or level-two terms in index sets, as well as classes of first-level terms not included in the Main List, e.g. commodity terms (List C), and fossils (List F).

For each list there are explanatory notes on indexing, online searching and searching in the Bibliography and Index of Geology (B.I.G.).

Under "Indexing," the current indexing practice is given, consisting of instructions on the construction of index sets, which needs to be read along with the instructions under the individual level-one terms involved, in the body of the Thesaurus.

The notes on searching attempt to guide the searcher in the use of the list. Searchers might also read the notes on Indexing for further clues but should be aware that these notes reflect current practice which in some cases differs from past. Further notes on specific terms are in the body of the Thesaurus.

LIST A - LEVEL-ONE TERMS

This List consists of the terms used as level-one index terms in GeoRef. Level-one terms are main topics in the papers they index. They are also broad terms. There are relatively few level-one terms and throughout the life-span of GeoRef they have been the most carefully controlled of the index terms. Level-one terms can also be used as subtopics on levels two and three and as supplemental index terms.

Level-one terms are entry points in the subject index of the five subfiles of GeoRef:

Bibliography and Index of Geology - 'B' in the List
Bibliography and Index of Geology Exclusive of North America - 'E' in the List
Geophysical Abstracts - 'G' in the List
Bibliography and Index of North American Geology - 'N' in the List
Bibliography of Theses in Geology - 'T' in the List

INDEXING

In the index sets used in GeoRef there are three levels. The second-level terms subdivide the citations on level-one, the third-level terms those on level-two.

e.g. Brazil	plate tectonics	igneous rocks
stratigraphy	mechanism	basalts
Cretaceous	ophiolite	textures

The rules for constructing index sets for each level-one term are found under that term in the body of the Thesaurus and under the appropriate List for that term.

ONLINE SEARCHING

In online searches, a term can be limited to its occurrences on level-one. Such a search will retrieve citations in which the term was a main topic as opposed to papers in which it was used on level-two or level-three as a subtopic or as a supplemental index term.

B.I.G. SEARCHING

The letters B, E, G, N, and T following terms stand for the subfiles in which the term was a level-one term. For subfile B, the Bibliography and Index of Geology, the years during which the term was valid are given as well.

Absolute age B72-
Absolute age, dates B69-71 E G N T
Absolute age, methods B69-71 E G N T
Absorption spectrophotometry N
Acoustical exploration N
Acoustical logging N
Actinium B69- N
Aden G
Adriatic Sea B69 E G
Agean Sea B77-
Aeronomy B75-
Afars and Issas G
Afars and Issas Territory B73-77
Afghanistan B69- E G
Africa B69- E G
Agnatha B81-
Alabama B69- G N T
Alaska B69- G N T
Albania B69- E
Alberta B69- G N T
Algae B69- E N T
Algal flora B81-
Algeria B69- E G
Alkali metals B69 E
Alluvial fans N
Alluvial plains N
Alpha activation analysis N
Alps B69- E G
Aluminum B69- E N T
Aluminum ores B81-
American Samoa N
Americium B69-
Amphibia B69- E N
Amphibians B81-
Amphineura N
Andes B69- E G T
Andesite N
Andorra B69- E
Angiosperms B69- E N
Angola B69- E G
Anhydrite N
Anion exchange-spectrochemical analysis N
Annelida B69-80 E N T
Antarctic Ocean B69- E G T
Antarctic region G
Antarctica B69- E G T
Anthozoa B69-71 E N T
Anticlines N
Antimony B69- E N T
Antimony ores B81
Apennines B69- E
Appalachian Basin N
Appalachians B69 G N T
Arabia B69 E G
Arabian Peninsula B69- E
Arabian Sea B77-
Arachnida B69-71 E N
Archaeocyatha B69- E N
Archean B78-
Arctic G N
Arctic America N
Arctic Ocean B69- E G N T
Arctic region B69- E T
Argentina B69- E G
Argon B69- E N T
Arid regions N
Arizona B69- G N T
Arkansas B69- G N T
Arsenic B69- E N
Arsenic ores B81-
Artesian waters and wells B69-71 E N T
Arthropoda B69- E N
Artifacts B69-71 E N
Asbestos B69-80 E N
Asbestos deposits B81-
Ascension Island B69-72 E G
Asia B69- E G T
Asia Minor B69 E
Asphalt N
Associations B69- E N
Astatine N
Asteroidea B69-71 E N T
Asteroids B70- G
Asteroyoa B69-71 E N
Astrophysics and solar physics B76-
Atlantic Coastal Plain B70- N
Atlantic Ocean B69- E G N T
Atlantic Ocean Islands B81-
Atlantic region B76-
Atmosphere B69- E G N T
Atomic energy N
Aurora B78-
Australasia B69- E
Australia B69- E G
Austria B69- E G T
Automatic data processing B69- E G N T
Autoradiography B70- N
Aves B69- E N
Azores B69- G
Bacteria B69- E N
Bahamas B69- G N T
Bahrain B77-
Balearic Islands B74-
Balkan Peninsula B69- E
Baltic region B69- E G
Baltic Sea B69- E G
Banda Sea G
Bangladesh B73-
Barbados B74- G N
Barents Sea G
Barite B69-80 E N T
Barite deposits B81-
Barium B69- E N T
Bars N
Basalt N
Base metals B78- N
Baselevel N

Basin and Range N
Basin and Range Province B78- G
Basin, structural N
Basins, structural B69-71 E N T
Basutoland B69 E
Batholiths B69-71 E N T
Bauxite B69- E N T
Bay of Biscay G
Beaches N
Bechuanaland G
Belgium B69- E G
Belize B 75-
Benin B69- E
Bentonite B69-80 E N
Bentonite deposits B81-
Bering Sea B69- E G N
Bermuda B69- G N T
Bermuda Islands G
Beryl N
Beryllium B69 E G N
Beryllium ores B81-
Bhutan B75-
Bibliography B69- E G N
Biogeochemical prospecting B69-71 E N
Biogeochemical surveys N
Biogeochemistry N
Biogeography B69- N
Biography B69- E N
Birds B81-
Bismark Islands G
Bismuth B69- E N
Bismuth ores B81-
Bitumens B69- E N
Bituminous sands N
Black Sea B69- E G
Blastoidea B69-71 E N
Bogs N
Bohemia B69 E
Bolivia B69- E G T
Book reviews B73-
Borates N
Borneo B69-73 E G
Boron B69- E G N T
Boron deposits B81
Botswana B69- E G
Boudinage B69 E N
Boulders N
Brachiopoda B69- E N T
Brachiopods B81-
Branchiopoda B69-71 E N
Brazil B69- E G T
Breccia B69-71 E N T
Brines B69- E N
British Columbia B69- G N T
British Guiana G
British Honduras B69-74 N T
British Isles G
Bromine B69- E N
Bromine deposits B81-
Brucite N
Bryophytes B69- E N
Bryozoa B69- E N T
Bryozoans B81-
Bulgaria B69- E G
Burma B69- E G
Burundi B69-
Cadmium B69- E N
Calcite B69- E N
Calcite deposits B81-
Calcium B69- E N
Calderas N
Caliche N
California B69- G N T
Californium B71- N
Cambodia B69- E G
Cambrian B69- E N T
Cameroon B69- E
Cameroun G
Canada B69- G N T
Canadian Shield B78
Canal Zone N
Canary Islands B74 G
Cape Verde Islands B69- G
Carbon B69- E G N T
Carbonate rocks N
Carbonates N
Carboniferous B69- E N T
Caribbean region B69- E G N
Caribbean Sea B69- E G N T
Caroline Islands B69 E
Carpathians B69- E
Cartography B69-71 E N T
Caspian Sea B69-
Catalogs B69- E N
Caves B69-71 E N T
Cayman Islands B81-
Celebes Sea B69- E
Celtic Sea B77
Cenozoic B69- E N T
Central African Republic B69- E
Central America B69- G N T
Cephalopoda B69-71 E N T
Ceramic materials B69- E N T
Cerium B69- E N
Cesium B69- E N T
Ceylon B69-73 E G
Chad B69- E G
Chad Republic G
Changes of level B69- E G N T
Chelicerata B69-70 E
Chemical analyses N
Chemical analysis B69- E N T
Chert N
Chesapeake Bay B69-72 N
Chile B69- E G T
China B69- E G
Chlorine B69- E N T
Chordata B69- N
Christmas Island B69- E
Chromite B69-80 E N T
Chromite ores B81-
Chromium B69- E N
Cirripedia B69-71 E N
Clay N
Clay mineralogy B69- E N T
Clays B69- E N T
Coal B69- E N T
Coal balls N
Coast Rica N
Cobalt B69- E N
Cobalt ores B81-
Coelenterata B69- E N
Collapse structures N
Collections B69-71 E N
Colombia B69- E G T
Colorado B69- E G N T
Colorado Plateau B78- G N
Colorimetric analysis N
Columbia G
Columbia Plateau B78-
Columbia river B69
Columbium B72 N
Comets G
Comoro Islands B77-
Compressibility N
Concretions B69-71 E N
Conglomerate N
Congo B69- E G
Congo Republic G
Connate water B69-71 N
Connecticut B69- G N T
Conodonta B81-
Conodonts B69- E N T
Conservation B78-
Construction materials B69- E N T
Continental drift B69- E G N
Continental margin G N
Continental shelf B69- E N T
Continental slope B69- E G N T
Continents B69-71 E G N
Cook Island G
Copper B69- E N T
Copper ores B81-
Coprolite N
Coprolites B69-- E N
Coral Sea B77- G

Corals B81-
Core B69- E G N
Correlation B69-71 E N
Corsica B69- G
Corundum B69-80 E N
Corundum deposits B81-
Cosmic dust B69-71 N
Cosmic-ray methods G
Cosmogeny G
Costa Rica B69- G N
Cratering B69-71 E G N
Craters N
Cretaceous B69- E N T
Crete B69
Crinoidea B69-71 E N T
Crust B69- E G N T
Crustacea B69-71 E N T
Cryogeology B69
Cryptoexplosion structures B69-71 E G N T
Cryptogam B69
Crystal chemistry B69- E N T
Crystal growth B71- N
Crystal structure B69- E N
Crystallography B69- E N T
Cuba B69- G N
Curium B69- N
Cyprus B69- E G
Cystoidea B69-71 E N
Czechoslovakia B69- E G
Dacite N
Dahomey B69-76 E G
Dams B78-
Deception Island G
Deformation B69- E G N T
Delaware B69- G N
Deltas B69-71 E N T
Denmark B69- E G
Density G
Desert pavement N
Deserts N
Deuterium B69-
Devonian B69- E N T
Diabase N
Diagenesis B69- E N T
Diamond N
Diamonds B69- E N
Diapirs B69-71 E N T
Diatomite B69- E N
Diatoms B69-71 E N
Differential thermal analysis B69-77 E N
Dikes B69-71 E N T
District of Columbia B69- G N
Djibouti B78-
Dolomite B69-76 E N
Dolomitization N
Dolostone B77-80
Dolostone deposits B81-
Domes N
Dominican Republic B72- N
Drainage changes N
Drainage patterns N
Dunes N
Dysporium N
Dysprosium B73-
Earth B69- E G N T
Earth current exploration N
Earth currents G N
Earth tides B69-71 E G N
Earth-current exploration G N
Earth-current methods B69-71 E G N
Earth-current surveys B69-71 E G N
Earthquakes B69- E G N T
East China Sea B78- G
East Germany B78-
East Pacific Ocean Islands B81-
Easter Island B69-72 E
Eastern Hemisphere B70-
Eastern U. S. B78-
Echinodermata B69- E N T
Echinoderms B81-
Echinoidea B69-71 E N T
Ecology B69- E N T
Economic geology B69- E N T
Ecuador B69- E G T
Education B69- E G N
Egypt B69- E G T
El Salvador B69- G N
Elastic properties B69-71 E G N
Elastic waves N
Elasticity N
Elba G
Electric exploration N
Electrical exploration G N
Electrical logging N
Electrical methods B69-71 E G N T
Electrical properties B69-71 E G N T
Electrical surveys B69-71 E G N T
Electron diffraction analysis B71 N
Electron-diffraction analysis N
Electron microscopy B69- E N T
Electron paramagnetic resonance N
Electron probe analysis N
Electron-probe analysis N
Electron spin resonance N
Elements B69-73 E G N
Emission spectroscopy N
Energy sources B69- E N
Engineering geology B69- E N T
England B69- E G
English Channel B69- E G
English Channel Islands B81-
Environmental geology B70-
Eocene B71- N
Epeirogenesis B69-71 E N
Epeirogeny B72-
Equatorial Guinea B78-
Erbium B78-
Eritrea G
Erosion B69-71 E N T
Erosion surfaces N
Estuaries B69-71 E N
Ethiopia B69- E G
Ethiopis G
Eurasia B69- E
Europe B69- E G
Europium B69- E N
Eurypterida B69-71 E N
Evaporite deposits B81-
Evaporites B69-80 E N T
Evolution B69-71 E N T
Explosion phenomena B69-71 E G N T
Explosions B78-
Extraterrestrial geology B69-71; 81- E T
Faeroe Islands B69- E G
Faeroes G
Falkland Islands B69- E
Far East B69- E
Faroe Islands G
Faults B69- E G N T
Fauna N
Feldspar B69-80 E
Feldspar deposits B81- N
Fennoscandia B69-71 E
Fernando Poo G
Ferns B81-
Fiji B69- G
Finland B69- E G
Fish B81-
Fjords N
Flame photometric analysis N
Flame photometry N
Flora B69 N
Florida B69- G N T
Florida keys B69
Fluid inclusions B69-71; 78- E N
Fluorescence N
Fluorine B69- E N
Fluorite N
Fluorometric analysis N
Fluorspar B69- E N T
Folds B69- E G N T
Foliation B69- E N T
Foliations N
Foraminifera B69- E N T
Foraminifers B81-
Formation pressure N

Formosa B69-76 E G
Fossil man B78-
Fossils, problematic B69-77 E N
Fossils, problematical N
Foundations B78-
Fractures B69- E G N T
France B69- E G
Francium N
Franz Josef Land G
French Guiana B69- E
Fuel resources B81-
Fulgurites B69 E N
Fuller's earth N
Fumaroles B69-71 E G N
Fungi B69- E N
Fusulinidae N
Gabbro N
Gabon B69- E G
Gadolinium B69- N
Galapagos Islands B69- E G
Gallium B69- E N T
Gambia B70-73
Garnet B71 N
Gas chromatographic analyses N
Gas, natural B69-77 E N T
Gastroliths N
Gastropoda B69-71 E N T
Gegenschein G
Gems B69- E N
General B69-71 E G N
Geobotanical prospecting N T
Geochemical exploration N
Geochemical methods B70
Geochemical prospecting B69-71 E N
Geochemical surveys B69-71 E N T
Geochemistry B69- E N T
Geochronology B69- E G N T
Geodes N
Geodesy B69- E G N T
Geologic barometry B69-71 E G N
Geologic cryometry B69 E
Geologic exploration N
Geologic hazards B78-
Geologic mapping N
Geologic thermometry B69-71 E G N
Geological barometry G
Geological exploration B69-71 E N T
Geology B72-
Geology as a profession N
Geomorphology B69- E N T
Geophysical exploration G N
Geophysical logging N
Geophysical methods B69- E G N T
Geophysical observations B69
Geophysical research G
Geophysical surveys B69-77 E G N T
Geophysics B69- E G N T
Georgia B69- G N T
Geosynclines B69- E G N
Geotectonics N
Geothermal energy B69- E G N
Geothermal gradient G N
Geothermal surveys N
Geothermonetry N
Germanium B69- E N
Germany B69- E G
Geysers B69-71 E G N
Ghana B69- E G
Glacial deposits N
Glacial features N
Glacial geology B72- N
Glacial lakes N
Glaciation B69-71 E G N T
Glaciers B69-71 E G N T
Glaciology N
Glauconite B69-80 E N
Glauconite deposits B81-
Glossaries B69- E N
Glossary N
Glossopterio flora B69-71
Goa B69-72 E
Gold B69- E N T
Gold ores B81-
Grabens N
Granite B69-80 E N
Granite deposits B81-
Granodiorite N
Graphite B69-80 E N
Graphite deposits B81-
Graptolites B81- N
Graptolithina B69- E N T
Gravel B69-80 E N
Gravel deposits B81-
Gravity anomalies N
Gravity exploration G N
Gravity field, Earth B69-71 E G N T
Gravity field, Moon B70-71 G
Gravity methods B69-71 E G N T
Gravity surveys B69-71 E G N T
Gravity tectonics N
Great Basin B78-
Great Britain B69- E G
Great Lakes B69-
Great Lakes region B69- G N
Great Plains B78- N
Greater Antilles B78-
Greece B69- E G
Greenland B69- G N T
Greenland Sea G
Groundwater T
Ground water B69- E N T
Guadalupe B69 E
Guadeloupe B69-
Guam G N
Guatemala B69- G N T
Guinea B69- E G
Guinea-Bissau B75-
Gulf Coastal Plain B69- G N
Gulf of Aden B69- E G
Gulf of Alaska G
Gulf of Bothnia G
Gulf of California B69- G N
Gulf of Maine N
Gulf of Mexico B69- G N T
Gulf of Saint Lawrence G N
Gulf of St. Lawrence G N
Guyana B69- E G
Guyots N
Gymnosperms B69- E N
Gypsum B69-80 E N T
Gypsum deposits B81-
Hafnium B69- E N
Haiti B69- N T
Halogens B69-71 E N
Hawaii B69- G N T
Heat flow B69- E G N T
Heat transfer G
Heavy metals N
Heavy minerals B69-80 E N T
Heavy mineral deposits B81-
Helium B69- E N
Helium gas B81-
Hemichordata B76-
High-pressure research N
Highways B78-
Himalaya B69- E
Himalayas B69- E
Historical geology N
History B69-71 E N
Holmium B78-
Holocene B71-
Holothuroidea B69-71 E N
Honduras B71- N
Hong Kong B69- E
Hungary B69- E G
Hydrocarbons N
Hydrogen B69- E N
Hydrogeology B69- E N T
Hydrology B75-
Hydrosphere N
Hydrothermal alteration B69-71 E N T
Hydrothermal solutions N
Hydrozoa B69-71 E N
Icarus G
Ice N
Ice ages (ancient) N
Ice ages, ancient B69-71 E N

Ice islands N
Ice, non-glacial B69-71 E N T
Ice, nonglacial G N
Iceland B69- E G
Ichnofossils B72-
Idaho B69- G N T
Igneous petrology N
Igneous rocks B69- E N T
Illinois B69- G N T
Illinois Basin N
Ilmenite N
Impact phenomena G N
Impact statements B78-
Impactite N
Impactites G N
Inclusions B69- E N T
India B69- E G T
Indian Ocean B69 E G T
Indian Ocean Islands B81-
Indiana B69- G N T
Indium B69- E N
Indochina B69- E
Indonesia B69- E G T
Industrial Materials B68
Industrial minerals B69- E N T
Infrared exploration G N
Infrared methods B69-71 E G N
Infrared spectroscopy N
Infrared surveys B69-71 G N
Insecta B69- E N
Interplanetary space B75-
Intrusions B69- E N T
Invertebrata B69- E N T
Invertebrates B81-
Iodine B69- E N
Iodine deposits B81-
Ionian Sea B69- E
Ionosphere B77-
Iowa B69- G N T
Iran B69- E G
Iraq B69 E G
Ireland B69- E G
Iridium B69- E N
Irish Sea B69- E G
Iron B69- E N T
Iron ores B81-
Island ores N
Isostasy B69- E G N
Isotopes B69- E G N T
Israel B69- E G
Italy B69- E G
Ivory Coast B69- E G
Jamaica B69- G N
Jan Mayen B69- E G
Japan B69- E G
Japan Sea B69- E G
Johnston Island N
Joints G N
Jordan B69- E G
Jupiter B69- G
Jurassic B69- E N
Kansas B69- G N T
Kaolin B69-80 E N
Kaolin deposits B81-
Karst N
Kashmir B69-71
Katanga B69-73
Kentucky B69- G N T
Kenya B69- E G
Kerguelen Islands G
Kermadec Islands G
Korea B69- E G
Krypton B69- E G N
Kuril Islands B69 E
Kuwait B69- E
Labrador B69- G N T
Laccoliths N
Lake Erie G
Lake Ontario G
Lake Superior G T
Lake Superior region B69-70 G N
Lakes B69-71 E N T
Lakes, extinct B69-71 E N T
Lamprophyre N
Land subsidence B78-
Land use B78-
Landforms N
Landslides N
Lanthanides B69-72 E
Lanthanum B69- E N
Laos B69- E
Lapland B69
Laser methods G N
Laser surveys N
Laterite N
Laterites B69-71 E N
Lava B69- E N T
Lava flows N
Lead B69- E G N T
Lead ores B81-
Lead-zinc deposits B78-
Lebanon B69- E G T
Lesotho B69-
Lesser Antilles B77-
Leveling G
Leveling Networks G
Lexicon N
Lexicons B69- E N
Liberia B69- E G
Libya B69- E G
Lichens B70-
Life G
Lignite B69- E N T
Limestone B69-80 E N T
Limestone deposits B81-
Lineaments N
Lineation B69- E N T
Lithium B69- E G N
Lithium ores B81-
Lithofacies N
Loess N
Louisiana B69- G N T
Low-temperature analysis N
Luminescence B69-71 E G N T
Lutetium B69-
Luxembourg B69- E G
Lycopoda B69-71
Macao G
Macedonia B69
Macquarie Island G
Madagascar B69-74 E G T
Madeira B77- G
Magma G
Magmas B69- E G N T
Magmas and magmatic differentiation N
Magnesite B69-80 E N
Magnesite deposits B81-
Magnesium B69- E N
Magnesium ores B81-
Magnetic anomalies N
Magnetic exploration G N
Magnetic field of the Earth N
Magnetic field, Earth B69-71 E G N T
Magnetic field, Jupiter G
Magnetic field, Mars B69 G
Magnetic field, Mercury B69 G
Magnetic field, Moon B69-71 G
Magnetic field, Sun G
Magnetic field, Venus B69 G
Magnetic field, asteroids B69
Magnetic field, interplanetary G
Magnetic field, planets G
Magnetic methods B69-71 E G N T
Magnetic properties B69-71 E G N T
Magnetic surveys B69-71 E G N T
Magnetite B71 N
Magnetosphere B75-
Magnetotelluric exploration G N
Magnetotelluric methods B69-71 E G N T
Magnetotelluric surveys B69-71 E G N T
Maine B69- G N T
Major-element analyses B 69-71 E N T
Malacostraca B69-71 E N
Malagasy Republic B74-
Malawi B69- E G
Malay Archipelago B71-

Malaya B69-72 E G
Malaysia B69- E G
Maldive Islands B69- E
Mali B69- E G
Malta B73-
Mammalia B69- E N T
Mammals B81-
Man, fossil B69-77 E N
Manganese B69- E N T
Manganese ores B81-
Manitoba B69- E G N T
Mantle B69- E G N T
Maps B72-
Marble B69-80 E N
Marble deposits B81-
Mariana Islands B69- E
Marine geology B69- E G N T
Marine installations B78-
Maritime Provinces B78-
Mars B69- E G
Marshall Islands B69- E
Martinique N
Maryland B69- G N T
Mass spectroscopy N
Mass wastage N
Massachusetts B69- G N T
Mathematical geology B71-
Mauritania B69- E G T
Mauritius B76- G
Medical geology B71
Mediterranean region B69- E G
Mediterranean Sea B69- E G T
Melanesia B74-
Mercury B69- E G N T
Mercury ores B81-
Mercury Planet B75-
Merostomata B69-70 E N
Meozoic B69- E N T
Metal B69 E
Metals B69- E N T
Metal ores B81-
Metamorphic petrology N
Metamorphic rocks B69- N T
Metamorphism B69- E N T
Metasomatic rocks B78-
Metasomatism B69- N T
Metazoa B69
Meteor craters B69- E G N
Metorites B69- E G N T
Meteorology B75-
Meteors G
Mexico B69- G N T
Mica B69-80 E N
Mica deposits B81-
Michigan B69- G N T
Micronesia B69- E
Micropaleontology B69- E N T
Microscope methods B69-71 E N
Microscope techniques N
Microscopic methods B69 E
Microseisms B69-71 E G N T
Microwave exploration N
Microwave methods G N
Microwave surveys G N
Middle East B69- E G
Midway Islands N
Midwest B78-
Migmatites N
Military geology B69-70 N
Mineragraphy B69-71 E N
Mineral collecting B69-71 E N
Mineral collections N
Mineral data B69-71 E N T
Mineral deposits, genesis B69- E N T
Mineral descriptions N
Mineral economics B69-71 E N T
Mineral exploration B69- E G N T
Mineral resources B69- E N T
Mineral zoning B69-71 E N
Mineralogy B69- E N T
Minerals B71
Mining geology B69- E N T
Minnesota B69- G N T
Minor-element analyses N
Miocene B69- E N
Mississippi B69- G N T
Mississippi Delta G
Mississippi River N
Mississippi Valley B78- G N
Mississippi embayment N
Mississippian B69- N T
Missouri B69- G N T
Models N
Mohorovicic discontinuity B69- E G N
Mollusca B69- E N T
Mollusks B81-
Molybdenum B69- E N T
Molybdenum ores B81-
Monaco B69- E
Monazite B69-80 E N
Monazite deposits B81-
Mongolia B69- E G T
Montana B69- G N T
Moon B69- E G N T
Morocco B69- E G
Mossbauer analysis N
Mossbauer effect G
Mozambique B69- E G
Mud volcanoes B69- E N
Mudflows N
Museums B69- E N
Myriapoda B69-71 E N
Namibia B81-
Natural bridges N
Natural gas B78-
Near East 69 E
Nebraska B69- G N T
Neodymium B71-
Neogene B78-
Neon B69- E G N
Neotectonics B78-
Nepal B69- E
Nepheline syenite N
Neptune B69- G
Neptunium B72-
Netherlands B69- E G
Netherlands Antilles B69- N
Neutron activation analysis N
Neutron diffraction analysis N
Nevada B69- G N T
New Britain G
New Britain Island G
New Brunswick B69- G N T
New Caledonia B69- E G
New England B69- G N
New Guinea B69- E G T
New Hampshire B69- G N T
New Hebrides B69- E G
New Jersey B69- G N T
New Mexico B69- G N T
New South Wales B69- E
New York B69- G N T
New Zealand B69- E G T
Newfoundland B69- G N T
Nicaragua B70- G N
Nickel B69- E N T
Nickel ores B81-
Niger B69- E
Niger Republic G
Nigeria B69- E G
Niobium B69- E N T
Niobium ores B81-
Nitrate deposits B81-
Nitrates B68-80 N
Nitrogen B69- E N
Noble gases B71- N
Nodules B69- E N T
Nonmetal deposits B81-
Nonmetals B74-
North Africa B69-71 E
North American B69- G N T
North Carolina B69- G N T
North Dakota B69- G N T
North Korea G
North Sea B69- E G T
North Vietnam G

Northern Hemisphere B69-
Northern Ireland B69- E G T
Northern Territory B69- E
Northwest Territories B69- G N T
Norway B69- E G
Norwegian Sea G
Nova Scotia B69- G N T
Nuclear explosions B69-71 E G N T
Nuclear facilities B78-
Nuclear magnetic resonance N
Nuclear science N
Nucleosynthesis G
Nutation G
Ocean basins B72- G N
Ocean circulation B78-
Ocean floors B72-
Ocean waves B72-
Oceania B69- E
Oceanography B69- EG N T
Oceans G N
Oceans, circulation B72-77
Ohio B69- G N T
Oil and gas fields B69- E N T
Oil sand B70-71
Oil sands B69; 72- N
Oil shale B69- E N T
Okhotsk Sea B69- E G
Oklahoma B69- G N T
Oligocene B69- E
Oman B69- E
Ontario B69- G N T
Oolites N
Optical data processing N
Optical mineralogy B69-71 N T
Optical properties B69 E
Ordovician B69- E N T
Oregon B69- G N T
Organic materials B69- E N T
Orogeny B69- E G N T
Osmium B69- E N
Ostracoda B69- E N T
Ostracods B81-
Oxygen B69- E N T
Ozone B73
Pacific Coast B78-
Pacific Islands G N
Pacific Ocean B69- E G N T
Pacific region B69-
Pakistan B69- E G T
Paladium N
Paleobiogeography N
Paleobotany B69- E N T
Paleocene B71-
Paleoclimatology B69- E G N T
Paleoecology B69- E N T
Paleogene B78-
Paleogeography B69- E N T
Paleomagnetism B69- E G N T
Paleontology B69- E N T
Paleosalinity G N
Paleotemperature G
Paleotemperatures N
Paleozoic B69- E N T
Palladium B69- E N
Palynology B69- E N
Palynomorphs B69- E N T
Panama B69- G N T
Pantellerite N
Paper chromatography N
Papua B69-75 E G
Papua New Guinea B75-
Paragenesis B69- E N T
Paraguay B69- E
Patterned ground B69-71 E N
Peat B69- E N
Pebbles B69-71 E N
Pediments N
Pegmatite B69- E N T
Pegmatites N
Pelecypoda B69-71 E N T
Peneplanes N
Pennsylvania B69- G N T
Pennsylvanian B69- N T
Peridotite N
Periglacial features N
Periglacial phenomena N
Perlite N
Permafrost B69-71; 78- E G N
Permeability B69-71 E G N
Permian B69- E G N T
Persian Gulf B69- E G
Peru B69- E G T
Petrofabrics B69-71 E G N T
Petrogenesis N
Petrography B69 E N
Petroleum B69- E N T
Petroleum engineering N
Petrology B69- E N T
Phanerozoic B72- N
Phase equilibria B69- E G N T
Philippine Islands B69- E T
Philippine Sea B78- G
Philippines G
Phosphate B69-80 E N T
Phosphate deposits B81-
Phosphorescence N
Phosphorous N
Phosphorus B69- E N
Photogeology B69-71 E G N T
Photomicroscopy N
Physical geology N
Physical properties G N
Phytoliths B69
Pisces B69- E N T
Placers B69- E N
Planetology B71-
Planets B69-71 E G
Plantae B72-
Plasticity N
Plate tectonics B76-
Platinum B69- E N
Platinum ores B81-
Playas N
Pleistocene B71- N
Pliocene B71-
Pluto B69- G
Plutonium B69- E G N
Plutons N
Poland B69- E G
Polar wandering G
Polarographic analysis N
Pollution B78-
Polonium B71- N
Polymetallic ores B69- E N
Polynesia B74-
Popular and elementary geology B69-72 E N
Porifera B69- E N
Porosity B69-71 E G N T
Portugal B69- E G
Portuguese Guinea B69-73 E
Portuguese Timor B76
Potash B69- E N
Potassium B69- E N T
Praesodymium B78-80
Praseodymium B81-
Precambrian B69- E N T
Primates N
Prince Edward Island B70- G N
Problematic fossils B78-
Problematical fossils N
Promethium B72-
Protactinium B69- N
Proterozoic B69; 78- E
Protista B69- E N T
Protozoa B69-71 E N
Pseudomorphs N
Pteridophytes B69 E N
Pterobranchia B70- N
Pteropoda B69-71 E N
Puerto Rico B69- G N T
Pumice B70-80 N
Pumice deposits B81-
Pyrenees B69- E
Pyrite B69-80 E N
Pyrite ores B81-
Qatar B74-

Quark G
Quartz crystal B69- E N
Quartzite N
Quaternary B69- E N T
Quebec B69- G N T
Queensland B69- E
Radar exploration N
Radar methods G N
Radar surveys G N
Radioactive-waste disposal N
Radioactivity B69-71 E G N T
Radioactivity exploration G N
Radioactivity logging N
Radioactivity methods B69-71 E G N T
Radioactivity surveys B69-71 E G N T
Radiochemical analysis N
Radiolaria B69- E N T
Radiolarians B81-
Radiowave methods N
Radiowave surveys G N
Radium B69- E N
Radon B69- E N
Rare earth deposits B81-
Rare earths B69- E N T
Rare gases B71 N
Reclamation B78-
Red Sea B69- E G T
Red Sea region B76-
Reefs B69- E N T
Reflectivity N
Refractometry N
Refractory materials N
Remote sensing B78-
Remote-sensing methods G N
Remote-sensing surveys G N
Reptiles B81-
Reptilia B69- E N
Reservoirs B78-
Reunion B69- E
Reunion Island G
Rhenium B69- E N
Rhode Island B69- G N T
Rhodesia B69-80 E G
Rhodium B69- N
Ripple marks N
Rivers B69-71 E N T
Rock creep N
Rock glaciers N
Rock mechanics B78- G N
Rocky Mountains B69- G N
Romania B69- E
Rubidium B69- E N T
Rumania G
Ruthenium B69- N T
Rwanda B69- E G
Ryukyu Islands B69-68 E
Sahara B69- E
Saint Helena G
Saint Pierre and Miguelon B81-
Saint-Pierre and Miquelon N
Salt B69- E N T
Salt tectonics B69- E G N T
Salt water intrusion N
Samarium B71- N
Samoa B69- E T
Samoa Islands B69- E
Sampling N
San Marino B81-
Sand B69-80 E N
Sand volcanoes B69- E
Sands B81-
Sandstone B69-80 E N
Sandstone deposits B81-
Sarawak B69-75 E
Sardinia B74- G
Saskatchewan B69- G N T
Saturn B69- G
Saudi Arabia B69- E G T
Scandinavia B69- E G
Scandiuon B69- E N
Scaphopoda B69-71 E N
Scolecodonts N
Scotia Arc G
Scotia Sea G
Scotland B69- E G T
Sea mounts N
Sea of Japan B69 E G
Sea water B69- E N T
Sea-floor spreading B69- E
Seamounts B69-71 G N
Sedimentary petrology B69- E N
Sedimentary rocks B69- E N T
Sedimentary structures B69- E N T
Sedimentation B69- E N T
Sediments B69- E N T
Seiche N
Seismic exploration G N
Seismic methods B69-71 E G N T
Seismic surveys B69-71 E G N T
Seismic waves N
Seismology B69- E G N T
Selenium B69- E N T
Senegal B69- E G
Serpentine N
Seychelles B78-
Shale N
Shatter cones N
Shetland Islands B69- E
Shore features N
Shorelines B69-71; 78- E N T
Sierra Leone B69- E
Sikkim B69-77 E
Silica N
Silicate rocks N
Siliceous earth N
Silicification N
Silicon B69- E N
Sills N
Silurian B69- E N T
Silver B69- E N
Silver ores B81-
Singapore B69- E
Slate B69-80 N
Slate deposits B81-
Slope stability B78-
Slopes N
Snow B69-71 E N T
Society Islands B69; 75- E
Sodium B69- E N
Sodium carbonate B71- N
Sodium sulfate B69- N
Soil mechanics B78-
Soils B69- E N T
Solar system B70-71 G
Solomon Islands B69- E G
Solubility N
Somali Republic B71-
Somalia B69 G
South Africa B69- E G T
South America B69- E G T
South Arabia B69 E
South Australia B69- E
South Carolina B69- G N
South China Sea B78-
South Dakota B69- G N T
South Korea G
South Sandwich Islands G
South Shetland Islands G
South West Africa G
South-West Africa B69-80 E G
Southern Hemisphere B69- E
Southern Rhodesia B69-72 E G T
Southern U.S. B81-
Southern Yemen B77-
Southwestern U. S. B78-
Spain B69- E G
Spanish Guinea B69-73 E
Spanish Sahara B69-76 E G
Spanish West Africa B69- E
Specific gravity B69-71 G N
Spectrochemical analysis N
Spectrophotometric analysis N
Spectroscopy B69- E G N T
Spermatophyta B81- N
Spitsbergen B69- E G
Spitzbergen G

Springs B69- E N T
Sri Lanka B74-
Standard materials B76-
Standard rocks B69-75
Statistical measures N
Statistical methods B69-71 E G N T
Statistics B N
Stocks B69-71 E N T
Stratigraphy B69- E N T
Stream capture N
Streams N
Strength G N
Stress N
Stromatolites B69-71 E N
Stromatoporoidea B69-71 E N T
Strontium B69- E N T
Structural analysis B72-
Structural geology B69- E N T
Styolites N
Submarine geology G N
Subsidence B69-71 E N
Sudan B69- E G T
Sulfides B76 N
Sulfur B69- E N T
Sulfur deposits B81-
Sun B69- G
Surinam B69- E G
Survey organizations B81-
Surveys B69-80 E N T
Swaziland B69- E G
Sweden B69- E G
Switzerland B69- E G
Symposia B69- E N
Syria B69- E G
Tahiti B69; 71- E
Taiwan B77- G
Talc B69-80 E N T
Talc deposits B81-
Tanganyika B69 E
Tantalum B69- E N T
Tantalum ores B81-
Tanzania B69- E G
Tasman Sea B77- G
Tasmania B69- E G T
Technetium B76-
Tectites G
Tectonics B69- E G N T
Tectonophysics B71-
Tektites B69- E G N T
Television logging N
Tellurium B69- E N
Temperature methods G
Temperature surveys G
Tenerife G
Tennessee B69- E G N T
Terbium B70-72; 78-
Terraces N
Tertiary B69- E N T
Tetrapoda B81-
Texas B69- G N T
Thailand B69- E G
Thallium B69- E N
Thallophytes B71-
Thermal analysis B78-
Thermal conductivity G N
Thermal methods G N
Thermal properties G N
Thermal springs B69 E G N
Thermal surveys G N
Thermal waters B69- E N
Thermodynamic properties B69-73 E G N
Thermogravimetric analysis B73-77 N
Thermoluminescence G N
Thorium B69- E N T
Thorium ores B81
Thoron B69
Throium B70
Thulium B71-72; 78- N
Till N
Tillites N
Timor B69-73 E
Timor Sea G
Tin B69- E N T
Tin ores B81-
Titanium B69- E N T
Titanium ores B81-
Togo B69- E
Tonga B69- E
Tonga Islands B69 E G
Trace elements N
Trace-element analyses B69-71 E N T
Tracks and trails B69-71 E N T
Traps N
Triassic B69- E N T
Trieste B69-71 E
Trilobita B69- E N T
Trilobites B81-
Trinidad B69-76 G N
Trinidad and Tobago B77-
Tristan da Cunha G
Tritium B69- E N
Trucial Coast B69-73 E
Trucial Oman B69-72
Tsunami G
Tsunamis B69-71 E G N T
Tuff N
Tungsten B69- E N T
Tungsten ores B81-
Tunisia B69- E G
Tunnels B78-
Turbidity currents N
Turkey B69- E G
Tyrrhenian Sea B69- E
U.S.S.R. G
Uganda B69- E G
Ultramafic rocks N
Ultraviolet spectroscopy N
Unconformities B69-71 E N T
Underground installations B78-
Uniformitarianism N
United Arab Emirates B74-
United Kingdom B74- G
United States B69- G N T
Uplifts B69-71 E N
Upper Volta B69- E G
Uranium B69- E N T
Uranium ores B81-
Uranus B70- G
Uruguay B69- E G
USSR B69- E
Utah B69- G N T
Valleys N
Vanadium B69- E N
Vanadium ores B81-
Varves N
Veins B69-71 E N T
Venezuela B69- E G T
Venus B69- E G
Vermes B69; 74
Vermiculite B69-80 E N
Vermiculite deposits B81-
Vermont B69- G N T
Vertebrata B69- E N T
Vertebrates B81-
Victoria B69- E
Viet Nam G
Vietnam B69- E G T
Virgin Islands G N T
Virginia B69- G N T
Volcanism B69-71 E G N T
Volcanoes B69-71 E G N T
Volcanology B72-
Wake Island N
Wales B69- E G
Washington B69- G N T
Waste disposal B78-
Water B69 E N
Water falls N
Water resources B71-
Waterways B78-
Weathering B69- E N T
Well logging G N
Well-logging B69- E T
Wells and drill holes B69-71 E N T
West Africa B69-72 G
West Germany B78-

West Indies B69- G N
West Virginia B69- G N T
Western Australia B69- E T
Western Hemisphere B69-
Western Interior B78-
Western Sahara B81-
Western U.S. B78-
Williston Basin N
Windward Islands T
Wind work N
Wisconsin B69- G N T
World B69 E N
Worms B69- E N
Wyoming B69- G N T
X-ray analysis B78-
X-ray diffraction N
X-ray diffraction analyses N
X-ray diffraction analysis B69-77 E N T
X-ray fluorescence analysis N
X-ray radiography N
X-ray spectrographic analysis N
Xenoliths N
Xenon B69- E G N
Yellow Sea B78- G
Yemen B69- E
Ytterbium B70; 77- N
Yttrium B69- E N
Yugoslavia B69- E G
Yukon G N T
Yukon Territory B69-
Zaire B72-
Zambian B69- E G T
Zimbabwe B81-
Zinc B69- E N T
Zinc ores B81-
Zircon B69-80 E N
Zircon deposits B81-
Zirconium B69- E N

LIST B - AREA SETS

Index entries for areas in the Bibliography and Index of Geology consist of an area term which is valid on the first-level (see List O), followed by one of the second- and third-level combinations in this List.

INDEXING
Use the following terms as second-level terms under area terms (List O). Third-level options are also given.

ONLINE SEARCHING
The second-level terms here are basically equivalent to the GeoRef categories and can be similarly used for general orientation in searching on a geographic area. Thus a search for iron ores in Nigeria would have:

iron (truncated) AND Nigeria AND economic geology

B.I.G. SEARCHING
The second-level terms here are used to group into broad categories the references treated under first-level area terms (see List O). Within these categories further subdivision is achieved by the topics on level three.

---

areal geology (2)
Used for entries that might properly be placed under three or more of the second-level headings such as geomorphology, stratigraphy, structural geology.

topic (3)
bibliography, explanatory text (for a map), guidebook, maps or locality or regional.

e.g. France
areal geology
Narbonne

economic geology (2)

commodity (List C) (3)

e.g. Germany
economic geology
coal

engineering geology (2)

topic (3)
See Main List, second-level terms or dams, earthquakes, foundations, geologic hazards, highways, land subsidence, marine installations, nuclear facilities, permafrost, reservoirs, rock mechanics, shorelines, slope stability, soil mechanics, tunnels, underground installations, waste disposal, waterways.

e.g. California
engineering geology
earthquakes

environmental geology (2)

topic (3)
See Main List, second-level terms, or conservation, geologic hazards, impact statements, land use, pollution, reclamation, waste disposal.

e.g. Idaho
environmental geology
land use

geochemistry (2)

isotopes or element or other terms from List D or material or topic (Main List). (3)

e.g. USSR
geochemistry
isotopes

geochronology (2)

topic (3)
absolute age or topic from Main List, second-level terms or age (List E).

e.g. United States
geochronology
absolute age

e.g. Maine
geochronology
Precambrian

geomorphology (2)

topic (3)
See Main List, second-level terms, or glacial geology, meteor craters, changes of level, weathering.

e.g. Sahara
geomorphology
eolian features

geophysical surveys (2)

topic (3)
Type of survey (see Main List under geophysical surveys) or geodesy, heat flow, remote sensing, well-logging.

e.g. Oklahoma
geophysical surveys
infrared surveys

general (2)

topic (3)
E.g. annual report, bibliography, current research, education, geology, maps, philosophy, research

e.g. United States
general
current research

hydrogeology (2)

topic (3)
ground water, hydrology, springs, thermal waters

e.g. Louisiana
hydrogeology
ground water

mineralogy (2)
Name of mineral group (List L). In the case of miscellaneous minerals, that term suffices here.

e.g. Italy
mineralogy
framework silicates, feldspar group

oceanography (2)

topic (3)
See Main List, second-level terms; or continental shelf, continental slope, estuaries, geophysical methods, geophysical surveys, marine geology, nodules, ocean basins, ocean circulation, ocean floors, ocean waves, reefs, sea ice, sea water, sedimentation, sediments or locality.

e.g. Pacific Ocean
oceanography
marine geology

paleobotany (2)

fossil group (List F, first-level terms) (3)
Repeat set if more than one equally-emphasized fossil group.

e.g. Arizona
paleobotany
gymnosperms

paleontology (2)
Also used for studies on both fauna and flora.

Fossil group (List F, first-level terms) (3)
Repeat set if more than one equally-emphasized fossil group.

e.g. California
paleontology
foraminifera

petrology (2)

topic (3)
fluid inclusions, igneous rocks, inclusions, intrusions, lava, magmas, metamorphic rocks, metamorphism, metasomatism, meteorites, phase equilibria, tektites, volcanism, volcanology.

e.g. Colorado
petrology
metamorphism

sedimentary petrology (2)

topic (3)
clay mineralogy, diagenesis, heavy minerals, reefs, sedimentary rocks, sedimentary structures, sedimentation, sediments, weathering.

e.g. New South Wales
sedimentary petrology
sedimentary rocks

seismology (2)

topic (3)
See Main List, second-level terms.

e.g. Washington
seismology
earthquakes

soils (2)

topic (3)
See Main List, second-level terms.

e.g. Alberta
soils
name of soil group

stratigraphy (2)

age (List E) or
archaeology, biogeography, changes of level, continental drift, paleogeography, paleomagnetism, paleoclimatology. (3)

Repeat set if more than one equally-emphasized age.

e.g. South Carolina
stratigraphy
Cretaceous

structural geology (2)

topic (3)
deformation, epeirogeny, faults, folds, foliation, fractures, geosynclines, lineation, orogeny, neotectonics, salt tectonics, structural analysis, tectonics.

e.g. India
structural geology
tectonics

tectonophysics (2)

topic (3)
See Main List, second-level terms; or continental drift, core, crust, heat flow, mantle, plate tectonics, sea-floor spreading.

e.g. North America
tectonophysics
plate tectonics

volcanology (2)
For Quaternary volcanic activity.

Name of volcano, or locality for several volcanoes (3)

e.g. Italy
volcanology
Mount Etna

LIST C - COMMODITIES

The terms in this list are used for earth materials of actual or potential commercial value, i.e. from the viewpoint of economics.

Beginning in 1981, for some commodities, new terms were adopted to distinguish between the economic and geochemical viewpoints. For example, beryllium ores was added, to be used for the economic, and beryllium, the existing term, was restricted to the geochemical sense. Where new terms have been added, this is noted below in the list and in the body of the Thesaurus. Prior to 1981, these commodity terms were also used for geochemical treatments.

INDEXING

Commodities designated as (1,2,3) are used on level three under area terms (1) and economic geology (2). Each such term set causes a cross reference to be generated in the Subject Index of the B.I.G., from the commodity to the area term.

e.g. Wyoming economic geology coal	coal see also under economic geology under Wyoming

Commodities are also used as first-level terms when no geography is emphasized. In this case the set options are:

commodity (1)
topic (2)
Such as affinities, economics, exploration, genesis, geochemistry, inventory, possibilities, production, properties, reserves, resources

Subtopic (3)
Such as one of the topics listed above.

e.g. gold ores
reserves
possibilities

In general, only the most specific commodity term applicable should be used. However, a general term like heavy mineral deposits or mineral resources can be used with one or two of the specific commodities which it includes where these commodities have received special attention in a discussion of the group.

ONLINE SEARCHING

For complete coverage, search both the specific and the general terms. For example, a thorough search on sillimanite deposits requires a search on ceramic materials, excluding andalusite deposits and kyanite deposits.

To isolate commodity-oriented papers from non-economic, it is advisable to coordinate pre-1981 commodity terms with the economic geology Category Codes or ore deposits (for metals) or deposits (for nonmetals) and to also search the 1981 form of the term.

B.I.G. SEARCHING

Most papers on commodities deal with the commodities in relation to geographic locations. To find commodity papers on an area, look under the area (1), economic geology (2), and the commodity (3). Another approach, which will show the areas related to a particular commodity, is to look under the commodity. There, see also references to each of these areas will appear.

e.g. Coal see also under economic geology under Alabama; Alberta; Australia; Colorado; England; etc.

For those papers on commodities which are unrelated to specific areas, look under the commodity. For papers on the genesis of commodities look under mineral deposits, genesis (1).

Other sets related to commodities can be found under sedimentary rocks (for coal), metamorphic rocks (for slate deposits or marble deposits), and igneous rocks (for granite deposits).

In the list which follows, the commodities are used in index sets on all three levels, except those followed by (3) which are only permitted in sets on level three.

Asterisks indicate terms introduced in 1981.

abrasives (3)
  see under industrial minerals
aluminum ores
andalusite deposits* (3)
  see under ceramic materials
anhydrite deposits* (3)
  see under gypsum deposits*
antimony ores*
arsenic ores*
asbestos deposits*
asphalt (3)
  see under bitumens
barite deposits*
base metals
bauxite
bentonite deposits*
beryl (3)
  see under beryllium ores* or gems
beryllium ores*
  includes beryl (3)
bismuth ores*
bitumens
  includes asphalt (3)
borate deposits* (3)
  see under boron deposits*
boron deposits*
  includes borate deposits*
brines
  includes bromine deposits*, iodine deposits* (3)
  see also salt
bromine deposits*
  see also brines, salt
brown coal
  use lignite
building stone (3)
  see under construction materials, granite deposits*, sandstone deposits*, limestone deposits* or marble deposits*
calcite deposits*
cement materials (3)
  see under limestone deposits* or construction materials
ceramic materials
  includes andalusite deposits* (3), kyanite deposits* (3), refractory materials (3), sillimanite deposits* (3)
chalk deposits* (3)
  see under limestone deposits*
chromite ores*
clays
  includes fuller's earth (3), and shale used as brick clay
coal
cobalt ores*
columbium ores
  use niobium ores*
construction materials
  includes perlite (3), gravel deposits*, building stone(3), cement materials (3), dimension stone (3)
copper ores*
corundum deposits*
cryolite deposits* (3)
  see under fluorspar
diamonds
diatomite
dimension stone (3)
  see under construction materials, granite deposits*, sandstone deposits*, limestone deposits*, marble deposits*
dolostone deposits*
energy sources
  a grouper for petroleum, natural gas, coal, uranium ores*, etc.
evaporite deposits*
feldspar deposits*
fluorite
  use fluorspar
fluorspar
fuel resources
  a grouper for petroleum, natural gas (1,2,3), oil shale, oil sands
fuller's earth (3)
  see under clays
garnet deposits* (3)
  see under industrial minerals
gems
  includes beryl (3)
geothermal energy
glauconite deposits*
gold ores*
granite deposits*
graphite deposits*
gravel deposits*
  includes sands when sand is used as a construction material
gypsum deposits*
  includes anhydrite deposits* (3)
heavy mineral deposits*
helium gas*
hematite (3)
  see under iron ores*
industrial minerals
  includes abrasives (3), garnet deposits (3)
iodine deposits* (3)
  see under brines, salt
iron ores*
  includes hematite (3), limonite (3), magnetite (3)
kaolin deposits*
  see also clays, bentonite deposits*
kyanite deposits* (3)
  see under ceramic materials
lead ores*
lead-zinc deposits*
  see also lead ores*, zinc ores*
lignite
limestone deposits*
  includes chalk deposits* (3)
  see also dolostone deposits*
limonite (3)
  see under iron ores*
lithium ores*
magnesite deposits*
magnesium ores*
magnetite (3)
  see under iron ores*
manganese ores*
marble deposits*
mercury ores*
metal ores*
mica deposits*
mineral resources
  for very general treatments
molybdenum ores*
monazite deposits*
  see also heavy mineral deposits*, rare earth deposits*, thorium ores*
natural gas
nickel ores*
niobium ores*
nitrate deposits*
nonmetal deposits*
  do not use for energy sources, fuel resources, or water resources
oil
  use petroleum
oil and gas fields*
  use only for detailed discussions of specific fields; otherwise use petroleum or natural gas
oil sands
oil shale
peat
pegmatite
perlite (3)
  see under construction materials
petroleum
phosphate deposits*
platinum ores*
polymetallic ores
potash
pumice deposits*
pyrite ores*
quartz crystal
rare earth deposits

refractory materials (3)
  see under ceramic materials
salt
sands
  includes silica (3) for glass, ceramic, chemical use, etc.
  Use gravel deposits for sand as a construction material.
sandstone deposits*
shale (3)
  use clays for shale as brick clay
  see under construction materials
silica (3)
  see sands
sillimanite deposits* (3)
  see under ceramic materials
silver ores*
slate deposits*
soapstone (3)
  see under talc deposits*
soda ash
  use sodium carbonate
sodium carbonate
  includes trona (3)
sodium sulfate
sulfur deposits*
talc deposits*
  includes soapstone (3)
tantalum ores*
tar sands
  use oil sands
thorium ores
  see also monazite deposits*
tin ores*
titanium ores*
trona (3)
  see under sodium carbonate
tungsten ores*
uranium ores*
vanadium ores*
vermiculite deposits*
water resources
zinc ores*
zircon deposits*

LIST D - ELEMENTS

All of the elements in the periodic table are index terms and are valid as first-, second-, and third-level terms. Prior to 1981, element terms were used for both economic and geochemical papers. Starting in 1981, other terms have been added for economic papers (see List C), and the element terms have been restricted to geochemical papers.

Terms related to elements:
  deuterium (valid on level 1, level 2, and level 3)
  tritium (1,2,3)
  metals (1,2,3)
  noble gases (1,2,3)
    Use for rare gases and inert gases.
  rare earths (1,2,3)
    Use for lanthanide series, inner transition elements, and lanthanoans. Includes cerium, dysprosium, erbium, europium, gadolinium, holmium, lanthanum, lutetium, neodymium, praesodymium, prometheum, samarium, scandium, terbium, thulium, ytterbium, and yttrium, all of which are (1,2,3)
  major elements (3)
  minor elements (3)
  trace elements (3)

INDEXING
  Elements can be used as first-, second- and third-level terms, as can the related terms above, except major elements, minor elements, and trace elements which are valid on the third level only. The set structures for elements are given below.

ONLINE SEARCHING
  For geochemical references entered through 1980, search the element AND (geochemistry OR the appropriate Category Code). From 1981 on, search the element term only. Also consider searching the terms related to elements, above.

  For economic references, see List C, Commodities.

B.I.G. SEARCHING
  For geochemical references, see the element as a first-level term. Also look under the terms isotopes, chemical analysis, geochemistry, X-ray analysis, spectroscopy, thermal analysis, and electron microscopy. Also consider using the terms related to elements, above. Note that a cross-reference will appear in the Subject Index from trace elements to each first-level term under which it is used from 1977 on.

  For economic references, see List C, Commodities.

Sets having elements or related terms on level one:

abundance (2)
  name of material (3)
    e.g. manganese
        abundance
          soils
    rare earths
        abundance
          granite

geochemistry (2)
  name of material (3)
    e.g. magnesium
        geochemistry
          magmas
    iron
        geochemistry
          sedimentary rocks

analysis (2)
  type of analysis (3) such as:

activation analysis	mass spectroscopy
alpha-ray spectroscopy	microwave spectroscopy
atomic spectroscopy	Mossbauer spectroscopy
autoradiography	neutron activation analysis
chemical analysis	neutron diffraction analysis
chromatography	nuclear magnetic resonance
colorimetry	optical spectroscopy
electrolytic analysis	polarography
electron diffraction analysis	radio-frequency spectroscopy
electron paramagnetic resonance	Raman spectroscopy
	spectroscopy
	ultraviolet spectroscopy
electron probe	vacuum fusion analysis
emission spectroscopy	volumetric analysis
flame photometry	x-ray analysis
gamma-ray spectroscopy	x-ray diffraction analysis
ion probe	x-ray fluorescence
laser probe	x-ray spectroscopy

  e.g. rare earths
      analysis
        neutron activation analysis

isotopes (2)
  name of isotope or isotope ratio (3)
  e.g. oxygen
      isotopes
        0-18/0-16
  lead
      isotopes
        PB-21Ø

LIST E - GEOLOGIC AGE (STRATIGRAPHIC) TERMS

The authority used in GeoRef for age and stratigraphic terms is the Geologic Time Table, compiled by F.W.S. Van Eysinga, 3rd edition, Amsterdam, Elsevier, 1975. This is followed for nomenclature except in a few cases where previous usage in GeoRef strongly supports alternate terms.

Throughout, the "stratigraphic" upper, middle, and lower rather than the "chronologic" early, middle, and late are used to modify age terms, e.g. Upper Cambrian instead of Late Cambrian. "Uppermost" and "lowermost" are not used.

Additional stratigraphic and rock-unit terms as well as the terms listed below appear in the main body of the Thesaurus.

INDEXING

Terms in the list with (1,2,3) are used as age terms on the first, second or third level in index sets. The other terms on the list are valid on level three only, but are usually used as supplemental index terms, and only when specified in the paper. In such a paper, only the specific term need be indexed if a set with an age term is not appropriate, since use of the specific term will result in automatic posting of its broader term(s).

The capitalization of "upper", "middle", and "lower" depends on whether the word is part of a formalized term in the Geologic Time Table as indicated in the list below.

ONLINE SEARCHING

All terms in the list may be searched. A high degree of specificity is often possible.

In searching for rock units such as Dakota Sandstone, it is advisable to search for the common variants of such names. Dakota Group OR Dakota Sandstone will give better results than Dakota Sandstone alone.

In searching for certain broad terms such as Cretaceous, it suffices to use the term as is and not worry about subdivisions, since those are automatically posted to Cretaceous. This applies to the Precambrian and the Paleozoic, Mesozoic, and Cenozoic subdivisions, (e.g. Tertiary or Paleocene), but not the Paleozoic, Mesozoic, and Cenozoic themselves. To find the latter, you must also search for their major subdivisions.

In current practice (post-1977) the search for the age terms found on level one will yield those papers that are not restricted to a certain geographic area. Thus a search for first-level occurrences of Precambrian in the September 1980 GeoRef update gives four hits, one title of which illustrates the point: "The problem of the Precambrian-Cambrian boundary". Whereas an unrestricted search for Precambrian for the same month gives some thirty hits, most of which are related to specific geographic areas, illustrated by the following title: "Precambrian age of Reguibat Ridge dolerites, Mauritania".

B.I.G. SEARCHING

The terms to search in the subject index of the B.I.G. are those marked as possibly occurring on level one, i.e. those followed by (1,2,3) in the list. Under Tertiary, Vol. 44, No. 9, Sept. 1980, p. I-198 is the entry:

> Tertiary see also under geochronology under France; USSR; see also under stratigraphy under Algeria; Benin; California; Cuba; Czechoslovakia; England; France; India; Indian Ocean; Indonesia....
>
> Tertiary - stratigraphy
> biogeography: Tertiary 33978

In this example, the see also reference is a guide to each area set in which Tertiary occurs on level three. Papers cited under Tertiary as a level one term are those not devoted to a specific area. In the example above the title of the paper happens to be "Tertiary".

Papers dealing with stratigraphy as a discipline, i.e. papers on its practice, philosophy, principles, etc., are found under the heading stratigraphy.

Many papers on fossils contain stratigraphic information. If this is not a major part of the paper it may not have a stratigraphy set. Such papers can be found under appropriate fossil terms (see List F). Note that as of 1981, a number of non-systematic terms have been added to index fossils in biostratigraphic papers, e.g. amphibians, birds, crustaceans and brachiopods. From 1981 on, these non-systematic fossil terms are used for stratigraphic papers while the systematic fossil terms are reserved for paleontologic papers.

In the list which follows, the terms followed by (1,2,3) are valid on all levels. The remaining terms are valid only on level three, and are used mostly as supplemental index terms. Dashed lines separate parallel stratigraphic terminology used in different areas.

- Phanerozoic (1,2,3)
  - Cenozoic (1,2,3)
    - Quaternary (1,2,3)
      - Holocene (1,2,3)
      - Flandrian
      - Tyrrhenian (marine)
      - Sicilian
      - Calabrian
      - \- - - - - - - - - - - -
      - Holocene (1,2,3)
      - Pleistocene (1,2,3) (continental)
        - upper Pleistocene
        - middle Pleistocene
        - lower Pleistocene
          - Villafranchian
    - Tertiary (1,2,3)
      - Neogene (1,2,3)
        - Pliocene (1,2,3)
          - upper Pliocene
          - middle Pliocene
          - lower Pliocene
        - Miocene (1,2,3)
          - upper Miocene
            - Meotian
            - Sarmatian
            - Tortonian
          - middle Miocene
            - Helvetian
          - lower Miocene
            - Burdigalian
            - Aquitanian
      - Paleogene (1,2,3)
        - Oligocene (1,2,3)
          - Stampian
          - upper Oligocene
          - middle Oligocene
          - lower Oligocene
        - Eocene (1,2,3)
          - upper Eocene
          - middle Eocene
          - lower Eocene
        - Paleocene (1,2,3)
          - Thanetian
          - Montian
          - Danian
  - Mesozoic (1,2,3)
    - Cretaceous (1,2,3)
      - Upper Cretaceous
        - Senonian
          - Maestrichtian
          - Campanian
          - Santonian
          - Coniacian
        - Turonian
        - Cenomanian
      - Lower Cretaceous
        - Albian
        - Aptian
        - Neocomian
        - Barremian
        - Hauterivian
        - Valanginian
        - Berriasian
    - Jurassic (1,2,3)
      - Upper Jurassic
        - Portlandian
          - Tithonian
        - Kimmeridgian
        - Oxfordian

- Middle Jurassic
  - Callovian
  - Bathonian
  - Bajocian
  - Aalenian
- Lower Jurassic
  - upper Liassic
    - Toarcian
  - middle Liassic
    - Domerian
    - Pliensbachian
  - lower Liassic
    - Sinemurian
    - Hettangian

Triassic (1,2,3)
- Rhaetian
- Upper Triassic
  - Norian
  - Carnian
- Middle Triassic
  - Ladinian
  - Anisian
- Lower Triassic

- - - - - - - - - - -

- Keuper
- Muschelkalk
- Bunter

Paleozoic (1,2,3)

Permian (1,2,3)
- Thuringian
- Saxonian
- Autunian

- - - - - - - - - - -

- Zechstein
- Rotliegendes

- - - - - - - - - - -

- Tatarian
- Kazanian
- Zungurian
- Artinskian
- Sakmarian

- - - - - - - - - - -

- Upper Permian
- Lower Permian

Carboniferous (1,2,3)
- Upper Carboniferous
  - Stephanian
  - Westphalian
  - Namurian
- Dinantian
  - Visean
  - Tournaisian

- - - - - - - - - - -

- Upper Carboniferous
  - Gzhelian
  - Kasimovian
  - Moscovian
  - Bashkirian
  - Namurian
- Dinantian
  - Visean
  - Tournaisian

- - - - - - - - - - -

- Pennsylvanian (1,2,3)
  - Upper Pennsylvanian
  - Middle Pennsylvanian
  - Lower Pennsylvanian
- Mississippian (1,2,3)
  - Upper Mississippian
  - Middle Mississippian
  - Lower Mississippian

Devonian (1,2,3)
- Upper Devonian
  - Strunian
  - Famennian
  - Frasnian
- Middle Devonian
  - Givetian
  - Eifelian
- Lower Devonian
  - Emsian
  - Siegenian
  - Gedinnian

Silurian (1,2,3)
- Upper Silurian
  - Ludlovian
  - Wenlockian
- Lower Silurian
  - Tarannon
  - Llandoverian

Ordovician (1,2,3)
- Upper Ordovician
  - Ashgillian
  - Caradocian
- Middle Ordovician
  - Llandeilian
  - Llanvirnian
- Lower Ordovician
  - Arenigian
  - Tremadocian

Cambrian (1,2,3)
- Upper Cambrian
- Middle Cambrian
- Lower Cambrian

Precambrian (1,2,3)
- Proterozoic (1,2,3)
  - upper Proterozoic
    - Vendian
  - middle Proterozoic
  - lower Proterozoic
- Archean (1,2,3)

LIST F - FOSSILS

The terminology in this List was developed in consultation with micropaleontologists at the American Museum of Natural History for whom we produce the monthly Bibliography and Index of Micropaleontology, and with vertebrate paleontologists at the Museum of Paleontology of the University of California-Berkeley with whom we produce the annual Bibliography of Fossil Vertebrates.

All terms in the List are systematic names except those on the right margin under the heading 'Common Names'. The latter were reserved in 1981 for biostratigraphy papers.

INDEXING

Terms on the left margin are used on all levels. Terms indented from the left are used on levels two or three only, except Graptolithina and Pterobrachia which are also valid on level one. Terms on the right margin (common names) which are oppsite terms on the far left margin are valid on levels one and three. The remaining common names are valid on level three only. Use the most specific term applicable.

Several codes are provided. The numbers from 8-11 stand for the section in the Bibliography and Index of Geology for that fossil group:

08 - general paleontology
09 - paleobotany
10 - invertebrate paleontology
11 - vertebrate paleontology

The letter 'V' under a term indicates that a citation with that index term is to be given a 'V' Special Bibliography Code to mark it for the Bibliography of Fossil Vertebrates.

A three character 'M' code under a term indicates that the citation of a paper in which that fossil group is the main topic should have that Special Bibliography Code for the Bibliography and Index of Micropaleontology sections:

M01	algae	M07	problematic microfossils
M02	scolecodonts		
M03	nannofossils; Coccolithophoraceae	M08	Ostracoda
		M09	palynomorphs
M04	Conodonta	M10	Protista
M05	diatoms	M11	Radiolaria
M06	foraminifera	M12	miscellanea

Fossil sets are constructed as follows:

fossil group (1)
  fossil subgroup (2)
    age (List E) (3)

e.g. Bryozoa
  Cheilostomata
    Miocene

foraminifera
  Globigerinacea
    Cretaceous

If the age is not known, an area term should be used:

fossil group (1)
  fossil subgroup (2)
    area (List O) (3)

e.g. Mollusca
  Cephalopoda
    France

If neither age nor area are known, use a topic, e.g. morphology, on level three. For a paper about more than one fossil subgroup, e.g. Buliminacea and Carterinacea, use an available broader subgroup on level two, e.g. Rotaliina.

For papers on more than one major subgroup, e.g. Rotaliina and Miliolina, construct the set as follows:

fossil group (1)
  topic (2)
  (biocenoses, biochemistry, biogeography, biostratigraphy (see below), distribution, ecology, evolution (including phylogeny), faunal studies, fossilization (including taphonomy), habitat, miscellanea, morphology, paleoecology, occurrence, ontogeny, taxonomy.
    topic (3)
    age (List E), benthonic taxa, bibliography, biometry, catalogs, faunal list, floral list, paleoclimatology, phylogeny, planktonic taxa, statistical analysis, taphonomy, thanatocenoses, zoning, or type of material, e.g. bones, fossil wood, plankton, shells, skeletons, skulls, teeth, or tests.

e.g. Mamalia
  faunal studies
    Cenozoic

Reptilia
  taxonomy
    morphology

Ostracoda
  paleoecology
    paleoclimatology

Beginning in 1981, terms on the right margin, common names, are used on level one instead of their systematic level-one equivalents for biostratigraphy papers.

e.g. ferns
  biostratigraphy
    Cretaceous

If there is no common name term on the List, use the equivalent systematic term in the set and enter a proposed common name as a supplemental index term.

e.g. Arthropoda
  biostratigraphy
    Cambrian

arthropods (supplemental index term)

For paleoecology and biogeography sets having fossil terms on level two, use systematic names only. These sets are not to be used for stratigraphy papers.

e.g. paleoecology
  Radiolaria
    Paleogene

biogeography
  Pisces
    Mesozoic

For papers which emphasize both paleontology and biostratigraphy, two fossil sets may be used.

e.g. Pisces
  morphology
    taxonomy

AND

fish
  biostratigraphy
    Tertiary

ONLINE SEARCHING

For paleontology papers on a fossil group, search the systematic name. To exclude stratigraphy papers from the search you can add NOT Stratigraphy to the strategy. Narrower terms in the List currently autopost each higher level broader term except in the case of Hemichordata which is not autoposted. With the exception of Hemichordata, a search for a term on the left margin of of the list should be complete without addition of its narrower terms to the search. This is true for current citations and for citations entered prior to the beginning of autoposting in 1978, since the set structure required use of these broad terms whenever a narrower term was used.

For stratigraphy papers on a fossil group, search the systematic name AND biostratigraphy through 1980. And to retrieve stratigraphy citations added from 1981 on, search the common name if one exists or search the systematic name and biostratigraphy.

For studies on specific taxa the species and/or genus name may be used in searching, keeping in mind that general studies discussing more than 10 taxa will rarely be indexed using the specific terms. A more general search is advisable for such studies.

B.I.G. SEARCHING

Both the systematic and, since 1981, the common names of fossil groups are entry terms in the subject index. The sets with these terms on level one are subdivided on level two by either narrower fossil terms or topics (see under Indexing, above).

Related headings include paleontology (for general treatments, discussions of evolution, etc.), biogeography, paleoecology, paleoclimatology, paleobotany, (the discipline), palynology (subdiscipline or technique), micropaleontology (subdiscipline or technique).

Fossil Group		Common Name
Agnatha		
11, V		
Algae		algal flora
09,(M01),		
(M03),(M05)	Chlorophyta	
	Charophyta	
	Chlorophyceae	
	Codiaceae	
	Dasycladaceae	
	Receptaculitaceae	
	Desmidiales	desmids
	Tasmanites	
	Crysophyta	
M03	Coccolithophoraceae	coccoliths
	Cyanophyta	
M05	diatoms	diatom flora
M03	nannofossils (incl. discoasters, Nannoconus)	nannoconids
	Phaeophyta	
	Pyrrhophyta	
	Rhodophyta	
	Corallinaceae	
	Gymnocodiaceae	
	stromatolites	
Amphibia	Labyrinthodontia (incl. Anthracosauria, Icthyostegalia, Temnospondyli)	amphibians
(11), V		
	Lepospondyli (incl. Aistropoda, Microsauria, Nectridea)	
	Lissamphibia (incl. Anura, Apoda, Gymnophiona, Caecilia, Proanura, Urodela)	
angiosperms		
09	Dicotyledoneae	
	Asteridae	
	Caryophyllidae	
	Dilleniidae	
	Hamamelididae	
	Magnoliidae	
	Rosidae	
	Monocotyledoneae	
	Alismidae	
	Arecidae	
	Commelinidae	
	Liliidae	
Archaeocyatha		
10		
Arthropoda		
10	Chelicerata	
	Arachnida	
	Merostomata	
	Mandibulata	
	Crustacea	crustaceans
	Branchiopoda	
	Cirripedia	
	Copepoda	
	Malacostraca	
	Myriapoda	
	Trilobitomorpha	
Aves	Archaeornithes	birds
11,V	Neornithes	
bacteria		
09		
Brachiopoda		brachipods
10		
	Articulata	
	Orthida	
	Dictyonellidina	
	Pentamerida	
	Rhynchonellida	
	Spiriferida	
	Strophomenida	
	Terebratulida	
	Thecideidina	
	Inarticulata	
bryophytes		
09	Hepaticae	
	Musci	
Bryozoa		bryozoans
10		
	Cheilostomata	
	Cryptostomata	
	Ctenostomata	
	Cyclostomata	
	Trepostomata	
Chordata	(incl. Protochordata, Vertebrata)	
11, V		
Coelenterata		corals
10	Anthozoa	
	Actiniaria	
	Ceriantipatharia	
	Corallimorpharia	
	Heterocorallia	
	Hexactiniaria	
	Octocorallia	
	Rugosa	
	Scleractinia	
	Tabulata	
	Zoanthiniaria	
	Hydrozoa	
	Scyphozoa	
	Conularida	
	Stromatoporoidea	
Conodonta		conodonts
08,M04		
coprolites		
08,11,(V)		
Echinodermata		echinoderms
10	Asteroidea	
	Blastoidea	
	Camptostromatoidea	
	Crinoidea	
	Cyclocystoidea	
	Cystoidea	
	Echinoidea	
	Edrioasteroidea	
	Edrioblastoidea	
	Eocrinoidea	
	Helicoplacoidea	
	Holothuroidea	
	Homoiostelea	
	Homostelea	
	Lepidocystoidea	
	Machaeridia	
	Ophiocistioidea	
	Ophiuroidea	
	Parablastoidea	
	Paracrinoidea	
	Somasteroidea	
	Stylophora	
Foraminifera		foraminifers
10, M06	Allogromiina	
	Fusulinina	
	Fusulinidae	
	Miliolina	
	Alveolinellidae	
	Miliolacea	
	Rotaliina	
	Buliminacea	
	Carterinacea	
	Cassidulinacea	
	Discorbacea	
	Globigerinacea	
	Nodosariacea	
	Orbitoidacea	
	Orbitoididae	
	Robertinacea	
	Rotaliacea	
	Nummulitidae	
	Spirillinacea	
	Textulariina	
	Ammodiscacea	
	Lituolacea	
fossil man:	Use only when fossil remains are discussed; otherwise use archaeology or anthropology under stratigraphy (2nd level) under area terms.	
11, V		
fungi		
09	Myxomycetes	

gymnosperms		
09	Bennettitales	
	Caytoniales	
	Coniferales	
	Cordaitales	
	Cycadales	
	Cycadofilicales	
	Ephedrales	
	Ginkgoales	
	Glossopteridales	
	Gnetales	
	Nilssoniales	
	Pentoxylales	
	Welwitschiales	
Hemichordata	Enteropneusta	
10	Graptolithinia (1)	graptolites
	Camaroidea	
	Dendroidea	
	Graptoloidea	
	Didymograptina	
	Diplograptina	
	Glossograptina	
	Monograptina	
	Stolonoidea	
	Tuboidea	
	Pterobranchia (1)	
	Cephalodiscida	
	Rhapdopleurida	
ichnofossils		
08,09,10,11, (V)	(Use lebensspuren for sedimentary structures)	
Insecta	Blattopteroida (Blattodea)	
	Coleopteroida (Coleoptera)	
	Dermapteroida (Dermaptera)	
	Ectotropha	
	Entotropha	
	Ephemeropteroida (Ephemeroptera)	
	Hemipteroida (Hemiptera)	
	Hymenopteroida (Hymenoptera)	
	Lepidopteroida (Lepidoptera)	
	Mecopteroida (Mecoptera)	
	Neuropteroida (Neuroptera)	
	Odonatopteroida (Odonata)	
	Orthopteroida (Orthoptera)	
	Palaeodictyopteroida (Palaeodictyoptera)	
	Psocopteroida (Psocoptera)	
	Siphonapteroida (Siphonaptera)	
	Thysanopteroida (Thysanoptera)	
Invertebrata		invertebrates
10		
lichens		
09		
Mammalia		mammals
11, V	Acreodi (incl. Mesonychoidea)	
	Anagalida	
	Arctocyonia	
	Artiodactyla	
	Ruminantia	
	Suiformes	
	Tylopoda	
	Astrapotheria	
	Carnivora	
	Fissipeda	
	Pinnipedia	
	Cetacea	
	Chiroptera	
	Condylarthra	
	Creodonta	
	Dermoptera	
	Desmostylia	
	Dinocerata	
	Diprotodonta	
	Docodonta	
	Edentata	
	Embrithopoda	
	Hyracoidea	
	Insectivora	
	Lagomorpha	
	Litopterna	
	Macroscelida	
	Marsupialia	
	Monotremata	
	Multituberculata	
	Nomarthra	
	Notoungulata	
	Pantodonta	
	Paucituberculata	
	Pantotheria (incl. Eupantotheria)	
	Perissodactyla	
	Ceratomorpha	
	Hippomorpha	
	Pholidota	
	Polyprotodontia	
	Primates	
	Hominidae	
	Homo sapiens	
	Neanderthal	
	pre-Neanderthal	
	Pongidae (incl. Apes)	
	Prosimii (incl. Lemuroidea, Tarsioidea, Pleisiadapoidea, Anthropoidea)	
	Simians	
	Cynomorpha	
	Platyrrhina	
	Proboscidea	
	Barytherioidea	
	Deinotherioidea	
	Elephantoidea	
	Mastodontoidea	
	Moeritherioidea	
	Proteutheria	
	Pyrotheria	
	Rodentia	
	Hystricomorpha	rodents
	Myomorpha	
	Sciuromorpha	
	Scandentia	
	Sirenia	
	Symmetrodonta	
	Taeniodonta	
	Theria	
	Tillodontia	
	Triconodonta	
	Tubulidentata	
	Xenarthra	
	Xenungulata	
Mollusca		mollusks
10	Aplacophora	
	Bivalvia	bivalves
	Actinodontida	
	Astartida	
	Carditida	
	Ctenodontida	
	Cyrtodontida	
	Arcina	
	Ostreacea	
	Pteriina	
	Inocerami	
	Pectinacea	
	Pholadomyida	
	Praecardiida	
	Rudistae	rudists
	Septibranchia	
	Solemyida	
	Venerida	
	Cephalopoda	cephalopods
	Tetrabranchiata	
	Ammonoidea	ammonoids
	Anarcestida	
	Bactritida	
	Ceratitida	
	Clymeniida	
	Desmoceratida	
	Goniatitida	
	Lytoceratida	
	Perisphinctida	
	Phylloceratida	
	Prolecanitida	
	Psiloceratida	
	Nautiloidea	
	Dibranchiata	
	Belemnoidea	

- Gastropoda — gastropods
  - Acoela
  - Archaeogastropoda
  - Basommatophora
  - Bellerophontina
  - Entomotaeniata
  - Mesogastropoda
  - Neogastropoda
  - Pteropoda
  - Sacoglossa
  - Stylommatophora
- Hyolithes
- Monoplacophora
- Polyplacophora
- Rostroconchia
- Scaphopoda
- Tentaculites

Ostracoda — ostracods
10,M08
- Archeocopida
- Bairdiomorpha
- Beyrichicopina
- Cladocopina
- Cypridocopina
- Cytherocopina
- Entomozocopina
- Eridostraca
- Kirkbyocopina
- Leperditicopida
- Myodocopina
- Platycopida

palynomorphs — palynomorphs
09,M09
- acritarchs — acritarch flora
- Chitinozoa
- Dinoflagellata — dinoflagellates
- Megaspores
- miospores

Pisces — fish
11, V
- Chondrichthyes
  - Elasmobranchii
  - Eubradyodonti
  - Holocephali
- Osteichthyes
  - Actinistia
  - Brachiopterygii
  - Chondrostei
  - Dipnoi
  - Halecostomi
  - Holostei
  - Rhipidistia
  - Teleostei
- Placodermi

Plantae
09

Porifera
10
- Calcispongea
- Demospongea
- Hyalospongea

problematic fossils
08,10,11,V, M07
- problematic microfossils

Protista
08,09,10,M10
- ebridians
- Silicoflagellata
- Thecamoeba
- Tintinnidae

pteridophytes — ferns
09
- Filicopsida
- Lycopsida
- Noeggerathiales
- Psilopsida
- Pteridophyllen
- Sphenopsida

Radiolaria — radiolarians
10,M11
- Acantharina
- Nassellina
- Phaeodarina
- Spumellina

Reptilia — reptiles
11, V
- Anapsida
  - Captorhinomorpha
  - Chelonia
  - Cotylosauria
  - Pareiasauria
- Archosauria
  - Crocodilia
  - Ornithischia
  - Pterosauria
  - Saurischia
  - Thecodontia
- dinosaurs
- Euryapsida
  - Araeoscelidia
    - Trilophosauria
  - Sauropterygia
    - Nothosauria
    - Plesiosauria
- Ichthyopterygia
  - Ichthyosauria
- Lepidosauria
  - Eosuchia
  - Rhynchocephalia
  - Squamata
- Synapsida
  - Mesosauria
    - Proganosauria
  - Pelycosauria
  - Placodontia
  - Therapsida

Spermatophyta (incl. angiosperms and gymnosperms)
09

Tetrapoda (incl. Aves, Amphibia, Reptilia, Mammalia)
11, V

thallophytes (incl. bacteria, algae, fungi, lichens)
09

Trilobita — trilobites
10
- Agnostida
- Corynexochida
- Lichida
- Odontopleurida
- Phacopida
- Ptychopariida
- Redlichiida

Vertebrata — vertebrates
11, V

worms
10
- Annelida
- Chaetognatha
- Echiurida
- Myzostomia
- Nematoida
- Nematomorpha
- Nemerta
- Oligochaetia
- Phoronida
- Pogonophora
- Polychaetia
- Priapulida
- M02 Scolecodonts
- Sipunculoida

- -

pollen analysis (results)
palynology (technique)

LIST G - GENERAL TERMS

INDEXING

The following general terms are to be used sparingly. They are permitted in certain sets on levels two and three.

ONLINE SEARCHING

Search for these terms in combination with specific topics, e.g. gravity surveys AND anomalies.

B.I.G. SEARCHING

Look for these terms on the second level under topics of interest.

---

age
annual report
anomalies
applications
causes
changes
composition
concepts
detection
distribution
effects
environment
evolution
experimental studies
general
genesis
history
identification
interpretation
mechanism
methods
objectives
observations
occurrence
origin, use genesis
patterns
philosophy
practice
preparation
principles
processes
properties
rates
regional
review
surveys (different from survey organizations)
techniques
theoretical studies
varieties

LIST H - IGNEOUS ROCKS

INDEXING

The following list indicates the second-level terms permitted in sets under the term igneous rocks. These terms are also permitted on the second level under phase equilibria.

The most specific term is to be used on the second level. Should the paper under consideration deal with several rocks belonging to different groups, a general topic should be selected (see igneous rocks in the Thesaurus).

Singular forms of terms such as granite may be used as index terms but not in igneous rocks sets.

e.g.
```
igneous rocks                 igneous rocks
  granites          not         granites
    textures                      granite
```

Should there be no need for an igneous rocks set (i.e. if these rocks are present but not emphasized in the paper) use specific rock names in the singular form,

e.g. alaskite, latite, microdiorite, etc.

ONLINE SEARCHING

Search for specific rocks by name, truncated at the end to retrieve singular and plural forms.

To retrieve all papers in which igneous rocks were a major topic, search igneous rocks. To add papers on phase equilibria, also search all the major groups such as volcanic rocks, truncated. Since only specific rock names may have been used in papers in which the rocks are minor topics, it is necessary to also search specific rock names for an exhaustive search for igneous rocks.

Search petrology as a level-one term for papers dealing with the discipline, such as papers on techniques, practice, etc.

Related topics such as intrusions, isotopes, metamorphic rocks, metasomatic rocks, magmas, metamorphism, metasomatism, phase equilibria, etc. can be searched separately.

B.I.G. SEARCHING

At the entry for igneous rocks, look under the appropriate second-level term(s) selected from the list below.

Look under related headings such as intrusions, isotopes, magmas, metamorphic rocks, metamorphism, metasomatic rocks, metasomatism, and phase equilibria.

Look under petrology for papers dealing with the discipline, such as papers on techniques, practice of, etc.

```
igneous rocks
  plutonic rocks (2)
  hypabyssal rocks (2)

  granites (2)
    microgranite
    alkali granites (2)
    granodiorites (2)
    charnockite

      aplite
  diorites (2)
    microdiorite
    alkali diorites (2)
    quartz diorites (2)
  syenites (2)
    microsyenite
    alkali syenites (2)
    monzonites (2)
  gabbros (2)
    microgabbro
    alkali gabbros (2)
      nephelinite
      leucitite
    anorthosite
    norite
  ultramafics (2)
    pyroxenite
    peridotites (2)
    kimberlite
  lamprophyres (2)
  carbonatites (2)
  diabase (2)

  volcanic rocks (2)
    rhyolites (2)
    rhyodacites (2)
    dacites (2)
    andesites (2)
    trachyandesites (2)
    trachytes (2)
      keratophyre
    phonolites (2)
    basalts (2)
      alkali basalts (2)
        trachybasalts (2)
    pyroclastics (2)
      ignimbrite
      perlite
      pumice
      tuff
      tuffite
    glasses (2)
      obsidian
```

LIST I - SEDIMENTARY ROCKS

INDEXING

The term sedimentary rocks is used as a first-level term in a set. Terms from the following list are used on the second and third levels.

e.g. sedimentary rocks
  clastic rocks
    sandstone

The most specific second-level term is used except when the paper discusses several rock groups, or general topics such as textures, in which case the second-level term should be a topic:

e.g. sedimentary rocks
  textures
    grain size

Other related sets include sediments, sedimentation, and sedimentary structures.

If the rocks are not emphasized, a set for sedimentary rocks is not needed, but the rock terms or rock group should be used as index terms.

ONLINE SEARCHING

To search for a specific rock, enter the rock name truncated at the end, e.g. limestone:

To be thorough, include variants, thus:
limestone: OR micrite: OR calcarenite: etc.....

For all sedimentary rocks search:
sedimentary rock: OR carbonate rock: OR ....
(list all groups)

Sedimentary petrology can be searched for papers dealing with the discipline, such as papers on techniques, methods, etc.

Related topics such as sedimentation, sedimentary structures, sediments and diagenesis can be searched separately.

B.I.G. SEARCHING

At the entry for sedimentary rocks, look under the appropriate second-level term(s) selected from the list below.

Look under related headings such as sedimentation, sedimentary structures, sediments, diagenesis, and weathering.

For papers dealing with the discipline, such as papers on techniques, practice of, etc., look under sedimentary petrology.

Look under specific sedimentary rocks which are commodities, e.g. limestone deposits (see List C, Commodities).

- sedimentary rocks
  - carbonate rocks (2)
    - algal limestone
    - beachrock
    - biocalcarenite
    - calcarenite
    - calcrete
    - chalk
    - coquina
    - dolomitic limestone
    - dolostone
    - limestone
    - micrite
    - oolitic limestone
    - travertine
  - chemically precipitated rocks (2)
    - chert
      - jasperoid
    - evaporites
      - anhydrite
      - gypsum
      - potash
      - salt
  - chemically precipitated rocks (cont.)
    - ferricrete
    - flint
    - iron formations
    - phosphate rocks
      - bone beds
    - silcrete
    - siliceous sinter
  - clastic rocks (2)
    - argillite
    - arkose
    - bauxite
    - bentonite
    - breccia
      - tectonic breccia
      - volcanic breccia
    - conglomerate
    - diatomaceous earth
    - flysch
    - graywacke
    - marl
    - molasse
    - radiolarite
    - red beds
    - sandstone
    - shale
    - siltstone
    - spongolite
    - tillite
    - tonstein
    - turbidite
  - organic residues (2)
    - coal
      - anthracite
      - bituminous coal
      - coke coal
      - lignite
      - sapropelite
      - steam coal
    - macerals
      - durain
      - exinite
      - inertinite
      - vitrain
      - vitrinite
    - oil sands
    - oil shale

LIST J - METAMORPHIC ROCKS

INDEXING

Metamorphic rocks is a first-level term with terms from the following list used on the second level. These terms can also be used on the second level under the first-level term phase equilibria.

Make a metamorphic rocks set only when the rocks are emphasized. Otherwise, add metamorphic rocks or a second-level term from the list below as a supplemental index term, without constructing a set.

For documents on petrology as a discipline, make a petrology set.

Make a metamorphism set for papers which stress the process.

A specific rock group should be used on the second level unless the paper discusses several groups and/or general topics such as mineral assemblages. For such papers, the second level should be a topic:

e.g. metamorphic rocks
  mineral assemblages
    hornblende schist

No singular form of the group names is allowed as an index term. Use amphibolites not amphibolite, and gneisses not gneiss.

e.g. metamorphic rocks
  gneisses
    chemical composition

ONLINE SEARCHING

Papers with major emphasis on these rocks can be retrieved by searching for metamorphic rocks as a level-one term.

For a thorough search of metamorphic rocks, use metamorphic rock: OR gneiss: OR schist: etc.... (include all major groups).

For specific searches on a certain rock, use the rock name, truncated.

For composite rock names, e.g. biotite gneiss, it is advisable to search:

biotite gneiss: OR (gneiss AND biotite).

B.I.G. SEARCHING

At the entry for metamorphic rocks, look under the appropriate second-level term(s) selected from the list below.

Look under related headings such as metamorphism, petrology, metasomatism, phase equilibria, and paragenesis.

For papers dealing with the study of metamorphic rocks, look under petrology.

- metamorphic rocks
  - metasedimentary rocks
  - metavolcanic rocks
  - metaplutonic rocks
  - metaigneous rocks
  - hornfels
  - slates
  - phyllites
  - schists
    - greenschist
    - blueschist
  - gneisses
    - orthogneiss
    - paragneiss
  - amphibolites
    - orthoamphibolite
    - para-amphibolite
  - granulites
  - marbles
  - mylonites
  - cataclasites
  - phyllonites
  - migmatites
    - embrechite
    - anatexite
  - quartzites
  - garnetite
  - eclogite
  - itabirite

LIST K - SEDIMENTARY STRUCTURES

INDEXING

Use specific terms from the following list. If a set needs to be constructed, use the general structure type on level two and the specific name on level three.

ONLINE SEARCHING

Search for specific or general structure type as need be.

B.I.G. SEARCHING

At the entry for sedimentary structures, look under the appropriate level-two and level-three term(s) selected from the following list.

Also under sedimentary structures, look for topics on level two, such as environmental analysis.

- sedimentary structures
  - bedding plane irregularities
    - antidunes
    - chevron marks
    - (current lineations use parting lineation)
    - current markings
    - (current partings use parting lineation)
    - flute casts
    - frost features
    - grooves
    - groove casts
    - ice wedges
    - load casts (also under soft sediment deformation and turbidity current structures)
    - megaripples
    - mounds
    - mudcracks
    - mud lumps
    - parting lineation
    - ripple marks (also under primary structures) (asymmetrical, interference, etc.)
    - sand waves
    - scour casts
    - scour marks
    - sole marks (also under turbidity current structures)
    - striations
    - tool marks
  - biogenic structures
    - algal banks
    - algal biscuits
    - algal mats
    - algal mounds
    - algal structures
    - banks
    - bioturbation
    - burrows
    - coprolites
    - girvanella
    - lebensspuren
    - oncolites
    - stromatolites
    - tracks
    - trails
  - cylindrical structures
  - planar bedding structures (i.e. bedding structures)
    - bars
    - bedding (also under primary structures)
    - channels
    - cross-bedding
    - cross-laminations
    - cross-stratification
    - cut and fill
    - cyclothems
    - flaser bedding
    - graded bedding (also under turbidity current structures)
    - imbrication
    - laminations
    - massive bedding
    - megacyclothems
    - rhythmic bedding
    - ripple drift-cross laminations
    - (ripple-cross-laminations use ripple drift-cross laminations)
    - sand bodies
    - stratification
    - varves
  - primary structures
    - bedding (also under planar bedding structures)
    - ripple marks (also under bedding plane irregularities) (asymmetrical, interference, etc.)
  - secondary structures
    - armored mud balls
    - concretions
    - cone-in-cone
    - geodes
    - microstylolites
    - septaria
    - stylolites

sedimentary structures (cont.)
  soft sediment deformation
    ball-and-pillows
    boudinage
    clastic dikes
    convoluted beds (also under turbidity current structures)
    (convoluted deformation use convoluted beds)
    flame structures
    flow structures (also under turbidity current structures)
    load casts (also under bedding plane irregularities and turbidity current structures)
    olistoliths (also under turbidity current structures)
    olistostromes (also under turbidity current structures)
    (pull apart structures use boudinage)
    sandstone dikes
    slump structures
  turbidity current structures
    convoluted beds (also under soft sediment deformation)
    (convoluted deformation use convoluted beds)
    flow structures (also under soft sediment deformation)
    graded bedding (also under planar bedding structures)
    load casts (also under bedding plane irregularities and soft sediment deformation)
    olistoliths (also under soft sediment deformation)
    olistostromes (also under soft sediment deformation)
    sole marks (also under bedding plane irregularities)

LIST L - MINERALS

INDEXING

The following group names (chemical for the non-silicates and structural for the silicates) are used on the second level under minerals, crystal chemistry, crystal structure, phase equilibria, and crystal growth. On the third level a specific mineral name (if the paper deals with one mineral primarily) or a topic such as composition or crystal chemistry (if the paper deals with more than one mineral) is used.

A mineral set is required whenever the paper deals with mineralogic aspects of a rock or with a specific mineral in a substantial manner. Papers discussing petrology with simple mention of mineral composition should not receive a mineral set. Rather, the mineral names involved in that case are used simply as index terms. Thus, a paper about the crystal structure of quartz will definitely have the following two sets:

minerals
  framework silicates, silica minerals
    quartz

crystal structure
  framework silicates, silica minerals
    quartz

But a paper discussing quartz distribution in a sandstone will simply have quartz as an index term.

In cases where the particular mineral species may be referred to more than one group (e.g. kainite belonging to both halides and sulfates), two minerals sets are made.

Papers dealing with the disciplines of mineralogy and crystallography are indexed using those terms on the 1st level.

e.g.	

mineralogy
  practice
    mineralogists
	crystallography
	  research
	    current research

For papers dealing with new mineral species, the term new minerals is used as an index term, and the chemical formula is given as an index term if available.

ONLINE SEARCHING

To search for a particular mineral species, the common name of that species must be used in keeping in mind possible varieties (agate, chalcedony... or titanite and sphene...) and that the American f is used for the British ph. To be thorough, a search should also be made of the chemical formula (or parts thereof) of the mineral species.

Some mineral species may not be treated from a mineralogical point of view. It is advisable therefore to search for "mineralogic" papers using the mineral species together with the mineral group and * minerals, and, for very specific searches, to use the mineralogy category as well.

For papers dealing with gemmological aspects, one should search for a mineral in combination with (gems OR collecting OR popular geology).

For papers dealing with the economic geology of minerals, see List C.

For papers dealing with a minerals group search for * minerals (i.e. minerals as a level-one term) and the name of the group.

Clay mineralogy as a discipline is treated separately. A search of the term clay mineralogy will retrieve papers dealing with all aspects of clays: mineralogy, structure, heat treatment, occurrence as constituents of rocks, etc.

B.I.G. SEARCHING

The headings minerals, mineralogy, crystal structure, crystal chemistry, crystal growth, phase equilibria, clay mineralogy, and crystallography are the main access points for mineralogical papers. Other related headings are igneous rocks, metamorphic rocks, sedimentary rocks, metasomatic rocks, sediments, geochemistry, and paragenesis.

For major mineral groups look on the second level under the headings in 1 above. Also cross-references are provided to direct your attention to these groups.

e.g.	

sulphur
  see sulfur
	sulfosalts
	  see under minerals

For the economic geology of minerals, look under the appropriate minerals selected from List C, Commodities.

---

minerals
  alloys (including carbides, nitrides, phosphides, silicides)
  antimonates
  antimonides
  antimonites
  arsenates
  arsenides
  arsenites
  bismuthides
  borates
  bromides
  carbides
  carbonates
  chlorides
  chromates
  fluoborates
  fluorides
  fluosilicates
  halides (includes bromides, chlorides, fluoborates, fluosilicates, iodides, and fluorides)
  iodates
  iodides
  miscellaneous minerals (used for several groups or for minerals of unknown affinity)
  molybdates
  native elements
  niobates
  niobotantalates
  nitrates

minerals (cont.)
- nitrides
- organic compounds
- oxalates
- oxides (including niobates, niobotantalates, tantalates)
- oxysulfides
- phosphates
- phosphides
- selenates
- selenides
- selenites
- silicates (use a narrower term below if dealing with specific mineral; otherwise larger group)
  - aluminosilicates
  - orthosilicates
    - sorosilicates
      - orthosilicates, epidote group
      - orthosilicates, melilite group
    - nesosilicates
      - orthosilicates, garnet group
      - orthosilicates, humite group
      - orthosilicates, olivine group
  - ring silicates
  - chain silicates (inosilicates)
    - chain silicates, amphibole group
      - chain silicates, alkalic amphibole
      - chain silicates, clinoamphibole
      - chain silicates, orthoamphibole
    - chain silicates, pyroxene group
      - chain silicates, alkalic pyroxene
      - chain silicates, clinopyroxene
      - chain silicates, orthopyroxene
  - sheet silicates (phyllosilicates)
    - sheet silicates, chlorite group
    - sheet silicates, clay minerals
    - sheet silicates, mica group
    - sheet silicates, serpentine group
  - framework silicates (tectosilicates)
    - framework silicates, feldspar group
      - framework silicates, alkali feldspar
      - framework silicates, plagioclase
      - framework silicates, barium feldspar
    - framework silicates, nepheline group
    - framework silicates, scapolite group
    - framework silicates, silica minerals
    - framework silicates, sodalite group
    - framework silicates, zeolite group
- silicides
- sulfates
- sulfosalts (including sulfantimonates, sulfantimonites, sulfarsenates, sulfarsenites, sulfobismuthites, sulfogermanates, sulfostannates, and sulfovanadates)
- sulfides (including antimonides, arsenides, bismuthides, oxysulfides, selenides, and tellurides)
- tantalates
- tellurates
- tellurides
- tellurites
- tungstates
- vanadates

NOTES

1. Papers on artificial minerals are usually not treated except when a mineralogic application is clearly indicated. The indexing term <u>synthetic</u> <u>materials</u> is then added.
2. Papers about mineral collecting will have the term <u>collecting</u> in the indexing.

LIST M - SOILS

INDEXING

The classification of soils has undergone several changes in American practice. What follows are two of the classifications using different criteria. The 1965 classification has been retained fairly intact since that time. In addition there follows a comparative table giving equivalent French and German terms for some of the basic soil terms in English. We do not attempt to standardize the soils. Use the soil terms given in the source documents, without attempting to interpret types of soil according to one of the available classifications.

The terms in the list below are used on level three under soils (1) and soil groups (2). In addition, terms such as <u>loam</u>, <u>laterites</u>, <u>paleosols</u>, and <u>volcanic</u> <u>ash</u> may be used on level three.

ONLINE SEARCHING

To search for a soil group, one must search for all possible varieties, including different names in different classifications and in different languages. For example, to search for <u>Latosols</u>, you must also search for <u>Sol-ferralitique</u>.

B.I.G. SEARCHING

All the soil entries may be found under the term <u>soils</u>. On the second level the term <u>surveys</u> followed by an area will occur, if the article discusses aspects of soils in a specific region. Or <u>soil</u> <u>groups</u> will occur on the second level with one of the terms from the list below on the third level. Otherwise, a topic such as <u>geochemistry</u> or <u>water</u> <u>regimes</u> will be found on the second level.

AMERICAN SOIL CLASSIFICATION SYSTEM - 1964

Zonal soils
- Tundra soils
- Subarctic brown forest soils
- Desert soils
- Red desert soils
- Sierozems
- Brown soils
- Reddish brown soils
- Chestnut soils
- Reddish chestnut soils
- Chernozems
- Brunizems
- Reddish prairie soils
- Noncalcic brown soils
- Podzols
- Brown podzolic soils
- Gray wooded soils
- Sols bruns acides
- Gray-brown podzolic soils
- Red-yellow podzolic soils
- Reddish-brown lateritic soils
- Yellowish-brown lateritic soils
- Latosols

Intrazonal soils
- Solonchak soils (saline soils)
- Solonetz soils
- Soloth soils
- Humic-gley soils
- Alpine meadow soils
- Bog soils
- Low-humic gley soils
- Planosols
- Ground-water podzols
- Ground-water laterite soils
- Brown forest soils
- Rendzinas
- Grumusols
- Calcisols

Azonal soils
- Lithosols
- Regosols
- Alluvial soils

AMERICAN SOIL CLASSIFICATION SYSTEM - 1965 et seq.

Alfisols
- Udalfs
- Boralfs
- Ustalfs
- Xeralfs

Aridisols
- Argids
- Orthids

Entisols
- Aquents
- Fluvents
- Orthents
- Psamments

Histosols
- Fibrists
- Folists
- Hemists
- Saprists

Inceptisols
- Andepts
- Aquepts
- Ochrepts
- Tropepts
- Umbrepts

Mollisols
- Albolls
- Aquolls
- Borolls
- Redolls
- Udolls
- Ustolls
- Xerolls

Oxisols
- Aquox
- Humox
- Orthox
- Torrox
- Ustox

Spodosols
- Aquods
- Humods
- Orthods

Ultisols
- Aquults
- Humults
- Udults
- Ustults
- Xerults

Vertisols
- Torrerts
- Uderts
- Usterts
- Xererts

AMERICAN	GERMAN	FRENCH
Acrisols	Acrisol	Sol-mediterraneen
Albolls		
Alfisols		
Alluvial soils	Auen-Boden	Sol-d'alluvions
Alpine meadow soils	Alpiner Wiesenboden	Sol-hydromorphe
Andepts		
Andosols	Andosol	Sol-peu-evolue roche-volcanique
Aqualfs		
Aquents		
Aquepts		
Aquods		
Aquolls		
Aquox		
Aquults		
Arctic tundra soils	Arktische Tundra Boden	Sol-de-toundra
Arenosols	Arenosol	Sol-brut sable
Arents		
Argids		
Aridisols		
Azonal soils	Roh-Boden	Sol-brut
Black earth use Chernozems	Schwarzerde	Chernozem
Bog soils	Moorboden	Tourbe
Boralfs		
Boreal frozen taiga soils		Sol-gele
Boreal taiga and forest soils		Sol
Borolls		Mollisol
Brown desert steppe soils	Burozem	Sierozem
Brown forest soils	Brauner Wald Boden	Sol-brun
Brown loam	Braunlehm	Couche-rouge
Brown podzolic soils	Podsoliger Braunerde	Podzol
Brown soils	Braunerde	Sol-brun
Brunizems	Brunizem	Brunizem
Calcisols		
Cambisols	Cambisol	Sol-brun
Caar	Uebergangsmoor	Tourbe
Chernozems	Chernozem	Chernozem
Chestnut soils	Kastannozem	Sol-chatain
Cryosols	Cryosol	Sol-gele
Desert soils	Wuesten-Boden	Sol-subdesetique
Desert raw soils	Wuesten-roh-Boden	Sol-de-desert
Dy	Dy	Vase
Entisols		
Fen		
Ferralites	Eisen-Silikat-Boden	Sol-fersialitique
Ferralsols	Ferralsol	Sol-ferralitique
Ferrods		
Ferruginous soils	Eisenhaltiger-Boden	Sol-ferrugineux
Fibrists		
Fluvents		
Fluviosols		Sol-d'alluvions
Folists		
Gleys	Gley	Gley
Glensols	Glensol	Glensol
Gray-brown podzolic soils	Fahlerde	Sol-brun lessivage
Gray podzolic soils	Podsolierter Grauer Boden	Podzol
Gray warp soils	Paternia	Sol-peu-evolue or sol-d'alluvions
Gray wooded soils		
Ground-water podzols	Gley-Podsol	Podzol
Ground-water laterite soils	Grundwasser-Laterite	Laterite
Grumosols	Grumosol	Vertisol
Half bog soils	Anmoor	Tourbe
Halomorphic soils	Salz-Boden	Sol-halomorphe
Halosols	Halosol	Sol-halomorphe
Hemists		
High moor	Hochmoor	Tourbe
Histosols		
Humic gley soils	Humus Gley Boden	Sol-hydromorphe
Humic soils	Humus-reiche-Boden	Sol-riche-en-humus
Humods		
Humox		
Hydromorphic soils	Hydromorpher-Boden	Sol-riche-en-humus
Inceptisols		
Intrazonal soils	Intrazonaler Boden	Sol
Kastanozems		
Krasnozems	Krasnozem	Krasnozem
laterites	Laterit-Boden	Sol-ferralitique
Latosols	Latosol	Sol-ferralitique
Lithosols	Gesteins-roh-Boden	Sol-squelettique
Low-humic gley soils		
Luvisols	Luvisols	Sol lessivage
Mediterranean soils	Mediterraner Boden	Sol-mediterraneen
Mollisols		
Mor	Auflagehumus	Humus
Mull	Mull	Humus
Mull soils	Mull-Boden	Sol-a-mull
Muck	Niedermoor	Tourbe
Nitosols	Nitosol	
Noncalcic brown soils		
Ochrepts		
Orthents		
Orthids		
Orthods		
Orthox		
Oxisols		
Parachernozems	Smonitza	Chernozem
Paramosols	Paramosol	Sol-chatain
Pararendzina	Borowina	Rendzine
Parasierozems	Parasierozem	Sierozem
Peat	Torf	Tourbe
Pelosols	Pelosol	Sol-hydromorphe
Pergelisols	Pergelisol	Sol-gele
Plaggenesch	Plaggenesch	Sol-riche-en-humus action-homme
Plaggepts		
Planosols	Planosol	Couche-rouge
Podzols	Podsol	Podzol
Podzoluvisols	Podzoluvisol	Sol lessivage
Poorly developed soils	Unreife-boden	Sol-peu-developpe
Prairie soils	Brunizem	Brunizem
Protopedons	Protopedon	Sol-brut or sol-d'alluvions
Psamments		
Pseudogleys	Pseudogley	Gley
Rambla	Rambla	Sol-brut or sol-d'alluvions

AMERICAN	GERMAN	FRENCH
Ranker	Ranker	Ranker
Red desert soils	Roter Wuestenboden	Sol-subdesertique
Reddish brown soils	Roetlich-brauner Halbwueste Boden	Sol-subdesertique
Reddish-brown lateritic soils		
Reddish chestnut soils	Roetlich-Kastanien-farbiger Boden	Sol-chatain
Reddish prairie soils	Roetlicher Prairie Boden	Brunizem
Red-yellow podzolic soils	Gelbig roterpod-soliger Boden	Sol-ferralitique
Reg	Reg	Sol-de-desert
Regosols	Regosol	Sol-peu-evolue
Regur	Regur	Vertisol
Rendolls		
Rendzinas	Rendzina	Rendzine
Rhegosols	Rhegosol	Sol-peu-developpe
Rigosols	Rigosol	Sol-brut
Rutmark	Raamark	Sol-brut
Saprists		
Sapropels	Sapropel	Sierozem
Sierozems	Sierozem	Sierozem
Solonchak soils	Solonchak	Sol-halomorphe
Solonetz soils	Solonetz	Sol-halomorphe
Soloth soils	Solod	Sol-halomorphe
Spodosols		
Stagnogleys	Stagnogley	Gley
Subarctic brown forest soils		
Subboreal desert soils		
Subboreal humid soils		
Subboreal steppe soils		
Subtropical desert soils		
Subtropical dry soils		
Subtropical humid soils		
Syrogleys	Syrogley	Gley
Syrozems	Syrozem	Sol-brut
Takyr	Takyr	Sol-de-desert
Terrae calcis	Terrae calcis	Couche-rouge
Terra rossa	Terra rossa	Couche-rouge
Terra fusca	Terra fusca	Couche-rouge
Tir	Tir	Vertisols
Torrox		
Torrerts		
Tropepts		
Tropical desert soils		
Tropical dry soils		
Tropical humid soils		
Tundra soils	Tundra-Boden	Sol-de-toundra
Udalfs		
Uderts		
Udolls		
Udults		
Ultisols		
Umbrepts		
Ustalfs		
Ustserts		
Ustolls		
Ustsults		
Vega	Vega	Sol-brun or sol d'alluvions
Vertisols		
Wet meadow soils	Wiesenboden	Sol-hydromorphe
Xeralfs		
Xerests		
Xerolls		
Xerosols	Xerosol	Sierozem
Xerults		
Yeltozems	Yeltozem	Sol-mediterraneen
Yermosols	Ermosol	Sol-de-desert
Zonal soils	Zonaler Boden	Sol

LIST N - SEDIMENTS

INDEXING

Under this general heading, entries are made for unconsolidated (Recent or Quaternary) materials. The second-level entry is clastic sediments, carbonate sediments, marine sediments, organic residues, or an appropriate topic such as textures, composition, or distribution.

Examples of level-three terms under clastic sediments are given in the list below. Other specific terms may be included on the third level.

ONLINE SEARCHING

To search for unconsolidated sediments, use the specific term such as sand or the general term sediments.

B.I.G. SEARCHING

Look under the appropriate second-level terms under the heading sediments.

Other related headings include: sedimentary rocks, sedimentary structures, sedimentation, soils, weathering, and the individual commodities such as kaolin deposits and clays.

- sediments
  - clastic sediments
    - alluvium
    - boulder clay
    - boulders
    - clay
    - colluvium
    - eluvium
    - erratics
    - flint clay
    - gravel
    - kaolin
    - loess
    - mud
    - ooze
    - pebbles
    - proluvium
    - residual clays
    - residuum
    - sand
    - silt
    - till
  - carbonate sediments
  - marine sediments
  - organic residues

LIST O - GEOGRAPHIC TERMS

This List includes terms for large regions, continents, countries, large subdivision of some countries, and large bodies of water. In addition to the terms on this List there are thousands of terms for smaller geographic areas. The more common of these terms have been added to the Thesaurus, based on frequency of use in GeoRef, and can be found listed under their broader terms. Other specific geographic terms infrequently used and not in the Thesaurus, are also in GeoRef. When one of these specific locations is indexed, a broader term from List O is also indexed.

Geographic locations are key factors in most geology papers. Since 1978, in addition to the relevant geographic terms, the coordinates of primary areas of papers have been added to GeoRef citations. These coordinates are in the form of the latitudes and longitudes defining the perimeters of the areas. This makes it possible to search for particular areas by both geographic terms and coordinates.

INDEXING

The terms in List O are the only geographic terms permitted on level one in area sets. No geographic term is used on level two. On level three, any geographic term may be used in an area set. The smallest applicable region is to be used in this case on level three.

e.g. Kansas
  areal geology
    Wichita County

In addition, on level three the following apply:

hydrology surveys name of river or lake or area	paleogeography age area
ground water surveys name of aquifer or area	clay mineralogy areal studies area
soils surveys area	geodesy surveys area

In the above six examples of area terms on level three, the terms permitted are those on List O except the ones followed by (3). The terms followed by (3) are permitted on level three in area sets and as supplementary terms.

ONLINE SEARCHING

Geographic searching is best achieved by using all variants of an area name including political, physiographic and geologic. It is possible to be as specific as town or county, and as general as country or continent. For oceans and continental margins, it is advisable to use the coordinates as well.

In searching online for a topic related to a particular geographic region on country or continent scale, search for that region using List O for reference, and combine it with the proper subject from List B to give a general subject orientation. Thus for gold ores in Argentina, search for Argentina AND economic geology AND all terms starting with gold. In a search for a List O term it is not necessary to also search its narrower terms.

B.I.G. SEARCHING

All terms on List O, except those designated (3), are headings in the Subject Index of the Bibliography and Index of Geology. It is our practice to make an area set (see List B, Area Sets) for each paper cited in GeoRef, excepting of course the relatively few geology papers which do not relate to geographic locations.

Note that the most specific term from List O which is applicable for a paper is used as the heading. Consequently a search for a broad region should include all terms in List O not designated (3), on that region. In each issue of the Bibliography, 'see also' references from broad terms to narrower terms used in that issue are included in the Subject Index, such as the following from the May 1981 issue:

Africa *see also* Algeria; Benin; Botswana; Cape Verde Islands; Djibouti; Egypt; Ethiopia;... Zambia

In the List which follows, the geographic terms are used in index sets on levels one and three, except those followed by (3) which are only permitted on level three.

Global (3)
- Western World (3)
- \- - - - - - - - -
- Eastern Hemisphere
- Northern Hemisphere
- Southern Hemisphere
- Western Hemisphere
- \- - - - - - - - - -
  - Africa
    - Sahara
    - East African Lakes (3) (Lake Albert, Lake Baringo, Lake Edward, Lake Kariba, Lake Kivu, Lake Magadi, Lake Malawi, Lake Natron, Lake Tanganyika, Lake Turkana, Lake Victoria)(3)
    - \- - - - - - - - - - - - -
    - Central Africa (3)
      - Angola
      - Burundi
      - Central African Republic
      - Congo
      - Equatorial Guinea
      - Gabon
      - Rwanda
      - Zaire
        - Bandundu (3)
        - Equatorial Zaire (3)
        - Kasai (3)
        - Kivu (3)
        - Lower Zaire (3)
        - Shaba (3)
        - Upper Zaire (3)
    - East Africa (3)
      - Djibouti
      - Ethiopia
      - Kenya
      - Malawi
      - Mozambique
      - Uganda
      - Somali Republic
      - Sudan
      - Tanzania
      - Zambia
    - North Africa (3)
      - Algeria
        - Ahaggar (3)
        - Algiers region (3)
        - Constantine region (3)
        - Oasis region (3)
        - Oran region (3)
        - Saoura region (3)
      - Egypt
        - Eastern Desert (3)
        - Middle Nile Valley (3)
        - Nile Delta (3)
        - Sinai (3)
        - Western Desert (3)
      - Libya
      - Morocco
        - Anti-Atlas (3)
        - Ceuta (3)
        - Melilla (3)
        - Moroccan Atlas Mountains (3)
        - Rif (3)
        - Western Sahara (1)
      - Tunisia
    - Southern Africa (3)
      - Botswana
      - Lesotho
      - Namibia
      - South Africa
        - Cape Province (3)
        - Natal (3)
        - Orange Free State (3)
        - Transvaal (3)

- (Southern Africa Contd.)
  - Swaziland
  - Zimbabwe
- West Africa (3)
  - Benin
  - Cameroon
  - Chad
  - Gambia
  - Ghana
  - Guinea
  - Guinea-Bissau
  - Ivory Coast
  - Liberia
  - Mali
  - Mauritania
  - Niger
  - Nigeria
    - East-Central Nigeria (3)
    - Northern Nigeria (3)
    - Southeastern Nigeria (3)
    - West-Central Nigeria (3)
    - Western Nigeria (3)
  - Senegal
  - Sierra Leone
  - Togo
  - Upper Volta

America
- Caribbean region
- Latin America (3)
- - - - - - - - - - - -
- Central America
  - Belize
  - Costa Rica
  - El Salvador
  - Guatemala
  - Honduras
  - Nicaragua
  - Panama
- North America
  - Appalachians
  - Atlantic Coastal Plain
  - Basin and Range Province
  - Canadian Shield
  - Colorado Plateau
  - Columbia Plateau
  - Great Basin
  - Great Lakes
  - Great Lakes region
  - Great Plains
  - Gulf Coastal Plain
  - Mississippi Valley
  - Rocky Mountains
  - Western Interior
  - - - - - - - - - - - -
  - Canada
    - Eastern Canada (3)
      - Maritime Provinces
        - New Brunswick
        - Nova Scotia
        - Prince Edward Island
      - Labrador
      - Newfoundland
      - Ontario
      - Quebec
    - Northwest Territories
      - District of Franklin (3)
        - Arctic Archipelago (3)
      - District of Keewatin (3)
      - District of Mackenzie (3)
    - Western Canada (3)
      - Alberta
      - British Columbia
      - Manitoba
      - Saskatchewan
      - Yukon Territory
  - Mexico
  - Saint Pierre and Miquelon
  - United States
    - Eastern United States
      - New England
        - Connecticut
        - Maine
        - Massachusetts
      - (New England Contd.)
        - New Hampshire
        - Rhode Island
        - Vermont
      - Delaware
      - District of Columbia
      - Florida
      - Georgia
      - Maryland
      - New Jersey
      - New York
      - North Carolina
      - Pennsylvania
      - South Carolina
      - Virginia
      - West Virginia
      - Midwest
        - Illinois
        - Indiana
        - Iowa
        - Kansas
        - Michigan
        - Minnesota
        - Missouri
        - Nebraska
        - North Dakota
        - Ohio
        - South Dakota
        - Wisconsin
      - Southern United States
        - Alabama
        - Arkansas
        - Kentucky
        - Louisiana
        - Mississippi
        - Tennessee
      - Southwestern United States
        - Arizona
        - New Mexico
        - Oklahoma
        - Texas
      - Western United States
        - Pacific Coast
          - California
          - Oregon
          - Washington
        - Alaska
        - Colorado
        - Hawaii
        - Idaho
        - Montana
        - Nevada
        - Utah
        - Wyoming
- South America
  - Andes
  - - - - - - - - - - - - - - - -
  - Argentina
    - Argentine Andes (3)
    - Pampas (3)
    - Patagonia (3)
  - Bolivia
  - Brazil
    - Acre (3)
    - Alagoas (3)
    - Amapa (3)
    - Amazonas (3)
    - Bahia (3)
    - Ceara (3)
    - Espirito Santo (3)
    - Goias (3)
    - Maranhao (3)
    - Mato Grosso (3)
    - Minas Gerais (3)
    - Para (3)
    - Paraiba (3)
    - Parana (3)
    - Pernambuco (3)
    - Piaui (3)
    - Rio de Janeiro (3)
    - Rio Grande do Norte (3)
    - Rio Grande do Sul (3)
    - Rondonia (3)

- (Brazil Contd.)
  - Roraima (3)
  - Santa Catarina (3)
  - Sao Paulo (3)
  - Sergipe (3)
- Chile
- Colombia
- Ecuador
- Falkland Islands
- Guyana
- French Guiana
- Paraguay
- Peru
- Surinam
- Uruguay
- Venezuela
  - Apure Guarico Plain (3)
  - Maracaibo Falcon Plain (3)
  - Venezuelan Andes (3)
  - Venezuelan Islands (3)
  - Eastern Venezuela (3)
  - Southeastern Venezuela (3)

- West Indies
  - Cayman Islands
  - Greater Antilles
    - Bahamas
    - Cuba
    - Dominican Republic
    - Haiti
    - Jamaica
    - Puerto Rico
  - Lesser Antilles
    - Barbados
    - Dominica (3)
    - Leeward Islands (3)
      - Guadeloupe
      - Virgin Islands (3)
        - British Virgin Islands (3)
        - U. S. Virgin Islands (3)
    - Netherlands Antilles
    - Trinidad and Tobago
    - Windward Islands (3)
      - Grenada (3)
      - Martinique (3)

- Asia
  - Himalayas
  - \- - - - - - - - - - - - - -
  - Arabian Peninsula
    - Bahrain
    - Kuwait
    - Oman
    - Qatar
    - Saudi Arabia
    - Southern Yemen
    - United Arab Emirates
    - Yemen
  - Far East
    - Borneo (3)
    - Indochina
    - \- - - - - - - - - - - - - -
    - Brunei (3)
    - Burma
    - Cambodia
    - China
    - Hong Kong
    - Indonesia
      - Celebes (3)
      - Irian Jaya (3)
      - Java (3)
      - Kalimantan (3)
      - Lesser Sunda Islands (3)
      - Moluccas (3)
      - Sumatra (3)
    - Japan
      - Hokkaido (3)
      - Honshu (3)
      - Kyushu (3)
      - Ryukyu Islands (3)
      - Shikoku (3)
    - Korea
      - North Korea (3)
      - South Korea (3)

- (Far East Contd.)
  - Laos
  - Malaysia
    - Sabah (3)
    - Sarawak (3)
  - Mongolia
  - Philippine Islands
  - Singapore
  - Taiwan
  - Thailand
  - Vietnam
- Indian Peninsula (3)
  - Afghanistan
  - Bangladesh
  - Bhutan
  - India
    - Andhra Pradesh (3)
    - Bengal Islands (3)
    - Bihar (3)
    - Gujarat (3)
    - Haryana (3)
    - Himachal Pradesh (3)
    - Jammu and Kashmir (3)
    - Karnataka (3)
    - Kerala (3)
    - Laccadive Islands (3)
    - Madhya Pradesh (3)
    - Maharashtra (3)
    - Northeastern India (3)
    - Orissa (3)
    - Punjab State (3)
    - Rajasthan (3)
    - Tamil Nadu (3)
    - Uttar Pradesh (3)
    - West Bengal (3)
  - Nepal
  - Pakistan
  - Sri Lanka
- Middle East
  - Cyprus
  - Iran
  - Iraq
  - Israel
  - Jordan
  - Lebanon
  - Syria
  - Turkey
    - Anatolia (3)
    - Pontic Mountains (3)
    - Sea of Marmara region (3)
    - Taurus Mountains (3)
    - Turkish Aegean region (3)

- Atlantic Ocean Islands
  - Azores
  - Bermuda
  - Canary Islands
  - Cape Verde Islands
  - Madeira
  - Sao Tome (3)
  - Saint Helena (3)
- Atlantic region
- Australasia
  - Australia
    - New South Wales
    - Northern Territory
    - Queensland
    - South Australia
    - Tasmania
    - Victoria
    - Western Australia
  - New Zealand
    - North Island (3)
    - South Island (3)
  - Papua New Guinea
- East Pacific Ocean Islands
  - Easter Island (3)
  - Galapagos Islands
- Eurasia
  - Baltic region
- Europe
  - Alps
    - Eastern Alps (3)
    - Western Alps (3)

- (Europe Contd.)
  - Balkan Peninsula
  - Carpathians
  - Dinaric Alps (3)
  - Jura Mountains (3)
  - Pyrenees
  - - - - - - - - - - -
  - Central Europe (3)
    - Austria
      - Austrian Vienna Basin (3)
      - Central Austrian Alps (3)
      - North Austrian Alps (3)
      - North Austrian Crystallines (3)
      - North Austrian Molasse (3)
      - South Austrian Alps (3)
      - South Austrian Molasse (3)
      - - - - - - - - - - - - - - - - -
      - Burgenland (3)
      - Carinthia (3)
      - Lower Austria (3)
      - Salzburg State (3)
      - Styria (3)
      - Tyrol (3)
      - Upper Austria (3)
      - Vorarlberg (3)
    - Czechoslovakia
      - Berounka System (3)
      - Bohemia-Moravia System (3)
      - Bohemian Basin (3)
      - Czech Bohemian Forest (3)
      - Czech Erzgebirge (3)
      - Czech Sudeten Mountains (3)
      - Slovakia (3)
        - Slovakian Carpathians (3)
        - Slovakian Pannonian Basin (3)
    - Germany
      - Alpenvorland (3)
      - Bavarian Alps (3)
      - Bavarian Massif (3)
      - Harz Mountains (3)
      - Hesse Basin (3)
      - Northeastern German Plain (3)
      - Northwestern German Plain (3)
      - Northern German Hills (3)
      - Rhenish Schiefergebirge (3)
      - Rhine Westphalian Basin (3)
      - Saar-Nahe Basin (3)
      - Saxonian Massif (3)
      - Southwestern German Hills (3)
      - Southwestern German Massifs (3)
      - Thuringian Hills (3)
      - Thuringian Massif (3)
      - Upper Rhine Valley (3)
      - - - - - - - - - - - - -
      - East Germany
        - Cottbus Bezirk (3)
        - Dresden Bezirk (3)
        - Erfurt Bezirk (3)
        - Frankfurt Bezirk (3)
        - Gera Bezirk (3)
        - Halle Bezirk (3)
        - Karl-Marx-Stadt Bezirk (3)
        - Leipzig Bezirk (3)
        - Magdeburg Bezirk (3)
        - Neubrandenburg Bezirk (3)
        - Potsdam Bezirk (3)
        - Rostock Bezirk (3)
        - Schwerin Bezirk (3)
        - Suhl Bezirk (3)
      - West Germany
        - Baden-Wurttemberg (3)
        - Bavaria (3)
        - Hesse (3)
        - Lower Saxony (3)
        - North Rhine Westphalia (3)
        - Rhineland-Palatinate (3)
        - Saarland (3)
        - Schleswig-Holstein (3)
    - Hungary
      - Alfold (3)
      - Budapest region (3)
      - Central Transdanubia (3)
      - Northeastern Hungarian Hills (3)
    - (Hungary Contd.)
      - Northwestern Transdanubia (3)
      - Southeastern Transdanubia (3)
      - Southern Transdanubia (3)
    - Liechtenstein (3)
    - Poland
      - Northeastern Polish Plain (3)
      - Northwestern Polish Plain (3)
      - Polish Carpathians (3)
      - Polish Sudeten Mountains (3)
      - Southeastern Polish Hills (3)
      - Swiety Krzyz Mountains (3)
    - Switzerland
      - Central Swiss Alps (3)
      - Eastern Swiss Alps (3)
      - Northern Swiss Alps (3)
      - Southern Swiss Alps (3)
      - Swiss Jura Mountains (3)
      - Swiss Rhine Graben (3)
      - Swiss Molasse Basin (3)
  - Southern Europe (3)
    - Albania
    - Bulgaria
      - Balkan Mountains (3)
      - Bulgarian Dobruja (3)
      - Bulgarian Rhodope Mountains (3)
      - Central Bulgaria (3)
      - North Bulgarian Hills (3)
      - West Bulgarian Hills (3)
    - Greece
      - Crete (3)
      - Epirus (3)
      - Euboea (3)
      - Greek Aegean Islands (3)
      - Greek Ionian Islands (3)
      - Greek Macedonia (3)
      - Greek Thrace (3)
      - Peloponnesus (3)
      - Sterea Ellas (3)
      - Thessaly (3)
    - Iberian Peninsula (3)
      - Portugal
      - Spain
        - Asturian Arc (3)
        - Betic Zone (3)
        - Cantabrian region (3)
        - Catalonian Coastal Ranges (3)
        - Duero Basin (3)
        - Ebro Basin (3)
        - Galaico Massif (3)
        - Guadalquivir Basin (3)
        - Hercinico Centro (3)
        - Hercinico Sur (3)
        - Iberica (3)
        - Spanish Pyrenees (3)
        - Tagus Basin (3)
        - - - - - - - - - - - - - -
        - Andalusia (3)
          - Almeria Province (3)
          - Cadiz Province (3)
          - Cordoba Province (3)
          - Granada Province (3)
          - Huelva Province (3)
          - Jaen Province (3)
          - Malaga Province (3)
          - Seville Province (3)
        - Aragon (3)
          - Huesca Province (3)
          - Saragossa Province (3)
          - Teruel (3)
        - Asturias (3)
        - Balearic Islands
        - Basque Provinces (3)
          - Alava (3)
          - Guipuzcoa (3)
          - Vizcaya (3)
        - Catalonia (3)
          - Barcelona Province (3)
          - Gerona Province (3)
          - Lerida (3)
          - Tarragona (3)
        - Extremadura (3)
          - Badajoz Province (3)
          - Caceres Province (3)

(Spain Contd.)
- Galicia (3)
  - La Coruna Province (3)
  - Lugo (3)
  - Orense Province (3)
  - Pontevedra Province (3)
- Leon region (3)
  - Leon Province (3)
  - Salamanca Province (3)
  - Zamora Province (3)
- Murcia region (3)
  - Albacete (3)
  - Murcia Province (3)
- Navarra (3)
- New Castile (3)
  - Ciudad Real (3)
  - Cuenca Province (3)
  - Guadalajara Province (3)
  - Madrid Province (3)
  - Toledo Province (3)
- Old Castile (3)
  - Avila Province (3)
  - Burgos Province (3)
  - Logrono Province (3)
  - Palencia Province (3)
  - Santander Province (3)
  - Segovia (3)
  - Soria Province (3)
  - Valladolid Province (3)
- Valencia region (3)
  - Alicante Province (3)
  - Castellon de la Plana Province (3)
  - Valencia Province (3)

Italy
- Apennines
- - - - - - - - - - -
- Abruzzi (3)
- Apulia (3)
- Basilicata (3)
- Calabria (3)
- Campania (3)
- Emilia-Romagna (3)
- Friuli-Venezia Guiulia (3)
- Latium (3)
- Liguria (3)
- Lipari Islands (3)
- Lombardy (3)
- Marches (3)
- Molise (3)
- Piemonte (3)
- San Marino
- Sardinia
- Sicily (3)
- Trentino-Alto Adige (3)
- Tuscany (3)
- Umbria (3)
- Valle d'Aosta (3)
- Veneto (3)

Malta

Romania
- Apuseni Mountains (3)
- Eastern Carpathians (3)
- Getic Nappe (3)
- Moldavian Platform (3)
- Romanian Dobruja (3)
- Romanian Pannonian Basin (3)
- Romanian Plain (3)
- Transylvanian Alps (3)
- Transylvanian Basin (3)

Yugoslavia
- Bosnia-Herzegovina (3)
- Croatia (3)
- Montenegro (3)
- Serbia (3)
- Slovenia (3)
- Yugoslav Macedonia (3)

Western Europe (3)
- Andorra
- Belgium
- France
  - Monaco
  - - - - - - - - - - -

(France Contd.)
- Aquitaine Basin (3)
- Armorican Massif (3)
- Central Massif (3)
- French Alps (3)
- French Pyrenees (3)
- Paris Basin (3)
- Saone-Rhone Basin (3)
- Vosges Mountains (3)
- - - - - - - - - - - - - - -
- Ain (3)
- Aisne (3)
- Allier (3)
- Alpes-de-Haute-Provence (3)
- Alpes-Maritimes (3)
- Ardeche (3)
- Ardennes Department (3)
- Ariege (3)
- Aube (3)
- Aude (3)
- Aveyron (3)
- Bas-Rhin (3)
- Belfort (3)
- Bouches-du-Rhone (3)
- Calvados (3)
- Cantal (3)
- Charente (3)
- Charente-Maritime (3)
- Cher (3)
- Correze (3)
- Corsica
- Cote-d'Or (3)
- Cotes-du-Nord (3)
- Creuse (3)
- Deux-Sevres (3)
- Dordogne (3)
- Doubs (3)
- Drome (3)
- Essonne (3)
- Eure (3)
- Eure-et-Loir (3)
- Finistere (3)
- Gard (3)
- Gers (3)
- Gironde (3)
- Haut-Rhin (3)
- Haute-Garonne (3)
- Haute-Loire (3)
- Hautes-Alpes (3)
- Haute-Marne (3)
- Haute-Saone (3)
- Haute-Savoie (3)
- Haute-Vienne (3)
- Hautes-Pyrenees (3)
- Hauts-de-Seine (3)
- Herault (3)
- Ille-et-Vilaine (3)
- Indre (3)
- Indre-et-Loire (3)
- Isere (3)
- Jura (3)
- Landes (3)
- Loir-et-Cher (3)
- Loire (3)
- Loire-Atlantique (3)
- Loiret (3)
- Lot (3)
- Lot-et-Garonne (3)
- Lozere (3)
- Maine-et-Loire (3)
- Manche (3)
- Marne (3)
- Mayenne (3)
- Meurthe-et-Moselle (3)
- Meuse (3)
- Morbihan (3)
- Moselle (3)
- Nievre (3)
- Nord Department (3)
- Oise (3)
- Orne (3)
- Paris (3)
- Pas-de-Calais (3)

- (France Contd.)
  - Puy-de-Dome (3)
  - Pyrenees-Atlantiques (3)
  - Pyrenees-Orientales (3)
  - Rhone (3)
  - Saone-et-Loire (3)
  - Sarthe (3)
  - Savoie (3)
  - Seine-Maritime (3)
  - Seine-et-Marne (3)
  - Seine-Saint-Denis (3)
  - Somme (3)
  - Tarn (3)
  - Tarn-et-Garonne (3)
  - Val-de-Marne (3)
  - Val-d'Oise (3)
  - Var (3)
  - Vaucluse (3)
  - Vendee (3)
  - Vienne (3)
  - Vosges (3)
  - Yonne (3)
  - Yvelines (3)
- Iceland
- Ireland
- Luxembourg
- Netherlands
- Scandinavia
  - Denmark
    - Faeroe Islands
  - Finland
  - Norway
    - Northern Norway (3)
    - Southern Norway (3)
  - Sweden
    - Gotland (3)
- United Kingdom
  - English Channel Islands
  - Great Britain
    - England
      - London Basin (3)
      - Pennines (3)
      - South-West England (3)
      - Welsh Borderland (3)
    - Scotland
      - Grampian Highlands (3)
      - Hebrides (3)
      - Midland Valley (3)
      - Northern Highlands (3)
      - Orkney Islands (3)
      - Shetland Islands
    - Wales
  - Isle of Man (3)
  - Northern Ireland

- Indian Ocean Islands
  - Comoro Islands
  - Malagasy Republic
  - Maldive Islands
  - Mauritius
  - Reunion
  - Seychelles
- Malay Archipelago
  - New Guinea
- Mediterranean region
- Oceania
  - Melanesia
    - Fiji
    - New Caledonia
    - New Hebrides
    - Solomon Islands
  - Micronesia
    - Caroline Islands (3)
    - Gilbert Islands (3)
    - Mariana Islands
    - Marshall Islands
    - Nauru Island (3)
  - Polynesia
    - Cook Islands (3)
    - French Polynesia (3)
      - Marquesas Islands (3)
      - Society Islands
        - Tahiti
      - Tuamotu Islands (3)

- (Polynesia Contd.)
  - Samoa
  - Tokelau (3)
  - Tonga
  - Tuvalu (3)
- Pacific region
- Polar regions (3)
  - Antarctica
    - Kerguelen Islands (3)
    - Scotia Sea Islands (3)
  - Arctic region
    - Greenland
    - Jan Mayen
    - Spitsbergen
- Red Sea region
- USSR
  - Russian Republic (3)
  - Siberia (3)
  - Soviet Arctic (3)
  - Ukraine (3)
  - \- - - - - - - - - - - -
  - Caucasus
    - Greater Caucasus (3)
    - Lesser Caucasus (3)
    - Northern Caucasus (3)
  - Central Asia (3)
    - Caspian Depression (3)
    - Central Kazakhstan (3)
    - Eastern Kazakhstan (3)
    - Fergana Basin (3)
    - Karakum (3)
    - Kopet-Dag Range (3)
    - Kyzylkum (3)
    - Pamirs (3)
    - Tien Shan (3)
    - Turgay Basin (3)
    - Ustyurt (3)
  - Northeastern USSR (3)
    - Amur region (3)
    - Chukchi Peninsula (3)
    - Dzhugdzhur region (3)
    - Kolyma Uplift (3)
    - New Siberian Islands (3)
    - Sikhote-Alin Range (3)
    - Stanovoy Range (3)
    - Verkhoyansk region (3)
  - Russian Platform (3)
    - Baltic Plain (3)
      - Estonia (3)
      - Latvia (3)
      - Lithuania (3)
    - Byelorussia (3)
    - Crimea (3)
    - Dnieper-Donets Basin (3)
    - Franz Joseph Land (3)
    - Moscow-Pechora Syneclise (3)
    - Soviet Carpathians (3)
    - Soviet Fennoscandia (3)
    - Timan Ridge (3)
    - Transcarpathia (3)
    - Ukrainian Shield (3)
    - Ukrainian Syneclise (3)
    - Voronezh-Volga Anteclise (3)
  - Siberian fold belt (3)
    - Baikal region (3)
    - Eastern Sayan (3)
    - Soviet Altai Mountains (3)
    - Tuva (3)
    - Western Sayan (3)
    - Western Transbaikalia (3)
  - Siberian Platform (3)
    - Anabar Shield (3)
    - Angara-Lena Basin (3)
    - Aldan Shield (3)
    - North Siberian Plain (3)
    - Tunguska (3)
    - Vilyuy Syneclise (3)
    - Yakutia region (3)
    - Yenisei Ridge (3)
  - Soviet Pacific region (3)
    - Kamchatka Peninsula (3)
    - Koryak Range (3)
    - Kuril Islands (3)
    - Sakhalin (3)

(USSR Contd.)
- Urals (3)
  - Central Urals (3)
  - Mugodzhar Hills (3)
  - Northern Urals (3)
  - Novaya Zemlya (3)
  - Polar Urals (3)
  - Southern Urals (3)
- West Siberia (3)
  - Kuznetsk Alatau (3)
  - Minusinsk Basin (3)
  - Severnaya Zemlya (3)
  - Siberian Lowland (3)
  - Taymyr (3)

World ocean (3)
- Antarctic Ocean
  - Bellingshausen Sea (3)
  - Mid-Antarctic Ridge (3)
  - Ross Sea (3)
  - Scotia Sea (3)
  - Weddell Sea (3)
- Arctic Ocean
  - Baffin Bay (3)
  - Barents Sea (3)
  - Beaufort Sea (3)
  - Chukchi Sea (3)
  - East Siberian Sea (3)
  - Greenland Sea (3)
  - Kara Sea (3)
  - Laptev Sea (3)
  - Mid-Arctic Ocean Ridge (3)
  - Norwegian Sea (3)
  - White Sea (3)
- Atlantic Ocean
  - North Atlantic (3)
    - Azores-Gibraltar Ridge (3)
    - Cape Verde Atlantic (3)
    - European Atlantic (3)
      - Baltic Sea
        - Gulf of Bothnia (3)
        - Gulf of Finland (3)
        - Gulf of Riga (3)
      - English Channel
      - Bay of Biscay
      - Irish Sea
      - North Sea
    - Gulf of Guinea (3)
    - Guyanese Atlantic (3)
    - North American Atlantic (3)
      - Caribbean Sea
      - Gulf of Mexico
      - Gulf of Saint Lawrence (3)
      - Hudson Bay (3)
      - Labrador Sea (3)
    - North Atlantic Ridge (3)
  - South Atlantic (3)
    - South American Atlantic (3)
    - South Atlantic Ridge (3)
    - Southeast Atlantic (3)
- Caspian Sea
- Indian Ocean
  - Andaman Sea (3)
  - Arabian Sea
    - Gulf of Aden
    - Gulf of Oman (3)
    - Persian Gulf
  - Bay of Bengal (3)
  - Carlsberg Ridge (3)
  - Central Indian Ridge (3)
  - Mozambique Channel (3)
  - Ninetyeast Ridge (3)
  - Red Sea
    - Gulf of Aqaba (3)
    - Gulf of Suez (3)
  - Southeast Indian Ridge (3)
  - Southwest Indian Ridge (3)
- Mediterranean Sea
  - East Mediterranean (3)
    - Adriatic Sea
    - Aegean Sea
    - Black Sea
    - Ionian Sea

(Mediterranean Sea Contd.)
- West Mediterranean (3)
  - Alboran Sea (3)
  - Gulf of Lion (3)
  - Tyrrhenian Sea

Pacific Ocean
- East Pacific Rise (3)
- North American Pacific (3)
  - Gulf of Alaska (3)
  - Gulf of California
- South American Pacific (3)
  - Carnegie Ridge (3)
  - Chile Ridge (3)
  - Cocos Ridge (3)
  - Nazca Ridge (3)
- South Polynesian Pacific (3)
- West Pacific (3)
  - Bering Sea
  - Coral Sea
  - East China Sea
  - Indonesian Seas (3)(Celebes Sea, Java Sea (3))
  - Japan Sea
  - North Australian Seas (3)(Arafura Sea (3), Timor Sea (3))
  - Okhotsk Sea
  - Philippine Sea
  - South China Sea
    - Gulf of Siam (3)
    - Gulf of Tonkin (3)
  - Tasman Sea
  - Yellow Sea

Pacific-Antarctic Ridge (3)